Methods in Enzymology

Volume 96

BIOMEMBRANES

Part J

Membrane Biogenesis:

Assembly and Targeting

(General Methods, Eukaryotes)

METHODS IN ENZYMOLOGY

EDITORS-IN-CHIEF

Sidney P. Colowick   Nathan O. Kaplan

Methods in Enzymology

Volume 96

Biomembranes

Part J
Membrane Biogenesis: Assembly and Targeting (General Methods, Eukaryotes)

EDITED BY

Sidney Fleischer

Becca Fleischer

DEPARTMENT OF MOLECULAR BIOLOGY
VANDERBILT UNIVERSITY
NASHVILLE, TENNESSEE

1983

ACADEMIC PRESS

A Subsidiary of Harcourt Brace Jovanovich, Publishers

New York London
Paris San Diego San Francisco São Paulo Sydney Tokyo Toronto

ACADEMIC PRESS, INC.
111 Fifth Avenue, New York, New York 10003

United Kingdom Edition published by
ACADEMIC PRESS, INC. (LONDON) LTD.
24/28 Oval Road, London NW1 7DX

LIBRARY OF CONGRESS CATALOG CARD NUMBER: 54-9110

ISBN 0-12-181996-5

PRINTED IN THE UNITED STATES OF AMERICA

83 84 85 86 9 8 7 6 5 4 3 2 1

Table of Contents

Contributors to Volume 96 xi

Preface xvii

Volumes in Series xix

Membrane Biogenesis: An Overview *by* George E. Palade xxix

Section I. Biogenesis and Assembly of Membrane Proteins

A. General Methods

1. Preparation of Rough Microsomes and Membrane-Bound Polysomes That Are Active in Protein Synthesis *in Vitro* — Sancia Gaetani, Julia A. Smith, Ricardo A. Feldman, and Takashi Morimoto — 3

2. Methods for the Preparation of Messenger RNA — Paul M. Lizardi — 24

3. Cell-Free Translation of Messenger RNA in a Wheat Germ System — Ann. H. Erickson and Günter Blobel — 38

4. Preparation and Use of Nuclease-Treated Rabbit Reticulocyte Lysates for the Translation of Eukaryotic Messenger RNA — Richard J. Jackson and Tim Hunt — 50

5. *In Vivo* and *in Vitro* Systems for Studying Bacterial Membrane Biogenesis — Pamela Green and Masayori Inouye — 74

6. Preparation of Microsomal Membranes for Cotranslational Protein Translocation — Peter Walter and Günter Blobel — 84

7. Methods for the Study of Protein Translocation across the RER Membrane Using the Reticulocyte Lysate Translation System and Canine Pancreatic Microsomal Membranes — George Scheele — 94

8. Immunoprecipitation of Proteins from Cell-Free Translations — David J. Anderson and Günter Blobel — 111

9. Use of Proteases for the Study of Membrane Insertion — Takashi Morimoto, Monique Arpin, and Sancia Gaetani — 121

10. Identifying Primary Translation Products: Use of *N*-Formylmethionyl-tRNA and Prevention of NH_2-Terminal Acetylation — Richard D. Palmiter — 150

11. Selection of Lectin-Resistant Mutants of Animal Cells — PAMELA STANLEY — 157

12. Two-Dimensional Polyacrylamide Gel Electrophoresis of Membrane Proteins — R. W. RUBIN AND C. L. LEONARDI — 184

13. Immunochemical Identification of Membrane Proteins after Sodium Dodecyl Sulfate–Polyacrylamide Gel Electrophoresis — ALBERT HAID AND MORDECHAI SUISSA — 192

14. Immunologic Detection of Specific Proteins in Cell Extracts by Fractionation in Gels and Transfer to Paper — JAKOB REISER AND GEORGE R. STARK — 205

15. Use of Fluorography for Sensitive Isotope Detection in Polyacrylamide Gel Electrophoresis and Related Techniques — WILLIAM M. BONNER — 215

16. Peptide Mapping in One Dimension by Limited Proteolysis of Sodium Dodecyl Sulfate-Solubilized Proteins — DON W. CLEVELAND — 222

17. Silver Staining Methods for Polyacrylamide Gel Electrophoresis — CARL R. MERRIL, DAVID GOLDMAN, AND MARGARET L. VAN KEUREN — 230

18. Slab Gel System for the Resolution of Oligopeptides below Molecular Weight of 10,000 — F. A. BURR AND B. BURR — 239

19. Two-Dimensional Immunoelectrophoresis of Membrane Antigens — CAROLYN A. CONVERSE AND DAVID S. PAPERMASTER — 244

B. Eukaryotic Membranes

Plasma Membrane

20. Methods for Study of the Synthesis and Maturation of the Erythrocyte Anion Transport Protein — HARVEY F. LODISH AND WILLIAM A. BRAELL — 257

21. Glycophorins: Isolation, Orientation, and Localization of Specific Domains — HEINZ FURTHMAYR AND VINCENT T. MARCHESI — 268

22. Glycophorin A: *In Vitro* Biogenesis and Processing — CARL G. GAHMBERG, MIKKO JOKINEN, KIMMO K. KARHI, OLLE KÄMPE, PER A. PETERSON, AND LEIF C. ANDERSSON — 281

23. Erythrocyte Membrane Proteins: Detection of Spectrin Oligomers by Gel Electrophoresis — JON S. MORROW AND WALLACE B. HAIGH, JR. — 298

METHODS IN ENZYMOLOGY

EDITORS-IN-CHIEF

Sidney P. Colowick   Nathan O. Kaplan

VOLUME VIII. Complex Carbohydrates
Edited by ELIZABETH F. NEUFELD AND VICTOR GINSBURG

VOLUME IX. Carbohydrate Metabolism
Edited by WILLIS A. WOOD

VOLUME X. Oxidation and Phosphorylation
Edited by RONALD W. ESTABROOK AND MAYNARD E. PULLMAN

VOLUME XI. Enzyme Structure
Edited by C. H. W. HIRS

VOLUME XII. Nucleic Acids (Parts A and B)
Edited by LAWRENCE GROSSMAN AND KIVIE MOLDAVE

VOLUME XIII. Citric Acid Cycle
Edited by J. M. LOWENSTEIN

VOLUME XIV. Lipids
Edited by J. M. LOWENSTEIN

VOLUME XV. Steroids and Terpenoids
Edited by RAYMOND B. CLAYTON

VOLUME XVI. Fast Reactions
Edited by KENNETH KUSTIN

VOLUME XVII. Metabolism of Amino Acids and Amines (Parts A and B)
Edited by HERBERT TABOR AND CELIA WHITE TABOR

VOLUME XVIII. Vitamins and Coenzymes (Parts A, B, and C)
Edited by DONALD B. McCORMICK AND LEMUEL D. WRIGHT

VOLUME XIX. Proteolytic Enzymes
Edited by GERTRUDE E. PERLMANN AND LASZLO LORAND

VOLUME XX. Nucleic Acids and Protein Synthesis (Part C)
Edited by KIVIE MOLDAVE AND LAWRENCE GROSSMAN

VOLUME XXI. Nucleic Acids (Part D)
Edited by LAWRENCE GROSSMAN AND KIVIE MOLDAVE

VOLUME XXII. Enzyme Purification and Related Techniques
Edited by WILLIAM B. JAKOBY

VOLUME XXIII. Photosynthesis (Part A)
Edited by ANTHONY SAN PIETRO

VOLUME XXIV. Photosynthesis and Nitrogen Fixation (Part B)
Edited by ANTHONY SAN PIETRO

VOLUME XXV. Enzyme Structure (Part B)
Edited by C. H. W. HIRS AND SERGE N. TIMASHEFF

VOLUME XXVI. Enzyme Structure (Part C)
Edited by C. H. W. HIRS AND SERGE N. TIMASHEFF

VOLUME XXVII. Enzyme Structure (Part D)
Edited by C. H. W. HIRS AND SERGE N. TIMASHEFF

VOLUME XXVIII. Complex Carbohydrates (Part B)
Edited by VICTOR GINSBURG

VOLUME XXIX. Nucleic Acids and Protein Synthesis (Part E)
Edited by LAWRENCE GROSSMAN AND KIVIE MOLDAVE

VOLUME XXX. Nucleic Acids and Protein Synthesis (Part F)
Edited by KIVIE MOLDAVE AND LAWRENCE GROSSMAN

VOLUME XXXI. Biomembranes (Part A)
Edited by SIDNEY FLEISCHER AND LESTER PACKER

VOLUME XXXII. Biomembranes (Part B)
Edited by SIDNEY FLEISCHER AND LESTER PACKER

VOLUME XXXIII. Cumulative Subject Index Volumes I–XXX
Edited by MARTHA G. DENNIS AND EDWARD A. DENNIS

VOLUME XXXIV. Affinity Techniques (Enzyme Purification: Part B)
Edited by WILLIAM B. JAKOBY AND MEIR WILCHEK

VOLUME XXXV. Lipids (Part B)
Edited by JOHN M. LOWENSTEIN

VOLUME XXXVI. Hormone Action (Part A: Steroid Hormones)
Edited by BERT W. O'MALLEY AND JOEL G. HARDMAN

VOLUME XXXVII. Hormone Action (Part B: Peptide Hormones)
Edited by BERT W. O'MALLEY AND JOEL G. HARDMAN

VOLUME XXXVIII. Hormone Action (Part C: Cyclic Nucleotides)
Edited by JOEL G. HARDMAN AND BERT W. O'MALLEY

VOLUME XXXIX. Hormone Action (Part D: Isolated Cells, Tissues, and Organ Systems)
Edited by JOEL G. HARDMAN AND BERT W. O'MALLEY

VOLUME XL. Hormone Action (Part E: Nuclear Structure and Function)
Edited by BERT W. O'MALLEY AND JOEL G. HARDMAN

VOLUME XLI. Carbohydrate Metabolism (Part B)
Edited by W. A. WOOD

VOLUME XLII. Carbohydrate Metabolism (Part C)
Edited by W. A. WOOD

VOLUME XLIII. Antibiotics
Edited by JOHN H. HASH

VOLUME XLIV. Immobilized Enzymes
Edited by KLAUS MOSBACH

VOLUME XLV. Proteolytic Enzymes (Part B)
Edited by LASZLO LORAND

VOLUME XLVI. Affinity Labeling
Edited by WILLIAM B. JAKOBY AND MEIR WILCHEK

VOLUME XLVII. Enzyme Structure (Part E)
Edited by C. H. W. HIRS AND SERGE N. TIMASHEFF

VOLUME XLVIII. Enzyme Structure (Part F)
Edited by C. H. W. HIRS AND SERGE N. TIMASHEFF

VOLUME XLIX. Enzyme Structure (Part G)
Edited by C. H. W. HIRS AND SERGE N. TIMASHEFF

VOLUME L. Complex Carbohydrates (Part C)
Edited by VICTOR GINSBURG

VOLUME LI. Purine and Pyrimidine Nucleotide Metabolism
Edited by PATRICIA A. HOFFEE AND MARY ELLEN JONES

VOLUME LII. Biomembranes (Part C: Biological Oxidations)
Edited by SIDNEY FLEISCHER AND LESTER PACKER

VOLUME LIII. Biomembranes (Part D: Biological Oxidations)
Edited by SIDNEY FLEISCHER AND LESTER PACKER

VOLUME LIV. Biomembranes (Part E: Biological Oxidations)
Edited by SIDNEY FLEISCHER AND LESTER PACKER

VOLUME LV. Biomembranes (Part F: Bioenergetics)
Edited by SIDNEY FLEISCHER AND LESTER PACKER

VOLUME LVI. Biomembranes (Part G: Bioenergetics)
Edited by SIDNEY FLEISCHER AND LESTER PACKER

VOLUME LVII. Bioluminescence and Chemiluminescence
Edited by MARLENE A. DELUCA

VOLUME LVIII. Cell Culture
Edited by WILLIAM B. JAKOBY AND IRA H. PASTAN

VOLUME LIX. Nucleic Acids and Protein Synthesis (Part G)
Edited by KIVIE MOLDAVE AND LAWRENCE GROSSMAN

VOLUME LX. Nucleic Acids and Protein Synthesis (Part H)
Edited by KIVIE MOLDAVE AND LAWRENCE GROSSMAN

VOLUME 61. Enzyme Structure (Part H)
Edited by C. H. W. HIRS AND SERGE N. TIMASHEFF

VOLUME 62. Vitamins and Coenzymes (Part D)
Edited by DONALD B. MCCORMICK AND LEMUEL D. WRIGHT

VOLUME 63. Enzyme Kinetics and Mechanism (Part A: Initial Rate and Inhibitor Methods)
Edited by DANIEL L. PURICH

VOLUME 64. Enzyme Kinetics and Mechanism (Part B: Isotopic Probes and Complex Enzyme Systems)
Edited by DANIEL L. PURICH

VOLUME 65. Nucleic Acids (Part I)
Edited by LAWRENCE GROSSMAN AND KIVIE MOLDAVE

VOLUME 66. Vitamins and Coenzymes (Part E)
Edited by DONALD B. MCCORMICK AND LEMUEL D. WRIGHT

VOLUME 67. Vitamins and Coenzymes (Part F)
Edited by DONALD B. MCCORMICK AND LEMUEL D. WRIGHT

VOLUME 68. Recombinant DNA
Edited by RAY WU

VOLUME 69. Photosynthesis and Nitrogen Fixation (Part C)
Edited by ANTHONY SAN PIETRO

VOLUME 70. Immunochemical Techniques (Part A)
Edited by HELEN VAN VUNAKIS AND JOHN J. LANGONE

VOLUME 71. Lipids (Part C)
Edited by JOHN M. LOWENSTEIN

VOLUME 72. Lipids (Part D)
Edited by JOHN M. LOWENSTEIN

VOLUME 73. Immunochemical Techniques (Part B)
Edited by JOHN J. LANGONE AND HELEN VAN VUNAKIS

VOLUME 74. Immunochemical Techniques (Part C)
Edited by JOHN J. LANGONE AND HELEN VAN VUNAKIS

VOLUME 75. Cumulative Subject Index Volumes XXXI, XXXII, XXXIV–LX
Edited by MARTHA G. DENNIS AND EDWARD A. DENNIS

VOLUME 76. Hemoglobins
Edited by ERALDO ANTONINI, LUIGI ROSSI-BERNARDI, AND EMILIA CHIANCONE

VOLUME 77. Detoxication and Drug Metabolism
Edited by WILLIAM B. JAKOBY

VOLUME 78. Interferons (Part A)
Edited by SIDNEY PESTKA

VOLUME 79. Interferons (Part B)
Edited by SIDNEY PESTKA

VOLUME 80. Proteolytic Enzymes (Part C)
Edited by LASZLO LORAND

VOLUME 81. Biomembranes (Part H: Visual Pigments and Purple Membranes, I)
Edited by LESTER PACKER

VOLUME 82. Structural and Contractile Proteins (Part A: Extracellular Matrix)
Edited by LEON W. CUNNINGHAM AND DIXIE W. FREDERIKSEN

VOLUME 83. Complex Carbohydrates (Part D)
Edited by VICTOR GINSBURG

VOLUME 84. Immunochemical Techniques (Part D: Selected Immunoassays)
Edited by JOHN J. LANGONE AND HELEN VAN VUNAKIS

VOLUME 85. Structural and Contractile Proteins (Part B: The Contractile Apparatus and the Cytoskeleton)
Edited by DIXIE W. FREDERIKSEN AND LEON W. CUNNINGHAM

VOLUME 86. Prostaglandins and Arachidonate Metabolites
Edited by WILLIAM E. M. LANDS AND WILLIAM L. SMITH

VOLUME 87. Enzyme Kinetics and Mechanism (Part C: Intermediates, Stereochemistry, and Rate Studies)
Edited by DANIEL L. PURICH

VOLUME 88. Biomembranes (Part I: Visual Pigments and Purple Membranes, II)
Edited by LESTER PACKER

VOLUME 89. Carbohydrate Metabolism (Part D)
Edited by WILLIS A. WOOD

VOLUME 90. Carbohydrate Metabolism (Part E)
Edited by WILLIS A. WOOD

VOLUME 91. Enzyme Structure (Part I)
Edited by C. H. W. HIRS AND SERGE N. TIMASHEFF

VOLUME 92. Immunochemical Techniques (Part E: Monoclonal Antibodies and General Immunoassay Methods)
Edited by JOHN J. LANGONE AND HELEN VAN VUNAKIS

VOLUME 93. Immunochemical Techniques (Part F: Conventional Antibodies, Fc Receptors, and Cytotoxicity)
Edited by JOHN J. LANGONE AND HELEN VAN VUNAKIS

VOLUME 94. Polyamines
Edited by HERBERT TABOR AND CELIA WHITE TABOR

VOLUME 95. Cumulative Subject Index Volumes 61–74, 76–80 (in preparation)
Edited by MARTHA G. DENNIS AND EDWARD A. DENNIS

VOLUME 96. Biomembranes [Part J: Membrane Biogenesis: Assembly and Targeting (General Methods, Eukaryotes)]
Edited by SIDNEY FLEISCHER AND BECCA FLEISCHER

VOLUME 97. Biomembranes [Part K: Membrane Biogenesis: Assembly and Targeting (Prokaryotes, Mitochondria, and Chloroplasts)]
Edited by SIDNEY FLEISCHER AND BECCA FLEISCHER

VOLUME 98. Biomembranes [Part L: Membrane Biogenesis (Processing and Recycling)] (in preparation)
Edited by SIDNEY FLEISCHER AND BECCA FLEISCHER

VOLUME 99. Hormone Action (Part F: Protein Kinases)
Edited by JACKIE D. CORBIN AND JOEL G. HARDMAN

VOLUME 100. Recombinant DNA (Part B)
Edited by RAY WU, LAWRENCE GROSSMAN, AND KIVIE MOLDAVE

VOLUME 101. Recombinant DNA (Part C)
Edited by RAY WU, LAWRENCE GROSSMAN, AND KIVIE MOLDAVE

VOLUME 102. Hormone Action (Part G: Calmodulin and Calcium-Binding Proteins)
Edited by ANTHONY R. MEANS AND BERT W. O'MALLEY

VOLUME 103. Hormone Action (Part H: Neuroendocrine Peptides) (in preparation)
Edited by P. MICHAEL CONN

VOLUME 104. Enzyme Purification and Related Techniques (Part C) (in preparation)
Edited by WILLIAM B. JAKOBY

VOLUME 105. Oxygen Radicals in Biological Systems (in preparation)
Edited by LESTER PACKER

Membrane Biogenesis: An Overview

By George E. Palade

General Characteristics of the Process

Membrane biogenesis, the synthesis of specific molecular components and their assembly into cellular membranes, is a critical, major activity in which cells engage throughout their existence; critical, because cell growth—leading to, and including cell division—depends stringently on the production of sufficient membrane to equip two daughter cells; major, because in logarithmic growth a cell nearly doubles in one generation time the amount of membrane inherited from its mother. Doubling amounts, on the average, to the production of $\sim$20 μm^2 in $\sim$30 min for a prokaryote and $\sim$10,000 μm^2 in $\sim$24 hr for a eukaryote. Per minute, a eukaryote produces, therefore, $\sim$10 times more membrane than a prokaryote.

In eukaryotes, membrane production is also a remarkably diversified operation, since such cells are equipped with a relatively large number of distinct types of membrane, each characterized by its own lipid and protein composition. At present we recognize $\sim$10 distinct types of cellular membranes in animal eukaryotes and $\sim$12 distinct types in their plant counterparts. Moreover, these numbers are expected to increase, since not all types of intracellular membranes have been characterized, and since differentiated domains or microdomains have already been identified within certain cellular membranes.

Net membrane production is not limited to cell growth: it occurs also in cell differentiation. Well-documented cases concern, for instance, the endoplasmic reticulum membrane in differentiating hepatocytes[1,2] and the plasmalemma in intestinal epithelia.[3] Even at zero net growth, however, production and assembly of membrane components continues as part of a general turnover process, which in some cell types proceeds at remark-

[1] G. Dallner, P. Siekevitz, and G. E. Palade, Biogenesis of endoplasmic reticulum membranes. I. Structural and chemical differentiation in developing rat hepatocyte. II. Synthesis of constitutive microsomal enzymes in developing rat hepatocyte. *J. Cell Biol.* **30**, 73–96, 97–117 (1966).

[2] A. Leskes, P. Siekevitz, and G. E. Palade, Differentiation of endoplasmic reticulum in hepatocytes. I. Glucose-6-phosphatase distribution *in situ*. II. Glucose-6-phosphatase in rough microsomes. *J. Cell Biol.* **49**, 264–287, 288–302 (1971).

[3] F. Moog, The differentiation and redifferentiation of the intestinal epithelium and its brush border membrane. *In* "Development of Mammalian Absorptive Processes" (K. Elliot and J. Whelan, eds.) (Ciba Foundation Symposium 70), p. 31. Pitman, London, 1979.

ably high rates.[4,5] Membrane biogenesis is therefore a continuous, massive, and, in eukaryotes at least, diversified cell activity.

General Chemistry of Cellular Membranes

In general terms, the chemistry of cellular membranes is reasonably well understood. The main molecular components are (a) polar lipids, i.e., phospholipids, glycolipids, and cholesterol, organized into a continuous bilayer, fluid at the temperature of the immediate ambient, and (b) proteins—many of them glycoproteins—interacting in different ways with the lipid bilayer. Nonpolar lipids and nucleic acids have been repeatedly reported as minor components in cellular membranes, but the significance (even the validity) of the corresponding findings is still uncertain.

The Fluid Mosaic Model. Our understanding of the physical state and molecular architecture of cellular membranes has been significantly advanced by the formulation of the fluid mosaic model of Singer and Nicolson.[6] The model recognizes two types of membrane proteins: (a) an integral type, characterized by hydrophobic domains buried in the lipid bilayer and hydrophilic domains projecting on one, or both sides of the latter; and (b) a peripheral type located on the cytoplasmic side of the lipid bilayer and bound to its components primarily by hydrophilic interactions. Since occupancy of the bilayer by integral membrane proteins appears to be limited, the model is reminiscent of an archipelago with potentially mobile islands scattered in a shallow sea of viscous, fluid lipid, bottomed by a mat of fibrillar proteins. Most integral membrane proteins are transmembrane entities with one (or more) endodomain(s) in the cytoplasm, one (or more) hydrophobic domain(s) in the bilayer, and one (or more) ectodomain(s) that protrude(s) in the extracellular medium. The ectodomains bear all the oligosaccharide chains of all the transmembrane glycoproteins so far studied.

Since the initial formulation of the fluid mosaic model, new findings have shown that protein interactions with a membrane are more diversified than originally assumed. Certain secretory glycoproteins, discharged into the extracellular matrix, e.g, fibronectin,[7] laminin,[8] and proteogly-

[4] R. T. Schimke, Turnover of membrane proteins in animal cells. *Methods Membr. Biol.* **3**, 201–237 (1975).

[5] T. Omura, P. Siekevitz, and G. E. Palade, Turnover of constituents of the endoplasmic reticulum membranes of rat hepatocytes. *J. Biol. Chem.* **242**, 2389–2396 (1967).

[6] S. J. Singer and G. L. Nicolson, The fluid mosaic model of the structure of cell membranes. *Science (Washington, D.C.)* **175**, 720–731 (1972).

[7] L. H. E. Hahn and K. M. Yamada, Isolation and biological characterization of active fragments of the adhesive glycoprotein fibronectin. *Cell* **18**, 1043–1051 (1979).

[8] H. L. Malinoff and M. S. Wicha, Isolation of a cell surface receptor for laminin from murine fibrosarcoma cells. *J. Cell Biol.* **96**, 1475–1479 (1983).

cans,[9] interact specifically with components accessible on the outer aspect of the bilayer. Accordingly, they can be considered as external counterparts of the peripheral membrane proteins found on the inner aspect of the bilayer. They appear to act not so much as supportive or stabilizing structures, but as integrative elements involved in the transition from the cellular to the tissular levels of biological organization. Some of them interact, however, with intracellular stress fibers,[10] presumably through the mediation of integral membrane proteins, and thereby become a factor in the stabilization of membranes and membrane infrastructures.

Protein complexes that behave in part like peripheral membrane proteins, but also interact hydrophobically with the corresponding bilayer, were recently identified, in the case of the signal recognition protein complex or particle.[11] In addition, glycoproteins synthesized initially as transmembrane proteins were found to be converted to secretory proteins during their intracellular life span.[12,13] Hence, the classification of membrane proteins is becoming more complex and more diversified. The distinction between peripheral membrane proteins, on the one hand, and cytosol and extracellular matrix proteins, on the other, becomes less clearly definable. Yet, the original categories remain valid.

General Function of Cellular Membranes

On account of their hydrophobic lipid bilayer, cell membranes function essentially as diffusion barriers primarily for electrolytes and hydrophilic micro- and macromolecular solutes. Most of the transmembrane proteins so far identified are modifiers of this permeability barrier in which they act as ion channels, ion pumps, metabolite (monosaccharide, amino acids) transporters, receptors for physiological ligands (hormones, neurotransmitters, complex metabolites), and signal transducers.

Most of these activities generate across the corresponding membranes electrical potentials and chemical or electrochemical gradients. These

[9] L. Kjellen, A. Oldberg, and M. Höök, Cell surface heparan sulfate: mechanism of proteoglycan-cell association. *J. Biol. Chem.* **255**, 407–413 (1980).

[10] I. I. Singer, The fibronexus: a transmembrane association of fibronectin-containing fibers and bundles of 5 nm microfilaments in hamster and human fibroblasts. *Cell* **16**, 675–685 (1979).

[11] P. Walter and G. Blobel, Purification of a membrane protein complex required for protein translocation across the endoplasmic reticulum. *Proc. Natl. Acad. Sci. U.S.A.* **77**, 7112–7116 (1980).

[12] L. C. Kuhn and J. P. Kraehenbuhl, The membrane receptor for polymeric immunoglobulin is structurally related to secretory component. Isolation and characterization of membrane secretory component from rabbit liver and mammary gland. *J. Biol. Chem.* **256**, 12490–12495 (1981).

[13] R. L. Rotundo and D. M. Fambrough, Synthesis, transport and fate of acetylcholinesterase in cultured chick embryo muscle cells. *Cell* **22**, 583–594 (1980).

gradients are homeostatically maintained and on them—or on their controlled fluctuations—critically depend essential cell functions, such as macromolecular (protein, RNA, DNA) synthesis, ATP production, lysosomal digestion, ligand–receptor interactions, contraction vs relaxation, and assembly vs disassembly of internal supportive structures, to give only a few examples. Membranes and the gradients they subtend appear, therefore, as sine qua non requirements for cellular existence.

Continuity Principles

A few decades of extensive structural and physiological studies at the cellular and subcellular level of biological organization have generated enough information to justify the conclusion that cellular membranes are endowed with spatial, temporal, and functional continuity.[14]

Spatial Continuity. Each membrane is continuous in three-dimensional space with itself, thereby defining a compartment with uninterrupted boundaries down to the limit of resolution practically attained in biological electron microscopy (i.e., ~1.5 nm). Cellular membranes have no free margins or exposed edges. Discontinuities—when encountered—can reasonably be traced to artifacts of preparation procedures. The spatial continuity of cellular membranes is, in fact, a corollary of the physical chemistry of polar lipid bilayers. Its final expression is a set of intracellular compartments, each defined by a specific membrane, with the entire set enclosed within a continuous plasmalemma.

Temporal Continuity. Cellular membranes are also endowed with temporal continuity. Each cell inherits from its mother a complete set of cellular membranes organized in a complete set of compartments. At present, there is no record of any type of membrane lost during cell division and "regenerated" postmitosis in any daughter cell. The nuclear envelope may look like an exception, since it appears to vanish at each cell division. But, in fact, it peels off the mother nucleus at the end of prophase as large cisternal elements, which reassemble into an envelope around each daughter nucleus at telophase. Spatial continuity is maintained in time, irrespective of the processes to which the cell subjects its membranes: growth, division, remodeling (as in differentiation), or turnover.

Functional Continuity. It can be logically assumed that spatial and temporal continuity exist to ensure functional continuity; that on their account permeability barriers, gradients, and membrane potentials are

[14] G. E. Palade, Membrane biogenesis. *In* "Molecular Specialization and Symmetry in Membrane Function" (A. K. Solomon and M. Karnovsky, eds.), pp. 3–30. Harvard University Press, Cambridge, 1977.

24. Isolation of the Chemical Domains of Human Erythrocyte Spectrin — WILLIAM J. KNOWLES AND MARCIA L. BOLOGNA 305

25. Proteins Involved in Membrane–Cytoskeleton Association in Human Erythrocytes: Spectrin, Ankyrin, and Band 3 — VANN BENNETT 313

26. Assembly of Histocompatibility Antigens — BERNHARD DOBBERSTEIN AND SUNE KVIST 325

27. Biosynthesis and Intracellular Transport of Acetylcholine Receptors — DOUGLAS M. FAMBROUGH 331

28. Acetylcholinesterase Biosynthesis and Transport in Tissue Culture — RICHARD L. ROTUNDO 353

29. Biosynthesis of Acetylcholine Receptor *in Vitro* — DAVID J. ANDERSON AND GÜNTER BLOBEL 367

30. Biosynthesis of Myelin–Specific Proteins — DAVID R. COLMAN, GERT KREIBICH, AND DAVID D. SABATINI 378

31. Sucrase–Isomaltase of the Small-Intestinal Brush Border Membrane: Assembly and Biosynthesis — JOSEF BRUNNER, HANS WACKER, AND GIORGIO SEMENZA 386

32. Structure and Biosynthesis of Aminopeptidases — SUZANNE MAROUX AND HÉLÈNE FERACCI 406

Enveloped Viruses

33. Use of the Heavy-Isotope Density-Shift Method to Investigate Insulin Receptor Synthesis, Turnover, and Processing — MICHAEL N. KRUPP, VICTORIA P. KNUTSON, GABRIELE V. RONNETT, AND M. DANIEL LANE 423

34. Biosynthesis of Myxovirus Glycoproteins with Special Emphasis on Mutants Defective in Glycoprotein Processing — HANS-DIETER KLENK 434

35. Methods for Assay of Cellular Receptors for Picornaviruses — RICHARD L. CROWELL, DAVID L. KRAH, JOHN MAPOLES, AND BURTON J. LANDAU 443

36. Transport of Virus Membrane Glycoproteins, Use of Temperature-Sensitive Mutants and Organelle-Specific Lectins — LEEVI KÄÄRIÄINEN, ISMO VIRTANEN, JAAKKO SARASTE, AND SIRKKA KERÄNEN 453

37. Immunoelectron Microscopy Using Thin, Frozen Sections: Application to Studies of the Intracellular Transport of Semliki Forest Virus Spike Glycoproteins — G. Griffiths, K. Simons, G. Warren, and K. T. Tokuyasu 466

38. Immunocytochemistry of Retinal Membrane Protein Biosynthesis at the Electron Microscopic Level by the Albumin Embedding Technique — Barbara G. Schneider and David S. Papermaster 485

39. Expression of Viral Membrane Proteins from Cloned cDNA by Microinjection into Eukaryotic Cell Nuclei — Beate Timm, Claudia Kondor-Koch, Hans Lehrach, Heimo Riedel, Jan-Erik Edström, and Henrik Garoff 496

40. Biosynthesis of Sindbis Virus Membrane Glycoproteins *in Vitro* — Stefano Bonatti 512

Endoplasmic Reticulum

41. Ribophorins I and II: Membrane Proteins Characteristic of the Rough Endoplasmic Reticulum — Gert Kreibich, Eugene E. Marcantonio, and David D. Sabatini 520

42. Biosynthesis of Hepatocyte Endoplasmic Reticulum Proteins — Gert Kreibich, David D. Sabatini, and Milton Adesnik 530

43. Membrane Induction by Drugs — Gustav Dallner and Joseph W. DePierre 542

44. Preparation of Microsomal β-Glucuronidase and Its Membrane Anchor Protein Egasyn — Aldons J. Lusis 557

Sarcoplasmic Reticulum

45. Biosynthesis of Sarcoplasmic Reticulum Proteins — David H. MacLennan and Stella de Leon 570

Plant Vacuoles

46. Plant Vacuoles — Clarence A. Ryan and Mary Walker-Simmons 580

Nuclear Membrane

47. Preparation of a Nuclear Matrix–Pore Complex–Lamina Fraction from Embryos of *Drosophila melanogaster* — Paul A. Fisher and Günter Blobel 589

48. Proteins of Pore Complex–Lamina Structures from Nuclei and Nuclear Membranes — Georg Krohne and Werner W. Franke — 597

Other Specialized Systems

49. Subcellular Fractionation and Immunochemical Analysis of Membrane Biosynthesis of Photoreceptor Proteins — David S. Papermaster — 609

50. Transducin and the Cyclic GMP Phosphodiesterase of Retinal Rod Outer Segments — Lubert Stryer, James B. Hurley, and Bernard K.-K. Fung — 617

51. Avian Salt Gland: A Model for the Study of Membrane Biogenesis — Russell J. Barrnett, Joseph E. Mazurkiewicz, and John S. Addis — 627

Section II. Targeting: Selected Techniques to Study Transfer of Newly Synthesized Proteins into or across Membranes (Eukaryotic Cells)

52. Control of Intracellular Protein Traffic — Günter Blobel — 663

53. Signal Recognition Particle: A Ribonucleoprotein Required for Cotranslational Translocation of Proteins, Isolation and Properties — Peter Walter and Günter Blobel — 682

54. Proteins Mediating Vectorial Translocation: Purification of the Active Domain of the Endoplasmic Reticulum Docking Protein — David I. Meyer and Bernhard Dobberstein — 692

55. Biosynthesis of Glyoxysomal Proteins — H. Kindl and C. Kruse — 700

56. *In Vitro* Processing of Plant Preproteins — F. A. Burr and B. Burr — 716

57. Biogenesis of Peroxisomal Content Proteins: *In Vivo* and *in Vitro* Studies — Paul B. Lazarow — 721

58. Studies of Lysosomal Enzyme Biosynthesis in Cultured Cells — Rachel Myerowitz, April R. Robbins, Richard L. Proia, G. Gary Sahagian, Christina M. Puchalski, and Elizabeth F. Neufeld — 729

59. Inhibitors of Lysosomal Function — Per O. Seglen — 737

60. Biosynthesis of Lysosomal Enzymes — Melvin G. Rosenfeld, Gert Kreibich, David D. Sabatini, and Keitaro Kato — 764

61. Applications of Amino Acid Analogs for Studying Co- and Posttranslational Modifications of Proteins — Glen Hortin and Irving Boime — 777

62. Quantitative Assay for Signal Peptidase — Robert C. Jackson — 784

63. Fatty Acid Acylation of Eukaryotic Cell Proteins — Milton J. Schlesinger — 795

64. Yeast Secretory Mutants: Isolation and Characterization — Randy Schekman, Brent Esmon, Susan Ferro-Novick, Charles Field, and Peter Novick — 802

65. Secretory Mutants in the Cellular Slime Mold *Dictyostelium discoideum* — Randall L. Dimond, David A. Knecht, Kevin B. Jordan, Robert A. Burns, and George P. Livi — 815

Author Index 829

Subject Index 867

Contributors to Volume 96

Article numbers are in parentheses following the names of contributors. Affiliations listed are current.

JOHN S. ADDIS (51), *Department of Anatomy, University of Mississippi School of Medicine, Jackson, Mississippi 39216*

MILTON ADESNIK (42), *Department of Cell Biology, New York University School of Medicine, New York, New York 10016*

DAVID J. ANDERSON (8, 29), *Institute of Cancer Research, Columbia University College of Physicians and Surgeons, New York, New York 10032*

LEIF C. ANDERSSON (22), *Department of Pathology and Transplantation Laboratory, University of Helsinki, 00290, Helsinki 17, Finland*

MONIQUE ARPIN (9), *Unité Biologie des Membranes, Department Biologie Moleculaire, Institut Pasteur, 75724 Paris Cedex 15, France*

RUSSELL J. BARRNETT (51), *Department of Cell Biology, Yale University School of Medicine, New Haven, Connecticut 06510*

VANN BENNETT (25), *Department of Cell Biology and Anatomy, The Johns Hopkins University School of Medicine, Baltimore, Maryland 21205*

GÜNTER BLOBEL (3, 6, 8, 29, 47, 52, 53), *Laboratory of Cell Biology, The Rockefeller University, New York, New York 10021*

IRVING BOIME (61), *Department of Pharmacology, Washington University School of Medicine, St. Louis, Missouri 63110*

MARCIA L. BOLOGNA (24), *Department of Pathology, Yale University School of Medicine, New Haven, Connecticut 06510*

STEFANO BONATTI (40), *Istituto di Biochimica Cellulare e Molecolare, II Facoltà di Medicina e Chirurgia, Via Pansini 5, 80131 Napoli, Italy*

WILLIAM M. BONNER (15), *Laboratory of Molecular Pharmacology, National Cancer Institute, National Institutes of Health, Bethesda, Maryland 20205*

WILLIAM A. BRAELL (20), *Department of Biochemistry, Stanford University School of Medicine, Stanford, California 94305*

JOSEF BRUNNER (31), *Laboratorium für Biochemie, ETH-Zentrum, CH-8092 Zurich, Switzerland*

ROBERT A. BURNS (65), *Department of Genetics, University of Wisconsin, Madison, Wisconsin 53706*

B. BURR (18, 56), *Department of Biology, Brookhaven National Laboratory, Upton, New York 11973*

F. A. BURR (18, 56), *Department of Biology, Brookhaven National Laboratory, Upton, New York 11973*

DON W. CLEVELAND (16), *Department of Physiological Chemistry, The Johns Hopkins University School of Medicine, Baltimore, Maryland 21205*

DAVID R. COLMAN (30), *Department of Cell Biology, New York University School of Medicine, New York, New York 10016*

CAROLYN A. CONVERSE (19), *Department of Pharmacy, University of Strathclyde, Glasgow G1 1XW, Scotland*

RICHARD L. CROWELL (35), *Department of Microbiology and Immunology, Hahnemann University School of Medicine, Philadelphia, Pennsylvania 19102*

GUSTAV DALLNER (43), *Department of Pathology at Huddinge Hospital, Karolinska Institutet Medical School, S-14186 Huddinge, Sweden*

STELLA DE LEON (45), *Banting and Best Department of Medical Research, C. H. Best Institute, University of Toronto, Toronto, Ontario M5G IL6, Canada*

JOSEPH W. DEPIERRE (43), *Department of Biochemistry, Arrhenius Laboratory, University of Stockholm, S-10691 Stockholm, Sweden*

RANDALL L. DIMOND (65), *Department of Bacteriology, University of Wisconsin, Madison, Wisconsin 53706*

BERNHARD DOBBERSTEIN (26, 54), *European Molecular Biology Laboratory, D-6900 Heidelberg, Federal Republic of Germany*

JAN-ERIK EDSTRÖM (39), *European Molecular Biology Laboratory, D-6900 Heidelberg, Federal Republic of Germany*

ANN H. ERICKSON (3), *Laboratory of Cell Biology, The Rockefeller University, New York, New York 10021*

BRENT ESMON (64), *Department of Anatomy, University of California, San Francisco, California 94720*

DOUGLAS M. FAMBROUGH (27), *Department of Embryology, Carnegie Institution of Washington, Baltimore, Maryland 21210*

RICARDO A. FELDMAN (1), *Department of Viral Oncology, The Rockefeller University, New York, New York 10021*

HÉLÈNE FERACCI (32), *Centre de Biochimie et de Biologie, Moléculaire-CNRS, 13402 Marseille Cedex 9, France*

SUSAN FERRO-NOVICK (64), *Department of Microbiology and Molecular Genetics, Harvard Medical School, Boston, Massachusetts 02115*

CHARLES FIELD (64), *Department of Biochemistry, University of California, Berkeley, California 94720*

PAUL A. FISHER (47), *Department of Pharmacological Sciences, State University of New York at Stony Brook, Stony Brook, New York 11794*

WERNER W. FRANKE (48), *Division of Membrane Biology and Biochemistry, Institute of Cell and Tumor Biology, German Cancer Research Center, D-6900 Heidelberg, Federal Republic of Germany*

BERNARD K.-K. FUNG (50), *Department of Radiation Biology and Biophysics, University of Rochester, Rochester, New York 14642*

HEINZ FURTHMAYR (21), *Department of Pathology, Yale University School of Medicine, New Haven, Connecticut 06510*

SANCIA GAETANI (1, 9), *Istituto Nazionale della Nutrizione, Rome, Italy*

CARL G. GAHMBERG (22), *Department of Biochemistry, University of Helsinki, 00170, Helsinki 17, Finland*

HENRIK GAROFF (39), *European Molecular Biology Laboratory, D-6900 Heidelberg, Federal Republic of Germany*

DAVID GOLDMAN (17), *National Institute on Alcohol Abuse and Alcoholism, Bethesda, Maryland 20205*

PAMELA GREEN (5), *Department of Biochemistry, State University of New York at Stony Brook, Stony Brook, New York 11794*

G. GRIFFITHS (37), *European Molecular Biology Laboratory, D-6900 Heidelberg, Federal Republic of Germany*

ALBERT HAID (13), *Institut für Genetik und Mikrobiologie der Universität, München, D-8000 München 19, Federal Republic of Germany*

WALLACE B. HAIGH, JR. (23), *Department of Protein Chemistry, Molecular Diagnostics, Inc., West Haven, Connecticut 06510*

GLEN HORTIN (61), *Department of Pharmacology, Washington University School of Medicine, St. Louis, Missouri 63110*

TIM HUNT (4), *Department of Biochemistry, University of Cambridge, Cambridge CB2 1QW, England*

JAMES B. HURLEY (50), *Biology Division, California Institute of Technology, Pasadena, California 91125*

MASAYORI INOUYE (5), *Department of Biochemistry, State University of New York at Stony Brook, Stony Brook, New York 11794*

RICHARD J. JACKSON (4), *Department of Biochemistry, University of Cambridge, Cambridge CB2 1QW, England*

ROBERT C. JACKSON (62), *Department of Biochemistry, Dartmouth Medical School, Hanover, New Hampshire 03756*

MIKKO JOKINEN (22), *Department of Biochemistry, University of Helsinki, 00170, Helsinki 17, Finland*

KEVIN B. JORDAN (65), *Department of Genetics, University of Wisconsin, Madison, Wisconsin 53706*

LEEVI KÄÄRIÄINEN (36), *Recombinant DNA Laboratory, University of Helsinki, SF-00380, Helsinki 38, Finland*

OLLE KÄMPE (22), *Department of Cell Research, University of Uppsala, Uppsala S-75122, Sweden*

KIMMO K. KARHI (22), *Department of Biochemistry, University of Helsinki, 00170, Helsinki 17, Finland*

KEITARO KATO (60), *Department of Biochemistry, Kyushu University, Fukuoka 812, Japan*

SIRKKA KERÄNEN (36), *Recombinant DNA Laboratory, University of Helsinki, SF-00380 Helsinki 38, Finland*

H. KINDL (55), *Biochemie, Fachbereich Chemie der Philipps-Universität, D-3550 Marburg, Federal Republic of Germany*

HANS-DIETER KLENK (34), *Institut für Virologie, Justus-Liebig-Universität, D-6300 Giessen, Federal Republic of Germany*

DAVID A. KNECHT (65), *Department of Biology, University of California at San Diego, La Jolla, California 92037*

WILLIAM J. KNOWLES (24), *Department of Pathology, Yale University School of Medicine, New Haven, Connecticut 06510*

VICTORIA P. KNUTSON (33), *Department of Physiological Chemistry, The Johns Hopkins University School of Medicine, Baltimore, Maryland 21205*

CLAUDIA KONDOR-KOCH (39), *European Molecular Biology Laboratory, D-6900 Heidelberg, Federal Republic of Germany*

DAVID L. KRAH (35), *Department of Virology, The Rockefeller University, New York, New York 10021*

GERT KREIBICH (30, 41, 42, 60), *Department of Cell Biology, New York University School of Medicine, New York, New York 10016*

GEORG KROHNE (48), *Division of Membrane Biology and Biochemistry, Institute of Cell and Tumor Biology, German Cancer Research Center, D-6900 Heidelberg, Federal Republic of Germany*

MICHAEL N. KRUPP (33), *Department of Metabolic Diseases, Pfizer Inc., Central Research Division, Groton, Connecticut 06340*

C. KRUSE (55), *Biochemie, Fachbereich Chemie der Philipps-Universität, D-3550 Marburg, Federal Republic of Germany*

SUNE KVIST (26), *European Molecular Biology Laboratory, D-6900 Heidelberg, Federal Republic of Germany*

BURTON J. LANDAU (35), *Department of Microbiology and Immunology, Hahnemann University School of Medicine, Philadelphia, Pennsylvania 19102*

M. DANIEL LANE (33), *Department of Physiological Chemistry, The Johns Hopkins University School of Medicine, Baltimore, Maryland 21205*

PAUL B. LAZAROW (57), *The Rockefeller University, New York, New York 10021*

HANS LEHRACH (39), *European Molecular Biology Laboratory, D-6900 Heidelberg, Federal Republic of Germany*

C. L. LEONARDI (12), *Department of Anatomy and Cell Biology, University of Miami School of Medicine, Miami, Florida 33101*

GEORGE P. LIVI (65), *Department of Bacteriology, University of Wisconsin, Madison, Wisconsin 53706*

PAUL M. LIZARDI (2), *Department of Cell Biology, The Rockefeller University, New York, New York 10021*

HARVEY F. LODISH (20), *Department of Biology, Massachusetts Institute of Technology, Cambridge, Massachusetts 02139*

ALDONS J. LUSIS (44), *Departments of Medicine and Microbiology, University of California, Los Angeles, California 90024*

DAVID H. MACLENNAN (45), *Banting and Best Department of Medical Research, C. H. Best Institute, University of Toronto, Toronto, Ontario M5G 1L6, Canada*

JOHN MAPOLES (35), *Department of Microbiology and Immunology, Hahnemann University School of Medicine, Philadelphia, Pennsylvania 19102*

EUGENE E. MARCANTONIO (41), *Department of Cell Biology, New York University School of Medicine, New York, New York 10016*

VINCENT T. MARCHESI (21), *Department of Pathology, Yale University School of Medicine, New Haven, Connecticut 06510*

SUZANNE MAROUX (32), *Centre de Biochimie et de Biologie, Moléculaire-CNRS, 13402 Marseille Cedex 9, France*

JOSEPH E. MAZURKIEWICZ (51), *Department of Anatomy, Albany Medical School, Albany, New York 12208*

CARL R. MERRIL (17), *Laboratory of General and Comparative Biochemistry, National Institute of Mental Health, Bethesda, Maryland 20205*

DAVID I. MEYER (54), *Cell Biology Program, European Molecular Biology Laboratory, D-6900 Heidelberg, Federal Republic of Germany*

TAKASHI MORIMOTO (1, 9), *Department of Cell Biology, New York University Medical Center, New York, New York 10016*

JON S. MORROW (23), *Department of Pathology, Yale University School of Medicine, New Haven, Connecticut 06510*

RACHEL MYEROWITZ (58), *Genetics and Biochemistry Branch, National Institute of Arthritis, Diabetes, Digestive and Kidney Diseases, National Institutes of Health, Bethesda, Maryland 20205*

ELIZABETH F. NEUFELD (58), *Genetics and Biochemistry Branch, National Institute of Arthritis, Diabetes, Digestive and Kidney Diseases, National Institutes of Health, Bethesda, Maryland 20205*

PETER NOVICK (64), *Department of Biology, Massachusetts Institute of Technology, Cambridge, Massachusetts 02139*

GEORGE E. PALADE (xxix), *Section of Cell Biology, Yale University School of Medicine, New Haven, Connecticut 06510*

RICHARD D. PALMITER (10), *Department of Biochemistry, University of Washington, Seattle, Washington 98195*

DAVID S. PAPERMASTER (19, 38, 49), *Department of Pathology, Yale University School of Medicine, New Haven, Connecticut 06510*

PER A. PETERSON (22), *Department of Cell Research, University of Uppsala, Uppsala S-75122, Sweden*

RICHARD L. PROIA (58), *Genetics and Biochemistry Branch, National Institute of Arthritis, Diabetes, Digestive and Kidney Diseases, National Institutes of Health, Bethesda, Maryland 20205*

CHRISTINA M. PUCHALSKI (58), *Genetics and Biochemistry Branch, National Institute of Arthritis, Diabetes, Digestive and Kidney Diseases, National Institutes of Health, Bethesda, Maryland 20205*

JAKOB REISER (14), *Institut für Molekularbiologie I, Universitat Zurich, 8093 Zurich, Hönggerberg, Switzerland*

HEIMO RIEDEL (39), *European Molecular Biology Laboratory, D-6900 Heidelberg, Federal Republic of Germany*

APRIL R. ROBBINS (58), *Genetics and Biochemistry Branch, National Institute of Arthritis, Diabetes, Digestive and Kidney Diseases, National Institutes of Health, Bethesda, Maryland 20205*

GABRIELE V. RONNETT (33), *Department of Physiological Chemistry, The Johns Hopkins University School of Medicine, Baltimore, Maryland 21205*

MELVIN G. ROSENFELD (60), *Department of*

Cell Biology, New York University School of Medicine, New York, New York 10016

RICHARD L. ROTUNDO (28), *Department of Embryology, Carnegie Institution of Washington, Baltimore, Maryland 21210*

R. W. RUBIN (12), *Department of Anatomy and Cell Biology, University of Miami School of Medicine, Miami, Florida 33101*

CLARENCE A. RYAN (46), *Institute of Biological Chemistry, Washington State University, Pullman, Washington 99164-6340*

DAVID D. SABATINI (30, 41, 42, 60), *Department of Cell Biology, New York University School of Medicine, New York, New York 10016*

G. GARY SAHAGIAN (58), *Genetics and Biochemistry Branch, National Institute of Arthritis, Diabetes, Digestive and Kidney Diseases, National Institutes of Health, Bethesda, Maryland 20205*

JAAKKO SARASTE (36), *Department of Virology, University of Helsinki, SF-00290 Helsinki 29, Finland*

GEORGE SCHEELE (7), *Laboratory of Cell and Molecular Biology, The Rockefeller University, New York, New York 10021*

RANDY SCHEKMAN (64), *Department of Biochemistry, University of California, Berkeley, California 94720*

MILTON J. SCHLESINGER (63), *Department of Microbiology and Immunology, Washington University School of Medicine, St. Louis, Missouri 63110*

BARBARA G. SCHNEIDER (38), *Department of Pathology, Yale University School of Medicine, New Haven, Connecticut 06510*

PER O. SEGLEN (59), *Department of Tissue Culture, Norsk Hydro's Institute for Cancer Research, The Norwegian Radium Hospital, Montebello, Oslo 3, Norway*

GIORGIO SEMENZA (31), *Laboratorium für Biochemie, ETH-Zentrum, CH-8092 Zurich, Switzerland*

K. SIMONS (37), *European Molecular Biology Laboratory, D-6900 Heidelberg, Federal Republic of Germany*

JULIA A. SMITH (1), *Peter Bent Brigham Hospital, Brookline, Massachusetts 02146*

PAMELA STANLEY (11), *Department of Cell Biology, Albert Einstein College of Medicine, Bronx, New York 10461*

GEORGE R. STARK (14), *Department of Biochemistry, Stanford University School of Medicine, Stanford, California 94305*

LUBERT STRYER (50), *Department of Structural Biology, Stanford University School of Medicine, Stanford, California 94305*

MORDECHAI SUISSA (13), *Department of Biochemistry, Biocenter, University of Basel, CH-4056 Basel, Switzerland*

BEATE TIMM (39), *European Molecular Biology Laboratory, D-6900 Heidelberg, Federal Republic of Germany*

K. T. TOKUYASU (37), *Department of Biology, University of California at San Diego, La Jolla, California 92093*

MARGARET L. VAN KEUREN (17), *Laboratory of General and Comparative Biochemistry, National Institute of Mental Health, Bethesda, Maryland 20205*

ISMO VIRTANEN (36), *Department of Pathology, University of Helsinki, SF-00290 Helsinki 29, Finland*

HANS WACKER (31), *Laboratorium für Biochemie, ETH-Zentrum, CH-8092 Zurich, Switzerland*

MARY WALKER-SIMMONS (46), *Institute of Biological Chemistry, Washington State University, Pullman, Washington 99164-6340*

PETER WALTER (6, 53), *Department of Biochemistry and Biophysics, University of California School of Medicine, San Francisco, California 94143*

G. WARREN (37), *European Molecular Biology Laboratory, D-6900 Heidelberg, Federal Republic of Germany*

Preface

Volumes 96 to 98, Parts J, K, and L of the Biomembranes series, focus on methodology to study membrane biogenesis, assembly, targeting, and recycling. This field is one of the very exciting and active areas of research. Future volumes will deal with transport and other aspects of membrane function.

We were fortunate to have the advice and good counsel of our Advisory Board. Additional valuable input to this volume was obtained from Drs. Vincent T. Marchesi, Harvey F. Lodish, and Keith Mostov. We were gratified by the enthusiasm and cooperation of the participants in the field whose contributions and suggestions have enriched and made possible these volumes. The friendly cooperation of the staff of Academic Press is gratefully acknowledged.

SIDNEY FLEISCHER
BECCA FLEISCHER

METHODS IN ENZYMOLOGY

EDITED BY

Sidney P. Colowick and Nathan O. Kaplan

VANDERBILT UNIVERSITY
SCHOOL OF MEDICINE
NASHVILLE, TENNESSEE

DEPARTMENT OF CHEMISTRY
UNIVERSITY OF CALIFORNIA
AT SAN DIEGO
LA JOLLA, CALIFORNIA

I. Preparation and Assay of Enzymes
II. Preparation and Assay of Enzymes
III. Preparation and Assay of Substrates
IV. Special Techniques for the Enzymologist
V. Preparation and Assay of Enzymes
VI. Preparation and Assay of Enzymes (*Continued*)
Preparation and Assay of Substrates
Special Techniques
VII. Cumulative Subject Index

maintained throughout the existence of each cell and transmitted without interruption through cellular and organismic generations. Extrapolating to the origin, it can be assumed that cellular membranes have functioned continuously through evolution in all successful (i.e., surviving) cases since the very beginning of cellular life.

Sites of Synthesis of Membrane Components

Polar Lipids. In animal eukaryotes, the lipid bilayers of all cellular membranes consist of the same classes and families of phospholipids. Differences from one type of membrane to another are essentially quantitative.[15] The only exception is diphosphatidylglycerol, present and synthesized only in the inner mitochondrial membrane. All other phospholipids are synthesized by enzymes located apparently exclusively in the membrane of the endoplasmic reticulum (ER)[16,17] from soluble precursors (glycerolphosphate, fatty acids, ethanolamine, etc.) imported into—or generated within—the cytosol. From this unique source, phospholipids must be transported to all other cellular membranes by means that are still poorly understood. The cytosol contains a number of specific phospholipid exchange proteins[18] which—under certain experimental conditions—may effect net transport,[19] as opposed to exchange, but their functional role in living cells is still uncertain. Experimentally, net transport of phospholipids can be demonstrated among polar lipid bilayers (organized in vesicles) in the absence of exchange proteins; in such cases, transport is presumably mediated by molecules or micelles in solution.[20] Transport of gangliosides from micelles to phospholipid vesicles does also occur *in vitro*.[21] Still another possibility is that polar lipids diffuse laterally from the ER membrane to other cellular membranes (i.e., Golgi membranes

[15] S. Fleischer and B. Fleischer, Membrane diversity in the rat hepatocyte. *In* "Membrane Alterations as Basis of Liver Injury" (H. Popper, ed.), pp. 31–47. MTP Press, Ltd., Lancaster, England, 1977.

[16] R. M. Bell and R. M. Coleman, Enzymes of glycerolipid synthesis in eukaryotes. *Ann. Rev. Biochem.* **48**, 47–71 (1980).

[17] R. M. Bell, L. M. Ballas, and R. A. Coleman, Lipid topogenesis. *J. Lipid Res.* **22**, 391–403 (1981).

[18] K. W. A. Wirtz, Transfer of phospholipids between membranes. *Biochim. Biophys. Acta* **344**, 95–117 (1974).

[19] D. B. Zilversmit and M. A. Hughes, Extensive exchange of rat liver microsomal phospholipids. *Biochim. Biophys. Acta* **469**, 99–110 (1977).

[20] M. A. Roseman and T. E. Thompson, Mechanism of the spontaneous transfer of phospholipids between bilayers. *Biochemistry* **19**, 439–444 (1980).

[21] P. L. Felgner, T. E. Thompson, Y. Barenholz, and D. Lichtenberg, Kinetics of transfer of gangliosides from their micelles to dipalmitoylphosphatidylcholine vesicles. *Biochemistry* **22**, 1670–1674 (1983).

and eventually the plasmalemma) when the corresponding bilayers become continuous as a result of membrane fusion. Lateral diffusion of polar lipids in cellular membranes is known to be rapid[22] and may occur even when membrane proteins are not mobile.[23] Yet, all the transport mechanisms so far envisaged cannot explain the differences known to exist in the relative concentration of different polar lipids among cellular membranes. And thus we are left with the assumption that, within the limits defined by the biosynthetic capabilities of each cell, membrane proteins control the lipid composition of the corresponding membranes either by protein–phospholipid interactions or by subsequent phospholipid modifications.

Cholesterol, the other common component of lipid bilayers, is synthesized through a long series of reactions by enzymes located either in the cytosol or the ER membrane. The key enzyme in sterol synthesis, 3-hydroxy-3-methyl glutaryl CoA reductase, is an integral ER membrane protein,[cf. 24] and the enzymes involved in squalene cyclization and in the final steps of cholesterol synthesis are associated with the same membrane. So, in this case we are again faced with the same problem: how is cholesterol transported from a unique site of production to a multiplicity of destinations,[24a] and how are characteristic differences in relative cholesterol concentration maintained among different cellular membranes. This is, in fact, a recurring theme in membrane biogenesis: the cells operate unique biosynthetic sites and channel the products to a multiplicity of differentiated destinations. In the case of membrane lipids, there are few exceptions to this pattern. Recent data suggest that sphingomyelin is produced mainly in the plasmalemma.[26] Enzymes that can modify phospholipids by transacylation are found in practically all cellular mem-

[22] J. Schlessinger, D. Axelrod, D. E. Koppel, W. W. Webb, and E. L. Elson, Lateral transport of a lipid probe and labelled proteins on a cell membrane. *Science (Washington, D. C.)* **195**, 307–309 (1977).

[23] E. L. Elson, Regulation of the lateral mobility of cell surface proteins by interactions with the cytoskeleton. *In* "New Perspective on Membrane Dynamics" (A. Waksman, ed.), pp. 84–87. Strasbourg, France, 1983.

[24] D. J. Chin, K. L. Luskey, R. G. W. Anderson, J. R. Faust, J. L. Goldstein, and M. S. Brown, Appearance of crystalloid endoplasmic reticulum in compactin-resistant Chinese hamster cells with a 500-fold elevation of 3-hydroxy-3-methylglutaryl coenzyme A reductase. *Proc. Natl. Acad. Sci. U.S.A.* **79**, 1185–1189 (1982).

[24a] A cholesterol exchange protein has also been found in the cytosol.[25]

[25] B. Bloj and D. B. Zilversmit, Rat liver protein capable of transferring phosphatidylethanolamine. Purification and transfer activity for other phospholipids and cholesterol. *J. Biol. Chem.* **252**, 1613–1619 (1977).

[26] D. R. Voelker and E. P. Kennedy, Cellular and enzymic synthesis of sphingomyelin. *Biochemistry* **21**, 2753–2759 (1982).

branes, but transmethylases that convert phosphatidylethanolamine to phosphatidylcholine (and thereby increase the fluidity of the bilayer) are active in the plasmalemma.[27] In plant cells in which galactolipids are major components of chloroplast (thylakoid) membranes, the cognate enzymes are located in the membranes of the chloroplast envelope.[28]

Membrane Proteins. In animal eukaryotes, the equipment involved in the synthesis of all cellular proteins, membrane proteins included, is concentrated essentially within a single cell compartment, the cytosol or cytoplasmic matrix. Up to 98% of the total protein production is carried out by cytoplasmic polysomes which, in the course of their activity, are found either free in the cytosol (free polysomes) or attached to the ER membrane (attached or bound polysomes).[29] The remaining ~2% is accounted for by the activity of mitochondrial ribosomes located in the mitochondrial matrix.[29a]

Membrane Glycoproteins. The carriers and the enzymic equipment involved in the synthesis of N-linked polymannose chains and in their conjugation to nascent polypeptide chains are concentrated in a single cellular membrane, that of the RER.[30] The enzymes that convert these polymannose chains into complex oligosaccharide chains (by removal of mannosyl residues and terminal glycosylation of the residual core) are restricted, however, to the Golgi complex,[31–34] but within the latter only

[27] F. Hirata and J. Axelrod, Enzymatic synthesis and rapid translocation of phosphatylcholine by two methyltransferases in erythrocyte membranes. *Proc. Natl. Acad. Sci. U.S.A.* **75**, 2348–2352 (1978).

[28] J. Joyard and R. Douce, Site of synthesis of phosphatidic acid and diacylglycerol in spinach chloroplasts. *Biochim. Biophys. Acta* **486**, 273–285 (1977).

[29] G. E. Palade, Microsomes and ribonucleoprotein particles. *In* "Microsomal Particles and Protein Synthesis" (R. B. Roberts, ed.), pp. 36–61. Pergamon, New York, 1958.

[29a] Plant eukaryotes have a third ribosomal population restricted to the chloroplast whose activity accounts for 10–20% of the total protein production. The relative output of their cytoplasmic polysomes is commensurately reduced.

[30] D. K. Struck and W. J. Lennarz, The function of saccharide lipid in synthesis of glycoproteins. *In* "The Biochemistry of Glycoproteins and Proteoglycans" (W. Lennarz, ed.), pp. 35–83. Plenum, New York, 1980.

[31] R. Kornfeld and S. Kornfeld, Structure of glycoproteins and their oligosaccharide units. *In* "Biochemistry of Glycoproteins and Proteoglycans" (W. Lennarz, ed.), pp. 1–34. Plenum, New York, 1980.

[32] B. Fleischer and S. Fleischer, Preparation and characterization of Golgi membranes from rat liver. *Biochim. Biophys. Acta* **219**, 301–319 (1970).

[33] R. Bretz, H. Bretz, and G. E. Palade, Distribution of terminal glycosyltransferases in hepatic Golgi fractions. *J. Cell Biol.* **84**, 87–101 (1980).

[34] H. Schachter and S. Roseman, Mammalian glycosyltransferases: their role in the synthesis and function of complex carbohydrates and glycolipids. *In* "Biochemistry of Glycoproteins and Proteoglycans" (W. Lennarz, ed.), pp. 85–160. Plenum, New York, 1980.

the localization of galactosyltransferase is well established.[35] Sulfation of glycoproteins, proteoglycans, and sulfatides is also limited to the Golgi complex.[36,cf. 37] Other similarly restricted modifications include the phosphorylation of mannosyl residues[38] in polymannose chains, and probably the synthesis of O-linked oligosaccharide chains for glycoproteins and proteoglycans.[34]

Consequences of Biosynthetic Monopolies

The existence of a protein biosynthetic monopoly (or near-monopoly) introduces a number of significant restrictions in our attempts to analyze and eventually understand the ways in which cells carry out membrane biogenesis.

- Traffic of membrane proteins must be controlled since from a common site of synthesis, the cytosol, proteins must be directed to a multiplicity of specific sites of final functional residence. Transport must be specifically and efficiently directed, since cellular membranes differ in their protein composition both quantitatively and qualitatively, a situation clearly different from that encountered in the case of membrane lipids. Traffic regulation must also apply to soluble proteins directed to certain intracellular compartments, such as those represented by the ER and Golgi cisternal space, and those found in peroxisomes, lysosomes, mitochondria, and (in plants) chloroplasts.
- Partial or complete translocation of proteins across the lipid bilayer has been successfully mastered by all cells, notwithstanding the thermodynamically unfavored nature of the operation. Translocation is incurred by practically all proteins synthesized in the cytosol, with the exception of those destined to the nucleoplasm or the cytosol itself.
- All glycoconjugates are expected to follow a common route that takes them from the ER (where they are proximally glycosylated) through the Golgi complex (for terminal glycosylation and O-linked glycosylation) to their final destinations.

[35] J. Roth and E. G. Berger, Immunocytochemical localization of galactosyltransferase activity in HeLa cells: codistribution with thiamine pyrophosphatase in *trans*-Golgi cisternae. *J. Cell Biol.* **93**, 223–229 (1982).

[36] B. Fleischer and F. Zambrano, Golgi apparatus of rat kidney. Separation and role in sulfatide formation. *J. Biol. Chem.* **249**, 5995–6003 (1974).

[37] R. W. Young, The role of the Golgi complex in sulfate metabolism. *J. Cell Biol.* **57**, 175–189 (1973).

[38] S. Kornfeld, M. L. Reitman, A. Varki, D. Goldberg, and C. A. Gabel, Steps in the phosphorylation of the high mannose oligosaccharides of lysosomal enzymes. *In* "Membrane Recycling" (Ciba Foundation Symposium 92) (D. Evered and G. M. Collins, eds.), pp. 138–156. Pitman, London, 1982.

- All sulfoconjugates—including proteoglycans—must follow a common route that takes them in transit through the Golgi complex.
- Traffic pathways may initially involve common steps reflecting common modifications, but may diverge thereafter for subsequent specific modifications to which different products (membrane proteins included) are subjected.

Mechanisms Involved in Traffic Regulation

Signal Sequences. In principle, traffic control can rely on mutual recognition between a signal on the protein to be transported and a signal-recognition element at the site of its initial destination. The existence of a signal sequence on secretory proteins was first postulated by Blobel and Sabatini,[39] but the first evidence suggestive of events and mechanisms involved in traffic control came from experiments carried out by Milstein *et al.*[40] in an *in vitro* system in which mRNA for light IgG chains was translated in the presence or absence of membranes. The primary translate was found to be smaller by ~2 kilodaltons than the known secretory product. This line of investigation was subsequently fully developed by Blobel and his collaborators[41,42] using *in vitro* experiments in which a variety of mRNAs were translated in heterologous reconstituted systems. The results led to the identification of a signal sequence of 15–30 amino acid residues at the N terminal of the original translate, and to the realization that the information it contains is not encoded in its primary structure (which characteristically comprises a core of hydrophobic residues), but probably in its tertiary structure. The signal may consist of specific residues disposed in space at critical distances from one another. Similar sequences have been found in prokaryotic cells[43] in which genetic manipulations are currently extensively used to define mandatory residues and

[39] G. Blobel and D. D. Sabatini, Ribosome-membrane interactions in eukaryotic cells. *In* "Biomembranes" (L. A. Mason, ed.), pp. 193–195. Plenum, New York, 1971.

[40] C. Milstein, G. G. Brownlee, T. M. Harrison, and M. B. Mathews, A possible precursor of immunoglobulin light chains. *Nature (London), New Biol.* **239**, 117–120 (1972).

[41] G. Blobel and B. Dobberstein, Transfer of proteins across membranes. I. Presence of proteolytically processed and unprocessed nascent immunoglobulin light chains on membrane-bound ribosomes of murine myeloma. II. Reconstitution of functional rough microsomes from heterologous components. *J. Cell Biol.* **67**, 835–851, 852–862 (1975).

[42] G. Blobel, P. Walter, C. N. Chang, B. M. Goldman, A. H. Erickson, and V. R. Lingappa, Translocation of proteins across membranes. The signal hypothesis and beyond. *In* "Secretory Mechanisms" (C. R. Hopkins and C. J. Duncan, eds.), pp. 9–36. Cambridge Univ. Press, Cambridge, 1979.

[43] S. D. Emr, M. N. Hall, and T. J. Silhavy, A mechanism of protein localization: the signal hypothesis in bacteria. *J. Cell Biol.* **86**, 701–711 (1980).

critical positions within the primary structure of the signals.[44] Signal sequences are recognized by competent membranes and their recognition leads to the translocation of the nascent polypeptides into the ER cisternal space. The signals are generally (but not obligatorily) removed cotranslationally by a signal peptidase as translocation proceeds.[41–43]

The same type of signal sequences is found in the primary translates of secretory proteins, membrane proteins[45,46] (including glycoproteins of enveloped virus[47,48]), and lysosomal proteins,[49,50] which can compete with each other for common recognition and translocation sites. These findings clearly establish that this lane of protein traffic is controlled in a succession of steps and that signal sequences recognize in the ER membrane the first step on an initially common pathway, irrespective of the final destination of the individual proteins involved in the process.

Signal Recognition Particle (SRP). More recent work carried out by Walter and Blobel[51] has identified a signal recognition protein (SRP) complex or particle that consists of 6 polypeptides and one small (7S) RNA molecule.[52] SRP requires low detergent concentrations for maximal activ-

[44] S. D. Emr and T. J. Silhavy, Molecular components of the signal sequence that function in the initiation of protein export. *J. Cell Biol.* **95**, 689–696 (1982).

[45] B. Dobberstein, H. Garoff, and G. Warren, Cell-free synthesis and membrane insertion of mouse H-2D^d histocompatibility antigen and β_2-microglobulin. *Cell* **17**, 759–769 (1979).

[46] D. J. Anderson and G. Blobel, *In vitro* synthesis, glycosylation and membrane insertion of the four subunits of *Torpedo* acetylcholine receptor. *Proc. Natl. Acad. Sci. U.S.A.* **78**, 5598–5602 (1981).

[47] V. R. Lingappa, F. N. Katz, H. F. Lodish, and G. Blobel, A signal sequence for the insertion of a transmembrane glycoprotein. *J. Biol. Chem.* **253**, 8667–8670 (1978).

[48] R. A. Irving, F. Toneguzzo, S. H. Rhee, T. Hofmann, and H. P. Gosh, Synthesis and assembly of membrane glycoproteins: presence of leader peptide in nonglycosylated precursor of membrane glycoprotein of vesicular stomatis virus. *Proc. Natl. Acad. Sci. U.S.A.* **76**, 570–574 (1979).

[49] A. H. Erickson and G. Blobel, Early events in the biosynthesis of the lysosomal enzyme cathepsin D. *J. Biol. Chem.* **254**, 11771–11774 (1979).

[50] M. G. Rosenfeld, G. Kreibich, D. Popov, K. Kato, and D. D. Sabatini, Biosynthesis of lysosomal hydrolases: their synthesis on bound polysomes and the role of co- and post-translational processing in determining their subcellular distribution. *J. Cell Biol.* **93**, 135–143 (1982).

[51] P. Walter and G. Blobel, Translocation of proteins across the endoplasmic reticulum. II. Signal recognition protein (SRP) mediates the selective binding to microsomal membranes of *in vitro* assembled polysomes synthesizing secretory proteins. III. The signal recognition protein (SRP) causes signal sequence dependent and site-specific arrest of chain elongation that is released by microsomal membranes. *J. Cell Biol.* **91**, 551–556, 557–561 (1981).

[52] P. Walter and G. Blobel, Signal recognition particle contains a 7S RNA essential for protein translocation across the endoplasmic reticulum. *Nature (London)* **299**, 691–698 (1982).

ity. It is distributed at equilibrium among ER membrane, cytosol, and ribosomes, but it has considerably higher affinity for ribosomes programmed by mRNAs for secretory proteins. In the absence of competent membranes, it arrests reversibly polypeptide elongation, thereby relieving the cell of the necessity of coordinating the synthesis of mRNAs with the availability of translocation sites.[52a] SRP recognizes both secretory and membrane proteins,[53] and it is the first cytoplasmic, small ribonucleoprotein particle to which a function can be ascribed. SRP interacts with a 72-kilodalton integral protein of the ER membrane (called "docking protein")[54] and the series of interactions mentioned above eventually leads to partial or complete translocation across the ER membrane of proteins provided with signal sequences.

Translocation. It is clear, however, that the translocation apparatus of the ER membrane must be more complex than the sum of the components thus far uncovered. The translocase itself has not yet been identified. Pore-forming proteins have been postulated[41,42] but not defined. Other components that may play a role in the series of recognition–translocation reactions include the ribophorins, integral glycoproteins known to be involved in the attachment of ribosomes to the ER membrane.[55]

Translocation amounts to the transfer of a predominantly hydrophilic macromolecule across a hydrophobic barrier, an operation that, in principle, requires considerable energy expenditure. In the cases so far investigated, energy demands are minimized, however, since the proteins negotiate the barrier as nascent peptides, i.e., at a stage when two of their dimensions are minimal (~0.5 nm). The energy input has been estimated[56] and found to be manageable. Moreover, it has been proposed that nascent peptides may be spontaneously inserted into membranes as helical hairpins, so that the solvation of hydrophobic sequences may provide the

[52a] It does not bind to ribosomes synthesizing cytosolic (and presumably other) proteins and does not arrest the elongation of the corresponding nascent polypeptides in the absence of microsomal membranes.[51]

[53] D. J. Anderson, P. Walter, and G. Blobel, Signal recognition protein is required for the integration of acetylcholine receptor subunit, a transmembrane glycoprotein, into the endoplasmic reticulum membrane. *J. Cell Biol.* **93**, 501–506 (1982).

[54] D. I. Meyer, E. Krause, and B. Dobberstein, Secretory protein translocation across membranes—the role of the "docking protein." *Nature (London)* **297**, 647–650 (1982).

[55] G. Kreibich, B. L. Ulrich, and D. D. Sabatini, Proteins of rough microsomal membranes related to ribosome binding. I. Identification of ribophorins I and II membrane characteristic of rough microsomes. *J. Cell Biol.* **77**, 464–487 (1978).

[56] G. von Heijne and C. Blomberg, Trans-membrane translocation of proteins (the direct transfer model). *Eur. J. Biochem.* **97**, 175–181 (1979).

energy needed for the transfer of hydrophilic residues.[57] We are at present still far from a satisfactory understanding of translocation even in its simplest form. The process seems to be controlled, rather than spontaneous, but the energy sources on which it depends remain, at least in part, undefined.[cf. 58]

Stop Sequences and Asymmetric Assembly. Integral membrane proteins are only partially translocated across the ER membrane, presumably because they have within their primary structure an internal, hydrophobic sequence (stop sequence, or stop transfer sequence) that blocks the operation half way through.[59,60] Signal sequence removal and initial glycosylation, usually by N-linked, mannose-rich oligosaccharide chains,[61] occur cotranslationally—more precisely cotranslocationally—since they affect only the translocated sequences of the nascent polypeptides. It is worthwhile pointing out that glycosylation, which can greatly increase the hydrophilicity of membrane proteins, is restricted to their ectodomains and is carried out only after the hydrophobic barrier of the bilayer has been negotiated.

Carried out as described, partial translocation gives an asymmetrically assembled transmembrane protein: its polypeptide chain traverses the bilayer in a single pass, its intracisternal domain (future ectodomain) contains its N terminal, and its C terminal remains within its endodomain in the cytoplasm. This type of asymmetry is shared by many—but not all—integral membrane proteins.[45,48,59,60,62] Many others are apparently assembled in reverse: their C terminal, rather than the N terminal, is in the ectodomain.[62] Moreover, many transmembrane proteins traverse the bilayer more than once: three times in the case of the anionic channel (band 3 protein) or erythrocytic membranes,[63] and seven times in the case of

[57] D. M. Engelman and T. A. Steitz, The spontaneous insertion of proteins into and across membranes: the helical hairpin hypothesis. *Cell* **23**, 411–422 (1981).

[58] B. D. Davis and P. C. Tai, The mechanism of protein secretion across membranes. *Nature (London)* **283**, 433–438 (1980).

[59] G. Blobel, Intracellular membrane topogenesis. *Proc. Natl. Acad. Sci. U.S.A.* **77**, 1496–1500 (1980).

[60] D. D. Sabatini, G. Kreibich, T. Morimoto, and M. Adesnik, Mechanisms for the incorporation of proteins in membranes and organelles. *J. Cell Biol.* **92**, 1–22 (1982).

[61] J. E. Rothman and H. F. Lodish, Synchronized membrane insertion and glycosylation of a nascent membrane protein. *Nature (London)* **269**, 775–780 (1977).

[62] W. Wickner, Assembly of proteins into biological membranes. The membrane trigger hypothesis. *Ann. Rev. Biochem.* **48**, 23–45 (1979).

[63] L. K. Drickamer, Orientation of band 3 polypeptide from human erythrocyte membranes. *J. Biol. Chem.* **253**, 7242–7248 (1978).

opsin[64,65] or bacterioopsin.[65,66] In addition, there are integral membrane proteins that do not seem to traverse the bilayer: they appear to be only partially embedded in it.[67]

Other Means of Anchoring Proteins into Lipid Bilayers. Certain membrane proteins are modified posttranslocationally by acylation of some of their (hydroxylated?) amino acid residues with palmitic (and possibly other fatty) acid(s). This modification was originally detected on viral glycoproteins,[68,69] but more recent work indicates its presence on bona fide eukaryotic plasmalemmal proteins, such as the human transferrin receptor,[70] murine glycophorins,[71] and probably many other integral membrane proteins. Fatty acylation may represent another means of anchoring a membrane protein into the bilayer, but the position of the acylated residues (on, or close to, the stop sequence?) is still uncertain.

Acylation of viral and plasmalemmal proteins occurs at a still undefined site early in transit through the Golgi complex.[38]

A different case is represented by certain transforming proteins, such as those of the Rous sarcoma virus, Harvey sarcoma virus, and Abelson virus. They are found in either soluble or membrane-bound form, the latter being fatty acylated.[72] The site where their acylation occurs is un-

[64] P. A. Hargrave, J. H. McDowell, D. R. Curtis, J. K. Wang, E. Juszczak, S. L. Fong, J. K. M. Roa, and P. Argos, The structure of bovine rhodopsin. *Biophys. Struct. Mech.* **9**, 235–244 (1983).

[65] Y. A. Ovchinhikov, Rhodopsin and bacteriorhodopsin: structure-function relationships. *FEBS Lett.* **148**, 179–191 (1982).

[66] R. Henderson and P. W. T. Unwin, Tridimensional model of purple membrane obtained by electron microscopy. *Nature (London)* **257**, 28–32 (1975).

[67] Y. Okada, A. B. Frey, T. M. Guenthner, F. Oesch, D. D. Sabatini, and G. Kreibich, Studies on the biosynthesis of microsomal membrane proteins. Sites of synthesis and modes of insertion of cytochrome b_5, cytochrome b_5 reductase, cytochrome *P*-450 reductase and epoxide hydrolase. *Eur. J. Biochem.* **122**, 393–402 (1982).

[68] M. F. G. Schmidt and M. J. Schesinger, Fatty acid binding to vesicular stomatitis virus glycoprotein: a new type of post-translational modification of the viral glycoprotein. *Cell* **17**, 813–819 (1979).

[69] M. G. F. Schmidt and M. J. Schlesinger, Relation of fatty acid attachment to the translation and maturation of vesicular stomatitis and Sinbis virus membrane glycoproteins. *J. Biol. Chem.* **255**, 3334–3339 (1980).

[70] O. B. Omary and I. S. Trowbridge, Covalent binding of fatty acids to transferrin receptor in cultured human cells. *J. Biol. Chem.* **256**, 4715–4718 (1981).

[71] E. D. Dolci and G. E. Palade, Biosynthetic labeling of murine erythroid cell sialoglycoproteins with [^{3}H]palmitic acid. *Fed. Proc.* **41**, 1140 (1982).

[72] B. M. Sefton, I. S. Trowbridge, and J. A. Cooper, The transforming proteins of Rous sarcoma virus, Harvey sarcoma virus and Abelson virus contain tightly bound lipid. *Cell* **31**, 465–474 (1982).

known. Such molecules may behave transiently as integral membrane proteins and their behavior may be subject to regulation.

Multiplicity of Assembly Procedures

It is clear, therefore, that cells use a multiplicity of procedures for inserting proteins in lipid bilayers. The one that gives a single pass transmembrane protein with the N terminal in the ectodomain appears to be the simplest and energetically the least expensive. This could be the reason all cells seem to use it for the translocation of secretory and lysosomal proteins, an operation which acquires high volume and proceeds at rapid rates in glandular and phagocytic cell types.

Procedures that result in reversed asymmetry of assembly or in multiple pass assembly of proteins into bilayers are perhaps variants of the single pass formula.[cf. 59,60] They might depend on the existence of repeated, alternating signal and stop sequences in the corresponding polypeptides. Proteolytic cleavage of the N terminal side of a stop sequence could generate a transmembrane protein with the C terminal in its intracisternal (future) ectodomain. Yet, it is difficult to understand how attached polysomes could carry through such a "syncopated" operation. Moreover, at present, there is no independent evidence to indicate or suggest that individual ribosomes within the same polysome detach from, and reattach to, the ER membrane serially and repeatedly during translation.

Other procedures appear to operate on partially different principles, since translocation across cognate membranes occurs posttranslationally, as clearly indicated by the detection of soluble precursors in the cytosol. This is the case with all mitochondrial[73,74] and chloroplast[73] proteins of cytoplasmic origin so far studied, irrespective of their final functional residence, i.e., membranes[75–77] or matrix.[78,79] It appears to be also the case

[73] G. W. Schmidt and N. H. Chua, Transport of proteins into mitochondria and chloroplasts. *J. Cell Biol.* **81**, 461–483 (1979).

[74] G. Schatz and R. A. Butow, How are proteins imported into mitochondria? *Cell* **32**, 316–318 (1983).

[75] G. W. Schmidt, S. G. Bartlett, A. R. Grossman, A. R. Cashmore, and N. H. Chua, Biosynthetic pathways of two polypeptide subunits of the light harvesting chlorophyll a/b protein complex. *J. Cell Biol.* **91**, 468–478 (1981).

[76] A. S. Lewin, I. Gregor, T. L. Mason, N. Nelson, and G. Schatz, Cytoplasmically made subunits of yeast mitochondrial F_1-ATPase and cytochrome *c* oxidase are synthesized as individual precursors, not as polyproteins. *Proc. Natl. Acad. Sci. U.S.A.* **77**, 3998–4002 (1980).

[77] K. Mihara and G. Blobel, The four cytoplasmically made subunits of cytochrome *c* oxidase are synthesized individually and not as a polyprotein. *Proc. Natl. Acad. Sci. U.S.A.* **77**, 4160–4164 (1980).

with certain viral coat proteins[80] in prokaryotic cells. Moreover, some prokaryotic periplasmic proteins seem to be translocated not seriatim (residue after residue) but a domain at a time, independent of elongation.[81] Under certain conditions, cytoplasmic polysomes, involved in the synthesis of yeast mitochondrial proteins, are found attached to mitochondrial outer membranes[82] but the attachment is neither general nor obligatory.[74]

The traffic of mitochondrial and chloroplastic proteins is efficiently controlled and—as in the case of ER-directed proteins—control is assumed to rely on mutual recognition between a signal sequence in the polypeptide and a signal recognition protein, or protein complex, on the cognate membrane. Many (but not all[77]) of these precursor proteins have removable sequences[78] usually long and generally hydrophilic,[83,84] but information about their primary structure is too limited to allow the identification of common, presumably characteristic features. Moreover, information about the corresponding signal recognition proteins or "import receptors"[74] is still indirect and limited.

In *in vitro* reconstituted systems and probably *in vivo*, posttranslational translocation is an energy-requiring process.[83,84] It is assumed to occur as a result of a change in conformation of the precursor protein triggered by its interaction with unknown membrane components,[62] presumably receptors. Removal of the signal sequence follows translocation[62,73,74] and renders it irreversible. In this case, translocation per se deserves proper attention, since the translocated species are indeed large

[78] J. G. Conboy and L. E. Rosenberg, Posttranslational uptake and processing of *in vitro* synthesized ornithine transcarbamoylase precursor by isolated rat liver mitochondria. *Proc. Natl. Acad. Sci. U.S.A.* **78**, 3073–3077 (1981).

[79] N. H. Chua and G. W. Schmidt, Posttranslational transport into intact chloroplasts of a precursor to the small subunit of ribulose-1,5-bisphosphate carboxylase. *Proc. Natl. Acad. Sci. U.S.A.* **75**, 6110–6114 (1978).

[80] K. Ito, G. Mandel, and W. Wickner, Soluble precursor of an integral membrane protein: synthesis of precoat protein in *E. coli* infected with bacteriophage M13. *Proc. Natl. Acad. Sci. U.S.A.* **76**, 1199–1203 (1979).

[81] L. L. Randall, Translocation of domains of nascent periplasmic proteins across the cytoplasmic membrane independent of elongation. *Cell* **33**, 231–240 (1983).

[82] M. Suissa and G. Schatz, Import of proteins into mitochondria. Translatable mRNAs for mitochondrial proteins are present in free as well as mitochondria-bound polysomes. *J. Biol. Chem.* **257**, 13048–13055 (1982).

[83] A. Grossman, S. Bartlett, and N. H. Chua, Energy-dependent uptake of cytoplasmically synthesized polypeptides by chloroplasts. *Nature (London)* **285**, 625–628 (1980).

[84] D. Kolansky, J. G. Conboy, W. A. Fenton, and L. E. Rosenberg, Energy dependent translocation of the precursor of ornithine transcarbamylase by isolated rat liver mitochondria. *J. Biol. Chem.* **257**, 8467–8471 (1982).

hydrophilic macromolecules. Translocation is carried through (notwithstanding apparently unfavorable thermodynamic conditions) without running down gradients that depend on low and controlled membrane permeability. In fact, proton gradients and membrane potentials appear to be needed or facilitating factors in translocation.[85,86]

Still another insertion formula may involve the initial production of cytosolic protein micelles (by interactions among their hydrophobic domains) followed by transport of the proteins from micelles to membranes. Transport of this type was demonstrated in *in vitro* experiments for cytochrome b_5. This integral (but apparently not transmembrane) protein of the ER has a C terminal located in a small hydrophobic domain buried in the bilayer; its N terminal is comprised in a large hydrophilic domain located in the cytosol.[87]

This special type of membrane assembly is shared by other ER membrane proteins, NADH–cytochrome b_5 reductase and NADPH–cytochrome *P*-450 reductase,[67] for instance. Such proteins have a relatively wide distribution. They are found in ER and Golgi membranes[88–92] and apparently in the plasmalemma.[92] The broad distribution as well as the fact that they do not have removable or recognizable signal sequences

[85] T. Date, J. M. Goodman, and W. Wickner, Precoat, the precursor of M13 coat protein, requires an electrochemical potential for membrane insertion. *Proc. Natl. Acad. Sci. U.S.A.* **77**, 4669–4673 (1980).

[86] S. M. Gasser, G. Daum, and G. Schatz, Import of proteins into mitochondria. Energy dependent uptake of precursors by isolated mitochondria. *J. Biol. Chem.* **257**, 13034–13041 (1982).

[87] L. Spatz and P. Strittmatter, A form of cytochrome b_5 that contains an additional hydrophobic sequence of 40 amino acid residues. *Proc. Natl. Acad. Sci. U.S.A.* **68**, 1042–1046 (1971).

[88] N. Borgese and J. Meldolesi, Localization and biosynthesis of NADH–cytochrome b_5 reductase, an integral membrane protein, in rat liver cells. I. Distribution of the enzyme activity in microsomes, mitochondria and Golgi complex. *J. Cell Biol.* **85**, 501–515 (1980).

[89] J. Meldolesi, G. Corte, G. Petrini, and N. Borgese, Localization and biosynthesis of NADH–cytochrome b_5 reductase, an integral membrane protein, in rat liver cells. II. Evidence that a single enzyme accounts for the activity in its various subcellular locations. *J. Cell Biol.* **85**, 516–526 (1980).

[90] N. Borgese, G. Petrini, and J. Meldolesi, Localization and biosynthesis of NADH–cytochrome b_5 reductase, an integral membrane protein, in rat liver cells. III. Evidence for the independent insertion and turnover of the enzyme in various subcellular compartments. *J. Cell Biol.* **86**, 38–45 (1980).

[91] K. E. Howell, A. Ito, and G. E. Palade, Endoplasmic reticulum marker enzymes in Golgi fractions. What does this mean? *J. Cell Biol.* **79**, 581–589 (1978).

[92] E. D. Jarasch, J. Kartenbeck, G. Bruder, A. Fink, J. Morré and W. W. Franke, β-type cytochromes in plasma membranes isolated from rat liver, in comparison with those of endomembranes. *J. Cell Biol.* **80**, 37–52 (1979).

may be taken as a suggestion that a signal sequence–signal recognition complex is not involved in their assembly. Yet large differences in their concentration among different cellular membranes suggest that their insertion does not occur at random.

Vesicular Transport

Upon insertion into a membrane by any one of the procedures mentioned above, some proteins already reach their ultimate destination. This is the case with the ER membrane proteins so far studied[67] and with the proteins of mitochondrial membranes, except that a second proteolytic step may be required for transport to the mitochondrial inner membrane and matrix.[74,84,92a] But for many other membrane proteins, the site of their primary insertion is not the site of their final functional residence. This appears to be the case with the integral membrane proteins of the Golgi complex, lysosomes and plasmalemma. The transport is effected by vesicular carriers assumed or proved to operate between the ER and the Golgi complex, in between different Golgi subcompartments, between the Golgi complex and lysosomes, and between the Golgi complex and different domains of the plasmalemma.[93] Vesicular traffic must also be effectively controlled, since every membrane protein initially inserted in the ER membrane eventually reaches its proper destination.

Vesicular transport is not limited to newly synthesized membrane proteins. Cells use it for the intracellular transport of secretory and lysosomal proteins, for exocytosis,[94] endocytosis,[95,96] transcytosis,[97] as well

[92a] A more complex processing is apparently involved in the case of some of the proteins of the mitochondrial intermembrane space.[74,86]

[93] M. G. Farquhar, Multiple pathways of exocytosis, endocytosis and membrane recycling: validation of a Golgi route. *Fed. Proc.* **42**, 2407–2413 (1983).

[94] G. E. Palade, Intracellular aspects of the process of protein secretion. *Science* **189**, 347–358 (1975).

[95] I. S. Mellman, Endocytosis, membrane recycling and Fc receptor function. *In* "Membrane Recycling" (Ciba Foundation Symposium 92) (D. Evered and G. M. Collins, eds.), pp. 35–58. Pitman, London, 1982.

[96] A. H. Helenius and M. Marsh, Endocytosis of enveloped animal viruses. *In* "Membrane Recycling" (Ciba Foundation Symposium 92) (D. Evered and G. M. Collins, eds.), pp. 59–76. Pitman, London, 1982.

[97] R. Rodewald and D. R. Abrahamson, Receptor mediated transport of IgG across the intestinal epithelium of the neonatal rat. *In* "Membrane Recycling" (Ciba Foundation Symposium 92) (D. Evered and G. M. Collins, eds.), pp. 209–232. Pitman, London, 1982.

as for membrane recycling[93,98] and at present it is still unknown to what extent these circuits overlap and to what extent common carriers are used for different proteins and different destinations. However, there are indications that in certain cell types secretory proteins and membrane proteins destined to the plasmalemma are transported by different vesicular carriers.[99] And, there is also evidence that vesicular carriers involved in recycling and probably biosynthetic transport have topographically defined ports of entry in cultured renal epithelia (MDCK) cells.[100]

Control of Vesicular Traffic and Control of Membrane Specificity

The mechanisms by which vesicular traffic is controlled are still unknown. But it can be assumed that they involve mutual recognition between signals and signal recognition complexes, as in the case of secretory and membrane proteins at the initial ER entry. In this case, however, each partner is probably affixed to one of the interacting membranes.

Wherever and for whatever purpose it operates, vesicular transport involves membrane fusion–fission that leads to continuity between the lipid bilayer of the carrier and that of either the donor compartment or the receiving compartment at the two termini of the transport shuttle or circuit.[98] Given the fluidity of the bilayers, established continuity is an invitation to randomization, by lateral diffusion, of the lipid and protein components of the interacting membranes. Randomization would be expected to lead to loss of membrane specificity, but since specificity is retained (as well documented in the literature), it follows that cells have developed means to control lateral diffusion. These means must include (a) stabilization of the chemistry of one or both membranes for the duration of the continuity of their bilayers, and (b) nonrandom removal of the membrane of the vesicular carrier from either the donor or the receiving compartment.[101,102] Stabilization followed by nonrandom removal would be ex-

[98] G. E. Palade, Problems in membrane traffic. *In* "Membrane Recycling" (Ciba Foundation Symposium 92) (D. Evered and G. M. Collins, eds.), pp. 1–14. Pitman, London, 1982.

[99] B. Gumbiner and R. B. Kelly, Two distinct intracellular pathways transport secretory proteins and membrane proteins through the Golgi complex. *Cell* **28**, 51–59 (1982).

[100] D. Louvard, Apical membrane aminopeptidase appears at site of cell-cell contact in cultured kidney epithelial cells. *Proc. Natl. Acad. Sci. U.S.A.* **77**, 4132–4136 (1980).

[101] J. H. Ehrenreich, J. J. M. Bergeron, P. Siekevitz, and G. E. Palade, Golgi fractions prepared from rat liver homogenates. I. Isolation procedures and morphological characterization. II. Biochemical characterization. *J. Cell Biol.* **59**, 45–72, 73–88 (1973).

[102] G. E. Palade, Interactions among cellular membranes. *In* "Pontificiae Academiae Scientiarum Scripta Varia. Semaine d'Etude sur le Théme Membranes Biologiques et Artificielees et la Désalinisation de l'Eau" (R. Passino, ed.), pp. 85–109. Pontifical Academy of Sciences, Vatican, Rome, 1976.

pected at both termini, if the vesicular membrane is different from that of the two termini; it would be needed at one terminus only, if the carrier derives its membrane from that of the other terminus.

Means of Controlling Membrane Specificity. If the need for stabilization and nonrandom removal is clear, the means by which these operations are carried out are still poorly understood. In principle, stabilization could be achieved by interactions among integral membrane proteins within the bilayer itself or within its immediate vicinity. 5′-Nucleotidase, localized by cytochemical procedures on isolated Golgi fractions, is found restricted to the distended rims of Golgi cisternae; notwithstanding bilayer continuity, the center region of the cisternae is nonreactive.[103] Another example, not connected this time with vesicular traffic, concerns the ribophorins, restricted in their distribution to the rough ER notwithstanding the latter's membrane continuity with the smooth ER.[55] Other, similar cases can be anticipated primarily in connection with proteins involved in signal recognition and nascent peptide translocation in the ER.

As an alternative, stabilization could be achieved by interactions among the endodomains of integral membrane proteins and peripheral membrane proteins that generate recognizable, often highly organized, infrastructures under differentiated domains of the plasmalemma and other cellular membranes. The most frequently encountered type of infrastructure is the geodetic cage formed by clathrin and clathrin-associated protein molecules on the cytoplasmic side of coated pits on the plasmalemma as well as on other cellular membranes (Golgi elements, transitional elements of the ER, secretion granules, endosomes, lysosomes). The clathrin cage is known to function as selector and stabilizer for various receptors.[104,105] But it may also act as a molder of the size and shape of the pits as well as an excision device which, upon cage completion, converts a coated pit into a coated vesicle.[106] Available evidence indicates that coated vesicles are a variety, in fact a major variety, of vesicular carriers. The coated pit could be viewed as the form of the carrier that interacts with the membrane of a terminus (plasmalemma, endosome, Golgi elements, etc.) and is stabilized for the duration of the interaction

[103] M. G. Farquhar, J. J. M. Bergeron, and G. E. Palade, Cytochemistry of Golgi fractions prepared from rat liver. *J. Cell Biol.* **60**, 8–25 (1974).

[104] B. M. F. Pearse and M. S. Bretscher, Membrane recycling by coated vesicles. *Ann. Rev. Biochem.* **50**, 85–101 (1981).

[105] M. S. Brown, R. G. W. Anderson, and J. L. Goldstein, Recycling receptors: the round trip itinerary of migrant membrane proteins. *Cell* **32**, 663–667 (1983).

[106] R. A. Crowther and B. M. F. Pearse, Assembly and packing of clathrin into coats. *J. Cell Biol.* **91**, 790–797 (1981).

by the infrastructure of the clathrin cage. At present, however, pertinent information is limited and fragmentary. The integral membrane proteins of coated pits and coated vesicles are still unknown and so is the identity of their partners and the nature of their interactions within the clathrin cage.

Morphological evidence suggests that secretion granule membranes are stabilized during exocytosis by a rapidly forming fibrillar infrastructure.[94] Its chemical nature is still unknown, but its structure is reminiscent of feltworks made by actin and actin-binding proteins on other domains of the plasmalemma.[107] Spectrins,[108] fodrins,[109] and other related proteins[110] may also be used for similar purposes. In fact, the best example of stabilizing infrastructure currently available is the spectrin–ankyrin–actin complex found under the plasmalemma of mature erythrocytes.[111] The interactions among its components and among the latter and the integral proteins of the erythrocytic plasmalemma are reasonably well understood, but the functional role of the resulting infrastructure is quite different from that considered here. The mature erythrocyte has lost all intracellular membrane systems and with them the need for vesicular transport. The infrastructure presumably confers tensile strength to the plasmalemma of a cell which is continuously stressed while squeezing through blood capillaries in the circulatory system.

The evidence so far obtained suggests that coated pits can differentiate between certain receptor–ligand complexes, which are retained and clustered, and unoccupied receptors, which are presumably allowed to move in and out.[105,112,113] The means by which this remarkable operation is

[107] J. Boyles and D. F. Bainton, Changing patterns of plasma membrane-associated filaments during the initial phases of polymorphonuclear leukocyte adherence. *J. Cell Biol.* **82**, 347–368 (1979).

[108] V. T. Marchesi, Spectrin: the present status of a putative cytoskeletal protein of the red cell membrane. *J. Membr. Biol.* **51**, 101–131 (1979).

[109] J. Levine and M. Willard, Fodrin: axonally transported polypeptides associated with the internal periphery of many cell types. *J. Cell Biol.* **90**, 631–643 (1981).

[110] J. R. Glenney, P. Glenney, and K. Weber, Erythroid spectrin, brain fodrin and intestinal brush border proteins (TW 260/240) are related molecules containing a common calmodulin-binding subunit bound to a variant cell-type specific subunit. *Proc. Natl. Acad. Sci. U.S.A.* **79**, 4002–4005 (1982).

[111] D. Branton, C. M. Cohen, and J. Tyler, Interactions of cytoskeletal proteins on the human erythrocyte membrane. *Cell* **24**, 24–32 (1981).

[112] R. G. W. Anderson, M. S. Brown, and J. L. Goldstein, Role of the coated endocytic vesicle in the uptake of receptor-bound low density lipoprotein in human fibroblasts. *Cell* **10**, 351–364 (1977).

[113] J. A. McKanna, H. T. Haigler, and S. Cohen, Hormone receptor topology and dynamics: morphological analysis using ferritin-labelled epidermal growth factor. *Proc. Natl. Acad. Sci. U.S.A.* **76**, 5689–5693 (1979).

carried out are still unknown. At first sight, the finding appears to contradict the premise that the clathrin cage of coated pits stabilizes the chemistry of the associated plasmalemma. But the situation can be rationalized by assuming that the clathrin cage can effect a differential stabilization in which elements responsible for selection and clustering remain fixed, whereas receptors move in and out and are retained only when complexed by their appropriate ligand. Perhaps a similar procedure is used in bringing newly synthesized proteins to the plasmalemma and other membranes: the carrier vesicle may release the transported proteins upon membrane fusion, while retaining its specific proteins in its own membrane. A comparable situation is encountered in the Golgi complex in which resident enzymes have fixed positions[35,114] while substrate proteins are allowed to diffuse through the entire membrane complex, as clearly indicated by work done on the G protein of vesicular stomatitis virus.[115]

At present, the distinction between stabilizing infrastructures and cytoskeletal elements (microtubules, intermediary filaments) is primarily a matter of dimensions over which the pertinent macromolecular assemblies exert their function, e.g., membrane microdomains vs cell regions or whole cell. The distinction is largely arbitrary, since we are dealing with a continuum of supportive or stabilizing structures that may interact with one another across our categories.

Peripheral Membrane Proteins

Information concerning the synthesis of peripheral membrane proteins is still limited to a few cases. Those so far studied belong to the erythrocytic plasmalemma, are synthesized on free polysomes, and are assembled in the membrane either directly or after proteolytic processing.[116,117]

Peripheral membrane proteins may have many other roles besides generating stabilizing infrastructures. They do participate in protein traffic control, as in the case of the signal recognition protein complex, and they are strategically positioned for a role in the control of the traffic of vesicular carriers.

[114] M. G. Farquhar and G. E. Palade, The Golgi apparatus (complex)—1954–1981—from artefact to center stage. *J. Cell Biol.* **91**, 77s–103s (1981).

[115] J. E. Bergmann, K. T. Tokuyasu, and S. J. Singer, Passage of an integral membrane protein, the vesicular stomatitis glycoprotein, through the Golgi apparatus en route to the plasma membrane. *Proc. Natl. Acad. Sci. U.S.A.* **78**, 1746–1750 (1981).

[116] H. F. Lodish, Biosynthesis of reticulocyte membrane proteins by membrane free polysomes. *Proc. Natl. Acad. Sci. U.S.A.* **70**, 1526–1530 (1973).

[117] H. F. Lodish and B. Small, Membrane proteins synthesized by rabbit reticulocytes. *J. Cell Biol.* **65**, 51–64 (1975).

Perspectives

This is for the moment the sum of our knowledge and the extent to which speculations can be reasonably advanced about the synthesis, assembly, and transport of membrane proteins, considered individually. Two main sets of findings set off important issues that will undoubtedly attract considerable attention in the future. The first concerns the traffic regulation of proteins in general and membrane proteins in particular. The second refers to the maintenance of membrane specificity under conditions which—in principle—favor chemical randomization. Both sets of findings can be considered as reasonably well established, but the means by which traffic is controlled and specificity is retained have been uncovered to a very limited extent only. A lot remains to be done, therefore, in this promising, challenging, and critically important research area.

An Integrated View of Membrane Assembly

Over the past decade, work on membrane biogenesis has been focused mostly on the synthesis and assembly of individual membrane components, primarily integral membrane proteins. The way in which eukaryotic cells process populations of various lipid and protein molecules during membrane biogenesis has received less attention. Hence, on this particular topic, one has to rely essentially on evidence generated in the late 1960s and early 1970s at a time when modes of assembly and topography of assembly of membrane components were considered in terms of molecular populations rather than individual molecules.[14]

Modes of Assembly. In principle, cells could assemble all the components of a given membrane in one step—after synthesizing them in synchrony or mobilizing them from undefined intracellular stores. The corollary of this "one-step" mode of assembly is constant biochemistry and full functional competence in the sense that all components are present at all times in constant, presumably optimal proportions. None is in excess over the others. The alternative is multiple-step assembly, a mode in which individual components or groups of components are assembled in their cognate membrane at different times and possibly at different rates.[14] In this case, the biochemistry of the membrane is continuously variable and the membrane apparently never reaches full competence, in the sense that certain components are present in excess, and as such inefficiently used.

In the few biological systems so far investigated, multiple step has been the mode of assembly consistently found, the order of appearance of enzymic activities or molecularly defined components being always the same for a given system. This is the case, for instance, with the ER membrane of hepatocytes in the rapid and extensive differentiation these cells incur in the newborn animal,[1] and this is also the case with the proteins and lipids of the same membrane in the ER proliferation induced by xenobiotics.[5] Activities appear in the ER membrane at different times even when the corresponding enzymes catalyze different steps in a common sequence of reactions; for instance, an increase in NADPH–cytochrome *P*-450 reductase precedes by a few hours the increase in cytochrome *P*-450 within the same membrane. Both enzymes are part of the microsomal mixed function oxygenase system.

On a broader scale, and considering all components of the membrane examined, the same mode of assembly was found in the case of the thylakoid membranes of the green alga *Chlamydomonas reinhardtii,* an organism that can be easily synchronized by repeated exposure to 12 hr light followed by 12 hr darkness,[118] the equivalent of a day–night cycle. Thylakoid lipids (primarily mono- and digalactosyldiglycerides) appear in the membranes at different times in the morning and accumulate at different rates.[119] Chlorophylls and carotenoids are synthesized and accumulate in the same membranes during the day, primarily around noon.[118,120] And proteins are synthesized and assembled through the entire cycle,[121] but each individual protein appears in the membrane singly (or in small groups) at specific, precise times in the cell cycle and accumulates at different rates. The first new proteins to be detected in membranes in the morning are in general those synthesized on chloroplastic ribosomes.[122] The biochemistry of the membrane follows, therefore, a well-defined pattern of repeated cyclic variations, clearly connected with the cell cycle, and as a result, at any given time some membrane components are in considerable excess over the others.

At present, it is not understood why cells use a multistep mode of assembly in membrane biogenesis. Their practice may reflect energetic or

[118] S. Schor, P. Siekevitz, and G. E. Palade, Cyclic changes in thylakoid membrane of synchronized *Chlamydomonas reinhardi. Proc. Natl. Acad. Sci. U.S.A.* **66**, 174–180 (1970).

[119] D. R. Janero and R. Barrnett, Thylakoid membrane biosynthesis in *Chlamydomonas reinhardtii* 137+. I. Cell cycle variations in the synthesis and assembly of polar glycerolipids. *J. Cell Biol.* **91**, 126–134 (1981).

[120] D. R. Janero and R. Barrnett, Thylakoid membrane biogenesis in *Chlamydomonas reinhardtii* 137+. II. Cell cycle variations in the synthesis and assembly of pigment. *J. Cell Biol.* **93**, 411–416 (1982).

[121] L. Y. W. Bourguignon and G. E. Palade, Incorporation of polypeptides into thylakoid membranes of *Chlamydomonas reinhardtii*—cyclic variations. *J. Cell Biol.* **69**, 327–344 (1976).

[122] A. Michaels and G. E. Palade, unpublished data.

biosynthetic stringencies. During the night, for instance, chloroplastic ribosomes are known to be inactive, or considerably less active, than during the day.[118] The obvious connection with the cell cycle suggests, however, that these recurrent waves of synthesis and assembly reflect an inherent pattern of gene activation within the overall program of cell replication.

Topography of Assembly. It can be assumed that newly synthesized membrane components could be assembled in a new membrane independently of any preexisting membranous structure. This "de novo" assembly was found so far in a single case, that of an inner membrane of the vaccinia virus[123] which shows free, apparently growing margins during its assembly. The membrane in question is an unusual, protein-rich structure. More importantly, vaccinia is a virus whose existence, unlike that of a cell, does not depend on its own metabolic continuity.

In all cellular systems so far investigated, new membrane components—lipids as well as proteins—are inserted into preexisting membranes. The result is membrane growth by continuous expansion of each preexisting structure. The conclusion is based on cytochemical studies carried out during hepatocyte differentiation in the preborn and newly born mammal: glucose-6-phosphatase, a new enzyme activity, appears in all ER elements at a set time in the differentiation program. Past that time, the cells do not have nonreactive ER elements that could represent residual ER, predating glucose-6-phosphatase production.[2] Moreover, the activity is evenly dispersed throughout the ER membrane since liver homogenization yields a population of uniformly reactive microsomal vesicles.[2]

The same conclusion was drawn from experiments in which thylakoid membranes were found to change their density as a result of the greening of a *Chlamydomonas* yellow mutant.[124] As the membranes acquired more chlorophylls, they became heavier in synchrony. No residual fraction of light membranes was detected in the extensively fragmented (vesiculated) preparations. Similar results were obtained in mutant bacteria unable to synthesize membrane lipids. Glycerol supplementation resulted in a uniform shift in the density of the membranes[124a] assessed upon extensive vesiculation.[125]

[123] S. Dales and E. H. Mosbach, Vaccinia as a model of membrane biogenesis. *Virology* **35**, 564–583 (1968).

[124] G. Eytan and I. Ohad, Biogenesis of chloroplast membranes. *J. Biol. Chem.* **245**, 4297–4307 (1970).

[124a] L. Mindich, Membrane biosynthesis in *Bacillus subtilis*. I. Isolation and properties of strains bearing mutations in glycerol metabolism. II. Integration of membrane proteins in the absence of lipid synthesis. *J. Mol. Biol.* **49**, 415–432, 433–439 (1970).

[125] Since the cells (*Bacillus subtilis*) studied have a single membrane, the interpretation of these results concerns both the mode and the topography of assembly (see later).

New molecules could be introduced either regionally or dispersively into a preexisting membrane. But, given the rapid lateral diffusion of lipids as well as proteins within a fluid bilayer, it may be difficult or impossible to detect regional insertion, because, by the time results are recorded, all components may appear randomly distributed even if some of them (or all of them) were initially regionally inserted. So far the converse, i.e., dispersive insertion, was demonstrated on thylakoid membranes in synchronized cultures of the green alga *Chlamydomonas*. During the light phase of the cycle, chloroplastic polysomes, varying in size from pentamers to octamers, were found attached to thylakoid membranes in intact cells and fragmented chloroplasts, as well as isolated thylakoids. They occurred widely distributed, rather than regionally clustered, on all these specimens.[126] The proteins synthesized and inserted into thylakoid membranes by these polysomes are products of the chloroplast genome.

The Biological Significance of Membrane Fluidity

The integration of the body of information so far secured allows a number of conclusions to be drawn and a few logical extrapolations to be considered.

The common structural denominator of all cellular membranes, their polar lipid bilayer, is fluid at the temperature of the cell's immediate environment. Given the generality of this property, it can be concluded that membrane fluidity is required for some common, basic process in membrane biology. Work done primarily in prokaryotic cells indicates convincingly that cells modify extensively the fatty acyl groups of their phospholipids in response to variations in environmental temperature, so as to maintain fluid at least a large fraction of the bilayer.[127] Mutants that cannot synthesize or modify fatty acids stop growing and eventually die when supplemented with fatty acids that have a melting point higher than the temperature of the environment.[128]

In a fluid bilayer, new lipid and protein molecules can be inserted, and thus the membrane can grow—by expansion of its preexisting matrix—without loss of permselectivity and without decay of gradients and potentials, which means without loss of functional continuity. Insertion of new

[126] N. H. Chua, G. Blobel, P. Siekevitz, and G. E. Palade, Periodic variations in the ratio of free to thylakoid-bound chloroplast ribosomes during the cell cycle of *Chlamydomonas reinhardtii*. *J. Cell Biol.* **71**, 497–514 (1976).

[127] J. E. Cronan, Jr., Thermal regulation of the membrane lipid composition in *Escherichia coli*. *J. Biol. Chem.* **250**, 7074–7077 (1975).

[128] J. E. Cronan, Jr., Molecular biology of bacterial membrane lipids. *Ann. Rev. Biochem.* **47**, 163–189 (1978).

molecules in a solid structure could not be carried out without transient but repeated loss of permselectivity. Accordingly, it can be postulated that cells generate fluid membranes because such membranes are a prerequisite for functional continuity and survival.

Membrane fluidity is also a prerequisite for membrane fusion–fission,[129] an event that occurs at the last step in cell division and at the first step in zygote formation. Moreover, membrane fusion–fission is involved in many other "minute-by-minute" cell activities, such as intracellular transport of macromolecules, exocytosis, endocytosis, membrane transport, and membrane recycling. With fluid membranes, all these operations are possible under conditions that ensure retention of the control of permeability for the plasmalemma and for each intracellular compartment.

On all these accounts, it can be concluded that membrane fluidity is a sine qua non condition for cellular existence. Cells could not exist without it.

Yet fluid membranes have potential drawbacks, of which the most important is the ever-present danger of randomization of lipids and proteins by lateral diffusion within a given membrane of in-between interacting membranes, when interaction involves continuity of the corresponding bilayers. Randomization means loss of membrane specificity, but cells need membrane specificity for the differentiated functions of their various compartments and microdomains of their membranes. If cells cannot exist without fluid membranes, and if they need membrane specificity, they must control the lateral diffusion of critical components among differentiated microdomains, or in-between interacting membranes. They achieve control of lateral diffusion by using apparently a variety of stabilizing mechanisms among which the most obvious are, at present, various stabilizing infrastructures. With such mechanisms a solution is provided to the dilemma, and both fluidity and specificity of cellular membranes are maintained. Control of fluidity so as to allow retention of membrane specificity should be introduced as an important modifying element in the fluid mosaic model.

Membranes Function as Their Own Templates

The intracellular traffic of proteins for whatever destination, including whatever membrane, is apparently effectively regulated by mutual recog-

[129] D. Papahadjopoulos, G. Poste, and B. E. Schaeffer, Fusion of mammalian cells by unilamellar lipid vesicles: influence of lipid surface charge, fluidity and cholesterol. *Biochim. Biophys. Acta* **323**, 23–42 (1973).

nition between specific signals or pilots and specific signal receptors, or ports of entry. This type of mechanism is known to operate for newly synthesized proteins initially directed to the ER membrane; it also operates most likely for proteins addressed to mitochondria and chloroplasts; and (by extension) it is assumed to operate for vesicular carriers plying between two specific termini. In the ultimate analysis, traffic control enables each membrane to recognize "like" proteins and to accept them as "self" in its existing matrix. Unlike proteins are not recognized as "self" and are not accepted. Therefore, it can be logically postulated that membranes act as their own templates by the interplay of specific signals and specific signal receptors. They recognize and incorporate like components, grow by expansion in two dimensions, and eventually divide into two sets of descendant membranes, one for each daughter cell. These sets are qualitatively identical but not necessarily quantitatively equal. They acquire this quantitative margin of freedom because membrane proteins are generally present in multiple copies. Two-dimensional templates of the type considered differ in some respects from the essentially linear templates represented by nucleic acid molecules. Like other templates, they function on the basis of the complementarity principle, but complementarity is probably involved only at the entry in the membrane matrix when specific pilots and specific ports of entry recognize one another. This recognition does not lead, however, to a permanent or lasting association of the two partners, so that, in principle, a limited number of signal receptors can control the entry of a large population of signal-bearing components. In this postulate, signal receptors or ports of entry emerge as the most important components of each membrane. Inherited by virtue of temporal continuity, they ensure maintenance of membrane specificity and make possible specific membrane differentiation.

Section I

Biogenesis and Assembly of Membrane Proteins

A. General Methods
Articles 1 through 19

B. Eukaryotic Membranes
Plasma Membrane
Articles 20 through 32

Enveloped Viruses
Articles 33 through 40

Endoplasmic Reticulum
Articles 41 through 44

Sarcoplasmic Reticulum
Article 45

Plant Vacuoles
Article 46

Nuclear Membrane
Articles 47 and 48

Other Specialized Systems
Articles 49 through 51

[1] Preparation of Rough Microsomes and Membrane-Bound Polysomes That Are Active in Protein Synthesis *in Vitro*

By SANCIA GAETANI, JULIA A. SMITH, RICARDO A. FELDMAN, and TAKASHI MORIMOTO

Rough microsomes (RM) and membrane-bound polysomes that are active in protein synthesis in an *in vitro* system are useful tools for the study of membrane–ribosome interactions, reconstitution of RM, membrane biogenesis, etc. It is particularly desirable to obtain preparations of undegraded RM from cells, such as rat hepatocytes, which have been extensively studied both biochemically and morphologically. In fact, difficulty of preparation primarily due to the high levels of endogenous RNase activity has greatly limited the use of rat liver microsomes for *in vitro* protein synthesis studies. The development of conditions for liver microsome preparation will, therefore, provide us with accessibility to an important system as well as valuable knowledge that could be used in other difficult systems, such as tissue culture cells.

In our studies, we have attempted to develop procedures that minimize polysome degradation based on the systematic analysis of existing conventional cell fractionation techniques.[1–5] The key observation is that pelleting of RM frequently causes polysome degradation, even if it is performed in the presence of an appropriate amount of rat liver high-speed supernatant (HSS) as an RNase inhibitor (Fig. 1). This is because RNase, which is present on the surface of microsomes, acts on the many more bound polysomes that are brought in its proximity within packed microsomes. We have therefore modified the procedure so that we could isolate with good recovery RM and bound polysomes that were relatively pure, in addition to being intact as evaluated by their ability to carry out *in vitro* protein synthesis. The major modifications we introduced are listed below.

1. Cell homogenization is performed in a slightly hypertonic sucrose solution lacking salts to prevent aggregation of organelles.
2. The postmitochondrial supernatant (PMS) was adopted not only as a satisfactory source of RM fractions, but also as a source of a

[1] G. Blobel and V. R. Potter, *J. Mol. Biol.* **26,** 279 (1967).
[2] M. A. Adelman, G. Blobel, and D. D. Sabatini, *J. Cell Biol.* **56,** 191 (1973).
[3] J. Kruppa and D. D. Sabatini, *J. Cell Biol.* **74,** 414 (1977).
[4] G. C. Shore and J. R. Tata, *J. Cell Biol.* **72,** 726 (1977).
[5] J. C. Ramsey and W. J. Steele, *Biochemistry* **15,** 1704 (1976).

ISBN 0-12-181996-5

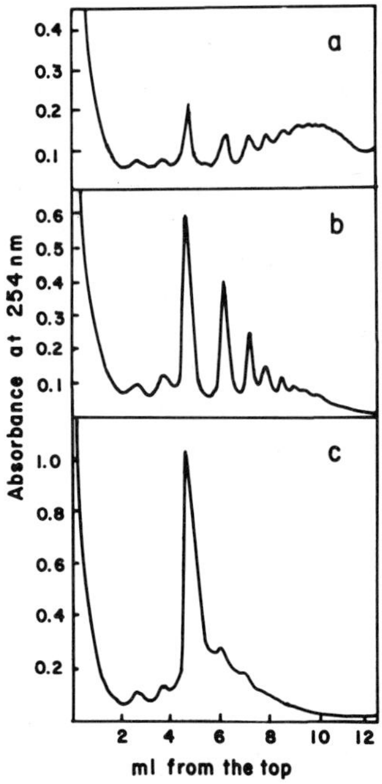

FIG. 1. Degradation of membrane-bound polysomes observed during the washing and concentrating of the rough microsomes (RM). The crude RM prepared by the procedure described in the text, were washed and concentrated in three different ways (a–c). Panel a: An aliquot of the crude RM fraction was diluted 3.5- to 4-fold with TKM containing 10% (v/v) high-speed supernatant (HSS), and layered on top of a step gradient consisting of 1 ml of 1.8 *M* STKM containing 10% (v/v) HSS, 2 ml of 1.3 *M* STKM containing 20% (v/v) HSS, and 3 ml of 1.0 *M* STKM containing 10% (v/v) HSS. The gradient was centrifuged at 40,000 rpm for 60 min in an SW41 rotor, and the microsomes banded at the 1.0/1.8 *M* STKM interface. Panel b: An aliquot of crude RM was diluted 3-fold with TKM containing 10% (v/v) HSS and centrifuged at 40,000 rpm for 30 min in a Ti50 rotor. The microsomal pellets were resuspended in a small volume of TKM containing 20% (v/v) HSS. Panel c: An aliquot of crude RM was diluted 3.5- to 4-fold with TKM and treated in the same way as in (a) except that none of the sucrose solutions used for making the gradient contained HSS.

Each sample, containing about 10 OD_{260} measured in the presence of deoxycholate (DOC), was solubilized with a mixture of DOC and Triton X-100 at final concentrations of 0.5% and 1%, respectively, and layered onto a 20 to 60% linear sucrose density gradient containing TKM. The gradient was centrifuged at 40,000 rpm for 4 hr in an SW41 rotor in the cold. After centrifugation, the absorbance at 254 nm was monitored by a Buchler Auto Densiflow coupled to an LKB UV monitor. The extent of polysome degradation was found to be greater when RM were kept longer in ice as pellets.

potent RNase inhibitor and as a source of endogenous salts for the preparation of the step gradient, so that microsomes may remain in contact with the supernatant throughout the preparation.

3. Pelleting of the RM fraction is avoided at all stages of the fractionation through the use of sucrose cushions.

These key points have been successfully applied to obtain undegraded RM from various sources such as HeLa cells, WI-38 human fibroblasts, MDCK cells, chicken embryonic muscle, and rat brain. In this chapter, we describe primarily a modified procedure for the cell fractionation of rat liver to obtain RM and bound polysomes that are active in cell-free protein synthesis. We also comment on the ways in which the procedure can be helpful for the fractionation of new materials.

Materials and Methods

Reagents

TKM: 50 m*M* Tris-HCl, pH 7.4, 50 m*M* KCl, 5 m*M* $MgCl_2$
STKM: Sucrose TKM
HSB: 50 m*M* Tris-HCl, pH 7.4, 500 m*M* KCl, 5 m*M* $MgCl_2$
Homogenizing solution: 0.5 *M* sucrose (S), 3 m*M* dithiothreitol (DTT)

Animals. Livers from unstarved rats were generally used for the preparation of microsomes and bound polysomes, but not for the preparation of free polysomes. Smooth microsome (SM) fractions prepared from livers of starved rats were found to be heavily contaminated with light rough microsomes (RM_L), perhaps because of the extensive ribosome runoff that may be caused by starvation.[6,7]

High-Speed Supernatant and a Sephadex G-100 Fraction. The liver of one rat was homogenized in 2 volumes (w/v) of 0.25 *M* sucrose, 3 m*M* DTT. The homogenate was centrifuged for 15 min at 15,000 rpm, and the supernatant was decanted and centrifuged in a Ti50 rotor for 60 min at 200,000 *g*. Four-fifths of this HSS was then gently removed and either stored at −70° for later use as an RNase inhibitor, or applied to a Sephadex G-100 column equilibrated with TKM. Fractions (1 ml) were collected by elution with TKM after adjusting the salt concentration of HSS to that of TKM. The most concentrated fractions (G-100 fraction), corresponding in volume to the amount applied, were pooled, divided into aliquots, and stored at −70°. The $A_{260}:A_{280}$ ratio of the G-100 fraction was about 0.8, and the protein concentration was about 10 mg/ml.

[6] S. Hiem Yap, R. K. Strair, and D. A. Schfritz, *J. Biol. Chem.* **253,** 4944 (1978).
[7] T. Staehelin, E. Verney, and H. Sidransky, *Biochim. Biophys. Acta* **145,** 105 (1967).

Translation Mixture. For each assay per 100 μl of incubation mixture there was 3.5–4 mM $MgCl_2$, 100 mM KCl [120 mM KCl for translation in the presence of 7-methylguanosine 5′-phosphate (7-mG5′p)], 3 mM DTT, 10 mM Tris-HCl, pH 7.5, 4 μg of creatine phosphokinase, 8 mM creatine phosphate, 0.2 mM GTP, 1 mM ATP, 30 μM each of unlabeled amino acids, 10 μCi of ^{3}H-labeled amino acids or [^{35}S]methionine, 30 μl of the G-100 fraction, and 1–2 OD_{260} units of microsomes or polysomes (it is recommended that the polysome pellets be suspended in TKM containing 10%, v/v, of the Sephadex G-100 fraction). The mixture was incubated for up to 2.5 hr at 28°. At various times, aliquots (5 μl) from each translation mixture were assayed for the kinetics of total radioactivity incorporated into hot trichloroacetic acid-insoluble material on Whatman 3 MM paper disks that were processed in the presence of 1% hydrolyzed amino acids.[8]

We have observed that amino acid incorporation by polysomes can last for up to 3 hr at 28° upon addition of the precipitates obtained from the HSS with 40 to 60% saturated ammonium sulfate at a final concentration of 1–1.2 mg/ml and bovine or yeast tRNA at a final concentration of 75 μg/ml.

A Sephadex G-100 fraction has been chosen for the reasons described below. For efficient RM translation, polysome degradation by endogenous microsomal RNase during incubation must be minimized to at least an extent that will allow continued peptide synthesis. We therefore inspected the stability of bound polysomes in the presence of microsomes that were stripped of ribosomes (stripped microsomes) in various *in vitro* cell-free translation systems including the reticulocyte lysate,[9] wheat germ extract,[10] pH 5 fraction,[11] and the Sephadex G-100 fraction from rat liver HSS, in conditions under which no elongation or termination occurred during incubation.

It was found that with incubation at 4° for 10 min, bound polysomes incubated with stripped RM in the wheat germ system were degraded into monomers and dimers. The polysome degradation was not prevented upon addition of higher concentrations of the wheat germ extract (up to 30% v/v). Incubation in the reticulocyte lysate or the pH 5 fraction, also resulted in degradation, but to a considerably lesser extent. Raising the temperature of incubation in these three systems to 38°, however, generally caused polysome degradation. In contrast, the polysomes and stripped RM incubated in a cell-free system containing the G-100 fraction

[8] R. J. Mans and G. D. Novelli, *Arch. Biochem. Biophys.* **94,** 48 (1961).
[9] H. R. E. Pelham and R. J. Jackson, *Eur. J. Biochem.* **67,** 247 (1976).
[10] K. Marcus and B. Duodock, *Nucleic Acids Res.* **1,** 1385 (1974).
[11] G. Blobel and B. Dobberstein, *J. Cell Biol.* **67,** 852 (1975).

were not degraded. Their profiles remained intact even after incubation at 28° for 90 min.

Based on our results, we have chosen the Sephadex G-100 fraction because (*a*) as opposed to the other systems, polysome stability was considerably more constant in the G-100 system; and (*b*) the Sephadex G-100 fraction can be used as the source of RNase inhibitor in sucrose cushions for collecting and concentrating RM that can then be used directly for translation. Neither the reticulocyte lysate nor the pH 5 fraction afford this advantage.

In this system (see Fig. 3) the radioactive amino acid incorporation by bound polysomes increased linearly for about 90 min, after which it gradually decreased and then stopped completely. This cessation of incorporation has been confirmed to be due to termination of protein synthesis, not to polysome degradation. Incorporation by RM increased linearly as well, but it ceased earlier than that by bound polysomes (Fig. 3); frequently it ceased in 15 min. Cessation with RM, in contrast to bound polysomes, has been found to be due neither to polysome degradation nor to termination of protein synthesis. It was also observed that when stripped microsomal membranes were added to the G-100 cell-free system containing bound polysomes, the overall amino acid incorporation decreased in proportion to the amount of stripped microsomes added, although, by sucrose density gradient analysis, the bound polysomes were found undegraded at the end of incubation.

As with the cessation of RM translation, other factors must be involved. These may include any of the following, as previously described in detail[12]: (*a*) alteration of the nature of the microsomal membrane; (*b*) loss of protein synthetic components by adsorption to microsomal membranes; (*c*) freshness of materials.

Considerations concerning the translation conditions and hydrolysis of the newly formed peptides by endogenous proteinase that may occur during translation (incubation) have been described previously.[12]

Chemical Analysis and Enzyme Assay. Protein concentrations were determined according to the procedure of Lowry *et al.*[13] RNA concentrations were determined using a slightly modified Fleck and Munro's procedure[14] as described by Blobel and Potter.[1] Cytochrome oxidase was assayed according to the procedure of Smith and Stotz[15] with reduced

[12] R. A. Feldman, S. Gaetani, and T. Morimoto, *in* "Cancer Cell Organelles" (E. Reid, G. Cook, and D. J. Morré, eds.), p. 263. Ellis Horwood Limited, Chichester, 1982.

[13] O. H. Lowry, N. J. Rosebrough, A. L. Farr, and R. J. Randall, *J. Biol. Chem.* **193,** 265 (1951).

[14] A. Fleck and H. N. Munro, *Biochim. Biophys. Acta* **55,** 571 (1962).

[15] F. G. Smith and E. Stotz, *J. Biol. Chem.* **179,** 891 (1949).

cytochrome *c*. Concentration of microsomes was routinely determined by measuring A_{280} and A_{260} after solubilization with 0.5% deoxycholate (DOC) or 0.2% sodium dodecyl sulfate (SDS).

Cell Fractionation

Preparation of Crude RM and SM Fractions

Livers were minced and homogenized in 4 volumes (w/v) (see *Note 1*) of homogenizing medium using a Potter-Elvehjem homogenizer with a motor-driven Teflon pestle (10 strokes). The homogenate was centrifuged in a Sorvall swinging-bucket HB-4 rotor at 8000 rpm for 10 min (see *Note 2*) and the supernatant was recentrifuged at 8000 rpm for 20 min. This PMS was then transferred with a syringe, taking care not to contact or disturb the pellet. The second centrifugation can be eliminated for preparation of bound polysomes, but it is necessary for the preparation of RM for translation and stripped RM for reconstitution experiments.

A three-step discontinuous gradient in Ti60 tubes was prepared: 5 ml of mixture A (1.8 *M* sucrose containing PMS, prepared by mixing 1 volume of PMS with 2 volumes of 2.5 *M* sucrose) on the bottom; 3 ml of mixture B (1.55 *M* sucrose containing PMS, prepared by mixing 1 volume of PMS with 1.1 volume of 2.5 *M* sucrose) as the middle layer; and 16–18 ml of mixture C (1.35 *M* sucrose containing PMS, prepared by mixing 1 volume of PMS with 0.7 volume of 2.5 *M* sucrose) on top (see *Note 3*). The tubes were overlaid with 1.0 *M* sucrose. This layer was intended to promote banding of the SM at 1.0/1.35 *M* sucrose interface in order to avoid exposure to air and to prevent tight packing at the top of the 1.35 *M* sucrose layer (it can be eliminated when the SM fraction is not needed) and centrifuged at 48,000 rpm (Beckman, Ti60 rotor) for a minimum of 6 hr (see *Note 4*). Crude heavy rough microsomes (RM_H) banded in the lower two-thirds of the mixture B layer, and crude light rough microsomes (RM_L) banded on top of the mixture B layer. The separation of light and heavy microsomal fractions as visible bands depends on the volume of mixture B and centrifugation conditions. Longer centrifugation at a high speed will cause less separation, and the crude RM appear as single band. In this case the upper and lower halves are defined as the RM_L and RM_H fractions, respectively. Crude SM banded on top of the mixture C layer.

Free and Bound Polysome Preparation

Free Polysomes. The PMS prepared from rat livers of animals starved overnight was adjusted to 1.35 *M* sucrose by addition of 0.7 volume of

2.5 *M* sucrose and was layered over a two-step discontinuous gradient consisting of 5 ml of 2 *M* STKM, 3 m*M* DTT, and 2–3 ml of mixture B. Centrifugation at 44,000 rpm (Beckman, Ti60 rotor) for 16 hr yielded a pellet of free polysomes whose surface was washed twice with TKM, drained well, and stored (see *Note 5*). Microsomal membranes, which frequently adhere to the inside wall of the centrifuge tube, may slide down to the free surface of the polysome pellet during transfer of the supernatant. Care must be taken to avoid this in order to prevent polysome degradation.

Bound Polysomes. Both light and heavy crude RM were mixed with an equal volume of two times concentrated TKM (20 m*M* $MgCl_2$) containing 10% (v/v) HSS (see *Note 6*), followed by addition of a mixture of DOC and Triton X-100 in final concentrations of 0.5% and 1%, respectively (see Recovery section under Properties). About 20 ml of this mixture was layered onto Ti60 tubes containing a two-cushion discontinuous gradient of 5 ml of 2 *M* STKM (20 m*M* $MgCl_2$), 3 m*M* DTT, and 2–3 ml of a mixture of 2 *M* STKM (20 m*M* $MgCl_2$) and solubilized RM fraction (3 : 1, v/v). Detergent was included in this layer so as to prevent reaggregation of membrane proteins that had been previously dissolved when these proteins reached this layer during centrifugation. The gradients were centrifuged at 48,000 rpm (Beckman, Ti60 rotor) for a minimum of 12 hr. The pellets of bound polysomes were washed twice with TKM, drained well (see *Note 7*) and, particularly for reconstitution experiments, used immediately.

Preparation of Rough Microsomes for Translation in a Cell-Free System

When the concentration of the crude RM_H is high enough for *in vitro* translation, the following washing and concentrating steps can be eliminated. Crude RM_H were combined with a solution containing 10% (v/v) HSS in TKM and layered over a three-step discontinuous gradient in SW41 tubes composed of 1 ml of 1.8 *M* STKM containing 10% (v/v) HSS on the bottom, 1 ml of 1.35 *M* STKM containing 20% (v/v) G-100 rat liver fraction (see *Note 8*) as the second layer, and 2 ml of 0.7 *M* STKM containing 10% (v/v) HSS as the third layer. After centrifugation of the gradients at 40,000 rpm (Beckman, SW41 rotor) for 60–90 min (60 min when the RM_H fraction was diluted about 4-fold, and 90 min when diluted 2.5-fold), the washed RM banded at the 1.35/1.8 *M* STKM boundary. These washed and concentrated RM were pipetted off and pooled or, if necessary, were further concentrated by mixing with a solution containing 20% (v/v) HSS in TKM, then layering over another two-step discontinuous gradient made in either SW41 or SW56 tubes (1.5 ml of 1.8 *M*

STKM containing 10% G-100 fraction and 1.5 ml of 1.0 *M* STKM containing 20% G-100 fraction, v/v, for SW41 tubes and 0.5 ml each of the two corresponding sucrose layers for SW56 tubes). These gradients were also centrifuged at 40,000 rpm (Beckman, SW41 or SW56 rotors, respectively) for 40 min, which produced a thick band of RM at the 1.0/1.8 *M* STKM boundary. These washed, concentrated RM were syringed off and shaken to eliminate aggregation, and the $A_{260} : A_{280}$ ratio was determined in 0.2% SDS or 0.5% DOC.

Notes on the Above Procedures

Note 1. Volume of Homogenizing Solution. Low recovery of RM seemed to be caused by loss to the nuclear and mitochondrial pellets, probably because of aggregation of subcellular organelles. It was found that, although membrane aggregation was reduced by lowering the salt concentration of the solutions used during fractionation, it occurred again if salts were eliminated entirely. The extent of aggregation was found to be optimally reduced when cells were homogenized in an appropriate volume of salt-free sucrose solution so that the endogenous salts from the cell homogenate would be just sufficient to prevent aggregation. In this way, the internal milieu of the cell could be conserved as much as possible during the steps of the fractionation, and the Mg^{2+} concentration could be maintained at the level required for polysome integrity.

With rat liver, the choice of volume was based not only on the concentration of endogenous salt, but also on the concentration of the endogenous RNase inhibitor. The homogenizing solution could be increased to 6 volumes, unless the increase caused handling problems. For example, when 25 g of rat livers are homogenized in 4 volumes of the homogenizing solution, the PMS obtained is just enough to make eight step gradients in centrifuge tubes for a Spinco Ti60 rotor.

Note 2. Centrifugation Conditions to Obtain the PMS. When the centrifugal force was decreased to 4000 *g* for 10 min in order to achieve a better recovery of RM, the resulting supernatant contained 74% of total RNA, but also 55% of the total cytochrome *c* oxidase activity of the homogenate, indicating greater contamination of mitochondria. This method of maximizing RNA recovery was therefore abandoned (see also Recovery section under Properties).

Note 3. The Cutoff Density of Sucrose in Step Gradients Used for the Separation of SM and RM. Since there is no discrete boundary between the buoyant density of SM and RM, we chose a cutoff density of 1.35 *M* sucrose to optimize their separation. This allowed the preparation of relatively pure RM, but the SM fraction was somewhat contaminated with

RM_L that floated up. However, the sucrose concentration should be changed according to (*a*) the aim of preparation, such as the preparation of RM_H or both RM_L and RM_H together; (*b*) the material used—for example, in rat brains a slightly higher sucrose concentration is required; (*c*) the salt concentration in the sucrose solution used for step gradients, as was the case for chicken embryonic muscle cells.[16]

Note 4. Centrifugation Conditions to Separate SM and RM. For gradients lacking a 2 *M* sucrose cushion at the bottom, a 6-hr centrifugation at 220,000–240,000 *g* in a Ti60 rotor is enough to sediment free ribosomes through 1.8 *M* sucrose. These gradients are designed to prepare only microsomes from the PMS.

For gradients containing a 2 *M* sucrose cushion, a 14-hr or longer centrifugation at 200,000–220,000 *g* is necessary to maximize the recovery of free polysomes. These gradients are designed to prepare both microsomes and free polysomes from the PMS. When rat livers are used for this purpose, the rats should be starved for about 16 hr, or the homogenate or PMS from unstarved rats should be treated with amylase[5]; otherwise the free polysome pellets will be heavily contaminated with glycogen.

Note 5. Storage and Activity of Rough Microsomes and Polysomes. Freshly prepared free and bound polysomes were used immediately for an *in vitro* translation, particularly when they were to be used for reconstruction of functional RM. These polysomes could, however, be kept at −20° for about a week without any diminution in the initial rate of radioactive amino acid incorporation, but in this case the duration of the linearity of the incorporation of radioactive amino acids into proteins was lower than that obtained using freshly prepared polysomes. Purified RM membranes to be used in an *in vitro* translation experiment could be stored in 50% glycerol for up to 3 weeks at −70°, but the level of overall activity was found to decrease. However, the use of fresh membrane preparations every time is recommended, so that membrane receptors, cofactors, and enzyme systems being assayed for have their highest activities. This is particularly true when RM are translated.

Note 6. Use of an Additional HSS and a Higher Concentration of $MgCl_2$. Although in most cases undegraded polysomes were obtained when the RM fraction was solubilized in the absence of an additional HSS, it is better to add it for safety (10% HSS can inhibit the RNase activity of microsomal suspension at a concentration of up to about 8 $OD_{280\ nm}$ units). For preparation of bound polysomes, $MgCl_2$ concentration of microsomal suspension and sucrose cushions is raised to 20 m*M*,

[16] T. Chyn, A. N. Martonosi, T. Morimoto, and D. D. Sabatini, *Proc. Natl. Acad. Sci. U.S.A.* **76,** 1241 (1979).

because this condition seemed to be better for obtaining undegraded polysomes.

Note 7. Quantitative Recovery of Polysomes as Pellets. Occasionally, the centrifugation temperature is lower by 1–2° than indicated; this causes a lower sedimentation rate for polysomes in the 2 *M* sucrose layer, in particular when the layer contains high-salt buffer. It is better, therefore, to save the lower portion of the 2 *M* sucrose layer and measure the absorbance at 260 nm and 280 nm, in order to make sure that all or most ribosomes are recovered as pellet.

Note 8. Reason for Inclusion of Rat Liver HSS or a G-100 Fraction (Prepared from the HSS) in the Step Gradients Used for Washing and Concentrating RM. We have observed that if RM are exposed to the medium lacking RNase inhibitors at any stage of their preparation, frequently membrane-bound polysomes are degraded (Fig. 1). For this reason, all solutions making up step gradients used for washing and concentrating RM are supplemented with 10–20% (v/v) rat liver HSS. In the last concentration step before *in vitro* translation, however, the Sephadex G-100 fraction is substituted for HSS as the source of RNase inhibitor for the specific sucrose layer in which RM band. The optimal final concentration range of the G-100 fraction with respect to RNase inhibitory activity is the same as that of HSS, but small components of the HSS that may interfere with subsequent *in vitro* translation of RM are removed during preparation of the G-100 fraction, making it more appropriate to use at this stage. It was also found that when RM were added to a translation mixture these microsomes absorbed some factors from the mixture, which might cause early cessation of overall incorporation. Suspending RM in a solution containing a Sephadex G-100 fraction could minimize the absorption.

The reason for supplementing the 1.35 *M* STKM layer with 20% v/v HSS or Sephadex G-100 fraction instead of 10% is that, after centrifugation, the RM are concentrated in this layer above the 1.8 *M* sucrose cushion, and extra protection against the endogenous RNase activity of this fraction is sought. Concentration of the salts in this layer can be changed so as to be easier to compensate to that of the *in vitro* translation mixture.

Properties of Isolated Rough Microsomes and Bound Polysomes

Recovery and Purity

About 60% of the total cellular RNA was recovered in the PMS, and this fraction contained about 9% cytochrome *c* oxidase activity. The RNA recovery of this fraction could be increased to 80% by washing and recentrifuging the nuclear and mitochondrial pellets. This step resulted, however, in significant mitochondrial contamination (18%). In addition, often polysomes obtained from RM prepared from this nuclear and mitochondrial wash were degraded even if the rehomogenization step was done in the presence of HSS. The crude RM fraction (both heavy and light combined) contained about 25% of the total cellular RNA. The crude SM fraction and the 1.35 *M* sucrose layer contained 4.7% and 8.2% of the total RNA, respectively. The RNA to protein ratios of the crude RM and SM fractions are about 0.195 and 0.043, respectively. The cytochrome *c* oxidase activity of the crude RM and SM fractions is only 1.3% and 3.2% of that of the homogenate, respectively.

The extent of free ribosome contamination was related to the duration of centrifugation at 48,000 rpm. A preliminary experiment to determine the minimum centrifugation time at that speed, using a ribosome suspension of mixture C, revealed that there were no ribosomes remaining in the mixture B layer (corresponding to the 1.55 *M* STKM layer) when centrifugation lasted 6 hr. It is still possible, however, that some free ribosomes contaminate this layer in the cell fractionation procedure, where rapidly banding microsomes may physically trap free ribosomes and interfere with their descent or decrease their sedimentation rate. The extent of free ribosome contamination into the crude RM fraction was about 10% of the total ribosomes in this fraction when examined using three different sucrose gradients as described before.[12] Here free ribosomes are defined as those ribosomes that are unattached to RM membranes in low salt buffer (TKM).

The free ribosomes contaminating this fraction can be almost completely removed by washing through a step gradient containing HSS. The crude RM fraction thus obtained, therefore, is relatively pure and free of both mitochondrial and free ribosomal contamination, which was also shown by a comparison of the translation products (see Fig. 4a).

With respect to the RM fraction, electron microscopic observation (electron micrographs are not shown here) has shown that the RM_L fraction is mainly composed of RM, with some contamination of smooth membrane vesicles, mitochondria, and other organelle fragments. The RM_H fraction is rather homogeneous, with typical RM consisting of rough-surfaced vesicles densely coated with ribosomes. The polysome profiles of the two subfractions of RM were indistinguishable, despite the fact that their buoyant densities are different, which led us to use both fractions for preparation of bound polysomes, and RM_H for translation and for the preparation of stripped RM.

Intactness of Isolated RM and Membrane-Bound Polysomes

Two criteria have been used to test for the intactness of isolated RM and membrane-bound polysomes: (*a*) intactness of mRNA evidenced by the presence of a wide molecular weight range of newly synthesized polypeptides, and (*b*) capability for protein synthesis evidenced by occurrence of reinitiation. For the latter purpose, albumin, a secretory protein, was chosen as a marker protein for the following reasons.

1. It is synthesized on membrane-bound polysomes in a precursor form (preproalbumin) that has an extra 18 amino acid residues at the N terminus of proalbumin[17]; preproalbumin is cotranslationally converted to proalbumin and vectorially transferred across the microsomal membrane into the lumen, where it is resistant to proteolysis by exogenously added proteinase.[18]
2. Since the size of preproalbumin is larger by about 2000 daltons, the two molecules can be resolved as separate bands, though very close to each other, on SDS–PAGE.[19]
3. Sodium dodecyl sulfate–polyacrylamide gel electrophoresis (SDS–PAGE) of the cyanogen bromide fragments of albumin permits evaluation of the extent of reinitiation of protein synthesis by bound polysomes or RM in an *in vitro* system. As shown in Fig. 5, the cyanogen bromide fragment I, which includes the N terminus, and fragment VI, which is located near the C terminus, are very well separated from each other on SDS–PAGE. There are no major intervening bands thus eliminating the possibility of spillover into either fragment by a contaminating band. Therefore, the radioactivity in the two bands representing fragments I and VI can be accurately measured. When mRNA is translated in the presence of a radioactive amino acid, e.g., [^{3}H]phenylalanine, all albumin molecules will be equivalently labeled starting from their N terminus. When bound polysomes are translated for one cycle, however, each nascent chain will be labeled from a different starting point: some will be labeled starting near the N terminus, whereas others will not show incorporation until the middle of the molecule. If the bound polysomes do not reinitiate protein synthesis in the cell-free system, the C-terminal fragment (VI) will be more heavily labeled than the N-terminal fragment (I). Therefore, the radioactivity ratio of fragment VI to fragment I of polysome translation products would be much larger than that of mRNA translation products. If the bound polysomes do reinitiate protein synthesis, the amount of

[17] A. W. Strauss, C. D. Bennet, A. M. Donohue, J. A. Rodkery, and A. W. Albert, *J. Biol. Chem.* **252,** 6846 (1977).

[18] A. W. Strauss, C. A. Bennet, A. M. Donohue, J. A. Rodkery, I. Boime, and A. W. Albert, *J. Biol. Chem.* **253,** 6270 (1978).

[19] J. M. Taylor and R. T. Schimke, *J. Biol. Chem.* **248,** 7661 (1973).

label in the N terminus (fragment I) will increase in proportion to the extent of reinitiation, and eventually the radioactivity ratio will approach to that of mRNA translation. The extent of reinitiation can, thus, be estimated by comparing the radioactivity ratio of fragment VI to fragment I between the mRNA and the bound polysome translation products.

Intactness of mRNA. Poly(A)$^+$ mRNA was isolated by oligo(dT) cellulose column chromatography[20] from total RNA extracted from bound polysomes with phenol.[21] The mRNA was translated in the reticulocyte lysate system,[9] plus or minus stripped RM from dog pancreas. Messenger RNA extracted from free polysomes was also translated. Autoradiography (Fig. 2) has shown a wide molecular weight range of labeled polypeptides with a distinct difference between the two classes.

The anti-albumin immunoprecipitates of bound polysomal mRNA translation products, in the absence of the stripped RM (Fig. 2, lane c), comigrate with a major band seen in Fig. 2, lane b, which is barely identifiable in the free polysomal mRNA translation products (Fig. 2, lane a). This band migrates slightly slower than both mature albumin and anti-albumin immunoprecipitates of bound polysomal mRNA translation products in the presence of stripped RM. This latter product was isolated from microsomes that were stripped and trypsinized after the translation had been terminated and, in Fig. 2, lanes d and e, it can be seen that this product comigrates with mature albumin. The immunoprecipitates of bound polysomal mRNA translation products, plus and minus stripped RM membranes, were confirmed by amino acid sequence analysis to be preproalbumin and proalbumin, respectively.

Capability for Protein Synthesis in Vitro: Occurrence of Reinitiation. In the present *in vitro* system, protein synthesis is linear for 90 min with bound polysomes and for a minimum of 30 min with RM (Fig. 3). The ability of the *in vitro* system to reinitiate protein synthesis after termination of growing nascent chains was tested using specific inhibitors of initiation, ATA[22,23] and 7-mG5′p.[24,25] These inhibitors led to a progressive decrease in amino acid incorporation, which stopped in 20 min (Fig. 3). These effects on translation contrast with those using puromycin or cycloheximide (inhibitors of elongation) in which incorporation halted immedi-

[20] H. Aviv and P. Leder, *Proc. Natl. Acad. Sci. U.S.A.* **69,** 1408 (1972).

[21] G. Brawerman, J. Mendecki, and Se Y. Lee, *Biochemistry* **11,** 637 (1972).

[22] M. L. Stewart, A. P. Grollman, and M. T. Huang, *Proc. Natl. Acad. Sci. U.S.A.* **68,** 97 (1971).

[23] M. B. Matthew, *FEBS Lett.* **15,** 201 (1971).

[24] E. D. Hickey, L. A. Weber, and C. Baglioni, *Proc. Natl. Acad. Sci. U.S.A.* **73,** 9 (1976).

[25] L. A. Weber, E. D. Hickey, and C. Baglioni, *J. Biol. Chem.* **253,** 178 (1978).

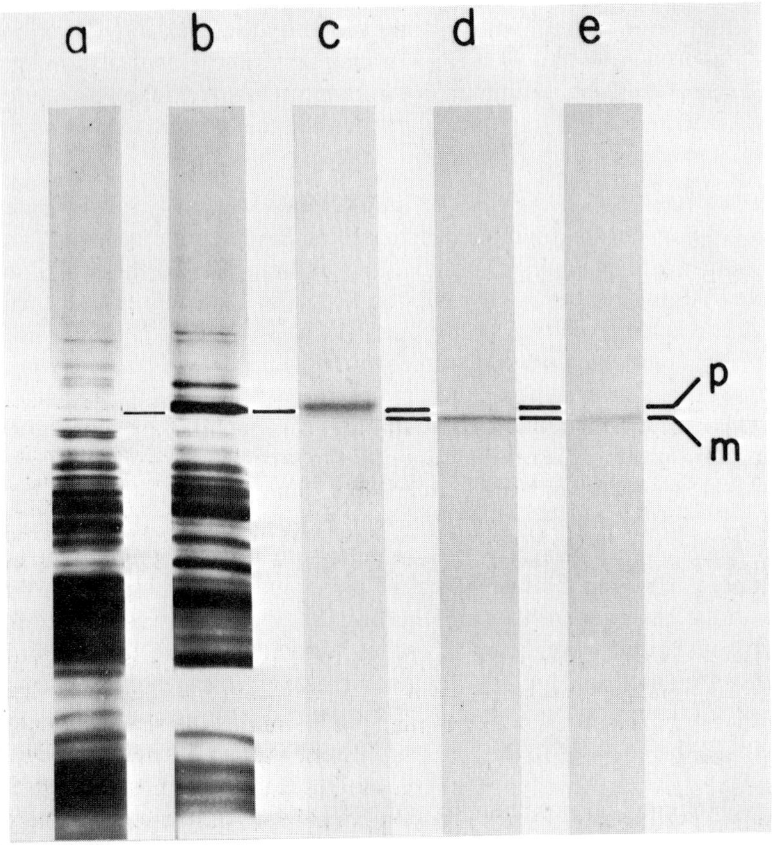

FIG. 2. Autoradiographs showing total translation products programmed with mRNA of free (lane a) and membrane-bound (lane b) polysomes, and the anti-albumin immunoprecipitates from the translation products of mRNA from membrane-bound polysomes in the absence (lane c) and in the presence (lane d) of stripped RM. Poly(A)-containing mRNA was prepared from free and membrane-bound polysomes as described in the text. *In vitro* protein synthesis was carried out using reticulocyte lysate prepared as described by Pelham and Jackson.[9] In this experiment, the stripped RM were prepared from dog pancreas as described by Blobel and Dobberstein.[11] The translation mixture (100 μl) with 0.1 OD_{260} unit of poly(A)-containing RNA and [^{35}S]methionine was incubated at 28° for 60 min either in the absence or in the presence of 0.4 OD_{280} unit of stripped RM. After incubation, the mixture was subjected to immunoprecipitation with anti-albumin antisera followed by SDS–PAGE analysis,[26] except for the sample in lane d, which was treated with exogenously added trypsin (0.1 mg/ml, 30 min at room temperature), followed by incubation with soybean trypsin inhibitor (0.5 mg/ml, 10 min, at room temperature). Lane e: ^{125}I-labeled mature rat serum albumin. p, Preproalbumin; m, mature and proalbumin.

[26] J. V. Maizel, Jr., *Methods Virol.* **5,** 179 (1971).

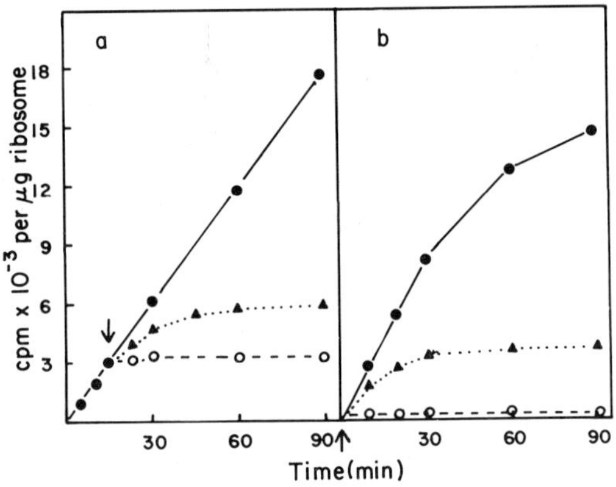

FIG. 3. Incorporation of [^{35}S]methionine into hot trichloroacetic acid-insoluble material. Membrane-bound polysomes (panel a, 2 OD_{260} units per 100 μl) and RM (panel b, 2 OD_{260} units per 100 μl) were translated in an *in vitro* translation system containing a Sephadex G-100 fraction as described in Materials and Methods. The incorporation profile in the presence of 7-methylguanosine 5′-phosphate (1 m*M* for bound polysome and 2 m*M* for rough microsome translation) was similar to that obtained in the presence of ATA. Arrows indicate the time of addition of the inhibitor. Control, ●——●; 10^{-4} *M* ATA, ▲···▲; 10^{-2} *M* cycloheximide, ○---○.

ately upon their addition. The effects of these various inhibitors were visualized by sucrose density gradient analysis as described in the legend to Fig. 4. A control sample, examined at 15 min of translation, contained very few free ribosomes (less than 10% of those contained in Fig. 4e). This amount was almost the same as that present at 0 time (figure not shown). During this period, 151,786 cpm of [^{35}S]methionine were incorporated into microsomes as hot trichloroacetic acid-insoluble precipitates associated with the membranes after high salt buffer treatment. Fifty-four percent of the radioactivity was resistant to proteolysis, and 6.5% of this radioactivity was recovered as immunoprecipitates with antiserum against albumin; 2–3% was recovered from untrypsinized microsomes, indicating that the newly synthesized albumin was vectorially discharged into the microsomal lumen, where it was protected from proteolysis. The control sample at 60 min of translation contained about the same amount of free ribosomes as observed in the 15-min sample, while hot trichloroacetic acid-insoluble radioactivity associated with the membranes was increased approximately threefold in the additional 45 min (Fig. 4a and d). The ATA-treated sample contained many free ribosomes (about 70% of those in Fig. 4e). Most of them were present as ribosomal subunits derived from inactive ribosomes that could not reinitiate protein synthesis after termination. Hot trichloroacetic acid-insoluble radioactivity incor-

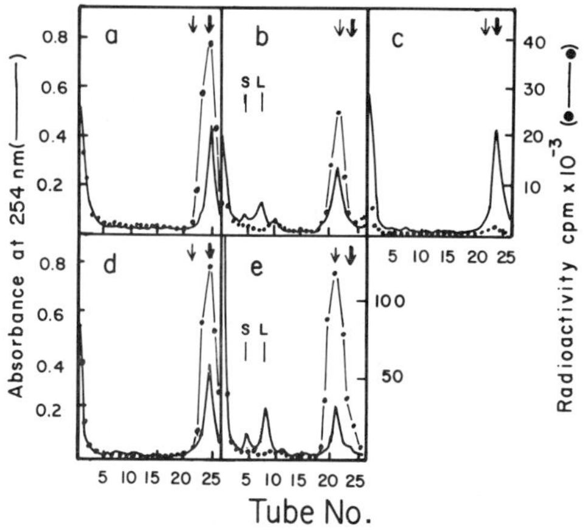

FIG. 4. Sucrose density gradient profiles of the translation mixture containing rough microsomes (RM) incubated at 28° for 15 min (a, b, and c) and for 60 min (d and e). A translation mixture (130 μl) containing RM (2.5 OD_{260} units, measured in the presence of 0.5% deoxycholate) was incubated in the presence of inhibitor of protein synthesis (panel b, 10^{-4} *M* ATA; panel c, 10^{-2} *M* cycloheximide) or in the absence of the inhibitor (panels a, d, and e). After incubation, samples were layered on 7.5 to 22.5% sucrose linear density gradients in HSB. Sample e was treated with 5×10^{-4} *M* puromycin at 37° for 10 min in the presence of HSB and then loaded onto the gradient. All gradients were made over two sucrose cushions (1.5 ml of 1.3 *M* STKM over 1.8 *M* STKM) and centrifuged at 39,000 rpm and 4° for 80 min in an SW41 rotor. Optical density at 254 nm was monitored as described in the legend to Fig. 1. Each gradient was fractionated into about 26 tubes. A 100-μl portion from each tube was pipetted onto a Whatman 3 MM filter disk and processed according to the procedure of Mans and Novelli.[8] S and L indicate the position of the small and large ribosomal subunits, respectively. Thin and thick arrows indicate the position at which stripped microsomes and reconstituted RM banded, respectively.

porated into microsomes was about half that of the control, and the buoyant density of the microsomes shifted to that of membranes treated with puromycin-KCl (Fig. 4b and e), indicating that addition of ATA led to a gradual decrease of amino acid incorporation with a concomitant increase in the number of ribosomes released by HSB treatment. The cycloheximide-treated sample at 15 min (Fig. 4c) contained very few free ribosomes, but a negligible amount of hot trichloroacetic acid-insoluble radioactivity was detected in the microsomes that retained the original buoyant density. In every case, no significant hot trichloroacetic acid-insoluble radioactivity was detected in the free ribosome region, other than the counts in the monomer region of the ATA-treated sample (Fig. 4b). These results, therefore, confirmed that the incorporation kinetics shown in Fig. 3 reflect the specific effect of the given inhibitor, and that initiation of protein synthesis occurs in this system with both membrane-bound polysomes and RM.

The extent of reinitiation of protein synthesis was examined by analyzing the anti-albumin immunoprecipitates as follows: The kinetics of radioactive amino acid incorporation in the presence of ATA and 7-mG5′p (Fig. 3) show that one cycle of translation requires approximately 20 min. Aliquots from bound polysome translation were sampled at 20 and 90 min after the start of incubation and subjected to immunoprecipitation with albumin antibodies. It was observed that, as predicted, bound polysome translation products contain two forms, one that comigrated with mRNA translation products, and another that comigrated with radioactive mature serum albumin. The immunoprecipitates from 20 min of translation contained mainly proalbumin, while those from 90 min of translation contained increasing amounts of preproalbumin. This indicates that most proalbumin nascent chains are completed during the first cycle of translation, followed by completion of newly initiated preproalbumin nascent chains. Furthermore, the immunoprecipitates from the translation products which had been labeled for 20 min after 20 min of incubation without radioactive amino acids contained preproalbumin in higher amounts than those obtained from the products that had been labeled for the first 20 min of incubation. These results demonstrate the occurrence of more than two cycles of reinitiation of protein synthesis in this cell-free system. This occurrence was corroborated by examination of the anti-albumin immunoprecipitates with CNBr fragmentation, followed by SDS–PAGE (Fig. 5). The CNBr fragments of the purified immunoprecipitates from the mRNA translation products were resolved as several bands on 15% polyacrylamide gels using SDS–PAGE. As described by Strauss *et al.*,[17] the radioactivity distribution patterns and the optical density distribution patterns of the fragments coincide, except at fragment I, where the radioactivity migrates more slowly than the optical density. The CNBr fragments of the immunoprecipitates purified from bound polysome translation products in the G-100 system (Fig. 5b) also show a pattern of radioactivity distribution coincident to the optical density, except that the radioactivity in fragment I is present in two distinct peaks. One of these peaks coincides with the optical density of this fragment I, and the other coincides with the radioactivity of fragment I from the mRNA translation. In contrast, the CNBr fragments of the immunoprecipitates purified from the RM translation in the G-100 system show total coincidence of the pattern of radioactivity and optical density. It should be noted, however, that the radioactive peak of the faster migrating component of fragment I from these RM does not precisely overlap with, but consistently migrates slightly slower than mature albumin fragment I. The same is true for this component from bound polysome translations. This may reflect the six amino acid difference between the mature and proalbumin molecules. These SDS–PAGE profiles of CNBr fragments demon-

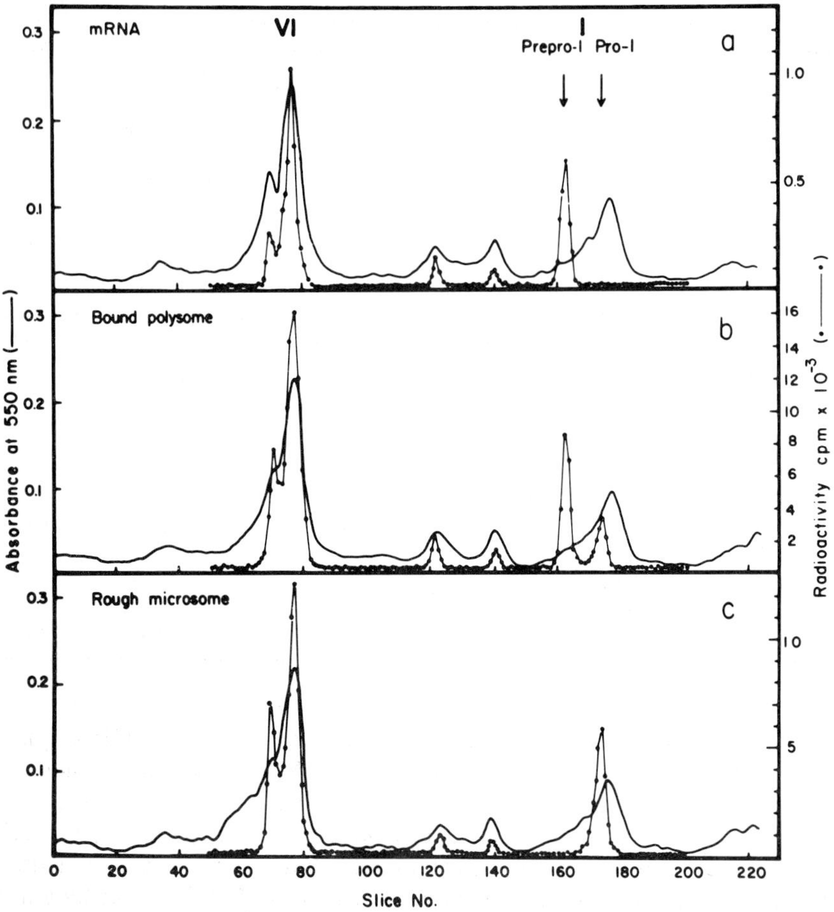

FIG. 5. The radioactivity distribution on SDS–PAGE of CNBr fragments of the anti-albumin immunoprecipitates from mRNA translation products in the reticulocyte lysate system (panel a), membrane-bound polysomes (panel b), and RM translation products (panel c), both in a Sephadex G-100 system. The labeled, purified immunoprecipitates containing about 200 μg of mature rat serum albumin as a carrier were dissolved in 70% formic acid and then subjected to CNBr treatment as described by Strauss *et al.*[17] I and IV indicate fragments I and IV, respectively.

strate that the purified immunoprecipitates of bound polysome translation products contain preproalbumin in addition to proalbumin.

As shown in the table, the molar ratios of leucine between fragments VI and I in the prepro molecule and between these fragments in the pro molecule are 2.13 and 2.90, respectively. The theoretical molar ratios of phenylalanine in these comparisons are 0.90 and 1.25, respectively. When

RATIO OF THE RADIOACTIVITY BETWEEN CNBr FRAGMENTS VI AND I

Immunoprecipitates from	Radioactivity (cpm)		Ratio of leucine (or phenylalanine) between two fragments		
			Calculated		
	Fragment VI	Fragment I	VI/I with prepro-albumin	VI/I with pro-albumin	Observed
mRNA translation products in wheat germ system using [^{3}H]leucine	3,504	1,955	2.13	—	1.75
Bound polysome translation products in G-100 system using					
[^{3}H]Leucine	75,421	38,956	2.13	2.90	1.93
[^{3}H]Phenylalanine	2,830	3,200	0.90	1.25	0.88
Rough microsome translation products in G-100 system using [^{3}H]leucine	67,913	21,582	—	2.90	3.15

mRNA is translated in the wheat germ system containing [^{3}H]phenylalanine, the molar ratio of the products is 1.75. When bound polysomes are translated for 90 min in the G-100 system with either [^{3}H]leucine or [^{3}H]phenylalanine, the ratio of fragment VI to fragment I is 1.93 ([^{3}H]leucine) and 0.88 ([^{3}H]phenylalanine). These values closely resemble those obtained with mRNA translation products. The fact that they are slightly higher, however, indicates that some bound polysomes did not reinitiate during the incubation. In addition, reinitiation in the RM translation system is emphatically supported by the observation that RM showed at all times only one product with the mobility expected from proalbumin, and that the radioactivity of the immunoprecipitates is close to the calculated value.

Concluding Remarks

Advantages and Disadvantages of the Modified Cell Fractionation Procedure

That this modified procedure produces undegraded RM and membrane-bound polysomes is substantiated by the proportion of large polysomes in the ribosome population (Fig. 1a) and the ability to reinitiate protein synthesis in an *in vitro* system (Figs. 3–5). Furthermore, it is a very simple and economical procedure with the least risk of polysome degradation and has allowed us to apply this method successfully to a large-scale preparation of undegraded RM using a Beckman Ti15 zonal rotor as described below. Two hundred gram livers were homogenized in 800 ml of homogenizing solution, and the PMS was prepared as described in the text. The supernatant was mixed with 2.5 *M* sucrose to prepare three different mixtures: 230 ml of mixture A, 140 ml of mixture B, and 1050 ml of mixture C. Mixture C was poured into a Beckman Ti15 rotor, which was set in the rotor chamber and then accelerated. When the rotor speed reached 3000 rpm, mixture B followed by mixture A were loaded from the edge of the rotor, and 250 ml of 1 *M* sucrose solution were loaded from the center to make a step gradient corresponding to that made in a Ti60 tube. The gradient was centrifuged at 35,000 rpm for 20 hr in the cold. In this method, about 210 ml of SM fraction and about 230 ml of RM fraction were obtained. The polysome profile of these RM examined by sucrose density gradient centrifugation and their ability to carry out protein synthesis in a cell-free system were the same as those of RM obtained by a small-scale preparation using a Ti60 rotor.

The new procedure, however, has the disadvantage of sacrificing material to the nuclear-mitochondrial pellet, which decreases the recovery of RM. For this reason, the other procedures should be employed for the quantitative recovery of free and membrane-bound polysomes for RNA extraction and for the quantitative recovery of RM and SM for physical, chemical, and biochemical analyses. As for the first case, there is a modification of Loeb *et al.* procedure[27] by Ramsey and Steele.[5] These free and membrane-bound polysomes were less contaminated and of good recovery, and they were found to be undegraded when the first nuclear, mitochondrial, and microsomal pellet was resuspended in a buffer containing 50% (v/v) rat liver HSS as soon as the centrifugation stopped. It is important to proceed with the next step immediately, since polysome degradation appears to take place in the pellet. It has been frequently observed that the extent of polysome degradation depends upon how long it had taken before the pellet was suspended. It is particularly true when a large amount of tissue is used. This procedure has been successfully used for various tissues, but several precautions similar to those described in the following chapter must be taken when the procedure is applied to the tissue containing many cytoskeletal elements. As for the second case, there are the procedures of Blobel and Potter[1] and Adelman *et al.*[2]

[27] J. N. Loeb, R. R. Howell, and G. M. Tomkins, *J. Biol. Chem.* **242,** 2069 (1967).

Collecting microsomes as a suspension in a sucrose layer not only prevents polysome degradation, but also preserves microsome intactness. For example, we have obtained reproducible results when microsomes were collected as suspensions after being incubated cotranslationally in the cell-free system, and were then subjected to proteinase digestion to determine whether newly synthesized secretory proteins were vectorially discharged into the microsomal lumen and newly synthesized membrane proteins were inserted into microsomal membranes. In contrast, microsomes collected in a pellet gave variable results.

The principle of avoiding the pelleting of subcellular fractions has been applied to the preparation of intact polysomes.[28] Polysomes were collected in 2.5 *M* sucrose buffer layer containing RNase inhibitor. In this manner, polysome degradation that could have taken place within a tightly packed pellet could be avoided by the continuous exposure of the polysomes to the RNase inhibitor. Furthermore, undesirable events such as the aggregation and damage of polysomes during pelleting and resuspension could be avoided. This method could be used for *in vitro* translation if conditions were established to obtain a high enough concentration of ribosomes in the sucrose layer (about 50 OD_{260} units/ml or higher) so that the polysome sample could be used directly for *in vitro* translation without any further steps, i.e., concentration of the polysomes and/or dialysis to reduce the sucrose concentration (the presence of sucrose at a concentration higher than 10% in the translation mixture has been found to reduce the rate of incorporation of amino acids).[12]

Precautions for Application of the Modified Procedure To Prepare RM and Membrane-Bound Polysomes from New Materials

The application of the procedure to the fractionation of other cells may require minor modifications adopted to characteristics of each cell type: RNase activity of the microsomes, RNase inhibitory activity of the cell sap, development of cytoskeleton, etc. In addition, a given RNase inhibitor may not always be effective against nonhomologous microsomal RNase; i.e., rat liver HSS does not inhibit the microsomal RNase of mouse liver as effectively as does that of rat liver. Therefore, the optimal concentration of rat liver HSS should be determined for each case, as should the endogenous RNase inhibitory activity of the cell sap and the endogenous RNase activity of microsomes, as described by Feldman *et al.*[12] Then, attempts should be made to minimize aggregation of subcellular organelles that otherwise decrease the purity of individual subcellular fractions and diminish RM recovery by loss into the nuclear

[28] J. P. Kraus and L. E. Rosenberg, *Proc. Natl. Acad. Sci. U.S.A.* **79,** 4015 (1982).

and mitochondrial pellets. For example, for the preparation of RM from chicken embryonic muscle cells, sucrose-buffer solutions containing a relatively high concentration of salt (250 m*M* NaCl) were used to avoid aggregation of subcellular organelles caused by cytoskeletal elements.[16] For the preparation of RM from rat brains, 6–8 volumes of the homogenizing solution were used, because the nuclear and mitochondrial pellets obtained from the 4-volume homogenate formed a large, fluffy layer that made it difficult to recover the supernatant quantitatively. Finally, for the preparation of RM from tissue culture cells whose cytosol does not contain sufficient amounts of RNase inhibitor, the rat liver HSS was added to the homogenizing solution (10–15%, v/v).[12]

[2] Methods for the Preparation of Messenger RNA

By PAUL M. LIZARDI

Currently available techniques permit the isolation of messenger RNA (mRNA) from virtually any organism, tissue, or subcellular fraction. Methods for RNA extraction have been reviewed in this series[1,2] and elsewhere.[3,4] The techniques described here were developed specifically for the study of eukaryotic mRNAs, but are largely adaptable to prokaryotes. The methods outlined below were chosen because of their simplicity and high yield and are especially suited for work with moderate or small amounts of biological material. A major section is dedicated to RNA size fractionation, including a detailed protocol for high-resolution preparative gel electrophoresis of mRNA in agarose gels and for mRNA recovery by electroblotting.

RNA Extraction Methods

Precautions against Exogenous Nucleases

To eliminate contamination by nucleases from exogenous sources, all glassware should be heat-treated for at least 2 hr at 170°. Solutions can be made RNase-free by treatment with 0.07% diethyl pyrocarbonate (DEP)

[1] M. Girard, this series, Vol. 12A, p. 581.
[2] K. S. Kirby, this series, Vol. 12B, p. 87.
[3] G. Brawerman, *Methods Cell Biol.* **7,** 1 (1973).
[4] J. M. Taylor, *Annu. Rev. Biochem.* **48,** 681 (1979).

ISBN 0-12-181996-5

followed by autoclaving for 30 min.[5] Sucrose-containing solutions treated with DEP may be heated in a boiling water bath for 30 min to avoid the caramelization that occurs on autoclaving. Solutions can be made free of particulates by filtration through a 0.45 μm (type HAWP) Millipore filter. Disposable pipette tips and plastic microcentrifuge tubes should be autoclaved in clean plastic containers. Vinyl or latex gloves should be worn for most manipulations.

Extraction of Whole-Cell or Tissue RNA

The ideal starting material for isolation of mRNA is a polyribosome or cytosol fraction. However, in many instances cell fractionation turns out to be laborious or may result in mRNA degradation. In such cases mRNA must be prepared from whole cells or tissues. The three techniques outlined below have been used successfully to isolate total RNA from a wide variety of sources. The method of choice for a particular application will often have to be determined empirically.

Phenol Extraction

This is the most widely used method for RNA extraction, and current versions are based on the phenol–sodium dodecyl sulfate (SDS) procedure described by Scherrer and Darnell.[2,6] While generally very effective, phenol extraction is rather sensitive to a number of variables, such as temperature, pH, ionic strength, presence of chloroform in the organic phase. These variables can have a dramatic effect on the recovery of poly(A)-containing mRNA, which can sometimes be trapped at the organic/aqueous interphase.[3,7]

Here I describe a version of the phenol/chloroform-SDS method that includes a preliminary digestion step in a buffer containing proteinase K and SDS. Digestion with this protease[8] results in a marked reduction in the size of the organic/aqueous interphase, regardless of whether phenol is used alone or in combination with chloroform. Besides increasing RNA yield, proteinase K digestion leads to rapid inactivation of most nucleases, which means that it is often unnecessary to add other ribonuclease inhibitors.

[5] DEP reacts with primary amino groups, including those of Tris buffer. The small drop in pH of DEP-treated solutions is usually of no consequence. It is important to destroy DEP by autoclaving to preclude modification of adenines in mRNA.

[6] K. Scherrer and J. E. Darnell, *Biochem. Biophys. Res. Commun.* **7,** 486 (1962).

[7] R. P. Perry, J. La Torre, D. E. Kelley, and J. R. Greenberg, *Biochim. Biophys. Acta* **262,** 220 (1972).

[8] H. Hilz, U. Wiegers, and P. Adamietz, *Eur. J. Biochem.* **56,** 103 (1975).

Reagents

Proteinase K (E. Merck, Darmstadt, Germany; U.S. distributor: E. M. Laboratories) is used as a powder or from a 6 mg/ml stock stored at −80° in small aliquots. It is stable for at least 6 months at −80°

Sodium lauryl sulfate (SDS): "specially pure" grade from BDH chemicals (U.S. distributor: Gallard Schlessinger); used as supplied

Phenol–chloroform–isoamyl alcohol. Commercial reagent grade phenol is colored and should be distilled by collecting the material that boils at 165–180° using an air-cooled condenser. Distilled water (18% of phenol volume) is added to the phenol and mixed thoroughly. The water-saturated phenol is mixed with 1 volume of chloroform and 0.02 volume of isoamyl alcohol and stored at 4° in the dark.

NETS buffer: 0.12 *M* NaCl, 10 m*M* ethylenedinitrilotetraacetic acid (EDTA), 25 m*M* Tris-HCl, pH 7.4, 1.5% SDS

pH 9 buffer: 100 m*M* Tris-HCl, pH 9, 0.5% SDS

ETS: 0.5 m*M* EDTA, 20 m*M* Tris-HCl, pH 7.4, 0.2% SDS

Procedure

1. Preparation of biological material. To obtain rapid inactivation of nucleases, tissue fragments must be solubilized in the protease–detergent mixture in the shortest possible time. For small, soft tissues this may be achieved relatively easily using a loose-fitting Dounce homogenizer. Other tissues may require preliminary disruption in a tissue press. An efficient method that minimizes nuclease activation is to freeze small fragments of tissue in liquid nitrogen, followed by grinding in a cold mortar on an electric coffee grinder to produce a coarse powder. Very hard tissues may be lyophilized after liquid nitrogen freezing in order to facilitate pulverization with a mortar. Slow thawing of frozen tissue should be avoided, as it often leads to RNA degradation. Cell culture monolayers can be digested *in situ* with protease–detergent mixture, but cells growing in suspension should be pelleted and resuspended in about 2.5 volumes of saline in order to produce a single-cell suspension.

2. Proteinase K–SDS digestion. The volume of protease–detergent mixture should be about 10 ml per gram of tissue or per milliliter of packed cells. Place an appropriate amount of NETS in homogenization vessel, bring to 37°, and add proteinase K (200 μg/ml for cells, 300 μg/ml for tissues). Then add cell suspension or tissue fragments and immediately begin homogenization with a plunger or a motor-driven spiral glass rod. Incubate for 30 min at 37° (some tissues require 40–60 min of diges-

tion) with intermittent agitation. Violent agitation (such as with a Waring blender) may partially inactivate proteinase K and should not be used for periods longer than 10 sec.

3. Phenol extraction. Transfer homogenate to tube or bottle containing an equal volume of phenol–chloroform–isoamyl alcohol at room temperature. Shake vigorously for 3–5 min, then centrifuge for 5 min at ≥2000 *g* (room temperature). Separate aqueous phase, and reextract organic phase with one-half volume of pH 9 buffer. Pool the first and second aqueous phases, and extract once more with fresh phenol–chloroform–isoamyl alcohol. Collect the final aqueous phase, adjust NaCl to about 0.2 *M*, add 2.5 volumes of ethanol, and incubate for 3 hr at −20°. Spin down nucleic acids at 3000 *g* for 15 min at 0°. Residual phenol may be removed by dissolving the nucleic acid pellet in ETS buffer and extracting twice with ethyl ether or by several ethanol precipitations from 0.2 *M* salt.

Ethanolic Sodium Perchlorate Extraction

This method utilizes a proteinase K–SDS digestion step, followed by a simple precipitation from 68% ethanol in the presence of a high concentration of sodium perchlorate.[9] The method is based on the observation[10] that sodium perchlorate greatly increases the solubility of proteins in ethanol. A subsequent precipitation step with isopropyl alcohol is included to remove a number of ethanol-insoluble contaminants, such as certain plant pigments. The yield of mRNA with this method is usually equal to or greater than with phenol extraction, since phase separation steps are not used.

Reagents

Proteinase K, SDS, NETS buffer, and ETS buffer are exactly as above, but SDS concentration in NETS should be increased to 2%.

Sodium perchlorate, 7 *M*. Commercial sodium perchlorate that gives a clear solution with neutral pH (6.0–7.5) is satisfactory. The salts obtained from Aldrich Chemical Company and British Drug Houses have been found to be of good quality. Solutions are routinely filtered through a Millipore filter (10 μm, type LC, or 5 μm, type SM) equipped with a glass prefilter, and then stored at room temperature.

EPR (ethanolic perchlorate reagent). This is simply saturated sodium perchlorate in approximately 85% ethanol solution. It is conveniently prepared by mixing 1 volume of saturated salt (~9 *M*) in

[9] P. Lizardi and A. Engelberg, *Anal. Biochem.* **98,** 116 (1979).

[10] J. Wilcockson, *Anal. Biochem.* **66,** 64 (1975).

H_2O (Millipore filtered) with 4 volumes of saturated salt in ethanol (unfiltered). EPR is stored at room temperature.

Procedure. The first step is a digestion using proteinase K in NETS buffer containing 2% SDS exactly as outlined above for the phenol extraction method. After incubation at 37° for 30 min, any excessive viscosity due to high molecular weight DNA can be reduced by several passages of the digest through a 16- or 18-gauge needle. Sodium perchlorate (7 *M*) is then added to give a final concentration of 1.4 *M*, and the solution is clarified by mixing for 3–4 min at 52°.[11] An equal volume of EPR is added to the warm lysate and mixed thoroughly to initiate nucleic acid precipitation. Another 3 volumes of EPR are added (final EPR concentration 80%), and the mixture is incubated for 30–45 min at 0° to 4° in a water bath. Precipitated nucleic acids are collected by centrifugation at 3000 *g* for 10 min (2°). The supernatant is decanted, and the pellet is dissolved in a volume of ETS equal to or larger than the original proteinase K digestion volume. Brief heating (50°) may be required for complete solubilization. The solution is brought up to 0.18 *M* NaCl, followed by the addition of 0.6 volume of isopropyl alcohol.[12] After 3 hr at −20°, nucleic acids are pelleted by centrifugation at 3000 *g* as before. Residual perchlorate may be removed by dissolving in a small volume of ETS followed by precipitation with 0.2 *M* salt and 2.5 volumes of ethanol. The final pellet contains both RNA and DNA.

Guanidinium Thiocyanate–CsCl Extraction

The guanidinium thiocyanate method is the method of choice for RNA isolation from tissues of very high RNase content or from tissues that are very difficult to homogenize. One of the two procedures outlined by Chirgwin *et al.*[13] involves a series of ethanol precipitations from guanidinium thiocyanate, and subsequently guanidine hydrochloride. This technique is efficient but rather time consuming. The alternative procedure, which is described below, involves the use of a CsCl centrifugation step[14]

[11] Most cells and tissues will yield a clear lysate at this step. Those that do not should be freed of debris by centrifugation or by filtration through a Millipore filter (5 μm, type SM, or 10 μm, type LC) equipped with a glass prefilter. This can be conveniently done using a reusable Swinnex-type syringe adaptor or millipore centrifuge tubes (Catalog No. XX6202550).

[12] Isopropanol will precipitate all the nucleic acids, except mononucleotides, at −20°, whereas at 18° tRNA will remain in solution. Quantitative precipitation of mRNA with isopropanol or with EPR requires an initial nucleic acid concentration of 80 μg/ml or higher when low-speed centrifugation (~3000 *g*) is used.

[13] J. M. Chirgwin, A. E. Przybyla, R. J. MacDonald, and W. J. Rutter, *Biochemistry* **18,** 5294 (1979).

[14] V. Glisin, R. Crkvenjakov, and C. Byus, *Biochemistry* **13,** 2633 (1974).

that yields a very pure RNA pellet after an overnight spin. Extracts from tissues rich in fat or polysaccharides are easily handled by this method because these substances float in dense CsCl solutions. Ribonuclease, which is completely inactivated by guanidinium thiocyanate in the presence of 2-mercaptoethanol, also remains at the top of the CsCl gradient.

Reagents

Guanidinium thiocyanate stock (4 *M*): Mix 50 g of Fluka purum grade guanidinium thiocyanate (Tridom, Inc.) with 0.5 g of sodium lauroylsarcosine (Sarkosyl NL-97, ICN, Inc.), 2.5 ml of 1 *M* sodium citrate, pH 7.0, 0.7 ml of 2-mercaptoethanol, and 0.33 ml of Sigma 30% Antifoam A. Add deionized H_2O, with some heating, to obtain 100 ml of solution at room temperature. Filter, adjust to pH 7, and store tightly closed for up to 5 weeks at room temperature.

CsCl solution: High-purity grade cesium chloride made up to 5.66 *M* and buffered with 60 m*M* sodium EDTA, pH 7.0. Sterilize with 0.07% diethyl pyrocarbonate and pass through a 0.45 μm Millipore filter. Technical grade CsCl may be used, but some batches will require prior purification through a bed of Chellex-100 (Bio-Rad Laboratories) to remove excess divalent metal ions.

Procedure. Set up homogenizing equipment in a fume hood (use a Tissumizer, Polytron, or Potter–Elvehjem homogenizer). Add 1 g of tissue per 16 ml of guanidinium thiocyanate solution, and homogenize for 30–60 sec at full speed. Layer homogenate on ultracentrifuge tubes (polyallomer) filled one-quarter (for swinging-bucket rotors) to one-third (for angle rotors) with CsCl solution. Spin at 36,000 rpm, 20°, for 12 hr in small rotors or for 18 hr in large rotors. Only certain rotors can be used for this procedure (i.e., SW50.1, or angle 55.2Ti) because the high density of CsCl (1.7 g/ml) and guanidine thiocyanate (1.15 g/ml) may impose speed limitations. At the end of the spin, remove the supernatant and wash the RNase-contaminated tube walls before reaching the bottom part of the solution. Wash the small, transparent RNA pellet with cold 70% ethanol to remove residual CsCl. Dissolve the RNA pellet in sterile 1 m*M* EDTA, pH 7.0, by brief heating (about 5 min) at 60°.

Comments on Alternative Extraction Methods

The reader may find that different combinations of the techniques outlined here can be used effectively for RNA extraction from "problem" tissues. For example, guanidinium thiocyanate homogenization followed by phenol extraction of the ethanol pellet has been used to isolate mRNA from corn leaves.[15] Digestion with proteinase K–sodium lauroylsarcosine followed by pelleting through CsCl yields excellent mRNA from the fat body of lepidopteran larvae.[16]

Removal of DNA and Other Contaminants

Although pelleting through CsCl yields RNA free of DNA or polysaccharide contamination, phenol or sodium perchlorate extraction do not. It has been reported[17] that proteinase K digestion can be performed in the presence of Ca^{2+} and DNase I to obtain simultaneous digestion of protein and DNA, but this promising method has not yet received widespread application. Removal of DNA from purified nucleic acid pellets may be achieved with RNase-free DNase. Commercial DNase can be made RNase-free by treatment with iodoacetate[18] or by chromatography in aminophenylphosphoryluridine 2′,3′-phosphate resin (Miles Yeda) as described by Maxwell *et al.*[19] DNase digestion is done in 10 m*M* MES (pH 6), 0.1 *M* NaCl, 5 m*M* $MgCl_2$, 2 m*M* $CaCl_2$. Incubation is performed using 15–50 μg of DNase I per milliliter at 25° for 30 min. DNA oligonucleotides can be removed by exclusion chromatography of the mRNA in Sephadex G-100 or Sephacryl S-200.

A simpler, alternative method to remove DNA and low molecular weight RNA, as well as certain polysaccharide contaminants, is precipitation from 2 *M* LiCl[20] or 3 *M* sodium acetate.[21] In a typical application, total nucleic acids are dissolved at a concentration of about 500 μg/ml, and any excessive DNA viscosity is reduced by several passages through a syringe needle. RNA is precipitated by adding 0.7 volume of 5 *M* LiCl followed by incubation in an ice-water bath for about 5 hr. The relatively loose pellet is collected by centrifugation at 16,000 *g* for 15 min at 0°. The supernatant contains DNA and tRNA and may be discarded or used as a source of endogenous tRNA for *in vitro* translation. The small amount of contaminating DNA that remains in the mRNA pellet can be removed by dT-cellulose chromatography or by size fractionation of the mRNA (see below).

Some tissues contain large amounts of particulate polysaccharides (such as glycogen), which are coextracted with RNA. A simple way to

[15] R. Broglie and N. H. Chua, personal communication (1982).

[16] P. Lizardi, unpublished observations (1981).

[17] R. H. Tullis and H. Rubin, *Anal. Biochem.* **107,** 260 (1980).

[18] S. B. Zimmerman and G. Sandeen, *Anal. Biochem.* **14,** 269 (1966).

[19] I. H. Maxwell, F. Maxwell, and W. E. Hahn, *Nucleic Acids Res.* **4,** 241 (1977).

[20] J. J. Barlow, A. P. Mathias, and R. Williamson, *Biochem. Biophys. Res. Commun.* **13,** 61 (1963).

[21] K. S. Kirby, *Biochem. J.* **96,** 266 (1965).

remove these is to dissolve the RNA in ETS buffer and centrifuge at 100,000 *g* for 30 min to pellet the polysaccharide.[22]

Isolation of mRNA from Subcellular Fractions

Methods for cell fractionation and for the preparation of free or bound ribosomes from various sources have been discussed in the preceding chapter. Optimal conditions for preserving mRNA integrity should be established for each application, as there is no truly universal method for inhibition of ribonucleases during cell fractionation. Certain commonly used ribonuclease inhibitors, such as heparin, may contaminate RNA preparations, but are easily removed by precipitating the RNA with 2 *M* LiCl or 3 *M* sodium acetate as described or by subsequent mRNA size fractionation steps (see below).

RNA from cytosol or from whole microsomes or polyribosomes can be extracted using any of the various methods described above. The sodium perchlorate extraction is particularly well suited for this purpose because it can be done very rapidly (20 min of proteinase K followed by 30 min of EPR precipitation) thanks to the absence of DNA viscosity.

The high RNA-to-protein ratio of polyribosomes has led some investigators to devise interesting abbreviated extraction methods. For example, Wiegers and Hilz[23] described the isolation of mRNA by digestion of polysomes with proteinase K–SDS followed by sucrose density gradient centrifugation. Digested protein, as well as proteinase K, remained at the top of the gradient. Krystosek *et al.*[24] have isolated poly(A)-containing mRNA by dT-cellulose chromatography of polysomes which were simply solubilized by treatment with SDS in high salt.

Methods for RNA Fractionation

Preparation of Poly(A)-Containing RNA

Methods for the isolation of poly(A)-containing mRNA using oligo(dT)-cellulose have been described elsewhere.[25–27] The yields of

[22] Polycarbonate ultracentrifuge tubes should not be treated with DEP and may be made RNase free with 10% hydrogen peroxide for 2 hr at 37°.

[23] U. Wiegers and H. Hilz, *FEBS Lett.* **23**, 77 (1972).

[24] A. Krystosek, M. L. Cawthon, and D. Kabat, *J. Biol. Chem.* **250**, 6077 (1975).

[25] H. Nakazato and M. Edmonds, this series, Vol. 29, p. 431.

[26] H. Aviv and P. Leder, *Proc. Natl. Acad. Sci. U.S.A.* **69**, 408 (1972).

[27] J. A. Bantle, I. H. Maxwell, and W. E. Hahn, *Anal. Biochem.* **72**, 413 (1976).

poly(A)-mRNA depend greatly on the average length and concentration of oligo(dT) chains, and it is advantageous to use commercial dT-cellulose where these parameters are well controlled (such as Collaborative Research type T3, or P-L Biochemicals type 7).

Binding of mRNAs with short poly(A) tails is most efficient with poly(U)-Sepharose[28] or poly(U)-Sephadex[29] owing to the greater stability of RNA–RNA hybrids. The Sephadex matrix is superior because all poly(U) chains are on the surface of the beads, equally accessible to all mRNAs, while Sepharose beads exclude mRNAs larger than 4000 nucleotides. The major disadvantage of poly(U)-containing affinity matrices is their relatively low stability compared to oligo(dT)-cellulose. However, it is possible to use poly(U)-Sephadex repeatedly over a period of many months by storing the beads at $-20°$ in loading buffer containing 35% glycerol as antifreeze or in mRNA elution buffer containing 50% formamide.

Size Fractionation of mRNA

Centrifugation and Gel Filtration Methods

The advantage of these two techniques lies in their relatively large capacity. Resolving power is poor, but considerable enrichment can be obtained for molecules smaller than 10 S or larger than 24 S, where fewer mRNA species are found. The choice of solvent system is very important in order to minimize RNA aggregation artifacts. The most widely used denaturing solvent for preparative sucrose gradients is formamide. Suzuki *et al.*[30] have described conditions for the fractionation of mRNA in gradients containing 70% formamide. Their method requires the use of a high-speed rotor (40,000 rpm) because RNA sediments very slowly in formamide–sucrose mixtures. It is possible to reduce centrifugation time while maintaining a high denaturation criterion by using gradients containing 60% formamide in an ultracentrifuge capable of maintaining a temperature of 32–35°. Gel filtration using agarose bead columns (Sepharose, BioGel A) has been used for mRNA size fractionation in aqueous solvents. Whereas some mRNAs can be substantially enriched by this method,[31] other species produce a smeared size profile due to aggregation artifacts. The availability of cross-linked agarose beads (Sepharose CL-2B, 4B) makes it possible to carry out mRNA chromatog-

[28] M. Adesnik and J. E. Darnell, *J. Mol. Biol.* **67,** 397 (1972).

[29] R. G. Deeley, J. I. Gordon, A. T. H. Burns, K. P. Mullinex, M. Binastein, and R. F. Goldberger, *J. Biol. Chem.* **22,** 8310 (1977).

[30] Y. Suzuki, L. P. Gage, and D. D. Brown, *J. Mol. Biol.* **70,** 637 (1972).

[31] Y. Suzuki and E. Suzuki, *J. Mol. Biol.* **88,** 393 (1974).

raphy under partially denaturing conditions (6 *M* urea, or 50% formamide), but this method has not received widespread application.

Gel Electrophoresis

The method of choice for high-resolution fractionation of mRNA is electrophoresis in agarose or polyacrylamide gels. Polyacrylamide can be used preparatively only for mRNAs of relatively low molecular weight (<20 S) and therefore agarose gels are of more general applicability. Agarose gel electrophoresis can be carried out preparatively under partially denaturing conditions using 6 *M* urea or 40% formamide. Fully denaturing conditions can be obtained by electrophoresis in the presence of methylmercury hydroxide.[32] Although this method yields excellent results, concerns about mercury toxicity present a problem for widespread laboratory use.

The method outlined here utilizes a buffer system containing 40% formamide and can be used with gels of 1.3–2% agarose content. The sample is heated in 50% formamide just before loading, which produces complete mRNA disaggregation in most applications. For example, poly(U) molecules hybridized to poly(A)-mRNA will be dissociated in this gel system. An example of electrophoretic separation of mRNAs from hen oviduct is shown in Fig. 1A. The mobility of ovalbumin mRNA (~1900 nucleotides) relative to the ribosomal RNA markers shows that migration is not directly proportional to molecular weight, as expected for a partially denaturing gel.

Reagents

Formamide: Commercial formamide (Fluka 47670, Eastman 565, or equivalent) must be deionized to remove ammonia, formate, and metal ion contaminants. Mix three ion-exchange resins as follows: 30 g of AG 501 X8 (20–50 mesh), 20 g of AG 50W X8 (20–50-mesh, hydrogen form), and 25 g of Chellex 100 (100–200 mesh, sodium form), obtained from Bio-Rad Laboratories, Richmond, California. Prewash mixture with 60 ml of formamide in order to dehydrate the beads. Filter to remove formamide, then add beads to 1 liter of formamide and stir for 2 hr at 4°. Filter in a large Büchner funnel, and store at −20° in the dark in 100-ml aliquots.

Agarose: Electrophoresis grade, from Bethesda Research Laboratories (BRL).

20× gel buffer: 0.7 *M* triethanolamine, 20 m*M* EDTA, 1% SDS, pH unadjusted

[32] J. Bailey and N. Davidson, *Anal. Biochem.* **70,** 75 (1976).

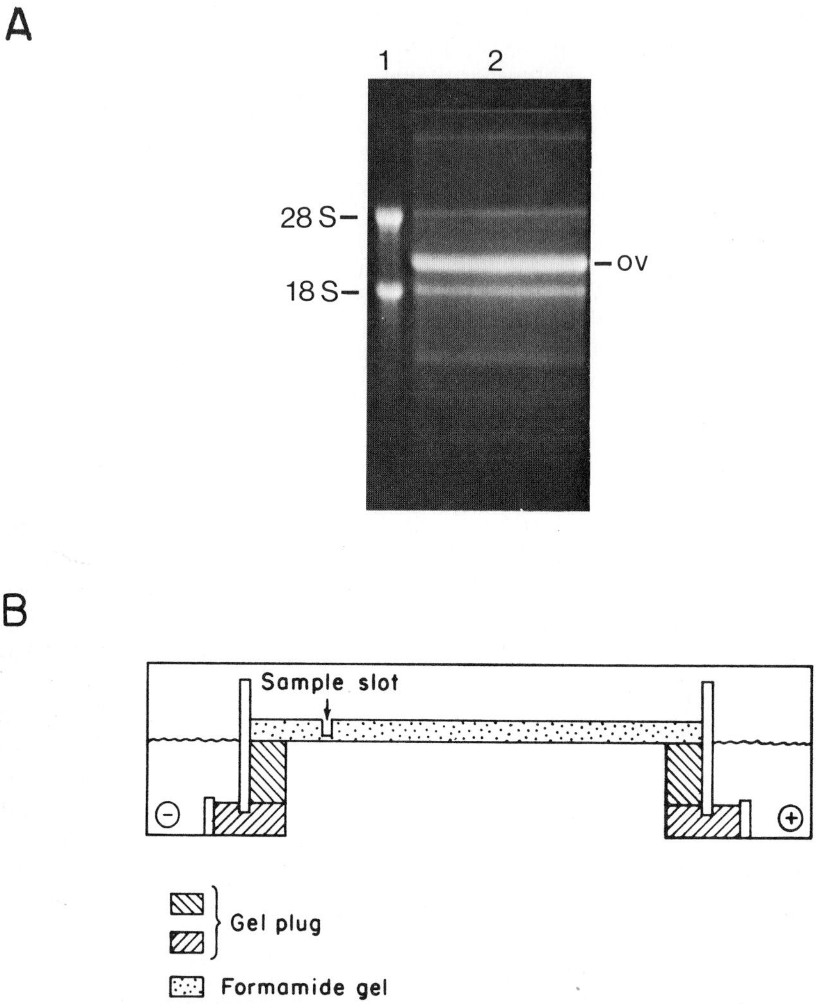

FIG. 1. Agarose gel fractionation of mRNA, and schematic of gel apparatus. (A) Hen oviduct poly(A)-RNA selected by two rounds of dT-cellulose chromatography was electrophoresed in a 1.8% agarose gel containing 40% formamide, as described in the text, and stained with ethidium bromide. Lane 1, ribosomal RNA markers; lane 2, poly(A)-RNA containing major band of ovalbumin mRNA (ov). (B) Schematic drawing of horizontal gel apparatus showing formamide gel and gel plugs.

Gel plug buffer: 35 mM triethanolamine, 1 mM EDTA, 0.05% SDS, 3 mM NaH_2PO_4, pH adjusted to 7.5 with phosphoric acid.

Electrode buffer: 35 mM triethanolamine, 1 mM EDTA, 8 mM NaH_2PO_4, pH adjusted to 7.5 with phosphoric acid. Can make 20× stock.

RNA sample buffer: 35 mM triethanolamine, 1 mM EDTA, 0.07% SDS, 50% formamide, 0.03% Bromphenol Blue, pH 7.5 (adjust with H_3PO_4 or NaOH). Store small aliquots at −20°.

Procedure. Use a horizontal gel apparatus with gel plug chambers, such as the type H2 from BRL. The plugs, which do not contain formamide, are cast in 1.2% agarose and are polymerized in two stages to reach the level of the gel platform as shown schematically in Fig. 1B. Gel solution, designed to produce a 1.8% agarose gel of 4 mm thickness, is made as follows: Mix 73 ml of deionized formamide, 9 ml of 20× gel buffer; adjust pH to 7.5, and bring volume to 85 ml with H_2O. Place 3.25 g of agarose in 97 ml of distilled H_2O and melt by autoclaving for 15 min or by boiling in a microwave oven (low setting). Heat the formamide solution to 50° and add to hot agarose solution (50–55°) using a strong magnetic stirrer to prevent agarose clumping. Once mixed, the formamide will prevent the agarose from gelling at room temperature. Pour the gel in a cold room (4°) and allow to polymerize for $1\frac{1}{2}$ hr. Return apparatus to room temperature, and add electrode buffer (1×) to a level about 6 mm below the gel surface. Fill slots with sample buffer (without RNA) and prerun for 1 hr at 75 V. Dissolve RNA in sample buffer (500–750 μg/ml), heat to 62° for 3 min, and quench for 10 sec on ice. Remove prerun buffer from slots, load RNA, and run at 2 V/cm (70 V) for 12 hr at 24°. For staining, gel strips are washed for 15 min in 0.12 M ammonium acetate, then incubated for 1 hr in the same buffer with 1 μg of ethidium bromide per milliliter. Fluorescent bands are examined on a 300-nm UV lamp.

Recovery of RNA from Gels

An efficient method for recovery of RNA from gels is continuous-flow electroelution. Electrophoresis units designed specifically for this purpose are available from commercial suppliers,[33] and the reader is referred to publications on the subject.[34,35] Another commonly used method for recovery of mRNA is mechanical crushing of gel fragments followed by

[33] Bethesda Research Laboratories, Gaithersburg, Maryland; Savant Instruments, Hicksville, New York.

[34] B. Wieringa, J. Mulder, A. van der Ende, A. Bruggeman, G. Ab, and M. Gruber, *Eur. J. Biochem.* **89,** 67 (1978).

[35] C. Auffray and F. Rougeon, *Eur. J. Biochem.* **107,** 303 (1980).

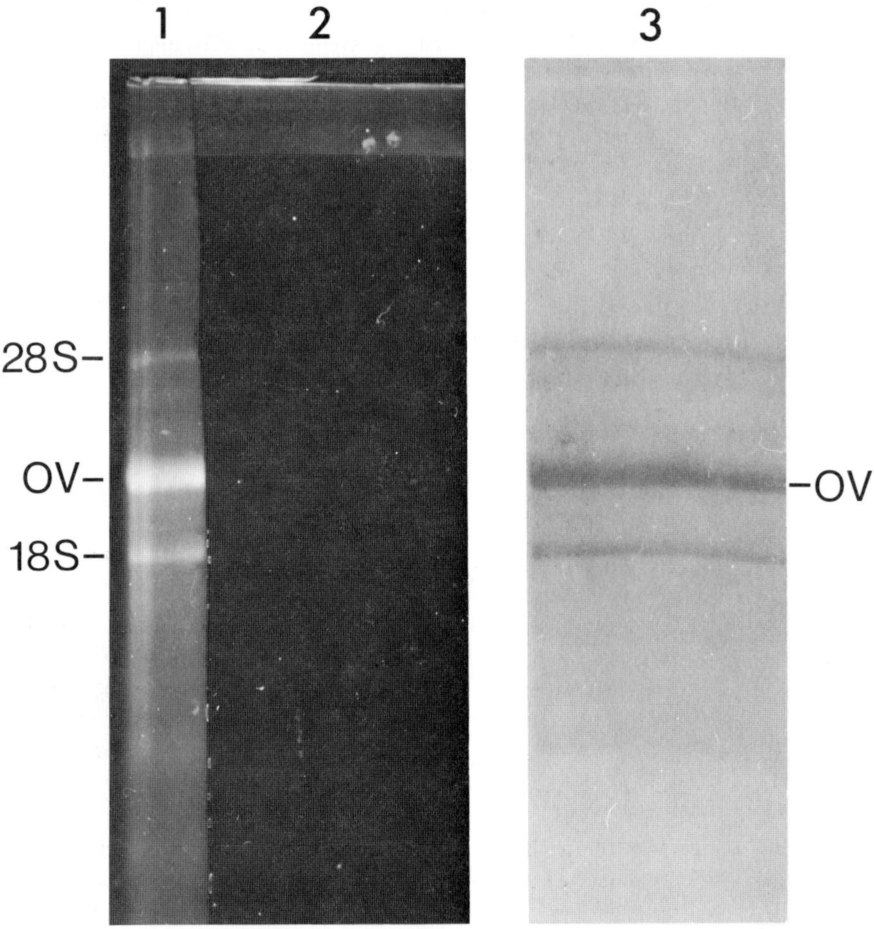

FIG. 2. Electroblotting of mRNA from agarose gel to DEAE membrane. Lane 1, stained RNA from control gel that was not blotted; lane 2, portion of same gel, stained after electroblotting; lane 3, DEAE membrane stained for 15 min with Stains-all (1) the staining solution consists of 0.01% 1-ethyl-2-[3-(1-ethylnaphtho[1,2-d]thiazolin-2-ylidene)-2-methylpropenyl]naphtho[1,2-d]thiazolium bromide (Eastman Organic Chemicals), 25% isopropyl alcohol, 15% formamide, 30 m*M* triethanolamine, pH 8.5. Destain for 30 min in 20% methanol, 30 m*M* triethanolamine, pH 8.5 [A. E. Dahlberg, C. W. Dingman, and A. C. Peacock, *J. Mol. Biol.* **41,** 139 (1969)].

soaking in high-salt buffer and dT-cellulose chromatography,[36] or phenol extraction from low-melting-point agarose.[37] These methods are relatively time consuming when a large number of gel fragments are to be processed at once.

[36] D. W. Cleveland, M. W. Kirschner, and N. J. Cowan, *Cell* **15,** 1021 (1978).

[37] L. Wieslander, *Anal. Biochem.* **98,** 305 (1979).

A third method, which is described in detail below, consists of electroblotting[38] fractionated mRNAs from an agarose gel to a membrane of (diethylaminoethyl)cellulose (DEAE membrane). mRNAs can then be recovered from membrane strips by a single one-step elution. Figure 2 shows an electrophoretic pattern of hen oviduct mRNA before and after blotting to a DEAE membrane. Transfer of mRNA is quantitative, and there is no apparent loss of resolution.

Materials and Reagents

Electroblotting apparatus: Commercial units with platinum electrodes are available from Bio-Rad Laboratories or Hoeffer Instruments.

DEAE membrane: Schleicher & Schuell NA-45 is pretreated by washing for 10 min in 10 m*M* EDTA, pH 7.6, then 5 min in 0.5 *M* NaOH, followed by several washes in distilled H_2O.

Electroblot buffer: 35 m*M* Tris, 1 m*M* EDTA, 3 m*M* NaH_2PO_4, adjusted to pH 7.5 with H_3PO_4 (do not use HCl).

RNA release buffer: 55% formamide, 30 m*M* HEPES, 2.0 *M* NaCl, 2 m*M* EDTA, 0.2% SDS, pH 7.5.

Procedure. Soak electroblot unit, Scotchbrite pads, Whatman 3 MM paper, and working trays with 0.1% SDS, 0.07% diethyl pyrocarbonate for about 2 hr, then rinse once with sterile H_2O. Using clean vinyl gloves, place prewashed NA-45 membrane over wet gel surface, remove any air bubbles, and cover on each side with wet 3 MM paper, then Scotchbrite pads to complete electroblot sandwich.[38] Place in electroblot apparatus that contains buffer precooled to about 10°, and turn on cooling circulation (ideally H_2O–methanol at −7°). Turn on voltage to 7 V/cm and electrophorese RNA toward anode for about 30 min (this time is adequate for a 1.8% agarose gel of 3.7 mm thickness). After blotting, mark membrane to define orientation by cutting arrow-shaped dents, then wash in 1 m*M* EDTA, pH 7, to remove agarose residues. Do not allow membrane to dry, as this causes irreversible mRNA binding. Membranes can be stored for months at −20° in 1 m*M* EDTA pH 7, 20% glycerol. Thin vertical strips containing part of a blotted RNA profile may be stained with Stains-all as shown in Fig. 2. The 28 S and 18 S ribosomal RNAs present as contaminants in mRNA preparations provide convenient size markers.

To recover mRNA, place unstained, wet membrane on a sterile glass surface and cut horizontal strips about 3 mm wide with a scalpel and a glass guide. Place wet strips in plastic microcentrifuge tubes containing 250 μl of release buffer and 8 μg of carrier tRNA and heat at 68° for 8 min with occasional vortexing. Transfer released RNA solution to clean mi-

[38] E. J. Stellwag and A. E. Dahlberg, *Nucleic Acids Res.* **8,** 299 (1980).

crocentrifuge tube, and add 0.75 ml of H_2O to reduce salt concentration. Centrifuge at 12,000 *g* for 5 min to remove debris, and transfer supernatant to clean siliconized glass or plastic tubes. Precipitate RNA with 2.5 volumes of ethanol (overnight at −20°). One or two additional ethanol precipitations from 0.2 *M* sodium acetate are required to remove residual NaCl, formamide, and SDS. RNA mass recovery is routinely 70–80%, while recovery of biological activity is higher than 60%.[39]

Saris *et al.*[40] have recently described the *in situ* translation of mRNA bound to ion-exchange paper, which expands the potential versatility of the blotting techniques described here.

[39] P. Lizardi and R. Binder, manuscript in preparation.

[40] C. J. M. Saris, H. J. Franssen, J. H. Heuyerjans, J. van Eenbergen, and H. P. J. Bloemers, *Nucleic Acids Res.* **10,** 4831 (1982).

[3] Cell-Free Translation of Messenger RNA in a Wheat Germ System[1]

By ANN H. ERICKSON and GÜNTER BLOBEL

Cell-free extracts of wheat germ embryos, containing ribosomes and soluble factors, support the *in vitro* translation of a wide variety of mRNAs into protein. The translation apparatus is sufficiently conserved so that a cell-free system can translate both prokaryotic and eukaryotic mRNAs with fidelity. The wheat germ system has been used widely because it is relatively easy to prepare and contains a relatively low amount of endogenous mRNA. The methods described below are modifications of those previously described by Marcus *et al.*[2] and Roberts and Paterson.[3] They have been successfully applied to the *in vitro* synthesis of proteins that each comprise several percent of total cellular protein, such as globin, prolactin, and the caseins, as well as to the synthesis of the lysosomal proteins cathepsin D,[4] β-glucuronidase,[5] and glucocerebrosidase,[5] which each comprise less than 0.1% of total cellular protein.

[1] Supported by Grant No. NP268B from the American Cancer Society.

[2] A. Marcus, D. Efron, and D. P. Weeks, this series, Vol. 30, p. 749.

[3] B. E. Roberts and B. M. Paterson, *Proc. Natl. Acad. Sci. U.S.A.* **70,** 2330 (1973).

[4] A. H. Erickson and G. Blobel, *J. Biol. Chem.* **254,** 11771 (1979).

[5] A. H. Erickson and G. Blobel, in preparation.

ISBN 0-12-181996-5

Source of Wheat Germ

Batches of fresh wheat germ are generally available without charge from large commercial mills, such as General Mills, Vallejo, California or Pillsbury Company, Minneapolis, Minnesota. Since the wheat strains in these batches are variable, it is best to request that several 1-lb samples be taken and sent over a period of time. The variability in the activity of the wheat germ is not well understood, but apparently relates to the wheat strains being processed by the mill. Mills frequently blend the germ from several types of wheat, and the composition of the blend varies with the strains of wheat available at the time. The germ retains activity for over a year when stored in a vacuum desiccator at 4° or in aliquots at −70°.

Flotation of Wheat Germ

Solvent flotation is used to enrich the wheat germ for viable, intact embryos.[6] Reagent-grade cyclohexane (240 ml) and carbon tetrachloride (600 ml) are stirred until no schlieren mixing lines are visible. About 40 g of wheat germ are added, and the mixture is stirred gently with a glass rod. The damaged embryos and contaminating endosperm fragments are allowed to settle away from intact, floating embryos for 2–3 min. The floating germ is collected in a large Büchner or sintered-glass funnel in a fume hood and allowed to dry by pulling air through the funnel for 30 min. About 20–40% of the wheat germ should float. If the recovery is much lower, the solvent ratio can be altered by increasing the amount of carbon tetrachloride added until about 30% of the germ floats. Although the altered solvent ratio allows collection of active germ, the most active preparations are generally obtained when the 1 : 2.5 ratio of cyclohexane to carbon tetrachloride floats approximately 30% of the germ. The organic solvent mixture may be reused for consecutive flotations. Floated germ should be stored in a vacuum desiccator at 4° or in aliquots at −70°.

Preparation of Wheat Germ Extract

The wheat germ extract to be used in the *in vitro* translation system is prepared in three operations: homogenization of the floated germ, centrifugation of the homogenate, and gel filtration of the supernatant. All these procedures should be carried out in the minimum amount of time in a 4° cold room and should employ sterilized buffers and glassware heat-treated at 150° overnight to minimize contamination of the extract by

[6] F. B. Johnston and H. Stern, *Nature* (*London*) **179,** 160 (1957).

RNases. These and all subsequent stock solutions should be made with double-distilled water and should be Millipore-filtered to maximize purity.

Preparation of Buffers

Homogenization buffer contains the following five solutes at the concentrations indicated:

- Potassium acetate, 100 m*M*
- Magnesium acetate, 1 m*M*
- Calcium chloride, 2 m*M*
- *N*-2-Hydroxyethylpiperazine-*N*′-2-ethanesulfonic acid (HEPES), 40 m*M*, pH 7.5
- Dithiothreitol (DTT), 4 m*M*

Column buffer contains the following four solutes at the concentrations indicated:

- Potassium acetate, 100 m*M*
- Magnesium acetate, 5 m*M*
- HEPES, 40 m*M*, pH 7.5
- Dithiothreitol (DTT), 4 m*M*

Each buffer is prepared from stock solutions without DTT, adjusted to pH 7.6 with KOH, and sterilized in an autoclave. Just before use, DTT powder is added and the buffer is cooled to 4°. Anions should be acetate rather than chloride. At concentrations higher than the cytoplasmic level of 70–80 m*M*, Cl^- severely inhibits the binding of mRNA to ribosomes to form initiation complexes.[7] Use of the acetate anion allows the K^+ concentration in the *in vitro* translation system to be as high as the physiological concentration.

Homogenization of the Wheat Germ

A mortar and a pestle are heat-treated, cooled to room temperature, and chilled with liquid N_2. Floated wheat germ (3 g) is added and ground (usually within 1–2 min) to a fine powder in liquid N_2.[8] The powdered wheat germ is transferred to a second heat-treated mortar resting in an ice bath and ground (again for 1–2 min) in homogenization buffer (10 ml added in three increments) until a thick paste is obtained. Powdering the wheat germ in liquid N_2 presumably minimizes the extent of enzymic degradation occurring between cellular disruption and exposure to buffer.

[7] L. A. Weber, E. D. Hickey, P. A. Maroney, and C. Baglioni, *J. Biol. Chem.* **252,** 4007 (1974).

[8] A. R. Grossman, S. G. Bartlett, G. W. Schmidt, J. E. Mullet, and N.-H. Chua, *J. Biol. Chem.* **257,** 1558 (1982).

Centrifugation of the Homogenate

The homogenate is scraped into a chilled centrifuge tube with a heat-treated spatula and spun at 4° for 10 min at 23,000 *g*. The supernatant is transferred with a heat-treated Pasteur pipette to a clean tube. Care is taken to avoid transferring the floating lipid layer into the tube. The centrifugation step is repeated. The supernatant recovered after the second centrifugation is traditionally known as the S23 wheat germ extract. Its volume is measured.

Gel Filtration of the Supernatant

Gel filtration of the wheat germ extract reduces the concentration of free amino acids, which compete with radiolabeled amino acids for incorporation into protein during *in vitro* translation of mRNA. Isotope dilution assays, which determine the rate of protein synthesis in aliquots of wheat germ extract that contain the same amount of radioactive amino acid but different amounts of added unlabeled amino acid, indicate that gel filtration does not completely deplete free amino acids from the wheat germ extract.[9] The reason for this is unknown. Free amino acids might conceivably be generated by proteolysis during the *in vitro* translation reaction.

A glass column (1.7 × 45 cm) containing a bed of Sephadex G-25 fine (1.7 × 33.3 cm) is equilibrated with sterile column buffer. The wheat germ supernatant is loaded onto the column, and the column is eluted with column buffer under gravity flow. The brownish solution eluting just behind the void volume is collected in approximately 15 fractions (each 2 ml). The slower-running yellow pigment is discarded. The most opaque fractions are pooled to provide a solution whose volume is approximately equal to the volume loaded onto the column. The final yield from 3 g of wheat germ should be about 7 ml. This solution is centrifuged at 4° for 10 min at 23,000 *g*. The scale of the whole procedure can be increased, using more wheat germ and a larger G-25 column. Autoclaved microfuge tubes containing 100–200-μl aliquots of gel-filtered wheat germ extract are immediately frozen in liquid N_2. The wheat germ extract is stable for at least a year when stored at −70°.

Components of the Translation Assay

Amino Acids and Energy-Generating System

Stock solutions are prepared at the following concentrations:

Adenosine 5′-triphosphate (ATP), 0.1 *M* (disodium salt)

[9] P. Walter, I. Ibrahimi, and G. Blobel, *J. Cell Biol.* **91,** 545 (1981).

Guanosine 5′-triphosphate (GTP), 0.02 *M* (sodium salt)
Creatine phosphate (CP), 0.6 *M* (disodium salt hydrate)
[^{35}S]Methionine or a desired tritiated amino acid
The 19 other amino acids, 1 m*M* each
Potassium hydroxide, 1.0 *M*
Creatine phosphokinase (CPK) from rabbit muscle, 8 mg/ml

The ATP, GTP, CP, and CPK stock solutions should be stored in small aliquots at −20°. All except the CPK solution can be refrozen. The stock solution of 19 amino acids is conveniently prepared from 20 m*M* stock solutions of each amino acid. All amino acids are soluble at 20 m*M* in distilled water except Phe, Asn, Ile, Trp, Val, Glu, and Asp, which require 0.01 *N* HCl, and Tyr, which is soluble in 0.1 *N* HCl. The stock solution of the 19 combined amino acids and the individual 20 m*M* amino acid solutions should be stored at −20° and may be refrozen. Translation-grade [^{35}S]methionine (1000 Ci/mmol, 1 mCi/110 μl) should be stored in small aliquots at −70° to minimize oxidation, which may occur on repeated freezing and thawing and on exposure to air. If the protein under study lacks methionine, [^{35}S]cysteine, available at a higher specific activity (500 Ci/mmol) than tritiated amino acids, may be useful. If the protein is unusually rich in a particular amino acid, even one available only at a low specific activity, this may be a useful choice for incorporation, as the background of labeling of total cellular protein should be low relative to incorporation into the particular protein under study.

A 50-μl mixture of the above components, which is enough for twelve 20-μl translation reactions, is prepared by mixing, in the order given, 3 μl of 0.1 *M* ATP, 1 μl of 0.02 *M* GTP, 4 μl of 0.6 *M* CP, 5 μl of the 19 amino acid solution, 9 μl of water, 25 μl of [^{35}S]methionine, and 1 μl of 1.0 *M* KOH. The pH of this mixture is checked by spotting 1 μl on pH paper. If the pH is less than 7.0, the solution is adjusted with additional KOH to pH 7–7.6. Finally, 2 μl of the CPK stock solution are added. This resulting solution can be stored at −70° and reused by addition of 2 μl of fresh CPK stock solution, but this generally results in a 10–20% decrease in amino acid incorporation. In the final translation reaction, the mixture is diluted 1:5, giving final concentrations of 1.2 m*M* ATP, 0.08 m*M* GTP, 9.6 m*M* CP, 20 μ*M* of each of the 19 amino acids, CPK at 64 μg/ml, and [^{35}S]methionine at 910 μCi/ml.

The amount of radioactivity incorporated into protein can be increased by substituting the radiolabeled amino acid solution for the portion of water. In addition, a desired quantity of the radiolabeled amino acid may be lyophilized in a microfuge tube and resuspended in 25 μl of distilled water. The other components of the mixture may be added directly to this tube. Concentration by lyophilization is especially important to maximize the incorporation of an amino acid that is available only at a dilute concentration. If the radioactive amino acid is supplied in a salt solution, the salt concentration should be taken into account when preparing the compensating buffer (see below). A methionine concentration of 2–3 mCi per milliliter of total translation mix is quite adequate for the *in vitro* synthesis of lysosomal enzymes,[4] which each comprise less than 0.1% of total cellular protein.

Wheat Germ Compensating Buffer

The K^+ and Mg^{2+} concentrations of the wheat germ translation system have dramatic effects on the efficiency of translation of particular mRNAs.[10,11] The S23 wheat germ extract prepared as described above contains 100 m*M* K^+ and 5 m*M* Mg^{2+}. The compensating buffer is used to adjust the ion concentrations of the total translation reaction to an optimum that must be determined for each mRNA being translated. The optimal concentrations are generally 130–140 m*M* for K^+ and 2.0–2.5 m*M* for Mg^{2+}. It is important to examine, by polyacrylamide gel electrophoresis, the effect of ion concentrations on the radiolabeled amino acid incorporated into the protein under study rather than be guided by the amount of radiolabeled amino acid incorporated into total protein.

Compensating buffer, which is diluted 1 : 10 in the final translation mixture, contains the following four solutes at the concentrations indicated:

- Potassium acetate, 1.0 *M*
- Magnesium acetate, 5 m*M*
- Spermine, 0.8 m*M*
- Dithiothreitol (DTT), 20 m*M*

This compensating buffer is used when the final translation mixture contains 40% of the S23 wheat germ extract. When the final translation mixture contains 50% wheat germ extract, the magnesium acetate is omitted from the compensating buffer and the potassium acetate concentration is decreased to 0.9 *M*. In both cases, the concentration in the final translation mixture will be 140 m*M* for K^+ and 2.5 m*M* for Mg^{2+}. Spermidine, at a concentration approximately 10 times that used for spermine,[11] can replace spermine in the compensating buffer and may improve translation of certain mRNAs.[12] The compensating buffer can be stored in aliquots at −70°.

[10] P. T. Lomedico and G. F. Sunders, *Science* **198,** 620 (1977).
[11] T. P. H. Tse and J. M. Taylor, *J. Biol. Chem.* **252,** 1272 (1977).
[12] K. E. Mostov and G. Blobel, this series, Vol. 98 [40].

mRNA

Total RNA or poly(A)-containing mRNA, prepared as described in this volume,[13] can be translated in a wheat germ system. When the mRNA under study is present at low concentration in a given tissue, it is advantageous to enrich its concentration by translating a particular size class of RNA, which is isolated from a sucrose gradient or a sizing gel.

The RNA is suspended in sterile distilled water at a concentration of approximately 1 mg/ml (20 A_{260}/ml). The optimal final RNA concentration in the translation reaction should be determined for each RNA preparation to be translated, but is generally 50–150 μg/ml (1–3 μl of RNA at 20 A_{260}/ml per 20 μl of total translation mixture). Higher RNA concentrations often result in decreased protein synthesis, probably due to inhibitors in the RNA preparation. Dilute RNA preparations may be concentrated by ethanol precipitation or lyophilization.

Wheat Germ Extract

The optimal concentration of S23 wheat germ extract should be determined for each RNA to be translated, but is generally either 40 or 50% of the total translation mix. Appropriate concentrations of K^+ and Mg^{2+} in the compensating buffer are used to adjust for the percentage of the wheat germ extract present.

The low concentration of endogenous mRNA present in the wheat germ extract competes with the added RNA for ribosomes and factors required for translation. It is advantageous to reduce the concentration of endogenous RNA by treating with nuclease.[14] Micrococcal nuclease from *Staphylococcus aureus* is dissolved in distilled water at a concentration of 1875 units/ml. Aliquots of the diluted enzyme are stored at −70° and can be refrozen. This nuclease solution (2 μl) is mixed with 100 μl of freshly thawed wheat germ extract and 2 μl of 0.1 *M* $CaCl_2$ and incubated for 5 min at 21° with occasional mixing. The reaction is terminated by transfer of the tube to an ice bath and addition of 4 μl of 0.1 *M* EGTA, which chelates the Ca^{2+} ion required for nuclease activity. Nuclease-treated wheat germ loses only 5–10% of its activity when refrozen immediately in liquid N_2. Alternatively, wheat germ may be nuclease-treated in large batches immediately after elution from the Sephadex G-25 column and prior to centrifugation and freezing in aliquots.

[13] P. M. Lizardi, this volume [2].

[14] H. R. B. Pelham and R. J. Jackson, *Eur. J. Biochem.* **67**, 247 (1976).

Optional Components

Transfer RNA. For some batches of wheat germ, the efficiency of translation of certain messenger RNAs can be increased by adding a mixture of heterologous transfer RNAs, such as commercial calf liver tRNA. The optimal concentration of added tRNA is determined by titration. It usually lies between 50 and 500 μg of tRNA per milliliter of total translation mixture. Use of higher concentrations can decrease the net incorporation of radiolabeled amino acids into protein. The effect on the net incorporation into the protein under study should be the guiding factor in choosing the optimal tRNA concentration. Depending on the amount of tRNA added, it may be necessary to alter the final Mg^{2+} concentration in the translation.

Ribonuclease Inhibitor. The addition of human placental RNase inhibitor to the wheat germ system can increase the yield of high molecular weight proteins.[15] The magnitude of this increase varies with different batches of wheat germ. This RNase inhibitor can be prepared by affinity chromatography on RNase A–Sepharose.[16] The optimal concentration of RNase inhibitor for a given translation system is determined by titration. It usually lies between 2 and 10 μg of RNase inhibitor per milliliter of total translation mixture.

Protease Inhibitors. Selected protease inhibitors that do not inhibit protein synthesis may be added at the following final concentrations: pepstatin A, 0.1 μg/ml; chymostatin, 0.1 μg/ml; antipain, 0.1 μg/ml; leupeptin, 0.1 μg/ml; and Trasylol, 10 units/ml.[9]

Microsomal Membranes. Microsomal membranes from the endoplasmic reticulum can translocate nascent secretory and lysosomal proteins, which usually results in cleavage of the amino-terminal signal sequence and addition of high-mannose carbohydrate chains, and can integrate newly synthesized membrane proteins. Preparation of these membranes and their use in the wheat germ system are described in this volume.[17]

Scale of the Reaction

The scale of the translation reaction required for detection of the protein under study depends upon the amount of its mRNA in the total RNA preparation. A major secretory protein, which usually comprises several percent of the total cellular protein, can normally be visualized by

[15] G. Scheele and P. Blackburn, *Proc. Natl. Acad. Sci. U.S.A.* **76,** 4898 (1979).

[16] P. Blackburn, *J. Biol. Chem.* **254,** 12484 (1979).

[17] P. Walter and G. Blobel, this volume [6].

polyacrylamide gel electrophoresis of a 20-μl reaction mixture containing [^{35}S]methionine, followed by autoradiography for 24 hr. In contrast, a lysosomal protein comprising less than 0.1% of the total cellular protein may require immunoprecipitation from a 100-μl reaction mixture and fluorography for 2–3 weeks.

Order of Component Addition

The components of the translation mixture are generally combined in a tube chilled in an ice bath in the following order: 4 μl of mixture of radiolabeled amino acid, 19 unlabeled amino acids, ATP, GTP, CP, and CPK; 2 μl of compensating buffer; distilled water; transfer RNA (if present); RNase inhibitor (if present); 8 μl of nuclease-treated S23 wheat germ extract; 1–3 μl of mRNA solution; microsomal membranes (if present).

The volume of distilled water added is chosen to bring the total volume of the translation mixture to 20 μl. Plastic microfuge tubes with conical bottoms are convenient reaction vessels. After addition of each reagent, the mixture is mixed thoroughly but gently. Brief centrifugation in a microfuge eliminates bubbles and ensures that no liquid remains as beads on the walls of the vessel. In order to maximize reproducibility when the amount of one reagent is being varied, the other reagents are combined, mixed, and divided equally among a series of tubes containing varying amounts of the last reagent.

Time and Temperature of the Reaction

At the beginning of the reaction period, a portion (2–5 μl) of the translation mixture is spotted on a filter disk to measure the background radioactivity (see below). The translation mixture is generally incubated at 25–27° for 60–90 min. Longer reaction periods often result in decreased net incorporation of radiolabeled amino acids into protein, presumably owing to degradation of newly synthesized polypeptide chains. At higher temperatures, protein synthesis proceeds faster but terminates sooner, which also decreases the net incorporation.

Termination of the Reaction

Cooling the reaction mixture in an ice bath is sufficient to terminate the translation reaction. Permanent termination is achieved by detergent denaturation or trichloroacetic acid precipitation of the protein in the reaction mixture.

Analysis of Translation Products

Incorporation of Radiolabel into Total Protein

The extent of incorporation of the radiolabeled amino acid into total protein can be determined by measuring the amount of radioactivity incorporated into acid-precipitable protein, using a modification of the method of Mans and Novelli.[18] This assay provides an overall estimate of the efficiency of the RNA translation. A small aliquot (2–5 μl) of the translation mix is spotted on a Whatman 3 M filter disk (2.5 cm in diameter) and dropped into a beaker of 10% trichloroacetic acid resting in an ice bath. Several filter disks can be treated in the same beaker, using 3 ml of trichloroacetic acid per filter. Handling is facilitated by insertion of a straight steel pin into the paper disk, which can be labeled with pencil. After 30 min for precipitation of protein, the 10% trichloroacetic acid is decanted and 5% trichloroacetic acid is added and heated to boiling on a hot plate in a fume hood for 15 min. If the filters are boiled too long, the concentration of the acid increases and protein hydrolysis can occur, which decreases the radioactivity in the precipitated protein. The hot 5% trichloroacetic acid is decanted, and the filters are rinsed three times in fresh 5% acid. The filters are immersed in a 1 : 1 solution of ethanol and ether for about 5 min. The solvent mixture is then decanted and ether is added. After 5 min, the ether is decanted and the filters are air-dried in a fume hood and counted in a toluene-based scintillation fluid. After counting and removal of the filter disks, the scintillation vial and fluid can be reused numerous times without significantly increasing the background radioactivity.

Definitions and Controls

Several control experiments are normally carried out to facilitate the interpretation of the translation results.

1. Background radioactivity is the percentage of trichloroacetic acid precipitable radioactivity at the beginning of the translation reaction. This control measures the binding of the free radiolabeled amino acid to protein in the wheat germ extract or to the filter disk.
2. Stimulation of the incorporation of radiolabeled amino acid into newly synthesized protein is defined as the ratio of the incorporation due to a specific source of mRNA divided by the background radioactivity.
3. Endogenous stimulation is the ratio of radioactivity incorporated

[18] R. J. Mans and G. D. Novelli, *Arch. Biochem. Biophys.* **94,** 48 (1961).

owing to the presence of endogenous mRNA in the wheat germ extract divided by the background radioactivity. This control is measured by omitting the mRNA under study from the translation mixture.

4. Efficiency of nuclease treatment is the percentage decrease in endogenous stimulation due to treatment of the S23 wheat germ extract with nuclease. It is measured by determining the incorporation using both nuclease-treated and untreated wheat germ extract.
5. Standard net stimulation is the net incorporation due to the presence of a standard mRNA minus the endogenous stimulation. This control is especially helpful when changing any components of the reaction. It is measured by subtracting the endogenous stimulation from the observed stimulation. It is useful to set aside aliquots of a particular mRNA preparation for use as a standard mRNA.
6. If a new RNA preparation proves to be inactive, it is helpful to translate a mixture of the new mRNA and the standard mRNA. If the observed stimulation is less than the standard stimulation, the new RNA preparation probably contains inhibitors of protein synthesis. Further purification of the RNAs should be helpful in this case.

Typical mRNA-Specific Stimulation of Protein Synthesis

The stimulation of protein synthesis varies greatly with the mRNA being translated. The average standard net stimulation is 40- to 100-fold but can be as high as 400-fold using certain RNAs under optimal conditions.

So many factors affect amino acid incorporation in the wheat germ system that it is difficult to generalize. Under our conditions, the background radioactivity using [^{35}S]methionine is about 5000 cpm per 2.5 μl of translation mixture. Endogenous stimulation using nuclease-treated wheat germ extract is normally 4- to 5-fold, or 20,000–25,000 cpm per 2.5 μl. Standard net stimulation by an exogenous mRNA preparation will be 40- to 100-fold, or 200,000–500,000 cpm per 2.5 μl. Highly efficient standard net stimulation can be 400-fold, or 2,000,000 cpm per 2.5 μl.

The secretory protein prolactin comprises several percent of the total protein of bovine pituitary glands. Total RNA from this tissue produces about a 40-fold standard net stimulation in radioactive amino acid incorporation in the wheat germ system. It has been estimated[19] that approximately 10 fmol of preprolactin are synthesized in a 25-μl wheat germ *in vitro* translation assay using bovine pituitary RNA.

[19] P. Walter and G. Blobel, *Proc. Natl. Acad. Sci. U.S.A.* **77,** 7112 (1980).

Polyacrylamide Gel Electrophoresis (PAGE) of the Translation Products in Sodium Dodecyl Sulfate (SDS)

The method most commonly used for examining the proteins produced by cell-free translation is SDS–PAGE. Various aspects of this technique are discussed elsewhere in this volume.[20,21] The products of *in vitro* synthesis are usually analyzed by autoradiography or fluorography because these sensitive techniques for detection of radiolabeled protein are usually necessary to visualize the small amounts of protein synthesized.

If the protein under study is coded for by an RNA comprising a large percentage of the total RNA population of the tissue, such as a secretory protein, it may be possible to detect that protein directly among the total translation products. It is important to run a sample of the proteins produced by endogenous mRNA on the same gel in order to distinguish which protein bands are due to endogenous mRNA in the wheat germ extract. Electrophoresing radiolabeled molecular weight standards allows determination of the molecular weights of particular protein products.

A small aliquot (5 μl) of translation mixture can be diluted with a loading buffer (25 μl) containing 3% SDS and loaded directly on the gel. Dithiothreitol may also be added at a concentration of 20 m*M*. Subsequent alkylation using iodoacetamide (100 m*M*, 30 min, 37°) is often required to prevent oxidation of the thiol groups and re-formation of disulfide bonds.

Large aliquots of translation mixture (>25 μl) may be precipitated with trichloroacetic acid and the pellets resuspended in the SDS loading buffer. Trichloroacetic acid precipitation also reduces the concentration of salts, which may distort the pattern of protein bands in the gels. Since the wheat germ extract contains nonradioactive proteins that are precipitated by trichloroacetic acid and loaded onto the gel along with the radiolabeled translation products, it is important not to precipitate too large a sample of the translation mixture. Loading large amounts of protein in a single gel slot produces distortion of the shape of the protein bands and aberrant electrophoretic migration, and thus incorrect estimates of the molecular weights of certain proteins.

Immunoprecipitation of Specific Translation Products

If the protein under study comprises a small percentage of the total protein of a particular tissue (<1–2%), it is necessary to immunoprecipi-

[20] R. W. Rubin and C. L. Leonardi, this volume [12].

[21] W. M. Bonner, this volume [15].

tate the protein prior to PAGE. Immunoprecipitation of *in vitro* translation products is discussed in this volume.[22]

Other Cell-Free Translation Systems

In principle it should be possible to prepare a cell-free extract for RNA translation from any cell type. In practice, relatively few cell-free systems have been developed for *in vitro* translation. In general, these systems are derived from cells engaged in a high rate of protein synthesis. They include (*a*) rabbit reticulocyte lysate[23]; (*b*) bacterial cell lysate[24]; (*c*) yeast cell lysate[25–27]; (*d*) HeLa, Chinese hamster ovary, or L-cell lysate[28,29]; (*e*) mouse Krebs ascites cell lysate.[30,31]

[22] D. J. Anderson and G. Blobel, this volume [8].
[23] R. J. Jackson and T. Hunt, this volume [4].
[24] P. Green and M. Inouye, this volume [5].
[25] E. Gasior, F. Herrera, I. Sadnik, C. S. McLaughlin, and K. Moldave, *J. Biol. Chem.* **254,** 3965 (1979).
[26] M. F. Tuite, J. Plesset, K. Moldave, and C. S. McLaughlin, *J. Biol. Chem.* **255,** 8761 (1980).
[27] B. Wolska-Mitaszko, T. Jakubowicz, T. Kucharzewska, and E. Gasior, *Anal. Biochem.* **116,** 241 (1981).
[28] M. J. McDowell, W. K. Joklik, L. Villa-Komaroff, and H. F. Lodish, *Proc. Natl. Acad. Sci. U.S.A.* **69,** 2649 (1972).
[29] L. A. Weber, E. M. Feman, and G. Baglioni, *Biochemistry* **14,** 5314 (1975).
[30] I. M. Kerr, N. Cohen, and T. S. Work, *Biochem. J.* **98,** 826 (1966).
[31] M. B. Mathews and A. Korner, *Eur. J. Biochem.* **17,** 328 (1970).

[4] Preparation and Use of Nuclease-Treated Rabbit Reticulocyte Lysates for the Translation of Eukaryotic Messenger RNA

By RICHARD J. JACKSON and TIM HUNT

In 1968 two research groups independently found that when rabbit reticulocyte lysates were supplemented with hemin they synthesized protein at the same rate as the intact cell for periods of about 60 min.[1,2] The remarkably high activity of this cell-free system has still not been equaled

[1] S. D. Adamson, E. Herbert, and W. Godschaux, *Arch. Biochem. Biophys.* **125,** 671 (1968).
[2] W. V. Zucker and H. M. Schulman, *Proc. Natl. Acad. Sci. U.S.A.* **59,** 582 (1968).

ISBN 0-12-181996-5

by extracts of other eukaryotic cells. Whereas these early experiments were concerned only with the translation of endogenous reticulocyte mRNA, it was subsequently shown that exogenous heterologous mRNA is also efficiently and faithfully translated in the lysate.[3] The disadvantage of such experiments is that the added mRNA is translated in competition with the endogenous reticulocyte mRNA, which makes quantitation of the template activity of the exogenous RNA difficult and tedious. The ideal translation system of which we used to dream would be a reticulocyte lysate from which the endogenous mRNA had been removed by fractionation or selective destruction without impairing the intrinsic high activity of the translation machinery. This proved to be easy in practice by using micrococcal endonuclease (EC 3.1.31.1) to destroy the endogenous mRNA.[4] This nuclease has an absolute requirement for Ca^{2+} as cofactor, so that globin mRNA can be destroyed by preincubating the lysate with nuclease and Ca^{2+}. EGTA is then added to chelate the Ca^{2+} and thereby inactivate the nuclease. Such is the specificity of EGTA that it is possible to lower the calcium levels below the level required for enzyme activity without significantly affecting the level of Mg^{2+}, which is critical for high-level protein synthesis. (After addition of EGTA, the concentration of free Ca^{2+} is about 10^{-7} *M*, whereas the Mg^{2+} concentration is reduced by less than 10%.) The resulting nuclease-treated lysate has a very low activity unless eukaryotic mRNAs are added. These are translated with remarkable efficiency, more or less whatever their origin (from yeast to higher plants or mammals). The efficiency of the system is indicated by the recovery of up to 70% of the original activity of the "parent" lysate when globin mRNA is added.[4] The failure to regain 100% activity may stem from the fact that after the nuclease treatment many ribosomes are blocked or stranded on short fragments of globin mRNA. One would not expect these ribosomes to be available for translating added mRNA, though preliminary experiments suggest that they do at least release their incomplete nascent chains during the incubation. We do not know whether they are also released from their mRNA fragments.

In spite of the fact that reticulocytes are highly specialized cells, the translation machinery shows little specialization or preference with regard to initiation of protein synthesis on different eukaryotic mRNAs. It is only with respect to the complement of tRNA species that the reticulocyte lysate shows special properties that may impair its ability to translate heterologous mRNA. The relative abundance of different types of tRNA in reticulocytes has been shown to be highly adapted to the synthesis of

[3] R. D. Palmiter, *J. Biol. Chem.* **248,** 2095 (1973).

[4] H. R. B. Pelham and R. J. Jackson, *Eur. J. Biochem.* **67,** 247 (1976).

globin,[5] and since the amino acid composition of globin is somewhat unusual (quite apart from questions of codon frequency), the tRNA population may not be optimal for the translation of mRNAs that code for proteins with more typical compositions. The consequences of this are most clearly seen when mRNAs (particularly viral RNAs) that code for long proteins are translated. Since the tRNA population is ill-adapted for their translation, it can take a very long time to complete the synthesis of the proteins specified by these messages; in extreme cases no full-length products are formed during a normal incubation lasting an hour.[6] This effect depends on the dose of added mRNA or viral RNA; at low input levels, full-length products are formed and the assembly time is not abnormally long, but as the amount of RNA is increased, the assembly time for the protein increases.[6] These observations are most consistent with the idea that although the lysate contains all species of tRNA necessary to read the 61 meaningful codons, some of them are present in rather low amounts. This tends to delay the translation of codons that call for them, and naturally the effect of their shortage becomes more and more significant as the concentration of mRNA or viral RNA increases, and with it the demand for the rare tRNA.

This deficiency is easily overcome by supplementing the lysate with tRNA from a less specialized cell; the rate of assembly of polypeptides specified by added mRNA is increased to near-normal levels, and the inverse dependence of the assembly time on the concentration of added mRNA is abolished.[4,6] For most mRNAs the source of the added tRNA does not make a significant difference, and we routinely use commercial liver tRNA or yeast tRNA.[6] However, if the translation of an mRNA specifying a protein of highly abnormal amino acid composition were under study, it might well be necessary to use tRNA from the same source as the mRNA in order to obtain efficient translation. Such has been shown to be the case for silk fibroin mRNA.[7]

The nuclease-treated rabbit reticulocyte lysate supplemented with heterologous tRNA is now by far the most widely used translation system for eukaryotic mRNAs. In this chapter we describe the procedures we use to make the lysate, and how to make it dependent on added mRNA. We also summarize what we have learned about the lysate during more than 10 years of study of its properties. Understanding the peculiar behavior of the lysate's regulatory systems is important for some aspects of troubleshooting and the design of translation assays. Finally, we put forward

[5] D. W. E. Smith, *Science* **190,** 529 (1975).

[6] T. Hunt and R. J. Jackson, unpublished observations (1975).

[7] P. M. Lizardi, V. Mahdavi, D. Shields, and G. Candelas, *Proc. Natl. Acad. Sci. U.S.A.* **76,** 6211 (1979).

some suggestions for methods of preparing gel-filtered nuclease-treated lysates. In most experiments lysates have been used without gel filtration or dialysis, since these procedures remove components that are essential for high activity. Some of these components have now been identified. Besides the obvious requirement for ATP and GTP, spermidine is needed for maximum rates of translation, and it is also necessary to add glucose 6-phosphate to prevent an early inhibition of protein synthesis, which otherwise shuts down after about 15 min.[8] The glucose 6-phosphate plays two roles: it acts as a direct cofactor enhancing the rate of initiation of protein synthesis, and it also serves as an NADPH-generating system. An adequate supply of NADPH and an active thioredoxin/thioredoxin reductase system are essential for the maintenance of high initiation rates in the lysate.[8]

Preparation of Rabbit Reticulocyte Lysates

Chemicals of Analar or equivalent grade are used when available. All solutions are made up in water of the highest available purity; we normally use glass-distilled deionized water. It has never been our practice to autoclave any of our solutions.

Reagents

Acetylphenylhydrazine (APH) (Sigma). A stock solution of 1.25% w/v in water is prepared on a heater-stirrer and stored at −20°. It requires gentle warming under a hot water tap (for example) to get it back into solution after freezing. The solutions are clear and colorless and should be discarded if they go yellow. However, they seem to keep well frozen. It is curious that it does not seem to be necessary to make up the APH in saline or to sterilize the solutions.

Hypnorm (Janssen Pharmaceuticals, 2340 Beerse, Belgium; obtained in the U.K. through the sole agents, Crown Chemical Company Ltd., Lamberhurst, Kent) is a mild veterinary analgesic containing, per milliliter, 0.2 mg of Fentanyl base and 10 mg of fluanisone. It seems to keep indefinitely at 2–4°. No other product seems to make rabbits so easy to handle. Unfortunately, Hypnorm is not available in the United States or Canada, but we have been informed that a similar formulation, Inovar Vet (obtainable from Pitman-Moore, Inc., P.O. Box 344, Washington Crossing, New Jersey 08560), should produce the same results.

[8] R. J. Jackson, P. Herbert, E. A. Campbell, and T. Hunt, *Eur. J. Biochem.* **131,** 313 (1983).

Nembutal or Sagatal is a commercially available solution of 60 mg of sodium pentobarbitone per milliliter in 10% v/v ethanol, 20% v/v propylene glycol.

Heparin (Evans Medical Ltd., Speke, Liverpool; Fisher Scientific). We normally add 5 ml of water to a bottle containing 100,000 units and keep the solution at 2–4°. The current batch contains 171 units/mg.

Unbuffered saline: 0.134 *M* NaCl, 5 m*M* KCl, 7.5 m*M* $MgCl_2$. We usually store this at 2–4° as a 10× stock.

Buffered saline: 0.134 *M* NaCl, 5 m*M* KCl, 7.5 m*M* $MgCl_2$, 5 m*M* D-glucose, 10 m*M* HEPES, pH 7.2 at room temperature. This solution rapidly grows molds and should be prepared freshly before use.

Brilliant Cresyl Blue. A 1% w/v solution is made in unbuffered saline and stored at room temperature.

Making Rabbits Anemic

We have been using the procedure described here for several years using New Zealand white rabbits of both sexes in the weight range of 3 to 4 kg (usually between 3 and 6 months old). The same procedures work on a variety of other kinds of rabbits; the only caution is that it is easy to overdose small rabbits. If more than 10% of the animals die, the dose of APH should be reduced. It is good practice to keep animals for a week or two before starting to inject them, to make sure that they are healthy and adjusted to their new diet and environment. We would also advise processing several rabbits at once; 5–10 give a lot of lysate (count on about 40 ml per rabbit), and processing does not take very much more time or effort than for 1 or 2 animals. We always pool the blood from all the rabbits, since this averages the potential variations between individuals, and makes for more consistent lysates.

Each animal is injected in the scruff of the neck with 4–5 ml of 1.25% APH solution each day on days 1, 2, and 3 of the schedule. The dose should be varied somewhat according to the age and size of the rabbit; only a very large rabbit would get as much as 15 ml, and we often give something like 4, 4, and 3 ml to a good-sized 6-month-old animal or 3, 3, and 2 ml to a 3-month-old animal. Unfortunately, weight does not seem to be an infallible guide to dosage, possibly because large caged animals are obese and the APH does not enter adipose tissue. Experience is the best guide, but if there are persistent doubts about the efficacy of the injection schedule, with animals either dying unexpectedly or not getting anemic at all it is probably best to make some hematological tests. Blood samples should be obtained from the marginal vein of the ear. They are stained by mixing with an equal volume of the Brilliant Cresyl Blue solution, followed by incubation for 20–30 min at room temperature. The stain is examined under the highest power of a microscope under oil, without phase. A sample stained with Brilliant Cresyl Blue should show virtually 100% Heinz bodies by day 3–4, and virtually none by day 8; it is very easy to tell a cell with a Heinz body in it from a real reticulocyte: the Heinz bodies are compact, round and deeply staining, and the cells containing them are smaller than reticulocytes. The reticulocytes themselves contain a wispy, bobbly blue network. The hematocrit should also be determined. It should fall below 20% by days 3–5 and then start rising back toward the normal value of about 40%; our rabbits usually have a hematocrit of around 25% when we bleed them, but it varies somewhat.

We normally bleed our rabbits on day 8 of the schedule, but a day sooner or later does not seem to make much difference. In our experience this schedule gives lysates of more consistent activity than the commonly described regime of five daily injections followed by bleeding on day 6 or 7. A short interval between the last injection and the bleeding makes for a low hematocrit, and hence low yield of lysate; such lysates do, on the other hand, have a high ribosome content and high initial activity. In our experience, however, they also have a fair probability of showing early shutdown of protein synthesis even in the presence of optimal levels of hemin. We think this is because there are still toxic by-products of APH around; by waiting 5 days between the last injection and bleeding the animal, this possibility is avoided. Our rabbits yield more lysate because they have a higher hematocrit, but with a lower ribosome content. We never get more than about 40% reticulocytosis as judged by staining, but it is very noticeable that almost all the cells are larger than mature erythrocytes, even though they do not stain.

Bleeding the Rabbits

We begin by giving an injection of 0.5 ml of Hypnorm into the thigh muscle (or the scruff of the neck if the rabbit is very hard to handle, in which case it takes longer for the effects of the drug to take). This mild analgesic makes the rabbits very docile and easy to handle.

Once the Hypnorm has taken effect, and the rabbit is rather dopey, the margin of one of the ears is shaved to expose the big marginal vein, which is injected with 2–2.5 ml of Nembutal containing 2000 units of heparin.

The rabbit should go out very quickly as the last of the syringe's contents go in. If the vein is missed, it is best to quickly try the other ear. It can be helpful to have an assistant pinch the vein between thumb and forefinger in order to engorge it with blood, letting go when the injection starts.

When the animal is completely unconscious (which can be tested if necessary by squeezing the back knee as hard as possible—there should be no reaction) the rabbit's chest is damped with 95% ethanol, and the skin is cut away with a pair of sharp scissors. When the ribcage is well exposed, make an incision from the bottom midline away from you, and up the midline toward the head, thereby making a triangular flap of flesh and bone with about 1-inch sides. Push the flap away from you, and cut either the heart itself or one of the great vessels leading from the heart. The chest cavity should rapidly fill with blood, which is removed with a 30–50 ml syringe attached to a 3-inch long piece of silicone rubber or Tygon tubing into a chilled beaker in an ice bucket. One should get an average of about 100 ml of blood from each rabbit; the most we ever get this way is about 130 ml.

Preparing the Lysate

The blood is filtered through cheesecloth or nylon mesh to remove hairs and debris. At this stage we take two drops for staining, mixing them with two drops of the Brilliant Cresyl Blue solution as described above. We do not, however, pay too much attention to the result, for the correlation between the reticulocyte count and the quality of the lysate is not absolute. Experience suggests that it is better to stain whole blood than washed cells.

The cells are harvested from the blood by centrifugation. We use 250-ml polycarbonate bottles in the Sorvall GSA rotor spinning at 2000 rpm for 10 min, but this is probably not very critical. Ideally one would use a swing-out rotor because removing the supernatant would be somewhat easier than with the angle-head. After removal of the supernatant by aspiration, the cells are resuspended in buffered saline containing 5 m*M* glucose (the addition of the sugar and the buffering seems to have given slightly more consistent results than our previous practice of using unbuffered saline) and spun again as before. The washes are repeated three times, i.e., four low-speed spins altogether; on the last wash the volume of cells is determined by resuspending them in a measured volume of saline and measuring the total volume. After the final spin, remove as much of the saline as possible and then add 1.5 volumes (with respect to the packed cell volume) of ice-cold distilled water. Mix the cells and the water thoroughly, and mix the contents of the different bottles with each other. Spin the lysate at 10,000 rpm for 20 min in the GSA rotor (about 15,000 *g*) at 2°. Pour the supernatant into a beaker through a fine nylon mesh (we use 53 μm Nitex) to prevent any detached lumps of the glutinous stroma getting into the lysate; it contains inhibitors of protein synthesis. Some people spin the lysate a second time to ensure complete clarification, but we do not bother; the filtration seems to work well enough.

At this point the lysate can either be frozen in its present state or treated with micrococcal nuclease before freezing, depending on the particular requirements of the user. It is advisable to freeze the lysate in liquid nitrogen to achieve as rapid cooling as possible. For storage, liquid nitrogen is to be preferred, and lysates stored in this way lose no activity over at least 3 years. Storage in a −70° freezer seems to be satisfactory for periods of at least several months, but we have not tested longer-term storage at this temperature. We and others have found that lysates lose activity very rapidly when stored at −20°. We generally store small aliquots of lysate in screw-cap plastic vials (e.g., Nunc N 1076-1), which hold 1 ml nicely, or in standard 1.5-ml conical capped Eppendorf centrifuge tubes (e.g., Sarstedt No. 72.690). Larger aliquots are stored in plastic scintillation vials. Since we do not normally refreeze aliquots of lysate once they have been thawed, we usually store lysate as 0.5–1 ml aliquots. In fact it is very unusual for a single round of thawing and refreezing to cause any loss of activity, but storage methods that require multiple cycles of freezing and thawing should be avoided.

Preparation and Use of Nuclease-Treated Lysate

It is convenient to carry out small-scale incubations in disposable plastic tubes: we use Sarstedt No. 72.698 or No. 72.699 for incubations between 10 and 100 μl and No. 72.690 for incubations between 100 and 1000 μl. We do not autoclave or wash the tubes before use, nor do we treat pipettor tips in any way. We just keep them in a safe, clean place and avoid touching them with bare hands.

Unless otherwise stated, solutions are made up in the highest quality water as explained previously, and chemicals should be of the best available grade. We store many solutions as small aliquots, but, unlike lysate, refreeze them repeatedly. No ill effects seem to result from this practice.

Reagents

Creatine kinase (Boehringer). A stock solution of 5 mg/ml in 50% v/v aqueous glycerol is kept (unfrozen) at −20°.

Creatine phosphate (Boehringer). A 0.2 *M* solution is made up in water and stored in 0.2-ml aliquots at −20°.

KM solution: 2 *M* KCl, 10 m*M* $MgCl_2$ stored in 0.2-ml aliquots at −20°

Unlabeled amino acid mixture. The stock solution contains 3 m*M* L-leucine and L-valine and 2 m*M* each of the 17 other L-amino acids;

the amino acid to be used as label is omitted. The solution is dissolved by warming on a heater-stirrer, neutralized with KOH to pH 7.2, made 1 m*M* in dithiothreitol, and stored in 0.2-ml aliquots at −20°.

[^{35}S]Methionine (Amersham International). The material delivered by the supplier is thawed and divided into 50-μl aliquots, which are stored in liquid nitrogen.

Hemin (Sigma or Eastman Kodak or isolated from whole blood by published procedures[9]), a 1 m*M* solution in 85% v/v ethylene glycol. For 10 ml of solution, 6.5 mg of hemin are dissolved in 0.5 ml of 1 *N* NaOH; then 0.5 ml of 1 *M* Tris-HCl, pH 7.5, is added, followed by 8.5 ml of ethylene glycol. The pH is the brought down to about 8.0 with 1 *N* HCl added slowly while stirring rapidly on a pH meter. The meter responds sluggishly in this medium, and it is important to go carefully. Finally, the concentration of the solution is checked by diluting a suitable sample in 50 m*M* KCN and reading the A_{540}; a 1 m*M* solution of the cyanide complex of hemin has an absorbance of 11.1 at this wavelength. This solution is extremely stable; we store it in the freezer.

Micrococcal (staphylococcal) nuclease (Boehringer), 15,000 units/ml. The contents of a bottle of 15,000 or 45,000 units are dissolved in water, and the solution is stored at −20°.

$CaCl_2$, 0.1 *M*

Ethylene glycol-bis(2-aminoethyl ether)-*N*,*N*′-tetraacetic acid (EGTA) (Sigma). A 200-m*M* solution, pH 7.5, is made by neutralizing a suspension of the free acid with KOH with stirring on a pH meter.

Calf liver tRNA (Boehringer). Water is added to the dry solid to give a solution of 10 mg/ml.

Decolorizing solution: to 1000 ml of 1 *N* NaOH are added approximately 1 g of solid DL-methionine and 50 ml of 30% w/v H_2O_2 ("100 volume"). This solution keeps surprisingly well at room temperature. However, after prolonged storage the hydrogen peroxide decays, and it is necessary to add more if the bleaching power of the solution is insufficient.

Trichloroacetic acid: 25% w/v and 8% w/v aqueous solutions

Nuclease Treatment of the Lysate

The procedure described below has been used for volumes as small as 0.1 ml and as great as 100 ml, using clean plastic or more usually glass containers for the incubation. The same procedure is used for both fresh and frozen lysate.

For 10 ml of lysate, add 0.2 ml of 1 m*M* hemin, 0.1 ml of 5 mg/ml creatine kinase solution, 0.1 ml of 0.1 *M* $CaCl_2$ and 0.1 ml of the micrococcal nuclease solution; mix thoroughly and incubate at 20° for 15 min (the time is not particularly critical, and if a large volume is being treated it may be advisable to lengthen the incubation to allow time for temperature equilibration). Digestion is stopped by adding 0.1 ml of 0.2 *M* EGTA; 60 μl of 10 mg/ml tRNA solution are added and mixed in well. The lysate is dispensed in suitable aliquots; we usually use about 0.5 ml.

This procedure is not very demanding, for there is considerable margin for error in the concentrations of most of the components. Failure is most likely caused by making up the $CaCl_2$ or EGTA solutions incorrectly; obviously, residual calcium would be fatal. It is reassuring that the lysate itself loses only about 5% of its activity per hour during incubations at 20° in the presence of hemin, and can tolerate at least five times the level of micrococcal nuclease specified above without showing significant inhibition, provided no free calcium is present. The incubation time is not particularly critical: suggested limits are 10–30 min, depending on the temperature, which may itself be varied over the range 15–25°. The precise level of added tRNA is also not important, though, depending on the particular preparation, very high amounts (above about 200 μg/ml) may be inhibitory.

Use of the Nuclease-Treated Lysate for Assay of mRNA

The basic incubation mix contains 0.8 volume of nuclease-treated lysate, prepared as above, and 0.05 (1/20th) volumes of each of the following four components: KM solution; creatine phosphate solution; unlabeled amino acids solution; labeled amino acid solution. Typically one would take a tube of nuclease-treated lysate (480 μl) from the liquid nitrogen and add 30 μl of each of these four solutions. After thorough mixing (very important), suitable aliquots of this mix are dispensed into plastic incubation tubes, and the appropriate amount of mRNA solution is added to each tube. All these operations should be conducted with the tubes embedded in ice. The most difficult part of this operation, the one that causes most trouble to beginners, is mixing: since the lysate is a very concentrated protein solution, it is viscous and requires fairly vigorous vortexing to mix it thoroughly, yet it is important to avoid frothing, which prevents good mixing and may also result in some inhibition of protein synthesis through oxidation or denaturation of proteins. If excessive foaming does occur, a quick microfuging usually clears it. With practice one can vortex-mix incubations efficiently, quickly, and without introducing bubbles.

We normally try to add the mRNA in a volume not greater than 10% of the volume of incubation mixture, and preferably less—i.e., while one might add 2 μl of mRNA solution to a 20-μl sample of incubation mix, 1 μl is preferable. In emergencies, as much as 20% by volume can be added, but it is important to remember that protein synthesis is rather sensitive to the ionic environment, and this dilution can change the concentrations of the K^+ and Mg^{2+} significantly; account has to be taken of any salts in the RNA preparation, and if necessary, compensatory adjustments must be made to the composition of the KM solution. For this reason, we always dissolve RNA in water if we want to translate it. Not only does it make the problems above easier to solve, but the RNA can always be concentrated by lyophilization if it should prove to be too dilute. Indeed, the solution of mRNA can be lyophilized in the same tube that will subsequently be used for the assay, and the 10–20 μl of assay mix added directly to the dry mRNA.

With a previously untested mRNA preparation it is a good idea to carry out a dose-response experiment, since the highest concentration does not necessarily give the most protein synthesis if (as often happens) it is contaminated with inhibitors of translation. The simplest way to construct a dose-response curve is to prepare an assay mix and set out one tube of 40 μl followed by a series of 20 μl aliquots. Add 4 μl of mRNA to the first tube, mix well, and transfer 20 μl to the next tube down the line. Continue this process down the line of tubes, mixing well by pipetting the mix up and down in the pipettor used for making the transfer. In this way a series of assays containing serial twofold dilutions of mRNA is set up.

Do not forget to have at least one control tube to which no mRNA is added (see below).

When all the tubes are ready, the rack is transferred from the ice bucket to a 30° water bath and left for at least an hour. At 37°, protein synthesis proceeds quicker but is more likely to stop earlier and at a lower final level of incorporation of radioactivity than at 30°. If the synthesis of very high molecular weight proteins is expected, it is worth leaving the tubes to incubate for 2 hr; since it takes ribosomes 30–40 min to assemble a polypeptide of 250,000 daltons, a fair proportion of the ribosomes would still be carrying uncompleted chains at 60 min. One of the fortunate features of the lysate is that it tends to stop initiating new chains before it loses the ability to elongate and complete chains that have already been started; thus, leaving incubations going for 2 hr tends to improve the yield of full-length translation products. We have not encountered problems of degradation (except in cases such as the translation of EMC RNA, where specific proteolytic processing of precursor polypeptides is normal[10]).

[10] H. R. B. Pelham, *Eur. J. Biochem.* **85,** 457 (1978).

Assay of Incorporation of Radioactivity into Products

After the incubation, the incorporation of labeled amino acid into protein can be determined by taking samples from the tubes and diluting them with water. We usually take 1–5 μl samples [using a Hamilton syringe, a Drummond (in Europe, Brand) digital pipettor, or an Oxford Micropettor] into 1 ml of water in 12 × 75 mm glass tubes. Having delivered the sample, the pipette is rinsed several times in this same water (the sample is very dense, sinks to the bottom, and leaves clean water above). This procedure achieves two things; it stops the reaction by dilution and cleans the sampling device ready for the next sample. When all the samples have been taken, 0.5 ml of decolorizing solution is added to each tube. This deacylates charged tRNA and bleaches the red color. The color should fade in less than 10 min at room temperature; if it does not, more H_2O_2 should be added.

When the samples are colorless, about 2 ml of 25% w/v trichloroacetic acid are squirted into each tube; the contents are mixed and then filtered through Whatman GF/C filters. The tube is rinsed with 8% w/v trichloroacetic acid, and the filters are washed with the rinsings. The filters are dried in an oven or in a hot airstream and counted in a scintillation counter.

Average preparations of mRNA usually give a stimulation over the zero-message control of about 10- to 20-fold, and really good preparations over 100-fold, but this is exceptional. If your mRNA gives little or no stimulation, it is still worthwhile to analyze the samples on a gel because the background can be suppressed by added mRNA, and specific products will often show up on the fluorogram, given a long enough exposure.

Controls

In any assay of mRNA with the nuclease-treated lysate it is vital to include proper controls. One is obviously the zero-message control mentioned above, which measures the background (we say more about this later). But another equally important, or at least very useful, test, which we include in almost every experiment, is an incubation containing a standard mRNA—an RNA that can be obtained easily in fairly large amounts and is efficiently translated into well-defined products. This serves to calibrate the system. Plant viral RNAs are very suitable for this purpose, and we use tobacco mosaic virus RNA (TMV RNA) at a final concentration in the assay of 50–100 μg/ml. This level of added TMV RNA should give a stimulation of at least 100 times the zero-message background with [^{35}S]methionine as label (the record is just over 500-fold stimulation). If it does not, something is wrong.

This standard control is very important. It is the only way to tell whether failure to obtain significant translation of a given mRNA preparation is due to the mRNA itself or to some fault in the assay system—bad lysate, decomposed label, etc. Usually the problem lies with the mRNA, but without a standard to test the lysate, it is impossible to know. When a message that ought to work well in the lysate does not, the commonest explanation is not that the preparation lacks active mRNA, but that strong inhibitors of translation are present as contaminants. The "standard" RNA can be used to reveal such inhibitors: an assay is set up containing the mRNA under test together with TMV RNA at 50 μg/ml, to see if the translation of the viral RNA is markedly inhibited.

TMV RNA is also useful for testing the tRNA preparation. At high levels (above 100 μg/ml) of added TMV RNA, the presence of liver tRNA at 50–100 μg/ml should stimulate protein synthesis by a factor of 4 or 5; at lower levels of TMV RNA, the stimulation will be somewhat less.

Troubleshooting

What can you do if a batch of nuclease-treated lysate does not work, that is, fails to translate a tried and trusted mRNA preparation? In a subsequent section we discuss various properties of lysates that are known to affect the efficiency of protein synthesis irrespective of whether such lysates have been treated with micrococcal nuclease. We confine ourselves here to the diagnosis of possible faults in the nuclease treatment procedure.

The first point to emphasize is that nuclease treatment only serves to destroy endogenous mRNA; it cannot transform a lysate with poor activity into an active translation system. Therefore the first step is to examine the activity of the parent lysate that has not been treated with nuclease. This is very simple; the hemin, creatine kinase, KM, creatine phosphate, amino acid mixture, and label are added to a tube of the lysate, but the $CaCl_2$, nuclease, EGTA, and tRNA are left out. After incubation at 30° for 60 min, the incorporation of label into protein is measured. If all is well, about half the added radioactivity should have been incorporated into protein when high-specific-activity [^{35}S]methionine is used. This is only a rough guide, and we would recommend keeping back a small number of samples of previous lysates so that the new batch can be compared with the old ones; in this way one can also catch mistakes in making up the standard master-mix components, bad batches of label, and so on.

Assuming that the parent lysate shows normal activity, the next step is to check the individual components of the nuclease preincubation mix. EGTA (2 m*M*) should not inhibit the parent lysate, nor should 1 m*M*

$CaCl_2$ plus 2 mM EGTA. The combination EGTA, $CaCl_2$, and micrococcal nuclease should not inhibit, either. If it does, test the nuclease with EGTA alone; if inhibition is observed, there must be something wrong with the nuclease. If nuclease plus EGTA does not inhibit, but the addition of calcium to them does, something is wrong with the Ca^{2+} and/or EGTA solutions, and they should be replaced.

Analysis of Translation Products by SDS–Polyacrylamide Gels

It is beyond the scope of this chapter to discuss different ways of running translation products on SDS–polyacrylamide gels; we actually use 0.8 mm thick gels run according to the recipe of Anderson *et al.*,[11] but other methods produce comparable results. The important point to stress for those who are not familiar with the reticulocyte lysate is that, since the lysate contains such a high concentration of protein (typically 50–60 mg/ml), it is essential that the sample be diluted in enough sample buffer containing enough SDS so that all the protein present is saturated with the detergent; horrible streaks can otherwise result. Our routine practice is to dilute one volume of standard translation mix with five volumes of an SDS–gel sample buffer that contains 2% w/v SDS, and to load the slot about 5–10 mm high. Resolution is seriously impaired if the dilution of the sample is less, but is fairly tolerant of the amount loaded per slot, despite the tremendous overloading with globin, which makes up about 90% of the total protein in the lysate.

With the recommended dilution and loading, resolution of proteins larger than globin is excellent, and that of proteins smaller than globin adequate. It is only with proteins of about the same mobility as globin that difficulties are encountered; for instance, the core histones are not very well resolved compared to most translation products.

The high protein content has consequences that make two further traps for the unwary. First, if there is an empty track next to a lysate sample lane, protein tends to spread from the one to the other, which is not very attractive to look at; the solution is to flank the labeled tracks with dummy samples containing unlabeled lysate and sample buffer so that all lanes contain the same protein loading. The second caution is that there is some distortion of mobility such that accurate estimates of molecular weight of products can be made only by adding labeled molecular weight markers to appropriate unlabeled dummy incubation samples. The marker proteins have to be radioactively labeled for them to show up on the autoradiographs, and we usually make our own by incubating suitable

[11] C. W. Anderson, P. R. Baum, and R. F. Gesteland, *J. Virol.* **12,** 241 (1973).

proteins with 5 mM [^{14}C]iodoacetic acid at pH 8.5 in 8 M urea overnight at room temperature in the dark, followed by dialysis or gel filtration into sample buffer. Alternatively, reductive methylation with [^{14}C]formaldehyde works too, and such labeled marker mixes are commercially available.

An aliquot of the zero-mRNA control should always be run on the gel along with the real assays containing mRNA. The control samples give some specific labeled bands that may interfere with the detection of the true translation products. The most prominent of these is a group of bands with an apparent M_r of 25,000–30,000, which are almost certainly labeled peptidyl-tRNA resulting from translation of fragments of globin mRNA (Fig. 1). Their labeling is completely suppressed by a variety of inhibitors of protein synthesis and also by the addition of efficiently translated RNAs to the assay, presumably as a result of competition for some limiting component of the translation machinery. They are therefore most troublesome when one is trying to detect synthesis programmed by very poor or scarce messages, as during mRNA selection assays; and addition of high concentration of label can sometimes result in their intensity far outweighing that of the "real" bands. Fortunately they are easy to eliminate, for they disappear if the sample is treated with 50–100 μg of RNase A per milliliter (after the protein synthesis incubation), and then incubated for a further 10 min at 30° prior to adding the SDS–gel sample buffer (Fig. 1).[12]

The other main background band seen when [^{35}S]methionine is used as label runs with an apparent M_r of 42,000 (Fig. 1). The intensity of this band is very variable among different lysates, and its presence is *not* eliminated by inhibitors of protein synthesis. Its labeling is thought to result from a tRNA-dependent but ribosome-*in*dependent addition of methionine to a preexisting protein.[12] If a different amino acid is used as label, this protein is not labeled (though others may be labeled instead, by a similar mechanism). If the 42,000 M_r band is troublesome, its mobility can be altered by omitting the customary heating step before loading the gel; it now runs with an apparent M_r of about 220,000 (Fig. 1). It makes very little difference to the resolution and mobility of the translation products if the samples are not heated before loading.

Finally in this section, a word about analysis by two-dimensional gel electrophoresis. It seems that good resolution of proteins in the isoelectric range 5–7.5 can be obtained; the globin is basic and runs off backward. We add incubation samples directly to weighed aliquots of solid urea and

[12] H. R. B. Pelham, Ph.D. Thesis, University of Cambridge (1978).

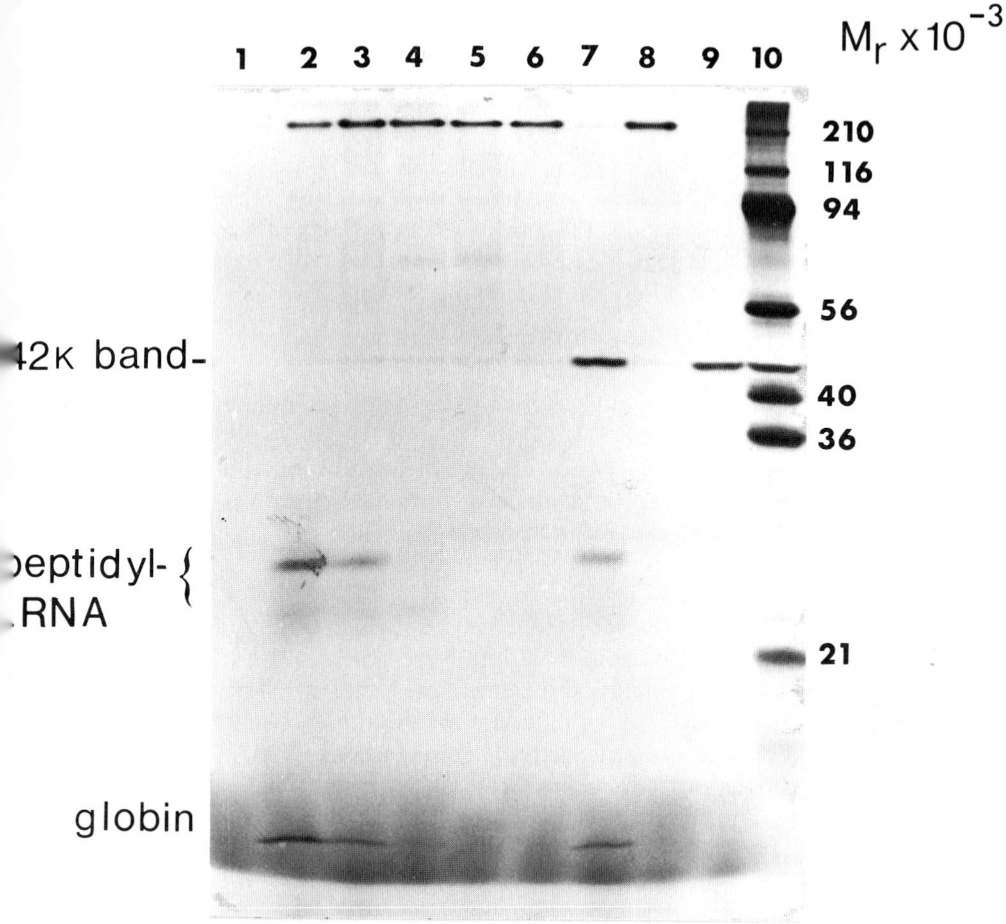

FIG. 1. Background incorporation in the nuclease-treated lysate. A nuclease-treated lysate was incubated with [^{35}S]methionine, but no added mRNA, for 1 hr, except for the samples displayed in track 1, 0.5-min incubation, and track 2, 30-min incubation. Some incubations contained inhibitors as follows: 0.5 m*M* m^{7}GTP (track 4); 5 μM edeine (track 5); or 1 m*M* emetine (track 6). The assays were diluted with sample buffer as described in the text, but with the exception of the samples in tracks 7 and 9 they were not heated before loading on a 15% gel. The samples in tracks 8 and 9 were incubated for 1 hr, made 0.1 mg/ml in pancreatic RNase, and incubated for a further 10 min before dilution in sample buffer. Track 10 contains the same material as track 9, but supplemented with radioactive marker proteins, whose molecular weights are marked on the autoradiogram. From Pelham[12] by courtesy of H. R. B. Pelham.

then add an equal volume of O'Farrell gel sample buffer before loading directly onto the first-dimension gels.[13]

Important Properties of Reticulocyte Lysates

Reticulocyte lysates contain a number of components necessary for their high protein synthesis activity. They can also be inhibited by a bizarre range of compounds that one might unwittingly or in all innocence add to translation assays. This section summarizes what is known about essential cofactors and potential inhibitors.

Composition of Reticulocyte Lysates

According to our measurements, lysates typically contain 25–40 m*M* K^+, 10 m*M* Na^+, 1.6–1.8 m*M* Mg^{2+}, 0.4 m*M* spermidine, less than 0.07 m*M* spermine, about 20 μM NADP(H), 1.5–2.0 m*M* glutathione, 0.5 m*M* ATP, and 0.1 m*M* GTP. In addition there are present, in unknown amounts, glycolytic intermediates, acetyl-CoA, some citric acid cycle intermediates, and 2,3-diphosphoglycerate. Some of these compounds are potential chelators of Mg^{2+}, and it is not surprising that the optimum Mg^{2+} concentration in gel-filtered lysates is somewhat lower than in normal lysates, because these compounds will have been removed.

The lysate also contains all the enzymes of the glycolytic and pentose phosphate pathways, and the latter are important for NADPH generation to maintain protein synthesis activity. Aconitase and NADP-linked isocitrate dehydrogenase are also present, so lysates can reduce NADP using citrate or isocitrate. The NADPH-dependent reducing systems present in the lysate are (*a*) glutathione reductase and (*b*) thioredoxin reductase together with thioredoxin.[8]

Mg^{2+} and K^+ Concentrations for Optimum Translation

When the incubation mix is made up by addition of the amount of KM solution specified above, it brings the *total* K^+ concentration to about 125 m*M*, and the *total* Mg^{2+} concentration to 2.0 m*M* in the regular system, or 1.4 m*M* in the gel-filtered system to be described later in this chapter. These concentrations have been found to be optimal for the translation of endogenous globin mRNA, and it turns out that they are optimal for the translation of most messages and viral RNAs. There are, however, some exceptions, of which the most extreme example is poliovirus mRNA, which requires a final concentration of only 60 m*M* K^+ and is translated

[13] P. J. Farrell, T. Hunt, and R. J. Jackson, *Eur. J. Biochem.* **89,** 517 (1978).

very poorly under the standard ionic conditions. It may therefore be worth testing the effects of varying the K^+ and Mg^{2+} concentrations on the efficiency of translation, looking not only at total incorporation but also the pattern of translation products on gels. It is possible for different initiation sites to have different Mg^{2+} optima, and termination can be suppressed to some extent by slightly superoptimal Mg^{2+} concentrations.[14]

Acetate or Chloride?

In the past few years it has become fashionable to use potassium acetate in the place of KCl. Whereas raising the final KCl concentration to 150–175 m*M* results in an inhibition of protein synthesis over that seen at 125 m*M*, no very great inhibition occurs if one adds similar high levels of potassium acetate.[15] This has led to claims that the use of potassium acetate is more physiological since it allows one to use K^+ levels closer to what is believed to be the normal intracellular level. The flaw in this argument is that the high acetate concentration is quite unphysiological; cells contain a far higher concentration of Cl^- than they do of acetate$^-$.

As far as we can see, the merits of potassium acetate have been rather exaggerated, for it does not in general stimulate the rate or extent of protein synthesis significantly over what can be achieved using KCl,[15] but merely allows the use of higher K^+ concentrations without suffering inhibition. It is also worth noting that high Cl^- tends to reduce the rate of initiation, while the elongation rate is relatively unaffected. Thus at the very worst, the use of excess KCl will only cause a reduction in yield of translation products, but will not change the product pattern. We are not really convinced by the arguments put forward in favor of the acetate ion.

Amino Acid Pools and Amino Acid Requirements

The concentration of unlabeled amino acids in the lysate varies according to the species. The methionine pool is very low (about 5 μM), and the concentration of cysteine, valine, leucine, isoleucine, phenylalanine, and tyrosine are also low, in the 10–15 μM range.[12] On the other hand, the pools of alanine, glycine, serine, threonine, lysine, and arginine are high—up to 0.5 m*M*—and the use of these amino acids as label in translation assays should be avoided unless absolutely necessary. If their use is essential, it is advisable to use a gel-filtered, and hence amino acid-de-

[14] H. R. B. Pelham, *Nature (London)*, **272,** 469 (1978).

[15] L. A. Weber, E. D. Hickey, P. A. Moroney, and C. Baglioni, *J. Biol. Chem.* **252,** 4007 (1977).

pleted, lysate. It is difficult to state an exact figure of the sizes of the various pools, for not only do they vary between different batches of lysate, but different methods of measurement give somewhat different values. Direct measurement by taking the trichloroacetic acid-soluble fraction of the lysate and analyzing it on an amino acid analyzer gives lower values than are obtained from isotope dilution experiments. Either the pools expand during the incubation (through proteolysis, for example), or part of the pool is cryptic to acid extraction.

Our lysates consume up to 2.5 nmol of leucine per milliliter per minute if they are not nuclease treated; this corresponds to about 8 nmol of globin per milliliter per hour, or about 1 chain per active ribosome per minute (the concentration of ribosomes in these lysates is about 2.5×10^{-7} *M*, of which roughly half are active). The nuclease-treated lysate will not achieve such a high rate even with a very good mRNA, but even so, it is clear that the endogenous pools of amino acids would be inadequate to sustain a reasonable rate of translation for a long time. It is therefore important to add amino acids, and the amounts recommended here should be more than adequate for all eventualities.

Attention must also be paid to the question of whether the amount of labeled amino acid is limiting. In any experiment in which a high proportion (say 70–80%) of the added label is incorporated into protein, it is likely that the amount of that amino acid will have limited synthesis, and that translation may have stopped before all the polypeptide chains have been completed. For this reason, it is as well to be aware what the maximum possible incorporation could be in a given size of sample.

When using [^{35}S]methionine, it is unlikely that all of the added label is available for protein synthesis, since some of it gets converted into other compounds, like methionine sulfoxide and possibly *S*-adenosylmethionine. It has also been known for labeled methionine solutions to go bad; since they are generally supplied at a concentration in the micromolar range, they easily get oxidized on repeated freezing and thawing, and should ideally be stored under liquid nitrogen vapor as small aliquots that are used in strict rotation; storage at −20° is very inadvisable.

If the amount of labeled amino acid added does prove to be limiting, there are two ways out; either more labeled amino acid or some unlabeled amino acid must be added to increase the pool size. In the former case it is generally necessary to concentrate the labeled amino acid by lyophilization just before use, and to add the assay mix directly into the tube used for the freeze-drying. If this course is impractical on the grounds of time or expense, a trick to get around the dilution of the label implicit in the second solution above is to add cold amino acid part way through the incubation; i.e., to get most of the label incorporated, and then complete the chains by the cold chase.

Fortunately these problems do not really arise when using [^{35}S]methionine (which must be by far the most widely used radioactive amino acid). It is extremely unusual to encounter a situation where the amount of methionine limits protein synthesis; that is, if another amino acid is used as label, it turns out to be almost impossible to starve a lysate for methionine. We think that a small pool is constantly maintained by some process, perhaps proteolysis. However, it is possible to get more incorporation of radioactivity into translation products by adding more methionine, up to a point. Concentration by lyophilization works well, but one needs to take care; the [^{35}S]methionine supplied by Amersham International contains potassium acetate at 0.02 *M*, and 14 m*M* mercaptoethanol as antioxidant. The salt is obviously concentrated by drying, and although the mercaptoethanol is probably removed, it is doubtful whether any oxidized mercaptoethanol is, and this could inhibit translation. It is advisable to add a final concentration of 1 m*M* dithiothreitol to the assay mix to guard against this potential problem. In practice we have found that it is always safe to resuspend dried-down [^{35}S]methionine in its original volume of assay mix, and generally possible to go as high as twice this concentration, i.e., about 20 mCi/ml, before trouble starts; we should report, however, that at higher concentrations we found that some batches of label completely inhibited protein synthesis. These are extremes that we would never go to in real life—for one thing, it would cost a fortune; but it is useful to know what are the limits.

Control of Initiation by Hemin, GSSG, etc.

If reticulocyte lysates are incubated without hemin the initial rate of protein synthesis is normal, but after 5–10 min the rate falls dramatically to 2–10% of its former value.[1,2,8] A similar shutoff is seen when lysates containing hemin are incubated with double-stranded RNA or with oxidized glutathione. Incubation at 40–42° or incubation of gel-filtered lysates in the absence of glucose 6-phosphate also causes this kind of shutoff.[8] In all cases, the inhibition is due to a reduced rate of initiation of protein synthesis, and all types of mRNA are more or less equally affected. It is believed that the inhibition occurs because protein kinases that phosphorylate the initiation factor eIF-2 are activated under all the disparate circumstances sketched above.[8,16]

[16] P. J. Farrell, K. Balkow, T. Hunt, R. J. Jackson, and H. Trachsel, *Cell* **11**, 187 (1977).

Because of this, there are certain precautions that must be observed when using lysates. The hemin should be added as soon as possible after thawing if the lysate was stored without it. Incubation at high temperature should be avoided. Oxidized thiols (e.g., oxidized mercaptoethanol, cystine) and other potential oxidizing agents should be excluded, because they may exhaust the capacity of the system to reduce them via NADPH, itself produced from the oxidation of endogenous glucose 6-phosphate.

One way of overriding these inhibitory effects is to add 1/20th volume of a neutralized 100 m*M* solution of 3′-5′ cyclic AMP (cAMP) to the assay mix; 5 m*M* cAMP inhibits the kinase and prevents shutoff.[16] Another approach is to add 2 m*M* GTP and 2 m*M* $MgCl_2$, which prevents activation of the hemin-regulated kinase.[17] If one experiences an unexpected early reduction in the rate of protein synthesis, it is always worth testing whether these compounds help, and it may be worth trying them in conjunction with added 1 m*M* dithiothreitol.

Double-Stranded RNA (dsRNA)

Of the many agents that inhibit initiation by causing the phosphorylation of eIF-2, dsRNA deserves special mention since it is a possible contaminant of mRNA and viral RNA preparations added to the translation system. RNA preparations containing as little as 0.1% dsRNA can still be inhibitory, since protein synthesis is sensitive to dsRNA levels as low as 1 ng/ml.[18] The inhibition of initiation by dsRNA can be prevented by the addition of 5 m*M* cAMP, as mentioned above.[16] It can also be prevented by the addition of high concentrations of dsRNA (20–50 μg/ml); surprisingly, high concentrations of dsRNA do not activate the protein kinase, and can actually reverse the activation of the kinase brought about by low levels of dsRNA.[16,18] Naturally occurring dsRNAs seem to be much more effective anti-inhibitors at high concentrations than synthetic dsRNAs like poly(rI : rC). We use *Penicillium chrysogenum* dsRNA for this purpose, but doubtless reovirus RNA would be equally good. If the translation of a given RNA preparation is significantly stimulated by the addition of 20 μg of natural dsRNA per milliliter, this is an absolutely unambiguous indication that the RNA preparation contains contaminating dsRNA.

Besides activating eIF-2 kinase, dsRNA has another effect: it promotes the synthesis of oligoisoadenylate (2–5 A), which in turn activates a nuclease latent in the lysate.[19,20] The result is that mRNA is nicked and

[17] K. Balkow, T. Hunt, and R. J. Jackson, *Biochem. Biophys. Res. Commun.* **67,** 366 (1975).
[18] T. Hunter, T. Hunt, R. J. Jackson, and H. D. Robertson, *J. Biol. Chem.* **250,** 409 (1975).
[19] B. R. G. Williams, C. S. Gilbert, and I. M. Kerr, *Nucleic Acids Res.* **6,** 1335 (1979).
[20] M. J. Clemens and B. R. G. Williams, *Cell* **13,** 565 (1978).

inactivated. Neither 5 mM cAMP nor high levels of dsRNA seem to prevent the synthesis of 2–5 A and the consequent endonuclease activation.[19] The serious problem about this type of inhibition is that because the mRNA gets nicked during the incubation, the synthesis of incomplete protein chains occurs, with ribosomes stranded at the nicked ends of the RNA fragments. Fortunately, we find that low concentrations of dsRNA inhibit translation mainly via the pathway of activation of eIF-2 kinase and inhibition of initiation, and that the activation of 2–5 A-dependent nuclease is of relatively minor significance in the overall inhibition of protein synthesis. The amounts of dsRNA added to translation assays as contaminants of mRNA are unlikely to lead to much nuclease activation. Nevertheless, one should watch out for incomplete translation products whenever there are suspicions of the presence of dsRNA, which can be diagnosed as explained above, by testing whether protein synthesis is stimulated by high concentrations of dsRNA.

Polysaccharides

When RNA is isolated by phenol extraction, any polysaccharides in the tissue or cell suspension will be coextracted with the RNA. Any heparin added as anticoagulant or as ribonuclease inhibitor will also be extracted and contaminate the RNA preparation. Whereas neutral polysaccharides do not seem to be particularly inhibitory (indeed, some workers recommend the addition of glycogen as carrier for small amounts of RNA), acidic polysaccharides are strong inhibitors of translation, and preparations of RNA that contain significant amounts of them may not be efficiently translated. We have known cases where good mRNA was present, but absolutely no translation could be detected until the polysaccharide had been removed. We know of no test that specifically diagnoses inhibition by polysaccharides, but such contamination would be detected by the mixing test described previously, in which one tests whether the translation of a standard mRNA like TMV RNA is inhibited by the suspect unknown RNA preparation. Unfortunately there are no known methods by which the inhibitory effects of acidic polysaccharides can be prevented or overcome; the only solution is to get rid of them by purification of the RNA. In our experience, precipitation of the RNA with 2 M LiCl or 3 M sodium acetate achieves this aim, but one gets good recovery of the RNA only if it is at a relatively high concentration, of the order of 1 mg/ml or higher.[21] Since the polysaccharides do not bind to oligo(dT)-cellulose this particular source of trouble should not be encountered when poly(A)$^+$ RNA is being translated.

[21] R. D. Palmiter, *Biochemistry* **13,** 3606 (1974).

The Preparation of Gel-Filtered Nuclease-Treated Lysates

We recommend using Sephadex G-25 for making gel-filtered lysates. Although we have had good results with G-50, the properties of these lysates are more variable, probably reflecting a variable degree of removal of thioredoxin, which is partially included on G-50. A further disadvantage of G-50 is that hemoglobin is slightly included, so that the hemoglobin concentration is not a good guide to tell which column fractions to pool.

A peculiar and little-publicized feature of Sephadex is that it binds ribosomes more or less irreversibly, so that, when a column is used for the first time, the emerging lysate is seriously depleted in ribosomes. The binding sites are easily saturated, and the second or third column runs give excellent activity in the gel-filtered lysate. We generally presaturate our columns before their first "real" use by passing about 0.5 column volumes of lysate through them. It is obviously preferable to use old or somewhat inactive lysate for this purpose.

Reagents

Buffer A: 25 m*M* KCl, 10 m*M* NaCl, 1.1 m*M* $MgCl_2$, 0.1 m*M* EDTA, 10 m*M* HEPES, pH 7.2 with KOH at 20°

Buffer B: 25 m*M* KCl, 10 m*M* NaCl, 1 m*M* $MgCl_2$, 0.1 m*M* EGTA, 10 m*M* HEPES, pH 7.2 with KOH at 20°

Sephadex G-25: superfine, fine, or medium grade according to the size of the column—as a rough guide, for 1–10 ml, 10–100 ml, and over 100 ml total bed volume columns, respectively.

ATP (Sigma), a 0.1 *M* stock solution neutralized with KOH, stored at −20°

GTP (Sigma or Boehringer), a 0.02 *M* solution neutralized with KOH, stored at −20°

Glucose 6-phosphate (Boehringer), a 0.1 *M* stock solution stored at −20°

Spermidine trihydrochloride (Sigma), a 0.1 *M* solution in water stored at −20°. This solution is odorless when fresh, but acquires a characteristic smell if it becomes oxidized, in which case it should be discarded.

5,5′-Dithiobis(2-nitrobenzoic acid) (DTNB) (Sigma or Boehringer). A 0.1 *M* solution in 96% ethanol is stored at −20°; 1 m*M* working solutions are made fresh by dilution of this stock in 50 m*M* Tris-HCl, pH 8.5.

Procedure

There are two alternative ways of making a nuclease-treated gel-filtered lysate: either the gel filtration or the nuclease treatment can be done first. Some people who have tried both have told us that they obtained very poor activity when gel filtration was done first, but we have obtained good results either way, though with a very slight margin in favor of nuclease treatment before the gel filtration.

To gel-filter a lysate prior to nuclease treatment, we first make the lysate 20 μM in hemin and 50 μg/ml in creatine kinase, and then apply it to a column of Sephadex G-25, equilibrated with buffer A, whose irreversible ribosome binding sites have already been saturated as described above. The total bed volume of the column should be about 5 times the volume of the load, and the ratio of height to diameter of the bed should be between 10 and 20. One can actually load up to about 30% of the total bed volume before overloading occurs. Overloading can be checked by measuring the separation between the hemoglobin peak (A_{540}) and the glutathione peak, which is detected by mixing a 50-μl sample of each column fraction with 1 ml of the 1 mM DTNB reagent and measuring the absorbance at 410 nm. The flow rate of the column does not seem to be very critical, and we normally pump it slightly faster than it would flow under its own hydrostatic head.

Fractions are pooled according to one of two possible criteria, either their hemoglobin content or their protein synthetic activity. If the A_{540} is used, only fractions containing more than 70% of the absorbance of the parent lysate should be taken, because protein synthesis is quite sensitive to dilution. The more rigorous but also more time-consuming method is to make a 10-min translation assay with samples of each red fraction and make a pool of the most active fractions.

The pool is made 0.5 mM in ATP, 0.1 mM in GTP, 0.4 mM in spermidine, and 0.2 mM in glucose 6-phosphate. From this stage on it can be treated exactly as if it were a normal lysate, already supplemented with hemin and creatine kinase. It can be stored in liquid nitrogen as is, or preincubated with $CaCl_2$ and micrococcal nuclease as described previously.

In the alternative approach, nuclease-treated lysate is prepared exactly as described earlier in this chapter and is loaded on a Sephadex G-25 column equilibrated with buffer B. The pooled fractions are supplemented with ATP, GTP, spermidine, and glucose 6-phosphate, as described above, and frozen in liquid nitrogen.

Dithiothreitol may be included in the column buffers at a concentration of 0.1–0.5 mM, but its presence rarely improves the activity of the lysate. On no account should spermidine be added to the column buffers, since the emergent lysate will have a spermidine content far above optimum, in the range that strongly inhibits translation. As the lysate passes down the column, macromolecular components evidently bind spermidine quite tightly, and they sweep it up from the buffer as they go through the column.

In a very few atypical preparations of gel-filtered lysates, glucose 6-phosphate alone is not enough to maintain high initiation rates and prevent an early shutoff of protein synthesis. In such cases, it is worth trying combinations of 0.2 m*M* glucose 6-phosphate and 0.5 m*M* dithiothreitol. In some cases, 5 m*M* cAMP together with these other two compounds may give better activity.

Since gel filtration eliminates most of the amino acid pools, an increase in the incorporation of labeled amino acids into proteins should occur, the exact magnitude of this increase depending on which amino acid is used as label. If, however, radioactive amino acids of high specific activity are used, one must bear in mind that the concentration may be so low (in relation to the K_m of the cognate activating enzyme) as to limit the *rate* of protein synthesis, quite apart from any limitation of the extent of protein synthesis as discussed previously. In experiments using [^{35}S]methionine in gel-filtered lysates, we found the rate of incorporation to be markedly enhanced by the addition of about 1 μM unlabeled methionine.

[5] *In Vivo* and *in Vitro* Systems for Studying Bacterial Membrane Biogenesis

By PAMELA GREEN and MASAYORI INOUYE

A valuable approach to the study of membrane biogenesis in bacteria is to uncouple the synthesis and assembly of membrane proteins in such a way that processing intermediates accumulate. This has been accomplished *in vivo* by altering cell growth conditions[1–4] or *in vitro* by using

[1] S. Halegoua and M. Inouye, *J. Mol. Biol.* **130,** 39 (1979).
[2] J. M. DiRienzo and M. Inouye, *Cell* **17,** 155 (1979).
[3] A. Hirashima, G. Childs, and M. Inouye, *J. Mol. Biol.* **79,** 373 (1973).
[4] T. Maeda, J. Glass, and M. Inouye, *J. Biol. Chem.* **256,** 4712 (1981).

ISBN 0-12-181996-5

purified components.[4–7] Semi-*in vitro* systems for membrane protein synthesis have also been developed that are dependent upon the addition of ATP.[1,8,9] Owing to the unique molecular events involved in the biogenesis of each membrane protein, no one method can be universally effective. At present, the precursors of many bacterial membrane proteins have been identified and characterized using a variety of techniques.

This chapter describes the methods that we have found to be most useful for the study of the biogenesis of the major outer membrane proteins of *Escherichia coli* K12: lipoprotein,[4–7,10,11] the matrix proteins,[1,9] and ompA protein.[1,2,9] The same methods should be applicable to the study of other bacterial membrane proteins.

In Vivo Systems

Membrane Perturbation

Treatment of growing cells with membrane perturbants such as phenethyl alcohol (PEA) causes the accumulation of specific membrane protein precursors.[1] This may be the result of increased membrane fluidity[1] or of a decrease in the membrane potential.[12] The effects of membrane fluidity can also be examined through the use of an unsaturated fatty acid auxotroph. In this mutant, the fatty acid composition of the membrane can be dictated by supplementing the cells with elaidate or oleate. Lowering the temperature gives rise to a crystalline lipid state and arrests the biogenesis of certain membrane proteins in cells grown on elaidate.[2,13] A comparison of the membrane proteins synthesized at 37° and 20° in the presence of elaidate is shown in Fig. 1. Restoring membrane fluidity by raising the temperature allows accumulated biosynthetic precursors to be chased into their mature forms and locations.[2]

PEA Treatment.[1] *Escherichia coli* K12 strain MX74T2 is grown at 37° in M9 glucose medium[14] containing, per milliliter, 2 μg of thiamin, 4 μg of

[5] A. Hirashima and M. Inouye, *Eur. J. Biochem.* **60,** 395 (1975).

[6] A. Hirashima, S. Wang, and M. Inouye, *Proc. Natl. Acad. Sci. U.S.A.* **71,** 4149 (1974).

[7] S. Halegoua, A. Hirashima, and M. Inouye, *J. Bacteriol.* **126,** 183 (1976).

[8] S. Halegoua, A. Hirashima, J. Sekizawa, and M. Inouye, *Eur. J. Biochem.* **69,** 163 (1976).

[9] J. Sekizawa, S. Inouye, S. Halegoua, and M. Inouye, *Biochem. Biophys. Res. Commun.* **77,** 1126 (1977).

[10] S. Halegoua, J. Sekizawa, and M. Inouye, *J. Biol. Chem.* **252,** 2324 (1977).

[11] M. Inukai, M. Takeuchi, K. Shimizu, and M. Arai, *J. Antibiot.* **31,** 1203 (1978).

[12] C. J. Daniels, D. G. Bole, S. C. Quay, and D. L. Oxender, *Proc. Natl. Acad. Sci. U.S.A.* **78,** 5396 (1981).

[13] K. Ito, T. Sato, and T. Yura, *Cell* **11,** 551 (1977).

[14] J. H. Miller, "Experiments in Molecular Genetics." Cold Spring Harbor Lab., Cold Spring Harbor, New York, 1972.

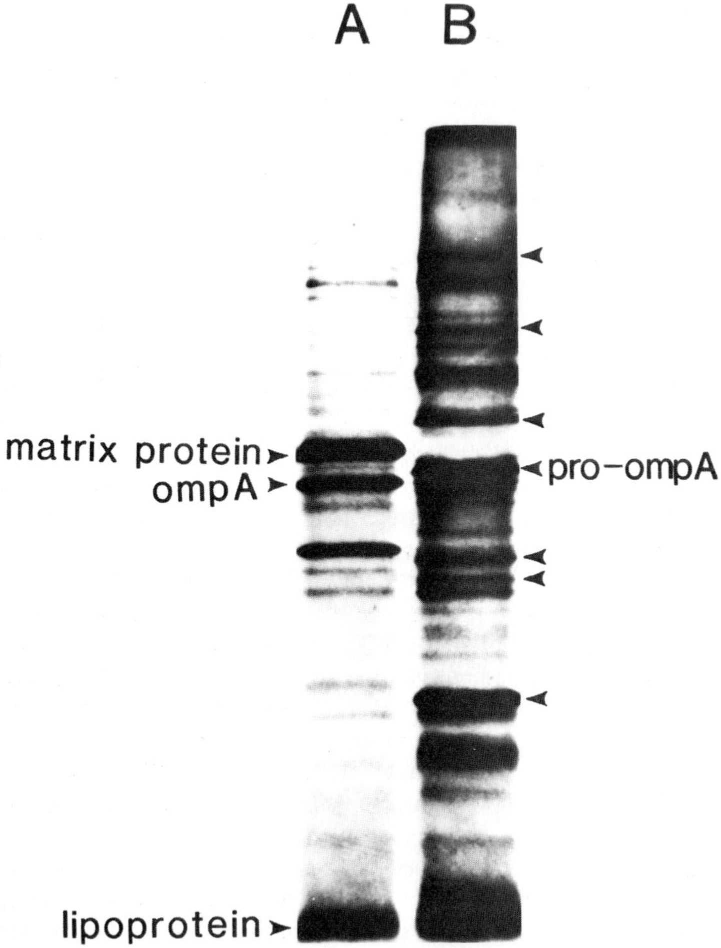

FIG. 1. Effects of lipid fluidity on the biosynthesis of membrane proteins. *Escherichia coli* strain K1060 grown on elaidate was pulse labeled at 37° and at 20°. Membranes were prepared by the freeze–thaw procedure and electrophoresed as described in the text. Lane A: Elaidate membranes pulse labeled at 37°; lane B: elaidate membranes pulse labeled at 20°. The major outer membrane proteins are indicated. In lane B the arrows show the positions of pro-ompA protein and other new membrane proteins synthesized at 20°. Adapted from Ref. 2. Copyright MIT Press.

thymidine, and 10 μg of methionine. When the culture reaches a density of about 2×10^8 cells/ml, PEA is added to 0.2% and incubation is continued for 30 sec, at which time 1 μCi of [^{35}S]methionine per milliliter (>400 Ci/mmol) is added. After a 2-min incubation, the culture is placed in an ice bath, formaldehyde is added to 0.4%, and membranes are prepared. In the presence of 0.2% PEA, cells will grow, but they form filaments. In the semi-*in vitro* system presented in the next section, the concentration of PEA can be increased from 0.3% to 1% to effect the processing and assembly of a variety of membrane proteins.

Membranes are prepared according to the method of Inouye and Guthrie.[15] All steps are carried out at 4°. The cells are collected by centrifugation, washed twice with 0.01 *M* $NaPO_4$ buffer, pH 7, and resuspended in the same buffer for sonication. After 3–5 min of sonication, the cells are centrifuged at 3000 *g* for 10 min to remove unbroken cells and debris. The supernatant is then centrifuged for 30 min at 100,000 *g*, and the pellet is washed once with the same buffer. The washed pellet is dissolved in a solution containing 80 m*M* Tris-HCl, pH 6.8, 2% sodium dodecyl sulfate (SDS), 10% glycerol, 1% 2-mercaptoethanol, and 0.002% Bromphenol Blue. The membrane preparations are heated at 100° for 7 min immediately prior to SDS–polyacrylamide gel electrophoresis (SDS–PAGE). The separating and stacking gel compositions are as described by Laemmli[16] and Anderson *et al.*[17] The gels are subjected to autoradiography to visualize the labeled membrane proteins.

Alteration of Lipid Fluidity.[2] *Escherichia coli* strain K1060 (F^-, *Thi fabB fadE*), an unsaturated fatty acid auxotroph,[18] is grown at 37° in medium E[19] containing 0.5% glycerol, 100 μg of elaidic acid (potassium salt) per milliliter, 4 μg of methionine per milliliter, and 0.2% Triton X-100, to mid-log phase. Half the culture is then shifted to 20° for 5 min, and 2 μCi of [^{35}S]methionine per milliliter (>700 Ci/mmol) are added. The other half of the culture is labeled with 0.1 μCi of [^{35}S]methionine per milliliter at 37°. After a 60-min incubation, 50 μg of chloramphenicol and 0.5 mg of methionine are added per milliliter to stop protein synthesis and [^{35}S]methionine incorporation. At this time, the cells can be chilled and membranes isolated. If a chase is desired, under conditions favoring formation of the fluid membrane state, the cultures grown at 20° are shifted to 37° after the addition of chloramphenicol and excess nonradioactive methionine. During the chase, aliquots are removed from the cultures at various time intervals, and membranes are isolated. We have found that the pro-ompA protein accumulated in the nonfluid state can be completely processed to ompA protein during a 2-min chase.[2]

Membranes can be prepared as described for PEA-treated cells; however, using this method, difficulty may arise in sedimenting proteins

[15] M. Inouye and J. P. Guthrie, *Proc. Natl. Acad. Sci. U.S.A.* **64,** 957 (1969).
[16] U. K. Laemmli, *Nature (London)* **227,** 680 (1970).
[17] C. W. Anderson, P. R. Baum, and R. F. Gesteland, *J. Virol.* **12,** 241 (1973).
[18] D. F. Silbert, M. Cohen, and M. E. Harder, *J. Biol. Chem.* **247,** 1699 (1972).
[19] H. J. Vogel and D. M. Bonner, *J. Biol. Chem.* **218,** 97 (1956).

loosely associated with the membrane.[2] Under these conditions, we have found it most useful to isolate the membranes using a modification[2] of the freeze-thaw procedure developed by Ron *et al.*[20]

Specific Inhibitors

As a whole, bacterial membrane proteins are more resistant to puromycin and kasugamycin but more sensitive to tetracycline and sparsomycin than are soluble proteins, for reasons that are not yet fully understood.[3,7] Often the synthesis of one membrane protein is much more sensitive or resistant to a particular antibiotic than are other membrane proteins. For instance, lipoprotein synthesis is inhibited by only 40% in the presence of 600 μg of puromycin per milliliter, while nearly all other membrane protein synthesis is completely inhibited.[3] Although information can be gained from the study of this differential inhibition, the development of specific inhibitors is an even more powerful tool in the study of bacterial membrane biogenesis. Presently we are using two such inhibitors to probe the mechanism of lipoprotein secretion and assembly. The first, benzyloxycarbonylalanine chloromethyl ketone (Z-Ala-CH_2Cl), inhibits either the cleavage or the modification step(s) required for the conversion of prolipoprotein to lipoprotein.[4] Prolipoprotein[11] and the precursor molecules of several other minor lipoprotein species[21] also accumulate when *E. coli* is grown in the presence of the cyclic peptide antibiotic globomycin. A more detailed description of globomycin and its mode of action is presented in Volume 97 [12] of this series.

Z-Ala-CH_2Cl.[4] Synthesis of Z-Ala-CH_2Cl is as follows: diazomethane is acylated using the mixed anhydride of benzyloxycarbonylalanine to form the diazomethyl ketone,[22] which is then treated with anhydrous hydrochloric acid to produce Z-Ala-CH_2Cl.[23,24]

To study the effect of this compound on membrane protein biogenesis, *E. coli* cultures are grown at 37° in M9[14] medium containing 0.2% glucose, 2 μg of thiamin and 4 μg of methionine per milliliter to a density of about 10^8 cells/ml. Z-Ala-CH_2Cl (dissolved in 95% ethanol) is added to the desired concentration, and incubation is continued for 5 min. The cells are labeled with 5 μCi of [^{35}S]methionine per milliliter (>700 Ci/mmol) at 37° for an additional 20 min, and then membrane fractions are prepared as described in the first section. We have found that processing of prolipo-

[20] E. Z. Ron, R. E. Kohler, and B. D. Davis, *Science* **153,** 1119 (1966).

[21] S. Ichihara, M. Hussain, and S. Mizushima, *J. Biol. Chem.* **256,** 3125 (1981).

[22] B. Penke, J. Czombos, L. Balaspiri, J. Peter, and K. Kovacs, *Helv. Chim. Acta* **53,** 1057 (1970).

[23] R. C. Thompson and E. R. Blout, *Biochemistry* **12,** 44 (1973).

[24] E. Shaw, this series, Vol. 11, p. 677.

protein is inhibited by 55% in the presence of 250 μg and by 35% in the presence of 120 μg of Z-Ala-CH_2Cl per milliliter.[4] At concentrations greater than 80 μg of Z-Ala-CH_2Cl per milliliter, no growth occurs. Lower concentrations do not severely inhibit growth during the first 30 min.[4] Z-[^3H]Ala-CH_2Cl may be used to label proteins that could be involved in the modification and assembly of membrane proteins.

Globomycin. Escherichia coli cultures are grown at 37° in M9 medium[14] containing 0.2% glucose and 2 μg of thiamin and 4 μg of methionine per milliliter to a density of about 2×10^8 cells/ml, at which time globomycin is added. After a 15-min incubation, the cells are labeled with 2–5 μCi of [^{35}S]methionine per milliliter (>700 Ci/mmol), and incubation is continued for 15–30 min. The cells are then chilled on ice and membranes are prepared.

The optimal concentration of globomycin varies from strain to strain. Generally, we add globomycin (from a 20 mg per milliliter of methanol stock solution) to a final concentration of about 100 μg per milliliter to inhibit the processing of lipoprotein in *E. coli* K12. This concentration is approximately 10 times the minimum inhibitory concentration (i.e., the lethal concentration on solid media) for *E. coli* K12. *Escherichia coli* B strains are about 30 times more sensitive than *E. coli* K12 strains.[25] The minimum inhibitory concentrations for a number of microorganisms has been determined.[26]

Semi-*in Vitro* Systems

The usefulness of certain membrane perturbants in the study of the biosynthesis and assembly of membrane proteins can sometimes be hampered by the inhibitory effect of the solvent on the cell's ability to synthesize proteins.[27,28] To overcome this limitation, semi-*in vitro* protein synthesizing systems have been established for use with toluene[7,8] or PEA[1]-treated cells. In these systems, protein synthesis is facilitated by (and is dependent upon) the external addition of ATP as an energy source. Varying the concentration and time of exposure to the perturbant dictates which membrane proteins are affected and to what extent. A precursor of lipoprotein accumulates in cells treated with 1% toluene for 10 min,[7,10] whereas optimal accumulation of the precursors of ompA protein and the matrix proteins occurs with milder treatment.[9] In PEA-treated cells, inhibition of outer membrane protein assembly seems to occur at concentra-

[25] M. Inukai, unpublished data (1978).

[26] M. Inukai, M. Nakajima, M. Osawa, T. Haneishi, and M. Arai, *J. Antibiot.* **31,** 421 (1978).

[27] R. L. Peterson, C. W. Radcliffe, and N. R. Pace, *J. Bacteriol.* **107,** 585 (1971).

[28] R. W. Jackson and J. A. Demoss, *J. Bacteriol.* **90,** 1420 (1965).

tions lower than that required to elicit precursor protein accumulation. The biogenesis of the matrix proteins is more sensitive to PEA than ompA protein biogenesis, and lipoprotein biogenesis is the most resistant.[2]

The system for semi-*in vitro* protein synthesis in PEA-treated *E. coli* cells described below offers distinct advantages over the analogous system using toluenized cells. First, the extent of exposure to the cells can be better controlled with PEA than toluene because PEA is more soluble in water. Second, the effects of PEA are reversible and, therefore, less damaging to membrane structure and macromolecular synthesis. This system has been modified for use with the gram-positive bacterium *Bacillus licheniformis*.[29]

Semi-in Vitro Protein Synthesis in PEA-Treated Cells[1]

Escherchia coli strain MX74T2 is grown to 2×10^8 cells/ml as described in the first section. The cells are pelleted by centrifugation at room temperature and resuspended in 10 m*M* Tris-HCl, pH 7.8, 50 m*M* NH_4Cl, 10 m*M* $Mg(CH_3COO)_2$, and 7 m*M* 2-mercaptoethanol at one third of the original volume. The cells are then diluted threefold to contain, in addition to PEA, 14.5 m*M* ATP, 0.2 m*M* GTP, 17.4 m*M* $Mg(CH_3COO)_2$, 60 m*M* NH_4Cl, 75 m*M* Tris-HCl, pH 7.4, 10 μCi of [^{35}S]methionine per milliliter (>700 Ci/mmol), and 0.2 m*M* each of the other 19 amino acids. Reaction mixtures are incubated for 15 min at 37° and then chilled on ice. Since the effects of PEA are reversible, a pulse-chase experiment can be performed as follows: The PEA-treated cells are labeled for 5 min at 37° as described above. After the pulse, half the culture is transferred to an ice bath and the other half is immediately centrifuged at room temperature. The cell pellet is resuspended in the original M9 growth medium supplemented with 200 μg of methionine per milliliter, incubated at 37° for one generation, and then chilled in an ice bath. This system can be used to examine the effects of a variety of PEA concentrations ranging from 0.3% to 1%. At higher PEA concentrations, the $Mg(CH_3COO)_2$ concentration can be lowered to 12 m*M* to obtain optimum incorporation of [^{35}S]methionine.

Total membrane fractions and separated inner and outer membranes can be analyzed. Two methods are generally used to separate the inner and outer membranes. One method employs sucrose gradient centrifugation[30] to facilitate the separation, and the other uses sodium sarcosinate.[31] In order to make a complete analysis, we feel that it is necessary to

[29] J. B. Nielsen, M. P. Caulfield, and J. O. Lampén, *Proc. Natl. Acad. Sci. U.S.A.* **78,** 3511 (1981).

[30] M. J. Osborn, J. E. Gander, E. Parisi, and J. Carson, *J. Biol. Chem.* **247,** 3962 (1972).

[31] C. Filip, G. Fletcher, J. L. Wulff, and C. F. Earhart, *J. Bacteriol.* **115,** 717 (1973).

compare the results obtained using both of these techniques. Periplasmic proteins can be isolated by the procedure described by Neu and Heppel[32] after semi-*in vitro* protein synthesis at 0.3% PEA with the following modification. Since the cells are sensitive to EDTA, to avoid lysis, EDTA should be omitted from the shock procedure and 1 m*M* $Mg(CH_3COO)_2$ should be used instead.

In Vitro Systems

An important approach to the study of the biosynthesis and assembly of bacterial membrane proteins is the development of RNA-dependent cell-free translation systems. Although bacterial membrane proteins have been translated in eukaryotic protein-synthesizing systems,[33] the use of a bacterial system is not as restrictive and can reveal more details about *in vivo* biosynthetic mechanisms. A prokaryotic system also allows the use of antibiotics[7] and specific inhibitors.[4] As outlined below, the mRNA can be introduced in one of several forms: in purified form,[6] as free polysomes,[5,34] or as membrane-bound polysomes.[4,35] Such tools can be used to probe how the translation of membrane proteins differs from the translation of cytoplasmic proteins.

Biosynthesis of Membrane Proteins on Free Polysomes[5,7]

The 0.5-ml protein synthesis reaction mixture contains 50 m*M* Tris-HCl, pH 7.8, 9 m*M* $Mg(CH_3COO)_2$, 80 m*M* NH_4Cl, 1 m*M* dithiothreitol, 4 m*M* phosphoenolpyruvate, 1 m*M* ATP, 0.2 m*M* GTP, 4 μCi of [^{35}S]methionine (>400 Ci/mmol), 0.05 m*M* of each of the other 19 amino acids, 17.1 μg of pyruvate kinase per milliliter, 2–5 A_{260} units of polysomes (see below), and 240 μg of the soluble enzyme fraction (see below). The reaction mixture is incubated at 37° for 45 min, after which 0.5 ml of 10% trichloroacetic acid is added and the mixture is boiled for 30 min. The precipitate is then washed three times with 1 ml of 5% trichloroacetic acid. After two washes with 1 ml of acetone, the products are analyzed by immunoprecipitation[4] and by SDS–PAGE.[16,17]

Isolation of Polysomes. Escherichia coli strain CP78 (RC^+, *Thi His Leu Thr Arg*) is grown at 37° in 1.5 liters of M9 glucose medium[14] containing, per milliliter, 2 μg of thiamin and 20 μg of each of the following amino acids: histidine, leucine, threonine, and arginine. When the culture

[32] H. C. Neu and L. A. Heppel, *J. Biol. Chem.* **240,** 3685 (1965).

[33] S. Wang, K. B. Marcu, and M. Inouye, *Biochem. Biophys. Res. Commun.* **68,** 1194 (1976).

[34] L. L. Randall and S. J. S. Hardy, *Mol. Gen. Genet.* **137,** 151 (1975).

[35] L. L. Randall and S. J. S. Hardy, *Eur. J. Biochem.* **75,** 43 (1977).

reaches a density of about 3.2×10^8 cells/ml, chloramphenicol is added to 50 μg/ml and the culture is quickly chilled by immersion in an acetone–dry ice mixture. All subsequent steps are performed at 0–4°. The cells are pelleted by centrifugation, and all the medium is removed. The pellet is quickly resuspended in 13.5 ml of a solution containing 0.5 *M* sucrose, 0.1 *M* Tris-HCl, pH 8, and 0.1 *M* NaCl. Next, 1.8 ml of lysozyme (1 mg/ml freshly dissolved in the above sucrose–salt solution) and 1.2 ml of 0.14 *M* EDTA, pH 8, are added; the suspension is left on ice for 10 min with occasional gentle swirling. After the addition of 0.25 ml of 1 *M* $Mg(CH_3COO)_2$, the protoplasts are centrifuged for 5 min at 5000 *g*. The pellet is resuspended in 4 ml of lysing medium containing 0.5% Brij 58 in buffer A (10 m*M* Tris-HCl, pH 7.8, 50 m*M* NH_4Cl, 10 m*M* $Mg(CH_3COO)_2$ and 7 m*M* 2-mercaptoethanol). DNase I is added to a final concentration of 20 μg/ml, and the lysate is centrifuged for 10 min at 10,000 *g* to remove cell debris. The supernatant is passed through a Sepharose 4B column (1.8×28 to 30 cm) equilibrated with buffer A, and 1.0–1.5-ml fractions are collected. Polysomes are eluted in the first A_{260} peak.

At this time, the polysome peak can be subdivided to separate larger and smaller polysomes. The first and last parts of the peak are analyzed by layering the respective pooled polysome fractions on 11 ml of 15 to 30% linear sucrose gradients in buffer A. The gradients are centrifuged at 37,000 rpm for 70 min in a Beckman SW41 rotor. By measuring the absorbance at 260 nm using the Gilford spectrophotometer flow cell system, the polysome distribution can be assessed. The smaller or larger polysome fraction can be used to enrich for synthesis of a particular membrane protein in the cell-free translation system.[5] The polysomes are stored in liquid nitrogen until used.

Preparation of the Soluble Enzyme Fraction. Escherichia coli strain, Q13 (*Hfr, RNaseI$^-$, Met Tyr*) is grown to mid-log phase in Penassay broth (Difco). A cell-free extract (S-30) is prepared by the method of Nirenberg and Matthaei[36] except that buffer A is used as the standard buffer. This S-30 extract is centrifuged at 150,000 *g* for 3 hr. The upper two-thirds of the supernatant are removed and stored at −90° until used.

Biosynthesis and Processing of Membrane Proteins on Membrane-Bound Polysomes[4]

The reaction mixture for *in vitro* protein synthesis is as described by Zubay[37] except that UTP, CTP, 3′,5′-AMP, *p*-aminobenzoic acid, $CaCl_2$,

[36] M. W. Nirenberg and J. H. Matthaei, *Proc. Natl. Acad. Sci. U.S.A.* **47,** 1588 (1961).

[37] G. Zubay, D. A. Chambers, and L. C. Cheong, *in* "The Lactose Operon" (J. R. Beckwith and D. Zipser, ed.), p. 375. Cold Spring Harbor Lab., Cold Spring Harbor, New York, 1970.

methionine and Ø80d*lac* DNA are omitted. The 50-μl reaction mixture also contains 20 μCi [^{35}S]methionine (>700 Ci/mmol) and 10 A_{260} units of the membrane-bound polysome fraction (see below). If desired, the membrane-bound polysome fraction can be pretreated with the specific inhibitor mentioned earlier, Z-Ala-CH_2Cl, by incubating the suspension at room temperature with 100–400 μg of the inhibitor in 0.5% ethanol for 10 min. The complete reaction mixture is incubated for 30 min at 37° and the products are analyzed as described in the previous section. The reaction products can also be further characterized by discontinuous sucrose density gradient centrifugation as described by Randall and Hardy.[35]

In this system, inclusion of the membrane fraction allows the synthesis of prolipoprotein and its subsequent processing to lipoprotein except in the presence of globomycin (10 μg/ml) or polysomes pretreated with Z-Ala-CH_2Cl (100–400 μg).[4]

Preparation of Membrane-Bound Polysomes. Membrane-bound polysomes are prepared by the method of Randall and Hardy,[35] with one slight modification. The final pellet is resuspended in 10 m*M* Tris-acetate, pH 8.2, 15 m*M* $Mg(CH_3COO)_2$, 60 m*M* NH_4Cl, and 1 m*M* dithiothreitol and stored at −70° until used.

Biosynthesis of Specific Membrane Proteins Using Purified mRNA[6]

This cell-free protein synthesis is carried out in a 100-μl reaction mixture containing 50 m*M* Tris-HCl, pH 7.8, 8.3 m*M* $Mg(CH_3COO)_2$, 85 m*M* NH_4Cl, 1 m*M* dithiothreitol, 2.1 m*M* phosphoenolpyruvate, 0.43 m*M* ATP, 0.043 m*M* GTP, 20 μCi of [^{35}S]methionine per milliliter (>700 Ci/mmol), 0.043 m*M* of each of the other amino acids, 0.053 m*M* palmitic acid, 85 μg of pyruvate kinase per milliliter, 68 A_{260} units per milliliter of preincubated S-30 extract,[35] and the purified mRNA of interest. The reaction mixture is incubated at 33° for 45 min, then 0.5 ml of 10% trichloroacetic acid is added. The reaction products are then precipitated by trichloroacetic acid and washed as described for the products of protein synthesis on free polysomes. The final precipitate is solubilized and examined by SDS–PAGE as described in the first section of this chapter. The amino acid sequence of the pure translation products can also be determined using Edman degradation.[38] The amount of mRNA to use must be determined empirically. In our laboratory, typically 60–150 μg of highly purified lipoprotein mRNA per milliliter is used in this system. The rate of incorporation of [^{35}S]methionine increases linearly with the amount of RNA added to the translation system. Time course experiments indicate

[38] S. Inouye, S. Wang, J. Sekizawa, S. Halegoua, and M. Inouye, *Proc. Natl. Acad. Sci. U.S.A.* **74,** 1004 (1977).

that incorporation of radioactivity into hot trichloroacetic acid-insoluble material continues for at least 15 min and then levels off.[6]

The most difficult step in this procedure is the isolation of highly purified mRNA. The lipoprotein mRNA is the only mRNA of *E. coli* that has been purified to homogeneity.[39] When the unique features of an mRNA cannot be used to facilitate its isolation from whole cells, perhaps total cellular RNA or RNA synthesized *in vitro* from a DNA template can be used (for an example, see Queen and Rosenberg[40]).

[39] K. Takeishi, M. Yasumura, R. Pirtle, and M. Inouye, *J. Biol. Chem.* **251,** 6259 (1976).
[40] C. Queen and M. Rosenberg, *Cell* **25,** 241 (1981).

[6] Preparation of Microsomal Membranes for Cotranslational Protein Translocation

By PETER WALTER and GÜNTER BLOBEL

Secretory, lysosomal, and many integral membrane proteins are translocated *across* or asymmetrically integrated *into* the membrane of the rough endoplasmic reticulum.[1,1a] The events of this translocation or integration process can be faithfully reproduced *in vitro*.[2–4] For this purpose, an *in vitro* protein translation system programmed with a suitable mRNA is supplemented with microsomal membranes, a fraction of closed vesicles derived from the rough endoplasmic reticulum.[2]

In all cases investigated, translocation is a *co*translational process; i.e., the nascent chain is vectorially translocated across the membrane as it emerges from the ribosome. Consequently, the microsomal membrane fraction has to be present *during* protein synthesis. This also implies that

[1] Abbreviations: DTT, dithiothreitol; EDTA, ethylenediaminetetraacetic acid; EGTA, ethylene glycol-bis(β-aminoethyl ether)-*N*,*N*'tetraacetic acid; IAA, iodoacetamide; NEM, *N*-ethylmaleimide; PAGE, polyacrylamide gel electrophoresis; PMSF, phenylmethylsulfonyl fluoride; RM, rough microsomes; SDS, sodium dodecyl sulfate; SRP, signal recognition particle; TCA, trichloroacetic acid; TEA, triethanolamine; TPCK, L-1-tosylamide-2-phenylethyl chloromethyl ketone.
[1a] G. Palade, *Science* **189,** 347 (1975).
[2] G. Blobel and B. Dobberstein, *J. Cell Biol.* **67,** 852 (1975).
[3] F. N. Katz, J. E. Rothman, V. R. Lingappa, G. Blobel, and H. F. Lodish, *Proc. Natl. Acad. Sci. U.S.A.* **74,** 3278 (1977).
[4] A. H. Erickson, G. Conner, and G. Blobel, *J. Biol. Chem.* **256,** 11224 (1981).

ISBN 0-12-181996-5

all the components added to study the translocation process have to be compatible with *in vitro* protein synthesis.[2] If the microsomal vesicles are added after protein synthesis is completed, a *post*translational translocation is not observed.

In most cases translocation is accompanied by the proteolytic removal of the signal peptide by signal peptidase, an endoprotease located on the luminal face or inside the lipid bilayer of the endoplasmic reticulum.[2,5,6] The conversion of the preprotein synthesized in the absence of microsomes to the processed form can therefore be taken as an assay for successful translocation. However, cleavage of the signal peptide (1500–3000 daltons) may be difficult to detect by mobility differences in SDS–PAGE in cases where the protein is larger than 60,000 daltons. Moreover, it is essential to determine the partial NH_2-terminal amino acid sequences of the processed chain to ascertain that cleavage did in fact occur at the signal peptidase site, not at some other nearby site by a protease other than signal peptidase. In addition to the signal peptide removal by signal peptidase, the translocated protein ends up on the inside of the microsomal vesicles and is therefore protected from degradation by exogenously added proteases.[2] This provides additional evidence that actual chain translocation occurred. In the case of proteins with uncleaved signal sequences, cosedimentation of the translocated protein with the vesicles under stringent conditions (high salt, high pH, urea, etc.) or the appearance of a coreglycosylated form of the translocated protein have been used as assays for translocation.[7,8] Occasionally, however, the translocation-coupled loss of the signal peptide and the gain of a single oligosaccharide compensate for each other so that the molecular weight difference between primary translation product and translocated chain is nil.[9]

The Translocation Machinery

We like to view the translocation machinery in the membrane of the endoplasmic reticulum as an assembly of proteins, representing the enzymic activities required for the specific polysome recognition and attachment, chain translocation, and the cotranslational modification of the nascent chain. A few of the activities of such a putative multienzyme

[5] R. C. Jackson and G. Blobel, *Proc. Natl. Acad. Sci. U.S.A.* **74,** 5598 (1977).

[6] V. R. Lingappa, A. Devillers-Thiery, and G. Blobel, *Proc. Natl. Acad. Sci. U.S.A.* **74,** 2059 (1977).

[7] V. R. Lingappa, D. Shields, S. L. C. Woo, and G. Blobel, *J. Cell. Biol.* **79,** 567 (1978).

[8] B. Goldman and G. Blobel, *J. Cell. Biol.* **90,** 236 (1981).

[9] D. J. Anderson and G. Blobel, *Proc. Natl. Acad. Sci. U.S.A.* **78,** 5598 (1981).

assembly have been isolated and characterized. Both specific polysome recognition[10] and binding[11] to the membrane are catalyzed by the signal recognition particle (SRP), a ribonucleoprotein consisting of six different polypeptide chains and one molecule of 7 S RNA of 300 nucleotides.[12,34] SRP appears to shuttle between a free, a ribosome/polysome-associated, and a membrane-bound state, and thus functions as adaptor between the cytoplasmic translation and the membrane-bound translocation machinery. It can be efficiently extracted from the membrane with solutions of high ionic strength.[13] On the membrane, SRP interacts with a SRP-receptor protein,[31] which was also termed docking protein.[32] SRP receptor is an integral membrane protein (72,000 daltons)[14] containing a 60,000-dalton cytoplasmic domain which can be removed by mild proteolytic digestion of the membrane. The 60,000-dalton fragment can be added back to proteolyzed membranes to reconstitute the translocation activity.[15–17,30,33]

Two integral membrane proteins (ribophorins)[18,19] have been identified on the basis of their apparent physical association with attached polysomes, but so far no evidence for their functional involvement in protein translocation has been presented.

The enzymic activities involved in cotranslational modifications of the nascent chain have not yet been fully characterized. Signal peptidase is an integral membrane protein that (or at least its active side) faces the luminal site of the endoplasmic reticulum membrane,[5,15] and, so far, there is no known inhibitor of its activity. If the microsomes are rendered translocation-inactive (e.g., by salt extraction, proteolysis, or *N*-ethylmaleimide treatment), the peptidase activity remains in a latent form in the membrane fraction and can be exposed by disrupting the lipid bilayer with detergents (this volume [62]). In detergent solutions it requires phospholipid for its activity.[20] The oligosaccharide transferase(s) that transfers the core oligosaccharide from the dolichol phosphate carrier to certain aspar-

[10] P. Walter, I. Ibrahimi, and G. Blobel, *J. Cell Biol.* **91,** 545 (1981).

[11] P. Walter and G. Blobel, *J. Cell Biol.* **91,** 551 (1981).

[12] P. Walter and G. Blobel, *Proc. Natl. Acad. Sci. U.S.A.* **77,** 7112 (1980).

[13] G. Warren and B. Dobberstein, *Nature (London)* **273,** 569 (1978).

[14] D. I. Meyer, D. Louvard, and B. Dobberstein, *J. Cell Biol.* **92,** 579 (1982).

[15] P. Walter, R. C. Jackson, M. M. Marcus, V. R. Lingappa, and G. Blobel, *Proc. Natl. Acad. Sci. U.S.A.* **76,** 1795 (1979).

[16] D. I. Meyer and B. Dobberstein, *J. Cell Biol.* **87,** 498 (1980).

[17] D. I. Meyer and B. Dobberstein, *J. Cell Biol.* **87,** 503 (1980).

[18] G. Kreibich, C. M. Freienstein, P. N. Pereyra, B. C. Ulrich, and D. D. Sabatini, *J. Cell Biol.* **77,** 464 (1978).

[19] G. Kreibich, B. C. Ulrich, and D. D. Sabatini, *J. Cell Biol.* **77,** 488 (1978).

[20] R. C. Jackson and W. R. White, *J. Biol. Chem.* **256,** 2545 (1981).

agine residues in the nascent polypeptide chain has been considerably enriched from detergent extracts of chicken oviduct rough microsomes.[21]

Some General Remarks on the Preparation of Microsomal Membranes

At present, canine pancreas is our main source of actively translocating microsomal membranes. Because the pancreas actively secretes digestive enzymes, it is essential to take a number of precautions. We work as fast as possible. It usually takes us less than 90 min from sacrificing the dog to the start of the final centrifugation step to pellet the microsomes. All steps are carried out at 0–4° to minimize degradation. We also add PMSF (a covalent serine protease inhibitor) and EDTA (to inhibit metalloproteases) to the homogenization buffer. Under these conditions, we obtain a rough microsomal fraction which is essentially unproteolyzed (as judged by the intactness of the three higher-molecular-weight polypeptides of SRP by immunological criteria).

The translocation activity of the microsomal membranes is dependent on free SH groups.[12,17,22] We therefore include 1 m*M* DTT in all buffers to keep the membranes under reducing conditions. Although there is no absolute requirement for the presence of DTT, we seem to obtain a better reproducibility as far as their translocation activity is concerned between different preparations if it is included. It should also be noted that, owing to the sulfhydryl requirement of the translocation activity, no reagents modifying SH groups (e.g., NEM, IAA, TPCK) can be added at any stage of the preparation.

Microsomal membranes can (and should) be rapidly frozen in small aliquots (<5 ml) in liquid N_2. We include 250 m*M* sucrose as a cryoprotectant. They can be stored at −80° for at least a year without a loss of activity. When stored at −20° (frozen or in 50% glycerol), they seem to lose activity more rapidly. For thawing, the tube is warmed up fast in a water bath at room temperature with rapid agitation; only after the contents are completely thawed is the tube placed on ice. We observe no detectable loss of activity upon at least three thawing and freezing cycles.

Solutions

A stock solution of 1.0 *M* triethanolamine was adjusted to pH 7.5 at room temperature with acetic acid and, as such, is referred to as TEA. A stock solution of 4.0 *M* KOAc was adjusted to pH 7.5 at room tempera-

[21] R. C. Das and E. C. Heath, *Proc. Natl. Acad. Sci. U.S.A.* **77**, 3811 (1980).

[22] R. C. Jackson, P. Walter, and G. Blobel, *Nature (London)* **284**, 174 (1980).

ture with acetic acid. A stock solution of 0.2 *M* EDTA was adjusted to pH 7.5 at room temperature with NaOH. Stock solutions of 2.5 *M* sucrose and 1.0 *M* $Mg(OAc)_2$ were not further adjusted. All stock solutions mentioned above were filtered through a 0.45-μm Millipore filter, except the sucrose solution, which was filtered through a 1.2-μm Millipore filter. Phenylmethylsulfonyl fluoride (PMSF) was freshly dissolved in ethanol or dimethyl sulfoxide to a concentration of 100 m*M* and diluted into the buffers immediately before use. Dithiothreitol was kept in small aliquots as a 1 *M* stock solution at −20° and diluted into the buffers immediately before use.

Buffer A: 250 m*M* sucrose, 50 m*M* TEA, 50 m*M* KOAc, 6 m*M* $Mg(OAc)_2$, 1 m*M* EDTA, 1 m*M* DTT, 0.5 m*M* PMSF

Buffer B: 250 m*M* sucrose, 50 m*M* TEA, 1 m*M* DTT

Buffer C: 50 m*M* TEA, 1.5 m*M* $Mg(OAc)_2$, 1 m*M* EDTA, 1 m*M* DTT, 0.5 m*M* PMSF

Preparation of Crude Rough Microsomes

Dogs weighing 10–25 kg, of either sex, were used. Although it appears to be an important parameter to check, the effect of fasting or feeding the animal prior to sacrifice has not been systematically investigated. Most (although not all) of our dogs were fed about 2 hr before sacrifice. Acepromacine maleate {[10-(3-methylaminopropylphenothiazin-2-yl)methyl ketone] maleate; Ayerst}, 1 ml of a 10 mg/ml solution, was injected intramuscularly as a sedative. After about 30–60 min the animals were anesthetized with an intraveneous injection of Nembutal (pentobarbital sodium salt; Abbott) (approximately 5 ml of a 50 mg/ml solution, the extract amount depending on the size of the dog). The dogs were bled by severance of the great vessels at the base of the heart or excision of the heart, and the pancreas was removed with scissors. The gland was immediately rinsed with 50 ml of ice-cold buffer A and immersed in another 50-ml aliquot of buffer A on ice. According to the size of the dog, the excised pancreas weighed 15–60 g. All subsequent steps were carried out at 0–4°.

The gland was freed of connective tissue and large vessels, repetitively rinsed with buffer A, and finally extensively minced with a razor blade. No tissue press was required if the tissue was well minced. Four milliliters of buffer A were then added per gram of tissue. The tissue was extensively homogenized with 5 strokes (10 sec down, 10 sec up) in a motor-driven Potter–Elvehjem homogenizer, avoiding foam formation and heating. Milder homogenization (1 stroke) yielded a microsome fraction with identical activity, but at a reduced yield. The homogenate was centrifuged for 10 min at 1000 g_{av}. Floating fatty material was removed by aspiration, and the supernatant was recentrifuged for 10 min at 10,000 g_{av}. The supernatant was immediately decanted from the pellet, taking care not to include the loose top layer of the pellet. Crude rough microsomes (RM) were collected by centrifugation of the 10,000 g_{av} supernatant for 2.5 hr at 140,000 g_{av} (Beckman Ti50.2 rotor at 40,000 rpm) through a cushion of 1.3 *M* sucrose in buffer A. The ratio of load to cushion was approximately 3 : 1. The supernatant, including the cushion and the membranous material at the interface, were removed by aspiration. The pellets were resuspended by manual homogenization in a Dounce homogenizer (A pestle; 2–3 strokes) in buffer B to a concentration of 50 A_{280} units/ml (determined in a 1% SDS solution). In a typical preparation 50 A_{280} units of RM were obtained from 1 g of tissue. The $A_{260}:A_{280}$ ratio was usually 1.84–1.92. The RM preparation obtained in this way can be used directly in an *in vitro* translation system.

Column Washing of RM

To remove adsorbed ribosomes and proteins we employed a column washing procedure.[23] Washed RM retained all their translocation activity and were consistently less inhibitory to protein synthesis; at most 20% inhibition of protein synthesis was observed at 3 Eq of washed RM per 25 μl of translation mix (see below for definition of Eq). The column washing buffer was of low ionic strength to avoid loss of membrane-bound SRP. The buffer also contained a low concentration of magnesium ions, enough to prevent unfolding of the membrane-bound ribosomes.

For a typical washing procedure a 20-ml portion of RM was loaded on a 200-ml Sepharose Cl-2B column (2.5 cm × 40 cm, flow rate 15 ml/hr) that was developed in upward flow in buffer C. The turbid fractions corresponding to the void volume of the column were pooled (about 40 ml), and the membranes were collected by centrifugation for 15 min at 50,000 g_{av}. The resulting washed RM were resuspended in 20 ml of buffer B.

EDTA Stripping of RM

The bulk of the membrane-bound ribosomes (all of the small subunits and at least half of the large subunits) and mRNA as well as many adsorbed proteins or peripheral membrane proteins can be removed from RM by an EDTA-extraction. The resulting membrane fraction has an approximately twofold decreased optical density and a considerably reduced endogenous mRNA activity. It should be noted, however, that the

[23] H. C. Hawkins and R. B. Freedman, *Biochim. Biophys. Acta* **558,** 85 (1979).

translocation activity of EDTA-stripped RM is not increased over that of the starting RM preparation; i.e., in spite of the fact that ribosomes have been unfolded and extracted, no new "translocation-active sites" are generated. SRP is not extracted by the EDTA treatment.

For a typical EDTA-stripping procedure, a 20-ml portion of RM was added to 20 ml of a solution of 50 m*M* EDTA (sodium salt) in buffer B. The mixture was incubated at 0–4° for 15 min. EDTA-stripped RM were collected by centrifugation of the mixture for 1 hr at 140,000 g_{av} through a cushion of 0.5 *M* sucrose in buffer B (without EDTA). The ratio of load to cushion was about 3 : 1. The resulting pellet was resuspended in 20 ml of buffer B.

Nuclease Treatment of RM

Rough microsomes can be treated with staphylococcal nuclease (EC 3.1.31.1)[24] to deplete them of endogenous mRNA activity, which, depending on the *in vitro* translation system used, might contribute more or less to the background.

To a 2-ml fraction of RM, washed RM, or EDTA-stripped RM, 20 μl of a 100 m*M* $CaCl_2$ solution were added. Staphylococcal nuclease (Boehringer, Catalog No. 107921) was added to a final concentration of 20 units/ml (here we added 8 μl of a stock solution of 5000 units/ml—which is stable for at least 2 years when stored in aliquots at −80°). Digestion was carried out for 10 min at 23°. It was stopped by the addition of 40 μl of a 100 m*M* EGTA solution (adjusted to pH 7.5 with NaOH). Membranes were pelleted at 100,000 g_{av} for 30 min and resuspended in 2 ml of buffer B.

No endogenous mRNA activity was detectable when the nuclease-treated RM were assayed in the reticulocyte lysate translation system. The translocation activity of the microsomes and the activity of membrane-associated SRP were not affected by the small concentrations of nuclease used.

Microsomes in Different Translation Systems

We have described here a rapid isolation procedure that reproducibly yields highly active microsomal membranes. We also described procedures for refining this crude RM fraction by column washing, EDTA stripping, or nuclease treatment. None of these procedures affects the translocation activity of the vesicles. The choice of procedure(s) will depend on the specific application—in particular, on which *in vitro* trans-

[24] H. R. B. Pelham and R. J. Jackson, *Eur. J. Biochem.* **67**, 247 (1976).

lation system will be used. For example, if one works in a wheat germ translation system, the readout of the RM membrane-bound polysomes is generally negligible, and therefore neither EDTA nor nuclease treatment will be required. A column wash, however, is advantageous, since otherwise microsomes tend to inhibit considerably.[25] In the reticulocyte lysate system readout of membrane-bound polysomes is very effective,[15] and therefore EDTA stripping and/or nuclease treatment will be required.

Rough microsomes can also be extracted with high-ionic-strength buffers and still retain their translocation activity in the reticulocyte lysate system. The same salt-extracted membranes, however, are completely inactive in the wheat germ translation system, indicating that a factor required for translocation (now identified as SRP) is removed from the membranes with the salt extract, but is present in the reticulocyte lysate.[32] Therefore, translocation in the wheat germ translation system is SRP dependent (and sensitive to salt extraction of the microsomes), whereas in the reticulocyte lysate it is independent of SRP added directly or with the microsomes (E. Evans and P. Walter, unpublished observation).

In summary: column-washed RM have in our hands almost no inhibitory effects on protein synthesis; EDTA-stripped RM have a high "specific activity" (translocation activity per milligram of protein), because of removal of ribosomes and proteins; nuclease-treated RM have essentially no endogenous mRNA activity. All these microsome preparations still contain SRP. They are therefore translocation-active in wheat germ as well as reticulocyte lysate.

Use of Microsomes in Translation Systems

We relate all concentrations of microsomes back to the original crude RM preparation, which has been adjusted to a concentration of 50 A_{280} units/ml. We refer to 1 μl of this suspension as 1 equivalent (1 Eq). Upon column washing, nuclease treatment, or EDTA extraction, UV-absorbing material is removed, but in theory the number of recovered vesicles should remain constant. We therefore resuspend these treated RM to the original volume to obtain the same (equivalent) vesicle concentration, rather than readjusting the concentration to the same optical density. As a result, all microsome preparations described here are at 1 Eq/μl.

Using bovine pituitary RNA translated in a wheat germ system, we obtain approximately 50% of the synthesized chains of preprolactin translocated (and processed to prolactin) with a microsome concentration of 1 Eq per 25 μl of translation mix. The optimal microsome concentration for different mRNAs ranges from 0.5–3 Eq per 25 μl of translation mix.

[25] D. Shields and G. Blobel, *J. Biol. Chem.* **253,** 3753 (1978).

When using the wheat germ system, translocation activity can be boosted to 100% at a fixed microsome concentration of 1 Eq/μl if an excess of purified SRP is added.

To add microsomes to a translation system, we first mix all other components except the microsomes and the mRNA. We than add the microsomes, mix, and add the mRNA. All our translations contain human placental RNase inhibitor at a concentration of 0.16 A_{280} units/ml.

It is not necessary to compensate for the buffer, sucrose, or DTT added with the membrane fraction. None of these components effects protein synthesis noticeably (unless the DTT is badly oxidized). The membrane fraction therefore simply displaces water.

Core Glycosylation Activity of RM

Glycoproteins containing asparagine-linked sugars are core-glycosylated by microsomal membranes *in vitro* cotranslationally. Most of our microsome preparations are glycosylation active, but there is a high variability from preparation to preparation in the extent of glycosylation obtained. Whereas essentially all translocated chains are cleaved by signal peptidase, usually only a fraction of them (ranging from 0 to 70%) will be glycosylated). There is evidence[26] that it is the dolichol-oligosaccharide intermediate that is limiting in the poorly glycosylating microsomal preparations, but no successful way to charge *in vitro* the microsomes with the corresponding core sugar precursors has been reported. In general, screening of a couple of different microsome batches with a suitable mRNA preparation (like rat mammary gland RNA, which codes for a number of major and readily detectable glycoproteins) is used to select a microsomal preparation which glycosylates satisfactorily.

After disruption of the lipid bilayer with detergents [e.g., 0.1% of the nonionic detergent Nikkol (Nikko Chem. Corp., Tokyo)], core glycosylation does not occur, but signal peptidase cleavage still takes place. This allows one to observe processing uncoupled from glycosylation.

Posttranslational Proteolysis Assay

Translocated proteins are protected from exogenously added proteases by the microsomal membrane. In practice, the choice of the proteolysis conditions will determine the successful outcome of these protection experiments. Too extensive protease digestion leads to degradation of even the segregated forms, presumably owing to breakdown of the

[26] D. D. Carson, B. J. Earles, and W. J. Lennarz, *J. Biol. Chem.* **256,** 11552 (1982).

membrane barrier. The reasons for this permeabilization are not clear and have not been systematically investigated.

We generally obtain good protection (80–100% of the translocated protein) by employing the following protocol. Immediately after translation the translation mix (25 μl, wheat germ or reticulocyte lysate) is cooled to 0° by placing the tube in an ice-water bath. A 2-μl aliquot of a 15-m*M* $CaCl_2$ solution is added. Calcium ions seem to stabilize the vesicles and improve recovery of the protected form. A 3-μl aliquot of a trypsin–chymotrypsin (Boehringer) solution (3 mg/ml each) is added. Digestion is allowed to proceed for 30–90 min on ice and is terminated by the sequential addition of 3 μl of 10 m*M* PMSF in DMSO and 5 μl of Trasylol (FBA Pharmaceuticals, New York), followed by PAGE sample buffer. The sample is placed in a boiling water bath immediately. The use of amphipathic molecules to further improve recovery of protected material is described elsewhere (this volume [7]). Controls should include a sample, where in addition to the proteases a detergent (Triton X-100 to 1% final) is added to destroy the lipid bilayer. For protection experiments we usually use a membrane aliquot that has not been more than once frozen and thawed.

Microsomal Membranes from Other Sources

Microsomal membranes from sources other than dog pancreas have also been employed in cotranslational studies. The most successfully used alternative systems are probably chicken oviduct RM[27] (which glycosylate very well) and adrenal microsomes.[28] Rat liver RM,[28] ascites RM,[28] bovine pituitary RM,[6] and an RM fraction of *Drosophila melanogaster* embryos[29] also have been reported to be translocation active *in vitro*.

[27] R. C. Das, S. A. Brinkley, and E. C. Heath, *J. Biol. Chem.* **255,** 7933 (1980).

[28] M. Bielinska, G. Rogers, T. Rucinsky, and I. Boime, *Proc. Natl. Acad. Sci. U.S.A.* **76,** 6152 (1979).

[29] M. D. Brennan, T. G. Warren, and A. P. Mahowald, *J. Cell Biol.* **87,** 516 (1980).

[30] R. Gilmore, G. Blobel, and P. Walter, *J. Cell Biol.* **95,** 463 (1982).

[31] R. Gilmore, P. Walter, and G. Blobel, *J. Cell Biol.* **95,** 470 (1982).

[32] D. I. Meyer, E. Krause, and B. Dobberstein, *Nature (London)* **297,** 647 (1982).

[33] D. I. Meyer and B. Dobberstein, this volume [54].

[34] P. Walter and G. Blobel, *Nature (London)* **299,** 691 (1982).

[7] Methods for the Study of Protein Translocation across the RER Membrane Using the Reticulocyte Lysate Translation System and Canine Pancreatic Microsomal Membranes

By GEORGE SCHEELE

Over the past 25 years the exocrine pancreas has represented a model system for the study of the synthesis and secretion of exportable protein. Recently, the first step in this process has been shown to involve the cotranslational transport of nascent polypeptide chains from their site of synthesis on membrane-bound ribosomes to the luminal space enclosed by the rough endoplasmic reticulum (RER).[1] This transport step is mediated by short-lived amino-terminal transport peptides that are proteolytically cleaved during the translocation event. This chapter summarizes the methods developed to elucidate this mechanism for the transport of pancreatic exocrine proteins across the RER membrane. The methods were optimized for the simultaneous study of 17 discrete pancreatic secretory proteins, which include individual members of several families of enzymes and proenzymes (serine protease zymogens, metalloenzymes represented by procarboxypeptidases A and B, glycosidases, and glycerol ester hydrolases). This approach allowed us not only to make internal comparisons among individual proteins, but also to monitor the fidelity of their synthesis *in vitro* compared to their synthesis *in vivo*. This latter comparison provided us with a sensitive assay for our success in reproducing *in vitro* the biogenetic events that occur *in vivo*.

Isolation of Rough Microsomes from Canine Pancreas

Dogs weighing 20–50 lb (9–23 kg) and fed *ad libitum* are sedated with 20-ml acepromazine, 10-[3-(dimethylamino)propyl]phenothiazin-2-yl-methyl ketone (Ayerst Laboratories, New York, New York) injected intramuscularly 1 hr prior to sacrifice. Several minutes prior to sacrifice sodium pentobarbital, 300–500 mg, is administered intravenously. When deep central nervous system depression results, the dog is placed on a surgical table, the thorax is opened, and the great vessels at the base of the heart are severed to allow rapid exsanguination into the chest cavity. The abdomen is opened by a midline incision, the pancreas (all of the tail and most of the head) is removed by excision of the surrounding mesen-

[1] G. Scheele, R. Jacoby, and T. Carne, *J. Cell Biol.* **87,** 611 (1980).

ISBN 0-12-181996-5

tery with scissors, and the gland is immersed in an ice-cold solution of 0.25 *M* sucrose, 50 m*M* Tris-HCl, pH 7.5, 25 m*M* KCl, and 5 m*M* $MgCl_2$. All subsequent steps are carried out at 3°. The pancreas is spread on a surface of Parafilm in the cold room and freed of connective tissue, fat, and large blood vessels. A razor blade is used to cut the pancreas into small pieces that are passed through a tissue press (stainless steel with perforations 1 mm in diameter) that removes connective tissue and large blood vessels. The resulting brei is mixed with 3 volumes of 0.25 *M* sucrose, 50 m*M* Tris-HCl, pH 7.5, 25 m*M* KCl, 5 m*M* $MgCl_2$, and 2 m*M* dithiothreitol and homogenized (3–4 strokes with a motor-driven Teflon pestle) in a Brendler tissue grinder (type C, Arthur H. Thomas Co., Philadelphia, Pennsylvania). The thiol reagent is added to preserve the endogenous ribonuclease inhibitor, which is present in canine pancreas at a 40-fold molar excess compared to ribonuclease.[1] Cell debris, nuclei, and mitochondria are removed by sedimentation for 10 min at 10,000 *g*. Visible lipid is removed by aspiration, and the postmitochondrial supernatant is layered over a discontinuous sucrose gradient containing 1.5 ml of 1.3 *M* sucrose and 2.0 ml of 2.25 *M* sucrose, both in the presence of 20 m*M* HEPES-KOH, pH 7.5, 1 m*M* $MgCl_2$, and 2 m*M* dithiothreitol. Sucrose gradients contained in 14-ml polyester tubes are centrifuged in an SW 40 rotor for 30–60 min at 40,000 rpm. Rough microsomes that sediment at the 1.3/2.25 *M* sucrose interface are pooled (absorbance at 260 nm ranges from 150 to 300 units/ml), aliquoted in 100-μl samples, and stored at $-80°$ after rapid freezing in liquid N_2. Aliquots are thawed for use only once.

Preparation of Stripped Microsomal Membranes

Micrococcal nuclease is used to strip rough microsomes of endogenous mRNA immediately prior to their use in reconstitution experiments. Rough microsomes, isolated as described in the preceding section, are treated with 1800 units of micrococcal (*Staphylococcus aureus*) nuclease per milliliter (Boehringer, Mannheim, West Germany) in the presence of 0.5 m*M* $CaCl_2$ for 30 min at 22°.[1] Ribosomes and tRNA remain functionally intact under these conditions.[2] Nuclease activity is terminated by the addition of 2.0 m*M* EGTA. After the addition of 1/10 volume of nuclease-treated membranes to translation mixtures, the final concentration of micrococcal nuclease during *in vitro* translation is 180 units/ml. Pelham and Jackson[3] used 80 units of micrococcal nuclease per milliliter for the degradation of globin mRNA in the rabbit reticulocyte lysate.

[2] P. Blackburn and G. Scheele, unpublished findings.

[3] R. B. Pelham and R. J. Jackson, *Eur. J. Biochem.* **67**, 247 (1976).

In the reticulocyte lysate translation system the use of micrococcal nuclease-treated microsomal membranes is superior to that of microsomal membranes stripped of endogenous mRNA by EDTA treatment.[1] Figure 1 shows a fluorogram of the translation products synthesized by EDTA-treated microsomal membranes in the absence and in the presence of dog pancreas mRNA and compares these to translation products synthesized

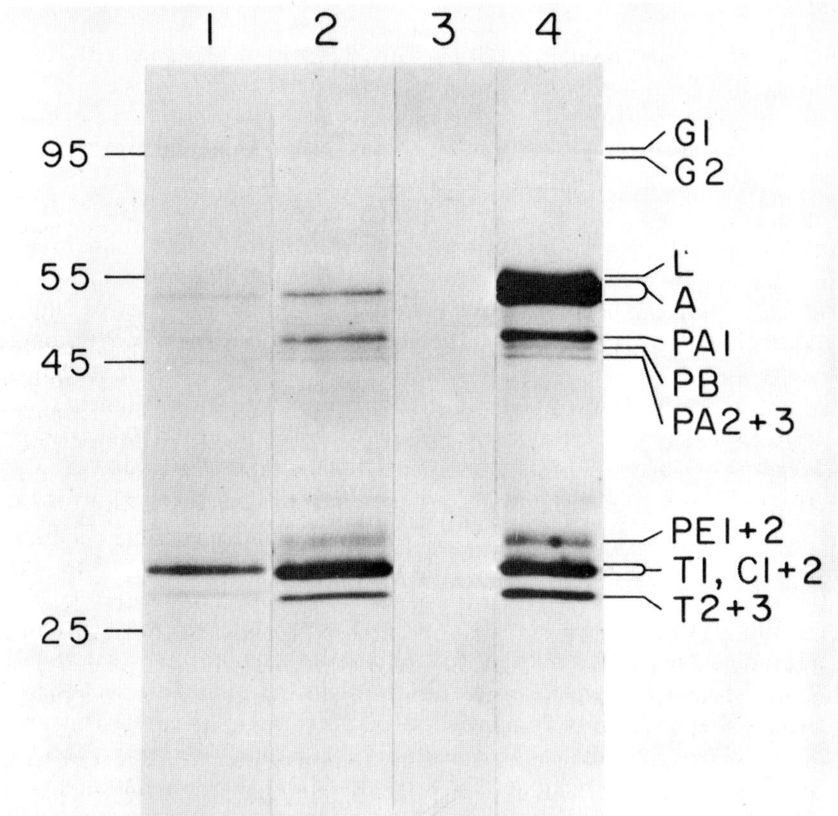

FIG. 1. Comparison of *in vitro* translation products directed by EDTA-treated (tracks 1 and 2) vs micrococcal nuclease-treated (tracks 3 and 4) dog pancreas microsomal membranes in the absence (tracks 1 and 3) and the presence (tracks 2 and 4) of dog pancreas mRNA. Translation products were separated by polyacrylamide gel electrophoresis in sodium dodecyl sulfate and analyzed by fluorography. Numbers to the left indicate molecular weight values $\times 10^{-3}$. Pancreatic exocrine proteins are identified on the right according to actual or potential enzyme activity by the following abbreviations: G, glycoprotein; L, lipase; A, amylase; PA, procarboxypeptidase A; PB, procarboxypeptidase B; PE, proelastase; T, trypsinogen; C, chymotrypsinogen. Numbers following the abbreviations indicate individual isoenzyme forms. Adapted in modified form from Scheele *et al.*[2]

PERCENTAGE DISTRIBUTION OF RADIOACTIVITY[a] AMONG DOG PANCREAS SECRETORY PROTEINS: COMPARISON OF CELL-FREE AND CELLULAR SYNTHESIS

Protein	mRNA	mRNA + EDTA-stripped membranes	mRNA + nuclease-stripped membranes	SP
Amylase, M_r range = 53,000–54,500	24.2	17.5	37.1	41.7
Procarboxypeptidases A + B, M_r range = 45,500–47,300	22.9	18.7	15.0	6.8
Serine proteases M_r range = 26,400–30,000	52.8	63.8	47.9	51.5

[a] Translation products or authentic secretory products contained in zymogen granule lysates (SP) were separated by one-dimensional polyacrylamide gel electrophoresis in sodium dodecyl sulfate, and analyzed by gel scanning after 2,5-diphenyloxazole fluorography. Peaks representing the groups of secretory proteins indicated were quantitated by gravimetric analysis.

by nuclease-treated microsomal membranes also in the absence and in the presence of mRNA. In the absence of added mRNA, EDTA-treated membranes synthesized identifiable dog pancreas secretory proteins (track 1) indicating incomplete removal of endogenous mRNA. Nuclease-treated membranes did not synthesize detectable levels of protein (track 3). In the presence of added mRNA, protein synthesis proceeds more efficiently in the presence of nuclease-treated microsomal membranes. Levels of protein synthesis were ~5-fold greater in the presence of nuclease-treated (track 4) than EDTA-treated (track 2) membranes.[1] In addition, the distribution of dog pancreas secretory proteins synthesized by the two membrane preparations was markedly different. Secretory proteins synthesized by dog pancreas mRNA in the presence of nuclease-treated microsomal membranes were similar to those synthesized *in vivo* (see the table). Secretory proteins synthesized in the presence of EDTA-treated membranes showed a progressive diminution in the larger molecular weight products, owing to a defect in polypeptide chain elongation caused by defective ribosomes associated with these membranes.[1] Synthesis of amylase, one of the larger molecular weight proteins, was 9.1 times greater in the presence of nuclease-treated membranes than in the presence of EDTA-treated membranes.

Isolation of Messenger RNA

In tissues containing favorable ribonuclease inhibitor to ribonuclease ratios (e.g., canine pancreas) undegraded RNA can be isolated using the sodium perchlorate (British Drug Houses, Poole, England)–proteinase K (EM Biochemicals, Darmstadt, Federal Republic of Germany) method of Lizardi and Engelberg.[4] This procedure as applied to the isolation of RNA from canine pancreas has been described in detail by Scheele *et al.*[1] In tissues that contain unfavorable inhibitor-to-ribonuclease ratios (e.g., rat or guinea pig pancreas), RNA should be isolated by the guanidinium thiocyanate (Fluka, AG, Basel, Switzerland)–guanidinium hydrochloride (Schwarz-Mann, Orangeburg, New York) procedure of Chirgwin *et al.*[5] In this latter procedure residual guanidinium hydrochloride can be removed by extraction of the purified RNA with chloroform butanol.[6] Polyadenylated mRNA is extracted from total RNA by chromatography on oligo(dT)-cellulose as described by Aviv and Leder.[7]

Isolation of the Ribonuclease Inhibitor from Human Placenta

Ribonuclease inhibitor present in the full-term human placenta ($K_i = 3 \times 10^{-10}$ *M* for bovine pancreatic ribonuclease A) was isolated and purified to homogeneity as described by Blackburn.[8] In this procedure, the inhibitor is extracted from a postmitochondrial supernate by precipitation with ammonium sulfate between 35 and 60% saturation. The precipitate is resuspended in a minimum volume of 45 m*M* potassium phosphate buffer, pH 6.4, 1 m*M* in EDTA and 5 m*M* in dithiothreitol. The suspension is dialyzed overnight against 20 volumes of this buffer and clarified by centrifugation at 48,000 *g* for 1 hr at 3°. The sample is then applied to a RNase-Sepharose affinity column prepared in the above buffer. Proteins adsorbed nonspecifically to the column are removed by washing with 0.5 *M* NaCl in the above buffer. Inhibitor is eluted with 0.1 *M* sodium acetate, pH 5.0, 3 *M* NaCl, 1 m*M* EDTA, 5 m*M* dithiothreitol, and 15% (v/v) glycerol. After elution of the inhibitor from the affinity column, it is dialyzed overnight against 20 m*M* HEPES–KOH, pH 7.4, 10 m*M* KCl, 5 m*M* dithiothreitol (DTT) and 15% glycerol.[9] Two milligrams of purified inhibitor can be obtained from a single placenta. When stored in excess thiol (5 m*M* DTT) the inhibitor at 100–200 μg/ml can be repeatedly frozen and thawed over a period of months during storage at either −20° or −80°. The purified inhibitor is added to the translation mixture at 10 μg/ml at 0° 5 min prior to the addition of mRNA as described by Scheele and Blackburn.[9] Figure 2 demonstrates the level of ribonuclease contamination

[4] P. Lizardi and A. Engelberg, *Anal. Biochem.* **98,** 116 (1979).
[5] J. M. Chirgwin, A. E. Przybyla, R. J. MacDonald, and R. W. Rutter, *Biochemistry* **18,** 5294 (1979).
[6] R. MacDonald, personal communication.
[7] H. Aviv and P. Leder, *Proc. Natl. Acad. Sci. U.S.A.* **69,** 1408 (1972).
[8] P. Blackburn, *J. Biol. Chem.* **254,** 12484 (1979).
[9] G. Scheele and P. Blackburn, *Proc. Natl. Acad. Sci. U.S.A.* **76,** 4898 (1979).

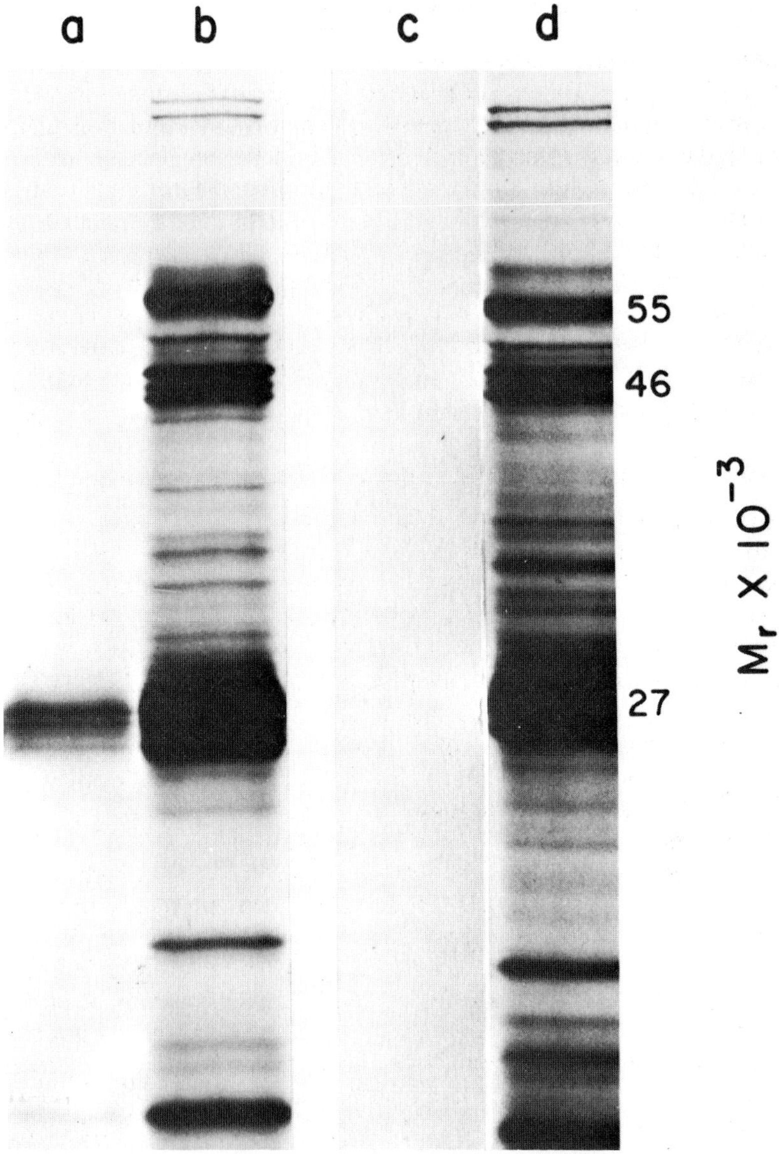

FIG. 2. Ribonuclease contamination in the translation components used for *in vitro* protein synthesis. The combination of translation components, excluding the wheat germ extract, was incubated for 30 min at 24° with dog pancreas polyadenylated mRNA, without (track a) and with (track b) placental RNase inhibitor. The residual mRNA was translated upon the addition of the wheat germ extract. Equal portions of the translation mixtures were prepared for SDS–gel electrophoresis and fluorography as described. Track c shows the effect of 250 ng of RNase per milliliter on the stability of dog pancreas mRNA in the absence (track c) and the presence (track d) of 10 μg of RNase inhibitor per milliliter. Similar results were obtained when reticulocyte lysate was used to translate mRNA. $M_r \times 10^{-3}$ is shown at the right.

present in the translation components used for *in vitro* protein synthesis. Microsomal membranes from a variety of sources introduce additional amounts of ribonuclease activity.[1] Freshly prepared rabbit reticulocyte lysate contains an excess quantity of endogenous ribonuclease inhibitor. However, during prolonged storage (>3 months) inhibitor activity disappears. In this circumstance, the addition of the purified ribonuclease inhibitor is critical.

Rabbit Reticulocyte Lysate Translation System

Preparation of Lysate. Circulatory reticulocytes (55–70% of circulating red blood cells) are stimulated in pathogen-free New Zealand white rabbits (≤2.3 kg) by five daily subcutaneous injections of freshly prepared and neutralized phenylhydrazine (Sigma Chemical Co., St. Louis, Missouri) at 10 μg/kg. Blood is collected by heart puncture on day 9. Blood cells are sedimented at 2000 rpm × 20 min, and the white buffy coat is removed by aspiration. Red blood cells are washed 5 times in a solution containing 130 m*M* NaCl, 5 m*M* KCl, and 7.5 m*M* $MgCl_2$. Cells are lysed in an equal volume of autoclaved double-distilled water (10 min at 0°) containing 2 m*M* DTT in order to prolong the stability of the endogenous ribonuclease inhibitor. Cell membranes are sedimented at 15,000 *g* for 10 min at 4°. Lysates from individual animals can be frozen in liquid N_2 and stored prior to further manipulation. Endogenous globin mRNA is degraded with 80 units of micrococcal nuclease per milliliter according to the procedure of Pelham and Jackson.[3] In order to remove endogenous amino acids, the nuclease-treated reticulocyte lysate is passed over a Sephadex G-25 column equilibrated and eluted with 50 m*M* KCl, 20 m*M* HEPES-KOH, pH 7.5, 1.0 m*M* $MgCl_2$, 0.5 m*M* EGTA, and 2 m*M* DTT. Fractions are taken which contain concentrations of hemoglobin (absorbance at 415 nm) greater than 60% of the loaded sample. One-half milliliter aliquots are frozen in liquid N_2, stored at −80°, and thawed for use only once.

In Vitro Translation Assay. Assays are carried out in a final volume of 25–100 μl with 40% (v/v) nuclease-treated and gel-filtered reticulocyte lysate and the following translation components: 1 m*M* ATP, 0.2 m*M* GTP, 6 m*M* creatine phosphate, 80 μg of creatine phosphokinase per milliliter, 70 m*M* KCl, 28 m*M* HEPES-KOH, pH 7.5, 3 m*M* DTT, 0.5 m*M* glucose 6-phosphate, 100 μg of bovine liver transfer RNA per milliliter, 10 μg of placental ribonuclease inhibitor per milliliter, 100 μ*M* each of 19 unlabeled amino acids, and 100 μCi of $[^{35}S]$methionine per milliliter (~1000 Ci/mmol). The optimum for $MgCl_2$ varies between 0.95 and 1.45 m*M* final concentration and the optimal concentration of spermine varies between 50 and 100 μ*M* depending upon the individual lysate preparation.

In reconstitution experiments, micrococcal nuclease-treated microsomal membranes are added at a final concentration of 20 A_{260} units/ml. *In vitro* translation is initiated with the addition of 0.2 A_{260} unit of polyadenylated mRNA per milliliter, and a temperature shift from 3° to 22° and incubations are carried out at 22° for 90 min in the absence of hemin. Incubations above RT should contain hemin at 20 μM. Incorporation of [^{35}S]methionine into protein is determined using 10-μl samples applied to Whatman 3 MM filter disks, which are then processed for hot trichloroacetic acid-insoluble radioactivity.[10]

Posttranslational Analysis for Segregation of Polypeptide Chains

After *in vitro* translation, the location of nascent polypeptide chains, whether intracisternal or extracisternal with respect to microsomal vesicles, is determined by their sensitivity to added proteases. Previous investigations have yielded ambiguous data, since protease activity resulted in the partial disruption of microsomal vesicles and therefore incomplete resistance of segregated polypeptide chains to added proteases. Using the conventional segregation assay, 17–19% of nascent amylase, synthesized with dog pancreas mRNA and nuclease-treated microsomal membranes, was resistant to a 1-hr incubation with 50 μg of trypsin and 50 μg of chymotrypsin per milliliter at 0° (Fig. 3a). Amylase resistance to posttranslational proteolysis for 1 hr at 22° was zero. Addition of divalent and trivalent cations including calcium resulted in marginal increases in protection from proteases. However, addition of membrane stabilizing agents, tetracaine and dibucaine (both obtained from Sigma Chemical Co., St. Louis, Missouri), prior to the addition of proteases, results in a dramatic increase in the protection of radioactive amylase to 95% (Fig. 3a). In the presence of 20 A_{260} units per milliliter of microsomal membranes from dog pancreas, tetracaine showed a biphasic effect on the protection of radioactive amylase. At 1–3 mM, tetracaine dramatically increased the resistance of amylase to the action of protease. At 10 mM the protective effect of this agent was markedly diminished. The protective effect of dibucaine as a function of concentration was similar to tetracaine, although concentrations greater than 1 mM could not be studied owing to the solubility characteristics of the compound. Tetracaine did not confer protease resistance to the proteins themselves, as judged by their complete degradation when protease treatment was carried out in the presence of 3 mM tetracaine and 1% sodium deoxycholate.

Figure 3b shows a kinetic analysis of the protection of amylase by microsomal membranes during a 60-min incubation in the presence of trypsin and chymotrypsin as a function of time, temperature, and the

[10] R. J. Mans and G. D. Novelli, *Arch. Biochem. Biophys.* **94,** 48 (1961).

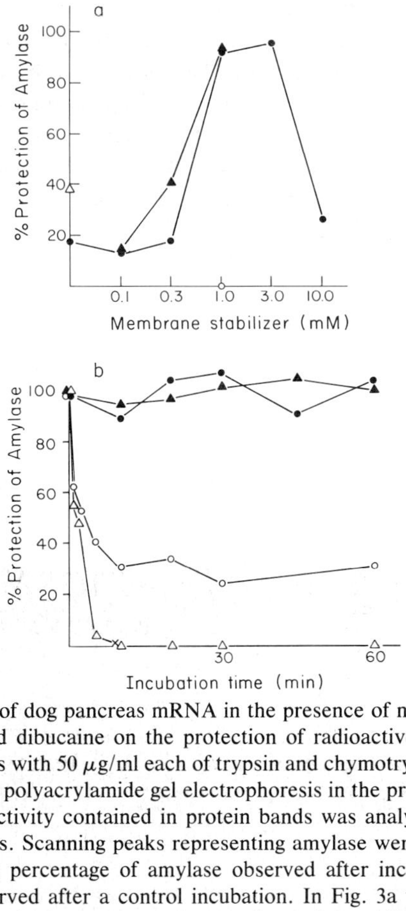

FIG. 3. Translation of dog pancreas mRNA in the presence of microsomal membranes; effect of tetracaine and dibucaine on the protection of radioactive amylase during post-translational proteolysis with 50 μg/ml each of trypsin and chymotrypsin. Translation products were separated by polyacrylamide gel electrophoresis in the presence of sodium dodecyl sulfate, and radioactivity contained in protein bands was analyzed by optical density scanning of fluorograms. Scanning peaks representing amylase were cut out and weighed. Data are expressed as percentage of amylase observed after incubation with proteases compared to that observed after a control incubation. In Fig. 3a the protective effect of microsomal membranes in the presence of tetracaine (–●–) or dibucaine (–▲–) is shown and compared to the effect of membranes plus calcium (–△–) or 1 m*M* tetracaine and 1% sodium deoxycholate (–○–) during a 60-min period of proteolysis at 0°. In Figure 3b resistance of sequestered amylase to proteolysis as a function of time is given for proteolysis at 0° (circles) or 22° (triangles) in the presence (filled symbols) or the absence (open symbols) of 3 m*M* tetracaine. Amylase protection after 10 min of trypsin–chymotrypsin treatment at 0° in the presence of 3 m*M* tetracaine and 1% deoxycholate is also indicated (–×–). Taken from Scheele *et al.*[1]

presence or absence of tetracaine. Proteolysis in the absence of tetracaine, either at 0° or 22°, showed either partial or complete degradation of nascent amylase. In contrast, proteolysis at either 0° or 22°, carried out in the presence of 3 m*M* tetracaine, resulted in the complete protection (95–102%) of radioactive amylase by microsomal membranes. The improvement in protection of amylase observed in Fig. 3b compared to 3a was due to the addition of 10 μg of cycloheximide (Sigma) per milliliter prior to proteolysis in order to inhibit, in the control mixture, further synthesis and segregation of small quantities of [^{35}S]-methionine-labeled proteins during the 1-hr incubation in the segregation assay.

Figure 4 shows, at the level of fluorography, the increase in protection of all dog pancreas secretory proteins as a function of increasing concentrations of tetracaine. The results for all secretory proteins including G1 and G2, at 97,000 and 92,000 M_r, respectively, parallel that seen for amylase.

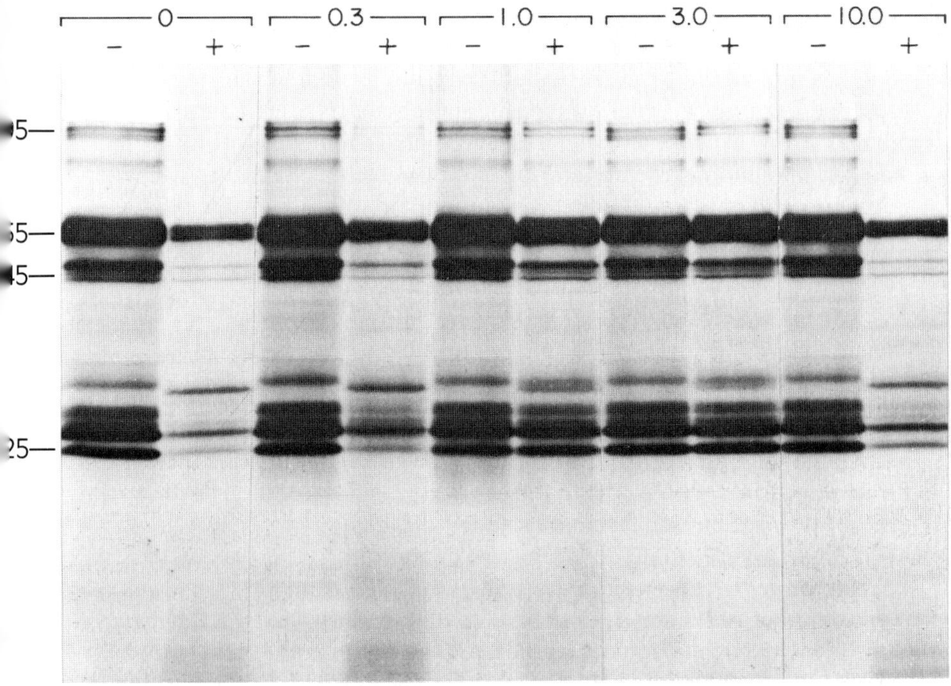

FIG. 4. Translation of dog pancreas mRNA in the presence of nuclease-treated microsomal membranes; effect of increasing concentrations of tetracaine (indicated in millimolar concentrations) on the protection of translation products during posttranslational proteolysis. Radioactive methionine-labeled products were synthesized in the reticulocyte lysate system in the presence of 0.2 A_{260} unit per milliliter of dog pancreas mRNA and 20 A_{260} units per milliliter of nuclease-treated dog pancreas microsomal membranes. Segregation of radioactively labeled proteins by microsomal membranes was monitored by posttranslational incubations in the absence (−) and in the presence (+) of proteases as described in the legend to Fig. 3. Translation products were then separated by sodium dodecyl sulfate–polyacrylamide gel electrophoresis and analyzed by fluorography. $M_r \times 10^{-3}$ is shown on the left.

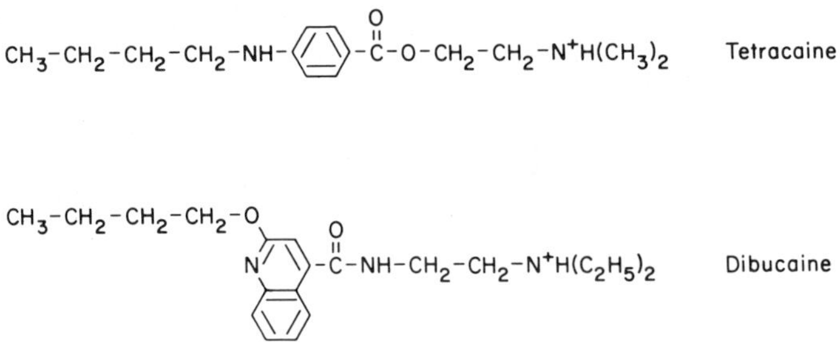

FIG. 5. Molecular structures of amphiphilic membrane stabilizers, tetracaine and dibucaine.

These findings indicate that microsomal membranes isolated from dog pancreas and added to the reticulocyte lysate system are unstable in the presence of 50 μg/ml each of trypsin and chymotrypsin. Tetracaine and dibucaine, both positively charged amphiphilic molecules (Fig. 5), may stabilize microsomal membranes by two possible mechanisms. The insertion of such agents into the lipid bilayer may inhibit, by charge repulsion, the degradative effect of trypsin and chymotrypsin on integral membrane proteins, and/or the insertion of these agents into the membrane may stabilize the lipid bilayer itself. In regard to the latter possibility, rough microsomes isolated from dog pancreas are associated with significant quantities of secretory prophospholipase A_2. Trypsin treatment during posttranslational proteolysis resulted in a 6-fold increase in the activity of phospholipase A_2 and a 23-fold increase in the quantity of lysolecithin associated with membranes. Tetracaine at 3 m*M* inhibited the increase in lysolecithin observed in microsomal membranes incubated in the presence of trypsin. In the absence of added proteases, dog pancreas microsomal vesicles are stable under *in vitro* conditions for periods up to 3 hr.[1]

Based on the above information, we recommend that analyses for segregation of polypeptide chains be conducted as follows. After *in vitro* protein synthesis in the presence of microsomal membranes, cycloheximide at 10 μg/ml and tetracaine at 3 m*M* are added to the incubation mixture and the resulting mixture is divided into two equal aliquots. Samples should be incubated with cycloheximide and tetracaine for 5 min at 22° before shifting to 0°. A fresh solution of 500 μg of trypsin and 500 μg of chymotrypsin per milliliter at 0° is added to one aliquot to achieve a final concentration of 50 μg/ml each,[11] and posttranslational proteolysis is carried out for 1 hr at 0°. The second aliquot, to which only buffer is added, serves as the control. Both tetracaine and trypsin–chymotrypsin stock

[11] D. D. Sabatini and G. Blobel, *J. Cell Biol.* **45,** 146 (1970).

solutions should contain 70 mM KCl and 1 mM $MgCl_2$. The stock solution of tetracaine, adjusted to pH 7.0 with HCl, can be repeatedly frozen and thawed. Posttranslational proteolysis is terminated by the addition of Trasylol (FBA Pharmaceuticals, New York, New York) at 1500 KIU/ml. Samples are incubated in Trasylol at 0° for 5 min before further processing. Using this procedure, one can successfully carry out posttranslational proteolysis for 1 hr at either 0° or 22°.

Gel Electrophoretic Analysis of Nascent Secretory Proteins

Proteolytic processing of presecretory proteins to secretory proteins by the transport peptidase associated with microsomal membranes can be followed by either one-dimensional polyacrylamide gel electrophoresis in SDS[12] or two-dimensional separation of proteins using isoelectric focusing in the first dimension and SDS–gel electrophoresis in the second dimension.[13,14] The first procedure measures apparent molecular weights. The second procedure determines not only the size and charge of nascent proteins, but also their conformational properties.

The procedure for separation of translation products according to apparent molecular weight (M_r) by one-dimensional polyacrylamide gel electrophoresis in SDS is as follows. Ten microliter samples of the translation mixture are prepared for electrophoresis in a final volume of 40 μl containing 5% SDS, 15% sucrose, 0.0025% Bromphenol Blue, Tris, and glycine in concentrations identical to those in electrode buffer, and 50 mM dithiothreitol. Samples are heated in boiling water for 3 min and maintained at 37° for 30 min. Freshly prepared iodoacetamide is added to a final concentration of 100 mM and samples are further incubated at 22° for 30 min. Forty microliters are applied to individual gel slots. The 1-mm slab electrophoresis gel contains a 12 to 17% acrylamide (British Drug Houses, Poole, England) gradient serving as a resolving gel and a 5% acrylamide stacking gel, both in 0.1% SDS and buffers as described by Maizel.[12] Gels are aged at 3° for ≥2 days prior to use. After electrophoresis, gels are stained in 50% methanol, 10% acetic acid, and 0.06% Coomassie Blue for 1 hr and destained in 40% methanol, 10% acetic acid. Gels are prepared for fluorography using dimethyl sulfoxide (DMSO) and 2,5-diphenyloxazole (PPO) as described by Bonner and Lasky[15] except that 3% glycerol is added to both the 20% PPO in DMSO solution and the final solution of water, and 5 min after the gel is immersed in water,

[12] J. V. Maizel, *in* "Fundamental Techniques in Virology" (K. Habel and N. P. Salzman, eds.), p. 334. Academic Press, New York, 1969.

[13] G. Scheele, *J. Biol. Chem.* **250,** 5375 (1975).

[14] W. Bieger and G. Scheele, *Anal. Biochem.* **109,** 222 (1980).

[15] M. Bonner and R. A. Laskey, *Eur. J. Biochem.* **46,** 83 (1974).

methanol is added to a concentration of 20%. Methanol at this concentration serves to restrict the gel's expansion in water without resolubilizing the PPO; glycerol serves as a stabilizer, and the combination prevents the gels from cracking during the vacuum drying procedure at 22° on Whatman 3 MM paper. Gels containing impregnated PPO and covered with Saran wrap are dried and then exposed for either fluorography (−80°) or autoradiography (22°) using medical X-ray film.

Figure 6 demonstrates, using polyacrylamide gel electrophoresis in SDS, that the addition of micrococcal nuclease-treated rough microsomes from dog pancreas results in complete proteolytic processing of pancreatic presecretory proteins (removal of transport peptides) to polypeptide chains with apparent molecular weights identical with authentic secretory enzymes and zymogens. However, analysis of these same proteins by two-dimensional isoelectric focusing (IEF)–SDS gel electrophoresis indicates that the conformations of nascent secretory proteins synthesized *in vitro* in a gel-filtered reticulocyte lysate supplemented with microsomal membranes are distinctly different than those of authentic secretory proteins. The inability of these nascent secretory proteins to separate into discrete spots during two-dimensional gel analysis has been shown to be secondary to either the lack of formation or incorrect formation of disulfide bonds during *in vitro* protein synthesis.[16] However, addition of oxidized glutathione (Sigma Chemical Co., St. Louis, Missouri), during *in vitro* translation results in nascent secretory proteins that comigrate in two dimensions with authentic secretory proteins, indicating similarity in isoelectric point, apparent molecular weight, and conformation properties including formation of the correct set of disulfide bonds (Fig. 7). In the presence of dithiothreitol in the translation mixture, oxidized glutathione (GSSG) is rapidly reduced (GSH), but small amounts of GSSG remain, owing to the equilibrium established by glutathione reductase present in the reticulocyte lysate. For each translation protocol it is necessary to optimize the concentration of GSSG in order to catalyze the correct formation of disulfide bonds without causing marked inhibition in protein synthetic activity.[16] In the presence of 3.25 mM dithiothreitol in the translation mixture, optimal concentration of added GSSG is 2–3 mM. In the absence of dithiothreitol, 0.5 mM GSSG is required to achieve correct formation of disulfide bonds. In this case, reduction of GSSG to GSH occurs primarily at the expense of cysteine and sulfhydryl groups attached to proteins present in the translation mixture.

In the absence or the presence of glutathione, pancreatic presecretory proteins, derived from the translation of mRNA in the absence of microsomal membranes, separate into only a few discrete two-dimensional

[16] G. Scheele and R. Jacoby, *J. Biol. Chem.* **257,** 12277 (1982).

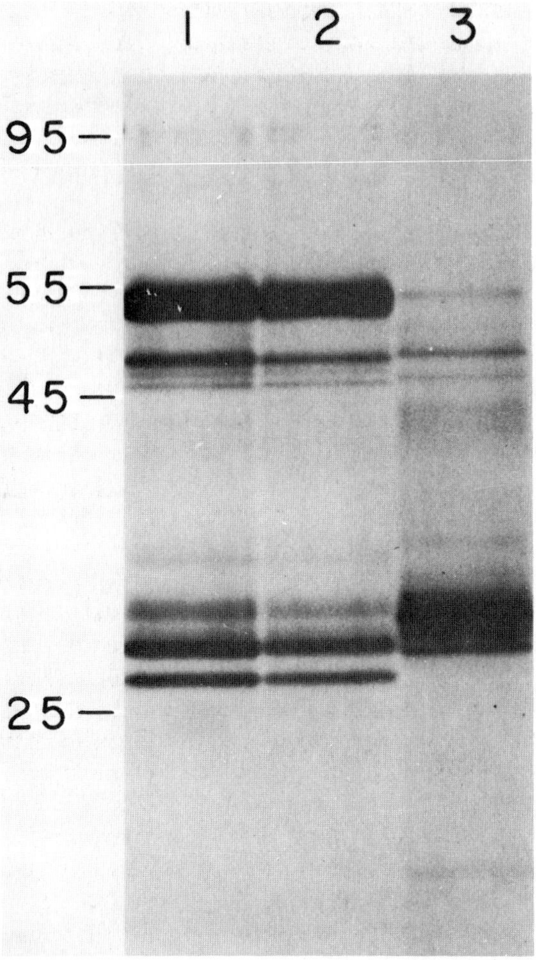

FIG. 6. Analysis of translation products directed by dog pancreas mRNA in the absence (track 3) and the presence (track 2) of micrococcal nuclease-treated microsomal membranes. In this case *in vitro* translation was carried out in the absence of added bovine liver tRNA. Track 1 shows authentic [^{35}S]methionine-labeled secretory proteins derived from a zymogen granule fraction obtained from dog pancreas. *In vitro* and *in vivo* translation products were separated by polyacrylamide gel electrophoresis in sodium dodecyl sulfate and analyzed by fluorography. Numbers to the left indicate molecular weight values $\times 10^{-3}$. Note the downward shift in apparent molecular weights of products translated in the presence of microsomal membranes and the similarity in size to authentic secretory protens.

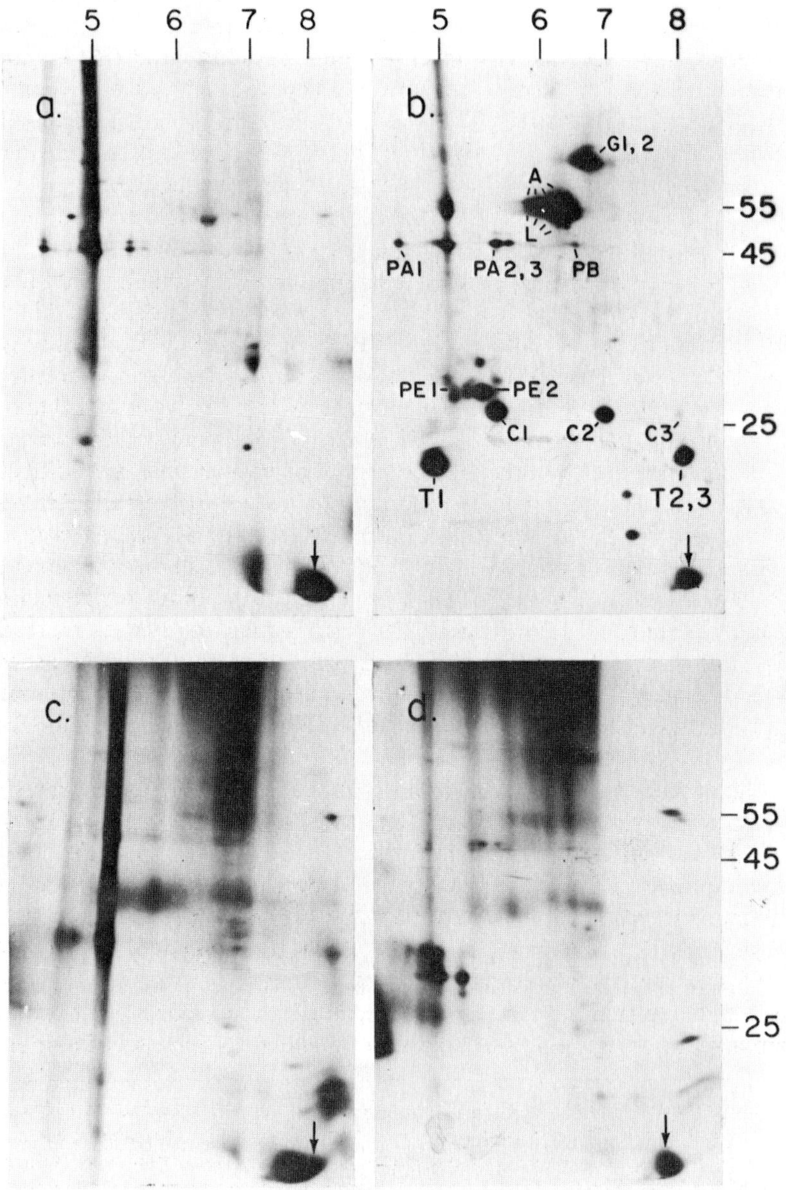

FIG. 7. Analysis of *in vitro* translation products by two-dimensional isoelectric focusing–sodium dodecyl sulfate–gel electrophoresis. [^{35}S]Methionine-labeled proteins were synthesized in a gel-filtered reticulocyte lysate translation system containing 3.25 m*M* dithiothreitol. Protein synthesis was directed in the presence of dog pancreas rough microsomes (panels a and b) or dog pancreas poly (A)$^+$ mRNA (panels c and d) in the absence (a,

spots (Fig. 7). Unlike the case with secretory proteins synthesized in the presence of microsomal membranes, little improvement in the migration of presecretory proteins was observed with the addition of GSSG (compare Fig. 7d with 7c). None of the observed spots comigrate with those of authentic secretory proteins, although several appear in the vicinity of secreted proteins. The conformational instability observed among presecretory proteins, containing amino-terminal transport peptides, is secondary to the formation of incorrect disulfide bands and nonspecific protein–protein interactions among nascent polypeptide chains. This latter interaction results in aggregations of protein in both dimensions of the gel procedure despite the presence of 8 *M* urea and 0.1% Triton X-100 in the first-dimension focusing gel and 0.1% SDS in the second-dimension gel. The conclusions reported here for conformational properties of presecretory and secretory proteins, as determined by two-dimensional gel analysis, has been confirmed by independent studies that analyzed the sensitivity of nascent (pre)secretory proteins to proteases and the solubility properties of these proteins in aqueous and organic solvents.[16] Removal of the transport peptide by the transport peptidase associated with the RER membrane, in the presence of an optimal redox potential, thus allows the proper folding of translocated secretory proteins, which include the correct set of disulfide bonds. The resulting molecules thus represent stable and discrete entities soluble in an aqueous environment. Nonsecretory proteins, e.g., hemoglobin, require neither microsomal membranes nor glutathione during their synthesis to fold into stable configurations (Fig. 7, vertical arrows).

The procedure for separation of translation products, which represent secretory proteins, by two-dimensional IEF–SDS-gel electrophoresis is as follows. Radioactive-labeled proteins are synthesized in a gel filtered reticulocyte lysate in the presence of optimal concentrations of glutathione. Translation samples containing membranes are treated with 1% Triton X-100, 25 m*M* Tris-HCl, pH 9.0, 100 KIU of Trasylol per milliliter, and 1 m*M* diisopropylfluorophosphate. Nonradioactive secretory proteins, either proteins secreted into the incubation medium or extracted from a secretory granule fraction, can be added as markers of authentic

c) or the presence (b, d) of 2 m*M* glutathione. Translation products were made 4 *M* in urea and 1.0% in Triton X-100 and separated in two dimensions by isoelectric focusing and SDS–gel electrophoresis. Radioactive proteins, applied in equal amounts, are shown after fluorography of dried gels. Numbers on the upper abscissa indicate isoelectric points. Numbers on the right ordinate indicate molecular weight values $\times 10^{-3}$ for nonreduced proteins. The majority of radioactive proteins separated by the two-dimensional gel technique in panel b comigrated with authentic secretory proteins detected by Coomassie Blue stain. These proteins are labeled according to their actual or potential activities by abbreviations described in the legend to Fig. 1. Vertical arrows indicate newly synthesized globin.

secretory proteins. Proteins are separated using the two-dimensional isoelectric focusing–SDS-gel electrophoresis procedures developed by Scheele[13] and modified by Bieger and Scheele.[14] These procedures utilize slab gel isoelectric focusing in the absence[13] or in the presence[14] of 8 *M* urea in the first dimension and SDS–slab gel electrophoresis in a polyacrylamide gradient (10 to 20%) in the second dimension. Isoelectric focusing and SDS-gel electrophoresis are carried out as previously described.[13] Second-dimension gels are stained with Coomassie Blue R, destained, impregnated with PPO,[15] dried on Whatman 3 MM paper, and exposed for fluorography as described above for one-dimensional gels. Alternatively, proteins separated in two-dimensional gels can be transferred electrophoretically to nitrocellulose paper,[17,18] and protein patterns contained on paper transcripts can be analyzed by amido black staining and autoradiography.

Analysis of Biological Activity Associated with Presecretory and Secretory Proteins

In order to demonstrate that translocation across the microsomal membrane is physiological, it is necessary to demonstrate that secretory proteins segregated within microsomal vesicles are biologically active. Until recently[19] it has not been possible to demonstrate significant biological activity among translocated secretory proteins. As described in the preceding section, nascent secretory proteins translocated into microsomal vesicles *in vitro* showed conformational properties distinctly different from those of authentic secretory proteins unless glutathione was added during translation. Addition of glutathione is also critical for the determination of biological activities in these molecules. The biological activities of two pancreatic exocrine proteins have been examined in detail: (*a*) the binding of nascent amylase to its substrate, glycogen; and (*b*) the binding of nascent trypsinogen to Sepharose-bound protease inhibitors, soybean trypsin inhibitor, and Trasylol. Secretory proteins translocated into microsomal vesicles showed significant activity only when glutathione was added during *in vitro* translation. The conditions optimal for determination of biological activity were identical to those given in the preceding section for the correct development of conformation, including formation of the appropriate set of disulfide bonds, in these polypeptide chains. Little or no activity was measured when presecretory proteins, synthesized in the absence of microsomal membranes, were assayed after

[17] H. Towbin, T. Staehelin, and J. Gordon, *Proc. Natl. Acad. Sci. U.S.A.* **76,** 4350 (1979).
[18] J. Symington, M. Green and K. Brackmann, *Proc. Natl. Acad. Sci. U.S.A.* **78,** 177 (1981).
[19] G. Scheele and R. Jacoby, *J. Biol. Chem.* **258,** 2005 (1983).

translation in either the absence or presence of glutathione. Thus, for secretory proteins with amino terminal transport peptide, these leader sequences must necessarily be removed to allow conformational development that not only stabilizes the molecule but also allows expression of authentic biological activity.

[8] Immunoprecipitation of Proteins from Cell-Free Translations

By DAVID J. ANDERSON and GÜNTER BLOBEL

Indirect immunoprecipitation has already become a widespread technique for the identification of low-abundance *in vitro* translation products. (By low abundance we mean comprising less than 1% of the total protein synthesized.) In at least one case,[1] it has been used to isolate a protein that has been estimated to constitute less than 0.01% of the total products of cell-free synthesis. With the advent of monoclonal antibody technology, it should be possible to study the *in vitro* synthesis of low-abundance polypeptides on a routine basis. In what follows we shall describe the basic steps in the indirect immunoprecipitation procedure, with specific examples drawn from the protocols currently followed in our laboratory.

Synthesis

In vitro synthesis techniques are described in detail elsewhere in this volume. It should be borne in mind that translation reaction mixtures will have to be scaled up for successful visualization of minor products, the degree depending upon the abundance of the mRNA of interest and the efficiency of cell-free translation. For example, a single polypeptide species containing ca 2000 dpm of [^{35}S]methionine can be visualized as a dark band on a sodium dodecyl sulfate (SDS) slab gel in a 3-day fluorographic exposure, using preflashed, highly sensitive film such as Kodak XAR-5. If this protein were 0.01% of total protein synthesis, to obtain ca 2000 dpm of it one would have to start with 2×10^7 dpm of total cell-free translation products. If the cell-free translation gave 5×10^5 dpm per 5 μl of trichloroacetic acid-precipitable radioactivity, then 2×10^7 dpm would correspond to 200 μl of translation reaction mixture. Since commercially avail-

[1] A. H. Erickson and G. Blobel, *J. Biol. Chem.* **254,** 11771 (1979).

ISBN 0-12-181996-5

able *in vitro* translation "kits" consider 25 μl to be a "standard assay," this represents a scaling up of 8-fold for material to be applied to a single lane on a gel. The above estimate could be reduced either by running very small gels or by allowing for longer exposure times.

Solubilization

Dispersing the translation reaction mixture with some sort of detergent is essential if one wishes to immunoprecipitate secretory or membrane proteins synthesized in the presence of rough microsomal membranes. Detergent is not absolutely necessary for proteins synthesized in the absence of membranes, but it does reduce the nonspecific adsorption background considerably and is recommended. The choice of detergent depends upon the conformational specificity of the antibody used. Sodium dodecyl sulfate is most desirable, as it gives the most complete solubilization and denaturation, but one must first determine independently that the antibody will recognize the SDS-denatured form of the protein. (This can be ensured by denaturing the antigen in SDS prior to immunization.) An added benefit of heat denaturation in the presence of SDS is that it is virtually guaranteed to inactivate any proteases in the cell-free system that might degrade the translation products during the immunoprecipitation procedure. After SDS, the next best thing seems to be a nondenaturing detergent such as Triton X-100 or Nonidet P-40 (NP-40). These detergents are less likely than SDS to destroy conformationally sensitive antigenic determinants, but they also give dirtier immunoprecipitates. Furthermore, cell-free products immunoprecipitated under these conditions are much more prone to proteolytic degradation, and a variety of protease inhibitors should be included in all buffers to minimize this. Deoxycholate (DOC) may also be used, but, like SDS, it must be complexed subsequently with another, weaker detergent to prevent interference with antigen–antibody binding.

We find it convenient to perform immunoprecipitations in the same 1.5-ml plastic Eppendorf tubes we use for translation. When SDS is used in the primary solubilization, we routinely add it to 1% final concentration from a 25% (w/v) stock solution and then heat the sample for 4 min in a boiling water bath. (The potassium in the cell-free synthesis reaction mixture will initially form a precipitate with the SDS, but this disperses upon heating and subsequent dilution.) While this provides for optimal solubilization, in some cases heating can cause a protein to behave anomalously in subsequent SDS gel electrophoresis and should be avoided.[2,3] Heating

[2] D. Henderson, H. Eibl, and K. Weber, *J. Mol. Biol.* **132,** 193 (1979).

[3] D. J. Anderson and G. Blobel, this volume [29].

should also be omitted when nondenaturing detergents are used, as the proteins will coagulate. When solubilizing samples synthesized in a reticulocyte lysate system,[4] higher concentrations of detergent must be used (e.g., 4% SDS) to compensate for the higher protein concentration. In the case of solubilizing radiolabeled proteins from tissue-culture cells, including a sonication step at this time to shear the DNA seems to reduce backgrounds considerably. It is imperative not to freeze cell-free translations prior to this solubilization step, as this increases substantially the background due to nonspecific adsorption.

Dilution

Detergent-solubilized translation reaction mixtures are next diluted with a buffered salt solution containing nonionic detergent, prior to addition of antibody. This step serves a twofold purpose. First, it decreases nonspecific adsorption by lowering the concentration of total protein while keeping the desired polypeptide in the range of its K_D for antibody binding. Second, the addition of nonionic detergent complexes the free SDS into mixed micelles where it can no longer disrupt antigen–antibody interactions. It is not clear to what extent substitution of nonionic detergent for protein-bound SDS occurs, and whether there is partial renaturation of proteins as a result of this. In cases where SDS has not been used, dilution is still helpful for the first reason.

Typically we add nonionic detergent so that its final concentration (percentage, w/v) is in fivefold excess over the SDS. Thus, to a wheat germ system reaction mixture previously denatured in 1% SDS, we add four volumes of a solution containing 1.25% Triton X-100 (or NP-40), 190 m*M* NaCl, 6 m*M* EDTA, 50 m*M* Tris-HCl, pH 7.4, and 10 units of Trasylol per milliliter as an inhibitor of proteolysis. This brings the final detergent concentrations to 1% Triton X-100, 0.2% SDS, and the EDTA to 4.8 m*M* (included as a metalloprotease inhibitor). In the case of the reticulocyte lysate system, after denaturing in 4% SDS, the sample is diluted first with an equal volume of water and then with four volumes of the aforementioned buffered salt solution containing 2.5% Triton X-100, so that the final detergent concentrations are 2% Triton, 0.4% SDS and the protein is twice as dilute as in the preceding example.

Incubation with Antibody

The amount of antibody added will depend upon its titer and affinity and upon the amount of cell-free product present. Since a typical 100-μl *in*

[4] T. Hunt and R. J. Jackson, this volume [4].

vitro translation reaction mixture may contain on the order of $(1-10) \times 10^{-15}$ mol of product, with an antiserum titer of 10^{-6} mol/liter, 1–20 μl of antiserum would be sufficient. In practice the exact amount of antibody needed for quantitative immunoprecipitation must be determined empirically by titration.

Since antibody–antigen interactions are tighter at low temperature, incubation is almost always performed at 4°. In one case,[5] preincubation for 1 hr at 37° was performed to speed up the reaction; however, this is not necessary. Although we usually incubate at least 12 hr at 4°, in some cases[6,7] the incubation time may be decreased to as little as 3 hr, although in these cases recovery was not quantitated. Agitation of the sample during incubation with antibody is unnecessary; we usually just leave the samples in the refrigerator.

Direct immunoprecipitation of cell-free translation products is practical only with the addition of carrier antigen, as the molar quantities of *in vitro* product made are vanishingly small. If this method of recovering immune complexes is chosen, unlabeled carrier antigen must be titrated against fixed amounts of translation reaction mixture and antibody until the region of equivalence is attained and macroscopic antibody–antigen complexes can be recovered, by high-speed centrifugation. In our opinion this method is more time-consuming and far less convenient than the solid-phase immunoadsorbent methods described below.

Incubation with Immunoadsorbent

Prior to this step, aggregated material is removed from the samples by centrifugation for 2–3 min in an Eppendorf microcentrifuge (or equivalent) followed by transfer of the supernatants to fresh tubes. This helps to reduce nonspecific contamination considerably, since the adsorbed antigen–antibody complexes are eventually washed by centrifugation and aggregated material would cosediment under these conditions.

The use of a second antibody to precipitate antigen–antibody complexes has declined since the advent of solid-phase immunoadsorbents. The latter have become the method of choice as they are easier to wash and to handle than are conventional second-antibody complexes. One class of immunoadsorbents is based upon protein A. This is a polypeptide found on the surface of *Staphylococcus aureus* that has a high affinity for the Fc portion of immunoglobulin G molecules from a variety of species.[8]

[5] B. M. Goldman and G. Blobel, *Proc. Natl. Acad. Sci. U.S.A.* **75,** 5066 (1978).
[6] B. Dobberstein, H. Garoff, G. Warren, and P. J. Robinson, *Cell* **17,** 759 (1979).
[7] W. A. Braell and H. F. Lodish, *Cell* **28,** 23 (1982).
[8] J. W. Goding, *J. Immunol. Methods* **20,** 241 (1978).

A popular, and inexpensive, protein A immunoadsorbent consists simply of heat-killed, formalin-fixed *S. aureus*.[9] Antigen–antibody complexes are bound to the surface of the bacteria and are separated from all other components by low-speed centrifugation. Although convenient, this method has the slight disadvantage that sonication of the pelleted bacteria is frequently necessary to achieve thorough washing. Since washing can involve as many as five steps (see below), this sonication can be cumbersome when processing many tubes at a time.

A more expensive but convenient alternative to *S. aureus* consists of purified protein A covalently coupled to Sepharose CL-4B beads. This can be obtained from Pharmacia (Uppsala, Sweden) as well as from other firms. It is not any cheaper to buy purified protein A and couple it to Sepharose yourself. However, we have found that the protein A on these beads is remarkably resistant to denaturants (probably because of its high α-helical content) and can thus be recycled simply by boiling in 5% SDS with 10 m*M* dithiothreitol followed by thorough washing in buffers containing 1% Triton X-100.[10] For adsorption of antibody–antigen complexes, we typically add three volumes of packed protein A–Sepharose beads per volume of crude rabbit antiserum, and agitate for at least 2 hr at room temperature. It is easier to pipette the Sepharose accurately as a 1 : 1 suspension in water. When working with monoclonal antibodies, it is important to determine independently that they bind to protein A before using this immunoadsorbent.

If the antibody of interest does not bind to protein A, a solid-phase immunoadsorbent can be made by purchasing an affinity-purified second antibody (Cappel laboratories has a wide selection) and coupling it to CNBr-activated Sepharose CL-4B. In general, a lower concentration of coupled antibody per unit volume of Sepharose should give lower nonspecific binding. However, in some cases this may necessitate using large volumes of beads (e.g., over 100 μl), which can be cumbersome when preparing samples for electrophoresis (see below). We have found that coupling antibody to a final concentration of 10 mg/ml Sepharose gives adequate results.

Washing Immunoprecipitates

A major source of contamination in immunoprecipitation is nonspecific adsorption of material to the solid-phase immunoadsorbent. It is to reduce this contamination that the washing procedures described below are used. Selecting the appropriate wash buffer is a compromise between

[9] E. O'Keefe and V. Bennett, *J. Biol. Chem.* **255,** 561 (1980).

[10] E. Ching and G. Blobel, unpublished observations.

conditions stringent enough to remove adsorbed contaminants, but not so harsh that they elute the desired antigen–antibody complexes. Nonspecific adsorption to the Sepharose itself is less a contributing factor than is adsorption directly to the antibody or protein A coupled to the Sepharose, or to the first antibody bound to these. Since different antisera contain different antibodies, wash conditions may have to be optimized in cases where a particularly "sticky" antiserum has been obtained. The different solutions described below have been used with success in a variety of cases in our laboratory.

Sodium dodecyl sulfate alone disrupts antigen-antibody complexes and should be used only as a last resort in cases where the desired product is of such low abundance that massive contamination results with any other procedure. Lysosomal proteins comprising 0.01% of total cell-free synthesis[1] have been washed with a buffer consisting of 0.1% SDS, 150 m*M* NaCl, 150 m*M* triethanolamine (TEA), pH 7.4, 5 m*M* EDTA, and 10 units of Trasylol per milliliter. As expected, however, the procedure results in some loss of specifically bound material.

More typical are wash buffer containing SDS complexed in a mixed micelle with nonionic detergent at a 1:5 ratio. For example, a buffer containing 0.1% Triton X-100, 0.02% SDS, 150 m*M* NaCl, 50 m*M* Tris-HCl, pH 7.4, 5 m*M* EDTA, and 10 units of Trasylol per milliliter has been used successfully with rabbit, rat, sheep, and goat antisera.[5] Increasing the ratio of SDS to Triton (or NP-40) gives more stringent washing, but is still safer than using SDS alone; thus a buffer consisting of 0.1% SDS, 0.05% Triton X-100 (SDS to Triton ratio, 2:1) in the above salts has been useful in some cases. In the event that more stringent conditions are desired without resorting to SDS, the salt concentration can be raised to 0.5 *M* NaCl without appreciable loss of material. Another wash buffer in common use is the so-called "RIPA" buffer used in radioimmunoassays, consisting of 1% Triton X-100, 1% DOC, 0.1% SDS, 150 m*M* NaCl, 10 m*M* Tris-HCl, pH 7.2, and 1% Trasylol.[11]

We have found it sufficient to wash adsorbed antigen–antibody complexes three or four times with 1 ml per wash of one of the above solutions. The beads are vortexed during each wash and, without further incubation, are recovered by centrifugation in a desk-top microcentrifuge. The supernatants are aspirated conveniently by suction through a drawn-out Pasteur pipette into a trap bottle. Care should be taken at this step to avoid aspirating any of the beads. In general, the greater the volume of translation reaction mixture being immunoprecipitated from, the greater

[11] Z. Gilead, Y. Jeng, W. Wold, K. Sugawara, H. M. Rho, M. L. Harter, and M. Green, *Nature (London)* **264**, 263 (1976).

the number of washes should be. If stubborn contaminants persist in the immunoprecipitates, however, they may reflect the presence of actual contaminating antibody species, which react with polypeptides other than the one of interest. In these cases other steps can be taken, such as immunoselecting the antiserum (see below).

After the above washes, a final wash is usually performed in an identical solution without detergent, so that complete solubilization in SDS prior to electrophoresis will not be interfered with by contaminating nonionic detergent. At this point, as much residual wash buffer as is possible should be removed from the pelleted adsorbent with a drawn-out Pasteur pipette, to avoid excessive dilution of the SDS–gel electrophoresis sample buffer. It is ideal to aspirate the beads to dryness, and for this the Pasteur pipette must be drawn very thin to avoid removing any of the beads. The washed material may be stored at −20° for a few days prior to electrophoresis without harmful effect on most proteins. Membrane proteins, however, should as usual be treated as guilty until proved innocent and freezing–thawing of them be avoided.

Elution of Sample from the Immunoadsorbent

If the volume of packed immunoadsorbent is sufficiently small (e.g., 50 μl or less), SDS gel sample buffer may be simply added directly to it. After heating under reducing conditions, the beads are spun out and an aliquot of the supernatant is applied to the gel. This method is convenient but has the disadvantage that it will elute large amounts of immunoglobulin from the beads when second antibody–Sepharose conjugates are used as the immunoadsorbent. This can be a problem if the radiolabeled protein of interest has a molecular weight in the range of 50,000; the large amounts (as much as 30 μg) of IgG heavy chain tend to obscure and distort minor labeled bands that have the same electrophoretic mobility. One way to avoid this problem is to use affinity-purified antibody, which results in a much smaller amount of IgG being applied to the gel. If this is impractical, the adsorbed antigen–antibody complexes can be preincubated in SDS–gel sample buffer lacking reducing agent. After spinning out the beads, the supernatant containing efficiently eluted antigen, but relatively little antibody, is transferred to a fresh tube and heated in the presence of reducing agent.

Eluting packed Sepharose beads with an equal volume of sample buffer tends to yield about 60% of the total bound radioactivity. For quantitative recovery, successive elutions may be performed (4 is maximal) and the eluates pooled. Alternatively, the adsorbed protein may be eluted from the beads into a relatively large volume (e.g., 0.5 ml) of 1%

SDS, 10 mM DTT, 10 mM Tris-HCl, pH 7.4, and then precipitated with ice-cold 10% trichloroacetic acid or 0.1 N HCl in acetone. This precipitate is easily taken up in a small volume of SDS–gel sample buffer with heating.

Checking the Specificity of Immunoprecipitation

It is useful but not sufficient to demonstrate the specificity of immunoprecipitation by performing a parallel incubation with nonimmune or preimmune sera. Antisera, even in the same animal (over time, as the animal produces new antibodies), can change their intrinsic stickiness. For this reason it is essential to establish specificity using the identical serum, by competition with excess unlabeled antigen. Ideally, the antigen used for competition should be purified by a different procedure than that used to prepare the original immunogen. If this is not possible, peptide mapping or partial radiosequence analysis can be used to prove the identity of the immunoprecipitated product.

Cleaning Up Immunoprecipitates

The following procedures may be useful in cases where certain contaminants are not removed by thorough washing, although they are not panaceata.

1. Pretreatment of solubilized reaction mixtures with the immunoadsorbent prior to addition of antisera.[6,7] By incubating with the adsorbent in the absence of antiserum for a few hours, many high-affinity contaminants can be adsorbed out of solution and removed by pelleting the beads. The supernatant is transferred to a new tube, and antiserum is added. This technique may be particularly useful where antibody–Sepharose rather than protein A–Sepharose is the adsorbent, since the former appears to be an appreciably stickier protein than the latter. A liability in this method is the possibility of losing some of the specific protein under study, if it is itself very sticky.

2. Cleaning up the antiserum. Making an IgG fraction from the antiserum by ammonium sulfate fractionation can help a bit, but by far the best approach is to affinity-purify the antibodies. This can be done with as little as 50 μg of pure antigen, since the antigen–Sepharose columns can be regenerated after use. Using such immunoselected antibodies gives immeasurably cleaner immunoprecipitates, since the amount of IgG added is often less than 1% of that added when crude serum is used. Furthermore, using such a selected reagent provides an additional level of confidence in the identity of the immunoprecipitated product, particularly if the antigen

used to affinity-purify the antibody is purer than that used to immunize the animal.

3. Reduction of the amount of antisera added. In cases where it is impractical to affinity-purify antibodies, immunoprecipitates can be cleaned up if a limiting rather than an excess amount of antiserum is used. The disadvantage is that recovery will not be quantitative, but, if this is not crucial to the experiment, this technique can reduce background considerably.

4. Reduction of the volume of the original synthesis reaction. For some reason unknown to us, it seems that the degree of nonspecific contamination is inversely proportional to the specific activity of protein labeling in the cell-free synthesis reaction. Thus, it is better to immunoprecipitate a product from 50 μl of reaction mixture than it is to do so from 200 μl containing the same amount of radioactivity. The specific activity of a translation can be increased by adding more label, adding more mRNA, and being sure to optimize ions properly.

5. Make a new antibody. Unfortunately, nonspecific contamination seems to depend upon the particular serum more than on any other variable. In our experience, for example, rat antisera yield consistently cleaner immunoprecipitates than rabbit antisera. The disadvantage is that the rat yields much less serum than the rabbit. If antigen is available in only limited quantities, it is probably better to use it to immunoselect an extant serum than to raise more antibodies in another immunization.

Lack of Specific Immunoprecipitation

There are a number of reasons for failure to immunoprecipitate a particular *in vitro* synthesized protein. Some are obvious, such as degradation of mRNA, poor incorporation of label into protein, and proteolysis. When studying membrane or secretory proteins, however, there is the additional possibility that the antiserum raised may not react with the primary translation product. This may be especially true if the antigenic determinants are on conformationally sensitive regions that are exposed only after certain posttranslational modifications of the protein have occurred. One way of getting around this is to include dog pancreas microsomal membranes[12] in the translation reaction mixture, so that signal sequence cleavage and core glycosylation can occur. It is possible that an antibody would react with this proteolytically processed and glycosylated product, but not with its unprocessed precursor. Another way to minimize this possibility is to prepare the original antibody against SDS-dena-

[12] P. Walter and G. Blobel, this volume [6].

tured antigen. In this way the antibodies raised are less likely to have conformational sensitivity. We have never observed an antibody raised in this way to fail to react with the corresponding primary translation product.

Summary

A typical procedure for immunoprecipitating a protein (abundance ca 0.5%) synthesized in the wheat germ cell-free system is summarized below.

1. Two microliters of 25% SDS are added to 48 μl of translation reaction mixture, and the sample is heated to 100° for 4 min.
2. Four volumes (i.e., 200 μl) of dilution buffer at 4° are added to the above sample. Dilution buffer is 1.25% Triton X-100, 190 m*M* NaCl, 60 m*M* Tris-HCl, pH 7.4, 6 m*M* EDTA, 10 units of Trasylol per milliliter.
3. Five microliters of appropriate antisera are added, and the sample is incubated for at least 12 hr at 4°.
4. The sample is spun for 2 min in a microcentrifuge, and the supernatant is transferred to a fresh tube.
5. Thirty microliters of a 1 : 1 suspension of protein A–Sepharose CL-4B (15 μl of packed beads) are added, and the sample is incubated with end-over-end mixing at room temperature for 2 hr.
6. The Sepharose beads are pelleted by a 10-sec centrifugation in the microcentrifuge, and the supernatant is aspirated.
7. The beads are washed four times in 1 ml, per wash, of 0.1% Triton X-100, 0.02% SDS, 150 m*M* NaCl, 50 m*M* Tris-HCl, pH 7.5, 5 m*M* EDTA, 10 units of Trasylol per milliliter at room temperature with vortexing at each wash.
8. The beads are given a final wash with the above solution not containing detergent, and the supernatant is aspirated as completely as possible with a drawn-out Pasteur pipette.
9. Forty microliters of SDS-gel electrophoresis sample buffer containing 50 m*M* DTT are added to the beads, and the sample is heated for 4 min in a boiling water bath.
10. Free —SH groups are blocked by adding 10 μl of 1.0 *M* iodoacetamide in sample buffer and incubating for 45 min at 37°.
11. The beads are centrifuged out, and the supernatant is applied to an SDS–polyacrylamide slab gel.

[9] Use of Proteases for the Study of Membrane Insertion

By TAKASHI MORIMOTO, MONIQUE ARPIN, and SANCIA GAETANI

Introduction

Cellular membrane proteins are synthesized in either membrane-bound or free polysomes. Those made in membrane-bound polysomes are inserted into the rough endoplasmic reticulum (RER) membranes in the course of protein synthesis (cotranslational insertion), after which they either stay in the endoplasmic reticulum (ER) membranes or are transferred to the site of their function, such as Golgi apparatus and plasma membranes. Nascent polypeptides synthesized in free polysomes are discharged into the cell sap and then inserted into the appropriate subcellular organelles (posttranslational insertion).

Proteases have been used for the study of these processes for various purposes which can be tentatively divided into two parts: (1) to study the mechanism by which newly synthesized membrane proteins recognize their destined subcellular organelles; and (2) to examine whether the newly synthesized polypeptides are inserted into the subcellular membranes properly.

The use of protease for this type of investigation has been initiated by Redman and Sabatini[1] to demonstrate that newly synthesized polypeptides of secretory proteins are transferred across the ER membranes into their lumen, where they are protected from digestion by exogenously added protease. Since then, chymotrypsin and trypsin have been most widely used for the above purposes. Organelle-specific proteases, such as mitochondrial peptidases, as will be described later, are apparently important for the study of the mechanism by which cytoplasmically synthesized mitochondrial proteins are inserted posttranslationally into the site of their function. Since these studies have just begun and very little information is available at the present time, we will primarily describe here the experiments carried out using trypsin and chymotrypsin. The following order will be used: (a) overview of the process of membrane protein insertion, based on the recent review by Sabatini *et al.*[2] focusing on the aim and step(s) in which proteases would be used, together with some further aspects of their use; and (b) proteolytic conditions and comments.

[1] C. M. Redman and D. D. Sabatini, *Proc. Natl. Acad. Sci. U.S.A.* **56,** 608 (1966).
[2] D. D. Sabatini, G. Kreibich, T. Morimoto, and M. Adesnik, *J. Cell Biol.* **92,** 1 (1982).

ISBN 0-12-181996-5

Overview

I. Cotranslational Insertion of Nascent Polypeptides

A. Segregation of Membrane-Bound Polysomes from Free Polysomes

In eukaryotic cells, cellular proteins are synthesized in either membrane-bound or free polysomes.[3–7] It is now widely acknowledged that nascent polypeptides contain specific information which plays a major role in determining their sites of synthesis.[2,8–11] Therefore, the polysomes translating specific classes of mRNA are associated with ER membranes, whereas others remain free in the cytoplasm. As with secretory proteins, their primary translation products synthesized *in vitro* in the absence of microsomal membranes contain amino-terminal hydrophobic segments (referred to as "signal segments") which are not present in the final secretory products or found within the microsomal lumen.[12–17] These signal segments have been identified in the primary translation products of all eukaryotic secretory proteins so far examined,[10,11,18] with the exception of ovalbumin,[19] which appears to have an uncleaved signal.[20,21]

[3] F. S. Rolleston, *Subcell. Biochem.* **3,** 91 (1974).

[4] R. W. Hendler, *Biomembranes* **5,** 147 (1974).

[5] D. D. Sabatini and G. Kreibich, "The Enzymes of Biological Membranes" (A. Martonosi, ed.), Vol. 2, p. 531. Plenum, New York, 1976.

[6] G. C. Shore and J. R. Tata, *Biochim. Biophys. Acta* **472,** 197 (1977).

[7] G. Kreibich, S., Bar-Nun, M. Czako-Graham, U. Czichi, E. Marcantonio, M. G. Rosenfeld, and D. D. Sabatini, *in* "International Cell Biology, 1980–1981" (H. G. Schweiger, ed.), p. 579. Springer-Verlag, Berlin and New York, 1981.

[8] P. N. Campbell and G. Blobel, *FEBS Lett.* **72,** 215 (1976).

[9] G. Blobel, *in* "International Cell Biology, 1976–1977" (B. R. Brinkley and K. E. Porter, eds.), p. 318. Rockefeller, New York, 1977.

[10] G. Blobel, *in* "Gene Expression: Proceedings FEBS Meeting, 11th, 1978" (B.F.C. Clark, A. H. Klenow, and J. Zeuthen, eds.), p. 99. Pergamon, Oxford.

[11] D. F. Steiner, P. S. Quinn, S. J. Chan, J. Marsh, and H. S. Tager, *Ann. N.Y. Acad. Sci.* **343,** 1 (1980).

[12] C. Milstein, G. G. Brownlee, T. M. Harrison, and M. B. Mathews, *Nature* (*London*) *New Biol.* **239,** 117 (1972).

[13] D. Swan, H. Aviv, and P. Leder, *Proc. Natl. Acad. Sci. U.S.A.* **69,** 1967 (1972).

[14] I. Schechter, *Proc. Natl. Acad. Sci. U.S.A.* **70,** 2256 (1973).

[15] B. Kemper, J. F. Habener, R. C. Mulligan, J. J. Potts, Jr., and A. Rich, *Proc. Natl. Acad. Sci. U.S.A.* **71,** 3731 (1974).

[16] G. Blobel and B. Dobberstein, *J. Cell Biol.* **67,** 835 (1975).

[17] G. Blobel and B. Dobberstein, *J. Cell Biol.* **67,** 852 (1975).

[18] B. M. Austin, *FEBS Lett.* **103,** 308 (1979).

[19] R. D. Palmiter, D. Labrecque, J. R. Duguid, R. J. Carroll, P. S. Keim, R. L. Heinrikson, and D. F. Steiner, *Proc. Natl. Acad. Sci. U.S.A.* **75,** 1260 (1978).

[20] V. R. Lingappa, D. Shield, S. L. C. Woo, and G. Blobel, *J. Cell Biol.* **79,** 567 (1978).

[21] V. R. Lingappa, J. R. Lingappa, and G. Blobel, *Nature* (*London*) **281,** 117 (1979).

Although their amino acid composition is quite variable, all signal segments are characterized by a high proportion of hydrophobic amino acids.

Binding of ribosomes that bear nascent chains of secretory proteins containing exposed signal segments is mediated by a "signal recognition particle" (SRP), a cytoplasmic factor which has been isolated from dog pancreas,[22-24] and the "docking protein," an SRP receptor present in the ER membrane which has been isolated from dog pancreatic microsomes.[25-28] The SRP appears to interact first with the signal segment in a nascent chain emerging from an unattached ribosome and then to establish conditions for insertion of the chain into the ER membrane by virtue of its subsequent association with the docking protein. Interaction of SRP with the signal segment in the absence of microsomal membranes leads to a block in further elongation of the peptide, ensuring that the signal segment does not become embedded within a folded domain of the nascent chain. This SRP block is relieved, however, and elongation leading to cotranslational insertion takes place when translocation component membranes are added. Mild proteolytic digestion (in order to examine if the signal segment would be recognized by microsomal proteins), followed by exposure to high salt medium (to remove cleaved peptides which may otherwise remain attached surface of the membrane) abolishes the translocation capacity of dog pancreas microsomal membrane.[26]

The nascent chains of secretory proteins thus bound to the ER membranes are transferred into the ER lumen where they are protected from proteolysis by exogenously added proteases. The process, known as "vectorial discharge of nascent polypeptides,"[1] operates during the elongation phase of protein synthesis. Proteases in this case are used to confirm whether the newly synthesized polypeptides are vectorially discharged across the microsomal membranes into their lumen. The signal segment is cleaved during the early stage of this process by a membrane-associated "signal peptidase,"[16,17] which, based on its resistance to exogenous proteases, is probably located on the luminal side of the ER membrane.[29,30] Such signal that is cleaved during cotranslational insertion is defined as "transient insertion signal."[2] Thus, not only the insertion of newly synthesized polypeptides, but also the topology of the preexisting

[22] P. Walter, I. Ibrahimi, and G. Blobel, *J. Cell Biol.* **91,** 545 (1981).

[23] P. Walter and G. Blobel, *J. Cell Biol.* **91,** 551 (1981).

[24] P. Walter and G. Blobel, *J. Cell Biol.* **91,** 557 (1981).

[25] G. B. Warren and B. Dobberstein, *Nature (London)* **273,** 569 (1978).

[26] D. I. Meyer and B. Dobberstein, *J. Cell Biol.* **81,** 498 (1980).

[27] D. I. Meyer and B. Dobberstein, *J. Cell Biol.* **87,** 503 (1980).

[28] D. I. Meyer, E. Krause, and B. Dobberstein, *Nature (London)* **297,** 647 (1982).

[29] R. C. Jackson and G. Blobel, *Proc. Natl. Acad. Sci. U.S.A.* **74,** 5598 (1977).

[30] P. Walter, R. C. Jackson, M. Marcus, V. R. Lingappa, and G. Blobel, *Proc. Natl. Acad. Sci. U.S.A.* **76,** 1795 (1979).

membrane components which are involved in the membrane insertion are examined based on their resistance to exogenous proteases.

Proteins that are synthesized in membrane-bound polysomes and released into the ER lumen are not all destined for secretion. Some of these proteins may remain as permanent residents of the ER cisternae,[31] while others may become associated with the luminal face of ER membrane as peripheral proteins, as appears to be the case with calsequestrin (a Ca^{2+} binding protein within the sarcoplasmic reticulum of muscle cells[32]). Other proteins, initially vectorially discharged into the ER lumen, may be later diverted from the secretory pathway for segregation within the membrane-bounded organelles.[33] Thus, the lysosomal enzyme cathepsin D of spleen cells[34,35] and β-glucuronidase of rat preputial gland[35] have been shown to be synthesized exclusively in membrane-bound polysomes. It has also been shown by their resistance to exogenous protease that these proteins are cotranslationally inserted into the ER vesicle.

As opposed to the proteins just described, several integral membrane proteins of the ER and plasma membrane have been shown to be synthesized in the membrane-bound polysomes and to be first inserted into the RER membrane before being transferred to their destination. These include the Ca^{2+}-ATPase of the sarcoplasmic reticulum[32,36], cytochrome *P*-450[37–39] and NADPH-cytochrome *P*-450 reductase,[40,41] integral membrane proteins of the smooth and rough ER, the envelope glycoprotein of vesicular stomatitis virus (VSV),[42–44] Sindbis and Semliki Forest virus,[45,46] cel-

[31] G. Kreibich and D. D. Sabatini, *J. Cell Biol.* **61,** 789 (1974).

[32] D. C. Greenway and D. H. MacLennan, *Can. J. Biochem.* **56,** 452 (1978).

[33] M. R. Natowitz, M. M-Y. Chi, O. H. Lowry, and W. S. Sly, *Proc. Natl. Acad. Sci. U.S.A.* **76,** 4322 (1979).

[34] A. H. Erikson and G. Blobel, *J. Biol. Chem.* **254,** 11771 (1979).

[35] M. G. Rosenfeld, G. Kreibich, D. Popov, K. Kato, and D. D. Sabatini, *J. Cell Biol.* **93,** 135 (1982).

[36] T. L. Chyn, A. N. Martonosi, T. Morimoto, and D. D. Sabatini, *Proc. Natl. Acad. Sci. U.S.A.* **76,** 1241 (1979).

[37] M. Negishi, Y. Fujii-Kuriyama, Y. Tashiro, and Y. Imai, *Biochem. Biophys. Res. Commun.* **71,** 1153 (1976).

[38] Y. Fujii-Kuriyama, M. Negishi, R. Mikawa, and Y. Tashiro, *J. Cell Biol.* **81,** 510 (1979).

[39] S. Bar-Nun, G. Kreibich, M. Adesnik, L. Alterman, M. Negishi, and D. D. Sabatini, *Proc. Natl. Acad. Sci. U.S.A.* **77,** 965 (1980).

[40] Y. Okada, A. B. Frey, T. M. Gunthner, F. Oesch, D. D. Sabatini, and G. Kreibich, *Eur. J. Biochem.* **122,** 393 (1982).

[41] F. J. Gonzalez and C. B. Kasper, *Biochem. Biophys. Res. Commun.* **93,** 1254 (1980).

[42] F. N. Katz, J. E. Rothman, D. M. Knipe, and H. F. Lodish, *J. Supramol. Struct.* **7,** 353 (1977).

[43] F. N. Katz, J. E. Rothman, V. R. Lingappa, G. Blobel, and H. F. Lodish, *Proc. Natl. Acad. Sci. U.S.A.* **74,** 3278 (1977).

lular plasma membrane proteins such as Band 3,[47,48] glycophorin,[49] and the β-subunit of Na^+,K^+-ATPase.[50,51] Among these proteins, nascent polypeptides of envelope glycoprotein and glycophorin have the transient insertion signal and their primary insertion into the ER membrane appeared to be carried out in the same way as the signal present in the secretory proteins.

As with cytochrome *P*-450[39] and NADPH-cytochrome *P*-450 reductase,[40,41] the amino-terminal sequence of the primary translation product synthesized *in vitro* in the absence of microsomal membranes was found to correspond to that of the mature polypeptide. The N-terminal segment is rich in hydrophobic amino acid residues and resembles the signal sequence present in the nascent chains of secretory protein. It is likely to function as a signal for cotranslational insertion, which indicates that insertion signals for membrane proteins need not always be transient and therefore may remain in the mature protein (such signal is defined as a "permanent insertion signal," and, in this case, it is called "N-terminal permanent insertion signal"[2] based on the location of the signal in the polypeptide).

Certain membrane polypeptides, such as the Band 3 protein of the erythrocyte membrane, have their N-terminal region exposed to cytoplasmic side.[52–54] This suggests that signals for cotranslational insertion need not always be at the amino-terminal end of the polypeptide as was the case for cytochrome *P*-450. In fact, it has been recently reported that Band 3 is inserted into the RER membrane in the middle stage of translation,[48] which indicates that the insertion signal is located in the middle of the nascent chain. This type of signal is defined as an "interior permanent insertion signal,"[2] which is probably inserted into the membrane by the same mechanisms as transient and N-terminal permanent insertion signals.

[44] J. E. Rothman and H. F. Lodish, *Nature* (*London*) **269,** 775 (1977).

[45] D. F. Wirth, F. Katz, B. Small, and H. F. Lodish, *Cell* **10,** 253 (1977).

[46] H. Garoff, K. Simons, and B. Dobberstein, *J. Mol. Biol.* **124,** 587 (1978).

[47] E. Sabban, V. Marchesi, M. Adesnik, and D. D. Sabatini, *J. Cell Biol.* **91,** 637 (1981).

[48] W. A. Braell and H. A. Lodish, *Cell* **28,** 23 (1981).

[49] C. G. Gahmberg, J. Jokinen, K. K. Karhi, I. Ulmanen, L. Kaarianen,and L. C. Andersson, *Rev. Fr. Transfus. Immunohematol.* **24,** 53 (1981).

[50] J. Sherman, D. D. Sabatini, and T. Morimoto, *J. Cell Biol.* **87,** 307a (1980).

[51] D. D. Sabatini, D. Colman, E. Sabban, J. Sherman, T. Morimoto, G. Kreibich, and M. Adesnik, *Cold Spring Harbor Symp. Quantitative Biol.* **46,** 807 (1982).

[52] T. L. Steck, R. Ramos, and E. Strapazon, *Biochemistry* **15,** 1154 (1976).

[53] L. K. Drickamer, *J. Biol. Chem.* **251,** 5115 (1976).

[54] G. Guidotti, *J. Supramol. Struct.* **7,** 489 (1977).

B. Segregation of Newly Synthesized Proteins Remained Associated with the ER Membranes from Those Discharged into the ER Lumen

As described, nascent polypeptides of some membrane proteins contain transient insertion signals and are inserted into the ER membrane by an interrupted vectorial discharge process.[2] In this case, the transient signal serves as temporary anchor of the nascent chains to the membrane, but it essentially functions in facilitating rather than halting translocation across the ER membrane. Since these membrane proteins are not vectorially discharged in the ER lumen, but remain associated with the ER membrane, it would be expected that in addition to the transient insertion signal the nascent polypeptide contains a "halt transfer signal,"[2] which interrupts the vectorial discharge of the nascent chains and leads to retention of the nascent chain in the phospholipid bilayer. An examination of the amino acid sequence of membrane protein segments which are membrane-associated and resistant to proteolysis[52,53,55] and comparison of the amino acid sequences[56,57] and nucleic acid sequence[58-60] of membrane-associated and nonassociated polypeptides and estimation of the membrane-embedded region from the distribution of hydrophobic amino acid residues determined from nucleic acid sequence analysis of the cloned DNA[61-65] have shown that the signal is represented by hydrophobic segments (24 or more amino acid residues) delimited on both sides by charged amino acid residues. Protease treatment of subcellular membranes has thus been used to search for the hydrophobic membrane-associated domain which may function as the "halt transfer signal."

[55] R. J. Winzler, E. D. Harris, D. J. Pekas, C. A. Johnson, and P. Weber, *Biochemistry* **6,** 2195 (1967).

[56] M. Kehry, S. E. Wald, R. Douglas, R. Sibley, C. Raschke, W. Fambrough, and L. Hood, *Cell* **21,** 393 (1980).

[57] P. B. Williams, R. T. Kubo, and H. M. Grey, *J. Immunol.* **121,** 2435 (1978).

[58] J. Rogers, P. Early, C. Carter, K. Calame, M. Bond, L. Hood, and R. Wall, *Cell* **20,** 303 (1980).

[59] Y. Yamawaki-Kataoka, S. Nakai, T. Miyata, and T. Honjo *Proc. Natl. Acad. Sci. U.S.A.* **79,** 2623 (1982).

[60] H. L. Cheng, F. R. Blattner, L. Fitzmaurice, J. F. Musinski, and P. W. Tucker, *Nature (London)* **296,** 410 (1982).

[61] H. L. Ploegh, H. T. Orr and J. L. Stominger, *Proc. Natl. Acad. Sci. U.S.A.* **77,** 6081 (1980).

[62] A. P. Zerinilofsky, A. D. Levinson, H. E. Varmus, J. M. Bishop, E. Tischer, and H. M. Goodman, *Nature (London)* **287,** 198 (1980).

[63] J. K. Rose, W. J. Welch, M. Sefton, F. S. Esch, and N. C. Ling, *Proc. Natl. Acad. Sci. U.S.A.* **77,** 3884 (1980).

[64] H. Garoff, A. M. Frischauf, K. Simons, H. Lehrach, and H. Delius, *Nature (London)* **288,** 236 (1980).

[65] W.Min Jou,G. Threlfall, M. Verhoeyen, R. Devos, E. Saman, R. Fang, D. Huylebroeck, W. Fiers, C. Barber, N. Carey, and S. Emtage, *Cell* **19,** 638 (1980).

C. Transfer from the Site of Insertion to the Site of Function (Intracellular Transport)

Among newly synthesized polypeptides on the RER, some remain associated with the ER membrane as the permanent membrane components, such as cytochrome *P*-450[39] and NADPH-cytochrome *P*-450 reductase,[40,41] whereas others are transferred to Golgi apparatus, lysosomes, and plasma membranes.[42,43,47–51,65–69] As with plasma membrane proteins of the polarized cells, domain-specific localization of plasma membrane proteins has been demonstrated.[66–69] For example, γ-glutamyltransferase[66] and sucrase-isomaltase[67] are localized in the apical surface (membrane) and influenza, Sendai, and Simian viruses bud from the apical surface of the plasma membrane.[68] On the contrary, Na^+,K^+-ATPase is localized on the basolateral membrane[69] and VSV buds from the basolateral membrane.[68] Very little is known, however, about the mechanisms by which newly synthesized polypeptides are sorted out to their destination from Golgi apparatus. Because of this, there has been little chance to use proteases for study of this process. It should be noted here that the use of proteases is a useful technique for the study of the relationship of the orientation of newly inserted polypeptides into the lipid bilayer to that of the intracellular itinerary of particular membrane proteins, as will be described.

Proteases have been successfully used to examine whether newly synthesized secretory proteins are vectorially discharged into the microsomal lumen. Since newly formed secretory proteins are protected by limiting membranes from protease attack, the occurrence of the vectorial discharge is demonstrated by the fact that the protein from the treated microsomes has the same mobility as those from control microsomes analyzed by SDS–PAGE. This method has also been used successfully in the case of transmembrane proteins whose major portions are exposed to the luminal side of the ER. Such proteins include glycophorin,[49] G protein of VSV,[44,70] and the β-subunit of Na^+,K^+-ATPase (N. Nabi and J. Sherman, unpublished). These proteins have short peptide domain near the C-terminus which is exposed to the cytoplasmic sidc and, therefore, the insertion of the protein has been corroborated by a slight shift of the band in the SDS–PAGE when treated with a peptidase.

[66] B. Nash and S. S. Tate, *J. Biol. Chem.* **257,** 585 (1982).

[67] H-P. Hauri, H. Wacker, E. E. Rickli, B. Bigler-Meier, A. Quaroni, and G. Semenza, *J. Biol. Chem.* **257,** 4522 (1982).

[68] E. Rodriquez-Boulan, and D. D. Sabatini, *Proc. Natl. Acad. Sci. U.S.A.* **75,** 5071 (1978).

[69] J. Kyte, *J. Cell Biol.* **68,** 287 (1976).

[70] R. A. Feldman, S. Gaetani, and T. Morimoto, *in* "Cancer Cell Organelles" (F. Reid, G. M. W. Cook, and D. J. Morre, eds.), p. 263. Horwood, London, 1982.

As with transmembrane proteins such as cytochrome *P*-450,[39] Ca^{2+}-ATPase,[36,71] and the α-subunit of Na^+,K^+-ATPase,[72] which seem to span several times the lipid bilayer, no one has succeeded in demonstrating that newly synthesized polypeptides which had been cotranslationally inserted into the RER are resistant to exogenous protease. Since the method used for the above studies was immunoprecipitation following digestion of the microsomal membrane with a protease, this failure could be considered to be due to technical problems with the detection of the proteolytic fragments and/or properties of the newly inserted polypeptides with respect to their orientation to the lipid bilayer. For example, Ca^{2+}-ATPase is cleaved into three fragments when sarcoplasmic reticulum is exposed briefly to trypsin,[73] but the newly synthesized polypeptide which *in vitro* appeared to be inserted cotranslationally into the microsomes was sensitive to trypsin digestion.[71] On the other hand, the polypeptide behaved like a transmembrane protein when the microsomes were subjected to stepwise solubilization with an increasing amount of DOC[36,39] or when subjected to high pH treatment.[71] Therefore, it seems that the polypeptide is oriented in a different way from that of the *in vivo* inserted one. Another example is the α-subunit of Na^+,K^+-ATPase for which we have not been able to determine as yet whether newly synthesized polypeptides are inserted into ER co- or posttranslationally because the newly inserted polypeptides were very sensitive to exogenous proteases. Since it has been observed (J. Sherman, unpublished) that the polypeptides are transferred to the plasma membrane through the Golgi apparatus, the α-subunit may combine with the β-subunit and form an active enzyme at any step(s) of the intracellular transfer. It is, therefore, conceivable that the newly inserted peptide may have a different orientation within the lipid bilayer of the ER than it does in the membranes of other subcellular organelles. Proteases will be necessary to examine whether the newly inserted polypeptides in each subcellular organelle are sensitive to proteolysis.

The involvement of proteases in the insertion and the subsequent intracellular transport of membrane proteins includes: (a) Cleavage of the transient insertion signal by the ER membrane-associated signal peptidase, as was the case for envelope glycoproteins of various viruses.[42–46] The significance of the cleavage has not been fully understood. (b) Removal of some amino acid residues from the N-terminus of the newly synthesized polypeptide which does not contain transient insertion signal.

[71] K. E. Mostov, P. DeFoo, S. Fleischer, and G. Blobel, *Nature* (*London*) **292,** 87 (1981).

[72] P. L. Jørgensen, *in* "Membranes and Transport" (A. N. Martonosi, ed.), Vol. 1, p. 537. Plenum, New York, 1982.

[73] P. S. Stewart, D. H. MacLennan, and A. E. Shamoo, *J. Biol. Chem.* **251,** 712 (1976).

The primary translation product and the mature polypeptide of α- and β-subunits of the Na^+,K^+-ATPase are indistinguishable in their size from each other in SDS–PAGE, while the N-terminus is not initiated from methionine. This indicates that some amino acid residues from the N-terminus must be cleaved in some step(s) of insertion and/or intracellular transport. Since not all membrane proteins that are inserted in the same way have amino acid residue cleaved from their N-terminus, the occurrence of the cleavage should have biochemical meaning. The role of the cleavage in the insertion and subsequent intracellular transport as well as the location and properties of the enzyme involved and the place where it occurs are not known. (c) Cleavage of the polypeptide by organelle-specific protease(s) which may occur during the step of intracellular transfer. For example, renal γ-glutamyltransferase,[66] a plasma membrane protein which consists of two polypeptides and is synthesized as a common precursor, seems to be cleaved in the Golgi apparatus, while sucrase-isomaltase,[67] a plasma membrane protein which is also synthesized as a common precursor, is cleaved after it reaches the plasma membrane. These two proteins are very similar to each other with respect to their site of biosynthesis, intracellular pathway, and precursor form, while one seems to be cleaved at the Golgi apparatus and the other at the plasma membrane. The biochemical properties of the enzyme(s) involved and the significance of the cleavage in the intracellular transport are not known. It is also known that Golgi membrane-associated proteases are involved in the conversion of presecretory proteins.[74]

The use of proteases affords one way of studying the above processes, but it is not technically feasible for an *in vitro* system. Finding specific protease inhibitors and introducing genetic approaches are the methods that will be feasible for this kind of study in the near future.

II. Posttranslational Insertion

A. *Mitochondria and Chloroplasts, Subcellular Organelles which Have Double Limiting Membranes*

Synthesis in membrane-bound polysomes and cotranslational insertion into a membrane are not the only mechanisms by which polypeptides can be transferred across or incorporated into the membrane of eukaryotic cells. It is now evident that many proteins synthesized on free polysomes and discharged in the cell sap are subsequently taken up posttranslationally into specific organelles such as mitochondria and chloro-

[74] A. J. Kenny, *in* "Proteinases in Mammalian Cells and Tissuess" (A. J. Barrett, ed.), p. 311. North-Holland, Amsterdam, 1977.

plasts. Therefore, newly synthesized free polypeptides must in some way pass through the limiting membranes[75–78] and become resistant to proteolysis. Such polypeptides must contain structural features, which may be designated as "primary addressing signal,"[2] that serve to determine or to stabilize an interaction with specific components of the receptor membrane. We have recently identified a peptide domain within apocytochrome *c* (from residue 66 to the carboxy terminal end) which appears to contain the primary addressing signal that determines the uptake of this protein into mitochondria.[79,80] The recognition of the addressing signal by outer mitochondrial membrane proteins has been suggested by the observation that the interaction between the signal and a hypothetical receptor was completely abolished by mild treatment of intact mitochondria with trypsin. The failure of the native holocytochrome, which contains the same domain, to be recognized by a hypothetical receptor on the mitochondrial surface suggests that the acquisition of the heme leads to sequestration of the signal segment within the protein molecule and that the final folding of the polypeptide mediated by heme binding cccurs within the mitochondria.

With the exception of cytochrome *c* and the ADP–ATP translocators of the inner mitochondrial membrane,[81] other mitochondrial polypeptides of cytoplasmic origin so far studied have been found to be synthesized as larger molecular weight precursors which are proteolytically processed upon entrance into the organelles.[78,82] A similar mechanism appears to be operating in chloroplasts.[77,83] For example, it has been demonstrated that the small subunit of ribulose-1,5-bisphosphate carboxylase, which is synthesized in cytoplasmic free polysomes as a large molecular weight precursor, has a large segment at its N-terminus[83] that is removed posttranslationally. The transient segment contains 44 amino acid residues in its length, but does not contain the cluster of hydrophobic amino acid residues characteristic of a signal for cotranslational insertion.

[75] M. A. Harmey, G. Hallermayer, H. Korb, and W. Neupert, *Eur. J. Biochem.* **81,** 533 (1977).

[76] G. Hallermayer, G. Zimmerman, and W. Neupert, *Eur. J. Biochem.* **81,** 523 (1977).

[77] N. H. Chua and G. W. Schmidt, *J. Cell Biol.* **81,** 461 (1979).

[78] W. Neupert and G. Schatz, *TIBS* **6,** 1 (1981).

[79] S. Matsuura, M. Arpin, C. Hamum, E. Margoliash, D. D. Sabatini, and T. Morimoto, *Proc. Natl. Acad. Sci. U.S.A.* **78,** 4368 (1981).

[80] M. Arpin, S. Matsuura, E. Margoliash, D. D. Sabatini, and T. Morimoto, *Eur. J. Cell Biol.* **22,** 152 (1980).

[81] R. Zimmermann, U. Paluch, M. Sprinzl, and W. Neupert, *Eur. J. Biochem.* **99,** 247 (1979).

[82] G. Schatz, *FEBS Lett.* **103,** 201 (1979).

[83] G. W. Schmidt, A. Devilliers-Thiery, H. Desruisseaux, G. Blobel, and H.-N. Chua, *J. Cell Biol.* **83,** 615 (1979).

Subunits IV and V of cytochrome oxidase (an integral protein of the inner mitochondrial membrane) are synthesized in cytoplasmic free polysomes as a large molecular weight precursor that is processed upon being transferred into mitochondria.[84,85] We have observed that a polypeptide fragment of apocytochrome *c* which contains an addressing signal competed with these subunits for their transfer into mitochondria.[80] Since denatured subunit IV competed with the newly formed subunit IV for its transfer into mitochondria,[86] the structural feature to be recognized by mitochondria must be present within the denatured polypeptide. Carbamol phosphate synthetase, a matrix protein of mitochondria, is known to be synthesized in cytoplasmic free polysomes as a large molecular weight precursor. We have observed that the precursor peptide failed to be taken up by mitochondria when the transient segment was cleaved by mitochondrial peptidase before incubation with intact mitochondria (M. Arpin and C. Lusty, unpublished). These observations suggest that the presence of a transient segment may be required to maintain a newly synthesized polypeptide in a conformation capable of interacting with the organelle receptor. A conformational difference between precursor or apoform and mature or holoform membrane proteins has been presented by their different susceptibilities to exogenously added proteases. For example, newly formed apocytochrome *c* is more sensitive to trypsin treatment than holocytochrome *c* (M. Arpin, unpublished) and the unprocessed precursor form of the leucin-binding protein of *E. coli* membrane is much more sensitive to proteolytic cleavage at sites located within the body of the molecule than the mature protein.[87]

Cytoplasmically synthesized mitochondrial and chloroplast proteins could be transferred to their destination through four steps, except for their outer membrane proteins: (1) interaction between newly synthesized polypeptides and outer mitochondrial membrane (recognition process); (2) transfer of newly formed proteins across the outer (for proteins in the intermembrane space and for some inner membrane proteins) and outer and inner mitochondrial membrane (for some of inner membrane proteins and all matrix proteins) (vectorial transfer process); (3) conformational change of incorporated polypeptide induced by the cleavage of the transient segment (most of mitochondrial proteins so far examined) or by binding with heme to the holoform (posttranslational processing); and (4)

[84] K. Mihara and G. Blobel, *Proc. Natl. Acad. Sci. U.S.A.* **77,** 4160 (1980).

[85] A. S. Lewin, I. Gregor, T. L. Mason, N. Nelson, and G. Schatz, *Proc. Natl. Acad. Sci. U.S.A.* **77,** 3998 (1980).

[86] E. Schmelzer and P. C. Heinrich, *J. Biol. Chem.* **255,** 7503 (1980).

[87] D. L. Oxender, J. J. Anderson, C. J. Daniels, R. Landicks, R. P. Gunsalus, G. Zurawski, and C. Yanofsky, *Proc. Natl. Acad. Sci. U.S.A.* **77,** 2005 (1980).

binding or insertion of processed polypeptides to the site of their function (disposition process).

Proteases have been used for the study of these four processes. For example, as with the study of posttranslational transfer of cytochrome *c* into mitochondria, which we have been presently carrying out, trypsin has been used in the following ways. (1) The recognition process with respect to an interaction between newly synthesized cytochrome *c* and outer mitochondrial membrane protein was examined by the change of the capability of mitochondria to take up polypeptides when they were treated with trypsin very briefly before incubation. (2) The occurrence of the vectorial transfer of newly synthesized cytochrome *c* was examined by its resistance to an exogenous protease when mitochondria were subjected to trypsin treatment after being incubated posttranslationally. (3) The occurrence of posttranslational processing was examined by the resistance of cytochrome *c* taken up by mitochondria to trypsin digestion after being treated with borohydrate (to make the reduced form). (4) The disposition process with respect to an interaction between newly synthesized cytochrome *c* and inner mitochondrial membrane was examined by the change of the capability of the mitoplasts to bind after they were briefly treated with trypsin.

Another important aspect of membrane protein insertion can be examined through the use of specific proteases. Parts of these studies are being carried out currently in our laboratory and other laboratories as well. They are as follows.

1. Subfragmentation by use of various proteases of a CNBr fragment of apocytochrome *c* (Frag. II, residues from 66 to C-terminus) which appears to contain an addressing signal. It has been observed that Frag. II generated by CNBr treatment from horse, rabbit, and yeast apocytochrome *c* (residues from 66 to C-terminus) competed with newly synthesized rat liver cytochrome *c* for its transfer into mitochondria, though there are only a few amino acid residues within the fragment that are different from one another. Cleavage of the fragment into two, even if present together, abolished its ability to compete with newly synthesized cytochrome *c* for its transfer. These observations suggest that the hypothetical addressing signal may be present in a narrow region of the CNBr fragment and also that the secondary and tertiary structures are important and function as an addressing signal. Since the amino acid sequence of the fragment is known, it is possible to obtain various sizes of the subfragments by digestion with a specific protease chosen by its catalytic specificity to search for the essential feature for its function as an addressing signal.

2. Digestion (cleavage) by the organelle-specific proteases of the larger molecular weight precursor polypeptide of mitochondrial proteins

which had been made in cytoplasmic free polysomes. As described, observations so far obtained indicate that the transient segment present in mitochondrial precursor proteins appears to have a different function from that present in secretory proteins. It has been postulated that the segment is required to maintain a newly formed polypeptide in a conformation capable of being recognized by the organelle receptor and of facilitating vectorial transfer as well. The enzyme (organelle-specific peptidase) participating in the cleavage seems to be common to transient segments, since a newly synthesized polypeptide containing the transient segment *in vitro* was taken up and processed by mitochondria prepared from the cells which do not synthesize the protein.[88] Success in isolating the enzyme and characterizing it will lead us to study the role of the transient segment in mitochondrial protein insertion. Such success may come soon since some characterization of the enzyme has recently been achieved by McAda and Douglas.[89]

3. Protease-catalyzed formation of peptide bonds to construct various chimeric polypeptides. In the early 1900s several observations were reported that proteases catalyzed the formation of peptide bonds when hydrolyzed materials were reincubated with the same protease under conditions different from those used for hydrolysis.[Cf. 90] In recent years, the success of protease-catalyzed synthesis of physiologically active angiotensin by Isowa *et al.*,[91] human insulin by Inouye *et al.*[92] and Morihara *et al.*,[93] and bovine pancreatic RNase A by Homandberg and Laskowski,[94] has led to a reassessment of this technique, which was recently reviewed by Fruton.[95] In this reaction the specificity of a protease in hydrolysis is utilized for peptide bond formation, since the same structural features of RCO-NHR that influence the rate of hydrolytic cleavage of CO—NH bonds are also involved in the formation of the bond. This technique will provide us with another aspect of the use of proteases in the study of protein insertion into membranes, particularly posttranslational insertion of membrane proteins that can be prepared in large quantities, such as cytochrome *c* and cytochrome b_5.

As described, Frag. II generated by CNBr treatment from horse apocytochrome *c* appears to contain the primary addressing signal to

[88] M. Mori, T. Morita, S. Miura, and Tatibana, *J. Biol. Chem.* **256,** 8263 (1981).

[89] P. C. McAda and M. G. Douglas, *J. Biol. Chem.* **257,** 3177 (1982).

[90] H. Wasteney and H. Borsook, *Physiol. Rev.* **10,** 110 (1930).

[91] Y. Isowa, N. Ohmori, M. Sato, and K. Mori, *Bull. Chem. Soc. Japan* **50,** 2766 (1977).

[92] K. Inouye, K. Watanabe, K. Morihara, Y. Tochino, T. Kanaya, J. Emura, and S. Sakakibara, *J. Am. Chem. Soc.* **101,** 751 (1979).

[93] K. Morihara, T. Oka, and H. Tsuzuki, *Nature (London)* **280,** 412 (1979).

[94] A. Homandberg and M. Laskowski, Jr., *Biochemistry* **18,** 586 (1979).

[95] J. S. Fruton, *Adv. Enzymol.* **53,** 239 (1982).

mitochondria. Direct evidence for a functional role of the segment as an addressing signal may be obtained by the construction of chimeric polypeptides using this method. Since protease-catalyzed peptide formation proceeds not only in aqueous solution but also in organic solvent, this technique will allow use of a peptide-containing hydrophobic domain. Thus, it would be possible to synthesize a chimeric polypeptide from a hydrophilic segment produced by trypsin treatment of cytochrome b_5 and a fragment of apocytochrome *c* containing the hypothetical signal, and another chimeric peptide from a hydrophobic fragment of cytochrome b_5 which may contain an addressing signal to ER and a fragment of cytochrome *c* extending from residues 1 to 65 (Frag. I). These chimeric peptides would be posttranslationally incubated with microsomes and mitochondria to examine whether they are inserted properly.

Newly synthesized cytochrome *c* is inserted into inner mitochondrial membranes through at least four steps, as previously mentioned. It is therefore conceivable that structural features in the fragment other than Frag. II, for example, Frag. I, may be involved in any of the four steps, most likely in the vectorial transfer and the ensuing modification processes. If so, a chimeric peptide containing Frag. II, for example, hydrophilic segment of cytochrome b_5–Frag. II, may be recognized by the outer mitochondrial membrane, but may not be transferred across the outer membrane into the intermembrane space. This possibility can be examined using the same chimeric peptide as above, but containing Frag. I at its N-terminus.

B. Peroxisomes and Endoplasmic Reticulum: Organelles having Single Limiting Membrane

Studies of the biosynthesis of the peroxisomal enzymes urate oxidase and catalase have shown that these proteins are synthesized in free polysomes.[96–99] The completed polypeptides must also cross an organellar membrane posttranslationally but it is yet to be demonstrated if they are incorporated directly into preexisting peroxisomes or developing ones budding from the ER. The occurrence of the posttranslational modification has not been documented in these cases.

[96] B. M. Goldman and G. Blobel, *Proc. Natl. Acad. Sci. U.S.A.* **75,** 5066 (1978).

[97] G. Kreibich, S. Bar-Nun, M. Czako-Graham, W. Mok, E. Nack, Y. Okada, M. G. Rosenfeld, and D. D. Sabatini, "Biological Chemistry of Organelle Formation," p. 147. Springer-Verlag, Berlin and New York, 1980.

[98] W. M. Becker, H. Reizman, E. M. Weir, D. E. Titus, and C. J. Leaver, *Ann. N.Y. Acad. Sci.* **386,** 329 (1982).

[99] H. Desel, R. Zimmermann, M. James, F. Miller, and W. Neupert, *Ann. N.Y. Acad. Sci.* **386,** 377 (1982).

Several ER integral membrane proteins such as cytochrome b_5 and NADH cytochrome b_5 reductase which are exposed on the cytoplasmic membrane face and do not appear to have a transmembrane disposition have recently been demonstrated to be synthesized in free polysomes[40,100–102] and must therefore be inserted posttranslationally into their membrane sites of function. These proteins contain hydrophobic segments located near the carboxyl terminal end[103–105] which are likely to serve as posttranslational insertion signals responsible for the permanent association of the polypeptides with the ER membrane.

Proteases have not been successfully used for the study of this type of membrane protein insertion, except for the study of the orientation and membrane domain of mature proteins such as cytochrome b_5.[103–105] As was the case for membrane proteins which appear to span several times the lipid bilayer, newly synthesized polypeptide inserted *in vitro* posttranslationally into the ER membrane was digested by trypsin.

While the mature protein was cleaved into hydrophilic and hydrophobic peptides under the same conditions, though the *in vitro* inserted cytochrome b_5 was not released by treatment with a buffer containing high salt (500 mM KCl),[40] this difference in behavior may be related to the lack of heme in cytochrome b_5 inserted *in vitro,* which may render it more susceptible to proteolysis. Alternatively, the spatial disposition may not be the same as in the native microsomes. Carboxypeptidase is successfully used for the study of the role of hydrophobic domain of the C-terminal region in the posttranslational insertion into microsomal membranes.[105]

Thus, the use of proteases cannot easily demonstrate that newly synthesized polypeptides are posttranslationally inserted into the RER membranes, as was the case for the transmembrane proteins which seem to span the membrane several folds and those whose major domains are exposed to the cytoplasmic side. This method, however, is still useful for the study of the posttranslational insertion of proteins into peroxisomes and ER with respect to the spatial disposition in the native organelles, as was done with cytochrome b_5,[103–105] for vectorial transfer of the newly synthesized proteins across the membranes into their lumen, as was done with some glyoxysomal enzymes[98,99] (peroxisomal enzymes), and for insertion of proteins whose domains are exposed to their luminal side.

[100] R. A. Rachubinski, D. P. S. Verma, and J. J. M. Bergeron, *J. Cell Biol.* **84,** 705 (1980).
[101] N. Borgese and S. Gaetani, *FEBS Lett.* **112,** 216 (1980).
[102] F. J. Gonzales and C. B. Kasper, *Biochemistry* **19,** 1790 (1980).
[103] J. Ozoles and C. Gerald, *Proc. Natl. Acad. Sci. U.S.A.* **74,** 3725 (1977).
[104] P. J. Fleming, H. A. Dailey, D. Corcoran, and P. Strittmatter, *J. Biol. Chem.* **253,** 5369 (1978).
[105] H. A. Dailey and P. Strittmatter, *J. Biol. Chem.* **253,** 8203 (1978).

Proteolytic Conditions and Comments

As described, proteolytic treatment has been most successfully used to examine whether newly synthesized secretory proteins are vectorially discharged into microsomal lumens, whether newly synthesized membrane proteins whose major portions are exposed to microsomal lumens are inserted into microsomal membranes, and whether newly synthesized mitochondrial and chloroplast proteins are posttranslationally taken up by corresponding subcellular organelles. Therefore, this discussion is focused on the proteolytic conditions used for the above studies. Comments described here are based on published and unpublished observations obtained by our laboratory, by S. Gaetani and J. Smith, with albumin, a secretory protein, by F. Waldman and R. Feldman with G protein, a membrane glycoprotein of vesicular stomatitis virus, by S. Matsuura and M. Arpin with mitochondrial proteins, by T. Chyn with Ca^{2+}-ATPase, and by J. Sherman and N. Nabi with Na^{+},K^{+}-ATPase.

Proteases

Chymotrypsin and trypsin have been most widely used. Both are relatively stable when solubilized in 0.001 *N* HCl (stable for several weeks at 5°). Extinction coefficients of α-chymotrypsin and trypsin at 280 nm are 20.4 and 14.3, respectively. Isoelectric points of chymotrypsin and trypsin are 8.1 and 10.1, respectively. These enzymes can be used immediately after being solubilized in distilled water or solubilized in 10 m*M* Tris-HCl, pH 7.5 (10 mg/ml), and kept at −80°.[30] Diphenyl carbamylchloride (DPCC) treated trypsin and N^{α}-*p*-tosyl-L-lysine chloromethyl ketone (TLCK) treated chymotrypsin are not necessary for this purpose. The basic techniques in the enzymic hydrolysis of peptide chains have been described in this series.[106] Commercially available proteases are listed with their pH optimum, specificity, and inhibitors in the table. Further information will be found in any of the reviews featuring proteases [this series, Vols. 19, 45, and 80; "The Enzymes" (P. D. Boyer, ed.), Vol. 3, Academic Press, New York (1960); S. Blackburn, "Enzyme Structure and Function," Dekker, New York (1974); "Structure–Function Relationships of Proteolytic Enzymes" (P. Desnuelle, H. Neurath, and M. Ottesen, eds.), Academic Press, New York (1970); "Proteinases in the Mammalian Cells and Tissues" (A. J. Barret, ed.), Elsevier/North-Holland, Amsterdam (1977); K. Morihara, *Adv. Enzymol.* **41,** 179 (1974); etc.].

[106] D. G. Smyth, this series, Vol. 11, p. 214.

Commercially Available Proteases[a]

Protease	Specificity	pH optimum	Inhibitors
I. Serine proteases			
Carboxypeptidase Y from yeast	Hydrolyzes C-terminal L-amino acids from proteins and peptides	5.5–6.5	Aprotinin, PMSF, DFP, ZPCK, *p*-hydroxymercuribenzoate
α-Chymotrypsin from bovine pancreas	Hydrolyzes amide bonds of proteins and peptides at the carboxyl side of aromatic amino acid residues	7.5–8.5	Aprotinin, PMSF, DFP, ZPCK
Endoproteinase lysine-C from *Lysobacter enzymogenes*	Very specifically hydrolyzes proteins and peptides at the C-terminal side of lysine residues	7.7	Aprotinin, PMSF
Proteinase K from *Tritirachium album*	Completely hydrolyzes both native and denatured proteins	8.0–11.5	PMSF
Proline-specific endopeptidase from *Flavobacterium meningosepticum*	Specifically hydrolyzes peptide bonds on the carboxyl side of proline residues	7.8	DFP
Protease, *S. aureus* V8	Specifically hydrolyzes peptide bonds on the carboxyl side of either aspartate or glutamate residues	4.0, 7.8	DFP
Subtilisin from *Bacillus subtilis*	Nonspecifically hydrolyzes proteins and peptides	7.0–8.0	Aprotinin, DFP, PMSF
Trypsin from bovine pancreas	Hydrolyzes proteins, peptides, and amino acid esters on the carboxyl side of arginine and lysine residues	7.5–8.5	Trypsin inhibitors from soybean and hen egg white, aprotinin, leupeptin
II. Thiol peptidases			
Bromelain from *Anans sativus*	Nonspecifically hydrolyzes proteins, peptides, and amino acid esters	8.0–8.5	Hg, TLCK, TPCK
Cathepsin C from bovine spleen	Cleaves N-terminal dipeptides from polypeptides and proteins	4.0–6.0	Iodoacetate, formaldehyde

(*continued*)

COMMERCIALLY AVAILABLE PROTEASES (*continued*)

Protease	Specificity	pH optimum	Inhibitors
Clostripain from *Clostridium histolyticum*	Hydrolyzes L-arginyl bonds in proteins, peptides, and amino acid esters	7.4–7.8	EDTA
Ficin from *Ficus carica*	Nonspecifically hydrolyzes proteins and peptides	6.4	Heavy metals, DFP
Papain from *Papaya carica*	Nonspecifically hydrolyzes proteins and peptides	6.2	Heavy metals, sulfhydryl reagents, TPCK, TLCK, leupeptin
III. Metalloproteases			
Aminopeptidase M from hog kidney	Sequentially cleaves N-terminal L-amino acids from proteins and peptides	7.0–7.5	EDTA, *o*-phenanthroline (only after pretreatment)
Aminopeptidase M from mouse submaxillaris glands	Specifically hydrolyzes L-arginyl bonds in proteins and peptides	7.5–8.0	Hg, Cu, Zn, DFP
Carboxypeptidase A from bovine pancreas	Cleaves C-terminal amino acids from proteins and peptides, with exception of arginine, lysine, and proline residues	7.0–8.0	Chelating agents of Zn, β-phenylpropionic acid
Carboxypeptidase B from porcine pancreas	Cleaves C-terminal arginine, lysine, and ornithine peptide bonds	7.0–9.0	Chelating agents of Zn
Leucine aminopeptidase from porcine kidney	Sequentially cleaves N-terminal amino acids from proteins and peptides with the exception of proline and hydroxyproline	7.5–9.0	EDTA
Protease, neutral (dispase) from *Bacillus polymyxa*	Hydrolyzes proteins and peptides, preferentially splitting leucyl and phenylalanyl bonds	7.0	
Protease, nonspecific, from *Streptomyces griseus*	Nonspecifically hydrolyzes proteins and peptides	7.0	EDTA, *o*-phenanthroline

COMMERCIALLY AVAILABLE PROTEASES (*continued*)

Protease	Specificity	pH optimum	Inhibitors
Thermolysin from *Bacillus thermoproteolyticus*	Hydrolyzes peptide bonds of hydrophobic L-amino acids	7.0–9.0	EDTA
IV. Acid proteases			
Pepsin from porcine gastric mucosa	Hydrolyzes the carboxyl end of aromatic, hydrophobic, and acidic amino acids	2.3	Pepstatin A

[a] This table is based on the product characteristics of Boehringer Mannheim, the catalogs of Sigma Chemical Corp. and Miles Laboratories Inc., and references cited in the Proteases section (page 136). Abbreviations: DFP, diisopropyl fluorophosphate; PMSF, phenylmethylsulfonyl fluoride; TPCK, L-1-tosylamide-2-phenylethyl chloromethyl ketone; TLCK, N^{α}-*p*-tosyl-L-lysine chloromethyl ketone; and ZPCK, *N*-CBZ-L-phenylalanine chloromethyl ketone.

General Procedure

Proteolytic analysis of translation products is ideally carried out through the steps described below. Although many procedures have been reported to date, the ones to be described can be simplified by omitting one or more of steps 2, 3, and 4 according to the aim of the experiment.

Step 1. Translate in the presence of rough microsomes (2–5 $OD_{280\ nm}$ units/ml).

Step 2. Add a protein synthesis inhibitor or RNase (10–20 μg/ml) to inhibit further protein synthesis.

Step 3. Add tetracaine-HCl (3 m*M*, pH 7)[107] to improve the efficiency of protease protection of dog pancreatic microsomes.

Step 4. Adjust to 8–10 m*M* $CaCl_2$ to retard autolysis of trypsin.

Step 5. Incubate an aliquot of the mixture at 0–4° or 20–27° with one or two different proteases (mostly trypsin alone or trypsin and chymotrypsin), and another aliquot with the same amount of protease, but in the presence of 1% Triton X-100, as in the control.

Step 6. Add protease inhibitors or trichloroacetic acid to stop further proteolysis.

Step 7. Add detergent to solubilize organellar membranes and then boil for 2 min for the subsequent steps of the immunoprecipitation or centrifuge to separate the organelles and the supernatant, and subject

[107] G. Scheele, R. Jacoby, and T. Carne, *J. Cell Biol.* **80**, 611 (1980).

both fractions to immunoprecipitation after being solubilized with detergent, followed by boiling for 2 min.

A. *Enzyme Concentration and Incubation Conditions*

A variety of enzyme concentration and incubation conditions are used depending upon the protein studied but, even with the same protein, they vary according to the investigator. With the secretory proteins, for example, the conditions reported include 60 μg/ml each of trypsin and chymotrypsin at 0–2° for 3 hr,[16,108] 50 μg/ml each of trypsin and chymotrypsin at 0° for 1 hr,[107] 250 μg/ml each of trypsin and chymotrypsin at 4° for 1 hr,[109] and 100 μg/ml trypsin at 20° for 30 min. With G protein of VSV, the conditions reported include 500 μg/ml trypsin at 23° for 20 min,[43] 600 μg/ml each of trypsin and chymotrypsin at 27° for 1 hr,[110] 250 μg/ml each of trypsin and chymotrypsin at 0° for 1 hr,[111] 1 mg/ml trypsin at 0° for 1 hr,[48] and with band 3 of the erythrocyte plasma membrane protein, 20 μg/ml of trypsin and chymotrypsin at 0° for 20 min,[48] and 62.4 μg/ml each of trypsin and chymotrypsin at 4° for 3 hr.[47]

Concerning the incubation temperature, it is better to carry out the incubation at a lower temperature so as to avoid any artifacts which may be caused by some unknown reasons during incubation at a higher temperature. This is particularly true with membrane proteins.

The conditions used for the posttranslational transfer of cytoplasmically made proteins into mitochondria and chloroplasts are also variable: with inner mitochondrial membrane proteins, the conditions reported include 3 mg/ml mitochondria, 70 μg/ml trypsin at 4° for 30 min,[79] 3–5 mg/ml mitochondria, 250 μg/ml each of trypsin and chymotrypsin at 23° for 20 min,[112] 100 μg/ml proteinase K at 4° for 60 min.[113] With intermembrane space protein, 3–5 mg/ml mitochondria, 100 μg/ml each of trypsin and chymotrypsin at room temperature for 15 min,[114] and with chloroplast proteins 1.5–2 mg/ml chloroplasts, 160 μg/ml each of trypsin and chymotrypsin at 0° for 30 min.[115] It is important to consider the sensitivity of the

[108] G. Scheele, B. Dobberstein, and G. Blobel, *Eur. J. Biochem.* **82,** 593 (1978).

[109] D. Shields, *Proc. Natl. Acad. Sci. U.S.A.* **77,** 4074 (1980).

[110] F. Toneguzzo and H. P. Ghosh, *Proc. Natl. Acad. Sci. U.S.A.* **75,** 715 (1978).

[111] V. R. Lingappa, A. N. Katz, H. F. Lodish, and G. Blobel, *J. Biol. Chem.* **253,** 8667 (1978).

[112] M. L. Maccecchini, U. Rudin, G. Blobel, and G. Schatz, *Proc. Natl. Acad. Sci. U.S.A.* **76,** 343 (1979).

[113] R. Zimmermann, B. Hennig, and W. Neupert, *Eur. J. Biochem.* **116,** 455 (1981).

[114] M. L. Maccecchini, Y. Rudin, and G. Schatz, *J. Biol. Chem.* **254,** 7468 (1979).

[115] A. R. Grossman, S. G. Bartlett, G. W. Schmidt, J. E. Mullet, and N.-H. Chua, *J. Biol. Chem.* **257,** 1558 (1982).

newly transported proteins to proteolysis, which may help to interpret the results; that is, (a) when mitochondria were treated with trypsin at a lower concentration, for example, 20 μg/ml after the posttranslational incubation, the newly synthesized cytochrome *c* taken up by the mitochondria was digested and recovered as two or three fragments by immunoprecipitation. However, when treated at a higher concentration, for example, 70 μg/ml, no degradation of the newly transported cytochrome *c* was observed and intact cytochrome *c* was recovered quantitatively. (b) After posttranslational transfer incubation, the mitochondria were separated from the supernatant by centrifugation and then subjected to proteolysis after being suspended in a buffer containing 0.3 *M* sucrose. The newly synthesized cytochrome *c* appeared to be less sensitive to trypsin digestion than when the mitochondria were digested together with the incubation mixture.

B. Termination of Proteolysis

Proteolysis is terminated by addition of either trichloroacetic acid (at a final concentration of 10%) or protease inhibitor(s). However, it is better to use protease inhibitor(s) for the subsequent treatment of membrane proteins primarily because the use of the inhibitor has less trouble with solubilization of membrane proteins. The following precautions should be kept in mind for the use of the inhibitor(s). (a) When protease inhibitor(s) is added the mixture should be incubated at 0–4° for at least 10 min so as to block the protease activity completely, and then it is solubilized; otherwise the target protein may be artificially digested because of incomplete inhibition of the protease when subcellular organelles are solubilized. (b) The washing solution of the organelles after the proteolytic treatment should contain a reasonable amount of inhibitor.

The conditions used for the termination of proteolysis are also variable, except for soybean trypsin inhibitor which is used at a concentration five times as much as that of trypsin. With Trasylol, 1500 units/ml was used to 50 μg/ml each of trypsin and chymotrypsin,[107] 20,000 units/ml to 250 μg/ml each of trypsin and chymotrypsin,[109] 8000 units/ml to 250 μg/ml each of trypsin and chymotrypsin,[111] and 2000 units/ml to 62.4 μg/ml each of trypsin and chymotrypsin.[47] Phenylmethyl sulfonyl fluoride (PMSF) is also used by adding directly to the incubation mixture at a final concentration of 0.5 m*M* to 100 μg/ml proteinase K[113] and by diluting 100 μl mixture with 5 ml of buffer solution containing 1 m*M* PMSF, 1 m*M* benzamide-HCl, and 5 m*M* ε-amino-*n*-capric acid (100 μl incubation mixture containing 160 μg/ml each of trypsin and chymotrypsin[115]).

C. Other Factors To Be Considered

1. Handling of Subcellular Organelles

When it is necessary to isolate organelles by centrifugation from a co- or posttranslational incubation mixture before proteolytic treatment, it is better to recover them as a band on a sucrose cushion because the results were found not to be reproducible when recovered as a pellet.

2. Endogenous Protease Activity

When *in vitro* protein synthesis was carried out using a reticulocyte lysate system we frequently observed unspecific (detected as an overall decrease of hot TCA-insoluble radioactivity as the incubation continued) and specific proteolysis (detected as the degradation of a particular newly synthesized protein, for example, when *N*-acetyl blocking agent was used,[116] the newly synthesized cytochrome *c* was degraded but the newly synthesized preproalbumin was not). We have noticed that the extent of the proteolysis varied depending upon batches of the reticulocyte lysate preparation. The proteolysis could be minimized upon addition of Trasylol (200 units/ml for unspecific proteolysis) and a mixture of protease inhibitors (chymostatin, pepstatin, and L-leucylleucylleucine for specific proteolysis).[79] Proteolysis of newly synthesized cytochrome *c* was occasionally observed during posttranslational incubation and the extent of the proteolysis varied according to the mitochondrial preparation, most likely due to its intactness. This proteolysis seems to be very specific to newly synthesized cytochrome *c* (and perhaps to other mitochondrial proteins) because the proteolysis of newly synthesized preproalbumin was not observed under the same conditions. Similar proteolysis was also observed when mitochondria were solubilized in the absence of protease inhibitors other than those for trypsin and chymotrypsin. This proteolysis could be minimized upon addition of chymostatin. For the reasons described above, it is important to know the endogenous protease activity present in the translation system and that derived from subcellular organelles.

Upon using protease inhibitors for inactivating endogenous protease activity, several considerations are required: (a) the effect on the incorporation of radioactive amino acid, (b) the effect on the co- and posttranslational processings, (c) the effect on the transfer of the newly synthesized proteins into their destined organelles, and (d) the effect on subsequent treatment with exogenously added proteases. We have observed that Trasylol (200 units/ml), *o*-phenanthroline (1 m*M*), L-leucylleucylleucine (0.5 mg/ml), pepstatin (0.2 mg/ml), and chymostatin (0.1 mg/ml) have no effect

[116] R. D. Palmiter, *J. Biol. Chem.* **252,** 8781 (1977).

on the incorporation and posttranslational transfer of newly synthesized cytochrome *c* into mitochondria (occasionally *o*-phenanthroline affects the rate of incorporation slightly). PMSF (1 m*M*) has no effect on posttranslational transfer of mitochondrial proteins, but TPCK (1 m*M*) seems to have a strong effect on their transport. Soybean trypsin inhibitor (300 μg/ml) has no effect on the transport of newly synthesized cytochrome *c*. Thus, each inhibitor has its own inhibitory pattern and its effect should be examined before use.

Conditions Used for the Study of the Interaction between Newly Synthesized Proteins and Organelle Membranes

A. *Translocation Ability of Rough Microsomes*

The translocation ability of rough microsomal membranes from dog pancreas was examined by limited trypsinization.[30] A total of 78 μl of stripped rough microsome suspension (50 $OD_{280\ nm}$/ml), 250 m*M* sucrose, 50 m*M* triethanolamine-HCl, pH 7.5, 1 m*M* dithiothreitol when stripped with EDTA, 40 $OD_{280\ nm}$/ml when stripped with EDTA, followed by KCl was used. The absorbance was measured in the presence of 2% SDS, and was incubated for 30 min on ice with 2 μl trypsin solution of the appropriate concentration so that the final concentration of trypsin was 5–60 μg/ml. A concentration of 5 μg/ml was used for preparation of the supernatant fraction to be used to examine whether the translocation ability of the trypsin-treated microsomes could be restored upon addition of the fraction, and a concentration of 60 μg/ml was used to obtain almost completely inactivated but fully restorable microsomes. Stripped rough microsomes which had been treated with 60 μg/ml trypsin lost their translocation ability almost completely but the ability was restored upon addition of an increasing amount of the supernatant fraction obtained from rough microsomes treated with 5 μg/ml trypsin. The digestion was terminated by addition of 10 μl of Trasylol(100 units). Aliquots of the mixture are used directly for assay of the translocation ability in the cell-free protein synthesis or used for separation by high-speed centrifugation into a membrane and a supernatant fraction for assay of the restoration of the translocation ability by combining both trypsinized membrane and the supernatant fraction in the cell-free system (see details, Ref. 30).

The procedure has been modified so as to obtain the supernatant fraction that has the highest activity to restore the translocation activity of the trypsinized membrane.[26] In this procedure 0.5 *M* KCl was used in conjunction with trypsin treatment to remove the active component from microsomal membranes. Microsomal membranes from dog pancreas suspended in 250 mM sucrose, 50 mM KCl, 20 mM HEPES, pH 7.4, to a concentration of 50 $OD_{280\ nm}$/ml were treated at 4° for 60 min with 0.2 μg/ml (for the preparation of the supernatant fraction) and 5 μg/ml (for the preparation of membranes which lost the translocation ability, but can be restored upon addition of the supernatant fraction). After the digestion was terminated by addition of PMSF, 0.5-ml aliquots of the digest were layered onto 0.5 ml cushion of 0.5 M sucrose in 1.5 ml Eppendorf tubes and centrifuged at 40,000 rpm for 90 min in a Beckman Ti75 rotor. The pellets were suspended in 0.5 M KCl, 20 mM HEPES, pH 7.5, and resedimented in an Eppendorf tube.

B. Role of the Hydrophobic Domain of Cytochrome b5 *in Its Binding to Microsomal Membranes*

Cytochrome b_5 is known to be synthesized in free polysomes[40,100–102] and inserted into endoplasmic reticulum posttranslationally. This protein contains hydrophobic segments located near the carboxy terminal end (residues from 96 to 134[103–105]) which is likely to serve as a posttranslational insertion signal to the ER. Since carboxypeptidase Y hydrolyzes X-His-Leu very slowly and carboxypeptidase A does not hydrolyze X-Pro, three derivatives of cytochrome b_5 are prepared by treatment with carboxypeptidase Y alone (derivative 1: residues from 1 to 127) or carboxypeptidase A alone (derivative 2: residues from 1 to 115) or a combination of the two (derivative 3: residues from 1 to 106).[105] These derivatives were incubated with [^{14}C]dimyristyl phosphatidylcholine vesicles at a molar ratio of 1 to 150 (cytochrome to lipid) overnight at 30° to examine whether the carboxy-shortened derivative can bind to the vesicles (for details, see Ref. 105). It was found that derivatives 1 and 2 could bind to the synthetic lipid vesicles, but derivative 3 failed. The proteolytic conditions used for these derivative preparations were: Derivative 1: 500 μM cytochrome b_5 in 0.1 M sodium acetate, pH 5.5, at a molar ratio of 500 to 1 (cytochrome to enzyme) at 23° for 60 min. The reaction was terminated by addition of phenylmethyl sulfonyl fluoride (50 μg/ml mixture) and then the mixture was subjected to a BioGel A 1.5-m column chromatography to separate the derivative from free amino acids and enzyme. Derivative 2: 700 μM cytochrome b_5 in 0.01 M Tris-acetate, pH 8.1, an equal volume of 0.2 M N-ethylmorpholine acetate, pH 8.5, and 0.02% sodium azide were mixed and then incubated at 23° for 16 hr with carboxypeptidase A at a molar ratio of 100 to 1 (cytochrome to enzyme). After digestion the mixture was subjected to a BioGel A 1.5-m column chromatography to purify the derivative. Derivative 3: 500 μM derivative 2 in 0.1 M N-ethylmorpholine acetate, pH 8.5, and 0.02% sodium azide were digested for 30 min at 23°

with carboxypeptidase Y at a molar ratio of 100 to 1 (cytochrome to enzyme), followed by digestion with carboxypeptidase A at a molar ratio of 100 to 1 (cytochrome to enzyme) for 16 hr at 23° in the presence of 50 μg/ml phenylmethyl sulfonyl fluoride. The derivative was separated from free amino acids and enzymes by passing it through a BioGel P 60 column.

C. Ability of Outer Mitochondrial Membrane to Recognize Cytoplasmically Synthesized Mitochondrial Proteins

Newly synthesized apocytochrome *c* molecule has a domain that appears to function as an addressing signal.[79,80] Whether or not the domain is recognized by outer mitochondrial membrane proteins was examined by limited proteolysis. Intact mitochondria (6 mg/ml in 70 m*M* sucrose, 220 m*M* mannitol, 2 m*M* HEPES, pH 7.4, 1 m*M* EDTA : H medium) were incubated at 0° for 30 min with trypsin (10 μg/ml). The digestion was terminated by addition of soybean trypsin inhibitor (50 μg/ml). The mitochondria were separated from the supernatant fraction by centrifugation through a 0.5 *M* sucrose cushion containing the same salts and inhibitors and the pellet was suspended in 1 ml of H medium containing protease inhibitor and used for assay in the posttranslational mixture. Alternatively, aliquots were directly used for assay in the posttranslational incubation mixture and another aliquot which contained the same amount of soybean inhibitor was also assayed as the control.

Alternative Methods to Verify *in Vitro* Membrane Insertion

In vitro insertion of membrane proteins which seem to span the lipid bilayer several times and whose major portions are exposed to the cytoplasmic side has not been successfully demonstrated by resistance to protease treatment for some unknown reasons. This may be due to technical limitation for identification of the proteolytic products (because the amount of newly synthesized protein is very little, the antibody which can react with newly synthesized protein cannot react with proteolytic fragments, and so on). The orientation of *in vitro* inserted protein may be different from that of the *in vivo* inserted one, which becomes more susceptible to proteolysis. The properties of microsomal membranes may be altered during cotranslational incubation, which may affect slightly the cotranslational translocation taking place in the beginning of the incubation, but may affect significantly the subsequent secondary and tertiary insertion of the growing nascent chains into microsomal membrane. Therefore the inserted chain is much more susceptible than those inserted *in vivo*. Membrane proteins requiring assembly on microsomes for their proper insertion have, in addition to the problems just described, another technical problem: in a cell-free system that is commonly used for translocation assay, the number of microsomal vesicles and number of mRNA are almost the same, and therefore there is very little probability that mRNAs encoded for each assembly unit are translocated on a vesicle simultaneously, unless purified mRNAs for each component are used.

Several treatments have been used as alternative methods to verify *in vitro* membrane protein insertion based on the observation that known integral membrane proteins in human erythrocyte membrane (band 3 and glycophorin) and rat liver ER (cytochrome *P*-450, cytochrome b_5, etc.) are resistant to release from these membranes when treated under conditions which do not destroy gross membrane structure. Such include treatment with alkali, high salt, chaotropic agent, or detergent.

A. *Insertion into Microsomal Membranes*

Alkali Treatment

In vitro insertion of newly synthesized Ca^{2+}-ATPase has been demonstrated by its resistance to alkali treatment[71] because it was unexpectedly found that several discrete fragments produced by trypsin digestion of *in vivo*-inserted enzyme were not the result of integration of polypeptide into lipid bilayer. Almost identical fragments were generated when the authentic enzyme was digested in the absence or presence of nonionic detergent Triton X-100. Furthermore, when the *in vitro* product was digested with trypsin these fragments were never found, whether or not the enzyme was synthesized in the presence or absence of microsomal membranes. Thus, resistance to trypsin digestion could not be used as a criterion for membrane insertion. That integral membrane proteins are not extractable by alkali treatment (one packed volume of erythrocyte membrane and 7–9 volumes of distilled water whose pH was adjusted to 12 or 7–9 volumes of 0.1 *M* NaOH were mixed and incubated for 15 min on ice) was used as the criterion.[117]

A 100-μl translation mixture (wheat germ system) programmed with total mRNA at 25° for 180 min in the presence of microsomal vesicles from dog pancreas was adjusted to a pH of 11.5 by addition of 5 μl of 1 *M* NaOH. After 15 min on ice, the whole sample (105 μl) was layered onto a step gradient made in a Beckman air-fuge tube (80 μl, 0.2 *M* sucrose and 20 μl, 2.0 *M* sucrose, both containing the same salts as the translation mixture) and the pH was adjusted to 11.5 with 1 *M* NaOH. After centrifugation at 160,000 *g* for 15 min microsomes were collected from 0.2/2.0 *M*

[117] T. L. Steck and J. Yu, *J. Supramol. Struct.* **1,** 220 (1973).

interface and diluted to 120 μl, with the solution containing the same salts as the translation mixture, also adjusted to 11.5. By this treatment, about half of the *de novo* synthesized enzyme was integrated into dog pancreas microsomal membranes, but only when these membranes were present cotranslationally.[71]

Stepwise Solubilization with Deoxycholate (DOC)

It has been shown that when rat liver microsomes are subjected to DOC treatment at 0° for 30 min most microsomal contents are released at a concentration of 0.1 mg DOC/mg protein in 0.3 ml, while membrane components remain associated. The microsomal membranes are, however, completely solubilized at a concentration of 0.5 mg DOC/mg protein/0.3 ml.[31,118] To examine the nature of the association between newly synthesized Ca^{2+}-ATPase *in vitro* and microsomal membranes, rough microsomes which had been translated were treated with 10 m*M* EDTA and 500 m*M* KCl to remove bound polysomes and peripheral proteins. These stripped membranes were recovered by centrifugation in a 10–50% linear sucrose gradient in 10 m*M* EDTA/500 m*M* KCl, and then subjected to deoxycholate treatment at two different concentrations (0.1 and 0.5 mg DOC/mg protein/0.3 ml) for 30 min on ice. About 20% of the newly synthesized Ca^{2+}-ATPase was released at a lower DOC concentration and the rest was released at only a higher DOC concentration.[36]

Other Treatments

A high salt treatment was used to verify the posttranslational insertion of the newly synthesized cytochrome b_5 to rat liver microsomal membranes because it was found that the *in vitro* inserted protein was digested by exogenously added trypsin (a 100-μl translation mixture containing 150 μg rough microsomal membranes stripped of ribosomes was treated at 0° for 60 min with 10 μg/ml trypsin[40]) while similar treatment of native rough microsomes led to release of the hydrophilic segment, as was previously reported.[119,120] In this case 500 m*M* KCl was used.

It has been reported that peripheral proteins, but not integral proteins, of human erythrocyte membranes are preferentially released by treatment with 1 *M* NaI at a neutral pH, 7.5.[121] The condition is as follows: 1 volume of membrane suspension (4 mg/ml in 7.5 m*M* sodium phosphate, pH 7.5) is mixed with 7 volumes of an aqueous solution of NaI, pH 7.5, to give a final concentration of 1.0 *M*. The mixture is then incubated for 30

[118] G. Kreibich, P. Debby, and D. D. Sabatini, *J. Cell Biol.* **58,** 436 (1973).
[119] P. Strittmatter, *J. Biol. Chem.* **235,** 2492 (1960).
[120] T. Omura, P. Siekevitz, and G. Palade, *J. Biol. Chem.* **242,** 2389 (1967).
[121] A. Kahlenberg, *J. Biol. Chem.* **251,** 1582 (1976).

min on ice. The supernatant and pellet fractions are separated by centrifugation. Thus, this treatment seems to be very effective to differentiate integral from peripheral membrane proteins, but has not been used as yet for the examination of *in vitro* membrane insertion.

B. Transfer to Mitochondria

It was found that cytoplasmically synthesized porin, a major outer mitochondrial membrane protein of *Saccharomyces cervisiae,* remained undigested when the posttranslational incubation mixture containing mitochondria (330 μg mitochondrial protein to 200 μl incubation mixture) was treated at 4° for 60 min with 250 μg/ml trypsin.[122] Another outer mitochondrial membrane protein from rat liver, a 35-kilodalton protein, remains associated with the membrane after sonication at 0° for six 10-sec periods (MSE microsonicator operating with a peak-to-peak amplitude of 12 μm) in the presence of 1.5 *M* KCl and 0.1% DOC, followed by incubation at 25° for 30 min.[123] As these proteins appear to be located in the intermembrane-space face of the outer membrane, proteolytic treatment or other harsh methods may be used to examine their insertion into the membrane. However, in the case of those proteins that may span the membrane several times or whose major portions are exposed on the cytoplasmic surface, the proteolytic treatment may not be used to examine their insertion into the outer membrane for the same reasons as described earlier for insertion of microsomal and plasma membrane proteins, which have a similar orientation to the lipid bilayer. High salt treatment of mitochondria has been utilized as an alternative method for demonstrating that *in vitro* uptake of newly synthesized protein by isolated mitochondria is not the result of electrostatic (nonspecific) binding of protein to the mitochondrial membrane, but is due to insertion. A posttranslational incubation mixture was chilled to about 0° and salt concentration was adjusted to 1 *M* KCl after the addition of an appropriate amount of protease inhibitor(s) (see the precaution in the following paragraph). The mixture was incubated for 15 min on ice, and then the mitochondria were separated from the supernatant by centrifugation through a 0.5 *M* sucrose cushion containing the same salt concentration and protease inhibitors. This treatment has also been used to examine whether cytoplasmically synthesized inner membrane and matrix proteins achieve the appropriate localization.

Digitonin, a nonionic detergent, in combination with differential centrifugation, has been used for the isolation of outer mitochondrial mem-

[122] K. Mihara, G. Blobel, and R. Sato, *Proc. Natl. Acad. Sci. U.S.A.* **79,** 7102 (1982).
[123] G. C. Shore, F. Powever, M. Bendayan, and P. Carignan, *J. Biol. Chem.* **256,** 8761 (1981).

branes, intermembrane space enzymes, and intact mitoplasts.[124] Application of an appropriate concentration of digitonin, leading to release of most enzymes of the intermembrane space but retention of most matrix proteins in the mitochondrial pellet, has been used to demonstrate that a cytoplasmically synthesized precursor matrix protein (ornithine transcarbamylase) is transferred into the matrix where it is processed to the mature form.[88] A total of 52 μl of posttranslational incubation mixture containing rat liver mitochondria (50 μg protein) was mixed with 450 μl of 10 m*M* potassium HEPES, pH 7.4, 0.25 *M* sucrose, 0.2 m*M* each of four protease inhibitors (antipain, leupeptin, chymostatin, and pepstatin), and 0.3 mg/ml digitonin. After incubation at 0° for 15 min, the mixture was centrifuged for 1 min in an Eppendorf Microfuge (Model 5412). The sedimented mitochondria and the supernatant were analyzed. It was found that about 80% of newly synthesized ornithine transcarbamylase taken up by mitochondria remained in the mitochondrial pellet while more than 90% of adenylate kinase activity (intermembrane space enzyme) was present in the supernatant fraction.

It should be noted that protease inhibitors must be added to the incubation mixture and washing solution to protect the newly transferred proteins from proteolysis by endogenous protease(s) which is released and/or activated during the treatment. For example, chymostatin (0.1 mg/ml) was used for study of the posttranslational transfer of cytochrome *c* and cytoplasmically synthesized subunits of cytochrome oxidase[79,80] and four protease inhibitors for that of ornithine transcarbamylase.[88] The choice of the inhibitor seems to depend upon the target protein.

Concluding Remarks

The study of mechanisms by which newly synthesized proteins are properly inserted into membranes, followed by intracellular transport to their final destination or site of function, is essential to the understanding of membrane biogenesis. The use of proteases is an important technique in an analysis of early events in membrane biogenesis. However, as we have mentioned, this technique can be successfully applied only to certain types of membrane proteins. Those conditions permiting the correct *in vitro* insertion of membrane proteins which span lipid bilayers several times or whose major portions are exposed to the cytoplasmic side have not been determined and hence, have not been discussed. Although proteases have been successfully used in combination with enzymic and affinity-labeling techniques in the analysis of membrane protein orientation, these methods were not for the study of the insertion of the newly synthe-

[124] C. Schnaitman and J. W. Greenawalt, *J. Biol. Chem.* **38,** 158 (1968).

sized proteins *in vitro* so that they have not been discussed here. These include Ca^{2+}-ATPase[125–127] cytochrome b_5,[103–105,119,120,128,129] cytochrome *P*-450 reductase,[130–133] α-subunit of Na^+,K^+-ATPase,[134–136] band 3,[52,53,137] and glycophorin.[138–140]

[125] D. A. Thorley-Lawson and N. M. Green, *Eur. J. Biochem.* **40,** 403 (1973).
[126] A. Migala, B. Agostini, and W. Hasselbach, *Z. Naturforsch.* **28,** 178 (1973).
[127] A. E. Shamoo, T. E. Ryan, P. S. Stewart, and D. H. MacLennan, *J. Biol. Chem.* **251,** 4147 (1976).
[128] L. Spatz and P. Strittmatter, *Proc. Natl. Acad. Sci. U.S.A.* **68,** 1042 (1971).
[129] P. Strittmatter, M. Rogers, and L. Spatz, *J. Biol. Chem.* **247,** 7188 (1972).
[130] S. D. Black, J. S. French, C. H. Williams, and M. J. Coon, *Biochem. Biophys. Res. Commun.* **91,** 1528 (1979).
[131] J. R. Gum and H. Strobel, *J. Biol. Chem.* **254,** 4177 (1979).
[132] J. S. French, F. P. Guengerich, and M. J. Coon, *J. Biol. Chem.* **255,** 4112 (1980).
[133] S. D. Black and M. J. Coon, *J. Biol. Chem.* **257,** 5929 (1982).
[134] R. A. Farley, D. W. Goldman, and H. Bayley, *J. Biol. Chem.* **255,** 860 (1980).
[135] J. Castro and R. A. Farley, *J. Biol. Chem.* **254,** 2221 (1979).
[136] P. L. Jørgenen, S. J. D. Karlish, and C. Gitler, *J. Biol. Chem.* **257,** 7435 (1982).
[137] T. J. Mueller and M. Morrison, *J. Biol. Chem.* **252,** 6573 (1977).
[138] M. S. Bretscher, *Nature (London), New Biol.* **231,** 229 (1971).
[139] J. P. Segrest, I. Kahane, R. L. Jackson, and V. T. Marchesi, *Arch. Biochem. Biophys.* **155,** 167 (1973).
[140] M. Tomita and V. T. Marchesi, *Proc. Natl. Acad. Sci. U.S.A.* **72,** 2964 (1975).

[10] Identifying Primary Translation Products: Use of *N*-Formylmethionyl-tRNA and Prevention of NH_2-Terminal Acetylation

By RICHARD D. PALMITER

The synthesis of all proteins is initiated with methionine; however, the NH_2 terminus of only a small percentage of mature proteins is methionine. Furthermore, in those cases where methionine is the NH_2-terminal residue, it may not have been derived from initiator Met-$tRNA_f$. The NH_2 terminus of a mature protein may be derived from a precursor by the action of methionine aminopeptidase as well as other aminopeptidases, or endoproteases, such as signal peptidase (in the case of most secreted proteins), or other processing enzymes (as in the case of many peptide hormones, viral and mitochondrial proteins).

Methionine aminopeptidase acts cotranslationally; in several cases the initiator methionine has been shown to be cleaved when the nascent chain

ISBN 0-12-181996-5

RELATIONSHIP BETWEEN PRIMARY TRANSLATION PRODUCTS AND NH_2 TERMINUS[a]

Primary translation product	Possible NH_2 terminus of mature protein
Met-Arg-Leu-	Met-Arg-Leu-
	Ac-Met-Arg-Leu-
Met-Gly-Leu	Gly-Leu-
	Ac-Gly-Leu-
	Met-Gly-Leu-
	Ac-Met-Gly-Leu-

[a] For these examples, assume that methionine aminopeptidase may act if the next amino acid is glycine, but not if it is arginine; also assume that acetylation (Ac) may occur if the NH_2 terminus is methionine or glycine.

is about 20 residues long. The specificity of this enzyme is such that methionine is usually cleaved when the adjacent amino acid is small and uncharged, e.g., glycine, alanine, and valine.

The original or the newly generated NH_2 terminus may also be modified. The most common modification is acetylation. Brown[1] has estimated that about 90% of the soluble proteins of mouse L cells are acetylated and there are numerous specific examples of NH_2-terminal acetylation.[2] Acetylation usually occurs cotranslationally, but there are examples of posttranslational acetylation as well. The enzyme involved, protein acetyltransferase, uses acetyl-CoA as the acetyl donor. The specificity requirements of this enzyme are not known. The only residues known to be acetylated are alanine, glycine, serine, threonine, valine, methionine, and aspartic acid, but these residues are not always acetylated.

The table shows the relationship between hypothetical primary translation products and the resulting mature NH_2 terminus based upon the specificities of the processing enzymes described above. In most cases, sequencing the mature protcin by Edman degradation would either be impossible owing to the blocked NH_2 terminus or would not provide definitive identification of the primary translation product.

Because of polypeptide cleavage and NH_2-terminal modification, special techniques are required to identify and sequence primary translation products. The most definitive approach for identifying the primary translation product of a mRNA is to use *N*-formyl[^{35}S]Met-tRNA$_f^{Met}$ in a cell-

[1] J. L. Brown, *J. Biol. Chem.* **254,** 1447 (1979).
[2] H. Bloemendal, *Science* **197,** 127 (1977).

free translation system.[3-10] This tRNA will label only the NH_2 terminus, and the formyl group prevents the action of both the eukaryotic methionine aminopeptidase and protein acetyltransferase. After isolation of the protein, the formyl group can be removed by mild acid hydrolysis for sequencing. The techniques described below assume the availability of active mRNA, an efficient cell-free translation system, a means of isolating the translation product of interest, and radiochemical microsequencing capability.

Preparation of *N*-Formyl[^{35}S]Met-tRNA$_f^{Met}$

Isolation of tRNA

There are many satisfactory methods of isolating tRNA. Some are described in earlier volumes in this series. The basic principles involved are (*a*) isolation of total nucleic acid by phenol–chloroform extraction; (*b*) isolation of low molecular weight RNA by ultracentrifugation, column chromatography, or salt precipitation; (*c*) degradation of DNA with RNase-free DNase I; (*d*) deacylation of tRNA by incubation at alkaline pH.

One method suitable for isolation of tRNA from rabbit liver or rabbit reticulocytes is outlined below. Frozen livers are available from Pel-Freeze Biologicals. Reticulocytes are prepared as described elsewhere[11]; they have the advantage of containing no DNA. A 5% (w/v) homogenate is prepared from rabbit liver or reticulocytes in SET buffer (1× SET = 1% SDS, 5 m*M* EDTA, 10 m*M* Tris-HCl, pH 7.5) containing 50 μg of proteinase K per milliliter and incubated at 45° for 1 hr. NaCl is added to 0.1 *M* along with 1/2 volume of phenol. After brief mixing, 1/2 volume of chloroform is added, and the mixture is shaken well. The mixture is centrifuged at room temperature to separate the phases; the lower organic phase is removed and replaced with an equal volume of chloroform, shaken, and

[3] A. E. Smith and K. A. Marcker, *Nature (London)* **226,** 607 (1970).

[4] N. K. Gupta, N. K. Chatterjee, K. K. Base, S. Bhaduri, and A. Chung, *J. Mol. Biol.* **54,** 145 (1970).

[5] D. Housman, M. Jacobs-Lorena, U. L. RajBhandary, and H. F. Lodish, *Nature (London)* **227,** 913 (1970).

[6] A. R. Hunter and R. J. Jackson, *Eur. J. Biochem.* **19,** 316 (1971).

[7] B. Kemper, J. F. Habener, J. T. Potts, Jr., and A. Rich, *Biochemistry* **15,** 20 (1976).

[8] R. D. Palmiter, J. Gagnon, L. Ericsson, and K. A. Walsh, *J. Biol. Chem.* **252,** 6386 (1977).

[9] R. Zemell, Y. Burstein, and I. Schechter, *Eur. J. Biochem.* **89,** 187 (1978).

[10] S. J. Chin, E. J. Ackerman, P. S. Quinn, P. B. Sigler, and D. F. Steiner, *J. Biol. Chem.* **256,** 3271 (1981).

[11] R. J. Jackson and T. Hunt, this volume [4].

centrifuged as above. The aqueous phase is carefully removed, avoiding any material at the interface, and precipitated with 2 volumes of ethanol at −20°. The precipitate is dissolved in 10 m*M* Tris-HCl, pH 7.5 (5 ml per gram of original tissue), and an equal volume of 4 *M* LiCl is added. After incubation for 1 hr at 0°, the precipitate is removed by centrifugation (10 min at 10,000 *g*).

The supernatant, containing tRNA, 5 S RNA, and DNA, is precipitated with 2 volumes of ethanol, collected by centrifugation, and washed once in 65% ethanol, 0.1 *M* NaCl. If DNA is present (e.g., if starting with liver), the precipitate is dissolved in 10 m*M* Tris-HCl, pH 7.5, adjusted to 5 m*M* $MgCl_2$ and 0.5 m*M* $CaCl_2$; then RNase-free DNase I is added to a final concentration of 5 μg/ml and the solution is incubated at 37° for 30 min. DNase I can be freed of RNase activity by treatment with iodoacetate,[12] passage through a UMP-agarose column as described by Brisson and Chambon,[13] or incubation with proteinase K in the presence of 10 m*M* $CaCl_2$.[14] Then Tris, pH 9.5, is added to 0.1 *M*, and the solution is incubated for 1 hr at 37° to deacylate tRNA. This step is followed by phenol chloroform extraction and ethanol precipitation as described above. The RNA is dissolved in 0.5 ml of 10 m*M* Tris-HCl (pH 7.5) per gram of original tissue, and the concentration is estimated spectrophotometrically assuming that a 1 mg/ml solution has an absorption of 20 at 260 nm. Expect a yield of 0.5–1.0 mg of tRNA per gram of liver or reticulocytes.

Preparation of tRNA Acylating and Formylating Enzymes

The method is described in detail by Stanley.[15] It involves harvesting *Escherichia coli* cells in mid log-phase growth, lysing the cells by grinding a cell paste with alumina, removing debris and ribosomes by ultracentrifugation, adsorption of the enzymes to DEAE-cellulose in low salt, and batch elution in 0.2 *M* KCl.

Preparation of [^{35}S]Met-tRNA$_f^{Met}$ and N-*Formyl[^{35}S]Met-tRNA$_f^{Met}$*

Rabbit tRNA (up to 20 mg/ml) is incubated with the *E. coli* extract (7 mg of protein per milliliter) in 50 m*M* cacodylic acid (pH 7.4), 10 m*M* ATP, 1 m*M* CTP, 15 m*M* $MgCl_2$, and 50 μ*M* [^{35}S]methionine (100–500 Ci/mmol) for up to 30 min at 37°. If N-formylation is desired, include 25 μ*M* 5-formyl-5,6,7,8-tetrahydrofolate (Leucovorin, Lederle Laboratories) in the above reaction. Perform a preliminary time course to ascertain the

[12] S. B. Zimmerman and G. Sandeen, *Anal. Biochem.* **14,** 269 (1966).
[13] O. Brisson and P. Chambon, *Anal. Biochem.* **75,** 402 (1976).
[14] R. H. Tullis and H. Rubin, *Anal. Biochem.* **107,** 260 (1980).
[15] W. M. Stanley, Jr., *Anal. Biochem.* **48,** 202 (1972).

optimal time for acylation. Isolate the product by phenol–chloroform extraction at pH 5 (to prevent alkaline deacylation) and ethanol precipitation. The product can be purified further by chromatography on Sephadex G-50 (in 0.1 *M* NaOAc, pH 5) or benzoylated DEAE-cellulose (20 m*M* NaOAc with a 0.4 to 0.8 *M* NaCl gradient). Identify the radioactive tRNA peak and concentrate it by ethanol precipitation. Aminoacyl-tRNA can be stored as an ethanol precipitate in pH 5 buffer at −20°. The extent of formylation can be ascertained by electrophoresis of methionyl oligonucleotides.[15] Nonformylated Met-$tRNA^{Met}$ can be specifically deacylated by Cu^{2+} hydrolysis.[10]

Preparation of [^{35}S]Met-$tRNA_m^{Met}$ and an Alternative Procedure for Preparation of [^{35}S]Met-$tRNA_f^{Met}$

Incubate a reticulocyte lysate for 20 min at 26° under conditions for protein synthesis, except omit mRNA and include 10 μM cycloheximide and [^{35}S]methionine (100–300 μCi/ml). For a greater yield, the lysate can be fortified with additional rabbit tRNA. Isolate tRNA by phenol extraction at pH 5, ethanol precipitation, LiCl fractionation; then separate [^{35}S]Met-$tRNA_f^{Met}$ from [^{35}S]Met-$tRNA_m^{Met}$ on benzoylated DEAE-cellulose with a linear 0.4 to 0.8 *M* NaCl gradient as described[7] (see Fig. 1). The fractions of interest can be conveniently concentrated with greater than 90% yield by diluting them 1 : 1 with distilled water and applying them to a small column of benzoylated DEAE-cellulose equilibrated with 0.4 *M* NaCl, 20 m*M* NaOAc, pH 5. After the tRNA is bound, it can be eluted with a few milliliters of 1.0 *M* NaCl, 20 m*M* NaOAc, pH 5, and precipitated with ethanol.

If only labeled Met-$tRNA^{Met}$ is desired, the [^{35}S]Met-$tRNA_f^{Met}$ can be specifically deacylated by using the *E. coli* extract,[15] thus omitting benzoylated DEAE-cellulose column.

Labeling the NH_2 Terminus with *N*-Formyl[^{35}S]methionine

The experiments are generally performed under normal translation conditions and scaled up to give the desired incorporation. The major problem is competition between endogenous, unlabeled Met-$tRNA_f^{Met}$ and added *N*-formyl[^{35}S]Met-$tRNA_f^{Met}$; thus, one should add as much labeled tRNA as possible without inhibiting translation significantly. Inhibition can be tested in small pilot experiments by measuring simultaneously the incorporation of [^{35}S]methionine and a tritiated amino acid into hot acid-precipitable material. Unlabeled methionine (about 100 μM) is generally included in the reaction mixture to prevent acylation of $tRNA_m^{Met}$ with [^{35}S]methionine.

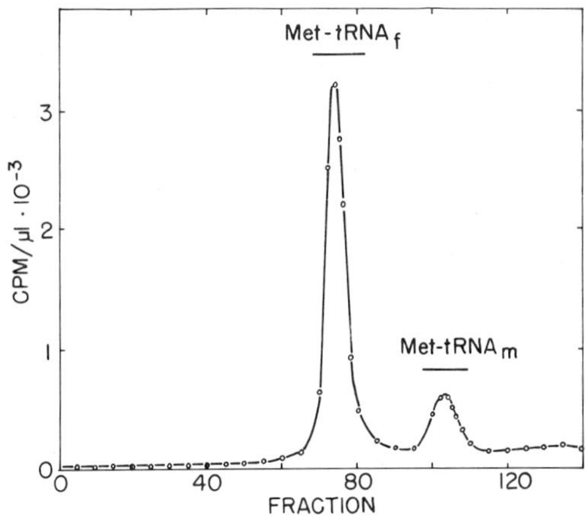

FIG. 1. Preparation of [35S]Met-tRNA$_f^{Met}$ and [35S]Met-tRNA$_m^{Met}$. A 2-ml translation reaction mixture was prepared with a nuclease- and Sephadex G-50-treated reticulocyte lysate including 200 μCi of [35S]methionine (500 Ci/mmol), 10 μM cycloheximide, and no mRNA. After a 20-min incubation at 26° to allow complete charging of methionyl-tRNAs, total RNA was isolated by phenol–chloroform extraction at pH 5, and the RNA was precipitated with ethanol. The precipitate was dissolved in 2 ml of 20 m*M* NaOAc, pH 5, and rRNA was precipitated with an equal volume of 4 *M* LiCl. The supernatant containing tRNA and 5 S RNA was precipitated with 2 volumes of ethanol, resuspended in 0.4 *M* NaCl, 8 m*M* $MgCl_2$, and 20 m*M* NaOAc, pH 5, and loaded onto a column (1 × 15 cm) of benzoylated DEAE-cellulose (Serva) equilibrated with the same buffer. The methionyl-tRNAs were eluted using a gradient of 0.4 to 0.8 *M* NaCl. The peak fractions were concentrated by diluting them 1 : 1 with distilled water and applying them to a small amount of benzoylated DEAE-cellulose in a centrifuge tube plugged with glass wool. The tRNA was eluted with 1 ml of the 0.8 *M* NaCl buffer with the aid of gentle centrifugation and precipitated with 2 volumes of ethanol. Total incorporation of [35S]methionine into aminoacyl-tRNA was 30 μCi, and the final yield of [35S]Met-tRNA$_f^{Met}$ was 7 μCi and of [35S]Met-tRNA$_m^{Met}$ was 1.6 μCi. Reprinted from Palmiter *et al.*[8]

For initial experiments the strategy will likely involve dual labeling with *N*-formyl[35S]Met-tRNA^{Met} and one tritiated amino acid followed by Edman degradation and analysis of radioactivity in sequential cycles.[16] For example, imagine that the primary translation sequence is Met-Ala-Arg-Leu and that the methionine residue is normally removed by methionine aminopeptidase. Thus, after translating the mRNA for this protein in the presence of *N*-formyl[35S]Met-tRNA^{Met} and [3H]leucine, the expected translation products would have the following NH_2-terminal sequences:

N-formyl[35S]Met-Ala-Arg-[3H]Leu. . .

and

Ala-Arg-[3H]Leu. . .

[16] S. N. Thibodeau, R. D. Palmiter, and K. A. Walsh, *J. Biol. Chem.* **253,** 9018 (1978).

After protein purification, direct Edman degradation of this mixture should give a tritium peak in cycle 3 and no ^{35}S in any cycle. After deformylation by mild acid hydrolysis (3 *M* HCl at 37° for 2 hr), Edman degradation should yield ^{35}S in the first cycle and tritium in cycles 3 and 4, the tritium ratio in these two cycles being proportional to the abundance of the two protein sequences. Acid hydrolysis can be performed on the sample before or after loading it into the spinning cup of the Sequenator. Another possibility worth considering is delaying acid hydrolysis several cycles. Thus, in the example above the tritium peaks could be clearly separated. Theoretically one could monitor only the primary translation product if one blocked the free NH_2 terminus of the Ala-Arg-Leu. . . sequence prior to deformylation, e.g., by treating the mixture with dansyl chloride,[10] phenylisocyanate,[17] acetic anhydride, or some other reagent that would be stable to mild acid hydrolysis.

Note that if the initiator methionine were not cleaved by methionine aminopeptidase, then tritium would be recovered in cycle 4 and the yield would increase after deformylation; ^{35}S would be recovered in cycle 1 only after deformylation. In this latter case, formylation of the methionine is not required to deduce the correct sequence of the primary translation product.

An experiment with *N*-formyl[^{35}S]Met-tRNA$_f^{Met}$ will provide information about the primary translation product unambiguously. Moreover, it precludes acetylation. However, it requires preparation of an acylating enzyme from *E. coli* and generally two sequenator runs—one with and one without deformylation. On the other hand, if initiator methionine is not cleaved or acetylated, then the answer can be derived with [^{35}S]Met-tRNA$_f^{Met}$, which can be readily prepared from reticulocyte lysates. In this case, a control with [^{35}S]Met-tRNA$_m^{Met}$ is often included to rule out the possibility of nonspecific loss of ^{35}S in cycle 1 of Edman degradation.[7]

Prevention of NH_2-Terminal Acetylation

The N-terminal acetyl group is derived from acetyl-CoA; thus, one strategy for prevention of acetylation is to reduce the pool of acetyl-CoA using a metabolic trap.[18] The metabolic trap takes advantage of the fact that acetyl-CoA can be effectively depleted by addition of oxaloacetate and citrate synthase to the lysate by the following reaction:

$$\text{Acetyl-CoA} + \text{oxaloacetate} \xrightarrow[\text{synthase}]{\text{citrate}} \text{citrate} + \text{CoA}$$

[17] A. Boosman, *in* "Methods in Peptide and Protein Sequence Analysis" (C. Birr, ed.), p. 513. Elsevier/North-Holland, Amsterdam, 1980.

[18] R. D. Palmiter, *J. Biol. Chem.* **252,** 8781 (1977).

A complete reaction mixture minus mRNA is prepared with the addition of 1 mM oxaloacetate and 20–30 units of citrate synthase (Sigma) per milliliter. The mixture is incubated for a few minutes at 26° to deplete the acetyl-CoA pool prior to addition of mRNA. At the end of the normal translation reaction the products are isolated for Edman degradation following usual procedures.

Some important considerations are to avoid acetate salts and to minimize the addition of $(NH_4)_2SO_4$ with the citrate synthase.[18] Citrate synthase from Sigma comes as an $(NH_4)_2SO_4$ precipitate. The simplest procedure for preparing it for translation is to place the appropriate amount of precipitated enzyme in a small microfuge tube, centrifuge, and remove as much of the supernatant as possible with a fine micropipette, then dissolve the enzyme in 10 mM Tris, pH 7.5. Because more than 20 mM $(NH_4)_2SO_4$ begins to inhibit translation, the volume of the enzyme pellet, assumed to be 4 M $(NH_4)_2SO_4$, must be diluted >200-fold in the final reaction mixture.

Rubenstein *et al.*[19] found that the competitive inhibitor acetonyl-CoA was not effective alone, but it significantly improved the yield of nonacetylated actin when it was added at a concentration of 20–100 μM along with oxaloacetate and citrate synthase. At the time of this writing acetonyl-CoA is not commercially available, but its synthesis has been described.[20]

[19] P. Rubenstein, P. Smith, J. Deuchler, and K. Redman, *J. Biol. Chem.* **256,** 8149 (1981).
[20] P. Rubenstein and R. Dryer, *J. Biol. Chem.* **255,** 7858 (1980).

[11] Selection of Lectin-Resistant Mutants of Animal Cells

By PAMELA STANLEY

The role of cell-surface carbohydrates in recognition phenomena of biological importance is currently of great interest. Biochemical studies have shown that the surfaces of a variety of animal cells contain molecules that specifically recognize particular carbohydrate moieties.[1,2] However, because of the complexity of the cell membrane, it is difficult to

[1] J. Lilien, J. Balsamo, J. McDonough, J. Hermolin, J. Cook, and R. Rutz, *in* "Surfaces of Normal and Malignant Cells" (R. O. Hynes, ed.), p. 389. Wiley, New York, 1979.
[2] E. Neufeld and G. Ashwell, *in* "Biochemistry of Glycoproteins and Proteoglycans" (W. J. Lennarz, ed.), p. 241. Plenum, New York, 1980.

ISBN 0-12-181996-5

determine the biological functions of carbohydrate–receptor interactions solely by a biochemical approach. Carbohydrate moieties of similar or identical structure are distributed among a multitude of different glycoproteins, proteoglycans, and glycolipids,[3–5] whereas specific receptor molecules may be present as only a minor proportion of total membrane components. Consequently, it is important to exploit the power of genetics in structure–function studies of cell surface molecules by isolating the appropriate mutants whenever possible. This chapter describes a general approach, which has proved to be highly successful, at isolating one family of mutants—those that express altered carbohydrate at the cell surface.

Selection of a wide range of such mutants has been made possible by the availability of a large number of different lectins. Lectins are proteins or glycoproteins of nonimmune origin that specifically recognize certain sugar conformations. The majority of the well-characterized lectins have been purified from plant seeds, although lectins have been obtained also from both invertebrates and animal tissues.[6,7] Although it is common practice to quote lectin binding specificities in terms of simple sugars (e.g., galactose, fucose, α-methylmannoside), it is now clear that this represents a gross oversimplification of the molecular structures for which any given lectin has the highest affinity. For example, concanavalin A (Con A) has been shown to bind carbohydrate moieties that contain two adjacent mannose residues, both unsubstituted at the 3, 4, and 6 positions.[8–10] However, the presence of such mannose residues does not guarantee Con A binding (Table I).[11–13] Likewise, pea lectin, which was originally thought to recognize mannose residues in a manner similar to Con A, has now been shown to require, in addition, a fucose residue linked α1,6 to the asparaginyl *N*-acetylglucosamine of complex carbohydrate moieties, for binding.[14] Evidence has also been provided that two galactose-binding lectins (ricin and L-PHA) are specific for different galactose

[3] L. Roden and M. I. Horowitz, *in* "Glycoconjugates" (M. I. Horowitz and W. Pigman, ed.), Vol. 2, p. 3. Academic Press, New York, 1978.

[4] H. Rauvala and J. Finne, *FEBS Lett.* **97,** 1 (1979).

[5] R. Kornfeld and S. Kornfeld, *in* "The Biochemistry of Glycoproteins and Proteoglycans" (W. J. Lennarz, ed.), p. 1. Plenum, New York, 1980.

[6] I. J. Goldstein and C. E. Hayes, *Adv. Carbohydr. Chem. Biochem.* **35,** 127 (1978).

[7] J. T. Powell, *Biochem. J.* **187,** 123 (1980).

[8] S. Ogata, T. Muramatsu, and A. Kobata, *J. Biochem.* (*Tokyo*) **78,** 687 (1975).

[9] T. Krusius, *FEBS Lett.* **66,** 86 (1976).

[10] S. Narasimhan, J. R. Wilson, E. Martin, and H. Schachter, *Can. J. Biochem.* **57,** 83 (1979).

[11] J. U. Baenziger and D. Fiete, *J. Biol. Chem.* **254,** 2400 (1979).

[12] N. Harpaz and H. Schachter, *J. Biol. Chem.* **255,** 4894 (1980).

[13] L. Hunt, *Biochem. J.* **205,** 623 (1982).

[14] M. L. Reitman, I. S. Trowbridge, and S. Kornfeld, *J. Biol. Chem.* **255,** 9900 (1980).

TABLE I
CARBOHYDRATE MOIETIES RECOGNIZED BY CON A[a]

Bound	Not bound

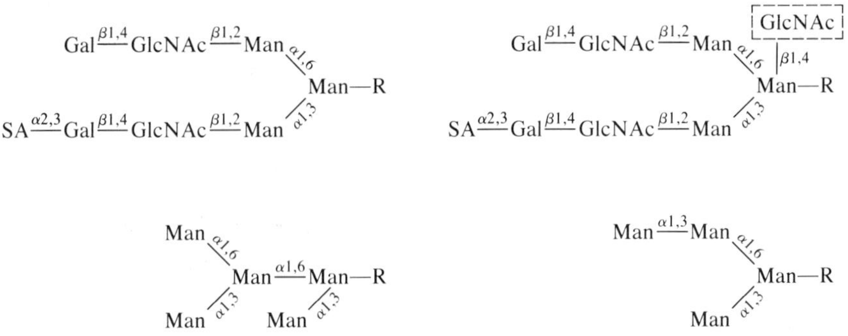

[a] SA, sialic acid; Gal, galactose; GlcNAc, *N*-acetylglucosamine; Man, mannose; R, GlcNAc $\xrightarrow{\beta 1,4}$ GlcNAc-asparagine peptide.

residues in a triantennary carbohydrate moiety containing at least three galactose-β-1,4-*N*-acetylglucosamine sequences.[15,16]

These findings suggest that all lectin sugar specificities may eventually be described by very precise carbohydrate configurations, making lectins exquisitely sensitive detectors of carbohydrate structure. It is for this reason that lectins make ideal reagents for selecting a wide range of mutants that express altered carbohydrate at the cell surface. However since lectin receptor specificities are still the subject of current research, it is not possible to predict accurately which lectins will give rise to particular carbohydrate alterations. In fact, the characterization of lectin-resistant mutants has, in a number of instances, revealed a great deal about the recognition properties of the lectins used in their selection.[14–17]

There are numerous ways in which lectin selections for cells with altered surface carbohydrate (i.e., altered lectin binding properties) might be devised. However, it happens that a number of lectins are cytotoxic. Although molecular mechanisms of toxicity are known only for ricin and abrin,[18] it is clear that the initial step in lectin toxicity is binding to cell surface carbohydrate. The most direct approach to isolating mutants that express altered surface carbohydrate, therefore, is to search for rare cells

[15] P. Stanley and T. Sudo, *Cell* **23,** 763 (1981).

[16] P. Stanley and P. H. Atkinson, *Proc. Int. Congr. Biochem., 12th, 1982* p. 343 (1982).

[17] P. Stanley, T. Sudo, and J. P. Carver, *J. Cell Biol.* **85,** 60 (1980).

[18] S. Olsnes and A. Pihl, *in* "The Molecular Actions of Toxins and Viruses" (P. Cohen and S. van Heyningen, eds.), pp. 51–105. Elsevier/North-Holland, Amsterdam, 1982.

that are resistant to toxic lectin concentrations. Using this strategy, our laboratory has isolated 15 genetically and/or phenotypically distinct cell surface carbohydrate mutants from Chinese hamster ovary (CHO) populations. [19–23] The methods that we have developed for selecting and partially characterizing lectin-resistant CHO cells form the basis of this chapter.

A number of other laboratories have also isolated lectin-resistant cell lines from a variety of different cell types.[24–27] For the most part, cytotoxic lectin selections have been performed. However, in some cases, an indirect approach has been used (e.g., binding of lectin followed by anti-lectin antibody and complement)[28] or lectin-resistant cell lines have been found to arise from selections that do not even include lectins.[29–31] The latter phenomenon is not difficult to understand since almost all lectin-resistant mutants described to date are glycosylation-defective. Therefore, any selection that enriches for a phenotype caused by altered glycosylation is likely to enrich for lectin-resistant mutants. Methods aimed at selecting specifically for glycosylation mutants, which include cytotoxic lectin selections, have been described in this series by Baker *et al.*[31]

The fact that lectin-resistant or, in a broader sense, lectin-binding mutants may be isolated not only by direct selection with cytotoxic lectins, but also by indirect selection with lectin, anti-lectin antibody, and complement, or potentially with nontoxic lectins conjugated to the A chain of ricin or diphtheria toxin[32,33] or by selection protocols that do not include the use of lectins at all, has led to a serious problem with nomen-

[19] P. Stanley, V. Caillibot, and L. Siminovitch, *Somatic Cell Genet.* **1,** 3 (1975).

[20] P. Stanley, V. Caillibot, and L. Siminovitch, *Cell* **6,** 121 (1975).

[21] P. Stanley and L. Siminovitch, *Somatic Cell Genet.* **3,** 391 (1977).

[22] P. Stanley, *Mol. Cell Biol.* **1,** 687 (1981).

[23] P. Stanley, *Somatic Cell Genet.,* in press (1983).

[24] R. M. Baker and V. Ling, *Methods Membr. Biol.* **9,** 337 (1978).

[25] P. Stanley, *in* "Biochemistry of Glycoproteins and Proteoglycans" (W. J. Lennarz, ed.), p. 161. Plenum, New York, 1980.

[26] J. A. Wright, W. H. Lewis, and C. L. J. Parfett, *Can. J. Genet. Cytol.* **22,** 443 (1980).

[27] E. B. Briles, *Int. Rev. Cytol.* **75,** 101 (1982).

[28] R. Hyman, M. Lacorbiere, S. Stavarek, and G. Nicolson, *J. Natl. Cancer Inst.* **52,** 963 (1974).

[29] I. S. Trowbridge, R. Hyman, and C. Mazauskas, *Cell* **14,** 21 (1978).

[30] C. B. Hirschberg, R. M. Baker, M. Perez, L. A. Spencer, and D. Watson, *Mol. Cell Biol.* **1,** 902 (1981).

[31] R. M. Baker, C. B. Hirschberg, W. A. O'Brien, T. E. Awerbuch, and D. Watson, this series, Vol. 83, pp. 444–458.

[32] T. Uchida, E. Mekada, and Y. Okada, *J. Biol. Chem.* **255,** 6687 (1980).

[33] D. G. Galliland, R. J. Collier, J. M. Moehring, and T. J. Moehring, *Proc. Natl. Acad. Sci. U.S.A.* **75,** 5319 (1978).

clature. This is compounded by the fact that the majority of these mutants exhibit altered interactions with one or more lectins of different sugar specificities.[19,20,22,23] The consequence is that mutants with identical phenotypes have been given different names because they were obtained from selections using different lectins, leading to confusion in the literature. For example, the CHO mutant that is characterized by the lack of a specific *N*-acetylglucosaminyltransferase activity (GlcNAc-T1) was isolated independently by this laboratory using L-PHA from *Phaseolus vulgaris* as the selective agent[19] and by Gottlieb *et al.*[34] using ricin (RIC) as the selective agent. Subsequently, we showed by complementation analysis in somatic cell hybrids, that the same phenotype could be isolated with at least four different lectins: L-PHA, RIC, WGA (the agglutinins from wheat germ), and LCA (the agglutinins from *Lens culinaris*).[20,21] We have shown by complementation analysis that the mutant isolated by Gottlieb *et al.*(15B) and one of our isolates (termed Wga^{R_I}) are affected in the same gene.[38] It would therefore seem important to define a nomenclature that may be uniformly applied to all lectin-resistant mutants. The nomenclature that this laboratory has recently devised, and that should be generally applicable to somatic cell mutants of this type, is described below.

Nomenclature

The lectin-resistant cell lines described to date arise from mutations that affect the activities of specific glycosyltransferases or glycosidases or enzymes involved in the biosynthesis of such intermediates as nucleotide-sugars or lipid-linked sugars.[24–27] In some mutants, all enzyme activities and substrates appear to be present in cell-free extracts, suggesting that altered intracellular compartmentalization might account for the observed surface carbohydrate changes.[14,35] In at least one mutant, the biochemical defect appears to affect internalization rather than altered lectin binding.[36] Although it seems likely that, in many cases, the altered enzyme activity that correlates with a particular phenotype actually reflects a mutation in the structural gene coding for that enzyme, in no case has this been proved. In fact, the term mutant as applied to lectin-resistant cells actually describes a stable phenotype that differs from the wild type. When proof of mutational sites becomes available, it will be possible to use gene symbols to refer to each mutation (e.g., Glt_1^- for GlcNAc-T1-deficient,

[34] C. Gottlieb, A. M. Skinner, and S. Kornfeld, *Proc. Natl. Acad. Sci. U.S.A.* **71,** 1078 (1974).

[35] E. B. Briles, E. Li, and S. Kornfeld, *J. Biol. Chem.* **252,** 1107 (1977).

[36] B. Ray and H. C. Wu, *Mol. Cell Biol.* **2,** 535 (1982).

lectin-resistant cells).[25] However, until that time, the broad generic symbols Lec (for recessive mutants) and LEC (for dominant mutants) are proposed to describe all animal cell mutants that exhibit at least a 2-fold increase in resistance or sensitivity to the toxicity of one or more lectins or that express significantly altered lectin-binding properties at the cell surface. This group would include mutants selected without the use of lectins that are subsequently found to be altered in their lectin-resistance or lectin-binding properties.

Since no standard nomenclature has been agreed upon for somatic cell mutants, we have adopted the general rules used by yeast geneticists.[37] Different mutant phenotypes that behave recessively in hybrids formed with parental cells will be designated Lec1, Lec2, etc., based on complementation between mutants. Mutants that behave dominantly in hybrids, will, by necessity, be assigned to different groups on the basis of a set of unique phenotypic characteristics and designated LEC10, LEC11, etc. Since an understanding of the alteration at the DNA level involved in any of the mutations affecting lectin resistance or lectin binding is lacking, a procedure for naming particular loci or alleles seems premature. Likewise, intragenic complementation has not been observed at this time and therefore has not been taken into account by the proposed nomenclature. Mutants will be routinely referred to by their Lec or LEC phenotypes. However, in discussing genotypes the symbols will be italicized (*lec* or *LEC*). Novel phenotypes that appear to fall into previously described complementation groups will be designated Lec1A, Lec1B, etc.[23] New mutants will be named according to their order of isolation, which will not necessarily reflect their frequency of occurrence. Individual cell lines will be identified by placing a period after the phenotypic designation followed by numerical and/or alphabetical symbols (e.g., $\mathrm{Pro^-5Wga^{R}{}_{I}3C}$ from our previous nomenclature becomes $\mathrm{Pro^-}$Lec1.3C). The new nomenclature as it applies to lectin-resistant mutants previously described by this laboratory is given in Table II.

Considerations Prior to Selection

The first requirement is to decide on the cell type from which mutants are to be selected. This laboratory has focused on CHO cells because they possess a pseudodiploid, stable karyotype, grow in suspension as well as substratum culture, and have a high colony-forming efficiency. In addition, they express a functional haploidy that has allowed the isolation of recessive mutants at frequencies of 10^{-5} to 10^{-7}.[39] Also, there are many

[37] F. Sherman and C. W. Lawrence, *in* "Handbook of Genetics" (R. C. King, ed.), Vol. 1, p. 359. Plenum, New York, 1974.
[38] P. Stanley, in preparation.
[39] L. Siminovitch, *Cell* **7**, 1 (1976).

TABLE II

NOMENCLATURE FOR LECTIN-RESISTANT CHINESE HAMSTER OVARY (CHO) CELLS[a]

Old nomenclature		Lectin-resistance properties[d]					New nomenclature	
Comp. Gp.	Phenotype	L-PHA (3)	WGA (2)	Con A (18)	RIC (0.005)	LCA (18)	Phenotype	Comp. Gp.
I	Pha^{R_I}, Wga^{R_I} Ric^{R_I}, Lca^{R_I}	R	R	S	R	R	Lec1	1
		>1000	30	6	100	>200		
II	Wga$^{R_{II}}$	(S)	R	—	S	S	Lec2	2
			11		100	2		
II,I	Wga$^{R_{II}}$Pha$^{R_{III}}$	R	R	S	S	R	Lec2.Lec1	1,2
		>1000	>40	6	10	>200		
III	Wga$^{R_{III}}$	(S)	R	—	S	S	Lec3	3
			5		10	2		
IV	Pha$^{R_{IV}}$	R	(R)	(S)	(S)	(S)	Lec4	4
		>1000						
V[b]	Con A^{R_I}	R	(R)	(R)	R	R	Lec5	5
		7			3	3		
I,VI	Pha^{R_I}Con A$^{R_{II}}$	R	R	R	R	R	Lec1.Lec6	1,6
		>1000	30	3	100	>200		
II,VII	Wga$^{R_{II}}$Ric$^{R_{III}}$	R	R	S	—	S	Lec2.Lec7	2,7
		5	>30	3		16		
VIII	Wga$^{R_{VIII}}$	R	R	(S)	R	S	Lec8	8
		10	100	—	2	10		
IX[b]	Ric$^{R_{IX}}$	(R)	(R)	—	R	(R)	Lec9	9
					10			
[c]	Ric$^{R_{II}}$	S	(S)	—	R	—	LEC10	*
		2			20			
[c]	Wga^{R} + Con A^{R} + Lca^{R}	R	R	—	S	R	LEC11	*
		4	8		25	3		
[c]	Wga^{R}	R	R	—	S	R	LEC12	*
		3	50		4	2		

[a] The old nomenclature is based on names published by this laboratory up to 1981.[15,17,19–22,25] The majority of these mutants are available as clones from independent selections of both Pro$^-$ and Gat$^-$ CHO populations. Phenotypic names were originally derived from the lectin used in each selection (i.e., Pha^{R_I} cells were selected with L-PHA). Those mutants which belong to two complementation groups (Comp. Gp.) were obtained following a second lectin selection of a cloned lectin-resistant cell line. By contrast, LEC11 cells were isolated from a selection using three lectins mixed in combination.[23]

[b] These mutants are temperature-sensitive for growth at 38.5°.

[c] These mutants behave dominantly for the expression of lectin resistance in somatic cell hybrids and therefore cannot be classified into complementation groups.

[d] Abbreviations: L-PHA, leukoagglutinin from *Phaseolus vulgaris;* WGA, agglutinins from *Triticum vulgaris;* Con A, lectin from *Canavalia ensiformis;* RIC, toxin from *Ricinus communis;* LCA, agglutinins from *Lens culinaris;* R, resistant compared to parental CHO cells; S, sensitive compared to parental CHO cells. —, not significantly different from parental cells. The values in parentheses directly under each lectin are the D_{10} values for parental CHO cells (in μg/ml) for that lectin. The values opposite each mutant correspond to the fold-difference ($\geq$2-fold) in D_{10} values of mutants compared with parental cells. The lectin-resistance properties of each mutant are important because they enable rapid phenotypic identification and the planning of specific selection protocols.[22,23]

genetic markers that have been studied in CHO cells.[40] However, as a transformed cell line, CHO cells express no specifically regulated or differentiative growth properties. Certain questions regarding structure–function relationships of cell surface carbohydrate must therefore be asked with other cell lines. Many of these cell types grow only in suspension or only in substratum culture. Modified selection protocols that may be useful in isolating surface carbohydrate mutants from these cell types are discussed at the end of this chapter.

Culture of CHO Cells. Two CHO clones (Pro^-5 and Gat^-2) are used as parental cell populations. Pro^-5 is a clone of a Pro^- cell population obtained from Raymond Baker in 1973. These cells originally came from W. C. Dewey and have been shown to differ in certain genetic respects from CHO-K1 cells.[41] Gat^-2 is an auxotrophic mutant that requires glycine, adenosine, and thymidine for growth. It is a Pro^+ revertant of the line AUXB1 isolated by Michael McBurney.[42] Each lectin-resistant phenotype is usually isolated from both Pro^-5 and Gat^-2 populations. In this manner it is shown that particular mutations may be obtained from independent clones. The fact that these clones carry complementary auxotrophic markers is extremely useful in selecting hybrids for complementation analysis.

Both parental lines are cultured in complete alpha medium that contains nucleosides and deoxynucleosides.[43] Any medium that fulfills the auxotrophic requirements should suffice. Fetal calf serum (FCS) at 10% v/v or horse serum (10% v/v) with 2% FCS are used to supplement the medium, but antibiotics are not usually included. Cells are routinely grown in 10-ml suspension cultures in sterile, round-bottom, plastic tissue culture tubes that are rotated on a roller drum (164-tube capacity) at approximately 50 rpm. Larger tubes may also be accommodated so that 30–40-ml cultures are obtained. For even larger cultures, spinner flasks of 100 ml or greater capacity are employed. Cells are inoculated at $(2–3) \times 10^4$ cells/ml and allowed to grow to a density of 10^6 cells/ml. They are used for experiments between 2×10^5 and 8×10^5 cells/ml when they are known to be in exponential growth. Under these conditions, the plating efficiency (i.e., ability of single cells to form colonies) is 70–90%, and generation times are 12–16 hr.

Cells may be passaged from monolayer to suspension and vice versa in the same culture medium. If cells are not growing well in suspension culture, often they can be rejuvenated by passage in monolayer under less

[40] P. N. Ray and L. Siminovitch, "Somatic Cell Genetics" (C. T. Caskey, ed.), NATO ASI Series, Ser. A, Life Sciences, Vol. 50, pp. 127–168. Plenum, New York, 1982.

[41] R. S. Gupta and L. Siminovitch, *Somatic Cell Genet.* **4,** 715 (1978).

[42] M. W. McBurney and C. F. Whitmore, *Cell* **2,** 173 (1974).

[43] C. P. Stanners, G. L. Elicieri, and H. Green, *Nature (London), New Biol.* **230,** 52 (1971).

stressful culture conditions. In doing this care must be taken not to select a minor cell population that may differ from the original cell line. Trypsin (0.25% w/v) in citrate saline pH 7.2 (0.015 *M* sodium citrate, 0.134 *M* potassium chloride) releases CHO cells from plastic after a few minutes at room temperature. Cells are routinely counted in a particle data counter.

To avoid the accumulation of mutants in parental cultures, they are carried no longer than 4–5 consecutive months from the original cloning. Monolayer cultures remain 50% viable if stored for up to 2 weeks in the dark at room temperature. Suspension cultures may be stored for a similar period at 4° with good retention of cell viability. In the cold, CHO cells do not adhere to glass or plastic culture vessels. For long-term storage, cells are frozen at 3×10^6 to 10^7 cells/ml in culture medium containing 10% dimethyl sulfoxide (DMSO) and placed at −70°. Representatives of important cell lines are subsequently transferred to liquid nitrogen. Cultures are routinely monitored for mycoplasma contamination soon after thawing. Mouse 3T6 cells obtained from G. J. McGarrity at the Institute for Medical Research in New Jersey and known to be mycoplasma-free are used as an indicator line. After cocultivation with CHO cells in antibiotic-free medium, they are stained with Hoechst 33258 and examined by fluorescence microscopy.[44,45]

Mutagenesis. Although many of our mutants have been selected from nonmutagenized cell populations, often it is desirable to search for mutants after mutagenesis or to examine the effects of mutagens on mutation frequency. Since, in our experience, different mutagens may give rise to different numbers of lectin-resistant mutants, we have routinely used three chemical mutagens—ethylmethane sulfonate (EMS), *N*-methyl-*N*-nitrosoguanidine (MNNG), and ICR-191. The mutagenesis protocol is similar in all cases except for the time of exposure to mutagen and the concentration of mutagen used. The latter is chosen from survival curves of parental lines plated in increasing concentrations of mutagen. A mutagen concentration that gives 50% to 80% cell survival is chosen. Mutagenesis resulting in a much lower survival rate is avoided because of the increased likelihood of obtaining mutants that carry numerous mutations.

Mutagens are stored at 4° in an airtight tin. They are handled with great care and disposed of after chemical inactivation. For CHO cells, EMS is routinely used at 100–200 μg/ml. A 10 mg/ml solution of EMS is made by adding 81 μl of EMS to 10 ml of medium. This mixture is vortexed for a full minute, and an aliquot is taken immediately for addition to the cell culture. The other mutagens are used at much lower concentrations—0.5 μg/ml for ICR-191 and 0.02–0.04 μg/ml for MNNG.

[44] T. R. Chen, *Exp. Cell Res.* **104,** 255 (1977).

[45] G. McGarrity, J. Sarama, and V. Vanaman, *In Vitro* **15,** 73 (1979).

Both these mutagens are weighed from the solid form and dissolved in distilled water (ICR-191) or DMSO (MNNG) before being added to the cell cultures. MNNG in DMSO may be stored frozen at −20°. The volume added to cells in culture is always 1% or less of the total volume of the culture.

The general protocol is to expose cells (at least 2×10^7) in suspension culture at approximately 2×10^5 cells/ml. Prior to the addition of mutagen, an aliquot of cells is diluted to 100 cells/ml and added to two 60-mm tissue culture plates containing 5 ml of medium for the determination of the plating efficiency of the untreated culture. Cells and mutagen are incubated in suspension culture at the appropriate temperature for 18 hr (EMS and ICR-191) or 2 hr (MNNG). The cells are centrifuged and washed once to remove mutagen. An aliquot is taken for plating at 100 and 1000 cells per duplicate 60-mm tissue culture dish. This is for the determination of cell survival following mutagen treatment. The mutagenized cells are returned to suspension culture and allowed to undergo approximately five doublings before being subjected to a lectin selection. Since expression times may differ for the optimal survival of different mutants, it may be wise to vary this time between 2 and 10 days. The time chosen is aimed at a compromise between allowing mutations to become fixed and mutant cell populations to expand, and ensuring that those mutants with slower generation times are not overgrown by parental cells.

Lectins. Many lectins are now commercially available from a variety of companies. For our selections we have used the following five lectin preparations: PHA (from *P. vulgaris*), the purified phytohemagglutinin (L-PHA) from Burroughs Wellcome; WGA (from *Triticum vulgaris*), the mixture of agglutinins from Sigma Chemical Company; Con A (from *Canavalia ensiformis*) from Pharmacia Fine Chemical Company; RIC (the toxin from *Ricinus communis*) prepared by affinity chromatography based on the method of Nicolson *et al.*[46]; and LCA, (the agglutinins from *Lens culinaris*) prepared by affinity chromatography based on the method of Sage *et al.*[47] These lectins have been shown to be comparatively pure (i.e., approximately 90–95%) by sodium dodecyl sulfate (SDS)–gel electrophoresis. However, every preparation contains visible contaminating bands. In addition, it is known that the WGA preparation contains three isolectins,[48,49] and that the LCA preparation is also a mixture of agglutinins.[47]

[46] G. L. Nicolson, J. Blaustein, and M. E. Etzler, *Biochemistry* **13,** 196 (1974).

[47] I. K. Howard, H. J. Sage, M. D. Stein, N. M. Young, M. A. Leon, and D. F. Dyckes, *J. Biol. Chem.* **246,** 1590 (1971).

[48] A. K. Allen, A. Neuberger, and N. Sharon, *Biochem. J.* **131,** 155 (1973).

[49] R. H. Rice and M. E. Etzler, *Biochemistry* **14,** 4093 (1975).

It seems likely that the toxic molecule(s) in each preparation is at least one of the lectin molecules. In support of this is the fact that mutants isolated from lectin selections almost always exhibit surface carbohydrate changes consistent with the loss of sugars that would be expected to be part of the selective lectin's receptor specificity. Also, more purified lectin preparations from other sources are not markedly altered in toxicity. In fact, purified L_4 from Felsted[50] was 2- to 3-fold less toxic for CHO cells than L-PHA from Burroughs Wellcome (P. Stanley, unpublished observations). Although this result might reflect the presence of a previously uncharacterized cytotoxin in Burroughs Wellcome L-PHA preparations, it is more likely that the toxic properties of different preparations vary somewhat depending on their state. Most lectins are prepared by affinity chromatography and eluted with high concentrations of the sugar for which they are most specific. Failure to remove sugar completely might account for some variation in toxicity. These considerations notwithstanding, we have found the lectins used in our experiments to be of remarkably uniform toxicity from batch to batch over many years. However, to maintain reproducibility, it is advisable to start with a large quantity of well-characterized lectin to be used in mutant selections and subsequent mutant characterization. Given a choice, the most toxic lectin preparation available should be used in selections. This is important because, at high lectin concentrations, a precipitate may be formed with serum glycoproteins. Also, if the selective lectin concentration is initially high, it may be difficult to determine the degree of resistance of putative mutants to even higher concentrations of lectin.

Lectins are stored at 4° in solution at 2 mg/ml in prerinsed, autoclaved glass vials with screw caps. They are dissolved in deionized, distilled water or phosphate-buffered saline, pH 7.2, depending on their solubility properties. For example, peanut agglutinin must be stored at room temperature since, when pure, it precipitates at 4°.[51] Nonsterile solutions are filtered through 0.22-μm filters, and aliquots are taken for measurements of protein concentration and SDS–gel electrophoresis. In cases where an extinction coefficient is not known, protein concentration is determined by a modification[52] of the method of Lowry *et al.* using bovine serum albumin as a standard. The specificity of new lectins is checked by the appropriate sugar inhibition assay of red blood cell agglutination.[6]

Sera. The fact that lectin preparations may vary somewhat in com-

[50] R. L. Felsted, R. D. Leavitt, C. Chen, N. R. Bachur, and R. M. K. Dale, *Biochim. Biophys. Acta* **668,** 132 (1981).

[51] M. Decastel, R. Bourrillon, and J.-P. Frenoy, *J. Biol. Chem.* **256,** 9003 (1981).

[52] M. A. K. Markwell, S. M. Haas, L. L. Bieber, and N. E. Tolbert, *Anal. Biochem.* **87,** 206 (1978).

TABLE III
EFFECTS OF DIFFERENT SERA ON LECTIN TOXICITY[a]

Lectin	Cell line	Lectin resistance (μg/ml)		
		FCS	HS/FCS	CS
Con A	Gat⁻	25–30	>50	>50
	Lec1	3	20	3
LCA	Gat⁻	20–25	50	50
	Lec2	15	15	25
	Lec8	3–5	10	3
WGA	Gat⁻	3–5	<1	<1
	Lec2	25–30	20	5–10
	Lec8	>150	75	75

[a] Lectin toxicities were determined by titration of cells against lectins dissolved in alpha medium supplemented with 10% fetal calf serum (FCS), 10% horse serum and 2% FCS (HS/FCS) or 10% γ-globulin-free calf serum (CS) in 96-well Microtiter dishes. The end point of the titration was determined by the concentration of lectin (μg/ml) at which cell growth was ≤10% that in control wells. All mutant cells carry the Gat⁻ auxotrophic marker.

position and toxicity is further complicated by the fact that different sera affect lectin toxicity in different ways. For example, WGA becomes approximately 3-fold more toxic in horse serum compared with FCS whereas the toxicities of Con A and LCA are reduced more than 2-fold under the same conditions (Table III). By contrast, L-PHA and RIC exhibit essentially identical toxicities in FCS or horse serum (data not shown). γ-Globulin free calf serum behaves very much like horse serum. Although the fold differences between parental and mutant types are similar in different sera, there is the problem of determining maximum degrees of resistance for both Con A and LCA. For this reason, we perform cell lectin selections and toxicity testing in alpha medium containing 10% FCS. Results from different batches of non-heat-inactivated FCS have been found to be extremely reproducible.

Lectin Toxicity: The Survival Curve. To determine which concentration of lectin is to be used in a selection, two experiments are performed. The first provides semiquantitative information on the concentration range over which lectin toxicity occurs, and the second provides quantitative data for a full survival curve.

In the first test, cells are plated in the presence of increasing lectin concentrations in the wells of a 96-well Microtiter culture dish. The lectin dilutions are added in 100 μl of alpha medium 10% FCS, and the cells (2×10^3 per well) are subsequently added in 100 μl of alpha medium 10% FCS. After 4 days in a humidified CO_2 incubator at the temperature to be used in selection, cell growth in wells containing lectin is compared with the growth in those that do not contain lectin. Confluency is determined by examination of each well by inverted microscopy. When wells containing no lectin are confluent, the medium is removed and the plates are stained with methylene blue (0.2% w/v in 50% methanol). The concentration of lectin at which $\leq$10% growth occurs is taken as the end point of the titration, which determines the range of lectin concentrations to be used in generating a survival curve. An example of a stained 96-well plate is given in Fig. 3. The results in Table III were obtained from this type of test.

The survival curve is generated by plating various numbers of cells in increasing lectin concentrations and counting surviving colonies after 8–10 days of undisturbed incubation. A typical result is shown in Fig. 1. To obtain such a curve, surviving colonies must be observed at all lectin concentrations so that relative plating efficiencies may be calculated. This is achieved by plating a range of cell numbers at each lectin concentration. The curves shown in Fig. 1 are comparatively ideal in the sense that both parental populations behave similarly and a distinct plateau occurs at a relative plating efficiency of approximately 10^{-5} for unmutagenized cells and 10^{-3} for EMS-mutagenized cells. Both plateaus are indicative of a subpopulation of PHA-resistant cells. In fact, colonies picked from plates over the plateau regions of the curves were found to be highly PHA-resistant and to retain this phenotype under nonselective conditions. This type of survival curve has been discussed in detail by Thompson and Baker.[53] Colonies that arise on the steep portion of the curve are often found not to breed true for resistance to the selective agent when cultured under nonselective conditions, suggesting that their initial survival was due to statistical chance rather than mutation.

Survival curves are essential in order to determine the minimal lectin concentration necessary to eliminate all parental cells. In curves with a marked plateau region, the minimal lectin concentration to achieve maximum killing is easy to determine. However, some survival curves do not have a clear plateau. In these cases, we have found that the lowest lectin concentration giving survival of 10^{-5} is appropriate for selection. Since lectin-resistant CHO cells exhibit differing degrees of resistance to different lectins (Table II), it is important to know the full range of lectin concentrations over which mutants might be selected.

[53] L. H. Thompson and R. M. Baker, *Methods Cell Biol.* **6,** 209 (1973).

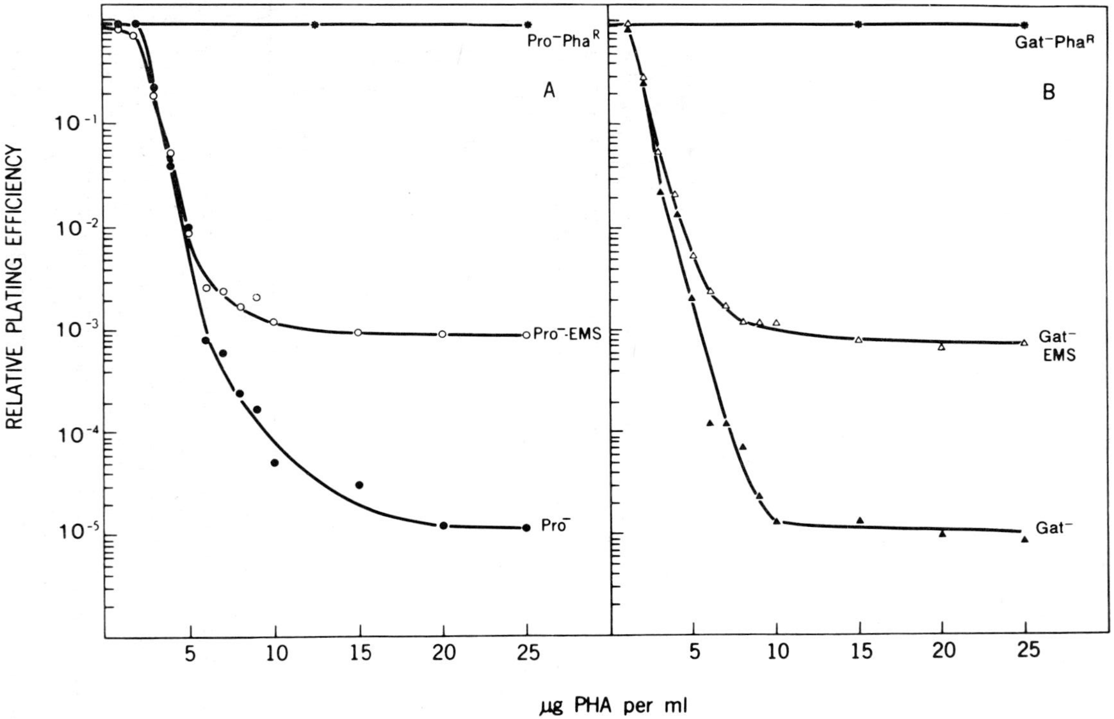

FIG. 1. Survival of mutagenized and unmutagenized Chinese hamster ovary (CHO) cells plated in the presence of phytohemagglutinin (PHA). Suspensions of exponentially growing Pro⁻ cells (A) and Gat⁻ cells (B) were divided, and half were treated with ethylmethane sulfonate (EMS) (150 μg/ml). After 18 hr, the cells were washed, resuspended in fresh medium, and returned to suspension culture. The survival of the Pro⁻-EMS cells was 75% that of the control population, and the survival of the Gat⁻-EMS cells was 50% that of the control. After a further 3 days, the mutagenized and unmutagenized cells were plated in various concentrations of PHA and incubated for 8 days (A) or 9 days (B) at 37°. The relative plating efficiencies of typical colonies that were selected from nonmutagenized cultures at PHA concentrations of 12 μg/ml (Pro⁻) and 15 μg/ml (Gat⁻), grown in the absence of PHA, and then replated in the drug are also shown (*).

Selection Protocols

From the properties of the lectin-resistant CHO cells described in Table II, two general rules regarding lectin selections emerge: (*a*) in using a single lectin to select mutants, many different mutant types might be expected among the survivors; i.e., a *single* lectin may select for a *variety* of lectin-resistant phenotypes; and (*b*) lectins of different sugar specificity may select for identical phenotypes; i.e., *different* lectins may select for *identical* lectin-resistant phenotypes. The latter observation follows directly from the fact that lectin-resistant cells almost always exhibit characteristic cross-resistances and/or hypersensitivities to two or more lectins of different sugar specificity. This property of lectin-resistant cells is extremely important to the design of selection protocols.

Many different categories of survivors may be sought from a lectin selection. From parental cells, a range of mutants resistant to different lectin concentrations may be obtained. From mutants already resistant to one or more lectins, revertants or double mutants may be selected. The hypersensitivity of most lectin-resistant mutants to one or more lectins provides an ideal approach to selecting revertants[54] or double mutants (Table II). Cells carrying multiple mutations are excellent material for studying genetic and biochemical relationships among mutations affecting glycosylation pathways. Revertants are also very useful in this way, as well as in providing evidence that a point mutation was probably responsible for the original mutant phenotype.

Three simple lectin selections are routinely used in this laboratory: (*a*) selections using a single lectin; (*b*) selections using combined lectins; and (*c*) selections using lectins added sequentially. To date only cytotoxic lectins have been employed. Cells and lectin are mixed in alpha medium containing 10% FCS on tissue culture plates and incubated until colonies 1–2 mm in diameter become visible (8–10 days). The cell population is exposed to lectin only during the incubation period. This is therefore called a single-step selection protocol. In fact, all our routine protocols (including sequential lectin selections), are single-step selections in that the mutants isolated have knowingly been subjected only to one selection pressure. The strategies developed for these selections are generally applicable to the use of lectin–toxin conjugates or, with minor modifications, to selections using anti-lectin antibody and complement.

Single Lectin Selections. Selections using a single cytotoxic lectin give rise to cells that are resistant to that lectin at the selective concentration and often at higher lectin concentrations as well. In a typical single-lectin selection, 10–12 ml of alpha medium, 10% FCS, is added to a 100-mm

[54] J. Finne, M. M. Burger, and J.-P. Prieels, *J. Cell Biol.* **92,** 277 (1982).

tissue culture dish, followed by 1 ml of the appropriate lectin concentration and 1 ml of cells (10^5 or 10^6) in alpha medium, 10% FCS. The number of cells added per plate depends on the frequency of mutants expected and on the effects of cell density on lectin toxicity. The toxic effects of many lectins are reduced at high cell density, giving rise to a significant background of survivors that do not breed true for lectin resistance. In cases of severe cell density effects, it may be more convenient to use 150-mm plates.

Cells should be evenly dispersed on the plate, and, during the incubation, the plates should not be jolted or moved in any way. If movement does occur, loosely attached (? mitotic) CHO cells are dislodged and form satellite colonies elsewhere on the plate. At worst, this can lead to a high background of cells, making individual colonies difficult to discern. At best, movement leads to "tailing" of colonies and increased cross-contamination between colonies. Since often more than one lectin-resistant mutant type is present on a single plate (Table II),[22] movement of selection plates during colony formation is highly undesirable.

The frequency of survivors in each selection is calculated from the relative plating efficiency of the untreated population at the time of plating. All cell dilutions are made from a cell suspension containing the highest number of cells to be plated. Although we routinely add lectin and cells together on day 1 of the selection, it is also possible to add lectin a day later than cells. In the few instances in which we have done this, lectin toxicity appears to be unchanged. Some authors have used brief exposure to lectin followed by culture in lectin-free medium to isolate mutants.[55] With CHO cells, it is probable that only the toxic effects of ricin or abrin occur quickly enough to make this approach feasible, although this point has not been systematically investigated.

Mixed Lectin Selections. Not only do single lectin selections give rise to different mutant types on the same selection plate, but often these mutants occur at very different frequencies. Consequently, even if a particular mutant is known to exist, it is not necessarily easy to reisolate that mutant if it occurs at a low frequency relative to other mutants resistant to the same lectin. For example, there are two CHO mutants (Lec1 and Lec4) that are highly resistant to L-PHA (Table II). However, out of 100 survivors in one L-PHA selection, 95% behaved like Lec1 cells and none appeared to be of the Lec4 type. Therefore, in order to reisolate the Lec4 mutation (originally discovered by chance in a large mutant selection), it was necessary to devise a selection that would eliminate survivors from complementation group 1. Lec1 cells are hypersensitive to Con A (Table II). Therefore, by mixing L-PHA and Con A a selection specific for Lec4

[55] Y. Wollman and L. Sachs, *J. Membr. Biol.* **10**, 1 (1972).

cells should theoretically be achieved. In fact, this combination proved successful. L-PHA was used at 15 μg/ml and Con A at 7.5 μg/ml which is too low to select for Con A-resistant mutants directly. From a mutagenized population of Gat^- cells, colonies expressing a Lec 4 phenotype were obtained at a frequency of about 10^{-5}.[15]

Another rationale behind using a mixture of lectins stems from the finding that a combination of lectins at low concentrations may be considerably more toxic than any one of those lectins used alone at the same low concentration. For example, a mixture of WGA, Con A, and LCA, each at 10 μg/ml, is extremely toxic to CHO cells,[23] whereas 10 μg/ml of any one of those lectins alone would allow many survivors (Table II). By using lectin combinations, therefore, it is hoped to isolate a different range of mutations than might be obtained from single lectin selections. In at least one case, this hope has been realized.[23]

The experimental approach to combined lectin selections is identical to that for performing single lectin selections. Alpha medium, 10% FCS, is added to plates followed by the combined lectins in 1 ml, followed by cells at the appropriate concentration in 1 ml. Plates are incubated undisturbed for 8–10 days at the selective temperature, and surviving colonies are picked. The concentrations of lectin to be used are arrived at somewhat empirically, depending on the ultimate aim of the selection. Determining survival curves with mixed lectins has not so far seemed helpful. However, it is conceivable that a survival curve in which the concentrations of all lectins are varied in constant ratio may be useful in deciding selection conditions. The main problems with the combined lectin approach are that the key selective agent in obtaining a particular mutant is necessarily obscure and only lectins that do not interact may be combined. For example, PHA and WGA or RIC and Con A cannot be combined because they form a precipitate. To overcome this limitation, sequential lectin selections were designed.[22,23]

Sequential Lectin Selection Protocols. The rationale behind sequential lectin selections is similar to that for combined lectin selections. Two complementary objectives are achieved—the elimination of known mutants so that new mutants might be uncovered and the development of specific selection protocols aimed at isolating particular mutants at will. As mentioned previously, the protocols that we use maintain the characteristics of a single step selection. They take advantage of the fact that degrees of lectin resistance and hypersensitivity vary widely among lectin-resistant mutants. For example, four WGA-resistant CHO mutants exhibit D_{10} values for WGA of ~15 μg/ml (Lec3), ~25 μg/ml (Lec2), ~75 μg/ml (Lec1), and ~240 μg/ml (Lec8).[20,22] Each mutant type exhibits a unique pattern of lectin hypersensitivity (Table II). We have been able to isolate specifically each mutation with high efficiency by selecting with an appropriate WGA concentration, allowing survivors to form barely visible colonies, and changing the lectin to one that is toxic to unwanted phenotypes.[22] After a few days of further incubation, those colonies that are not inhibited by the second lectin are large whereas those that are sensitive to it are small or have disappeared from the plate (Fig. 2). The use of two or more lectins consecutively in the same selection is termed a sequential lectin selection. A selection protocol typical of that used specifically to generate Lec8 mutants (which correspond to the large colonies shown in Fig. 2) may be outlined as follows:

Parental cells (10^6 per plate) are added to 12 ml of alpha medium, 10% FCS, containing 20 μg of WGA per milliliter (selects for Lec1 and Lec8 mutants, but is toxic to Lec2 and Lec3 mutants) in 20 plates 100 mm in diameter. After 6 days at 37°, one plate is stained so that the number of initial WGA-resistant survivors may be observed. The culture medium is removed from the remaining selection plates, and they are washed twice with 10 ml of warm alpha medium. Fresh medium containing 7.5 μg of Con A per milliliter (inhibits growth of Lec1 colonies) is added, and the plates are returned to the incubator. After 2 days, the control plates, established on the first day of the experiment to determine the plating efficiency of the starting population, are stained. After 2 more days (4 days from the addition of Con A), the selection plates are examined for very large colonies. These are picked into nonselective medium and cultured for genetic and biochemical characterization.

A major advantage of the sequential lectin approach is that no mixing of lectins is required. A single lectin is used as the selective lectin while the counterselective lectin(s) serve mainly to identify desired phenotypes. It is possible to extend the number of lectins employed sequentially to three, allowing 4–6 days of incubation between changes of medium. However, more than three consecutive incubations would probably be difficult because colonies would become very large and the background generated by washing the plates extremely heavy. It is also possible to combine lectins in the counterselective step(s).[22] An obvious disadvantage of this approach is that new phenotypes may be unknowingly selected against in the search for new mutants. However, there is no way to avoid this problem. Since the strategy has already been successful in isolating two new complementation groups and at least four novel phenotypes, it is clearly worth pursuing to its logical limit. The alternative approach of picking and testing all survivors of a lectin selection for new phenotypes does not seem wise in view of the predominance of certain phenotypes in both mutagenized and unmutagenized CHO populations.[22]

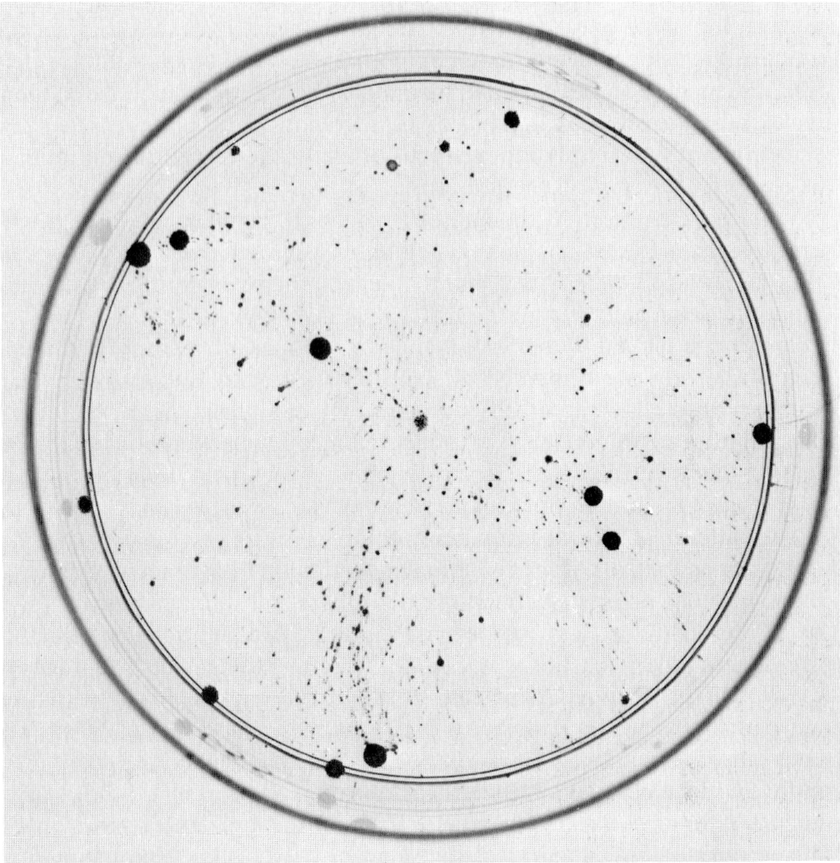

FIG. 2. Specific selection of Lec8 cells. Pro^-5 cells were subjected to selection with 20 μg/ml of wheat germ agglutinin for 6 days followed by 7.5 μg/ml of Con A for 4 days. The large, darkly stained colonies are typical of those shown subsequently to exhibit a Lec8 phenotype and to belong to complementation group 8.

Identification of Isolates as Lectin-Resistant Mutants. Performing a selection is only the first step in isolating a mutant. Once surviving colonies are growing on the selection plates, a number of decisions must be made. If there are few survivors, all should be picked and tested. If, however, there are many survivors (e.g., 10–20 per plate), we usually pick a total of 20–30 colonies (1 or 2 from each plate) taking care to include colonies that cover a representative range of colony sizes and morphologies. To ensure against unconscious biases, it may be wise in addition to pick all the colonies from two or three plates because this may provide the best chance of observing the range of surviving types. It is always important to select mutants from different plates so that special circumstances peculiar to one plate do not skew the results of the selection. This may also help in defining the characteristics of identical mutant isolates. However, isolation of colonies from different plates in the same selection does not imply independent origin of the respective mutations, as is often suggested in the literature.

Picking Colonies. At the end of a selection, colonies are circled with grease pencil. Lectin-containing medium is aspirated from the plates, and they are washed once with alpha medium to remove floating cells. Fresh alpha medium containing 10% FCS is added back to the dish and the colonies are picked by scraping while aspirating into a 1-ml Pasteur pipette. The cells are dispersed by pipetting vigorously into a small, clear, sterile plastic tube that already contains 1 ml of alpha medium, 10% FCS. The tube is capped loosely and incubated in a CO_2 incubator at the selective temperature. The cells from the colony grow in a clump on the bottom of the tube but do not adhere to the plastic, since it is not treated for tissue culture. This means of picking colonies is rapid and is therefore not rate-limiting for the number of colonies that might be examined. After 4–6 days, the cells are resuspended and 1.5 ml are transferred from the tubes to T25 flasks or to 10-ml suspension cultures. Medium is added back to the colony tube in case it is necessary to return to cells at that stage.

Phenotype Testing. As soon as possible after picking, cells are frozen and phenotype tested (P-tested). It is extremely important to freeze the cells quickly because of the likelihood that each colony of cells contained more than one cell type at the time of picking. Depending on the relative growth rates of its various members, the culture may become more or less mixed during continued expansion. Consequently, the culture should be frozen at the earliest time feasible after picking so that access to a population closely resembling that of the original colony is possible. For colonies of interest, it is this population that will be thawed for cloning.

Phenotype testing is accomplished by the semiquantitative test described previously to measure lectin toxicity. The number of lectins and cell lines to be compared varies with the aim of the selection. In searching for new mutants, all lectins are used and a bank of well-characterized mutants is included as a control. In this case, the lectin concentration ranges covered are very broad. However, in cases where specific mutants are being sought, fewer lectins over a more restricted concentration range may be used.

This type of test can reveal, in a comparatively short time (4–5 days) which colonies are of interest to pursue further. It will also reveal mixed colonies such as that shown in Fig. 3. It is clear from this P-test result that

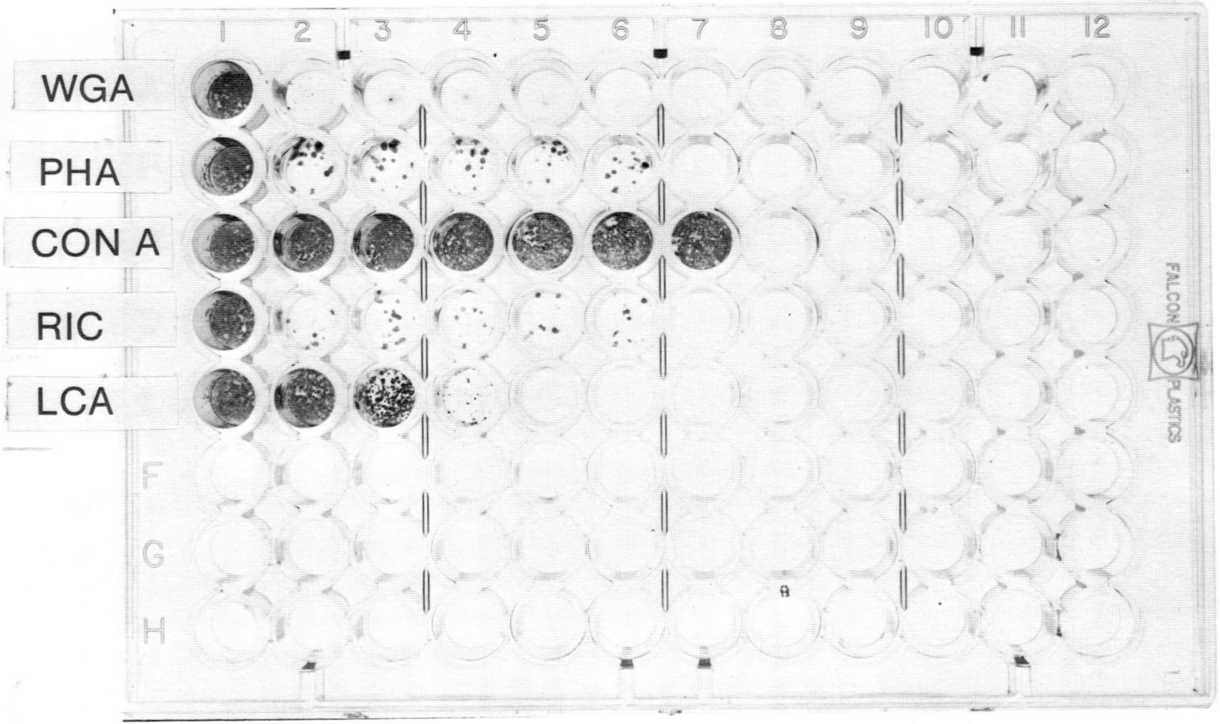

FIG. 3. Phenotype testing performed as described in the text showing a culture that contains more than one cell type. The concentrations of lectins were as follows:

WGA (μg/ml):	0	50	100	150	200	
L-PHA (μg/ml):	0	10	15	20	25	30
Con A (μg/ml):	0	10	15	20	35	
RIC (ng/ml):	0	5	10	25	50	75
LCA (μg/ml):	0	0.5	1	2	4	5

the cell population in the colony contained at least two types of cell. The predominant type is sensitive to WGA, L-PHA, RIC, and LCA. However a minor cell population is present that is clearly resistant to both L-PHA and RIC. This population would not have been detected if only resistance to WGA, Con A, and LCA had been compared. The necessity to purify cell populations by cloning is obvious.

Cloning. Once a colony is identified for further study, the earliest frozen sample is thawed into a T25 flask and a 10-ml suspension culture. As soon as the cells in suspension are growing exponentially, they are cloned and refrozen. If the cells do not grow well in suspension, those in the T25 flask are trypsinized, cloned, and refrozen.

Cloning is achieved by limiting dilution. Cells are diluted to 2 cells/ml, aliquoted as 200 μl per well of a 96-well Microtiter dish, and incubated in a CO_2 incubator. After 8 days, the wells are scanned for those containing single colonies. If the plates are moved during the incubation, very few, if any, wells will appear to contain single colonies. If different morphologies are observed among the clones, enough clones are picked to encompass the range of phenotypes. Usually 5–10 clones are picked depending on the expected frequency of the desired phenotype. As soon as possible the clones are P-tested for lectin resistance. Those that exhibit the desired phenotype are expanded and frozen in a large number of vials at $-120°$ (10–15 vials) and in liquid nitrogen (3 vials). If there is any indication of a mixed cell culture after cloning, a second cloning is performed. In almost all cases, this has not been found to be necessary with our CHO clones.

Determination of D_{10} Values. Once purified clones are available, their lectin-resistance phenotypes are determined quantitatively. A D_{10} value, which is defined as the lectin concentration at which 10% of the cells survive, is determined from survival curves performed with each lectin. For this determination, a full curve is not generated, but survival to 10^{-3} relative plating efficiency is examined. Cells are plated in increasing lectin concentrations in 24-well tissue culture dishes. Lectin dilutions are added in 0.5 ml of alpha medium, 10% FCS. Cells are added subsequently in 0.5 ml of alpha medium, 10% FCS, so that 50 or 500 cells are present per well. Plates are incubated for 8–10 days for colonies to form, and relative plating efficiencies are calculated. By comparing D_{10} values from different experiments in which eight independently isolated Lec1 mutants were examined, we have shown that the error in these measurements is ±30%.[56] It is preferable to use values from 10% survival rather than 50% survival because the survival curve is more variable at 50% relative plating efficiency.

[56] P. Stanley and J. P. Carver, *Adv. Exp. Med. Biol.* **84,** 265 (1977).

Genetic Characterization. Before beginning a biochemical investigation of any new isolate, it is essential to determine whether it is, in fact, representative of a new mutation or whether it belongs to a previously described mutant group. If a purified clone expresses a novel lectin-resistance phenotype, it is, more often than not, a new mutant. Similarly, new isolates that express a phenotype typical of a previously described mutant type have usually been found to fall into the appropriate complementation group. However, exceptions to these generalizations have been observed.[22,23] For example, there is a mutant termed Lec1A that belongs to complementation group 1, but expresses a novel lectin-resistance phenotype. Also there is a mutant (Lec2A) that belongs to complementation group 2, but expresses a novel WGA-binding phenotype. The genetic basis of these cell types is not known. They could represent rare double mutants or alternative mutations of the gene affected in the other members of their respective complementation groups. However, in the absence of complementation analysis, they would be viewed as having sustained mutations in new genes affecting glycosylation pathways.

These two examples serve to illustrate the absolute necessity of performing complementation analysis on new isolates. Although the genetic characterization of all mutants is important, it is of particular importance with respect to lectin-resistant mutants because of their pleiotropic phenotypes. Thus different laboratories studying mutants with similar phenotypic properties may conclude that two isolates represent identical mutations, having missed the crucial biochemical test that reveals their uniqueness. Provided the mutations behave recessively in somatic cell hybrids, complementation analysis between them will quickly reveal whether they are, in fact, affected in the same gene. This argument applies equally to the assumption that two isolates represent different mutations when, in fact, only different phenotypic properties have been compared. For example, a number of laboratories have described CHO cell lines that express properties typical of the Lec1 phenotype, but few have examined hybrids formed between Lec1 cells and their own isolates. It is important that, at the least, glycosylation-defective cell lines are examined for altered lectin-resistance or lectin-binding properties. Those that express changes in these parameters should be subjected to complementation analysis. Only in this way will isolates that actually represent new mutations be quickly identified for further study.

In this laboratory all cloned isolates are subjected to complementation analysis. For new phenotypes, the dominant or recessive behavior of hybrids formed with parental cells is determined. Those that behave dominantly cannot be pursued further by this approach. However, those that behave recessively are hybridized with other mutants to define different complementation groups. In performing complementation analyses, both the population of hybrids formed as well as individual hybrid colonies are tested for their lectin-resistance properties. Isolates that exhibit a previously defined phenotype are crossed with a member of the complementation group expressing the same phenotype and with a member of one other complementation group that exhibits a phenotype similar, though not identical to, the one expressed by the new clone. New isolates that exhibit a novel, recessive phenotype are crossed with members of most complementation groups. The lectin chosen for determining the lectin-resistance properties of hybrids is the one to which both parental cells are most resistant. A typical experiment in which hybrids are selected and tested from the cross $Pro^-Lec1 \times Gat^-Lec8$ may be summarized as follows:

Day 1: Cells from each of the two lines to be hybridized are plated separately at 10^2 cells per 60-mm plate in duplicate to determine their relative plating efficiencies.

Each cell line is also plated at $(1–3) \times 10^5$ cells per 100-mm plate in duplicate in hybrid medium (alpha medium lacking glycine, adenosine, thymidine, and proline), 10% dFCS (dialyzed FCS) to determine the frequency of reversion of the Pro^- and Gat^- markers. Hybrid medium contains penicillin (1 mg/liter), streptomycin (1 mg/liter), and Fungizone (2.5 mg/liter).

For the hybridization itself, parental cells are mixed in equal numbers (1.5×10^6 cells of each) in three 35-mm plates to form mixed monolayers of cells for polyethylene glycol (PEG) treatment on day 2.

Day 2: Two of the 35-mm mixed parental cell monolayers are treated with PEG, and the third is mock-treated. The latter is used as a measure of spontaneous hybrid formation and reversion frequency among the mixed parental cell cultures.

Eight grams of PEG, M_r 1000 (Koch-Light Chemical Co.) is weighed into a glass test tube, loosely capped, and autoclaved. Before it has cooled completely, 10 ml of alpha medium at 45° are added, and the solution is well mixed. A 1.0-ml aliquot is added to the two mixed monolayers for exactly 1 min, then aspirated; the cells are washed 4 times with 1.0 ml of alpha medium, 10% DMSO, and once with alpha medium, 10% FCS. Fresh alpha medium, 10% FCS, is added to the monolayers, and the plates are returned to the incubator for 24 hr.

Day 3: PEG-treated and untreated monolayers are trypsinized and resuspended as single cells in hybrid medium, 10% dFCS.

Cells which were not treated with PEG are plated in duplicate at $(1–3) \times 10^5$ cells per 100-mm dish containing hybrid medium, 10% dFCS.

PEG-treated cells are plated in duplicate at (1–3) × 10^5 cells per 100-mm dish in hybrid medium, 10% dFCS, alone or in the same medium containing 1 μg, 2 μg, or 3 μg of WGA per milliliter. The cell number plated here and throughout the experiment depends on the frequency of viable hybrid formation. We aim for approximately 200 colonies per plate from PEG-treated cells plated in the absence of WGA. Plating too many cells tends to inhibit viable hybrid formation.

Day 8: Stain control plates plated on day 1.

Days 11–12: Pick 6 large, diffuse-looking colonies from PEG-treated cells on plates that do not contain WGA. These are used to check lectin resistance of hybrids by P-test and for karyotype analysis.

Stain remainder of plates.

Count colonies on all plates, and calculate relative plating efficiencies for PEG-treated and untreated cells in the presence and absence of WGA. This will give the WGA resistance of the hybrid cell population.

Days 17–20: P-test and karyotype hybrids picked on days 11–12. This allows degree of lectin-resistances and cross-resistances of hybrids to be explored and confirms the validity of the result obtained with the mass culture. Karyotyping confirms the hybrid nature of each isolate.

In addition to performing complementation analysis, it is important to show that subclones of each mutant possess identical phenotypic properties and that these properties are stable and hereditable characteristics of the cell line. If a new mutant is identified, it is important to isolate it independently from a different parental line and to show that it exhibits noncomplementation with the original mutant in somatic cell hybrids. Karyotypes are routinely performed for new mutants. In theory, a point mutation should not give rise to a change in modal chromosome number. However, it is important to show that a new isolate is pseudodiploid like Pro^- and Gat^- CHO cells. It is not uncommon to isolate pseudotetraploid cells in lectin selections.[22,23] Ultimately at least two independently derived genetically characterized clones should be available for in depth biochemical studies of each phenotype identified.

Biochemical Characterization. Since most lectin-resistant or lectin-binding mutants are glycosylation-defective, their biochemical characterization involves localizing carbohydrate structural alterations at the cell surface and determining the molecular basis of each alteration. Detailed discussion of the techniques involved in these investigations is beyond the scope of this chapter.[25,27] In summary, animal viruses that carry one or two glycoproteins (e.g., vesicular stomatitis virus or Sindbis virus) have been very useful tools in identifying asparagine-linked carbohydrate lesions expressed by lectin-resistant cells. With the advent of ^1H nuclear magnetic resonance spectroscopy to determine the carbohydrate structures of purified viral glycopeptides,[57] it is now possible to identify carbohydrate structural lesions directly from virus grown in mutant cells[16] and to use these structurally characterized molecules as substrates in assays to determine the enzymic lesion that led to their production.

Lectin-Resistant Mutants from Cells Other Than CHO

Lectin-resistant mutants have been selected from a number of cell lines that exhibit very different growth properties from CHO cells; e.g., mouse lymphoma cells that grow only in suspension[29] and rat myoblast cells that grow only in monolayer.[58] For cell types with these growth properties, certain precautions must be taken in applying the experimental approach described for CHO cells. However, the principles of selection and genetic characterization remain the same.

Cells that grow only in suspension cannot be plated in liquid medium to form colonies, and therefore survival curves and P-tests cannot be generated by calculating relative plating efficiencies or by comparing confluency of monolayers. Likewise, selections cannot be performed so that surviving colonies may be identified and picked. The major disadvantage of this situation is that single-step selection protocols are not easily achieved. The criterion of lectin toxicity must be cessation of cell division, which is determined by monitoring cell numbers in cultures containing increasing concentrations of lectin. The growth curves generated can be misleading, since cells that have stopped dividing are not necessarily dead. Therefore selection pressure must be maintained so that only resistant cells will grow. This usually means continuous addition of selective agent during prolonged culture or, alternatively, proceeding through several cycles of selection. Both these protocols result in multistep selections increasing the chance that the final mutant type will be the product of more than one mutation. However, this risk is unavoidable. Provided that the final lectin-resistant culture is cloned at least once and preferably twice, bona fide mutant clones should be derived.

Cells that grow only in monolayer can more easily be pursued by the methods described for CHO cells. The main concern with these lines is that, prior to plating, they must be released from the substratum by trypsinization, which grossly affects carbohydrate structures at the cell surface. Consequently, it is conceivable that if they are exposed to medium containing lectin immediately after trypsinization, many cells might behave as lectin-resistant, leading to a high background of wild-type sur-

[57] J. Hakimi, J. Carver, and P. H. Atkinson, *Biochemistry* **20,** 7314 (1981).
[58] C. L. J. Parfett, J. C. Jamieson, and J. A. Wright, *Exp. Cell Res.* **136,** 1 (1981).

vivors from the selection. This problem may be overcome by plating the cells the day before lectin is added.

The final consideration in using cells other than CHO is that many cell lines do not carry genetic markers which might be used in selecting cell hybrids. This creates a problem for classifying mutants into recessive and dominant types and subsequently into complementation groups. The easiest approach to generating the appropriate situation for forming hybrids between parental and mutant or two mutant cell lines is to make one parent line deficient in hypoxanthine-guanine phosphoribosyltransferase ($HGPRT^-$) by selection for thioguanine resistance[59] as well as making it resistant to ouabain (OUA^R).[60] Hybrids may then be formed between this line and any unmarked line and selected for in HAT medium (hypoxanthine, aminopterin, and thymidine supplemented) containing ouabain. Hybrids survive in this medium because the $HGPRT^-$ marker behaves recessively while the OUA^R market behaves codominantly in hybrid cells. On the other hand, both parental lines are selected against in this medium because $HGPRT^-$ cells are killed by aminopterin and the unmarked line will be killed by ouabain.

Hyman and Trowbridge[61] have classified their $Thy1^-$ and lectin-resistant mouse lymphoma mutants into five complementation groups (A–E) by this approach. However, it is not known how these groups relate to the complementation groups that we have defined for CHO cells (Table II). Unfortunately, the interpretation of complementation analyses between these mutants and the CHO mutants might be difficult because of the different lectin-resistance and growth properties of the cells and the problem of rapid chromosome loss that occurs in interspecies hybrid cells.[62] Comparisons of biochemical phenotypes suggest that some of the lymphoma mutants exhibit glycosylation lesions typical of certain CHO mutants.[63] However, a number of the lymphoma mutants exhibit unique glycosylation defects that do not appear to be represented among the CHO isolates.[14,63,64] By contrast, we have shown that a number of lectin-resistant CHO lines isolated by other laboratories all fall into one or other of the complementation groups described in Table II.[38]

[59] O. P. van Diggelen, T. F. Donahue, and S.-I. Shin, *J. Cell. Physiol.* **98,** 59 (1979).

[60] R. M. Baker, D. M. Brunette, R. Mankovitz, L. H. Thompson, G. F. Whitmore, L. Siminovitch, and J. E. Till, *Cell* **1,** 9 (1974).

[61] R. Hyman and I. Trowbridge, *Cold Spring Harbor Conf. Cell Proliferation* **5,** 741 (1978).

[62] U. Franke, P. A. Lalley, W. Moss, J. Ivy, and J. D. Minna, *Cytogenet. Cell Genet.* **19,** 57 (1977).

[63] I. S. Trowbridge, R. Hyman, T. Ferson, and C. Mazauskas, *Eur. J. Immunol.* **8,** 716 (1978).

[64] A. Chapman, K. Fujimoto, and S. Kornfeld, *J. Biol. Chem.* **255,** 4441 (1980).

Acknowledgments

The writing of this chapter was supported by PHS grant CA30645 awarded by the National Cancer Institute, DHHS, and grant PCM80-23672 from the National Science Foundation. The author thanks Lisa Youkeles and Barbara Dunn for excellent technical assistance and Dr. Christine Campbell for critical comments on the manuscript. P. S. is a recipient of a senior faculty award from the American Cancer Society.

[12] Two-Dimensional Polyacrylamide Gel Electrophoresis of Membrane Proteins

By R. W. RUBIN and C. L. LEONARDI

We have been studying the polypeptide content of membranes obtained from a variety of sources including erythrocytes,[1] lymphoblasts,[2] and subfractions of cultured bovine adrenal chromaffin cells.[3] The ultrahigh resolution of two-dimensional polyacrylamide gel electrophoresis (2-D PAGE) in combination with new ultrasensitive staining procedures and electroblot transfer techniques have permitted us and other laboratories to ask questions in cell biology previously not technically possible. When these methods are employed in combination with lectin-binding studies and specific immunolocalization of polypeptides initially separated on two-dimensional gels, the full potential of these methods becomes apparent (Fig. 1). The 2-D PAGE system we employ in our laboratory is based on the one originally described by O'Farrell.[4] However, many specific methods and procedures have evolved since that time. In the following description we will point out some of the pitfalls we have encountered, suggest possible remedies, and identify those procedures that have been successfully applied to the system in our laboratory.

Specific Methods and Procedures

Sample Preparation. Along with other laboratories, we have found sample preparation to be a critical step in high-resolution 2-D PAGE.[1,5]

[1] R. W. Rubin and C. Milikowski, *Biochim. Biophys. Acta* **509,** 100 (1978).

[2] R. W. Rubin, M. Quillen, J. J. Corcoran, R. Ganapathi, and A. Krishan, *Cancer Res.* **42,** 1384 (1982).

[3] S. J. Suchard, J. J. Corcoran, B. C. Pressman, and R. W. Rubin, *Cell Biol. Int. Rep.* **5,** 953 (1981).

[4] P. H. O'Farrell, *J. Biol. Chem.* **250,** 4007 (1975).

[5] D. L. Wilson, G. E. Hall, G. C. Stone, and R. W. Rubin, *Anal. Biochem.* **83,** 33 (1977).

ISBN 0-12-181996-5

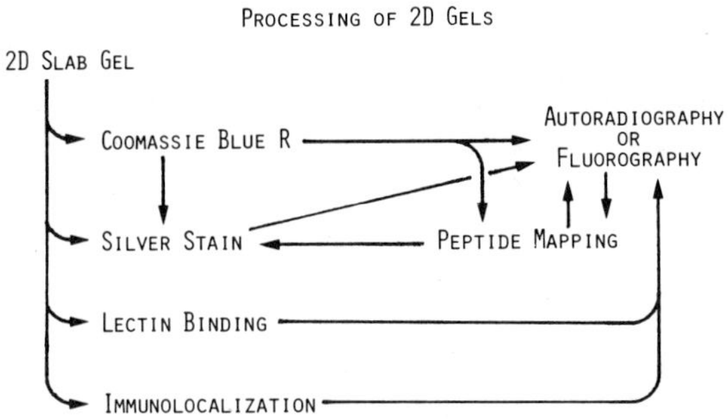

FIG. 1. Processing of two-dimensional gels.

Frequently a given system will be highly valuable for the analysis of membranes of cultured cells but inadequate for resolving the membrane proteins obtained from intact tissues, where collagen and other relatively insoluble proteins tend to induce protein aggregation during the first-dimension isofocusing run. In most cases, the key to success is the initial solubilization of the sample in ultrapure SDS and the avoidance of heating, which can produce artifactual spots via secondary charge modifications of the proteins.[4,5] We utilize a procedure developed for eukaryotic tissues that maximizes the protein-to-volume ratio yet maintains a constant sodium dodecyl sulfate (SDS) concentration for increased reproducibility.[5] Usually, a membrane pellet obtained by ultracentrifugation is transferred at room temperature to a low-volume (0.3 ml) precision-ground glass homogenizer. Here, a minimal amount of stock grind solution (0.1 g of SDS, 1.0 ml of 2-mercaptoethanol, and 5.7 g of urea to make 10.0 ml, pH 7.4) is added with homogenization until no light scattering is observed in the preparation. A protein assay of this material can then be run (1–3 μl usually suffice) using the trichloroacetic acid filter method of Schaffner and Weissmann.[6] The volume of stock grind used is noted, and the remaining solubilization solutions [lysis supplement: 4 ml of 10% NP-40, and 0.4 ml of Ampholines; and lysis buffer: 2.0 ml of 10% Nonidet P-40 (NP-40), 0.2 ml of Ampholines, and 5.7 g of urea to make 10 ml] are added sequentially with homogenization to yield a final ratio of 3 parts stock grind, 1 part lysis supplement, and 4 parts lysis buffer. In lieu of the glass homogenizer, the above procedure may be conveniently performed by resuspending the pellet with a microliter syringe of appropriate size. However, the glass homogenizer is recommended for tissue or whole-cell homogenates.

[6] W. Schaffner and C. Weissmann, *Anal. Biochem.* **56,** 502 (1973).

At this point, the homogenate is centrifuged at high speed (100,000 *g*, 60 min, 20°) to remove unsolubilized material (DNA, collagen, etc.), which frequently makes the sample too viscous for accurate measurement.[7] Furthermore, such material cannot migrate properly in the first dimension, "clogs" the top of the isofocusing gel, and can produce horizontal streaking artifacts in the final stained pattern. Solubilization efficiency may be empirically gauged at this stage by noting the size of the pellet resulting from clarification. Typically, a 0.05-ml membrane pellet is dissolved in 0.16 ml of two-dimensional solubilization buffers with at least 95% recovery of the sample in the supernatant. An aliquot of this sample is then loaded on the first-dimensional gel, and the remainder is immediately frozen at −70°. When required, frozen samples are thawed at room temperature, loaded on the gels, and quickly refrozen. Samples prepared in this manner may be stored for several months and will survive numerous freeze–thaw cycles without apparent loss of resolution.

Occasionally it is necessary to concentrate a dilute sample prior to solubilization for 2-D PAGE (e.g., the contents of exocytotic granules may be examined by collecting the supernatant after hypotonic lysis of a granule fraction). For such cases, the sample is dialyzed against 10 m*M* ammonium acetate buffer (pH 7.0) to remove nonvolatile salts, which, in high concentration, interfere with solubilization and may distort the pH gradient in the first dimension. After dialysis, the sample is lyophilized to dryness and the residue is prepared for 2-D PAGE as described above.

Samples that possess contaminating proteases frequently exhibit highly variable spot pattern artifacts. One of the obvious symptoms of proteolysis in membrane preparations is the appearance of numerous, clearly resolved spots running just above the tracking dye. These tend to spread out over a wide pH range at the bottom of a 10 or 15% gel. We have had some success in dealing with this problem by isolating membranes in the presence of 0.5 m*M* phenylmethylsulfonyl fluoride (PMSF), 1 m*M* EGTA, and 0.1% soybean trypsin inhibitor. For such cases, we find it necessary to keep this cocktail within the 2-D PAGE solubilization buffers, as many proteases are fully active in SDS.[8]

Electrophoresis. Isofocusing gels are prepared as described by O'Farrell with a number of modifications that we feel permits greater resolution of the high molecular weight membrane proteins. The acrylamide concentration of the first dimension was reduced to 3%. The Ampholine concentration is 4% (v/v) and consists of a mixture of 3/10, 4/6, and 5/7 Ampho-

[7] A Beckman Airfuge is routinely used.

[8] D. W. Cleveland, S. G. Fischer, M. W. Kirschner, and U. K. Laemmli, *J. Biol. Chem.* **252,** 1102 (1977).

lines (2 : 1 : 1) required to generate a linear pH gradient[9] from approximately pH 4 to 8.9. Owing to variation between new batches of Ampholines, the exact proportions are adjusted using the positions of actin, troponin, and tropomyosin from rabbit skeletal muscle homogenate as protein standards. For this reason, Ampholines are purchased in bulk quantity, divided into 1.0-ml aliquots, and stored at −70° for up to 6 months without effect.

Ten milliliters of isofocusing gel solution are prepared from 5.5 g of ultrapure urea,[10] 1.00 ml of acrylamide stock (30% acrylamide, 0.8% bis-acrylamide), 2.00 ml of 10% NP-40, 2.54 ml of H_2O, 0.4 ml of Ampholines, 10 μl of 10% (w/v) ammonium persulfate, and 10 μl of TEMED. In cases where a narrow pH range is required to increase protein separation in the first dimension, the mixture of Ampholines may be replaced by one of a single type (i.e., 4/6 only).

The isofocusing gel solution is poured to a height of 12 cm into 3 mm (i.d.) by 15-cm glass tubes that are sealed off at the end with Parafilm. The solution is overlayered with 8 *M* urea. After polymerization, the gel is equilibrated for 1 hr with lysis buffer. The Parafilm is pierced repeatedly with a needle to allow even current flow and prevent the gel from falling out of the tube during the long isofocusing run. The tubes are then placed in the isofocusing chamber, the lower reservoir is filled with 0.012 *M* H_3PO_4 and the upper reservoir with 0.02 *M* NaOH.

The gels are then prerun under conditions of constant current, where the voltage is allowed to rise continuously from 175 V until it reaches 400 V. At that time, the prerun is stopped and lysis buffer and NaOH are removed from the tubes. The samples are loaded and overlaid with sample overlay (4 *M* urea, 1% Ampholines). The tubes and upper reservoir are again filled with NaOH, and the gels are run at 400 V for 19 hr. After completion of the isofocusing run, the gels are extruded by carefully applying pressure (using a water-filled syringe and a Tygon tubing connector) to the top of the tube.

The isofocusing gels are equilibrated for 30 min in 5 ml of sample buffer (10% glycerol, 5% 2-mercaptoethanol, 2% SDS, 0.0625 *M* Tris-HCl, pH 6.8), after which they are placed on beveled slab gel plates and anchored to the second-dimension stacking gel with sample buffer containing 1% agarose. A small amount of Bromphenol Blue is usually added to this agarose solution for use as a tracking dye.

Slab gels typically consist of a 3% acrylamide stacking gel and a 10%

[9] The pH is measured directly with a Micro Combination pH Probe (MI-410) available from Microelectrodes Inc. (Londonderry, New Hampshire 03053).

[10] Available from Schwarz-Mann.

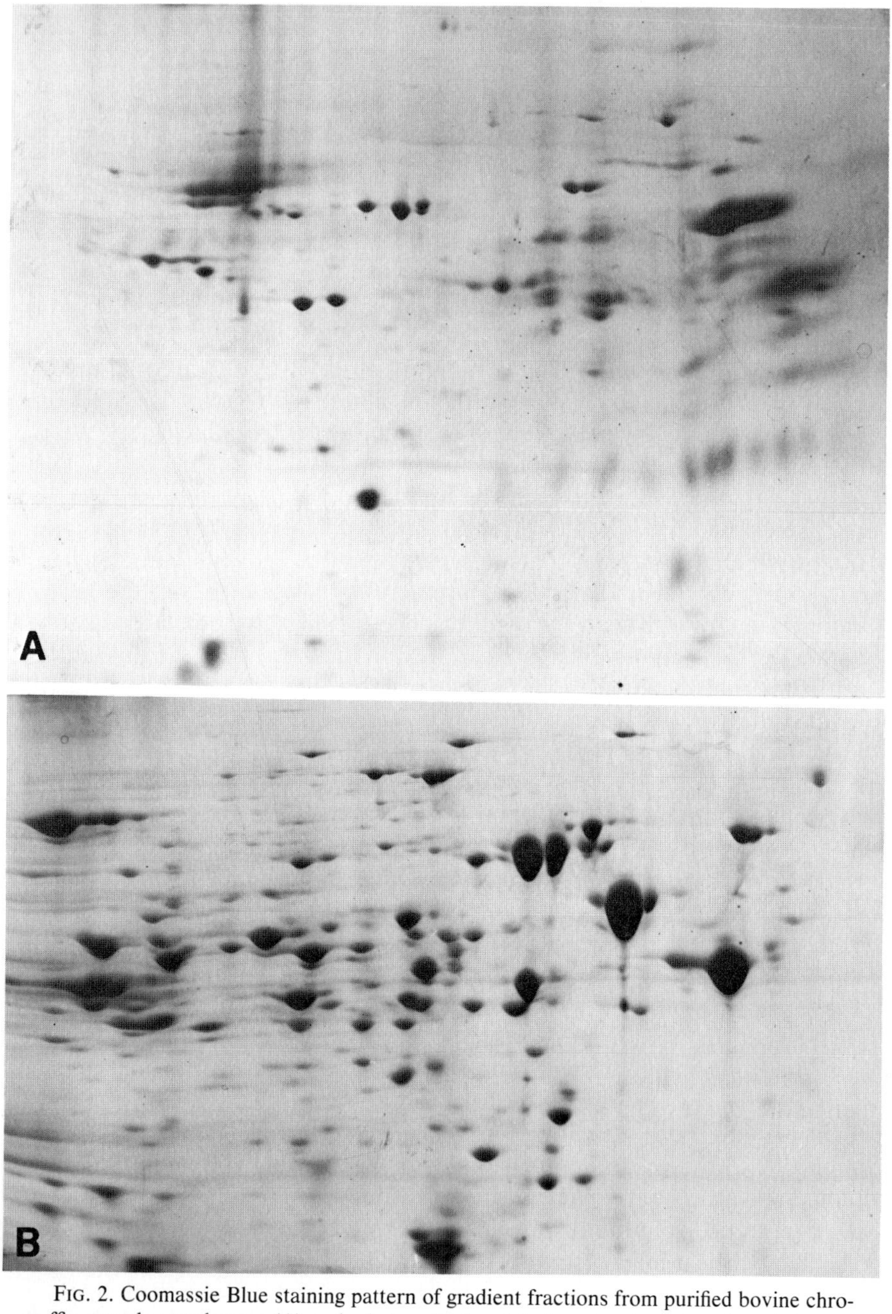

FIG. 2. Coomassie Blue staining pattern of gradient fractions from purified bovine chromaffin granule membranes (A) and whole mitochondria from bovine adrenal medulla (B). Note the total lack of protein spot homology between the two fractions as expected. This indicates the absence of cross-contamination and illustrates the utility of the two-dimensional electrophoretic technique. pH gradient: 7.4 (left) to 4.0 (right). M_r range: 240,000 (top) to 12,000 (bottom).

acrylamide separation gel. This permits resolution of proteins with molecular weights ranging from 12,000 to 250,000. The concentration of acrylamide in the separation gel may be increased or decreased to allow resolution of proteins with higher or lower molecular weights. Slab gels are run at 20 mA constant current for 3–4 hr (see Fig. 2).

Analysis Techniques

Protein Staining. Slab gels are fixed and stained with 0.1% Coomassie Brilliant Blue R-250 (Eastman Kodak) in 25% trichloroacetic acid for 1 hr, and destained overnight in 7% acetic acid. Destaining efficiency is greatly increased by placing a small square of inert sponge foam rubber in the destaining solution during the overnight step. The foam regenerates the destaining solution by removing Coomassie dye from the acetic acid. Slab gels treated in this fashion have extremely low background and are comparable to gels destained for 2 days with several changes of destain solution.

Photography. Destained gels are placed on an X-ray film-viewing box and photographed with 35 mm Kodak Technical Pan Film (2415) through Kodak Wrattan gelatin filters Nos. 15 and 57A, or an equivalent yellow-green combination. A reusable label (acetate strip, marking pen) is conveniently placed along the gel to create a permanent photographic record.

Silver Stain. Recent improvement in silver stain methodology[11] has led to the routine use of this ultrasensitive protein stain (roughly 100× more sensitive than Coomassie Blue R) in our laboratory. The most recent version of the stain is photochemically based (as opposed to earlier histochemical versions), permits detection of 0.01 ng of protein per square millimeter and can be performed in less than 1 hr at low cost. The disadvantages of the silver stain for protein detection are related to the great sensitivity of the method. All glassware must be protein free from electrophoresis (suggestion: chromic acid and/or Chem-Solve[12] followed by methanol and distilled H_2O), and extreme care must be exercised in handling the gels without touching them. We have even been plagued by a protein contaminant (ca. M_r 65,000) that appears in some commercial sources of 2-mercaptoethanol. This contaminant appears as a broad band extending across the isofocusing dimension. The reader is directed to Merril *et al.*[11] for a detailed silver stain protocol.

Fluorography. After Coomassie staining and photography, slab gels are prepared for fluorography using the 2,5-diphenyloxazole–dimethyl

[11] C. R. Merril, D. Goldman, and M. L. Van Keuren, this volume [17].

[12] Mallinckrodt Chemical Works.

sulfoxide (PPO-DMSO) method described by Bonner.[13] Here, gels are dehydrated during two DMSO washes (1 hr each), and impregnated with fluorochrome by shaking in 22% PPO (w/v) in 50 ml of DMSO. The PPO is precipitated by washing the gel in several changes of distilled water (30 min each). The gels are then dried down on Whatman No. 3 filter paper at 80° using a commercially available slab gel drier. Exposure of the fluorograph proceeds at −70° using Kodak XAR film (5 × 7 inches) in a suitable lighttight container (recommended: Kodak X-Ray Film holder, 5 × 7 inches). As an alternative to the above procedure, "one step" solutions for impregnating slab gels with fluorochrome are available.[14] While the convenience of these products is unsurpassed, the products are not as fluorographically efficient in our system as the method described above.[15]

Autoradiography. Samples containing proteins radiolabeled with ^{125}I are stained, photographed, and dried down as described above. For samples of low specific activity, exposure of the autoradiograph is enhanced by the use of rare earth image intensifying screens[16] and Kodak XAR film at −70°. As the screens are extremely sensitive, sheets of lead are usually placed between adjacent autoradiographs while in the freezer, since as few as 10 disintegrations per minute above background can be detected. Samples with high specific activity are exposed directly at room temperature without intensifying screens. When only a small amount of membrane is available, we have had success in using an iodogen[17] to ^{125}I-radiolabel samples solubilized in H_2O or 0.5% SDS. Here, 100 μCi/0.1 ml of sample is added to an iodogen-coated glass tube, and the reaction is allowed to proceed for 30 min with frequent shaking. At the end of the iodination period, stock grind, lysis supplement, and lysis buffer solutions are added in the proportions described above. These samples are directly loaded on the first-dimension gel and run as described above in a radiation hood. We have found that unbound counts migrate through the gel and into the lower chamber. In avoiding the lengthy dialysis and lyophilization steps normally employed, more sample is recovered and secondary charge artifacts are significantly reduced. Furthermore, the method produces less radioactive waste for disposal.

Since ^{125}I has a significant (90%) β emission, we have further enhanced our ability to detect this isotope by combining the PPO-based fluorographic and rare-earth screen autoradiographic techniques.

Lectin Binding. Several methods for performing lectin-binding studies

[13] W. M. Bonner, this volume [15].
[14] ENHANCE, available from New England Nuclear.
[15] D. Rulli and D. Wilson, personal communication.
[16] DuPont Cronex Quanta III intensifying screen.
[17] P. J. Fraker and J. C. Speck, Jr., *Biochem. Biophys. Res. Commun.* **80,** 849 (1978).

in polyacrylamide gels have been published.[18,19] These methods are similar in that they involve an initial "soft" fixation of the slab gel with acetic acid–alcohol–water solutions followed by incubation in buffered saline. After pH neutralization, iodinated and affinity-purified lectin is applied to the gel with gentle shaking. Control gels should be preincubated in the appropriate saccharide. We have successfully applied the method of Horst *et al.*[19] to our gel system.

Electroblot and Immunolocalization. Antibodies prepared against membrane proteins can be localized to two-dimensional gel patterns after transfer to a nitrocellulose sheet[20–22] which is then incubated in antiserum, thereby avoiding the problem of poor gel penetration by the antibody when attempting to stain the gel directly.[23] After completion of electrophoresis, two-dimensional gels are briefly washed (15 min) in transfer buffer (20 m*M* Tris base, 150 m*M* glycine, and 20% methanol, pH 8.3[21]) to remove excess SDS, placed in intimate contact with a sheet of nitrocellulose,[24] and electrophoretically transferred in an inexpensive destaining device.[25] After transfer, the nitrocellulose replica is washed in a solution containing bovine serum albumin and incubated with antiserum. Antibody is visualized after binding ^{125}I-labeled Protein A[24] followed by rare earth screen autoradiography (described above). Transfer efficiency is greatly increased by presoaking the nitrocellulose for several hours in transfer buffer and taking great care to ensure a strong intimate contact between gel and nitrocellulose sheet.

Peptide Mapping. Peptide mapping by limited proteolysis[8] is a powerful means of comparing proteins isolated by 2-D PAGE. In general, individual spots are punched out of two-dimensional gels and digested with *Staphylococcus aureus* protease in a second 15% SDS slab gel. Partial digestion products are visualized by the silver stain method described above. We have extended this technique by using dried gels as a source of protein for digestion. Here, protein spots are removed from the gel along with the filter paper backing, and rehydrated in Cleveland's "gel slice incubation buffer" (0.125 *M* Tris-HCl, pH 6.8, 0.1% SDS, and 1 m*M* EDTA).[8] After incubation, the filter paper is removed with forceps and the gel slice inserted into the stacking portion of a second 15% gel for

[18] K. Burridge, this series, Vol. 50, p. 54.

[19] M. N. Horst, S. Mahaboob, M. Basha, G. A. Baumbach, E. H. Mansfield, and R. M. Roberts, *Anal. Biochem.* **102,** 399 (1980).

[20] H. Towbin, T. Staehelin, and J. Gordon, *Proc. Natl. Acad. Sci. U.S.A.* **76,** 4350 (1979).

[21] W. N. Burnette, *Anal. Biochem.* **112,** 195 (1980).

[22] G. R. Stark and J. Reiser, this volume [14].

[23] K. Olden and K. M. Yamada, *Anal. Biochem.* **78,** 483 (1977).

[24] Schleicher & Schuell (Keene, New Hampshire), grade 470.

[25] Canalco gel destainer is available from Miles Laboratories (Elkhart, Indiana).

digestion and subsequent analysis. We have found the method to be equally successful with proteins punched out of fluorographs and with gels that had been dried and stored for several years. Thus, it is possible to compare digestion products of individual proteins after autoradiography and fluorography, or even several years after electrophoresis. If the proteins were originally radiolabeled, the Cleveland digest itself may be fluorographed or autoradiographed.

[13] Immunochemical Identification of Membrane Proteins after Sodium Dodecyl Sulfate–Polyacrylamide Gel Electrophoresis

By ALBERT HAID and MORDECHAI SUISSA

Electrophoresis in the presence of sodium dodecyl sulfate (SDS) has become a key method for analyzing membrane proteins. Yet it can at best provide information on the size and the amount of these molecules. The unambiguous identification of a given protein or a comparison between related proteins requires additional tests that respond to the amino acid sequence. This may be important if, e.g., homologous proteins differ in electrophoretic mobility because of a mutational change in the sequence or if proteins of different size are related by a precursor–product relationship.

Assays utilizing specific antibodies as a detecting probe greatly facilitate the identification and characterization of individual membrane proteins. Until recently, a method of choice was to immunoprecipitate the radioactive protein under study from crude cell extracts or cellular subfractions followed by analysis of the immunoprecipitate by electrophoresis in SDS–polyacrylamide gels.[1] Although immunoprecipitation has proved to be very useful, it is rather time-consuming and requires biosynthetic labeling of membrane proteins with radioactive tracers. An even more serious problem is proteolytic degradation of the protein under study during the lengthy procedure and nonspecific coprecipitation of hydrophobic contaminants. These shortcomings are circumvented by procedures in which unlabeled proteins are first separated on SDS–polyacrylamide gel slabs and then identified by a radioimmune assay. These procedures are based on the transfer of proteins from the polyacrylamide

[1] G. Schatz, this series, Vol. 56, p. 40.

ISBN 0-12-181996-5

gel slab to an insoluble matrix, such as chemically reactive cellulose or, more conveniently, commercially available nitrocellulose sheets.[2–8,8a,8b] This transfer, which can be induced either by electrophoresis or by diffusion, yields a faithful replica of the original gel pattern. The immobilized proteins on the replica are readily accessible to antibodies or other specific probes (e.g., lectins); this makes possible the *in situ* localization and characterization of individual proteins in complex mixtures.

The method outlined here is a modification of that originally developed by Towbin *et al.*[3] We describe the procedure in detail and discuss its various possible applications in the study of membrane proteins.

Method

Special Equipment

Nitrocellulose sheets (Millipore, pore size 0.22 μm or 0.45 μm; nitrocellulose sheets from Sartorius or from Schleicher & Schuell are also satisfactory)
Perforated plastic plates [poly(vinyl chloride) or Perspex], 15 cm × 10 cm
Sponge pads, 15 cm × 10 cm
Filter paper (Whatman 3 MM)
Electrophoretic destaining chamber for slab gels, 20 cm × 20 cm × 8 cm, buffer capacity 2.5 liters (a chamber specially designed for electrophoretic transfer is commercially available from Bio-Rad)
Electrophoretic destaining power supply (40 V, 1.5 A)

Reagents

Newborn calf serum (Gibco)
Bovine serum albumin (BSA), fraction V (Sigma)
Na ^{125}I (Amersham)

[2] J. Renart, J. Reiser, and G. R. Stark, *Proc. Natl. Acad. Sci. U.S.A.* **76,** 3116 (1979).
[3] H. Towbin, T. Staehelin, and J. Gordon, *Proc. Natl. Acad. Sci. U.S.A.* **76,** 4350 (1979).
[4] H. A. Erlich, J. R. Levinson, S. N. Cohen, and H. O. McDevitt, *J. Biol. Chem.* **254,** 12240 (1979).
[5] B. Bowen, J. Steinberg, U. K. Laemmli, and H. Weintraub, *Nucleic Acids Res.* **8,** 1 (1980).
[6] E. J. Stellwag and A. E. Dahlberg, *Nucleic Acids Res.* **8,** 299 (1980).
[7] M. Bittner, P. Kupferer, and C. F. Morris, *Anal. Biochem.* **102,** 459 (1980).
[8] W. N. Burnette, *Anal. Biochem.* **112,** 195 (1981).
[8a] J. G. Howe and J. W. B. Hershey, *J. Biol. Chem.* **256,** 12836 (1981).
[8b] W. Lin and H. Kasamatsu, *Anal. Biochem.* **128,** 302 (1983).

Protein A from *Staphylococcus aureus* (Pharmacia)
Horseradish peroxidase-conjugated antibodies directed against a primary antibody, e.g., swine anti-rabbit immunoglobulin or rabbit anti-mouse immunoglobulin (DAKO)
4-Chloro-1-naphthol, 0.3% in methanol
Hydrogen peroxide, 30%
Transfer buffer: 25 m*M* Tris, 192 m*M* glycine, 20% (v/v) methanol, 0.02% SDS, pH 8.3
Transfer indicator: 1% methyl green in SDS–gel sample buffer[9]
Phosphate-buffered saline (PBS): 0.14 *M* NaCl, 10 m*M* NaP_i, pH 7.4, 0.1% NaN_3
Serum buffer: 5% newborn calf serum in PBS
BSA buffer: 2.5% BSA in PBS
Staining solution A: 0.1% amido black in 45% methanol, 10% acetic acid
Staining solution B: 0.25% Coomassie Blue in 45% methanol, 10% acetic acid
Destaining solution: 45% methanol, 10% acetic acid

Procedures

Gel Electrophoresis. One-dimensional SDS–polyacrylamide gel electrophoresis is performed according to conventional techniques[9,10] in slab gels. Subsequent transfer of proteins to nitrocellulose sheets is possible with acrylamide concentrations between 5% and 20%. This is also valid for gradient gels with top and bottom concentrations of acrylamide within this range. It is convenient to add 5 μl of transfer indicator either to the sample prior to electrophoresis or on top of the stacking gel when the Bromphenol Blue tracking dye front has moved to within 1–2 cm of the bottom of the separating gel. In the latter case it is important to know that methyl green migrates in the stacking gel about twice as fast as the dye front in the separating gel. The methyl green should enter the top of the separating gel before the dye front runs off the gel. Methyl green binds tightly to nitrocellulose and can therefore serve as a readily visible indicator of distortion-free and efficient transfer. It also allows the easy identification of individual gel lanes on the replica so that they can be excised and treated separately, if desired. However, methyl green competes with proteins for binding sites on the nitrocellulose; in order not to distort the protein replica, it is therefore advisable to limit the band of methyl green to either the top or the bottom of the separating gel.

Electrophoretic Transfer. Plastic gloves should be worn during all

[9] U. K. Laemmli, *Nature (London)* **227,** 680 (1970).
[10] M. Douglas, D. Finkelstein, and R. A. Butow, this series, Vol. 56, p. 58.

manipulations. After SDS–gel electrophoresis, the gel slab is briefly soaked in transfer buffer and then incorporated into a "sandwich" as illustrated in Fig. 1. To prepare the sandwich a perforated plastic plate (a in Fig. 1) is laid flat on a support and overlayered successively with (b) a sponge pad, (c) a 14 cm × 10 cm sheet of filter paper (Whatman 3 MM), (d) the polyacrylamide gel slab (12.5 cm × 9 cm) without the stacking gel, (e) the nitrocellulose sheet cut to the size of the gel (optionally one or more additional nitrocellulose sheets can be used, depending on the experiment), (f) a second sheet of Whatman filter paper, (g) a second sponge pad, and finally, (h) a second perforated plastic plate. All filter papers and nitrocellulose are prewetted in the transfer buffer prior to assembly. All air bubbles that may be trapped between the different layers must be painstakingly removed since they block the flow of current and thereby seriously distort the replica pattern on the nitrocellulose sheet. The sandwich is finally fixed into position with thick rubber bands and immersed in an electrophoretic transfer chamber (Fig. 1) filled with transfer buffer, in such a way that the nitrocellulose sheet faces the anode (in order to avoid errors, it is convenient to mark the topmost perforated plastic plate of the sandwich). The electrodes of the transfer chamber consist of two stainless steel wire grids placed 5 cm apart. Electrophoresis for 1.5–2 hr at 40 V and 1 A is usually sufficient to ensure efficient transfer of the protein from the gel slab to the nitrocellulose sheet. To avoid excessive heating of the buffer during the transfer, the entire chamber can be placed into a container filled with ice water. Stirring or recirculating the buffer is not necessary.

After transfer, the polyacrylamide gel is fixed and stained in staining solution B for 30–60 min and then destained to assess the efficiency of the protein transfer. For direct visualization of transferred proteins, the nitrocellulose sheet is stained for 1–5 min in staining solution A, followed by rapid destaining in destaining solution.

Immunological Detection of Transferred Polypeptides. The nitrocellulose sheet is incubated with gentle shaking in serum buffer for 2 hr at room temperature to saturate all remaining protein binding sites on the nitrocellulose. The sheet is then briefly rinsed with PBS and, if desired, cut into vertical 3- to 8-mm-wide strips, using the methyl green bands as indicator for the transferred gel lanes. (Nitrocellulose filters containing transferred polypeptide antigens can be stored in a desiccator at room temperature for at least several months.) The nitrocellulose sheet (or one of the strips derived from it) is incubated with gentle shaking (for 3–16 hr) with antiserum diluted into fresh serum buffer. We routinely use 25 ml or 3 ml of serum buffer solution and 5–20 μl or 0.5–2 μl of antiserum for immune decoration of sheets or strips, respectively. After washing out excess antibody with serum buffer (3 times for 30 min each at room temperature),

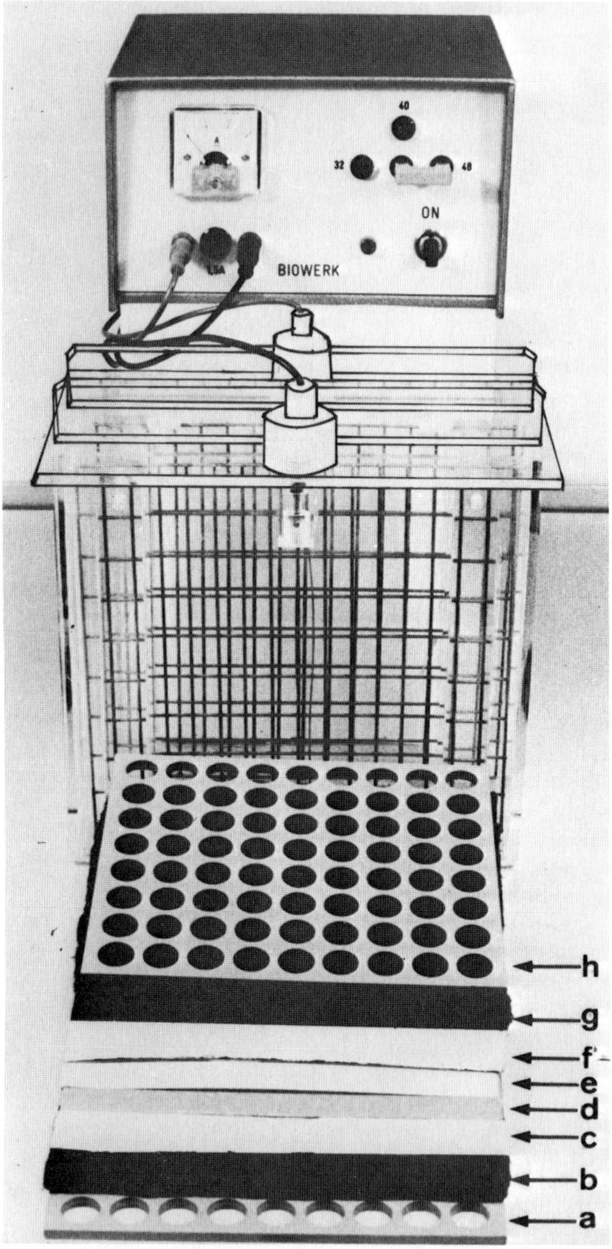

FIG. 1. Setup for electrophoretic transfer. Components necessary for preparing a replica are shown. The electrophoretic transfer setup consists of a power supply, an electrophoretic transfer chamber, and the sandwich. This sandwich is composed of perforated plastic plates (a, h), sponge pads (b, g), filter paper (c, f), polyacrylamide gel slab (d), and a nitrocellulose sheet (e).

sheets are incubated for 1–3 hr at room temperature with 25 ml of serum buffer containing ^{125}I-labeled protein A (5×10^5 cpm to 1×10^6 cpm/ml, 2.2×10^7 cpm/μg protein A) or with ^{125}I-labeled antibodies directed against the first antibody [(2–5) $\times 10^5$ cpm/ml, 2×10^6 cpm/μg IgG]. For strips, all volumes and amounts are lowered 8-fold. Nonspecifically adsorbed iodinated protein A or immunoglobulins are then removed by 3–5 further washes with PBS for 30 min each at room temperature. The signal-to-background ratio can be increased by including 1% Triton X-100 in the final washing solution. The nitrocellulose membrane is dried between two sheets of filter paper with a stream of warm air, wrapped in a thin plastic foil (e.g., Saran wrap) and exposed to Kodak XS-5 or SB-5 film either at room temperature or at −70° utilizing a Kyokko high-speed intensifying screen.

Immune complexes on nitrocellulose sheets can also be detected by a color reaction, using a second antibody conjugated to horseradish peroxidase.[3] The protein antigen is localized on the replica by the peroxide reaction product, which is deposited at the site of the antigen–antibody complex. This staining method is faster and less expensive, but also somewhat more prone to interference than the autoradiographic method. If the staining method is to be used, azide should be omitted from all solutions, since it inhibits peroxidase. The nitrocellulose sheets are incubated with the first antibody, washed as described earlier (except for the absence of azide), and incubated for 2 hr at room temperature in 25 ml of serum buffer containing 10–20 μl of peroxidase-conjugated secondary immune globulin. Excess of second antibody is removed by washing with PBS as above. For staining, the nitrocellulose sheet is immersed in 25 ml of a freshly prepared solution of 0.02% 4-chloro-1-naphthol, 0.006% hydrogen peroxide in PBS and incubated until staining of individual bands no longer increases. Since the color is not stablc, thc stained sheets can be photographed or xerographed.

Reuse of Solutions and Filter Replicas. Most solutions used for incubating the nitrocellulose sheets (e.g., solutions for blocking excess protein binding sites, for immune reaction or for decoration of the immunocomplexes) can be reused several times without any apparent decrease in the sensitivity of the immunodetection. They can be stored frozen at −20° or, if no peroxidase staining is planned, at 4° in the presence of 0.1% NaN_3. Even a filter replica can be used several times. For example, it can be sequentially treated with different antisera either with[11] or without[12] the removal of the preceding antibodies.

[11] R. P. Legocki and D. P. S. Verma, *Anal. Biochem.* **111,** 385 (1981).

[12] R. T. M. J. Vaessen, J. Kreike, and G. S. P. Groot, *FEBS Lett.* **124,** 193 (1981).

Efficiency of Protein Transfer to Nitrocellulose. Several recent articles deal with the problem of transferring proteins from SDS–polyacrylamide gels to nitrocellulose sheets.[8,8a,8b,11,12] Basically, transfer efficiency depends on the elution rate of the individual proteins from the polyacrylamide gel and their ability to adsorb onto nitrocellulose. With most proteins the velocity of electrophoretic elution is inversely related to the apparent molecular weight, in other words, large proteins are eluted less effectively than smaller ones. Elution of large proteins can be improved by resolving proteins in SDS–polyacrylamide gradient gels[10] (5% to 20% acrylamide) or by limited proteolysis of proteins during electrophoretic transfer in order to reduce their size.[12a] Even some relatively small mitochondrial membrane proteins (such as the adenine nucleotide translocator) are eluted only very slowly, perhaps because they contain strongly hydrophobic regions. However, virtually all proteins can be eluted if 0.02% SDS is added to the transfer buffer. This modified buffer system does not interfere with the immobilization of proteins onto the nitrocellulose sheets or with their immunodetection.

Not all proteins adsorb equally well to nitrocellulose. Some proteins, particularly small ones, are bound less efficiently and tend to leak through the membrane. This can be readily detected by conducting the electrophoretic elution with two nitrocellulose sheets placed next to each other and checking for appearance of antigens on the sheet closest to the anode. If the protein of interest "leaks" through, conditions of transfer should be modified, e.g., by using nitrocellulose sheets with smaller pore size, sheets of material other than nitrocellulose [e.g., (diazobenzyloxymethyl) cellulose,[2,4,6,7] Zeta bind[13]], different pH, different composition of the transfer buffer, decreased electrical field strength during elution, or shorter elution times.[12]

Applications of the Filter Transfer Method

Since its introduction by Towbin *et al.*,[3] the electrophoretic protein transfer to nitrocellulose sheets has become an extremely valuable tool in many biological fields. The power of the method will be illustrated by showing how it has been applied to a variety of problems encountered in studying mitochondrial biogenesis.

Detection and Quantitation of Antigens in Complex Mixtures

The combination of SDS–polyacrylamide gel electrophoresis with a radioimmune assay enables the detection of individual proteins even in complex mixtures of tightly interacting membrane proteins. Because the

[12a] W. Gibson, *Anal. Biochem.* **118,** 1 (1981).

[13] J. M. Gershoni and G. E. Palade, *Anal. Biochem.* **124,** 396 (1982).

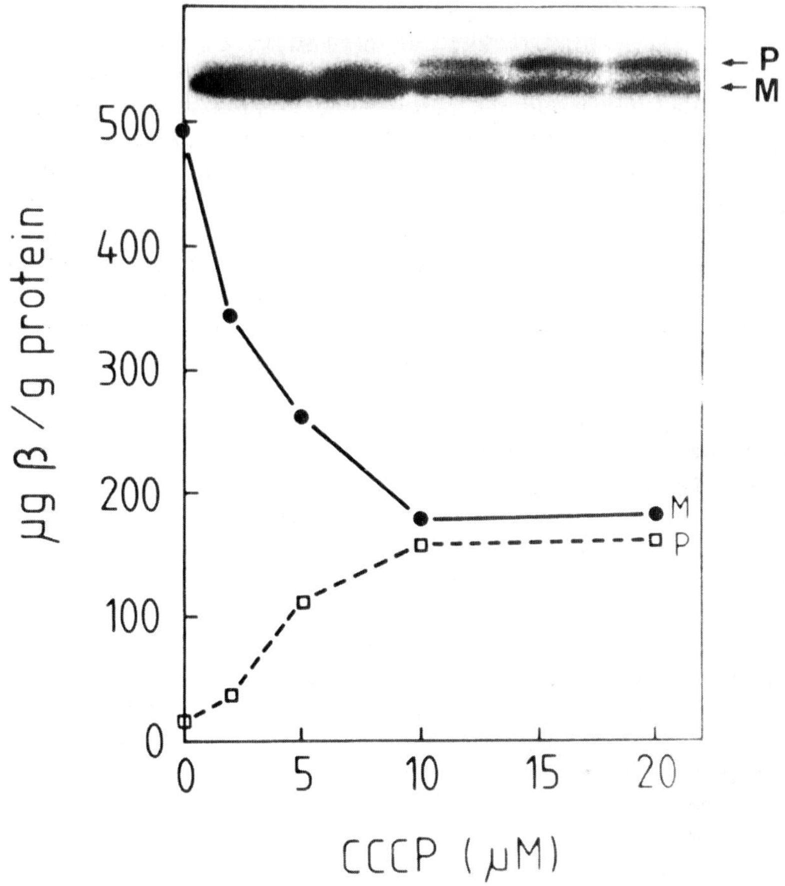

FIG. 2. Identification and quantitation of the mature and precursor form of the β-subunit of F_1-ATPase in yeast cells after growth in the presence of carbonyl cyanide *m*-chlorophenyl hydrazone (CCCP). A *rho*⁻ mutant was grown in the presence of CCCP, and cell extracts were prepared as outlined by Ohashi *et al.*[15] Aliquots of the cell extracts containing 200 μg of protein each were electrophoresed in a 10% SDS–polyacrylamide gel slab, and the separated bands were transferred to a nitrocellulose sheet. The replica was allowed to react with antiserum directed against the β-subunit of F_1-ATPase and decorated with ^{125}I-labeled protein A. The bands corresponding to the mature (M) and the precursor (P) forms in the five lanes (upper part of the figure) are from cells grown in the presence of (left to right) 0, 2, 5, 10, and 20 *μM* CCCP, respectively. They were located on the immune replica by autoradiography, excised, and counted in a gamma counter; the specific radioactivity was calculated from a β-subunit standard that was treated in the same manner. The results of these calculations are plotted as a function of the CCCP concentration in the bottom part of the figure.

antigen is immobilized on the nitrocellulose, the antibody need only bind to the antigen; it is not required to form an immune precipitate. This allows detection of even very small amounts of antigen. Also, within certain limits, the darkening of the autoradiogram is proportional to the amount of ^{125}I-labeled protein A bound to the immune complex and, hence, to the amount of antigen present on the filter. Both criteria make it possible to determine the steady-state concentration of proteins present in complex mixtures such as whole-cell lysates. The sensitivity of the method is very high; detection of as little as 100 pg of a protein has been reported.[12]

A typical example is the detection and quantitation of precursors to imported mitochondrial proteins in intact yeast cells.[14] This is shown in Fig. 2 for the β-subunit of F_1-ATPase. Intact yeast cells that had been grown in the presence of the uncoupler carbonyl cyanide *m*-chlorophenyl hydrazone (CCCP) were rapidly denatured in 20% trichloroacetic acid to avoid proteolysis and broken with glass beads.[15] After dissociation in SDS, the cell extracts were resolved by SDS–polyacrylamide gel electrophoresis, and the gel slab was then subjected to the immune replica technique using antiserum against the F_1-ATPase β-subunit. By coelectrophoresing calibrated amounts of purified β-subunit with graded amounts of cell extracts, the concentration of mature and precursor forms was measured by cutting the corresponding bands out of the nitrocellulose sheet and counting them for ^{125}I. The only assumptions in these measurements are that the SDS-denatured form of the precursor binds the same amount of antibody as the SDS-denatured mature form and that denatured mature and precursor forms transfer to the nitrocellulose filter with equal efficiency.

In addition to quantitative studies, the technique has also been used for (*a*) the immunological characterization of mitochondrial subfractions by marker antigens[16] (Fig. 3); (*b*) the screening of nuclearly inherited yeast mutants defective in the formation of a functional cytochrome c_1 (Fig. 4) and the analysis of evolutionary relationships between proteins from various sources.[17,18]

Characterization of Antisera

Immunoreplication has also proved to be valuable for checking the purity and the titer of antisera. It readily detects antibodies exhibiting low

[14] G. A. Reid and G. Schatz, *J. Biol. Chem.* **257,** 13056 (1982).

[15] A. Ohashi, J. Gibson, I. Gregor, and G. Schatz, *J. Biol. Chem.* **257,** 13042 (1982).

[16] G. Daum, P. C. Böhni, and G. Schatz, *J. Biol. Chem.* **257,** 13028 (1982).

[17] B. Ludwig, *Biochim. Biophys. Acta* **594,** 177 (1980).

[18] R. Rott and N. Nelson, *J. Biol. Chem.* **256,** 9224 (1981).

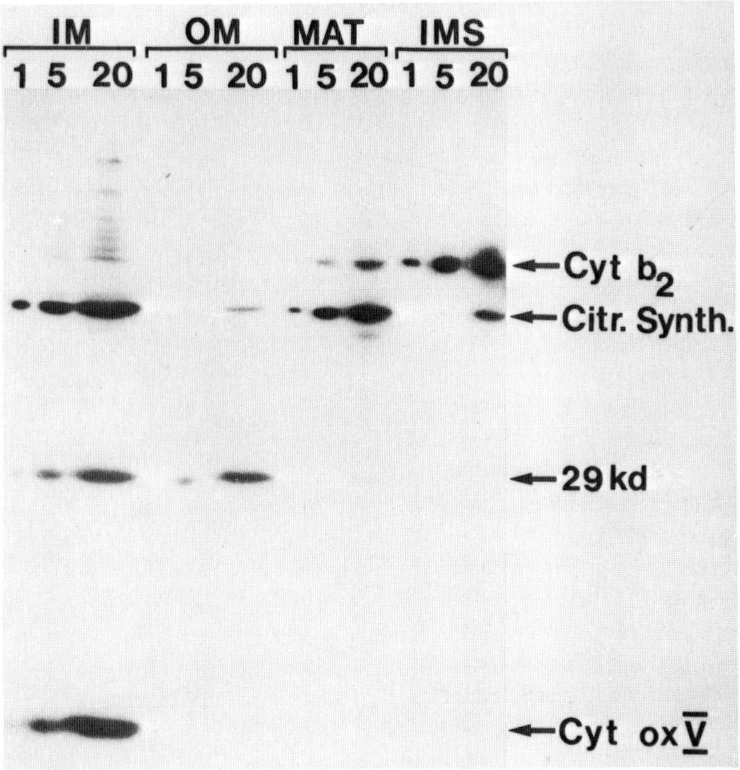

FIG. 3. Immunological test for purity of mitochondrial subfractionation. Yeast cells grown on lactate were converted to spheroplasts. The mitochondria were isolated and separated into the following subfractions: inner membrane (IM), outer membrane (OM), matrix space (MAT), and intermembrane space (IMS).[16] Aliquots of each fraction equivalent to 2, 10, and 40 μg of whole mitochondria (ratio 1 : 5 : 20) were electrophoresed in a 12.5% SDS–polyacrylamide gel, and the resolved proteins were subsequently transferred to a nitrocellulose sheet. The immune replica was prepared after treating the replica with a mixture of antisera against cytochrome b_2 (Cyt b_2; an IMS protein), citrate synthase (Citr. Synth., a MAT protein), the major 29,000 dalton (29K polypeptide of the OM), and cytochrome *c* oxidase subunit V (Cyt oxV, an IM protein). Cross-contamination in the different fractions can be assessed as described in Fig. 2.

affinities or antibodies directed against a single antigenic site, such as monoclonal antibodies produced by hybridomas. For testing antisera against yeast proteins, we routinely use nitrocellulose test strips with electrophoretically resolved polypeptides of total yeast cells or total isolated yeast mitochondria. For preparing these strips, yeast cells are quickly denatured in 20% trichloroacetic acid, broken with glass beads,

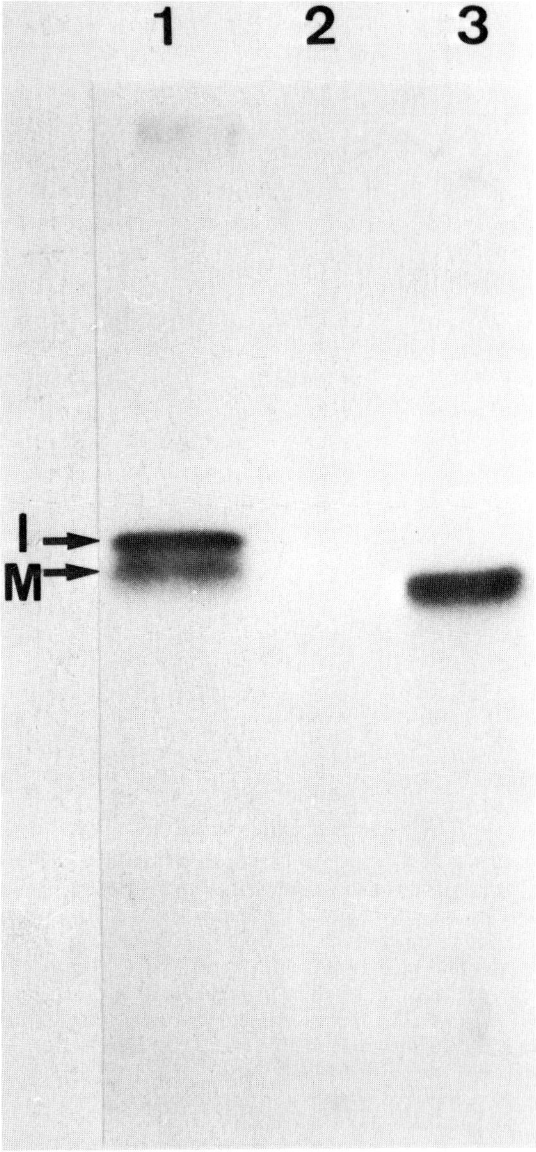

FIG. 4. Immunological screening of yeast mutants lacking functional cytochrome c_1. Wild-type yeast cells and nuclearly inherited yeast mutants lacking functional cytochrome c_1 were grown in 1-ml cultures, converted to spheroplasts, and dissociated in 5% SDS at 95°.[14] Aliquots of the extracts containing 50 μg of protein each were electrophoresed in a 15% SDS–polyacrylamide gel slab, followed by transfer to a nitrocellulose sheet, which was subsequently allowed to react with antiserum against cytochrome c_1. Lanes 1 and 2, two different cytochrome c_1-defective mutants; lane 3, wild-type cells. M and I represent the mature and the intermediate form of cytochrome c_1, respectively.[15]

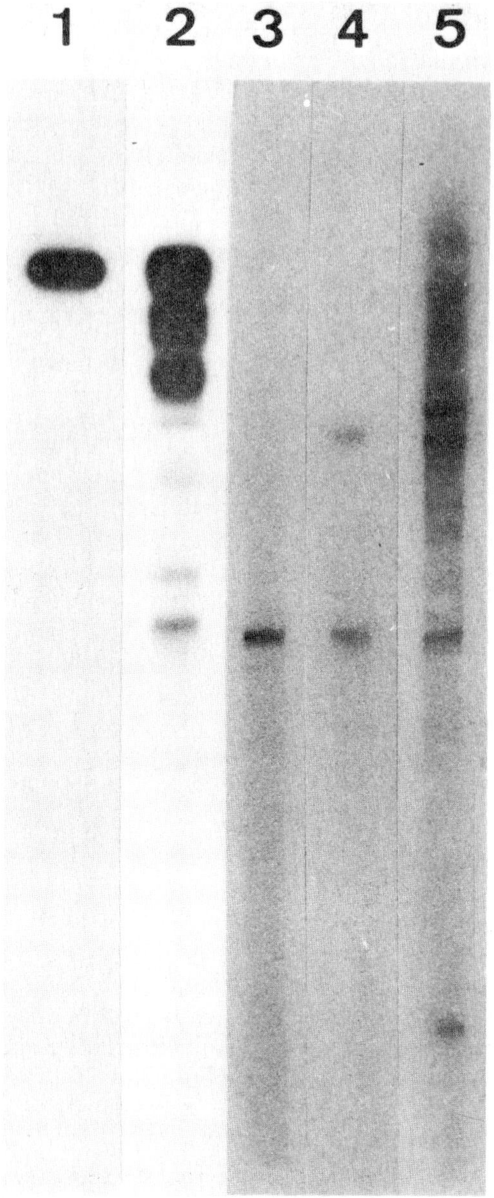

FIG. 5. Testing the specificity of antisera by immune replication. Rabbit antisera against yeast hexokinase and the γ-subunit of F_1-ATPase were allowed to react against SDS-dissociated, electrophoretically resolved polypeptides of yeast spheroplasts or yeast mitochondria by the immune replica method, and the immune complexes on the nitrocellulose sheet were visualized with ^{125}I-labeled protein A and autoradiography. The spheroplasts had been dissociated in 5% SDS for 3 min at 95° either in the presence (lane 1) or the absence (lane 2) of protease inhibitors.[19] The mitochondria (lanes 3–5) were dissociated as described above in

and extracted with hot SDS.[15] Alternatively, yeast mitochondria are isolated from spheroplasts in the presence of protease inhibitors[19] and immediately dissociated in hot SDS. The extracts are centrifuged as described by Reid.[20] An aliquot containing approximately 1 mg of proteins is layered over the entire width of an SDS–polyacrylamide gel and subjected to electrophoresis. The resolved bands are transferred to a nitrocellulose sheet; the sheet is saturated with serum buffer and cut into strips (3–8 mm wide). For testing an antiserum, strips are incubated with 0.05–10 μl of antiserum in 3 ml of serum buffer and the immune complex is visualized either with iodinated protein A or the peroxide reaction, as described earlier.

This test may present some problems. If the tester antigen has been proteolytically degraded, multiple bands appear (Fig. 5, lane 2). However, all "extra" bands have a lower apparent molecular weight than the authentic antigen. This feature usually differentiates proteolytic breakdown products from genuine contaminants (Fig. 5, lane 5). The appearance of proteolytic fragments can often be prevented by adding protease inhibitors during the isolation of the tester antigen (Fig. 5, lane 1).

Another problem is the appearance of immunoreactive bands exhibiting a lower mobility than the antigen. These bands are often aggregates, reflecting incomplete dissociation of the test mixture prior to electrophoresis (Fig. 5, lane 4). If this is suspected, dissociation of the extract in SDS should be repeated at a higher SDS : protein ratio or at a lower absolute concentration of protein. Additionally, heating in SDS should be avoided, even though this may entail the risk of generating proteolytic fragments because of incomplete denaturation of cellular proteases.

A third problem is the failure of some antisera to react with antigen present in total cell proteins, even though adequate reaction is observed with purified antigen or with antigen partially enriched in subcellular fractions. Although we have not investigated this phenomenon in detail, we suspect that it results from the presence of excess proteins in total cell extracts. If these proteins have the same electrophoretic mobility as the antigen under study, they would compete with it for the limited binding

[19] C. Côté, M. Solioz, and G. Schatz, *J. Biol. Chem.* **254,** 1437 (1979).
[20] G. A. Reid, this series, Vol. 97 [31].

the presence of protease inhibitors, except that the SDS concentration was lowered to 1% with the sample run in lane 4. The electrophoresed samples were transferred to a nitrocellulose sheet, and five strips corresponding to each of the five lanes were cut out. Strips 1 and 2 were allowed to react with antiserum against hexokinase, strips 3 and 4 with antiserum against the γ-subunit of F_1-ATPase, and strip 5 with an antiserum against the γ-subunit of F_1-ATPase containing various contaminating antibodies.

sites on the nitrocellulose sheet. This problem can often be overcome simply by preparing nitrocellulose test strips with a lower amount of total cell protein.

Identification of Glycoproteins

Polypeptides transferred to nitrocellulose sheets can also be identified with specific probes other than antibodies.[21] For example, glycoproteins can be visualized by reaction with radiolabeled lectins, such as concanavalin A or wheat germ agglutinin. Transfer of proteins and saturation of the filter is done essentially as described above, except that serum buffer is replaced by BSA buffer during all reaction steps. (The serum, present in serum buffer contains glycoproteins.) After saturation, the nitrocellulose sheet is incubated for 3–4 hr at room temperature in 25 ml of BSA buffer containing radioiodinated lectins (in the case of concanavalin A, 1.3×10^7 cpm; specific activity 1.8×10^7 cpm/μg). For strips, all amounts and volumes are reduced 8-fold. After several washes with BSA buffer the sheet is processed as described earlier for immunodetection. Since proteins may bind lectins through interaction not involving sugar groups, a second identical replica should always be treated with the appropriate hapten carbohydrate prior to and during incubation with the radiolabeled lectin; with ^{125}I-labeled concanavalin A, α-methyl mannoside (40 mg/ml in BSA buffer) is used as the competing hapten. Only bands that react in the absence, but not in the presence, of the hapten sugar can be presumed to be glycoproteins.

[21] J. C. S. Clegg, *Anal. Biochem.* **127,** 389 (1982).

[14] Immunologic Detection of Specific Proteins in Cell Extracts by Fractionation in Gels and Transfer to Paper

By JAKOB REISER and GEORGE R. STARK

Transfer of Proteins from Gels to Paper

Background and General Principles

Specific proteins can be characterized and quantified in cell extracts even if enzymic or biological assays are not available for them. For example, high-resolution two-dimensional gel electrophoresis[1] makes it possi-

[1] P. Z. O'Farrell, H. M. Goodman, and P. H. O'Farrell, *Cell* **12,** 1133 (1977).

ISBN 0-12-181996-5

ble to separate more than 1000 different cellular proteins in a single gel. However, since minor proteins are difficult to detect in this way, immunoprecipitation with specific antisera is used to enrich for them before analysis on gels. But immunoprecipitation has an inherent ambiguity: It is impossible to distinguish proteins that react directly with the antibody from those that simply are bound to the specific antigen. To circumvent such complications, alternative approaches were developed in which the proteins are first separated by gel electrophoresis and then detected with radioactive antibody after transfer from the gel to a solid support. Renart *et al.*[2] first described a system in which the proteins were separated in polyacrylamide–agarose composite gels, with or without SDS. The polyacrylamide cross-links were cleaved with periodate or alkali to facilitate transfer of the proteins by diffusion to diazobenzyloxymethyl (DBM) paper, where they coupled covalently. Using a similar approach, Bowen *et al.*[3] transferred proteins by diffusion from standard SDS–polyacrylamide gels to sheets of nitrocellulose. Towbin *et al.*[4] were the first to introduce a simple electrophoretic technique for transferring proteins from polyacrylamide gels containing SDS or urea to nitrocellulose. Transfer of proteins from polyacrylamide gels by electrophoresis is much more rapid than transfer by diffusion, which is slow and fairly inefficient. But electrophoretic transfer is still incomplete if standard polyacrylamide gels containing SDS are used.[4] Stellwag and Dahlberg[5] described an electrophoretic system similar to the one of Towbin *et al.*,[4] but they used DBM paper. Transfer from urea gels was efficient, but transfer from SDS gels was not. Bittner *et al.*[6] claimed to transfer proteins quantitatively from standard SDS gels to DBM paper, but they estimated the efficiency by comparing the intensity of radioactive bands visually before and after transfer of ^{35}S-labeled proteins. Since quenching causes the signals obtained from ^{35}S-labeled proteins in dried gels to be less intense than from labeled proteins attached to DBM paper, visual comparison of autoradiograms leads to an overestimate of transfer efficiency. Also, soaking the gel before transfer under the conditions described results in substantial loss of protein. Using the exact conditions of Towbin *et al.*,[4] Burnette[7] obtained good transfer efficiencies for proteins with molecular weights up to 97,000 by electrophoresis onto nitrocellulose for 22 hr. Disadvantages are that long times are required and that distortion of the bands due to unequal expansion of the gel and the nitrocellulose may occur during long-term electrophoresis.

[2] J. Renart, J. Reiser, and G. R. Stark, *Proc. Natl. Acad. Sci. U.S.A.* **76,** 3116 (1979).

[3] B. Bowen, J. Steinberg, U. K. Laemmli, and H. Weintraub, *Nucleic Acids Res.* **8,** (1980).

[4] H. Towbin, T. Staehelin, and J. Gordon, *Proc. Natl. Acad. Sci. U.S.A.* **76,** 4350 (1979).

[5] E. J. Stellwag and A. E. Dahlberg, *Nucleic Acids Res.* **8,** 299 (1980).

[6] M. Bittner, P. Kupferer, and C. F. Morris, *Anal. Biochem.* **102,** 459 (1980).

[7] W. N. Burnette, *Anal. Biochem.* **112,** 195 (1981).

The system of Reiser and Wardale,[8] which allows efficient transfer of proteins from SDS–polyacrylamide gels to diazophenyl thioether (DPT) paper with only 1 hr of electrophoresis, is described in detail below.

Preparation of Aminophenyl Thioether (APT) Paper

The following procedure was worked out by Brian Seed, California Institute of Technology, and is given here with his permission. APT paper is also available commercially.

Materials and Reagents

Paper for derivatization: Whatman 50 or 540, or Schleicher & Schuell 589 Red Ribbon or 507-C
Heat-sealable plastic bags
Apparatus for rotating papers during reaction
Sodium borohydride
1,4-Butanediol diglycidyl ether (Aldrich)
Ethanol, 95%
2-Aminothiophenol (Aldrich)
Solutions
NaOH, 0.5 *M*
NH_4OH, 1 *M*
Hydrochloric acid, 0.1 *M*

Procedure. Place seven sheets of paper (14 cm by 25 cm each, about 20 g) in a heat-sealable freezer bag. Add 100 ml of 0.5 *M* sodium hydroxide containing 2 mg/ml of sodium borohydride to wet all the paper, then add 45 ml of 1,4-butanediol diglycidyl ether. Since this reagent may be carcinogenic, carry out the entire procedure in a hood and wear gloves. Expel air from the bag, seal it, and rotate it slowly end-over-end for 16 hr at room temperature. Gas evolved during the first hour should be released by making a small hole in one corner of the bag, which should then be resealed. Pour the liquid into 1 *M* NH_4OH and leave for 24 hr before disposing down the drain. Wash the papers for 1 hr in 150 ml of 0.5 *M* NaOH : ethanol (1 : 1). Drain. Add a mixture of 75 ml of 0.5 *M* NaOH and 75 ml of 5% 2-aminothiophenol in ethanol. Seal the bag again and rotate it end-over-end for another 2 hr. Discard this solution into a container for chemical wastes. Wash the papers with a large volume of ethanol, then with 0.1 *M* hydrochloric acid and repeat these two washes. Wash the papers with running distilled water, dry them in air, and store them over silica gel at 4°, or at room temperature in the dark. Carry out all operations up to the water wash in a fume hood.

[8] J. Reiser and J. Wardale, *Eur. J. Biochem.* **114,** 569 (1981).

SDS–Polyacrylamide Gel Electrophoresis

We have used the discontinuous borate–sulfate buffer system of Neville[9] because diazonium groups are inactivated rapidly by the glycine present in the more widely used electrophoresis buffer of Laemmli.[10] Gradient gels are recommended for proteins with a molecular weight (M_r) greater than 100,000, since the efficiency of transfer is quite substantially affected by the concentration of acrylamide in the gel and the degree of cross-linking. Gradient gels (14 cm by 10.5 cm by 0.1 cm) have been used with acrylamide ranging from 7% to 20% and with bisacrylamide ranging from 0.09% to 0.15%. We give only one example here: preparation of a 10% nongradient gel at pH 9.18.

Solutions

Upper reservoir buffer, pH 8.64: boric acid, 0.04 *M*; Tris base, 0.04 *M* (approximately); SDS, 0.1%. Prepare 5 times concentrated stock solution.

Buffer for stacking gel, pH 6.1: Tris base, 0.054 *M*; H_2SO_4, 0.027 *M*. Prepare 8 times concentrated stock solution.

Lower reservoir and running gel buffer, pH 9.18: Tris base, 0.42 *M*; HCl, 0.031 *M*. Prepare 6 times concentrated stock solution.

Acrylamide stock solution for resolving gel, 3 times concentrated (3×): 30 g of acrylamide, 0.30 g of bisacrylamide, 73 ml of quartz-distilled water.

Acrylamide stock solution for stacking gel, 5 times concentrated (5×): 3.2 g of acrylamide, 0.2 g of bisacrylamide, 20 ml of quartz-distilled water.

Procedure. To prepare a 10% acrylamide gel (13.5 cm × 13.5 cm × 0.15 cm), mix and deaerate 6× buffer, pH 9.18, 5.0 ml; 3× acrylamide stock solution for resolving gel, 10.0 ml; H_2O, 14.2 ml. Then add 10% SDS, 0.30 ml; ammonium persulfate, 0.30; TEMED, 0.05. Swirl the mixture gently and pour the gel immediately. Carefully layer isobutanol on top and let the gel polymerize. Remove the isobutanol and wash the gel surface with water before adding the stacking gel.

To prepare the stacking gel, mix and deaerate 8× buffer, pH 6.1, 1.0 ml; 5× acrylamide stock solution, for stacking gel, 1.6 ml; H_2O, 5.2 ml. Then add 10% SDS, 0.10 ml; 10% ammonium persulfate, 0.05; TEMED, 0.012. Swirl to mix, pour, and introduce the well spacer immediately. Allow the gel to polymerize for 30 min.

[9] D. M. Neville, Jr., *J. Biol. Chem.* **246,** 6328 (1971).

[10] U. K. Laemmli, *Nature (London)* **227,** 680 (1970).

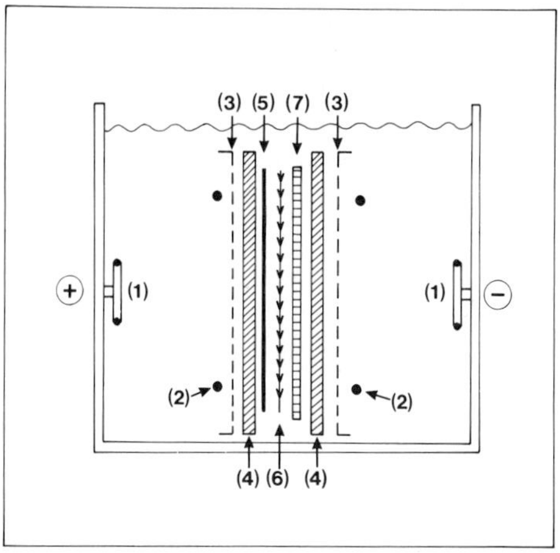

FIG. 1. Assembly for electrophoretic transfer of proteins. 1, Circular platinum wire electrodes; 2, rubber bands; 3, disposable pipette-tip trays; 4, foam pads; 5, Whatman 50 paper; 6, diazophenyl thioether paper; 7, SDS–polyacrylamide gel. The tank is constructed of 0.6-cm Plexiglas and has inner dimensions of 19.7 × 20.4 × 23.2 cm. The circular platinum electrodes are 12.5 cm apart. Each electrode has a circumference of 12.5 cm. (Note that larger electrodes give less voltage.)

Preparation of Samples. Mix 4 volumes of cell extract and 1 volume of 5 times concentrated sample buffer (5% SDS, 0.5 g; 250 m*M* dithiothreitol, 0.38 g; 50 m*M* sodium phosphate, pH 7.0; 1 ml of 0.5 *M* stock solution; 50% glycerol, 5 ml; Bromphenol Blue, 3 mg; water to 10 ml). Store 0.5-ml aliquots at −20°. Then incubate the mixture at 90° for 5 min.

Electrophoretic Transfer at pH 9.2 and 0° Using High Voltage

Materials and Reagents

Transfer apparatus (Fig. 1). Note that the size of the electrode is critical for getting high enough voltage.
Foam pad, 5 mm thick
Stiff plastic grid (e.g., disposable micropipette tray, Gilson)
Whatman 50 paper
Aminophenyl thioether (APT) paper
Enamel or glass trays
$NaNO_2$
Buffers and solvents

Wash buffer: sodium borate, 50 m*M*, pH 9.2
Transfer buffer: sodium borate, 10 m*M*, pH 9.2
Hydrochloric acid, 1.2 *M*
Inactivation buffer: 0.1 *M* Tris-HCl, pH 9.0, 10%
Ethanolamine (optional—see below)
Gelatin, 0.25%

Procedure. Prepare the gels for transfer by incubating them at room temperature three times for 10 min each with 250 ml of distilled water and once for 10 min with 50 m*M* sodium borate, pH 9.2, all with gentle rocking. Begin diazotizing the APT paper and washing the gel at the same time. Use a sheet of APT paper that exceeds the size of the gel by about 15% in each dimension. Prechill 100 ml of 1.2 *M* HCl in an enamel or glass tray on ice. Add 2.7 ml of $NaNO_2$ solution (10 mg/ml), made immediately before use. Add the paper and occasionally shake gently on ice for 20–40 min. The paper should become bright yellow, with uniform density of color. Wash the DPT paper with several changes of 100 ml each of ice-cold water and ice-cold transfer buffer. Time this so that the paper is in the ice-cold buffer for only 1–2 min before final assembly with the gel.

Put the foam pad, soaked in transfer buffer, on top of a stiff plastic grid, and put the gel on top of the foam pad. Place freshly prepared DPT paper on the gel, followed by a sheet of Whatman 50 paper that has been soaked briefly in transfer buffer. Take care to remove all air bubbles between the gel and the DPT paper. Add a second foam pad and plastic grid, and string rubber bands around all the layers. (The same kind of assembly can be used for transferring proteins from two gels simultaneously. The second gel is put on top of the second foam pad, and DPT paper and Whatman 50 paper are added, followed by an additional foam pad and plastic grid.) Put the assembly into the tank, which contains 5 liters of ice-cold transfer buffer.

Perform the transfer for 1 hr at 400 V and 300–400 mA (32 V/cm). The temperature in the tank will increase by 10–20° during the electrophoresis. This does not affect the quality of the transfer. Note that there is no physical separation between the positive and negative chambers and that current is allowed to go through the gel as well as through the buffer surrounding it. After transfer, incubate the paper at room temperature for 15 min with gentle rocking in 100 ml of a buffer containing 0.1 *M* Tris-HCl, pH 9.0, 10% ethanolamine, 0.25% gelatin (or for 2–16 hr in the same buffer without ethanolamine) to inactivate any remaining diazonium groups. The paper responds to high pH by becoming brick-red. The paper can be stored in a sealed container at room temperature or 4° after this step for at least 6 months without affecting the results.

Electrophoretic Transfer at pH 6.5 and Room Temperature Using Low Voltage

This procedure gives results quite comparable to those Reiser and Wardale[8] have obtained using pH 9.2 and high voltage, with transfer efficiencies greater than 50% for most proteins (E. A. Swyryd and G. R. Stark, unpublished results).

Materials and Reagents

Transfer apparatus. Several are available commercially. We have obtained excellent results with the Transphor (Hoefer Scientific Instruments).
Whatman 3 MM paper
APT paper
Enamel or glass trays
$NaNO_2$
Buffers and solvents
  Wash and transfer buffer: 15 m*M* sodium phosphate, pH 6.5
  Hydrochloric acid, 1.2 *M*
  Inactivation buffer (see preceding section).

Procedure. Displace the running buffer from the gel with three 5-min washes in 300 ml of 15 m*M* sodium phosphate buffer at room temperature with good agitation.

Timing is important. The following protocol is used:

At −30 min: Fill the tank with 15 m*M* sodium phosphate buffer pH 6.5 and begin to circulate cold tap water through the heat exchanger, to lower the starting temperature to about 18°.
At −22 min: Begin diazotizing the APT paper (see procedure in preceding section)
At −15 min: Begin gel wash 1.
At −10 min: Change to gel wash 2.
At −5 min: Change to gel wash 3.
At −2 min: Wash DPT paper and place in buffer on ice.
At time 0: Assemble transfer apparatus without delay.

Lift the gel from the bottom of the wash tray with a piece of Whatman 3 MM paper soaked in buffer by inverting the tray. Place the gel and paper onto the pad of the apparatus, with the gel surface up. Remove the activated paper from the cold buffer and place it on top of the gel. Finally, place another sheet of paper soaked in buffer and a second pad on top. Place the whole stack into the tank of buffer in the following orientation:

cathode–gel–DPT paper–anode. Attach electrodes and begin the transfer without delay. Run at 2 A (about 50 V) for 1 hr. The temperature will rise about 12° by the end of the run. Continue as in the preceding procedure.

Identification of Specific Proteins Bound to DPT paper

Background and General Principles

Renart *et al.*[2] identified specific proteins bound to DBM paper by sequential incubation with unfractionated unlabeled specific antiserum and ^{125}I-labeled protein A from *Staphylococcus aureus*. (If desired, an intermediate incubation with unlabeled anti-antibody can also be used, to increase sensitivity or to provide a bound γ-globulin that has high affinity for protein A.) This general approach has been used subsequently for detecting specific proteins bound to DBM paper[11] or sheets of nitrocellulose.[7,12] Provided that an excess of labeled affinity probe is used, the amount of film darkening in the autoradiogram is a relative measure of the steady-state level of the specific protein detected. Towbin *et al.*[4] have used ^{125}I-labeled second antibody, fluorescein-labeled second antibody or horseradish peroxidase-conjugated second antibody. Affinity probes other than antibody can also be used. Labeled DNA can serve to identify DNA binding proteins,[3] and the possibility of detecting RNA binding proteins or proteins capable of forming complexes with other ligands has also been discussed by Bowen *et al.*[3] J. Reiser and Karlsson (unpublished) used ^{125}I-labeled lectins to identify specific red cell membrane glycoproteins bound to DPT paper, and a similar procedure has been published for detecting glycoproteins bound to nitrocellulose.[13] J. Reiser and J. Wardale (unpublished) used ^{125}I-labeled avidin to detect proteins labeled with *N*-hydroxysuccinimide-biotin or glycoproteins tagged with biotin hydrazide.

Method

Materials

Specific antibody: It is best to prepare this in rabbits, hamsters, or guinea pigs because of the good binding between γ-globulins of these species and protein A.[14] Otherwise, use a sandwich, with unlabeled rabbit antiglobulin as the second stage.

[11] J. Symington, M. Green, and K. Brackmann, *Proc. Natl. Acad. Sci. U.S.A.* **78,** 177 (1981).

[12] R. P. Legocki and D. P. S. Verma, *Anal. Biochem.* **111,** 385 (1981).

[13] W. F. Glass, II, R. C. Briggs, and L. S. Hnilica, *Anal. Biochem.* **115,** 219 (1981).

[14] J. J. Lagone, this series, Vol. 70, p. 356.

^{125}I-labeled protein A: Prepare this at 3- to 6-month intervals, or more often as needed, using the Iodogen (Pierce Chemicals) technique.[15] Labeled protein A is also sold commercially.

Ten times concentrated (10×) NET buffer: 1.5 *M* NaCl, 0.05 *M* EDTA, 0.50 *M* Tris-HCl, pH 7.4

NET–gel–NP-40 buffer: 1× NET buffer, 0.25% gelatin, 0.05% Nonidet P-40 (NP-40)

NET–Sarkosyl buffer: 1× NET, 0.5 *M* NaCl, 0.4% Sarkosyl

Procedure. The procedure of Renart *et al.*[2] is followed, with minor modifications. Wash the inactivated paper in a tray in NET–gel–NP-40 for 20–30 min with a couple of changes of solution. There will be a slight change of color. Rinse with water between and after washes.

Blot excess moisture from the paper and place it in a close-fitting, heat-sealable plastic bag. For a 14-cm^2 sheet, add 5 ml of NET–gel–NP-40 to the bag and 10–100 μl of antiserum, depending on the titer. Seal the bag and place it on a rocking device for 1–2 hr at room temperature. Rinse the paper with distilled water, place it in a rocking tray in NET–gel–NP-40 for 2 hr at room temperature.

Rinse the paper with distilled water, blot it, replace it in a fresh, close-fitting bag. Add 5 ml of buffer plus ^{125}I-labeled protein A [(1–2) × 10^6 cpm]. Seal and place in a rocking device for 1 hr at room temperature. (The antiserum and protein A can also be mixed before adding to the paper all at once, but, in this case, enough protein A must be used to saturate the binding sites of the excess free γ-globulin.) Rinse the paper well with distilled water, wash for 1 hr at room temperature with two or three changes of NET–Sarkosyl buffer, rinse again, and blot damp-dry. Autoradiograph using Kodak SB-5, presensitized Kodak XRP5, or Fuji RX X-ray film and an Ilford fast tungstate or DuPont Cronex Lightning-plus intensifying screen at −70°.

Reuse of Transfers

Renart *et al.*[2] showed that antibody and protein A bound to DBM paper could be removed and that the same paper could be reprobed again with a different antiserum. The potential for reusing the same blot several times becomes important when precious antigen is used or when different antigens in two-dimensional gels have to be compared with great accuracy. A similar procedure has been described for reprobing nitrocellulose blots.[12]

A solution containing 10 m*M* Tris-HCl, pH 7.5, and 3 *M* potassium thiocyanate can be used to remove the antibody and protein A from the

[15] M. A. K. Markwell and C. F. Fox, *Biochemistry* **17,** 4807 (1978).

paper, with incubation for 1 hr at 37°, followed by incubation for 30 min at room temperature in 0.01 *M* hydrochloric acid. (Alternatively, use a solution of 5 *M* guanidinium thiocyanate, 50 m*M* Tris-hydrochloride, pH 7.0, and 10 m*M* dithiothreitol for 2 hr at room temperature.) After rinsing with distilled water and NET–gel–NP-40 buffer, the paper is ready for incubation with a second antiserum. If the paper is to be reused, it is best to store it moist at 4° sealed in a plastic bag.

Some antisera give high backgrounds. Large immune complexes, which may be responsible, can be removed by centrifugation or by passing the serum through GF/C glass fiber filters[16] or Millipore filters. In some instances the backgrounds can be lowered by incubating the paper in a buffer containing 50 m*M* sodium phosphate, pH 7.5, 3 *M* urea, 1 *M* NaCl, 0.25% NP-40, and 0.25% Sarkosyl for 30 min at room temperature after incubation with the antiserum and also after incubation with protein A.

Discussion

When electrophoresis in the presence of SDS or urea is employed, the protein transfer technique is limited by the requirement that at least some antigenic determinants are either insensitive to denaturation and dissociation of subunits or are capable of refolding after transfer. This is especially critical when the technique is used in conjunction with monoclonal antibodies,[17] since only one determinant is available. Some proteins are difficult to transfer out of the gel at a particular pH, and others do not bind well to the solid support. For example, small proteins do not bind well to nitrocellulose. Although a small fraction of such proteins might end up on the paper, they will be underrepresented and quantitation is not possible. Dr. Brian Sauer (personal communication) has obtained good results with large proteins by using agarose–acrylamide composite gels cross-linked with DATD[2] and electrophoretic transfer at pH 6.5. He has also successfully used the gel system of Laemmli[10] for electrophoresis in conjunction with transfer at pH 6.5 by substituting a Tris-borate reservoir buffer for the usual Tris-glycine buffer. Transfer efficiency may be improved in some cases if a low concentration of SDS (0.05% or less) is included in the transfer buffer.

In a related technique, specific affinity ligands, such as antibodies, lectins, or antigens, have been coupled to DBM paper, which is then

[16] L. V. Crawford and D. P. Lane, *Biochem. Biophys. Res. Commun.* **74,** 323 (1977).

[17] N. Hogg, M. Slusarenko, J. Cohen, and J. Reiser, *Cell* **24,** 875 (1981).

placed on top of the gel containing the separated proteins.[18] Proteins that interact with the affinity ligand are specifically transferred from the gel to the paper. This procedure is useful for analysis of both radiolabeled and unlabeled proteins. More recently, two groups[19,20] have used paper activated by cyanogen bromide to bind proteins transferred by diffusion from SDS–polyacrylamide gels. Quantitation of the transfer efficiency was not reported. Paper activated by cyanogen bromide has not given good results in electrophoretic transfers at pH 6.5 (E. A. Swyryd and G. R. Stark, unpublished results).

[18] H. A. Erlich, J. R. Levinson, S. N. Cohen, and H. O. McDevitt, *J. Biol. Chem.* **254,** 12240 (1979).

[19] B. Ratzkin, S. G. Lee, W. J. Schrenk, R. Roychoudhury, M. Chen, T. A. Hamilton, and P. P. Hung, *Proc. Natl. Acad. Sci. U.S.A.* **78,** 3313 (1981).

[20] D. J. Kemp and A. P. Cowman, *Proc. Natl. Acad. Sci. U.S.A.* **78,** 4520 (1981).

[15] Use of Fluorography for Sensitive Isotope Detection in Polyacrylamide Gel Electrophoresis and Related Techniques

By William M. Bonner

Many radioactive isotopes are most efficiently detected by trapping some of the energy released from their radioactive disintegrations and converting it to light. This conversion is accomplished by placing compounds called scintillants or fluors in the path of the radioactive disintegration. The light emitted by the fluors can then be detected by photomultipliers, as in the case of scintillation counting, or by film in the case of fluorography. Because the wavelength of the emitted light depends on the fluor, not on the energy of the radioactive disintegration, the detectors, including films, can be designed to detect a small range of photoenergies more efficiently than would be possible if the detectors had to interact directly with the whole range of energies from many kinds of radioactive isotopes.

Fluorographic procedures fall into two categories, depending on the energy of the radioactive disintegration. For the higher energy β-emitters or γ-emitters such as ^{32}P and ^{125}I, much of the radiation passes through the film without exposing it. The scintillant, in the form of an intensifying screen, is placed next to the film on the side opposite the gel. Radiation passing through the film interacts with the intensifying screen, generating

ISBN 0-12-181996-5

light, which is then detected by the film. The increase in sensitivity over autoradiography is approximately 10-fold for ^{32}P and 15-fold for ^{125}I.

For lower energy β-emitters, such as ^{3}H, ^{14}C, ^{33}P, and ^{35}S, the scintillant must be placed inside the gel, since most of the radiation from these isotopes never emerges from the gel to expose the film. The difficulty is that most scintillants are hydrophobic and not soluble in the hydrophilic solvents necessary for keeping polyacrylamide gels solvated. Dimethyl sulfoxide (DMSO) is one solvent found to have the necessary properties, and 2,5-diphenyloxazole (PPO) dissolved in DMSO can be easily and efficiently introduced into polyacrylamide gels. For ^{14}C, ^{33}P, and ^{35}S, the increase in sensitivity over autoradiography is approximately 15-fold. For ^{3}H, the enhancement in sensitivity is in excess of 1000-fold.

Gel Preparation

For easier fluorography, certain factors should be considered even before gels are prepared. One factor is elasticity. Gels should be elastic rather than brittle for ease in handling and resistance to cracking during drying. For a given acrylamide concentration, decreasing the bisacrylamide concentration increases the elasticity of the gel. Blattler *et al.*[1] found that if the final acrylamide percentage (A) times the final bisacrylamide percentage (B) satisfies the equation $AB = 1.3$, the gels will have satisfactory elasticity throughout the range from 4% acrylamide to 40%. This laboratory has routinely used the formula $AB = 1.5$ for gels from 5% to 60% acrylamide and found that these gels have good elasticity and resistance to cracking. A sodium dodecyl sulfate (SDS) gel with 15% acrylamide and 0.1% bisacrylamide follows this formula and separates proteins greater than 10,000 daltons. Stacking gels that are removed before fluorography need not follow this formula.

A second factor to be considered is the thickness of the gels. According to Fick's second law, the time required for a substance to diffuse a certain distance is proportional to the square of that distance. Therefore, by using thinner gels, one can greatly decrease the time necessary for infusing the gels with scintillants. In practice, gels 0.8 mm thick satisfying the formula $AB = 1.5$ can be handled easily and can be infused with scintillant in less than 1 hr instead of the 3 hr required for 1.5-mm gels. Spacers and slot combs for these gels can be easily cut with scissors or a razor blade from skived Teflon tape (0.8 mm = 0.030 inch) available from most plastic supply houses.

[1] D. P. Blattler, F. Garner, K. Van Slyke, and A. Bradley, *J. Chromatogr.* **64,** 147 (1972).

Fluorographic Procedures

Although the compounds used in fluorography have no known mutagenicity or carcinogenicity, they are not without potential hazard, and appropriate precautions should be used. DMSO may cause skin irritation. It readily penetrates the skin and may facilitate the entry of dissolved substances such as PPO. The procedures described below should be carried out in a hood, and gloves should be worn.

Original PPO–DMSO and Related Procedures

The following procedures are for weak β-emitters such as ^{14}C, ^{35}S, ^{33}P, and ^{3}H. For an overnight exposure using unflashed film, sensitivities are approximately 20 dpm/mm^2 for the first three and 300 dpm/mm^2 for ^{3}H.[2] After electrophoresis, the gel may be stained or may be processed directly for fluorography. Small proteins and peptides may be fixed in the gel with formaldehyde.[3]

All soaking steps should be performed with gentle shaking in closed containers large enough to contain the gel without folding. Plastic freezer boxes with lids can be obtained in a variety of sizes ideal for this purpose. The times presented are for 0.8-mm-thick gels; for 1.5-mm-thick gels, times should be increased 3-fold.[2] Gels are soaked (*a*) in two changes of DMSO for 10 min each to remove water; (*b*) then in four volumes of a 22% (w/v) solution of PPO in DMSO for 1 hr; (*c*) followed by two 10-min changes of water to precipitate the PPO in the gels. The gels are now ready to dry. Drying an 0.8-mm gel on a commercial gel dryer set to 80° generally takes about 1 hr. In a modification of this procedure, glacial acetic acid can be substituted for DMSO as the solvent. Then the gels can be immersed directly in 22% PPO (w/v) in glacial acetic acid, thereby eliminating the first two 10-min washes. Possibly more important is that the staining pattern of a gel is retained better when acetic acid is used instead of DMSO. PPO can be recovered from either solvent for reuse by adding water and collecting the precipitated PPO on a filter.[4]

Several other procedures and solutions for fluorography of polyacrylamide gels have been developed since the original procedure using PPO in DMSO. One such solution is a commercial preparation EnHance (New England Nuclear). This solution is acetic acid based, so gels can be added directly in EnHance for 1 hr, followed by two 10-min changes of water.

[2] W. M. Bonner and R. A. Laskey, *Eur. J. Biochem.* **46,** 83 (1974).

[3] G. Steck, P. Leuthard, and R. R. Bürk, *Anal. Biochem.* **107,** 21 (1980).

[4] R. A. Laskey, A. D. Mills, and J. S. Knowland, Appendix in R. A. Laskey and A. D. Mills, *Eur. J. Biochem.* **56,** 335 (1975).

The gel dryer should not be set above 80°. EnHance is considerably less expensive than PPO containing solutions if one does not recover the PPO for reuse.

Another procedure uses the water-soluble scintillant sodium salicylate.[5] This method is quicker than, and as sensitive as, the methods just described, but it has significantly less resolution because the resulting bands are rather fuzzy. Stained gels are soaked for 30 min in water to remove acetic acid, then in 10 volumes of 1 *M* (16% w/v) sodium salicylate for 30 min (for 1.5-mm-thick gel). Gels at neutral or alkaline pH can be added directly to the sodium salicylate. The gels are then dried directly on paper under an acetate sheet. The usual Mylar sheets should not be used, since the dried gel becomes glued to them. The gel dryer should be set for about 80°.

Gel Dryers

Commercial gel dryers do an excellent job of drying gels, particularly when connected to a freeze dryer. Generally the gel dryer should not be set above 80°, as some fluors are volatile at higher temperatures. (This is less important for PPO than the other compounds.) Most commercial gel dryers have a rheostat, accessible through a hole in the bottom or after removing the bottom, which can be used to adjust the temperature. A 0.8-mm-thick 15% polyacrylamide gel can be dried in about 1 hr. Generally, drying a gel on thin paper (i.e., Whatman 1) rather than heavy paper results in a more manageable dried gel. Do not be tempted to release the vacuum to see how well a gel is drying as this can cause the whole gel to fragment immediately into small pieces. The dried gels tend to curl up so they should be put in a cassette or otherwise flattened. If gels generally crack while drying, then a recipe with less bisacrylamide, as discussed previously, should be tried.

Exposing the Gel to Film

For fluorography the gel is placed in a cassette with the correct film and exposed at −70°.[6,7] The appropriate type of film is essential for good results, as the sensitivities of different types of film can easily vary by a factor of ten.[2] Kodak X-Omat AR or the equivalent from other manufacturers should be used.

[5] J. P. Chamberlin, *Anal. Biochem.* **98,** 132 (1979).
[6] K. Randerath, *Anal. Biochem.* **34,** 188 (1970).
[7] U. Lüthi and P. G. Waser, *Nature (London)* **205,** 1190 (1965).

Flashing Film To Linearize Its Response[8]

The response of film to light is nonlinear. At low light intensities, the density of the resultant image is approximately proportional to the square of the radioactive density. If it is desired to quantitate data by microdensitometry, the response of the film can be made more linear by exposing it to a hypersensitizing light flash of less than 1 msec duration. A single flash from an electronic flash unit is suitable, but it is usually necessary to decrease its intensity with a Kodak Wratten 21 or 22 (orange) filter, and to diffuse the light by covering the flash unit with filter paper. In addition to the hypersensitized grains, some grains in the film are also exposed by the flash, resulting in an increased background fogging density. Usually a fogging density of 0.15 OD results in a linear film response, and the distance between the film and flash unit is varied until that density is obtained. The film is placed on a yellow background and flashed, and its flashed face is placed against the gel. Since electronic flash units produce a high current drain, it is a great convenience to choose a unit that uses rechargeable batteries or plugs directly into the mains.

Flashed film is useful in two situations even when accurate quantitation is not desired. When the spots of interest are near the level of detection, flashing the film results to a 2- to 3-fold increase in relative density for these spots.[8,9] This is because the nonlinear response of unflashed film underrepresents very faint spots. The second situation is when no $-70°$ storage is available. ^{3}H and ^{14}C fluorography is possible with flashed film, but not unflashed film, at ambient temperature. However, the sensitivity is about half that at $-70°$.

Radioactive Ink

Scribing a few doodles of radioactive ink (about 1 μCi ^{14}C/ml) in the corners or around the edge of a dried gel enables one to line up the final fluorograph exactly with the gel. This ability is essential for locating radioactive spots on the dried gel when quantitation by scintillation counting is desired.

Quantitation

As discussed above, film images can be quantitated using microdensitometry. Another method particularly useful for small numbers of spots is merely to cut out the spots from the dried fluorographed gel. If the gel is dried on thin paper and spotted around the edges with radioactive ink, the

[8] R. A. Laskey and A. D. Mills, *Eur. J. Biochem.* **56,** 335 (1975).

[9] W. M. Bonner and J. D. Stedman, *Anal. Biochem.* **89,** 247 (1978).

fluorograph can easily be placed in register over the dried gel and the desired spots marked with pencil on the paper backing while holding the film-gel sandwich in front of a strong light. Spots are cut out and put in a scintillation vial. Pieces of dried gel infused with fluor will give counts in a scintillation counter, but the geometry must be kept constant for accurate quantitation. It is generally more accurate to digest the pieces of dried gel overnight at 37° in 1 ml of a fresh solution of 95 parts of 30% H_2O_2 and 5 parts of concentrated NH_4OH. The procedure is performed in capped scintillation vials—preferably plastic, as occasionally glass ones explode. The next morning, the vials are cooled and a water miscible scintillation fluid is added. Since this is a homogeneous system, internal standards can be added to calculate counting efficiency and hence the absolute amount of radioactivity in the spot. This method also works well for doubly labeled spots (i.e., ^{3}H and ^{14}C).

Intensifying Screens[10]

For strong β-emitters (^{32}P) and γ-emitters (^{125}I), most of the radiation passes through the film without exposing it. By means of an intensifying screen, some of this radiation is captured and converted into light, which then exposes the film. Film cassettes can be purchased with intensifying screens permanently attached to one or both faces. In general, calcium tungstate intensifying screens are most suitable. Dried gels should be exposed at −70°. Wet gels are better exposed to 0°, because freezing may crack them. Caution should be exercised in combining flashed film and intensifying screens since some screens contain enough endogenous radioactivity to blacken the film during a long exposure. After an overnight exposure, sensitivities are approximately 1 dpm/mm^2 for ^{125}I and 0.5 dpm/mm^2 for ^{32}P.

Other Potentially Useful Techniques

Fluorography of Agarose Gels[11]

Agarose gels can be soaked in a 3% (w/v) solution of PPO in ethanol for 3 hr after they have been dehydrated in several changes of absolute ethanol. PPO is precipitated by immersing the gel in water, and the gel is dried. During drying, the gel should not be heated or should be heated only gently so that it does not melt.

[10] R. A. Laskey and A. D. Mills, *FEBS Lett.* **82,** 314 (1977).
[11] R. A. Laskey, A. D. Mills, and N. R. Morris, *Cell* **10,** 237 (1977).

Fluorography of Solid Porous Supports (Thin-Layer Plates, Nitrocellulose Filters, etc.)[9]

Efficient fluorography of ^{3}H and ^{14}C on solid supports such as thin-layer plates, paper chromatograms, and nitrocellulose filters can be obtained by dipping the completely dried support into a slightly warm solution (about 37°) of 0.4% PPO in 2-methylnaphthalene (Aldrich Chemical Co.). The support is stood or hung vertically to allow the excess to run off. As the solution cools to ambient temperature, it solidifies. The support can then be exposed to flashed or unflashed film at −70°. One caution is that plates should not be left uncovered overnight, as the 2-methylnaphthalene is quite volatile. The levels of detection are similar to those for polyacrylamide gels.

Spark Chamber

A high-resolution spark chamber connected to a Polaroid camera (Beta Camera, Beta Analytical, Inc., Coraopolis, Pennsylvania) is also a very sensitive detector of radioactivity on solid supports and has been used with thin-layer plates. A radioactive disintegration generates a spark of light that can be recorded on film; 100 dpm of $^{3}H/mm^{2}$ and 20 dpm of $^{14}C/mm^{2}$ can be detected in 15 min. Fluorography would require about 6 hr to detect that density of ^{3}H and 2.5 hr for that density of ^{14}C, whereas autoradiography would require about 36 hr for that density of ^{14}C. However, this differential in sensitivity may not hold for ^{3}H and ^{14}C in polyacrylamide gels, since the geometry is different.

Intensifying X-Ray Films[12]

In this unique procedure, the exposed silver grains of the film image are made radioactive with [^{35}S]thiourea. The now radioactive film is then exposed to X-ray film at −70°.

A great deal of washing is required. First the film is washed (1 liter each for an 18 × 24 cm sheet) in two changes of distilled water (2 min each), then 20% methanol (2 min), 50% methanol (2 min), 20% methanol again (2 min), and finally two changes of distilled water (10 min each). After the film is dry, it is soaked in a solution of [^{35}S]thiourea in 0.1 *N* ammonia, pH 11, for 1 hr. The solution should be approximately 2 μCi/ml, and one should allow 1 μCi of [^{35}S]thiourea per square centimeter of film. Lower backgrounds can be obtained by adjusting the radioactive solution to pH 8 with HCl and soaking for 2 hr.

[12] Amersham Research News No. XIII (1981).

Again the film is washed very thoroughly in two changes of distilled water (2 min each), 20% methanol (5 min), 50% methanol (2 min), 20% methanol (5 min), and two changes of distilled water (5 min each).

After the film is dry, it is exposed to X-ray film (such as Kodax X-Omat R film) at −70°. Intensification of at least 10-fold is claimed after an overnight exposure. This method could be useful in those instances when the original gel is no longer available, when the isotopes in the original gel have a very short half-life and have all decayed, or when it took a very long time to get the original exposure. It is likely to be more useful with unflashed rather than flashed film, since the background fogging will also be intensified.

[16] Peptide Mapping in One Dimension by Limited Proteolysis of Sodium Dodecyl Sulfate-Solubilized Proteins

By DON W. CLEVELAND

Electrophoresis in polyacrylamide gels containing sodium dodecyl sulfate (SDS) is a powerful tool for separation of polypeptide chains in complex biological samples. Often, however, the unambiguous identification of relationships among specific proteins cannot be made on the basis of electrophoretic mobility alone. This is the case, for example, when larger and smaller peptides are related by a precursor–product relationship, when related proteins differ in mobility as the result of artifactual proteolysis during preparation, or when closely related proteins differ in mobility owing to a chemical modification such as glycosylation. In addition, the identity or near-identity of proteins of indistinguishable mobilities but that have been isolated using different assays or protocols cannot be established by simple electrophoresis. In each of these instances it is necessary to subject individual polypeptides to further biochemical analysis.

I report here on a procedure for peptide analysis of proteins that is especially suitable for analysis of proteins that have been isolated from SDS gels. The method involves the partial digestion of the protein of interest with any of several proteases in a buffer containing SDS. Under these conditions surprisingly stable partial digests are produced whose peptides have sufficiently large molecular weights that their resolution on 15% polyacrylamide–SDS gels is possible. The pattern of peptides so generated is characteristic of the protein substrate and the proteolytic

ISBN 0-12-181996-5

enzyme and is highly reproducible.[1] Moreover, the technique can be completed in a matter of hours and requires as little as 50 ng of protein or as few as 1000 cpm of a radiolabeled substrate.

I shall discuss three general protocols for production of one-dimensional peptide fingerprints by partial proteolysis in the presence of SDS. The first procedure is applicable to polypeptides that have been purified either by classical biochemical techniques or by elution from preparative polyacrylamide gels. The second method is suitable for substrates contained in gel slices which have been excised from an initial (hydrated) SDS–polyacrylamide gel. The third method is suitable for substrates contained in gel slices excised from *dried* polyacrylamide gels. This last protocol is especially suitable for radioactively labeled substrates.

General Principles and Procedures

The principle by which peptide fingerprints are generated with all three procedures is identical. By addition of an exogenous protease to a protein sample in the presence of SDS, relatively stable limited proteolysis products are produced. Digestion may be achieved either in solution (using Method I) or *directly in the 4.5% stacking gel* of a 15% resolving gel (using Methods I, II, or III). In either case, the peptide fragments produced are then separated by molecular weight in the resolving portion of a 15% gel.

Gel electrophoresis is performed in slab gels with the discontinuous system described by Laemmli.[2] Typical gel mixtures are given in Table I. Gels of thicknesses between 0.75 and 1.5 mm have been used successfully, although for Method II (using hydrated gel slices) thicker gels (1.5 mm) have been found to be easier to handle. Similarly, for Method III using dried gel slices, most satisfactory results have been obtained with thinner gels (0.75 mm).

For gels 1.5 mm thick, electrophoresis is at 60 V until the Bromphenol Blue clears the stacking gel, at which time the voltage is increased to 150 V. For 0.75-mm gels, these voltages may be increased to 100 and 250 V, respectively.

When using protocols in which proteolysis is achieved directly in the stacking gel of the 15% acrylamide gel, it should be remembered that for a given amount of protease, greater proteolysis may be obtained (*a*) by using longer stacking gels (3–5 cm); (*b*) by electrophoresing the samples more slowly through the stacking gel (i.e., at lower voltages); or (*c*) by

[1] D. W. Cleveland, S. G. Fischer, M. W. Kirschner, and U. K. Laemmli, *J. Biol. Chem.* **252,** 1102 (1977).

[2] U. K. Laemmli, *Nature (London)* **227,** 680 (1970).

TABLE I
ELECTROPHORESIS SOLUTIONS

1. Acrylamide stock: 30 g of acrylamide, 0.8 g of bisacrylamide, to 100 ml final volume
2. Resolving gel buffer: 18.1 g of Tris base, titrated to pH 8.8 with HCl, to 100 ml final volume
3. Stacking gel buffer: 6.0 g of Tris base, titrated to pH 6.8 with HCl, to 100 ml final volume
4. Gel reservoir buffer: 6.0 g of Tris base, 28.8 g of glycine, 1.0 g of SDS, to 1000 ml final volume
5. Proteolytic enzymes (stored at −20° as 1–5 mg/ml stocks in 0.125 *M* Tris-HCl, pH 6.8)
   a. *Staphylococcus aureus* V8 (Miles Laboratories, 36-900-1)
   b. Chymotrypsin (Worthington-Millipore, CDI-1450)
   c. Papain (Sigma, A-4762)
   d. Subtilisin (Sigma, P-5380)
6. Solution for 15% resolving gel (50 ml): 25 ml of acrylamide stock, 12.5 ml of resolving gel buffer, 0.5 ml of 10% SDS, 0.1 ml of 0.5 *M* EDTA, 11.4 ml of distilled H_2O, 50 μl of *N′*,*N′*,*N′*,*N′*-tetramethylethylenediamine, 0.5 ml of 10% ammonium persulfate
7. Solution for 4.5% stacking gel (10 ml): 1.5 ml of acrylamide stock, 2.5 ml of stacking gel buffer, 0.1 ml of 10% SDS, 20 μl of 0.5 *M* EDTA, 5.8 ml of distilled H_2O, 10 μl of *N′*,*N′*,*N′*,*N′*-tetramethylethylenediamine, 0.1 ml of 10% ammonium persulfate
8. Gel slice equilibration solution/protease dilution solution (10 ml): 2.5 ml of stacking gel buffer, 0.1 ml of 10% SDS, 1.0 ml of glycerol, 20 μl of 0.5 *M* EDTA, 30 μl of 2-mercaptoethanol, 6.3 ml of distilled H_2O, trace of Bromphenol Blue
9. Gel slice overlay solution (10 ml): 2.5 ml stacking gel solution, 0.1 ml of 10% SDS, 2.0 ml of glycerol, 20 μl of 0.5 *M* EDTA, 30 μl of 2-mercaptoethanol, 5.4 ml of distilled H_2O

turning off the voltage for 15–30 min after the samples have completely stacked in the stacking gel.

Method I. Digestion Protocol for Purified Proteins

Proteins purified by standard biochemical techniques or by gel electrophoresis followed by elution of the protein from gel slices are dissolved or diluted into solution 8 (see Table I). The samples are denatured by heating to 100° for 2 min. Proteolytic digestions are performed at 37° by addition of given amounts of any of several proteases (see Table II for suggested amounts). After addition of 2-mercaptoethanol and SDS to final concentrations of 5% and 2%, respectively, and addition of a trace of bromphenol blue, proteolysis is stopped by boiling for 2 min. Each sample is then loaded into a sample well of the stacking gel of a 15% polyacrylamide gel. Electrophoresis is then conducted as detailed above.

Alternatively, proteolysis may be achieved directly in the stacking gel portion of the 15% gel. In this instance the samples are dissolved/diluted

TABLE II
USEFUL RANGES OF PROTEASES[a]

Protease	Proteolysis in solution[b] (μg/ml)	Proteolysis in stacking gel[c]
Staphylococcus aureus V8	50	10 ng–300 ng
Chymotrypsin	50	100 ng–3 μg
Papain	3–30	50 ng–1 μg
Subtilisin	50	50 ng–1 μg

[a] Proteases are stored at 1–5 mg/ml in 0.125 *M* Tris, pH 6.8. Aliquots are thawed just prior to use, and appropriate working dilutions into solution 8 are made.

[b] Final concentration of protease

[c] Total amount of protease overlayed onto gel slice.

into gel slice overlay solution (solution 9), boiled, and loaded into sample wells of the 15% gel. Each sample is then overlayed with protease diluted into solution 8, and electrophoresis is conducted as given above.

An example of fingerprint patterns generated by solution proteolysis using protease *Staphylococcus aureus* V8 to digest samples of tubulin purified from porcine brain or from chicken brain are presented in Fig. 1. Each sample consists of equimolar amounts of α-tubulin and β-tubulin polypeptides. The comparative fingerprints are displayed in pairs with the first pair (part A) showing the undigested hog and chicken samples, the second pair (part B) showing the proteolysis products following digestion with 6 μg of protease per milliliter, and parts C and D showing the products of digestion with increasing concentrations of protease. Since hog and chicken α-tubulins are known to differ in only 2 of 450 amino acids[3,4] (and the β-tubulin chains are expected to be equally homologous), the pattern of peptides produced from either sample should be similar. In fact, the patterns are identical with the exception of a single extra peptide found in the chicken samples (marked with an arrow). Hence, the fingerprinting technique confirms the known similarity in tubulins from the two species, yet retains sufficient sensitivity to distinguish differences between proteins whose sequence divergence is less than 1%.

[3] H. Ponstingl, E. Krauhs, M. Little, and T. Kempf, *Proc. Natl. Acad. Sci. U.S.A.* **78,** 2757 (1981).

[4] P. Valenzuela, M. Quiroga, J. Zaldivar, W. J. Rutter, M. W. Kirschner, and D. W. Cleveland, *Nature (London)* **289,** 650 (1981).

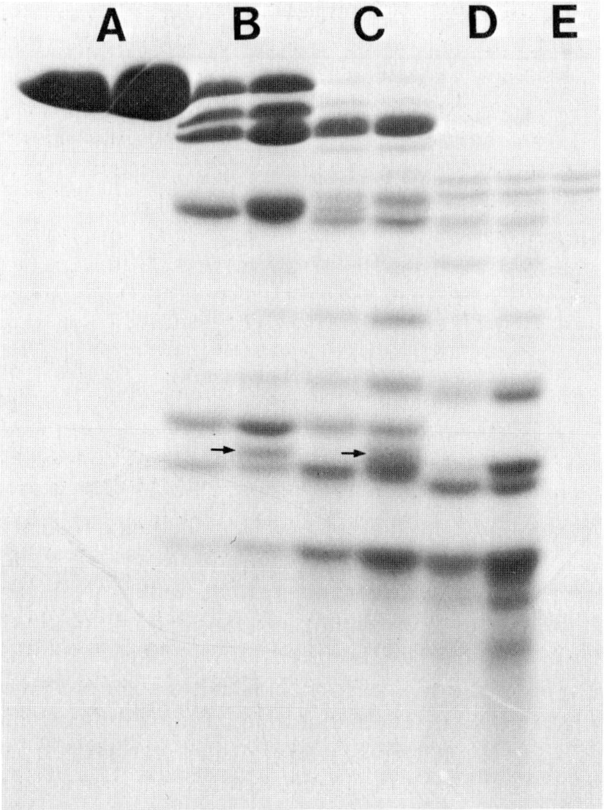

FIG. 1. Peptide fingerprints of porcine and chicken tubulins. Peptide fingerprints were obtained by solution proteolysis using protease *Staphylococcus aureus* V8 to digest samples of tubulin purified from hog brain or from chicken brain. The comparative fingerprints as visualized by Coomassie Blue staining of a 15% acrylamide gel are displayed in pairs with the first pair (part A) showing the undigested hog (part A, slot 1) and chicken (part A, slot 2) proteins. Equivalent aliquots of these samples were digested at 37° for 30 min with increasing concentrations of protease. The hog and chicken samples in part B were incubated with 6 μg of enzyme per milliliter; those in part C with 20 μg/ml; and those in part D with 60 μg/ml. Approximately 15 μg of tubulin substrate were digested in each slot. Slot E represents the staining pattern of a mock incubation of 60 μg of protease per milliliter only. The arrows in the figure mark the position of a peptide found only in the chicken tubulin.

Method II. Digestion Procedure for Proteins Contained in Hydrated Gel Slices

Protein bands isolated from SDS gels following staining with Coomassie Blue are conveniently fingerprinted without the laborious and inefficient procedure of elution from the acrylamide. The initial gels are stained in 0.1% Coomassie Blue, 50% methanol, and 10% acetic acid and are destained by diffusion in 5% methanol, 10% acetic acid. [Protein bands containing more than 3–4 μg of protein can be visualized by brief staining (10 min) in 1% Coomassie Blue in water followed by 5–10 min of destaining in water.] Note that when destaining gels in solutions containing acetic acid, care should be used not to heat the gels to speed destaining, since this may cause partial hydrolysis of the protein. Radioactive samples may also be isolated using this method if the mobility of the radioactive polypeptide is known with respect to a stained marker protein.

After sufficient destaining so that the bands can be visualized, the gel can be placed on a light box and individual bands excised with a razor blade. The slices should be trimmed at this point to the approximate width of the slot-former to be used on the subsequent fingerprinting gel. The slices are then equilibrated against 10 ml of gel slice equilibration buffer (solution 8) for at least 60 min. Failure to remove all residual acetic acid will lead to streaking on the fingerprinting gel. When it is desirable, the slices may be stored at −20° after this equilibration step. Next, each gel slice is pushed to the bottom of a sample well of the 15% acrylamide gel. This is conveniently accomplished by construction of a Plexiglas tool that is slightly thinner than the gel thickness and whose width is slightly less than the width of a sample well. This tool can be used to push intact slices into wells without fragmentation. Spaces around the slices are filled by overlaying each slice with slice overlay solution (solution 9). Finally, 10 μl of freshly diluted protease in solution 8 is overlayed over each slice, and electrophoresis is performed (see above for voltages, etc.).

Method III. Digestion Procedure for Proteins Contained in Dried Gel Slices

Proteins initially electrophoresed on an SDS–polyacrylamide gel may also be peptide-fingerprinted by excision of the appropriate bands *after* drying the gels. The method, first demonstrated by Oppermann and Levinson[5] (Genentech, Inc.) is particularly useful when the proteins to be mapped are radioactively labeled. The location of the appropriate band is determined by autoradiography of the dried gel. Slices are then carefully cut as intact bands from the dried gel using an autoradiogram as a guide. (This is most easily accomplished using radioactive marking spots on the dried gel for alignment, cutting the band of interest out of the film to create a mask, and then using a scalpel to excise the gel slice through the cut in the film.) Although good results have been obtained with gels dried

[5] H. Oppermann, A. D. Levinson, and H. E. Varmus, *Virology* **108**, 47 (1981).

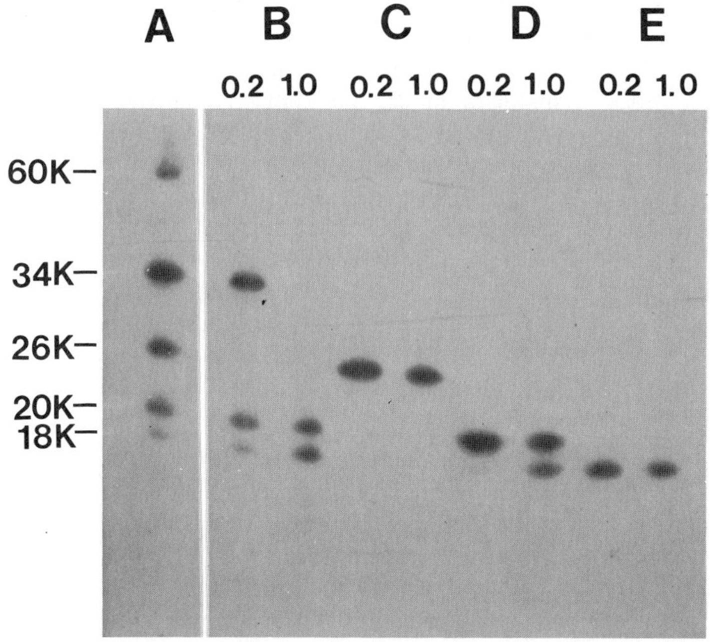

FIG. 2. Peptide fingerprints from proteins recovered in slices from dried gels. ^{32}P-labeled src protein (from Rous sarcoma virus) was localized on an initial dried gel by autoradiography. The appropriate band was excised and peptide fingerprinted by Method III using protease *S. aureus*. An autoradiograph of the peptide patterns is given in part A. Parts B–E represent autoradiographs of redigestions with *S. aureus* of the individual 34K, 26K, 20K, and 18K peptides seen, respectively, in part A. These redigestions were also performed using Method III after autoradiography and excision of each peptide from the initial mapping gel. The pairs of slots in parts B–E represent digestion with 0.2 μg or 1.0 μg of protease. This experiment is from unpublished data provided by Hermann Oppermann (Genentech, Inc.).

after standard staining and fixation, consistently good recoveries have been obtained only with unstained, unfixed gels that were dried directly after washing in water for 10 min. Reasonable care is used to isolate the slices as free of bound gel drying paper as possible, although small amounts of paper that remain adhered to individual gel slices seem to have little deleterious effect. The slices are then placed into sample wells (about 2 cm deep) of a 15% gel. Slices are hydrated *in situ* in the well for at least 15 min in 20 μl of solution 8. Sample wells are then filled with 10 μl of protease in solution 8 and electrophoresed in the standard manner (see above).

An example of the use of this method is shown in Fig. 2. This experiment, generously provided by Herman Oppermann (Genentech, Inc.) was used to map two distinct sites of phosphorylation in *src*, the transforming gene of Rous sarcoma virus. ^{32}P-labeled *src* protein was localized on an initial dried gel by autoradiography and subjected to mapping with *S. aureus* protease. This mapping gel was then dried and autoradiographed and individual peptide fragments were subjected to a second round of proteolysis and analysis. Slot A shows the five ^{32}P-containing peptides produced by partial digestion of *src*. These include (*a*) the 60K (kilodalton) undigested protein; (*b*) a 34K fragment; (*c*) a 26K fragment; (*d*) a 20K fragment, and (*e*) an 18K fragment. Part B shows that redigestion of the 34K fragment results in appearance of both the 20K and 18K fragments. Similarly, the isolated 20K fragment is converted into the 18K fragment upon extensive redigestion (part D). The 26K fragment, however, is distinct from the others and does not yield additional labeled digestion fragments (part C). One may conclude from this experiment that src must contain two distinct sites for phosphorylation, i.e., one on the 26K fragment and one on the 34K fragment. This latter site is also contained on the 20K and 18K subfragments. It is noteworthy in this example that the protease most easily digests the 60K protein into 34K and 26K halves, less easily converts the 34K fragment to 20K, and only with great difficulty reduces the 20K fragment to 18K. Such site preference is commonly seen in these analyses.

Final Comments

The patterns of peptides produced with any of the procedures detailed above may be visualized on the 15% resolving gels by Coomassie Blue staining, by silver staining (see Merril *et al.*, this volume [17]; for the protocol) or for radioactive samples by autoradiography or fluorography (see Bonner [15], this volume). Silver staining allows as little as 50 ng of substrate protein to be fingerprinted. Fluorographic techniques allow analysis of as few as 1000 cpm of ^{35}S- or ^{14}C-labeled protein.

The partial proteolytic patterns are surprisingly stable, particularly with protease *S. aureus* V8, which has become the protease of first choice. Nonetheless, with any protease to maximize the ability to compare different samples, it is best to perform multiple digestions with increasing levels of proteolysis (as is demonstrated in Fig. 1 and 2). The sensitivity of the technique can be further increased by use of additional proteases or by further proteolysis of individual peptide fragments with the same or additional enzymes (as is shown in Fig. 2).

[17] Silver Staining Methods for Polyacrylamide Gel Electrophoresis

By CARL R. MERRIL, DAVID GOLDMAN, and MARGARET L. VAN KEUREN

Image formation with silver was discovered in the mid-seventeenth century. It was used first as a light-sensitive agent in photography and within a decade to visualize cellular structures. Early cellular silver staining techniques were refined and developed through extensive empirical observations. However, the chemical basis of the selective reduction of ionic to metallic silver in most histological silver stains is still unknown. Nevertheless, adaptation of these stains to the detection of proteins in polyacrylamide matrices resulted in an approximate 100-fold increase in sensitivity over organic dyes such as Coomassie Blue. Unfortunately, these histologically derived silver stains take hours to perform,[1-3] give variable results,[4] and use large quantities of silver or other expensive reagents.[4-6]

Many photographic processes also rely on the selective reduction of silver from an ionic to a metallic form. By soaking a PAGE gel containing separated proteins in a silver nitrate solution, in the dark, followed by the use of most common photographic developers (e.g., Eastman Kodak D76 in 1 : 10 dilution), a negative image is obtained. This negative image contains clear regions in portions of the electrophoretogram containing protein and background regions that are brown or gray. Production of a negative image may indicate a Donnan equilibrium type of exclusion of silver ions from gel regions containing charged biopolymers. Exposure of this image to light results in some photoreversal unless the remaining silver salts are removed by rinsing the gel for 5 min with photographic fixer followed by three 20-min washes with water. This photochemical method is not very sensitive. However, by the utilization of chemical photoreversal procedures and formaldehyde as the developer, a highly sensitive and simple to perform silver stain was developed.[7-10]

[1] R. C. Switzer, C. R. Merril, and S. Shifrin, *Anal. Biochem.* **98,** 231 (1979).
[2] C. R. Merril, R. C. Switzer, and M. L. Van Keuren, *Proc. Natl. Acad. Sci. U.S.A.* **76,** 4335 (1979).
[3] B. R. Oakley, D. R. Kirsch, and R. Morris, *Anal. Biochem.* **105,** 361 (1980).
[4] P. Verheecke, *J. Neurol.* **209,** 59 (1975).
[5] L. Kerenyi and F. Gallyas, *Clin. Chim. Acta* **38,** 465 (1972).
[6] D. Karcher, A. Lowenthal, and G. Van Soom, *Acta Neurol. Belg.* **79,** 355 (1979).
[7] C. R. Merril, M. L. Dunau, and D. Goldman, *Anal. Biochem.* **110,** 201 (1981).

METHODS IN ENZYMOLOGY, VOL. 96

ISBN 0-12-181996-5

In this procedure either oxidizing agents, reducing agents, or a strong light source can be used to obtain photoreversal. By soaking the gel in one of these reagents (we prefer dichromate) prior to exposure to silver nitrate, full photoreversal can be obtained during image development. The image developing solution contains sodium carbonate and formaldehyde. The formaldehyde serves as the silver-reducing agent and is converted to formic acid during image development. Sodium carbonate makes the solution alkaline so that formaldehyde can be oxidized.[11]

Photochemical Silver Stain

To stain PAGE gels the following steps are executed.

Step 1. PAGE gels may be fixed in either 20% w/v trichloroacetic acid or 50% methanol–12% acetic acid (v/v) for at least 1 hr (gels that are thinner than 0.5 mm should be fixed in trichloroacetic acid).

Step 2. Gels are washed twice for a total of at least 30 min with 10% ethanol–5% acetic acid (v/v). This allows the gels to swell back to normal size. Gels more than 1 mm thick require more than twice as much time in 10% ethanol–5% acetic acid.

Step 3. Gels are soaked for 5 min in 3.4 m*M* potassium dichromate containing 3.2 m*M* nitric acid.

Step 4. Gels are soaked in 12 m*M* silver nitrate for 20 min.

Step 5. Gels are rinsed with agitation in 0.28 *M* sodium carbonate with 0.5 ml of formaldehyde (37% commercial formaldehyde) per liter. This step requires at least two changes of the solution to prevent precipitated silver salts from adsorbing to the surface of the gel. The pH of the gel is made alkaline so that formaldehyde can reduce silver from the ionic to the metallic form as described.

Step 6. When a slightly yellowish background appears, development is stopped by making the gel acidic by placing it in 3% v/v acetic acid (for 5 min). Development is carried out for 20 min for maximum sensitivity for gels 1 mm thick. Thicker gels require more time.

Gels are washed twice with water before storage. They may be stored in water or soaked in 3% glycerol for 5 min and dried between dialysis membrane (Bio-Rad, Richmond, California) under vacuum at 80–82° for 3 hr. This latter procedure results in a transparency that is relatively permanent and easy to store. For autoradiography or fluorography, gels are dried down onto Whatman 3 MM filter paper.

[8] C. R. Merril, D. Goldman, S. A. Sedman, and M. H. Ebert, *Science* **211,** 1437 (1981).
[9] C. R. Merril, D. Goldman, and M. Ebert, *Proc. Natl. Acad. Sci. U.S.A.* **78,** 6471 (1981).
[10] C. R. Merril, D. Goldman, and M. L. Van Keuren, *Electrophoresis* **3,** 17 (1982).
[11] G. Ehrenfried, *Photogr. Sci. Tech.* **18B,** 2 (1952).

TABLE I
EFFECTS OF RECYCLING ON PROTEIN SPOT AND BACKGROUND DENSITIES

Cycle	Mean protein spot[a] density (optical density × mm²)	Mean background density (optical density)
1	10.23	0.43
2	20.21	0.58
3	25.63	1.11

[a] Mean densities were determined by measuring the densities of 20 *Escherichia coli* lysate proteins separated by two-dimensional electrophoresis.[12] Densities were calculated using a microdensitometer and a computer program that finds the model density in the vicinity of each protein spot, subtracts this background from the average density, and multiplies this net density by the spot area.

Recycling for Increased Sensitivity

With gels 1 mm thick, the density of the silver deposited in each band or spot reaches a maximum in about 20 min in the sodium carbonate–formaldehyde solution (step 5). Addition of fresh sodium carbonate–formaldehyde will not enhance the density of the spots at this point; however, if the silver nitrate is replaced, additional density can be achieved.[10] To achieve this additional sensitivity, gels are stained as described; however, development during the first cycle is stopped just prior to the appearance of a yellowish background, at step 6, with 3% acetic acid (v/v). The gel is then rinsed twice with additional 3% acetic acid and recycled through steps 4–6. If proteins require further staining, another recycling procedure (steps 4–6) can be conducted. However, background darkening becomes a problem with continued recycling (Table I).

Other Image Intensification and Destaining Procedures

The image obtained by silver staining a polyacrylamide gel is similar to a developed black and white photograph. Numerous image intensification methods have been developed for photography.[13] These procedures are

[12] P. H. O'Farrell, *J. Biol. Chem.* **250,** 4007 (1975).

[13] E. J. Wall, F. I. Jordan, and J. S. Carrol, *in* "Photographic Facts and Formulas," p. 168. Am. Photogr. Book Publ. Co., New York, 1976.

directly applicable to silver-stained electrophoretograms; however, some modifications in chemical concentrations and procedure timing are usually necessary because most polyacrylamide gels are thicker and have different diffusion properties than photographic emulsions. Image intensification methods increase the optical density of the image by adding more metal to the image, as illustrated by the use of additional silver in the recycling procedure described above. They may also make use of other metals, such as copper, mercury, chromium, or uranium. Some of these procedures increase all densities proportionally whereas others, such as copper iodide and mercuric chloride, are superproportional and increase contrast by building up optically dense areas more rapidly. Intensifiers that act selectively on optically less dense areas, such as uranium, mercuric iodide, and chromium, are called subproportional.

Methods of destaining the silver image from electrophoretograms may also be adapted from the photochemical reagents known as reducers.[13] Again by varying reagents proportional, sub-, and superproportional effects can be achieved. It is difficult to control the speed of the reduction process and even more difficult to stop it at a defined end point. For these reasons photographs should be made of the gel prior to and during image reduction. The reducing method we have used most frequently is performed by dissolving 37 g of sodium chloride and 37 g of cupric sulfate in 850 ml of deionized water. Concentrated ammonium hydroxide is added until the precipitate first formed is completely dissolved, and then the deep blue solution is diluted to 1 liter with deionized water. A second solution is made by dissolving 436 g of sodium thiosulfate in 1 liter of deionized water. Equal volumes of these solutions are mixed and used directly if extensive destaining is required. They may be diluted (1 : 10) with water to remove light silver deposits. The use of photography to preserve images is essential when performing destaining because it is difficult to stop the destaining process at a precise point. To restain destained gels, repeat steps 2–6.

Sensitivity of Silver Staining

A silver stain derived from histological methods detected 0.38 ng/mm^2 of albumin in PAGE gels.[1] The recently developed photochemical silver stain without recycling has a similar sensitivity[8,10] and with recycling can detect 0.01 ng/mm^2. Sensitivity varies for different proteins, and higher sensitivities can be achieved with thinner gels. As with Coomassie Blue staining, the ability to detect particular proteins depends both on their migration characteristics and on their affinity for the stain. Sensitivity with silver stains is generally more than 100-fold greater than with Coomassie Blue stains.

TABLE II
EFFECT OF SILVER STAINING ON FLUOROGRAPHY AND AUTORADIOGRAPHY[a]

Staining method	Fluorography of ^{3}H-labeled proteins (% density of unstained proteins)	Autoradiography of ^{14}C-labeled proteins (% density of unstained proteins)
Coomassie Blue	85 ± 3	93 ± 4
Histological silver stain	ND	53 ± 3
Photochemical silver	ND	98 ± 6
Histological silver stain (silver removed)	ND	48 ± 7
Photochemical silver stain	43 ± 9	81 ± 10

[a] Mouse cell lysates were separated by two-dimensional electrophoresis.[11] Densities from 13 protein spots were measured on each electrophoretogram using computerized microdensitometry.[14] ND, not detectable.

Effects of Silver Staining on Autoradiography and Fluorography

The use of a histological silver stain almost completely quenches the detection of ^{3}H-labeled proteins by fluorography.[14] The photochemical silver stain, which uses less silver nitrate per gel, also causes quenching; however, some ^{3}H-labeled proteins may be faintly visualized by fluorography. Removal of silver with photographic reducers restores much of the intensity of fluorography in photochemically stained gels, but not in gels stained with the histological silver stain. Quenching of autoradiography with ^{14}C-labeled proteins was moderate with the histological stain but barely perceptible with the photochemical stain. The effect of silver staining on autoradiographic and fluorographic detection of ^{14}C and ^{3}H, respectively, is illustrated in Table II.

^{14}C quenching with the photochemical stain is small enough so that it is not a practical hindrance. However, silver staining makes fluorographic detection of ^{3}H-labeled proteins more difficult. Detection of ^{3}H-labeled proteins can be improved by destaining photochemically silver-stained gels after their images have been recorded by photography.

Quantitative Analysis of Silver-Stained Proteins

Measurements with three ^{14}C-labeled proteins demonstrated that a histologically derived silver stain maintained a linear relationship between

[14] M. L. Van Keuren, D. Goldman, and C. R. Merril, *Anal. Biochem.* **116,** 248 (1981).

protein concentration and density over a 10-fold range in concentration (from 0.1 to 1 μg).[1] The photochemical stain[8] showed a similar linearity for eight purified proteins. The range of linearity may be extended by photographing the image during development so that protein images are recorded before they have begun to approach saturation or nonproportional staining. The linearity for these proteins is shown in Fig. 1. Above 2 ng/mm^2 the stain becomes nonlinear for the proteins studied, but the range of linearity can be extended by photographing at earlier times during development. In this way, protein images are recorded before staining has become nonproportional or saturating.

The slope of silver protein staining (density per nanogram of protein) varies and is probably characteristic for each protein. The slope for ovalbumin is almost nine times that for carbonate dehydratase (EC 4.2.1.1; carbonic anhydrase). Even with different methods of silver staining, similar slope ratios are obtained. The data of Poehling and Neuhoff[15] give a ratio of the albumin slope to carbonate dehydratase slope of 4.1 when analyzed for the linear part of their curve. Our ratio of the albumin to carbonate dehydratase slope was 3.9 when density was plotted against protein concentration/band area (ng/mm^2) or 3.0 when plotted against protein concentration (ng). It is significant that, even with different methods of silver staining, similar slope ratios are obtained.

Quantitative use of the stain is possible if constitutive or marker proteins are present on each gel so that densities can later be normalized to these and if care is taken to work within the linear range of the stain.

For comparing gels quantitatively, one should load equivalent amounts of protein. However, since image intensity can be controlled by varying the length of time the gel is maintained in the sodium carbonate–formaldehyde solution, a 10-fold variation in initial protein loading can be tolerated (Table III). Intergel comparisons require known standards or constitutively synthesized proteins to serve as markers for normalization.

Staining of DNA in Polyacrylamide Gels

DNA restriction fragments separated in 5% polyacrylamide slab gels can be stained using silver stain as described.[10,16–18] Gels should be washed free of buffer with distilled water and need not be fixed if stained immediately. Gels previously stained with ethidium bromide may be subsequently stained with silver. Optimal image development occurs over

[15] H. Poehling and V. Neuhoff, *Electrophoresis* **2,** 141 (1981).
[16] L. L. Somerville and K. Wang, *Biochem. Biophys. Res. Commun.* **10,** 53 (1981).
[17] D. Goldman and C. Merril, *Electrophoresis* (in press).
[18] T. Boulikas and R. Hancock, *J. Biochem. Biophys. Methods* **5,** 219 (1981).

TABLE III
DENSITY VARIATION WITH VARIABLE PROTEIN LOADING AND DIFFERENTIAL STAINING

Protein loaded[a] (μg)	Mean spot density[b]	Correlation coefficients (μg) 10	25	50	100	200
10	27.3	1.00	—	—	—	—
25	43.0	0.94	1.00	—	—	—
50	51.6	0.95	0.96	1.00	—	—
100	69.8	0.90	0.94	0.97	1.00	—
200	72.5	0.60	0.73	0.74	0.84	1.00

[a] Human lymphocyte lysate proteins, electrophoresed according to O'Farrell.[12]

[b] The densities of 45 protein spots were measured after differential staining of each gel with silver. An attempt was made to stain each gel to the same intensity despite the difference in the amount of protein loaded by limiting the image development time. Correlation coefficients were computed using linear regression analysis.

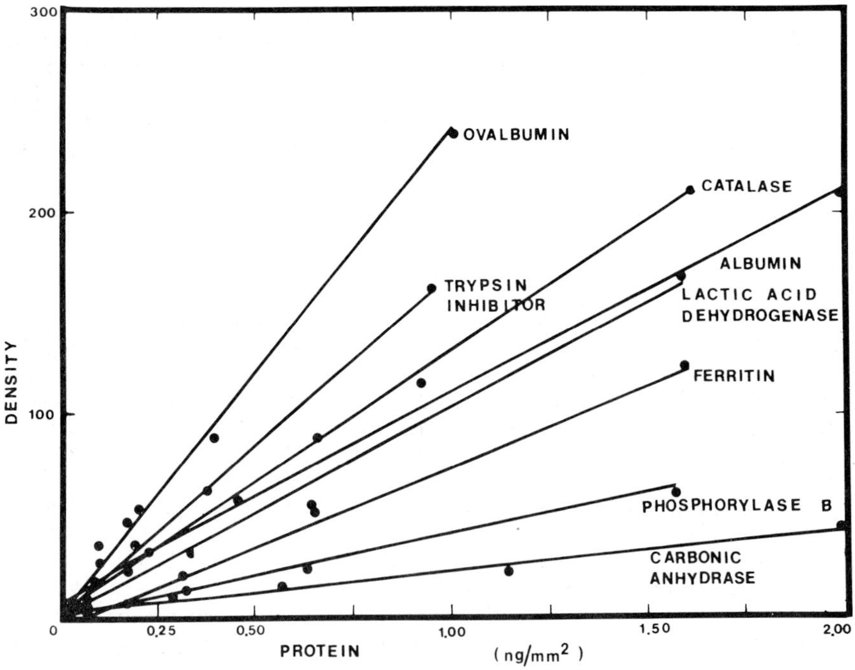

20–25 min in sodium carbonate–formaldehyde, and gels are placed on a light box during this step. Sensitivity is superior to that observed with ethidium bromide using fluorescent detection. The silver stain was capable of detecting a DNA band at 0.03 ng/mm^2 whereas ethidium bromide was useful to detect bands at 0.1 ng/mm^2. The relationship of total band density to nanogram of DNA was linear over a 10-fold range.[17]

Remarks

Loss of Stain Sensitivity. The purity of the water used in the stain and rinsing solutions is critical for maximum sensitivity. Deionized water is required. Formaldehyde can present a problem because it is sometimes

FIG. 1. Density versus concentration for eight proteins after silver staining. Different concentrations of eight proteins were electrophoresed in a single gel and stained to determine the linearity of the protein concentration–density relationship. Purified proteins [albumin, ovalbumin, phosphorylase *b*, carbonate dehydratase (carbonic anhydrase), ferritin, trypsin inhibitor, and lactate dehydrogenase] were purchased as molecular weight standards from Pharmacia, diluted in 20% glycerol, placed in wells, and electrophoresed at 20 mA/gel in a 10% polyacrylamide gel. The gel was fixed in 50% methanol–12% acetic acid (v/v) and stained with sequential dichromate, silver nitrate, and sodium carbonate–formaldehyde steps as described, except that the gel was left on a light box during development in sodium carbonate–formaldehyde so that it could be photographed at intervals. Data shown were obtained after 9 min of gel image development. Measurements were made using computerized microdensitometry. A band area for each protein was measured so that concentrations could be expressed as nanograms per square millimeter. Density units are optical density × mm^2. Linearity of staining was observed over the ranges (tabulated below) of proteins electrophoresed.

Protein	Total protein range (ng)	Protein concn./band area, range (ng/mm^2)	Number of concentrations measured	Slope[a]	Correlation coefficient
Albumin	5.4–270	0.05–2.27	5	101.7	0.995
Ovalbumin	7.2–360	0.02–1.01	5	231.5	0.998
Trypsin inhibitor	8.0–200	0.04–0.96	5	165.8	0.996
Ferritin	25.0–250	0.06–1.60	4	83.3	0.990
Carbonate dehydratase	4.0–200	0.06–2.85	5	26.0	0.993
Phosphorylase *b*	6.4–640	0.06–6.24	5	45.8	0.998
Catalase	3.6–180	0.03–1.59	5	127.0	0.990
Lactate dehydrogenase	9.6–240	0.06–1.57	5	107.5	0.990

[a] Slopes, *Y* intercepts, and correlation coefficients were determined by performing linear regression analysis.

not actually a 37% solution. Other causes of loss of sensitivity are inadequate fixation or the presence of riboflavin or glycine (which may cause the gel to turn black).

Precipitate on Gel Surfaces. If the carbonate–formaldehyde solution is not changed rapidly during initial image development or agitated continuously, a precipitate forms that adheres to the surface of the gel, causing a black surface discoloration. Pressure or exposure to air during processing or the presence of borate will also cause surface deposition.

Colored Protein Patterns. Most silver stains produce some colored bands or spots. In a study of human cerebrospinal fluid proteins[19] utilizing the histochemical silver stain,[1,2] some liproproteins stained blue whereas glycoproteins were yellow, brown, or red. The new photochemical stains also produce colored spots. By modifying the chemistry, color effects can be enhanced.[20] However, saturation and negative staining effects that occur with color enhancement make quantitation more difficult.

Protein Detection. Silver stains used for PAGE gels detect most proteins in crude cell lysates that can be detected with ^{14}C-labeled proteins by autoradiography. However, a few proteins are poorly stained or not stained at all. In this regard, the staining of individual proteins varies among the specific techniques of silver staining. In *Escherichia coli* lysates a major acidic protein does not stain with a single cycle of the photochemical silver stain, but it appears on recycling. Highly specific histological silver stains have been developed empirically to stain tissue differentially. When used in PAGE gels, one of these stains, the Bodian silver method, has been shown to stain mainly neurofilament proteins.[21]

Gel Thickness. Some staining parameters must be adjusted for gel thickness. Reagent concentrations and procedure times given above have been optimized for 1-mm-thick PAGE gels. Thicker gels require more time for each step, particularly step 2.

Artifacts with Silver Staining. Usually, two horizontal lines are visible upon staining at approximately M_r 60,000 and 67,000. These can also be visualized, at times, by Coomassie Blue staining. These lines appear on SDS–PAGE gels even when no protein sample has been applied. In two-dimensional gels where sodium dodecyl sulfate was used in isoelectric focusing, there is an anodal region that does not stain.

If the gel is dried, folded, or compressed, artifacts will occur. Storage of incompletely fixed gels in contact with one another may result in the transfer of proteins from one gel to the surface of another.

[19] D. Goldman, C. R. Merril, and M. H. Ebert, *Clin. Chem.* (*Winston-Salem, N.C.*) **26,** 1317 (1980).

[20] D. W. Sammons, L. D. Adams, and E. E. Nishizawa, *Electrophoresis* **2,** 135 (1981).

[21] P. Gambetti, Autilio-Gambetti, and S. C. Papasozonenos, *Science* **213,** 1521 (1981).

The most concentrated proteins may show greater density at their periphery and diminished staining at the center of the spot or band, making density measurements inaccurate.

Fingerprints. Touching gels prior to or during staining generally results in surface marks. If it is necessary to touch or move a gel during staining, it is useful to use a portion that will not be adversely affected by surface marks. Surface stains may be reduced by gently rubbing them with a cotton ball soaked in photographic reducer followed by rinsing with water. Considerable care must be exercised to minimize damage of the primary silver image by the reducer. Dilution of the reducer by about 10-fold is usually helpful.

Water. Deionized water with a conductivity of less than 1 μmho is required.

[18] Slab Gel System for the Resolution of Oligopeptides below Molecular Weight of 10,000

By F. A. Burr and B. Burr

The systems of polyacrylamide gel electrophoresis that are in current use for the molecular weight determinations of conventional proteins are not well suited for resolving proteins below 10,000.[1-4] In fact abnormal migration characteristics can frequently be detected in the range of M_r = 15,000.[4] Swank and Munkres[5] developed a popular method for tube-cast polyacrylamide gels that gives reproducible separations of polypeptides in the M_r 1000 to 10,000 range. Unfortunately, for reasons that are not apparent, their formulation does not give comparable results with slab gels.

Described here are our modifications of the basic Swank and Munkres procedure that allow the gels to be slab-cast. The system has been found to be particularly favorable for the resolution of small oligopeptides that are between M_r 1500 and 10,000 (Figs. 1 and 2). The gels can therefore be

[1] U. K. Laemmli, *Nature (London)* **227,** 680 (1970).

[2] D. M. Neville, *J. Biol. Chem.* **246,** 6328 (1971).

[3] D. F. Summers, J. V. Maizel, Jr., and J. E. Darnell, *Proc. Natl. Acad. Sci. U.S.A.* **54,** 505 (1965).

[4] K. Weber and M. Osborn, *in* "The Proteins" (H. Neurath and R. Hill, eds.), 3rd ed., Vol. 1, p. 179. Academic Press, New York, 1975.

[5] R. T. Swank and K. D. Munkres, *Anal. Biochem.* **39,** 462 (1971).

ISBN 0-12-181996-5

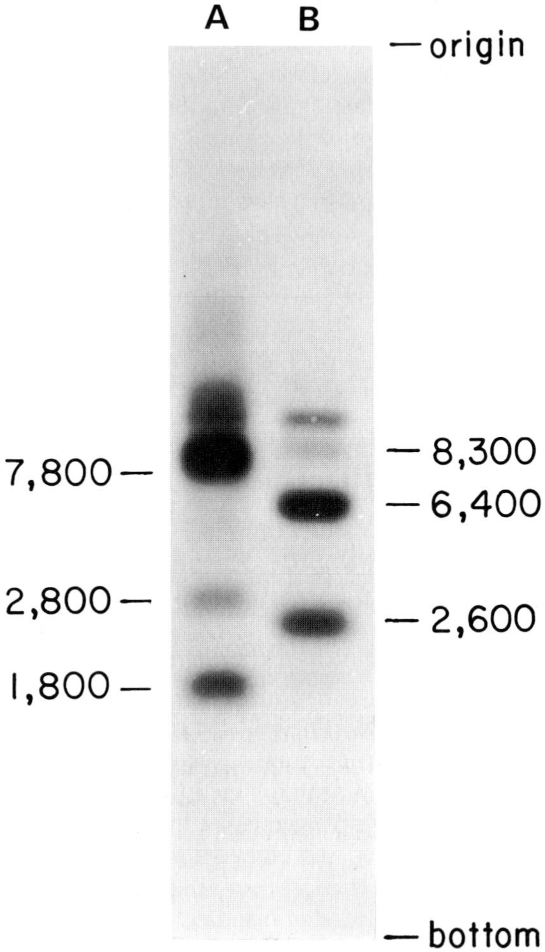

FIG. 1. Cyanogen bromide cleavage products of (A) cytochrome *c* and (B) myoglobin electrophoresed on low molecular weight resolving gel and stained with Coomassie Brilliant Blue. Direction of migration is from top to bottom (anode). Molecular weights of the fragments are indicated at the sides.

used to best advantage in the identification of preprotein cleavage peptides and peptides produced by proteolytic or chemical cleavage.[6] While the upper limit of the system has not been established, linearity is preserved to at least M_r 25,000 (Fig. 2). The bands produced by separation in this gel are considerably thicker (Fig. 1) than those obtained by conven-

[6] F. A. Burr and B. Burr, *J. Cell Biol.* **90,** 427 (1981).

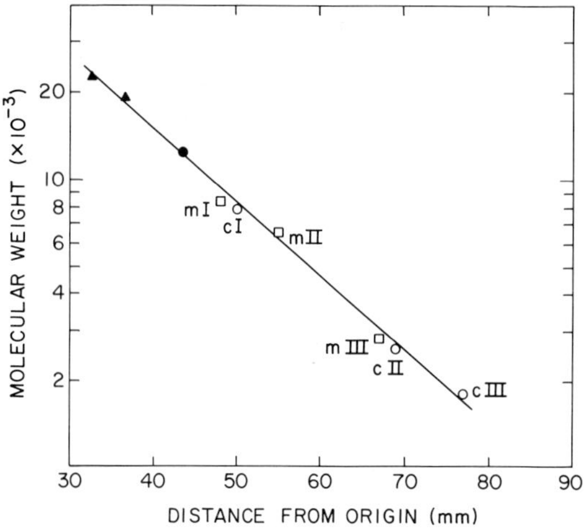

FIG. 2. Plot of the relative electrophoretic migration of molecular weight markers: ▲, zein (22,500 and 19,000); ●, uncleaved cytochrome *c* (12,300); □, mI, mII, mIII = cyanogen bromide cleavage peptides of myoglobin (8270, 6420, 2550, respectively); ○, cI, cII, cIII = cyanogen bromide cleavage peptides of cytochrome *c* (7760, 2780, 1810, respectively).

tional sodium dodecyl sulfate (SDS)–polyacrylamide gel electrophoresis, but reasonable accuracy is assured by always selecting the midpoint of a band.

Theoretical considerations in the estimation of a molecular weight of a protein by polyacrylamide gel electrophoresis have been discussed at length by previous authors.[4,7] It will suffice simply to reiterate here that factors such as the shape, final net charge, and side-chain modifications can greatly influence the migratory behavior of a polypeptide, and therefore molecular weights determined by this method should be regarded as provisional.

A similar procedure has been devised by Bethesda Research Laboratories, Inc., which appeared in their publication *Focus*.[8] Their method differs principally from the one reported here in having a higher bis : acrylamide ratio (0.8 : 30), less urea (6 *M*), a different buffer (0.1 *M* phosphate, pH 7.2), and no stacking gel.

The molecular weight standards that we have found to be the most convenient for our purposes are the cyanogen bromide cleavage peptides of myoglobin (horse muscle) and cytochrome *c* (horse heart): mI = 8270, mII = 6420, mIII = 2550; cI = 7760, cII = 2780, cIII = 1810 (Fig. 1). The average variability based on four recent tests where both cleavage

[7] A. L. Shapiro and J. V. Maizel, Jr., *Anal. Biochem.* **29,** 505 (1969).

[8] *Focus* **2,** 6 (1980).

TABLE I
GEL PREPARATION

Final concentration	Solution or reagent	Quantity
A. Running gel		
12.5% Acrylamide	Acrylamide, solid[a]	3.13 g
0.11% Bisacrylamide	1% Bisacrylamide[a] (w/v)	2.75 ml
8 *M* Urea	Ultrapure urea, solid[b]	12.0 g
0.34 *M* Tris-H_3PO_4, pH 8.0	2 *M* Tris[c]-H_3PO_4, pH 8.0	4.25 ml
0.1% SDS	20% SDS[a] (w/v)	125 μl
0.05% TEMED	TEMED[a]	12.5 μl
0.05% NH_4perSO_4[d]	10% NH_4perSO_4 (w/v)[a]	125 μl
	Water	4.4 ml
		25.0 ml
B. Stacking gel		
5% Acrylamide	30% Acrylamide (w/v)	1.67 ml
0.07% Bisacrylamide	1% Bisacrylamide (w/v)	0.7 ml
8 *M* Urea	Ultrapure urea, solid	4.8 g
0.125 *M* Tris-H_3PO_4, pH 6.8	1 *M* Tris-H_3PO_4, pH 6.8	1.25 ml
0.1% SDS	20% SDS (w/v)	50 μl
0.05% TEMED	TEMED	5 μl
0.05% NH_4perSO_4	10% NH_4perSO_4 (w/v)	50 μl
	Water	2.5 ml
		10.0 ml

[a] Bio-Rad Laboratories.
[b] Schwarz/Mann.
[c] Sigma.
[d] NH_4perSO_4, ammonium persulfate.

products were electrophoresed in the same gel were as follows: mI(±2.8%), mII(±5.4%), mIII(±9.3%); cI(±7.6%), cII(±11.1%), cIII(±3.3%). This compares favorably with the ±18% cited by Swank and Munkres for their method.[5] Uncleaved myoglobin (17,200) and cytochrome *c* (12,300) and insulins A (2300) and B (3500) are additional markers that are frequently used. Glucagon (3500) has not proved to be satisfactory.

Apparatus

The slab gel apparatus used was designed by Studier[9] and is available from the Aquebogue Machine and Repair Shop (Box 205, Main Road, Aquebogue, New York 11931). The 14-cm-high × 16-cm-wide glass plates

[9] F. W. Studier, *Science* **176,** 367 (1972).

(final gel size = 11 × 14.5 cm) are acid washed in a concentrated sulfuric acid–ammonium persulfate bath before use.

Preparation of the Gel

The composition of the gel and the protocol are given in Table I. The buffers and all other stock solutions, except the ammonium persulfate, are filtered through Whatman No. 1 filter paper. All solutions, except for the 20% SDS, are stored at 4° in dark bottles. The 10% ammonium persulfate solution remains active for several months if the salt from which it was made had been kept well desiccated and under vacuum. (When the persulfate is dissolved in water, a good active solution should be heard to crackle if the bottle is held up close to the ear.)

In preparing the gel, add everything except the SDS, TEMED, and ammonium persulfate to a small beaker; place the beaker in a 50–55° water bath, and swirl gently until solid material is completely dissolved. Degassing is not necessary. The SDS, TEMED, and ammonium persulfate are then added and mixed in before pouring. Polymerization should occur in <5 min.

Sample Preparation

The sample solution given in Table II should be made up just prior to use. Heat the solution gently to dissolve urea before adding SDS and mercaptoethanol. Samples are boiled for 3–5 min after dissolution and centrifuged briefly. Remains of samples may be stored at 4° and reboiled before use.

After the gel has been mounted on the gel apparatus and the buffer added, the sample wells must be washed out thoroughly just before application of the samples. (If the gel is allowed to stand before loading, the urea will diffuse into the wells and make it difficult to layer the sample.)

TABLE II
SAMPLE SOLUTION

Final concentration	Solution or reagent	Quantity
8 *M* Urea	Ultrapure urea, solid	2.4 g
0.02 *M* Tris-H_3PO_4, pH 6.8	1 *M* Tris-H_3PO_4, pH 6.8	0.1 ml
0.008% Bromphenol Blue	0.4% Bromphenol Blue (w/v)	0.1 ml
1% SDS	20% SDS (w/v)	0.25 ml
0.02% 2-Mercaptoethanol	2-Mercaptoethanol	0.1 ml
	Water	to 5.0 ml

The samples are loaded into the wells with a Hamilton syringe that has a length of polyethylene cannula fitted onto the needle.

Electrophoresis Conditions

The electrophoresis buffer is 0.1 *M* H_3PO_4-Tris, pH 6.8, containing 0.1% SDS: for 1 liter, use 7.0 ml of H_3PO_4; adjust to pH 6.8 with solid Tris; add 5.0 ml of 20% SDS.

The gel is electrophoresed at room temperature at 50 V until the tracking dye is 1 cm from the bottom edge of the gel (with our apparatus, this is 8.5 cm from the origin). The length of a run is generally about 15 hr.

Staining

The gel is stained in 0.25% Coomassie Brilliant Blue R250 (Bio-Rad) 0.1% Crocein Scarlet (Bio-Rad), 10% acetic acid, 10% methanol for 1–2 hr on an oscillating rocker; 0.5% $CuSO_4$, which reportedly aids in the retention of especially small oligopeptides, may be included. In this case, the $CuSO_4$ should be dissolved in water before adding the acetic acid and methanol. Destaining is carried out in 10% acetic acid, 10% methanol until the background is sufficiently clear.

Excess urea must be removed if the gel is to be dried back for autoradiography without prior staining or if it is to be treated for fluorography. This can be accomplished by first fixing the gel with 12.5% trichloroacetic acid, then washing with several changes of 7–10% acetic acid, 10% methanol, using 30–45 min for each step. These steps are unnecessary if the gel is stained.

[19] Two-Dimensional Immunoelectrophoresis of Membrane Antigens

By CAROLYN A. CONVERSE and DAVID S. PAPERMASTER

Antibodies to cell membrane proteins have been useful reagents for the localization of these constituents in cells[1–4] and for studies of their biosynthesis[5] and their subunit structure.[6,7] Since the original introduction

[1] D. S. Papermaster, B. G. Schneider, M. A. Zorn, and J. P. Kraehenbuhl, *J. Cell Biol.* **77**, 196 (1978).

[2] D. S. Papermaster, B. G. Schneider, M. A. Zorn, and J. P. Kraehenbuhl, *J. Cell Biol.* **78**, 415 (1978).

ISBN 0-12-181996-5

of this two-dimensional immunoelectrophoretic technique,[8] it has been modified for numerous special purposes[6,7,9] and has continued to be a useful approach despite the introduction of techniques dependent on immunochemical identification of proteins by electrophoretic blotting. The two-dimensional technique has the advantage of being relatively rapid and uniquely has the ability to determine cross-reactions when proteins exist in multimeric forms or form tight intermolecular associations with other proteins that result in the distribution of the antigen in several bands on a polyacrylamide gel. In the second dimension the immunoprecipitate forms a long rocket, based on the technique originally devised by Laurell.[10] Adjacent rockets fuse when the antigens have determinants in common, and the rockets cross when the antigens are distinct. In addition, the peak height is a linear function of the quantity of antigen in the first-dimension gel, when care is taken to avoid overloading the system. This chapter provides complete directions for the preparation of the first- and second-dimension gels and their recording by photography.

Briefly, the procedure consists of a first-dimension separation of membrane proteins by sodium dodecyl sulfate (SDS)–polyacrylamide gel electrophoresis, followed by electrophoretic elution of the proteins from this gel, in the second dimension, through a two-layer agarose gel containing Lubrol PX (ICI, Sigma) in the first layer and specific antibodies in the second layer, where precipitin arcs form. The principles of the procedure are shown diagrammatically in Fig. 1.

The unique feature of this technique is the intermediate Lubrol PX-agarose layer. We have shown, using the SDS–stain Pinacryptol Yellow (ICN, K & K), that interposition of the Lubrol layer considerably retards the migration of the bulk of the SDS from the polyacrylamide gel. Presumably, mixed micelles, of reduced electrophoretic mobility, are formed.[11–14]

---

[3] D. S. Papermaster and B. G. Schneider, *in* "Cell Biology of the Eye" (D. S. McDevitt, ed.), p. 475. Academic Press, New York, 1982.

[4] D. Louvard, H. Reggio, and G. Warren, *J. Cell Biol.* **92,** 92 (1982).

[5] D. S. Papermaster, C. A. Converse, and J. Siu, *Biochemistry* **14,** 1343 (1975).

[6] N.-H. Chua and F. Blomberg, *J. Biol. Chem.* **254,** 215 (1979).

[7] G. Piperno and D. J. L. Luck, *J. Biol. Chem.* **252,** 383 (1977).

[8] C. A. Converse and D. S. Papermaster, *Science* **189,** 469 (1975).

[9] F. H. Kirkpatrick and D. J. Rose, *Anal. Biochem.* **89,** 130 (1978).

[10] C.-B. Laurell, *Anal. Biochem.* **10,** 358 (1965).

[11] P. Becher, *in* "Nonionic Surfactants" (M. J. Schick, ed.), p. 478. Dekker, New York, 1967.

[12] R. J. Williams, J. N. Phillips, and K. J. Mysels, *Trans. Faraday Soc.* **51,** 728 (1955).

[13] M. J. Schick and D. J. Manning, *J. Am. Oil Chem. Soc.* **43,** 133 (1966).

[14] Not all of the SDS appears to be removed from the protein, however. Even a protein with an isoelectric point above pH 7.4 (IgG) still moves toward the anode.

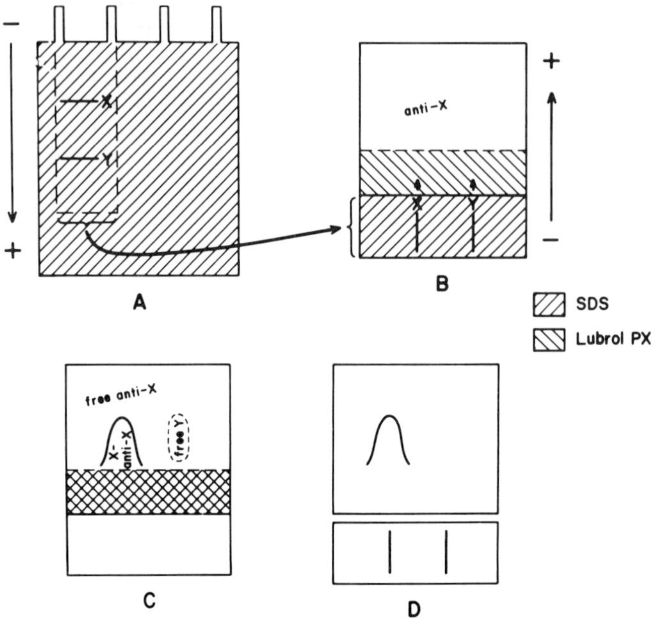

FIG. 1. Schematic diagram of the technique of two-dimensional immunoelectrophoresis. (A) Electrophoretic separation of two membrane antigens, X and Y, on SDS–polyacrylamide gel. (B) The two antigens are electrophoretically eluted from the polyacrylamide strip into an agarose gel. They pass through a layer of Lubrol PX (which removes excess SDS), into a gel containing anti-X. (C) Antigen X forms a precipitin arc (rocket) with anti-X. (D) After washing in saline to remove unreacted Y, the gel is stained. A stained polyacrylamide strip, identical to the strip that was eluted, is shown below the agarose gel, for comparison.

As a consequence, free SDS is essentially eliminated, an SDS–antiserum precipitin line is avoided, and well-formed precipitin rockets are obtained, similar to those usually seen with soluble proteins in the complete absence of SDS.

Outline of Procedure

First Dimension. The first-dimension electrophoresis is based upon the SDS–5.8% polyacrylamide recipe of Fairbanks *et al.*[15] (see the table), but the buffer is used at one-fourth the usual ionic strength to increase the mobility of proteins in the second dimension. The SDS concentration is reduced from 1% to 0.1%, to further diminish nonspecific precipitation of the antiserum. The gel is cast as a slab 157 mm wide × 114 mm long × 1.5

[15] G. Fairbanks, T. L. Steck, and D. F. H. Wallach, *Biochemistry* **10,** 2606 (1971).

SOLUTIONS FOR TWO-DIMENSIONAL IMMUNOELECTROPHORESIS

A. 10X Buffer (Fairbanks *et al.*[15])[a]
  Tris(hydroxymethyl)aminomethane, 96.8 g
  Sodium acetate · $3H_2O$, 54.4 g (or anhydrous sodium acetate), (32.8 g)
  Disodium EDTA (ethylenediamine, tetraacetic acid), 14.9 g

B. First-dimension buffer
  10X Buffer, 25 ml
  SDS (BDH No. 44215), 1 g
  Diluted to 1000 ml with distilled water

C. Second-dimension buffer
  10X Buffer, 400 ml
  Diluted to 2000 ml with distilled water

D. 10% Lubrol
  Lubrol PX (Sigma No. L3753), 2g
  2-D Electrophoresis buffer, to 20 ml
  Dissolve by adding Lubrol to the buffer gradually. Adding buffer to Lubrol is ineffective.

E. SDS–polyacrylamide slab gels (modified from Fairbanks *et al.*[15])[b]

	5.8%	4%	3.2%
N,N'-Methylene bisacrylamide (Eastman)	0.105 g	0.075 g	0.048 g
Acrylamide (BDH No. 15212)	2.815 g	1.930 g	1.570 g
SDS (BDH No. 44215)	0.050 g	0.050 g	0.050 g
Ammonium persulfate	0.075 g	0.075 g	0.075 g
10× Buffer	1.25 ml	1.25 ml	1.25 ml

F. 10% SDS denaturing solution[c]
  Disodium EDTA, 0.037 g
  SDS (BDH No. 44215), 0.50 g
  2-Mercaptoethanol, 0.50 ml
  42% w/w Sucrose, 0.50 ml
  Distilled H_2O added to 5.0 ml

G. Naphthol Blue Black stain
  Solvent:
    Absolute methanol, 400 ml
    Distilled water, 400 ml
    Glacial acetic acid, 80 ml
  Stain:
    1.25 g Naphthol Blue Black (C.I. No. 20470, Sigma) in 880 ml of solvent

[a] Made up to 1900 ml with distilled water, titrated to pH 7.4 with glacial acetic acid, then diluted to 2000 ml final volume.

[b] Distilled water is added to 45 ml and stirred until dissolved. Then 13 μl TEMED (*N,N,N',N'*-tetramethylenediamine, Eastman) are dissolved in 5 ml of H_2O and mixed into the other ingredients by swirling in a 150–250-ml beaker, just before pouring into the gel mold.

[c] One volume of denaturing solution should be added to every three volumes of sample solution to give 2.5% SDS final concentration.

mm thick, using a slot-forming comb with sample slots of a convenient width (6 mm, 10 mm, or wider when several identical strips are to be cut).

A preliminary gel should be run to establish the mobility of the proteins under investigation and to compare the banding pattern with that obtained under standard conditions of electrophoresis (see Modifications).

Samples are denatured by dissolving in the SDS denaturing solution (see the table) and incubating at 20° for 15 min. Since heat may induce aggregation of some membrane glycoproteins and is needed to disaggregate other proteins, the effect of heat should be evaluated by comparing the distribution of the proteins with and without heat denaturation. Gel slots are then loaded with up to 30 μg of protein (0.1–5 μg per band), optimally, in a volume of 20 μl or less, for use in the two-dimensional gels. Reference slots of the first dimension, however, may require more protein (up to 50 μg) for adequate visualization of the bands.[8] Electrophoresis is started at 25 V, then increased to 100 V (about 35 mA) after the samples have entered the gel. After the tracking dye has traveled the desired distance, the power is turned off, the slab is removed from the tank, and one of the glass plates is carefully removed. The gel is covered with Saran (Dow) plastic wrap to keep it moist while preparing the second-dimension gel.

Modifications of First-Dimension Conditions. The 5.8% polyacrylamide gel recipe (see the table) may be used for proteins of apparent M_r 15,000–150,000. When studying proteins of apparent M_r >200,000, we found that they traveled too slowly out of the 5.8% gel into the second-dimension gel, but excellent results were obtained when a 4% polyacrylamide gel was substituted in the first dimension.[16] To verify that a protein has actually left the first-dimension gel, the polyacrylamide strips used in the second dimension should be saved after the second-dimension run and stained in the usual way.

When samples are too dilute, and give diffuse bands in the first dimension, we have found that we can sharpen the bands adequately by introducing a 3.2% polyacrylamide stacking gel above the standard 5.8% (or 4%) resolving gel.

Other investigators have chosen to use the SDS–polyacrylamide gel recipes developed by Laemmli,[17] Maizel,[18] or Neville[19] instead of the Fairbanks procedure. In particular, Chua and Blomberg,[6] in adapting our technique for Neville gels, introduce an extra deoxycholate–agarose gel layer, behind the polyacrylamide gel strip in the second dimension, to accelerate the migration of proteins out of the first-dimension gel.

Preparation of Second-Dimension Gel. Molds are assembled from two acid- and alcohol-washed 83 × 102 × 1 mm projector slide cover glasses (Kodak) separated by three 25 × 75 × 1.5 mm-thick slides (Clay Adams Gold Seal No. A1454) and held together with three large (No. 100) binder clips (Fig. 2), so that the width of the gel (distance between the side spacer

[16] D. S. Papermaster, C. A. Converse, and M. Zorn, *Exp. Eye Res.* **23,** 105 (1976).

[17] U. K. Laemmli, *Nature (London)* **277,** 680 (1970).

[18] J. V. Maizel, *in* "Fundamental Techniques in Virology" (K. Habel and N. P. Salzman, eds.), p. 334. Academic Press, New York, 1969.

[19] D. Neville, *J. Biol. Chem.* **246,** 6328 (1971).

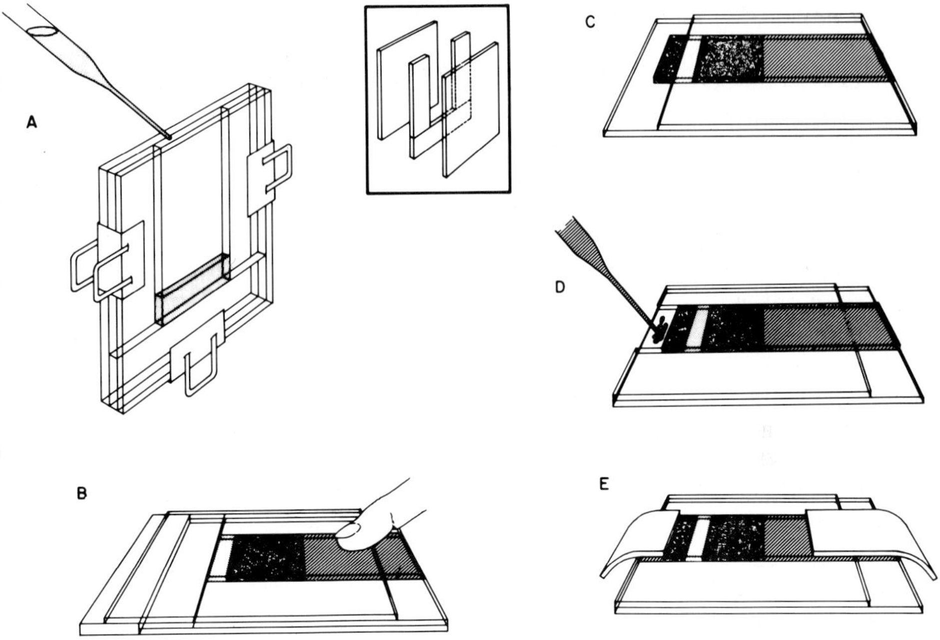

FIG. 2. Preparation of the second-dimension gel. (A) The mold is assembled from glass slides held together with large binder clips (shown smaller than actual size). The mold is then filled to a depth of 7 mm with warm Lubrol–agarose solution. After being allowed to gel, this is followed by layers of antiserum–agarose, then plain agarose to the top. (B) The top plate is slid off, using a 2.3-mm-thick piece of glass (window glass, on the left) as a stopper. (C) The bottom spacer is removed, and the polyacrylamide strip is placed snugly against the Lubrol layer. (D) The side spacers are slid to the left, and the remaining space is filled with plain agarose, to provide electrical contact to the wick. (E) The diagram shows the completed second-dimension gel, with the wicks in place on top of the plain agarose bridges at each end. Drawings were adapted from Converse and Papermaster.[8]

slides) is 5.0 cm. A sheet of Silastic of the same thickness cut into a U shape can substitute for the three glass slides. A line is marked on the molds 7 mm above the bottom spacer (for the Lubrol layer) and the molds, Pasteur pipettes, and a 5-ml disposable pipette are all warmed in a 60° oven. Test tubes are warmed in a heating block, also set at 60°. Agarose (Electrophoresis Grade, ICN Pharmaceuticals) is made up to a concentration of 1.1% w/v in second-dimension buffer (see the table) and stirred on a hot plate (being careful not to let it boil over) until it clears; then it is maintained at about 70–80° on the hot plate.

About 1–2 ml of Lubrol–agarose are required per mold. Seventeen volumes of hot 1% agarose are added to 3 volumes of prewarmed 10%

Lubrol solution in a prewarmed tube; using a warm Pasteur pipette, molds are filled to the 7-mm line, avoiding formation of bubbles (a stiff wire or thin plastic ruler cut to a point should be available, to burst bubbles before they harden). After this layer sets, the molds are returned to the oven.

The antiserum layer can be 1.5–3.0 cm deep (2 ml make a 2.5-cm layer) and usually contains 4–12% (v/v) antiserum (or 30–150 μg of purified antibodies per milliliter). The ratio of antiserum to antigen that will give the best precipitin rockets should be determined by testing a range of protein loadings on the first-dimension gel, and antiserum concentrations on the second-dimension gel.

To cast the antiserum layer, 1.1% agarose solution is measured into the prewarmed tubes in the heating block, using the heated 5-ml pipette (if it is to be reused, it should be immediately rinsed out in warm water). Serum is added to each agarose tube, and the solution is immediately transferred into the molds using preheated Pasteur pipettes. Typical quantities for a 4% layer are 80 μl of antiserum mixed with 1.92 ml of 1.1% agarose, and so forth. When this layer has set, the mold is filled to the top with buffered 1.1% agarose and allowed to cool to room temperature. The clips are then removed and the top plate is slid off (Fig. 2; or it may be gently lifted off), taking care not to tear the gel. The bottom spacer is removed. The top glass plate may be siliconized to ease its removal at this step.

Meanwhile, strips 6–8.5 mm wide by 5.0 cm long are cut out of the SDS–polyacrylamide gel, using a new, alcohol-cleaned razor blade (it is convenient to mark the strips on the covering Saran wrap first, using a sharp, indelible, felt-tip pen). A buffer-wetted, flat microspatula is used to pick up each gel strip and place it along the lower edge of the Lubrol–agarose layer, with the top of the strip to the left. It should be tightly abutted against the Lubrol–agarose without bubbles in between. The space on the other side of the polyacrylamide strip is then filled with plain 1.1% agarose in buffer (the slide may be tilted up at the bottom edge, while applying the agarose, to make certain this agarose layer is as thick as the acrylamide strip). The junction between the Lubrol–agarose and polyacrylamide strips is marked with a needle dipped in 1% Bromphenol Blue, then sealed with a minimal amount of hot buffered 1.1% agarose, which is applied at one end with a Pasteur pipette and allowed to run along the crack. The agarose should not be allowed to cover the acrylamide strip.

Modifications to Second-Dimension Gel. Many variations of the second-dimension gel are possible, depending on the application. The antiserum layer may be split vertically into segments, for assay of multiple sera,[20] or may contain purified IgG or specifically purified antibodies (we have not investigated the use of monoclonal antisera, which may present particular problems unless the antigen is multivalent for the unique determinant). When very high concentrations of serum (12% v/v or more) are used, the stock agarose concentration (normally 1.1%) also should be increased, to maintain the strength of the gel.

If only a few bands on the acrylamide gel are of interest, and these can be located fairly accurately on the unstained gel, then several short polyacrylamide segments (from different slots, e.g., different loadings) may be arranged in tandem and run into the same second-dimension gel. Quantities of antigen can also be varied by varying the width of the polyacrylamide strips; for proteins of M_r <200,000, the precise width of the strip does not seem to be critical to the success of the technique.

If Lubrol PX is not available, it is possible to use other nonionic detergents: in particular, we have found that Triton N-101 (Rohm and Haas) and Emulphogene BC 720 (also known as Mulgofen BC 720, GAF) are equally effective, and Triton X-100 is slightly less effective.[21] The width of the Lubrol (or other detergent)–agarose gel strip is very critical. A 5-mm width will often suffice, but to be certain of trapping all the SDS, we now routinely use a 7–8-mm width. If the strip is too narrow, a wavy SDS–antiserum line will appear along the top of the strip (see Fig. 4).

Second-Dimension Electrophoresis. Synthetic sponge wicks (Orion Diagnostica; LKB No. 217-202) cut to 5 × 14 cm are rinsed several times in distilled water and finally in second-dimension buffer. The completed second-dimension gels are placed in a horizontal electrophoresis tank (we use a tank 32 × 46 × 11 cm high, which will hold up to six gel plates), and the wicks are placed carefully on the agarose "bridges" (Fig. 2), taking care not to pull the acrylamide and agarose layers apart. The ends of the wicks on the agarose are held down with plastic rods.

The second-dimension gels are subjected to a constant voltage of 90 V for 2 hr, or until the Bromphenol Blue tracking dye passes from the antiserum–agarose layer into the agarose bridge, whichever occurs first. The current will be about 25 mA/gel and may increase, so a power supply with high current output is required, such as Model LPD-425A-FM or LPD-424A-FM (Lambda Electronics, Melville, New York). The gels should be checked occasionally for separation of the layers; if separation occurs, the gaps should be resealed with 1.1% agarose after shutting off the power. Power should be restarted after sealing.

[20] C. A. Converse and M. A. Nutley, *Biochem. Soc. Trans.* **10,** 458 (1982).

[21] D. S. Papermaster, C. A. Converse, and S. S. Coppock, *Science* **192,** 616 (1976).

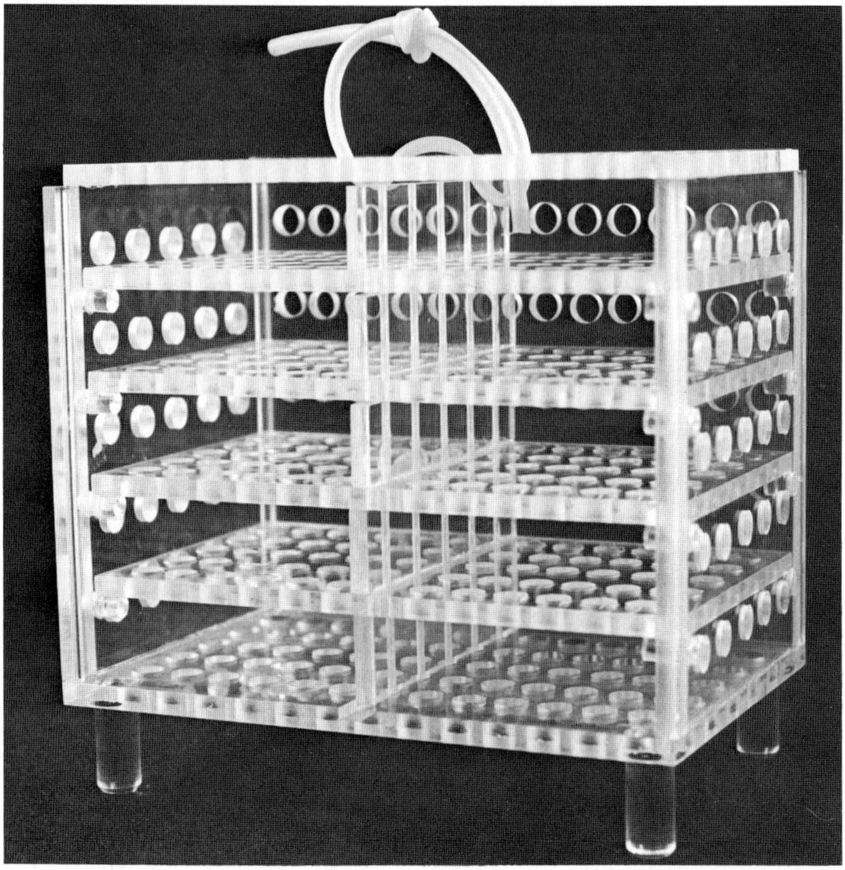

FIG. 3. Fenestrated gel washer, shown with front panel removed. The outside dimensions, minus feet, are 12 × 18.5 × 15 cm high, and the washer rests in a tank 19 × 25 × 21 cm high (capacity 7 liters).

At the end of the run, after turning off the power supply and removing the gel from the tank, the Bromphenol Blue front is marked with a needle dipped in India ink, and the agarose bridges are cut off and discarded. The acrylamide strip is marked with India ink in the lower right-hand corner, then removed to stain 1 of Fairbanks *et al.*[15] A blank portion of the first-dimension gel is also marked with India ink at 1 cm intervals (to determine the extent of swelling during staining) and subjected to the Fairbanks *et al.*[15] staining procedure (this first-dimension gel should also contain reference slots duplicating, perhaps with greater quantities, the strips removed for the second dimension).

Photography and Staining. Precipitin arcs are often visible on the gels immediately after electrophoresis, but they are more easily photographed after washing. Washing removes soluble proteins and nonspecific precipitation and is necessary if the gels are to be stained. A fenestrated gel washer (Fig. 3), constructed of Plexiglas (Perspex), contains 10 compartments each 83 × 102 × 22 mm high. Gel plates are first marked with a diamond marking pen or placed in identifiable positions, as saline will remove "waterproof" felt-tip markings. The washer is lowered into a tank of 0.075 *M* NaCl that is slowly stirred magnetically overnight, preferably with thin foam insulation on the stirring plate to prevent heating by the magnetic stirrer.

For photography, unstained gels (Fig. 4, top) are suspended over the center of a circular fluorescent lamp to give lighting from all sides. A black paper background is centered under the lamp, and black paper strips provide masking around the gel. The gel is photographed in a darkened room with a Polaroid MP-3 (or MP-4) Land camera and type 665 black and white film, a setting of f/16 or f/11, and $\frac{1}{4}$-sec shutter speed.

Before staining (Fig. 4, middle; Fig. 5), gels are first dried onto their glass plates. Two or three gels may be accommodated on one glass plate if desired: gels in a puddle of saline may be moved by pushing gently with a flat edge. Excess saline is blotted from the edges, the gels are then covered with pieces of Whatman No. 1 paper that are cut slightly larger than the gel, placed rough side down, and weighted with an inverted "finger" ("peg forest") rack and a 100–500 g weight. They are dried overnight at room temperature. The paper is carefully removed, then the gels are stained in Naphthol Blue Black stain (see the table) for 5–10 min and destained in the same solvent. They are air-dried and photographed, on a light box in a darkened room, through a yellow (Kodak Wratten No. 15) filter, using the same camera, film, and settings as for the unstained gels. The dried gels may then be autoradiographed.[5]

Interpretation of Results

To assign precipitin arcs to polyacrylamide gel bands, a polyacrylamide strip matching the one electrophoretically eluted (but containing more protein) may be cut out and placed along the bottom of the second-dimension gel. However, if the 1-cm markings show that the reference polyacrylamide gel strip has changed size during staining (Fairbanks gels usually swell to 105–110% of their original length), then the alignment will be only approximate. It may be necessary to determine the prestaining positions of the bands and draw them in, or to photograph the reference

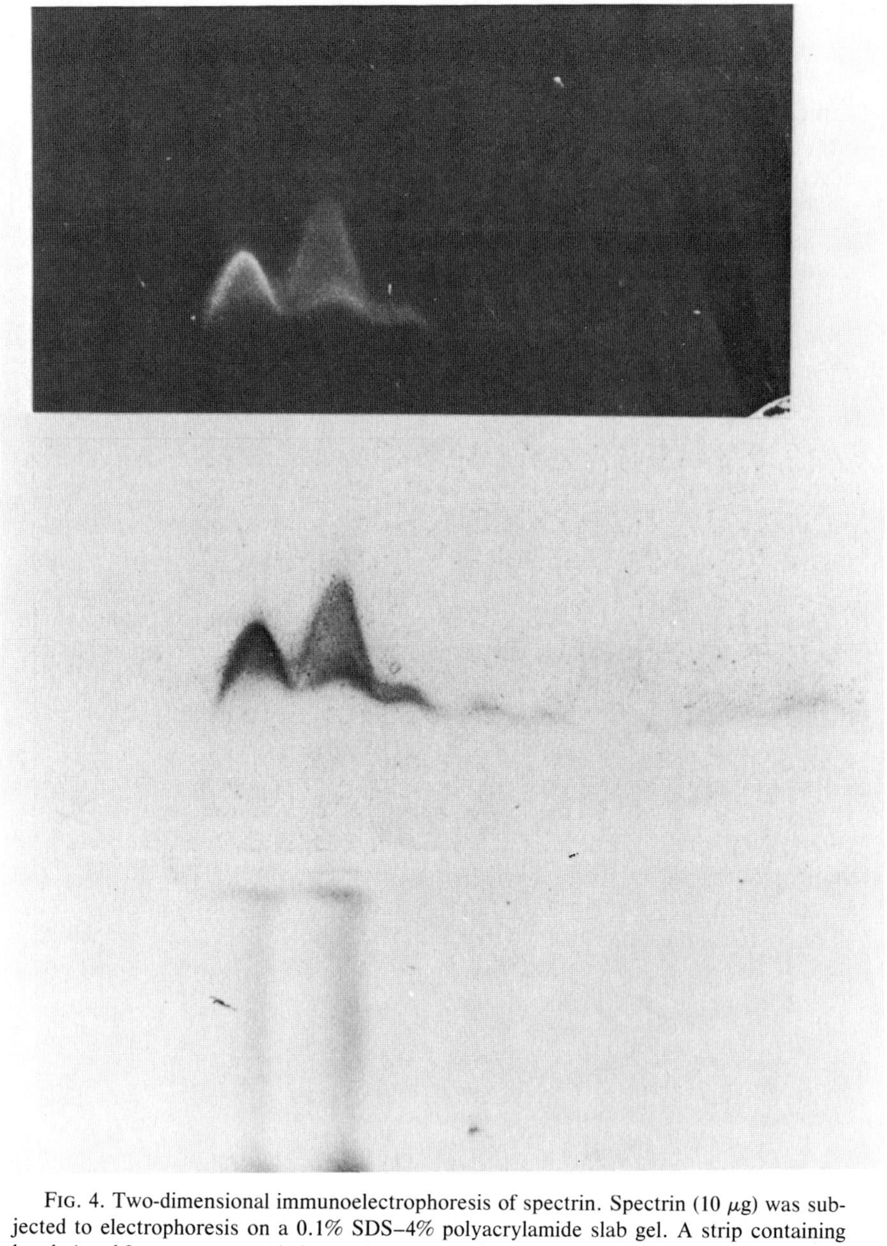

FIG. 4. Two-dimensional immunoelectrophoresis of spectrin. Spectrin (10 μg) was subjected to electrophoresis on a 0.1% SDS–4% polyacrylamide slab gel. A strip containing bands 1 and 2 was cut out and electrophoresed, at right angles, through the Lubrol–agarose layer and into an 8% anti-spectrin (rabbit 2)–agarose layer. *Top:* The second-dimension gel, photographed after washing, but before staining. *Middle:* The same gel, stained. *Bottom:* A stained polyacrylamide strip identical to the one used for the second-dimension experiment. Band 1 is on the left.

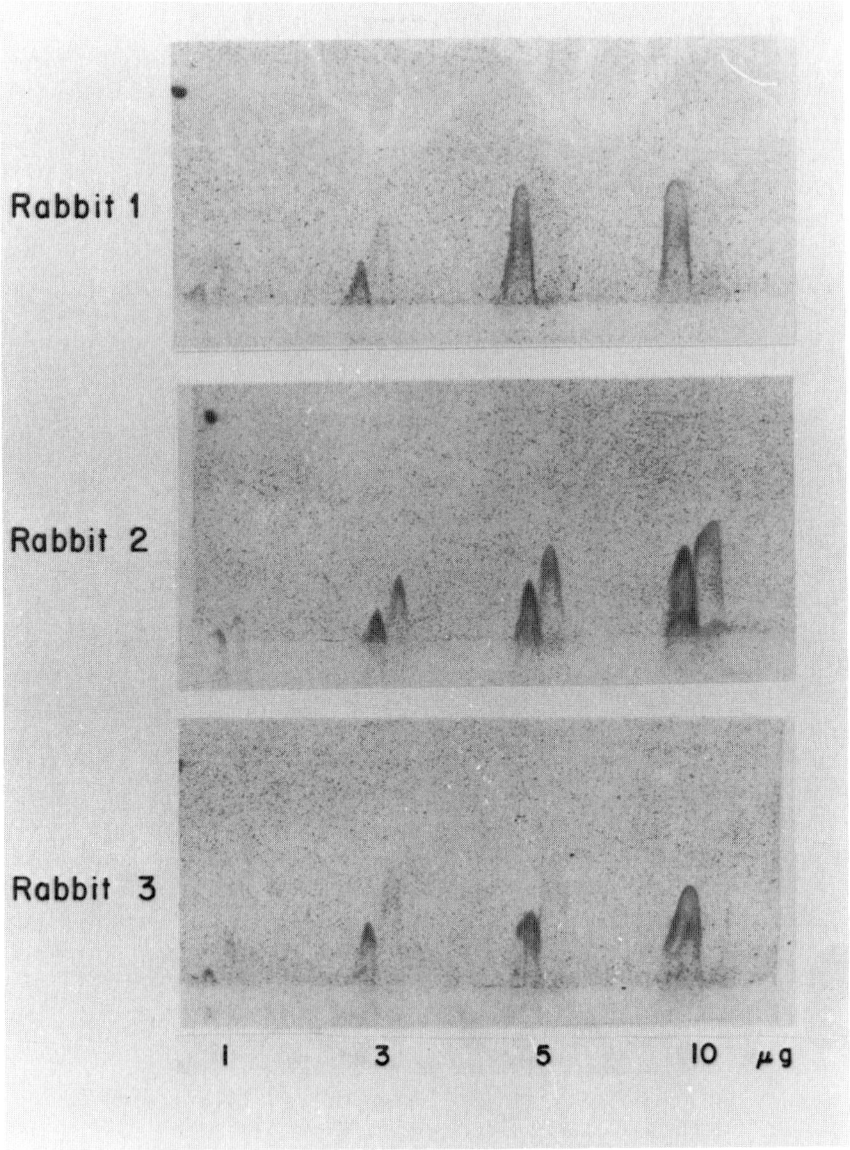

FIG. 5. Two-dimensional immunoelectrophoresis of spectrin, comparing the immune response to band 1 and band 2 in three rabbits. Conditions and concentrations as in Fig. 4.

gel and print it by reducing the image to its calculated prestaining size (as was done in Fig. 4).

One problem occasionally encountered is that rockets do not correspond with bands, but instead appear in pairs on either side of a band. This usually results when too much antigen has been used, and the center and tip of the rocket, being at antigen excess, fail to precipitate. The problem is usually remedied by repeating the second-dimension gel, using much less antigen in the first-dimension strip.

Some investigators have expressed fears that antigenic determinants are lost upon SDS denaturation and not recovered in the second dimension when the Lubrol removes the excess SDS. This is a valid concern, but surprisingly, it has not turned out to be a serious problem in any of the dozen or more antigen–antibody systems we have investigated, and we have heard of few instances where this has been a serious obstacle to the use or interpretation of the method (some problems encountered could have been due to the use of low-titer, or low-affinity antisera). Indeed, Weber and Kuter[22] have found that many enzymes regain their activity after removal of SDS; it is perhaps not unreasonable to expect that many antigenic determinants will be recovered as well. The fact that there are over 120 reports using our method would seem to support its applicability for many systems.

Finally, some two-dimensional immunoelectrophoretic gels[23] (Figs. 4 and 5) illustrate the results obtainable with this method. When polyacrylamide strips containing spectrin bands 1 and 2 were electrophoresed into anti-spectrin antisera,[24] only one of the three antisera tested gave a strong reaction with band 2, whereas all three rabbit antisera reacted with band 1 (Fig. 5). This seemed to indicate that bands 1 and 2 are antigenically distinct. This hypothesis was strengthened by the demonstration of lack of fusion, and apparent crossing, between the rockets for bands 1 and 2 in serum from rabbit 2 (note especially the 5- and 10-μg rockets in Fig. 5). Our preliminary conclusions have been substantiated by later workers, who have shown that bands 1 and 2 differ antigenically, chemically, and functionally.[25–29]

[22] K. Weber and D. J. Kuter, *J. Biol. Chem.* **246,** 4504 (1971).

[23] C. A. Converse, Y.-S. Wong, and D. S. Papermaster, unpublished experiments (1975).

[24] Spectrin and the anti-spectrin sera were the kind gifts of H. Furthmayr and V. T. Marchesi.

[25] M. P. Sheetz, R. G. Painter, and S. J. Singer, *Biochemistry* **15,** 4486 (1976).

[26] F. H. Kirkpatrick, D. J. Rose, and P. LaCelle, *Arch. Biochem. Biophys.* **186,** 1 (1978).

[27] C. J. Hsu, A. Lemay, Y. Eshdat, and V. T. Marchesi, *J. Supramol. Struct.* **10,** 227 (1979).

[28] S. E. Zweig and S. J. Singer, *Biochem. Biophys. Res. Commun.* **88,** 1147 (1979).

[29] D. Litman, C. J. Hsu, and V. T. Marchesi, *J. Cell Sci.* **42,** 1 (1980).

Thus our two-dimensional immunoelectrophoresis method is quick, and gives useful, relatively unambiguous information about relationships among proteins that can be separated on SDS–polyacrylamide gels.

Acknowledgments

Research was supported in part by the NIH Grant EY-00845 and postdoctoral fellowships (to C. A. C.) GM-2140 and EY-01360, an American Cancer Society Grant GC 129, and the Veterans Administration.

[20] Methods for Study of the Synthesis and Maturation of the Erythrocyte Anion Transport Protein

By HARVEY F. LODISH and WILLIAM A. BRAELL

The major integral membrane protein of the human erythrocyte is the anion transport protein, band III. The NH_2-terminal 40,000–45,000 daltons of the protein comprises a hydrophilic domain on the cytoplasmic face of the erythrocyte membrane (Fig. 1). This region is the membrane binding site for the cytoskeletal protein ankyrin, as well as for several glycolytic enzymes. The COOH-terminal 52,000 daltons comprise a membrane-bound domain that spans the membranes two or more times and forms the anion channel or pore.[1–3] The carboxyl-terminal segment of band III contains an asparagine-linked oligosaccharide with the repeating disaccharide Gal ($\beta1 \rightarrow 4$)GLcNAc($\beta1 \rightarrow 3$).[4] Thus, the transmembrane configuration of band III is quite different from that of well-studied membrane proteins like the vesicular stomatitis virus glycoprotein (VSV G) or glycophorin (Fig. 1). There the COOH-terminal 20–30 amino acid residues project into the cytoplasm, while the NH_2-terminal region is exposed on the extracytoplasmic face of the membrane.[1]

This difference in topography raises the question of whether the mechanism of biogenesis of band III is at all related to that of proteins of the VSV G class. G protein is cotranslationally inserted into the rough endoplasmic reticulum, where it receives two high-mannose Asn-linked oligo-

[1] H. F. Lodish, W. A. Braell, A. L. Schwartz, G. J. A. M. Strous, and A. Zilberstein, *Int. Rev. Cytol., Suppl.* **12,** 247 (1981).

[2] W. A. Braell and H. F. Lodish, *J. Biol. Chem.* **256,** 11337 (1981).

[3] W. A. Braell and H. F. Lodish, *Cell* **28,** 23 (1982).

[4] T. Tsuji, T. Irimura, and T. Osawa, *Biochem. J.* **187,** 677 (1980).

ISBN 0-12-181996-5

saccharides. It then matures to the Golgi complex, where occurs addition of fatty acids and also extensive modifications of the oligosaccharides. Subsequently it arrives at the cell surface.[1,5]

We have shown that membrane insertion of band III is cotranslational; in particular, microsomes can be added until the time when the nascent chain is half complete and still permit subsequent insertion of the polypeptide. Thus, an internal sequence is probably used for membrane insertion.[3] Insertion of band III is distinct from all other glycoproteins studied to date, such as the VSV G,[1,4] and all secretory proteins, including ovalbumin,[6] that utilize an NH_2-terminal sequence for membrane insertion. Band III follows a route of maturation to the cell surface that is similar in pathway and kinetics to that of cotranslationally inserted proteins such as VSV G and HLA-A: rough endoplasmic reticulum → Golgi → plasma membrane.[2]

As mammalian erythrocytes do not synthesize protein, we have utilized spleens of anemic mice as a source of cells that synthesize the band III protein.[2] Polyadenylated RNA prepared from these cells will direct the synthesis of band III polypeptide in cell-free systems from rabbit reticulocytes or wheat germ.[3] When supplemented with dog pancreatic microsomes, these extracts will catalyze glycosylation and transmembrane insertion of the band III polypeptide.[3]

Preparation of Erythropoietic Spleen Cells

Reagents

RBB solution: 0.14 *M* NaCl, 0.0015 *M* Mg $(CH_3COO^-)_2$, 0.005 *M* KCl, 100 μg of heparin per milliliter.

PBS (phosphate-buffered saline)

Solution A: 5 liters: NaCl, 50 g; KCl, 1.25 g; Na_2HPO_4, 7.18 g; KH_2PO_4, 1.25 g. Dissolve in 5 liters and check that pH is 7.2. Dispense in 400-ml bottle and autoclave.

Solution B: $CaCl_2 \cdot 2H_2O$, 6.67 g/5 liters or 1.35 g/liter. Dissolve in water and autoclave.

Solution C: $MgCl_2 \cdot 6H_2O$, 5 g/5 liters or 1 g/liter. Dissolve in water and autoclave. Before use for complete PBS, add 50 ml of solutions B and C per 400 ml of PBS solution A.

Procedure. Female CD-1 mice were made anemic by daily bleeding (0.5 ml of blood per day) from the tail over a 5-day period. Before each bleeding incubate the mouse at 37° for 10 min or longer in order to dilate

[5] D. D. Sabatini, G. Kreibich, T. Morimoto, and M. Adesnik, *J. Cell. Biol.* **92,** 1 (1982).

[6] W. A. Braell and H. F. Lodish, *J. Biol. Chem.* **257,** 4578 (1982).

the blood vessels so that the blood will flow more easily. Pour about 30 ml of ether in a beaker with a cushion of cheesecloth on top of a layer of cotton and cover the beaker with a petri dish. Prepare an ether mask by soaking with ether a piece of cotton placed near the tip of a 12-ml conical centrifuge tube and gently corking the tube. Place the mouse in the ether beaker immediately after removing it from the heat. Do not overanesthetize: remove mouse immediately after it loses consciousness. Place the etherized mouse on a paper surface and place the ether mask over its snout. With a sharp razor slice off about 1 cm of the tail for the first bleeding and just a tiny sliver for subsequent bleedings. Massage the tail until it has released about 15 drops or until the blood flow slows down dramatically. If collection of the blood is desired, it is dripped into RBB solution. After bleeding, cauterize the tip of the tail (far away from the ether!) with a hot glass rod or preferably a miniature soldering iron. The mouse will recover on its own. One can warm the mouse under a heat lamp if the blood flow is too slow.

Spleens were excised from animals made anemic by bleeding for 5 days as described above. The spleens were removed surgically after cervical dislocation and were minced with a razor blade in minimal Eagle's medium (MEM, GIBCO) containing 20% fetal calf serum (Microbiological Associates). The cells were then forced through a 20-gauge nickel mesh, aspirated with a 21-gauge needle, and passed through 35-μm Nitex cloth (TETKO, Inc.) to yield monodisperse cells. Cells were washed by centrifugation in medium, and then washed in PBS containing 20% dialyzed fetal calf serum.

Labeling of Erythropoietic Spleen Cells with [^{35}S]Methionine. Cells prepared as described above were incubated at 5×10^7 cells/ml with 500 μCi of [^{35}S]methionine per milliliter in minimal Eagle's spinner medium lacking methionine (Flow Laboratories), supplemented with 20% dialyzed fetal calf serum and 3% dialyzed bovine serum albumin (Miles Laboratories). Culture was performed at 37°, except for experiments where the elongation rate of protein synthesis needed to be slowed, where culture was at 25°.

Preparation of Methionine-Labeled Erythrocytes

Female CD-1 mice (Charles River Breeding Laboratories) were injected intraperitoneally with 1–5 mCi of [^{35}S]methionine (Amersham) in 0.5–1.0 ml of 0.9% saline. At 96 hr after injection, peripheral blood was removed by cardiac puncture, and washed by centrifugation three times with 10 volumes of PBS for 5 min at top speed in a clinical table-top centrifuge. The buffy coat layer was removed from the top of the cell pellet during the washing procedure.

Preparation of Erythrocyte Band III Protease Fragments

Defined proteolytic fragments of band III are required for two types of studies: as a standard for membrane insertion of *in vitro*-synthesized protein, and as immunogen for induction of antibodies. Most rabbit antibodies generated with total band III protein are, in our experience, directed toward the NH_2-terminal cytoplasmic domain. Thus, in order to generate antibodies that bind to the COOH-terminal membrane-protected region, it was necessary to use as immunogens the COOH-terminal fragments TR52 and CH38 (Fig. 1).

Reagent

Sodium phosphate, 5 m*M*, pH 8.0 (5P8)

Procedure. To prepare CH38, erythrocytes were washed by centrifugation three times in PBS, removing the top buffy coat after the first and second centrifugations. Chymotrypsin-digested membranes were prepared by digestion of washed, intact erythrocytes with 10 volumes (*re* packed cell volume) of 1 mg/ml α-chymotrypsin (Worthington Biochemicals) for 10 min at 37° in PBS supplemented with 1 mg of glucose per milliliter as described by Knipe *et al.*[7] Digestion was terminated by addition of 200 m*M* phenylmethylsulfonyl fluoride (PMSF, Sigma), in absolute ethanol, to a final concentration of 2 m*M*. The cells were incubated for 5 min at 37°, then washed three times in PBS.

Red cell ghosts were prepared from washed, chymotrypsin-digested or undigested erythrocytes by several washes in 5P8, as described by Steck *et al.*[8] Erythrocytes were washed first by centrifugation three times in RBB (5000 *g* for 5 min at 4°). The washed cell pellet was diluted with at least 40 volumes of cold 5P8; the resulting membranes (ghosts) were pelleted at 13,000 *g* for 20 min, then washed three times to remove residual hemoglobin.

To prepare TR52, undigested ghosts were first depleted of spectrin and then digested with trypsin.[9] For depletion of spectrin, membranes were incubated in 0.1 m*M* EDTA, 1 m*M* 2-mercaptoethanol, pH 8.0, at 37° for 15 min, and the membranes were isolated by centrifugation in a Sorvall SS34 rotor for 30 min at 16,000 rpm at 4°. Membranes were resuspended in 4 volumes of 25 m*M* Tris-HCl buffer, pH 8.0, and digested with 15 μg of TPCK-treated trypsin per milliliter (Worthington Biochemicals) for 60

[7] D. Knipe, H. F. Lodish, and D. Baltimore, *J. Virol.* **21,** 1121 (1977).

[8] T. L. Steck, J. J. Koziarz, M. K. Singh, G. Reddy, and H. Kohler, *Biochemistry* **17,** 1216 (1978).

[9] M. Fukuda, Y. Eshdat, G. Tarone, and V. T. Marchesi, *J. Biol. Chem.* **253,** 2419 (1978).

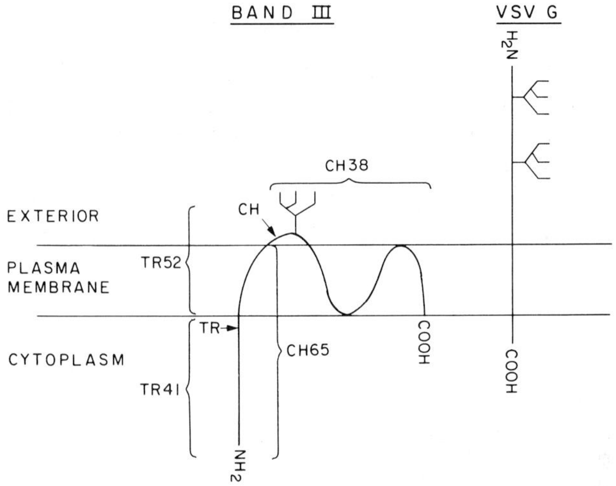

FIG. 1. The configurations of band III protein and vesicular stomatitis virus glycoprotein (VSV G) in the plasma membrane. The configuration of band III presented is that obtained from a compilation of structural information for the human erythrocyte protein.[1] The sites of cleavage by chymotrypsin on the exterior and trypsin on the interior of the cells are indicated by CH and TR, respectively. The characteristic cleavage fragments CH65 and CH38 are produced by chymotrypsin digestion, whereas TR52 and TR41 are produced by trypsin digestion. The numbers associated with each fragment refer to the molecular weight (×1000) of the fragment, as determined by ourselves for the mouse erythrocyte band III protein in the Laemmli gel electrophoresis system. These molecular weight values are similar to those obtained by other investigators working with the human protein in other gel electrophoresis systems. The structure of the VSV G protein is also based on information compiled by Lodish *et al.*[1]

min at 0°. Digestion was terminated by adding 50 μg of soybean trypsin inhibitor per milliliter (Sigma).

Both chymotrypsin-digested and trypsin-digested membranes were stripped of peripheral proteins by washing in ice-cold 0.1 *N* NaOH as described by Steck *et al.*[8] The membranes were resuspended in 7 volumes of cold 0.1 *N* NaOH, followed immediately by centrifugation for 30 min at 16,000 rpm in Sorvall SS34 rotor at 0–2° in polyethylene tubes. Pellets were washed once with phosphate buffer, 10 m*M*, pH 7.4, in the same volume as used for stripping: as the membrane pellet shrinks about 3-fold after NaOH treatment, this is about 20 volumes relative to the pellet. This brings the membranes to pH 7–8. This treatment removes the peripheral proteins and thereby enriches band III and its membrane-protected fragments TR52 and CH38. These stripped membranes were solubilized in sample buffer prior to SDS–gel electrophoresis.

Gel Electrophoresis and Autoradiography

Reagent

Sample buffer: 2% (w/v) sodium dodecyl sulfate, 10% (v/v) glycerol, 20 m*M* dithiothreitol, 0.7 *M* 2-mercaptoethanol, 0.01% Bromphenol Blue dye, and 80 m*M* Tris-HCl buffer, pH 6.8

Procedure. SDS–gel electrophoresis analysis was performed by the method of Laemmli[10] in 0.1% SDS–Tris/glycine buffer, using slab gels with 3.75% (w/v) stacking gels and 10% (w/v), or 7 to 15% (w/v) gradient, running gels. Samples for electrophoresis were heated to 100° for 5 min in sample buffer.

Purification of Band III for Use as Immunogen

Reagents

SDS buffer: 2% sodium dodecyl sulfate, 0.2 *M* Tris-HCl, pH 8.0, 1 m*M* NaEDTA

Thiol Sepharose column buffer: 0.1% sodium dodecyl sulfate, 0.1 *M* Tris-HCl, pH 8.0, 0.1 *M* NaCl, 1 m*M* NaEDTA

Sepharose column buffer: 0.2% sodium dodecyl sulfate, 20 m*M* Tris-HCl, pH 8.0, 1 m*M* NaEDTA, 20 m*M* 2-mercaptoethanol

Procedure. The antisera against total band III (anti-band III serum) was raised against mouse band III prepared by the method of Fukuda *et al.*[9] To prepare the band III immunogen, mouse erythrocyte ghosts were depleted of spectrin by incubation for 15 min at 37° in 20 volumes of 0.1 m*M* NaEDTA, 1 m*M* 2-mercaptoethanol, pH 8.0, recovered by centrifugation in the Sorvall SS34 rotor at 16,000 rpm for 30 min, and stripped of peripheral proteins by incubation for 20 min at 4° in the dark after the addition of 2 volumes of 50 m*M* lithium 3,5-diiodosalicylate (Eastman Kodak; final concentration = 33 m*M*) in 10 m*M* Tris-HCl, pH 8.0. The residual membranes were pelleted by centrifugation at 20,000 rpm for 1 hr in an SS34 rotor, and dissolved to a final protein concentration of 2 mg/ml in SDS buffer at 25° for 30 min. The solution was treated with 10 m*M* dithiothreitol for 1 hr at 25°, and dialyzed overnight against 500 volumes of thiol-Sepharose column buffer. Particulates were removed by centrifugation for 5 min at 2000 *g*, and the supernatant was applied to a 1 × 20 cm thiol Sepharose 4B column equilibrated in the same buffer. After washing with this buffer, the column was eluted with Thiol Sepharose column buffer containing 7 m*M* dithiothreitol. Band III was detected by gel electrophoresis of the eluate fractions. The eluate was lyophilized and

[10] U. K. Laemmli, *Nature (London)* **227,** 680 (1970).

resuspended in a small volume of Sepharose column buffer. This was made 2% in sodium dodecyl sulfate, incubated for 15 min at 25°, then heated for 5 min at 80°. The solution was applied to a 1.5 × 60 cm column of Sepharose 6B equilibrated with the column buffer, with a flow rate of 6 ml/hr. The pure fractions of band III were located by gel electrophoretic analysis, pooled, and dialyzed extensively against 0.1 m*M* NaEDTA, pH 8.0. The pure band III was concentrated by ultrafiltration on an Amicon PM-10 filter.

Purification of TR52 and CH38 Immunogens

To prepare antiserum against the carboxyl-terminal portions of band III (anti-TR52 serum), samples of [^{35}S]methionine-labeled and protease-digested band III, which contained the CH38 and TR52 protease fragments, were subjected to SDS–gel electrophoresis in slab gels by the Laemmli method, fixed in isopropanol–acetic acid–water (25 : 10 : 65), and dried from water. The CH38 and TR52 bands were located by autoradiography, and the bands were excised from the gel. Gel slices were hydrated in 0.1% deoxycholate in PBS, aspirated through a series of needles from 19 to 23 gauge, and dialyzed extensively against PBS. The adherent PBS was included with the gel fragments in the antigen emulsion.

Production of Antisera

In both cases, antigen was emulsified in 0.9% saline 1 : 1 with Freund's adjuvant (GIBCO). Complete adjuvant was used initially, and incomplete adjuvant was used for subsequent boosts. New Zealand white rabbits (Charles River Breeding Laboratories) were injected at multiple sites along the back and in the footpads with about 0.1 mg of each peptide, and boosted at 4-week intervals with similar dosages until the serum was immunoreactive. Sera were heat-inactivated at 56° for 60 min and clarified by centrifugation at 100,000 *g* for 60 min prior to use.

Preparation of Spleen Cell Membranes

Reagents

Buffer A: 1% Nonidet P-40, 0.50 *M* NaCl, 1 m*M* EDTA, 0.1 m*M* PMSF, 10 m*M* Tris-HCl buffer, pH 8.0, plus 1 mg of bovine serum albumin per milliliter

Procedure. Spleen cells that had been labeled with [^{35}S]methionine were washed in PBS, swelled in 100 volumes of 5P8 at 0° for 5 min, and homogenized by 50 strokes with a tight-fitting Dounce homogenizer. Total membranes were prepared by layering the homogenate over a discontinuous sucrose gradient with layers of 15% and 55% (w/w) sucrose in 5P8. Membranes were isolated from the 15/55% interface after 90 min of centrifugation at 4° and 250,000 *g* in an SW41 rotor of a Beckman ultracentrifuge. A fraction of membranes enriched in plasma membranes was obtained by centrifuging the homogenate for 90 min at 250,000 *g* in the SW41 rotor on a discontinuous sucrose gradient with layers of 15, 30, 40, and 55% (w/w) sucrose in 5P8, and collecting those membranes at the 30/40% interface. The recovered membranes were stripped in ice-cold 0.1 *N* NaOH as described above, and dissolved in 100 volumes of buffer A for immunoprecipitation, or dissolved in sample buffer for SDS–gel electrophoresis as required.

Preparation of Spleen Cell Microsomes

Reagents

Buffer B: 50 m*M* KCl, 2 m*M* $MgCl_2$, 25 m*M* HEPES buffer, pH 7.5

Procedure. Spleen cells labeled with [^{35}S]methionine were washed in PBS, and swelled at 0° in buffer B. Cells were homogenized by 50 strokes of a tight-fitting Dounce homogenizer and centrifuged at 1000 *g* for 5 min to remove nuclei and large plasma membrane fragments. The membranes in the remaining supernatant fluid constituted a crude microsomal fraction. As an assay for proper orientation of labeled band III in microsomal membranes,[2] these preparations were treated with trypsin: Treatment was for 60 min at 0° with 10–15 μg of trypsin-TPCK per milliliter, and digestion was halted with soybean trypsin inhibitor, as described above. Microsomes to be immunoprecipitated were dissolved in 20 volumes of buffer A.

Chymotrypsin Digestion of Pulse-Labeled Spleen Cells

This assay measures appearance of band III on the cell surface. Band III in the plasma membrane is digested by chymotrypsin to yield two fragments—CH38 and CH65 (Fig. 1). Intracellular band III is resistant to extracellular protease.

Spleen cells labeled with [^{35}S]methionine for 5- to 10-min periods as described above were chased by addition of excess unlabeled methionine to a concentration of 1.0 m*M*, sufficient to inhibit further incorporation of label. Incubation was continued at 37°, aliquots of cells being taken at the indicated periods of chase. Each aliquot was washed in PBS, split, and incubated with or without 1 mg of chymotrypsin per milliliter in PBS plus

1 mg of glucose per milliliter for 10 min at 37°, as described above. Digestion was stopped with PMSF, and total cell membranes were prepared as described.

Immunoprecipitation

Reagents and Materials

Buffer C: 1% sodium dodecyl sulfate, 1% 2-mercaptoethanol

Staphylococcus aureus (New England Enzyme Center, IgGSorb). The material is resuspended in 10 volumes of PBS using a water-bath sonicator to disperse the cells. It is then washed twice in 10 volumes of buffer A by centrifugation for 10 min and is resuspended to a 10% w/v solution in buffer A.

Procedure. In those experiments where it was desired to immunoprecipitate the carboxyl-terminal portions of band III, the antiserum against such fragments (anti-TR52 serum) was used; otherwise, the antiserum against total band III (anti-band III serum) was used.

Immunoprecipitation from solubilized spleen cells membranes was performed with formalin-treated *S. aureus* bacteria, essentially as described by Owen *et al.*[11] Typically, 1 ml of a lysate in buffer A was preincubated for 30 min at 0° with 100 μl of a 10% (w/v) solution of *S. aureus* in buffer A, and centrifuged at 12,000 *g* for 5 min to remove bacteria and debris. The supernatant fluid was incubated for 30 min at 25° with 10 μl of antiserum and then shaken for 60 min at 4° with 100 μl of a 10% (w/v) solution of *S. aureus* in buffer A. Recovery of antigen under these conditions was found to be quantitative. Sedimented bacteria were washed sequentially with buffer A, buffer A plus 0.1% sodium dodecyl sulfate, and 0.1% (w/v) Nonidet P-40 in 10 m*M* Tris-HCl buffer, pH 7.4. Resuspended, washed bacteria were heated for 5 min at 100° in 60 μl of sample buffer to release the immunocomplexes, and the bacteria were removed by centrifugation for 5 min at 12,000 *g*. Alternatively, release was accomplished by heating in 60 μl of buffer C for samples to be treated with endoglycosidase H.

Immunoprecipitation from cell-free translation extracts was performed in a modified manner. Extracts were made up in buffer A supplemented with 7.5 m*M* methionine (unlabeled) to inhibit nonspecific retention of label. Lysates were pretreated as above, but in the actual immunoprecipitation, 40 μl of antiserum per milliliter were used and incubation at 25° for 30 min was followed by incubation at 4° for 3 hr prior to *S. aureus* addition. Treatment with *S. aureus,* using 10 volumes of 10% (w/v) bacteria per volume of antiserum, and subsequent washing were performed as described above.

[11] M. J. Owen, A.-M. Kissonerghis, and H. F. Lodish, *J. Biol. Chem.* **255,** 9678 (1980).

Endoglycosidase H Digestion of Immunoprecipitations

Solutions of immunoprecipitated material in buffer C were made 0.25 *M* in sodium citrate buffer, pH 6.0, 0.01% NaN_3, and 1 m*M* PMSF. Each aliquot was split and incubated for 18 hr at 37° with 0.1 volume of buffer C or with 0.1 volume of endoglycosidase H (30 μg/ml), essentially as described by Zilberstein *et al.*[12] Reactions were precipitated with acetone, and the pellets were lyophilized and dissolved in 60 μl of sample buffer, with heating at 100° for 5 min, in preparation for SDS–gel electrophoresis.

Preparation of Spleen Cell Messenger RNA

This procedure yielded RNA that directed synthesis of band III protein in conventional cell-free systems prepared from wheat germ or rabbit reticulocytes.[3]

Reagents

Homogenization buffer: 5 *M* guanidinium thiocyanate, 1% sodium *N*-lauryl sarcosinate, 25 m*M* sodium citrate, pH 7.0, 0.1 *M* 2-mercaptoethanol, and 0.1% antifoam A (Sigma); filtered
Guanidine-HCl buffer: 7.5 *M* guanidinium-HCl, 5 m*M* dithiothreitol, 25 m*M* sodium citrate, pH 7.0
RNA extraction buffer: 0.5% sodium dodecyl sulfate, 0.1 *M* NaCl, 50 m*M* sodium acetate, pH 5.2, m*M* NaEDTA
Oligo(dT) binding buffer: 0.1% sodium dodecyl sulfate, 0.4 *M* sodium acetate, pH 7.5, 5 m*M* NaEDTA, treated with 0.1% diethyl pyrocarbonate (DEP)
Potassium binding buffer: 0.4 *M* potassium acetate, pH 7.5, 5 m*M* NaEDTA
Elution buffer: 10 m*M* sodium acetate, 1 m*M* NaEDTA

All solutions not containing guanidine in which pure RNA is to be dissolved are first treated with 0.1% DEP and then autoclaved.

Procedure. This was a modification of that of Chirgwin *et al.*[13] for the preparation of RNA from tissues enriched in ribonuclease. Individual spleens from anemic mice were excised and rapidly homogenized for 60 sec in 15 ml of homogenization buffer with a Brinkmann Polytron homogenizer on its highest setting. It was critical that the tissue be entirely dispersed within the first few seconds of the homogenization. The pooled homogenates were centrifuged for 10 min at 8000 rpm in a Sorvall HB-4

[12] A. Zilberstein, M. D. Snider, M. Porter, and H. F. Lodish, *Cell* **21,** 417 (1980).
[13] J. M. Chirgwin, A. E. Pryzbyla, R. J. MacDonald, and W. J. Rutter, *Biochemistry* **18,** 5294 (1979).

rotor to remove cell debris. The homogenate was acidified with 1/40 volume of 1 *M* acetic acid, and ethanol was precipitated with 1 volume of ethanol overnight at −20°.

The RNA was pelleted by centrifugation as described above, and resuspended in a minimal volume of guanidine-HCl buffer, heating at 68° if necessary. This solution was acidified and ethanol precipitated as described above, using 0.75 volume of ethanol. Resuspension in guanidine-HCl buffer and ethanol precipitation was repeated, and the pellet was now resuspended in 0.1% sodium dodecyl sulfate. To this was added an equal volume of RNA extraction buffer, and this aqueous phase was extracted with equal volumes of phenol–chloroform–isoamyl alcohol (25 : 24 : 1) a total of three times. The bulk of a black-brown heme-containing material will be removed into the phenol phase. The aqueous phase was extracted with chloroform, made 0.3 *M* in sodium acetate, pH 5.2, and ethanol precipitated with two volumes of ethanol overnight at −20°. The RNA was pelleted as before, resuspended in water, and mixed with an equal volume of ice-cold 4 *M* LiCl. This was allowed to precipitate overnight at 0° before pelleting the RNA at 12,000 rpm for 10 min in the Brinkmann microcentrifuge. The pellet was washed by centrifugation with 2 *M* LiCl and resuspended in water. This solution was made 0.4 *M* on sodium acetate, pH 7.5, and ethanol precipitated as before.

The RNA pellet was resuspended in water, made up in oligo(dT) binding buffer, and then passed twice over a 2 × 1-cm column of oligo(dT) cellulose (Collaborative Research). The column was washed successively with 10 column volumes of binding buffer, 2 volumes of binding buffer without sodium dodecyl sulfate, 2 volumes of potassium binding buffer and eluted with 5 column volumes of elution buffer. This solution was made 0.3 *M* in potassium acetate at pH 7.5 and was ethanol-precipitated overnight at −20° with 2 volumes of ethanol. The RNA was pelleted by centrifugation at 24,000 rpm for 60 min in an SW27 rotor at 0°, lyophilized, and resuspended in water to about 1 mg/ml. Typically the RNA was used in translation at 40–160 μg/ml.

[21] Glycophorins: Isolation, Orientation, and Localization of Specific Domains

By HEINZ FURTHMAYR and VINCENT T. MARCHESI

Glycophorins are relatively small and heavily glycosylated proteins that are found in plasma membranes of many, possibly all, cell types.[1] This class of proteins was initially discovered and studied extensively in human erythrocyte membranes.[2] More recently, similar proteins have been isolated from pig[3] and horse erythrocytes,[4] from plasma membranes of murine T lymphocytes,[5] rat hepatoma cells,[6] or platelets.[7] From these studies it became clear, that in most species and at least in some cell types the glycophorins represent a family of related or similar sialoglycoproteins, which are difficult to separate. In human red cells three distinct molecules, termed glycophorin A, B, and C,[8] have been identified, and similar heterogeneity is observed for platelet membranes.

Despite a considerable body of information on the primary structure of the protein and carbohydrates of human glycophorins,[9–13] in general the function of these proteins is unknown. Genetically determined changes in the amount or even lack of one or the other of the various proteins expressed at the cell surface of the mature erythrocyte appear to be of little biological consequence,[2] and mutational events that lead to structural alterations[14,15] within the glycosylated domain are of interest mainly

[1] H. Furthmayr, *Protides Biol. Fluids* **29,** 49 (1982).

[2] H. Furthmayr, *in* "Biology of Complex Carbohydrates" (V. Ginsburg and P. Robbins, eds.), p. 123. Wiley, New York, 1981.

[3] K. Honma, M. Tomita, and A. Hamada, *J. Biochem.* (*Tokyo*) **88,** 1679 (1980).

[4] J. Murayama, K. Takeshita, M. Tomita, and A. Hamada, *J. Biochem.* (*Tokyo*) **89,** 1593 (1981).

[5] W. R. A. Brown, A. N. Barcley, C. A. Sunderland, and A. F. Williams, *Nature* (*London*) **289,** 456 (1981).

[6] S. Nakajo, K. Nakaya, and Y. Nakamura, *Biochim. Biophys. Acta* **579,** 88 (1979).

[7] T. Okumura and G. A. Jamieson, *J. Biol. Chem.* **251,** 5944 (1976).

[8] H. Furthmayr, *J. Supramol. Struct.* **9,** 79 (1978).

[9] M. Tomita, H. Furthmayr, and V. T. Marchesi, *Biochemistry* **17,** 4756 (1978).

[10] H. Furthmayr, *Nature* (*London*) **271,** 519 (1978).

[11] W. Dahr, K. Beyreuther, E. Bause, and M. Kordowicz, *Protides Biol. Fluids* **29,** 57 (1982).

[12] H. Yoshima, H. Furthmayr, and A. Kobata, *J. Biol. Chem.* **255,** 9713 (1980).

[13] R. Prohaska, T. A. W. Koerner, I. M. Armitage, and H. Furthmayr, *J. Biol. Chem.* **256,** 5781 (1981).

[14] H. Furthmayr, M. N. Metaxas, and M. Metaxas-Bühler, *Proc. Natl. Acad. Sci. U.S.A.* **78,** 631 (1981).

ISBN 0-12-181996-5

because of effects on the expression of blood group antigens. Thus it remains unknown which genetic events compatible with life are required to affect red cell function severely in humans. In platelets, the expression of these proteins appears to be more tightly linked to the biological function of these particles, since different sialoglycoprotein profiles on polyacrylamide gels are correlated with a number of disorders.[16] Yet, the molecular basis of deficient function is not understood. Various ideas have been put forward, suggesting that the cell surface domain of these proteins serves as receptors for viruses and other ligands, and that the cytoplasmic domain of these transmembrane protein provides attachment sites for cytoskeletal proteins.[17] These observations are compatible with the idea that these proteins mediate functions via the cytoskeleton.[18] The function of these proteins could, however, be related to more general cellular requirements—to provide negatively charged molecules that may be required to prevent cell fusion. Although it is known, that glycophorins are expressed in early precursor cells during erythropoiesis,[19] we do not know whether the function of these membrane proteins in the early precursor cells is related to the postulated function in the mature red cells, which are much more readily accessible to study.

In this chapter the isolation procedures and properties of intact glycophorins are described, as well as simplified procedures to obtain preparative amounts of the glycosylated domains of the three glycophorin molecules.

Isolation Procedures

Preparation of the Crude Glycophorin Fraction

A crude sialoglycoprotein fraction can be readily prepared by the lithium diiodosalicylate (LIS)/phenol extraction procedure originally described by Marchesi and Andrews.[20] The methods outlined here include modifications and are currently used in the laboratory.

---

[15] O. O. Blumenfeld, A. M. Adamani, and K. V. Puglia, *Proc. Natl. Acad. Sci. U.S.A.* **78,** 747 (1981).

[16] A. T. Nurden, D. Dupuis, D. Pidard, T. Kunicki, and J. P. Caen, *Ann. N.Y. Acad. Sci.* **370,** 72 (1981).

[17] T. J. Mueller and M. Morrison, *Prog. Clin. Biol. Res.* **56,** 95 (1981).

[18] R. G. Painter and M. Ginsberg, *J. Cell Biol.* **92,** 565 (1982).

[19] P. D. Yurchenco and H. Furthmayr, *J. Supramol. Struct.* **13,** 255 (1980).

[20] V. T. Marchesi and E. P. Andrews, *Science* **174,** 1247 (1971).

Reagents

Lithium 3,5-diiodosalicylate (DIS) is prepared from twice recrystallized 2-hydroxy-3,5-diiodosalicylate (Eastman Kodak No. 2166, M_r 389.91).

Three hundred grams of DIS are dissolved in 4 liters of methanol at 38°, the solution is filtered through two layers of Whatman No. 1 filter paper on a Büchner funnel under suction, and the solution is kept at 4° (light protected) for 1–2 days. Crystals are collected on a Büchner funnel and dried *in vacuo* (yield approximately 50%). These steps are repeated once more. The mother liquor from this second crystallization step can be stored and used in the first step for future preparation.

LIS is prepared by dissolving DIS in water to an end concentration of 0.6 *M* by adding dropwise a concentrated solution of lithium hydroxide ($LiOH \cdot H_2O$, M_r 41.96) to a final concentration of 0.6 *M* or to neutrality. The solution can be warmed to 45°. After filtration of the almost clear solution through two layers of Whatman No. 1 filter paper, the solution is kept for 1–2 days at 4° (light protected), and crystals are collected and dried as above (end product 3 × crystallized LIS). Phenol (Mallinckrodt, loose crystals).

LIS/Phenol Extraction

1. Freshly drawn or outdated human blood is washed free of plasma by centrifugation and resuspension in phosphate-buffered saline. Hemoglobin-free ghosts are then prepared by lysis of the packed red cells in 20-fold excess (v/v) of 5 m*M* sodium phosphate, pH 8.0, containing 0.3 m*M* phenylmethylsulfonyl fluoride (PMSF) and stirring at ice temperature for 10 min. The red cell ghosts are pelleted by centrifugation at approximately 30,000 *g* for 30 min and are then washed repeatedly. The last wash should be done with 25 m*M* Tris-HCl, pH 8.0, and the pellet is then lyophilized. During the washing procedure and centrifugation of the membranes a small darker pellet is formed, which is carefully removed by suctioning with a Pasteur pipette.

2. Three grams of lyophilized membranes (white to pink) are dispersed by brief homogenization with a homogenizer (Tekman Co., Cincinnati, Ohio) in 100 ml of 0.3 *M* LIS (12 g/100 ml) in 50 m*M* Tris-HCl, pH 8.0. After stirring for 10 min at room temperature (light protected), 200 ml of cold distilled water is added and the solution is stirred for an additional 10 min on ice.

3. Undissolved material is removed by centrifugation for 1 hr at 40,000–50,000 *g* in a refrigerated centrifuge.

4. Fifty percent aqueous phenol is prepared by mixing 150 g of phenol and distilled water at room temperature to a final volume of 300 ml in a closed vessel. The solution is stored at 4° until use.

5. Equal volumes of supernatant from step 3 and 50% phenol are mixed, and the mixture is stirred at room temperature for 20 min (light protected). After centrifugation at 4000 rpm for 1 hr, in a refrigerated centrifuge, two phases are formed; the upper (water) phases are collected by suction into a sidearm flask (occasionally this phase is cloudy, but will become clear as the solution warms up). The interphase and phenol phase are discarded.

6. The water phase is extensively dialyzed in the cold against distilled water and is then lyophilized. After lyophilization the material is suspended in excess ice-cold ethanol by stirring for 30–60 min to extract excess LIS; the protein precipitate is recovered by centrifugation at 4000 rpm. The ethanol extraction is repeated once more. The final precipitate is dissolved in distilled water, then dialyzed against distilled water and lyophilized. The solution should be clear at this step and a precipitate indicates contamination with other membrane proteins. Removal of the precipitate by centrifugation before lyophilization is then required, and this may result in some loss of sialoglycoprotein. Excess amount of precipitate at this step indicates problems with the procedure and usually results in considerably lower yields of material still contaminated by other proteins.

The approximate yield of ethanol extracted protein is 30 mg/unit of blood.

Isolation of Glycophorin A

The various glycoproteins contained in the crude glycophorin fraction can be separated by analytical polyacrylamide electrophoresis in the presence of sodium dodecyl sulfate and staining of the gels with the periodic acid–Schiff's reagent. Depending on the gel system used, different protein patterns are obtained. A minimum number of four bands are seen on gels prepared according to Fairbanks,[21,22] and eight to ten bands are observed on gels prepared according to Laemmli.[8,23]

The major glycoprotein, glycophorin A, can be separated from the other sialoglycopeptides by gel filtration in the presence of detergents. We

[21] H. Furthmayr, M. Tomita, and V. T. Marchesi, *Biochem. Biophys. Res. Commun.* **65,** 113 (1975).

[22] G. Fairbanks, T. L. Steck, and D. F. H. Wallach, *Biochemistry* **10,** 2606 (1971).

[23] U. K. Laemmli, *Nature (London)* **227,** 680 (1970).

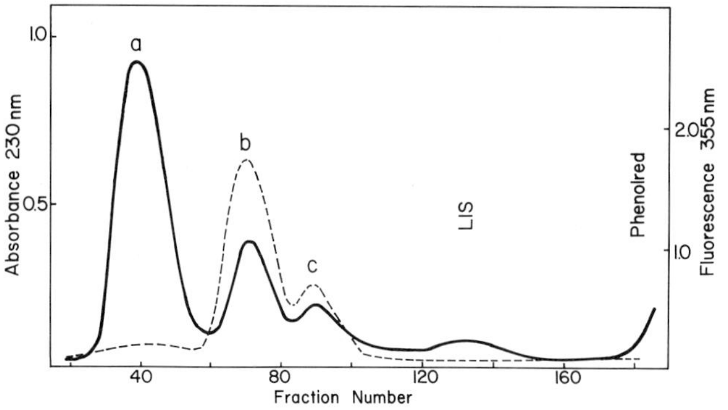

FIG. 1. Elution profile of human erythrocyte membrane glycophorins on agarose A-1.5 in the presence of the detergent Ammonyx-Lo. A maximum load of 100 mg of a crude sialoglycoprotein fraction is placed onto an 85 × 5 cm column. ——, Absorbance 230 nm; ---, fluorescence of tryptophan (excitation, 290 nm; emission, 355 nm); LIS: position at which excess lithium diiodosalicylate will elute. Phenol red is used as marker dye. Peak a contains nearly pure glycophorin A; peaks b and c contain mixtures of several glycophorins.

have determined that the Zwitterionic detergent Ammonyx-Lo (Onyx Chem. Co., Jersey City, New Jersey; stock solution 30%) is ideally suited for preparative separation, since column effluents can be monitored in the far UV range and the detergent does not interfere with measurements of protein, sialic acid, or fluorescence. An agarose A-1.5 (Bio-Rad Laboratories Richmond, California) column (90 × 5 cm) is equilibrated with a buffer containing 0.1% Ammonyx-Lo, 25 m*M* NaCl, 5 m*M* sodium phosphate, pH 8.0, and 0.01% NaN_3. The size of this column allows one to separate 100 mg of lyophilized crude glycophorin into three distinct peaks (Fig. 1). The first, and largest, peak contains almost pure glycophorin A, and the second and third peaks contain mixtures of two minor glycoproteins, glycophorins B and C (in addition to variable amounts of glycophorin A).[8] Figure 2 shows the complex patterns observed on SDS–polyacrylamide gels. It has become clear from a number of studies that some of the bands represent oligomeric species of the glycophorins that either require certain conditions for dissociation[2,24] or cannot be disaggregated once they have been formed.[8] The proteins from appropriately pooled fractions can be separated from salt and most of the detergent by extensive dialysis against distilled water. After lyophilization, residual detergent and lipid that is still contained in the material can be extracted by chloroform–methanol (2 : 1, v/v). The proteins will still be soluble in aqueous solutions (see below).

[24] H. Furthmayr and V. T. Marchesi, *Biochemistry* **15,** 1137 (1976).

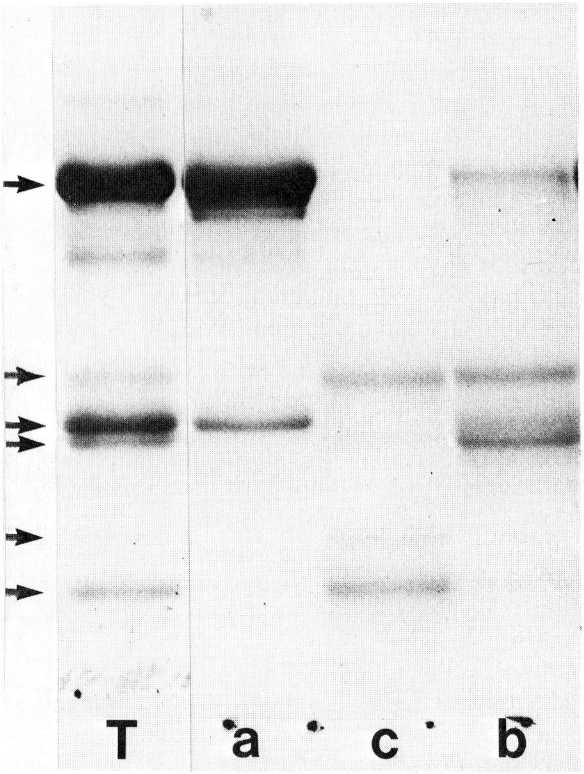

FIG. 2. Sodium dodecyl sulfate (SDS–polyacrylamide gel) electrophoresis of glycophorins separated by gel filtration. T: crude glycophorin fraction; a, b, c: protein pools from Fig. 1. Arrows and arrowheads indicate positions in the gel at which specific glycophorins are found: ⇡ glycophorin A, ↑ glycophorin B, ▲ glycophorin C.

Isolation of Glycophorin B and Glycophorin C

The two minor glycophorins could not be separated by column chromatography. However, larger amounts of intact glycophorins B and C can be isolated by elution from gels after preparative SDS–gel electrophoresis.[8]

The proteins contained in the minor peaks in Fig. 1 are labeled with dansyl chloride or fluorescamine in the presence of SDS and mixed with unlabeled material at a ratio of 1 : 9 before electrophoretic separation. Ten milligrams of protein are dissolved in 1 ml of 0.02 *M* sodium phosphate, pH 8.0, 1% SDS; 0.5 ml of freshly prepared dansyl chloride at a concentration of 0.5% in acetone is added. After incubation at 37° for 30 min, acetone is added and the precipitate is collected by centrifugation. The precipitate is dissolved in 20 m*M* Tris-HCl, pH 6.8, 4 *M* urea, 6% SDS at a concentration of 10 mg/ml, and aliquots are mixed with unlabeled protein dissolved in the same buffer. After incubation at 37° for 30 min and 2 min at 100°, 20 mg per 2 ml are loaded onto a slab gel (17 × 15 × 0.5 cm) prepared according to Laemmli,[23] and electrophoresis is done for 12 hr at 50 mA or until pyronine Y, a dye included into the sample before loading, has moved 9–10 cm into the gel bed.

The labeled protein bands are visualized under UV light, and the fluorescent gei strips are cut with razor blades. After homogenization of the gel with a Teflon pestle the elution buffer, containing 0.05% SDS, 50 m*M* sodium bicarbonate, and 0.02% sodium azide, is added. The protein is extracted into excess buffer at 37° by shaking overnight. The extract is dialyzed against water and then lyophilized. Since it was found to be difficult to remove excess glycine and acrylamide by dialysis, the lyophilized material is subjected to column chromatography on LKB Ultrogel AcA-54 in 0.05% SDS, 5 m*M* sodium phosphate, pH 8.0 (column dimensions 2.5 × 90 cm). The protein-containing fractions are pooled, dialyzed against distilled water and lyophilized. To remove residual SDS, the lyophilized material is extracted twice with ethanol, then solubilized in distilled water and lyophilized.

Preparation of the Glycosylated (Cell Surface) Domains of Glycophorins

The glycosylated domains of glycophorins A and C can be obtained in larger amounts by a rather simple procedure.[8] To 200 ml of packed and carefully washed (isotonic phosphate-buffered saline, pH 7.2) red blood cells, 400 ml of 0.02 *M* sodium phosphate, pH 8.0, 0.2 *M* NaCl is added. Then 30 mg of Tos-PheCh_2Cl-treated trypsin (Worthington) are added and the suspension is gently shaken at 37° for 2 hr. The cells are removed by centrifugation at 3000 *g* for 20 min and washed once with phosphate-buffered saline; both supernatants are combined, mixed with an equal volume of 50% phenol (see above). After stirring for 20 min at ice temperature, the solution is centrifuged in a Sorvall RC-3 centrifuge using a swinging-bucket rotor for 1 hr at 5000 rpm. The upper (water) phase is extensively dialyzed against water at 4° and then lyophilized. The glycopeptide material is then extracted twice with chloroform–methanol (2 : 1, v/v) at 4° and the peptides are lyophilized again. Separation of the glycopeptides is done by gel filtration on a column (2.5 × 150 cm) of LKB Ultrogel AcA-54 equilibrated with 100 m*M* sodium acetate, pH 6.8, containing 0.02% sodium azide (Fig. 3). Two peptide peaks are separated, which contain sialic acid. The first peak contains tryptophan, indicating

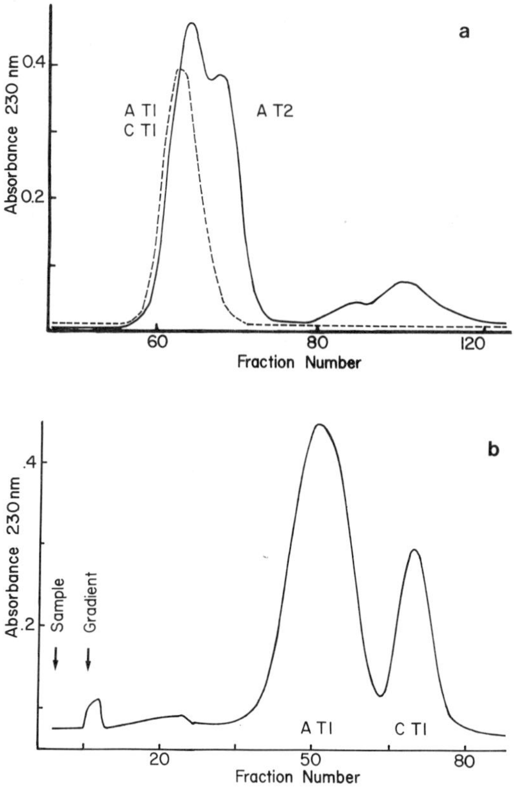

FIG. 3. Gel filtration of cell surface glycopeptides obtained by tryptic digestion of intact human erythrocytes. Peptides recovered from the water phase after phenolic partitioning of a tryptic digest of intact cells are separated by gel filtration on AcA-54 (panel A). Two major sialic acid-containing peaks are obtained (——) absorbance 230 nm), the first of which yields fluorescence indicative of tryptophan (excitation, 290 nm; emission, 355 nm; ---). CT1, AT1, and AT2 indicate the major glycopeptides derived from glycophorins A and C.[8,9] Panel b: Separation of peptides AT1 and CT1 by DEAE-cellulose chromatography. The peak indicated as CT1 yields tryptophan fluorescence (not shown).

the presence of the glycosylated peptide CT1 derived from glycophorin C. Fractions are pooled, concentrated by evaporation, and desalted on a BioGel P2 column equilibrated with 0.05 *N* acetic acid. After lyophilization, the fractions are chromatographed on a DEAE-cellulose (Whatman DE-52) column (15 × 1.5 cm) equilibrated with 50 m*M* sodium formate, pH 6.1, and eluted with a linear salt gradient from 0 to 0.3 *M* sodium chloride over a total volume of 400 ml.[9] This procedure separates peptides CT1 from AT1 (the larger amino-terminal glycopeptide derived from glycophorin A) as indicated in Fig. 3B.

Trypsin does not release the glycosylated fragment of glycophorin B from intact erythrocytes.[8] Thus, an alternative procedure is proposed to obtain this fragment. Glycophorins B and C are contained in peaks b and c after gel filtration of the crude glycophorin fraction (Figs. 1 and 2) in addition to traces of glycophorin A and possibly other glycoproteins. After trypsin digestion of peak b or peak c material in 50 m*M* Tris-HCl,

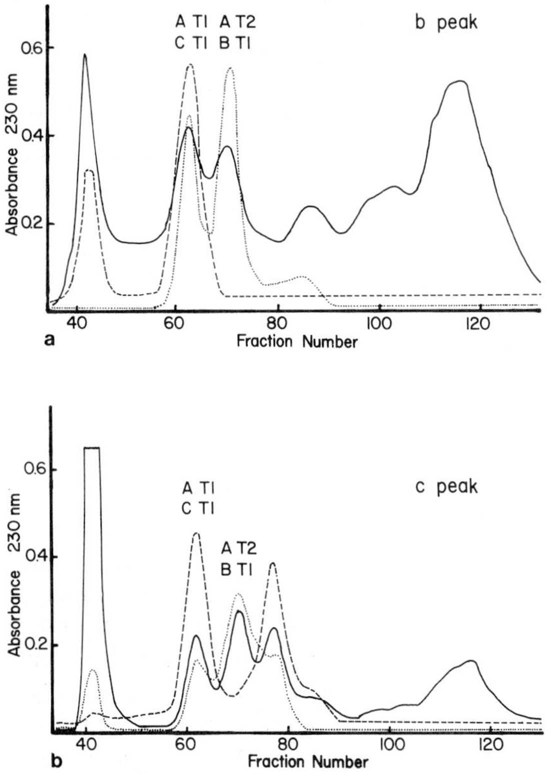

FIG. 4. Separation of tryptic fragments of glycophorin mixtures. (a) After tryptic digestion of protein pools b and c, peptides are separated by gel filtration on AcA-54. Positions in the chromatogram at which specific sialoglycopeptides elute are indicated by AT1, AT2, BT1, and CT1. (b) These peptides are further purified by DEAE-cellulose ion-exchange chromatography as in Fig. 3. ——, Absorbance at 230 nm; ····, sialic acid at 549 nm; ---, fluorescence (excitation 290 nm; emission, 355 nm).

pH 8.0, at an enzyme-to-substrate ratio of 1 : 30 for 20 hr at 37°, the digest is brought to pH 4.5 by dropwise addition of 1 *N* HCl. The precipitate is removed by centrifugation, and the peptides in the soluble fraction are separated by gel filtration on AcA-54 as described above (Fig. 4a, b). The peptides in the first two peaks are rechromatographed by ion-exchange chromatography on DEAE-cellulose equilibrated with 50 m*M* sodium formate, pH 6.1, as above to purify CT1 and BT1, the sialoglycopeptide fragments of glycophorins B and C.

Purity

Glycophorin A is heterogeneous with regard to carbohydrate composition and amino acid sequence. The latter is related to the MN blood group antigens,[10,25] and a homogeneous product can be obtained by using typed blood of individuals homozygous for these antigens. For corrections of

[25] K. Wasniowska, Z. Drzeniek, and E. Lisowska, *Biochem. Biophys. Res. Commun.* **76,** 385 (1977).

the amino acid sequence proposed earlier[9] in positions 11 and 17, it is referred to Dahr *et al.*[26] Heterogeneity of the oligosaccharides attached to threonine or serine has been inferred in the structural studies[9] and has been documented for the asparagine-linked oligosaccharide.[27]

Similar heterogeneity in amino acid sequence exists for glycophorin B. Depending on the Ss-blood type in position 29 a methionine or threonine is present.[26]

All of the glycophorins contain a hydrophobic region, but structural studies have been done only on glycophorin A.[28] Glycophorin preparations contain various contaminants that interact with this region tenaciously and are difficult to remove unless detergents such as SDS are used. Invariably, phospholipids, phosphatidylinositol,[29] heme, and possibly glycolipids can be detected in the preparations obtained by procedures as outlined above. In addition, it is difficult to exclude the possibility that minute amounts of other glycoproteins are present as contaminants.

Chemical and Physicochemical Properties

Glycophorins may be associated to form dimeric structures in the membrane. This is suggested on the basis of data on the protein in solution. When erythrocyte membranes or the isolated glycophorin A preparation are dissolved in sodium dodecyl sulfate-containing solutions and the proteins are analyzed by polyacrylamide gel electrophoresis, glycophorin A migrates as a homodimer that is in equilibrium with the monomeric form. The equilibrium is sensitive to ionic strength, concentration, temperature, SDS concentration, and other parameters. The site of interaction resides within the hydrophobic domain of the molecule.[24,30,31] Circular dichroism[32] and proton nuclear magnetic resonance studies[33] suggest differences in the conformation of the three domains of glycophorin A—the extracellular glycosylated, the intramembranous hydrophobic, and the cytoplasmic carboxy-terminal peptide regions. The hydrophobic domain assumes an α-helical conformation that is stable in a wide range of

26 W. Dahr, K. Beyreuther, H. Steinbach, W. Gielen, and J. Krüger, *Hoppe-Seyler's Z. Physiol. Chem.* **361,** 895 (1980).

27 H. Yoshima, H. Furthmayr, and A. Kobata, *J. Biol. Chem.* **255,** 9713 (1980).

28 H. Furthmayr, R. E. Galardy, M. Tomita, and V. T. Marchesi. *Arch. Biochem. Biophys.* **185,** 21 (1978).

29 I. M. Armitage, D. L. Shapiro, H. Furthmayr, and V. T. Marchesi, *Biochemistry* **16,** 1317 (1977).

30 M. Silverberg, H. Furthmayr, and V. T. Marchesi, *Biochemistry* **15,** 1448 (1976).

31 M. Silverberg and V. T. Marchesi, *J. Biol. Chem.* **253,** 95 (1978).

32 T. H. Schulte and V. T. Marchesi, *Biochemistry* **18,** 275 (1979).

33 J. A. Cramer, V. T. Marchesi, and I. M. Armitage, *Biochim. Biophys. Acta* **595,** 235 (1980).

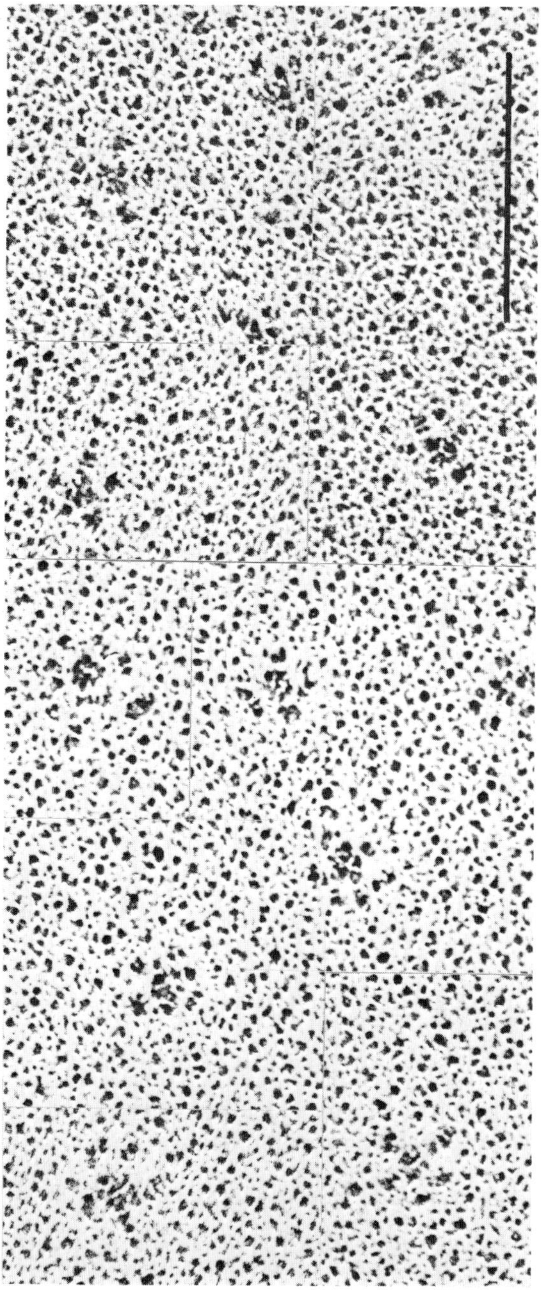

FIG. 5. Rotary shadowing of glycophorin A with carbon platinum. Glycophorin A was dissolved in 0.1 *M* ammonium acetate, pH 7.5, containing 70% glycerol, and the solution was sprayed onto freshly cleaved mica. After drying in a chamber at 2×10^{-5} Torr, shadowing was done at an angle of 9° with carbon–platinum and of 90° with carbon. The replica was floated off the mica and viewed in a Phillips 300 electron microscope. ×480,000; bar indicates 1000 Å.

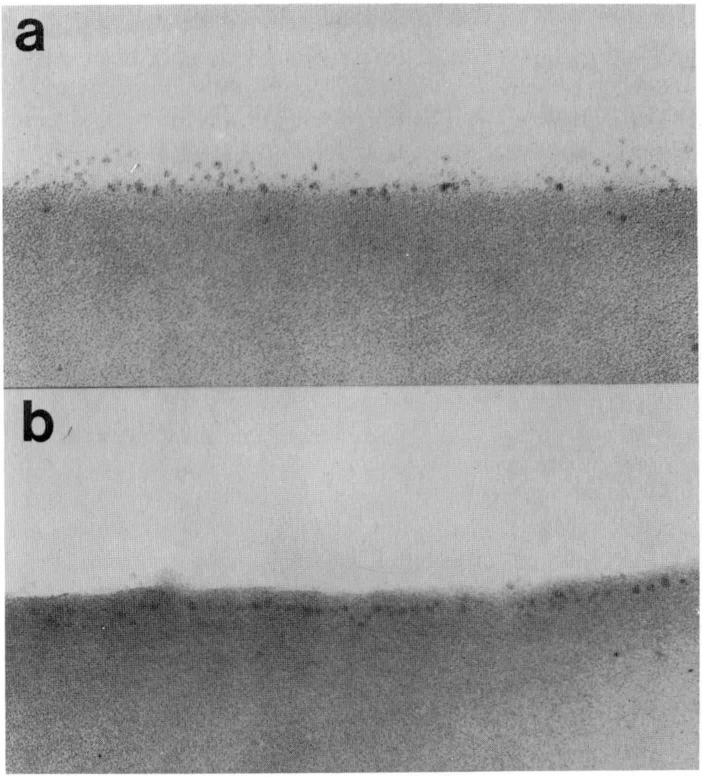

FIG. 6. Frozen thin sections of intact human red blood cells incubated with ferritin–wheat germ agglutination conjugates (a) or with specific ferritin–antibody conjugates (b). ×264,000. Reproduced from Cotmore *et al.*[36]

solvents, including SDS, guanidine, urea, but is disrupted by incubation at 100° for 3 min or in trifluoroacetic acid.

The proteins can aggregate to even larger molecular weight forms.[8] In aqueous buffers particles are formed that appear to consist of 5 or 6 dimeric structures or 10–12 molecules consistent with a molecular weight of about 400,000.[1] As shown in Fig. 5, glycophorin A in preparations obtained after metal shadow-casting with carbon–platinum gives starlike structures. These particles apparently are formed by association within the hydrophobic domain. The hydrophilic glycosylated portions appear to extend from the center of the particle. These images suggest that the glycosylated domains of glycophorin A may be fairly rigid rods, although it is not clear that the technique of metal shadow-casting visualizes the entire structure with sufficient detail.

Orientation

Glycophorin A is a transmembrane protein, and the linear arrangement of the three domains—extracellular, intramembranous, and cytoplasmic—is evident not only from studies on the primary structure, but also from experiments performed with cells. The strongly negative charge of sialic acid was utilized to label cells with colloidal iron. Sialic acid can be removed from intact cells with neuraminidase, and the oligosaccharides on glycophorin can interact with a variety of lectins (Fig. 6). Similarly, blood group antigens of the MNSs system are recognized by specific antibodies on intact cells.[2] The intramembranous domain can be labeled with chemical probes that partition into the hydrophobic environment of the lipid bilayer.[34] Finally, the cytoplasmic segment has been shown to be phosphorylated in the mature red cell, albeit to a minor extent, suggesting that it can act as a substrate for intracellular and possibly membrane-bound phosphokinases.[35] The location of the cytoplasmic domain of glycophorin A at the inner surface of the membrane has been demonstrated ultrastructurally by labeling with antibodies to a specific antigenic determinant of ultrathin frozen sections of intact cells[36] as shown in Fig. 6. There is no information available on the mode of insertion or transmembrane nature of glycophorins B and C.

It is interesting to note that red cells of a small number of individuals were discovered, by virtue of serological abnormalities during blood group testing or in population studies, to lack one or the other of the glycophorins.[1,10] These genetically variant red cells appear to be structurally and functionally normal. It remains to be seen whether these genetic defects are expressed only in the mature red cell and not also in erythroid precursor cells.

[34] D. W. Goldman, J. S. Pober, G. White, and H. Bayly, *Nature* (*London*) **280,** 841 (1982).
[35] D. L. Shapiro and V. T. Marchesi, *J. Biol. Chem.* **252,** 508 (1977).
[36] S. F. Cotmore, H. Furthmayr, and V. T. Marchesi, *J. Mol. Struct.* **153,** 539 (1977).

[22] Glycophorin A: *In Vitro* Biogenesis and Processing

By Carl G. Gahmberg, Mikko Jokinen, Kimmo K. Karhi, Olle Kämpe, Per A. Peterson, and Leif C. Andersson

The major sialoglycoprotein of the human red cell membrane, glycophorin A, is one of the best-studied mammalian integral membrane proteins.[1,2] Its amino acid sequence has been established,[3,4] and it contains three functional domains. The NH_2 terminus carries all the carbohydrate and is located on the external surface of the cell. The protein spans the membrane and has a hydrophilic cytoplasmic portion of about 30 amino acids, which probably interacts with the anion transport protein, band 3.[5] The intramembrane portion is a hydrophobic stretch of 23 nonpolar amino acids.

Glycophorin A contains 60% carbohydrate distributed in two main types of oligosaccharides. There are 15 serine/threonine-linked O-glycosidic oligosaccharides, most of which have the structure *N*-acetylneuraminic acid ($\alpha 2 \rightarrow 3$)galactose ($\beta 1 \rightarrow 3$)[*N*-acetylneuraminic acid ($\alpha 2 \rightarrow 6$)]*N*-acetylgalactosamine.[6] A single N-glycosidic oligosaccharide chain with a complex-type structure[7] is located at asparagine-26.[3]

Detailed information available on the molecular structure of glycophorin A would make it a good model for studies on the biosynthesis and assembly of membrane proteins. Mature human red cells, however, are not useful, since the synthesis of integral membrane proteins ceases already at the normoblast stage of erythropoiesis.

The continuous cell line K562 was originally established by Lozzio and Lozzio[8] from the pleural effusion of a patient with chronic myeloid leukemia in terminal blast crisis. The K562 cells have been reported to carry the Philadelphia chromosome marker and were considered to represent the outgrowth of a chronic myeloid leukemia clone.[8] The cell line has attained widespread use as a highly sensitive *in vitro* target in the natural killer cell assay.[9]

[1] D. J. Anstee, *in* "Immunobiology of the Erythrocyte," p. 67. Alan R. Liss, Inc., New York, 1980.

[2] H. Furthmayr, *in* "Biology of Carbohydrates" (V. Ginsburg, ed.), Vol. 1, p. 123. Wiley, New York, 1981.

[3] M. Tomita and V. T. Marchesi, *Proc. Natl. Acad. Sci. U.S.A.* **72,** 2964 (1975).

[4] M. Tomita, H. Furthmayr, and V. T. Marchesi, *Biochemistry* **17,** 4756 (1978).

[5] E. A. Nigg, C. Brown, M. Girardet, and R. J. Cherry, *Biochemistry* **19,** 1887 (1980).

[6] D. B. Thomas and R. J. Winzler, *J. Biol. Chem.* **244,** 5943 (1969).

[7] H. Yoshima, H. Furthmayr, and A. Kobata, *J. Biol. Chem.* **255,** 9713 (1980).

[8] C. B. Lozzio and B. B. Lozzio, *Blood* **45,** 321 (1975).

[9] E. Saksela, T. Timonen, A. Ranki, and P. Häyry, *Immunol. Rev.* **44,** 71 (1979).

ISBN 0-12-181996-5

During our previous analysis of the surface glycoprotein patterns of a large panel of human hematopoietic cell lines and of freshly isolated populations of normal leukocytes and of leukemic cells, we noticed that the surface glycoprotein profile of the K562 cells was completely different from those obtained with malignant and benign cells representing various stages of the myeloblast to granulocyte maturation sequence (for a review, see Gahmberg and Andersson[10]). Moreover, we observed that rabbit anti-K562 antiserum, after extensive absorptions with established cell lines and blood leukocytes to nonreactivity against normal myeloblasts and blood leukocytes, still showed strong reactivity with erythrocytes and erythroid precursor cells.[11]

Using glycophorin A antiserum, we could show that the K562 cells contained glycophorin A.[12] Because we knew that glycophorin A in normal human bone marrow is specific for the erythroid lineage of cells,[13] its presence in K562 cells strongly indicated their erythroid nature. Definite proof for the erythroid derivation was obtained after induction to hemoglobin synthesis and formation of erythroid particles.[14,15] The availability of the K562 cell line and anti-glycophorin A antiserum made biosynthetic studies of erythrocyte components possible.

We here summarize our studies on the synthesis of the glycophorin A polypeptide in the K562 cell line and its intriguing posttranslational modifications, including N- and O-glycosylation, phosphorylation, and sulfation. Furthermore its cell-free synthesis is described using glycophorin A mRNA obtained from K562 cells. The reader is referred to the original articles[16–18] and a previous review[19] for further detailed information.

Methods

Cells

The K562 cell line is readily available from many laboratories, especially because it has been extensively used as a target cell in natural killer

[10] C. G. Gahmberg and L. C. Andersson, *Biochim. Biophys. Acta, Rev. Cancer* **651,** 65 (1982).

[11] L. C. Andersson, K. Nilsson, and C. G. Gahmberg, *Int. J. Cancer* **23,** 143 (1979).

[12] C. G. Gahmberg, M. Jokinen, and L. C. Andersson, *J. Biol. Chem.* **254,** 7442 (1979).

[13] C. G. Gahmberg, M. Jokinen, and L. C. Andersson, *Blood* **52,** 379 (1978).

[14] L. C. Andersson, M. Jokinen, and C. G. Gahmberg, *Nature (London)* **278,** 364 (1979).

[15] T. R. Rutherford, J. B. Clegg, and D. J. Weatherall, *Nature (London)* **280,** 164 (1979).

[16] M. Jokinen, C. G. Gahmberg, and L. C. Andersson, *Nature (London)* **279,** 604 (1979).

[17] C. G. Gahmberg, M. Jokinen, K. K. Karhi, and L. C. Andersson, *J. Biol. Chem.* **255,** 2169 (1980).

[18] M. Jokinen, I. Ulmanen, L. C. Andersson, L. Kääriäinen, and C. G. Gahmberg, *Eur. J. Biochem.* **114,** 393 (1981).

[19] C. G. Gahmberg, M. Jokinen, K. K. Karhi, I. Ulmanen, L. Kääriäinen, and L. C. Andersson, *Blood Transfus. Immunohaematol.* **24,** 53 (1981).

cell assays. Our cells were obtained from K. Nilsson, Uppsala and G. Klein, Stockholm. The cells are cultivated in RPMI-1640, culture medium in the presence of 10% newborn calf serum. Normal and En(a-) human red cells were obtained from the Finnish Red Cross Blood Transfusion Service, Helsinki.

Radioactive Cell Surface Labeling of Erythrocytes

Reagents

Sodium metaperiodate (Merck AG, Darmstadt, Federal Republic of Germany)

Tritiated sodium borohydride, 8–15 Ci/mmol (The Radiochemical Centre Ltd., Amersham, United Kingdom)

Procedure. Normal erythrocytes are labeled with ^{3}H using the periodate–NaB^3H_4 surface labeling technique.[20] Red cells are incubated in the dark at 0° for 10 min with 1 m*M* periodate, washed, and reduced with 1 mCi of NaB^3H_4 for 30 min at room temperature. The cells are washed and the erythrocyte membranes are isolated as described elsewhere.[21]

Radioactive Metabolic Labeling of K562 Cells

Reagents

Tunicamycin (G. Tamura, University of Tokyo, Japan; R. Hamill, Eli Lilly Co., Indianapolis, Indiana). In preliminary experiments no difference has been found between the two tunicamycin preparations, and both can be used in this type of work. The preparations are weighed, dissolved in alkaline ethanol, and stored at −20°. The concentrations are confirmed by spectrometric analysis using the E_{260} extinction coefficient of 110.

[^{14}C]Galactose, [^{32}P]phosphate, [^{35}S]sulfate, and [^{35}S]methionine (The Radiochemical Centre Ltd., Amersham, United Kingdom)

Bovine trypsin, 3.5 units/mg (Merck AG, Darmstadt, Federal Republic of Germany)

Soybean trypsin inhibitor (Sigma, St. Louis, Missouri)

Triton X-100 (British Drug Houses Ltd., Poole, United Kingdom)

Phenylmethylsulfonyl fluoride (Sigma)

Procedure. K562 cells are washed three times in $NaCl/PO_4$ (0.15 *M* NaCl, 0.01 *M* sodium phosphate, pH 7.4) and, when indicated, preincubated for 3 hr in 5 ml of RPMI-1640 medium containing 10% newborn calf serum and 10–20 μg of tunicamycin per milliliter. Labeling with [^{14}C]galactose is performed in Eagle's minimum essential medium (MEM) con-

[20] C. G. Gahmberg and L. C. Andersson, *J. Biol. Chem.* **252,** 5888 (1977).

[21] C. G. Gahmberg and S.-i. Hakomori, *J. Biol. Chem.* **248,** 4311 (1973).

taining 10% newborn calf serum with or without 10 μg/ml of tunicamycin. The cells are incubated for 6 hr with 20 μCi of [^{14}C]galactose. Labeling with [^{32}P]phosphate is performed in phosphate-free MEM by using 10 mCi of [^{32}P]phosphate and incubation for 3 hr. For labeling with [^{35}S]sulfate, cells are incubated for 2 hr in MEM with 2 mCi [^{35}S]sulfate. For pulse–chase experiments the K562 cells that had been preincubated as above or with 20 μg of tunicamycin per milliliter are labeled for 5 min with 0.5 mCi of [^{35}S]methionine, rapidly washed with normal MEM at 37°, and transferred to culture bottles in normal MEM with or without tunicamycin. At indicated times, cell aliquots are removed and immediately cooled to 0°, and the cells are washed in NaCl/PO_4. For determination of surface exposure of glycophorin A the cell samples are divided into two equal aliquots, and one of these is treated for 10 min at 37° with 0.1 mg of trypsin per milliliter in NaCl/PO_4. After incubation, the cells are immediately cooled to 0°, a fivefold excess (over trypsin) of soybean trypsin inhibitor (Sigma) is added, and the cells are washed three times in NaCl/PO_4. All incubated cells are lysed in buffer A (NaCl/PO_4 containing 1% Triton X-100, 1% ethanol, and 2 m*M* phenylmethylsulfonyl fluoride). After centrifugation at 1000 *g* for 10 min, the supernatants are recovered for subsequent analysis.

Isolation of mRNA from K562 Cells

Reagents

Sodium dodecyl sulfate (Sigma)
Cycloheximide (Sigma)
Oligo(dT) polymer (Sigma)
Polyvinyl sulfate (Sigma)

Procedure. Two milliliters of packed K562 cells are washed twice in NaCl/PO_4 containing 0.1 mg of cycloheximide per milliliter, resuspended in 0.01 *M* Tris, pH 7.4–0.01 *M* NaCl–0.0015 *M* $MgCl_2$, supplemented with 30 μg of polyvinyl sulfate per milliliter, and left to swell for 15 min at 0°. The cells are broken with a Dounce homogenizer and centrifuged at 1000 *g* for 10 min. The supernatant is centrifuged at 1000 *g* for 1 hr in Sorvall HB-4 rotor at 4° to obtain membranes. Sodium dodecyl sulfate is added to the pellet to a concentration of 2%, and the solution is warmed to 60°. RNAs are extracted with phenol–chloroform–isoamyl alcohol (25 : 24 : 1) and precipitated with 2.5 volumes of ethanol at −20° overnight.

Size fractionation of RNAs is performed by centrifugation in a 15% to 30% sucrose gradient made in 0.01 *M* Tris, pH 7.4–0.015 *M* NaCl–0.001 *M* EDTA–0.1% sodium dodecyl sulfate using a Beckman SW27 rotor for 21 hr at 25,000 rpm. Fractions are precipitated twice with ethanol, washed, dried, and dissolved in sterile water for chromatography on oligo(dT) columns.[22] The RNAs are stored at $-70°$ until used.

Translation of mRNA in Vitro

Reagents

Micrococcal nuclease (EC 3.1.4.7, now EC 3.1.31.1, micrococcal endonuclease, P-L Biochemicals, Milwaukee, Wisconsin)
Dog pancreatic microsomal membranes

Procedure. Messenger RNAs from K562 cells are used for *in vitro* translation in a cell-free lysate made from rabbit reticulocytes as described[23] and containing 0.2 mCi of [^{35}S]methionine per milliliter. The lysate is made dependent on exogenous mRNA by micrococcal nuclease treatment.[24] The detailed conditions for the translation assays are described elsewhere.[25] When indicated, microsomal membranes (6–8 A_{260} units/ml) are added to the lysate to allow processing and glycosylation of membrane proteins.[26] The lysates are incubated at 30° for 2 hr.

Preparation of Lectin–Sepharose Columns

Reagents

Lentils (obtained locally)
Wheat germ (obtained locally)
Ovomucoid (Sigma)
Cyanogen bromide (Eastman Kodak Co., Rochester, New York)
Helix pomatia lectin–Sepharose (Pharmacia, Uppsala, Sweden)
D-Glucose, *N*-acetyl-D-glucosamine, *N*-acetyl-D-galactosamine (Sigma)

Procedure. We would like to emphasize that the use of different lectin–Sepharose columns with known specificities are excellent tools for studying the glycosylation of cellular proteins. Although indirect, these techniques give valuable information. The specific activity of commercially available [^{35}S]methionine preparations exceeds the specific activities of available radioactive monosaccharides at least with an order of

[22] H. Aviv and P. Leder, *Proc. Natl. Acad. Sci. U.S.A.* **69,** 1408 (1972).
[23] L. Villa-Komaroff, M. McDowell, D. Baltimore, and H. F. Lodish, this series, Vol. 30, p. 709.
[24] H. R. B. Pelham and R. J. Jackson, *Eur. J. Biochem.* **67,** 247 (1976).
[25] P. Lehtovaara, I. Ulmanen, L. Kääriäinen, S. Keränen, and L. Philipson, *Eur. J. Biochem.* **112,** 461 (1980).
[26] D. Shields and G. Blobel, *J. Biol. Chem.* **253,** 3753 (1978).

magnitude. The high specific activity makes the study of minor components possible in considerable detail.

Lentil lectin (*Lens culinaris*) is isolated by affinity chromatography on Sephadex G-50 in NaCl/PO_4 and elution with 0.05 *M* D-glucose essentially as previously described.[27] Wheat germ lectin is isolated on ovomucoid–Sepharose 4B as described elsewhere[28] by using 0.1 *N* acetic acid as the eluting agent. Purified lectins are coupled to Sepharose 4B (2 mg of lectin per milliliter of packed gel) activated with cyanogen bromide in the presence of a 0.1 *M* concentration of the appropriate sugar hapten.[29]

Preparation of Anti-glycophorin A Antiserum

Reagents

Freund's adjuvant (Difco, Detroit, Michigan)
Sodium azide (Merck AG, Darmstadt, Federal Republic of Germany)

Procedure. A crude preparation of sialoglycoproteins is obtained after chloroform–methanol extraction of erythrocyte membranes and partition with water.[30] Rabbits are immunized three times with 1 mg of protein of this preparation emulsified in Freund's adjuvant. Ten days after the last injection the rabbits are bled and the sera are collected. Two milliliters of antiserum are absorbed twice for 12 hr at 4° with 1 ml of packed red cell membranes of the En(a-) blood group, which lack glycophorin.[31–33] After centrifugation at 12,500 rpm for 30 min in a Sorvall SS-34 rotor, the supernatant is collected and stored at 4° in the presence of 0.2% sodium azide.

Lectin–Sepharose Affinity Chromatography and Immune Precipitations

Reagents

Staphylococcus aureus strain Cowan I. The *S. aureus* Cowan I strain is obtained by cultivation as described[34] and fixed, after heat treatment at 80° for 10 min, in 0.5% formaldehyde for 3 hr at 22°.

[27] H. J. Sage and R. W. Green, this series, Vol. 28, p. 332.
[28] V. T. Marchesi, this series, Vol. 28, p. 354.
[29] P. Cuatrecasas, *J. Biol. Chem.* **245,** 3059 (1970).
[30] H. Hamaguchi and H. Cleve, *Biochem. Biophys. Res. Commun.* **47,** 459 (1972).
[31] C. G. Gahmberg, G. Myllylä, J. Leikola, A. Pirkola, and S. Nordling, *J. Biol. Chem.* **251,** 6108 (1976).
[32] W. Dahr, G. Uhlenbruck, J. Leikola, W. Wagstaff, and K. Landfried, *J. Immunogenet.* **3,** 329 (1976).
[33] M. J. A. Tanner and D. J. Anstee, *Biochem. J.* **155,** 701 (1976).
[34] P. Landwall, Ph.D. Thesis, Karolinska Institute, Stockholm (1977).

Rabbit anti-mouse immunoglobulin serum

Rabbit anti-glycophorin A serum

Procedure. To determine the presence of different carbohydrates in glycophorin A and to reduce nonspecific radioactivity often observed during immune precipitation, [^{35}S]methionine-labeled or [^{32}P]phosphate-labeled cell or membrane extracts are passed in buffer A through 2 ml of lectin–Sepharose columns, and the absorbed material is eluted with buffer A containing a 0.05 *M* concentration of the appropriate sugar hapten. Aliquots are counted for radioactivity, and the radioactive peak fractions eluted with sugar are pooled and subjected to immunoprecipitation. The samples are preincubated with 5 μg of rabbit anti-mouse immunoglobulin G. The immunoglobulins are prepared by standard techniques. After 1 hr on ice, 0.2 ml of a 10% suspension of protein A-containing *Staphylococcus aureus* cells is added, and after 1 hr the samples are centrifuged at 2000 *g* for 5 min at 4° in a Wifug table centrifuge. After centrifugation the supernatants are recovered and identical aliquots are incubated for 2 hr at 0° with 5 μl of either anti-glycophorin A antiserum, glycophorin A antiserum plus 10 μl isolated red cell glycophorin A, or rabbit preimmunization serum. Then 0.2 ml of the staphylococcal suspension is added, and, after incubation for 1 hr at 0°, the staphylococci are centrifuged and washed three times in buffer A. Proteins are eluted by boiling in 1% sodium dodecyl sulfate for 1 min, and the samples are centrifuged. The supernatants are lyophilized, boiled in electrophoresis sample buffer, and electrophoresed. Control experiments using more anti-glycophorin A antiserum and staphylococci are done to check that no additional glycophorin A molecules can be precipitated from the different samples.

Treatment of Glycophorin A with Endoglycosidase H

Reagent

Endoglycosidase H (*Streptomyces griseus,* Seikagaku Kogyo Co., Tokyo, Japan)

Procedure. Glycophorin A precursor (GP_a) is immunoprecipitated with anti-glycophorin A antiserum from lentil lectin–Sepharose eluates of proteins labeled with [^{35}S]methionine in the cell-free system in the presence of membranes, or from [^{32}P]phosphate-labeled eluates from lentil lectin–Sepharose or wheat germ lectin Sepharose columns and eluted from the staphylococci with 0.05 ml of 0.3% sodium dodecyl sulfate. The samples are then diluted with 0.1 ml of H_2O and 0.02 ml of 0.125 *M* sodium citrate buffer, pH 5.0, and 2 mU of endoglycosidase H are

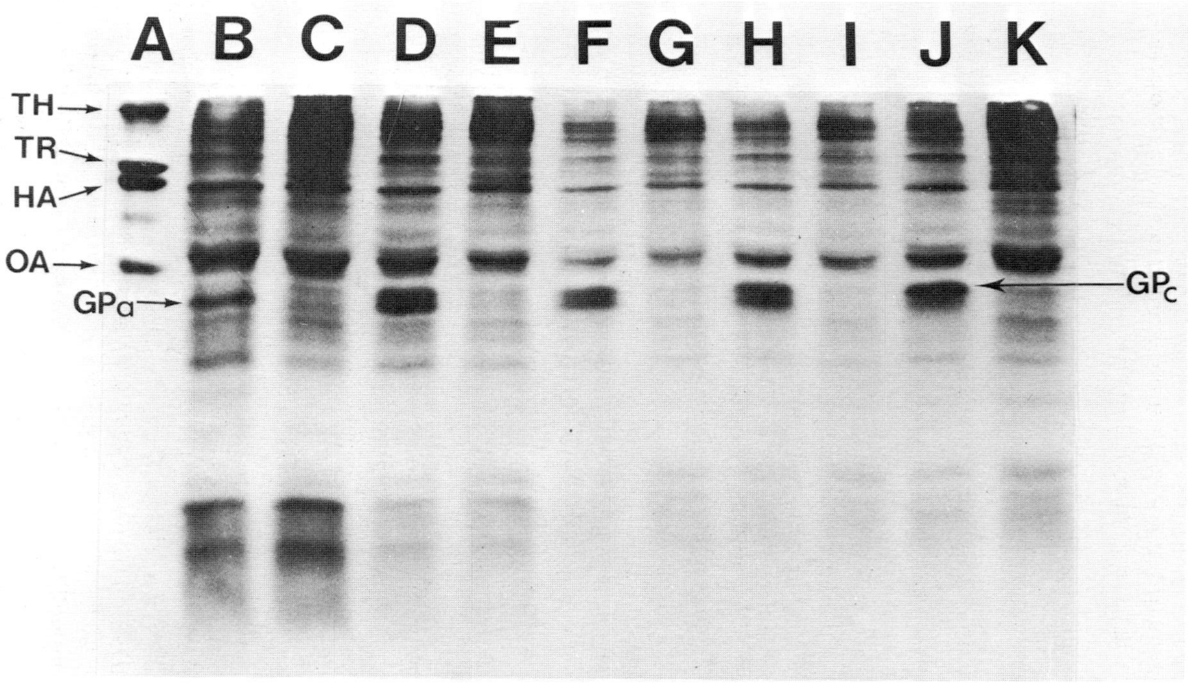

FIG. 1. Fluorography of a polyacrylamide slab gel of immune precipitates obtained from pulse-chased [^{35}S]methionine-labeled cells with anti-glycophorin A antiserum and control serum. (A) ^{14}C-Labeled standard proteins; TH, thyroglobulin; TR, transferrin; HA, human albumin; OA, ovalbumin. (B) Immune precipitates obtained with antiserum from cells labeled for 5 min; (C) with preimmune serum; (D) immune precipitates obtained with antiserum from cells labeled for 5 min followed by chase for 10 min; (E) with preimmune serum; (F) immune precipitates obtained with antiserum from cells after 25 min chase; (G) with preimmune serum; (H) immune precipitates obtained with antiserum from cells after 45 min chase; (I) with preimmune serum; (J) immune precipitates obtained with antiserum from cells after 60 min chase; (K) with preimmune serum. GP_a, glycosylated glycophorin A precursor; GP_c, final glycophorin A.

added.[35] After incubation at 37° for 15 hr the samples are boiled in electrophoresis sample buffer and subjected to electrophoresis.

Polyacrylamide Slab Gel Electrophoresis

Reagents

Acrylamide and *N,N'*-methylenebisacrylamide (Eastman Kodak Co., Rochester, New York)

^{14}C-Labeled protein standards (The Radiochemical Centre Ltd., Amersham, United Kingdom)

Procedure. Polyacrylamide slab gel electrophoresis in the presence of sodium dodecyl sulfate is performed according to Laemmli[36] using an acrylamide concentration of 12%. The gels are treated for fluorography as described elsewhere.[37] The apparent molecular weights of the proteins are determined using the ^{14}C-labeled standards.

Illustrative Data and Interpretation

Pulse–Chase Labeling of K562 Cells with [^{35}S]Methionine and Identification of a Glycophorin A Precursor

When K562 cells are labeled for 5 min with [^{35}S]methionine, immunoprecipitated with anti-glycophorin A antiserum, and analyzed by polyacrylamide gel electrophoresis, a protein with an apparent M_r of 37,000 is specifically precipitated (GP_a, Fig. 1B). After chase, two protein bands become visible (Fig. 1D), and finally only the GP_c molecule with an identical apparent molecular weight as normal glycophorin A is seen (Fig. 1J).

Inhibition of N-Glycosylation of Glycophorin A

Tunicamycin is known to inhibit N-glycosylation of proteins.[38–40] The inhibitory concentration can be determined by studying the adsorption of glycophorin A to lentil lectin–Sepharose. Lentil lectin interacts primarily with mannose/glucose residues, and in glycophorin A such are found in the single N-glycosidic oligosaccharide. Figure 2 shows that 10–20 μg of tunicamycin per milliliter completely inhibit the adsorption to lentil lec-

[35] A. L. Tarentino, T. H. Plummer, Jr., and F. Maley, *J. Biol. Chem.* **249,** 818 (1974).
[36] U. K. Laemmli, *Nature (London)* **227,** 680 (1970).
[37] W. M. Bonner and R. A. Laskey, *Eur. J. Biochem.* **46,** 83 (1974).
[38] A. Takatsuki, K. Kohno, and G. Tamura, *Agric. Biol. Chem.* **39,** 2089 (1975).
[39] J. E. Rothman, R. N. Katz, and H. F. Lodish, *Cell* **15,** 1447 (1978).
[40] E. Li, J. Tabas, and S. Kornfeld, *J. Biol. Chem.* **253,** 7762 (1978).

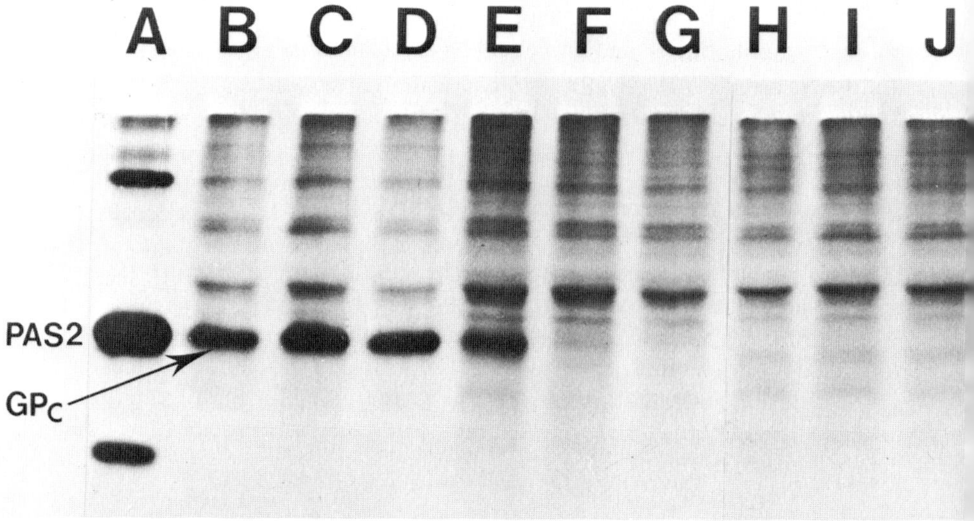

FIG. 2. Fluorography of a polyacrylamide slab gel of immune precipitates obtained with anti-glycophorin antiserum from eluates of lentil lectin–Sepharose columns of [^{35}S]methionine-labeled K562 cells grown in the presence of tunicamycin. Cells were preincubated with tunicamycin for 3 hr and labeled with [^{35}S]methionine for 1 hr with or without tunicamycin. After labeling, the cells were dissolved in buffer A and passed through lentil lectin–Sepharose columns. The adsorbed glycoproteins were eluted with α-methyl mannoside and immunoprecipitated with anti-glycophorin A antiserum or preimmune serum. (A) Pattern of periodate/NaB^{3}H$_4$-labeled erythrocyte membranes; PAS2, predominantly red cell glycophorin A monomer. (B) Pattern obtained with anti-glycophorin A antiserum from cells labeled in the absence of tunicamycin; GP$_c$, K562 cell glycophorin A monomer (for nomenclature, see Ref. 16). (C–G) Patterns obtained with antiserum from cells cultivated in the presence of tunicamycin; (C) 0.5 μg/ml; (D) 2 μg/ml; (E) 5 μg/ml; (F) 10 μg/ml; (G) 20 μg/ml. (H) Pattern obtained with preimmunization serum from cells cultured in the absence of tunicamycin. (I) Pattern obtained with preimmunization serum from cells cultured in the presence of 0.5 μg/ml of tunicamycin. (J) Pattern obtained with preimmunization serum from cells cultured in the presence of 2 μg/ml of tunicamycin.

tin–Sepharose. But in tunicamycin-treated cells glycophorin A is readily synthesized, as seen after labeling with [^{14}C]galactose (Fig 3). The apparent molecular weight is decreased to about 37,000, corresponding well with the absence of a single N-glycosidic oligosaccharide. The surface exposure of glycophorin A is possible to determine by labeling with [^{35}S]methionine in the presence or the absence of tunicamycin, followed by chase and trypsin treatment of intact cells to degrade surface-located glycophorin A. Tunicamycin does not affect the intracellular migration of O-glycosylated glycophorin A (Fig. 4).

Demonstration of Phosphate in Glycophorin A

K562 cells incubated with [^{32}P]phosphate are washed by centrifugation and lysed in buffer A. Half of the extract is passed over a lentil lectin–Sepharose column and half over a wheat germ lectin–Sepharose column. After elution with α-methyl mannoside and *N*-acetylglucosamine, respectively, labeled glycophorin A is immunoprecipitated (Fig. 5). The labeled molecules have identical apparent molecular weights as normal red cell glycophorin A (Fig. 5B, F) and endoglycosidase H has no effect on either the lentil lectin- or wheat germ lectin-adsorbed molecules. This shows that their carbohydrate chains are processed.

Demonstration of Sulfate in Glycophorin A

From K562 cells incubated with [^{35}S]sulfate a weakly labeled glycophorin A molecule is obtained by immunoprecipitation (Fig. 6C). The low radioactivity does not permit further analysis.

Cell-Free Synthesis of Glycophorin A

A 16–18 S mRNA fraction from K562 cells is translated in the rabbit reticulocyte cell-free protein synthesizing system. Immune precipitation with anti-glycophorin A antiserum gives a protein that evidently is a precursor of glycophorin A (GP-P, Fig. 7B). When 50 μg of red cell gly-

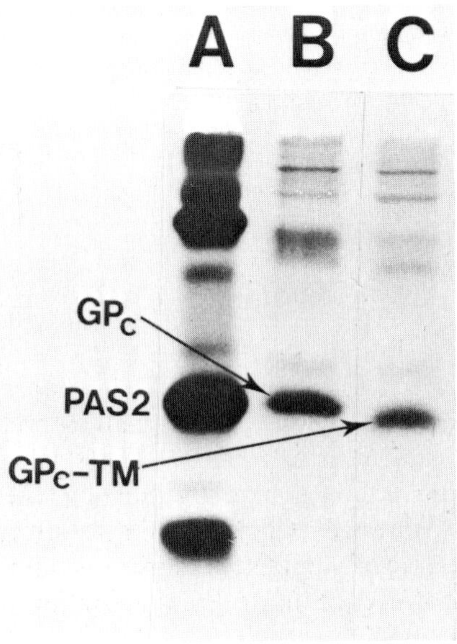

FIG. 3. Glycophorin A synthesized in the presence of tunicamycin. K562 cells were labeled with [^{14}C]galactose for 6 hr. (A) Pattern of periodate/NaB^{3}H$_4$-labeled erythrocyte membranes. (B) Pattern obtained with antiserum from cells cultivated in the absence of tunicamycin. (C) Pattern obtained with antiserum from cells cultivated in the presence of 10 μg/ml of tunicamycin. PAS2, red cell glycophorin A monomer. GP$_c$-TM, K562 cell glycophorin A synthesized in the presence of tunicamycin.

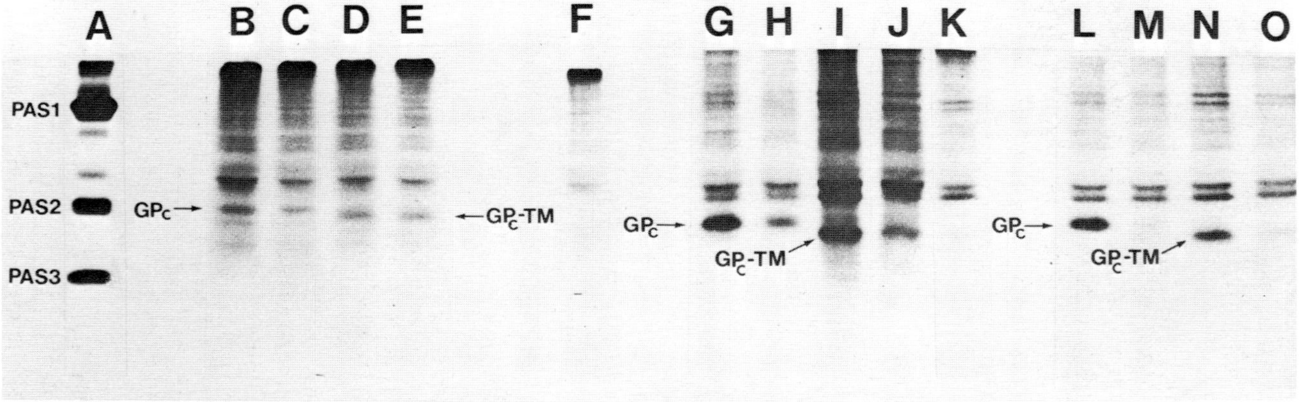

FIG. 4. Cell surface exposure of glycophorin A synthesized in the absence and presence of tunicamycin. K562 cells were pulse-chase labeled with [^{35}S]methionine in the absence or presence of 20 μg/ml of tunicamycin and half of the cell samples were treated with trypsin. After solubilization with buffer A, the extracts were applied to wheat germ lectin columns and identical aliquots of the D-glucosamine-eluted samples were immune precipitated with anti-glycophorin A antiserum or preimmunization serum and subjected to polyacrylamide slab gel electrophoresis and fluorography. (A) Pattern of periodate/NaB^3H_4-labeled erythrocyte membranes; (B) pattern obtained with antiserum from cells labeled for 5 min with [^{35}S]methionine with no tunicamycin; (C) pattern obtained with antiserum from trypsin-treated cells labeled for 5 min with [^{35}S]methionine and no tunicamycin present; (D) pattern obtained with antiserum from cells labeled for 5 min with [^{35}S]methionine in the presence of tunicamycin; (E) pattern obtained with antiserum from trypsin-treated cells labeled for 5 min with [^{35}S]methionine in the presence of tunicamycin; (F) pattern obtained with preimmunization serum from an identical sample as in (C); (G) pattern obtained with antiserum from cells after a 10 min chase with no tunicamycin; (H) pattern obtained with antiserum from trypsin-treated cells after a 10 min chase with no tunicamyçin; (I) pattern obtained with antiserum from cells after a 10 min chase in the presence of tunicamycin; (J) pattern obtained with antiserum from trypsin-treated cells after a 10 min chase in the presence of tunicamycin; (K) pattern obtained with preimmunization serum from an identical sample as in (H); (L) pattern obtained with antiserum from cells after a 25 min chase without tunicamycin; (M) pattern obtained with antiserum from trypsin-treated cells after a 25 min chase without tunicamycin; (N) pattern obtained with antiserum from cells after 25 min chase in the presence of tunicamycin; (O) pattern obtained with antiserum from trypsin-treated cells after a 25 min chase in the presence of tunicamycin.

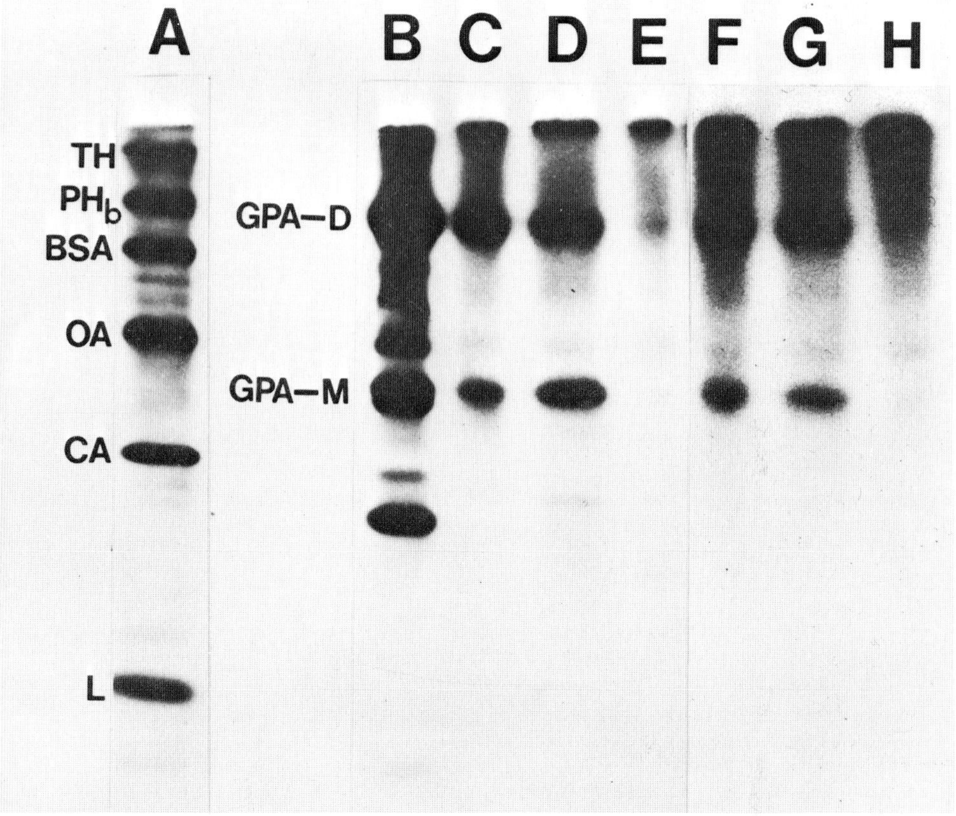

FIG. 5. Fluorography of a polyacrylamide slab gel of [^{32}P]phosphate-labeled glycophorin A. (A) ^{14}C-Labeled standard proteins: TH, thyroglobulin; PH_b, phosphorylase b; BSA, bovine serum albumin; OA, ovalbumin; CA, carbonic anhydrase; L, lysozyme. (B) Pattern of periodate/NaB^3H_4-labeled erythrocyte membranes. GPA-D, glycophorin A dimer; GPA-M, glycophorin A monomer. (C) Pattern obtained with anti-glycophorin A antiserum from the eluate of a lentil lectin–Sepharose column of K562 cells labeled with [^{32}P]phosphate. (D) Identical sample as in (C) but treated with endoglycosidase H. (E) Pattern obtained as in (C) but immune precipitated in the presence of 10 μg of nonradioactive red cell glycophorin A. (F) Pattern obtained with anti-glycophorin A antiserum from the eluate of a wheat germ lectin–Sepharose column of K562 cells labeled with [^{32}P]phosphate. (G) Identical sample as in (F) but treated with endoglycosidase H. (H) Pattern obtained as in (F) but immune precipitated in the presence of 10 μg of nonradioactive red cell glycophorin A.

cophorin A added to compete in the immune precipitation, no protein is specifically precipitated (data not shown). The apparent molecular weight of GP-P of 19,500 exceeds that of the glycophorin A apoprotein, with about 5000 corresponding to approximately 45 amino acid residues.

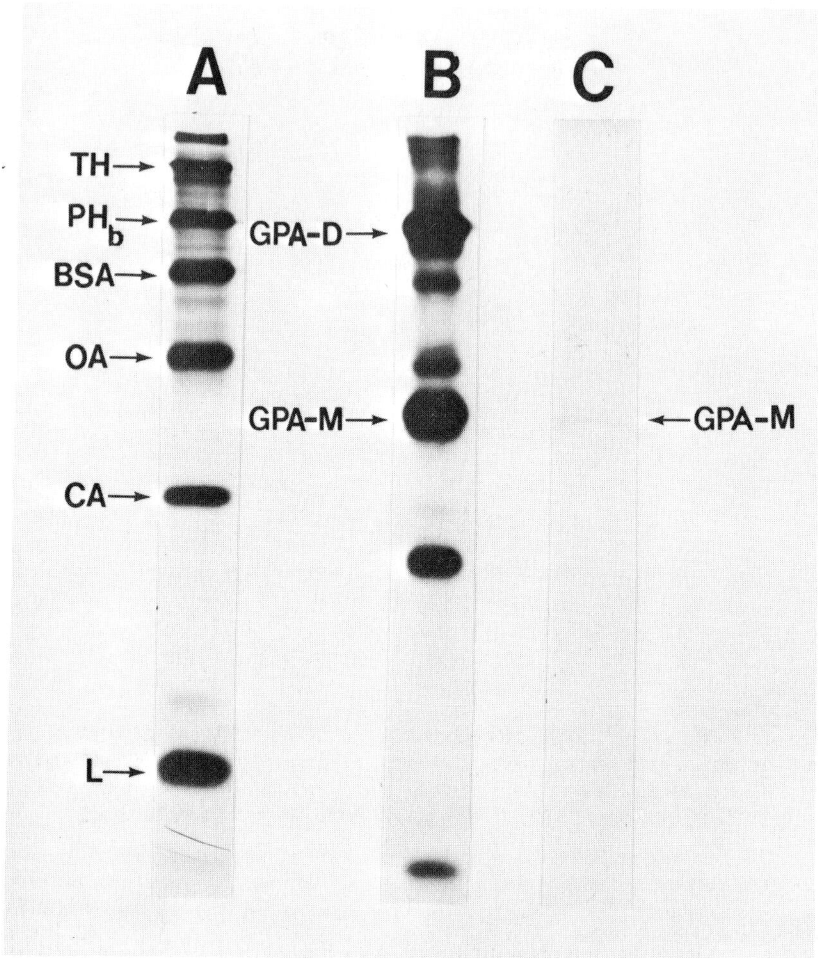

FIG. 6. Fluorography of a slab gel of glycophorin A labeled with [^{35}S]sulfate. Lane: (A) ^{14}C-labeled standard proteins as in Fig. 5; (B) pattern of periodate/NaB^3H_4-labeled erythrocyte membranes; (C) pattern obtained with anti-glycophorin A antiserum from [^{35}S]sulfate-labeled K562 cells. GPA-M, glycophorin A monomer.

When ribosome-stripped dog pancreatic microsomal membranes are included in the cell-free protein-synthesizing system, a glycoprotein is immune precipitated that corresponds in electrophoretic mobility to cellular GP_a (Fig. 8C). It is N-glycosylated because it absorbs to lentil lectin–Sepharose, and its molecular weight decreases with 2000 after endoglycosidase H digestion (Fig. 8D). Its apparent molecular weight is identical to

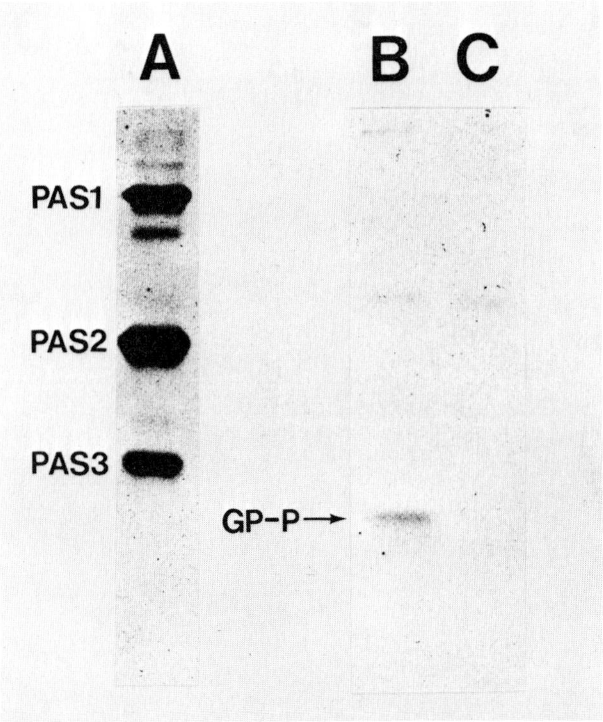

FIG. 7. Cell-free synthesis of glycophorin A in the absence of membranes. Lane: (A) fluorography pattern of polyacrylamide slab gel of periodate–NaB$^3H_4$-labeled erythrocyte membranes; (B) immune precipitates obtained with anti-glycophorin A antiserum; and (C) control with preimmune serum. GP-P = nonglycosylated glycophorin A precursor.

GP_a from K562 cells and substantially higher than that of the glycophorin A apoprotein. This indicates that it is also O-glycosylated. However, it cannot be immunoprecipitated from the eluate of a *Helix pomatia*–Sepharose affinity column specific for *N*-acetylgalactosamine residues.[18]

The biosynthesis of very few mammalian integral membrane glycoproteins has been studied in any detail; among these, the best studied are the major transplantation antigens.[41–44] This is due to the great difficulties when studying a membrane component not comprising more than a small

[41] H. L. Ploegh, L. E. Cannon, and J. L. Strominger, *Proc. Natl. Acad. Sci. U.S.A.* **76,** 2273 (1979).

[42] B. Dobberstein, H. Garoff, G. Warren, and P. J. Robinson, *Cell* **17,** 759 (1979).

[43] M. S. Krangel, H. T. Orr, and J. L. Strominger, *Cell* **18,** 979 (1979).

[44] I. D. Algranati, C. Milstein, and A. Ziegler, *Eur. J. Biochem.* **103,** 197 (1980).

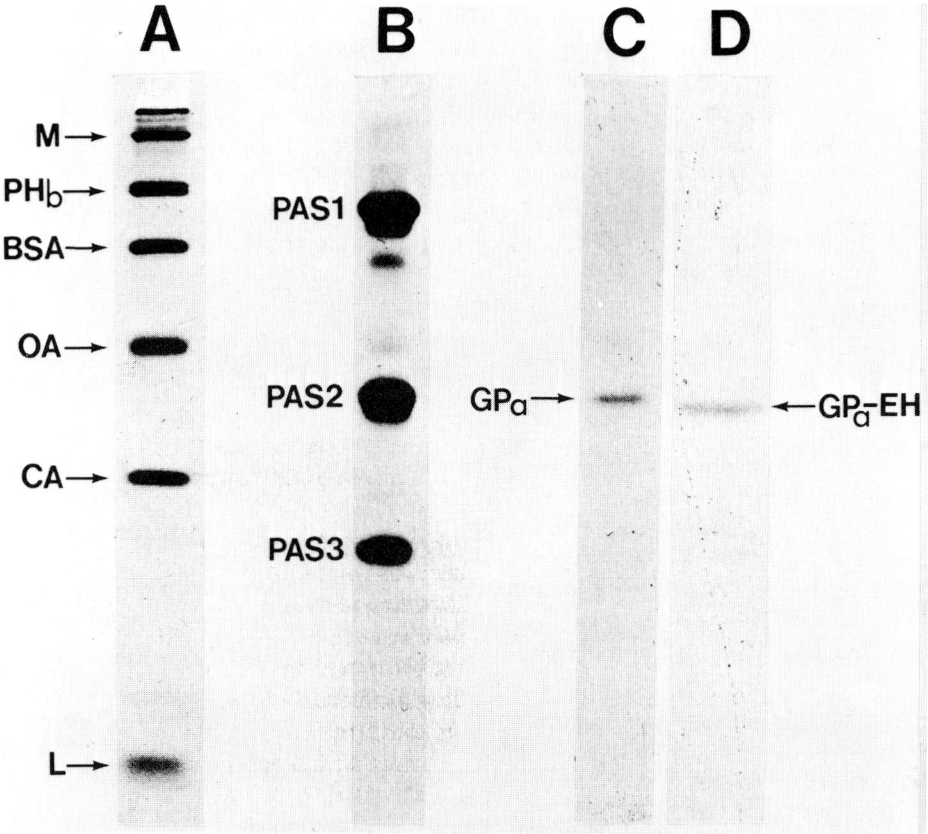

FIG. 8. Cell-free synthesis of glycophorin A in the presence of dog pancreatic membranes. Lane: (A) fluorography pattern of polyacrylamide slab gel of ^{14}C-labeled standard proteins (see legend to Fig. 5); (B) pattern of periodate–NaB^3H_4-labeled red cell membranes; (C) immune precipitates obtained with anti-glycophorin A antiserum from K562 cell mRNA-directed glycoprotein fraction isolated by lentil lectin–Sepharose affinity chromatography; (D) pattern of immune precipitate obtained as in (C) and then treated with endoglycosidase H. GP_a-EH, endoglycosidase H-treated glycophorin A glycosylated precursor.

fraction of the membrane. In addition, unless detailed structural information is available, meaningful conclusions are impossible.

The understanding of the erythroid characteristics of the K562 cells and the availability of specific anti-glycophorin A antiserum make biosynthetic experiments on glycophorin A possible. A great difficulty, however, is the limited amount of radioactivity attainable in glycophorin A molecules because of the multitude of proteins simultaneously synthesized. Using the methods described above, it is possible to study the synthesis of mammalian integral membrane glycoproteins in considerable detail. As an example, the synthesis of glycophorin A is described and a summary of this follows. The polypeptide is synthesized initially with a signal sequence that subsequently is cleaved. Early during synthesis, the NH_2 terminus becomes segregated into microsomes.[16] Cotranslationally it is N-glycosylated at the single asparagine-26 site. That it is also O-glycosylated early during synthesis is obvious for the following reasons. The apparent molecular weight of the GP_a molecule seen after 5 min of synthesis *in vivo,* or after synthesis *in vitro* in the presence of membranes, is about 37,000. The molecular weight of the sugar-free apoprotein is about 14,500 as calculated from the amino acid sequence. The contribution of the N-glycosidic oligosaccharide to the molecular weight is about 2000. Mild alkaline treatment lowered the apparent molecular weight of glycophorin A much below the 37,000 value (not shown), as did treatment of desialylated glycophorin A with endo-*N*-acetylgalactosaminidase (not shown). Even treatment of normal glycophorin A (containing the N-glycosidic oligosaccharide) with neuraminidase in the presence of sodium dodecyl sulfate gave an apparent molecular weight of about 37,000.[45] Interestingly, the newly synthesized GP_a molecule did not adsorb to *Helix pomatia* lectin–Sepharose (D-*N*-acetylgalactosamine specific) or to *Ricinus communis* lectin–Sepharose (*d*-galactose specific).[18]

These results indicate that early during synthesis of GP_a there is alkali-labile material added to serine/threonine residues that may not be *N*-acetylgalactosamine. The addition of *N*-acetylgalactosamine is perhaps a later event, probably taking place in the Golgi region, where UDP-*N*-acetylgalactosaminyltransferases[46] are thought to be located.

The addition of the peripheral sugars to glycophorin A occurred after a lag period of about 10 min. Then galactose, *N*-acetylglucosamine, and sialic acids were rapidly added. This was clearly evident by the use of *Ricinus communis* lectin- and wheat germ lectin–Sepharose columns and the effect of neuraminidase on the electrophoretic mobility of glycophorin A. The molecule was completely glycosylated before it appeared at the cell surface, which took about 25 min.

Red cell glycophorin A is known to contain phosphate, evidently at serine residue 102 or 104.[47] The K562-derived glycophorin A was also phosphorylated. Because the phosphorylated molecule adsorbed both to lentil lectin and to wheat germ lectin and was insensitive to endoglycosidase H, it was evidently completely glycosylated. The presence of sulfate in glycophorin A is interesting. Sulfate is known to be bound to proximal

[45] C. G. Gahmberg and L. C. Andersson, *Eur. J. Biochem.* **122,** 581 (1982).
[46] E. J. McGuire and S. Roseman, *J. Biol. Chem.* **242,** 3745 (1967).
[47] D. L. Shapiro and V. T. Marchesi, *J. Biol. Chem.* **252,** 508 (1977).

N-acetylglucosamines of some N-glycosidic oligosaccharides (see Gahmberg[48] for a review). Unfortunately we could not determine the location of sulfate in glycophorin A because of the low radioactivity obtained.

The tunicamycin experiments clearly showed that the absence of the N-glycosidic oligosaccharide did not affect the intracellular migration of glycophorin A. This is in contrast to the behavior of some other integral membrane proteins.[48] It must, however, be pointed out that glycophorin A still contained O-glycosidic carbohydrate. The O-glycosidic oligosaccharides alone may give the polypeptide essentially the conformation needed for intracellular migration.

Very few integral membrane glycoproteins are both O- and N-glycosylated. The presence of both types of oligosaccharides in glycophorin A makes the molecule an excellent candidate for more extensive studies of complex glycosylations. Work along this line is in progress.

Acknowledgments

The original work reported herein was supported by The Academy of Finland, The Association of Finnish Life Insurance Companies, and National Cancer Institute Grant 5 R01 CA26294-02. We thank Barbara Björnberg for secretarial work.

[48] C. G. Gahmberg, *in* "Membrane Structure" (J. B. Finean and R. H. Michell, eds.), p. 127. Elsevier/North-Holland Biomedical Press, Amsterdam, 1981.

[23] Erythrocyte Membrane Proteins: Detection of Spectrin Oligomers by Gel Electrophoresis

By JON S. MORROW and WALLACE B. HAIGH, JR.

Spectrin is the principal structural protein of the erythrocyte membrane skeleton. It is composed of two chemically distinct polypeptides, an α-subunit of 240,000 daltons and a β-subunit of 220,000 daltons. Although a fully functional spectrin molecule requires a single copy of each chain, the functional unit formed by the noncovalently joined α- and β-subunits has historically been referred to as the spectrin dimer ($\alpha\beta$).[1] The combination of two dimers forms a tetramer [$(\alpha\beta)_2$], and the addition of a third dimer yields a hexamer [$(\alpha\beta)_3$]. Higher association states are named accordingly. The remarkable self-associating ability of spectrin has been

[1] V. T. Marchesi, *J. Membr. Biol.* **51,** 101 (1979).

ISBN 0-12-181996-5

appreciated only in the past few years. The significance of this process arises from its implications for understanding the structure of the erythrocyte membrane skeleton and from the likelihood that many, if not all, cells possess large asymmetric structural proteins similar to spectrin. These proteins presumably act by linking F-actin to a membrane receptor and may also self-associate. Thus the ability to analyze noncovalent associations in these proteins is of critical importance.

The self-association of spectrin has been measured by a number of techniques. Most, such as gel filtration or sedimentation velocity, have readily detected the dimer and tetramer species,[2,3] but have proved to be less useful in differentiating oligomeric forms of pure spectrin from the high molecular weight complexes formed from spectrin and several other proteins. A particularly sensitive technique well suited to the measurement of spectrin oligomers has proved to be electrophoresis into low-percentage polyacrylamide gels at low temperature and under native conditions.[4,5] By this technique a number of previously unrecognized spectrin oligomeric states may be precisely measured; although poorly resolved in sedimentation experiments, these oligomers possess sedimentation values between 12 and 20 S.

A number of features of the spectrin molecule and its oligomerization make it particularly well suited for electrophoretic analysis. The protein is acidic, with an isoelectric point near pH 5.2. Thus the protein will migrate briskly toward the anode at physiological pH, whereas some common contaminating proteins, such as hemoglobin ($pI = 7.2$), will possess little charge under these conditions and can be made to migrate cathodically. A second feature of spectrin is the high activation energy associated with the self-association process.[3] This has been estimated to be approximately 250 kJ mol^{-1}. The consequence of this activation barrier is that at low temperatures the equilibrium is kinetically trapped. Typical equilibration times at 37° are 15–30 min, whereas at 0° equilibration requires weeks. Thus by conducting the electrophoretic analyses at 0–4° the oligomer distributions under a variety of conditions may be sampled without concern for changes occurring during the electrophoretic experiment itself. It has been our experience that no combination of salt, pH, or other added proteins has altered the ability of low temperature to trap kinetically the spectrin oligomer populations. Caution must be exercised when extending this technique to other proteins or to the interaction of spectrin with other

[2] G. B. Ralston, J. Dunbar, and M. White, *Biochim. Biophys. Acta* **491,** 345 (1977).

[3] E. Ungewickell and W. Gratzer, *Eur. J. Biochem.* **88,** 379 (1978).

[4] J. S. Morrow and V. T. Marchesi, *J. Cell Biol.* **88,** 463 (1981).

[5] J. S. Morrow, W. B. Haigh, Jr., and V. T. Marchesi, *J. Supramol. Struct. Cell. Biochem.* **17,** 275 (1981).

peptides and proteins, since not all associations will be so stable, even at 0°.

A second advantage inherent in nondenaturing polyacrylamide gel electrophoretic analysis of protein–protein associations is the ease with which the technique lends itself to two-dimensional studies. By analyzing a nondenaturing gel containing noncovalently associated proteins in a second dimension with sodium dodecyl sulfate (SDS), additional information may be obtained on the identity of the associating species. This approach has proved to be quite powerful in distinguishing specific noncovalently associated peptides from complex peptide mixtures.[6]

Procedure: Nondenaturing Polyacrylamide Gel Electrophoresis of Spectrin Oligomers

Apparatus and Glassware. Either standard disc or slab gel electrophoresis chambers may be used. Unless two-dimensional analysis is planned, the slab gel configuration is preferable. Owing to the extreme fragility of the nonstabilized low-percentage polyacrylamide gels, very thin gels (<1.5 mm) are impractical. In this laboratory, 160 × 160 × 3 mm slab gels have become standard and work well. These are used in conjunction with a form that provides eleven 10-mm sample wells across the top of the gel. Electrophoresis is performed in a simple homemade vertical slab gel chamber. Cooling is provided by conducting the experiment at 4° in a cold box and limiting the maximum applied potential to 50 V.

Electrophoresis Buffer. The buffer and acrylamide gel compositions are similar to those popular in SDS–gel electrophoresis except for the absence of all denaturants.[7] The gel and the running buffer are identical. Its final composition is 40 m*M* Tris–20 m*M* sodium acetate (approximate)–2 m*M* EDTA, pH 7.4.

It is conveniently prepared and stored as a 10× stock solution:

10× Buffer[7]: 1 *M* Tris base, 800 ml; 2 *M* sodium acetate, 200 ml; 0.2 *M* EDTA, 200 ml; glacial acetic acid, adjusted to pH 7.4; double glass-distilled water (DDW), adjusted to 2000 ml. Variation in pH between 6.5 and 8.0 may be made without degrading the resolution of the system.

Preparation of the Gel. The choice of acrylamide percentage depends on the nature of the experiment planned. To resolve oligomeric species with molecular weights in excess of several million, very low acrylamide concentrations are necessary. An optimal choice for slab gel experiments

[6] J. S. Morrow, D. W. Speicher, W. J. Knowles, C. J. Hsu, and V. T. Marchesi, *Proc. Natl. Acad. Sci. U.S.A.* **77,** 6592 (1980).

[7] G. Fairbanks, T. L. Steck, and D. F. H. Wallach, *Biochemistry* **10,** 2606 (1980).

appears to be a gradient ranging from 2.0 to 4.0% acrylamide. The acrylamide/bisacrylamide ratio is 25 : 1. The 2% acrylamide at the top of the gel allows entry of very large complexes, and the gradient to 4% bestows some mechanical stability and sharpens the bands. Alternatively, a straight 3% gel may be used with sacrifice of some resolution. When disc gels are prepared for use in two-dimensional separations, the gradient is impractical and straight 3, 4, or 5% acrylamide gels are used, the choice depending on the size of the peptide complexes to be separated. The following solutions are used to prepare the gel.

Acrylamide stock solution[7]: 40 g of acrylamide monomer (Bio-Rad, electrophoretic purity), 1.5 g of *bis-N,N'*-methylenebisacrylamide (Bio-Rad). Dilute to 100 ml of DDW. This solution is deionized for 1 hr using Amberlite MB-3 (6 g/100 ml) and then stored at 4° protected from light.

Ammonium persulfate (APS) (Bethesda Research Labs): 1.5% in DDW, prepared fresh weekly.

TEMED (*N,N,N'N'*-tetramethylethylenediamine; Bio-Rad): 5% in DDW

Pouring a 2–4% Slab Gel. A gradient 160 × 160 × 3 mm slab gel is prepared from 2% and 4% acrylamide solutions.

	2%	4%
Stock acrylamide solution (ml)	2	4
10× Stock buffer (ml)	4	4
DDW (ml)	29.8	22.8
TEMED, 5% (ml)	0.2	0.2
Sucrose, 2.5 *M* in DDW (ml)	0	5
	36 ml	36 ml

Thirty-four milliliters of each solution are placed in the appropriate chamber of a gradient former, 3.75 ml of 1.5% APS are mixed with each, and the gel is then immediately poured at a rate of 3–4 ml/min. Faster filling rates should be avoided, since they cause excessive mixing and distortion of the gradient. Typically the gel requires 1 hr at room temperature to polymerize fully. After an hour the comb used to form the sample wells may be removed; the 2% acrylamide at the top of the gel is very fragile and requires careful handling. After removal of the comb the sample wells are covered with 1× buffer, and the gel is allowed to stabilize at 4° for at least 6 hr. Degassing of the gel solutions or special precautions to exclude oxygen during polymerization[8] have not been necessary despite the low percentages of acrylamide employed.

[8] R. F. Peterson, this series, Vol. 25, p. 178.

Loading and Running the Gel. Aliquots containing approximately 10–70 μg of protein are used for analysis. For concentrated solutions of spectrin (5–20 mg/ml) it is often difficult to remove such small amounts of material accurately. In these cases a convenient procedure is first to dilute the protein sample to a more manageable concentration with cold (0°) buffer so that 50–100 μl may be loaded into each sample well. If precautions are taken always to maintain the sample at low temperatures, spectrin may be diluted to workable concentrations without detectable changes in the distribution of oligomers. Immediately prior to loading, the sample is diluted 20% with a cold solution of 10% sucrose containing 0.1% Bromphenol Blue. The sucrose produces a good layering of the sample atop the gel, and the Bromphenol Blue provides a convenient marker of the buffer front during electrophoresis. Excellent results have been obtained with sample volumes between 10 and 200 μl and for protein loads up to 70 μg. The ionic strength of the protein sample does not materially influence the results; our spectrin samples are typically in 10 m*M* Tris, 130 m*M* KCl, 20 m*M* NaCl, and 1 m*M* 2-mercaptoethanol.

After loading the gel at 4°, electrophoresis is conducted at 4° in the conventional manner. Two potential problems arise during the electrophoresis. If too much power is applied, the resultant heating facilitates the interconversion of the oligomeric species. While the amount of power that is tolerable will be a function of the cooling efficiency of the apparatus employed, we typically limit the applied potential to 50 V and electrophorese for 24–48 hr. Higher voltages yield more rapid separations, but one must be vigilant for the onset of sufficient heating in the gel to cause interconversion of oligomeric forms.

A second problem is encountered when the pH of the running buffer drifts due to electrolysis. While small changes in pH are without effect, larger changes will severely distort the electrophoretic mobilities and produce clearing of the top of the gel with concentration of the protein into a single sharp band sweeping down the gel. This problem may be avoided by recirculating the buffer or by changing the running buffer frequently. Our buffer reservoirs hold approximately 1 liter in each chamber. When the gels are electrophoresed at 50 V, it is sufficient to change the buffer once every 12 hr.

Staining the Gel. Staining follows published methods.[7] Removal of the gels from the glass plates must be done with utmost caution owing to the gel's extreme fragility. After the top plate is removed, a sheet of Saran wrap is placed over the gel, and the gel is inverted to remove the other plate. No attempt should ever be made to lift the unsupported gel directly. Once free of the glass plates, the gel is lowered on the Saran wrap into

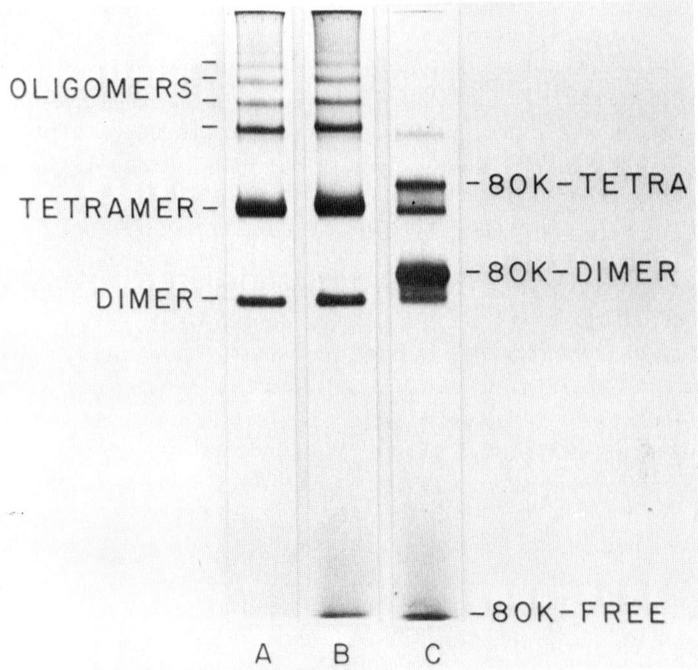

FIG. 1. Coomassie Blue stained 2–4% nondenaturing polyacrylamide gel analysis of spectrin oligomers and of an 80,000-dalton (80K) spectrin-binding spectrin peptide. The conditions of electrophoresis were as described in the text (50 V, 48 hr, 4°). Each lane contains 60 μg of protein. Lane A: Purified human erythrocyte spectrin (6 mg/ml) incubated at 30° for 90 min before loading gel. At this concentration, spectrin reversibly self-associates to oligomeric states larger than dimer.[4] The molecular weight of the largest oligomer notes is 2.8×10^6 daltons. Lane B: The same spectrin solution shown in A, after the addition at 0° of a purified 80K proteolytic fragment of spectrin (7 mg/ml). This peptide binds to spectrin in a temperature-dependent manner at the spectrin self-association site.[4] Without incubation, it has little influence on the observed oligomer distribution. Most of the unbound 80K peptide has electrophoresed off the bottom of the gel. Lane C: The same solution as shown in B, but after incubation for 90 min at 30°. The 80K peptide competitively reduces the amount of oligomers. The binding of the peptide to the dimer and tetramer forms of spectrin is apparent.

Coomassie Blue staining solution and stained for 6–12 hr at room temperature.

Coomassie Brilliant Blue stock solution: Combine 24 g of Coomassie Brilliant Blue R-250 and 300 ml of methanol; stir until dissolved. Add 60 ml of glacial acetic acid. Dilute to 600 ml with DDW.

Coomassie Blue working stain solution (0.025%): Combine 50 ml of Coomassie Blue stock solution, 2000 ml of methanol, and 400 ml of glacial acetic acid. Dilute to 8 liters with DDW.

Destaining of the gel is done in acetic acid–methanol–water (10 : 15 : 75, v/v/v) and requires 24–46 hr. The process is facilitated by adding a 50–100-cm length of virgin wool to the destaining solution to absorb the free stain. If the staining solution is too concentrated, artifactual bubbles and stain spots may appear near the top of the gel.

Handling, Drying, and Quantitating the Gel. In the wet state the gel is extremely fragile and is best handled by carefully pouring it from one container to another. For photography it may be spread in a shallow water-filled transparent tray. Alternatively, a particularly convenient way to preserve the gels and to also facilitate densitometric analysis of the bands is to dry them on cellophane paper. The resulting dried gel is crystal clear and durable and may be photographed, cut, and scanned without difficulty. To dry the gels, spread a 30 × 30 cm sheet of cellophane paper (obtainable in 36-inch-wide roles from Daimaru-Finjii Co., Sapporo, Japan) on a 20 × 25 cm glass plate. Prior to being placed on the plate, the cellophane paper is softened for 1 min in a 5% glycerol solution. Trapped bubbles should be removed; the gel is floated in approximately 50 ml of 5% glycerol and spread atop the cellophane paper. After centering, an additional 30 ml of 5% glycerol is poured over the gel and it is overlaid with a second softened sheet of cellophane paper. After the bubbles and excess fluid are forced out, the cellophane sheets are sealed on all sides, using plastic spacers and spring clamps. The gels are then allowed to dry overnight under a stream of cool air.

Results of a typical analysis by the procedure described here are shown in Fig. 1.

Detection of Other Protein–Protein Associations

A number of other protein–protein interactions may be detected using nondenaturing gels as described above. We have detected the binding of various spectrin proteolytic peptides, band 2.1, and band 4.1 by this method. These binding interactions may be quantitated by employing radiolabeled proteins; the dried gels are suitable for autoradiography, and the radiolabeled bands may be cut out for isotope counting.

[24] Isolation of the Chemical Domains of Human Erythrocyte Spectrin

By WILLIAM J. KNOWLES and MARCIA L. BOLOGNA

Spectrin is the primary component of the erythrocyte membrane skeleton.[1,2] Structural studies indicate that spectrin is composed of nine unique chemical domains and subdomains that range in molecular weight from 80,000 to 12,000.[3,4] Specific domains contain sites for oligomerization and membrane binding and are involved in the formation of noncovalent associations between the α- and β-subunits.[5–7] These functions, as well as spectrin interactions with other membrane skeletal proteins, play an important role in determining the red cell's shape and viscoelastic properties.[2,8]

Several reports indicate that spectrin is altered in at least two hemolytic anemias.[9,10] In one disease, hereditary pyropoikilocytosis, a structural change has been found in the αI-T80 domain. This change is responsible for altering the shape and deformability of the red cells and impairs spectrin oligomerization *in vitro*.[9] Precise structural changes within specific domains are, however, difficult to determine because of the complexity of the spectrin molecule (heterodimer molecular weight = 480,000) and the difficulties in isolating the individual domains by conventional purification procedures. Blood from patients with hemolytic anemias can usually be obtained in only small quantities, thereby compounding these difficulties. In order to compare the structural and functional properties of individual spectrin domains from different donors we have developed procedures that allow the isolation of all nine structural domains and their

[1] V. T. Marchesi, *J. Membr. Biol.* **51,** 101 (1979).
[2] S. Lux, *Semin. Hematol.* **16,** 21 (1979).
[3] D. Speicher, J. Morrow, W. Knowles, and V. Marchesi, *Proc. Natl. Acad. Sci. U.S.A.* **77,** 5673 (1980).
[4] D. Speicher, J. Morrow, W. Knowles, and V. Marchesi, *J. Biol. Chem.* **257,** 9093 (1982).
[5] J. Morrow, D. Speicher, W. Knowles, C. Hsu, and V. Marchesi, *Proc. Natl. Acad. Sci. U.S.A.* **77,** 6592 (1980).
[6] J. Morrow and V. Marchesi, *J. Cell Biol.* **88,** 463 (1981).
[7] J. Morrow, W. Haigh, and V. Marchesi, *J. Supramol. Struct. Cell Biochem.* **17,** 275 (1981).
[8] S. Lux and L. Wolfe, *Pediatr. Clin. North Am.* **27,** 463 (1980).
[9] W. Knowles, J. Morrow, D. Speicher, H. Zarbowsky, N. Mohandas, S. Shohet, and V. Marchesi, *J. Clin. Invest.* June (1983).
[10] H. Tomaselli, K. John, and S. Lux, *Proc. Natl. Acad. Sci. U.S.A.* **78,** 1911 (1981).

ISBN 0-12-181996-5

subfragments from less than 20 ml of whole blood. To accomplish this, we have used preparative two-dimensional electrophoresis [isoelectric focusing (IEF) and sodium dodecyl sulfate (SDS)] to resolve all peptides generated from spectrin by mild tryptic digestion. Each domain can be identified by comparison to analytical two-dimensional separations of spectrin peptides[3,4] and can be eluted rapidly from the gel.

Isolation of Spectrin and Generation of Intermediate-Sized Peptides

Spectrin is isolated from human red cell ghosts by low-ionic-strength extraction as previously described.[11] Intermediate-sized peptides consisting of the unique structural domains of the α- and β-subunits and their subfragments are generated by incubating spectrin (8 mg at 1 mg/ml) for 16 hr at 0° in 20 m*M* Tris-HCl, pH 8.0, 1 m*M* 2-mercaptoethanol (Eastman Kodak) containing TPCK-trypsin (Worthington; 217 units/mg) at an enzyme-to-substrate ratio of 1 : 200 (w/w). Different lots of trypsin have ranged from 180 units/mg to 517 units/mg necessitating adjustment of the enzyme concentration or time of the reaction. The extent of spectrin digestion is critical, since too mild a digest will result in a significant proportion of the peptides having molecular weights greater than 80,000. Large amounts of these higher molecular weight peptides interfere in subsequent electrophoretic separations and decrease the yield of the intermediate-sized peptides. Digestion is terminated by adding diisopropyl fluorophosphate (Aldrich Chemical Company) to 1 m*M* final concentration followed 30 min later by shell-freezing and lyophilization. The effect of different conditions on the generation of intermediate-sized peptides from spectrin has been previously studied.[4]

Preparative Two-Dimensional Electrophoresis

Preparative isoelectric focusing is performed using a modification of O'Farrell's analytical isoelectric focusing procedures and formulas.[12,13] Gels are 4.0% acrylamide (Bio-Rad), 0.013% *N,N'*-methylenebisacrylamide (Bio-Rad), 9 *M* urea (Ultrapure, Bethesda Research Laboratories), 2% Triton X-100 (New England Nuclear), 2.4% Ampholines (0.8% pH 4–6, 0.8% pH 5–7, 0.8% pH 3.5–10; LKB) and are polymerized by adding 1.5% ammonium persulfate and TEMED (*N,N,N',N'*-tetramethylethylenediamine; Eastman Kodak) to final concentrations of

[11] D. Litman, C. Hsu, and V. Marchesi, *J. Cell Sci.* **42,** 1 (1980).
[12] P. H. O'Farrell, *J. Biol. Chem.* **250,** 4007 (1975).
[13] G. Ames and K. Nikaido, *Biochemistry* **15,** 616 (1976).

0.009% and 0.0007%, respectively. Gels are cast 12.5 cm in length in 6 mm i.d. × 14.5 cm long Pyrex tubes and overlaid with 8 *M* urea for 5–6 hr.

The gels (which occasionally slip out of the Pyrex tubing during electrofocusing) are held in the glass tube with one layer of cheesecloth (Curity) secured on the lower end of the gel tube with a snugly fitting section of latex rubber tubing. After polymerization the 8 *M* urea is removed from the upper surface of the gel and replaced with solubilization buffer (9 *M* urea, 2.4% Ampholines, 5% 2-mercaptoethanol, and 2% Triton X-100), and the gels are preelectrophoresed for 60 min at a constant 340 V. The cathode and anode buffers are 0.02 *M* NaOH and 0.01 *M* H_3PO, respectively, and are prepared with degassed distilled water.[12]

The spectrin digest (8 mg) is solubilized at room temperature in 250 μl of a freshly prepared solution of solubilization buffer–9 *M* urea (1 : 1) and applied to the upper surface of the gel after removal of the upper electrode buffer. The sample tube is rinsed with 100 μl of an 8 : 2 dilution of 9 *M* urea, 1% Triton X-100, 1.2% Ampholines, and 2.5% 2-mercaptoethanol–H_2O, and this solution is used as an overlay.[12]

After replacing the upper electrode buffer, the digests are electrophoresed for 19 hr at a constant 340 V. Gels are removed from the tubes by rimming with a fine needle and are dansylated (see below) and/or equilibrated in gel storage solution for 10–15 min before freezing in a dry ice–ethanol bath.[12] The gel storage solution contains 10 m*M* Tris-HCl, pH 7.0, 10% glycerol, 3% SDS (Bio-Rad), 2% 2-mercaptoethanol, 4 m*M* EDTA, and 0.002% Bromphenol Blue (BPB). The equilibrated gels are stored at −80°.

A 10 to 15% acrylamide gradient gel (6 mm thick, 14 cm long, 16 cm wide) prepared with the SDS–discontinuous buffer system of Laemmli[14] is used for the second dimension. The stacking gel of 3.75% acrylamide, 0.1% *N,N'*-methylenebisacrylamide is 2.0 cm high. Immediately after thawing, the IEF gel is placed on the surface so that the entire surface of the IEF gel is in contact with the upper surface of the stacking gel. The second dimension is electrophoresed until the BPB reaches the bottom of the separating gel (16–18 hr at a constant 65 V). Care must be taken that the cathode buffer does not leak, since replenishment with fresh buffer will make it difficult if not impossible to identify the separated peptides by cold-KCl precipitation.

Identification of Peptides

Cold KCl. The most frequently used means of identifying spectrin peptides after two-dimensional electrophoresis is by immersing the gel in

[14] U. K. Laemmli, *Nature (London)* **227,** 680 (1970).

1.0 *M* KCl at 0–4°.[15,16] Within 2–5 min the SDS–peptide complexes form white precipitates against the frosted appearance of the nonpeptide-containing portions of the gel. The main advantage of this technique is that it does not modify the peptides. The disadvantages are that it is less sensitive than the dansylation reaction (below) and results in an approximate 5% decrease in the yield during electrophoretic elution when compared to peptides identified by dansylation.

Dansylation. One of the common procedures for identifying proteins after single-dimension SDS–polyacrylamide gel electrophoresis has been by prelabeling with dansyl chloride.[17,18] In most cases dansylation does not modify the apparent molecular weight of proteins as determined by SDS electrophoresis. However, dansylation before isoelectric focusing (IEF) modifies the p*I* of the protein by varying degrees, depending on the extent of the reaction. Adding 2–5% of dansylated protein to the unlabeled protein before IEF would not be useful since the fluorescent peptides would have a different p*I* than the unlabeled protein. In order to circumvent this problem, we have developed methods to dansylate the peptides after IEF and before separation in the second dimension. To dansylate the peptides, the entire IEF gel is rinsed three times (20 ml each) over a period of 10 min in 10% glycerol, 1.4% SDS, 0.5 *M* $NaHCO_3$ titrated to pH 9.5 with NaOH. All solutions used in this procedure must be at 30° or SDS precipitation will occur. Each gel is placed in a small beaker (50 ml) with 10 ml of the above solution, and 200 μl of a 10% dansyl chloride in acetone (Pierce Chemical Co.) are forcibly injected into the solution using a disposable tuberculin syringe with a 30-gauge needle. The resulting solution should be turbid, containing a suspension of dansyl chloride. Inadequate injection will result in precipitation of dansyl chloride and poor dansylation of peptides. The gel and dansyl solution are intermittently agitated for 15 min. Extending this time will result in increased labeling of the peptides, giving greater sensitivity, but it also more extensively modifies the peptides. The gel is rinsed 2–3 times in H_2O and equilibrated for 5 min in gel storage solution; it can be frozen and stored as above. The dansylated peptides are visible under longwave UV light shortly after they enter the separating gel in the second dimension. A large fluorescent band comigrates with the BPB dye front and presumably contains unreacted dansyl and low molecular weight dansylated peptides.

The two-dimensional electrophoretically separated peptides visualized by KCl precipitation or by dansylation form a pattern that is identical

[15] R. Wallace, P. Yu, J. P. Dieckert, and J. W. Dieckert, *Anal. Biochem.* **61,** 86 (1974).

[16] D. Peterson, *J. Biol. Chem.* **256,** 6975 (1981).

[17] D. Talbot and D. Yphantis, *Anal. Biochem.* **44,** 246 (1971).

[18] R. Stephens, *Anal. Biochem.* **65,** 369 (1975).

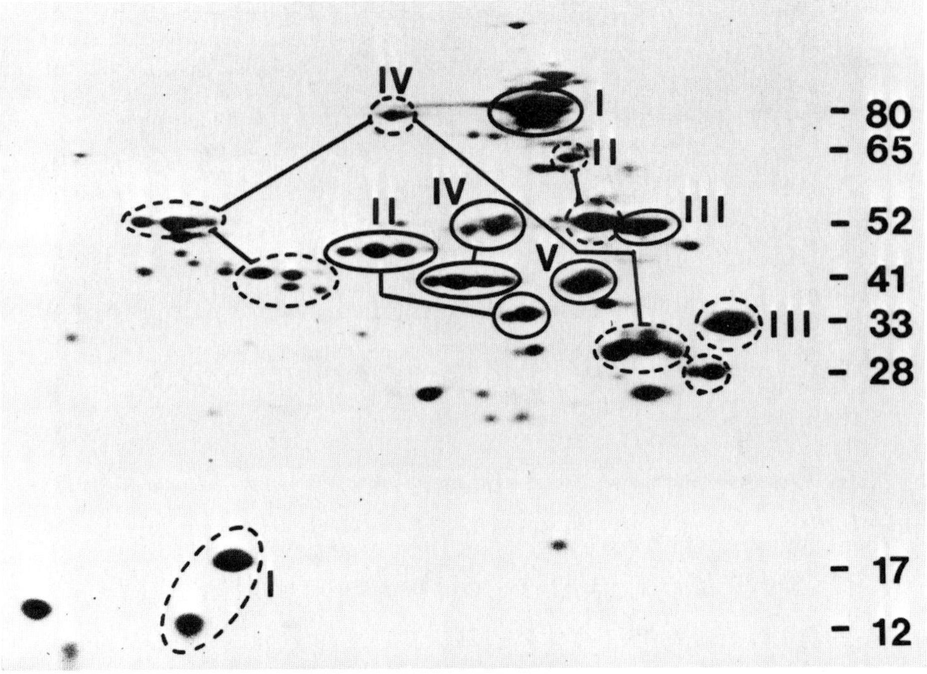

FIG. 1. Analytical two-dimensional separation of the intermediate-sized peptides of human erythrocyte spectrin. Peptides circled with a solid line are from the α-subunit; those circled with a dashed line are from the β-subunit. Numbers indicate specific domains within the subunit.[4] Circles connected by lines indicate that the lower molecular weight peptide is a fragment of the higher molecular weight peptide. This analytical gel was prepared by procedures identical to those used for the preparative separations except that both IEF and SDS were 3 mm thick. The gel was loaded with 200 μg of a spectrin digest and was stained with Coomassie Blue.

to that obtained by analytical two-dimensional electrophoresis (Fig. 1). All peptides can therefore be identified as belonging to an α- or β-subunit and to a specific domain within that subunit.[4] Once identified, the peptides are cut from the gel and stored at −20° in 12.5 m*M* Tris-glycine, pH 8.6, 0.01% SDS, 0.03 m*M* phenylmethylsulfonyl fluoride (PMSF) until electroelution.

Electrophoretic Elution

The electrophoretic elution of ^{125}I-labeled spectrin peptides from preparative two-dimensional gels was used to evaluate the effectiveness of different buffers, SDS concentrations, temperature, and conditions (volt-

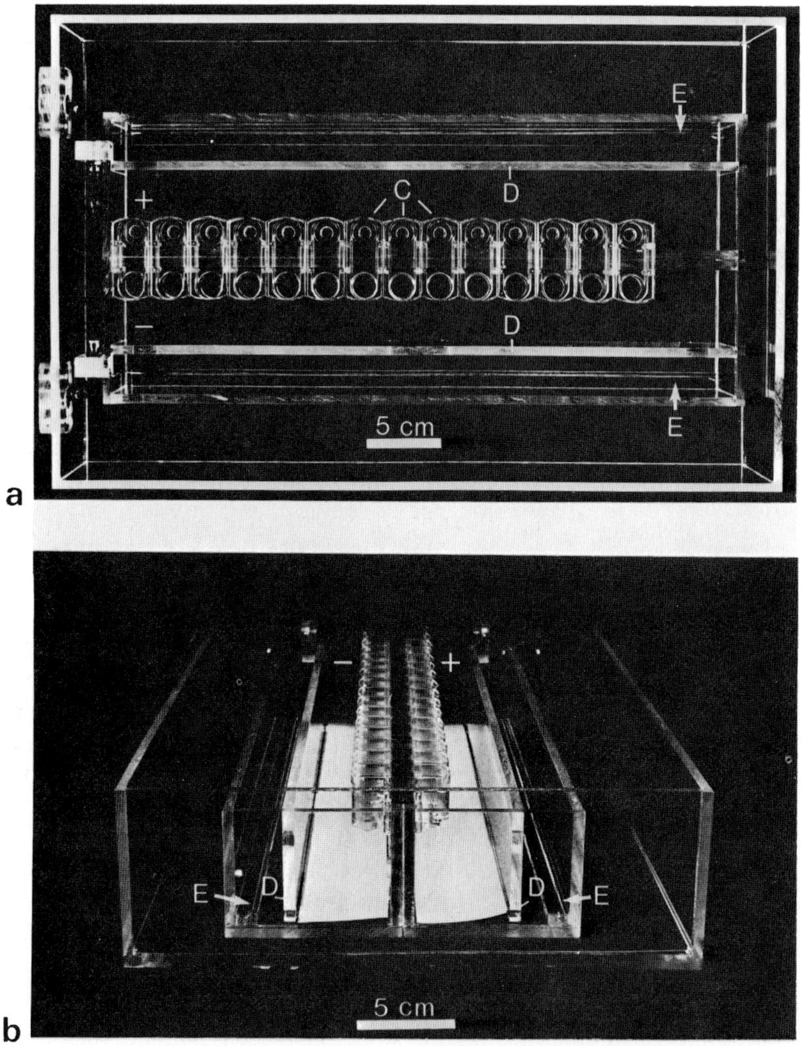

FIG. 2. Chamber used for the electrophoretic elution of spectrin peptides. All Plexiglas is $\frac{1}{4}$ inch thick. C indicates ISCO cups; E, electrodes; and D, a divider that partially divides the electrode side of the buffer chamber from the side containing the sample cups and prevents the accumulation of gas bubbles under the dialysis membrane. This divider extends the full depth of the buffer chamber except for the lower 2 mm. (a) Top view; (b) side view.

age and time) for peptide elution. The procedure described below is simple, allowing us to elute 16 peptides in a minimum amount of time, and it consistently produces the highest yield.

Apparatus. The peptides are eluted from the gel and concentrated in a small volume using ISCO sample concentrator cups (Instrumentation Specialities Co., Lincoln, Nebraska). The cup dimensions and their previous applications have been published.[19,20] These cups have been used with the ISCO sample concentrator buffer tanks or with a conventional thin-layer electrophoresis chamber shown in Fig. 2a,b. This chamber holds 16 sample cups and 4 liters of buffer, which serves as an effective coolant during the elution procedure.

The sample cups have a reusable membrane with a molecular weight cutoff of 3500.[20] If cross-contamination is a concern or if the membranes become damaged they can be replaced with sections of washed dialysis membrane[21] Spectrapor (45 mm width) with a molecular weight cutoff of 3,500. Thicker membranes will not fit and will result in breaking the plastic rings used to hold the membranes. After replacing the membrane, it is necessary to check for leakage by filling the cup with buffer and placing it on a dry surface for 5–10 min. Leaks, other than flow of buffer through the membrane, will be evident.

Procedure. All solutions must be prepared with the highest quality of water available. Any exogenous proteins (in the worst case protease) placed in the sample cup will be concentrated with the protein eluted from the gel. The buffers are prepared immediately before use with distilled water that has been subsequently purified by sequential passage through 1 charcoal filter, 2 ion-exchange filters, 1 organex Q filter, and a 0.22-μm filter (Milli Q System, Millipore Corporation).

Immediately before elution the gel sections are thawed and placed in the large sample well in the cathode buffer chamber. The sample cups and buffer chambers are filled with 12.5 m*M* Tris-glycine, pH 8.6, 0.01% SDS, and 0.03 m*M* PMSF, and the samples are electrophoresed at a constant 400 V for 100 min at room temperature.[22] The temperature rise of the buffer solutions is approximately 1.5° during this period. After electrophoresis all buffer except that in the concentration well is removed from the sample cup using a vacuum aspirator. The cup is immediately re-

[19] W. Allington, A. Cordry, G. McCullough, D. Mitchell, and J. Nelson, *Anal. Biochem.* **85,** 188 (1978).

[20] J. Nelson, A. Cordry, G. McCullough, and G. Meakin, *ISCO Appl. Res. Bull.* No. 32 (1980).

[21] A. Brown, J. Mole, F. Hunter, and J. Bennett, *Anal. Biochem.* **103,** 184 (1980).

[22] R. Stephens, *Anal. Biochem.* **84,** 116 (1978).

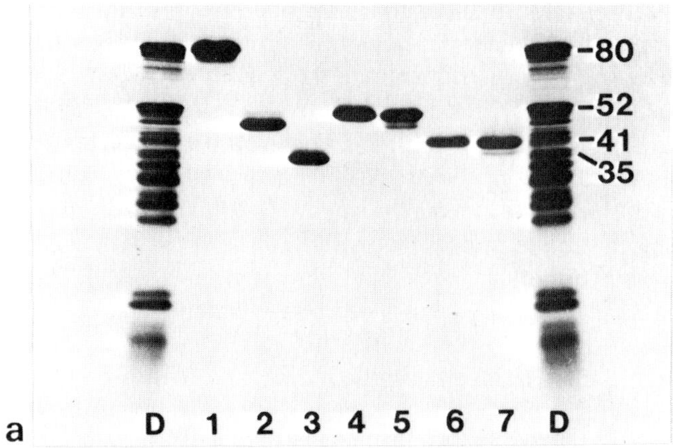

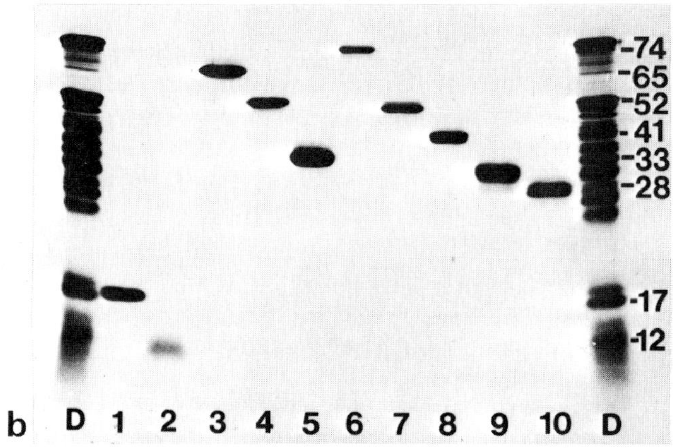

FIG. 3. Eluted peptides electrophoresed on a 10–15% acrylamide Laemmli gel. Peptides correspond to those identified in Fig. 1. (a) α-subunit peptides; 1, αI-T80; 2, αII-T46; 3, αII-T35; 4, αIII-T52; 5, αIV-T52; 6, αIV-T41, 7, αV-T41. (b) β-subunit peptides; 1, βI-T17; 2, βI-T12; 3, βII-T65; 4, βII-T52; 5, βIII-T33; 6, βIV-T74; 7, βIV-T52; 8, βIV-T41; 9, βIV-T30; 10, βIV-T28. D is 50 μg of a spectrin digest containing both α- and β-subunits.

moved from the buffer chamber, and the concentrated eluted protein is removed with a siliconized pipette (Prosil-28; PCR Research Chemicals). The membrane is rinsed twice with 200 μl of 12.5 mM Tris-glycine, pH 8.6, 0.001% SDS, and 0.03 mM PMSF. Approximately 85–95% of the total radioactivity in the original gel slice is recovered from the concentration well. The twice-rinsed dialysis membrane has <1% of the total radioactivity. Depending upon the subsequent experiments, the SDS concentration can be reduced by prolonged dialysis, electrodialysis,[23] or solvent extraction.[24,25]

The spectrin peptides isolated by this procedure are shown in Fig. 3. The total amount of each peptide recovered (from 8 mg of spectrin), ranges from 1 nm to several nanomoles. Each peptide can be compared from donor to donor using a variety of peptide mapping procedures.[26,27]

Acknowledgments

This work was supported by grants to Dr. Vincent T. Marchesi from the U.S. Public Health Service, No. GM21714 and by a National Service Award to William J. Knowles from the National Institutes of Health. We thank Drs. Marchesi, Atkinson, Speicher, Madri, and Morrow for critically reviewing this manuscript.

[23] G. Tuszynski and L. Warren, *Anal. Biochem.* **67,** 55 (1975).
[24] A. Weiner, T. Platt, and K. Weber, *J. Biol. Chem.* **247,** 3242 (1972).
[25] L. Henderson, S. Oroszlan, and W. Koningsberg, *Anal. Biochem.* **93,** 153 (1979).
[26] S. Stein, *in* "The Peptides" (E. Gross and J. Meienhofer, eds.), Vol. 4, p. 185. Academic Press, New York, 1981.
[27] J. Fishbein, A. Place, I. Ropson, D. Powers, and W. Sofer, *Anal. Biochem.* **108,** 193 (1980).

[25] Proteins Involved in Membrane–Cytoskeleton Association in Human Erythrocytes: Spectrin, Ankyrin, and Band 3

By VANN BENNETT

The major structural proteins of the human erythrocyte membrane have been purified, and details of the association of these proteins with each other have been established.[1,2] The erythrocyte contains a membrane-associated assembly of proteins referred to as the cytoskeleton or membrane skeleton. The cytoskeleton is composed principally of spectrin, which is complexed with erythrocyte actin and band 4.1 at the same region of the spectrin molecule. Spectrin is associated with the cytoplasmic surface of the membrane by attachment to ankyrin. Ankyrin, which is a peripheral membrane protein, is linked to the membrane by association with the cytoplasmic domain of band 3, a major membrane-spanning integral protein. The purpose of this chapter is to focus on the proteins in-

[1] D. Branton, C. Cohen, and J. Tyler, *Cell* **24,** 24 (1981).
[2] V. Bennett, *J. Cell Biochem.* **18,** 49 (1982).

ISBN 0-12-181996-5

volved in association of spectrin with the membrane. Methods are described for preparation of erythrocyte ghosts and inside-out vesicles, and for purification of spectrin, ankyrin, a spectrin-binding fragment of ankyrin, and a proteolytic fragment containing the cytoplasmic domain of band 3. Conditions are also described for measurement of binding of spectrin and ankyrin to inside-out vesicles.

Preparation of Erythrocyte Ghosts and Inside-out Vesicles

Protease activity is potentially a serious problem in purification of membrane proteins from erythrocyte ghosts. The source of the protease activity is both from cells of the buffy coat (neutrophils, platelets, lymphocytes, monocytes) and from protease endogenous to erythrocytes. The following procedure is designed to minimize protease activity by complete removal of buffy coat prior to lysis of erythrocytes, and by use of protease inhibitors during the lysis steps. In addition to these precautions, it also is important to work rapidly and to maintain carefully the temperature of buffers below 4°.

Blood. Freshly drawn blood anticoagulated with acid citrate–dextrose (see below) is used whenever possible; granulocytes aggregate during storage of blood for more than 24 hr, and these aggregates are difficult to remove. The amount of buffy coat varies with individuals and their white cell count. In general, blood should not be used from a donor with a cold, a fever, or a known infection. Patients with hemolytic anemias have elevated levels of reticulocytes that are enriched in protease activity. Samples of dextran-sedimented erythrocytes (see below) from these individuals can be pretreated with phenylmethylsulfonyl fluoride (100 μg/ml; 3–12 hr at 4°) or diisopropylfluorophosphate (0.05% v/v 2 hr at 4°) prior to lysis.

Solutions

Acid–citrate–dextrose (ACD): trisodium citrate, dihydrate (2.2 g), citric acid, monohydrate (0.8 g), and dextrose (2.5 g) dissolved in 100 ml of H_2O. Blood is anticoagulated with 1 volume of ACD per 7 volumes of blood.

Sedimenting buffer: 150 mM NaCl, 5 mM sodium phosphate, pH 7.5, 0.75% (w/v) Dextran 500 (Pharmacia)

Saline: 150 mM NaCl

Lysing buffer: 7.5 mM sodium phosphate, 1 mM sodium EDTA, pH 7.5

Spectrin extraction buffer: 0.2 mM sodium EDTA, pH 7.5

KI extraction buffer: 1 M KI, 7.5 mM sodium phosphate, 1 mM sodium EDTA, 1 mM dithiothreitol, pH 7.5

Stock solutions of phenylmethylsulfonyl fluoride and pepstatin A. These are dissolved in dimethyl sulfoxide at 200 mg/ml and 20 mg/ml, respectively.

Procedure. Erythrocytes free of other cell types are isolated from whole blood by mixing, at 4°, 1 volume of blood with 4 volumes of sedimenting buffer in a graduated cylinder.[3] The erythrocytes settle in 1–2 hr, and the supernatant is removed. The process is repeated with another 4 volumes of sedimenting buffer, and the settled erythrocytes are washed three times with 10 volumes of saline by centrifugation for 5 min at 2000 *g*. Any remaining buffy coat is carefully removed.

The washed erythrocytes are lysed by rapid addition of a minimum of 10 volumes of ice-cold lysing buffer with fresh additions of phenylmethylsulfonyl fluoride (20 μg/ml final) and pepstatin A (2 μg/ml final). Pepstatin A is included since erythrocyte membranes have been reported to contain an acid protease that is inhibited by this inhibitor.[4] Phenylmethylsulfonyl fluoride should be added no more than 5 min before lysis, since this reagent is relatively unstable in water. The lysed cells are immediately centrifuged at 2° for either 10 min at 18,000 rpm (37,000 *g*) in a JA-20 (Beckman) or SS34 (Sorvall) rotor, or for 20 min at maximum speed in a JA-14 (Beckman) or GSA (Sorvall) rotor. Volumes of cells up to 24 ml are processed in the JA-20 or SS34 rotors, where approximately 40-ml capacity tubes are used with a maximum of 3 ml of cells per tube. Larger volumes of cells are isolated in the JA-14 or GSA rotors, which accommodate 250-ml tubes and a maximum of 24 ml of cells per tube.

A viscous white button of cellular debris may be present underneath the pelleted ghosts if the buffy coat was not removed prior to lysing the cells. This button contains protease activity and should be removed carefully. The pelleted ghosts are washed once more with lysing buffer (and freshly added protease inhibitors), and then are washed with lysis buffer alone until white. Lysis buffer lacking EDTA can be substituted at this point. Ghosts are stored at 2–4° in lysis buffer with additions of 1 m*M* sodium azide and 0.2 m*M* dithiothreitol.

Spectrin-Depleted Inside-out Vesicles[5]

Ghosts are washed once with ice-cold spectrin extraction buffer, and then incubated for 30 min at 37° in at least 10 volumes of extraction buffer plus 20 μg of phenylmethylsulfonyl fluoride per milliliter and 0.2 m*M* dithiothreitol. This treatment causes spectrin to dissociate and the ghosts

[3] V. Bennett and P. Stenbuck, *J. Biol. Chem.* **255,** 2540 (1980).
[4] T. Murokami, Y. Suzuki, and T. Murachi, *Eur. J. Biochem.* **96,** 221 (1979).
[5] V. Bennett and D. Branton, *J. Biol. Chem.* **252,** 2753 (1977).

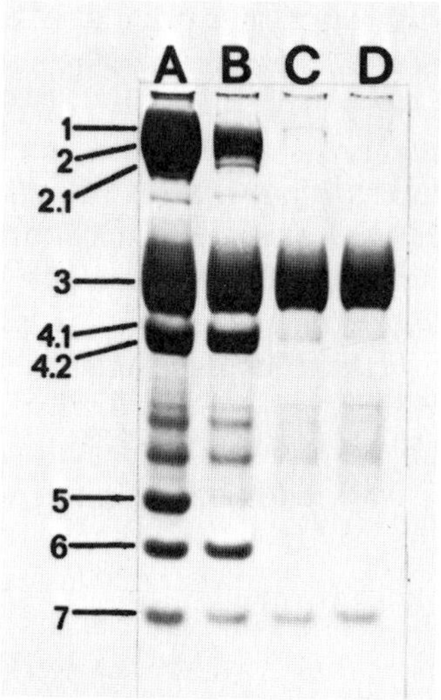

FIG. 1. Sodium dodecyl sulfate–polyacrylamide gel electrophoresis of erythrocyte ghosts (A), spectrin-depleted vesicles (B), KI-extracted vesicles (C), and 3 *M* guanidine-extracted vesicles (D). Reprinted, with permission, from Bennett and Stenbuck.[6]

to endovesiculate to form vesicles that are 80–90% inside out.[5] The vesicles are pelleted by centrifugation for 25 min at 19,000 rpm in a JA-20 or SS34 rotor or for 90 min at maximum speed in a JA-14 or GSA rotor, and washed once with extraction buffer and finally with lysis buffer. This treatment normally extracts over 95% of the spectrin and actin, and the procedure should be monitored by SDS–PAGE (Fig. 1). Vesicles are stored in the same buffer as ghosts.

KI-Extracted Inside-Out Vesicles.[6] Spectrin-depleted vesicles prior to the final wash with lysis buffer are incubated for 30 min at 37° in a minimum of 10 volumes of KI extraction buffer with 20 μg of phenylmethylsulfonyl fluoride per milliliter. The vesicles are collected by centrifugation for 40 min at 19,000 rpm in JA-20 or SS34 rotors or for 120 min at full speed in JA-14 or GSA rotors, and are washed once in the extraction buffer and finally in lysis buffer. The resulting membranes are depleted of 90% or more of ankyrin (band 2.1), degradation products of ankyrin (bands 2.2, 2.3), and bands 4.1, 4.2, and 6 (Fig. 1). The proteins that remain are operationally integral membrane proteins, since further extraction of these vesicles with 3 *M* guanidine does not elute any polypeptides (Fig. 1).

Purification of Spectrin

Spectrin (present at about 1 mg/ml of erythrocyte ghosts) is extracted by incubation of ghosts at low ionic strength and can then be purified by

[6] V. Bennett and P. Stenbuck, *J. Biol. Chem.* **255,** 6424 (1980).

gel filtration[7] or by sedimentation on preparative sucrose gradients. The sucrose gradient procedure[5,8] is preferable for small amounts of spectrin (3–4 mg), whereas gel filtration is practical for larger quantities.

Small-Scale Purification. Erythrocyte ghosts from 5 ml of packed cells are washed once with spectrin extraction buffer at 2°. The membranes are resuspended in a final volume of 7.5 ml with extraction buffer containing 20 μg of phenylmethylsulfonyl fluoride per milliliter and 0.2 mM dithiothreitol. The suspension is incubated for 30 min at 37° and then centrifuged for 15 min at 200,000 g. The resulting supernatant (A_{280} = 0.4–0.6) contains 60–80% of the erythrocyte spectrin, almost all the erythrocyte actin as well as some band 4.1 and numerous minor polypeptides. Care should be taken in aspiration of the supernatant so as not to disturb a loose layer of vesicles overlaying the membrane pellet. KCl is added to the supernatant to a final concentration of 20 mM, and 0.8–1 ml portions are layered onto 12.5-ml linear sucrose gradients (5–20%, w/v, in 20 mM KCl, 5 mM sodium phosphate, pH 7.5, 1 mM NaN_3, 0.2 mM dithiothreitol) in SW40 tubes. After centrifugation at 2° for 18 hr at 40,000 rpm in an SW40 rotor, fractions (0.8 ml) are collected. The major peak of protein has an A_{280} of 0.14–0.25 and contains only spectrin heterodimer. The yield is about 0.4 mg/ml of ghosts, or 40%. Spectrin is stable for several weeks in sucrose. Spectrin can also be quick-frozen and stored at −80° for a period of months. Frozen spectrin should be centrifuged (60 min at 200,000 g) after thawing to remove aggregated protein.

The same procedure can be used to isolate spectrin tetramers, with the modification that the low-ionic-strength extract is warmed for 3 hr at 30° after addition of KCl to a final concentration of 20 mM. The tetramer form is stable at 2°,[9] and can be isolated separately from dimer on sucrose gradients.

Large-Scale Purification. Erythrocyte ghosts from 200 ml of packed cells (20 ml of cells per 250-ml centrifuge tube) are washed once with 2.5 liters of spectrin extraction buffer at 2°. The membranes are resuspended to a final volume of 350 ml with extraction buffer containing 20 μg of phenylmethylsulfonyl fluoride per milliliter and 0.2 mM dithiothreitol and are incubated 30 min at 37°. The suspension is centrifuged 60 min at 40,000 rpm in a 45 Ti rotor, or, if this rotor is not available, for 180 min at full speed in a JA-14 or GSA rotor. Solid ammonium sulfate (31.3 g/100 ml of solution) is added to the supernatant on ice to give a final solution 50% saturated in ammonium sulfate. The precipitated protein is collected after 1–2 hr by centrifugation, dissolved in 30 ml of 0.1 M NaCl, 10 mM

[7] V. Marchesi, this series, Vol 32, p. 275.

[8] V. Bennett, *Life Sci.* **21,** 433 (1977).

[9] E. Ungewickell and W. Gratzer, *Eur. J. Biochem.* **88,** 379 (1978).

$NaPO_4$, 1 m*M* sodium EDTA, 1 m*M* NaN_3, 0.2 m*M* dithiothreitol, pH 7.5, and dialyzed overnight against this buffer. The dialyzed solution is centrifuged for 90 min at 200,000 *g* to remove aggregated protein and is applied to a Sephacryl S-500 column (5 cm × 90 cm) equilibrated with resuspension buffer. Spectrin heterodimer elutes in a major peak at about 2 V_0; the fractions are monitored for purity by SDS–polyacrylamide electrophoresis. The spectrin is dialyzed against 10% (w/v) sucrose, 10 m*M* sodium phosphate, 1 m*M* sodium EDTA, 1 m*M* sodium azide, 0.2 m*M* dithiothreitol. Spectrin prepared by this procedure may still contain minor associated proteins (e.g., protein kinase). These proteins can be minimized by running the gel filtration step in 1 *M* NaCl instead of 0.1 *M* NaCl.

Purification of Ankyrin

Ankyrin (band 2.1) comprises 5–6% or erythrocyte membrane protein and can be purified in milligram quantities from 200 ml of erythrocytes. Ankyrin is selectively extracted by first extracting ghosts with Triton X-100 and then incubating the Triton X-100-extracted ghosts in 1 *M* KCl in the absence of detergent. Ankyrin is preferentially solubilized under these conditions and can then be purified by DEAE-chromatography and sedimentation on preparative sucrose gradients.[3] Ankyrin is particularly susceptible to proteolysis; to minimize protease activity, washed erythrocytes are incubated overnight at 4° with 5 volumes of 150 m*M* NaCl, 5 m*M* sodium phosphate, 1 m*M* sodium EDTA, and 100 μg of phenylmethylsulfonyl fluoride per milliliter.

Erythrocyte ghosts from 240 ml of packed cells are incubated for 15 min on ice in 10 volumes of a buffer containing 100 m*M* KCl, 7.5 m*M* sodium phosphate, 1 m*M* sodium EDTA, 0.4 m*M* dithiothreitol, 20 μg of phenylmethylsulfonyl fluoride per milliliter, 0.5% (v/v) Triton X-100, pH 7.5, and centrifuged for 20 min at full speed in a JA-14 or GSA rotor. The supernatant is discarded and the pellet is washed once with buffer, and again with the same buffer except that detergent is omitted. If Triton X-100 is present during subsequent steps, ankyrin will be coextracted with band 3 as a stable ankyrin–band 3 complex.[10] Ankyrin is solubilized by resuspending the final pellet to the original ghost volume at 2° with 1 *M* KCl, 7.5 m*M* sodium phosphate, 1 m*M* sodium EDTA, 0.2 m*M* dithiothreitol, 20 μg of phenylmethylsulfonyl fluoride per milliliter, 1 μg of pepstatin A per milliliter, pH 7.5. After a 30-min incubation on ice, the suspension is centrifuged for 30 min at 19,000 rpm in a JA-20 or SS34 rotor. The supernatant, which contains about 40% of the ankyrin, is dialyzed overnight at 4° against 4 liters of 7.5 m*M* sodium phosphate, 0.1 m*M*

[10] V. Bennett, *Biochim. Biophys. Acta* **689,** 475 (1982).

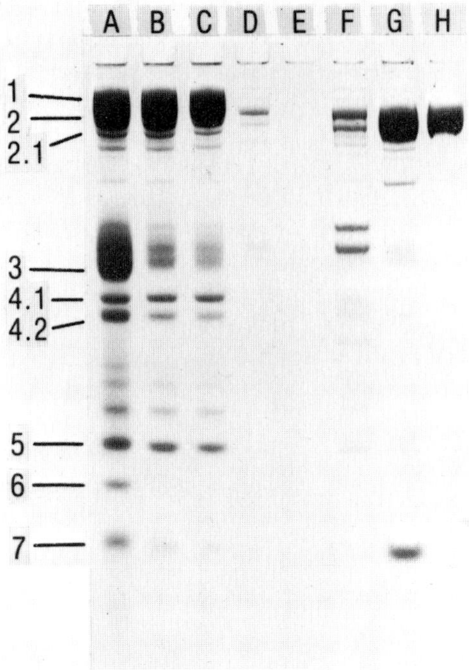

FIG. 2. Purification of human erythrocyte ankyrin. Samples from various stages of purification were analyzed by SDS–PAGE on a 7% polyacrylamide slab gel. Lanes: A, Erythrocyte ghosts; B, Triton X-100-extracted ghosts; C, 1 *M* KCl-extracted pellet from B; D, 1 *M* KCl-extracted crude ankyrin from B; E, breakthrough from DE-52 column; F, peak 1 from DE-52; G, ankyrin peak from DE-52; H, ankyrin peak from preparative sucrose gradient. Reprinted, with permission, from Bennett and Stenbuck.[6]

sodium EDTA, 1 m*M* sodium azide, 0.2 m*M* dithiothreitol (dialysis buffer), and 20 μg of phenylmethylsulfonyl fluoride per milliliter, pH 7.5. The dialysis bag is transferred in the morning to 1 liter of fresh buffer containing in addition 10% (w/v) sucrose and is dialyzed for another 2 hr. The purpose of this step is to reduce the volume of the extract. The dialyzed material is applied to a column (0.75 × 12 cm) of DE-52 cellulose equilibrated with dialysis buffer. The column is washed with 50 ml of dialysis buffer and then eluted with 180 m*M* KCl followed by 300 m*M* KCl, both dissolved in dialysis buffer with 2% (w/v sucrose). Ankyrin elutes with 300 m*M* KCl and is nearly pure except for a contaminating polypeptide of M_r = 29,000 (band 7) (Fig. 2). The 300 m*M* peak is pooled and concentrated to about 10 ml by dialysis against a slurry of polyethylene glycol 6000. The concentrated material is dialyzed further for 1 hr against dialysis buffer with 5% (w/v) sucrose, and then 3-ml portions are layered over 31-ml linear gradients of 10–30% (w/v) sucrose dissolved in dialysis buffer. The gradients are centrifuged for 6 hr at 48,000 rpm in a Sorvall TV-850 vertical rotor, and 2-ml samples are collected from the bottom of the tube. An SW40 rotor can also be used if a vertical rotor is not available, with a run of 22 hr at 40,000 rpm. Ankyrin elutes in a symmetrical peak separated from the M_r = 29,000 contaminant. Purity of the fractions is monitored by SDS–PAGE prior to pooling. A single band at M_r = 215,000 comprises about 96% of the Coomassie Blue staining on gels. A polypeptide of M_r = 155,000 also is present at 1–2% of the total protein. Occasionally, degradation products of ankyrin of M_r = 200,000 and 170,000 also are present or appear during storage. About 3 mg of ankyrin is obtained per unit of blood, which represents a recovery of 8% of the protein present as band 2.1 in ghost membranes. Losses of 60% occur during solubilization and dialysis, and of about 50% after DEAE chromatography and after the sucrose gradient.

Ankyrin is stored in the same solution as eluted from the sucrose gradient. If more concentrated solutions are required, sucrose is removed by dialysis and the protein is concentrated against polyethylene glycol. The concentrated protein is then dialyzed against 20% sucrose in dialysis buffer. Purified ankyrin stored in sucrose and dithiothreitol remains unaggregated based on sedimentation on sucrose gradients and retains ability to reassociate with spectrin and band 3 for at least 4 weeks. Manipulations with ankyrin are limited by the tendency of this protein to adsorb to surfaces. Protein losses and nonspecific interactions are minimized in the presence of 10–20% sucrose or glycerol, 0.1–0.2% nonionic detergent, or 0.2–0.5 mg of bovine serum albumin per milliliter.

Purification of a Proteolytic Fragment of Ankyrin Containing the Spectrin-Binding Site[11]

A 72,000 M_r fragment of ankyrin that contains the binding site for spectrin is released during digestion of inside-out vesicles with α-chymotrypsin. This fragment offers advantages over the intact molecule in certain studies, since the fragment does not adsorb to surfaces to the extent that ankyrin does and has lower levels of nonspecific binding.

Spectrin-depleted inside-out vesicles from 200 ml of erythrocytes are washed an additional time with 7.5 m*M* sodium phosphate buffer, pH 7.5, and resuspended at 2° to 200 ml with the phosphate buffer and 0.2 m*M* dithiothreitol. The vesicles are digested with 1 μg of α-chymotrypsin per milliliter on ice, a treatment that solubilizes about 15% of membrane protein and about 80–90% of the spectrin binding sites. After 45 min, phenylmethylsulfonyl fluoride is added at a final concentration of 200 μg/ml. The polypeptides solubilized by the digestion are collected by adding directly to the suspension 20 ml of DE-52 cellulose equilibrated with 7.5 m*M* sodium phosphate, pH 7.5. The slurry is mixed gently for 15 min, and the cellulose is collected by centrifugation (1 min at 500 *g*). The cellulose is washed 3 times with 100 ml of phosphate buffer with 100 μg of phenylmethylsulfonyl fluoride per milliliter and 0.2 m*M* dithiothreitol to

[11] V. Bennett, *J. Biol. Chem.* **253,** 2292 (1978).

remove trapped vesicles. The batchwise adsorption of protein allows rapid concentration of material from large volumes as well as removal of α-chymotrypsin, a basic protein that is not adsorbed to DEAE-cellulose at pH 7.5. The cellulose is poured into a column and washed with 100 ml of phosphate buffer, 0.2 m*M* dithiothreitol, and 50 μg of phenylmethylsulfonyl fluoride per milliliter, followed by 50 ml of 75 m*M* KCl in phosphate buffer, 0.2 m*M* dithiothreitol. The 72,000 M_r fragment is eluted with 200 m*M* KCl in the same buffer. The fractions are monitored by SDS–PAGE and pooled. The fragment at this stage is about 70% pure, with numerous contaminating polypeptides of lower M_r. Occasionally, polypeptides of M_r ~110,000 also are present, which can be removed by a second digestion with 1 μg of α-chymotrypsin per milliliter followed by DEAE chromatography. The fragment can be concentrated by precipitation with solid ammonium sulfate at 60% saturation. Further purification can be achieved by gel filtration on a AcA-44 Ultrogel column, where the fragment elutes at about 1.5 V_0. The fragment is stored frozen at $-20°$ at 1–1.5 mg of protein per milliliter in 7.5 m*M* sodium phosphate, 1 m*M* sodium EDTA, 1 m*M* sodium azide, 0.2 m*M* dithiothreitol. The yield of fragment is about 8–12 mg from a unit of blood after one step of DEAE chromatography and about 4–7 mg after gel filtration.

Purification of the Cytoplasmic Domain of Band 3

Steck and co-workers have reported that mild digestion of erythrocyte membranes with trypsin or papain releases an M_r = 41,000 water-soluble polypeptide that is derived from the portion of band 3 expressed selectively on the cytoplasmic surface of the membrane.[12] This fragment has been purified in the presence of SDS,[13] and a related fragment of M_r = 36,000 has been purified from SDS-purified band 3.[14] This section describes a simple procedure for isolation under nondenaturing conditions of large amounts of the band 3 fragment that retains ability to bind to ankyrin.[6]

Spectrin-depleted vesicles from 200 ml of erythrocytes are stripped of peripheral membrane proteins by extraction in 1.2 liters of 1% (v/v) acetic acid for 20 min at 24°. The acid-stripped vesicles (400 mg of membrane protein) are pelleted, washed once with 0.1 *M* sodium phosphate, pH 7.5, and resuspended by mild sonication in a 200-ml volume of 7.5 m*M* sodium phosphate, 0.2 m*M* dithiothreitol, pH 7.5. The vesicles are then digested for 30 min on ice with 1 μg of α-chymotrypsin per milliliter. The vesicles and released polypeptides are then processed just as described for isola-

[12] T. Steck, B. Ramos, and E. Strapazon, *Biochemistry* **15,** 1154 (1976).
[13] T. Steck, J. Koziarz, J. Singh, G. Reddy, and H. Kohler, *Biochemistry* **17,** 1216 (1978).
[14] M. Fukuda, Y. Eshdat, G. Tarone, and V. Marchesi, *J. Biol. Chem.* **253,** 2419 (1978).

tion of the ankyrin fragment (see above). The DE-52 cellulose column is eluted at 20 ml/hr with a linear gradient of 100 ml of 7.5 mM sodium phosphate, 1 mM sodium azide, 0.2 mM dithiothreitol, pH 7.5, versus 100 ml of 500 mM KCl dissolved in the same buffer. The peak fractions (32 mg of protein) eluting between 0.25 and 0.32 M KCl are pooled, and the protein is precipitated at 0° by addition of solid ammonium sulfate at 60% saturation. The pelleted protein is resuspended in 10 ml of 25 mM sodium phosphate, 1 mM sodium azide, 1 mM sodium EDTA, 0.2 mM dithiothreitol, pH 7.5, and applied to an Ultrogel AcA-44 column (2 × 100 cm) equilibrated with resuspension buffer. Peak fractions elute at 1.2 V_0 and contain about 20 mg protein. The yield is about 20% based on the amount of band 3. The fragment can be stored frozen at −20°. SDS–PAGE of the fragment (Fig. 3) indicates that about 90% of the protein is a polypeptide of M_r = 43,000, the remainder being composed of a polypeptide of M_r = 41,000. The amount of M_r = 41,000 polypeptide varies among preparations and is increased with more extensive digestion. The M_r = 41,000 polypeptide cross-reacts with antibody raised against the M_r = 43,000 band cut from SDS gels and thus is most likely a subfragment of the M_r = 43,000 polypeptide.

Radiolabeling of Proteins for Binding Studies

A variety of procedures are available for covalent incorporation of radioactive labels into proteins including iodination with ^{125}I, reductive formylation, and metabolic labeling. ^{125}I radiolabeling with Bolton–Hunter reagent[15] is an excellent method for labeling proteins that allows high specific activity, occurs under mild conditions, and does not alter the biological activity of the labeled protein. This method has been used to label spectrin, ankyrin, and the 43,000 M_r fragment of band 3.[3,6] A procedure is described here that is applicable to all these proteins.

^{125}I-labeled Bolton–Hunter reagent, *N*-succinimidyl-3-(4-hydroxy-[5-^{125}I]iodophenyl) propionate (1 mCi, 1500–2000 Ci/mmol purchased from New England Nuclear or Amersham), is dried down from benzene in the delivery vial with a stream of nitrogen at 24°. The gas leaving the vial contains ^{125}I and should be filtered across Norite. The vial is then transferred to ice and allowed to cool while nitrogen is still flowing. The cap is removed, and 0.1 ml of reaction mixture is added containing 50–80 μg of protein, 40 mM sodium phosphate, pH 8.5, Sucrose, EDTA, and azide may also be present, but dithiothreitol should not, since sulfhydral groups are potentially reactive with the ^{125}I-labeled reagent. Dithiothreitol is rou-

[15] A. Bolton and W. Hunter, *Biochem. J.* **133,** 529 (1973).

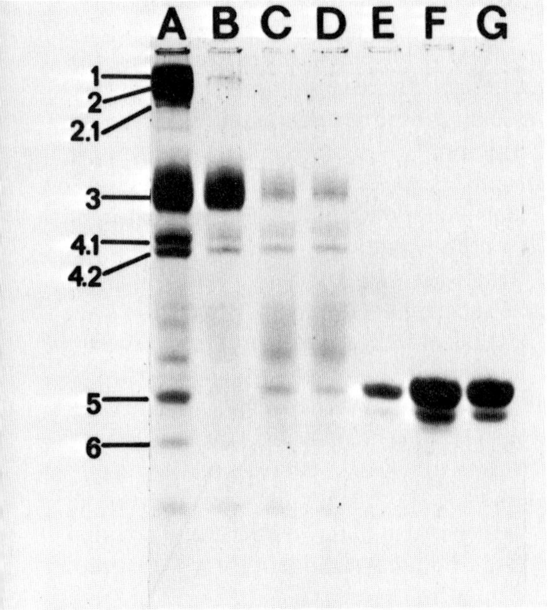

FIG. 3. Purification of an M_r = 43,000 water-soluble fragment of band 3. Samples from various stages of the purification were analyzed by SDS-electrophoresis on an 8% polyacrylamide slab gel. Lanes: A, starting ghosts; B, acetic acid-stripped vesicles; C, vesicles after digestion, which were washed and resuspended in buffer; D, a sample of the total digest after the DE-52 cellulose had settled; E, the protein released during digestion of the vesicles; F, the pooled fraction from DE-52 chromatography; G, the pooled fraction from the AcA-44 column. Reprinted, with permission, from Bennett and Stenbuck.[6]

tinely included in the preparation of proteins, but its reactivity is minimized to an acceptable level by the alkaline pH of the reaction and by gently blowing oxygen over the protein sample for 1–2 min in order to oxidize the dithiothreitol.

The incubation of protein and reagent is continued for 90 min on ice, and the reaction mixture is then diluted with 0.2 ml of 7.5 m*M* phosphate buffer, 0.2 m*M* sodium EDTA, 1 m*M* sodium azide (and 20% sucrose in the case of spectrin and ankyrin) and dialyzed overnight against this buffer. The specific activity and efficiency of the reaction are determined by measuring the amount of radioactivity that is precipitable by trichloroacetic acid. A 5-μl aliquot of the diluted reaction mixture is diluted to 5 ml with 50 m*M* glycine (to quench the reaction) and 1 mg of bovine serum albumin per milliliter; the radioactivity in a 25-μl portion is determined. One milliliter of 50% (w/v) trichloroacetic acid is added, and the tube is incubated on ice for 15 min. The protein is pelleted (10 min at 900 *g*), and a 30-μl aliquot of the supernatant is counted. The percentage incorporation of radioactivity into protein (i.e., TCA-precipitable material) is 1-(cpm/30 μl after TCA/cpm/25 μl before TCA). The specific activity (cpm/μg) is equal to (percentage of incorporation) × (total cpm in reaction)/microgram of protein in the reaction. Radiolabeling usually proceeds with 30-50% incorporation of ^{125}I-labeled reagent. Specific activity of labeled proteins is adjusted by mixing with unlabeled protein.

Binding of ^{125}I-Labeled Ankyrin and Spectrin to Inverted Vesicles

^{125}I-labeled spectrin reassociates with inside-out vesicles depleted of spectrin, and the binding interaction can be measured in a simple assay.[5] Similarly, ^{125}I-labeled ankyrin also will bind to inside-out vesicles depleted of ankyrin (KI-extracted vesicles).[6] ^{125}I-labeled spectrin or ankyrin (5–100 μg/ml; 100,000–300,000 cpm/μg) is incubated in plastic tubes (12 × 75 mm) in a 0.22-ml volume containing 20–50 μg of membrane protein, 0.1 *M* KCl, 7.5 m*M* sodium phosphate, 0.2 m*M* sodium EDTA, 0.2 m*M* dithiothreitol, 10% (w/v) sucrose, pH 7.5. The mixture is incubated for 90 min at 0° in the case of spectrin or 90 min at 24° with ankyrin. Free and membrane-bound ^{125}I-labeled protein are separated by layering 0.2 ml of the samples over 20% sucrose dissolved in assay buffer in polyethylene (Eppendorf, hard) microfuge tubes (0.4 ml capacity) followed by centrifugation for 30 min at 18,000 rpm in a SS34 or JA-20 rotor. Adapters are available for Microfuge tubes that accommodate 4 tubes each (Sorvall, Catalog No. 408) and allow 32 tubes to be processed per run. The microfuge tubes are frozen in crushed dry ice; the tips containing the membrane pellets are cut off, and their radioactivity is measured. The top of the tube can also be counted to determine directly the amount of free ligand. It is important to include controls for nonspecific binding in these assays. For example, binding of heat-denatured (10 min, 70°) ligand can be measured for each experimental condition, and these values can be subtracted. Such nonspecific binding is normally no more than 20% of the total.

These assays can be used to measure levels of activity of unlabeled ankyrin or spectrin by competitive displacement of binding of ^{125}I-labeled ligand as in a radioimmunoassay. Such measurements are useful for screening activities of spectrin and ankyrin in patients with abnormal erythrocytes.[16] The advantage of competitive displacement as opposed to direct binding is that the proteins do not have to be purified and that multiple samples can be tested. For example, spectrin in the crude low ionic strength extract of ghosts and ankyrin from the 1 *M* KCl extract of cytoskeletons can be analyzed. If a difference is noted, then further detailed studies can be performed with purified proteins.

Acknowledgments

This work has been supported in part by grants to V. B. from the National Institutes of Health (1 R01 AM29808-1, Research Career Development Award) and the Muscular Dystrophy Association.

[16] P. Agre, E. Orringer, D. Chui, and V. Bennett, *J. Clin. Invest.* **68,** 1566 (1981).

[26] Assembly of Histocompatibility Antigens

By BERNHARD DOBBERSTEIN and SUNE KVIST

Histocompatibility antigens are oligomeric proteins expressed on the plasma membrane. Two classes of these antigens can be distinguished: transplantation, or class I, antigens, and immune associated, or class II, antigens.[1] Class I antigens are composed of a glycosylated, membrane-spanning heavy chain that is noncovalently associated with β_2-microglobulin, a peripheral membrane protein.[1,2] They are called H-2, K, D, and L antigens in mouse and HLA-A, B, C antigens in man. Class II antigens consist of three glycosylated membrane-spanning proteins that are noncovalently associated.[3] In mouse they are called Ia antigens; in man, HLA-DR antigens.[1] Their polymorphic subunits are the α (35 kilodaltons) and β (29 kilodaltons) chains, and the invariant one is the I or γ (33 kilodaltons) chain.[4]

Both types of histocompatibility antigens are synthesized on membrane-bound ribosomes and cotranslationally inserted into the membrane of the endoplasmic reticulum.[5–10] Here they are proteolytically processed and glycosylated, and their subunits, which are synthesized on separate mRNAs, are assembled into an oligomeric complex. During intracellular transport their carbohydrate portions are modified.[5,9,10] In the case of the HLA-DR antigens, DRγ chains detach from the oligomeric complex before it reaches the cell surface.[10]

Biosynthesis and cell surface expression of these antigens can be studied *in vitro* and *in vivo*.[5–10] *In vitro* the events occurring in the endoplasmic reticulum can be reconstructed, i.e., cotranslational insertion, and even oligomeric assembly.[10] Selected methods to study biosynthesis and membrane insertion of histocompatibility antigens are described here.

[1] J. Klein, *Science* **203,** 516 (1979).

[2] H. L. Ploegh, H. T. Orr, and J. L. Strominger, *Cell* **24,** 287 (1981).

[3] R. J. Winchester and H. G. Kunkel, *Adv. Immunol.* **28,** 221 (1979).

[4] P. P. Jones, D. B. Murphy, D. Hewgwill, and H. O. McDevitt, *Immunochemistry* **16,** 51 (1978).

[5] B. Dobberstein, H. Garoff, G. Warren, and P. J. Robinson, *Cell* **17,** 759 (1979).

[6] J. S. Lee, J. Trowsdale, and W. F. Bodmer, *J. Exp. Med.* **152,** 3s (1980).

[7] M. S. Krangel, H. T. Orr, and J. L. Strominger, *Cell* **18,** (1979).

[8] M. J. Owen, A. M. Kissonerghis, and H. F. Lodish, *J. Biol. Chem.* **255,** 9678 (1980).

[9] M. J. Owen, A. M. Kissonerghis, H. F. Lodish, and M. J. Crumpton, *J. Biol. Chem.* **256,** 8987 (1981).

[10] S. Kvist, K. Wiman, L. Claesson, P. A. Peterson, and B. Dobberstein, *Cell* **29,** 61 (1982).

ISBN 0-12-181996-5

Choice of Cells to Study Biosynthesis of Histocompatibility Antigens

H-2 antigens are present in largely varying amounts in different tissues or cells.[11] Thus the proper choice of the cell type used for the study is important. Thymoma cells like the SL2 cells (H-2^d haplotype) express relatively large amounts of H-2 antigens on their cell surface. They are grown in ascites form in DBA/2 mice (about 0.5 to 1 $\times$ 10^9 cells per mouse). The human lymphoblastoid cell line Raji is a rich source of HLA-DR antigens. Both cell types contain low amounts of ribonuclease and can be efficiently biosynthetically labeled.[5]

Isolation of mRNA Coding for Histocompatibility Antigens

To study biosynthesis of histocompatibility antigens *in vitro,* a fraction enriched in mRNA coding for H-2 or HLA antigens must be isolated. This can be done by purifying messenger RNA (mRNA) from rough microsomes. The same procedure can be used for both cell types.

SL2 cells are grown in ascites form in DBA/2 mice (Bomholtgord, Denmark) and harvested from the peritoneal cavity 8–10 days after inoculation with 2 $\times$ 10^6 cells. The cells are washed 3 times in ice cold 50 m*M* Tris-HCl, pH 7.5, 100 m*M* KCl, 10 m*M* $MgCl_2$, 10 m*M* dithiothreitol (DTT), and 100 μg of cycloheximide per milliliter, swollen in 20 m*M* Tris-HCl, pH 7.5, for 10 min, and then broken using a Dounce homogenizer. Isotonic conditions are established by adding an equal volume of 20 m*M* Tris-HCl, pH 7.5, 300 m*M* KCl, 5 m*M* $MgCl_2$, 100 μg of cycloheximide per milliliter. Nuclei and mitochondria are removed by centrifugation at 8000 *g* for 10 min. From the resulting supernatant, crude rough microsomes are pelleted by centrifugation for 1 hr at 20,000 rpm in a Sorvall SS34 rotor. RNA is extracted from the pelleted microsomes by the phenol–chloroform–isoamyl alcohol method, and poly(A)$^+$ RNA is purified on oligo(dT)-cellulose.[12] mRNA coding for H-2 antigens is thus enriched about 10-fold.[5] Further purification can be conveniently achieved by centrifugation in an 8 to 20% aqueous sucrose gradient.[5] Such a separation is usually sufficient to obtain a further 5- to 10-fold enrichment.

Immunocharacterization of H-2 and HLA-DR Antigens

The major problem in the characterization of histocompatibility antigens synthesized *in vitro* or *in vivo* is that they represent only a very

[11] J. Klein, "Biology of the Mouse Histocompatibility Complex." Springer-Verlag, Berlin and New York, 1975.

[12] H. Aviv and P. Leder, *Proc. Natl. Acad. Sci. U.S.A.* **69,** 1408 (1972).

minor proportion of cellular proteins. Furthermore, antigenic determinants present on the mature protein might not be found on the one synthesized in a cell-free system. Like many membrane proteins, histocompatibility antigens are modified during intracellular transport. The signal sequence is removed, and they become glycosylated and assembled into an oligomeric complex. Some of these modifications are essential for antigenic determinants to become expressed and must be considered in the immunocharacterization.

Immunoprecipitation Protocol

After translation of SL2 mRNA in a cell-free system (25 μl) supplemented with dog pancreas microsomes, the following operations are performed at 4°. Ribosomes are dissociated by adding 75 μl of 10 m*M* EDTA, 150 m*M* KCl, 20 m*M* Tris-HCl, pH 7.5, and the membranes are pelleted by centrifugation at 30 psi for 10 min in a Beckman Airfuge. Antigens present in the pelleted membranes are then solubilized in 50 μl of Nonidet P-40 (NP-40) buffer [1% NP-40, 10 m*M* Tris-HCl, pH 7.5, 0.15 *M* NaCl, 2 m*M* EDTA, 10 μg of phenylmethylsulfonyl fluoride (PMSF) per milliliter] and transferred to a 1.5-ml Eppendorf tube. One microliter of a nonimmune serum and 50 μl of a 1 : 1 slurry of protein A–Sepharose (Pharmacia, Uppsala) in NP-40 buffer are added. Incubation is performed for 30 min under slow rotation of the tube on a Denley Mixer (England). Beads and nonsolubilized material are then removed by centrifugation for 10 min in an Eppendorf centrifuge. The supernatant is incubated for 30 min with the appropriate immune serum (5 μl in the case of an alloantiserum and 1 μl when a heteroserum is used), and then 50 μl of a 1 : 1 slurry of protein A–Sepharose NP-40 buffer are added. This is allowed to react for 2 hr as above. Beads are then washed 3 times with a 1-ml portion of NP-40 buffer, once with NP-40 buffer containing 0.5 *M* NaCl, and once with 20 m*M* Tris-HCl, pH 7.5. After all supernatant fluid has been carefully removed, 25 μl of sample buffer are added, and proteins are characterized by PAGE[13] and fluorography.[14]

H-2 antigens are synthesized in a cell-free system (derived from reticulocytes) as higher molecular weight precursors that are not recognized by most of the alloantisera used to identify H-2 antigens from different loci or haplotypes. When, however, H-2 antigens are inserted into microsomal membranes from dog pancreas, glycosylated and their signal sequence cleaved, the H-2D^d molecule can be specifically detected by an anti-H-2D^d alloantiserum.[5]

[13] U. K. Laemmli, *Nature* (*London*) **227,** 680 (1970).

[14] W. M. Bonner and R. A. Laskey, *Eur. J. Biochem.* **46,** 83 (1974).

Characterization of Oligomeric HLA-DR Antigens

In the case of HLA-DR antigens the DRγ chain is not recognized by an HLA-DR antiserum even after insertion of the chain into microsomal membranes. For its detection, assembly with DRα and DRβ chains is required.[10]

To achieve the oligomeric assembly of HLA-DR chains *in vitro*, it is necessary for the three different subunit chains to be inserted into the same microsomal vesicle. An alternative is fusing the microsomal vesicles together after translation is completed so that the subunit chains can laterally diffuse in the membrane and assemble. We tested membrane concentrations between 0.06 and 0.25 A_{280} unit of microsomal membranes in a 25-μl assay system. Assembly was observed only at the lowest microsome concentration. This concentration might be different for each antigen and mRNA studied, as it depends largely on the amount of specific mRNA coding for each subunit. Furthermore, in the case of HLA-DR antigens, a very favorable situation exists, as the DRγ chain is synthesized in large excess over DRα and DRβ chains.

Often it is desirable to obtain the separated subunits of an oligomeric protein complex in order to test the antigen specificity of an antiserum. In the *in vitro* system containing large amounts of membranes, exclusively native monomeric subunit proteins are synthesized. Thus it can be conveniently used in the characterization of antibodies raised against subunit chains. The advantage in this procedure is that oligomeric protein complexes need not be disassembled, and hence the subunits are not denatured. This could be particularly useful in the characterization of monoclonal antibodies to determine the subunit chain with which they react.

Assembly of HLA-DR Antigens *in Vivo*

H-2 and HLA-DR antigens are assembled from subunits synthesized in largely different amounts. β_2-Microglobulin is synthesized in excess over the H-2 heavy chain, and the DRγ chain in excess over the DRα and DRβ chains. In each case the unassembled subunits, β_2-microglobulin or the DRγ chain, remain in the endoplasmic reticulum and the oligomeric protein complex travels to the cell surface.

To follow oligomeric assembly of HLA-DR antigens during intracellular transport, two types of pulse–chase experiments can be performed: one in which unlabeled protein is allowed to be synthesized during the chase period, and one in which it is stopped, but after 100 min allowed to proceed again. The assembly process is followed by an antiserum that precipitates the DRγ chain only after it has assembled with DRα and DRβ chains.[10] Such an analysis will allow one to determine requirements for oligomeric assembly and intracellular transport of DR chains.

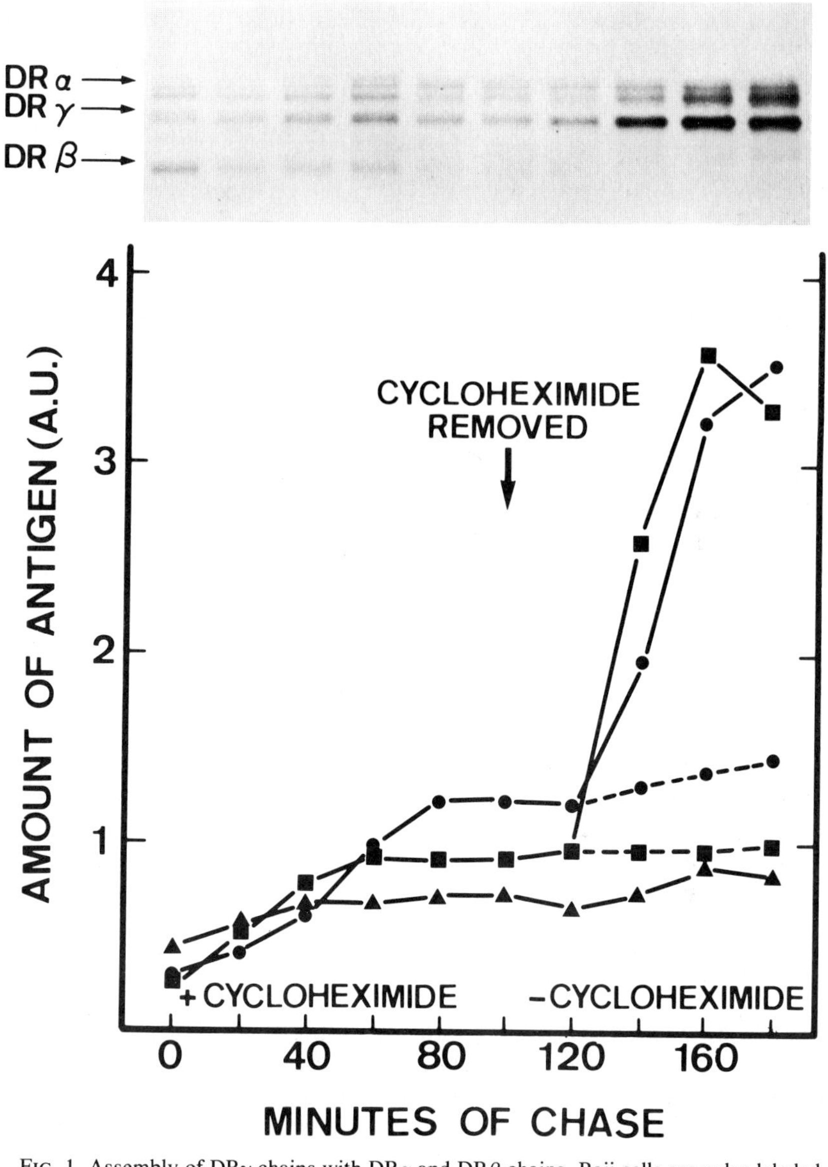

FIG. 1. Assembly of DRγ chains with DRα and DRβ chains. Raji cells are pulse-labeled for 10 min and then chased for 100 min in the presence of cycloheximide. Intracellular transport of membrane proteins proceeds under these conditions. After 100 min, cycloheximide is removed and unlabeled proteins are allowed to be synthesized again. At the time points indicated, [^{35}S]methionine-labeled antigens are immunoprecipitated with the anti HLA-DR antiserum, which precipitates the DRγ chains only after it has complexed with DRα and DRβ chains. Antigens are characterized by SDS–polyacrylamide gel electrophoresis and quantitated by densitometry. ▲——▲, DRβ chains; ■——■, DRγ chains; ●——●, DRα and processed DRγ chains. The dashed lines show the amounts of antigen precipitated when cycloheximide was not removed after 100 min of chase.

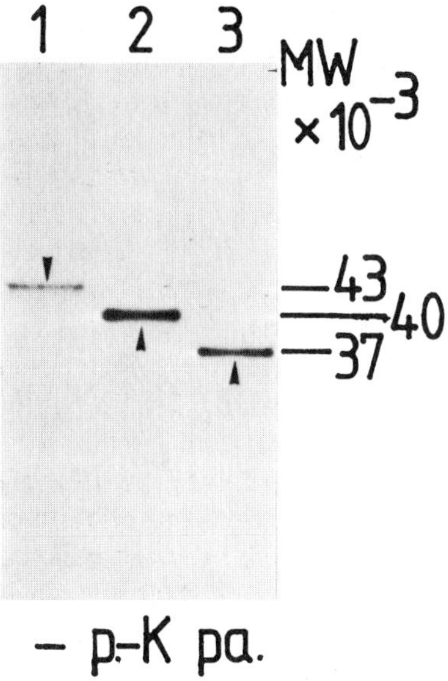

FIG. 2. Determination of the size of the membrane-spanning segment of an H-2D antigen. H-2D antigens, pulse labeled *in vivo*, are characterized by SDS–polyacrylamide gel electrophoresis and autoradiography after no proteolytic digestion (lane 1), after treatment of microsomes with proteinase K (lane 2), and after papain treatment of solubilized antigens (lane 3).

Procedure. The human lymphoblastoid cell line Raji is cultured in RPMI-1640 medium, 10% fetal calf serum. Cells are washed 3 times in methionine-free RPMI-1640 medium, and 1×10^7 cells/ml are incubated for 15 min at 37° to lower the endogenous methionine pool. [^{35}S]Methionine (250 μCi; 1000 Ci/mmol) is then added, and the cells are incubated for 10 min. Then 10 ml of RPMI-1640 complete culture medium and 10 m*M* methionine are added, and aliquots of 1 ml (1×10^6 cells) are removed after various times of incubation. To stop synthesis of unlabeled proteins after the 10-min pulse, cycloheximide is added to a final concentration of 100 μg/ml. To remove cycloheximide after the 90 min of chase, cells are washed twice at 0° with 10 ml of RPMI-1640 medium.

Cells removed at various time points are pelleted and solubilized in 100 μl of 50 m*M* Tris-HCl, pH 7.5, 0.15 *M* NaCl, 1% Triton X-100, and 40 μg of PMSF per milliliter. Particulate material is removed by centrifugation at 100,000 *g* for 15 min at 4° in an Airfuge. Antigens are characterized from the supernatant by immunoprecipitation and PAGE (Fig. 1). It can be concluded from this experiment that DRγ chains are synthesized in excess over DRα and DRβ chains.

Determination of the Size of the Membrane-Spanning Segment

Proteins that span the membrane often can be proteolytically cleaved close to the membrane. A papain cleavage site exists in the case of H-2,

K, D, L[2], and HLA-DR[3] antigens that removes the N-terminal portion in a soluble form. This cleavage can be performed either on the intact cell or on the detergent-solubilized proteins. In each case the same size fragment is liberated. The size of the cytoplasmically disposed portion of a membrane-spanning protein can be estimated by cleaving it with proteinase K. Comparing the sizes of the proteins generated after each of these cleavages will allow one to estimate the approximate length of the membrane-spanning portion. The size thus estimated was in good agreement with the stretch of uncharged residues found by sequence analysis (Fig. 2).

Procedure. SL2 cells (5 × 10^6) are labeled in 1 ml of RPMI medium for 10 min with 200 μCi of [^{35}S]methionine, washed, and broken as described above. Nuclei are removed by centrifugation at 4000 *g* for 10 min, and the resulting 1-ml supernatant is incubated at 4° for 90 min with 200 μg of proteinase K per milliliter. The reaction is terminated by the addition of PMSF (40 μg/ml) and serum albumin (10 μg/ml). Membranes are pelleted by centrifugation at 100,000 *g* for 20 min. After solubilization of the pelleted microsomes, antigens are immunoprecipitated as described above.

For papain cleavage of H-2 antigens, microsomes from SL2 cells (5 × 10^6) labeled for 10 min with [^{35}S]methionine are solubilized in 200 μl of 1% NP-40, 100 m*M* NaCl, and 0.25 mg of cysteine per milliliter and treated for 30 min at 37° with 0.5 mg of papain per milliliter. Proteolysis is terminated by the addition of neutralized iodoacetic acid to a final concentration of 0.5 mg/ml. Antigens are immunoprecipitated and characterized by PAGE and fluorography (Fig. 2).

[27] Biosynthesis and Intracellular Transport of Acetylcholine Receptors

By DOUGLAS M. FAMBROUGH

The acetylcholine receptors of vertebrate skeletal muscles and the electric organs of certain fish are multisubunit, integral membrane proteins that become cation-selective ion channels in response to the binding of acetylcholine. In normal adult muscles and electric organs, the acetylcholine receptors are located almost exclusively in the postsynaptic membrane at points of functional contact with nerves. The regulation of acetylcholine receptor number and distribution has been of considerable interest to neurobiologists for several decades in relation to denervation

ISBN 0-12-181996-5

hypersensitivity and more recently in relation to the receptor deficit in myasthenia gravis. The neuromuscular junction has also been studied extensively as a model for synaptic connections in the central nervous system. The production and organization of acetylcholine receptors during myogenesis and formation of neuromuscular junctions serve as paradigms for CNS synaptogenesis as well as being of major interest in their own right. The biosynthesis of acetylcholine receptors and their intracellular transport play important roles in all the processes mentioned above: the development of denervation hypersensitivity, the loss of receptors in myasthenia gravis, and the genesis of synaptic connections.

Extensive reviews focused upon acetylcholine receptor structure,[1] function,[2,3] biosynthesis and bioregulation,[4–6] and involvement in myasthenia gravis[7,8] have appeared. This chapter is focused on techniques and strategies that we have used to elucidate events in the biosynthesis and turnover of acetylcholine receptors.

Methods of Measurement of Acetylcholine Receptors

Electrophysiological techniques developed for the study of acetylcholine receptors were of pivotal importance both in the first studies of acetylcholine receptor distribution on skeletal muscle fibers and in continuing studies on the mechanism of receptor function. While few of these studies have been concerned directly with acetylcholine receptor biosynthesis and turnover, many of the changes in acetylcholine sensitivity described in these studies are explainable in terms of acetylcholine receptor metabolism. A principal technique is intracellular recording of transmembrane potential of a muscle fiber combined with very localized stimulation of acetylcholine receptors by electrophoretic ejection of acetylcholine ions from a micropipette positioned by the investigator at various points on the surface of the fiber. Relative acetylcholine sensitivities of membrane areas are calculated by dividing the magnitude of the response (millivolts depolarization of the muscle fiber) by the amount of charge ejected from

[1] A. Karlin, *in* "The Cell Surface and Neuronal Function" (C. W. Cotman, G. Poste, and G. L. Nicholson, eds.), p. 191. Elsevier/North-Holland, New York, 1980.

[2] C. F. Stevens, *Annu. Rev. Physiol.* **42,** 643 (1980).

[3] P. R. Adams, *J. Membr. Biol.* **58,** 161 (1981).

[4] D. M. Fambrough, *Physiol. Rev.* **59,** 165 (1979).

[5] J. Patrick and P. W. Berman, *in* "The Cell Surface and Neuronal Function" (C. W. Cotran, G. Poste, and G. L. Nicholson, eds.), p. 157. Elsevier/North-Holland, New York, 1980.

[6] D. W. Pumplin and D. M. Fambrough, *Annu. Rev. Physiol.* **44,** 319 (1982).

[7] A. Vincent, *Physiol. Rev.* **60,** 756 (1980).

[8] D. B. Drachman, *Annu. Rev. Neurosci.* **4,** 195 (1981).

the micropipette tip during application of the acetylcholine (measured in nanocoulombs). Values of acetylcholine sensitivity ranging over about six orders of magnitude can be measured.[9] The technique offers several important advantages compared with other techniques for measuring numbers of receptors. It is a nondestructive method for sampling acetylcholine sensitivities of individual cells or areas of a cell; it is compatible with gathering other physiological data on the same cells; individual measurements can be made fairly rapidly; and the technique can be upgraded to obtain data on the open times and conductances of single acetylcholine receptor ion channels.[10-12] The disadvantages of electrophysiological techniques for acetylcholine receptor measurement are several. While correlations between acetylcholine sensitivity and numbers of acetylcholine receptors per unit area of muscle surface have been made, the relations are highly nonlinear[13,14] and the absolute values of sensitivity are influenced not only by packing density of acetylcholine receptors in the membrane, but also by the resting potential of the cell, the ionic composition of the bathing medium, the varying characteristics of single receptor function, and also factors related to the functioning of the acetylcholine containing micropipettes, and so on. Furthermore, some particularly interesting portions of the muscle surface, the neuromuscular junctions and myotendonous junctions, are more difficult to reach than others. Fine-scale clustering of acetylcholine receptors may also result in problems of sampling.

Quantitative analyses of acetylcholine receptor number and distribution generally involve the use of very high affinity ligands whose binding sites on the receptor molecule overlap the binding sites for acetylcholine. By far the most widely used of these ligands is α-bungarotoxin.[15,16] Conjugates of α-bungarotoxin with fluorescein and tetramethylrhodamine and horseradish peroxidase as well as acetylated and iodinated derivatives remain highly toxic. The iodinated derivatives retain the potency of the native toxin. α-Bungarotoxin and radiolabeled iodo-α-bungarotoxin have been employed in nearly every study of acetylcholine receptor metabolism.

[9] R. Miledi, *J. Physiol. (London)* **151,** 1 (1960).

[10] B. Katz and R. Miledi, *J. Physiol. (London)* **224,** 665 (1972).

[11] C. R. Anderson and C. F. Stevens, *J. Physiol. (London)* **235,** 655 (1973).

[12] E. Neher and B. Sakmann, *Nature (London)* **260,** 799 (1976).

[13] H. C. Hartzell and D. M. Fambrough, *J. Gen. Physiol.* **60,** 248 (1972).

[14] B. R. Land, T. R. Podleski, E. E. Salpeter, and M. M. Salpeter, *J. Physiol. (London)* **269,** 155 (1977).

[15] C. Y. Lee, L. F. Tseng, and T. H. Chiu, *Nature (London)* **215,** 1177 (1967).

[16] C. Y. Lee, *in* "Neurotoxins: Tools in Neurobiology" (B. Ceccarelli and F. Clementi, eds.), p. 1. Raven, New York, 1979.

α-Bungarotoxin is a 74 amino acid polypeptide that represents roughly 30% of the protein in the venom of *Bungarus multicinctus,* the Formosan banded krait. α-Bungarotoxin contains five intrachain disulfide bridges that are important for toxicity. A number of properties of α-bungarotoxin make it especially suitable for studies of acetylcholine receptor metabolism.

1. The purified toxin is commercially available at moderate cost and is even available in radioiodinated form.
2. The toxin is stable for years in sterile solution with or without freezing and is relatively stable to heat and proteases.
3. The affinity of α-bungarotoxin for acetylcholine receptors is exceedingly high because the dissociation rate of toxin–receptor complexes is almost immeasurably slow.
4. α-Bungarotoxin is unable to cross cell membranes.
5. Acetylcholine receptors solubilized in nonionic detergent solutions retain their binding sites for α-bungarotoxin.
6. The physicochemical properties of α-bungarotoxin and the acetylcholine receptor are very different, facilitating separation of unbound ligand from toxin–receptor complexes by gel filtration, velocity sedimentation, isoelectric focusing, or ion-exchange chromatography.
7. At concentrations that result in saturation of acetylcholine receptors, α-bungarotoxin does not seem to have any side effects. In fact, local application of α-bungarotoxin *in vivo* has been used to tag the acetylcholine receptors in individual muscles of experimental animals, resulting only in local blockade of neuromuscular transmission.
8. α-Bungarotoxin does not seem to alter the rates of acetylcholine receptor biosynthesis or turnover appreciably.

These properties of α-bungarotoxin have made possible the development of simple strategies for determining many of the major events in the life history of acetylcholine receptor molecules. These strategies are described together with experimental results below.

Cell and Organ Cultures

Most of our studies of acetylcholine receptor biosynthesis have been done on tissue-cultured chick skeletal muscle. Myogenic cells, obtained by mechanical dissociation of embryonic muscle tissue at a stage when the population of myoblasts is highest, are grown in serum-containing medium in dishes coated with collagen. The myoblasts divide and locomote in the dish for about 2 days and then aggregate and fuse to form multinucleate myotubes. The ability to fuse is acquired approximately synchronously with the onset of synthesis of various muscle-specific contractile proteins and of acetylcholine receptors. (Evidence for a small number of nonfunctional acetylcholine receptors on dividing myoblasts is weak, and metabolic labeling studies have demonstrated that the receptors appearing during the period of cell fusion are newly synthesized.) Studies on acetylcholine receptor biosynthesis usually have been carried out on early postfusion myotubes that have several hundred receptor sites per square micrometer of cell surface.

To a much more limited extent, we have used organ-cultured rat and mouse muscles. Experiments with these muscles suggest that the metabolism of acetylcholine receptors in these systems is basically like that in the myogenic cultures, and to a pleasing extent the kinetics of production and turnover of extrajunctional receptors resemble those measured for chick muscle *in vitro*. Results obtained with adult muscles are briefly mentioned in the section on regulation below.

Metabolic Labeling with Heavy Isotopes

Among the most common strategies for determining the kinetics of biosynthesis and turnover of molecules in living systems, the metabolic labeling strategy involves supplying a tagged precursor to the system either continuously (pulse labeling) or for a defined period (pulse-chase labeling) and measuring the amount of precursor incorporated into the molecule of interest (product) as a function of time. Typically, radioisotopically labeled precursors are used, and the incorporation into the product is assessed by counting the radioactivity in the purified product. A key word here is "purified." Unless the product can be purified to homogeneity, the incorporation of radioisotope into it cannot be estimated confidently. And unless such purification can be accomplished with quantitatively reproducible yield, a careful kinetic analysis of labeling is not possible. Many of the most interesting membrane proteins have never been purified to homogeneity. Most are present in very small quantity. The acetylcholine receptors of skeletal muscle, for example, represent only about 1 part in 20,000 to 100,000 of the total protein in skeletal muscle cultures. Thus, the radioisotopic labeling strategy for study of acetylcholine receptor metabolism has proved to be difficult. Heavy isotopic labeling, on the other hand, has been quite useful. In early experiments we used as precursors a mixture of amino acids labeled with ^{13}C or at the nonexchangeable positions with deuterium.[17] Later we used triple-labeled (^2H, ^{13}C, ^{15}N) amino acids, as described below.[18–20]

The unique feature of heavy isotope labeling is that the assay for incorporation of the precursor is measured by the degree of change in the density of the product. In the case of proteins, this physical property can be assessed easily by equilibrium centrifugation in density gradients or by velocity sedimentation. In either case purification of the protein product is not involved. It is merely necessary to determine the protein's position in the gradient after centrifugation relative to that of the same, unlabeled protein. In the case of acetylcholine receptors, the normal and the density-shifted molecules are easily detected as complexes with iodinated α-bungarotoxin (which could be used either before or after the centrifugation, but when used before centrifugation eliminates the necessity of performing a separate binding assay on each fraction from the density gradient). It should be stressed that the heavy labeling technique is suitable for use in analysis of the biosynthesis and turnover of any protein that can be quantified as a soluble entity. The technique has been used to measure the turnover of insulin receptors[21,22] and the Na^+, K^+-ATPase heavy polypeptide chains.[23] In the case of insulin receptors, an insulin binding assay was used on each gradient fraction after equilibrium centrifugation in cesium chloride gradients. In the case of the ATPase, the heavy chain was covalently labeled in a partial enzymic reaction that transferred isotopically labeled phosphate from the γ position of ATP to an aspartyl carboxyl in the active site of the membrane-bound enzyme, after which the enzyme was solubilized in ionic detergent solution.

Heavy isotope-labeled amino acids were obtained from Merck, Sharp and Dohme Ltd. of Canada. This product is an acid hydrolyzate of protein isolated from algae grown in a medium containing $^{15}NO_3^-$ and deuterium oxide, gassed with $^{13}CO_2$. The amino acid mixture contains approximately 98% substitution of ^2H for ^1H in the nonexchangeable positions and 80% substitution of ^{15}N and ^{13}C for the normal ^{14}N and ^{12}C isotopes. This algal protein hydrolysate also contains some brown material, presumably caramelized sugar and, if used directly in culture medium, is toxic to the cells. The toxicity is nearly eliminated by filtration of the amino acid mixture through an ultrafiltration membrane (Amicon UM-05) with molecular

[17] P. N. Devreotes and D. M. Fambrough, *Proc. Natl. Acad. Sci. U.S.A.* **73,** 161 (1976).

[18] P. N. Devreotes, J. M. Gardner, and D. M. Fambrough, *Cell* **10,** 365 (1977).

[19] D. C. Linden and D. M. Fambrough, *Neuroscience* **4,** 527 (1979).

[20] J. M. Gardner and D. M. Fambrough, *Cell* **16,** 661 (1979).

[21] B. C. Reed and M. D. Lane, *Proc. Natl. Acad. Sci. U.S.A.* **77,** 285 (1980).

[22] B. C. Reed, G. V. Ronnett, P. R. Clements, and M. D. Lane, *J. Biol. Chem.* **256,** 3917 (1981).

[23] L. R. Pollack, E. A. Tate, and J. S. Cook, *Am. J. Physiol.* **241,** C173 (1981).

weight cutoff around 500. Residual toxicity results in a very slow decline in biosynthetic capacity of cultured chick skeletal muscle. However, a greater toxic effect on HeLa cells has been attributed tentatively to the deuterium.[23]

For the labeling of chick skeletal muscle (normally grown in Eagle's minimum essential medium supplemented with 10% horse serum and 2% chick embryo extract), the serum and embryo extract are exhaustively dialyzed against balanced salt solution, and the usual complement of amino acids is replaced with the heavy amino acid mixture (500 mg/liter) plus the normal quantities of cysteine, tryptophan, and glutamine. Myoblasts cultured in this medium fuse and differentiate as usual, but the resulting myotubes are somewhat fewer and thinner than normal. The adverse effects of the medium could be reversed by switching to normal medium; and myotubes that had differentiated in normal medium remained healthy for more than 48 hr in heavy labeling medium.

Separation of normal and heavy isotope-labeled acetylcholine receptors on the basis of density has been accomplished by equilibrium density gradient centrifugation in gradients of metrizamide[24] and D_2O and by velocity sedimentation[25] in gradients of sucrose in D_2O. Attempts to use gradients of CsCl, RbCl, and chloral hydrate were unsuccessful owing to denaturation of the acetylcholine receptors and loss of capacity to bind α-bungarotoxin. Velocity sedimentation proved superior to equilibrium centrifugation because toxin–receptor complexes were slightly unstable in higher concentrations of metrizamide and because the metrizamide gradients were quite viscous near the bottom and thus could not be dripped easily.

The sedimentation constant s of a protein is related to the protein's density (ρ_p), molecular weight (M), and frictional coefficient (f), and to the viscosity (η_s), and average density (ρ_s) of the solution, by Eq. (1).

$$s = (KM/f\eta_s)[1 - (\rho_s/\rho_p)] \quad (1)$$

where K is a constant.[25] The introduction of heavy isotopes into a protein molecule is formally identical to the addition of extra neutrons to the nuclei of the carbon, nitrogen, and deuterium atoms. There is negligible change in molecular size. Thus the frictional coefficient remains the same while M and ρ_p increase, each adding to the sedimentation constant of the protein.

Conditions for separation of heavy isotope-labeled (henceforth referred to simply as heavy) and normal acetylcholine receptors were determined empirically, using populations of heavy and normal receptors dif-

[24] A. Hüttermann and G. Wendlberger, *Methods Cell Biol.* **13,** 153 (1976).

[25] R. G. Martin and B. W. Ames, *J. Biol. Chem.* **236,** 1372 (1961).

ferentially tagged with radioiodinated α-bungarotoxins. The normal acetylcholine receptors were labeled by incubation of muscle cultures in medium containing ^{131}I-labeled α-bungarotoxin. Heavy receptors were generated by allowing differentiation of myogenic cells in medium containing heavy amino acids and were tagged by incubation of these cultures in medium containing ^{125}I-labeled α-bungarotoxin. After removal of unbound toxin, the labeled receptors were solubilized in 1% Triton X-100, 10 m*M* Tris-HCl, pH 7.8, with 1 m*M* EDTA and 0.5 m*M* phenylmethylsulfonyl fluoride as protease inhibitors. More than 95% of the toxin–receptor complexes were solubilized. The mixed extracts were layered onto sucrose–deuterium oxide gradients and centrifuged in a Beckman SW41 rotor at 40,000 rpm at 4°. The sucrose gradients were formed in cellulose nitrate tubes by layering solutions of the following sucrose concentrations in deuterium oxide: 4 ml of 40% (w/v), 4 ml of 32.5%, and 3.5 ml of 25%. The function of the gradient is to stabilize the solution against convection currents. It has negligible effects upon sedimentation velocity of the proteins, and the slight discontinuities in density probably dissipate during centrifugation. The overall high concentration of sucrose and the use of deuterium oxide were to increase solution density so that a small difference in the density between heavy and normal protein would result in large differences in sedimentation velocity [see Eq. (1)].

After centrifugation, sucrose gradients were dripped into vials and radioactivity was measured in a two-channel gamma counter or in a scintillation spectrometer after addition of scintillation cocktail. Data were corrected for isotope spillover. A typical separation of heavy and normal acetylcholine receptors is illustrated in Fig. 1.

The magnitude of the change in sedimentation constant resulting from isotopic substitution depends upon the extent of isotopic substitution of the precursor amino acids, the utilization of these amino acids versus utilization of amino acids resulting from protein breakdown and from *de novo* amino acid synthesis, and the amino acid composition of the protein. The maximum theoretical density shift for the acetylcholine receptor was calculated[26] from the amino acid composition of the eel acetylcholine receptor, using the values of isotopic substitution established by Merck for the amino acid mixture. This value was 8.5% increase in density.

The magnitude of our observed density shifts was determined as follows. Normal and heavy receptors labeled with iodinated α-bungarotoxins were separated by velocity sedimentation on sucrose gradients with average density varied by using 100%, 50%, or zero deuterium oxide as solvent. Sedimentation velocities for the normal and the heavy receptors were plotted as a function of solution density, and extrapolations were

[26] A. S. L. Hu, R. M. Bock, and H. O. Halvorson, *Anal. Biochem.* **4,** 489 (1962).

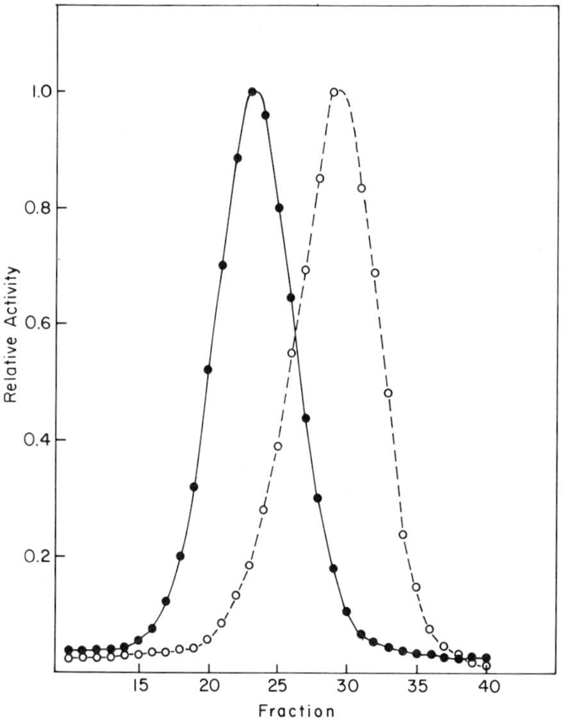

FIG. 1. Separation of heavy (●) and normal (○) acetylcholine receptors by velocity sedimentation on sucrose–deuterium oxide gradients. Fractions, 0.2 ml, from only the portion of the gradient containing the receptors (labeled differentially with radioiodinated α-bungarotoxins) are depicted. From Gardner and Fambrough.[20]

made to the solution densities at which sedimentation velocity would be zero: in other words, the proteins would be buoyant (Fig. 2). The resulting values of buoyant density were corrected for bound detergent (estimated by Meunier *et al.*[27] to be 20% of the weight of purified, solubilized acetylcholine receptors from the eel), and a very small correction was made for bound α-bungarotoxin (ρ = 1.33 g/cm^3), assuming 1 mol of bungarotoxin (M_r = 8000) was bound per 100,000 *g* of receptor protein.[28] The corrected values were 1.32 g/cm^3 for the normal receptor and 1.41 g/cm^3 for the heavy receptor. This is a density shift of 6.8%, or approximately 80% of the theoretically maximal shift. The difference is probably due primarily to dilution of the heavy amino acids with normal amino acids derived from protein breakdown, while *de novo* amino acid biosynthesis and the occur-

[27] J. L. Meunier, R. W. Olsen, and J. P. Changeux, *FEBS Lett.* **24,** 63 (1972).
[28] A. Karlin, *Life Sci.* **14,** 1385 (1974).

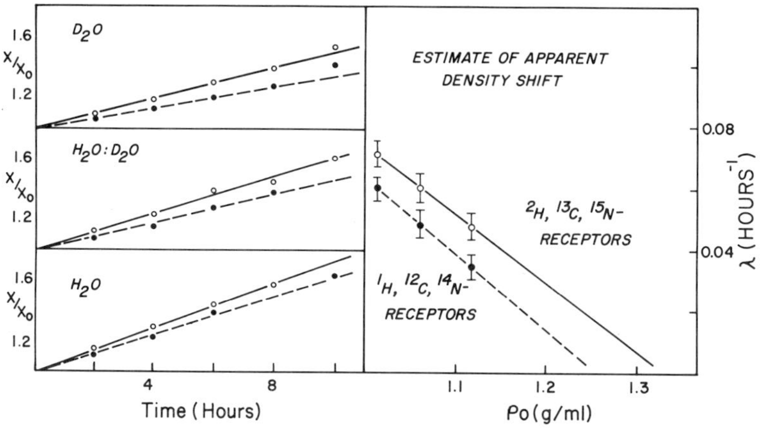

FIG. 2. Estimation of apparent densities of heavy (○) and normal (●) acetylcholine receptors by analysis of velocities of sedimentation in sucrose gradients of varying overall density. For details see text and Devreotes *et al.*[18]

rence of oligosaccharide chains in the receptor structure probably make additional contributions.

In the metabolic labeling studies described below, complexes of ^{131}I-labeled α-bungarotoxin with normal acetylcholine receptors were added to detergent extracts from each time point of the pulse and pulse-chase labeling with heavy amino acids. These extracts contained ^{125}I-labeled α-bungarotoxin complexed with the heavy and normal receptors for density analysis. The ^{131}I marker served to define the position and shape of the sedimentation profile or normal toxin–receptor complexes. With the aid of a DEC PDP-8 computer, the proportions of heavy and normal receptors in the experimental sample were estimated as follows: Data representing the amount of normal (^{131}I) and experimental (^{125}I) material in each gradient fraction were entered into the computer. An initial estimate of the percentage of heavy receptor in the ^{125}I profile was entered by the computer operator, and the computer proceeded to manipulate numerically the position and peak height of the marker ^{131}I profile, constantly comparing the resulting template to the experimental profile. Each iteration of the computer consisted of a revised estimate of the percentage of heavy receptors, and the computer continued to iterate until the best fit was found, as judged by numerical criteria. The program could be operated in a manual mode, so it was possible to compare the computer-obtained best fit with best fits obtained by visual inspection. Simulated profiles were made by mixing known quantities of heavy and normal receptors complexed with ^{125}I-labeled α-bungarotoxin and a marker of

^{131}I-labeled α-bungarotoxin–normal receptor complexes. These mixtures were subjected to velocity sedimentation, and the profiles were analyzed as described above. The computer-estimated percentages of heavy and normal receptors in the samples were generally within 1% of the actual values.[20]

Biosynthesis and Intracellular Transport

Three populations of acetylcholine receptors have been defined in cultures of skeletal muscle. The largest population, comprising about 75–80% of all the receptors, exists on the cell surface with binding sites oriented toward the outside of the myotubes. This population interacts readily with α-bungarotoxin added to the culture medium: the population of binding sites is saturated by incubation with 25 n*M* α-bungarotoxin at 37° for 20 min, and the cells become insensitive to acetylcholine. Two other populations of acetylcholine receptors are unavailable for interaction with extracellular α-bungarotoxin, and these have been shown by electron microscope autoradiography (see below) to occur on intracellular membranes. Each of the internal populations represents about 10–12% of total acetylcholine receptors.

One of these intracellular populations consists of newly synthesized molecules, most of which are destined to join the surface population. There is near quantitative transfer of this population to the surface in the absence of protein synthesis (see below). This population is rapidly labeled by heavy amino acids (Fig. 3). In order to observe the kinetics of labeling of this population, muscle cultures were switched to heavy labeling medium for various periods of time. Then the surface receptor population was blocked with unlabeled α-bungarotoxin, unbound toxin was washed away, and the acetylcholine receptors were solubilized in nonionic detergent solution. ^{125}I-labeled α-bungarotoxin was added to a final concentration of 1.25 n*M*, and the binding sites were saturated by incubation for 20–30 min at 37°. Free ^{125}I-labeled α-bungarotoxin was removed by gel filtration on BioGel P-60, using 1% sodium cholate in 10 m*M* Tris-HCl, pH 7.8, as solvent. Toxin–receptor complexes were recovered in the void volume and analyzed by velocity sedimentation in sucrose–deuterium oxide gradients. A marker of ^{131}I-labeled α-bungarotoxin complexed with normal acetylcholine receptors was added to each sample before sedimentation in order to define the position of normal complexes in the gradients.

The kinetics of density shift in the intracellular receptor population show that heavy receptors are synthesized and assembled in less than 30 min. Radioisotopic labeling experiments indicate that there is virtually

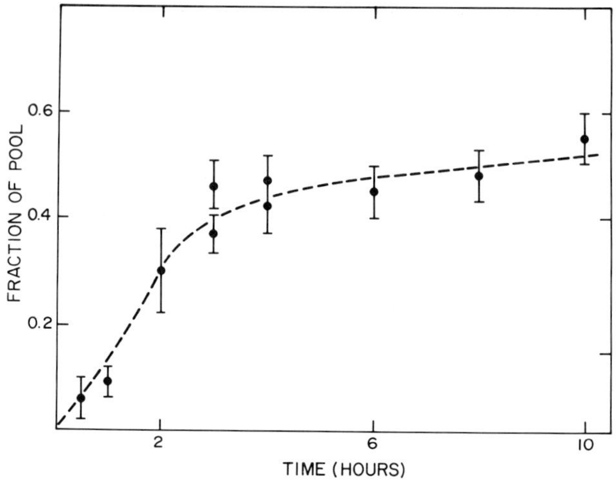

FIG. 3. Kinetics of appearance of heavy acetylcholine receptors in the intracellular compartment during continuous labeling from time zero in heavy isotope labeling medium. From Devreotes *et al.*[18]

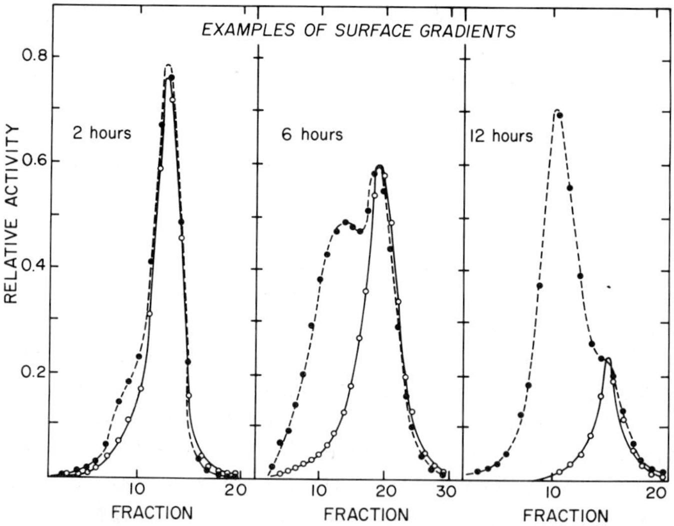

FIG. 4. Sucrose gradient velocity sedimentation profiles of acetylcholine receptors appearing at the cell surface during continuous labeling of the cells from time zero in heavy labeling medium. The surface receptor population had been blocked at time zero with unlabeled α-bungarotoxin. Newly appearing receptors were tagged with ^{125}I-labeled α-bungarotoxin (●). A marker of ^{131}I-labeled α-bungarotoxin complexed with normal receptors was added to the samples to mark the position of normal receptors in the gradients (○). From Devreotes *et al.*[18]

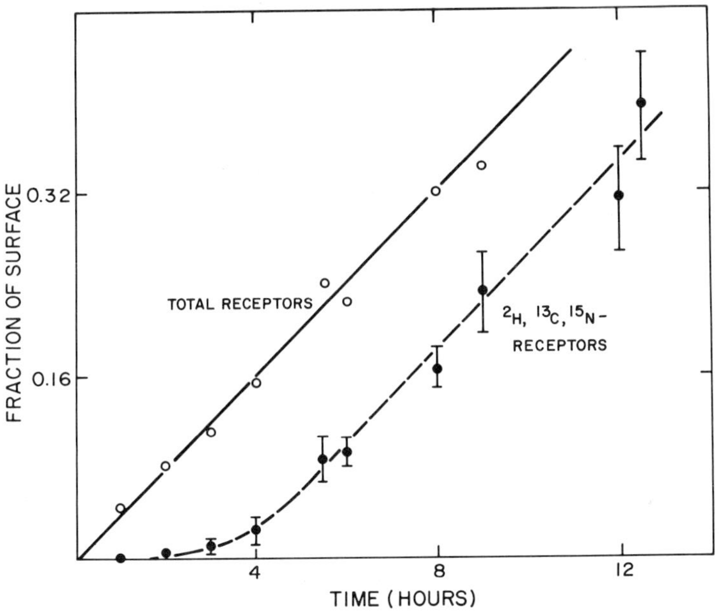

FIG. 5. Kinetics of appearance of total (○) and heavy (●) acetylcholine receptors at the cell surface during continuous labeling from time zero in heavy labeling medium. The data were calculated from analysis of sucrose gradient profiles such as those shown in Fig. 4. From Devreotes *et al.*[18]

instantaneous equilibration of intracellular amino acid pools with exogenous amino acids. Thus, the lag in initial labeling of the receptors after addition of heavy amino acids probably reflects the time course of polypeptide synthesis and assembly to form the toxin-binding units.

In contrast to the fast time-course of labeling of the intracellular pool of new receptors, the population of cell surface receptors receives labeled receptors only after a lag of several hours. The first heavy labeled receptors are available for interaction with exogenous ^{125}I-labeled α-bungarotoxin about 2 hr after a switch of the cultures to heavy labeling medium. After this lag period, the appearance of heavy receptors on the cell surface accounts for all the receptors arriving at the cell surface (Figs. 4 and 5). This observation confirms the validity of our previous conclusion that the appearance of acetylcholine receptors on the cell surface after blockade of the surface population with bungarotoxin reflects the biosynthesis and incorporation of newly synthesized receptors into the plasma membrane.[29,30] This observation is also part of a body of evidence against the

[29] H. C. Hartzell and D. M. Fambrough, *Dev. Biol.* **30,** 153 (1973).

[30] P. N. Devreotes and D. M. Fambrough, *J. Cell Biol.* **65,** 335 (1975).

existence of a population of old acetylcholine receptors slowly recycling out of and back into the plasma membrane (see below).

The intracellular location of the newly synthesized acetylcholine receptors has been determined by electron microscope autoradiography.[31] After blockade of cell surface receptors with unlabeled α-bungarotoxin, the intracellular receptors can be exposed for interaction with ^{125}I-labeled α-bungarotoxin either by solubilization, as described above, or by permeabilization of the cells. Fixation and permeabilization conditions were devised for maximal preservation of cell morphology and number of specific α-bungarotoxin binding sites. Cultured chick muscle cells were fixed for 1 hr at room temperature in the fixative devised by McLean and Nakane.[32] Fresh fixative was prepared by hydrolysis of paraformaldehyde in 20 m*M* sodium phosphate buffer, pH 7.2, at 80–85°. This solution was cooled to room temperature, and the other components of the fixative were added to give the final composition: 2% formaldehyde, 100 m*M* lysine, 60 m*M* sucrose, 10 m*M* sodium periodate, and 20 m*M* sodium phosphate, pH 7.2. After this fixation, the cultures were returned to culture medium containing 18 m*M* tricine or HEPES buffer and could be kept for more than a week at 4° without change in the number of α-bungarotoxin binding sites. Among techniques for permeabilization, treatment of the fixed cultures with 0.5% saponin in 150 m*M* KCl gave best preservation of ultrastructure while revealing the entire population of intracellular binding sites for α-bungarotoxin. After saponin treatment, the cultures were rinsed in HEPES-buffered culture medium and incubated for 90 min at room temperature in medium supplemented with 4 n*M* ^{125}I-labeled α-bungarotoxin, which saturated the intracellular binding sites. Nonspecific binding was defined as binding not blocked by 1.5×10^{-4} *M* *d*-tubocurarine (added 15 min before addition of α-bungarotoxin and kept in the rinse solutions during removal of unbound toxin).

The population of newly synthesized acetylcholine receptors was defined as the fraction of intracellular receptors that could be chased to the cell surface by incubation of live muscle cultures under conditions of no protein synthesis (10 or 20 μg of puromycin per milliliter; 37°). The decline in the intracellular populations of receptors was readily measured on fixed, saponin-treated cultures; it corresponded to the loss of about half of the receptors from the intracellular population and represented 10–14% of total receptors in the cultures, as expected (Fig. 6). The intracellular population, so defined, was identified by electron microscope autoradiography as consisting of binding sites in the Golgi apparatus and in other membranous elements in the perinuclear regions of the myotubes (Fig. 7).

[31] D. M. Fambrough and P. N. Devreotes, *J. Cell Biol.* **76,** 237 (1978).

[32] I. W. McLean and P. K. Nakane, *J. Histochem. Cytochem.* **22,** 1077 (1974).

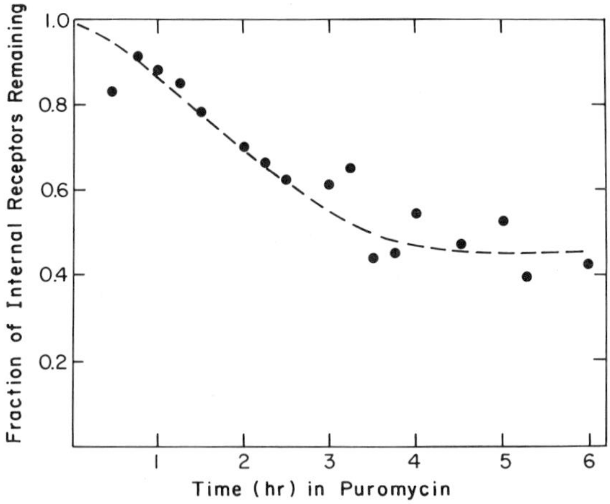

FIG. 6. Decline in intracellular acetylcholine receptor pool during inhibition of protein synthesis by puromycin. For determination of intracellular pool sizes, surface receptors were blocked with α-bungarotoxin, the cultures were fixed and permeabilized and incubated with 4 n*M* ^{125}I-labeled α-bungarotoxin to saturate the internal sites. From Fambrough and Devreotes.[31]

The other saponin-revealed binding sites for α-bungarotoxin were distributed through the myotubes and represent receptors associated with some poorly defined intracellular membrane system. This system could correspond to the developing T-tubule system.

The intracellular transport of acetylcholine receptors from sites of biosynthesis to points of exteriorization apparently involves vesicular transport. The newly synthesized receptors occur on internal membranes, which, in cell homogenates, are mostly sealed with the binding sites for α-bungarotoxin sequestered in their lumens. Thus, homogenization of myotubes that have been pulse-labeled for several hours with dense amino acids fails to reveal more than about 15% of the heavy-labeled receptor sites for interaction with added ^{125}I-labeled α-bungarotoxin. However, if such homogenates are solubilized with nonionic detergent solution, the full complement of heavy labeled receptors becomes accessible to bungarotoxin.[33]

The mechanism of intracellular transport is not established. However, the sensitivity of the process to various commonly used inhibitors has been explored.[30,34,35] The production and incorporation of receptors into

[33] J. M. Gardner, *Year Book—Carnegie Inst. Washington* **77,** 12 (1978).

[34] D. M. Fambrough, H. C. Hartzell, J. E. Rash, and A. K. Ritchie, *Ann. N.Y. Acad. Sci.* **228,** 47 (1974).

[35] R. L. Rotundo and D. M. Fambrough, *Cell* **22,** 595 (1980).

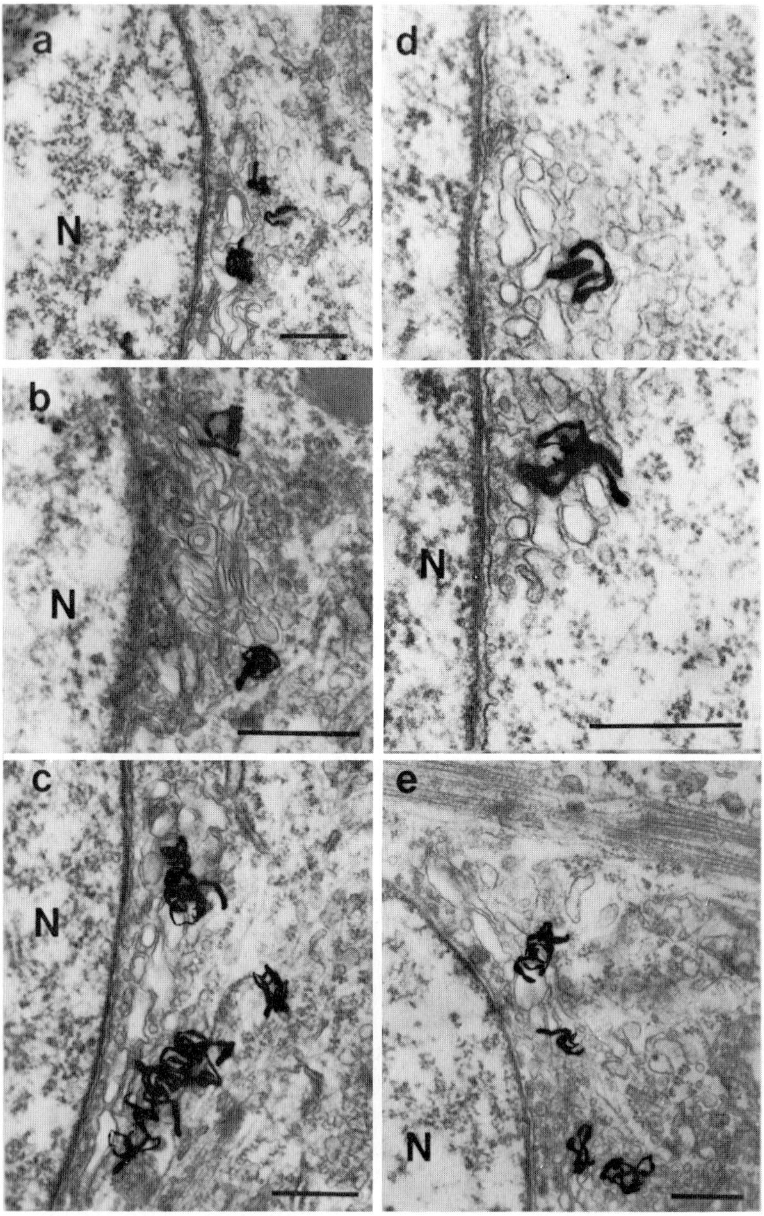

FIG. 7. Electron microscope autoradiographs demonstrating location of saponin-revealed binding sites for ^{125}I-labeled α-bungarotoxin in the Golgi apparatus in cultured chick myotubes. Magnification bars, 0.5 μm.

plasma membrane is unaffected by short-term incubation of cultures (up to 8 hr) in solutions containing 10^{-6} M to 100 mM calcium ions or in solutions that result in changes in transmembrane potential between 0 and −40 mV. Intracellular transport is also insensitive to inhibitors of protein synthesis (puromycin and cycloheximide) and protein glycosylation (tunicamycin), although each of these three inhibitors results in emptying of the intracellular pool of newly synthesized receptors onto the cell surfaces over a 2–3-hr period. The ionophores nigericin and monensin cause marked swelling of the Golgi apparatus[36,37] and slow the production of new receptors and their transport from Golgi to cell surface in a dose-dependent manner. Uncouplers of oxidative phosphorylation rapidly halt the exteriorization of receptors, leaving the intracellular pool filled. Colchicine, but not lumicolchicine, slows the incorporation of receptors into the plasma membrane. This effect occurs with immediate onset and saturates at about 50% slowing of incorporation rate at a colchicine concentration around 5 μM. None of the cytochalasins (B, D, or E) affect intracellular transport. Exteriorization is strongly temperature dependent and does not occur below around 25° in cultured chick muscle.

An interesting aspect of intracellular transport of newly synthesized acetylcholine receptors is that the process is difficult or impossible to distinguish from that involved in production and secretion of secretory proteins. This matter remains controversial. Smilowitz[37] offered evidence that the secretion of acetylcholinesterase by cultured muscle was separable from exteriorization of newly synthesized acetylcholine receptors by differential inhibition with monensin or nigercin. Rotundo and Fambrough[35] were unable to demonstrate this differential effect when the intracellular transport of acetylcholinesterase and acetylcholine receptors was studied in the absence of protein synthesis; and they offered an alternative explanation for earlier findings. They also showed that various other inhibitors, such as colchicine and tunicamycin, had quantitatively identical effects upon the time courses and magnitude of exteriorization of acetylcholinesterase and acetylcholine receptor molecules. Gardner and Fambrough[38] demonstrated that the kinetics of biosynthesis and secretion of fibronectin and various collagenous peptides by chick skeletal muscle *in vitro* followed the same slow time-course characteristic of acetylcholine receptors, whereas biosynthesis and secretion of apparently identical fibronectin molecules by chick muscle fibroblasts followed the rapid kinetics characteristic of these processes in most tissues studied to date.

[36] A. Tartakoff, P. Vassalli, and M. Detraz, *J. Cell Biol.* **79,** 694 (1978).

[37] H. Smilowitz, *Cell* **19,** 237 (1980).

[38] J. M. Gardner and D. M. Fambrough, *J. Cell Biol.* **96,** 474 (1983).

Early Events in Acetylcholine Receptor Biosynthesis

In the past year some progress has been made in several other laboratories toward defining the events in polypeptide synthesis and assembly of receptor subunits. A major problem has been the extreme protease sensitivity of several of the subunits of the receptor, either reducing them all to a common molecular weight or nicking them so much that they fall apart into small polypeptides upon denaturation. Nevertheless there is some evidence that the biosynthesis of receptor subunits occurs on membrane-bound ribosomes[39] and that assembly occurs over a 15- to 30-min period.[40,41] The acquisition of α-bungarotoxin binding sites heralds the completion of the primary assembly process. All the acetylcholine receptor subunits are glycosylated. The exact site of glycosylation and the site of assembly of the receptor are not known. Both the paucity of rough endoplasmic reticulum in myotubes and the rapid onset of depletion of receptors from the Golgi apparatus in the absence of protein synthesis argue for a very small compartment before the Golgi apparatus in the biosynthetic process.[31] Assembly of receptor units may actually occur at the level of the Golgi apparatus.

Sequence analysis of receptor subunits of *Torpedo* revealed remarkable homologies between subunits at their N-terminal ends.[42] The current studies of cDNA clones corresponding to mRNA sequences for the receptor subunits should provide further insight into the biosynthesis-processing mechanism by revealing the sites of possible signal sequences and defining the exact nature of homologies between subunits. Such studies may also shed some light on the mechanism of coordination of production of the subunits.

Evidence against Involvement of Acetylcholine Receptors in Membrane Recycling

Over the years we have kept in mind the possibility the acetylcholine receptors might participate in membrane recycling between the plasma membrane and some internal membrane system, and we have carried out a number of experiments to test the possibility. Most of our results argue against any significant recycling of acetylcholine receptors in skeletal muscle. Many of these experiments address the question: Is there a pool of intracellular receptors that are not newly synthesized but can become

[39] J. P. Merlie, J. G. Hoffer, and R. Sebbane, *J. Biol. Chem.* **256,** 6995 (1981).

[40] J. P. Merlie and R. Sebbane, *J. Biol. Chem.* **256,** 3605 (1981).

[41] D. Anderson and B. Blobel, *Proc. Natl. Acad. Sci. U.S.A.* **78,** 5598 (1981).

[42] M. A. Raftery, M. W. Hunkapiller, C. D. Strader, and L. E. Hood, *Science* **208,** 1454 (1980).

exteriorized? The population of intracellular receptors that is not part of the newly synthesized pool is described to some extent above. As shown in Fig. 3, for example, the labeling of these sites with heavy amino acids is a very slow process, actually suggesting a turnover rate similar or identical to that of the surface acetylcholine receptor population. Figure 6 shows that this intracellular population is not chased to the cell surface in the absence of protein synthesis; electron microscope autoradiographic studies indicate that the intracellular sites are located throughout the cytoplasm; and biochemical studies demonstrate that these sites are indistinguishable from surface receptors in their physicochemical properties and that they exist as membrane-bound receptor units.[18] If these intracellular receptors were part of a recycling system, then they should cycle from their protected position inside the myotubes to sites accessible for binding α-bungarotoxin added to the culture medium. Cultures maintained in the presence of α-bungarotoxin for days, however, contain the same number of intracellular sites as control cultures.[20] Moreover, within experimental error, the number of ^{125}I-labeled α-bungarotoxin binding sites saturated by long-term exposure of cultures to a saturating concentration of bungarotoxin at 4° or 25° is the same as the number of binding sites measured by binding experiments done at 37°. This number of sites is the same as the number measured on formaldehyde-fixed but not permeabilized cells. When exposed sites are saturated with α-bungarotoxin at 25° or 4°, the unbound toxin is removed, and the cultures are returned to 37°, the subsequent appearance of new binding sites is exactly that expected from incorporation of newly synthesized receptors into the plasma membrane (there is no burst of new binding sites that might represent the completion of a round of recycling suspended by the low-temperature treatment). Heavy isotope labeling experiments also show that all of the receptors appearing at the surface after several hours of incubation in heavy labeling medium are of the heavy (i.e., newly synthesized) variety.[18]

These results do not completely rule out all participation of acetylcholine receptors in recycling. However, they place major constraints on the possible number of receptors that could be involved and on the time-course of the hypothetical receptor recycling. Not excluded by the data is a very slow recycling involving a small fraction of receptors. Since the median lifetime of acetylcholine receptors is 22 hr (see below), most receptors could not participate in a very slow mechanism. Also not excluded is an extremely rapid recycling mechanism involving occasional interiorization of every acetlycholine receptor molecule, but extremely short intracellular residence time, and hence a minuscule population of internalized receptors in the recycling mode. However, the sort of total plasma membrane recycling described for some other cells, involving a sizable internal pool and a moderate recycling rate, is inconsistent with our data. It remains possible that there is a membrane recycling system in skeletal muscle, perhaps involved in uptake of specific macromolecules, which is selective in that acetylcholine receptors are excluded from the system.

Intracellular Transport and the Degradation Process

While the emphasis of this volume is upon biosynthetic and cycling aspects of membrane proteins, it should be mentioned that the turnover process for the acetylcholine receptors involves internalization and transport to secondary lysosomes, where proteolytic degradation to amino acids occurs rapidly. This process was first studied by studying the fate of ^{125}I-labeled α-bungarotoxin bound to surface acetylcholine receptors.[30] While dissociation of toxin from receptor did not appear to occur, nevertheless bound radioactivity was lost from the cells under normal culture conditions. This loss was shown to represent the diffusion of [^{125}I]iodotyrosine from the cells after proteolytic destruction of the bound bungarotoxin. Electron microscope autoradiographic analysis demonstrated that ^{125}I-labeled α-bungarotoxin–receptor complexes are transferred from the plasma membrane to secondary lysosomes. Blockade of proteolysis in the lysosomes caused by long-term incubation of cultured cells in medium containing trypan blue resulted in accumulation of labeled complexes in the lysosomal compartment (Fig. 8). Transport to the lysosomes precedes the appearance of iodotyrosine in the culture medium.[43] The degradation process was also studied using the heavy isotopic labeling strategy.[20] The degradation rate for acetylcholine receptors estimated from observing the rate of degradation of bound ^{125}I-labeled α-bungarotoxin was shown to be slightly slower than that measured directly by pulse–chase labeling experiments. The labeling experiments also showed that the normal and the heavy labeled receptors turned over at the same rate. Degradation is a first-order exponential process and the half-time is about 17 hr. The degradation process was shown to be insensitive to various inhibitors, including the cholinergic against cabachol, the antagonist *d*-tubocurarine, colchicine, the cytochalasins, protease inhibitors, tunicamycin, and variations in calcium ion and potassium ion concentrations in the culture medium.[20,30,34,35] The process was very sensitive to temperature, pH of the culture medium, and uncouplers of oxidative phosphorylation.[30] Libby *et al.*[44] showed that leupeptin, an inhibitor of proteolytic destruction of ace-

[43] D. M. Fambrough, P. N. Devreotes, D. J. Card, J. Gardner, and K. Tepperman, *Natl. Cancer Inst. Monogr.* **48,** 277 (1978).

[44] P. Libby, S. Bursztajn, and A. Goldberg, *Cell* **19,** 481 (1980).

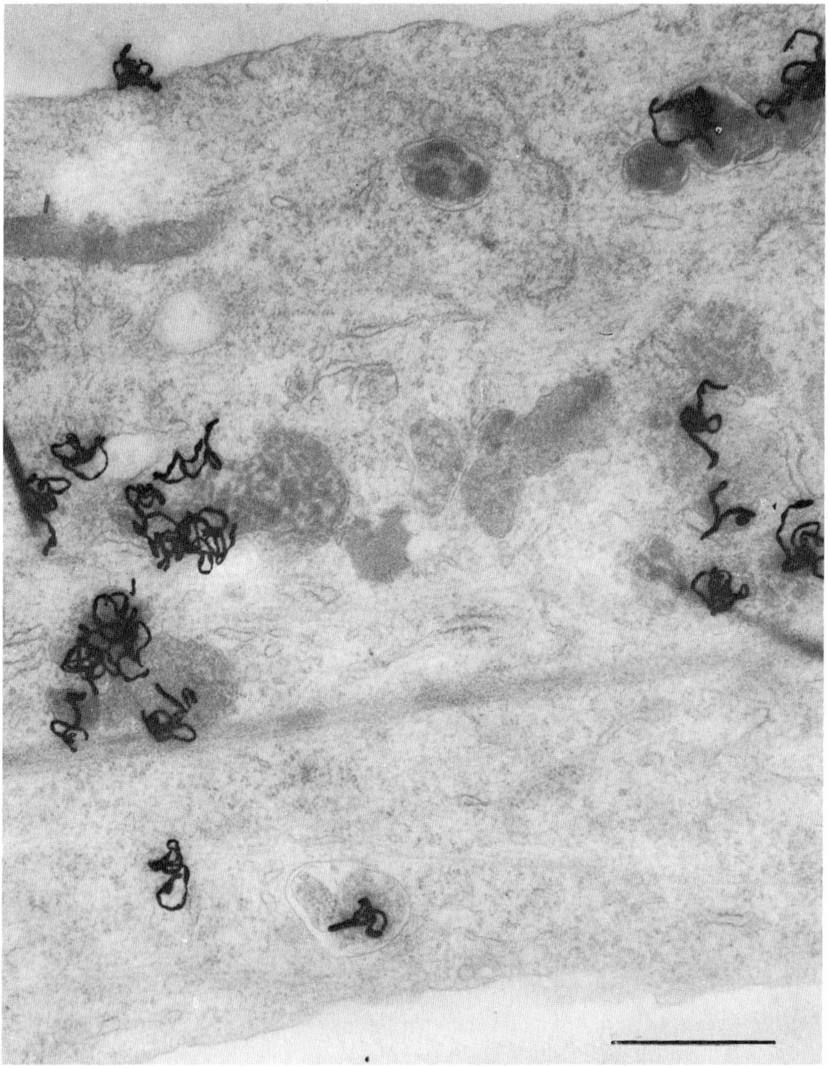

FIG. 8. Electron microscope autoradiograph demonstrating accumulation of ^{125}I-labeled α-bungarotoxin–receptor complexes in swollen lysosomes of cultured chick myotubes preincubated for 24 hr with trypan blue. Approximately 20% of silver grains are located over lysosomal profiles in specimens fixed 6 hr after brief exposure (5 min) of cells to ^{125}I-labeled α-bungarotoxin. Magnification bar, 1.0 μm.

tylcholine receptors in cell culture, failed to interfere with internalization of receptors. They also reported an increase in the number of coated vesicles seen in thin sections of leupeptin-treated myotubes, and they suggested that coated vesicles might mediate the transfer of acetylcholine receptors from surface to lysosomes.

Regulation of Acetylcholine Receptor Metabolism

Natural and pathological mechanisms have been shown to regulate receptor metabolism by accelerating biosynthesis, suppressing biosynthesis, accelerating degradation, and slowing degradation.[4–6,8] Biosynthesis is suppressed in myogenic cells until the onset of differentiation. At this time acetylcholine receptor biosynthesis becomes maximally activated and receptors accumulate to a packing density approaching 1000 sites per square micrometer of cell surface in some cases. Patches of receptors with a 10-fold higher packing density also form. *In vivo* the major patches are at sites of nerve–muscle interaction during formation of neuromuscular junctions. As the neuromuscular system matures, the production of acetylcholine receptors declines until very few new receptors are added to each area of membrane per unit time. The continued rapid degradation ($t_{1/2} \simeq 20$ hr) results in a net loss of receptors until the noninnervated surface membrane has few or even no receptors. Innervated areas maintain a packing density of about 20,000 receptor sites per square micrometer. In these areas the turnover rate changes, possibly abruptly, from the rapid rate ($t_{1/2} \simeq 20$ hr) to a much slower rate ($t_{1/2} \simeq 1$–2 weeks). The suppression of receptor biosynthesis is mediated in large part by muscle activity driven by the motor nerve. Following denervation, receptor biosynthesis is reactivated and the nonsynaptic areas of muscle plasma membrane become repopulated with receptors that have a fast turnover rate. This repopulation is the molecular basis for denervation hypersensitivity in skeletal muscle. In denervated muscle there is also a switch to a fast turnover rate of receptors present in the former postsynaptic membrane, although, paradoxically, the packing density of about 20,000 sites per square micrometer is maintained in this area for a very long time.

Myasthenia gravis is an autoimmune disease in which autoantibodies against acetylcholine receptors participate in diminution of the receptor populations at neuromuscular junctions, resulting in failure of neuromuscular transmission. The mechanisms resulting in a deficit of functional acetylcholine receptors include acceleration of turnover of acetylcholine receptors following cross-linking by antibodies.

[28] Acetylcholinesterase Biosynthesis and Transport in Tissue Culture

By RICHARD L. ROTUNDO

For the vast majority of proteins whose synthesis, transport, and assembly have been studied, one prerequisite has been the ability to obtain the protein in pure form. Whether by immunoprecipitation, gel electrophoresis, and/or direct labeling techniques, the molecule of interest must be unambiguously detected and quantified at all stages of metabolism and in all subcellular locations. For membrane-bound proteins this has not been an easy task. The inaccessibility of these molecules to externally applied labeling agents during their biosynthesis and assembly has made isotopic labeling the method of choice for pursuing their metabolic fate. This method has relied heavily on the use of immunological techniques for isolating the labeled proteins of interest and subsequent studies on their structure and metabolism.

Acetylcholinesterase (AChE) is unique among membrane and secretory proteins insofar as a wide variety of specific reversible and irreversible active site-directed ligands exist that can be used, either singly or in combination, to inactivate all the cell-associated enzyme or particular subpopulations of molecules.[1] Furthermore, the availability of highly sensitive assays of enzyme activity[2] have provided researchers with exquisitely fine tools with which to study the metabolism of this fascinating family of molecules. In this chapter we discuss in detail several methods employed in our laboratory to study the synthesis, transport, and assembly of this family of membrane-bound and secreted enzymes, using direct measurement of active enzyme molecules.

Acetylcholinesterase (EC 3.1.1.7) is the enzyme that hydrolyzes the neurotransmitter acetylcholine released at the neuromuscular junction and other cholinergic synapses in the nervous system. This enzyme exists as a complex family of molecular forms distinguished by their subunit composition, solubility characteristics, and hydrodynamic properties (see review by Massoulié[3]). The most commonly occurring forms consist of monomers, dimers, and tetramers of a common subunit, plus a unique asymmetric form comprising three tetramers covalently linked via disul-

[1] G. B. Koelle, ed. (1963). "Handbuek der experimentellen Pharmakologie," Vol. 15. Springer-Verlag, Berlin and New York.

[2] C. D. Johnson and R. L. Russell, *Anal. Biochem.* **64,** 229 (1975).

[3] J. Massoulié, *Trends Biochem. Sci.* **5,** 160 (1980).

ISBN 0-12-181996-5

fide bonds to a three-stranded collagen-like tail. This tail is believed to be involved in attaching these asymmetric molecules to the extracellular matrix surrounding muscle cells at their sites of highest concentration in the regions of nerve muscle contact.[3–5] The appearance and localization of this asymmetric form of AChE in muscle is in part under neural control, hence its interest to researchers studying the interactions between nerves and muscle.

In addition to the oligomeric forms of AChE, this enzyme also exists as membrane-bound molecules, such as those found on the outer membranes of erythrocytes and muscle[6,7] as well as secreted molecules released by muscle,[8,9] nerves,[10] and adrenal chromaffin cells.[11] In tissue culture, chick embryo muscle cells synthesize primarily the 11.5 S tetrameric form of the 7 S dimeric form of AChE.[12] Both of these forms occur as membrane-bound and secreted molecules.

The approaches described in this chapter for studying the synthesis, transport, and assembly of acetylcholinesterase are unique in that they permit the investigator to study the metabolism of functional enzyme molecules, using a variety of highly specific reversible and irreversible inhibitors. Taken together, these various inhibitors provide a very useful set of tools with which to selectively inhibit subpopulations of AChE molecules in tissue cultured cells. With some modifications, and with the appropriate tests and controls, these methods should also have wide applicability to the study of AChE *in vivo* and in organ culture. The following sections describe actual experiments applying the techniques described.

Subcellular Distribution of Acetylcholinesterase

Assay of Cell Surface AChE Activity

In order to study the subcellular distribution of AChE, a method was required to distinguish between intracellular and extracellular enzyme molecules. To this end we have adapted the radiometric assay procedure of

[4] Z. W. Hall, *J. Neurobiol.* **4,** 343 (1973).
[5] C. B. Weinberg and Z. W. Hall, *Dev. Biol.* **68,** 631 (1979).
[6] P. Ott, B. Jenny, and U. Brodbeck, *Eur. J. Biochem.* **57,** 469 (1975).
[7] R. L. Rotundo and D. M. Fambrough, *Cell* **22,** 583 (1980).
[8] R. L. Rotundo and D. M. Fambrough, *Cell* **22,** 595 (1980).
[9] B. W. Wilson, P. S. Nieberg, C. R. Walker, T. A. Linkhart, and D. M. Fry, *Dev. Biol.* **33,** 285 (1973).
[10] T. H. Oh, J. Y. Chyu, and S. R. Max, *J. Neurobiol.* **8,** 469 (1977).
[11] I. W. Chubb and A. D. Smith, *Proc. R. Soc. London, Ser. B* **191,** 263 (1975).
[12] R. L. Rotundo and D. M. Fambrough, *J. Biol. Chem.* **254,** 4790 (1979).

DISTRIBUTION OF LABELED SUBSTRATE AND PRODUCT IN MUSCLE CULTURES DURING ASSAY OF CELL SURFACE AChE[a]

	[^{14}C]ACh		[^{14}C]Acetate	
	Cpm	% Total	Cpm	% Total
Assay medium	666,360	86.7	99,150	13.0
Cultured cells	30	0	2,800	0.4
	666,390		101,950	

$$\frac{\text{Total CPM recovered}}{\text{Total CMP added}} = \frac{765{,}340}{767{,}760} = 0.997$$

[a] Muscle cells were incubated for 1 hr with HBSS containing 1.2 mM [^{14}C]ACh. The amount of [^{14}C]acetate and [^{14}C]acetylcholine in the medium and cells was determined using sodium tetraphenylboron as described by Fonnum.[13] See text and Rotundo and Fambrough[7] for details. Copyright MIT, with permission.

Johnson and Russell[2] to measure AChE activity associated with the surface plasma membrane of tissue-cultured muscle cells as well as the intracellular pools of these same cells.[7] This assay takes advantage of the fact that acetylcholine is a charged molecule and does not readily cross the plasma membrane. This can be shown by incubating intact muscle cells in a medium consisting of Hank's balanced salt solution (HBSS) at pH 7 containing 1.2 mM [^{14}C]acetylcholine (ACh). After allowing some time for hydrolysis of ACh to occur, the medium is removed and the cells are washed with HBSS followed by solubilization in detergent-containing buffer (buffer A: 50 mM Tris, pH 7, 150 mM NaCl, 0.25 mM EDTA, and 0.5% Triton X-100). The amount of [^{14}C]acetylcholine ([^{14}C]ACh) and [^{14}C]acetate in the medium and cell extract is then determined using sodium-tetraphenylboron as described by Fonnum.[13] The results of such an experiment, presented in the table, show that essentially all the substrate ([^{14}C]ACh) and more than 97% of the reaction product ([^{14}C]acetate) remain in the medium. Thus the hydrolysis of acetylcholine observed under these conditions must be due to enzyme molecules located external to the plasma membrane.

For routine determinations of cell surface AChE, enzyme activity is assayed directly in the culture dish. Muscle cells grown in 35-mm Falcon plastic tissue culture dishes are first rinsed with 3 × 2 ml of HBSS to

[13] F. Fonnum, *Biochem. J.* **115,** 465 (1969).

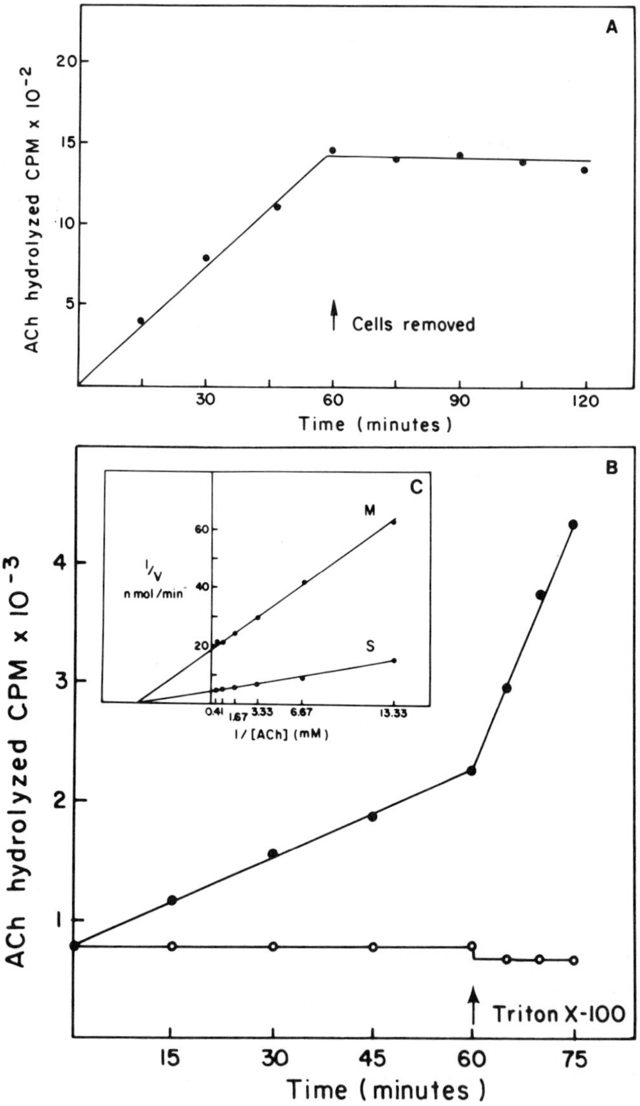

FIG. 1. Assay of cell surface acetylcholinesterase (AChE). (A) Hydrolysis of acetylcholine (ACh) by cell-associated AChE. Muscle cultures were washed with Hank's balanced salt solution (HBSS) and incubated with 1 ml of HBSS containing 1.2 m*M* [^{14}C]ACh added directly to the culture dish. At the end of 60 min the assay medium was removed to a collagen-coated culture dish and incubated for an additional hour in the absence of cells. At 15-min intervals, 20-μl aliquots of assay medium were removed and the [^{14}C]acetate was counted as described by Johnson and Russell.[2] Each value is the mean of three cultures with background subtracted, and the standard errors are less than 5%. (B) Effects of detergent

remove any esterase activity contributed by horse serum and embryo extract in the culture medium, then incubated on ice with 1 ml of 10 m*M* HEPES-buffered HBSS, pH 7, containing 1.2 m*M* [^{14}C]ACh per dish. The enzyme is assayed with the cells on ice to block the release of secretory AChE forms. Acetylcholinesterase is remarkably insensitive to temperature and still retains about 30% of its activity at 0–5°. At the appropriate time intervals 10–20-μl aliquots of incubation medium are removed and transferred to 7-ml scintillation vials containing 1.5 ml of 50 m*M* glycine–HCl buffer, pH 2.5, and 2 *M* NaCl. The reaction product, [^{14}C]acetate, is extracted for counting directly in the vial by adding 5 ml of scintillation cocktail containing 17.2 g PPO, 1 g of POPOP, 400 ml of *n*-butanol, and toluene to make 4 liters.[2] Blanks consist of collagen-coated culture dishes incubated in parallel with the muscle cells.

Figure 1A illustrates the results obtained with the cell surface AChE assay indicating that the hydrolysis of ACh is linear with time. The inclusion of specific inhibitors in the incubation medium, such as tetraisopropylpyrophosphoramide (Iso-OMPA; an irreversible inhibitor of butyrylcholinesterase) or 1,5-bis(4-allyldimethylammoniumphenyl)pentan-3-one dibromide (BW284c51; a specific inhibitor of AChE), indicate that the hydrolysis of ACh is due to true acetylcholinesterase. If the cells are removed from contact with the incubation medium (Fig. 1A) the hydrolysis of ACh ceases, indicating that the enzyme is associated with the muscle cell surface rather than released into the medium. Previous studies have shown that in these cultures only the muscle cells contain AChE and that this protein is tightly associated with the cell membrane.[8,14]

Cell Surface versus Intracellular AChE

An additional modification of the assay permits the measurement of both cell surface and intracellular AChE in the same culture dish. After a 1-hr incubation to determine surface enzyme activity an aliquot of Triton

[14] P. B. Taylor, F. Rieger, M. L. Shelanski, and L. A. Greene, *J. Biol. Chem.* **256,** 3827 (1981).

added during cell surface AChE assay. Cell surface AChE was assayed as described in (A) except that after 60 min a 100-μl aliquot of 5% Triton X-100 in HBSS was added to each culture to expose the intracellular AChE. Subsequent aliquots of medium were sampled at 5-min intervals. ●, Muscle cultures; ○, blank. (C) Kinetics of membrane-bound and solubilized AChE. Aliquots of cultured muscle cell homogenate were mixed with buffer or buffer plus Triton X-100 to give 0.5% final concentration and incubated with varying concentrations of [^{14}C]ACh as described previously.[17] The K_m was calculated by linear repression analysis of the reciprocal plots. M, Membrane-bound AChE; S, detergent-solubilized AChE. From Rotundo and Fambrough.[7] Copyright MIT, with permission.

X-100 (usually 100 μl of a 5% solution in HEPES-buffered HBSS, pH 7) is added to give 0.5% final concentration and restore the original assay volume. This procedure solubilizes all the enzyme associated with the cells and renders the intracellular AChE accessible to the substrate, resulting in an increase in enzyme activity (see Fig. 1B). This effect is not due to an activation of preexisting enzyme molecules (Fig. 1C) or to an effect on turnover number,[15] but rather reflects the exposure of new enzyme molecules to the substrate. After adding the detergent, several aliquots of incubation medium are extracted and counted as described above. The slope of this line represents the total AChE activity in the culture dish, and the value for intracellular AChE is obtained as the difference between the slopes of the cell surface and intracellular enzyme activity.

When assaying large numbers of cultures we routinely use an abbreviated version of this assay, which necessitates taking only two samples from each culture dish. The first aliquot is obtained at the end of 1 hr of incubation in the absence of detergent to determine surface AChE, usually 100 μl. A 100-μl aliquot of 5% Triton X-100 in HBSS is then added, incubation is continued for an additional 10 min, and a second 100-μl aliquot is removed for counting. The difference between the two values, corrected for background hydrolysis and time, represents the intracellular AChE activity. The percentage of enzyme activity located on the cell surface can be calculated by the following formula:

$$\% \text{ surface AChE} = \frac{(S - b)/t_1}{[T - (1 - s)\, S]/t_2} \times 100$$

where S is the counts per minute (cpm) (or nanomoles of ACh hydrolyzed) in the first sample, T is the CPM in the sample obtained after the addition of detergent, b is the value of the blank at the end of the first incubation, s is the size of the sample used to determine surface cpm expressed as its fraction of the total incubation volume, t_1 is the duration of the surface assay in hours, and t_2 is the duration of the assay in the presence of detergent, also in hours. Using this procedure for tissue-cultured chicken embryo muscle cells, we routinely find that approximately one-third of the AChE activity is localized on the external plasma membrane; the remaining two-thirds are found inside the cell associated with a rapidly turning over pool destined for secretion or incorporation into the external plasma membrane (see later sections).

This assay, used in conjunction with a wide variety of metabolic inhibitors and active site-directed ligands, has provided us with a valuable tool

[15] M. Vigny, S. Bon, J. Massoulie, and F. Leterrier, *Eur. J. Biochem.* **85,** 317 (1978).

with which to study the subcellular distribution of AChE molecules under different experimental conditions as well as evaluating the effects of specific inhibitors on the cell surface and intracellular enzyme.

Orientation of Newly Synthesized AChE

Since acetylcholinesterase in muscle cells is destined either for secretion or incorporation into plasma membrane,[7–9] these molecules should be synthesized on and sequestered within membrane-bound organelles.[16,17] This hypothesis can be tested directly by treating cultured cells with DFP, an irreversible inhibitor of AChE, and allowing the cells to synthesize AChE molecules for 2 hr. Under these conditions the cell surface AChE is completely inhibited, yet the cells contain a full intracellular pool of newly synthesized active enzyme (see later section of AChE recovery for details of this procedure). A microsomal fraction is then prepared from these cells and assayed for AChE activity by adding aliquots of microsomes to a buffer solution containing [^{14}C]ACh and varying concentrations of detergent (Fig. 2), in this case the nonionic detergent octylglucoside. Only a small amount of enzyme activity is detected in the absence of detergent. However, with increasing detergent concentration more enzyme activity is revealed, indicating that the newly synthesized active enzyme molecules are sequestered within the lumen of the microsomes. Identical results are obtained with other nonionic detergents, such as Triton X-100.

Use of Active Site-Directed Inhibitors for Studying AChE

There exists a vast number of specific active site-directed acetylcholinesterase inhibitors (for an extensive list, see Koelle[1]) that can be useful in the study of AChE metabolism. For our purpose, they can be classified into three categories. First are the lipid-soluble irreversible inhibitors such as diisopropyl fluorophosphate (DFP) and methane sulfonyl fluoride (MSF), which readily cross the plasma membrane and thus inactivate both the intracellular and extracellular pools of enzyme. These inhibitors react covalently with the catalytic site of AChE, and the enzyme remains inactive even after removal of the unreacted inhibitor and solubilization of the protein. These inhibitors can readily be washed out of the cells by simply rinsing several times with an isotonic salt solution. The net result is a population of cells that are functionally devoid of AChE activity. Since the recovery of enzyme activity requires *de novo* synthesis of AChE molecules, we can now study the metabolic pathway of transport and assembly using enzyme activity alone (see next section).

[16] G. Blobel, *Proc. Natl. Acad. Sci. U.S.A.* **77**, 1496 (1980).
[17] G. Palade, *Science* **189**, 347 (1975).

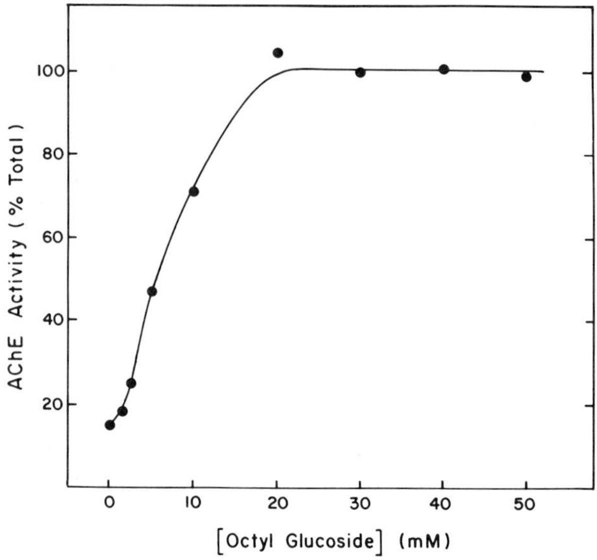

FIG. 2. Orientation of newly synthesized AChE in microsomes. Muscle cultures were treated with 10^{-4} *M* diisopropyl fluorophosphate, washed, covered with culture medium, and incubated at 37° for 2 hr to allow recovery of the intracellular pool of enzyme molecules. The cells were then homogenized in 0.25 *M* sucrose, 10 m*M* Tris, pH 7, and 1 m*M* EDTA using a Dounce homogenizer. Crude microsomes were prepared by pelleting a low speed supernatant onto a 60% sucrose cushion. Aliquots of the microsomes were mixed with an equal volume Tris buffer, pH 7, containing octyl glucoside to give the desired final concentration and assayed for AChE activity.

A second important class of compounds are the reversible water-soluble inhibitors, exemplified by the specific AChE inhibitor BW284c51. Inhibitors of this type were first used by McIsaac and Koelle[18] to "protect" the extracellular AChE from inhibition by DFP. Using the cell surface AChE assay we were able to demonstrate that for a monolayer of cells in tissue culture an initial incubation with 10^{-4} *M* BW284c51, followed by treatment with 10^{-5} *M* DFP resulted in complete inhibition of the intracellular enzyme while sparing over 90% of the cell surface AChE.[7] This technique is useful for studying the cell surface enzyme activity alone or in experiments where direct labeling of cell surface AChE is required. In this case incubation of the cells with [^{3}H]DFP after the above treatment labels only the extracellular enzyme molecules.

A third class of inhibitors that are extremely useful for metabolic studies are those that are irreversible and relatively impermeable to the

[18] R. S. McIsaac and G. B. Koelle, *J. Pharmacol. Exp. Ther.* **126,** 9 (1959).

plasma membrane, such as echothiophate or phospholine. Under the appropriate conditions of time, temperature, and concentration, echothiophate can be used to inhibit completely the extracellular AChE without affecting the intracellular pools. Convincing proof of its effectiveness has been worked out in tissue-cultured muscle cells. Using the cell surface AChE assay, Emmerling[19] could clearly show that incubation of a cell monolayer at 5° with 10^{-5} *M* echothiophate for 15 min inactivated essentially all the extracellular enzyme, without appreciably inhibiting the intracellular pool.

Synthesis of Acetylcholinesterase

Recovery of Intracellular AChE after DFP Treatment

This method was first used by Wilson and colleagues[9,20,21] with tissue cultured muscle cells, and it has been used extensively in our laboratory and by others to study the metabolism of AChE in culture.[7–9,19–22] Muscle cultures are rinsed with HBSS to remove esterases contributed by the horse serum and embryo extract followed by treatment at room temperature with 10^{-4} *M* DFP in HBSS. The DFP is conveniently kept as a 100 m*M* stock solution in isopropanol and diluted immediately before use. After a 10-min incubation, the cultures are washed four times with HBSS. This procedure inactivates more than 98% of the cell-associated enzyme, and the unreacted inhibitor is completely removed during the subsequent washes. The culture medium is then replaced, and the cells are returned to the incubator at 37°. The recovery of AChE in muscle cells after DFP treatment is illustrated in Fig. 3A. In this cell type, the rate of AChE synthesis is very high, approximately 20% of total cell contents of enzyme per hour. The plateau of cell-associated enzyme activity after 2–3 hr reflects the filling of the intracellular enzyme pool and the start of secretion (see next section). That this recovery requires *de novo* enzyme synthesis is shown in Fig. 3C; there is no recovery of AChE activity in the presence of protein synthesis inhibitors. There is no effect of the DFP treatment on the synthesis, transport, and secretion of AChE or the metabolism of other membrane-bound proteins, such as the acetylcholine receptor.

The use of DFP to inactivate total cell AChE has been useful in studying the time course of assembly of the different molecular forms of AChE

[19] M. R. Emmerling, Ph.D. Thesis, University of Wisconsin, Madison (1981).

[20] C. R. Walker and B. W. Wilson, *Neuroscience* **1,** 509 (1976).

[21] B. W. Wilson and C. R. Walker, *Proc. Natl. Acad. Sci. U.S.A.* **71,** 3194 (1974).

[22] F. Rieger, A. Faivre-Bauman, P. Benda, and M. Vigny, *J. Neurochem.* **27,** 1059 (1976).

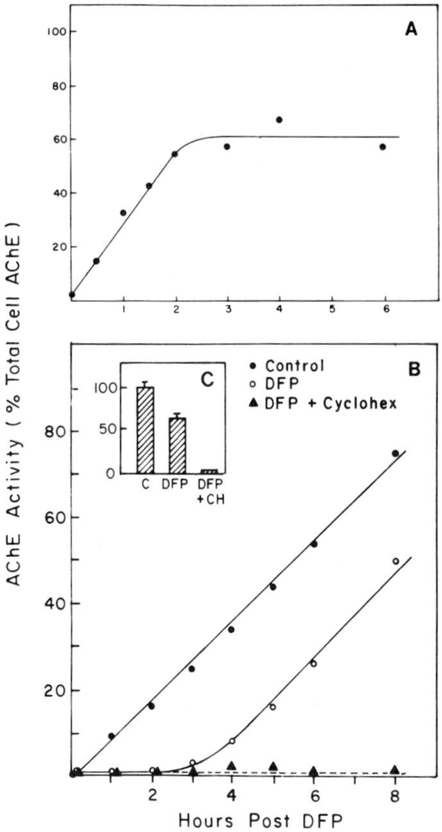

FIG. 3. Recovery of intracellular and secreted acetylcholinesterase following diisopropyl fluorophosphate (DFP) treatment. (A) Recovery of cell-associated AChE. Muscle cultures were treated with 10^{-4} *M* DFP, washed with HBSS, and returned to the incubator after adding 1 ml of DFP-treated medium per dish. At the indicated time intervals, three cultures were extracted with 10 m*M* Tris, pH 7, 0.5% Triton X-100, 1 m*M* EDTA, and 150 m*M* NaCl and assayed for AChE activity. Enzyme activity is expressed as percentage of untreated controls, and the standard errors are less than 10%. (B) Recovery of AChE secretion. Muscle cultures were treated with 2×10^{-4} *M* DFP, or HBSS alone, and incubated with 1 ml of DFP-treated medium in a 37° incubator. A third group of cultures were incubated with DFP-treated medium containing 100 μg of cycloheximide per milliliter after DFP treatment. At the indicated times 10-μl aliquots of medium were removed and stored at 5° until assayed at the end of the experiment. ●, Untreated cultures; ○, DFP-treated cultures; ▲, DFP-treated cultures incubated in the presence of 100 μg of cycloheximide per milliliter. Enzyme activity is expressed as percentage of the total AChE in control cells. (C) Cell-association AChE after an 8-hr recovery from DFP.

At the end of the experiment described in (B) the cultures were washed, extracted, and assayed as described in (A). C-control; DFP-DFP-treated; DFP + CH-DFP-treated and subsequent incubation with cycloheximide. From Rotundo and Fambrough[7]; Copyright MIT, with permission.

found in these cells (see Fig. 5). Used in conjunction with inhibitors of protein synthesis, pretreatment with DFP permits the use of pulse–chase studies of active enzyme molecules throughout their intracellular residence time. For example, cultures can be treated with DFP and allowed to synthesize AChE molecules for periods as short as 15 min before adding an inhibitor of protein synthesis. The fate of the population of newly synthesized AChE molecules, such as changes in subcellular location, oligomeric state, or appearance in the medium and/or on the plasma membrane, can then be followed over time.

Transport and Fate of Intracellular AChE

Secretion of AChE

Wilson and co-workers have shown that muscle cells release substantial amounts of AChE into the medium, and that this release is dependent upon protein synthesis.[9] That this release is due to secretion can be demonstrated using drugs that interfere with secretory processes, lowered temperature, or experiments employing direct labeling of the intracellular AChE molecules with [^{3}H]DFP.[7] In this latter experiment the intracellular AChE molecules were shown to be quantitatively transferred to the medium without any apparent residence time on the cell surface. An alternative strategy for studying secretion of AChE employs DFP-treated muscles and DFP-treated culture medium as described by Wilson *et al.*[9] The use of DFP-treated medium is necessary in order to detect the enzyme released by the cells; before use the treated medium is dialyzed against fresh medium to remove any unreacted inhibitor.

Figure 3B shows that muscle cells secrete at least 10% of their total cell contents of AChE per hour. If the cells are first treated with DFP and then allowed to recover in esterase-free medium, there is a 3-hr lag before active enzyme molecules appear in the medium. This secretion of active enzyme is dependent upon protein synthesis, as it is completely blocked by puromycin or cycloheximide. Once secretion has resumed, it continues at the same rate as control (untreated) cultures. The 3-hr lag period reflects the time from synthesis to release of the molecules at the plasma membrane and coincides with the time required to replenish the intracellular pool of active enzyme molecules (see Figs. 3A and C).

The converse experiment, measuring the time required to deplete the intracellular pool of secretable AChE, gives a result identical to that described above (Fig. 4). In this case untreated muscle cultures were incubated in DFP-treated medium in the presence or the absence of cycloheximide, and the appearance of AChE in the medium was monitored by

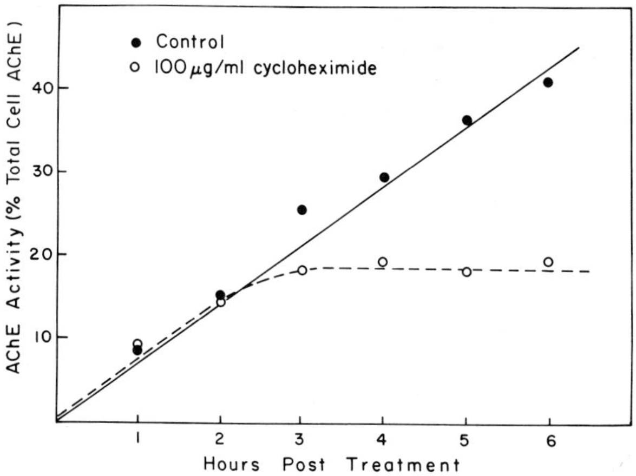

FIG. 4. Secretion of acetylcholinesterase in the presence of cycloheximide. Muscle cultures were washed with Hank's balanced salt solution and incubated at 37° with DFP-treated medium in the presence or the absence of 100 μg of cycloheximide per milliliter. At the indicated times 10-μl aliquots of medium were removed and kept on ice until assayed at the end of the experiment. Each point is the mean of three cultures, and enzyme activity is expressed as percentage of the cell-associated AChE in control cultures. From Rotundo and Fambrough[7]; copyright MIT, with permission.

repeatedly sampling 10-μl aliquots for enzyme activity. Inhibiting protein synthesis has no effect on the secretion of AChE for approximately 3 hr; thereafter the rate of release drops abruptly to zero. This indicates that the cells contain an intracellular pool of secretable AChE molecules equivalent to that released over a 3-hr period and that the time from synthesis to appearance at the plasma membrane is about 3 hr. Furthermore, these types of studies also indicate that the intracellular transport (and/or processing) of this protein is linear with time.

Transport and Accumulation of Cell Surface AChE

Experiments similar in design to those described for secretory AChE can also be used to study the cell surface enzyme.[7] In one study, muscle cultures were treated with DFP under sterile conditions and allowed to recover for periods as long as 48 hr. The appearance of newly synthesized AChE was monitored using the cell surface enzyme assay. Like the secretable forms of AChE, the cell surface membrane-bound enzyme also exhibits a 2–3-hr lag before its appearance, suggesting a similar transport mechanism. However, once externalized, it continues to accumulate on the plasma membrane at the rate of about 2–3% of total surface AChE per hour. Using BW284c51 to protect the surface AChE, DFP to inhibit irreversibly the intracellular enzyme, and [^{3}H]DFP to label the surface enzyme molecules, we can show that the cell surface AChE is not sloughed off into the medium, but rather is degraded by a process exhibiting first-order decay kinetics with a half-life of about 50 hr.

Localization and Assembly of Acetylcholinesterase Molecular Forms

Localization of Acetylcholinesterase Molecular Forms in Culture

Tissue-cultured chicken embryo muscle cultures contain two predominant forms of AChE with sedimentation coefficients of about 7 S and 11.5 S (Fig. 5). These forms correspond to dimers and tetramers[12] and occur in a ratio of about 4 : 1. In order to determine which forms are located on the external plasma membrane, muscle cultures are rinsed with HBSS to remove serum esterases and preincubated for 5 min at room temperature with 1 ml of 10^{-4} *M* BW284c51 in HBSS to inhibit reversibly

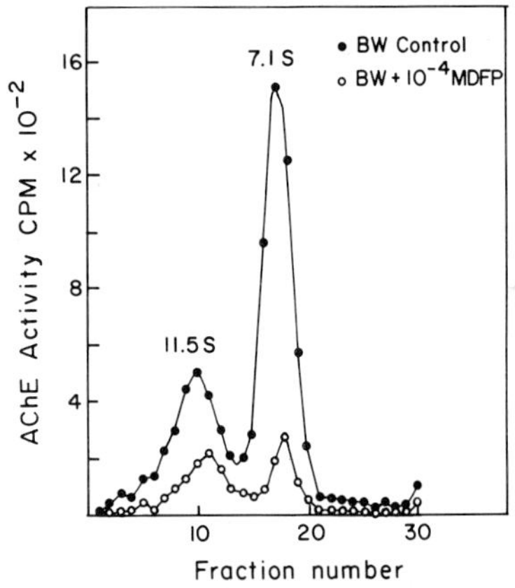

FIG. 5. Molecular forms of acetylcholinesterase on muscle cell surfaces. Muscle cultures were washed with Hank's balanced salt solution (HBSS) and covered with 1 ml of HBSS alone or containing 9×10^{-4} *M* BW284c51 to protect the cell surface acetylcholinesterase (AChE). Each set of control or protected cultures received in addition either 1 ml of HBSS alone or HBSS with 2×10^{-4} DFP. After incubation for 15 min at room temperature, the cultures were washed with 6×2 ml of HBSS, extracted in 500 μl of buffer A per three culture dishes, and centrifuged. The AChE molecular forms in the supernatant were analyzed by centrifuging 200-μl aliquots on 5 to 20% sucrose gradients made up in 50 m*M* Tris, pH 7, 1 *M* NaCl, 1 m*M* EDTA, and 0.5% Triton X-100. Centrifugation was for 15 hr at 41,000 rpm in an SW50.1 rotor at 4°. Fractions were collected from the bottom of tube (fraction 1). Ten-microliter aliquots of each fraction were assayed for 10 min. In this experiment 25% of the total AChE was protected by BW284c51, distributed about equally between the 11.5 S and 7 S molecular forms.

or "protect" the cell surface enzyme. A 100-μl aliquot of 10^{-4} *M* DFP in HBSS is then added to each dish, and incubation is continued for an additional 15 min, at which time both inhibitors are washed out with several changes of HBSS. The cells are then extracted in buffer A and centrifuged briefly to remove insoluble material; 200-μl aliquots of supernatant are analyzed by velocity sedimentation. A control set of cultures are run in parallel where DFP is omitted from the second HBSS treatment. The results of this experiment, illustrated in Fig. 5, indicate that both the 11.5 S and 7 S AChE forms are found on the external plasma membrane in approximately equal amounts. The difference between control and DFP-treated cultures represents the AChE forms located inside the cells.

Conversely, the cultures can be treated for 15 min on ice with 10^{-5} *M* echothiophate in HBSS to inhibit only the extracellular enzyme. In this case the difference between control and experimental cultures represents the molecular forms found on the cell surface, since with this soluble irreversible inhibitor the intracellular enzyme is spared.

Assembly of Acetylcholinesterase Molecular Forms

The rapid rate of AChE synthesis by cultured muscle cells and the sensitivity of the radiometric enzyme assay are such that the AChE forms synthesized during a 15-min period can easily be analyzed by velocity sedimentation and quantified. In one experiment muscle cell cultures are treated with DFP followed by incubation in complete culture medium at 37° for periods ranging from 15 min to 2 hr. Extraction of the cell-associated AChE and quantitation of each molecular form after velocity sedimentation indicates that the 7 S enzyme is synthesized at about 6 times the rate of the 11.5 S form. The rates of synthesis of both dimers and tetramers are linear with time and extrapolate to zero suggesting that assembly occurs at or near the time of synthesis. The possibility of interconversion of forms can be tested directly by using what is essentially a pulse–chase study of active enzyme molecules. After DFP treatment, the cells are allowed to synthesize AChE molecules for 1–2 hr, at which time the culture medium is replaced with medium containing 10 μg of puromycin per milliliter. The amounts of each molecular form of AChE are then determined over the next several hours using a combination of velocity sedimentation followed by enzymatic assay. Under these conditions there is no interconversion of molecular forms. Thus it appears unlikely that the dimeric form in these cells is the precursor of the tetrameric form. Once synthesized, each molecule retains its identity throughout its intracellular residence time and subsequent appearance either on the plasma membrane or in the external milieu.

Summary

The enzyme acetylcholinesterase consists of a family of molecular forms differing in subunit composition, solubility properties, and subcellular location. The use of a variety of reversible or irreversible active site-directed ligands with different membrane permeability properties permits the selective inactivation of separate pools of enzyme molecules. The application of these inhibitors together with standard biochemical techniques has permitted a detailed characterization of the synthesis and metabolism of the secretory and membrane-bound acetylcholinesterase in tissue-cultured cells. These techniques, with minor modifications and appropriate controls, can also be applied to the study of AChE metabolism in organ culture and *in vivo*.

Acknowledgments

This work was supported by a grant from the Muscular Dystrophy Association to Douglas M. Fambrough, a Postdoctoral Fellowship from the Muscular Dystrophy Association to R. L. R., and NSF Grant BNS-8015778 to R. L. R.

[29] Biosynthesis of Acetylcholine Receptor *in Vitro*

By DAVID J. ANDERSON and GÜNTER BLOBEL

The studies of acetylcholine receptor (AChR) structure and biosynthesis have until recently been dichotomous; the former utilizing the electric organs of rays or eels (for review, Heidmann and Changeux[1]), the latter skeletal myotubes grown in tissue culture[2]. While large quantities of material have made the electroplax AChR especially amenable to biochemical and immunological analysis, skeletal myotubes constitute a well-defined cell culture system in which the kinetics of AChR biosynthesis and turnover can be carefully measured under a variety of controlled physiological conditions. Unfortunately, muscle AChR comprises essentially a trace amount of the total cellular protein, making parallel biochemical studies extremely difficult in this system. In making the transition to AChR molecular biology, we and others have compromised physiological manipulability in favor of biochemical accessibility and have opted to study AChR synthesis in the electric organ. Since there are currently no

[1] T. Heidmann and J.-P. Changeux, *Ann. Rev. Biochem.* **47,** 317 (1978).
[2] D. M. Fambrough, this volume [27].

ISBN 0-12-181996-5

published *Torpedo* culture systems, we have had to confine our approach to the use of cell-free systems. The following methods are addressed to this *in vitro* approach to AChR biosynthesis.

RNA Isolation

Torpedo californica electroplax tissue contains extremely small quantities of RNA, about 1.0 A_{260} unit (40 μg) per gram of tissue. This value has been obtained consistently either from direct measurement in crude homogenates or on the basis of yields of a variety of independent extraction procedures. Part of the reason for this is that the tissue is almost 98% water by weight; the other part is that it is a terminally differentiated organ with very low biosynthetic activity (the constituent cells are multinucleated electrocytes that have ceased to divide). Nonetheless, this RNA can be extracted efficiently if the appropriate modifications are introduced into the isolation procedure to compensate for the low RNA to tissue ratio.

We obtain our electroplax tissue frozen in liquid nitrogen from Pacific Biomarine (Venice, California), and have stored it for up to 4 months at −70° without adverse effect upon RNA content. The various extraction procedures used have been discussed extensively elsewhere in this volume; rather than detail them we will simply mention the modifications we have introduced. The two best procedures among those tested so far appear to be the SDS–phenol–proteinase K method, and the guanidinium thiocyanate–CsCl method. For the former, we have scaled up the initial homogenate slightly so that four volumes of buffer per unit weight of frozen tissue are used. After combining the aqueous phases from one back-extraction and two reextractions, glycogen is removed by centrifugation at 30,000 *g* for 45 min. The nucleic acids are next precipitated with 0.6 volume of isopropanol, rather than with ethanol. This yields a precipitate that is easier to dissolve and eliminates a yellow pigment that contaminates ethanol precipitates. Care should be taken to use siliconized glassware, as the RNA concentration is quite low and considerable losses due to adsorption can otherwise occur. The sample should be left at −20° for at least 24 hr to ensure complete precipitation of the RNA.

After resuspending this material in water, it is wise to perform one last phenol extraction prior to LiCl precipitation of the RNA. If there is residual protein in the redissolved isopropanol precipitate, it will immediately "salt out" when LiCl is added to 2 *M*, appearing as a cloudy yellow precipitate. Should this occur, remove the precipitate immediately by centrifugation at 10,000 *g* for 20 min at 15. This will not remove any RNA, which requires at least 6 hr to precipitate, but eliminates the protein that would otherwise make it virtually impossible to redissolve the RNA.

We have found that the guanidinium thiocyanate–CsCl procedure[3] can be scaled up without adverse effects for electroplax RNA isolation. Thus, 50 g of frozen, pulverized tissue are added gradually to 40 ml of 5 *M* guanidinium thiocyanate (in 50 m*M* Tris, pH 7.6, 10 m*M* EDTA, 4% Sarkosyl NL-97, and 5% 2-mercaptoethanol) at 65°, with gentle swirling. After a 10-min incubation, the suspension is homogenized in a Dounce apparatus with a loose-fitting pestle and returned for another 10 min to the 65° incubation bath. This homogenate is centrifuged at 30,000 *g* for 20 min to remove insoluble material, 0.15 g of solid CsCl are added to each milliliter of supernatant, and the resulting solution is layered over a cushion of 5.7 *M* CsCl in 0.1 *M* EDTA for sedimentation of the RNA.[3]

Although all the AChR subunit mRNAs are polyadenylated,[4] it is not necessary to poly(A)-select the total electroplax RNA for efficient *in vitro* translation. Indeed, because of the small amounts of RNA obtained from the tissue and the losses that can occur during oligo(dT)-cellulose chromatography, we have in general avoided this enrichment procedure when the RNA is to be used exclusively for cell-free synthesis experiments.

Cell-Free Synthesis

Acetylcholine receptor subunits (sAChRs) are efficiently translated in both the rabbit reticulocyte lysate system[4] and the wheat germ system.[5] Both systems are discussed in this volume [3,4]. Although in general each RNA preparation should be titrated to achieve optimal incorporation of labeled amino acids, a good standard condition is 5 A_{260} units of total cellular RNA per milliliter of translation mixture. The pattern of total translation products obtained in the wheat germ system is shown in Fig. 1 (lane B). Each *in vitro* synthesized sAChR constitutes about 0.5% of the total radioactivity incorporated into newly synthesized protein. With a good RNA preparation, we are routinely able to achieve incorporations of 4 to 5×10^5 dpm of trichloracetic acid-insoluble radioactivity per 5 μl of total translation mixture. Thus, an individual sAChR immunoprecipitated from 50 μl of such a translation mixture will yield a dark band after 2 days of fluorographic exposure. Since each sAChR is synthesized from a separate mRNA, there are different Mg^{2+} optima for the synthesis of these separate polypeptides: 3.5 m*M* for α and γ, and 2.5 m*M* for β and δ. A good compromise is therefore to work at 3.0 m*M*. Under these conditions (with 150 m*M* KOAc) fairly comparable levels of all four chains are synthesized in the same reaction mixture.

[3] P. Lizardi, this volume [2].

[4] B. Mendez, P. Valenzuela, J. A. Martial, and J. D. Baxter, *Science* **209,** 695 (1980).

[5] D. J. Anderson, and G. Blobel, *Proc. Natl. Acad. Sci. U.S.A.* **78,** 5598 (1981).

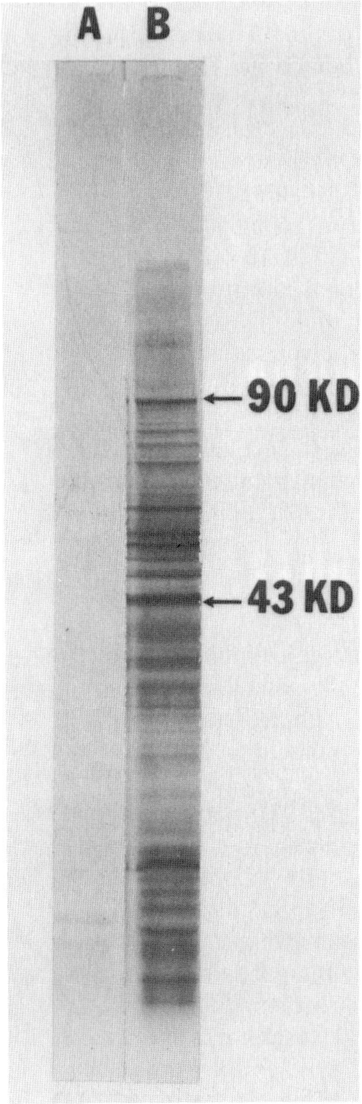

FIG. 1. Total translation products of *Torpedo californica* electroplax mRNA, synthesized in the wheat germ system. Five microliters of total reaction mixture incubated either with (B) or without (A) added mRNA were applied to a 10–15% linear gradient polyacrylamide slab gel. The dried gel was autoradiographed for 24 hr. Molecular mass calibration is indicated in the right-hand margin. KD, kilodalton.

Immunoprecipitation of *in Vitro* Synthesized AChR Subunits

All four sAChRs are transmembrane glycoproteins that are synthesized, in the absence of microsomal membranes, as nonglycosylated precursors that have different electrophoretic mobilities than their mature counterparts. Furthermore, under some conditions, these sAChR precursors have multiple electrophoretic forms that further confuse their analysis (see below). In order to distinguish clearly between these various species it is essential to use subunit-specific antisera for immunoprecipitation.

Anti-sAChR antibodies have been prepared in guinea pigs,[6] rabbits,[7] and rats.[7] In our own analysis,[5] we have used rat antibodies prepared against SDS-denatured sAChRs eluted from preparative polyacrylamide gels. An advantage of using rat antibody is that the IgG heavy chain has a sufficiently different molecular weight from the AChR β chain that the small radiochemical amounts of the latter are not obscured by the microgram quantities of the former upon electrophoresis. Rabbit IgG heavy chain, in contrast, comigrates (on 7.5 to 15% gradient gels) with AChR-β and obscures it almost completely.

The *in vitro* synthesized sAChRs are solubilized in SDS prior to immunoprecipitation. The rationale for this procedure is discussed in this volume.[8] In the case of the α chain, it is imperative that SDS solubilization be carried out at room temperature. Heating of α causes it to migrate anomalously in gel electrophoresis (see below).

All four sAChRs exhibit some degree of sequence homology in the first 54 N-terminal amino acid residues.[9] Consistent with this, "subunit-specific" antisera, including monoclonal antibodies,[10] show varying degrees of intersubunit cross-reactivity. It is possible, however, to render sAChR immunoprecipitates subunit-specific by using sufficiently low amounts of antibody such that the contribution from the cross-reacting species becomes negligible. A simple titration experiment will define such ideal conditions. When establishing these parameters it is essential to identify the sAChR primary translation products by competition experiments with unlabeled, mature subunit, as several of the species migrate quite close to one another in SDS–gel electrophoresis (see below).

Electrophoretic Analysis of *in Vitro* Synthesized sAChR Primary Translation Product

The electrophoretic behavior of *in vitro* synthesized sAChRs is quite sensitive to preparation conditions for SDS gel samples, and unless the

[6] T. Claudio and M. A. Raftery, *Arch. Biochem. Biophys.* **181,** 484 (1977).

[7] J. Lindstrom, B. Einarson, and J. Merlie, *Proc. Natl. Acad. Sci. U.S.A.* **75,** 769 (1978).

[8] D. J. Anderson and G. Blobel, this volume [8].

[9] M. A. Raftery, M. W. Hunkapiller, C. B. D. Strader, and L. E. Hood, *Science* 208, 11454 (1980).

[10] S. J. Tzartos and J. M. Lindstrom, *Proc. Natl. Acad. Sci. U.S.A.* **77,** 755 (1980).

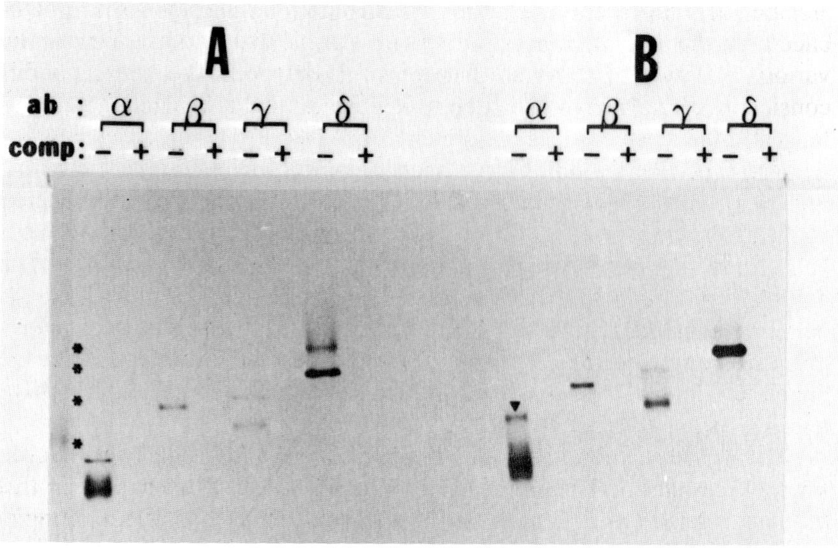

FIG. 2. Effect on alkylation on the electrophoretic behavior of sAChR precursors synthesized *in vitro*. Group A: Samples applied to gel after reduction alone (100 mM dithiothreitol, 4 min at 100°). Group B: Samples applied to gel with alkylation following reduction (500 mM iodoacetamide, 45 min at 37°). ab indicates the antibody used for immunoprecipitation (either anti-α, β, γ or δ); comp + indicates that, as a control, 5 μg of the corresponding mature sAChR (obtained by elution from a preparative SDS gel[5]) was included to inhibit competitively the immunoprecipitation of all *in vitro* synthesized species related to that particular subunit. The four asterisks mark the M_r values of the mature α-, β-, γ-, and δ-subunits (reading from the lowest to the uppermost, respectively). All samples in group B migrate slightly slower than their counterparts in group A owing to the higher salt concentration in the samples, which retards sample entry into the gel. Arrowhead (B, α) indicates alkylated form of the subunit.

parameters described below are adhered to strictly, a different pattern will be obtained. The standard sample solubilization buffer consists of 5% Pierce (Sequenal grade) sodium dodecyl sulfate (SDS), 15% (w/v) sucrose, 0.2 M Tris-HCl, pH 8.8, 1 mM EDTA, and 0.01% Bromphenol Blue. To effect reduction of disulfide bonds, dithiothreitol (DTT) is added directly to the above buffer at the concentrations indicated. Where subsequent alkylation is called for, iodoacetamide (IAA) is freshly prepared as a 1.0 M stock solution in the same buffer and added to achieve the appropriate final concentration. All four sAChRs are conveniently displayed on a 7.5 to 15% linear gradient polyacrylamide gel with a 5% stacking gel. Pierce (Sequenal grade) SDS is used in the running buffer. The brand of SDS should not be varied from experiment to experiment, as results will be difficult to compare. When reduction at 100 mM DTT is followed by alkylation at 500 mM IAA, the salt concentration in the sample is raised significantly. Since the rate at which the sample enters the gel is influenced by the salt concentration in the sample, we indicate below the various buffers that are compatible with this protocol. The running buffer consists of 50 mM Tris base, 250 mM glycine, and the stacking gel contains 50 mM Tris-HCl, pH 6.8, and 15% sucrose. When running samples that have been reduced but not alkylated, the salt concentration in the sample should be increased by adding Tris buffer to a final concentration of 0.4 M (pH 8.8). Nonetheless, samples prepared differently should not be run immediately adjacent to one another on the slab gel, as distortion will occur.

Of the four sAChRs, the β chain seems to be the best behaved. It migrates with an M_r of 50,000 (Fig. 2A,B, lane β) whether reduction alone or reduction-alkylation is used. The sample can be heated to 100° without altering its electrophoretic mobility.

In the case of the α chain, heating the sample above 25° causes it to run as a doublet, consisting of a sharp band at M_r 39,000 with a smeared band beneath it (Fig. 2A, lane α, Fig. 3, H). If, on the other hand, heating is omitted, almost all the material runs as the sharp, M_r 39,000 species (Fig. 3, nH). This sharp band, furthermore, runs at M_r 39,000 only if the sample is prepared with reduction alone; if alkylation is performed, α increases its apparent M_r by several thousand (see Fig. 2B, lane α) as shown originally by Froehner and Rafto.[11] Thus, we avoid heating and alkylation for α.

The γ- and δ-subunits can be heated to 100°. If alkylation is omitted, however, both of these subunits run as doublets (see Fig. 2A, lanes γ and δ). The upper band of these doublets is converted to the lower species by efficient alkylation of the sample (Fig. 2B), suggesting that it arises as the result of reoxidation of intramolecular disulfide bonds during electrophoresis. In both cases, reduction with 100 mM DTT followed by alkylation with 500 mM IAA is required; even under these stringent conditions, conversion to the lower band is occasionally incomplete (see Fig. 2B, lanes γ and δ). Reduced and alkylated pre-AChR δ migrates at M_r 59,000, whereas pre-AChR γ migrates at M_r 48,000–49,000. Note that pre-γ actually migrates *faster* than pre-β, whereas in mature AChR the converse is true.

The optimal sample preparation conditions for each sAChR are summarized below:

α chain: 100 mM DTT; 20 min at 25°, no alkylation
β chain: 100 mM DTT; 4 min at 100°, alkylation not necessary.
γ and δ chains: 100 mM DTT; 4 min at 100°; alkylation with 500 mM IAA, 45 min at 37°

[11] S. C. Froehner and S. Rafto, *Biochemistry* **18,** 301 (1979).

FIG. 3. Effect of heating on the electrophoretic behavior of pre-AChR α synthesized *in vitro*. A translation reaction mixture was solubilized, prior to immunoprecipitation, in 1% SDS for 20 min at 25°. After immunoprecipitation with anti-α antibodies, samples were solubilized in SDS gel sample buffer containing 100 m*M* dithiothreitol at either 100° (H) for 4 min, or at 25° (nH) for 20 min. The arrow indicates the M_r 39,000 species referred to in the text.

Integration of *in Vitro* Synthesized sAChRs into Microsomal Membranes

When synthesized in the presence of pancreatic microsomal membranes (ca. 2 A_{280} units/ml), sAChRs undergo both proteolytic removal of a signal sequence and core glycosylation.[12] Glycosylation of the γ- and δ-

[12] D. J. Anderson, P. Walter, and G. Blobel, *J. Cell Biol.* **93,** 501 (1982).

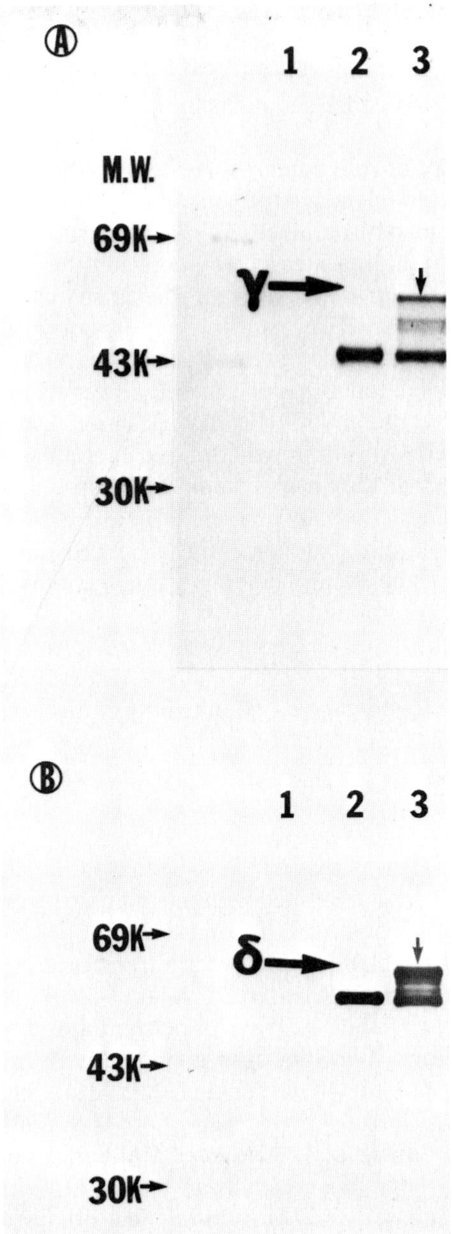

FIG. 4. Effect of dog pancreas microsomal membranes on forms of the γ (A) and δ (B) sAChRs synthesized *in vitro*. Lanes: 1, M_r of the mature subunit is marked by an arrow; 2, sAChR synthesized *in vitro* without microsomal membranes; 3, sAChR synthesized in the presence of 2 A_{280} units per milliliter of microsomal membranes. Arrows indicate (lanes 3) the fully glycosylated forms of the subunit. M.W., molecular weight markers (K = 1000). From Anderson and Blobel.[5]

subunits increases their apparent molecular weight to ca. 60,000 and 65,000, respectively (Fig. 4A,B, lanes 3). The α-subunit is shifted up only slightly when glycosylated, to about M_r 40,000 (Fig. 5A, lane 3). In contrast, the molecular weight of glycosylated β-subunit is the same as that of its precursor (Fig. 5B, lane 3).

Since the glycosylation capacity of rough microsomes is variable and limited,[13] partially glycosylated or nonglycosylated, but signal peptidase-cleaved, forms of the sAChRs can also be obtained. The γ- and δ-subunits show partially glycosylated forms that migrate midway between the precursor and fully glycosylated forms (Fig. 4A,B, lanes 3). They also exhibit signal peptidase-cleaved, nonglycosylated forms, which are poorly resolved from residual, unprocessed precursor in this gel system.[12] The signal peptidase-cleaved, nonglycosylated forms of α and β, however, are well resolved from their respective precursors (Fig. 5A,B, lanes 3, upward arrows). Partially glycosylated forms of these latter two subunits are not observed, which may indicate that they each contain one Asn-linked core oligosaccharide, whereas γ and δ contain several. We have experienced difficulty in consistently obtaining batches of rough microsomes with high glycosylation capacity. The factors affecting this variability have not as yet been systematically explored.

The existence of partially glycosylated and nonglycosylated forms of the sAChRs can be demonstrated using the enzyme endo-β-*N*-acetylglucosaminidase H (endo H). This enzyme cleaves N-linked core oligosaccharide moieties at the penultimate *N*-acetylglucosamine residue and eliminates the contribution of carbohydrate to the electrophoretic mobility of the polypeptide(s). Thus, all glycosylation variants of a particular sAChR will be converted to a single band by treatment with endo H.

To perform digestion with endo H, washed protein A-Sepharose beads containing immunoprecipitated sAChRs are eluted with one volume (typically 20 μl) of 1 m*M* Tris-HCl, pH 7.4, 50 m*M* DTT, and 1% SDS, at 100° for 3 min. (The α chain is treated at 25°.) The eluates are transferred to fresh tubes, the beads are rinsed once with an equal volume of 0.3 *M* sodium citrate, pH 5.5 (containing 100 units of Trasylol per milliliter and 0.1 m*M* PMSF as protease inhibitors), and this rinse is pooled with the original eluate. The sample, at this point in 40 μl, contains 0.15 *M* sodium citrate, 0.5 m*M* Tris-HCl, 25 m*M* DTT, and 0.5% SDS. 2 μl of a 0.66 units per milliliter endo H solution in 10 mg of bovine serum albumin per milliliter are added, so that the final enzyme concentration is 0.033 units per milliliter. The samples are incubated at 37° for 16 hr. (We obtained endo H as a gift from Dr. Phillips Robbins of Massachusetts Institute of

[13] P. Walter and G. Blobel, this volume [6].

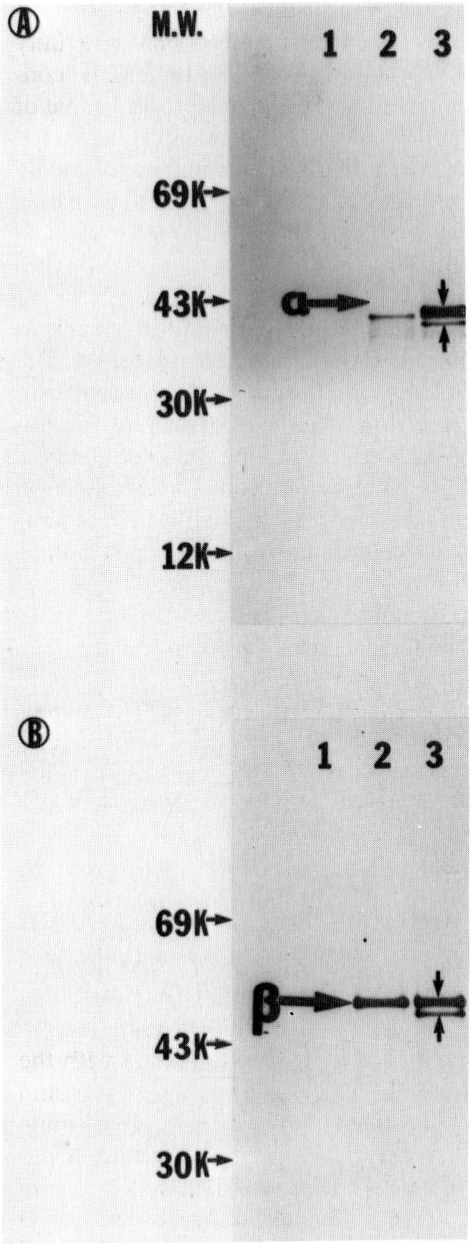

FIG. 5. Effect of dog pancreas microsomal membranes on forms of the α (A) and β (B) sAChRs synthesized *in vitro*. Lanes: 1, M_r of the mature subunit is marked by an arrow; 2, sAChR synthesized without microsomes; 3 sAChR synthesized with microsomes. Downward arrows (lanes 3) indicate the glycosylated forms; upward arrows, the nonglycosylated (but signal-peptidase-cleaved) forms of the subunits. M.W., molecular weight markers (K = 1000). From Anderson and Blobel.[5]

Technology; although it can be purchased from Miles-Yeda, Inc., the activity of this commercially available preparation is apparently variable.) Preparation of the endo H-digested samples for electrophoresis is performed by adding 10 μl of a solution containing 1.0 *M* sucrose, 1.0 *M* Tris-HCl, pH 8.8, 10 m*M* EDTA, 15% SDS, 0.05% Bromphenol Blue, and 100 m*M* DTT. If desired, alkylation can be achieved by subsequently adding 5 μl of 1.0 *M* IAA in 0.2 *M* Tris base and incubating as described earlier.

Summary

Torpedo acetylcholine receptor is accessible to analysis by cell-free protein synthesis, like other membrane proteins studied previously.[14] The factors deserving particular attention in this system are the small amounts of messenger RNA contained in electroplax tissue, the inherent complexity of an oligomeric membrane protein (necessitating the use of subunit-specific antibodies), the anomalous electrophoretic behavior of the primary translation products (attributable in part to the high —SH content of sAChRs), and the multiple glycosylated forms that appear upon processing by microsomal membranes. Although not all these complexities can be satisfactorily explained, they have been empirically characterized and can be well controlled if the appropriate procedures are followed.

[14] F. N. Katz, J. E. Rothman, V. R. Lingappa, G. Blobel, and H. F. Lodish, *Proc. Natl. Acad. Sci. U.S.A.* **74,** 3278 (1977).

[30] Biosynthesis of Myelin-Specific Proteins

By David R. Colman, Gert Kreibich, and David D. Sabatini

A simple procedure requiring one centrifugation through a discontinuous sucrose density gradient has been devised for the preparation of purified rough microsomes, free polysomes, and myelinated axon fragments from rat brainstems. Several other fractionation schemes for brain have been published,[1-5] and the one described here incorporates some

[1] C. E. Zomzley-Neurath and S. Roberts, *Methods Neurochem.* **1,** 95 (1972).
[2] J. M. Gilbert, *Biochem. Biophys. Res. Commun.* **52,** 79 (1973).
[3] J. M. Gilbert, *Biochim. Biophys. Acta* **340,** 140 (1974).
[4] L. Lim, J. O. White, C. Hall, W. Berthold, and A. N. Davison, *Biochim. Biophys. Acta* **361,** 241 (1974).
[5] J. Ramsey and W. J. Steele, *J. Neurochem.* **28,** 517 (1977).

ISBN 0-12-181996-5

features of these methods. It utilizes a high-salt medium for homogenization to facilitate removal of adsorbed free polysomes from rough microsome membranes,[6] and sedimentation of the rough microsomes into a pellet is avoided. Because ribosomes are rapidly removed from the sample layer in which rough microsomes slowly sediment, trapping of free polysomes in the rough microsome fraction is reduced. Myelin fragments obtained by this procedure may be used immediately for RNA extraction or may be further treated by osmotic shock to separate a myelin membrane preparation from which the major myelin proteins may be purified.

Solutions

Buffer A: 250 m*M* sucrose, 10 m*M* HEPES, 50 U of Trasylol per milliliter, 3 m*M* DTT

Buffer B: 500 m*M* KCl, 25 m*M* HEPES, pH 7.4, 5 m*M* $MgCl_2$, 50 U of Trasylol per milliliter, 3 m*M* DTT

Buffer C: 2.5 *M* sucrose, 25 m*M* HEPES, 5 m*M* $MgCl_2$, 50 U of Trasylol per milliliter, 3 m*M* DTT

NOTE: All solutions are autoclaved without DTT and Trasylol, which are added just prior to use.

Preparation of Subcellular Fractions from Brainstem for Extraction of RNA

Subcellular fractions are prepared (Fig. 1) from young rats (10–25 days old) in which the medulla, pons, and ventral midbrain are undergoing rapid myelinogenesis.[7]

In a typical preparation, six rats are decapitated, and the brains are removed and placed in ice-cold buffer A. Brainstems are dissected, and the tissue (approximately 4 g) is transferred to a Potter–Elvehjem glass homogenizer (type C, Thomas Instruments) containing 27 ml of ice-cold buffer B. Seven strokes of a Teflon pestle rotating at 500–800 rpm are usually sufficient for homogenization. During each stroke, care must be taken not to remove the spinning pestle from the homogenate, since this will cause extensive foaming. After homogenization, nuclei are removed by centrifugation in a Sorvall HB4 rotor (2000 rpm, 2 min). It is imperative that this step be performed immediately after homogenization, since a precipitate will form rapidly under the high-salt conditions. The postnuclear supernatant (approximately 25 ml) is collected and immediately di-

[6] J. Ramsey and W. J. Steele, *Biochemistry* **15,** 1704 (1976).

[7] N. Banik and M. Smith, *Biochem. J.* **162,** 247 (1977).

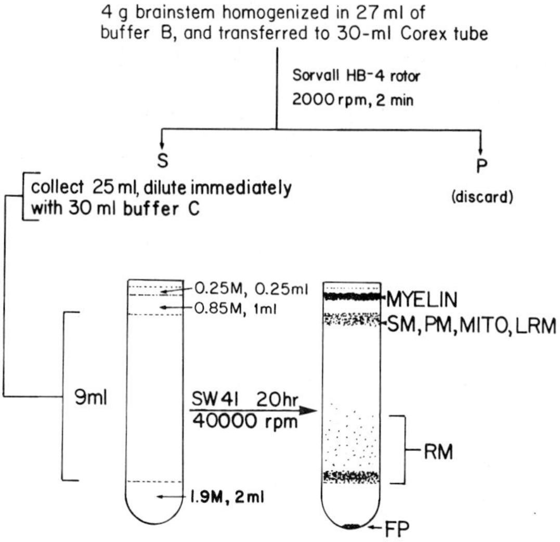

FIG. 1. Flow chart for the fractionation procedure. The sucrose concentration (*M*) and volume (ml) of each layer in the discontinuous sucrose gradient are indicated. Solutions in the gradient also contain 25 m*M* HEPES, 250 m*M* KCl, 5 m*M* $MgCl_2$, 50 U of Trasylol per milliliter, and 3 m*M* dithiothreitol. SM, Smooth membranes including Golgi membranes; PM, plasma membrane; MITO, mitochondria; LRM, "light" rough microsomes; RM, rough microsomes; FP, free polysomes.

luted with enough buffer C (30 ml) to adjust the final sucrose concentration to 1.4 *M*. Aliquots (9 ml) of this suspension are layered in six SW41 tubes, above cushions of 1.9 *M* sucrose (2 ml) and then overlaid with 0.85 *M* sucrose (1 ml) and 0.25 *M* sucrose (0.25 ml), each with identical salt concentrations as the diluted homogenate (25 m*M* HEPES, 250 m*M* KCl, 5 m*M* $MgCl_2$, 3 m*M* DTT, 50 U of Trasylol per milliliter). After centrifugation (40,000 rpm for 20 hr in a Beckman SW41 rotor), fractions are recovered from the 0.25 *M*/0.85 *M* and 1.4 *M*/1.9 *M* sucrose interfaces. These fractions consist of a pellicle of myelinated axon fragments (Fig. 2A), which is removed with a spatula from the 0.25 *M*/0.85 *M* interface, and a zone containing rough microsomes (Fig. 2B), which is aspirated from the 1.4 *M*/1.9 *M* interface with a Pasteur pipette. Free polysomes (Fig. 2C) are recovered in the pellets.

Rough microsome suspensions may be directly frozen in liquid nitrogen and stored at −80° or diluted with 2 volumes of sterile water (4°), sedimented in an SW41 rotor (40,000 rpm, 3 hr), and used immediately for RNA extraction.

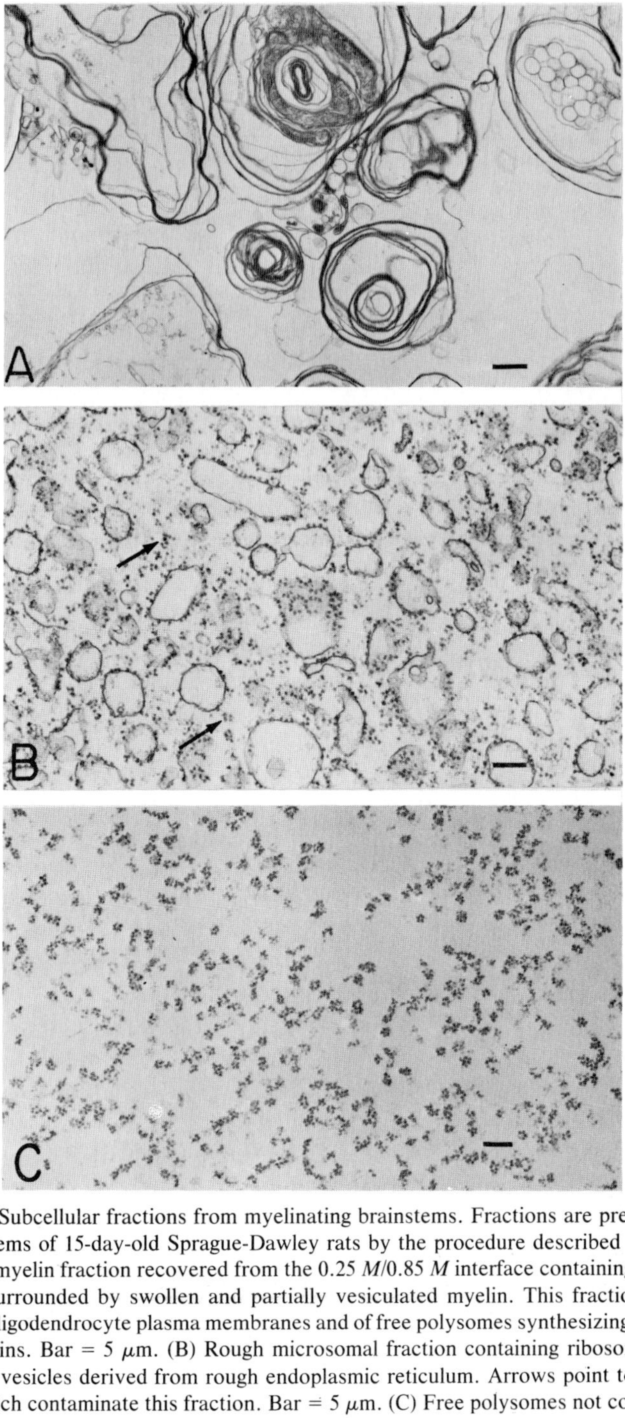

FIG. 2. Subcellular fractions from myelinating brainstems. Fractions are prepared from the brainstems of 15-day-old Sprague-Dawley rats by the procedure described in the text. (A) Crude myelin fraction recovered from the 0.25 *M*/0.85 *M* interface containing fragments of axons surrounded by swollen and partially vesiculated myelin. This fraction is a rich source of oligodendrocyte plasma membranes and of free polysomes synthesizing the myelin basic proteins. Bar = 5 μm. (B) Rough microsomal fraction containing ribosome-studded membrane vesicles derived from rough endoplasmic reticulum. Arrows point to free polysomes, which contaminate this fraction. Bar = 5 μm. (C) Free polysomes not contaminated by membranes, recovered in the pellet. Bar = 3 μm.

For the extraction of total RNA, each subcellular fraction is dissolved in a 6 *M* reagent grade guanidine hydrochloride solution (11 ml)[8] containing 25 m*M* sodium acetate, pH 5, and 20 m*M* DTT. The samples are then centrifuged over 1.7 ml of 5.7 *M* $CsCl_2$ with 100 m*M* EDTA, pH 6, in a Beckman SW41 rotor (35,000 rpm) for 20 hr at 25° to sediment total RNA. The glassy pellets are dissolved in 2–3 ml of water, and 2 *M* sodium acetate, pH 5, is added to a final concentration of 100 m*M*. Total RNA is precipitated by addition of 2.2 volumes of cold (−20°) 95% ethanol followed by overnight incubation at −20°. The pellet recovered after centrifugation (10,000 rpm, 20 min in a Sorvall HB4 rotor) is dissolved in 0.5% SDS, 500 m*M* NaCl, and poly(A)$^+$ mRNA is then prepared by affinity chromatography on an oligo(dT)-cellulose column.[9]

Purification of Myelin Basic Proteins (MBPs) and Proteolipid Protein (PLP) from Myelin Fragments

The myelinated axon fragments obtained by the above procedure may be subjected to osmotic shock[10] to partially remove nonmyelin contaminants, which include axonal and glial constituents. This step also releases oligodendrocyte free polysomes, which are found in this fraction and contain a high proportion of polysomes engaged in the synthesis of MBPs.[11] Myelinated axon fragments obtained as above by discontinuous sucrose gradient centrifugation are subjected to osmotic shock by homogenization of the pellicle in 20 volumes of 10 m*M* HEPES, pH 7.4, containing 50 U of Trasylol per milliliter. The suspension is then adjusted to 0.85 *M* sucrose by the addition of 2.5 *M* sucrose (in 10 m*M* HEPES, pH 7.4, containing 50 U of Trasylol per milliliter), and 12-ml aliquots of this suspension are placed in SW41 tubes and overlaid with 0.25 *M* sucrose (0.5 ml) in 10 m*M* HEPES. After centrifugation (40,000 rpm for 4 hr) a pellicle at the 0.25 *M*/0.85 *M* sucrose interface is collected and subjected two more times to the osmotic shock and centrifugation procedure. The final pellicle, consisting of purified myelin membranes, is used immediately for extraction of myelin proteins (Fig. 3) or may be frozen in liquid N_2 and stored at −80°.

Myelin membranes prepared as above are partially delipidated by suspension and incubation (−20°, 16 hr) in 20 volumes of cold (−20°) acidified (5 drops of 12 *N* HCl per liter) acetone. After centrifugation (10,000

[8] C. P. Liu, D. Slate, R. Gravel, and F. H. Ruddle, *Proc. Natl. Acad. Sci. U.S.A.* **76,** 4503 (1979).
[9] H. Aviv and P. Leder, *Proc. Natl. Acad. Sci. U.S.A.* **69,** 1408 (1972).
[10] W. T. Norton and S. E. Poduslo, *J. Neurochem.* **21,** 749 (1973).
[11] D. R. Colman, G. Kreibich, A. B. Frey, and D. D. Sabatini, *J. Cell Biol.* **95,** 598 (1982).

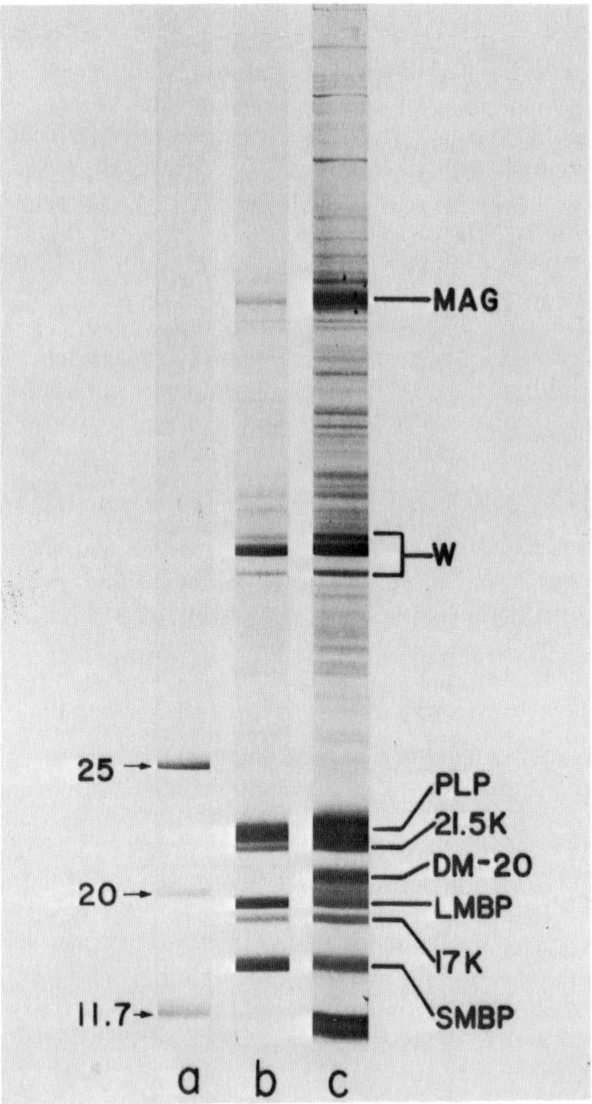

FIG. 3. Electrophoretic patterns of purified myelin membrane proteins. Purified myelin membranes were prepared according to the discontinuous sucrose gradient and osmotic shock procedures described in the text. Samples (75 μg protein) of delipidated myelin were analyzed by electrophoresis in a slab gel together with molecular weight standards (a): cytochrome *c* (11,700), soybean trypsin inhibitor (20,000), chymotrypsinogen A_4 (25,000). All samples were solubilized in sample buffer containing 10% SDS, heated to 100° and received 500 m*M* dithiothreitol after cooling. After electrophoresis (20 mA for 20 hr) the gel was fixed in acetic acid–methanol, stained with Coomassie Blue, photographed (b), and then silver stained and rephotographed (c) [W. T. Wray, T. Boulikas, V. Wray, and R. Hancock, *Anal. Biochem.* **118,** 97 (1981)]. MAG, Myelin-associated glycoprotein; W, Wolfgram region; PLP, proteolipid protein; LMBP, large myelin basic protein; SMBP, small myelin basic protein.

rpm, 20 min in a Sorvall HB4 rotor), the supernatant is discarded and the sediment is washed three times by suspension and resedimentation from ice cold water (20 volumes each). The washed material is suspended in 5 volumes of 0.5 *N* acetic acid or 0.5 *N* HCl[4] and is kept on ice for 1 hr, after which a supernatant and a sediment (pellet A) are separated by centrifugation (as above). A crude basic protein fraction is precipitated from the supernatant by the addition of 10 volumes of cold acidified acetone and incubated at −20° overnight. After centrifugation, the sediment is dried under nitrogen and solubilized for electrophoresis.

For extraction of the major myelin proteolipid protein, pellet A is suspended in 10 volumes of 2:1 chloroform–methanol[12] and maintained at 37° for 1 hr. After centrifugation (as above), 4 volumes of ether are added to the clear yellowish supernatant, which is kept overnight (−20°) to precipitate a proteolipid fraction that is recovered by centrifugation (as above). The proteolipid pellet is treated briefly with cold (−20°) 100% ethanol to remove the ether, washed by suspension and recentrifugation in several changes of cold H_2O, and finally solubilized for electrophoresis. At no point is the proteolipid pellet allowed to dry.

Preparative Electrophoresis

Samples are dissolved in a buffer containing 10% sucrose, 10% SDS, 50 m*M* Tris-HCl, pH 8, 6 m*M* EDTA, and Bromphenol Blue at a protein concentration of about 0.4 mg/ml. The dissolved mixtures are heated in a boiling water bath for 2 min and allowed to cool to room temperature. Dithiothreitol is then added to a final concentration of 500 m*M*. Each sample (2 ml) is loaded onto a single wide slot (30 × 2 cm) of an SDS (0.1%)–polyacrylamide gradient (7 to 17%) slab gel (2 mm thick)[13] immediately after thioglycolacetic acid (0.03% v/v) is added to the upper buffer reservoir. Electrophoresis is carried out for 20 hr at 20 mA. Gels are lightly stained with Coomassie Blue, and the bands corresponding to the small basic protein of myelin (SMBP; M_r 14,000) and PLP (M_r 23,000) are excised from the gels. Each excised piece is frozen slowly by overnight storage in a (−20°) freezer. The crust of frozen electrophoresis buffer is removed by a brief rinsing in water; the gel pieces are blotted dry, placed in Corex tubes, and pulverized by hand with a glass rod. A 1% SDS solution is added to the Corex tube (3 ml per gram of gel), which is then heated to 95° for 2 hr. The gel fragments are removed by centrifugation (10,000 rpm, 10 min, Sorvall HB4), and the eluted protein in the superna-

[12] J. Folch-Pi and M. Lees, *J. Biol. Chem.* **191,** 807 (1951).

[13] G. Kreibich, B. L. Ulrich, and D. D. Sabatini, *J. Cell Biol.* **77,** 464 (1978).

tant is precipitated by overnight incubation, with 10 volumes of cold (−20°) acetone.

Preparation of Antibodies

The PLP and SMBP purified by preparative gel electrophoresis are dissolved (50 μg each) in 250 μl of 0.1% SDS with 150 m*M* NaCl, to which equivalent volumes of complete Freund's adjuvant are added. Aliquots (50 μl) of these suspensions are injected into popliteal lymph nodes of young New Zealand rabbits (2.5 kg each) under barbiturate anesthesia. After 2 weeks a total of 100 μg of each antigen suspended in 0.5 ml of incomplete Freund's adjuvant (IFA) is administered to each rabbit by intradermal injections in several places across the back. Animals may be bled weekly thereafter, and booster injections of 10 μg of purified protein suspended in IFA are given monthly. The specificity of antisera may be characterized by immunoprecipitation[14] with detergent-solubilized ^{125}I-labeled myelin fragments.[15] Antisera raised against PLP yield a single band, which comigrates with the purified antigen. Antisera raised against SMBP recognize four polypeptides, corresponding to the M_r 21,500, 18,500 (LMBP), 17,000, and 14,000 (SMBP) proteins described for mouse myelin.[16]

All the procedures described here have been used in studies on the biosynthesis of the major myelin membrane proteins.[11] Experiments in which mRNAs derived from brainstem fractions are translated *in vitro* revealed that PLP and the MBPs are synthesized on membrane-bound ribosomes and free polysomes, respectively. Furthermore, it was found that mRNAs recovered in the myelin fraction are greatly enriched in those sequences coding for the myelin basic proteins, but not for PLP.

[14] B. M. Goldman and G. Blobel, *Proc. Natl. Acad. Sci. U.S.A.* **75,** 5066 (1978).
[15] A. Bolton and W. Hunter, *Biochem. J.* **133,** 529 (1973).
[16] E. Barbarese, P. Braun, and J. Carson, *Proc. Natl. Acad. Sci. U.S.A.* **74,** 3360 (1977).

[31] Sucrase–Isomaltase of the Small-Intestinal Brush Border Membrane: Assembly and Biosynthesis

By JOSEF BRUNNER, HANS WACKER, and GIORGIO SEMENZA

Among the constituents of the small-intestinal brush border membrane, the dimeric enzyme complex sucrase–isomaltase (sucrose α-D-glucohydrolase, EC 3.2.1.48 and oligo-1,6-glucosidase, EC 3.2.1.10, respectively) is the most abundant protein, accounting for approximately 10% of the intrinsic proteins. This enzyme complex accounts for all of the sucrase activity, for approximately 90% of the isomaltase activity, and for 70–80% of the maltase activity of the small intestine.[1,2]

Information concerning the topological arrangement of sucrase–isomaltase was first derived from studies involving the action of proteases, in particular of papain, on the luminal surface of sealed, right-side-out brush border membrane vesicles. Thus controlled treatment of the vesicles with papain was found to release quantitatively sucrase–isomaltase (as well as other hydrolases) from the membrane. Since papain can be assumed to act from the outside only (as shown by the unchanged permeability properties of the vesicles after papain treatment[3,4]), most of the protein mass, including the enzymically active sites, is located on the outside of the brush border membrane. In addition to the papain-solubilized, water-soluble sucrase–isomaltase, a different form of the enzyme exhibiting amphipathic properties is released by detergent treatment.[5,5a] For example, only the latter is capable of forming stable lipid–protein complexes with nascent liposomes.[6] Sodium dodecyl sulfate (SDS)–polyacrylamide gel electrophoresis of papain- and detergent-solubilized sucrase–isomaltase revealed differences solely in the electrophoretic mobility of the isomaltase subunit. The apparent molecular weight of this subunit was estimated to be approximately 140,000 in the detergent form and approximately 120,000 in the papain form.[7,7a] In contrast, the sucrase subunit

[1] G. Semenza, *in* "The Enzymes of Biological Membranes" (A. Martonosi, ed.), Vol. 3, p. 349. Plenum, New York, 1976.

[2] G. Semenza, *in* "Carbohydrate Metabolism and Its Disorders" (P. J. Randle, D. F. Steiner, and W. J. Whelan, eds.), Vol. 3, p. 425. Academic Press, New York, 1981.

[3] C. Tannenbaum, G. Toggenburger, M. Kessler, A. Rothstein, and G. Semenza, *J. Supramol. Struct.* **6,** 519 (1977).

[4] A. Klip, S. Grinstein, and G. Semenza, *J. Membr. Biol.* **51,** 47 (1979).

[5] H. Sigrist, P. Ronner, and G. Semenza, *Biochim. Biophys. Acta* **406,** 433 (1975).

[5a] Y. Takesue, R. Tamura, T. Akaza, and Y. Nishi, *J. Biochem., Tokyo* **74,** 415 (1973).

[6] J. Brunner, H. Hauser, and G. Semenza, *J. Biol. Chem.* **253,** 7538 (1978).

[7] J. Brunner, H. Hauser, H. Braun, K. J. Wilson, H. Wacker, B. O'Neill, and G. Semenza, *J. Biol. Chem.* **254,** 1821 (1979).

[7a] M. Spiess, H. Hauser, J. P. Rosenbusch and G. Semenza, *J. Biol. Chem.* **256,** 8977 (1981).

ISBN 0-12-181996-5

seemed to be identical in both types of sucrase–isomaltase preparations (apparent molecular weight, 120,000).

These studies strongly suggested that solubilization by papain results from cleavage of one or more peptide bonds that connect the "body" of sucrase–isomaltase (including the hydrolytic active sites) with a hydrophobic region that itself is inserted in the lipid core of the bilayer. That papain solubilization is accompanied by a change in the primary structure of the isomaltase subunit was confirmed by analyses of the NH_2-terminal amino acid (which changed after papain treatment) and of the amino acids released by carboxypeptidase Y from the COOH termini (which did not change).[7] Accordingly, papain solely cleaves a segment from the NH_2 terminus of the native (detergent-solubilized) form of sucrase–isomaltase, and this segment must be responsible for anchoring the enzyme complex to the membrane fabric. Indeed, Edman degradation of the isomaltase subunit showed a long hydrophobic sequence (see Fig. 6).[8,9]

The present chapter describes the application of chemical labeling reagents that have been developed in our laboratory to gain a better understanding of how this anchoring peptide is structured and folded within the membrane. Knowledge of the folding pattern is an important step toward formulating a possible mechanism of biosynthesis and membrane insertion of this enzyme complex. It is now known that sucrase–isomaltase complex is synthesized as a large one-chain precursor (pro-sucrase–isomaltase).[10–13] Cell-free translation of this precursor or even of a pre-form should provide another important tool to obtain information on the biosynthesis of this membrane protein. Conditions for the cell-free synthesis of the precursor[14] and a possible explanation for the particular topological arrangement of sucrase–isomaltase are presented in the second part of this chapter.

[8] G. Frank, J. Brunner, H. Hauser, H. Wacker, G. Semenza, and H. Zuber, *FEBS Lett.* **96,** 183 (1978).

[9] H. Sjöström, O. Norén, L. Christiansen, H. Wacker, M. Spiess, B. Bigler-Meier, E. E. Rickli, and G. Semenza, *FEBS Lett.* **148,** 321 (1982).

[10] G. Semenza, *in* "Processing and Turnover of Protein and Organelles in the Cell" (S. Rapoport and T. Schewe, eds.), 12th FEBS Meet., Vol. 53, p. 21. Pergamon, Oxford, 1979.

[11] H. Sjöström, O. Norén, L. Christiansen, H. Wacker, and G. Semenza, *J. Biol. Chem.* **255,** 11332 (1980).

[12] H.-P. Hauri, A. Quaroni, and K. J. Isselbacher, *Proc. Natl. Acad. Sci. U.S.A.* **77,** 6629 (1980).

[13] H. P. Hauri, H. Wacker, E. E. Rickli, B. Bigler-Meier, A. Quaroni, and G. Semenza, *J. Biol. Chem.* **257,** 4522 (1982).

[14] H. Wacker, R. Jaussi, P. Sonderegger, M. Dokow, P. Ghersa, H.-P. Hauri, P. Christen and G. Semenza, *FEBS Lett.* **136,** 329 (1981).

Identification by Hydrophobic Photolabeling of the Subunit of Sucrase–Isomaltase That Is Embedded in the Lipid Bilayer

Principle

Labeling of membranes from within the lipid core aims at identifying those polypeptide segments of membrane proteins that are buried in the lipid bilayer. This technique, therefore, represents a valuable complement to surface-labeling techniques of membranes.

Since most of the polypeptide segments that penetrate the lipid core are nonpolar and chemically inert, their modification requires a highly reactive reagent. Such considerations have led to the development of various types of compounds that are unreactive in the dark but generate highly reactive intermediates when exposed to ultraviolet irradiation. Some of the intermediates thus generated are, in principle, capable of reacting with the full range of amino acid side chain residues, including aliphatic ones. The principal elements of photolabeling techniques have been reviewed by Bayley and Knowles,[15] by Chowdhry and Westheimer,[16] and, with emphasis on hydrophobic labeling, in more recent articles.[17,18] To the selective labeling of the membranous segment of sucrase–isomaltase, the carbene-generator, 3-trifluoromethyl-3-(*m*-[^{125}I]iodophenyl)diazirine, has been applied successfully.[19] Since this reagent combines a number of favorable properties,[20] it should be similarly useful in topological studies of other integral membrane proteins.

Reagents and Equipment

3-Trifluoromethyl-3-(*m*-[^{125}I]iodophenyl)diazirine ([^{125}I]TID) was prepared by chemical synthesis from trifluoroacetophenone according to a procedure described in detail elsewhere.[20] It was stored as an ethanolic solution in the dark at −20° at a concentration of approximately 5 mCi/ml (specific radioactivity: 10 Ci/mmol). The apparatus used for photoactivation of the reagent is shown schematically in Fig. 1. Irradiation of [^{125}I]TID generates the actual labeling reagent trifluoromethyl-(*m*-[^{125}I]iodophenyl)carbene as shown below.

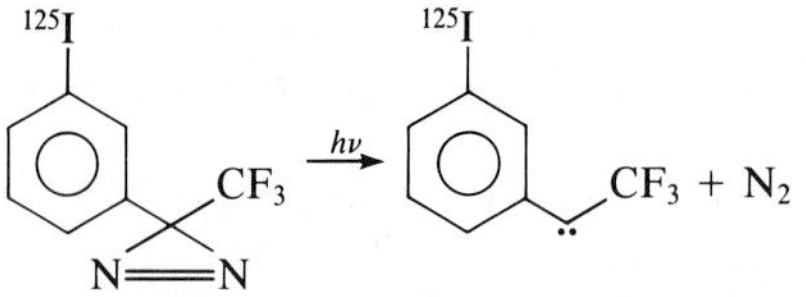

[15] H. Bayley and J. R. Knowles, this series, Vol. 46, p. 69.

[16] V. Chowdhry and F. H. Westheimer, *Annu. Rev. Biochem.* **48**, 293 (1979).

[17] J. Brunner, *Trends Biochem. Sci.* **6**, 44 (1981).

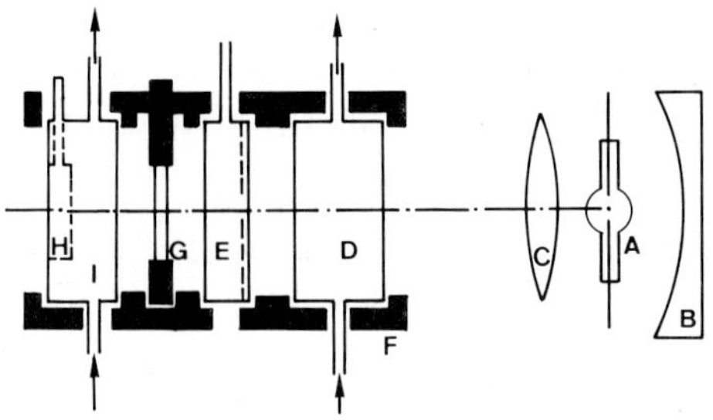

FIG. 1. Scheme of the photolysis apparatus: A, light source (medium- or high-pressure mercury lamp); B, reflector; C, lens; D, infrared filter (circulating cold water); E, G, combination of filters to screen out radiation of wavelengths shorter than that used for photoactivation of the reagent; F, tube; H, sample cuvette; I, thermostatted cell. From J. Brunner, *in* "Membrane Proteins" (A. Azzi U, Brodbeck, and P. Zahler, eds.), p. 170. Springer-Verlag, Berlin and New York, 1981.

Small-intestinal brush border membrane vesicles were prepared according to the procedure of Schmitz *et al.*[21] as modified by Kessler *et al.*[22] Buffer components and other chemicals were from commercial sources and of highest purity available.

Procedure

Brush border membrane vesicles (1–2 mg of protein per milliliter) in 50 m*M* sodium phosphate, pH 7.5, were flushed with a stream of nitrogen for 30 min at 0–5°. Then, 10 μl of the ethanolic solution of [^{125}I]TID was injected into 1 ml of the vesicle dispersion. After equilibration at 0–5° for 30 min in the dark, the vesicles were transferred into the photolysis cuvette and photolyzed for 15 sec at 0–5°. Subsequently, the vesicles were washed four times with sodium phosphate (50 m*M*, pH 7.5) containing 1% of bovine serum albumin and twice with albumin-free buffer. For immunoprecipitation of the sucrase–isomaltase complex, [^{125}I]TID-labeled and washed membrane vesicles were treated for 15 min at 4° in the same buffer to which Triton X-100 had been added to a final concentration of 2% (w/v). After pelleting non–solubilized material at 15,000 *g* for 10 min,

[18] H. Bayley, *in* "Membranes and Transport" (A. Martonosi, ed.), Vol. 1, p. 185. Plenum, New York, 1982.

[19] M. Spiess, J. Brunner, and G. Semenza, *J. Biol. Chem.* **257,** 2370 (1982).

[20] J. Brunner and G. Semenza, *Biochemistry* **20,** 7174 (1981).

[21] J. Schmitz, H. Preiser, D. Maestracci, B. K. Ghosh, J. J. Cerda, and R. K. Crane, *Biochim. Biophys. Acta* **323,** 98 (1973).

[22] M. Kessler, O. Acuto, C. Storelli, H. Murer, M. Müller, and G. Semenza, *Biochim. Biophys. Acta* **506,** 136 (1978).

the supernatant was transferred into a conical vial and sucrase–isomaltase complex was precipitated by means of an anti-sucrase–isomaltase antiserum raised in guinea pigs. The immunoprecipitate was washed by repeated sedimentation (3000 rpm for 90 sec in an MSE table centrifuge) with 150 m*M* NaCl, 10 m*M* sodium phosphate, pH 7.5, containing 1% (w/v) Triton X-100 until no further radioactivity was found in the supernatant. Alternatively to solubilizing labeled and washed vesicles with Triton X-100, vesicles were subjected to papain digestion according to a procedure described previously.[5] The supernatant containing papain-solubilized protein was dialyzed against 150 m*M* NaCl, 10 m*M* sodium phosphate, pH 7.5, containing 1% (w/v) Triton X-100 and sucrase–isomaltase immunoprecipitated as described above. Subsequently, immunoprecipitates derived from papain- and Triton X-100-solubilized sucrase–isomaltase were analyzed by SDS–polyacrylamide gel electrophoresis. The Coomassie Blue-stained slab gel was dried and autoradiographed (Fig. 2). For quantitation of the radioactivity associated with the individual subunits of sucrase–isomaltase complex, the dried gel was cut into 1-mm slices that were used for γ-counting.

Comments

Sucrase–Isomaltase Is Anchored via an N-Terminal Segment of the Isomaltase Subunit. SDS–polyacrylamide gel electrophoresis of the subunits of Triton X-100 and papain-solubilized sucrase–isomaltase complex has provided the first evidence that papain solubilization cleaves a polypeptide segment from the intact isomaltase subunit to yield an isomaltase with a slightly decreased apparent molecular weight (Fig. 2A). Subsequent analyses of the amino- and carboxyl-terminal amino acids of the individual subunits demonstrated that this polypeptide segment originates from the N terminus of the (intact) isomaltase subunit.[7] Obviously, its function is to "anchor" the enzyme complex to the brush border membrane. Such a mode of association is reflected in the distribution of radioactivity among the subunits of sucrase–isomaltase isolated from [^{125}I]TID labeled brush border membranes (Fig. 2). Evidently, radiolabel is present only in the isomaltase subunit of Triton X-100-solubilized sucrase–isomaltase. Therefore, this supports the presumption that papain solubilization in fact results from removal of a polypeptide segment that deeply penetrates the lipid bilayer and thus can be labeled by the highly lipid-soluble photoreagent [^{125}I]TID.

Choice of the Labeling Reagent. Based on current concepts of the insertion of integral proteins, polypeptide segments penetrating the lipid bilayer are arranged in such a way that their surface areas in contact with

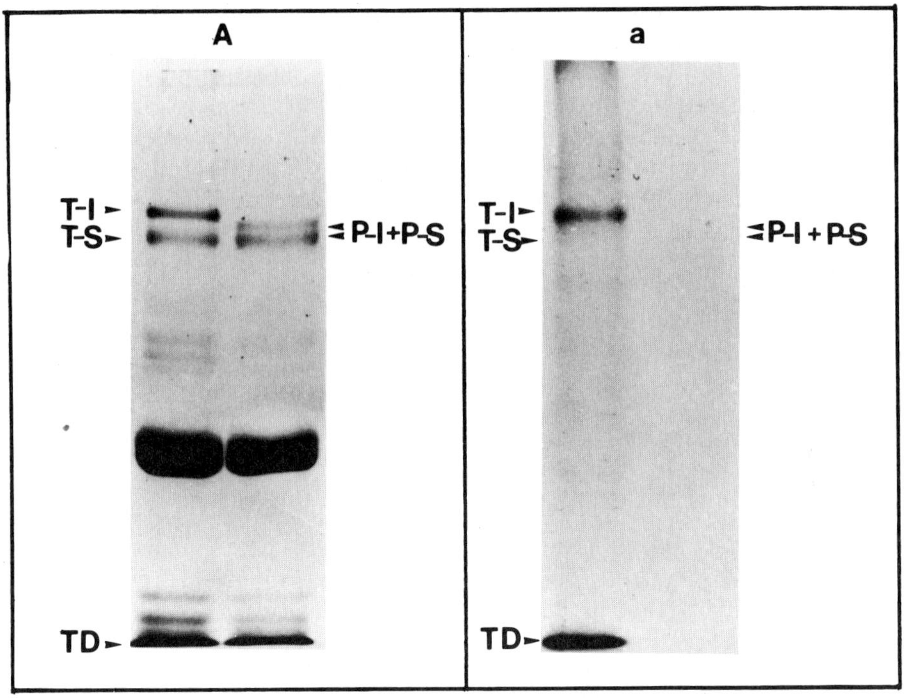

FIG. 2. Distribution of radiolabel in [^{125}I]TID-labeled sucrase–isomaltase complex. Labeled and albumin-washed brush border membrane vesicles were either solubilized by Triton X-100 or treated with papain. Solubilized sucrase–isomaltase complex was immunoprecipitated, and the precipitates were subjected to SDS–polyacrylamide gel electrophoresis. (A) Left lane: Immunoprecipitate from Triton X-100 solubilized sucrase–isomaltase complex (Coomassie Blue stained); right lane: immunoprecipitate from papain-solubilized sucrase–isomaltase (Coomassie Blue stained). (a) Autoradiography of the gel A. Positions of the subunits of detergent (T-I and T-S)- and papain-solubilized (P-I and P-S) sucrase–isomaltase complex are marked by arrows. TD, Tracking-dye front. From Spiess *et al.*[19]

the hydrocarbon matrix of the lipids are extraordinarily rich in amino acids with hydrophobic and chemically inert side chains. For instance, the membrane-associated segment of sucrase–isomaltase contains a succession (residues 12–31 in Fig. 6) that is likely to form a highly unpolar and inert surface.[8,9] Chemical modification of such areas obviously requires extremely reactive reagents capable of inserting even into C—H bonds of aliphatic residues. Mainly such considerations have led to the development of precursors of carbenes that are more reactive than the structurally related nitrenes.[15,16] TID offers a palette of properties that, in view of the general problems and difficulties related with photolabeling techniques,[15,16] seems to be suitably balanced.[20] That reagents which generate less reactive intermediates (arylnitrenes) may also be utilized successfully to investigate similar problems has been demonstrated by Norén and Sjöström,[23] who probed the insertion of pig microvillous aminopeptidase into the membrane by [^{125}I]iodonaphthylazide. The same reagent has also been used by Sigrist-Nelson *et al.*[24] to label the intrinsic proteins of the rat intestinal microvillous membrane. Another application of a hydrophobic labeling reagent, 1-azido-4-[^{125}I]iodobenzene, to the analysis of kidney microvillous membrane hydrolases has been reported by Booth *et al.*[25] [^{125}I]Iodonaphthylazide, which has also been examined for the selective labeling of the hydrophobic anchor peptide of the isomaltase subunit, gave a less clear labeling pattern (unpublished results) in that the isomaltase subunit was labeled only approximately twice as efficiently as the sucrase subunit. The high selectivity of [^{125}I]TID for labeling intrinsic domains of membrane proteins has been confirmed by labeling γ-glutamyltranspeptidase[26] and other intrinsic membrane proteins, such as glycophorin.[20] It should be emphasized that still little information is available concerning the detailed mechanisms underlying photochemical labeling processes. Nonetheless, evidence is accumulating that labeling by arylazides involve long-lived electrophilic species that react with a narrow range of functionalities present in amino acid side chains.[27]

Orientation of the N-Terminal Amino Acid (Alanine) of the Isomaltase-Anchoring Segment

Principle

Treatment of sealed, right-side-out brush border membrane vesicles with amino group-specific reagents, such as imidoesters, esters of *N*-hydroxysuccinimide, may convert a substantial fraction of the (total) amino groups to corresponding derivatives. If the modifying reagent is essentially impermeant and if all the vesicles are sealed and have the same orientation, derivatization of the amino groups should be restricted to the outer surface of the vesicles (in the case of brush border membrane vesicles, this corresponds to the luminal surface). Furthermore, if it is then possible to determine precisely to what extent the N terminus of a given

[23] O. Norén and H. Sjöström, *Eur. J. Biochem.* **104,** 25 (1980).
[24] K. Sigrist-Nelson, H. Sigrist, T. Bercovici, and C. Gitler, *Biochim. Biophys. Acta* **468,** 163 (1977).
[25] A. G. Booth, L. M. L. Hubbard, and J. Kenny, *Biochem. J.* **179,** 397 (1979).
[26] T. Frielle, J. Brunner, and N. Curthoys, *J. Biol. Chem.* **257,** 14979 (1982).
[27] J. Staros, *Trends Biochem. Sci.* **5,** 320 (1980).

polypeptide has been derivatized by the amino group-specific reagent, one should be able to correlate these data with a certain topological orientation of the N terminus.

Reagents and Equipment

3-Dimethyl-2-[(acetimidoxyethyl)ammonio]propanesulfonic acid (DAP) was synthesized as previously described in detail.[28] The white crystals were stored in a desiccator at room temperature. Solutions of DAP in sodium borate (pH 8.5) were prepared immediately before use. [G-^{3}H]Dansyl chloride and [*N-methyl*-^{14}C]dansyl chloride were from Amersham. Other chemicals and solvents were of reagent grade. Micropolyamide thin-layer sheets F 1700 were obtained from Schleicher & Schuell. Radioactivity (^{3}H and ^{14}C) was determined by scintillation counting and ^{125}I by γ-counting.

Procedure

Brush border membrane vesicles were prepared according to the procedure of Kessler *et al.*[21] Sucrase activity was determined by measuring the amount of glucose liberated from sucrose (33 m*M*) at 37° in 33 m*M* sodium maleate buffer, pH 6.8. D-Glucose was determined using the glucose dehydrogenase kit from Merck.[29] Radioiodination of purified Triton X-100-solubilized sucrase–isomaltase complex was performed by the chloramine-T method.[30]

Amidination with DAP of Brush Border Membrane Vesicles. Brush border membrane vesicles were dispersed in 100 m*M* sodium borate buffer (pH 8.5) at a concentration of 30 mg of protein per milliliter; 100 μl of this vesicle suspension were added to 2 ml of 100 m*M* sodium borate-buffered (pH 8.5) solution of DAP (100 m*M*). This solution was prepared immediately before use by adding DAP in small portions to sodium borate; the pH value of the solution was kept at pH 8.5 with NaOH. The mixture was incubated at 25° for 10 min, and then amidination was quenched by the addition of 21 ml of an ice-cold solution of 70 m*M* ethanolamine dissolved in 150 m*M* NaCl, 10 m*M* sodium phosphate, pH 7.5. The vesicles were pelleted by centrifugation at 4° (30 min at 80,000 *g*) and washed by resuspending in phosphate-buffered saline, pH 7.5, and pelleting again. Following the same procedure, amidination by DAP was also performed on brush border membrane vesicles that had been made

[28] R. Bürgi, J. Brunner, and G. Semenza, *J. Biol. Chem.* (in press).

[29] D. Banauch, W. Brümmer, W. Ebeling, H. Metz, H. Rindfrey, H. Lang, K. Leyold, and W. Rick, *Z. Klin. Chem. Klin. Biochem.* **13,** 101 (1975).

[30] F. C. Greenwood, W. M. Hunter, and J. S. Glover, *Biochem. J.* **89,** 114 (1963).

leaky by pretreatment with sodium deoxycholate (0.5 mg per milligram of protein).[31] This treatment allows the imidoester to have free access to both sides of the membrane.

Isolation of Sucrase–Isomaltase from DAP-Labeled Brush Border Membranes. Washed and sedimented membranes (corresponding to 3 mg of protein) were dispersed in 0.5 ml of PBS containing 2% (w/v) Triton X-100, and the dispersion was slowly rotated (end over end) in a small test tube at 4° for 2 hr. Solubilized sucrase–isomaltase was recovered in the supernatant after centrifugation at 4° at 100,000 *g* for 1 hr. Approximately 0.01 μCi of ^{125}I-labeled sucrase–isomaltase was added as internal standard for subsequent quantitation of sucrase–isomaltase complex (after denaturation). The enzyme complex was then precipitated by anti sucrase–isomaltase antiserum, and the precipitate was washed four times with PBS containing 0.1% of Triton X-100. After each extraction, the precipitate was sedimented in an MSE table centrifuge at 2500 rpm for 90 sec. Supernatants were discarded.

Quantitation of the Amidination Reaction at the N-Terminal Amino Groups of Sucrase–Isomaltase Complex. Immunoprecipitates of amidinated and nonamidinated sucrase–isomaltase (corresponding to 250 μg of sucrase–isomaltase) were denatured by boiling each sample for 2 min in 100 μl of aqueous SDS (4%, w/v). Each tube was then supplemented with 100 μl of *N*-ethylmorpholine (Pierce, Sequenal Grade) and 150 μl of a solution (12.5 mg/ml) of ^{3}H-labeled (300 μCi) dansyl chloride dissolved in dimethyl formamide. The mixtures were then allowed to react at room temperature for 2 hr. Dansylation should block free amino groups present in the polypeptide chains (see Comments). Each tube was then supplemented with 30-μl aliquots of a mixture obtained from dansylation (as described above) of sucrase–isomaltase in SDS with ^{14}C-labeled dansyl chloride. Acetone was added to a final concentration of 80% (v/v) which caused the dansylated protein to precipitate. The sedimented protein was washed once with 80% aqueous acetone, and the supernatants were discarded. The yellow-fluorescent pellet was dried with a gentle stream of nitrogen. As determined by γ-counting, this precipitation was quantitative ($\geq$98%) with respect to the sucrase–isomaltase. The dry residues were then subjected to acid hydrolysis (6 *M* HCl at 108° for 4 hr). After evaporation of the acid, the residues were dissolved in approximately 50 μl of 50 m*M* acetic acid, and dansylamino acids were extracted with ethyl acetate (3 times 100 μl for each sample). Two-dimensional thin-layer chromatography (see Fig. 3) on polyamide sheets using solvent systems 1 and 2 of Wood and Wang was used to separate the dansylamino acids. These were visualized under an ultraviolet lamp and identified by their R_f values.

[31] A. Klip, S. Grinstein, and G. Semenza, *Biochim. Biophys. Acta* **558,** 233 (1979).

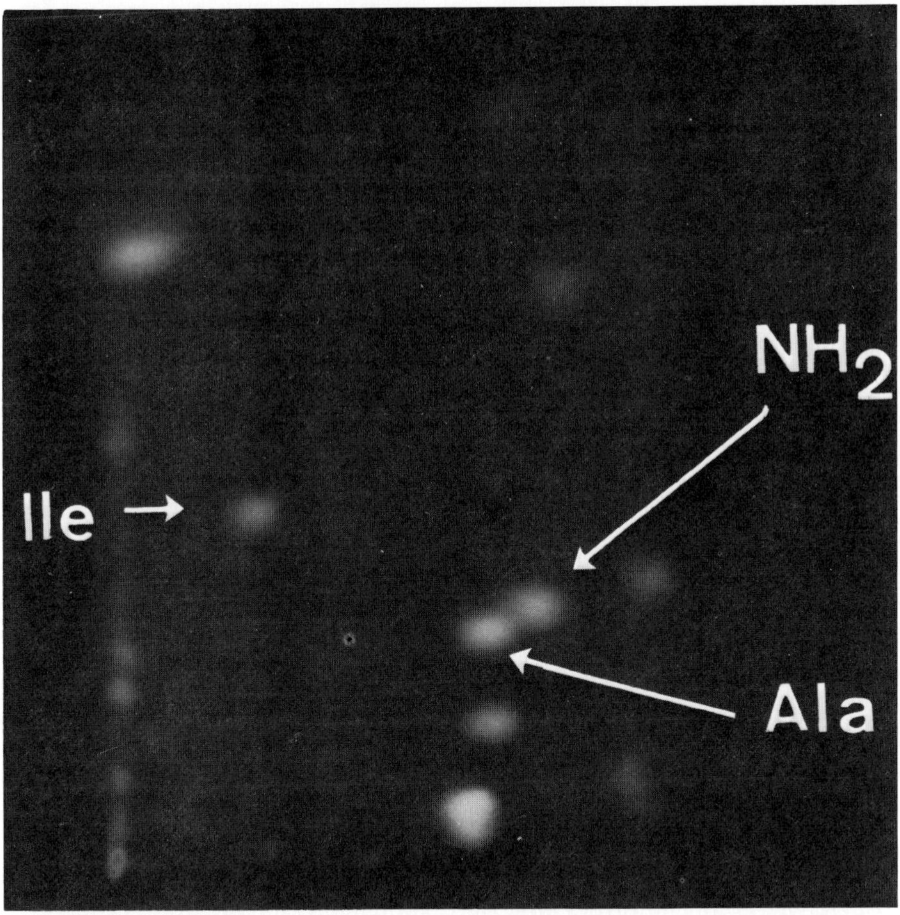

FIG. 3. Separation of dansylamino acids by polyamide thin-layer chromatography. Triton X-100-solubilized sucrase–isomaltase complex was precipitated by anti-sucrase–isomaltase antiserum, and the precipitate was washed extensively with 150 m*M* NaCl, 10 m*M* sodium phosphate, pH 7.5, 0.1% Triton X-100. The precipitate was dansylated and then subjected to acid hydrolysis (6 *M* HCl at 110° for 4 hr). Dansylamino acids were extracted by ethyl acetate and separated on polyamide sheets (10 × 10 cm). Chromatography in horizontal direction, first solvent: water–formic acid (100 : 1.2, v/v); in vertical direction, second solvent: benzene–acetic acid (9 : 1; v/v). Abbreviations: dansylalanine, Ala; dansylisoleucine, Ile; dansylamide, NH_2. From Bürgi *et al.*[28]

Fluorescent spots of dansylalanine (N terminus of isomaltase) and dansylisoleucine (N terminus of sucrase) were excised, and the radioactivity was determined by scintillation counting. From the ratios of $^3H/^{14}C$ radioactivity (appropriate corrections for background radioactivity, spillover,

AMINO ACID AMIDINATION[a,b]

Amidination by DAP	Extent of amidination (%) of N-terminal amino acid of	
	Sucrase (isoleucine)	Isomaltase (alanine)
Intact brush border membrane vesicles (sealed)	58(±8 SD; n = 4)	33(±4 SD; n = 4)
Deoxycholate-treated (leaky) brush border membranes	46(±8 SD; n = 4)	35(±6 SD; n = 4)

[a] From Bürgi *et al.*[28]

[b] Amidination was performed at 23° with DAP at an initial concentration of 40 m*M*. The pH was 8.5, and the reaction time 10 min.

and quenching were made), the percentage of N-terminal amino acid amidination was calculated according to the equation

$$\% \text{ of amidination} = [(Y - X)/Y]\,(100\%)$$

X is $^3H/^{14}C$ radioactivity from amidinated sucrase–isomaltase; Y is $^3H/^{14}C$ radioactivity from nonamidinated (reference) sucrase–isomaltase.

Some of the results thus obtained are shown in the table. Amidination of both N termini of sucrase–isomaltase with the sparingly permeant reagent DAP was the same with intact and leaky bush border membranes. This led us to conclude that the N terminus of the isomaltase subunit must be located on the outer (luminal) surface of the vesicle.

Comments

General. In contrast to procedures used in previous studies, the methodology described here does not include proteolytic degradation of the protein being investigated. Therefore, it is applicable to proteins of which suitable fragmentation cannot be obtained, as may be the case for integral proteins (e.g., isomaltase), the N terminus of which is a few amino acid residues away from a membrane-embedded domain. Although the procedure has given meaningful data in the case of intestinal sucrase–isomaltase, it is necessary to point out potential difficulties that could arise in attempts to adapt the procedure to other membranes and proteins.

Permeability Measurements. Proper interpretation of experimental data requires knowledge of the permeability properties to the modifying reagent used. For instance, permeability measurements can be carried out by following the time course by which a target entrapped into the vesicles is derivatized upon addition of the (impermeant) reagent to the aqueous buffer of the vesicle dispersion. The time course must then be compared with that obtained when the slightly permeant reagent is substituted for a highly permeant one with comparable reactivity. It should also be mentioned that the permeability properties of biological membranes to small charged molecules may vary by several orders of magnitude. Choice of the (impermeant) amidinating reagent (or any other amino group specific reagent, if available with suitable properties) as well as the method(s) for determining its (their) time course(s) of permeation must be evaluated carefully. Several chemical labeling reagents have been developed to probe the asymmetric organization of membranes.[32–36]

Isolation of the Protein from Modified (Amidinated) Membranes. For determining the extent to which the N-terminal amino acid residue of protein polypeptide chain has been derivatized, it is first necessary to isolate the protein from the membrane vesicles. Immunoprecipitation utilized in the present study should be of wide applicability, since amidination of proteins, in general, does not affect immunogenicity. Additional advantages might derive from the fact that immunoprecipitation allows essentially complete separation of protein from buffer components or detergents that could interfere with the subsequent dansylation reaction. Furthermore, it yields proteins of high purity in a single operation. It should be pointed out, however, that this technique alone is adequate only when the presence of antibodies during dansylation does not interfere with the N-terminal amino acid analysis of the membrane protein under investigation. Specifically, neither antibodies nor any possible contaminants should have N-terminal amino acids identical with those of the antigens being investigated. To evaluate this problem, N-terminal amino acid analysis of specific (affinity purified) antibodies alone or of immunoprecipitates obtained from fully amidinated antigen can be performed.[28] Alternatively to immunoprecipitation, the amidinated membrane protein might be isolated by affinity chromatography on a porous support coupled with antibodies or by SDS–polyacrylamide gel electrophoresis according to established techniques.

Quantitation of the Extent of N-Terminal Amino Group Derivatization in a Protein. The approach used relies on the assumption that dansylation of the (partially) derivatized and SDS-denatured protein quantitatively blocks amino groups left free. Although it is difficult to provide rigorous evidence for this claim, the following evidence supports this. If a several-

[32] M. S. Bretscher, *J. Mol. Biol.* **58,** 775 (1971).

[33] W. W. Bender, H. Garan, and H. C. Berg, *J. Mol. Biol.* **58,** 783 (1971).

[34] N. M. Whiteley and H. C. Berg, *J. Mol. Biol.* **87,** 541 (1974).

[35] P. P. Nemes, G. P. Miljanich, D. L. White, and E. A. Dratz, *Biochemistry* **19,** 2067 (1980).

[36] R. E. Abbott and D. Schechter, *J. Biol. Chem.* **251,** 7176 (1976).

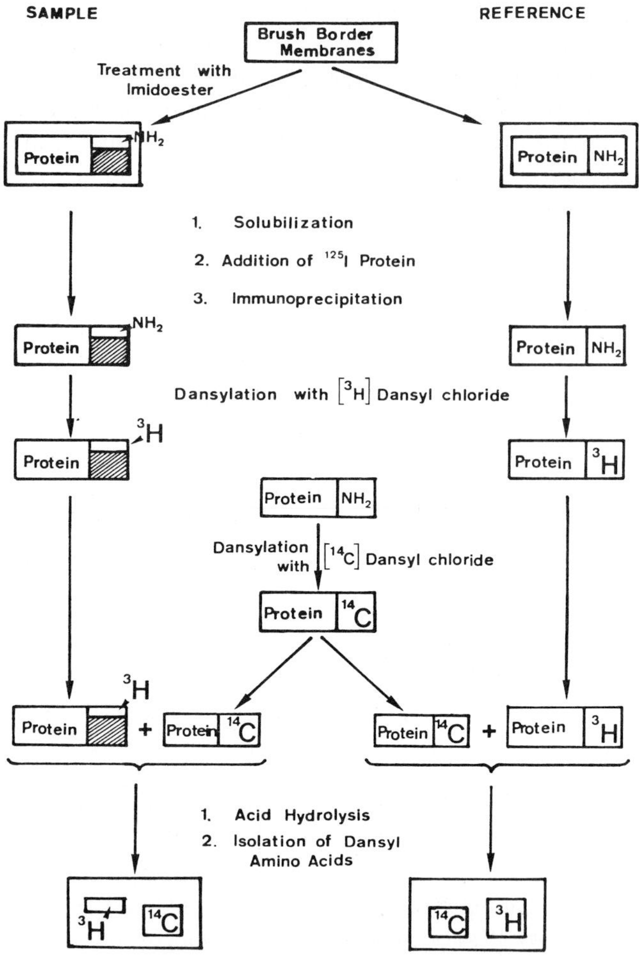

FIG. 4. Scheme of the procedure used for quantitation of the extent of amidination of NH_2-terminal amino acids in sucrase–isomaltase complex. From Bürgi *et al.*[28]

fold excess of dansyl chloride is present, the extent of dansylation of amino acids depends, among other factors, on the initial concentration of dansyl chloride. Thus, Gray demonstrated that at a concentration of dansyl chloride of 5 m*M* most of the amino acids are quantitatively dansylated.[37] In the case of sucrase–isomaltase, we have compared the extent of dansylation of N-terminal amino groups as a function of the initial dansyl chloride concentration. Since it was not possible to increase the extent of

[37] W. R. Gray, this series, Vol. 25, p. 121.

dansylation further by using dansyl chloride concentrations higher than 15 m*M*, this suggests that 15 m*M* is sufficient to give virtually complete dansylation of the protein.[28] In order to determine the extent of N-terminal amino group amidination, the double-labeling approach outlined in Fig. 4 has been devised. Although the procedure following the dansylation step includes various operations, this method for quantitation gives reproducible data.[28]

Assembly of Sucrase–Isomaltase Complex

As outlined above, a variety of methodologies have been applied to probe the topological organization of small-intestinal sucrase–isomaltase complex. These studies have allowed us to propose the model illustrated in Fig. 5. As indicated in this figure, the membranous segment has highly helical structure as concluded from measurements of circular dichroisms of the hydrophobic polypeptide segment in detergents and in liposomes.[19]

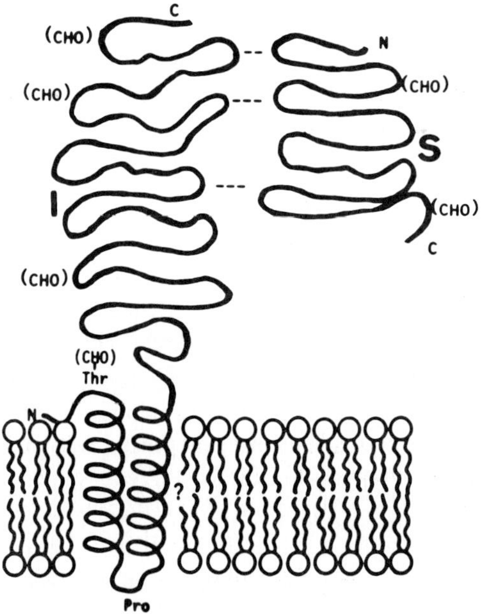

FIG. 5. Positioning of sucrase–isomaltase complex in small-intestinal brush border membrane. Note: The drawing is not to scale and is intended to convey the major features only. CHO, carbohydrate chains; I, isomaltase subunit; S, sucrase subunit. From Semenza.[2]

The "One Chain, Two Active Sites Precursor" (Pro-SI)

In order to explain within a single framework the particular arrangement of the sucrase–isomaltase complex, i.e., the peripheral position of S and the anchoring to the membrane via the N-terminal position of the isomaltase subunit (I, Fig. 5), the homology between the two subunits and their related or common control of biosynthesis, the "one chain, two active sites precursor" hypothesis was suggested.[10] According to it, the two subunits have arisen from a common ancestor gene (coding for an isomaltase with maltase activity) by partial gene duplication giving rise to one polypeptide chain carrying two identical active sites; subsequent mutation(s) changed the substrate specificity of one of the sites: maltose would still be accepted and hydrolyzed, but isomaltose would no longer be, and sucrose would now be hydrolyzed. This "one chain, two active sites sucrase–isomaltase" (pro-SI) would be inserted into the membrane during synthesis and then split into the two chains of the "final" SI complex by extracellular (e.g., pancreatic) proteases.

A one-chain pro-SI in the 260,000 M_r range was indeed isolated from small intestines that had not been exposed to pancreatic proteases,[11] which in all likelihood is identical with the large M_r precursor found to migrate from the Golgi membranes to the brush border in pulse–chase experiments.[12] Pro-SI is indistinguishable from "final" SI both enzymically and immunologically; this shows that the prospective S and I subunits are preformed in this precursor. The identity of the NH_2-terminal sequences of the I subunit and pro-SI shows that the I portion is at the N-terminal segment of pro-SI (Fig. 6) and indicates that the mode of anchoring of the precursor is the same as for SI.[9,13] This does not rule out the possible presence of additional hydrophobic segment(s) prolonging the C terminus of the S portion.

Cell-Free *in Vitro* Translation of Pro-Sucrase–Isomaltase (Pro-SI)

Pro-SI is an unusually large polypeptide chain (apparent M_r 270,000). Its *in vitro* translation has been achieved starting from total RNA. To the best of our knowledge this is the largest reported polypeptide synthesized *in vitro* from total RNA.[14]

Procedure

Total RNA was extracted with 4 *M* guanidinium thiocyanate (see this volume [3]) from fresh mucosal scrapings of rabbit small intestines, purified by repeated precipitation from 6 *M* guanidinium-HCl with ethanol, and translated in a rabbit reticulocyte lysate treated with micrococcal

```
                   ?                    CHO
HOG ProSI   Ala-Arg-Lys-Ser/Thr-Phe-Ser-Gly-Leu-Glu-Ile—X—|Leu-Ile-Val-Leu-Phe-Ala-Ile-Val-
                                        CHO                                                                         ?   ?
HOG I       Ala-Arg-Lys-Lys-Phe-Ser-Gly-Leu-Glu-Ile—X—|Leu-Ile-Val-Leu-Phe-Ala-Ile-Val-Leu-Ser-Ile-Ala-Ile- Ala-Leu-Val-Val-Val—X—Ala|Ser-Lys—X—Pro-Ala-Val
                        ?               CHO
RAT ProSI^a Ala-Lys-Lys-Lys-Phe-Arg-Ala-Leu-Glu-Ile—X—|Leu-Ile-Val-Leu-Phe-Ile-Ile-
                                        CHO
RAT I^a     Ala-Lys-Lys-Lys-Phe-Ser-Ala-Leu-Glu-Ile—X—|Leu-Ile-Val-Leu-Phe-Ile-Ile-Val-Leu-Ala-Ile-Ala-Ile- Ala-Leu-Val-Leu-Val-
                                        CHO
RABBIT I^b  Ala-Lys-Arg-Lys-Phe-Ser-Gly-Leu-Glu-Ile-Thr|Leu-Ile-Val-Leu-Phe-Val-Ile-Val-Phe-Ile-Ile-Ala-Ile- Ala-Leu-Ile-Ala-Val-Leu-Ala|—X—X—X—Pro-Ala-Val
HOG S       Ile-Lys-Leu-Pro-Ser-Asp-Pro-Ile-Pro-Thr-Leu-Arg-Val/Met-Glu-Val/Met-Lys/Thr-Tyr-His-Lys/Thr-Asp-Tyr-Val/Met-Leu-Glu- Phe—X—Arg-Tyr-Asp-Pro-Glu-Arg-
                                                                    ?           ?
RAT S^a     Ile-Lys-Leu-Pro-Ser-Asn-Pro-Ile-Arg-Thr-Leu-Arg-Val-Glu-Val-Thr-Tyr—X—Thr-Asn-Arg-Val-Leu-Glu/Gln- Phe-Arg-Ile-Tyr-Arg-Ala-Glu—X—X—Gly-
RABBIT S    Ile-Asn-Leu-Pro-Ser-Glu/Thr-Pro-Glu/Thr-

            1           5           10          15          20          25          30          35
```

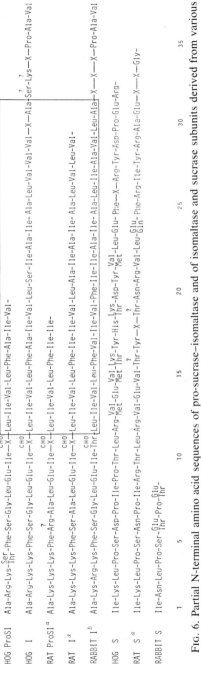

FIG. 6. Partial N-terminal amino acid sequences of pro-sucrase–isomaltase and of isomaltase and sucrase subunits derived from various species. The highly hydrophobic sequence 12–31 is indicated; one more hydrophobic sequence is likely to begin at res. 36. [a] From ref. 13; [b] from refs. 8 and 9; the other sequences are from ref. 9.

nuclease (see this volume [5]). Translations were performed in the presence of 0.5–1.5 mCi of [^{35}S]methionine per milliliter of lysate, 1.5–3 mg of total RNA, and 70 μg of tRNA from calf liver in the presence or the absence of dog pancreas microsomes.

The *in vitro* synthesized SI was purified by immunoprecipitation using anti-SI antiserum and *Staphylococcus aureus*. The translation mixture (up to 1 ml) was made 1% in Triton X-100 and diluted four times with 50 m*M* Tris-HCl, 150 m*M* NaCl, 5 m*M* EDTA, 1% Triton X-100, pH 7.2 (Triton buffer). A proteinase inhibitor mixture (30 μl) containing per milliliter, 2.8 mg of aprotinin, 1 mg each of pepstatin leupeptin, antipain, and chymostatin, respectively, was added. After centrifugation at 100,000 *g* for 1 hr, the pellet (100,000 *g* pellet) was resuspended in a very small volume of water. SDS and 2-mercaptoethanol were added to a final concentration of 3% and 1%, respectively, and the mixture was boiled for 2 min. After centrifugation as above, the supernatant was lyophilized, taken up in Triton buffer and diluted to 0.1% SDS. After another centrifugation, 100 μl of goat anti-denatured SI antiserum was added and incubated at 4° overnight. After addition of 100 μl of a 10% *S. aureus* suspension the incubation continued for 2 hr. After centrifugation the supernatant was incubated with another 100 μl of *S. aureus* suspension. The combined pellets were washed 3 times with 0.2 ml of Triton buffer containing 0.1% SDS, resuspended in a small volume of water, and heated in 5% SDS, 1% mercaptoethanol. The pellets were reextracted three times with water, and the combined supernatants were lyophilized, dissolved in Triton buffer to a final SDS concentration of 0.1%; immunoprecipitation was repeated with 100 μl of goat antiserum. The final staphylococcal pellet was resuspended in 50 μl of SDS–PAGE sample buffer and heated for 5 min in boiling water, the *S. aureus* cells were reextracted twice, and the combined supernatants were examined by SDS–PAGE. For immunoprecipitation from the supernatant of the 100,000 *g* pellet, 100 μl of anti-native SI-antiserum for guinea pig were added. The treatment with *S. aureus* was carried out as above; for the second precipitation, goat antiserum was used.

SDS–PAGE electrophoresis was performed using a discontinuous sulfate–borate system modified[14] from Neville; bands were detected by fluorography (see this volume [15]). Peptide maps were obtained by treatment of gel pieces containing the band with 75% (v/v) formic acid for 40 hr at 37°. The fragments were separated with a SDS 10% gel according to Laemmli.[38] The translation system used produced some high M_r bands, of

[38] U. K. Laemmli, *Nature (London)* **227,** 680 (1970).

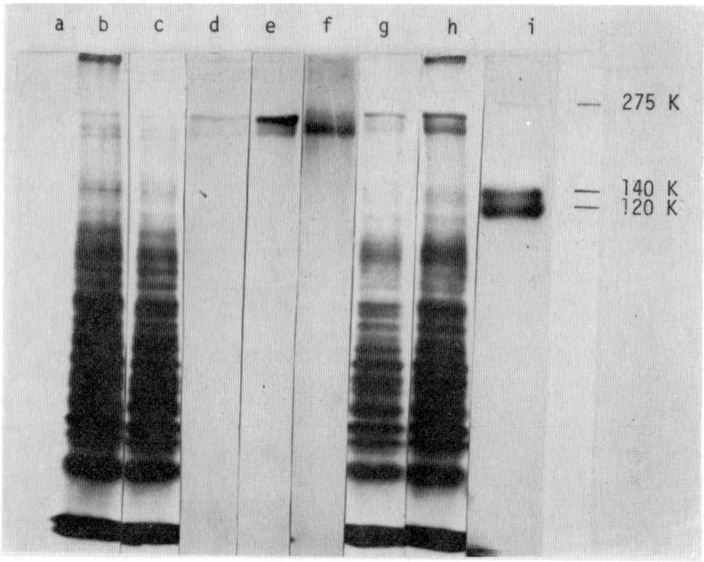

FIG. 7. *In vitro* synthesis of pro-SI (or a precursor thereof). Fluorograph of ^{35}S-labeled polypeptides synthesized in a reticulocyte lysate pretreated with nuclease in the presence or the absence of dog pancreas microsomes, in response to total RNA extracted from rabbit intestinal mucosa. After translation, the synthesized polypeptides were precipitated with anti-SI antiserum. The SDS–polyacrylamide gels (6%) were exposed for fluorography for 4 days (a–d, g, h) or 2 days (e, f). Lanes: a, control (no RNA, no microsomes); b, translation mixture without microsomes; c, as in b, with centrifugation after incubation (100,000 *g* supernatant); d, immunoprecipitate from lane c; e, mixture of the following 3 immunoprecipitates; d + immunoprecipitate from the 100,000 *g* pellet (see lane c) + immunoprecipitate from lane g (see below); f, translation mixture with microsomes, spun at 100,000 *g* after the incubation (immunoprecipitate of the pellet); g, translation mixture with microsomes, spun at 100,000 *g* after the incubation (supernatant, no immunoprecipitation); h, translation mixture with microsomes; i, [^{125}I]SI (M_r 120,000 sucrase subunit; M_r 140,000, isomaltase subunit). From Wacker *et al.*[14]

which two (M_r by SDS–PAGE, 270,000 and 240,000, Fig. 7, lanes d–f) were precipitated by anti-SI-antiserum. If preimmune serum was used instead, no band could be detected in the fluorogram, even after a six-times longer exposure (data not shown).

Comments

When translation was carried out in the presence of dog pancreas microsomes (Fig. 7, lane h) the immunospecific bands were much more intense than after translation in the absence of microsomes (lanes b, c, and d). In addition, the band with higher mobility was almost lacking in

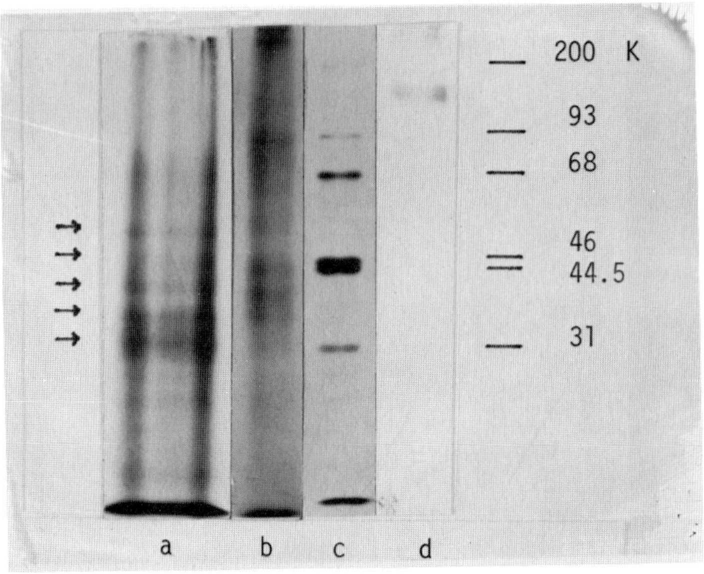

FIG. 8. Peptide map of the immunoprecipitated translation product. A gel piece containing the faster moving protein band (as in Fig. 7, f) was subjected to treatment with 75% formic acid and reelectrophoresed on SDS–polyacrylamide (10%) gel according to Laemmli.[38] Fluorography was for 4 days. Lanes: a, peptides from the translation product, [^{35}S]methionine labeled; b, peptides from [^{125}I]SI; c, ^{14}C-labeled standard M_r proteins (myosin, 200,000; glycogen phosphorylase, 93,000; bovine serum albumin, 68,000; cytosolic aspartate aminotransferase, 46,000; mitochondrial aspartate aminotransferase, 44,500; carbonic anhydrase, 31,000); d, [^{125}I]SI. The arrows indicate the peptide bands appearing in both lanes a and b. From Wacker *et al.*[14]

the supernatant of the 100,000 *g* pellet (lane g), from which it could be extracted with SDS and specifically precipitated (lane f). At the moment it cannot be decided whether this indicates the processing of a (still hypothetical) pre-pro-SI or whether the higher-mobility band arises from the other one by a proteolysis unrelated to the processes of translation and membrane insertion. Interpretation of this mobility change is complicated, at this stage, by the possibility that these polypeptides might be glycosylated, the sugar moieties of rabbit SI accounting for approximately 15% of its weight.

The similarity in the peptide patterns of [^{125}I]SI and of the [^{35}S]methionine-labeled *in vitro* translation product (Fig. 8) confirms that the one-chain large-M_r translation product(s) are (is) indeed identical with, or a precursor of, pro-SI.

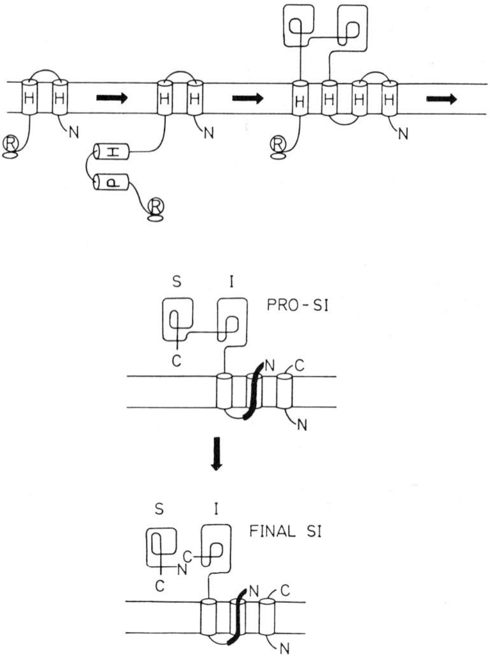

FIG. 9. A possible sequence of events in the biosynthesis and membrane insertion of prepro-SI and of its processing into pro-SI and SI. This model follows the scheme of the "helical hairpin" hypothesis.[39–42] For details see the text. R, ribosome; H, hydrophobic helix; P, polar (hydrophilic) helix; S, sucrase portion in pro-SI (or subunit in the SI complex); I, isomaltase portion in pro-SI (or subunit in the SI complex). The heavy line in the two lower figures corresponds to the left portion of the isomaltase anchor in Fig. 5, the amino acid sequence of which is known (see Fig. 6). From Sjöström *et al.*[9]

A Possible Biosynthetic Mechanism

Various authors[39–42] have suggested in various forms that the gliding of the nascent polypeptide chain across the membrane of the rough endoplasmic reticulum starts with the insertion of a hydrophobic-hydrophilic "loop" or "helical hairpin" into the membrane bilayer (the N-terminal of the pre-piece would at all times remain at the cytosolic side of the membrane). Subsequent insertion of more "hairpins" and/or the action of proteases (including signal protease) would lead to a variety of positionings of intrinsic membrane proteins.

[39] B. M. Austen, *FEBS Lett.* **103,** 308 (1979).
[40] M. Inouye and S. Halegoua, *CRC Crit. Rev. Biochem.* **10,** 339 (1980).
[41] G. von Heijne and C. Blomberg, *Eur. J. Biochem.* **97,** 175 (1979).
[42] D. M. Engelman and T. A. Steitz, *Cell* **23,** 411 (1981).

Figure 9 shows a possible model for pro-SI biosynthesis and insertion. It assumes that pro-SI is synthesized with a leader peptide—an assumption justified by the fact that the N-terminal of I is located on the *extracellular* side.[28] A first hairpin would establish the first insertion into the membrane, and a short hydrophilic sequence (including residue 11, a threonine to be glycosylated) could thus be extruded until the following hydrophobic sequence (left part of the membranous segment in Fig. 5) terminates extrusion. A second hydrophobic hydrophilic hairpin would follow immediately and the rest of pro-sucrase–isomaltase could be synthesized and extruded either completely, or up to another hydrophobic segment situated closely to the C terminus. "Signalase" would split the loop between the two helices of the first hairpin thus creating a secondary (extracellular) N terminus (i.e., the N terminus of isomaltase).

The major events in the glycosylation of pro-SI and in its transfer from the endoplasmic reticulum into the Golgi and finally into the brush border membrane have been studied and clarified by various authors in pulse-chase experiments.[12,43–45]

[43] E. M. Daneilsen, H. Sjöström, and O. Norén, *FEBS Lett.* **127,** 129 (1981).
[44] E. M. Danielsen, H. Skovbjerg, O. Norén, and H. Sjöström, *FEBS Lett.* **132,** 197 (1981).
[45] E. M. Daneilsen, *Biochem. J.* **204,** 639 (1982).

[32] Structure and Biosynthesis of Aminopeptidases

By SUZANNE MAROUX and HÉLÈNE FERACCI

Activity against β-naphthylamide derivatives of amino acids has been demonstrated in the apical part of the plasma membrane of several types of epithelial cells.[1,2] The enzymes responsible have been purified and characterized in hepatocytes,[3] proximal tubule kidney cells and intestinal absorbing cells.[4] They are aminopeptidase N (EC 3.4.11.2, microsomal) and aminopeptidase A (EC 3.4.11.7, aspartate aminopeptidase), respectively, which are active on neutral and acidic N-terminal amino acids in peptides and synthetic derivatives such as β-naphthylamides and *p*-nitroanilides.[5,6]

[1] B. Monis, H. Wasserkrug, and A. M. Seligman, *J. Histochem. Cytochem.* **13,** 503 (1965).
[2] Z. Loyda and R. Gossran, *Histochemistry* **67,** 267 (1980).
[3] H. G. Little, W. L. Starnes, and F. J. Behal, this series, Vol. 45, p. 495.
[4] J. A. Kenny and S. Maroux, *Phys. Rev.* **62,** 91 (1982).
[5] H. Feracci and S. Maroux, *Biochim. Biophys. Acta* **599,** 448 (1980).
[6] H. Feracci, A. Benajiba, J. P. Gorvel, C. Doumeng, and S. Maroux, *Biochim. Biophys. Acta* **658,** 148 (1981).

METHODS IN ENZYMOLOGY, VOL. 96

ISBN 0-12-181996-5

In a given species, the homology of aminopeptidases from different organs is high enough so that antibodies raised against aminopeptidase from one organ will precipitate the aminopeptidases from all the other organs.[7-9] Relationship between the structure of these enzymes and their mode of integration in the membrane, and consequently their biosynthesis, has been most extensively studied on kidney and intestinal brush-border aminopeptidases.[4]

Structure of Aminopeptidases in Relation with Their Mode of Integration in the Brush-Border Membrane

Aminopeptidases are integral membrane proteins that can be solubilized without loss of activity either by limited proteolysis or by neutral detergent extraction.[10,11] The proteinase and detergent forms thus obtained have molecular characteristics of hydrophilic and amphipathic molecules, respectively.[12] Limited proteolysis transforms the detergent form into the proteinase form. Simultaneously, a highly hydrophobic peptide, responsible for the hydrophobic properties of detergent form, is released.[13] It is called the anchor peptide.

Purification of Protease and Detergent Forms of Rabbit Aminopeptidase N

The main steps of the purification procedure are summarized in Scheme 1.

Common Steps. Frozen mucosa (600 g) was cut into pieces and gently stirred for 30 min in 2400 ml of buffer A (10 m*M* potassium phosphate, pH 6.0). Part of the mucus was removed by gauze filtration and a 15-min centrifugation at 3000 *g*. A 3-hr centrifugation at 30,000 *g* of the 3000 *g* supernatant yielded a second pellet that was suspended in 1 liter of buffer A containing 3% Emulphogen. The suspension was stirred overnight at 4° and centrifuged for 1 hr at 30,000 *g*. The resulting solution was brought up to 0.5 saturation ammonium sulfate, and the precipitate containing the aminopeptidase activity was dissolved in 100 ml of buffer A containing 1% Emulphogen. The gel thus obtained was dialyzed against 5 liters of buffer

[7] C. Vannier, D. Louvard, S. Maroux, and P. Desnuelle, *Biochim. Biophys. Acta* **455,** 185 (1976).

[8] T. Ito, K. Hiwada, and T. Kokubu, *Clin. Chim. Acta* **101,** 139 (1980).

[9] H. Feracci, A. Bernadac, S. Hovsépian, G. Fayet, and S. Maroux, *Cell Tissue Res.* **221,** 137 (1981).

[10] S. Maroux, D. Louvard, and J. Baratti, *Biochim. Biophys. Acta* **321,** 282 (1973).

[11] D. Louvard, S. Maroux, C. Vannier, and P. Desnuelle, *Biochim. Biophys. Acta* **375,** 236 (1975).

[12] A. Helenius and K. Simons, *Proc. Natl. Acad. Sci. U.S.A.* **74,** 529 (1977).

[13] S. Maroux and D. Louvard, *Biochim. Biophys. Acta* **419,** 189 (1976).

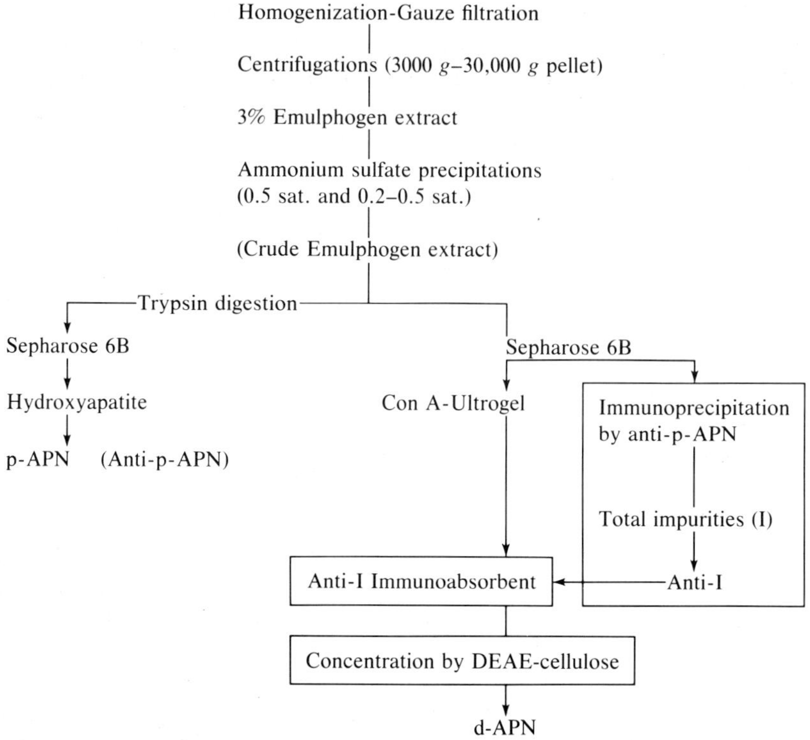

SCHEME 1. Flow sheet of the purification of the protease (p-APN) and detergent (d-APN) forms of rabbit intestinal aminopeptidase N. Trypsin hydrolysis was performed in 0.1 *M* Tris-HCl buffer, pH 8.0, at 4° for 2.5 hr with 0.25 mg of trypsin per milliliter. A Sepharose 6B column (3 cm × 400 cm) was equilibrated and eluted in 10 m*M* phosphate buffer, pH 6.0. A hydroxyapatite column (3 cm × 6 cm) was equilibrated with 10 m*M* monopotassium phosphate, pH 6.0, containing 0.2 *M* KCl. The column was washed with the same buffer, then eluted by a linear phosphate concentration gradient (2 × 450 ml) from 10 m*M* to 200 m*M*. A concanavalin A–Ultrogel column (6 ml) was equilibrated and eluted with a 10 m*M* Tris-HCl buffer, pH 7.2, containing 1 m*M* $CaCl_2$, 1 m*M* $MgCl_2$, 1 m*M* $MnCl_2$, and 1% Emulphogen. Column dimensions were for 600 g of mucosa as starting material. From Feracci and Maroux.[5]

A. The resulting solution was cleared up by centrifugation and brought to 0.2 saturation ammonium sulfate. After a 30-min centrifugation at 30,000 *g*, the upper gel phase was discarded and the lower phase was brought to 0.5 saturation ammonium sulfate. The new precipitate was dissolved in minimum volume of buffer A to give a solution (crude Emulphogen extract) that was dialyzed against 5 liters of buffer A. Beyond this point, the two forms were purified independently.

Purification of the Protease Form. The above solution made 0.1 *M* in Tris-HCl (pH 8.0) was incubated at 4° for 2.5 hr with trypsin (0.25 mg/ml). Hydrolysis was stopped by 1 m*M* phenylmethylsulfonyl fluoride (PMSF). After dialysis against buffer A, the solution was filtered through a Sepharose 6B column (3 cm × 400 cm) equilibrated with the same buffer. The aminopeptidase N activity emerged as a symmetrical peak at 2.15 void volumes. The fractions with specific activities higher than 6000 were pooled and dialyzed against buffer A containing 0.2 *M* KCl. The aminopeptidase A that emerged with the void volume of the column due probably to persisting aggregation was readily separated at this stage from aminopeptidase N.

The last purification step was a chromatography on a hydroxyapatite column (3 cm × 6 cm) equilibrated with a 10 m*M* monopotassium phosphate buffer (pH 6.0) containing 0.2 *M* KCl. The column was washed with the buffer and eluted as a symmetrical peak by a linear phosphate concentration gradient (2 × 450 ml) from 10 m*M* to 200 m*M*.

Purification of the Detergent Form. For this purification, the trypsin digestion of the crude Emulphogen extract was omitted (Scheme 1), and the detergent form was maintained in solution by 1% (v/v) Emulphogen in all buffers. The buffers also contained benzamidine (1 m*M*) and PMSF (1 m*M*) to keep to a minimum undesirable degradations. The crude Emulphogen extract was directly filtered through Sepharose 6B in the same type of column as before. The detergent form emerged at approximately 1.95 void volumes. It was again well separated from the detergent form of aminopeptidase A that emerged at 1.7 void volumes.

The pooled fractions containing the aminopeptidase N activity were loaded onto a 60-ml concanavalin A–indubiose column equilibrated with a 10 m*M* Tris-HCl buffer (pH 7.2) containing 1 m*M* $CaCl_2$, 1 m*M* $MgCl_2$, and 1 m*M* $MnCl_2$. Approximately 70% of the total activity migrated unretarded when the column was eluted with the same buffer.

Beyond this point, no conventional method could be found for increasing the specific activity of the detergent form above 12,000 compared to 15,000 for the most purified protease form. Anti-aminopeptidase immunoabsorbent chromatography was unsuccessful because of heavy activity losses during elution, and an entirely new technique must be worked out. The principle of this technique was to precipitate aminopeptidase selectively by the corresponding antibody, thus leaving the impurities in solution. In a second step, "anti impurities" (anti I) antibodies were raised and then used for reverse immunoabsorbent chromatography under suitable conditions.

In practice, a solution containing all the contaminants but devoid of enzyme was prepared from pooled active fractions emerging from

Sepharose 6B by specific precipitation of aminopeptidase using antibodies directed against the protease form. An antiserum against total impurities was raised in a goat. The resulting antibodies were separated and coupled with Ultrogel ACA-22 to give an immunoabsorbent through which the aminopeptidase migrates freely while the impurities are retained.

The resulting detergent form of aminopeptidase was freed from traces of remaining antibodies and concentrated by passage through a short (2 ml) DEAE-cellulose column equilibrated with 0.01 *M* potassium phosphate buffer, pH 6.0, containing 0.1% Emulphogen. After washing with the same buffer, enzyme was eluted by this buffer containing 0.15 *M* NaCl.

Obtaining Specific Antibody

Specific antibody is a very useful tool in the study of biosynthesis of a protein (see Section I,A of this volume, General Methods, in particular Chapter [8]. Intestinal aminopeptidases bear sugar sequences immunologically identical to human blood group A antigen.[14,15] Consequently, antibodies raised against highly purified aminopeptidase were found to cross-react with human blood group-like substances of mucus and other glycoproteins of the plasma membrane of enterocytes also bearing such antigenic determinants. To obtain antibodies strictly specific for the enzyme, depletion of antibodies with human A erythrocytes was necessary. Serum or antibody preparation at 0.6 mg/ml was incubated three times with an equal volume of packed human blood group A erythrocytes. Depletion was controlled by labeling of erythrocytes using the indirect immunofluorescence technique.

Preparation of the Anchor Peptide

The detergent form of aminopeptidases were labeled with ^{125}I to facilitate subsequent detection of the split product, and 5–6 mg/ml solutions in a 0.1 *M* Tris-HCl buffer, pH 7.8, were digested for 4 hr at 4° by 0.5 mg of Sepharose-bound trypsin per milliliter. The anchor peptides were purified from the hydrolyzate as summarized in Scheme 2B by successive chromatographies on a Sepharose G-50 column (1.5 × 60 cm) equilibrated and eluted with 10 m*M* phosphate buffer, pH 6.0, 0.1% in Emulphogen and on a DEAE-cellulose column (2 ml) equilibrated and eluted with the same 10 m*M* phosphate buffer, pH 6.0, 0.1% in Emulphogen.[13] The hydrophobic anchor emerged unretarded from these two columns, whereas small

[14] H. Feracci, A. Bernadac, J. P. Gorvel, and S. Maroux, *Gastroenterology* **82,** 317 (1982).
[15] J. P. Gorvel, A. Wisner-Provost, and S. Maroux, *FEBS Lett.* **143,** 17 (1982).

TABLE I

COMPARISON OF MOLECULAR PARAMETERS OF RABBIT AND PIG AMINOPEPTIDASES (DETERGENT FORM) AND THEIR RESPECTIVE HYDROPHILIC AND HYDROPHOBIC DOMAINS (p-FORM AND ANCHOR PEPTIDE, RESPECTIVELY) DETERMINED BY DIFFERENT TECHNIQUES

	Apparent molecular weight estimated by				
Aminopeptidases (AP) from intestinal brush border	High-speed sedimentation equilibrium	Gel filtration in presence of 1% neutral detergent	SDS–gel electro-phoresis	Isotopic dilution technique	NH_2-terminal residues
Rabbit APN[5]					
d-form		200,000	120,000	—	Tyr
p-form	125,000	125,000[a]	120,000	—	Ser
Anchor peptide		70,000	8,500	3800	Tyr
Pig APA[17]					
d-form		380,000	120,000	—	Leu
p-form	247,000	247,000[a]	120,000	—	Asx
Anchor peptide		70,000	8,500	4500	Leu
Pig APN[10,13,16]					
d-form		380,000 }	130,000	—	Ala, Ser
		}	95,000	—	
p-form	245,000	245,000[a] }	49,000	—	Val, Ser
Anchor peptide		70,000	8,500	3500	Ala

[a] Used as calibration protein.

degradative peptides were retarded on the Sepharose column and the two forms of the enzyme were bound on a DEAE-cellulose column.

Molecular Weight Determination and Detergent Binding of Aminopeptidases and Their Hydrophilic and Hydrophobic Parts

The molecular weight of the detergent form and its hydrophilic and hydrophobic domains in the case of intestinal rabbit aminopeptidase N and pig aminopeptidases A and N determined by several techniques, are compared in Table I. Among these techniques, only the unusual isotopic dilution method will be described in detail.

The exact molecular weight value of hydrophilic part has been determined by high-speed sedimentation equilibrium. Comparison of the values thus obtained with those found by electrophoresis in the presence of SDS, clearly shows that the rabbit aminopeptidase N is a monomer,

A. Principle

$$\text{d-AP} + \text{Pep}^* \xrightarrow{\text{Ti}} \text{p-AP} + (\text{Pep}^* + \text{Pep}) + \text{d-AP}$$

		d-AP	Pep*	p-AP	(Pep* + Pep)	d-AP
cpm		0	b	0	b	0
nmol	1	a	x	y	$(x + y)$	$+ (a - y)$
	2	a	x	y	$(x + 2y)$	$+ (a - y)$

1. AP is a monomer $\dfrac{\text{Pep}^* + \text{Pep}}{\text{Pep}^*} = \dfrac{x + y}{x}$

2. AP is a symmetrical dimer $\dfrac{\text{Pep}^* + \text{Pep}}{\text{Pep}^*} = \dfrac{x + 2y}{x}$

B. Determination of parameters

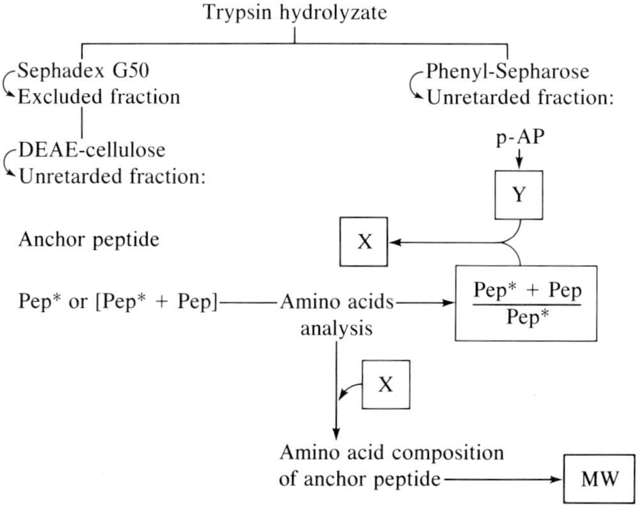

SCHEME 2. Molecular weight (MW) determination of another peptide by the isotopic dilution technique.

whereas pig aminopeptidases A and N are dimers. In this last case the native subunits are transformed into two smaller ones by proteolysis.[16]

Isotopic Dilution Technique. It is difficult to estimate the exact values of the molecular weights of the anchor peptides and detergent forms on account of the high affinity of the hydrophobic anchors for detergents and their aggregation in the absence of detergent. An isotopic dilution technique summarized in Scheme 2, which yields the amino acid composition of the anchor peptide, has been used to determine the molecular weight by summation.[5]

[16] A. Benajiba and S. Maroux, *Biochem. J.* **197**, 573 (1981).

TABLE II
AMINO ACID COMPOSITION OF THE ANCHOR PEPTIDE FROM RABBIT AMINOPEPTIDASE N AFTER 24 HR HYDROLYSIS[a]

Amino acid	Residues per b cpm		Dilution factor	Residues/mol	
	P*	P* + P	(P* + P)/P*	Calculated values	Next integer
Asp	10.6	74	7	2.9	3
Thr	5.45	29.8	5.4	1.5	1–2
Ser	11.0	81.3	7.4	3.0	3
Glu	15.0	101	6.7	4.1	4
Pro	3.6	22.7	6.3	0.99	1
Gly	2.1	15.0	7.1	5.8	6
Ala	12.9	89.7	6.9	3.5	3–4
Val	9.8	61	6.2	2.7	3
Ile	8.6	56	6.5	2.4	2
Leu	16.0	107	6.7	4.4	4
Tyr	4.4	30.4	6.9	1.2	1
Phe	4.7	34.6	7.4	1.3	1
His	2.3	16.2	7.1	0.6	1
Lys	7.5	50.5	6.7	2.0	2
Arg	3.3	22.9	6.9	0.9	1
Number of residues					36–38
Molecular weight					3710–3880

[a] From Feracci and Maroux.[5]

In a typical assay, the radioactivity b of P* was 4.2×10^6 cpm and a was equal to 50 nmol. The amount of protease form generated (y nmol) was exactly evaluated by chromatography on a phenyl–Sepharose column from which this hydrophilic form is excluded with a yield of 100%, whereas the detergent form and peptide were bound. In the present case y was found equal to 25 nmol (conversion yield, 50%). The amino acid composition of the peptides P* and (P* + P) calculated for the same radioactivity (b cpm) are indicated in Table II. The dilution factor (P* + P)/P* = $(x + y)/x$ was derived from the average of the ratio for each amino acid and found equal to 6.75. Then, $x = 4.3$ nmol was calculated. This value enabled the amino acid analysis of P* to be interpreted in terms of the number of residues per mole, and hence, the molecular weight of the peptide was computed.

The same technique has been very useful in showing that each subunit of dimeric aminopeptidases bears a hydrophobic part.[16,17] The very low

[17] A. Benajiba and S. Maroux, *Eur. J. Biochem.* **107,** 381 (1980).

value (about M_r 4000) thus obtained, explains why subunit molecular weight determinations from detergent and proteinase forms were found to be identical by SDS–gel electrophoresis.

In contrast, filtration on Sepharose 6B in the presence of 1% neutral detergent shows different molecular weights for the two native forms of the enzymes. This is due to the binding of a detergent micelle around the hydrophobic part(s),[12,18] the molecular weight of which was estimated to be 70,000.

Position of the Anchor Peptide in the Molecule

The position of hydrophobic anchorage peptide in the N-terminal region of each subunit was first strongly suggested by determinations of the N-terminal residues (Table I). The isolated hydrophobic part has always the same N-terminal residue as the detergent subunit. This residue is different from that found in the proteinase subunit.[5,13,16,17]

In the case of rabbit aminopeptidase N, the N-terminal sequence of the two forms has been determined.[19] Sixty nanomoles of protein were precipitated by 10% trichloracetic acid. In the case of the detergent form, the precipitate was washed by 98% acetone to eliminate the detergent. The proteins were dissolved in 300 μl of formic acid to be introduced in the cup of a Beckman liquid-phase Sequencer Model 890 C. The 0.1 *M* quadrol program was used. As shown in Fig. 1, the N-terminal sequence of the proteinase form is essentially composed of hydrophilic residues, whereas in the case of detergent form, after the first four residues, all the amino acids are hydrophobic. All results described until now strongly suggest that the anchor peptide released by trypsin contains this hydrophobic N-terminal sequence. Considering the amino acid composition of the peptide, the hydrophobic sequence determined can elongate by a maximum of 10 additional hydrophobic residues (3 Gly, 3 Ala, 3 Val, and 1 Leu). Then the 15 remaining polar residues form a highly hydrophilic sequence that constitutes the beginning of the hydrophilic domain.

Localization of Hydrophilic Part on the Membrane Surface by a Quantitative Immunochemical Technique

The brush-border membrane spontaneously forms closed right-side-out vesicles.[20] The hydrophilic part of the aminopeptidase or protease

[18] A. Helenius and K. Simons, *Biochim. Biophys. Acta* **415,** (1975).
[19] H. Feracci, S. Maroux, J. Bonicel, and P. Desnuelle, *Biochim. Biophys. Acta* **684,** 133 (1982).
[20] D. Louvard, S. Maroux, J. Baratti, P. Desnuelle, and S. Mutafschiev, *Biochim. Biophys. Acta* **291,** 747 (1973).

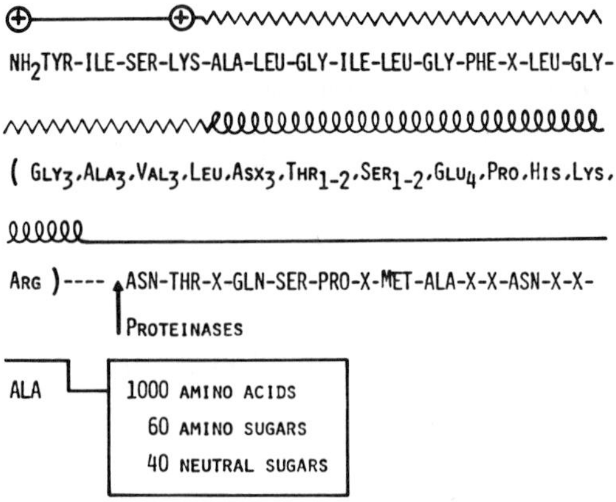

FIG. 1. N-terminal sequence of the detergent and proteinase forms of rabbit intestinal aminopeptidase N. The residues given in parentheses are contained in anchor peptide released by trypsin from the detergent form, minus the 14 first residues of the N-terminal sequence of the detergent form, indicated on the first line. The protease form begins on the right of the arrow that indicates the cleavage that releases the anchor peptide from the detergent form. Serrated line, hydrophobic sequence in anchor peptide; looped line, hydrophilic sequence in anchor peptide; dashed line, polypeptide chain of protease form that contains 1000 amino acids and 100 sugars.

form can be released from these membrane preparations by papain digestion without affecting the lipid bilayer limiting the vesicles.[11] This observation was the first evidence for its location on the external side of the membrane.

The extent of integration in the membrane of pig aminopeptidase N has been evaluated by a quantitative immunological technique.[21,22] The method is based on the observation that the maximum number of antibody molecules that can simultaneously bind to an antigen corresponds to the total covering of its surface.[21] Besides, the integration of a protein antigen in a membrane matrix can be expected to bury part of this antigen surface.

This technique involved the utilization of antibodies purified to homogeneity in a single step by affinity chromatography. Rabbit antiserum, raised against pure protease form of pig aminopeptidase was run through an immunoabsorbent column prepared by coupling this antigen to glutaraldehyde-activated Ultrogel ACA-22. Antibodies were labeled by iodination and reisolated on an immunoabsorbent column.[23] Then the maxi-

[21] D. Louvard, S. Maroux, and P. Desnuelle, *Biochim. Biophys. Acta* **389,** 389 (1975).

[22] D. Louvard, C. Vannier, S. Maroux, J. M. Pagès, and C. Lazdunski, *Anal. Biochem.* **76,** 83 (1976).

[23] J. M. Pagès, D. Louvard, and C. Lazdunski, *FEBS Lett.* **59,** (1975).

mum numbers of antibody molecules that can bind to membrane-bound (vesicles) and solubilized aminopeptidase are compared as follows.

A constant amount of antigen (60 pmol) was incubated (2 hr at 37° and overnight at 4°) with varying amounts of labeled antibodies (240–1320 pmol) sufficient for complete precipitation of the antigen. After centrifugation, the radioactivity measured in the supernatant corresponds to the antibody excess. This value is plotted versus the antibody-to-antigen molar ratio. No antibody was found in the supernatant as long as this molar ratio was lower than the value corresponding to the saturation of the antigen surface by the antibodies. Above this critical value the amounts of antibodies added in excess were found in the supernatant. The corresponding points were on a line that is parallel to that obtained by plotting the total amounts of antibody added to the antigen. The extrapolation of this line to the abscissa gives the value of the ratio corresponding to the saturation of the surface antigen by the antibody.

One can bind 12 molecules of antibody on 12 antigenic determinants at the surface of the solubilized pig intestinal aminopeptidase N whereas only 8 molecules can be attached simultaneously to the membrane-bound enzyme.

To determine the exact number of buried determinants, a more suitable technique is to use antibodies specific for these determinants for their titration on the soluble antigen. To obtain such antibodies, the total antibody population was incubated with a large excess of membrane-bound enzyme (1 mol of antigen per mole of antibody). In these conditions, there is a statistical occupancy of only one accessible determinant per antigen molecule, and steric hindrance that can occur in the direct method is avoided. The unbound antibodies recovered in the supernatant after centrifugation are specific for determinants really masked by integration. In experiments where these depleted antibodies were used, one has to operate in two steps. A first incubation is carried out in the presence of the specific ^{125}I-labeled antibody, then nonradioactive total antibody is added to induce precipitation. By this method it was found that only 2 determinants out of the 12 are inaccessible in the membrane-bound enzyme.

Integration of the Anchor Peptide in the Membrane

To show that the anchor peptides of aminopeptidases span or do not span the lipid bilayer, the cytoplasmic face of the membrane must be specifically labeled. The spontaneous vesiculization of the brush-border membrane offered the opportunity to trap, inside closed right-side-out vesicles formed, a high molecular weight reagent initially added in the medium. A photogenerated reagent obtained by coupling 4-fluoro-3-nitrophenylazide with a ^{125}I-labeled protein has been used to show that anchor peptide of pig intestinal aminopeptidase spans the membrane.[24] After extensive washing, irradiation induced the cross-linking of the trapped radioactive reagent to accessible proteins. The membrane proteins were solubilized by detergent or papain treatment. Aminopeptidase was then specifically precipitated by its antibody. The extent of the aminopeptidase photolabeling is determined by the radioactivity in the precipitate. The specific labeling of the hydrophobic part can be easily estimated by comparing the labeling of the detergent and protease forms. The extent of labeling of the protease form was found equal to the unspecific labeling from the outside due to a small amount of reagent that cannot be displaced by extensive washing. It has been directly estimated in a control experiment in which reagent was added after vesiculization. The detergent form is significantly more labeled than the papain form. Consequently, we can conclude that the anchor peptide has been significantly labeled from the inside of the vesicles. In other words it spans the membrane. The enzyme of pig kidney microvillous membrane has also been shown to be a transmembrane protein.[25]

Considering that a sequence of 15 or 20 hydrophobic amino acids forming a 3_{10} helix or an α helix is required to traverse the 30 Å hydrophobic core of the membrane,[26] the aminopeptidases span the membrane only once.

In the case of rabbit enzyme the first four residues of the detergent form could constitute the cytoplasmic hydrophilic segment. The intramembranous domain is constituted by the following hydrophobic sequence of 15–20 residues. Then the high molecular weight hydrophilic domain, located on the membrane surface, begins with a highly hydrophilic sequence of 15 residues.

Relationship between the Mode of Integration of Aminopeptidases in the Membrane and Their Biosynthesis

The monomeric structure of rabbit aminopeptidase seems to be an exception, the most general case being the dimeric structure of pig enzymes and also in the case of man[27] and rat[28] enzymes. In the case of pig enzymes it is clear that each subunit is the homolog of the rabbit mono-

[24] D. Louvard, M. Sémériva, and S. Maroux, *J. Mol. Biol.* **106,** 1023 (1976).
[25] A. Booth and A. J. Kenny, *Biochem. J.* **187,** 31 (1980).
[26] D. M. Engelman and T. A. Steitz, *Cell* **23,** 411 (1981).
[27] K. Hiwada, T. Ito, M. Yokoyama, and T. Kokubu, *Eur. J. Biochem.* **104,** 155 (1980).
[28] Y. S. Kim and E. J. Brophy, *J. Biol. Chem.* **251,** 3199 (1976).

mer. In particular, each subunit is anchored by its N-terminal sequence. These observations argue in favor of an independent biosynthesis of the two subunits, then subsequent association. In addition, *in vitro* pig aminopeptidase N is synthesized as an M_r 115,000 polypeptide corresponding to one nonglycosylated subunit.[29] By contrast, synthesis of a large precursor, which is subsequently split into two associated chains, has been demonstrated for the rat brush-border sucrase–isomaltase, a disymmetric dimeric complex anchored in the membrane by only one of its subunits.[30]

Another important structural character of aminopeptidases in regard to their biosynthesis is their high content of sugars (15–20% in weight). Considering the localization of the glycosylation reaction,[31,32] the asymmetric insertion of aminopeptidase must occur in the rough endoplasmic reticulum membrane, where the first steps of glycosylation of the extracytoplasmic domain block its insertion. The glycosylation is completed in the Golgi apparatus, where the presence of immunoreactive protein has been shown[14] before its final transfer into the plasma membrane.

In the absence of synthesis *in vitro,* it is impossible to say whether aminopeptidases are or are not synthesized with a transient leader sequence.[33-35] But, considering its mode of integration in the membrane, it is tempting to propose that its N-terminal transmembrane peptide could be considered to be a leader sequence not cleaved by a leader peptidase. However, no significant sequence homology can be found between this peptide and that of true signal sequences. By contrast, we have noticed[19] a high sequence homology between residues 11–24 and 5–18 of the proteolipids from, respectively, *Neurospora crassa* and the mitochondria of *Saccharomyces cerevisiae,* which are highly hydrophobic proteins synthesized without leader sequences.[36] Comparison of the amino acid compositions of the nondetermined sequence between residues 15–38 of the detergent form of aminopeptidase (Fig. 1) and sequences 25–52 and 19–46 of the proteolipids strongly suggest that homology will also exist in this part of the molecule.

Several mechanisms proposed for protein assembly into biological

[29] M. E. Danielsen, O. Norén, and H. Sjöström, *Biochem. J.* **204,** 323 (1982).

[30] H. P. Hauri, A. Quaroni, and K. Isselbacher, *Proc. Natl. Acad. Sci. U.S.A.* **76,** 5183 (1979).

[31] J. Molnar, *in* "The Enzymes of Biological Membranes" (A. Martonosi, ed.), Vol. 2, p. 385. Plenum, New York, 1976.

[32] F. N. Katz, J. E. Rothman, V. R. Lingappa, G. Blobel, and H. F. Lodish, *Proc. Natl. Acad. Sci. U.S.A.* **74,** 3278 (1977).

[33] G. Blobel, *Proc. Natl. Acad. Sci. U.S.A.* **77,** 1476 (1980).

[34] W. Wickner, *Science* **210,** 861 (1980).

[35] D. M. Engelman and T. A. Steitz, *Cell* **23,** 411 (1981).

[36] A. Tzagoloff and G. Macino, *Annu. Rev. Biochem.* **48,** 419 (1979).

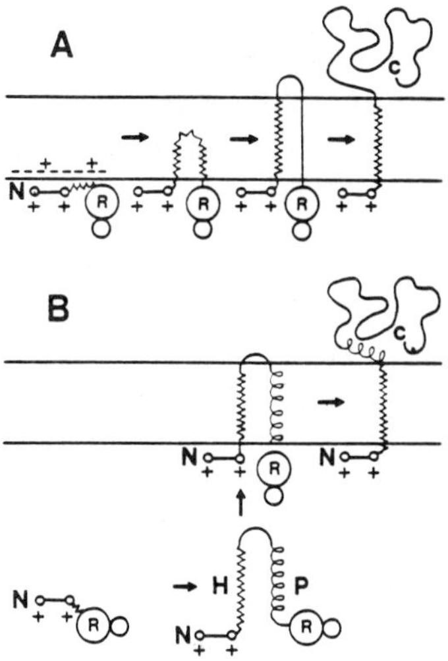

FIG. 2. Model of insertion of brush-border aminopeptidase by the N-terminal hydrophobic sequence during a cotranslational insertion according to the loop mechanism in A and helical hairpin hypothesis in B. Same symbols as in Fig. 1.

membranes can account for the insertion of aminopeptidase in the brush-border membrane. Figure 2 shows, in particular, that the mode of integration of aminopeptidase fits very well with the cotranslational translocation according to the "loop mechanism"[37] and "helical hairpin hypothesis."[35] In all these mechanisms the anchor region of aminopeptidase plays the role of a leader peptide not cleaved after translocation of the protein. Only cotranslational translocation processes have been considered because, according to Blobel,[33] the rough endoplasmic reticulum membrane is competent only for this particular type of insertion of the newly synthesized protein in the membrane.

Direct Transfer from the Golgi to the Brush-Border Membrane or via the Lateral Membrane?

Ultrastructural[38] as well as biochemical[39] techniques have generally shown that the brush-border glycoproteins, in particular sucrase–isomaltase[30] and aminopeptidase N,[15] two of the major brush-border glycoproteins, follow the same intracellular route as most membrane surface glycoproteins. They are synthesized in the rough endoplasmic reticulum and pass through the Golgi apparatus before insertion into the plasma membrane. The question open now is how the brush-border hydrolases

[37] J. M. Di Rienzo, K. Nakamura, and M. Inouye, *Annu. Rev. Biochem.* **47,** 481 (1978).
[38] G. Bennet, C. P. Leblond, and A. Haddad, *J. Cell Biol.* **60,** 258 (1974).
[39] A. Quaroni, K. Kirsh, and M. M. Weiser, *Biochem. J.* **182,** 203 (1979).

arrive in the brush-border membrane. Indeed, the use of different methods have led to different results.

Microscopy techniques, not described here, generally argue in favor of direct transfer from the Golgi to the apical brush-border membrane of the enterocytes. By ultrastructural autoradiography, Leblond's group[38] followed the intracellular route of the glycoproteins of the enterocytes (essentially composed of the brush-border hydrolases and glycocalyx) labeled during a short pulse with [^{3}H]fucose. They showed the presence, in the apical part of the cytoplasm, of vesicles labeled after the Golgi apparatus and before the plasma membrane. Their ultrastructure could correspond to brush-border membrane vesicles closed inside out, and sometimes their fusion with this membrane has been observed. Using ultrathin frozen sections of jejunum to improve the sensitivity of immunofluorescence labeling as proposed by Tokuyasu *et al.*,[40] we have revealed aminopeptidase N only in the brush-border membrane and the Golgi apparatus. On the same sections, histochemistry revealed active molecules in the brush-border membrane, and also in the apical region of the lateral membrane and in a structure located under the terminal web.[14]

By contrast, subcellular fractionation techniques generally argue in favor of a transit of brush-border glycoproteins by lateral membrane.

Presence of Brush-Border Hydrolases in the Enterocyte Basolateral Membranes

To follow the presence of newly synthesized proteins labeled during a short pulse in different membranes of the cell, it is better to prepare the different membrane fractions from the same starting material. However, it is always difficult to obtain with a good yield more than two purified membranes from the same homogenate. Simultaneous isolation of Golgi and basolateral membranes[41] and basolateral and brush-border membranes[42] from intestinal absorbing cells have been described. These three types of membranes characterized by their specific marker—glycosyltransferases for the Golgi, Na^+,K^+-ATPase for basolateral membrane, and hydrolases such as aminopeptidases for the brush-border membrane—can be easily separated from a microsomal fraction by centrifugation through sucrose gradients. Indeed, they have very different density: 1.11 for Golgi, 1.14 and 1.16 for two basolateral membrane fractions, and 1.19 for the brush-border membrane.

[40] K. T. Tokuyasu and S. J. Singer, *J. Cell Biol.* **71,** 894 (1976).

[41] M. M. Weiser, M. Malcolm Neumeier, A. Quaroni, and K. Kirsh, *J. Cell Biol.* **77,** 722 (1978).

[42] B. Colas and S. Maroux, *Biochim. Biophys. Acta* **600,** 406 (1980).

Precipitation of the Rabbit Enterocyte Basolateral Membranes by Specific Anti-Aminopeptidase N Antibodies.[42] All the enterocyte basolateral membrane preparations contained small amounts of brush-border hydrolases, in particular aminopeptidases. If this result is not due to contamination by brush-border membrane, anti-aminopeptidase N antibodies must coprecipitate aminopeptidase and Na^+,K^+-ATPase activities since they are borne by the same membrane.

A sample of basolateral membrane (1.16 of density) containing 50 units of aminopeptidase N was incubated for 2 hr at 37° in a total volume of 200 μl of 10 m*M* phosphate buffer (pH 7.4)–0.15 *M* NaCl with 50 μg of goat immunoglobulin anti-aminopeptidase N. Then precipitation was induced by addition of 200 μg of rabbit anti-goat γG-antibodies. After incubation (1 hr at 37° followed by 16 hr at 4°), the unsolubilized material was separated by a 2-min centrifugation in an IEC bench centrifuge operated at full speed (about 600 *g*). It was checked that native membranes were not sedimented under these conditions. The proportion of aminopeptidase N and Na^+,K^+-ATPase activities present in the sediment and in the supernatant was determined: 65% of the aminopeptidase N and 40% of Na^+,K^+-ATPase activities contained in the preparation were precipitated.

This result shows that a rather extended area of the basolateral membrane preparation contained brush-border hydrolases (about 10% of the brush-border amount). It cannot be excluded that these hydrolases come from the apical part of lateral membrane and brush border by diffusion during cell homogenization. Indeed, redistribution of the brush-border hydrolases over the whole cell surface often occurs immediately after

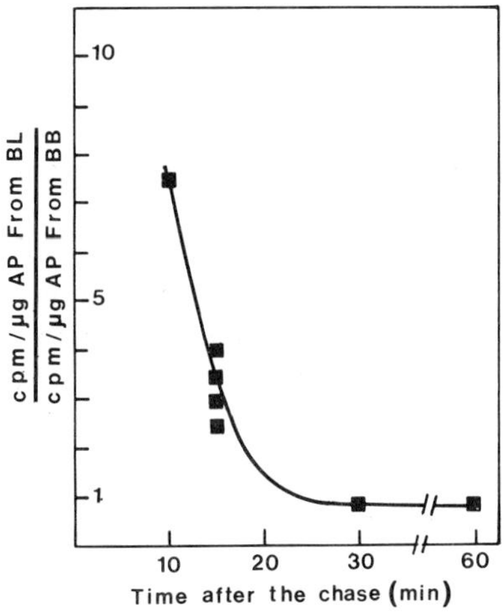

FIG. 3. Ratio of specific radioactivity of aminopeptidase (AP) from basolateral membrane fraction (BL) and from brush-border membrane fraction (BB), at various time intervals after the chase.

disruption of the tight junctions.[9,43] However, we are going to see, a strict relationship may well exist between the presence of brush-border hydrolases in the basolateral membrane and their intracellular transport if the newly synthesized hydrolases do not reach the brush border directly from the Golgi, but transit through the basolateral membrane.

Labeling the glycoproteins by [^{3}H]fucose intraperitoneal injection in rat, Quaroni *et al.*[39] demonstrated that at least 60% of these glycoproteins are first incorporated into the basolateral membrane fraction before reaching their final destination in the brush border. Hauri *et al.*[30] obtained similar evidence for the sucrase–isomaltase individualized by immunoprecipitation.

Kinetics of Incorporation of Newly Synthesized Aminopeptidase in the Basolateral and Brush-Border Membrane Fractions

Labeling with [^{35}S]Methionine. Rabbits (1.2–1.7 kg) were fasted overnight and anesthetized intravenously with the analgesic Imalgene. A jejunum loop about 30 cm long was isolated, and 1 mCi of [^{35}S]methionine, specific radioactivity about 1100 Ci/mmol in 2 ml of phosphate-buffered saline (PBS), was injected in the lumen. After 10 min of incorporation, radioactivity was chased with 1 ml of 50 m*M* methionine in PBS. Rabbits were sacrificed at different times after the chases. The radioactive loop was removed and washed with cold PBS. Basolateral and brush-border membranes were immediately prepared from the mucosal scraping.[42]

Solubilization and Immunoprecipitation of Aminopeptidase N. Aminopeptidase present in total homogenates or in purified membranes was solubilized by adding Triton X-100 to a final concentration of 2% and incubating overnight at 4°. After centrifugation at 18,000 *g* for 30 min in the angle rotor of a Haereus-Chris centrifuge, aminopeptidase activity is entirely in the supernatant. Generally, 6 μg of enzyme were specifically and quantitatively immunoprecipitated with 18 μl of guinea pig serum containing antibodies raised against pure enzyme.[14] A control experiment involved the same quantity of nonimmunized guinea pig serum. Each assay was done twice. After incubation for 2 hr at 37° and overnight at 4°, the immunoprecipitates were centrifuged for 3 min in an IEC bench centrifuge operated at full speed (about 600 *g*). The precipitates were washed 5 times with 0.5 ml of PBS containing 1% Triton X-100 and 0.5% sodium deoxycholate and twice with 10 m*M* Tris-HCl, pH 7.4. Immunoprecipitates were solubilized in 100 μl of Soluene (Packard) and 4 ml of Dimilume-30 (Packard) for scintillation counting.

[43] C. A. Ziomek, S. Schulman, and E. Edidin, *J. Cell Biol.* **86,** 849 (1980).

Results. Figure 3, which summarizes the results obtained, clearly shows that the newly synthesized enzyme is first integrated in the basolateral membrane fraction, then migrates to the brush-border membrane fraction. However, it has not been definitely demonstrated that the membrane containing newly synthesized brush-border hydrolases is the lateral membrane. Indeed, the criteria of purity, such as microscopy or absence of different markers in a membrane fraction, have very low sensitivity and do not permit exclusion of the presence of other type of membranes such as, in our case, transporting vesicles. To conclude, the basolateral membrane must be more specifically purified by immunoprecipitation with antibody specific for one of its characteristic proteins such as Na^+,K^+-ATPase. Coprecipitation or not of the newly synthesized brush border hydrolases will definitely answer the question.

[33] Use of the Heavy-Isotope Density-Shift Method to Investigate Insulin Receptor Synthesis, Turnover, and Processing

By MICHAEL N. KRUPP, VICTORIA P. KNUTSON, GABRIELE V. RONNETT, and M. DANIEL LANE

The responsiveness of a target cell to a peptide hormone, such as insulin, depends on the number of functional insulin receptors the cell possesses. It is recognized[1–10] that cells have the capacity to modulate

[1] J. R. Gavin, III, J. Roth, D. M. Neville, Jr., P. DeMeyts, and D. N. Buell, *Proc. Natl. Acad. Sci. U.S.A.* **71,** 84 (1974).
[2] R. S. Bar, P. Gorden, J. Roth, C. R. Kahn, and P. DeMeyts, *J. Clin. Invest.* **58,** 1123 (1976).
[3] B. C. Reed, S. H. Kaufmann, J. C. Mackall, A. K. Student, and M. D. Lane, *Proc. Natl. Acad. Sci. U.S.A.* **74,** 4876 (1977).
[4] B. C. Reed, G. V. Ronnet, P. R. Clements, and M. D. Lane, *J. Biol. Chem.* **256,** 3917 (1981).
[5] G. V. Ronnett, V. P. Knutson, and M. D. Lane, *J. Biol. Chem.* **257,** 4285 (1982).
[6] W. Blackard, P. Guzelian, and M. Small, *Endocrinology* **103,** 548 (1978).
[7] J. M. Livingston, B. J. Purvis, and D. H. Lockwood, *Metab., Clin. Exp.* **27,** Suppl. 2, 2009 (1978).
[8] M. N. Krupp and M. D. Lane, *J. Biol. Chem.* **256,** 1689 (1981).
[9] I. G. Fantus, J. Ryan, N. Hizuka, and P. Gorden, *J. Clin. Endocrinol. Metab.* **52,** 953 (1981).
[10] C. R. Kahn, I. D. Goldfine, D. N. Neville, Jr., and P. DeMeyts, *Endocrinology* **103,** 1054 (1979).

ISBN 0-12-181996-5

their insulin receptor level in response to various physiological stimuli, such as differentiation,[3,4,11] exposure to heterologous hormones,[12] or a change in the ambient concentration of insulin.[5,8] To understand the mechanisms that underlie these alterations in functional receptor level, it is useful to localize and identify the regulated step(s) in receptor metabolism that is affected by each of these stimuli. The heavy-isotope density-shift technique has been of particular value in localizing the steps in the metabolic pathway of the insulin receptor that are modulated. This approach has also been employed successfully with the acetylcholine,[13–15] triiodothyronine,[16] and EGF[17] receptors.

The number of active insulin receptors a cell possesses in the steady state is determined by the relative rates of receptor synthesis and decay,

$$R = k_s/k_d$$

where R is receptor number, k_s is the zero-order rate constant for receptor synthesis, and k_d is the first-order rate constant for receptor decay. To determine which of these parameters, i.e., k_s or k_d, is affected by a physiological perturbation that alters the cellular level of active[17a] insulin receptors, the heavy-isotope density-shift method may be used. The density-shift approach entails exposing cells to medium containing amino acids enriched in ^{15}N, ^{13}C, and ^{2}H. After the density shift, newly synthesized "heavy" and previously synthesized "light" insulin receptors are extracted from total cellular membranes with a nonionic detergent and are then separated by isopycnic banding on a CsCl density gradient.[4,8,11] The heavy and light receptor peaks are located and quantitated by determining the ^{125}I-insulin binding activity of each fraction of the density gradient. Measurement of the amount of heavy and light receptor present at increasing times of exposure of the cells to heavy amino acids provides the results for computing the rate of synthesis of new heavy receptor and the rate of decay of old light receptor.

It should be emphasized that the heavy-isotope density-shift method as applied here follows the rate at which the insulin binding activity of

[11] B. C. Reed and M. D. Lane, *Proc. Natl. Acad. Sci. U.S.A.* **77,** 285 (1980).

[12] V. P. Knutson, G. V. Ronnett, and M. D. Lane, *Proc. Natl. Acad. Sci. U.S.A.* **79,** 2822 (1982).

[13] P. N. Devreotes and D. M. Fambrough, *Proc. Natl. Acad. Sci. U.S.A.* **73,** 161 (1976).

[14] P. N. Devreotes, J. M. Gardner, and D. M. Fambrough, *Cell* **10,** 365 (1976).

[15] J. M. Gardner and D. M. Fambrough, *Cell* **16,** 661 (1979).

[16] B. M. Raaka and H. H. Samuels, *J. Biol. Chem.* **256,** 6883 (1981).

[17] M. N. Krupp, D. T. Connolly, and M. D. Lane, *J. Biol. Chem.* **257,** 11489 (1982).

[17a] Active insulin receptors are defined as receptors capable of binding insulin; functional receptors are defined as active receptors that have been inserted into the plasma membrane, and thus, are accessible to extracellular insulin.

receptors if formed or lost, i.e., the formation or loss of active receptor. These are the physiologically relevant steps in the synthesis or inactivation–degradation of functional receptor.

In this chapter chick liver cells in monolayer culture are used to illustrate the heavy-isotope density-shift method[8]; however, in principle, any cell or tissue preparation that can be maintained in culture for 24 hr could be used. Other cell types, including 3T3-C2 fibroblasts,[4,12] human skin fibroblasts,[18] undifferentiated 3T3-L1 cells,[4,11] differentiated 3T3-L1 adipocytes,[4,5,11] and human A431 epidermoid carcinoma cells,[17] have been used successfully to study receptors in heavy-isotope density-shift experiments.

Reagents and Materials

Heavy (^{15}N-, ^{13}C-, ^{2}H-Labeled) Amino Acids. Heavy amino acids were either purchased from Merck Sharp and Dohme Canada Ltd., Montreal, or were prepared[18a] from lyophilized *Chlorella pyrenoidosa* cells provided by Dr. Thomas Whaley (Stable Isotope Resource, Los Alamos Scientific Laboratory, Los Alamos, New Mexico 87545). The algae were disrupted by adding 230 ml of 10% (w/v) trichloroacetic acid to 15 g of lyophilized cells. The mixture was warmed to 70°, stirred for 30 min, cooled to 4°, and centrifuged for 10 min at 10,000 *g*. The supernatant was discarded and trichloroacetic acid precipitation and centrifugation were repeated twice. Chlorophyll and other pigments were removed by extracting the residue three times with 250 ml of ethanol–ether (2 : 1, v/v) followed by five extractions with 75 ml of ether. After each extraction, the residue was recovered by vacuum filtration through Whatman No. 42 filter paper. After drying under vacuum at 40° for 1 hr, the residue weight was approximately 10 g.

The residue was hydrolyzed with 200 ml of 6 *N* HCl at 100° for 16 hr in a sealed, heavy-wall glass vessel. After hydrolysis, the mixture was cooled to 5° and centrifuged for 20 min at 7000 *g*. The insoluble residue was washed with 200 ml of water, and the washings were added to the hydrolyzate. The HCl was removed from the hydrolyzate by repeated (3 to 5 times) concentration with a rotary evaporator attached to a KOH pellet trap. The last traces of HCl were removed from the syrup by drying in a vacuum oven at 110° for approximately 3 hr. The weight of the syrup after drying was 10 g.

[18] G. V. Ronnett, G. Tennekoon, V. P. Knutson, and M. D. Lane, *J. Biol. Chem.* **258,** 283 (1983).

[18a] The procedure for the isolation of heavy amino acids from lyophilized algae is a modification of that communicated to us by Thomas Whaley.

The syrup was diluted with 125 ml of water, filtered, and applied to a freshly washed, 5 cm × 50 cm column of Dowex 50-X2, 100–200 mesh, in the H^+ form. The column was washed with 1.3 liters of water at a flow rate of 1 ml per min. Elution of amino acids was performed with 3 *N* ammonium hydroxide. The void volume (approximately 700 ml) was discarded, and the following 700–800 ml were collected. The eluate was repeatedly evaporated to dryness with the rotary evaporator to remove the last traces of ammonia. The syrup was diluted to 150 ml with water and lyophilized; the dry weight of amino acids was 6 g.

The lyophilized amino acids were dissolved in phosphate-buffered saline at a concentration of 20 mg/ml. This solution was filtered through an Amicon UM-2 filter under pressure. The filtrate was then filter-sterilized using a Millipore 0.22-μm filter and was stored at −70°.

The composition of the heavy algal amino acid mixture expressed as milligrams of each amino acid per gram of total mixture was 38.2 mg of lysine, 13.0 mg of histidine, 63.4 mg of arginine, 87.6 mg of aspartic acid, 51.5 mg of threonine, 42.0 mg of serine, 111.1 mg of glutamic acid, 29.9 mg of proline, 43.2 mg of glycine, 85.6 mg of alanine, 2.6 mg of cysteine, 59.8 mg of valine, 21.8 mg of methionine, 50.2 mg of isoleucine, 91.1 mg of leucine, 43.6 mg of tyrosine, and 43.2 mg of phenylalanine.

Cesium Chloride Stock Solution. Density gradient grade Beckman CsCl was dissolved in 50 m*M* Tris-Cl^-, pH 7.4, containing 400 units of Trasylol per milliliter and was brought to a final density of 0.568 g/ml. The refractive index of the CsCl solution was determined with a Bausch and Lomb Abbe refractometer and was brought to a final value of 1.3735 at 20°.

^{125}I-Labeled Insulin. Crystalline bovine insulin from Eli Lilly and Co. was iodinated by the chloramine-T method using sodium [^{125}I]iodide (Amersham) and then purified as described by Gavin *et al.*[19] The ^{125}I-labeled insulin had a specific activity of 1 μCi/pmol and was about 98% precipitable with 10% (w/v) trichloroacetic acid. Prior to use, an aliquot of the stock ^{125}I-labeled insulin was further purified by gel filtration on a column (0.5 × 15 cm) of Sephadex G-50 in 50 m*M* phosphate buffer, pH 7.4, containing 0.1% (w/v) bovine serum albumin.

Procedures

Liver Cells in Monolayer Culture. Livers were removed from chicks (10–20 days of age), and hepatocytes were prepared under aseptic condi-

[19] J. R. Gavin III, J. Roth, D. M. Neville, Jr., P. DeMeyts, and D. N. Buell, *Proc. Natl. Acad. Sci. U.S.A.* **71**, 84 (1974).

tions by external digestion with collagenase as previously described.[20] Cells were plated in 60-mm culture dishes at a density of 10^7 cells per dish in Eagle's basal medium supplemented with amino acids, glucose, bicarbonate, antibiotics, and 2% rooster serum.[8,20] Cells were maintained under a 90% air, 10% CO_2 atmosphere at 37° during incubation. After 4 hr and every 24 hr thereafter, the cells were fed with serum-free medium. Under these conditions a confluent monolayer containing 6 to 7 × 10^6 cells per dish was obtained after 48 hr in culture.

Extraction–Solubilization of Total Cellular Insulin Receptors for Density-Gradient Centrifugation. To remove any bound insulin from the cell monolayers prior to extraction of receptors for isopycnic banding, monolayers were washed five times with Krebs–Ringer phosphate (KRP) buffer, pH 7.4, containing 1% (w/v) albumin and 10 m*M* glucose.[8] The first two washes were done in rapid succession, and each of the last three washes were separated by 20-min incubations at room temperature.

Ten cell monolayers (60-mm dishes) were scraped into 5 ml of 50 m*M* Tris-HCl, pH 7.5, buffer containing 400 units of Trasylol per milliliter and then were homogenized with 20 strokes of a motorized Teflon pestle. The resultant homogenate was centrifuged at 100,000 *g* at 4° for 1 hr. The pellet containing all cellular membranes was resuspended by homogenization in 1.0 ml of 50 m*M* Tris-HCl, pH 7.5, with Trasylol and 1% (v/v) Triton X-100, and then incubated at room temperature for 30 min. The suspension was centrifuged at 100,000 *g* for 1 hr at 4°, and the supernatant was stored at −70°. This procedure consistently extracts >90% of the ^{125}I-labeled insulin-binding activity present in the total cellular membrane pellet, leaving <10% in insoluble form.

Determination of ^{125}I-Labeled Insulin Binding to Cell-Surface and Triton X-100-Solubilized Insulin Receptors. Prior to conducting binding experiments with ^{125}I-labeled insulin, cell monolayers were washed 5 times with Krebs–Ringer phosphate (KRP) buffer, pH 7.4, containing 1% (w/v) albumin and 10 m*M* glucose to remove any medium and adherent hormone. The first two washes were done in rapid succession, whereas 20-min incubations at room temperature were interposed between the following three washes.

To determine cell-surface insulin-binding capacity,[8] cell monolayers were incubated overnight at 4° with 4 ml of KRP buffer containing the appropriate concentration, usually 0.5 n*M*, of ^{125}I-labeled insulin or ^{125}I-labeled insulin plus 5 μg of unlabeled insulin per milliliter (for nonspecific binding). After aspirating the medium, the cells were quickly washed 5

[20] D. M. Tarlow, P. A. Watkins, R. E. Reed, R. S. Miller, E. E. Zwergel, and M. D. Lane, *J. Cell Biol.* **73,** 332 (1977).

times with 5 ml of ice-cold phosphate-buffered saline, pH 7.4, to remove unbound insulin. The resultant monolayers were solubilized with 1.5% SDS and transferred to a 12 × 75 mm polystyrene tube for the determination of cell-associated ^{125}I activity.

The binding of ^{125}I-labeled insulin (0.5 n*M*) to soluble insulin receptor was determined by the polyethylene glycol precipitation method of Cuatrecasas[21] as modified by Krupp and Livingston.[22]

Isopycnic Banding of Solubilized Insulin Receptor in CsCl Density Gradients. One-half milliliter of Triton X-100-solubilized insulin receptor was mixed with 1.7 ml of the stock CsCl solution (0.568 g/ml, $\eta_D^{25°}$ = 1.3735) in 50 m*M* Tris-HCl, pH 7.4, containing 400 units of Trasylol per milliliter and overlaid with paraffin oil. The samples were then centrifuged for 18 hr at 4° in a Beckman SW56 Ti rotor (or SW60 Ti rotor) at 50,000 rpm (246,000 *g*). Fifty-microliter fractions were collected from the bottom of the tube. Each fraction was assayed for ^{125}I-labeled insulin-binding activity at 0.5 m*M* labeled insulin using the polyethylene glycol precipitation method. The refractive index of each fraction was determined with a Bausch and Lomb Abbe 3l refractometer.

Figure 1 shows a typical banding profile of unlabeled receptor in an isopycnic CsCl gradient.[8] Specific binding of ^{125}I-labeled insulin by receptor in the gradient (Fig. 1B) was determined by subtracting nonspecific insulin binding from the total ^{125}I-insulin binding (Fig. 1A). It should be noted that the positions of the light and heavy receptors in the gradient ($\eta_D^{25°}$ = 1.3633 and 1.3688, respectively) are nearly identical with those of the light and heavy receptors of 3T3-L1 adipocytes.[11]

To quantitate the distribution of light and heavy insulin receptor in CsCl gradients, it is necessary to estimate the total amount of ^{125}I-labeled insulin-binding activity (receptor) in each peak.[8] Insulin binding to receptor was determined at a subsaturating concentration of insulin (0.5 n*M*); therefore, comparisons of receptor levels from CsCl gradients results reflect relative rather than absolute receptor number. The amount of receptor at subsaturating concentrations of insulin has been shown to be proportional to receptor concentration.[11,21] It has been demonstrated[11] that CsCl carried over into the assay from the CsCl gradient does not affect ^{125}I-labeled insulin binding.

An empirical relationship exists between peak area and peak height. Thus, when varying amounts of soluble insulin receptor from chick liver cells were applied to and isopycnically banded on CsCl gradients, a linear relationship between peak area and peak height was obtained. The recip-

[21] P. Cuatrecasas, *Proc. Natl. Acad. Sci. U.S.A.* **69,** 318 (1972).
[22] M. N. Krupp and J. N. Livingston, *Proc. Natl. Acad. Sci. U.S.A.* **75,** 2593 (1978).

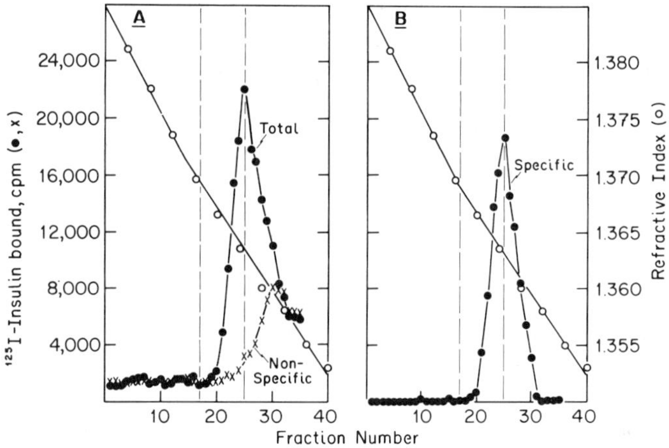

FIG. 1. Isopycnic banding of unlabeled insulin receptor. Panel A shows the total (●——●) and nonspecific binding (×——×) to the soluble receptor from 8 plates (~56 × 10^6 cells) banded on a CsCl gradient when 50-μl fractions are assayed. 5 × 10^{-10} *M* ^{125}I-labeled insulin was used in the soluble receptor assay. Nonspecific binding was determined as given in Procedures. Panel B is the result of the subtraction of total and nonspecific binding (panel A). The specific binding profile of light receptor peaks at a refractive index for CsCl of $\eta_D^{25°}$ = 1.3633. The position of the heavy receptor peak is shown by the left vertical dashed line and corresponds to a refractive index for CsCl of 1.3688. From Krupp and Lane.[8]

rocal of the slope of the line yields a factor of 5.4, which when multiplied by the peak height gives peak area.[8] In density shift experiments, light receptor contributed little to insulin binding at the heavy peak position ($\eta_D^{25°}$ = 1.3688, Fig. 1). Therefore, the relative amount of heavy receptor in the mixture was determined by multiplying the heavy peak height by 5.4. The relative amount of light receptor was then calculated from the difference between total insulin binding and the calculated binding to heavy receptor.

Heavy-Isotope Density-Shift Protocol. It is essential that the cells achieve a steady state with respect to total cellular level of insulin receptor, i.e., $k_s = k_d \cdot [R = \text{constant}]$, before the density-shift protocol is initiated. That a steady-state level of receptor has been attained is indicated by the constancy of total ^{125}I-labeled insulin binding activity per cell over a 12- to 24-hr period.

Cell monolayers are then shifted from medium containing light (^{14}N, ^{12}C, ^{1}H) amino acids to medium containing heavy (>95% ^{15}N, ^{13}C, ^{2}H) amino acids.[8] Just prior to the density shift, the cell monolayers were washed quickly three times with amino acid-free medium, and then 3 ml of the "heavy" medium was added. Each 100 ml of heavy medium contained the regular culture medium components minus the usual amino acid mixture, but also contained 0.8 mg of tryptophan, 3 mg of cystine, 6 mg of glutamine, 0.8 mg of histidine, 2 mg of lysine, and 100 mg of heavy amino acids (>95% enriched in ^{15}N, ^{13}C, ^{2}H).

The amino acid composition of the heavy amino acid-containing medium differed somewhat from the standard culture medium.[8] Experiments were conducted to assess the effect of this new amino acid mixture on the cell's insulin-binding properties. A synthetic mixture of light amino acids with an identical amino acid composition to that of the heavy medium was substituted for the normal culture medium, and the cells were maintained in this medium for several days. Cells maintained in this medium had insulin-binding characteristics and receptor down-regulation properties identical to those of cells maintained in normal culture medium.[8]

At various times (0, 3, 5, 7, 9, 12, 16, and 24 hr) after the shift to medium containing heavy amino acids, cells were scraped from the culture dish and homogenized; a total cellular membrane pellet was prepared, and the pellet was extracted with Triton X-100 as described above. The solubilized receptor can be stored at −70° for several months without significant loss of insulin binding activity or a change in isopycnic banding characteristics. Isopycnic density gradient centrifugation is carried out as described above.

Figure 2 shows the gradient profiles of heavy and light receptors from chick liver cells that had been maintained in insulin-containing (8×10^{-8} *M* insulin) or insulin-free medium for 24 hr.[8] During the 16-hr duration of the experiment, there was a continuous rise in peak area attributable to newly synthesized heavy receptor and a concomitant fall in peak area of old light receptor. It is evident (Fig. 2B) that active newly synthesized heavy receptor (as judged by insulin-binding activity) is detected as early as 3 hr after the shift to heavy medium. By 16 hr (Fig. 2G), most of the previously synthesized light receptor had turned over. These findings correlate well with the kinetic pattern of receptor synthesis and turnover obtained with 3T3-L1 adipocytes.[11] During the 16-hr duration of the experiment shown, there was a continuous rise in peak area attributable to newly synthesized receptor and a simultaneous decrease in the amount of old light receptor.

The relative amounts of heavy and light receptors at each time point are quantitated from heavy peak height and total peak area by using the peak area to peak height ratio constant of 5.4 as described above, and the results are plotted as progress curves for receptor synthesis and decay (Fig. 3A); the data for light receptor decay are plotted semilogarithmically (Fig. 3B) for an accurate assessment of the half-life of the receptor. From these results the synthetic rate constant (k_s), the inactivation rate constant

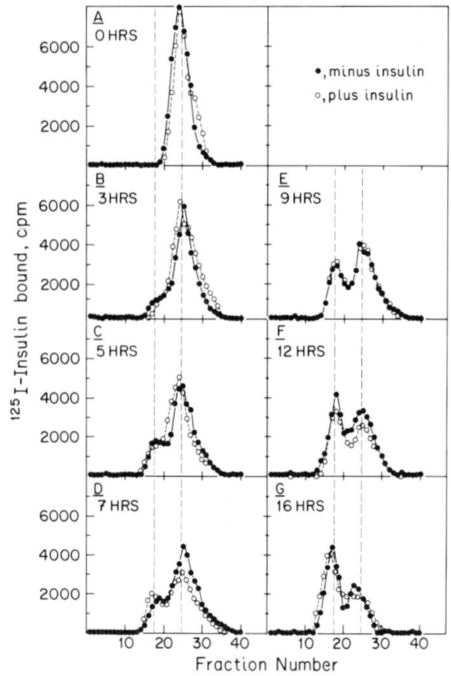

FIG. 2. Specific ^{125}I-insulin binding profiles of soluble insulin receptor on CsCl gradients after shifting chick liver cells to "heavy" amino acid-containing medium. Chick liver cells, which were either treated with insulin-containing or insulin-free medium, were labeled for various times in medium containing, per liter, 1 g of heavy (^{15}N, ^{13}C, ^{2}H) amino acids supplemented with lysine, histidine, cysteine, glutamine, and tryptophan. Cells were then homogenized, the total cellular membrane pellet was obtained, and the pellet was extracted with Triton X-100. The soluble receptor was banded on CsCl gradients as given in Procedures. The data shown represent specific ^{125}I-labeled insulin binding, i.e., the difference between total binding and nonspecific binding. Heavy receptor peaks at fractions 17–18; whereas, light receptor peaks at fractions 23–24. From Krupp and Lane.[8]

(k_d), and the half-life ($t_{1/2}$) for the receptor can be calculated. The k_d is calculated from the half-life using the relationship

$$k_d = 0.69/t_{1/2}$$

where $t_{1/2}$ is the half-life of receptor in hours and k_d is the first-order rate constant for receptor inactivation in hr^{-1}. The synthetic rate constant, k_s, can be calculated from the initial slope of the receptor synthetic progress curve (Fig. 3A) and from $k_s = k_d$ R.

Use of the Heavy Isotope Density-Shift Method to Determine the Transit Time of Insulin Receptors to and from the Cell Surface. The

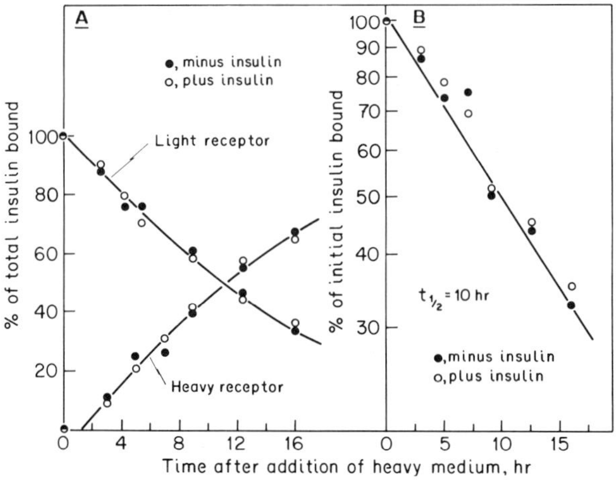

FIG. 3. Kinetics of heavy receptor synthesis and light receptor inactivation–decay in chick liver cells. Panel A shows the progressive rise in the fraction of total receptor that has become labeled with heavy amino acids and the concomitant fall in the fraction of light receptor. These data are calculated from the relative peak areas of the CsCl gradient binding profiles given in Fig. 2. Panel B is a semilogarithmic plot of the data presented in panel A for the degradation of light receptor. The slope of the line through the points of receptor decay yields a half-life ($t_{1/2}$) of 10 hr and a rate constant for the decay (k_D) of 0.069 hr^{-1}. From Krupp and Lane.[8]

heavy isotope density-shift method can be used in conjunction with a procedure for specifically labeling cell-surface insulin receptors[18] to determine (*a*) the rate of transit of newly synthesized receptors from their site of translation in the rough endoplasmic reticulum to the plasma membrane; and (*b*) the rate of net removal of active receptors from the plasma membrane.

To label cell-surface receptors, ^{125}I-labeled insulin was bound to cells at 4° and then covalently cross-linked to the receptors with disuccinimidyl suberate (DSS). The identity of the surface-labeled product as insulin receptor was established by immunoprecipitation with anti-receptor antibody and SDS–polyacrylamide gel electrophoresis.[18] Triton X-100-solubilized surface receptors to which ^{125}I-labeled insulin had been covalently cross-linked band at a position in CsCl density gradients identical to that of native solubilized insulin receptors that have not been cross-linked.

Cells were first shifted to medium containing heavy (^{15}N-, ^{13}C-, ^{2}H-labeled) amino acids; at time intervals following the density shift, cell monolayers were subjected to the ligand debinding protocol to remove cell-associated insulin and then were cooled to 4°. At this temperature the processes of exocytosis and endocytosis are blocked.[23]

To determine the amounts of heavy and light receptors on the cell surface, ^{125}I-labeled insulin was bound to the cell monolayers at 4°, after which bound insulin was covalently cross-linked to surface receptors with DSS. The ^{125}I-insulin labeled receptors were then extracted from total cellular membranes with Triton X-100, and the extract was applied to CsCl gradients to resolve and quantitiate heavy and light receptors. The gradients were fractionated, and the heavy and light peak fractions were subjected to SDS–polyacrylamide gel electrophoresis to quantitate the ^{125}I-labeled surface insulin receptor subunits. In the same experiment, but with different cell monolayers, rates of synthesis and decay of total cellular receptor were also determined by the heavy isotope density-shift method for comparison to rates accrual and decay of surface receptors.

It was found that after the shift of cells to medium containing heavy amino acids about 3 hr were required for newly translated heavy receptors to reach and become inserted into the plasma membrane when they become functional;[17a] heavy receptors then continued to accrue on the cell surface for at least 15 hr. After a 3-hr lag, which probably represents the time required for light receptor to clear the transit–processing pathway to the cell surface, a net loss of light surface receptor begins to occur. Half-replacement of light receptors by heavy receptors on the cell surface following the 3-hr lag required about 7 hr in cells exposed to insulin-containing medium.

Acknowledgments

This investigation was supported by grants from the National Institutes of Health (AM-14574) and the Kroc Foundation. G.V.R. was supported in part by the National Institutes of Health Medical Scientist Training Program (GM-07309) and the Juvenile Diabetes Foundation. V.P.K. was supported by a postdoctoral fellowship (AM-06609) from the National Institutes of Health.

[23] S. C. Silverstein, R. M. Steinman, and Z. A. Cohn, *Annu. Rev. Biochem.* **46,** 669 (1977).

[34] Biosynthesis of Myxovirus Glycoproteins with Special Emphasis on Mutants Defective in Glycoprotein Processing

By HANS-DIETER KLENK

The surface glycoproteins of myxoviruses, i.e., the hemagglutinin and the neuraminidase of influenza viruses and the hemagglutinin–neuraminidase and the fusion protein of paramyxoviruses are integral membrane glycoproteins with their carbohydrate side chains attached to the polypeptide by N-glycosidic linkages. These glycoproteins play essential roles in the infection process by promoting adsorption of the virus to the host cell surface and penetration of the viral genome into the cytoplasm. Furthermore, the surface glycoproteins are the viral antigens that induce and react with neutralizing antibodies and are therefore the prime target of the immune response of the infected organism. Finally, the glycoproteins of myxoviruses are suitable models to study structure, function, and biosynthesis of membrane proteins in general.

Synthesis of the myxovirus glycoproteins involves translation at membrane-bound ribosomes, insertion into the membrane of the rough endoplasmic reticulum, and transport to the plasma membrane from which the virus is released in a budding process. Insertion into the rough endoplasmic reticulum is fairly well understood with the influenza hemagglutinin, where it is mediated by a signal sequence at the amino terminus that is removed by cotranslational proteolytic cleavage.[1] Proteolytic cleavage at the posttranslational level has been observed with the influenza hemagglutinin,[2,3] the fusion protein,[4,5] and the hemagglutinin–neuraminidase of paramyxoviruses.[6] Proteases of trypsin specificity play an essential role in the cleavage reaction, and it depends on the presence of an appropriate enzyme, whether or not the glycoproteins are cleaved in a given cell. Posttranslational cleavage is necessary for the biological activities of the glycoproteins and thus for virus infectivity. It has also been found to be an important determinant for the spread of infection and for the pathogenicity of these viruses.[6,7]

[1] J. McCauley, J. J. Skehel, K. Elder, M.-J. Gething, A. Smith, and M. Waterfield, *in* "Structure and Variation in Influenza Virus" (W. G. Laver and G. Air, eds.), p. 97. Elsevier/North-Holland, Amsterdam, 1980.

[2] H.-D. Klenk, R. Rott, M. Orlich, and J. Blödorn, *Virology* **68,** 426 (1975).

[3] S. G. Lazarowitz and P. W. Choppin, *Virology* **68,** 440 (1975).

[4] M. Homma and M. Ohuchi, *J. Virol.* **12,** 1457 (1973).

[5] A. Scheid and P. W. Choppin, *Virology* **57,** 475 (1974).

[6] Y. Nagai, H.-D. Klenk, and R. Rott, *Virology* **72,** 494 (1976).

[7] F. X. Bosch, M. Orlich, H.-D. Klenk, and R. Rott, *Virology* **95,** 197 (1979).

ISBN 0-12-181996-5

In addition to proteolytic cleavage, processing of the myxovirus glycoproteins involves also covalent attachment of fatty acids[8] and glycosylation. Glycosylation is initiated by the en bloc transfer of mannose-rich oligosaccharides from dolichol pyrophosphate to the asparagine residues of the polypeptides. This occurs in the rough endoplasmic reticulum. Upon arrival of the glycoproteins in the Golgi apparatus, some of the side chains undergo extensive modifications involving removal of mannose and attachment of galactose and fucose residues. Thus, two major types of carbohydrate side chains are found on the glycoproteins: the mannose-rich type containing only mannose and glucosamine, and the complex type that contains in addition galactose and fucose.[9]

Viruses

Most studies devoted to the biosynthesis of influenza virus glycoproteins,[2,3,10–15] have been carried out on two strains of influenza A viruses: the Rostock strain of fowl plague virus [A/FPV/Rostock/34 (H7N1)][16] and strain A/WSN/33 (H1N1).[17] These strains have been selected, because in tissue cultures they grow to relatively high titers of infectious virus [$\sim 10^9$ plaque-forming units (PFU) per milliliter of tissue culture medium], form plaques, and suppress the synthesis of host cell proteins to allow the analysis of viral protein synthesis by metabolic labeling. Valuable information on the mechanism of proteolytic activation has been derived from the analysis of the primary structure of several hemagglutinins. These include again the hemagglutinins of fowl plague virus,[18] of the WSN strain,[19] of strain A/Chick/Germany/49 (H10N7),[20] and of a variety of hemagglutinins of the serotypes H2 and H3.[21] For the X-ray crystallo-

[8] M. F. G. Schmidt, *Virology* **116,** 327 (1982).
[9] H.-D. Klenk and R. Rott, *Curr. Top. Microbiol. Immunol.* **90,** 19 (1980).
[10] S. G. Lazarowitz, R. W. Compans, and P. W. Choppin, *Virology* **46,** 830 (1971).
[11] H.-D. Klenk, R. Rott, and H. Becht, *Virology* **47,** 579 (1972).
[12] H.-D. Klenk, W. Wöllert, R. Rott, and C. Scholtissek, *Virology* **57,** 28 (1974).
[13] H. J. Hay, *Virology* **60,** 398 (1974).
[14] R. W. Compans, *Virology* **51,** 56 (1973).
[15] K. Nakamura and R. W. Compans, *Virology* **93,** 31 (1979).
[16] W. Schäfer, *Z. Naturforsch., B: Anorg. Chem., Org. Chem., Biochem., Biophys., Biol.* **10B,** 81 (1955).
[17] T. Francis, Jr. and A. E. Moore, *J. Exp. Med.* **72,** 717 (1940).
[18] A. G. Porter, C. Barber, N. A. Carey, R. A. Hallewell, G. Threlfall, and J. S. Emtage, *Nature (London)* **282,** 471 (1979).
[19] A. C. Hiti, A. R. Davis, and D. P. Nayak, *Virology* **111,** 113 (1981).
[20] W. Garten, F. X. Bosch, D. Linder, R. Rott, and H.-D. Klenk, *Virology* **115,** 361 (1981).
[21] C. W. Ward, *Curr. Top. Microbiol. Immunol.* **94,** 1 (1981).

graphic analysis of the tertiary structure, also a H3 hemagglutinin has been used.[22]

The biosynthesis of paramyxovirus glycoproteins has been analyzed primarily with Sendai virus[23] and Newcastle disease virus.[24-26]

Seed stocks of myxoviruses are usually grown in the allantoic sac of 11-day-old embryonated chicken eggs that have been inoculated with about 10^4 PFU. Allantoic fluid is harvested 24–48 hr after infection, centrifuged at 2000 *g* for 15 min to remove debris, and stored at −80°.

Assays of Biological Activities of Viral Glycoproteins

Plaque Assay.[2] Virus infectivity is usually titrated by plaque assays in cultures of chick embryo cells, MDBK cells, or BHK cells (see below). Confluent monolayers on plastic petri dishes (5 cm in diameter) are rinsed with phosphate-buffered saline, pH 7.4, and infected with 0.2 ml of virus suspension diluted in the same buffer. After a 60-min adsorption at 37° the cultures are overlaid with 4 ml of medium 199 (Flow Laboratories) containing 0.08% bicarbonate and 0.7% Difco Bactoagar. Alternatively, reinforced Eagle's medium (REM)[27] with 0.7% Difco Bactoagar is used. When the viral glycoproteins are not activated by endogenous cellular proteases, trypsin is added to the overlay at final concentrations between 0.1 and 10 μg/ml.[2,3,5,6] Cultures are kept for 3 days at 37° and are then stained with 4 ml of phosphate-buffered saline (PBS), pH 7.4, containing 0.005% neutral red and 0.7% Difco Bactoagar. After 4 hr of incubation at 37°, plaques are counted.

Hemagglutination Titration. Serial twofold dilutions of virus in a volume of 0.05 ml of PBS are prepared in plastic hemagglutination trays. A 0.05-ml amount of a 1% suspension of chicken red blood cells in PBS is added to each egg, and the HA titer is read after 45 min. With influenza viruses the assay is carried out at room temperature; with Sendai and Newcastle disease virus, at 4°.

Hemadsorption. Cell cultures in plastic petri dishes are infected with virus. After the adsorption time, the inoculum is removed. The cultures are washed with warm PBS and incubated in culture medium. After an appropriate incubation period, the medium is removed and the cultures are washed with PBS. To each petri dish an equal amount of a suspension

[22] I. A. Wilson, J. J. Skehel, and D. C. Wiley, *Nature (London)* **289,** 366 (1981).

[23] R. A. Lamb and P. W. Choppin, *Virology* **81,** 371 (1977).

[24] Y. Nagai, H. Ogura, and H.-D. Klenk, *Virology* **69,** 523 (1976).

[25] Y. Nagai and H.-D. Klenk, *Virology* **77,** 125 (1977).

[26] W. Garten, T. Kohama, and H.-D. Klenk, *J. Gen. Virol.* **51,** 207 (1980).

[27] R. H. Bablanian, H. J. Eggers, and I. Tamm, *Virology* **26,** 100 (1965).

of erythrocytes (0.5% in PBS) is added. After 10 min at room temperature, the suspension is removed and the cultures are thoroughly rinsed with PBS. The cell monolayers are then inspected under the microscope for adsorbed erythrocytes.

Neuraminidase Assay. The enzyme is incubated for 30 min at 37° in 0.5 ml of buffer with fetuin at a final concentration of 100 μg of *N*-acetylneuraminic acid.[28] Free neuraminic acid is measured by the thiobarbituric acid method.[29] One neuraminidase unit is defined as the amount of enzyme that releases 1 nmol of *N*-acetylneuraminic acid in 1 min at 37°.

Hemolysis Assay. The capacity of a myxovirus to induce membrane fusion can be measured by its hemolytic activity. With paramyxoviruses, the virus sample in 1 ml of PBS is mixed with 2 ml of a 2% suspension of chicken erythrocytes in PBS. After incubation for 60 min at 37°, the erythrocytes are removed by centrifugation, and the optical density of the supernatant is measured at 540 nm.[6] The fusion capacity of influenza viruses has an acidic pH optimum. With these viruses, the hemolysis assay is therefore carried out at pH 5.0–6.0 depending on the virus strain used.[30]

Virus Replication in Cell Cultures

Cell Cultures. To analyze the individual steps in virus replication and assembly, myxoviruses are grown in cell cultures that are propagated as monolayers in plastic petri dishes or tissue culture flasks. Permanent cell lines frequently used for the propagation of influenza and paramyxoviruses are Madin–Darby bovine kidney (MDBK) and Madin–Darby canine kidney (MDCK) cells.[31] The BHK21 line of baby hamster kidney cells[32] is another permanent cell line that has been successfully used to study the biosynthesis of Sendai and Newcastle disease virus proteins. Such cultures are grown in REM containing 10% calf serum or, preferentially, fetal calf serum. Primary cultures of chicken embryo cells are also used for the growth of myxoviruses. Such cultures are prepared from 11-day-old embryos after trypsinization and growth in tissue culture medium supplemented with calf serum. Confluent cell sheets form after 24–48 hr of incubation at 37° in an atmosphere of 5% CO_2 in air.

Infection of Cell Cultures. Monolayers in petri dishes are washed once with warm sterile PBS and are then exposed to the virus at a multiplicity

[28] R. Drzeniek, J. T. Seto, and R. Rott, *Biochim. Biophys. Acta* **128,** 547 (1966).
[29] D. Aminoff, *Biochem. J.* **81,** 384 (1961).
[30] R. T. C. Huang, R. Rott, and H.-D. Klenk, *Virology* **110,** 243 (1981).
[31] S. H. Madin and N. B. Darby, Jr., *Proc. Soc. Exp. Biol. Med.* **98,** 574 (1958).
[32] M. G. P. Stoker and I. A. McPherson, *Virology* **14,** 359 (1961).

of 10–50 PFU/cell. Adsorption is allowed to proceed for 40–60 min at 37°. After adsorption the inoculum fluid is removed, fresh growth medium is added, and the dishes are incubated at the appropriate growth temperature.

Radioactive Labeling of Infected Cells. Conditions for metabolic labeling of viral proteins are optimal when the rate of viral protein synthesis has reached a maximum and shutoff of host protein synthesis approximates a minimum. Depending on the virus and the host cell used, it takes 4–10 hr after inoculation to reach this stage. To label viral proteins with radioactive amino acids, the culture medium is removed, and the cells are washed three times with medium lacking the unlabeled amino acids to be substituted in radioactive form. The radioactive amino acids are then added in the same medium, and the cultures are incubated for an appropriate time. Radioactive amino acids are used at the following concentration: ^{3}H-labeled amino acids (e.g., leucine, valine, tyrosin), 10 μCi/ml; ^{14}C-labeled protein hydrolyzate, 2 μCi/ml; [^{35}S]methionine, 10 μCi/ml. If the radioactive pulse (5–20 min) is followed by a chase period (1–2 hr), the radioactive medium is removed and the cells are washed twice with complete medium and incubated in the same medium.

The carbohydrate side chains of the viral glycoproteins can be labeled with radioactive sugars. Specific carbohydrate markers that are not converted, or only little, into other radioactively labeled compounds, are the following sugars: [6-^{3}H]glucosamine, [2-^{3}H]mannose, [1-^{3}H]galactose, [1-^{3}H]fucose, and [1-^{14}C]glucosamine. The sugars must be added in quite high concentrations (100 μCi/ml) to the culture medium, because the intracellular pools of their metabolic intermediates in the glycosylation reactions are high and cannot be lowered by simple washing procedures. For the same reason, labeling periods must be quite long (at least 2 hr) in certain host cells, particularly in primary chick embryo cells, and it is difficult to perform pulse–chase experiments. Because glucose and glucosamine have the same uptake mechanism, the incorporation of radioactive glucosamine is significantly increased when glucose in the medium is replaced by fructose (10 mM) as an energy source.[33]

Radioactive Labeling of Released Virus. For metabolic labeling of the proteins incorporated into mature virions, the radioactive isotopes are usually added immediately after inoculation and left in the cell culture medium until virus harvest (about 24 hr postinfection). When radioactive amino acids are used for labeling, the respective unlabeled amino acids should be reduced in the serum-free medium to 10–30% of their normal concentrations. When radioactive glucosamine is used, glucose is again

[33] C. Scholtissek, R. Rott, and H.-D. Klenk, *Virology* **63,** 191 (1975).

replaced in the medium by fructose (10 mM). ^{3}H-Labeled sugars and amino acids are added to the medium at a concentration of 2 μCi/ml, ^{14}C-labeled sugars and amino acids at 0.5 μCi/ml. To prepare fatty acid-labeled virus particles, REM with 2% fetal calf serum and 10–50 μCi/ml [9,10-^{3}H]palmitic acid is used.[8]

Localization of Viral Proteins in the Infected Cell. The compartmentalization of viral glycoproteins has been analyzed by a variety of different approaches, notably cell fractionation and immune electron microscopy.

Preparation of Cytoplasmic Membranes. The procedure of Caliguiri and Tamm[34] has been used to isolate the cytoplasmic membrane of cells infected with influenza viruses[12,14] and paramyxoviruses.[23,24] The cells from 20 petri dishes (10 cm in diameter) are washed with PBS and scraped in cold reticulocyte standard buffer (RSB) (0.01 M KCl, 0.0015 M $MgCl_2$, 0.01 M Tris-HCl, pH 7.4), held for 20 min in an ice bath, and disrupted with approximately 20 strokes of a tight-fitting Dounce homogenizer. Nuclei and cell debris are removed by centrifugation at 950 g for 10 min, and the supernatant is mixed with an equal volume of 60% sucrose in RSB. The resulting extract in 30% sucrose is included in a discontinuous gradient of 0, 25, 30, 40, 45, and 60% sucrose in RSB solutions and centrifuged at 24,000 rpm for 19 hr in a SW27 rotor. Generally six bands can be discriminated that are located (1) on the top of the gradient; (2) in the 25% sucrose layer; (3) in the interphases between the 25% and the 30% layers; (4) between the 30% and 40% layer; (5) between the 40% and 45% layers; and (6) in the 60% bottom layer. As revealed by electron microscopy and marker enzymes, fractions 2 and 3 contain smooth membranes derived from the Golgi apparatus and the smooth endoplasmic reticulum, fraction 5 contains rough membranes, and fraction 6 contains free ribosomes. With the exception of the first one, all bands are collected with a 10-ml syringe by puncturing from the side of the tube. After collection, each sample is diluted with RSB and pelleted at 100,000 g for 2 hr. The cell fractions can then be analyzed by polyacrylamide gel electrophoresis for radioactively labeled viral proteins and by the assays described above for the various biological activities of the virus.

Isolation of Plasma Membranes. Plasma membranes can be isolated from a variety of cells by the fluorescein mercuric acetate (FMA) method.[35] The procedure developed for the infected BHK cells is the following.[36] Monolayers of 15–20 petri dishes (10 cm in diameter) are washed three times with cold PBS and then placed in 0.003 M EDTA. After 5 min at room temperature, the cells are pipetted off the plastic,

[34] L. A. Caliguiri and I. Tamm, *Virology* **42,** 100 (1970).

[35] L. Warren, M. C. Glick, and M. K. Nass, *J. Cell. Comp. Physiol.* **68,** 269 (1967).

[36] H.-D. Klenk and P. W. Choppin, *Virology* **38,** 255 (1969).

pelleted, and washed once with PBS and twice with 0.16 *M* NaCl. The cells are suspended in 2 ml of 0.16 *M* NaCl and 2 ml of water, and allowed to swell with agitation for 10 min at room temperature. Then 6 ml of a saturated solution of FMA in 0.02 *M* Tris buffer (pH 8.0) is added, the suspension is kept at room temperature for 15 min and in an ice bath for another 15 min. After 15–25 strokes of gentle homogenization in a Dounce homogenizer, an equal volume of 60% (w/w) sucrose is added, and the homogenate is layered over 10 ml of a 45% sucrose solution. After centrifugation at 350 *g* for 40 min, the 30% layer, which contains mainly whole-cell ghosts or large pieces of membranes, is taken and pelleted by centrifugation at 2000 *g* for 2 hr. This membrane preparation is finally spun in a discontinuous sucrose gradient for 1 hr at 23,000 rpm in an SW27 rotor. Plasma membranes are found on the top and bottom of the 60% layer.

Immune Electron Microscopy by the Indirect Ferritin-Antibody Labeling Technique.[37] Monolayer cultures in plastic petri dishes (3.5 cm in diameter) are infected with virus and incubated at 37°. Cells are sampled at appropriate times after infection and prefixed *in situ* with cold 0.25% glutaraldehyde in PBS for 15 min. After several PBS washings, the cell sheet is covered with 2% bovine serum albumin solution for 15 min. The excess albumin solution is aspirated, and after two PBS washings the cells are covered with 0.2 ml of the appropriate antibody preparation for 15 min at 25°. The excess antibody solution is washed off with PBS, and the cells are covered with 0.2 ml of the ferritin-antibody label for 15 min at 25°. Untreated ferritin label is thoroughly washed with PBS. The cells are fixed with 2.5% glutaraldehyde in PBS for 1 hr, scraped, and pelleted by centrifugation at 1000 *g* for 10 min. The cell pellet is postfixed in 1% osmic acid, dehydrated in ethanol, and embedded in Epon 812.

Mutants of Influenza Virus with Defects in Glycoprotein Transport

Transport, glycosylation, and proteolytic cleavage are tightly coupled events in the biosynthesis of myxovirus glycoproteins. This has been demonstrated by the analysis of mutants of influenza virus that have temperature-sensitive defects in the transport of the hemagglutinin.[38,39]

Chemical Mutagenesis of Influenza A Virus. The mutants with a defective hemagglutinin transport have been derived from the Rostock strain of fowl plague virus.[40–42] The chemical mutagenesis by 5-fluoroura-

[37] J. Lohmeyer, L. T. Talens, and H.-D. Klenk, *J. Gen. Virol.* **42,** 73 (1979).

[38] J. Lohmeyer and H.-D. Klenk, *Virology* **93,** 134 (1979).

[39] H.-D. Klenk, W. Garten, W. Keil, H. Niemann, F. X. Bosch, R. T. Schwarz, C. Scholtissek, and R. Rott, *ICN-UCLA Symp. Mol. Cell. Biol.* **21,** 193 (1981).

[40] C. Scholtissek, R. Kruczinna, R. Rott, and H.-D. Klenk, *Virology* **58,** 317 (1974).

cil (5-FU) was performed by the following protocol originally described by Simpson and Hirst.[43] Chick embryo cell cultures infected with fowl plague virus at low multiplicity (~0.04) were incubated at 33° for 24 hr in virus growth medium containing 3 m*M* 5-FU, and virus released into the medium was harvested.

Isolation of ts Mutants. Treated or untreated fowl plague virus was appropriately diluted to give about 25 PFU per plate, and a sufficient number of monolayer cultures were inoculated to assure a minimum screening of at least 1000 plaques. Virus was adsorbed at room temperature for 30 min, after which agar overlay medium lacking neutral red was added. Infected cultures were incubated at 33° for 3–4 days before a second agar overlay containing neutral red was added. Distinct plaques were counted, and their borders were carefully traced on the plastic plate; the plates were then shifted to a 40° incubator for 18–24 hr. These plaques showing a minimal size increase after temperature shift were provisionally classified as foci containing ts virus and were selected for further testing. Plaque-derived virus was stored at 4° in 2-ml volumes of growth medium containing 20% normal chicken serum.

To verify the ts character of provisional ts clones, 0.2 ml of undiluted plaque suspension was inoculated into each of two culture dishes (6 cm in diameter), which were subsequently overlaid with agar medium lacking neutral red. These cultures were incubated at 33° and 40°, respectively. Under these conditions most *ts* mutants produced confluent plaques in the low-temperature plating, whereas in the plate incubated at 40°, they formed no plaques or a few, usually small, foci. Two separate stocks of each ts mutant were prepared by passage of virus in chick embryo cell cultures for 2 days at 33° using as inoculum an aliquot of the original plaque suspension (temperature-shift screening) and virus derived from the plaques of the second 33° plating. These "master stocks" were assayed for PFU content and plating efficiency at 40° as described in the following sections.

Preparation of ts Mutant Stocks. The *ts* mutant was first recloned under agar overlay in chick embryo cell cultures incubated at 33°. Virus was picked from two or more well-isolated distinct plaques and suspended in 1 ml of a 2% albumin solution; 0.2 ml of this suspension was injected into embryonated eggs, which were incubated at 33° for 48 hr.

Infectivity Assays and Determination of Plaquing Efficiency at 40°. Fluids to be assayed for plaque-forming titer were diluted, and cell mono-

[41] C. Scholtissek and A. C. Bowles, *Virology* **67,** 576 (1975).

[42] I. Koennecke, Thesis, Justus-Liebig-Universität, Giessen (1981).

[43] R. W. Simpson and G. K. Hirst, *Virology* **35,** 41 (1968).

layers were inoculated with 1-ml volumes of diluted material. Inocula were adsorbed at room temperature for 30 min, after which agar overlay medium lacking neutral red was added. The assay plates were incubated at 33° and 40°, respectively. A second agar medium containing neutral red was added on day 3 or 4, and the plates were incubated for an additional 24 hr at the same temperatures previously employed. Distinct plaques were counted. All mutants chosen for study showed plaquing efficiencies at the restrictive temperature (40°) of 10^{-4} or less, as determined by dividing the 40° PFU titer by the 33° titer.

Protease Activation Mutants of Paramyxoviruses

Mutants with alterations in glycoprotein processing have also been isolated with paramyxoviruses. The fusion protein of these mutants has an altered sensitivity to proteolytic activation. Cells that do not contain the appropriate enzyme to cleave the fusion protein, and thus do not allow multiple replication cycles of the virus, provide a suitable system for the isolation of such mutants.

Isolation of Sendai Virus Mutants Activated by Chymotrypsin and Elastase.[44] Unlike wild type, which can be activated by trypsin, the fusion protein of these mutants is cleaved by proteases of different specificities substituted to the cell culture medium. Sendai virus was mutagenized by treatment with nitrous acid.[45] Undiluted egg-grown virus, 0.5 ml, was incubated with 0.25 ml of 1 *M* sodium acetate, pH 4.4, and 0.25 ml of freshly prepared 4 *M* sodium nitrite. This treatment caused a decrease in virus infectivity, as determined by plaque assay with trypsin, of 1.5 logs in 1 min, 3.5 logs in 2 min, and 5.5 logs in 3 min. Virus treated for 1 min was used for mutant selection. Thirteen MDBK monolayers in 6-cm petri dishes received 0.5 ml of inocula containing ca. 1×10^4 PFU. After the adsorption period, the cells were incubated with REM containing chymotrypsin at a concentration of 1 μg/ml to activate and permit multiple cycle growth of any mutant virus susceptible to this enzyme. After 5 days, 0.5 ml of the medium from each dish was used to inoculate a fresh MDBK monolayer. Within 3 days, two of these 13 cultures developed cytopathic effects typical of Sendai virus infection, and only these two cultures produced hemagglutinin. The harvests from these cultures were then tested for plaque formation with chymotrypsin, 1 μg/ml, in the agar overlay. In contrast to wild-type virus, which does not form plaques under these conditions, these two inocula produced plaques. Individual plaques were

[44] A. Scheid and P. W. Choppin, *Virology* **69,** 265 (1976).

[45] K. W. Mundry and A. Gierer, *Z. Vererbungsl.* **89,** 614 (1958).

selected and subjected to a second plaque purification, and stocks of virus were propagated in MDBK cells with chymotrypsin, 1 μg/ml, added to the medium. Mutants activated by elastase were prepared by essentially the same procedure.

Isolation of Mutants of Newcastle Disease Virus with Altered Protease Sensitivity of the Fusion Protein.[46] The fusion protein of these mutants is cleaved by an endogenous protease of the host cell that is unable to activate wild-type virus. Mutants were induced in egg-grown virus (strains LaSota and Ulster) by treatment with 1 *M* sodium nitrite at pH 4.4 as described above. For mutant selection, monolayer cultures containing 10^6 MDBK cells were inoculated at a multiplicity of 10^{-3} and incubated with 5 ml of serum-free medium. After 3 days, 0.5 ml of the medium was used to inoculate a fresh culture, which was then incubated with serum-free medium for another 3 days. With both strains, in 1 out of about 500 twofold passages carried out in this way, virus production could be observed in the second passage as indicated by a distinct cytopathogenic effect and the release of hemagglutinin into the medium (128 HAU/ml). This virus was then subjected to three plaque passages in the same cell line.

[46] W. Garten, W. Berk, Y. Nagai, R. Rott, and H.-D. Klenk, *J. Gen. Virol.* **50,** 135 (1980).

[35] Methods for Assay of Cellular Receptors for Picornaviruses

By Richard L. Crowell, David L. Krah, John Mapoles, and Burton J. Landau

The picornaviruses are small (24–28 nm), single-stranded RNA-containing, nonenveloped viruses comprising over 170 immunologically distinct human viruses and a multitude of animal viruses. The several genera include the human enteroviruses (polioviruses, coxsackieviruses A and B, and echoviruses) and rhinoviruses, the murine cardioviruses, and the viruses of foot-and-mouth disease (FMDV).[1] Specific receptors for each of the several virus species exist on various cells in culture.[2–4] For example, the three poliovirus immunotypes compete for a receptor that is

[1] R. R. Rueckert, *Compr. Virol.* **6,** 131 (1976).
[2] R. L. Crowell, *J. Bacteriol.* **91,** 198 (1966).
[3] K. Lonberg-Holm, R. L. Crowell, and L. Philipson, *Nature (London)* **259,** 679 (1976).
[4] B. Baxt and H. L. Bachrach, *Virology* **104,** 42 (1980).

ISBN 0-12-181996-5

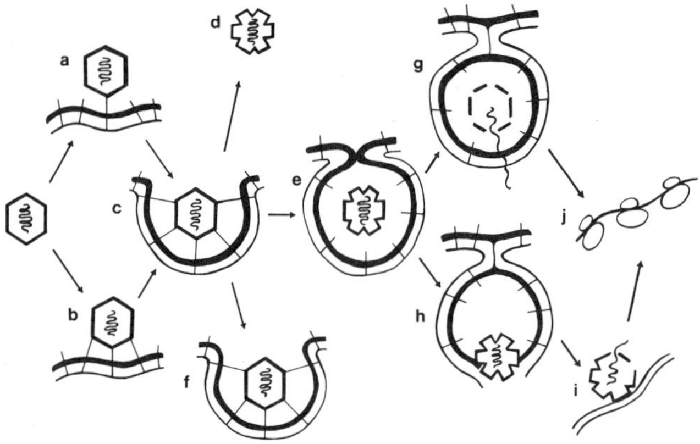

FIG. 1. Scheme for early events in picornavirus infection at 37°. a, Virion attached by virion attachment protein (VAP) to cellular receptor unit (CRU) (loose binding, reversible with EDTA); b, virion attached by multiple VAPs to CRUs (cellular receptor site) (tight binding, reversible at pH 2); c, virion penetration (virion inaccessible to neutralizing antibodies, but recoverable by pH 2 or SDS); d, eluted virus, "A" particle (virus eclipsed and not recoverable by pH 2 or SDS); e, engulfed A particle (virus eclipsed and not recoverable by pH 2 or SDS); f, virion that will remain static at the cell surface (recoverable by pH 2 or SDS); g, disassembly of A particle to release RNA (uncoating), which passes across membrane of vesicle (RNA susceptible to hydrolysis by RNase following disruption of cells by sonication or by SDS); h, A particle transported across membrane of vesicle, since it is more lipophilic than virions; i, A particle may attach to endoplasmic reticulum prior to disassembly and release of RNA (uncoating); j, RNA combines with ribosomes for translation.

distinct from the one that binds the six immunotypes of the group B coxsackievirus.[2] Receptor specificity based on virus attachment competition experiments has sorted these viruses into the same groups in which they were originally placed. The original grouping was based on the pattern of disease and histopathology produced in humans and animals, indicating that receptors are important determinants of virus tropism and pathogenesis.[5]

Attachment of a virion to its specific cellular membrane receptor is the initial event in the virus life cycle.[6–8] This interaction is mediated by a site on the virion, referred to as the virion attachment protein (VAP), occurring as multiple copies, probably at the vertices of the icosahedral-shaped capsid (Fig. 1,a,b). It is possible that these sites may reside on two or more of the capsid polypeptides for some picornaviruses. A single virion

[5] R. L. Crowell, B. J. Landau, and J.-S. Siak, *Recept. Recognition, Ser. B* **8,** 169 (1981).
[6] K. Lonberg-Holm and L. Philipson, *Monogr. Virol.* **9,** 1 (1974).
[7] C. Howe, J. E. Coward, and T. W. Fenger, *Compr. Virol.* **16,** 1 (1980).
[8] R. L. Crowell and B. J. Landau, *Compr. Virol.* **18,** 1 (1983).

binding site is considered to attach to a complementary structure on the external surface of the plasma membrane. This structure is defined as a cellular receptor unit (CRU) and is likely to be the receptor protein. The cellular receptor site (CRS) is composed of one or more copies of the CRU and has the capacity to bind one virion.[9] Presumably, multiple interactions occur between the VAP sites and the CRUs in the plasma membrane as the virion penetrates the cell surface (viropexis, endocytosis, or phagocytosis) (Fig. 1,c). The sequence of events that follow attachment of the virion to the CRS can be divided into stages (penetration, eclipse, and uncoating of the genomes) based on operational criteria. A scheme depicting these events is summarized in Fig. 1. Penetration is signaled by a decrease in the ability of virus type-specific neutralizing antibodies to prevent virus infection, while the virus–receptor complex (VRC) may be dissociated from cells by detergents with recovery of infectious virus.[10] Virus eclipse (Fig. 1,d,e) is defined as an irreversible event that occurs at 37° and is identified by a decrease in the amount of infectious virus that can be recovered from cells artificially disrupted by detergents or other reagents that dissociate the VRC.[11,12] Approximately 20% of the cell-associated virions will remain in their native form (Fig. 1,f). Uncoating (Fig. 1,g–i) of the viral genome is assessed by the ability of nucleases to hydrolyze the viral nucleic acid.[13] The transition from photosensitivity to photoresistance of virus labeled with acridine dyes also has been used to signal uncoating.[14,15]

Assays of Receptors for Virus Attachment

Attachment of virions to their receptors is primarily an electrostatic interaction that is strongly influenced by the charge distribution on both the ligand and the receptor. Additionally, van der Waals forces, hydrogen bonds, and hydrophobic bonds play a role in firm binding of viruses to receptors. Binding of viruses does not occur in the absence of ions, and some rhinoviruses, coxsackieviruses A9 and A13, and FMDV require divalent cations for attachment.

Factors influencing the attachment rate of virions to receptors include both cell and virion concentration, pH, temperature, ion species and con-

[9] K. Lonberg-Holm, *Recept. Recognition, Ser. B* **8,** 1 (1981).
[10] B. Mandel, *Virology* **17,** 288 (1962).
[11] J. J. Holland, *Virology* **16,** 163 (1962).
[12] M. L. Fenwick and P. D. Cooper, *Virology* **18,** 212 (1962).
[13] W. K. Joklik and J. E. Darnell, *Virology* **13,** 439 (1961).
[14] F. L. Schaffer and A. J. Hackett, *Virology* **21,** 124 (1963).
[15] H. J. Eggers, B. Bode, and D. T. Brown, *Virology* **92,** 211 (1979).

centration, viscosity, presence of sulfhydryl groups, proteolytic enzymes, and assorted inhibitors.[16] Whereas changes in cell concentration affect attachment kinetics in a directly proportional way, the virion concentration over a wide range (multiplicity of infection of 0.01 to 50) has relatively little effect on the attachment rate. As the higher limit is approached, saturation of receptors begins to occur, and the reduced rate of attachment reflects competition for receptor sites.[2]

Cells to be assayed for virus receptors are grown either in suspension or monolayer cultures.[17] Growth medium is removed, and the cells are washed (3 times) with phosphate-buffered saline (PBS) without calcium and magnesium and dispersed with 0.02% ethylenediaminetetraacetic acid in PBS. The cells are collected by centrifugation (350 *g*, 5 min), resuspended in PBS, and counted with a hemacytometer or electronic cell counter. Cell samples are distributed to tubes and centrifuged; the fluid phase is discarded, and the cells are resuspended in virus. The virus stock should be diluted in the desired buffered medium at the chosen temperature, to give a virus:cell ratio of about 0.1 and a final cell concentration of 5×10^6/ml or 1×10^7/ml. Virus attachment is allowed to proceed at 0–20°. At intervals, samples of virus–cell suspensions are diluted 1 : 100 in buffered medium at 4° effectively to stop virus attachment. The cells are removed by centrifugation, and the supernatant fluid is frozen for plaque assay of unattached virus. Alternatively, cells can be separated from unattached virus on membrane filters.[18] The rate of virus attachment is expressed as the percentage of input infectivity remaining in the supernatant fluid. Incorporation of 2–3% animal serum or 0.1% BSA (free from viral inhibitors) into the medium stabilizes the diluted virus preparations when frozen. Virus samples incubated without cells are used to provide a measure of virus stability and to monitor for nonspecific virus attachment to the vessel walls. Additional controls should include a cell devoid of receptors for the virus being assayed, e.g., L cells, which have no receptors for polioviruses and group B coxsackieviruses.

A theoretical treatment of the kinetics of virus attachment to cellular receptor sites has been presented by Lonberg-Holm.[9] Initial reaction rates are linear and first order, and under optimum conditions the rates are proportional to the number of cells (CRS) present. The velocity constant, K, for attachment of enteroviruses to HeLa cells lies between 1×10^{-8} and 2×10^{-9} ml min^{-1} cell^{-1}. Very slow rates of up to 1×10^{-12} ml min^{-1}

[16] R. L. Crowell, *in* "Cell Membrane Receptors for Viruses, Antigens, and Antibodies, Polypeptide Hormones, and Small Molecules" (R. F. Beers, Jr. and E. G. Basset, eds.), p. 179. Raven, New York, 1976.

[17] R. L. Crowell, *J. Bacteriol.* **86,** 515 (1963).

[18] K. Lonberg-Holm and N. M. Whiteley, *J. Virol.* **19,** 857 (1976).

$cell^{-1}$ have also been found for some viruses attaching to susceptible cells.[16]

Although plaque assays of virus infectivity (PFU) are more sensitive, radioactively labeled virus is often used to provide a more rapid assay and to facilitate measurement of both unattached and attached virus, thereby keeping track of the total amount of virus in the system. The experimental design is similar to that described for PFU, except that virus radioactivity is monitored. Caution must be exercised in the interpretation of results when attachment of labeled virus is determined at temperatures above 22°, since a large percentage (approximately 50%) of attached virions will elute or slough from the receptors in a noninfectious form.[12,13,19,20]

The number of receptor sites on cells is determined by virus saturation experiments.[2,3] Using a constant cell concentration (5×10^6/ml) in replicate tubes, increasing amounts of virus are added and allowed to attach under optimum conditions for a given amount of time (1 hr) at 20°. The cells are separated from the fluid phase by centrifugation, and the amount of virus that has become cell associated (CAV) is determined (PFU or cpm). To determine the number of virions attached per cell at saturation, one must know the number of virions in the purified virus preparation under study. This value is estimated from the absorbance at 260 nm. A value of 9.4×10^{12} virions per OD_{260} unit has been determined for picornaviruses.[1] The number of receptor sites per cell varies from about 3×10^3 for poliovirus,[6] to approximately 10^5 for coxsackievirus B3,[2] and 10^4–10^6 sites for human rhinoviruses.[21,22] These values are comparable to those found for several polypeptide hormones and represent a relatively small amount of specific receptor protein compared to the total protein of the cell.[23] Another method that has been used to determine the relative amounts of receptors for strains of FMDV used competition between different amounts of unlabeled virus and a constant amount of labeled virus for binding to cells.[4]

A number of selected strains of picornaviruses are known to agglutinate erythrocytes. However, the relationship between the erythrocyte receptor and the host cell receptor for attaching virus is unknown.[8] Extensive studies of the attachment of the murine cardioviruses to erythrocytes have revealed that the receptor is glycophorin A and contains sialic acid.[24]

[19] L. Philipson and S. Bengtsson, *Virology* **18,** 457 (1962).

[20] R. L. Crowell, B. J. Landau, and L. Philipson, *Proc. Soc. Exp. Biol. Med.* **137,** 1082 (1971).

[21] K. Lonberg-Holm and B. D. Korant, *J. Virol.* **9,** 29 (1972).

[22] L. Medrano and H. Green, *Virology* **54,** 515 (1973).

[23] M. D. Hollenberg and E. Nexø, *Recept. Recognition, Ser. B* **11,** 1 (1981).

[24] A. T. H. Burness, *Recept. Recognition, Ser. B* **8,** 63 (1981).

Assays of virus binding to erythrocytes are frequently measured in hemagglutination units.[19] It is recommended, however, that virus binding to erythrocytes be quantitated as for host cells described above. The binding of encephalomyocarditis (EMC) virus to host cells occurs very rapidly.[25]

Assays for Virus Attachment to Subcellular Fractions and to Solubilized Membranes

Subcellular membrane fractions are typically prepared by causing cells to swell in a low-salt buffer followed by Dounce homogenization. The preparation is fractionated on discontinuous sucrose gradients. Measurement of virus binding to subcellular fractions is performed with methods similar to those outlined for cells. It should be noted, however, that greater centrifugal force is needed to separate membrane-bound virus from free virus than is required for sedimenting cells. Attempts to measure virus binding to homogenized or minced animal or human tissues are fraught with problems in quantitation of receptors. Nevertheless, such studies have given rough comparative estimates of receptor presence on selected tissues.[26–29]

The anionic detergent sodium deoxycholate (DOC) has been used to solubilize the receptors for the group B coxsackieviruses and polioviruses from HeLa cell plasma membranes.[30,31] A solid-phase assay has been developed to measure virus binding to solubilized receptor preparations.[31] In these studies, plasma membrane preparations were made according to the method of Roesing *et al.*,[32] and solubilized in 0.2% DOC (final concentration) with vigorous agitation on a Vortex mixer. The DOC solution was prepared immediately prior to use by diluting an aliquot of a 10% stock solution (w/v in distilled water) in PBS. After 15 min for solubilization, the DOC-treated membranes were transferred to a cellulose nitrate or polycarbonate tube fitted in a type-50 fixed-angle rotor (Beckman Instruments, Inc.) and centrifuged for 60 min at 100,000 *g*. The clear supernatant fluid containing the soluble receptors can be stored in small volumes in the vapor phase of a liquid nitrogen freezer.

For the solid phase virus binding assay, serial 2-fold dilutions in PBS of solubilized membrane preparations (50 μl/well) are immobilized by passive adsorption (20 hr, 6°) on 96-well, flat-bottom polystyrene Microti-

[25] P. R. McClintock, L. C. Billups, and A. L. Notkins, *Virology* **106,** 261 (1980).

[26] J. J. Holland, *Virology* **15,** 312 (1961).

[27] C. M. Kunin and W. S. Jordan, *Am. J. Hyg.* **73,** 245 (1961).

[28] C. M. Kunin, *Bacteriol. Rev.* **28,** 382 (1964).

[29] D. H. Harter and P. W. Choppin, *J. Immunol.* **95,** 730 (1965).

[30] R. L. Crowell and J.-S. Siak, *Perspect. Virol.* **10,** 39 (1978).

[31] D. L. Krah and R. L. Crowell, *Virology* **118,** 148 (1982).

[32] T. G. Roesing, P. A. Toselli, and R. L. Crowell, *J. Virol.* **15,** 654 (1975).

ter plates (Micro ELISA substrate plates, Dynatech Division of Cooke Industries, Alexandria, Virginia). Plates are sealed with self-adhering acetate film (Dynatech) during all incubations to minimize sample evaporation. The receptor-coated wells are washed twice with PBS (300 μl/wash) to remove unadsorbed or weakly adsorbed receptors. Wells receive 5×10^6 PFU of virus in 50 μl of 0.05 *M* phosphate–citrate buffered saline (0.14 *M* NaCl), pH 4.5 (optimum for coxsackie B3), containing 0.5% bovine serum albumin, and the plates are incubated at 6° for an additional 48 hr to permit virus to attach to the immobilized receptors. We have found that the incubation time can be reduced to 6 hr at 23° or to 1 hr at 37°, since coxsackievirus B3 was not eclipsed under these conditions. Unattached virus is removed by five washes with PBS, and the plates are blotted dry on paper towels. The amount of attached virus is determined by plaque assays following dissociation from the receptors with 200-μl volumes of 0.05 *M* glycine-HCl buffer, pH 1.5 (2 min, 6°). The pH of the dissociated virus is adjusted to neutrality by a 10-fold dilution in balanced salt solution supplemented with 3% calf serum and 0.05 *M* *N*-2-hydroxyethylpiperazine *N*′-2-ethanesulfonic acid (HEPES), pH 7.2. Wells coated with solubilized L-cell membranes, which lack receptors for the human enteroviruses, serve as controls for nonspecific virus adsorption. The titer of receptor activity is defined as the reciprocal of the highest dilution of the receptor preparation that binds 5-fold more virus than the L-cell preparation. This assay is about 400-fold more sensitive than the virus eclipse assay for solubilized receptors.[30] Concentrations of DOC greater than 0.05% inhibit adsorption of solubilized receptors to the wells. Specificity of virus attachment to receptors is established by receptor competition experiments using saturating amounts of heterologous viruses, which are in the same receptor family.[2,3,16,31] The solid phase assay was found to provide a 2500-fold increase in sensitivity,[31] compared to the assay of coxsackievirus B3 binding to homogenates of mouse brain.[33] It is anticipated that partial purification of the specific cellular receptors for binding virus prior to assay in the solid phase system will permit the use of radioactively labeled virus instead of the virus infectivity assay. The PFU measurements of binding are used because the specific activity of labeled virus is too low for use on solubilized membrane preparations.

Assay of solubilized receptors by competition with cells for binding coxsackievirus B3 as done for solubilized receptors for adenovirus type 2,[34,35] was found to be about 100 times less sensitive than the solid-phase virus binding assay.[31]

[33] C. M. Kunin, *J. Immunol.* **88,** 556 (1962).

[34] B. Hennache and P. Boulanger, *Biochem. J.* **166,** 237 (1977).

[35] U. Svensson, R. Persson, and E. Everitt, *J. Virol.* **38,** 70 (1981).

Assay of Virus Penetration

Several studies have clearly demonstrated that virus penetration of the cell follows attachment and precedes virus uncoating.[12,36–38] To study virus penetration, cell-associated virus (CAV) at a multiplicity of infection of 0.1–1.0 is prepared at 6°, as described previously. At intervals during incubation at 37°, aliquots of the virus–cell suspension are removed, cooled below room temperature to stop penetration, and treated with neutralizing antiserum for 1 hr. The virus-specific antiserum should have a neutralization titer greater than 1000. The cells are washed 3 times to remove excess antibody, resuspended in growth medium, and enumerated; dilutions are plated on monolayers of susceptible cells (indicator cells) for assay of infectious centers. With time of incubation at temperatures above 20°, the CAV becomes progressively more resistant to neutralization by the antiserum (virus penetration), since these antibodies do not neutralize intracellular virus.[39] Virus that has penetrated the cell (resistant to neutralizing antibodies), however, can be recovered as infectious virus after dissociation of the virus–cell complex by use of low pH[40] or SDS.[10] Incubation of a sample of CAV in the cold serves as a control of virus neutralization by the antiserum (no plaques should be obtained when plated on indicator cells).

Assay of Virus Eclipse and Virus Uncoating

Eclipsed virus is considered to be that amount of virus infectivity that is determined initially to be cell-associated and is not released from cells after treatment of washed cells for 1 min at 2° with 0.05 *M* glycine buffer, pH 1.5,[20] or by use of SDS.[10] These reagents have been shown to dissociate the virus–receptor complex, SDS having a slight advantage for following the disappearance of virus infectivity at intervals of incubation of the CAV at 37°.[41]

Transition from photosensitivity to photoresistance of neutral red-labeled virus also has been used to herald virus disassembly or uncoating following virus interaction with cells.[14,15,36] Neutral red-labeled virus is prepared as follows. Cells are grown for 2 days in growth medium, which is exchanged with fresh medium containing neutral red at a concentration

[36] B. Mandel, *Virology* **31,** 702 (1967).
[37] J. N. Wilson and P. D. Cooper, *Virology* **21,** 135 (1963).
[38] S. Dales, *Bacteriol. Rev.* **37,** 103 (1973).
[39] B. Mandel, *Compr. Virol.* **15,** 37 (1979).
[40] I. Zajac and R. L. Crowell, *J. Virol.* **3,** 422 (1969).
[41] R. L. Crowell and B. J. Landau, unpublished observations (1970).

of 10 μg/ml.[36] After an additional 24-hr incubation, cells are infected with virus that has been passaged once or twice in the presence of dye. Virus growth medium with neutral red (10 μg/ml) is added, and the culture is incubated overnight at 37° in 5% CO_2 atmosphere. The virus is recovered following alternate freeze–thaw (3 times) of cells. All operations involving neutral red-labeled virus are conducted under red light (shown to be non-inactivating).

The photodynamic inactivation (5 logs) of neutral red-labeled virus is produced by exposing virus for 10 min at room temperature at a distance of 3 inches from two 15-watt "white" tubular fluorescent bulbs. At intervals of incubation at 37°, samples of photosensitive virus with cells are removed for assay of virus photosensitivity. Results of such experiments revealed that acquisition of light resistance (uncoating) and inability to recover infectious virus by SDS were chronologically indistinguishable.[36]

In following the early events of virus–cell interaction using labeled virus, about 50–80% of the CAV radioactivity elutes (sloughs) from the cells at 37°.[12,13,21,42] The eluted virus is equivalent to the "A" particle that has lost the native infectivity (eclipsed) associated with virions. "A" particles have lost VP4 (the smallest capsid polypeptide), contain the viral genome, which is resistant to applied ribonuclease, have a slower sedimentation rate (135 S) than virions (150 S), are less stable in CsCl than are virions, and become sensitive to proteolytic enzymes. In addition, poliovirus and coxsackievirus B3 "A" particles do not attach to the cellular receptors.[42–44]

"A" particles are generated by interaction of virions with cells or plasma membrane preparations.[32] [^{32}P]- or [^{3}H]uridine-labeled CAV or MAV (membrane associated virus) is prepared as described above. Preparations of CAV or MAV are incubated at 37° for up to 1 hr in HEPES-buffered saline, pH 7.0.[40] A replicate sample is kept at 0° to serve as a negative control (no A-particle formation). After incubation, the preparations are centrifuged for 20 min at 3700 *g* at 4° to pellet the MAV (less for CAV). The supernatant fluids (1 ml) are decanted into tubes, to which are added 1 ml of 10% trichloroacetic acid and a few drops of BSA (5 mg/ml) to serve as a carrier to aid precipitation of virus label (1–2 hr at 4°). After centrifugation (5000 *g*, 5 min, 4°), the trichloroacetic acid pellets and supernatants are separated and assayed for radioactivity. Virus labeled with ^{32}P can be counted in 10 ml of 0.1 *M* NaOH by Cerenkov radiation by using the ^{3}H-channel setting in a liquid scintillation spectrometer,[45]

[42] R. L. Crowell and L. Philipson, *J. Virol.* **8,** 509 (1971).
[43] M. L. McGeady and R. L. Crowell, *J. Virol.* **32,** 790 (1979).
[44] M. L. McGeady and R. L. Crowell, *J. Gen. Virol.* **55,** 439 (1981).
[45] T. Clausen, *Anal. Biochem.* **22,** 700 (1968).

whereas, ^{3}H-labeled virus is counted by standard scintillation methods. Radioactive counts remaining in the CAV or MAV pellet are considered to be attached virions that were neither eluted nor uncoated. Counts in the trichloroacetic acid pellet are considered to be A particles that remain intact, since they are not hydrolyzed by RNase. Counts in the trichloroacetic acid supernatant fluids are considered as being "uncoated" virus because they are susceptible to digestion by RNase. The sensitivity of the poliovirus A particle to disruption by 1.0% SDS also has been used to distinguish A particles from virions.[46]

HeLa cell plasma membranes solubilized by 0.2% DOC have been shown to generate A particles of coxsackievirus B3 after incubation at 37°.[47] The A particles were recovered on sucrose gradients. With increasing time of incubation at 37°, the A particles spontaneously disrupted to liberate the labeled RNA, which was hydrolyzed by applied RNase. Comparable experiments with poliovirus were not possible, since DOC dissociates the poliovirus–receptor bond.[12] Incubation of labeled poliovirus or coxsackievirus B3 with plasma membranes at 37° has revealed the formation of A particles and "C" particles (subviral particles that sediment more slowly than A particles).[44,48,49] The C particles of coxsackievirus B3 were sensitive to applied RNase and could not be detected by immunoprecipitation in washed and disrupted HeLa cells at intervals of incubation at 37°.[44] Although proteolytic enzymes have been used to aid the *in vitro* uncoating of poliovirus A particles,[46,48,50,51] no evidence was found for the formation of C particles on the infectious pathway of coxsackievirus B3 and proteolysis did not appear to be necessary for virus uncoating by HeLa cells.[44] Further studies are needed to determine the fate of the A particle on the pathway to virus infection and their acquired lipophilic nature.[52] Whether clathrin-coated pits and receptosomes serve to transport A particles to the Golgi and lysosomes for liberation of viral RNA is not known.[53]

It is hoped that the methods and comments presented herein will stimulate others to explore further the role of receptors in the molecular events in viral infections and in the cellular function(s) that receptors serve.

[46] N. Guttman and D. Baltimore, *Virology* **82,** 25 (1977).

[47] J-S. Siak, M. L. McGeady, and R. L. Crowell, *Abstr., Int. Congr. Virol. 4th, 1978* p. 182 (1978).

[48] J. DeSena and B. Mandel, *Virology* **78,** 554 (1977).

[49] J. DeSena and B. Torian, *Virology* **104,** 149 (1980).

[50] J. J. Holland and B. H. Hoyer, *Cold Spring Harbor Symp. Quant. Biol.* **27,** 101 (1962).

[51] J. DeSena and B. Mandel, *Virology* **70,** 470 (1976).

[52] K. Lonberg-Holm, L. B. Gosser, and E. J. Shimshick, *J. Virol.* **19,** 746 (1976).

[53] I. H. Pastan and M. C. Willingham, *Science* **214,** 505 (1981).

[36] Transport of Virus Membrane Glycoproteins, Use of Temperature-Sensitive Mutants and Organelle-Specific Lectins

By LEEVI KÄÄRIÄINEN, ISMO VIRTANEN, JAAKKO SARASTE, and SIRKKA KERÄNEN

The envelope glycoproteins of lipid-containing animal viruses have been used as models for the synthesis and transport of cellular membrane glycoproteins, since they are produced in large amounts during the virus infection. Virus mutants with defects in the transport of envelope glycoproteins have helped to elucidate the intracellular pathway of the membrane glycoproteins.[1] Here we describe the isolation and screening of transport-defective, temperature-sensitive mutants of Semliki Forest virus (SFV), and the use of organelle-specific lectins for the rapid localization of the arrested virus glycoproteins using double fluorescence.

Abbreviations: BSA, bovine serum albumin; CEF, chick embryo fibroblasts; Con A, concanavalin A; CPE, cytopathic effect; FCCP, carbonyl cyanide-*p*-trifluoromethoxyphenylhydrazone; FITC, fluorescein isothiocyanate; HA, hemagglutination; HBSS, Hank's balanced salt solution; HEPES, *N*-2-hydroxyethylpiperazine *N'*-2-ethanesulfonic acid; HI, hemagglutination inhibition; HPA, *Helix pomatia* agglutinin; medium 199, Parker's medium 199; MEM, Eagle's minimum essential medium; NTG, *N*-methyl-*N*-nitronitrosoguanidine; PBS, Dulbecco's phosphate-buffered saline without Ca^{2+} and Mg^{2+}; PFU, plaque-forming unit; SDS, sodium dodecyl sulfate; SFV, Semliki Forest virus; TRITC, tetramethylrhodamine isothiocyanate; WGA, wheat germ agglutinin.

Isolation of the Mutants

Reagents

Actinomycin D (a generous gift from Merck, Sharp & Dohme)
Bactoagar, 0.9% in medium 199[1a]
Chicken serum, inactivated (30 min at 56°)
Dulbecco's phosphate-buffered saline[1a]

[1] L. Kääriäinen and M. Pesonen, *in* "The Glycoconjugates" (M. I. Horowitz, ed.), Vol. 4, p. 191. Academic Press, New York, 1982.

[1a] Formulas for the tissue culture media and solutions appear in the GIBCO catalog "Products for Tissue Culture."

ISBN 0-12-181996-5

Dulbecco's phosphate-buffered saline without Ca^{2+} and Mg^{2+} (PBS)
Dulbecco's phosphate-buffered saline supplemented with 0.2% bovine serum albumin (BSA)
Eagle's minimum essential medium[1a] (MEM)
Eagle's minimum essential medium supplemented with 0.2% bovine serum albumin (BSA)
Hank's balanced salt solution[1a] (HBSS)
Medium 199[1a]
Newborn calf serum, inactivated (30 min at 56°)
N-Methyl-*N*-nitronitrosoguanidine (NTG)
Semliki Forest virus, prototype strain
Sodium dodecyl sulfate (SDS)
Trypsin, 0.25% in PBS
Trypsin, 0.125%, EDTA 0.01% in PBS
Tryptose phosphate broth (Difco)
[5-^{3}H]Uridine, 10–20 Ci/mmol (Radiochemical Centre, Amersham, United Kingdom)

Propagation of the Cells

Secondary chicken embryo fibroblasts (CEF) are used for virus propagation and isolation of the mutants. In spite of the fact that the production of SFV in CEF cells is considerably lower (1000 PFU/cell) than, e.g., in baby hamster kidney cells (5000 PFU/cell), the chicken cells have the advantages for isolation of temperature-sensitive mutants that (*a*) they tolerate high temperatures up to 41° for prolonged times; (*b*) they remain stable during long incubations at the low temperature (28°) required for plaque formation by the *ts* mutants; (*c*) they do not contain passenger viruses when obtained from a proper flock; (*d*) they are reasonably easy to prepare in large quantities.

Our primary cells are prepared from 10-day-old embryos from a leukosis-free flock. The flock is regularly tested for several avian pathogens.[2] The muscle and connective tissues of the embryos are chopped and trypsinized,[3] and 8×10^8 cells are seeded per 90-mm plastic dish in medium 199 supplemented with 7.5% fetal calf serum, 1.5% chicken serum, and 10% tryptose broth. The cells are grown in a humidified incubator flushed constantly with 5% CO_2 at 35°; on day 4 the monolayers are confluent and the secondary cells can be prepared. The medium is removed, and the cell sheet is rinsed once with 10 ml of PBS and once with 5 ml of 0.125% trypsin, 0.01% EDTA in PBS. After few minutes at 37° the cells are

[2] K. Sandelin and T. Estola, *Acta Vet. Scand.* **16,** 341 (1975).
[3] E. Hunter, this series, Vol. 58, p. 379.

collected and resuspended in the growth medium (MEM + 5% calf serum + 10% tryptose phosphate broth, Difco). Usually 15×10^6 cells are obtained from a dish. For cultures to be confluent in 2 days, 1.5×10^6 cells are seeded per 50-mm dish and 5×10^6 cells per 250-ml plastic flask.

Preparation and Storage of the Virus Stocks

To avoid accumulation of defective interfering virus particles, all virus stocks are prepared with low multiplicity of infection, 0.01 PFU per cell. The virus stocks are stored at −70° divided into small aliquots (1–2 ml) and never refrozen. When properly stored, the PFU titers of the stocks remain unchanged for several years.

First the virus strain is plaque purified several times and a primary stock is prepared and stored at −70°. The working stocks are prepared starting from the primary stock and thus represent the same second-passage level after the plaque purification. By this we try to avoid the accumulation of nucleotide changes in the viral RNA that occurs during continuous passaging.

Confluent monolayers of secondary CEF in 250-ml culture flasks containing 10^7 cells are used 2 days after seeding. The medium is removed, and the cells are rinsed once with HBSS. The virus inoculum (0.01 PFU/cell) is added in 1 ml of MEM supplemented with 0.2% BSA. The adsorption period is 1 hr at 37° (wild-type virus) or at 28° (*ts* mutants), after which the inoculum is removed and 10 ml of MEM + 0.2% BSA are added per flask. The replication of the virus is monitored by taking aliquots for hemagglutination (HA) titration[4] starting at 18 hr for wild-type virus and at about 42–44 hr for the *ts* mutants. When the HA titer has leveled off or extensive cytopathic effect (CPE) is observed, the virus is harvested. The culture fluid is collected and cell debris is removed by centrifugation for 30 min at 10,000 *g* at 4°. Previously an equal volume of Dulbecco's PBS containing 0.5% BSA was added to the supernatant. This is a good method for storing the SFV stocks. During 7 years of storage, the PFU titer of a *ts* mutant stock had been reduced from 1.2×10^9 to 7.8×10^8. However, the infectivity of the Sindbis virus stocks is reduced rapidly under these conditions. Presently we store our viral stocks in 20 m*M* HEPES, pH 7.2, and 5% glycerol. The PFU titers of the virus stocks are about 10^9 PFU/ml.

Cloning of the Wild-Type Virus

Before mutant isolation, the wild-type virus is cloned several (3–5) times by plaque purification. For isolation of the *ts* mutants, the cloning

[4] D. H. Clarke and J. Casals, *Am. J. Trop. Med. Hyg.* **7**, 561 (1958).

should be done by alternating the permissive and restrictive temperatures (selected for the *ts* mutants) during the successive plaque purifications to assure that the wild-type virus replicates equally well at both temperatures. Proper virus dilutions (in Dulbecco's PBS + 0.2% BSA) to give a few plaques only, are inoculated on 50-mm dishes. After 1 hr of adsorption the inoculum is removed and 4 ml of agar overlay medium, 0.9% agar in medium 199 supplemented with 2% (at 39°) or 5% (at 28°) calf serum. When the plaques have developed in 2 days at 39° and in 4–5 days at 28°, a block of agar on top of a well separated plaque is collected with a Pasteur pipette and transferred to 1 ml of virus growth medium. The virus is allowed to elute from the agar either by incubating for a few hours in an ice-water bath and vigorous vortexing or by freezing at −70°. The second cloning is done directly from the plaque eluate. About 10^4 to 10^5 PFU are obtained from a plaque. From the final plaque eluate a primary stock is grown with low multiplicity, as described above, and stored at −70° in aliquots. The working stock is prepared, and using this the virus growth curve is determined at the temperatures selected as permissive and as restrictive for the *ts* mutants.

The difference between the permissive and restrictive temperature is selected as large as possible. The upper limit of the restrictive temperature is determined by inactivation of the virus. The lower limit for the permissive temperature is mainly determined by the growth rate of the virus. For convenience, incubation periods for plaque formation should not exceed 1 week. We selected 39° as the restrictive, and 28° as the permissive, temperature for our mutants. The growth cycle of the virus is 8 hr at 39° and 16 hr at 28°.[5]

Mutagen Treatment and Selection of Mutants

Several mutagens have been used to induce *ts* mutants of alphaviruses.[5–7] Our *ts* mutants were selected after 30 min of treatment at 20° with NTG (100 g/ml). Among 16 mutants isolated 2 out of 7 RNA$^+$ mutants had a defect in envelope protein transport. Sindbis virus transport-defective mutants *ts*-10 and *ts*-23[8] have been isolated after NTG and HNO_2 treatment, respectively.[6] After treatment, the virus is diluted extensively and plaque assay is performed under agar overlay at the permis-

[5] S. Keränen and L. Kääriäinen, *Acta Pathol. Microbiol. Scand., Sect. B* **82,** 810 (1974).

[6] B. W. Burge and E. R. Pfefferkorn, *Virology* **30,** 204 (1966).

[7] E. G. Strauss and J. H. Strauss, *in* "Togaviruses" (R. W. Schlesinger, ed.), p. 393. Academic Press, New York, 1980.

[8] J. Saraste, C.-H. von Bonsdorff, K. Hashimoto, S. Keränen, and L. Kääriäinen, *Cell Biol. Int. Rep.* **4,** 279 (1980).

sive temperature. Alternatively, the treated sample can be dialyzed extensively against Dulbecco's PBS + 0.5% BSA. NTG treatment of SFV followed by immediate dilution to 10^{-7} reduced the infectivity by 93%. Dialysis overnight resulted in 97% reduction in the PFU titer.

Several hundred well separated plaques formed under agar overlay at 28° after mutagen treatment are randomly picked up, and the virus is eluted as described above. To screen out the wild-type plaques, the eluates were tested for replication at 39°. CEF cells on plastic well trays (1×10^6 cells/well) are inoculated with 0.2 ml of the plaque eluate, and CPE is recorded at 24 hr. Care should be taken to use low multiplicity of infection (0.01–0.05 PFU/cell), since the RNA-positive mutants do cause CPE at high multiplicity and would be discarded. If the virus does not cause CPE, any method to measure the released virus may be used.

Only those plaque eluates that do not show CPE or virus replication at 39° are similarly tested for replication at 28°, and CPE is recorded at 48 hr. For those eluates that show virus replication at 28°, but not at 39°, plaque assays are performed at both temperatures. Those eluates that have at least 100-fold higher PFU titer at 28° than at 39° are characterized further. Three plaques are picked up for each eluate from the low-temperature plaque assay and stored at −70°. This constitutes the second cloning of the mutants.

A primary stock is grown from one of the three plaques. The PFU titers of the primary stocks are determined at 28° and at 39°. The efficiency of plating at the restrictive temperature (PFU/ml at 39° : PFU/ml at 28°) gives an estimate for the reversion frequency of the putative mutants. Our mutants were selected to have a plating efficiency 39°/28° 10^{-4} or lower. Owing to the missense nature of the *ts* mutants all *ts* mutants are able to replicate to some extent at the elevated temperatures. This leakiness of the mutants is measured by infecting the cell cultures with 10 PFU/cell at both temperatures. After a 1-hr adsorption period, the inoculum is removed, the cells are washed three times with prewarmed HBSS, and medium is added. At 4 hr the medium is removed, the cells are washed once more with HBSS, and new medium is added. The virus released to the medium is harvested at 8 hr at 39° and at 16 hr at 28°. A plaque assay for both samples is performed at 28°.

$$\text{Leak yield} = \frac{\text{yield at 39° titrated at 28°}}{\text{yield at 28° titrated at 28°}}$$

Our mutants were selected to have leak yield values 10^{-3} or below. We deliberately selected stable mutants with low leakiness and low reversion frequencies to facilitate the biochemical studies.

Identification of Mutants with Defects in the Structural Proteins

Screening for alphavirus mutants with defects in the structural proteins is easily done by determining their ability to synthesize RNA at 39°. The mutants with defects in the nonstructural proteins are not able to induce viral RNA synthesis, whereas those with defects in structural proteins should exhibit normal RNA synthesis. The RNA phenotype of the mutants is determined as follows: Cells on 50-mm plates are infected with 50 PFU/cell at 39° in the presence of actinomycin D (1 μg/ml). At 3 hr the medium is removed and replaced with 2 ml of fresh medium containing 10–20 μCi of [^{3}H]uridine (10–20 Ci/mmol), and incubated for 2 hr. The medium is removed, and the cells are rinsed once with PBS (39°) and lysed in 1 ml of 2% SDS in water (39°). The cell lysate is collected using a rubber policeman. The DNA is sheared by passing the lysate four times through a 25-gauge needle. Duplicate samples are precipitated with trichloroacetic acid to determine the acid-insoluble radioactivity.

Our RNA positive (RNA^+) mutants synthesize 40–100% of RNA compared to the wild-type virus, whereas the synthesis by the RNA negative (RNA^-) mutants is less than 5%.

Screening for Transport-Defective Mutants

The RNA^+ mutants that synthesize virus-specific proteins in normal amounts[9] can be screened for the presence of envelope proteins at the cell surface using immunofluorescence[10] or radioimmune assay based on the binding of iodinated protein A of *Staphylococcus aureus*.[11] Mutants with transport defects should show reduced amounts of envelope proteins at the cell surface when grown at the restrictive temperature throughout infection.

Reagents

Cycloheximide, 10 mg/ml in PBS

Dulbecco's phosphate-buffered saline without Ca^{2+} and Mg^{2+} (PBS)[1a]

Eagle's minimum essential medium[1a] supplemented with 0.2% bovine serum albumin (BSA)

Fluorescein isothiocyanate (FITC)-coupled goat anti-rabbit IgG (Wellcome, Beckenham)

[9] S. Keränen and L. Kääriäinen, *J. Virol.* **16,** 388 (1975).

[10] J. Saraste, C.-H. von Bonsdorff, K. Hashimoto, L. Kääriäinen, and S. Keränen, *Virology* **100,** 229 (1980).

[11] L. Kääriäinen, K. Hashimoto, J. Saraste, I. Virtanen, and K. Penttinen, *J. Cell Biol.* **87,** 783 (1980).

Hank's balanced salt solution[1a] (HBSS)
Iodinated (^{125}I) protein A of *Staphylococcus aureus*
Mounting medium (50% glycerol, 100 m*M* NaCl, 50 m*M* Veronal, buffer, pH 8.6)
Paraformaldehyde 4% in 0.1 *M* sodium phosphate buffer, pH 7.2
Rabbit antiserum against virus membrane glycoproteins
Triton X-100, 0.05% in PBS

Immunofluorescence

Secondary CEF are grown on glass coverslips in 35-mm plastic dishes for 1–2 days to obtain nonconfluent monolayers in which the individual cells can be discerned. The cultures are washed with HBSS and infected with the mutants at 39° using 20–50 PFU/cell. After 60 min of absorption at 39° the cultures are washed three times with HBSS and thereafter MEM 0.2% BSA and 20 m*M* HEPES, pH 7.2, but no actinomycin is added. At the time of harvest (usually 5 hr after infection with SFV and 6 hr with Sindbis virus) the cultures are washed with PBS and fixed with 4% paraformaldehyde in 0.1 *M* sodium phosphate buffer, pH 7.2, for 10 min at 39°. Fixation with paraformaldehyde preserves the plasma membrane intact so that the antibody molecules cannot penetrate into the cell.[12] After fixation, the cultures are washed twice with PBS and stored at 4° in Dulbecco's PBS + 0.5% BSA. The fixed cultures can be stored at 4° for several days without affecting the result.

For intracellular immunofluorescence the fixed cells are made permeable for antibodies by incubating them in 0.05% Triton X-100 for 30 min at 20°, followed by washing as above. The permeabilized cells are a useful control to assure that all cells have been infected, since at this time all infected cells show strong fluorescence.[8,10]

Antibodies were prepared in rabbits against the envelope protein complex,[10] which can be isolated from the virus after Triton X-100 treatment. The solubilized envelope protein trimers are complexed to octamers (29 S) during sucrose gradient centrifugation.[13] These lipid and detergent free 29 S structures have SFV E1, E2, and E3 glycoproteins in roughly equimolar ratios and elicit hemagglutination inhibiting and protective antibodies.[14]

The cells are treated with anti-envelope serum in dilutions 1:80 to 1:320, respective to 8 to 2 HI units, for 30 min at 37°, followed by three

[12] P. Laurila, I. Virtanen, J. Wartiovaara, and S. Stenman, *J. Histochem. Cytochem.* **26,** 251 (1978).

[13] A. Helenius and C.-H. von Bonsdorff, *Biochim. Biophys. Acta* **436,** 895 (1976).

[14] B. Morein, A. Helenius, K. Simons, R. F. Pettersson, L. Kääriäinen, and V. Schirrmacher, *Nature (London)* **276,** 715 (1978).

washes with PBS and briefly with distilled water before addition of anti-rabbit IgG-FITC conjugate (Wellcome, Beckenham) in a dilution of 1:20 to 1:40. Incubation is for 30 min at 37° followed by three washes with PBS. Care must be taken that the cells are not dried during the above procedures. The coverslips are placed on a drop of mounting medium (50% glycerol, 100 m*M* NaCl in 50 m*M* Veronal buffer, pH 8.6).

Radioimmune Assay with ^{125}I-Labeled Protein A

Confluent monolayers of CEF in 35-mm plastic dishes infected as described above are incubated at 39°, and the cells are fixed with 4% paraformaldehyde 5–6 hr after infection. Treatment with anti-envelope serum or antisera prepared against E1, E2, or E3 is for 30 min at 37° in 300 μl (in Dulbecco's PBS + 0.5% BSA).

After incubation the cells are washed three times with 3 ml of Dulbecco's PBS + 0.5% BSA followed by addition of ^{125}I-labeled protein A in 250 μl (Pharmacia Fine Chemicals). The iodination we have made according to Dorval *et al.*[15] Incubation is at room temperature for 30 min under constant slow shaking. After removal of the unadsorbed material, the cultures are washed four times with Dulbecco's PBS + 0.5% BSA, with shaking for 5 min during each wash. The cells are solubilized by addition of 500 μl of 2% SDS (60°) in distilled water, incubated at 37° for 20 min, and scraped carefully with a rubber policeman. The radioactivity is assayed.

The dilution of anti-envelope serum is determined using wild-type SFV-infected cells fixed at 5 hr after infection. Serum dilutions giving a linear decrease of ^{125}I-labeled protein A binding are preferred to avoid oversaturation with antibodies. The amount of ^{125}I-labeled protein should also be determined to give a maximum binding with a minimum input.[11] We have used regularly 0.1–0.2 μg of ^{125}I-labeled protein A per dish, which corresponds to 100,000 to 300,000 cpm.

As controls we use mock-infected cultures treated exactly like the infected ones, and infected cultures that are treated with normal rabbit serum. Both controls give the same values, 1–2% of the added radioactivity, which is subtracted as background. By this criterion we found one SFV mutant (*ts*-1) that expressed less than 10% of the radioactivity at 39° as compared to the wild type at 5 hr after infection. The same mutant was also negative according to surface immunofluorescence. Another mutant, *ts*-7, which was regarded as ambiguous (±) in the surface immunofluorescence, showed 30% radioactivity (^{125}I-labeled protein A) of that bound to the wild-type infected cells.

[15] G. K. Dorval, I. Welsh, and H. Wigzell, *J. Immunol. Methods* **7**, 237 (1975).

Reversibility of the Transport Defect

To test the reversibility of the transport defect of the *ts*-1 mutant of SFV, the infected cultures grown at 39° for 5 hr are shifted to the permissive temperature (28°). In order to prevent protein synthesis at 28°, 100 μg of cycloheximide per milliliter are added 5 min before the shift down. Thus it is possible to follow the fate of envelope proteins that have been synthesized at the restrictive temperature. When the amount of envelope proteins at the cell surface is measured by the ^{125}I-labeled protein A binding assay, there is a lag of about 20 min followed by an almost linear increase for about 60–70 min. During this transport period, the amount of envelope proteins is increased about fivefold.

Organelle-Specific Lectins in Intracellular Localization of Virus Glycoproteins

Principle

Saccharide moieties of glycoproteins and glycolipids appear to be strictly compartmentalized to distinct cell organelles. Mannose residues can be detected already in rough endoplasmic reticulum and nuclear membranes, whereas the more terminal saccharides, such as galactose, *N*-acetylgalactosamine, and sialic acids, can be found only in the Golgi apparatus–lysosomal complex in addition to their presence in cell surface membranes.[16–18] Such a compartmentalization of saccharide moieties can be utilized to visualize the cell organelles using lectins as probes in fluorescence microscopy.[11,18] Lectins, such as concanavalin A (Con A), binding to mannose and glucose residues, reveal the endoplasmic reticulum as a wide perinuclear area in cultured cells (Fig. 1) whereas wheat germ agglutinin (WGA), binding to terminal sialic acid residues and, e.g., *Helix pomatia* agglutinin (HPA), binding to *N*-acetylgalactosamine residues, decorate intracellularly only the Golgi apparatus as a reticular juxtanuclear organelle in cultured cells (Figs. 2 and 3). The specific decoration of the Golgi is apparently due to a high content of sialic acids as, e.g., latex-phagolysosomes bind WGA only weakly in comparison to the Golgi complex.[19]

[16] J. M. Sturgess, M. Moscarello, and H. Schachter, *Curr. Top. Membr. Transp.* **11,** 15 (1978).

[17] E. Rodriguez Boulan, G. Kreibich, and D. D. Sabatini, *J. Cell Biol.* **78,** 874 (1978).

[18] I. Virtanen, P. Ekblom, and P. Laurila, *J. Cell Biol.* **85,** 429 (1980).

[19] I. Virtanen, P. Ekblom, P. Laurila, S. Nordling, K. O. Raivio, and P. Aula, *Pediatr. Res.* **14,** 1199 (1980).

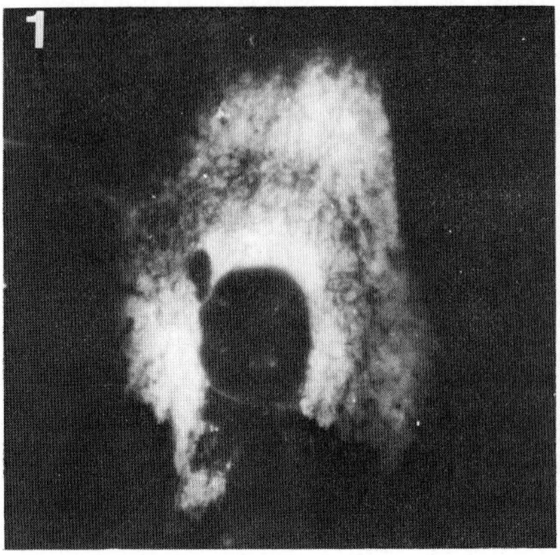

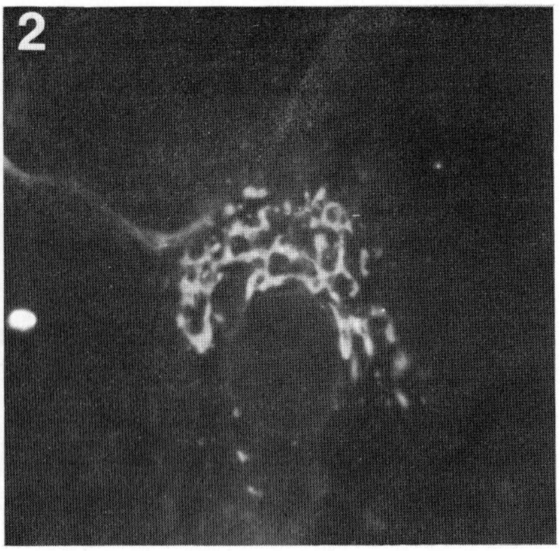

FIG. 1. Double-fluorescence microscopy of human embryonal fibroblasts with FITC–Con A. Note the wide perinuclear cytoplasmic fluorescence obtained with FITC–Con A distinctly differing from the reticular juxtanuclear staining obtained with the WGA conjugate (Fig. 2). ×700.

FIG. 2. Double-fluorescence microscopy of human embryonal fibroblasts with TRITC–WGA. ×700. See legend for Fig. 1.

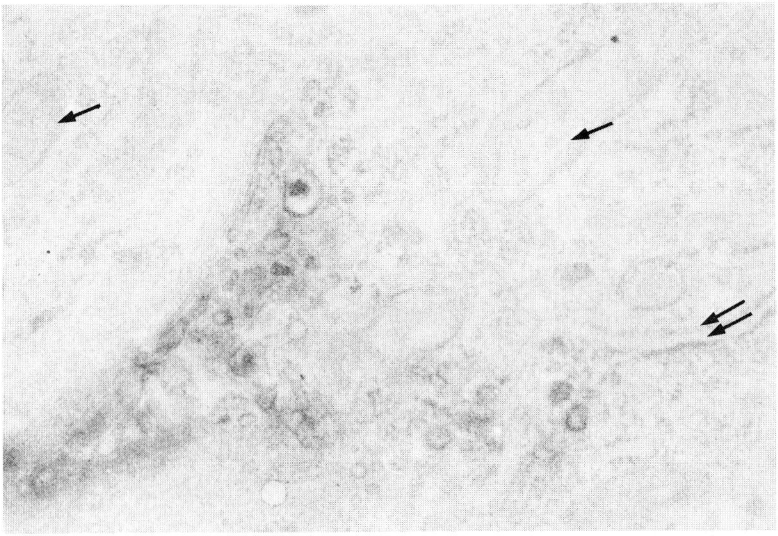

FIG. 3. Electron microscopy of human embryonal fibroblasts labeled with peroxidase-coupled WGA. Note the staining reaction in lamellar and vesicular parts of the Golgi apparatus and the lack of labeling of mitochondria (arrow) and rough endoplasmic reticulum (double arrow). ×24,000.

Reagents

Carbonyl cyanide-*p*-trifluoromethoxyphenylhydrazone (FCCP)
Dulbecco's phosphate-buffered saline without Ca^{2+} and Mg^{2+} (PBS)[1a]
Fluorescein isothiocyanate (FITC)-coupled goat anti-rabbit IgG
Monensin (a gift from Eli Lilly & Co.)
Mounting medium (50% glycerol, 100 m*M* NaCl, 50 m*M* Veronal buffer, pH 8.6)
Paraformaldehyde 4% in 0.1 *M* phosphate buffer, pH 7.2
Rabbit antiserum against virus membrane glycoproteins
Tetramethylrhodamine isothiocyanate (TRITC)-coupled concanavalin A, 100 g/ml in PBS (E-Y or Vector Laboratories)
Tetramethylrhodamine isothiocyanate-coupled wheat germ agglutinin (TRITC–WGA) (E-Y or Vector Laboratories)
Tetramethylrhodamine isothiocyanate-coupled *Helix pomatia* agglutinin (TRITC-HPA) (E-Y or Vector Laboratories)
Triton X-100, 0.05% in PBS.

Double Fluorescence with Antibodies and Organelle-Specific Lectins

For double-fluorescence experiments the cells are cultured on small glass coverslips, infected with the virus, and after a suitable period of time, fixed in 4% paraformaldehyde made in 0.1 *M* sodium phosphate buffer, pH 7.2, for 10 min. Thereafter the cells are washed in PBS and permeabilized in 0.05% Triton X-100 or Nonidet P-40 (NP-40) for 30 min at 22°.[11] After washing, the cells are first exposed to virus membrane antibodies in a humidified atmosphere for 30 min, washed in PBS, and then allowed to react with FITC-coupled goat anti-rabbit IgG (Wellcome) for 30 min. Thereafter the cells are allowed to react with tetramethylrhodamine isothiocyanate (TRITC)-coupled Con A (100 μg/ml) in Dulbecco's PBS for 30 min to visualize the endoplasmic reticulum in double fluorescence, or with TRITC–WGA or –HPA to visualize the Golgi apparatus. We have successfully used lectin conjugates both from Vector Laboratories (Burlingame, California) and from E-Y Laboratories (San Mateo, California) for fluorescence microscopy. After washing, the specimens are mounted in sodium-Veronal–glycerol buffer, pH 8.6, and examined in a Zeiss Universal microscope equipped with filters for FITC- and TRITC-fluorescence.

In cells infected with the wild-type Semliki Forest virus, a typical cytoplasmic fluorescence corresponding to the localization of ER can be found (Fig. 4) in addition to the pronounced juxtanuclear staining corresponding to the region of Golgi apparatus (Figs. 4 and 5). In double-

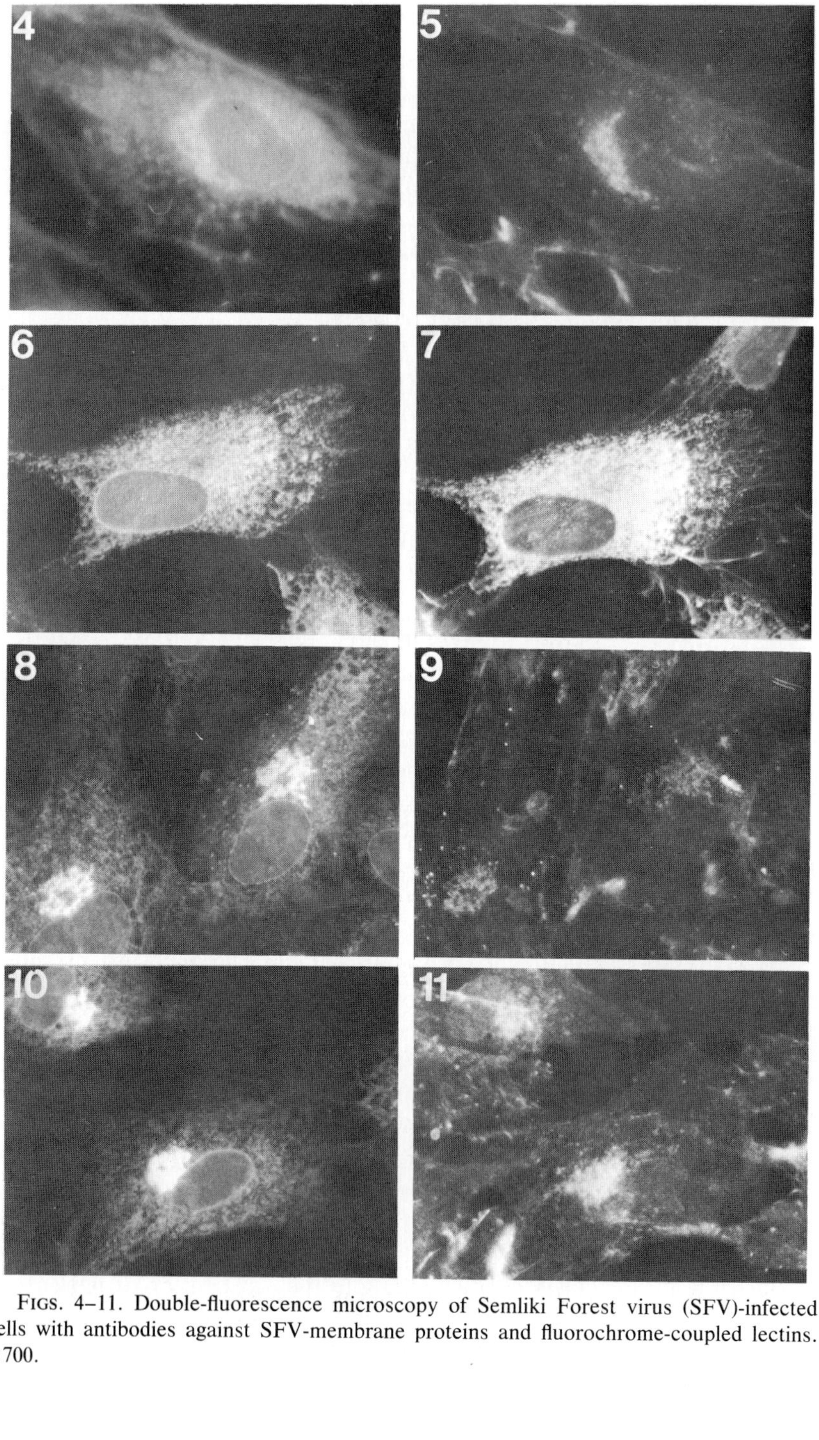

FIGS. 4–11. Double-fluorescence microscopy of Semliki Forest virus (SFV)-infected cells with antibodies against SFV-membrane proteins and fluorochrome-coupled lectins. ×700.

fluorescence experiments, the lectin conjugates (Con A, WGA, HPA) do not appear to bind immunoglobulins or vice versa.[10]

When the cells are infected with a temperature-sensitive mutant, *ts*-1, at the restrictive temperature, a typical cytoplasmic immunofluorescence corresponding to staining with Con A-coupled fluorochrome is seen, suggesting that the viral glycoproteins are arrested in ER (Figs. 6 and 7). This interpretation is compatible with the results obtained by immunoelectron microscopy[20] and glycan analysis of the virus-specific proteins.[21] A substantial proportion of the arrested virus glycoproteins migrate to the Golgi complex and plasma membrane when the infected cultures are shifted to 28° in the presence of cycloheximide (100 μg/ml) and are revealed by double staining with TRITC–WGA[10] (Figs. 8 and 9). When the shift to 28° is carried out in the presence of the monovalent ionophore, monensin (1–10 μM for 15 min to 4 hr), the transport of virus glycoproteins is arrested into a juxtanuclear organelle corresponding to the Golgi complex as judged by double fluorescence with TRITC–WGA and the lack of surface immunofluorescence[11] (Figs. 10 and 11).

When the shift to the permissive temperature is carried out in the presence of FCCP, an uncoupling agent of oxidative phosphorylation (10 μM), the virus glycoproteins remain in ER as shown by double fluorescence with FITC–Con A and lack of immunofluorescence at the juxtanuclear region and at the cell surface.[11]

The pathway of *ts*-1 glycoproteins from ER to plasma membrane via the Golgi complex as suggested by double fluorescence with organelle-specific lectins has been established by immunoelectron microscopy[20] and biochemical studies.[21] Thus we conclude that at the light microscopic level the organelle-specific lectins can be used for rapid intracellular localization of membrane glycoproteins of transport-defective mutants.

[20] J. Saraste, Thesis, University of Helsinki, Helsinki, Finland (1981).

[21] M. Pesonen, J. Saraste, K. Hashimoto, and L. Kääriäinen, *Virology* **109,** 165 (1981).

FIG. 4. In human fibroblasts infected with the wild-type virus, both a wide perinuclear virus-specific staining and a reticular juxtanuclear fluorescence are seen.

FIG. 5. The reticular staining codistributes with that obtained with TRITC–WGA, representing thus virus membrane glycoproteins in the Golgi apparatus.

FIG. 6. When the cells are infected with the temperature-sensitive virus mutant *ts*-1 at the restrictive temperature, a bright cytoplasmic immunofluorescence is seen.

FIG. 7. Staining obtained with FITC-Con A.

FIG. 8. When the mutant-infected cells are shifted to the permissive temperature, a bright juxtanuclear immunostaining is again seen.

FIG. 9. Staining obtained with the WGA conjugate.

FIG. 10. When the mutant-infected cells are shifted to the permissive temperature in the presence of monensin, a bright juxtanuclear immunostaining is seen.

FIG. 11. Staining obtained with the WGA-conjugate.

[37] Immunoelectron Microscopy Using Thin, Frozen Sections: Application to Studies of the Intracellular Transport of Semliki Forest Virus Spike Glycoproteins

By G. GRIFFITHS, K. SIMONS, G. WARREN, and K. T. TOKUYASU

Enveloped animal viruses have proved to be useful experimental systems for studying the structure and assembly of cell surface glycoproteins.[1] Of those studied, Semliki Forest virus (SFV) is one of the simplest known. It comprises a spherical nucleocapsid surrounded by a lipid bilayer containing virally coded proteins. The nucleocapsid contains a single RNA molecule (12.7 kilobases in length) complexed with about 180 molecules of a lysine-rich capsid protein (molecular weight 29,700).[2] The membrane contains about 180 molecules of spike protein, each comprising two spanning glycoproteins (E1 and E2, both having a molecular weight of about 50,000) and a peripheral protein (E3, with a molecular weight of about 10,000). E3 and most of the mass of E1 and E2 are on the outer surface of the viral particle.[3]

Cells infected with SFV synthesize only viral proteins. A 26 S viral RNA (4.2 kilobases) is translated from a single initiation site, and the proteins appear in the following order: nucleocapsid protein, glycoprotein precursor p62 (E2 and E3), and E1.[4,5] The nucleocapsid protein is cleaved from the nascent chain immediately after synthesis. The amino-terminal end of p62 functions as a signal sequence and causes the 26 S RNA–ribosome complex to bind to the endoplasmic reticulum (ER), where further synthesis of the glycoproteins is coupled to transfer across the membrane and glycosylation. The proteolytic cleavage that releases the p62 chain from the nascent E1 chain takes place when the p62 chain has been completed. Both p62 and E1 span the bilayer and form a complex, which then migrates from the endoplasmic reticulum through the Golgi complex to the cell surface.[6] The p62 protein is cleaved to E2 and E3 probably just after leaving the Golgi complex. After reaching the plasma

[1] K. Simons and H. Garoff, *J. Gen. Virol.* **50,** 1 (1980).

[2] H. Garoff, A. M. Frischauf, K. Simons, H. Lehrach, and H. Delius, *Proc. Natl. Acad. Sci. U.S.A.* **77,** 6376 (1980).

[3] H. Garoff, A. M. Frischauf, K. Simons, H. Lehrach, and H. Delius, *Nature (London)* **288,** 236 (1980).

[4] I. C. S. Clegg, *Nature (London)* **254,** 454 (1975).

[5] H. Garoff, K. Simons, and B. Dobberstein, *J. Mol. Biol.* **124,** 535 (1978).

[6] A. Ziemiecki, H. Garoff, and K. Simons, *J. Gen. Virol.* **50,** 111 (1980).

ISBN 0-12-181996-5

membrane, the mature spike proteins are recruited by underlying nucleocapsids, previously assembled in the cell cytoplasm, and this results in the eventual budding of the virus from the plasma membrane.[1]

The key to the use of SFV as an experimental model for studying plasma membrane assembly lies in the fact that the virus must make use of those cellular components that normally assemble the host cell plasma membrane. The SFV genome codes only for viral structural proteins, and there is no evidence that it can code for any of the proteins needed for membrane assembly.

Understanding membrane biogenesis demands that the precise route taken by proteins during intracellular transport be known. This in turn requires a high-resolution microscopic technique that can locate the transported protein in an identifiable membrane. One generally useful method is to follow the movement of newly synthesized proteins by immunolocalization at the ultrastructural level. We will describe here the practice and application of the thin frozen section technique to the biogenesis of the SFV membrane glycoproteins. We have also included a general procedure for making high-titered antiserum to amphiphilic membrane proteins.

Antibody Labeling of Intracellular Antigens

It is technically straightforward to localize extracellular and cell surface antigens, but there are a number of problems when one attempts to label the internal components of cells. The major difficulty lies in making the antigens accessible to the antibody reagents without destroying the fine structure of the cell. In conventional ultramicrotomy there are two basic approaches to this problem. The first is to permeabilize the fixed cells or tissues with agents such as Triton X-100 or acetone, which destroy or partially destroy membranes, and then the antibody labeling and embedding are carried out (pre-embedding techniques). The second approach is to embed the fixed tissues, to section them, and then to treat them with antibodies (postembedding techniques).

For pre-embedding techniques, the structural damage resulting from the use of solvents or detergents can be considerable. For cytoskeletal elements, which are usually stable to these agents, this is usually not important, and many techniques are available that give excellent results at the light and electron microscopic levels. For localization of antigens in or on membranes, however, this approach is fraught with difficulties, since the membrane structure is destroyed by the solvent or detergent. In addition, one is generally restricted to using small soluble markers in the horseradish peroxidase (HRP) technique (though there are exceptions;

see Willingham *et al.*[7]). The sensitivity is high because the reaction product is enzymically amplified, but there are three major disadvantages. First, the reaction product can diffuse away from the site at which it was produced. Second, a negative reaction is not necessarily significant since the reagents may not have had access to all intracellular sites. Third, the labeling cannot be quantitated.

Postembedding techniques offer a number of advantages over pre-embedding techniques. These are listed below.

1. Accessibility. The entire surface of the section is accessible for reaction with the antibody reagents.

2. Quantitation. This can be done reliably only using particular markers,[8–10] such as ferritin or colloidal gold. Whereas ferritin has been quantitated successfully on sections, most workers are now working with colloidal gold preparations.[11,12] Colloidal gold is more electron dense than other markers used. It can be easily and rapidly "conjugated" to a wide variety of proteins, the most useful probably being protein A.[13] The gold can be prepared in a variety of size classes that can be easily separated on sucrose gradients.[13a] These different sizes can then be used for double-labeling experiments.[14] There are other markers that, unlike colloidal gold, are chemically conjugated to the protein. These are theoretically more attractive for double-labeing experiments, e.g., ferritin and imposil,[15] although the tissue contrast must then be relatively low in order to see them.

3. Fine-structural preservation. There are a number of approaches for embedding that give very good fine-structural preservation. The mere absence of lipid-destroying agents, which are used in pre-embedding techniques, must *a priori* improve the fine-structural preservation of membranes.

There are three kinds of postembedding techniques that we consider useful.

[7] M. C. Willingham, S. S. Yamada, and I. Pastan, *J. Histochem. Cytochem.* **28,** 453 (1980).
[8] J. P. Kraehenbuhl, L. Racine, and J. D. Jamieson, *J. Cell Biol.* **72,** 406 (1977).
[9] J. P. Kraehenbuhl, E. R. Weibel, and D. S. Papermaster, "Immunofluorescence and Related Staining Techniques." North-Holland Publ., Amsterdam, 1978.
[10] J. P. Kraehenbuhl, L. Racine, and G. W. Griffiths, *Histochem. J.* **12,** 317 (1980).
[11] W. P. Faulk and G. M. Taylor, *Immunochemistry* **8,** 1081 (1971).
[12] M. Hornsberger and J. Rosset, *J. Histochem. Cytochem.* **25,** 295 (1977).
[13] E. L. Romano and M. Romano, *Immunochemistry* **14,** 711 (1977).
[13a] J. W. Slot and H. J. Geuze, *J. Cell Biol.* **90,** 533 (1981).
[14] H. J. Geuze, J. W. Slot, P. A. van der Ley, R. C. T. Scheffer, and J. M. Griffith, *J. Cell Biol* **89,** 653 (1981).
[15] B. Geiger, A. H. Dutton, K. T. Tokuyasu, and S. J. Singer, *J. Cell Biol.* **91,** 614 (1981).

Epon Embedding. In a few cases, conventional Epon sections have proved to be useful for antibody labeling.[16,17] In general, however, loss of antigenicity due to the dehydration and embedding means that the sensitivity is usually very poor and the electron-dense marker of choice, colloidal gold, tends to stick nonspecifically to the Epon giving high background labeling. Another serious difficulty in many applications is the problem of antigen accessibility: unless the antigen is abundant and exposed on the surface of the section, significant labeling cannot be expected.

Water-Soluble and Low-Temperature Embedding Media. Until recently, BSA embedding was the most useful technique in this category,[8] although the damage to fine structure was often considerable.[18] The introduction of the low-temperature embedding medium Lowicryl K4-M[19] has resulted in sections having good fine-structural preservation, very impressive specific labeling, and negligible background.[20,21] This method is clearly more sensitive than the labeling of Epon sections, but it is still less sensitive than the labeling of frozen sections.

Frozen Sections. In our opinion, the frozen-thin section technique[22,23] offers the best method available for intracellular localization of membrane antigens. The fine-structural preservation has improved considerably since the introduction of the methyl cellulose post-embedding technique[23,24] and variations on this.[25] The technique has now been successfully applied to a wide range of different antigens and gives significant labeling with negligible background. An extra advantage is that semithin sections cut from the same frozen block as that used for thin sections can be labeled with fluorescent markers for a low-resolution overview of the labeling pattern. In addition, the technique is very rapid; it takes less than 1 day from the fixation of the tissue to the printed micrograph. Since we consider this approach to be the best, it is the one we have chosen to describe in detail below.

[16] J. Roth, M. Bendayan, and L. Orci, *J. Histochem. Cytochem.* **26,** 1074 (1978).

[17] M. Ravazzola, A. Perrelet, J. Roth, and L. Orci, *Proc. Natl. Acad. Sci. U.S.A.* **78,** 566 (1981).

[18] G. W. Griffiths and B. M. Jockusch, *J. Histochem. Cytochem.* **28,** 969 (1980).

[19] E. Carlemalm, M. Garavito, and W. Villinger, *J. Microsc.* (*Oxford*) (in press).

[20] J. Roth and E. Berger, *J. Cell Biol.* **93,** 223 (1982).

[21] B. Thorens, J. Roth, A. W. Norman, A. Perrelet, and L. Orci, *J. Cell Biol.* **94,** 115 (1982).

[22] K. Tokuyasu, *J. Cell Biol.* **57,** 551 (1973).

[23] K. Tokuyasu, *J. Ultrastruct. Res.* **63,** 287 (1978).

[24] K. Tokuyasu, *Histochem. J.* **12,** 381 (1980).

[25] G. W. Griffiths, R. Brand, B. Burke, D. Louvard, and G. Warren, *J. Cell Biol.* **95,** 781 (1982).

The Frozen-Section Technique and Antibody Labeling

The different steps in this procedure will be dealt with in sequence.

Fixation. For immunocytochemistry, fixation can be considered a compromise between the preservation of structural detail and the retention of antigenic activity. In general, glutaraldehyde is the best fixative for structural preservation at the electron microscope (EM) level and is the preferred fixative for immunocytochemistry. Unfortunately, at the concentrations of glutaraldehyde needed for optimal cross-linking, the antigenicity of many proteins is often severely and adversely affected. Hence, for sensitive antigens, it may be necessary to treat the specimen briefly with a mixture of ethyl acetimidate and formaldehyde before fixing with glutaraldehyde.[15,26] The same fixation procedure may also improve the accessibility of antigens to the antibodies, particularly when the antigens are covered by a dense matrix of other proteins.[27]

For a wide range of cellular antigens we have found that treatment of the cells with 0.5% glutaraldehyde alone in 0.1 *M* PIPES buffer, pH 7.0, and 5% (w/v) sucrose, for 30–60 min gives good structural preservation and adequate retention of antigenicity. This fixation procedure was in fact used in the present application. It should be noted that a concentration of glutaraldehyde as low as 0.1% can be effectively employed in certain cases.[28]

Infusion. Infusion with sucrose was introduced in 1973 and represents the best method for freezing and sectioning specimens.[22] If sucrose should for any reason be avoided, other sugars or polyethylene glycol (M_r 300–500) can also be used (K. T. Tokuyasu, unpublished data). After a brief wash in buffer containing 10% (w/v) sucrose, the specimen is transferred to 0.6–2.3 *M* sucrose in the same buffer. The optimal sucrose concentration needed to cut good thin sections depends on the tissue itself. In general, the hardness of the frozen tissue is inversely related to the sugar concentration.

Following a recent innovation[29] we routinely infuse with 2.3 *M* sucrose, which has greatly increased the plasticity and improved the quality of sectioning. This necessitates cutting at colder temperatures (usually in the range of −100° to −110°). We must emphasize, however, that, when very thin sections are required, it is necessary to use significantly lower concentrations of sucrose to make the block harder. If a cryomicrotome

[26] K. T. Tokuyasu, A. H. Dutton, B. Geiger, and J. S. Singer, *Proc. Natl. Acad. Sci. U.S.A.* **78,** 7619 (1981).

[27] K. T. Tokuyasu and J. S. Singer, *J. Cell Biol.* **71,** 894 (1976).

[28] K. T. Tokuyasu, R. Scheckman, and S. J. Singer, *J. Cell Biol.* **80,** 481 (1979).

[29] H. J. Geuze and J. W. Slot, *Eur. J. Cell Biol.* **21,** 93 (1980).

could be stably operated below −110°, 2.3 *M* sucrose-infused blocks should, in principle, be successfully sectioned to 50 nm thickness.

Freezing and Mounting. For sucrose-infused tissues, a simple immersion in liquid nitrogen will be adequate to freeze the blocks without ice-crystal damage. However, if low concentrations of sucrose are used, liquid Freon 12 or 22 may be needed to minimize ice damage. The tissue is mounted on the copper chuck in a small amount of the sucrose infusion solution. For thin sections the tissue piece should be small (1–2 mm) and should be pretrimmed or mounted in such a way as to reduce or obviate trimming in the microtome itself. To cut cross sections of tubular structures, such as muscle fibers, a modified copper holder is recommended. This has a small hole drilled into the top into which the fibers can be inserted. A useful alternative is to cut a vertical slice through the top of the specimen holder to a depth of a few millimeters and to use the vertical wall to mount a tubular or rod-shaped specimen onto it.

Sectioning. Having mounted the specimen into the cryochamber of the cryoultramicrotome, it is of critical importance to select the correct temperature for cutting and to ensure that this temperature is accurately maintained. For any tissue infused with a given sucrose concentration, there exists a narrow temperature range within which a desired section thickness can be cut. The higher the sucrose concentration, the "softer" the tissue becomes and the colder the cutting temperature required for any given section thickness. Conversely, for any one temperature from −10° to −110° (or even colder on the newer microtomes), only a certain thickness of section can be cut optimally. In general, the thinner the section required, the colder the temperature must be. For 0.5-μm sections for immunofluorescence, the useful temperature range is usually between −30° and −60°. For thin sections the range is −70° to −110°. Using 2.3 *M* sucrose, as we now routinely do, the optimal temperature is between −100° and −110°.

The best sections are those cut with a very slow cutting speed. If the temperature is stable, sections can be cut with the automatic advance of the specimen. However, inadequate insulation of the cryochamber often results in heat transfer from the specimen holder through the mechanical arm. The specimen may continuously advance or retract even with the mechanical advance mechanism set to zero. Both of these problems can to some extent be compensated for by using a rapid return stroke of the specimen. In the case of continuous advance of the specimen, such action will reduce the effective thickness of the section, and in the opposite case it allows the operator to "chase" the retracting specimen while increasing the thickness setting on the mechanism.

The Knife. Undoubtedly, when one has overcome the basic initial problems, the *most critical factor* in obtaining good thin sections for electron microscopy is the sharpness of the knife. Both glass and diamond knives may be used for sectioning, but glass tends to be more useful. Delicate manipulations of the specimen block in the cryochamber can easily result in damage to the diamond knife, and the glue used to mount the diamond in its holder often cracks with regular freezing and thawing. In addition, diamond, much more often than glass, tends to build up an electrostatic charge that often results in the sections sticking to the knife and being more difficult to handle. It must also be remembered that frozen tissues are much softer than Epon blocks, and an optimal glass knife can cut thin sections for many hours.

The "Optimal" Glass Knife. The LKB knife-maker and, to a lesser extent, the competing machines from DuPont/Sorvall and Reichert are now standard equipment in most EM laboratories. The average glass knife prepared on these machines by following the manufacturer's instructions is adequate for cutting semithin frozen sections, but is simply not sharp enough for cutting ultrathin frozen sections.

In theory the procedure we adopt to prepare the "optimal" glass knife is simple. First, a perfectly square piece of glass is cut, and this is then broken diagonally into two. The closer this final break is to being exactly diagonal, the sharper is the knife.[30] In practice, however, it is difficult to prepare a perfectly square piece of glass, and thus we can not break it diagonally at exactly 45 degrees. The importance of this can be seen by considering Fig. 1, which shows a square piece of glass that has been broken approximately diagonally. The aim is to minimize the width (w) of the complementary face. The simplicity of the principle is in fact deceptive. If the corner edge is not quite perpendicular to the square, then, as the final diagonal break is made, two knives are produced each with a partially usable edge and each with a part of the complementary face. The first requirement, therefore, is that the corner edge should be as near perpendicular to the plane of the square as possible. To achieve this a rectangular piece of glass is prepared that is exactly twice the length of the required square. If squares are simply cut from a longer piece of glass, then the unbalanced weight of the extra length of glass prevents fracture occurring exactly perpendicularly to the glass plate, and hence the corner edge of the square does not fulfill the above requirement. Furthermore, it is important that the forces that are applied to fracture the rectangular piece of glass, and even more so to produce the final fracture of the square, act symmetrically about the fracture. This places stringent requirements on the knife-making machine. First, the score line must be

[30] K. Tokuyasu and S. Ukamara, *J. Biophys. Biochem. Cytol.* **6,** 305 (1959).

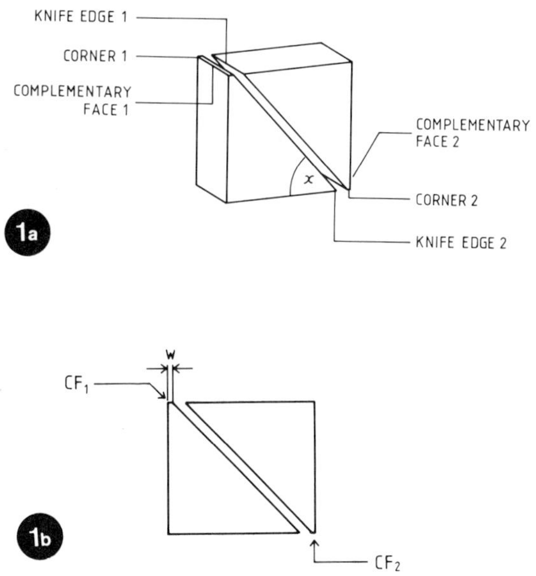

FIG. 1. Schematic drawings (a and b) of the knife-making process.

exactly perpendicular to the edges of the glass plate and centrally positioned with the same free-breaking length on each side. Second, the score line must lie exactly above the line joining the pair of breaking pins (refer to the manual for the instrument) and perpendicular to the line joining the support studs. Top and bottom edges of the rectangle as well as the ends of the score line should be equidistant from the midpoint of the pins (see diagram 2 of Fig. 3). Optimal results are obtained only from a perfectly symmetrical system.

When a perfectly square piece of glass with perpendicular corner edges is obtained, the final diagonal break is made following the same ideas of symmetry. A score line is marked along the vertical diagonal, and the glass square is supported by the pins and the studs symmetrically positioned along the diagonals (Fig. 2). The fracture is made by applying the least possible force and allowing the crack to develop slowly.

In practice, with the LKB machine, the position of the scoring line with respect to the pins is preset during manufacture and is not readily adjusted, as are the positions of the pins. Furthermore, owing to the mechanism for holding the glass, when preparing the "double squares" and squares it is difficult to ensure that the glass is scored accurately perpendicular to its long axis. Similarly, it is difficult to make the final score exactly along the diagonal. The mechanism for adjusting this is

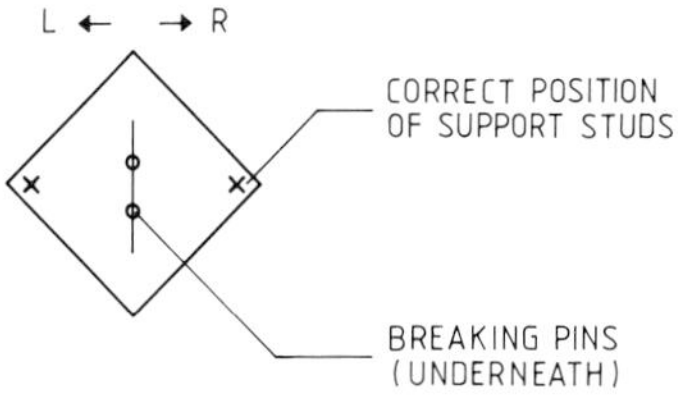

FIG. 2. Diagram to show the symmetrical positioning of the support studs and breaking pins. The line above the breaking pins represents the score. The vertex can be moved to the left (L) or right (R) in order to "aim" the break for the corner.

reproducible only for perfect squares with perfectly flat sides, and there is no way of telling prior to breaking the glass how close the score line is to being diagonal.

These inadequacies thus make glass knife-making an art. With the present machines, all one can do is to learn how to recognize the optimal glass knife and persevere by following the guidelines presented above. We emphasize that this effort is more than compensated for by the ease and speed of obtaining good-quality sections compared with the time wasted with inferior knives.

Practical Procedure for Making the Optimal Glass Knife Using the LKB Knife-Maker. This is shown diagrammatically in Fig. 3.

"Double squares" are made from longer glass rods. Ideally a new "arresting stud" should be mounted to define the length of the "double square," but simply marking the machine with a piece of tape may be adequate. Ensure that the glass strip is scored centrally and perpendicular to its long axis. Use the longest score setting on the machine, and break the glass gently.

Squares are made from the "double-squares" as above. The squares may be stored, but the final fracture to make the knives should be made just prior to sectioning.

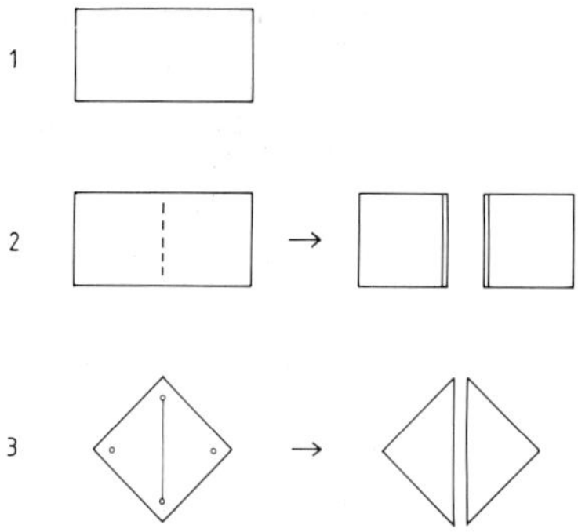

FIG. 3. Representation of the complete knife-making process.

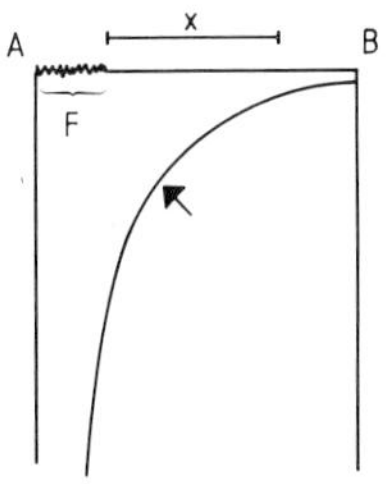

FIG. 4. Schematic drawing of the knife as viewed from the front. The hillock is indicated by the arrow, since it curves to the right. In this case, the knife would be designated a "right-handed knife." F indicates the region where knife marks can be seen using fiber optic illumination. In an optimal knife, these may be completely absent. X indicates the sharpest part of the knife.

Note which is the better corner produced by the preceding step, and place the glass between the corner holders (tensioned according to the LKB manual). Position it so that the studs are arranged symmetrically along the diagonals, score (still on the longest setting), and break slowly. The rubber damping pad is not used.

Remove the two knives and examine them, preferably under a stereomicroscope (see Fig. 1 and the next section). Although the corner selected in diagram 3 should produce the better knife, often, owing to incorrectly set dials or a nonperfect square of glass, the other corner may provide a perfectly usable knife. Note that both a left-handed and a right-handed knife can be obtained, depending on which side of the corner fracture occurs.

Repeat until a good knife is obtained.

Checking the Knife. Knives are most conveniently checked under a low-power stereomicroscope or in the microtome. We find that a portable fiber-optic illuminating system is a very useful addition to the microtome both for examining knives and for seeing interference colors in the cryosections.

First examine the cutting edge of the knife. Along the edge of a poor knife, fault lines (F in Fig. 4) give a jagged appearance and will clearly cause knife marks in any sections cut. Next examine the line that curves either to the left-hand or right-hand edge of the knife (indicated by the arrow in Fig. 4). This indicates the presence of a hillock and means that the actual angle of the knife is greater close to this line, and in practice this is the bluntest part of the knife [the right-hand part of the edge (B) in Fig. 4]. The important point is that the sharpest part of the knife is the region marked X in Fig. 4. The extent to which surface faults spoil this region depends directly on the proximity of the final fracture to the corner of the square. The smaller the complementary face (Fig. 1), the fewer will be the surface marks and the sharper the knife. In practice, when the complementary face is wider than about 0.2 mm, the surface faults can be seen to extend some half way across the knife. As the width (w) of the complementary face decreases, (0.1 mm or less), so the extent of the surface faults decreases, until in an optimal knife they are almost completely absent. In such a knife the best sections, in our experience, are cut from the center to the left-hand side of the knife.

Section Thickness. Frozen sections, like Epon sections, show interference colors that are related to their thickness. For routine EM work we find that sections giving gold-blue interference colors give optimal results. Assuming that the tissues infused with 2.3 *M* sucrose have a similar refractive index to Epon, this would give a thickness of 900–1200 Å. When these sections thaw out and stretch, they obviously become thinner. The best estimate of thickness we have at present comes from using a technique routinely employed for estimating the thickness of Epon sections.[31,32] Areas are found where the section has folded and where such folds are perpendicular to the electron beam, two electron-dense lines are apparent. This is assumed to be twice the section thickness. Although clear images of such folds are far more difficult to see in frozen sections, fortuitous images such as that shown in Fig. 5 indicate that the frozen section thickness giving blue interference color is of the order of 800 Å after thawing.

Section Retrieval. An eyelash probe is used to move sections away from the knife edge and to ensure that sections are not lying on top of each other. A platinum or tin-coated copper loop is used to pick up the sections. The loop is dipped into 2.3 *M* sucrose in PIPES buffer (or PBS), and the drop of sucrose is brought quickly into the cryochamber. The sections are touched onto the bottom surface of the drop. This is a critical and often difficult step. The sucrose must still be in a liquid state when the sections are touched, otherwise they will not flatten onto the sucrose surface. At $-100°$, for example, one has only about 1–2 sec to touch the sections before the sucrose freezes. For picking up thick (0.5 μm) sections, the problem is the opposite one—the loop must be held in the chamber long enough for the sucrose to become sufficiently viscous to ensure that the sucrose does not run onto the knife surface. About 30 sec or longer are needed for this when one cuts at $-30°$ to $-35°$. In this temperature range a reduction of the sucrose concentration to between 1.6 *M* and 1.8 *M* is often an advantage.

There are two sizes of loops for picking up sections. A small (no more than 1 mm in diameter) platinum loop has the disadvantage of freezing quickly but the advantage that, when it is touched to the surface of the grid, the sections are certain to be on the grid. A large (up to 2 mm) loop

[31] J. V. Small, *Proc. Eur. Congr. Electron Microsc. 4th, 1968* Vol. 1 (1968).

[32] E. R. Weibel, "Stereological Methods," Vol. 1. Academic Press, New York, 1979.

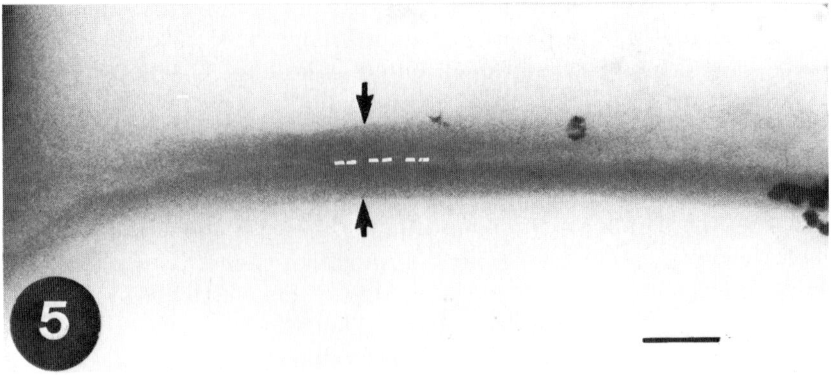

FIG. 5. Micrograph showing a fold in a frozen section that showed blue interference color. The white dotted line indicates the center of the fold. The arrows indicate the distance estimated to be twice the section thickness—approximately 160 nm. The bar = 160 nm.

has the advantage of allowing a valuable extra second or two before the sucrose freezes, but one must examine this loop under the binocular microscope with reflected light in order to locate the sections. The sections can then be centered on the grid. If the sections touch the grid (or glass slide surface), they will immediately attach.

For electron microscopy we use Formvar/carbon-coated, 100–200 mesh, hexagonal-lattice grids. These are stronger than the regular square-lattice grids. They are ionized by a glow-discharge apparatus that helps to reduce the hydrophobicity of the grids and allows the sucrose with the sections to "flow" onto the grid. It also helps to contrast the sections uniformly after immunolabeling.

Immunolabeling Procedure

1. Store the grids either on PBS drops or more conveniently on a layer of agarose (0.3%) plus gelatin (1%) plates covered by a thin film of PBS. This step washes away the sucrose.
2. Place the grids on drops of 2% gelatin in PBS for 10 min. The aim of this step is to block the nonspecific binding sites and hence lower the background labeling.
3. Float on PBS droplets containing 0.02 *M* glycine, pH 7.4 (2–3 min total). The glycine binds to free aldehyde groups from the fixative, which could lead to nonspecific binding. This process is conveniently carried out on a sheet of Parafilm in large puddles, and the grids can be transferred from one to the other with a 2-mm bacterial loop.

4. Using forceps, transfer the grids onto droplets of the antibody solution (that has been centrifuged for 1–2 min at 15,000 *g* in an Eppendorf bench-top centrifuge). Often as little as 3–4 μl per grid will suffice. The grids should be covered with a petri dish in a moist atmosphere. The routine incubation times are 15–45 min.
5. Rinse 5 times with PBS (15 min).
6. Transfer to the gold–protein A solution. The dilution of this is empirical but very critical; solutions that are too concentrated always give high background labeling that cannot be washed away. The incubation time is 15–20 min (for double-labeling procedures see references cited in footnotes 13a, 14, 15, 33, 34).
7. Rinse 5 times with PBS (30 min total).
8. Rinse 4 times with distilled water (4–5 min).

Contrasting and Embedding. Assuming good sections have been put on the grid, this final two-step procedure is the most critical for good fine structural preservation. The aim of the embedment is to infiltrate the section with a solution of a large molecular weight polymer that dries to form a "skeleton" for the section and prevents cellular structures, especially membranes, from collapsing. After simple air-drying, collapse of the structures is inevitable, and the introduction of this embedment process radically advanced the frozen section immunolabeling technique. Before this time, the fine-structure preservation after antibody labeling was very poor. For details of the process, including hydrophilic versus hydrophobic embedment, see Tokuyasu.[23,24] For routine studies, hydrophilic embedment is now always used, and the best compound for this is methyl cellulose, which is available in a range of molecular weights and hence viscosities. Following the advice of the Utrecht group (J. W. Slot, personal communication), we always use Tylose MH300 from Fluka, A.G. in the range of 0.4–1.5% (w/v). Methyl cellulose is more soluble in cold water than in warm water, and we routinely embed on ice.

The high density of colloidal gold has made it possible to use negative staining for contrasting. Good fine-structural preservation of frozen sectioned tissue has been in fact possible with negative staining since the late 1960s, but the contrast was too high to allow ferritin to be visualized. With negative staining, the heavy-metal stain itself helps to support the structures. In conjunction with gold, which is much more dense then ferritin, it is even possible to use negative staining without embedment and get reasonable fine-structural preservation. However, in the presence of

[33] A. H. Dutton, K. T. Tokuyasu, and S. J. Singer, *Proc. Natl. Acad. Sci. U.S.A.* **76,** 3392 (1979).

[34] J. Roth and M. Binder, *J. Histochem. Cytochem.* **26,** 163 (1978).

methyl cellulose embedment, structural preservation is improved. We have found empirically, when using gold, that a subtle mixture of positive and negative staining is possible that is satisfactory. The ratio of positive to negative staining varies with, for example, variations in the thickness of methyl cellulose, in the extent of staining and, more important, of destaining, when one puts the grid onto water or, as we now prefer, directly onto methyl cellulose. The procedure that we use at present is outlined below.

1. Transfer the grid to neutral uranyl acetate oxalate solution (pH 7–7.5) for 5 min.
2. Wash briefly with distilled water three times (1–2 min total).
3. Stain with aqueous uranyl acetate (2%, pH 4) for 3–5 min.
4. Transfer to the surface of large droplets of 1.5% methyl cellulose solution (Tylose M3400, Fluka) (three changes) and leave for a total of 20 sec. Using a 4-mm platinum loop, the grids are taken out and excess solution is carefully removed with filter paper. The final film should give a gold-blue interference color after air drying.
5. Sections can now be observed in the electron microscope. For clear visualization of membranes and gold it is essential that the sections be relatively thin and that the methyl cellulose film be of optimal thickness. Contrast can be improved by reducing the operating voltage and by reducing the size of the objective aperture.

This staining method was used in the present application using 5–8 nm gold particles.[13a] For using ferritin, Imposil,[33] or minute gold particles 3–5 μm in diameter as the marker, positive staining may be more appropriate. The "adsorption staining" method devised for this purpose[24] follows the same initial two steps as described above. The grids are then transferred to the surface of a droplet of a mixture of 1.8% Carbowax (M_r 1540), 0.2% methyl cellulose (400 cps), and 0.1–0.01% aqueous uranyl acetate. The sections are left for 10–15 min, then embedded in the same mixture without further washing in the same manner as described in the latter half of step 4 (above).

Labeling Thin Frozen Sections of SFV-Infected BHK Cells

Isolation of SFV. BHK21 cells are grown in ten 850-cm^2 roller bottles (Falcon Plastics) at 37° in minimal essential medium Glasgow (MEM-G) supplemented with 5% (v/v) fetal calf serum, 2 mM glutamine, 10% (v/v) tryptose phosphate broth, 100 μg of pencillin per milliliter, and 100 μg of streptomycin per milliliter. Stock virus for infection is prepared as described by Kääriäinen and Gomatos.[35] The monolayer of BHK21 cells must be subconfluent (about 1.5×10^8 cells per bottle), and each roller bottle is inoculated with 1.5×10^6 plaque-forming units of SFV in 10 ml of MEM-G. One hour after inoculation, 50 ml of MEM-G are added to each bottle, and infection is continued for 20 hr at 37°. The cell medium is collected on ice (600 ml). All subsequent operations are done at 4°. Cell debris is removed by centrifugation for 20 min at 8000 rpm in the GSA rotor in the Sorvall centrifuge. The virus in the supernatant is sedimented by centrifugation in the SW27 rotor for 150 min at 25,000 rpm using a Beckman ultracentifuge. The supernatant is poured off immediately after the rotor has stopped, and the walls of the centrifuge tube are dried carefully with sterile gauze. To each virus pellet, TN buffer (0.5 ml of 0.05 *M* Tris-HCl, pH 7.4, 0.1 *M* NaCl) is added and left overnight on ice to resuspend the virus. The pellets are resuspended by careful pipetting with a Pasteur pipette, and the suspension is further homogenized by five strokes in a loose Dounce homogenizer. Final purification is achieved by density gradient centrifugation on a 5 to 50% (w/w) potassium tartrate gradient. Two milliliters of virus suspension are layered on each gradient. Centrifugation is for 150 min at 25,000 rpm in the SW27 rotor. The opalescent virus bands are collected using a Pasteur pipette from above, and the virus is concentrated by centrifugation as before at 25,000 rpm for 150 min. The pellet is resuspended in TN buffer. Aliquots are frozen in liquid nitrogen and stored at −70°. The amount of virus protein is determined by the Lowry procedure using 0.1% SDS in the reaction mixture. The purity of the virus can be checked by SDS–polyacrylamide gel electrophoresis. The yield of virus protein is normally between 4 and 7 mg from ten roller bottles.

Preparation of Protein Micelles from the SFV Spike Glycoproteins. To prepare antibodies against the SFV spike proteins, they have to be separated from the nucleocapsids. This can be done by solubilizing the virus in Triton X-100 and separating the SFV spike proteins from the nucleocapsids by ultracentrifugation. The spike proteins are solubilized as a 4.5 S complex consisting of one spike monomer and 80 molecules of Triton X-100.[36] However, a much more immunogenic form of the membrane proteins can be prepared by forming so-called protein micelles from the SFV spike proteins. The protein micelles are 29 S complexes containing eight spike protein monomers essentially free of lipid and detergent.[37] Two injections of 1 μg of the 29 S protein micelles (without adjuvant) into

[35] L. Kääriäinen and P. J. Gomatos, *J. Gen. Virol.* **5,** 251 (1969).

[36] K. Simons, A. Helenius, and H. Garoff, *J. Mol. Biol.* **80,** 119 (1973).

[37] A. Helenius and C.-H. von Bonsdorff, *Biochim. Biophys. Acta* **436,** 895 (1976).

mice give rise to a specific antibody response of about 0.7 mg of IgG antibody per milliliter of serum. For a comparison of the immunogenicity of the different physical forms of the SFV spike glycoproteins see Balcarova *et al.*[38] The method described below seems generally applicable to amphiphilic proteins with a large hydrophilic domain and a small hydrophobic domain attaching the protein to the lipid bilayer.[39] The optimal centrifugation times and gradient designs will vary depending on the sedimentation coefficients of the protein detergent complexes and of the protein micelles. In all cases so far studied, the protein micelles are practically devoid of lipid and detergent. The protein micelles probably form by association of the hydrophobic domains by hydrophobic interactions.[39,40] Once formed, they are not easily dissociated without denaturing agents.

One milligram of SFV (containing [^{35}S]methionine-labeled virus) is solubilized in 4 mg of Triton X-100 in a volume of 0.2 ml of 0.025 *M* Tris-HCl, 0.05 *M* NaCl, pH 7.4. This is layered on top of a 12-ml sucrose gradient comprising a 20 to 50% (w/w) sucrose gradient and a layer of 0.3 ml of 15% sucrose with 1% Triton X-100 in TN buffer. Centrifugation is carried out at 40,000 rpm in the SW40 rotor for 24 hr. Fractions of 0.5 ml are collected from below, and the protein is located by measuring the radioactivity. The nucleocapsid is in the pellet, and the viral spike proteins are all found in the middle of the gradient. Fractions containing the spike protein are pooled, dialyzed against 0.005 *M* Tris-HCl, 0.015 *M* NaCl, pH 7.4, and finally lyophilized.

Immunization and Purification of the Antibodies. Each rabbit is immunized with 10–20 μg of the spike protein dissolved in 0.5 ml of H_2O to which the same volume of Freund's complete adjuvant is added. The rabbits are injected directly in the popliteal lymph nodes,[41] boosted subcutaneously 3 weeks later with the same amount in Freund's incomplete adjuvant, and bled after 10 days.

The antibodies against the spike proteins are purified on an immunoadsorbent prepared by attaching the 29 S complex to an acrylamide-agarose gel. HMD-Ultrogel Ac-34, carrying hexamethylene diamine spacers (Industrie Biologique Française, Paris), is activated with glutaraldehyde and then incubated overnight with 29 S complexes (0.3 mg of protein per milliliter of gel). After coupling, the immunoadsorbent is reduced with sodium borohydride and the protein is further cross-linked using dimethyl

[38] J. Balcarova, A. Helenius, and K. Simons, *J. Gen. Virol.* **53,** 85 (1981).

[39] K. Simons, A. Helenius, K. Leonard, M. Sarvas, and M. J. Gething, *Proc. Natl. Acad. Sci. U.S.A.* **75,** 5306 (1978).

[40] W. G. Laver and R. C. Valentine, *Virology* **38,** 105 (1969).

[41] R. B. Goudie, L. H. W. House, and P. C. Wilkinson, *Lancet* **2,** 1224 (1966).

suberimidate as described by Coudrier *et al.*[42] The washed immunoadsorbent is then used to affinity purify the antibodies to the spike protein from the crude rabbit serum by the procedure of Ternynck and Avrameas.[43]

Antibody Labeling of SFV-Infected BHK Cells. BHK21 cells, infected with SFV for 4–5 hr, are washed in ice-cold Dulbecco's PBS and released from the petri dish by a 5-min treatment with 1 ml of ice-cold proteinase K (50 μg/ml; Serva) in Dulbecco's PBS. The protease is then inhibited by addition of phenylmethylsulfonyl fluoride (40 mg/ml in ethanol) to a final concentration of 40 μg/ml, the mixture is centrifuged (1000 *g*, 2 min, 4°), and the pellet is fixed for 30 min at room temperature with 0.5–1% (v/v) glutaraldehyde (Polysciences, Inc.) in 1 m*M* PIPES buffer (pH 7.0) containing 5% (w/v) sucrose. The pellets of fixed cells are then washed briefly twice in 1 m*M* PIPES (pH 7.0) containing 10% (w/v) sucrose. For cryosectioning, small pieces of the fixed pellet are infused with 2.3 *M* sucrose in 1 m*M* PIPES (pH 7.0) for 10–15 min at room temperature. After freezing in liquid nitrogen, the sectioning, labeling, and contrasting procedures described above are used. The quantitation of gold and ferritin particles on micrographs of frozen sections is described by Griffiths *et al.*[25] and Green *et al.*,[44] respectively (see Fig. 6).

Summary of the Results

Antibody labeling of thin frozen sections has allowed us to follow the precise fate of newly synthesized spike proteins after their insertion into the ER membrane. Biochemical experiments (pulse–chase, cell fractionation) were carried out in parallel and together they have allowed us to draw the following conclusions.

1. The spike proteins pass through the Golgi stacks. Antibodies to the spike proteins specifically labeled all the membrane compartments on the intracellular transport pathway. All cisternae of the Golgi stacks were labeled, and the labeling was fairly uniform with no indication of a gradient of any kind. Cycloheximide stopped synthesis but not transport of the spike proteins, and, by quantitating the label over sections of ER and Golgi membranes, the spike proteins were seen to leave the ER and then pass through the Golgi stacks on their way to the cell surface.[45]

2. Complex sugars are added in trans Golgi cisternae. Parallel biochemical experiments showed that the spike proteins acquired complex

[42] E. Coudrier, H. Reggio, and D. Louvard, *J. Mol. Biol.* **152,** 49 (1981).

[43] T. Ternynck and S. Avrameas, *Scand. J. Immunol., Suppl.* **3,** 29 (1976).

[44] J. Green, G. W. Griffiths, D. Louvard, P. Quinn, and G. Warren, *J. Mol. Biol.* **152,** 663 (1981).

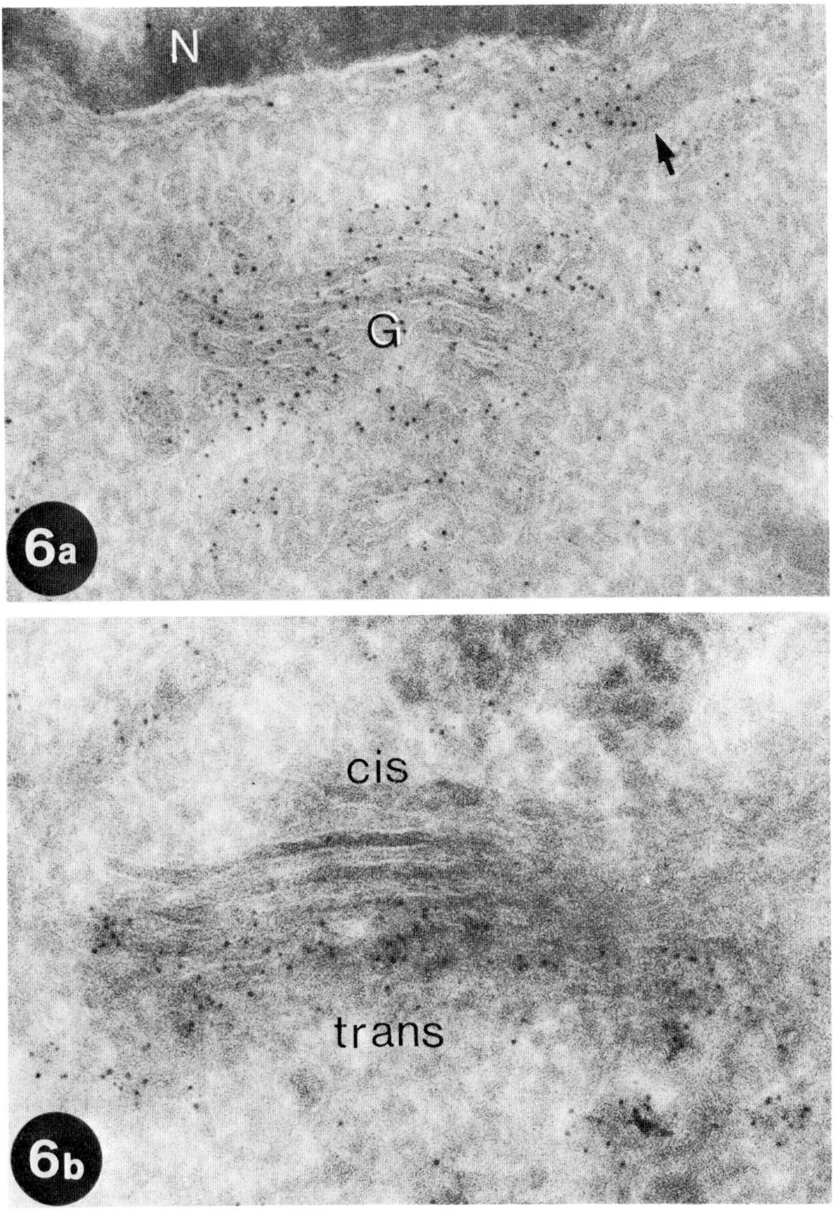

FIG. 6. (a) Frozen thin section of BHK cells infected with Semliki Forest virus and incubated with an affinity-purified antibody against the spike protein complex followed by protein A–colloidal gold. Labeling is seen uniformly over the Golgi (G) cisternae, as well as some over the nuclear envelope (N, nucleus). The labeled structure indicated with the arrow is a "naked capsid structure."[25] × 89,000. (b) Similar to (a) but labeled with *Ricinus communis* lectin followed by an antibody to the lectin and then protein A gold. The labeling is exclusively over the trans cisternae of the Golgi. ×105,000.

sugars in the Golgi stacks, but the precise location could not be determined.[44] One of the complex sugars added to the spike proteins is galactose. This sugar is specifically recognized by one of the plant lectins, *Ricinus communis* agglutinin I (RCA). Thin, frozen sections of infected cells labeled with RCA, followed by anti-RCA antibodies and protein A gold, gave only partial labeling of the intracellular transport pathway. The ER was unlabeled; the plasma membrane was heavily labeled. The transition in labeling occurred abruptly within the Golgi stack. Those cisternae in the trans half of the stack were labeled; the cis cisternae were unlabeled. After treatment with cycloheximide, resulting in loss of spike proteins from intracellular membranes, the level of RCA labeling dropped by about 50%. This shows that about half of the galactose revealed by RCA is on the transported spike proteins. We can also conclude that at least one of the complex sugars is added in trans Golgi cisternae.[25] Galactose is added by galactosyl transferase, which should also be present in trans Golgi cisternae. This has been confirmed using specific antibodies.[20]

3. The spike proteins move from cis to trans Golgi cisternae. The spike proteins pass through the Golgi stacks during intracellular transport and are present in all cisternae of the stack. RCA labels trans Golgi cisternae and approximately half of the lectin is bound to viral membrane proteins that have acquired complex sugars. If the movement of these proteins was from trans to cis Golgi, it would be difficult to explain why the proteins bearing complex oligosaccharides would suddenly lose their ability to bind RCA as they passed to the cis cisternae but subsequently regain the ability to bind at the plasma membrane. It is therefore reasonable to conclude that they move from cis to trans Golgi,[25] the direction generally assumed for other transported proteins.

4. The concentration of spike proteins in ER and Golgi membranes is extremely low. By radiolabeling infected cells with defined specific-activity amino acids, it is possible to show that a typical infected cell, 4–6 hr after infection, is synthesizing about 120,000 spike proteins per minute.[45] Quantitation on thin, frozen sections shows that the spike proteins spend on average about 15 min in both the ER membrane and Golgi stacks.[44] Hence at any time there are about 1.8×10^6 spike proteins in each of these membrane systems. The surface area of these membranes in infected cells has also been determined morphometrically so that the density of spike proteins can be calculated.[46] This amounts to about 90 and 750 spikes/μm^2 for ER and Golgi membranes, respectively. Since a typical concentration

[45] P. Quinn, G. Griffiths, and G. Warren, manuscript in preparation.

[46] G. Griffiths, G. Warren, P. Quinn, O. Mathieu, and H. Hoppeler, manuscript in preparation.

of spanning proteins in many biological membranes is about 30,000/μm^2,[45] it is clear that at no time during intracellular transport do the spike proteins constitute more than 0.14% and 1.14% of the total integral membrane proteins in ER and Golgi membranes, respectively. The completed viral particles contain about 20,000 spikes/μm^2. The low concentration of spike proteins in intracellular membranes ensures that budding does not take place there. For a typical plasma membrane protein, the low concentration would ensure that the ER and Golgi membranes did not behave at all like a plasma membrane.

Acknowledgment

We would like to thank Frank Booy for helping to write the section on the glass knife.

[38] Immunocytochemistry of Retinal Membrane Protein Biosynthesis at the Electron Microscopic Level by the Albumin Embedding Technique

By BARBARA G. SCHNEIDER and DAVID S. PAPERMASTER

Pathways of membrane protein biosynthesis and transport may be investigated at the electron microscopic level using antibodies as highly specific probes. Localization of intracellular antigens with antibodies and electron-opaque tags may be done on thin sections to eliminate tissue barriers to antigen exposure since all intracellular and extracellular compartments are accessible to antibodies applied to the surface of the section. In our studies of membrane biosynthesis in photoreceptor cells[1-3] we have applied antibodies against opsin and other membrane proteins to thin sections of retinas embedded in bovine serum albumin (BSA), a technique exploited by McLean and Singer[4] and Kraehenbuhl and Jamieson[5] in the study of intracellular antigens. The BSA penetrates between

[1] D. S. Papermaster, B. G. Schneider, M. A. Zorn, and J. P. Kraehenbuhl, *J. Cell Biol.* **77,** 196 (1978).

[2] D. S. Papermaster, B. G. Schneider, M. A. Zorn, and J. P. Kraehenbuhl, *J. Cell Biol.* **78,** 415 (1978).

[3] D. S. Papermaster and B. G. Schneider, *in* "Cell Biology of the Eye" (D. McDevitt, ed.), p. 475. Academic Press, New York, 1982.

[4] J. D. McLean and S. J. Singer, *Proc. Natl. Acad. Sci. U.S.A.* **65,** 122 (1970).

[5] J. P. Kraehenbuhl and J. D. Jamieson, *Int. Rev. Exp. Pathol.* **12,** 1 (1974).

ISBN 0-12-181996-5

cells, forming a supporting matrix but does not enter cells. Once infiltration is completed, the BSA is cross-linked to provide mechanical support of the fixed tissue for thin sectioning.

Fixation of Tissue

Frogs or toads are decapitated and pithed. Eyecups are immediately opened by slicing down with a razor blade on top of the eye along a line parallel to but slightly behind the corneal–scleral junction. The lens and anterior third of the eye are removed and the eyecup is flooded with fixative, either with 4% paraformaldehyde, 2% glutaraldehyde or a mixture of 4% paraformaldehyde and 2% glutaraldehyde in 0.1 *M* phosphate buffer, pH 7.4. After 1 hr of fixation at 20°, the eyecups are removed from the head and cut into crescents.

Embedding Protocol

The fixed tissue is rinsed thoroughly with several changes of 0.1 *M* phosphate buffer, pH 7.4, at 4°, then transferred to 30% (w/v) BSA (Armour Reheis, fraction V) in phosphate buffer. Commercial sources of albumin vary considerably in suitability for this technique. The tissue is incubated with a rocking motion in the albumin solution overnight for 16 hr at 4° to allow adequate infiltration. For dehydration, the tissue is transferred to a small open cylinder with its bottom covered with wet dialysis membrane. A plastic scintillation vial with the bottom cut away will serve. The dialysis membrane is held in place by a ring of rubber tubing. To embed cell suspensions or other samples of very small volume, a 400-μl microfuge tube with the tip sliced away is convenient. The tissue is placed on the dialysis membrane forming the bottom of the cylinder, and fresh BSA solution is added, 1.3 ml in a cylinder 1.1 cm in diameter. The cylinder, resting directly on the dialysis membrane, is placed upon Aquacide II (Calbiochem, La Jolla, California) and left overnight at room temperature in a desiccator over anhydrous calcium sulfate (Drierite). During desiccation, the BSA becomes further concentrated. After dehydration the top of the BSA layer may be liquid or firm, but it should not be overdried to the point of cracking.

The bulk of the dried Aquacide is gently scraped away from the bottom of the cylinder, without puncturing the dialysis membrane. The cylinder is then transferred to the second fixative to cross-link the BSA. It is submerged 1 cm into either 0.5% glutaraldehyde or freshly prepared 4% paraformaldehyde in 0.1 *M* phosphate buffer, pH 7.4 (a lower concentration of paraformaldehyde may be suitable and should be tested if antigens are destroyed by high concentrations of aldehydes). The cylinder can be supported by taping it to the side of the beaker containing the fixative. It may be prevented from floating by adding fixative above the BSA in the inside of the cylinder. The fixative beneath the cylinder is stirred to cross-link the BSA for 16 hr at 20°. The disk of cross-linked BSA containing the tissue is rimmed with a syringe needle to release it from the cylinder wall and moistened with phosphate buffer for ease in trimming. Excess BSA is discarded, and the tissue is cut into 1 mm^3 blocks and placed on filter paper to dry in a dessiccator. Blocks are mounted on Epon dummy blocks with cyanoacrylate glue (Krazy glue, Krazy Glue Inc., Chicago, Illinois).

Sectioning

Material embedded in BSA is sectioned nearly the same way as Epon-embedded material except that the water level in the boat of the diamond knife is kept very low. A wetting agent, such as saliva, applied to the edge of the diamond knife with a tapered pithwood stick will keep the knife edge wet despite the lower water level. After applying the wetting agent while the boat is full of water, the water is withdrawn carefully and the boat is rinsed vigorously three or four times by a stream of distilled water applied at the back of the boat so that the water flows over the knife edge. Only distilled H_2O stored in glass containers and aspirated with glass Pasteur pipettes should be used, since plastic bottles may leach components that form dark spots on some tissues. The boat is immediately refilled for sectioning at the low water level. Care must be taken that the block face does not become wet during sectioning because a wet block will immediately expand. If the block face becomes wet, sectioning must stop; a new block can be substituted while the wet block dries again in a desiccator. Block faces are trimmed to very small size, 0.5 mm or less across. A fast cutting speed, 20 mm/sec, is often helpful in reducing chatter. Curiously, the blocks change with age (weeks to months) so that sections of older blocks are less fragile. Sections are usually collected at thicknesses of 60–100 nm. Interference colors resemble Epon sections in the boat. Thinner sections have low contrast but simplify locating labeled regions; thicker sections reveal tissue structure more readily for illustration.

When the boat contains enough sections, the knife is retracted and water is added to the boat to form a convex surface. This simplifies positioning of sections for easier retrieval on carbon-coated Formvar grids. Grids with wide rims are employed (Pelco grids, Tustin, California), in order to reduce the tearing of Formvar with forceps tips during subsequent extensive handling in the staining steps. Thick sections (1 μm) may be heat-dried on glass slides and stained as one would prepare Epon-embedded material, so that orientation of the block may be discerned. Serial thin sections may be obtained. Before a labeling experiment is performed, a few sections should be examined in the electron microscope to check their quality and thickness. Once exposed to the electron beam, sections should not be used for labeling purposes.

Preparation of Reagents

If possible, antibodies should be affinity-purified to reduce nonspecific staining of tissues.[6] When care is taken in the preparation of reagents, background labeling is negligible (<4 ferritins/μm^2). When insufficient antigen is available or the antigen is not purified enough to prepare an affinity column, IgG fractions of antisera may be isolated by ion-exchange chromatography on QAE-Sephadex or DEAE-Sephadex (Pharmacia, Piscataway, New Jersey). Antibodies may be conjugated with biotin after isolation.[1,7,8] Freezing of antibody solutions at −70° in buffer containing 1% BSA is the preferred method of long-term storage. After thawing, antibody solutions are centrifuged for 15 min at 20° at 90,000 g_{av} in an Airfuge (Beckman Instruments, Palo Alto, California) or an ultracentrifuge (SW-50.1 rotor, Beckman). BSA solutions (4%) are centrifuged for 4 min at 4° at 12,000 rpm in a Microfuge (Beckman). Avidin-ferritin (AvF) is stored at 4° at 2 mg/ml in ferritin, but is centrifuged for 20 min at 18,000 rpm (JA-20 rotor, Beckman or SS34, Sorvall) and diluted to 0.03 mg/ml before use. Solutions of ferritin-containing reagents should never be frozen.

Labeling Protocol

One method of labeling employs two stages[1]: (1) first stage antibody, usually prepared in rabbits, (2) ferritin-sheep anti-rabbit IgG (or suitable antibody complementary to the first-stage antibody); or (1) biotinyl-antibody, (2) AvF. Amplification of label may be obtained with three stages[9]: (1) rabbit antibody (biotin conjugation optional), (2) biotinyl-sheep anti-rabbit IgG or anti-$F(ab')_2$, and (3) AvF. Sections are picked up on carbon-coated Formvar grids. Grids with torn Formvar are to be avoided because copper can react with labeling reagents. Grids are passed down lanes of

[6] D. S. Papermaster, this series, Vol. 81, p. 240.

[7] H. Heitzmann and F. M. Richards, *Proc. Natl. Acad. Sci. U.S.A.* **71,** 3537 (1974).

[8] E. A. Baer, E. Skultelsky, D. Wynne, and M. Wilchek, *J. Histochem. Cytochem.* **24,** 922 (1976).

[9] F. J. Roll, J. A. Madri, J. Albert, and H. Furthmayr, *J. Cell Biol.* **85,** 597 (1980).

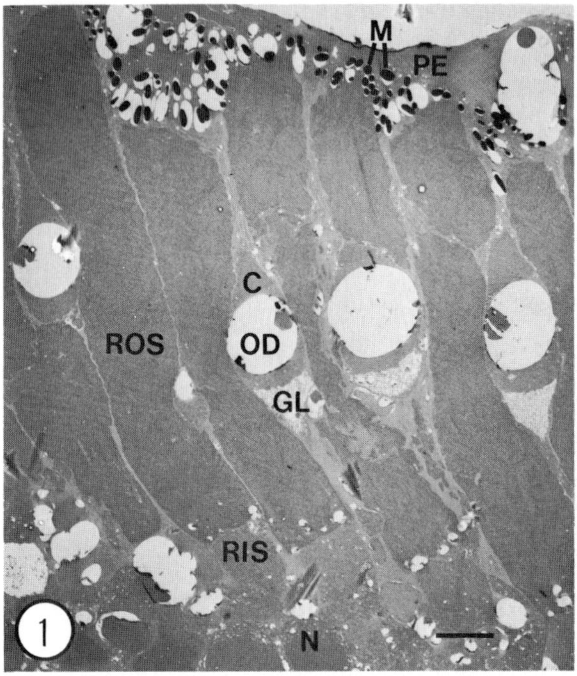

FIG. 1. *Xenopus laevis* tadpole retina embedded in bovine serum albumin (BSA). Outer segments of rods (ROS) and cones (C) appear dark because of the high density of ferritin bound to the anti-opsin labeled by the three-stage technique. Melanosomes (M) are retracted into the pigment epithelial cells (PE). Oil droplets (OD) of cone inner segments are not preserved by the aldehyde fixative and are lost during processing, leaving large holes. Cone inner segments also contain glycogen deposits (GL); both cone and rod inner segments (RIS) contain mitochondria, rough endoplasmic reticulum and Golgi (not resolved at this low magnification), and nuclei (N). This figure illustrates only a small portion of a section that contained over 40 rods. Thus quantitative data could be gathered from sufficient numbers of cells to evaluate variability of labeling. Higher magnification images are chosen to demonstrate the average labeling density in each cellular domain rather than the occasional exceptional result. ×1500; bar = 5 μm.

drops of reagents (30 μl per drop for three grids) on Parafilm strips at 20°. During incubations of 10 min or more, the drops are covered by small petri dishes to minimize evaporation. During incubation on antibody drops, grids are rotated several times on top of each drop so that sections on the edges of grids contact the labeling solutions thoroughly. Uniformity of label across the entire grid should be examined by quantitating the labeling density of sections in all locations on the grid so that inhomogeneous labeling is avoided in the analysis (Fig. 1).

Three-Stage Labeling Sequence

Tris-HCl buffer, 0.1 *M*, pH 7.4, 5 min
4% BSA in Tris-HCl buffer, 10 min
Glycine, 0.02 *M*, in Tris-HCl buffer, 15 min
First-stage antibody containing 1% BSA, 15 min
Tris-HCl buffer wash, 6 times, 1 min
Second-stage antibody: affinity-purified biotinyl-sheep anti-rabbit $F(ab')_2$ or anti-rabbit IgG, 0.1 mg/ml, containing 1% BSA, 15 min
Tris-HCl wash, 6 times, 1 min
Third stage: AvF, 0.03 mg/ml (in ferritin), 30 min
Tris-HCl wash, 6 times, 1 min
Phosphate buffer wash, 0.1 *M*, pH 7.4
Glutaraldehyde, 2% in phosphate buffer
H_2O, 6 times, 1 min

Sections may be stained with uranyl acetate and ferritin density be enhanced with bismuth subnitrate.[10] Lead citrate may also be used, but it may produce an excessively grainy image. Two-stage labeling is conducted in the same way except that the first- and second-stage reagents are chosen as described above and the third stage is eliminated.

Control of Background Labeling

We have found the following conditions to be advantageous in lowering background labeling.

1. Incubation of grids in 4% BSA solution prior to incubation with antibodies.
2. Presence of 1% BSA in each antibody solution.
3. Centrifugation of reagents prior to use to remove aggregates.
4. Adequate dilution of antibody solutions. With each new antibody, labeling patterns of sections stained with serial dilutions of antibody solutions should be compared. The antibody concentration that generates the highest labeling density with lowest background labeling is selected. It is important to note, however, that, as a polyclonal antibody solution is diluted, antibodies to minor components or contaminants may become diluted beyond the point of detection by the second- and third-stage reagents. The relative labeling density of "specific" and "nonspecific" sites should be compared quantitatively before drawing conclusions concerning the specificity of reaction. One approach to this evaluation is to compare the relative affinity of binding at each site with Langmuir adsorption

[10] S. K. Ainsworth and M. J. Karnovsky, *J. Histochem. Cytochem.* **20,** 225 (1972).

isotherm double-reciprocal plots of the percentage of maximum labeling density vs the antibody dilution.[11]

Conditions found to yield high levels of background at the electron microscope level include the following: (*a*) lyophilization of antibodies; (*b*) use of whole antiserum for labeling; (*c*) isolation of antibodies by $(NH_4)_2SO_4$ precipitation; (*d*) the use of aggregated antibodies.

Quantitation

Areas of labeled sections are selected either randomly or systematically,[12] and images are recorded in an electron microscope at 60 kV with a small aperture to enhance contrast at a minimal primary magnification of 10,000×. Calibration of magnification is achieved by photographing a carbon replica grating calibration grid (E. F. Fullam, Schenectady, New York). We gather data on 35 mm film at 10,000× and calibrate each film strip at the beginning in order to calculate the exact enlargement of the final image in the viewer. The use of 35 mm film greatly lowers the cost and eases the gathering and recording of data since printing on paper is virtually eliminated. The 35 mm negative strips or large-format negatives are contact printed on the same film using a long (6 ft) contact printer. The contact positive is viewed either with a rear-image projector[13] or on a Bellco plaque viewer (Bellco Glass, Inc.; Vineland, New Jersey), which has the capability of magnifying the image 6.5-, 9-, 13-, 17.5-, or 50-fold. The rear image projection screen is overlaid with a transparent acetate sheet lined with a double lattice grid system (Fig. 2). With the plaque viewer the image is viewed on a photocopy of the test grid. Ferritin grains falling in every 16th test square (marked on the lattice by a $\frac{3}{4}$ circle) are recorded. Ferritin density/μm^2, $N_{F(A,c)} = N_F/d^2P_c$, where N_F is the sum of ferritin counts in each of the small squares, d is the lattice spacing of the small square (in μm corresponding to the magnification of the image), and P_c is the number of lattice points falling in the compartment of interest.[14,15] Direct recording of the particles per square on an Apple II Plus computer with the VisiCalc program (VisiCorp, Sunnyvale, California) or Hewlett–Packard HP67 simplifies data storage and computation.

[11] D. S. Papermaster, P. Reilly, and B. G. Schneider, *Vision Res.* **22,** 1417 (1982).

[12] E. R. Weibel, "Stereological Methods," Vol. 1. Academic Press, New York, 1979.

[13] E. R. Weibel, *in* "Principles and Techniques of Electron Microscopy" (M. A. Hayat, ed.), Vol. 3, p. 237. Van Nostrand-Reinhold, Princeton, New Jersey, 1973.

[14] J. P. Kraehenbuhl, E. R. Weibel, and D. S. Papermaster, "Quantitative Immunocytochemistry at the Electron Microscope" (W. Knapp, K. Holubar, and G. Wick, eds.), p. 245. Elsevier/North-Holland Biomedical Press, New York, 1978.

[15] J. P. Kraehenbuhl, L. Racine, and G. Griffiths, *Histochem. J.* **12,** 317 (1980).

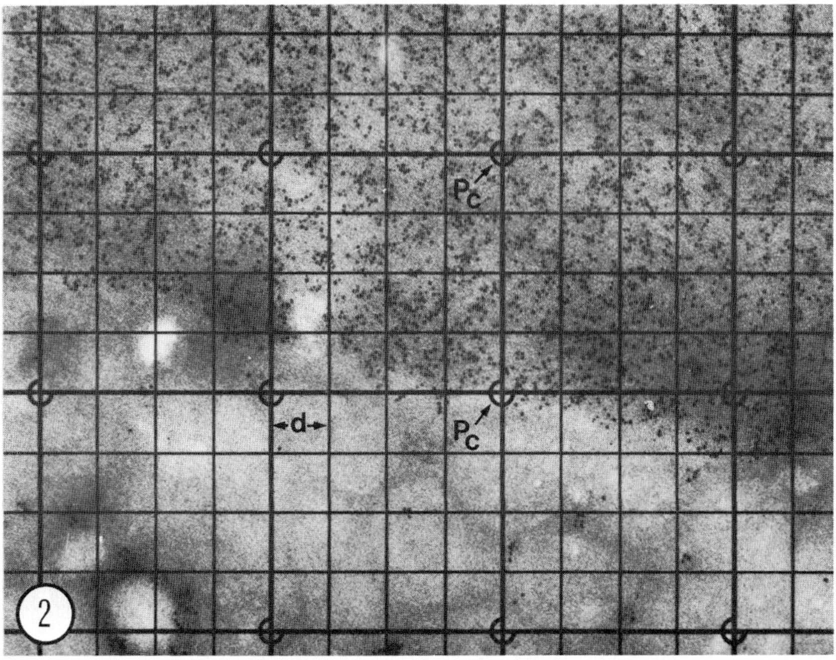

FIG. 2. Junction of rod outer segment and inner segment in a *Xenopus laevis* retina labeled with anti-opsin and ferritin conjugates by the three-stage technique. The micrograph is overlaid with a double-lattice test system for quantitation of ferritin densities. The $\frac{3}{4}$ circles indicate both the small squares within which ferritin grains are counted and the lattice points, P_c, for estimating the area occupied by the compartments of interest in the cell. The small square has a length, d, in micrometers corresponding to the magnification of the micrograph. Note that our definition of d is $\frac{1}{16}$th the d value of Kraehenbuhl *et al.*[14,15] This value can be selected for convenience of counting so that sufficient area is included to avoid squares that are empty of ferritin. ×64,000.

A comparison of labeling densities of anti-opsin detected by avidin-ferritin on tissue embedded in BSA and Lowicryl K4M and in BSA and frozen sucrose indicated a remarkable concordance of labeling levels. For example, a comparison of anti-opsin labeling over thin sections of rod outer segments labeled by the two-stage protocol generated the following results: BSA, 989 ± 43, n = 13; frozen sucrose, 964 ± 24, n = 8.

Choice of Method

Cross-linked BSA is one of several hydrophilic embedding media currently used for postembedding labeling. Tissues may also be embedded in

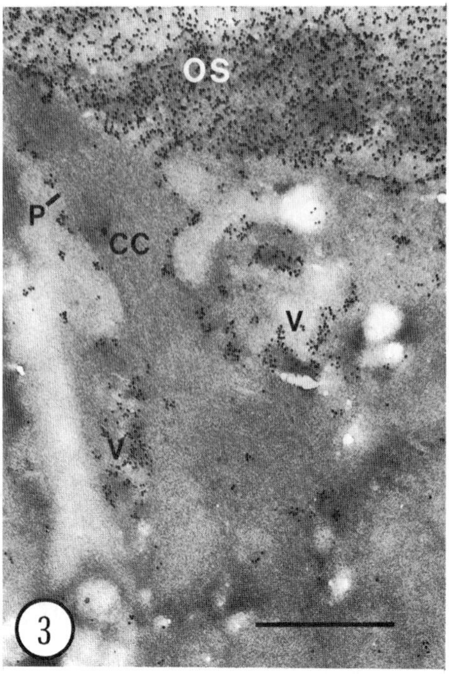

FIG. 3. A connecting cilium (CC) links the rod outer segment (OS) to the inner segment in a *Xenopus laevis* tadpole retina labeled with anti-opsin and ferritin conjugates by the three-stage technique. Opsin-bearing vesicles (V) cluster beneath the base of the CC. The plasma membrane (P) of the CC as well as the disk membranes of the OS are labeled. ×37,000; bar = 0.5 μm.

frozen sucrose[15–17] or in Lowicryl K4M (Polysciences, Warrington, Pennsylvania), a methacrylate resin.[18] Factors to be considered in choosing a method include the sensitivity of the antigen to the effects of fixation and embedding, degree of preservation of structure, cost of the method, and type of antibody to be used. With the increased use of monoclonal antibodies specific to single antigenic determinants, false negatives in labeling experiments at the electron microscope level have occurred more frequently than in experiments employing polyclonal antibodies. The use of monoclonal antibodies may impose upon the antigen a much lower tolerance of denaturation, cross-linking, or chemical modification by the

[16] K. T. Tokuyasu and S. J. Singer, *J. Cell Biol.* **71,** 894 (1976).
[17] G. Griffiths, K. Simons, G. Warren, and K. T. Tokuyasu, this volume [37].
[18] E. Carlemalm, M. Garavito, and W. Villinger, *J. Microsc.* (*Oxford*) **126,** 123 (1982).

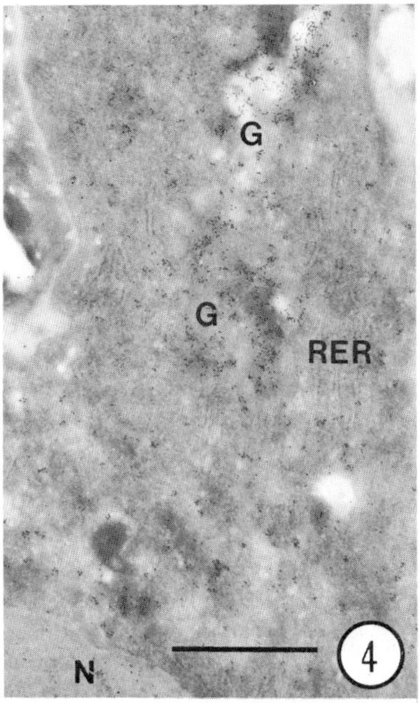

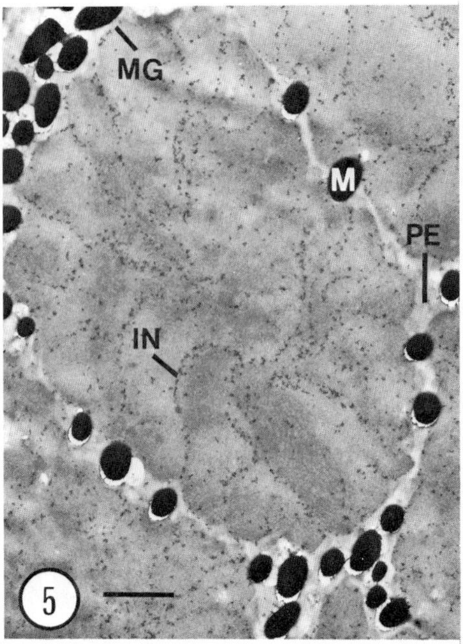

FIG. 4. Golgi region (G) in the inner segment of a *Xenopus laevis* rod cell is heavily labeled with anti-opsin by the three-stage technique. The Golgi apparatus of this cell lies near the nucleus (N) and is elongated axially in a pale gray area from which rough endoplasmic reticulum (RER) is excluded. The RER membranes are outlined by the negatively stained dense ribosomes. In sections of inner segments with aligned RER, anti-opsin is bound along the membranes of the cisternae.[3] ×37,000; bar = 0.5 μm.

FIG. 5. Cross section of bovine serum albumin-embedded rod outer segment of *Rana pipiens* retina labeled with antibodies to the large protein (M_r 290,000) by the three-stage technique. Antibody is bound along the deep incisures (IN) and the disk margins (MG). The interspersed processes of the pigment epithelium (PE) that contain melanosomes (M) are labeled at background levels (<4 ferritins/μm^2). ×9200; bar = 1 μm.

fixative and embedding medium than the tolerance that is acceptable in experiments with polyclonal antibodies.

One of the major advantages of this technique is that sections and blocks appear to be stable indefinitely. Consequently, sections from numerous blocks can be labeled in a single experiment if desired. Moreover, large numbers of grids can be prepared in advance and stored for subsequent staining or for use in comparative studies. Additional advantages include high retention of antigenicity with numerous cytoplasmic and membrane antigens tested so far, ease of orientation of blocks, ease of production of serial sections, relatively rapid production of sections and exposure of tissue only to aldehydes and drying but to no organic chemicals or heating. These advantages make possible complex labeling experiments involving, for example, sections from many different blocks handled simultaneously under identical labeling conditions.

Disadvantages of the BSA embedding technique include the time required for processing, fragility of sections, especially from fresh blocks, a tendency for sections from freshly prepared tissue blocks to contain chatter, poor delineation of some membranes when cytoplasm abutting the membrane has equivalent density, and the inability to examine extremely lightly fixed tissue.

Application in Analysis of Opsin Biosynthesis in the Retina

The vertebrate photoreceptor cell is a highly compartmentalized sensory neuron; light capture is a function of the opsin-laden membranes of the rod outer segment (ROS) while the biosynthetic apparatus is confined to the adjacent compartment, the inner segment. Labeling of BSA-embedded retinas with affinity purified anti-opsin antibodies[6] demonstrates the high density of opsin in the ROS, as expected (Figs. 1 and 3). In the inner segment, which is connected to the outer segment by a connecting cilium, the mitochondria are unlabeled, but small vesicular profiles between them and clustered beneath the cilium are labeled at high density. The upper part of the ciliary plasma membrane is also labeled (Fig. 3). Closer to the nucleus, the Golgi apparatus is also densely labeled (Fig. 4). At the dilution used in this study, the rough endoplasmic reticulum is only lightly labeled, but at higher concentration, it too is labeled along the edges of the cisternae.[3] These results have been correlated with the sequential appearance of radiolabeled protein after short exposure to labeled amino acids or mannose to support a proposed pathway of transport of opsin from RER to Golgi, then on vesicles to the apical plasma membrane of the inner segment. Subsequently, the plasma membrane of the cilium becomes the conduit to the emerging disks at the base of the outer segment.[3,19] Antibodies to the other intrinsic membrane protein of ROS (M_r 290,000) have localized it to the incisures and margins of rod disks and cone lamellae (Fig. 5). Since there are between 1000 and 3000 molecules of this large protein on each disk whose surface area is 40 μm^2 on each side (cf. rhodopsin, 10^6 molecules per disk), the technique clearly has the sensitivity to localize minor membrane components.[2,11]

Acknowledgments

We wish to thank Jean Pierre Kraehenbuhl and Ewald Weibel and L. Racine for their continuing interest and collaboration in applying this technique. Research was supported in part by the NIH Grants EY-03239 and EY-00845 and by the Veterans Administration.

[19] B. G. Schneider and D. S. Papermaster, *Invest. Ophthalmol. Visual Sci.* **20,** Suppl. 1, 76 (1981).

[39] Expression of Viral Membrane Proteins from Cloned cDNA by Microinjection into Eukaryotic Cell Nuclei

By BEATE TIMM, CLAUDIA KONDOR-KOCH, HANS LEHRACH, HEIMO RIEDEL, JAN-ERIK EDSTRÖM, and HENRIK GAROFF

The relationship between a protein's structure and function can be explored in detail by *in vitro* mutagenesis of a cloned DNA molecule encoding the protein and expression of its mutagenized form in a eukaryotic cell.[3,4] This approach should be ideal to study the biosynthesis and intracellular transport of viral and plasma membrane proteins, especially as several of the genes encoding these proteins have been cloned.[1,2,5-10]

For instance, the function of the signal sequence can be studied by adding a signal peptide to the amino terminus of a cytoplasmic protein (e.g., viral capsid proteins and hemoglobin molecules) through genetic manipulation. If the cytoplasmic protein becomes translocated when equipped with the signal sequence, it would unequivocally show that this peptide is *alone* sufficient for the segregation of a polypeptide chain.

The membrane-binding requirements for a polypeptide can be analyzed by mutating the membrane-spanning portion of a membrane protein. The effect of deletions and point mutations in this region should characterize the "anchoring" peptide in detail.

Does the signal for correct intracellular routing of spanning membrane proteins reside in their cytoplasmic or extracytoplasmic domain? This question can be answered by following the cellular distribution of a membrane protein that through gene manipulation is modified so that it is lacking the cytoplasmic domain either completely or partially, or this

[1] H. Garoff, C. Kondor-Koch, and H. Riedel, *Curr. Top. Microbiol. Immunol.* **99,** 1 (1982).

[2] D. D. Sabatini, G. Kreibich, T. Morimoto, and M. Adesnik, *J. Cell Biol.* **92,** 1 (1982).

[3] R. C. Mulligan, B. H. Howard, and P. Berg, *Nature (London)* **277,** 108 (1979).

[4] M.-J. Gething and J. Sambrook, *Nature (London)* **293,** 620 (1981).

[5] A. G. Porter, C. Barber, N. H. Carey, R. A. Hallewell, G. Threlfall, and J. S. Emtage, *Nature (London)* **282,** 471 (1979).

[6] W. Min Jou, M. Verhoeyen, R. Devos, E. Saman, R. Fang, D. Huylebroeck, W. Fiers, G. Threlfall, C. Barber, N. Carey, and S. Emtage, *Cell* **19,** 683 (1980).

[7] H. Garoff, A.-M. Frischauf, K. Simons, H. Lehrach, and H. Delius, *Nature (London)* **288,** 236 (1980).

[8] C. Kondor-Koch and H. Garoff, *J. Gen. Virol.* **58,** 443 (1982).

[9] J. K. Rose, W. J. Welch, B. M. Sefton, F. S. Esch, and N. C. Ling, *Proc. Natl. Acad. Sci. U.S.A.* **77,** 3884 (1980).

[10] M.-J. Gething, J. Bye, J. Skehel, and M. Waterfield, *Nature (London)* **287,** 301 (1980).

ISBN 0-12-181996-5

domain has been exchanged with that of a protein belonging to another membrane.

To address a question through this approach, a series of mutants must be constructed and analyzed for expression. Therefore the expression system must be both fast and easy to use. Further, the system should be applicable to a large variety of cell types including differentiated ones. We describe here an expression system that meets many of the requirements. The system has been tested with the cDNA encoding the structural proteins of Semliki Forest virus (SFV).[11]

The Expression System

The expression system is shown schematically in Fig. 1. It involves (*a*) the engineering of the cDNA encoding the membrane protein into a eukaryotic expression vector; (*b*) cloning of the recombinant DNA molecule in *Escherichia coli;* (*c*) extraction of the plasmid molecules from the bacteria; (*d*) introduction of the DNA into the nucleus of the cell by injection with a micropipette; and (*e*) analysis for viral proteins expressed from the cDNA using indirect immunofluorescence.

The expression vector we are using has been developed by Mulligan and Berg.[12] It consists of a 2.3-kilobase *Eco*RI-*Pvu*II restriction endonuclease fragment of the *E. coli* plasmid pBR 322 containing the origin for plasmid replication and the gene for ampicillinase resistance, and fragments of the SV40 genome. The *Pvu*II-*Hin*dIII fragment of simian virus 40 DNA (SV40) (coordinates 0.65–0.71) contains the promoter for the early SV40 transcripts. A second composite fragment of SV40 DNA, between the *Bgl*II and the *Eco*RI site, contains the small T intron (SV40 coordinates 0.44–0.56) and the polyadenylation signal for SV40 early transcripts (coordinates 0–0.19). The DNA between the *Hin*dIII and the *Bgl*II contains cDNA of rabbit β-globin, which will be replaced by that encoding the membrane protein. The plasmid sequences direct replication of the DNA molecule in the prokaryotic cell. The SV40 promoter ensures efficient transcription of the cDNA into an RNA molecule that contains necessary signals for its processing into a stable translatable form when introduced into the nucleus of the eukaryotic cell.

The cDNA is most conveniently cloned between a *Hin*dIII and a *Bgl*II site in any suitable cloning vector (including the expression vector) with its 5′ end at the *Hin*dIII site. After cleavage of the cloning vector with these enzymes, the cDNA molecule can be isolated and inserted directly

[11] C. Kondor-Koch, H. Riedel, K. Söderberg, and H. Garoff, *Proc. Natl. Acad. Sci. U.S.A.* **79,** 4525 (1982).

[12] R. C. Mulligan and P. Berg, *Science* **209,** 1422 (1980).

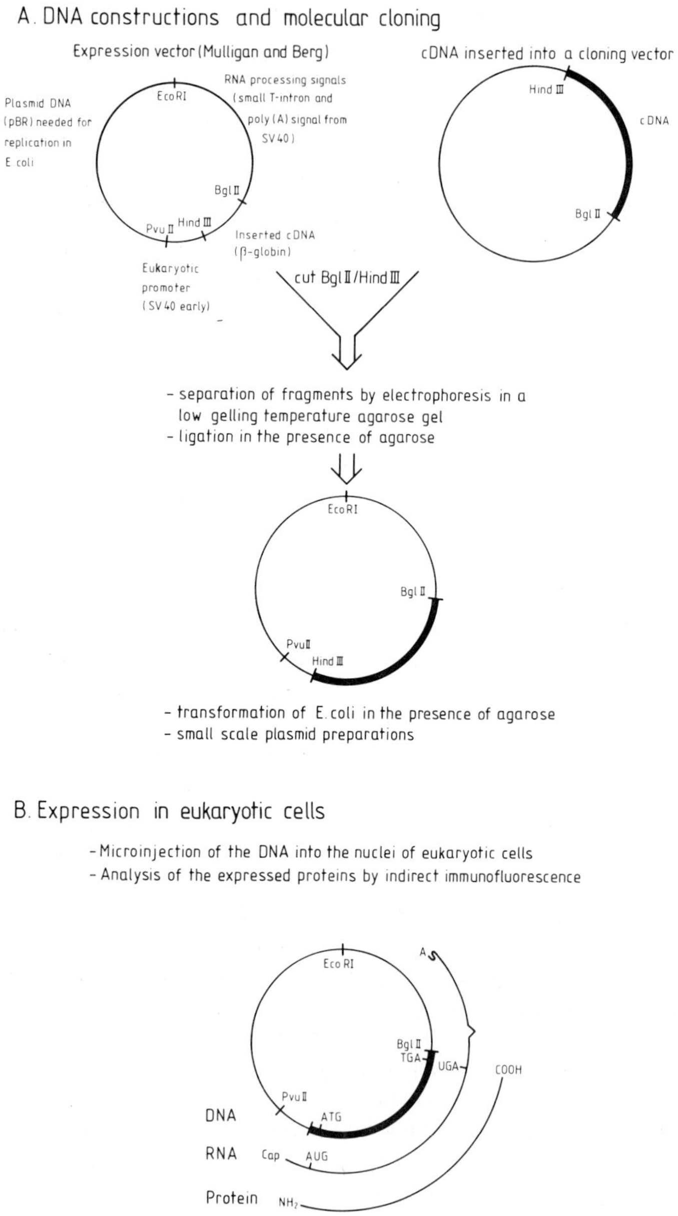

FIG. 1. Scheme of expression system.

in the sense orientation with respect to the SV40 promoter in the similarly cleaved expression vector molecule. We have found that both the ligation of the cDNA and the expression vector molecules as well as the subsequent transformation of *E. coli* can be done efficiently in the presence of melted low gelling temperature agarose, thus avoiding tedious DNA isolation steps (see Procedures, below). For the expression analysis the plasmid DNA is extracted from small bacterial cultures using our modification of the simple boiling procedure described by Holmes and Quigley.[13] This procedure allows up to 30 samples to be processed during a single working day.

Introduction of the DNA directly into the nucleus of the eukaryotic cell with the micropipette has several advantages.[14] First, the DNA is present in the nucleus in a high copy number. Second, the technique allows the expression to be studied probably in any cell that can be injected. We have so far demonstrated the SFV proteins in baby hamster kidney (BHK) cells, Madin–Darby canine kidney cells, and pig kidney cells that have been microinjected with the expression vector containing SFV cDNA. Third, the procedure is very rapid; some 100 cells can be injected in about 10 min.

The most convenient analysis for expressed proteins in the microinjected cells is indirect immunofluorescence. This technique is very sensitive, and it shows whether the protein is located in the cell cytoplasm, endoplasmic reticulum, or Golgi complex or on the cell surface. Figure 2a–e shows the structural proteins of SFV when expressed from cDNA that has been microinjected into BHK cells. The capsid protein is seen as a diffuse "cytoplasmic" staining in the cell cytoplasm (Fig. 2a,b) whereas the membrane proteins are associated with the endoplasmic reticulum (Fig. 2c), the Golgi complex (Fig. 2d) and the plasma membrane (Fig. 2e). The SFV proteins are detectable already 1 hr after microinjecting the DNA; they reach their highest concentration after about 10 hr, and they can gradually decrease in amount so that they are barely detectable some 24 hr after injection.[11]

Procedures

Ligation Reaction and Transformation in the Presence of Low Gelling Temperature Agarose

Principle

Low gelling temperature agarose has the property of melting already at 68° and remaining in the melted form at room temperature (22°) when

[13] D. S. Holmes and M. Quigley, *Anal. Biochem.* **114,** 193 (1981).
[14] M. R. Capecchi, *Cell* **22,** 479 (1980).

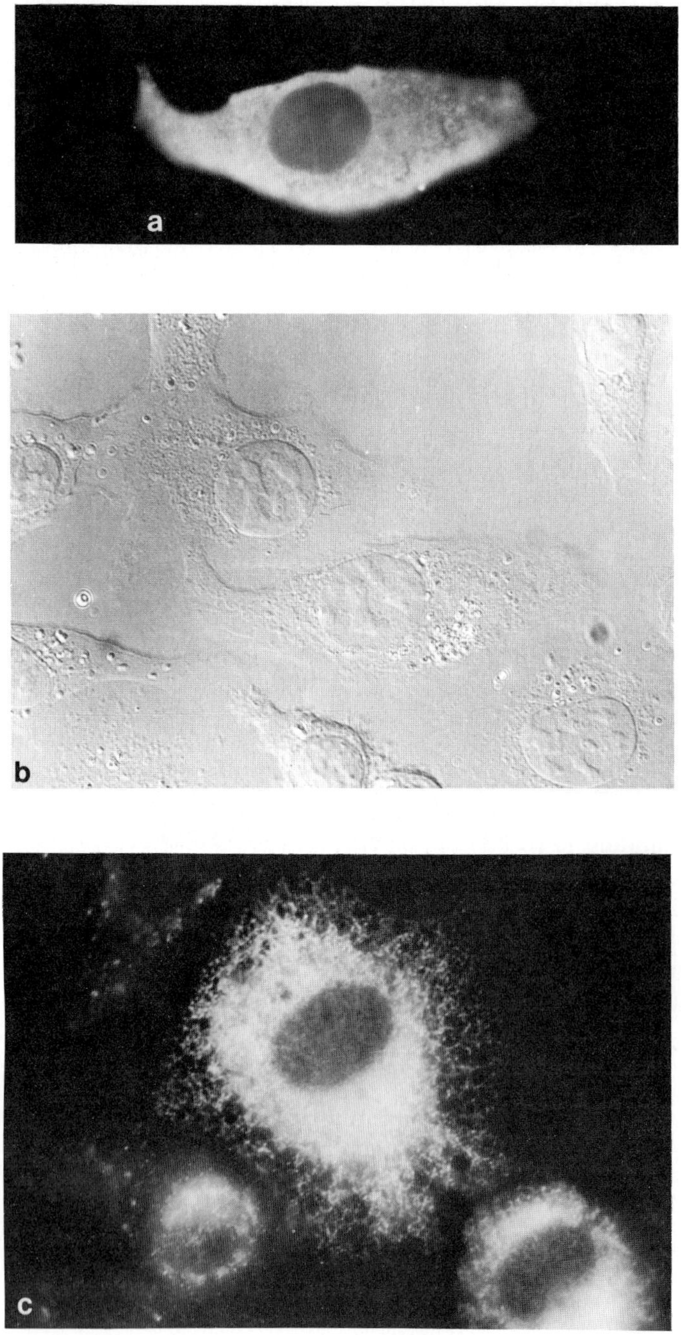
a
b
c

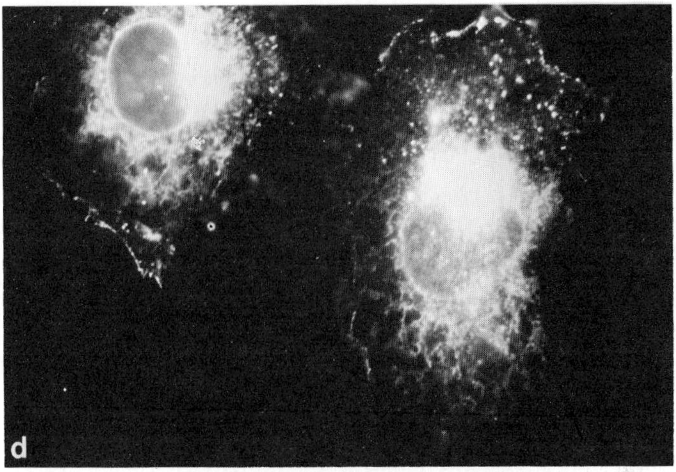

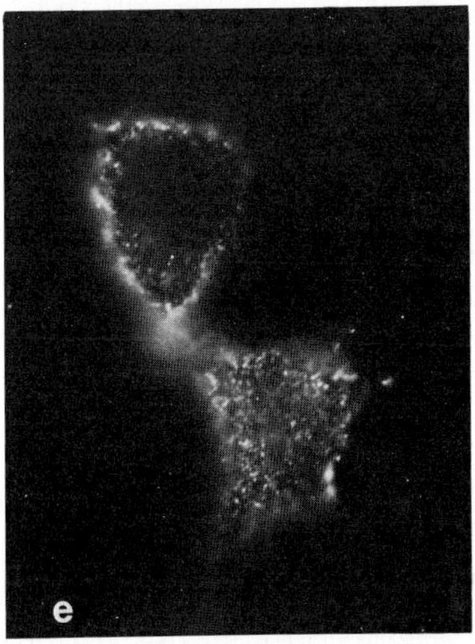

FIG. 2. Intracellular staining for (a) capsid protein, (c) E1 membrane protein, and (d) p62 membrane protein (the intracellular precursor for membrane proteins E2 and E3) in cells that have been microinjected with cDNA encoding the three structural proteins of SFV. (e) Surface staining for the E2 protein. All cells have been incubated for 6 hr after injection. (b) The cells in (a) when observed with Nomarsky optics. The capsid protein is seen as a diffuse cytoplasmic staining suggesting that this protein is in the cell cytoplasm. The membrane proteins are associated with membranes; E1 shows up in the reticular network of the rough endoplasmic reticulum; p62 is seen in addition as a strong perinuclear fluorescence, which probably represents the Golgi region, and on the plasma membrane. ×1300.

diluted to concentrations below 0.2% (w/v). These characteristics make it possible to use this agarose for normal electrophoretic separation of DNA molecules, cut out pieces of the gel with relevant DNA fragments, melt the agarose, dilute it, and perform further enzymic DNA-processing steps directly in the presence of the melted agarose.

Materials and Solutions

Low gelling temperature agarose (Bio-Rad Laboratories)
Low-salt buffer for agarose gel: 40 m*M* Tris, pH 7.8, 5 m*M* sodium acetate, 1 m*M* EDTA
Agarose gel: 0.8 g of agarose in 100 ml buffer molded into a 120 mm × 110 mm gel with 15 slots (20 μl volume)
TE: 10 m*M* Tris, pH 7.4, 1 m*M* EDTA
ATP solution: 100 m*M* in water
Restriction endonucleases *Hin*dIII and *Bgl*II (New England BioLabs)
T4 DNA ligase (Bethesda Research Laboratories)
Ligase buffer 5 times concentrated: 250 m*M* Tris, pH 7.6, 50 m*M* magnesium chloride, 50 m*M* dithiothreitol

Separation and Ligation of DNA fragments in Low Gelling Temperature Agarose

About 1 μg of the expression vector and the cDNA-cloning vector DNA are each cut with restriction endonucleases *Hin*dIII and *Bgl*II. The restricted DNA of the expression vector is applied into one slot of a 0.8% low gelling temperature agarose gel containing 0.5 μg of ethidium bromide per milliliter and run for about 1 hr at 5 V/cm before the other DNA sample is applied into the same slot (use 366-nm UV light to visualize the ethidium bromide–DNA complexes). The run is continued until the smaller cDNA molecule catches up to the larger vector molecule in the gel. If the cDNA molecule is longer than the vector molecule, then the samples are applied in the reverse order. The single band containing both molecules are carefully cut out from the gel with a scalpel (volume of gel piece is approximately 40 μl), diluted in a 1.5 ml Eppendorf tube with 4 volumes of water (200 μl), heated to 68° for 5 min, mixed by vortexing, and put into a 37° incubator. The agarose is now completely melted, and the DNA molecules can be joined together with T4 DNA ligase.

Forty-five microliters of concentrated ligase buffer, 2.5 μl of ATP solution, and 3 units of T4 DNA ligase are added, and the mixture is incubated first for 2 hr at 37° and then at room temperature for 8–10 hr. The ligation mixture is still liquid at room temperature. A 5% aliquot of the sample is analyzed for ligation products on an agarose gel after incubation.

Transformation in the Presence of Agarose

For transformation, about one-third (about 80 μl) of the ligation mixture is precipitated in ethanol, dried briefly, and taken up in 50 μl of TE. The precipitation is done by adding sodium acetate (pH 6.0) to 0.3 *M* and 3 volumes of ethanol, and then cooling in Dry Ice for 15 min before spinning in a Microfuge (approximately 12,000 *g*) for 10 min. Most of the agarose coprecipitates with the DNA; however, all material goes again into solution in TE when heated at 68° for 1–2 min with intermittent mixing using a vortex. The mixture is put into a 37° incubator, where it remains liquid. Two hundred microliters of freshly thawed competent cells are added to the sample, and it is incubated first in ice for 30 min and then further at 37° for 90 sec before plating.

Small-Scale Plasmid Preparation

Materials and Solutions

- Disposable loops (1 μl) to collect bacteria (Nunc, Roskilde, Denmark)
- STET buffer: 50 m*M* Tris, pH 8.0, 50 m*M* EDTA, 8% (w/v) sucrose, 5% (w/v) Triton X-100
- Lysozyme solution: 20 mg per milliliter of water, prepared fresh each time
- RNase solution, 20 mg per milliliter of water, boiled for 5 min
- DEPC solution, 10% (w/v) in ethanol
- Ammonium acetate solution, 5 *M* in water

Extraction of Plasmid DNA

Each transformed colony to be analyzed is streaked out on one half of a petri dish containing growth medium in agarose and grown to semiconfluency (usually overnight). The bacteria are carefully collected using a disposable loop and suspended in 1 ml of ice cold STET buffer in a 1.5-ml Eppendorf tube by briefly mixing. Ten microliters of lysozyme solution are added, and the bacteria are lysed by incubation at 95° for 7 min. A large clot of insoluble bacterial debris and genomic DNA is formed upon cooling of the sample, and it is removed by centrifugation for 30 min in a microfuge. The supernatant (0.8–1 ml), which contains most of the plasmid DNA together with tRNA, is transferred to a 2.5-ml tube for digestion

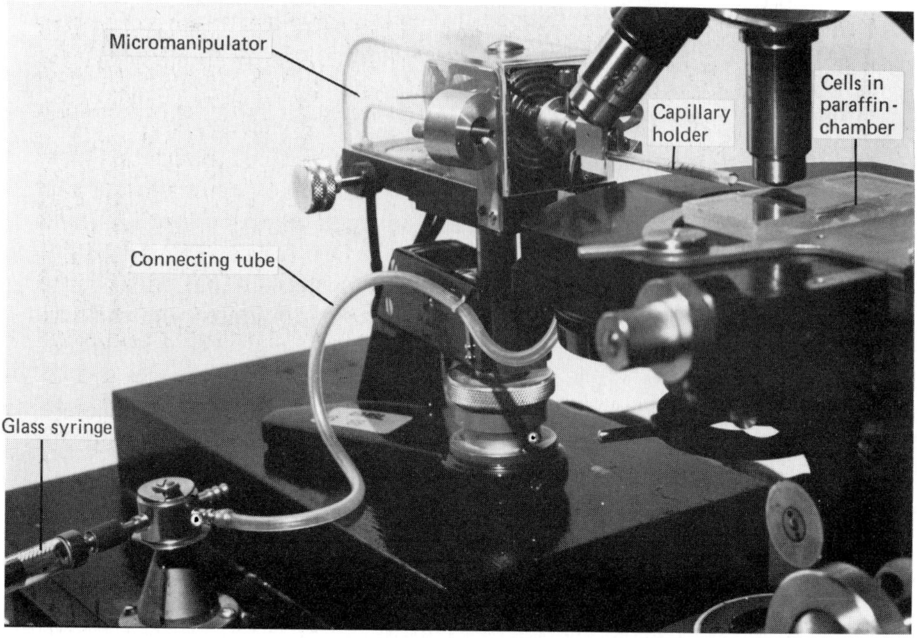

FIG. 3. The setup for microneedle injection of cell nuclei.

of RNA. One microliter of RNase solution is added, and the mixture is incubated for 15 min at 37°. The RNase is inactivated by adding 1 μl of DEPC solution and incubating for 15 min at 37°. The DEPC is destroyed by heating the sample for 10 min at 65°. Ammonium acetate (5 M; 0.5 ml) is mixed with the supernatant followed by 1.0 ml of isopropanol, and the DNA is precipitated by keeping the sample in Dry Ice for 15 min. The precipitate is recovered by centrifugation in the microfuge and taken up into 50 μl of TE.

Microinjection with a Micropipette

Principle

Microinjection of cell nuclei is done under a phase contrast microscope with a glass capillary, the tip of which forms a 45-degree angle with the body part (Figs. 3 and 4). The cells lie upside down on a coverslip in a paraffin oil chamber and are injected from below by moving the capillary upward against the cells (see Figs. 6 and 7). A micromanipulator is used to

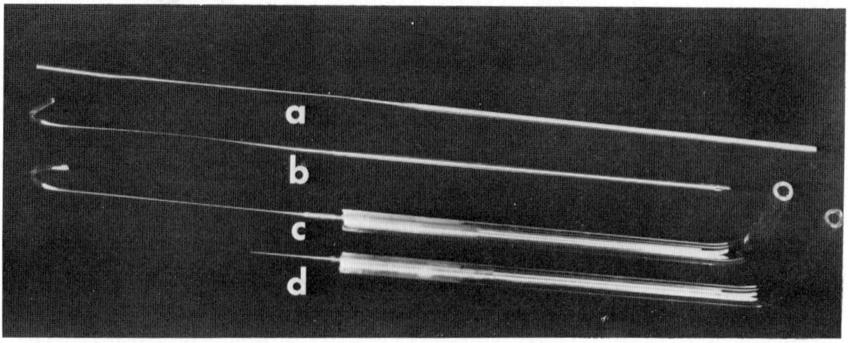

FIG. 4. Steps in preparing a microneedle. (a) Drawn-out microcapillary; (b) the hook for the weight is made; (c) the microcapillary is inserted in its holder and fixed in place with paraffin; (d) the ready microcapillary with its holder. The tip is not visible at this magnification.

control capillary movements and a 5-ml glass syringe to inject the fluid (Fig. 3). The procedure is essentially that of de Fonbrune.[15]

The cells (covered with a thin layer of medium) and the sample droplet are effectively protected from both evaporation and infection by paraffin oil during the whole microinjection procedure. Furthermore the system offers a superior visual control of the cells and all micromanipulations, as these are inspected directly through the thin coverslip with the cells.

Instruments, Materials, and Solutions

Microscope, phase contrast, any model in which focusing is performed by monitoring the microscope tube and that can be equipped with a long-distance focus phase-contrast condenser. We have been using a Zeiss Standard microscope.

Micromanipulator and Microforge of the de Fonbrune model[15] (Bachofer Laboratorium Geräte and Glas Technik, D-7410 Reutlingen, Postfach 7089, Federal Republic of Germany).

Syringe for injections, A 2-ml Inaltera glass syringe connected with a Luer-lock attachment to a three-way stopcock (both obtained from Henke-Sass Wolf GmbH, Kronenestrasse 16, No. 2, Tuttlingen, Federal Republic of Germany). The stopcock plug is fastened onto a cone-shaped brass stand fixed to a wooden platform on which both

[15] P. de Fonbrune, "Technique de micromanipulation." Monographes de l'Institut Pasteur, Masson, Paris, 1949.

the microscope and the micromanipulator are placed. When the syringe is moved sideways, different outlets will be connected with the syringe, one for a rubber tube connected to the micropipette, another one for intake or expulsion of air (Fig. 3).

Glass capillary (10 cm long, outer diameter 1 mm, wall thickness 0.1 mm) for preparation of micropipette (Clark Electromedical Instruments, P.O. Box 8, Pangbourne, Reading RG8 7HU, England)

Glass capillary (10 cm long, outer diameter 4 mm, wall thickness 1 mm) for micropipette holder

Coverslip, 24 mm × 32 mm for cells; and 6 mm × 32 mm for sample (AB Termoglas, Box 14137, S-40020 Göteborg, Sweden)

Oil chamber, a piece of glass, 6 mm × 35 mm × 70 mm, with a 3 mm deep and 25 mm wide slit (Bischoff Glastechnik, Alexanderstr. 2, 7518 Bretten, Federal Republic of Germany)

Paraffin for injection chamber: Paraffin, flüssig für Spectroscopie, Uvasol (Merck, Postfach 4119, 6100 Darmstadt 1, Federal Republic of Germany)

Paraffin for the microcapillary holder: Paraffin erstarrt für Histologie, melting point 61° (Merck)

10× Injection buffer: 0.48 *M* K_2HPO_4; 0.14 *M* NaH_2PO_4; 0.045 *M* KH_2PO_4, pH 7.2

Growth medium for BHK cells: G-Mem supplemented with 5% fetal calf serum, 10% Tryptose phosphate broth, 10 m*M* HEPES (Gibco-Biocult, Nunc GmbH, Göethestr. 5, 6200 Wiesbaden, Federal Republic of Germany)

Cell medium during injection: G-MEM without $NaHCO_3$

Preparation of a Micropipette

A glass capillary is heated above a small flame approximately 2 cm from one end and drawn out so much that it is flexible when moved (Fig. 4). Care should be exercised not to increase the relative wall thickness by prolonged heating. The short end is bent into a hook on the flame, and the capillary is then inserted into a glass holder from the long end and fixed in place using liquid paraffin as shown in Fig. 4. The holder with the capillary is mounted with the hook straight down on a microforge, and a weight of 0.15 g is added to the hook. The capillary is tilted 45 degrees and, with the aid of the heating coil, is bent just above (0.5 mm) its most narrow position (Fig. 5). The heating coil is shifted down to the thinnest point, and here the tip is formed by heating the coil and letting the weight slowly extend the capillary until it breaks. Using the 0.15 g weight, our capillaries have an outside diameter of approximately 1 μm at the tip. Tips with a

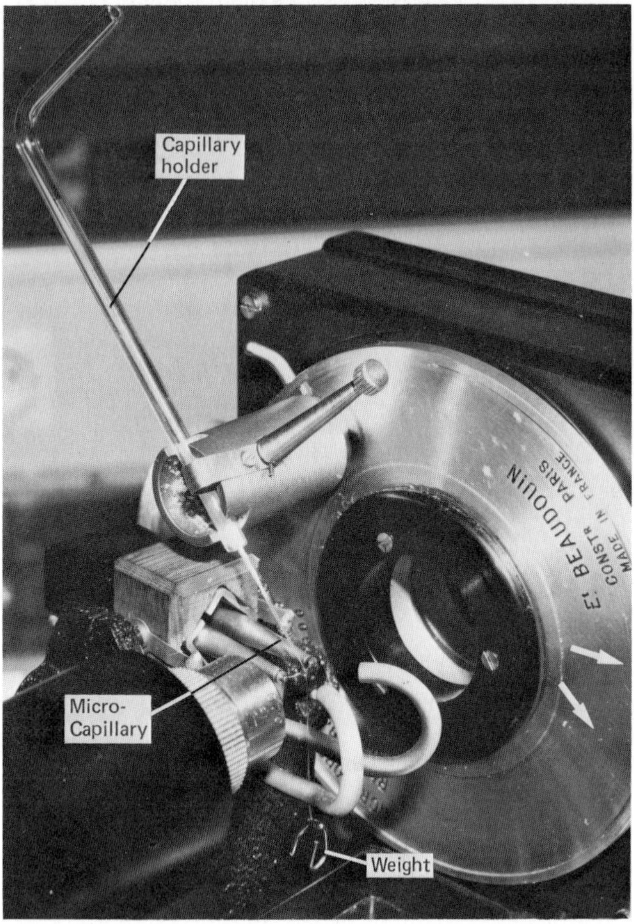

FIG. 5. The microcapillary is assembled in the microforge to make the fine tip of the microcapillary.

smaller or larger orifice can easily be made using lighter or heavier weights.

The Sample and the Cells

The nucleic acid sample is taken up in the injection buffer at a concentration of approximately 1 μg/μl and spun in a bench centrifuge for 5 min to remove material that could clog the micropipette. The cells are grown to subconfluency on a coverslip that has been marked with numbers at the

FIG. 6. The glass chamber with the coverslips for sample and cells.

cell side by scratching the glass with a diamond pen before plating the cells (Fig. 6). Prior to microinjection of the cells, the growth medium is changed to one lacking $NaHCO_3$.

The Injection Chamber

The chamber is formed by a small coverslip with the sample, a large coverslip with the cells and their glass holder (Fig. 6). About 0.5 μl of the DNA sample is pipetted on the small coverslip, and this is laid sample-side down on the holder. The space below the glass slide is immediately filled with paraffin oil. The coverslip with the cells is next taken out from the growth medium, dried on the back side with paper, and positioned on the holder very close to the sample coverslip, the cells being down. More paraffin oil is added until the chamber is filled.

Microinjection of Cell Nuclei

The microscope is focused on cells growing around one of the numbers scratched on the glass, using an 80-fold magnification. Keeping the focus, the stage of the microscope with the coverslips is withdrawn and the tip of the microcapillary is micromanipulated into focus position. The capillary is then moved 0.5–1.0 mm down and positioned under the droplet of sample on the small coverslip simply by returning the microscope stage (Fig. 7). With the aid of the micromanipulator, the capillary tip is inserted into the sample, filled with liquid by sucking with the syringe, and moved out again, now keeping a small positive pressure in the capillary. Next, the cells to be injected are repositioned under the microscope, and the magnification is changed to ×320. The injection of the nuclei is done by first manipulating the tip of the capillary just beneath the nucleus to be injected and then moving it upward into the nucleus (and into focus) where pressure is applied to expel some liquid (Fig. 8). This is seen as an immediate refractory change of the nuclei. The tip is then moved out from the cell, and the injection is repeated on the next cell. When there is no more sample left in the capillary, it can easily be refilled at the droplet of sample. After the desired number of cells have been injected, the capillary is positioned under the sample coverslip, the paraffin is carefully pipetted out from under the cell coverslip, and the coverslip is put into a petri dish and washed free from paraffin oil with complete growth medium before incubating at 37°.

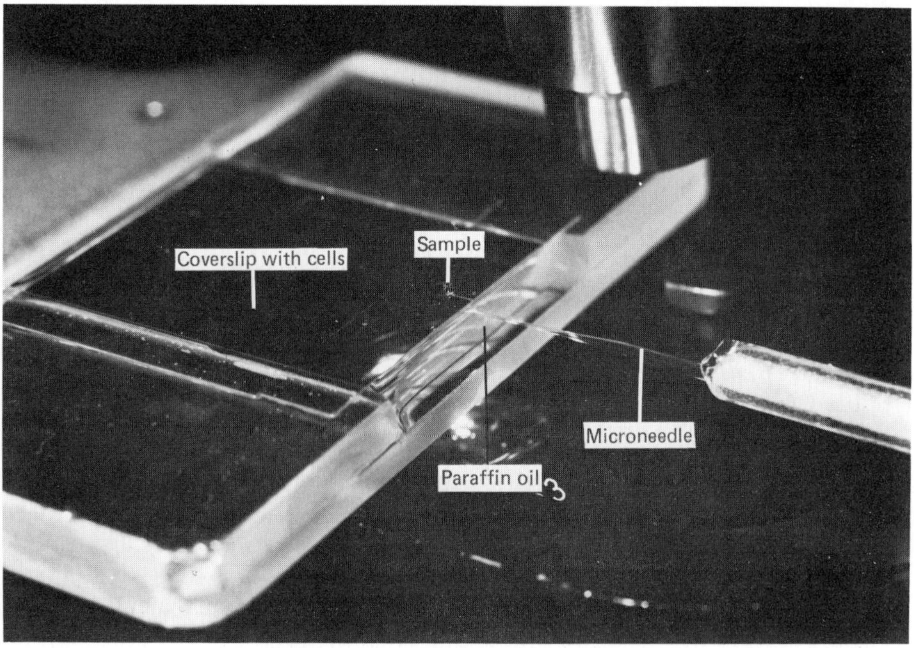

FIG. 7. Preparation for microinjection. The sample droplet is under the front coverslip, and the cells are at the down side of the back one. The microneedle is inserted into the paraffin oil chamber.

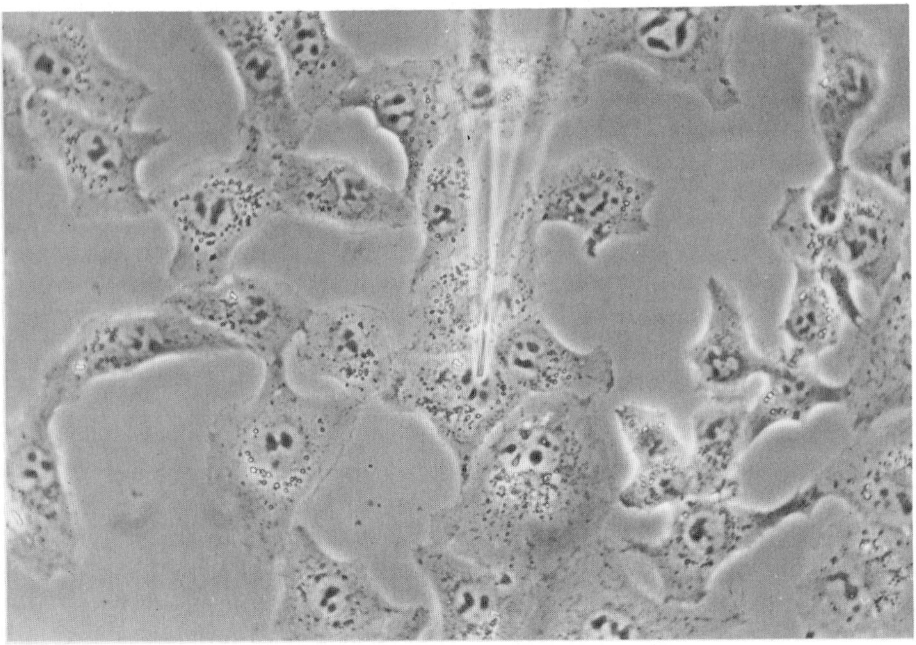

FIG. 8. Microinjection of nuclei.

Analysis of Microinjected Cells Using Indirect Immunofluorescence

Principle

The immunofluorescent staining procedures have been outlined in detail by others.[16,17] In the indirect staining technique, the cells are first treated with the specific antibody; then, in a second treatment, the antibody is bound to a fluorescently labeled anti-antibody.

Solutions and Reagents for Indirect Immunofluorescence Staining

Microscope: Zeiss photomicroscope equipped for epifluorescence
Dish, 55 mm; Falcon
Phosphate-buffered saline: 1.86 m*M* $NaH_2PO_4 \cdot H_2O$, 12.64 m*M* $Na_2HPO_4 \cdot 2\ H_2O$, pH 7.4, 150 m*M* NaCl
Formaldehyde, 3% (w/v) in PBS containing 0.1 m*M* $CaCl_2$ and 0.1 m*M* $MgCl_2$. Store in aliquots at −20°; use only once.

[16] J. F. Ash, D. Louvard, and S. J. Singer, *Proc. Natl. Acad. Sci. U.S.A.* **74,** 5584 (1977).
[17] K. S. Matlin, H. Reggio, A. Helenius, and K. Simons, *J. Cell Biol.* **91,** 601 (1981).

Gelatin 2% (w/v) in PBS as a 10× stock solution. Dilute with PBS before use.
NH_4Cl, 50 m*M*, in PBS
Glycerol 90% (w/v) in PBS
Triton X-100, 0.1% (w/v) in PBS
Affinity-purified antibodies against the structural proteins of SFV[11]

Staining of Cells

The injected cells on the coverslip are put into a Falcon dish, washed three times with 5 ml of PBS, and then fixed by adding 1 ml of paraformaldehyde solution for 20 min at room temperature. When washed three times with 5 ml of PBS, the fixative is quenched by adding 5 ml of 50 m*M* NH_4Cl–PBS solution and incubating for 10 min. After a further three washes with 5 ml of PBS, the cells are ready to be stained for surface antigens. For internal staining the coverslips are placed into 5 ml of 0.1% (w/v) Triton X-100 in PBS for 4 min and then washed three times with 5 ml of PBS.

The specific antibody is diluted to proper concentration in PBS–gelatin and centrifuged briefly in a microfuge; 100 μl of it is transferred to a strip of Parafilm laid out on a wet filter paper in a closable box. The coverslip with the cells is washed with 5 ml of PBS–gelatin (three times) and overlayed with 5 ml of PBS for 5 min and then put carefully on the top of the antibody droplet with the cell side down. After incubation for 20 min at room temperature, 200 μl of PBS are pipetted beneath the coverslip to facilitate its removal, and the cells are then washed with 5 ml of PBS–gelatin (twice) and overlayed twice with 5 ml of PBS for 5 min each time. Next, the second antibody is transferred to the Parafilm, and the treatment of the cells is repeated. When washed twice with PBS–gelatin and PBS, the coverslip is dried on the glass side and mounted on a microscope slide on top of a small drop of 90% glycerol–PBS. The slide is then observed in a fluorescence microscope. First, the area containing the injected cells is located at low magnification by identifying the guide number scratched at the cell side of the coverslip. After switching to epifluorescence and a high power objective, the fluorescence analyses are done.

Acknowledgments

We thank K. Matlin and K. Simons for helpful criticism and W. Moses for typing the manuscript.

[40] Biosynthesis of Sindbis Virus Membrane Glycoproteins *in Vitro*

By STEFANO BONATTI

A single virus-specific mRNA, 26 S RNA, synthesizes the three structural proteins present in Sindbis virus[1]: a 30,000-dalton capsid protein, C, which interacts with the genomic 42 S RNA to form the nucleocapsid, and two 50,000-dalton glycoproteins, E_1 and E_2, both integral membrane proteins of the viral envelope. E_2 is initially synthesized as a larger precursor, the 60,000-dalton PE_2,[2] which is converted to E_2 during its intracellular pathway to the plasma membrane by the removal of a 10,000-dalton aminoterminal portion. Another 26 S RNA translation product has been identified in infected cells, the "6K peptide,"[3] which will not end up in the extracellular virus and is probably membrane associated. The 26 S RNA is mainly found on membrane-bound polyribosomes[4,5] and represents the most abundant virus-specific RNA present in infected cells.[1] The gene order of 26 S RNA is C–PE_2–6K–E_1[6]; an identical situation is also found for the closely related Semliki Forest virus.[7] Because the 26 S RNA has only one initiation site for protein synthesis,[8] all its translation products are generated by proteolytic cleavages. Evidence obtained by *in vivo* pulse-labeling of infected cells suggests that these cleavages take place cotranslationally[9]; furthermore, proteolytic treatment of membrane fractions obtained from infected cells shows that newly synthesized C protein is degraded, whereas PE_2 and E_1 are mostly protected.[10] These findings indicate that the amino terminal portion of the nascent chain that comprises C protein is discharged into the cytosol, whereas the other two major products of the nascent chain, PE_2 and E_1, are integrated into the rough endoplasmic reticulum membrane, leaving the bulk of the molecule

[1] J. H. Strauss and E. G. Strauss, *in* "The Molecular Biology of Animal Viruses" (D. P. Nayak, ed)., p. 111. Dekker, New York, 1977.
[2] M. J. Schlesinger and S. Schlesinger, *J. Virol.* **10,** 925 (1972).
[3] W. J. Welch and B. M. Sefton, *J. Virol.* **33,** 230 (1980).
[4] G. Martire, S. Bonatti, G. Aliperti, G. De Giuli, and R. Cancedda, *J. Virol.* **21,** 610 (1977).
[5] S. Bonatti, A. Cerasuolo, R. Cancedda, N. Borgese, and J. Meldolesi, *Eur. J. Biochem.* **103,** 53 (1980).
[6] C. M. Rice and J. H. Strauss, *Proc. Natl. Acad. Sci. U.S.A.* **78,** 2062 (1981).
[7] H. Garoff, A. M. Frischauf, K. Simons, H. Lehrach, and H. Delius, *Nature (London)* **288,** 236 (1980).
[8] R. Cancedda, L. Villa-Komaroff, H. F. Lodish, and M. J. Schlesinger, *Cell* **6,** 215 (1975).
[9] J. C. S. Clegg, *Nature (London)* **254,** 454 (1975).
[10] D. F. Wirth, F. Katz, B. Small, and H. F. Lodish, *Cell* **10,** 253 (1977).

ISBN 0-12-181996-5

on the luminal side of the membrane. To study in more detail the mechanism by which a single mRNA is able to direct the synthesis of one cytosolic protein and at least two membrane glycoproteins, *in vitro* cell-free protein synthesis systems able to translate efficiently the 26 S RNA have been set up.[11,12] By supplementing these systems with dog pancreas microsomal membranes, it has been possible faithfully to reconstitute *in vitro* all the events coupled to the translation of Sindbis 26 S RNA *in vivo*. Subsequently, by using radiosequencing methods, the presence of an uncleaved signal sequence and the transmembrane orientation of glycoprotein PE_2 have been established.[12] In this chapter, I focus on the biosynthesis, membrane insertion, and core glycosylation *in vitro* of Sindbis glycoproteins.

General Methods

Preparation of 26 S RNA

Chick embryo fibroblasts ($\sim 1 \times 10^9$ cells) grown as monolayers in a roller cell apparatus are infected at 37° with Sindbis virus at a multiplicity of infection of 50 plaque-forming units per cell in phosphate-buffered saline containing calcium, magnesium, and 1% fetal calf serum. After 1 hr the unadsorbed virus is removed and minimum essential medium containing 2% fetal calf serum, 1 μg of actinomycin D per milliliter and 4 μCi of [^{3}H]uridine per milliliter are added. Six hours later the cells are washed three times with phosphate-buffered saline, scraped with a rubber policeman, and collected by centrifugation. The cells are resuspended in TNE buffer [50 m*M* Tris-HCl, pH 7.4, 150 m*M* NaCl, 1 m*M* ethylenediaminetetraacetic acid (EDTA)], and the RNA is extracted after addition of SDS to 1% and 1 volume of phenol–chloroform–isoamyl alcohol (50 : 50 : 1). The organic phase is reextracted once with phenol–chloroform–isoamyl alcohol, and the pooled aqueous phases are finally extracted with chloroform saturated with TNE buffer. After two ethanol precipitations the RNA pellet is dissolved in H_2O and the single-stranded RNA is precipitated twice by overnight incubations at 4° in 2 *M* LiCl. After repeated ethanol precipitations, Sindbis 26 S RNA is partially purified by sedimentation in 5 to 20% sucrose gradients made in TNE buffer containing 0.5% SDS (SW27, 14 hr at 19,000 rpm). About 4 A_{260} units of RNA are loaded on each gradient. The sedimentation of Sindbis and cellular RNA is determined by monitoring the radioactivity and the optical density content of each fractions, respectively. The pooled fractions containing 26 S RNA

[11] S. Bonatti, R. Cancedda, and G. Blobel, *J. Cell Biol.* **80,** 219 (1979).

[12] S. Bonatti and G. Blobel, *J. Biol. Chem.* **254,** 12261 (1979).

are ethanol precipitated twice, and finally the RNA is dissolved in H_2O at 20–50 A_{260} units/ml. The preparations are stored at −80° in small aliquots.

Preparation of Rabbit Reticulocyte Lysate and Wheat Germ S_{23} Translation Systems

Rabbit reticulocyte lysates are prepared and nuclease-treated as described[13] (see also this volume [4]). Wheat germ S_{23} extracts are prepared as detailed elsewhere[14] (see also this volume [3]), except that glass powder is used for the grinding step.

Preparation of Dog Pancreas Microsomal Membranes

Dog pancreas rough microsomes are prepared, and EDTA is stripped as reported[15] (see also this volume [6]).

Conditions for Protein Synthesis in the Translation Systems

Rabbit Reticulocyte. Reaction mixtures (25 μl) contain 10 m*M* Tris-HCl, pH 7.4, 1 m*M* ATP, 0.2 m*M* GTP, 10 m*M* phosphocreatine, 50 μg of creatine phosphokinase per milliliter, 100 m*M* potassium chloride, 2 m*M* magnesium acetate, 20 μ*M* hemin, 40 μ*M* amino acids minus methionine, 6–15 μCi of [^{35}S]methionine (600–1000 Ci/mmol), 12.5 μl of nuclease-treated rabbit reticulocyte lysate, 0.02 A_{260} unit of 26 S RNA, 2–6 A_{280} units per milliliter of EDTA-stripped dog pancreas microsomal membranes. Incubations are performed for 60 min at 29°.

Wheat Germ. Reaction mixtures (25 μl) contain 20 m*M* HEPES, pH 7.4, 2.5 m*M* ATP, 380 μ*M* GTP, 10 m*M* creatine phosphate, 80 μg of creatine phosphokinase per milliliter, 140 m*M* potassium acetate, 2.0 m*M* magnesium acetate, 80 μ*M* spermidine, 4 m*M* dithiothreitol (DTT), 40 μ*M* amino acids minus methionine, 6–15 μCi of [^{35}S]methionine (600–1000 Ci/mmol), 40 A_{260} units/ml of wheat germ extract, 0.01 A_{280} unit human placental ribonuclease inhibitor,[16] 0.02 A_{260} unit of 26 S RNA. Incubations are performed for 60 min at 24–27°.

Labeling of Sindbis Structural Proteins in Vivo

Chick embryo fibroblasts (~3 × 10^6 cells) grown as monolayers are infected with Sindbis virus as above except that the addition of [^{3}H]uridine is omitted. At 5 hr after infection the medium is replaced with amino acids-free minimum essential medium supplemented with 2% fetal calf

[13] H. R. B. Pelham and R. J. Jackson, *Eur. J. Biochem.* **67,** 247 (1976).
[14] M. McMullen, P. Shaw, and T. M. Martin, *J. Mol. Biol.* **132,** 679 (1979).
[15] D. Shields and G. Blobel, *J. Biol. Chem.* **253,** 3753 (1978).
[16] P. Blackburn, *J. Biol. Chem.* **254,** 12484 (1979).

serum. At 6 hr after infection the cells are pulsed for 3 min with 100 μCi of [^{35}S]methionine (600–1000 Ci/mmol) per milliliter, then quickly chilled, washed with cold PBS, scraped with a rubber policeman, and collected by centrifugation. After resuspension in 0.4 ml of 10 m*M* Tris-HCl, pH 7.4, 150 m*M* sodium chloride, the cells are homogenized in a Dounce homogenizer with 20 strokes of a B pestle. A postnuclear supernatant is prepared by centrifugation at 500 *g* and stored at −20°.

Analysis by Sodium Dodecyl Sulfate–Polyacrylamide Gel Electrophoresis (SDS–PAGE) of in Vitro and in Vivo Synthesized Translation Products

Twenty microliters of translation products are mixed with 15 μl of PAGE solution I (6% SDS, 0.5 *M* Sucrose, 20 m*M* Tris, 180 m*M* DTT, 0.06% Bromphenol Blue) and incubated for 3 min at 100°. Then 15 μl of PAGE solution II (2.5% SDS, 0.2 *M* sucrose, 0.5 *M* Tris, 1 *M* iodoacetamide, 0.06% Bromphenol Blue) are added and the incubation proceeds for 30–90 min at 37°. The samples are spun for 2 min in an Eppendorf centrifuge to remove any particulate and then are loaded onto 10 to 15% polyacrylamide gradient gels.[17]

After the run, the gels are processed for fluorography as described[18] (see also this volume [15]).

Analysis on SDS–PAGE of Translation Products Synthesized in Vitro and in Vivo

On the basis of its electrophoretic mobility, C protein is easily identified as a major product that is synthesized *in vivo* (Fig. 1, panel B, lane 7), and *in vitro* either in the absence or in the presence of pancreatic membranes (Fig. 1, panels A and B, lanes 2 and 4). Another prominent polypeptide is synthesized in both cell-free systems in the absence of microsomal membranes (Fig. 1, panel A, lanes 2 and 4; panel B, lane 2). This polypeptide has an apparent molecular weight of 110,000 and comigrates with a minor protein synthesized by the infected cells (Fig. 1, panel B, lane 7): this protein, designated B_1 by Schlesinger and Schlesinger[19] contains both PE_2 and E_1,[19] is not glycosylated and is located in the cytosol of the infected cells (S. Bonatti, unpublished results). The equivalence of the *in vivo* and *in vitro* synthesized B_1 forms is demonstrated by analysis of their tryptic peptides.[12] It should be noted that several major products in between C and B_1 are synthesized in the absence of microsomal mem-

[17] G. Blobel and B. Dobberstein, *J. Cell Biol.* **67,** 835 (1975).

[18] W. M. Bonner and R. A. Laskey, *Eur. J. Biochem.* **43,** 83 (1974).

[19] M. J. Schlesinger and S. Schlesinger, *J. Virol.* **11,** 1013 (1973).

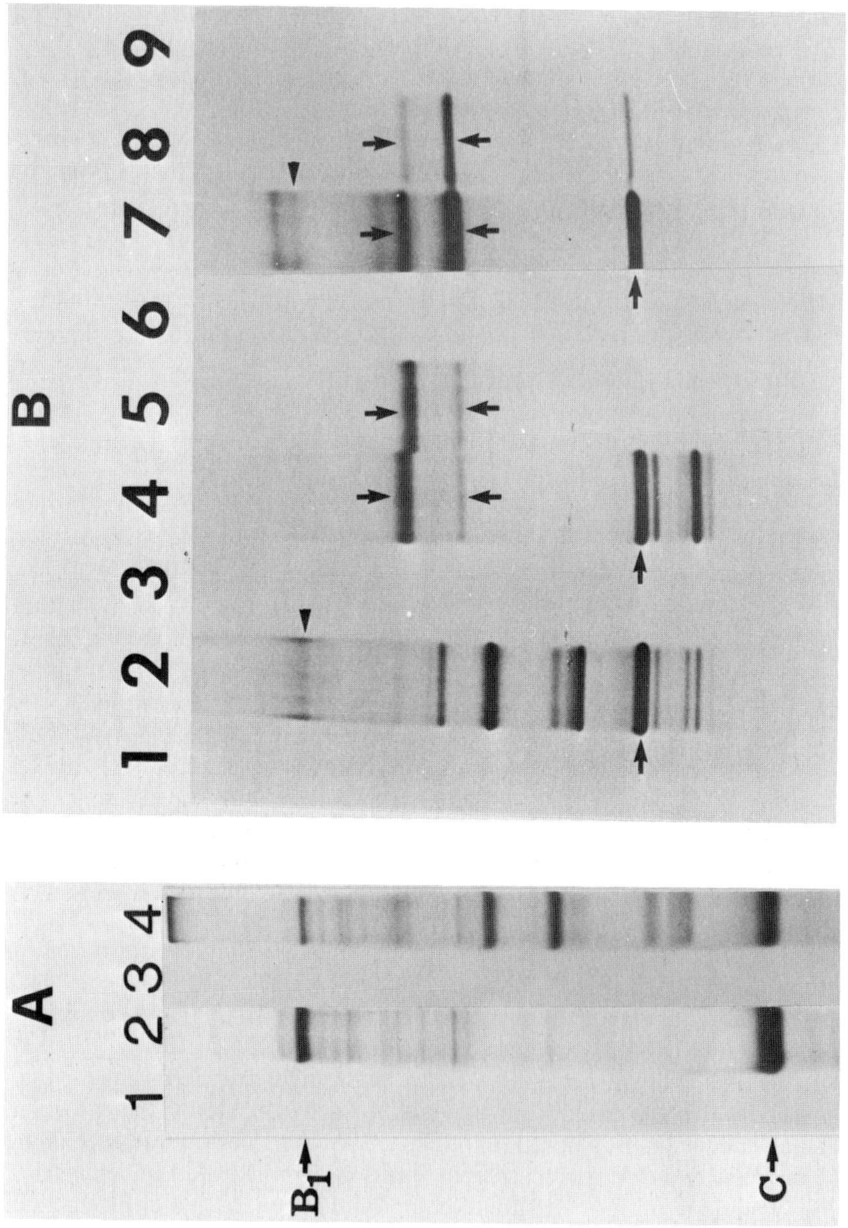

FIG. 1

branes if the translation is performed in the rabbit reticulocyte cell-free system (Fig. 1, panel A, compare lanes 2 and 4). All these polypeptides contain tryptic peptides characteristic of C protein,[11] and therefore they most likely represent incompleted nascent chain from which the C portion has not been cleaved. The addition of microsomal membranes to the translation mixtures programmed with 26 S RNA results in a drastic change of the synthesized products. Two new polypeptides are clearly detectable which comigrate with PE_2 and E_1 synthesized *in vivo* (Fig. 1, panel B, lanes 4 and 7); analysis of their tryptic peptides proves the identity of *in vitro* and *in vivo* synthesized forms of PE_2 and E_1.[11] The synthesis of both proteins is strictly dependent on the cotranslational presence of microsomal membranes.[20]

Posttranslational Assays

Proteolytic Protection

Method. Twenty microliters of translation products are adjusted to a final concentration of 10 m*M* $CaCl_2$ and kept on ice. Trypsin and chymotrypsin (Boehringer, Mannheim, Federal Republic of Germany) (stock solutions, 6 mg/ml, stored in small aliquots at −80°) are added to a final concentration of 300 μg/ml for each enzyme, and the samples are carefully mixed. Incubation is stopped after 20–60 min by the addition of 750 units of Trasylol (Mobay Chemical Corporation, New York) and PAGE solution I. The samples are immediately boiled for 5 min and prepared for PAGE analysis as detailed above. As a control, duplicate samples are treated with the same procedure, but in the presence of 1% Triton X-100.

Results. As shown in Fig. 1, panel B, lane 5, C protein is almost totally degraded, whereas E_1 is apparently protected. PE_2 instead is converted to

[20] H. Garoff, K. Simons, and B. Dobberstein, *J. Mol. Biol.* **124,** 587 (1978).

FIG. 1. [^{35}S]Methionine-labeled Sindbis proteins synthesized *in vivo* and *in vitro* and analyzed by polyacrylamide slab gel electrophoresis. Panel A: Products synthesized in the wheat germ cell-free system (lanes 1 and 2) or in the rabbit reticulocyte system (lanes 3 and 4) either in the absence (lanes 1 and 3) or in the presence (lanes 2 and 4) of 26 S RNA (S. Bonatti, unpublished results). Panel B: Products synthesized in the reticulocyte cell-free system (lanes 1 to 6) programmed with 26 S RNA (lanes 2–6) or obtained from pulse-labeled infected cells (lanes 7–9). Lanes 4–6: Products synthesized in the presence of dog pancreas microsomal membranes. Proteolytic treatment with trypsin and chymotrypsin was performed before the electrophoretic analysis either in the absence (lanes 3, 5, 8) or in the presence (lanes 6 and 9) of Triton X-100, respectively. Rightward-pointing arrows indicate C protein. Arrows pointing downward and upward indicate the position of PE_2-PE_2' (see text) and E_1 forms, respectively. Arrowheads indicate B_1 protein. Reproduced from Bonatti *et al.*,[11] *J. Cell Biol.* with permission of the Rockefeller University Press.

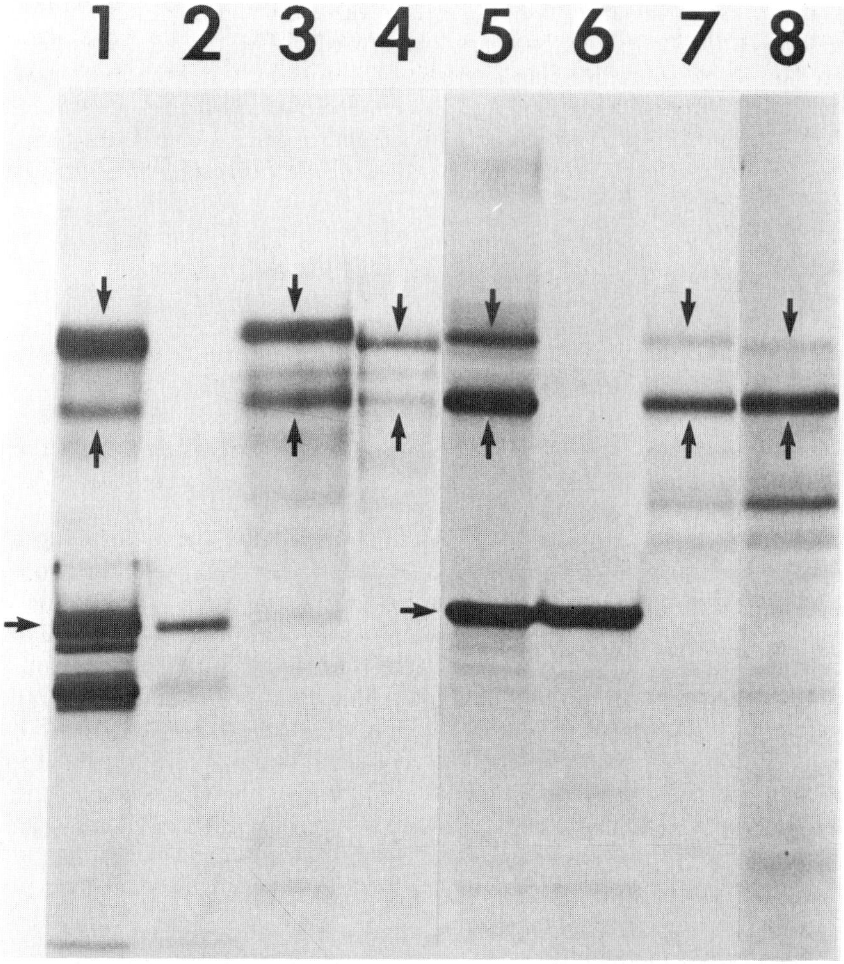

FIG. 2. Core glycosylation of [^{35}S]methionine-labeled Sindbis proteins synthesized *in vivo* and *in vitro*. Lanes 1–4: Products synthesized in the rabbit reticulocyte system programmed with 26 S RNA in the presence of dog pancreas microsomal membranes. Lanes 5–8: Products obtained from pulse-labeled infected cells. *In vivo* and *in vitro* synthesized products were subjected to concanavalin A–Sepharose affinity chromatography and subsequently analyzed by polyacrylamide slab gel electrophoresis. Lanes 1 and 5: Total products as applied to the Con A columns. Lanes 2 and 6: Unbound material. Lanes 3 and 7: Bound material eluted by α-methyl mannoside. Lanes 4 and 8: Bound and α-methyl mannoside eluted material of samples previously digested with trypsin and chymotrypsin. Arrows as in Fig. 1. Reproduced from Bonatti *et al.*,[11] with permission of the Rockefeller University Press.

a slightly smaller derivative, PE_2': however, if the proteolytic treatment is performed in the presence of detergent (Fig. 1, panel B, lane 6), all synthesized products are degraded. Similar results are obtained performing the treatment on postnuclear supernatant aliquots derived from *in vivo*-labeled infected cells (Fig. 1, panel B, lanes 8 and 9), thus strongly suggesting that *in vitro*-synthesized PE_2 and E_1 are correctly assembled into the membranes.

Concanavalin A (Con A)–Sepharose Affinity Chromatography

Method. Fifty microliters of translation products are adjusted to a final concentration of 2% SDS, incubated at 100° for 2 min, and diluted 40-fold with solution A (20 m*M* Tris-HCl, pH 7.4, 150 m*M* NaCl, 1 m*M* MgCl, 1 mm DTT, 12 μg of bovine serum albumin per milliliter). The sample is loaded onto a column of Con A–Sepharose CL-4B (Pharmacia Fine Chemicals, Uppsala, Sweden), 100 μl bed volume, equilibrated with solution B (solution A containing 0.05% SDS). The unbound fraction is collected and concentrated by precipitation with 2 volumes of ethanol–ether (50:50) at 0° for 1 hr. The column is washed with 10 ml of solution B and then 0.5 ml of solution B containing 0.4 *M* α-methyl mannoside (Sigma Chemical Company, St. Louis, Missouri) is applied. The column is sealed for an overnight incubation, and then the eluate is collected and concentrated as above but using 25 μg of tRNA as a carrier. The precipitated unbound and bound fractions are collected by centrifugation and prepared for PAGE analysis by the addition of 20 μl of H_2O and PAGE solutions as detailed above.

Results. As shown in Fig. 2, lanes 3 and 7, both *in vitro* and *in vivo* synthesized PE_2 and E_1 bind to Con A–Sepharose and are specifically eluted by α-methyl mannoside. Conversely, C protein does not bind to Con A–Sepharose (Fig. 2, lanes 2 and 6). These data indicate the presence of core sugars in both the *in vitro* and *in vivo* synthesized PE_2 and E_1. Moreover, PE_2' and E_1 obtained by posttranslational proteolytic treatment still bind to Con A–Sepharose (Fig. 2, lanes 4 and 8). This finding is consistent with the notion that the core-glycosylating enzymes are located on the luminal side of the microsomal vesicles.[21]

Acknowledgments

I thank R. Cancedda and G. Blobel for helpful discussions. This work was supported in part by Grant 80.00587.84 from Progetto Finalizzato Virus, Consiglio Nazionale delle Ricerche, Rome, Italy.

[21] H. Schachter, *Biochem. Soc. Symp.* **40,** 57 (1974).

[41] Ribophorins I and II: Membrane Proteins Characteristic of the Rough Endoplasmic Reticulum

By GERT KREIBICH, EUGENE E. MARCANTONIO, and DAVID D. SABATINI

Biochemical analysis of rough and smooth hepatic microsomes first revealed the existence of compositional differences between the membranes that appear to be related to the function of the rough endoplasmic reticulum (RER) in protein synthesis.[1,2] Purified rough microsomes (RM) were found to contain two transmembrane glycoproteins[3] that are not found in smooth microsomes (SM). These integral membrane proteins, which have been designated ribophorins (ribophorin I, M_r 65,000; ribophorin II, M_r 63,000) because of their relationship to membrane-bound ribosomes,[1,2,4] are solubilized when RM are treated with ionic detergents, but are quantitatively recovered as a sedimentable complex in association with polysomes when neutral detergents are used. In native RM the ribophorins can be cross-linked to ribosomes with glutaraldehyde or with reversible bifunctional reagents.[2] These experiments suggest a possible role for these characteristic components of RM in mediating the attachment of ribosomes to the membranes. In support of this idea it was found that the ribophorin content of microsomal fractions is stoichiometrically related to their ribosome content.

Ribophorins have now been identified in microsomal fractions from several other tissues and animal species. Although individual ribophorins appear to vary slightly in electrophoretic mobility, within a given species different tissues contain apparently identical ribophorin polypeptides.[5] Immunoprecipitation with specific antibodies raised against ribophorins I and II from rat liver showed that there is no immunological cross-reactivity between the two ribophorins. On the other hand, antibodies against each of the rat liver ribophorins cross-reacted with the corresponding proteins identified in RM from other sources.[5] The presence of ri-

[1] G. Kreibich, B. L. Ulrich, and D. D. Sabatini, *J. Cell Biol.* **77,** 464 (1978).

[2] G. Kreibich, C. M. Freienstein, B. N. Pereya, B. L. Ulrich, and D. D. Sabatini, *J. Cell Biol.* **77,** 488 (1978).

[3] E. Rodriguez-Boulan, D. D. Sabatini, N. Pereyra, and G. Kreibich, *J. Cell Biol.* **78,** 894 (1978).

[4] G. Kreibich, M. Czako-Graham, R. C. Grebenau, and D. D. Sabatini, *Ann. N.Y. Acad. Sci.* **343,** 17 (1980).

[5] E. E. Marcantonio, R. C. Grebenau, D. D. Sabatini, and G. Kreibich, *Eur. J. Biochem.* **124,** 217 (1982).

ISBN 0-12-181996-5

bophorins as a characteristic feature of RER membranes suggests that these proteins play an important functional role in the process by which nascent polypeptides are inserted into ER membranes.

Within the membranes, ribophorins appear to have a strong tendency to interact with each other, and it seems likely that an intramembranous network of these polypeptides may control the distribution of ribosome binding sites in the plane of the ER membrane. The presence of ribophorins may also be a determinant of the characteristic flattened sac appearance of RER cisternae.[1,4]

Solutions

TKM: 25 mM KCl; 50 mM Tris-HCl, pH 7.6; 5 mM $MgCl_2$
LSB: 50 mM KCl; 50 mM Tris-HCl, pH 7.6; 5 mM $MgCl_2$
Binding buffer: 100 mM KCl, 50 mM Tris-HCl, pH 7.6; 5 mM $MgCl_2$
HSB: 500 mM KCl; 50 mM Tris-HCl, pH 7.6; 5 mM $MgCl_2$
RSB: 10 mM Tris-HCl, pH 7.6; 10 mM KCl; 1.5 mM $MgCl_2$

Isolation of Ribophorins in Association with Membrane-Bound Polysomes after Partial Solubilization of Microsomes from Different Species

Animals are starved overnight before sacrifice. Rat and rabbit liver RM are prepared by the procedure of Adelman *et al.*[6] from male or female albino rats (150 g) of the Sprague–Dawley strain or from New Zealand white rabbits (3 kg). Rat lacrimal gland RM are isolated according to Herzog and Miller,[7] and dog pancreas RM by the procedure of Shields and Blobel.[8] The following protease inhibitors are included in all solutions during cell fractionation; Trasylol (10 units/ml, FBA Pharmaceuticals), chymostatin (1 μg/ml), leucylleucylleucine (1 μg/ml), *o*-phenanthroline (1 mM, Sigma Chemical Company). If not stated otherwise, all treatments of microsomes are carried out at 3°.

Rough microsomes from exocrine glands, such as the pancreas or lacrimal gland, which are characterized by high protein synthetic activity, bear more ribosomes on the vesicle surface then liver RM. Ribosome concentrations are estimated using an absorption coefficient of A_{260} = 233 cm^{-1}.[9] Variations in ribosome content are reflected in the RNA : protein ratios, which typically are about 0.17 for rat and rabbit liver RM, 0.23 for

[6] M. R. Adelman, G. Blobel, and D. D. Sabatini, *J. Cell Biol.* **56,** 191 (1973).
[7] V. Herzog and F. Miller, *J. Cell Biol.* **53,** 662 (1972).
[8] D. Shields and G. Blobel, *J. Biol. Chem.* **253,** 3753 (1978).
[9] Y. Tashiro and P. Siekevitz, *J. Mol. Biol.* **11,** 149 (1965).

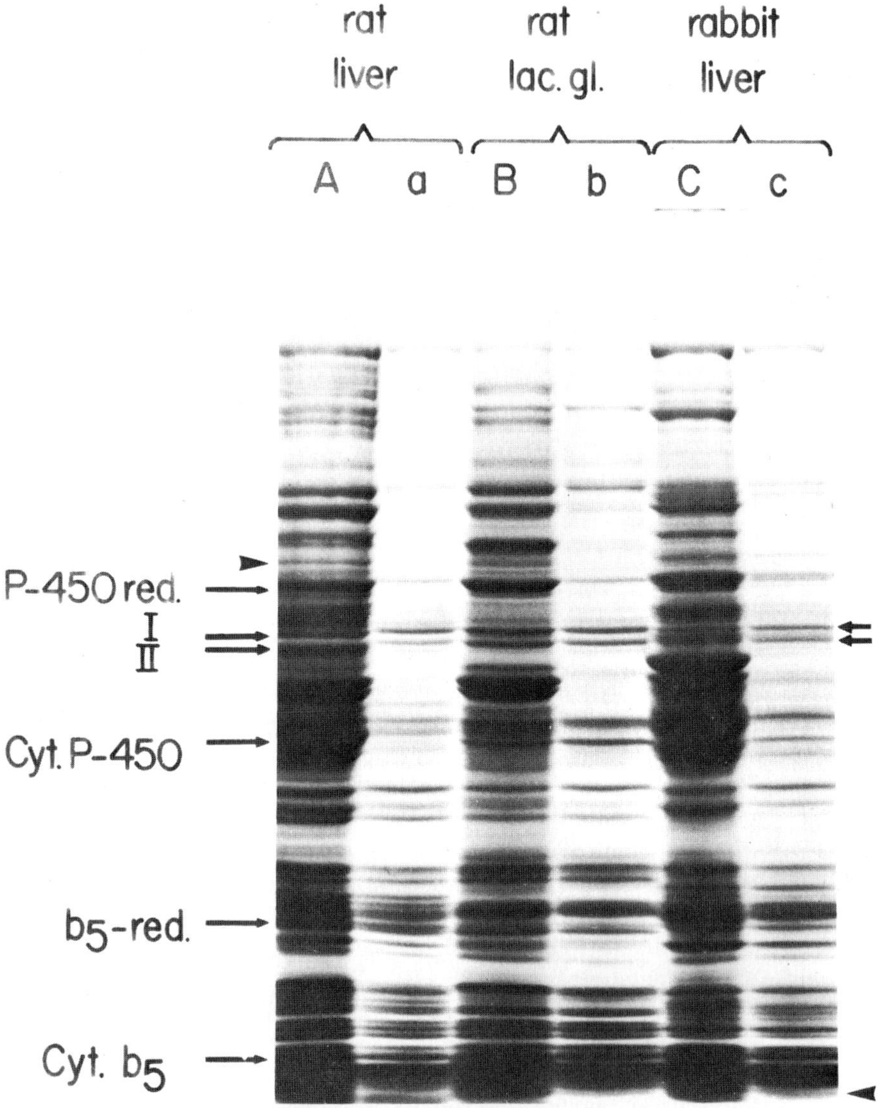

FIG. 1. Presence of ribophorin-like polypeptides in rough microsomes from various sources. Rough microsomes (600 μg of protein) from rat liver (A), rat lacrimal gland (B), and rabbit liver (C), and sedimentable subfractions (lanes a, b, and c) obtained from them after treatment with 1.5% Emulgen 913 were analyzed on SDS–polyacrylamide gradient (6 to 11%) gels and stained with Coomassie Blue. The positions of ribophorins in rat liver microsomes and of equivalent polypeptides in microsomes from other sources are marked I and II with arrows. Proteins identified in rat liver rough microsomes are indicated in lane a.

rat lacrimal gland, and 0.42 for dog pancreas RM. For storage, freshly prepared microsomes are resuspended in 0.25 *M* sucrose containing LSB, mixed with 2 volumes of glycerol, and kept at −70°. Before use, microsomes stored in glycerol are diluted four times with LSB-buffer containing 0.25 *M* sucrose and are recovered by sedimentation (20 min at 30,000 rpm in the Ti60 Beckman rotor). Rough microsomal pellets are resuspended in LSB (3.5 mg of protein per milliliter) by hand using a Potter–Elvehjem homogenizer. An amount of nonionic detergent equal to one-ninth the volume of the suspension is added while vortexing. Either a 15% stock solution of Emulgen 913 (Kao Atlas, Tokyo, Japan) or 10% Kyro EOB (a gift from D. D. Hughes, Miami Valley Laboratory, Proctor and Gamble Co.) may be used. After 30 min of incubation on ice, a supernatant and a sedimentable fraction are obtained by centrifugation (60 min at 40,000 rpm in a Ti50 Beckman rotor).

After RM are treated with Kyro EOB or Emulgen 913, ribophorins are quantitatively recovered in sedimentable fractions (Fig. 1). Electrophoretic patterns of RM from different organ sources vary widely, reflecting major differences in the protein composition of the membranes. However, sedimentable residues obtained after treatment with neutral detergents strikingly resemble each other. In addition to the presence of ribosomal proteins of low molecular weight, ribophorins are major constituents of these residues. Electron microscopic examination shows that the sedimentable residues consist mainly of curved membrane remnants bearing tightly packed ribosomes bound to their convex faces.[1,5] This suggests that a direct association between the ribosomes and the ribophorins is maintained after most other membrane components are solubilized by the nonionic detergent.

Recovery of Ribophorins in Inverted Rough Microsomes

When RM are treated with concentrations of Triton X-100 that are not sufficient to solubilize the membranes, extensive aggregation of bound ribosomes on the microsomal surface results, with a concomitant change in the configuration of the vesicles.[10] This observation led to the development of a procedure for the subfractionation of RM into two types of vesicles, derived from different regions of the rough microsomal membrane. One microsomal subfraction is obtained that consists largely of inverted rough vesicles, which contain most of the ribophorins and bound ribosomes, and another of lower density, from which bound ribosomes and ribophorins have been excluded (Fig. 2).

[10] G. Kreibich, G. Ojakian, E. Rodriguez-Boulan, and D. D. Sabatini, *J. Cell Biol.* **93,** 111 (1982).

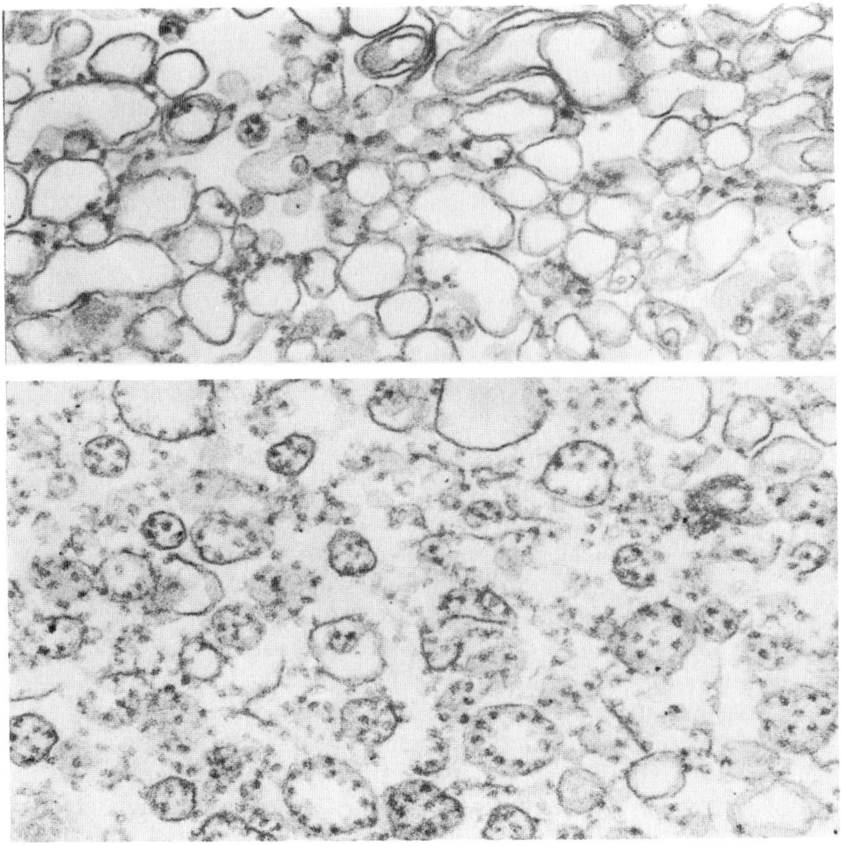

FIG. 2. Preparation of "inverted rough" vesicles. A suspension of rough microsomes treated with Triton X-100 (0.16% final) was layered on top of a sucrose step gradient. The material found after centrifugation on the interface between the 10% and 30% sucrose layers consisted largely of smooth-surfaced vesicles (upper panel) while the pellet contains inverted rough vesicles (upper panel). ×66,000.

To prepare inverted RM, rat liver RM are resuspended in LSB solution (3.5 mg of protein per milliliter) and one-tenth volume of 1.6% Triton X-100 is added while vortexing gently. Aliquots (1.5 ml) of this suspension are layered on top of step gradients (2 ml of 30% sucrose and 1.5 ml of 10% sucrose, both in LSB), which are centrifuged (60 min at 40,000 rpm) in an SW56 Beckman rotor. Inverted RM are recovered in the pellet (Fig. 2), and smooth-surfaced vesicles in a band found at the 10/30% sucrose interface. Luminal proteins and partially solubilized microsomal membrane proteins remain in the loading zone. The protein, RNA, and phos-

TABLE I
BIOCHEMICAL CHARACTERIZATION OF ROUGH INVERTED AND SMOOTH VESICLES OBTAINED FROM RM TREATED WITH 0.16% TRITON X-100[a]

	Control RM	Solubilized proteins	Smooth vesicles	Rough inverted vesicles	Recovery [μg (%)]
$OD_{260/280}$	1.57	0.78	1.00	1.76	—
Protein (μg)	3070	670	460	1520	2650 (86)
RNA (μg)	680	110	50	660	820 (120)
Phospholipid (μg)	1120	400	280	360	1040 (93)
RNA: protein	0.183	0.164	0.108	0.423	—
Lipid: protein	0.364	0.597	0.608	0.230	—
Cytochrome *P*-450 (μg)	103	31	16	42	89 (86)
Cytochrome b_5 (μg)	38	5	11	20	36 (95)
NADPH cytochrome *P*-450 reductase (μg)	12	6	2	3	11 (92)

[a] RM suspended in LSB and treated with 0.16% Triton X-100 were subfractionated into rough inverted vesicles, smooth surfaced vesicles, and soluble proteins. Values given are averages from two independent experiments. In each case individual values were within 10% of the average.

pholipid content, as well as the distribution of components of the microsomal electron transport chain in the three fractions yielded by this procedure, are presented in Table I.

Preparation of Antibodies against Ribophorins I and II

The sedimentable residue obtained from rat liver RM treated with Emulgen 913 (1.5% final) is solubilized in electrophoresis sample buffer[11] (4 mg of protein per milliliter) and aliquots (10 mg of protein) are loaded onto preparative SDS–polyacrylamide (6–11%) gels (26 cm × 22 cm × 2 mm).[11] After electrophoresis (20 hr, 20 mA) and brief staining (5 min) with Coomassie Blue, gel slices containing ribophorins are excised from several gels, pooled, and homogenized in sample buffer (omitting SDS) with a Potter–Elvehjem homogenizer. SDS (1% final) and sucrose (10% final) are then added and the slurry is loaded onto another SDS–acrylamide (6–11%) gel with a single wide sample well. Electrophoresis is carried out at

[11] J. V. Maizel, Jr., *Methods Virol.* **5**, 179 (1971).

5–10 mA for 16 hr until the tracking dye enters the separating gel, then the current is increased to 20 mA and electrophoresis is continued for 20 hr more until the tracking dye begins to enter the lower buffer reservoir.

Acrylamide gel slices containing each ribophorin are homogenized in a small volume of phosphate-buffered saline, mixed with complete Freund's adjuvant (CFA), and sonicated. Initially, for each antigen a total of 0.8 mg protein was injected intradermally at multiple sites along the dorsal midline of one goat (Pocono Rabbit Farm, Pennsylvania). At 3-week intervals, booster injections of 0.4 mg of protein in incomplete Freund's adjuvant were applied in a similar manner. Significant antibody titers were observed after three boosts. Goats were bled at regular intervals and IgG fractions were prepared by $(NH_4)_2SO_4$ precipitation.

In order to test for the specificity and cross reactivity of the antibodies, sedimentable residues obtained after Emulgen 913 treatment of rough microsomes (see above) are resuspended in phosphate-buffered saline (5 mg of protein per milliliter), and 0.5-mg aliquots are iodinated using the chloramine-T procedure.[12] The residues containing labeled proteins are recovered by sedimentation (60 min at 100,000 *g* through a 20% sucrose–LSB cushion), resuspended in phosphate-buffered saline, and solubilized in SDS (2% final) before immunoprecipitation by an indirect procedure[13] using protein A-Sepharose beads (Sigma Chemical Company). Immunoprecipitates were analyzed by SDS–polyacrylamide (6–11%) gel electrophoresis followed by autoradiography using a Cronex intensifying screen (DuPont, Photo Products).

Figure 3 shows that the two rat ribophorins are immunologically distinct; that is, each ribophorin was precipitated specifically from the SDS-solubilized iodinated residues by the corresponding antibodies. Polypeptides electrophoretically indistinguishable from rat liver ribophorins I and II were immunoprecipitated from residues obtained from other rat tissues, such as the lacrimal gland,[5] as well as from liver microsomes of other species, such as rabbit.

Correlation of the Ribophorin Content of Microsomal Fractions with Their Ribosome Binding Capacity

Specific sites on rough microsomal membranes, which are exposed only after the membranes are stripped of native ribosomes, are capable of rebinding 80 S ribosomes in media of physiological ionic strength, even in the absence of nascent polypeptide chains.[14] These sites recognize eu-

[12] W. M. Hunter and F. C. Greenwood, *Nature (London)* **194,** 495 (1962).
[13] B. M. Goldman and G. Blobel, *Proc. Natl. Acad. Sci. U.S.A.* **75,** 5066 (1978).
[14] N. Borgese, W. Mok, G. Kreibich, and D. D. Sabatini, *J. Mol. Biol.* **88,** 559 (1974).

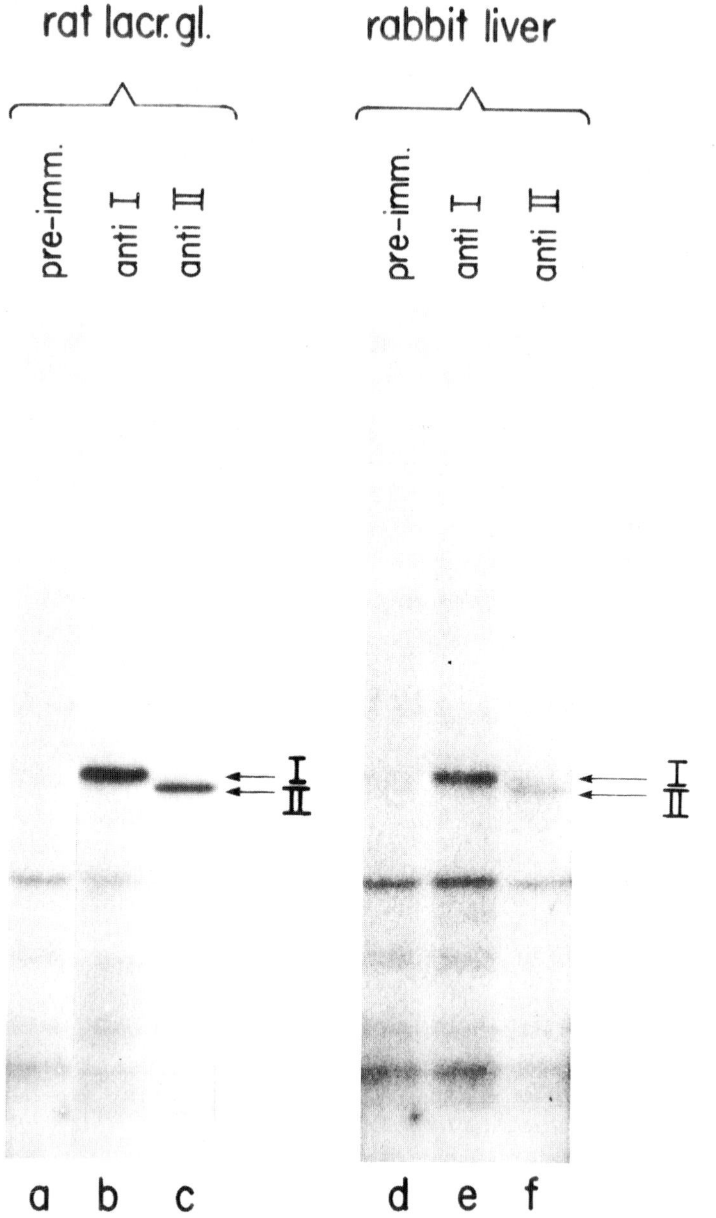

FIG. 3. Immunoprecipitation of ribophorins from rat lacrimal gland and rabbit liver with anti-rat ribophorin antibodies. Sedimentable residues obtained from rough microsomes of rat lacrimal gland (a–c) and rabbit liver (d–f) were labeled with ^{125}I and incubated for immunoprecipitation. Immunoprecipitates recovered after incubation with preimmune IgG (a, d), and anti-rat liver ribophorin I (b, e) or II (c, f) were analyzed by electrophoresis on SDS–polyacrylamide (6 to 11%) gels followed by autoradiography for 4 days.

karyotic (60 S) but not prokaryotic (50 S) large ribosomal subunits.[15] The capacity of the membranes to bind ribosomes is abolished by mild trypsinization, which suggests the involvement of membrane proteins in ribosome recognition.[14]

Preparation of ^{3}H-Labeled Ribosomes from Myeloma Cells

MPC 11 mouse myeloma cells grown in suspension (5 to 10 × 10^8 cells) are labeled for 2 days with [^{3}H]uridine (500 μCi). The cells are recovered by sedimentation (15 min, at 2000 rpm in the Sorvall 276 rotor) and are transferred with 20 ml of RSB to a tight Dounce homogenizer. The suspension is left on ice for 20 min and then homogenized with 20 strokes. The homogenate is centrifuged (15 min at 2000 *g*), and the supernatant is decanted and saved. The nuclear pellet is washed with TKM, and the resulting supernatant is combined with the first. The combined supernatants are transferred to a tube for the Ti60 rotor, underlaid with 5 ml of 1.5 *M* sucrose containing TKM, and centrifuged (12 hr at 40,000 rpm in a Ti60 Beckman rotor) to sediment the microsomes. Pellets are resuspended in 10 m*M* Tris-HCl, pH 7.6 (80 OD_{260}/ml), and an equal volume of 2 × 0.25 *M* sucrose HSB containing 10^{-3} *M* puromycin is added.[16] The suspension is incubated (25°, 45 min), centrifuged (15 min at 40,000 rpm in the Beckman Ti50 rotor), and the supernatant containing the dissociated ribosomes is decanted. After fivefold dilution with TKM, the supernatant is transferred to Ti50 tubes, underlaid with 1 ml of TKM containing 1.6 *M* sucrose, and centrifuged (12 hr at 40,000 rpm in a Ti50 rotor). The final ribosomal pellets are rinsed with binding buffer and stored at −70°. The specific activity of the labeled ribosomes ranges from 800 to 1700 cpm per microgram of protein.

In Vitro Binding of 80 S Ribosomes to Microsomes

Rat liver microsomal fractions of different isopycnic densities and RNA-to-protein ratios are isolated according to Kruppa and Sabatini.[17] RNA-to-protein ratios are generally 0.24 for heavy RM (RM_H), 0.115 for light RM (RM_L), and 0.05 for smooth microsomes (SM).

After removal of ribosomes by incubation in HSB containing 2.5 m*M* $MgCl_2$ and puromycin (for details, see Adelman *et al.*[16]) stripped microsomal membranes are sedimented (15 min at 37,000 rpm in a Ti60 rotor) and washed after resuspension and sedimentation in binding buffer. *In vitro* ribosome binding assays are carried out essentially according to

[15] W. Mok, C. Freienstein, D. D. Sabatini, and G. Kreibich, *J. Cell Biol.* **70** (2), 393a (1976).
[16] M. R. Adelman, D. D. Sabatini, and G. Blobel, *J. Cell Biol.* **56,** 206 (1973).
[17] J. Kruppa and D. D. Sabatini, *J. Cell Biol.* **74,** 414 (1977).

TABLE II
In Vitro REBINDING OF ^{3}H-LABELED 80 S RIBOSOMES TO MICROSOMAL MEMBRANES STRIPPED OF RIBOSOMES

Membrane fraction[a]	RNA : protein ratios of microsomal fractions: Unstripped	Stripped[b]	^{3}H-Labeled 80 S ribosomes[c] rebound[d]	Affinity constant K^e
RM_H	0.25	0.06	0.130 (68%)	$8.7 \times 10^7\ M^{-1}$
RM_L	0.12	0.03	0.054 (60%)	$10.3 \times 10^7\ M^{-1}$
SM	0.05	0.01	0.012 (30%)	$4.4 \times 10^7\ M^{-1}$

[a] Microsomal membranes were prepared according to Kruppa and Sabatini.[17]
[b] Stripping of ribosomes was done as described by Adelman *et al.*[16]
[c] Calculated from Scatchard plots by extrapolation to zero ordinate values.
[d] For the calculation of the percentage of ribosomes rebound, the values for the RNA : protein ratio of the stripped microsomal fraction was subtracted from that of the corresponding unstripped microsomes and taken as 100%.
[e] Affinity constant calculated from the slope of Scatchard plots.

Borgese *et al.*[14] The incubation mixtures contain stripped membranes (usually 0.2–0.3 mg of protein) and ^{3}H-labeled ribosomes (7–10 μg) in 0.12 ml of binding buffer. After a 30-min incubation at 0°, each sample receives 1.08 ml of binding buffer containing 2.3 *M* sucrose, so that the final sucrose concentration is about 2.1 *M*. The samples are thoroughly mixed, and 0.8-ml aliquots are layered under a continuous sucrose gradient (1.3–1.9 *M*, 3.8 ml) containing binding buffer that is covered with 0.2 ml of binding buffer. After centrifugation (60 min at 50,000 rpm in a Beckman SW56 rotor), the following subfractions are recovered using an Eppendorf pipette: (a) top 2 ml containing floated microsomal membranes; (b) middle 1 ml; (c) bottom 1.8 ml containing unbound ribosomes. Pellets are resuspended in water and added to fraction (c). All fractions receive 1 mg of bovine serum albumin, and volumes are adjusted to 2.25 ml with water. After addition of 50% trichloracetic acid (5% final), tubes are mixed, incubated at 0° for 2 hr, and centrifuged (45 min at 2000 *g*). Supernatants are discarded. After brief recentrifugation (5 min) and removal of remaining trichloroacetic acid, pellets are dissolved in 0.5 ml of Protosol [New England Nuclear (NEN)], and 16 ml of toluene-liquifluor are added for scintillation counting. Counting efficiency for tritium is 45–50%, and backgrounds are subtracted. The radioactivity recovered in fraction (a) corresponds to ribosomes bound to microsomal membranes, while unbound ribosomes are found in the pellet and in fraction (c). Data are presented in Scatchard plots[14] from which affinity constants and number of binding sites can be derived.

As can be seen from Table II, about 65% of ribosomal binding sites uncovered by stripping of rough microsomes with puromycin are active in rebinding 80 S ribosomes. The binding constant for this reaction, determined from the slope of Scatchard plots, is 10^{-8} mol^{-1}. The binding capacity of smooth microsomes is less than 10% of that observed for stripped membranes from heavy rough microsomes, and the binding constant was significantly lower. The ribophorin content of different microsomal fractions (RM_H, RM_L, and SM) determined from Coomassie Blue-stained gels is proportional to the RNA-to-protein ratios before stripping and the ribosome binding capacity after stripping.

[42] Biosynthesis of Hepatocyte Endoplasmic Reticulum Proteins

By GERT KREIBICH, DAVID D. SABATINI, and MILTON ADESNIK

The rat hepatocyte serves as an excellent model for the study of the biosynthesis of endoplasmic reticulum (ER) membrane proteins. The prominent role of the liver in the synthesis and secretion of serum proteins[1] accounts for the presence in the cytoplasm of the hepatocyte of a very well-developed system of rough ER cisternae to which a major fraction (50–70%) of the ribosome population is attached.[2,3] In addition, the participation of ER membrane enzymes in many metabolic functions characteristic of the liver is reflected in an extensive development of a tubular network of smooth ER membranes, which penetrates most regions of the hepatocyte cytoplasm.

In the hepatocyte, as in all cell types, the biosynthetic machinery for the synthesis of secretory and membrane proteins includes both rough and smooth ER membrane-associated components which affect a variety of co- and posttranslational modifications of polypeptides synthesized in

[1] D. D. Sabatini, G. Kreibich, T. Morimoto, and M. Adesnik, *in* "Plasma Protein Secretion by the Liver" (H. Glauman, T. Peters, Jr., and C. Redman, eds.), p. 563. Academic Press, New York, 1981.

[2] J. C. Ramsey and W. J. Steele, *Biochemistry* **15,** 1704 (1976).

[3] J. C. Ramsey and W. J. Steele, *Anal. Biochem.* **92,** 305 (1979).

ISBN 0-12-181996-5

bound polysomes.[1,4,5] Although only a few of them have been biochemically well defined, the existence of specific ER enzyme systems that carry out proteolytic cleavages,[6] transfer and trimming of core oligosaccharides[7] formation of disulfide bridges,[8] and other modifications of specific amino acid residues such as hydroxylation,[9] acetylation,[10,11] and phosphorylation,[12] is well established. Other components of ER membranes play a prominent role in the biosynthesis of lipids.[13] In the hepatocyte, phospholipids and cholesterol synthesized by ER membrane enzymes not only serve as constituents of cell membranes but, together with triglycerides synthesized in the ER and proteins synthesized in membrane-bound ribosomes, are exported in large amounts into the bloodstream as components of several classes of lipoproteins secreted by the liver.[14]

The rich enzymic endowment of hepatocyte ER membranes also includes many activities which participate in other specific aspects of liver metabolism such as the conjugation of bilirubin and steroids with glucuronic acid moieties[15,16] or the release of glucose from glucose 6-phosphate[17,18] produced by glycogen degradation or gluconeogenesis. In addition, hepatocyte ER membranes contain two distinct but interrelated electron transport chains which transfer reducing equivalents from NADH or NADPH through cytochromes b_5 and cytochrome P-450, respectively, to small molecular weight substrates.[19] These electron transport chains function in the desaturation of fatty acids[20] and the oxidative

[4] G. Kreibich, S. Bar-Nun, M. Czako-Graham, U. Czichi, E. Marcantonio, M. G. Rosenfeld, and D. D. Sabatini, *in* "International Cell Biology 1980–1981" (H. G. Schweiger, ed.), p. 579. Springer-Verlag, Berlin and New York, 1981.

[5] D. D. Sabatini, G. Kreibich, T. Morimoto, and M. Adesnik, *J. Cell Biol.* **92,** 1 (1982).

[6] R. C. Jackson and G. Blobel, *Proc. Natl. Acad. Sci. U.S.A.* **74,** 5598 (1977).

[7] S. C. Hubbard and R. J. Ivatt, *Ann. Rev. Biochem.* **50,** 555 (1981).

[8] F. DeLorenzo, R. F. Goldberger, E. Steers, Jr., D. Girol, and C. B. Anfinsen, *J. Biol. Chem.* **241,** 1562 (1966).

[9] B. R. Olsen, R. A. Berg, Y. Kishida, and D. J. Prockop, *Science* **182,** 825 (1973).

[10] R. D. Palmiter, J. Gagnon, and K. A. Walsh, *Proc. Natl. Acad. Sci. U.S.A.* **75,** 94 (1978).

[11] A. Pestana and H. C. Pitot, *Biochemistry* **14,** 1404 (1975).

[12] R. S. Sharma, C. A. McLaughlin, and H. C. Pitot, *Eur. J. Biochem.* **65,** 577 (1976).

[13] R. M. Bell and R. A. Coleman, *Ann. Rev. Biochem.* **49,** 459 (1980).

[14] P. Siuta-Mangano, D. R. Janero, and M. D. Lane, *J. Biol. Chem.* **257,** 11463 (1982).

[15] H. Beaufay, A. Amar-Costesec, D. Thines-Sempoux, M. Wibo, M. Robbi, and J. Berthet, *J. Cell Biol.* **61,** 213 (1974).

[16] J. P. Gorsky and C. B. Kasper, *J. Biol. Chem.* **252,** 1336 (1977).

[17] R. C. Nordlie, *Curr. Top. Cell. Reg.* **8,** 33 (1974).

[18] M. A. Zoccoli, R. R. Hoopes, and M. L. Karnovsky, *J. Biol. Chem.* **257,** 3319 (1982).

[19] R. F. Sato and T. Omura, "Cytochrome *P*-450." Kodansha, Tokyo and Academic Press, New York (1978).

[20] P. Strittmatter, L. Spatz, D. Corcoran, M. J. Rogers, B. Setlow, and R. Redline, *Proc. Natl. Acad. Sci. U.S.A.* **71,** 4565 (1974).

modification of a large number of endogenous and exogenous substances.[21]

The sets of protein constituents found in rough and smooth ER membranes appear to be very similar, the major differences apparently resulting from the presence in rough ER of components involved in protein synthetic functions, such as the recognition of signal segments in nascent polypeptide chains,[22–24] the binding of ribosomes to the membranes, [cf.1,25] and the modification of nascent polypeptides.[1,6,7] Detectable differences between rough and smooth regions of the ER in the relative content of some proteins, which are not directly related to the function of bound ribosomes, have also been reported.[cf.26] However, it has been proposed[26] that at least some of these are an indirect result of steric effects caused by the presence of ribosomes and their associated membrane proteins on the cytoplasmic aspect of rough ER membranes.

Studies of hepatocyte ER membrane biogenesis have been facilitated by the fact that continuous synthesis of membrane components occurs in adult animals to replace molecules destroyed during normal turnover.[27] In addition, a general stimulation in the synthesis of ER membrane proteins may be experimentally induced during liver regeneration after partial hepatectomy[28] or a selective increase in the rate of synthesis of certain proteins can be produced by the administration of specific drugs.[29]

Several general conclusions have been drawn from recent work on the biosynthesis of ER membrane proteins.[30–37] It has been found that both

[21] A. Y. H. Lu and S. B. West, *Pharm. Rev.* **31,** 277 (1980).

[22] D. I. Meyer, E. Krause, and B. Dobberstein, *Nature* (*London*) **297,** 647 (1982).

[23] R. Gilmore, G. Blobel, and P. Walter, *J. Cell Biol.* **95,** 463 (1982).

[24] R. Gilmore, P. Walter, and G. Blobel, *J. Cell Biol.* **95,** 470 (1982).

[25] G. Kreibich, E. E. Marcantonio, and D. D. Sabatini, this volume [41].

[26] H. Beaufay, A. Amar-Costesec, and C. deDuve, *in* "Mitochondria and Microsomes" (C. P. Lee, G. Schatz, and G. Dallner, eds.), p. 629. Addison-Wesley, London, 1981.

[27] A. L. Hubbard, *in* "Transport of Macromolecules in Cellular Systems" (S. C. Silverstein, ed.), p. 363. Dahlem Konferenzen, Berlin 1978.

[28] O. A. Scornik, *J. Biol. Chem.* **249,** 3876 (1974).

[29] J. W. DePierre, J. Seidegard, R. Morgenstern, L. Balk, J. Meijer, and A. Åstrôm, *in* "Mitochondria and Microsomes" (C. P. Lee, G. Schatz, and G. Dallner, eds.), p. 585. Addison-Wesley, London, 1981.

[30] S. Bar-Nun, G. Kreibich, M. Adesnik, L. Alterman, M. Negishi, and D. D. Sabatini, *Proc. Natl. Acad. Sci. U.S.A.* **77,** 965 (1980).

[31] E. Bresnik, M. Brosseau, W. Levine, L. Reik, D. E. Ryan, and P. E. Thomas, *Proc. Natl. Acad. Sci. U.S.A.* **78,** 4083 (1981).

[32] M. G. Rosenfeld, E. E. Marcantonio, V. M. Harnik, D. D. Sabatini, and G. Kreibich, *J. Cell Biol.* **91,** 404a (1981).

[33] Y. Okada, A. B. Frey, T. M. Guenthner, F. Oesch, D. D. Sabatini, and G. Kreibich, *Eur. J. Biochem.* **122,** 393 (1982).

[34] F. J. Gonzalez and C. B. Kasper, *Biochemistry* **19,** 1790 (1980).

co- and posttranslational insertion mechanisms are utilized for the incorporation of newly synthesized ER polypeptides into the membranes. In particular, certain integral membrane proteins such as cytochrome b_5 and NADH-cytochrome b_5 reductase, which are anchored to the membranes solely by a small carboxy terminal segment, are incorporated posttranslationally into the ER membrane after their discharge from free polysomes into the cell sap.[33,36,37] Other ER membrane proteins, however, only two of which have been demonstrated to be glycoproteins[32,38] and to span the membrane,[38] are synthesized in membrane-bound ribosomes and are inserted into the membrane during their synthesis.[32] Furthermore, except for the two glycoproteins, all other ER membrane polypeptides studied to date appear to retain in their mature form the signals which during synthesis mediate their cotranslational insertion into the membrane.[30,31,33–35]

Polypeptides Inserted Posttranslationally into ER Membranes

Cytochrome b_5 (M_r 18 × 10^3) and NADH-cytochrome b_5 oxidoreductase (M_r 33 × 10^3) are part of an electron transport chain which is involved in the desaturation of fatty acids.[19,20,39,40] Both components are concentrated in the ER but their presence in other cell fractions enriched in Golgi elements, plasma membrane fragments, mitochondria, and peroxisomes has also been demonstrated.[40–45] The accessibility of these proteins to nonpenetrating labeling probes,[46] to antibody ferritin complexes,[41] and the arrest of electron flow caused by the addition of specific antibodies to intact microsomal vesicles[44,47,48] suggest that both polypeptides are exposed on the cytoplasmic face of the ER. In addition, it was found that

[35] F. J. Gonzalez and C. B. Kasper, *Biochim. Biophys. Res. Commun.* **93,** 1254 (1980).
[36] N. Borgese, and S. Gaetani, *FEBS Lett.* **112,** 216 (1980).
[37] R. A. Rachubinski, D. P. S. Verma, and J. J. M. Bergeron, *J. Cell Biol.* **84,** 705 (1980).
[38] E. Rodriguez-Boulan, D. D. Sabatini, B. N. Pereyra, and G. Kreibich, *J. Cell. Biol.* **78,** 894 (1978).
[39] L. Spatz and P. Strittmatter, *J. Biol. Chem.* **248,** 793 (1973).
[40] J. Ozols and C. Gerard, *Proc. Natl. Acad. Sci. U.S.A.* **74,** 3725 (1977).
[41] S. Fowler, J. Remacle, A. Trouet, H. Beaufay, J. Berthet, M. Wibo, and P. Hauser, *J. Cell Biol.* **71,** 535 (1976).
[42] Y. Okada and T. Omura, *J. Biochem.* **3,** 1039 (1978).
[43] E.-D. Jarasch, J. Kartenbeck, G. Bruder, A. Fink, J. Morre, and W. W. Franke, *J. Cell Biol.* **80,** 37 (1979).
[44] N. Borgese and J. Meldolesi, *J. Cell Biol.* **85,** 501 (1980).
[45] N. Borgese, G. Pietrini, and J. Meldolesi, *J. Cell Biol.* **86,** 38 (1980).
[46] O. S. Nilsson, J. W. DePierre, and G. Dallner, *Biochim. Biophys. Acta* **511,** 93 (1978).
[47] J. Remacle, S. Fowler, H. Beaufay, A. Amar-Costesec, and J. Berthet, *J. Cell Biol.* **71,** 551 (1976).
[48] S. Kuwahara, Y. Okada, and T. Omura, *J. Biochem.* **83,** 1049 (1978).

polypeptides (12×10^3 for cytochrome b_5 and 28×10^3 for the reductase) representing major portions of cytochrome b_5[40,49] and its reductase[39,50,51] can be released from microsomal vesicles by proteolytic digestion. An analysis of the amino acid sequence of cytochrome b_5 and its hydrophilic and hydrophobic segments indicates that the native protein is anchored to the phospholipid bilayer via its hydrophobic carboxy terminal segments.[40] The limited structural data available suggest a similar disposition for NADH-cytochrome b_5 reductase.[52]

Cell fractionation experiments have demonstrated that translatable mRNAs for cytochrome b_5 and NADH-cytochrome b_5 reductase are recovered exclusively in free polysomes.[33,36,37] It was estimated that each messenger activity accounts for approximately 0.01% of the total [^{35}S]methionine incorporated into *in vitro*-produced polypeptides. When proper precautions were taken to prevent proteolytic degradation of the *in vitro* products during immunoprecipitation, their electrophoretic mobility corresponded to that of the mature proteins purified from detergent-treated microsomes. The *in vitro*-synthesized polypeptides appeared to be unusually sensitive to proteolysis. During incubation with antibodies for more than 15 min at 20° followed by protein A-Sepharose for 30 min, cytochrome b_5 was degraded to a product which migrated as a sharp band with apparent molecular weight of 11×10^3. This degradation could not be prevented by the addition of a mixture of protease inhibitors (Trasylol, pepstatin, leupeptin, chymostatin, or antipain) which, however, was sufficient to preserve the integrity of *in vitro*-synthesized NADH-cytochrome b_5 reductase.[33]

The localization of messenger RNAs for cytochrome b_5 and NADH-cytochrome b_5 reductase in free polysomes implies that *in vivo* the completed polypeptides are released into the cell sap before they are incorporated into the membranes. The *in vitro*-synthesized polypeptides were, indeed, incorporated posttranslationally into microsome membranes (rough as well as smooth) and into the membranes of mitochondria when the corresponding cell fractions were added to translation mixtures after synthesis was completed.[33] Since this association was not disrupted when the membranes were subsequently incubated in a medium of high salt concentration (0.5 *M* KCl), it is reasonable to presume that it reflects a partial insertion of the polypeptides into the hydrophobic phase of the membranes, probably through the same carboxy terminal segments that anchor the mature proteins to the ER membranes. These findings are in

[49] L. Spatz and P. Strittmater, *Proc. Natl. Acad. Sci. U.S.A.* **68,** 1042 (1971).
[50] S. Takesue and T. Omura, *J. Biochem.* **67,** 267 (1970).
[51] K. Mihara, K. Sato, R. Sakakibara, and H. Wada, *Biochemistry* **17,** 2829 (1978).
[52] S. Tajima, T. Mihara, and R. Sato, *Arch. Biochem. Biophys.* **198,** 137 (1979).

agreement with the results of *in vivo* labeling experiments[42,44,45] which showed the rapid and simultaneous incorporation of newly synthesized reductase into ER mitochondria and Golgi membranes. It has also been shown that the mature forms of cytochrome b_5 and its reductase purified by detergent treatments are readily incorporated into microsomes[53,54] or into synthetic phospholipid vesicles,[55] where they form functional complexes. The fact that the concentrations of cytochrome b_5 and its reductase are much higher in the ER than in other membranes may reflect the affinity of these proteins for other components of the microsomal electron transport chains, such as those which are synthesized on bound polysomes.

ER Polypeptides Inserted Cotranslationally into the Membranes

Recent work has shown that three nonglycosylated integral membrane proteins of the ER, cytochrome *P*-450[30,56] its reductase,[33,34] and epoxide hydrolase[33,35] are synthesized on membrane-bound ribosomes and inserted into the membranes without cleavage of peptide segments that may serve as cotranslational insertion signals. Cytochrome *P*-450 polypeptides (M_r 48–58 × 10^3) constitute a family of microsomal hemoproteins which in their reduced form bind CO and show characteristic difference absorption spectra with maxima near 450 nm. Hepatocyte cytochromes *P*-450 serve as terminal oxidases in an electron transport chain which includes a specific flavoprotein, cytochrome *P*-450 reductase (NADPH-cytochrome *P*-450 oxidoreductase; M_r 78 × 10^3). This chain functions in the oxidation of endogenous substrates as well as in the modification of drugs, carcinogens, and other xenobiotics.[cf.19,57,58] Distinct forms of cytochrome *P*-450 are induced in response to different drugs and can amount to more than 5% of the total microsomal membrane proteins.[21]

Epoxide hydrolase (M_r 49 × 10^3) is an enzyme that functions in association with the cytochrome *P*-450 system, being able to modify products of the terminal oxidase and to generate new substrates for oxidation by cytochrome *P*-450.[cf.59] Both cytochrome *P*-450 and its reductase have been shown to be exposed on the cytoplasmic face of ER membranes.

[53] P. Strittmatter, M. J. Rogers, and L. Spatz, *J. Biol. Chem.* **247,** 7188 (1972).

[54] M. J. Rogers and P. Strittmatter, *J. Biol. Chem.* **249,** 5565 (1974).

[55] M. J. Rogers and P. Strittmatter, *J. Biol. Chem.* **250,** 5713 (1975).

[56] M. Adesnik, S. Bar-Nun, F. Maschio, M. Zunich, A. Lippman, and E. Bard, *J. Biol. Chem.* **256,** 10340 (1981).

[57] A. H. Conney, *Pharm. Rev.* **19,** 317 (1967).

[58] J. F. Gillette, D. C. Davis, and H. A. Sasame, *Ann. Rev. Pharm.* **12,** 57 (1972).

[59] F. Oesch, *in* "Enzymatic Basis of Detoxication" (W. B. Jacoby, ed.), p. 277. Academic Press, New York, 1980.

Thus, anticytochrome *P*-450 antibodies added to intact microsomal vesicles inhibit the terminal oxidase function[cf.60] and proteolysis experiments show that a large portion of cytochrome *P*-450 reductase (72 kilodaltons), containing its active site, can be cleaved off from intact microsomes.[61] In direct contrast to cytochrome b_5 and its reductase, NADPH-cytochrome *P*-450 reductase is anchored to the ER membrane by the amino terminal segment of the polypeptide.[62] The location of the anchoring segments of cytochrome *P*-450 and of epoxide hydrolase within the polypeptides have not been determined. In fact, there is no direct evidence that any portion of these polypeptides is exposed on the luminal side of the ER membrane and therefore that the proteins have a true transmembrane disposition.

NADPH-Cytochrome *P*-450 Reductase

Treatment of rats with phenobarbital leads to a striking proliferation of ER membranes in hepatocytes[63] and to markedly increased levels of specific forms of cytochrome *P*-450,[cf.21] as well as to a less-pronounced increase in NADPH-cytochrome *P*-450 reductase.[60,64] Levels of translatable mRNA for NADPH-cytochrome *P*-450 reductase in rat liver increase severalfold during phenobarbital treatment for 3 days, while mRNA levels for cytochrome b_5, its reductase, or serum albumin remain unaffected.[33,34] *In vitro* translation experiments using free and bound polysomes obtained from livers of phenobarbital-treated rats showed that the reductase mRNA is found exclusively in bound polysomes. The electrophoretic mobility of NADPH-cytochrome *P*-450 reductase polypeptides synthesized *in vitro* in mRNA-dependent systems was indistinguishable from that of the mature protein under conditions of electrophoresis which allowed the resolution of preproalbumin (M_r 69 × 10^3) as a distinct band from proalbumin (M_r 67 × 10^3). This suggests that the nascent polypeptide of NADPH-cytochrome *P*-450 reductase does not undergo removal of a transient amino terminal signal. Partial amino terminal sequence analysis of the *in vitro*-synthesized reductase[33] showed that there are 7 leucine residues among the first 15 amino acids. It therefore seems likely that the amino terminal portion of the polypeptide is hydrophobic and may serve to initiate insertion of the nascent chain into the ER membrane. The sequence of the *in vitro*-synthesized product, however, could not be compared directly to that of the native rat protein, which contains a blocked

[60] Y. Kuriyama, T. Omura, P. Siekevitz, and G. E. Palade, *J. Biol. Chem.* **244,** 2017 (1969).
[61] B. S. S. Masters and R. J. Okita, *Pharm. Ther.* **9,** 227 (1980).
[62] S. D. Black and M. J. Coon, *J. Biol. Chem.* **257,** 5929 (1982).
[63] W. Staubli, R. Hess, and E. R. Weibel, *J. Cell Biol.* **42,** 92 (1969).
[64] L. Ernster and S. Orrenius, *Fed. Proc.* **24,** 1190 (1965).

amino terminus and has not yet been sequenced. On the other hand, amino acid sequence data for rabbit liver reductase[62] indicate that in this species the first 15 amino acids do not constitute a particularly hydrophobic segment, although the following 17 are very hydrophobic and are followed by 6 successive charged amino acid residues. The partial sequence data suggest that a segment corresponding to the amino terminal portion of the rat reductase can be aligned wth residues 13 to 31 in the rabbit sequence.[65] Thus, it is possible that whereas in the rat a signal for cotranslational insertion is located in the extreme amino terminal position, an equivalent signal in the rabbit protein is preceded by a short peptide segment which does not impair its signaling function. This would be consistent with a "loop" model[5,66,67] for the transmembrane disposition of the portion of the nascent chain initiating vectorial insertion into the membrane. The concentration of charged residues after the hydrophobic segment would be expected to prevent sequences immediately following it from penetrating the membrane. This would confer to the amino terminal segment the combined properties of insertion and halt-transfer signal and would explain the exposure of the remaining portion of the protein on the cytoplasmic side of the membrane.

Epoxide Hydrolase

Substantially increased levels of translatable epoxide hydrolase mRNA in rat liver can be obtained after treatment of animals with the carcinogens *trans*-stilbene oxide[35] or 2-acetylaminofluorene,[33] which lead to the formation of hyperplastic nodules.

Epoxide hydrolase mRNA was exclusively found on bound polysomes and the product of its *in vitro* translation had the same electrophoretic mobility as the mature protein, suggesting that the polypeptide is not processed by removal of an amino terminal insertion signal.[33,35] The retention of the original amino terminal segment in the mature protein was demonstrated directly[33] by comparing a partial amino terminal sequence of the immunoprecipitated translation product with the amino terminal sequence of mature epoxide hydrolase.[68] As is the case for cytochrome *P*-450, epoxide hydrolase has an amino terminal methionine and a glutamic

[65] S. D. Black, G. E. Tarr, and M. J. Coon, *in* "Cytochrome *P*-450, Biochemistry, Biophysics, and Environmental Implications" (E. Hietanen, M. Laitinen, and O. Hanninen, eds.), p. 277. Elsevier Biomedical, Amsterdam, 1982.

[66] M. Inouye and S. Halegoua, *CRC Crit. Rev. Biochem.* **7,** 339 (1979).

[67] G. Kreibich, M. Czako-Graham, R. C. Grebeneau, and D. D. Sabatini, *Ann. N.Y. Acad. Sci.* **343,** 17 (1980).

[68] L. H. Botelho, D. E. Ryan, and W. Levin, *J. Biol. Chem.* **254,** 5635 (1979).

acid close to its amino terminus. In addition, the amino terminal sequence of the mature epoxide hydrolase contains an uncharged, largely hydrophobic region of 15 amino acids and therefore resembles other signal sequences[cf.69] characteristic of polypeptides synthesized on bound polysomes.

Recently, cDNA clones for epoxide hydrolase[70] and NADPH-cytochrome *P*-450 reductase[71] as well as recombinant bacteriophages containing the natural gene for the latter protein[72] have been obtained and therefore the complete amino acid sequence of these proteins may be available in the near future.

Cytochrome *P*-450

Phenobarbital administration represents a classical treatment to increase hepatic microsomal levels of cytochrome *P*-450.[cf.57] In the rat this drug induces 30- to 50-fold higher levels[73] of two or more immunologically indistinguishable cytochrome *P*-450 forms.[74,75] Other treatments, such as administration of polycyclic hydrocarbons (3-methylcholanthrene), synthetic steroids (pregnenolone 16-α-carbonitrile), and natural (isosaffrole) and synthetic (Araclor) aromatic compounds lead to the accumulation of other forms of cytochrome *P*-450 which differ structurally and immunologically from each other and from the phenobarbital-induced forms.[cf.76,77] The various forms of *P*-450, however, have characteristic but broad and overlapping substrate specificities, and some show slight differences in spectroscopic properties.[cf.21,77] Both *in vitro* translation experiments and hybridization to a cloned cDNA for a phenobarbital-induced form of *P*-450 demonstrate that phenobarbital treatment leads to an increase in levels of the corresponding mRNA, which takes place soon after administration of the inducer.[30,56] Translatable cytochrome *P*-450 mRNA has only been found in membrane-bound polysome preparations[30] and quantitative

[69] D. F. Steiner, P. S. Quinn, S. J. Chan, J. Marsh, and H. S. Tager, *Ann N.Y. Acad. Sci.* **343,** 1 (1980).

[70] F. J. Gonzalez and C. B. Kasper, *J. Biol. Chem.* **256,** 4697 (1981).

[71] F. J. Gonzalez and C. B. Kasper, *J. Biol. Chem.* **257,** 5962 (1982).

[72] F. J. Gonzalez and C. B. Kasper, *J. Biol. Chem.* **258,** 1363 (1983).

[73] P. E. Thomas, D. Korzeniowski, D. Ryan, and W. Levin, *Arch. Biochem. Biophys.* **192,** 524 (1979).

[74] F. P. Guengerich, *Pharm. Ther.* **6,** 99 (1979).

[75] G. P. Vlasuk, J. Ghrayeb, D. E. Ryan, L. Reik, P. E. Thomas, W. Levin, and F. G. Waltz, Jr., *Biochemistry,* **21,** 789 (1982).

[76] G. P. Vlasuk, D. E. Ryan, P. E. Thomas, W. Levin, and F. G. Walz, *Biochemistry* **21,** 6288 (1982).

[77] D. E. Ryan, P. E. Thomas, L. M. Reik, and W. Levin, *Xenobiotica* **12,** 727 (1982).

cDNA hybridization experiments demonstrated that more than 97% of *P*-450 mRNA sequences are recovered with rough microsomes.[56] The *in vitro*-synthesized polypeptide has been cotranslationally incorporated into microsomal membranes and, as is the case with the native product, it could only be released by detergent concentrations much higher than those required to open the microsomal vesicles and permit escape of the luminal content.[30] Like its reductase and epoxide hydrolase, cytochrome *P*-450 does not undergo cotranslational removal of an insertion signal.[30] This was first suspected for the amino terminal sequence of the mature rabbit phenobarbital-induced cytochrome *P*-450.[78] The primary translation product does not change in size when inserted into membranes and its partial amino terminal sequence is in agreement with that of the mature protein.[30] It may be speculated that the amino terminal segment represents a composite insertion and halt-transfer signal,[cf.5] since a group of charged residues immediately follows the hydrophobic region.[79,80]

Recently, cDNA clones or phenobarbital-induced forms of rat liver cytochrome *P*-450 have been obtained and the complete sequences of two closely related phenobarbital-induced forms were derived from their nucleotide sequences.[80,81] These data are nearly in complete agreement with amino acid sequence information covering approximately 75% of the two related polypeptides.[82]

No segment of the *P*-450 polypeptide sequence long enough to span the membrane is as hydrophobic as the amino terminal region (residues 1–20), although two uncharged segments (residues 167–186 and 445–462) are present within the protein. Neither of these two segments, however, appears sufficiently hydrophobic to function as classical cotranslational insertion or halt transfer signals. It may therefore be proposed that the portion of the molecule following the putative amino terminal permanent-insertion half-transfer signal remains exposed on the cytoplasmic aspect of the ER membrane. It cannot be excluded, however, that the relatively uncharged segments of the polypeptide could become inserted into the membrane posttranslationally, in analogy to the carboxy terminal segments of cytochrome b_5 and its reductase. Lateral interactions between membrane-embedded segments could stabilize the structure.

[78] D. A. Haugen, L. G. Armes, K. T. Yasunobu, and M. J. Coon, *Biochem. Biophys. Res. Comm.* **77,** 967 (1977).

[79] D. J. Waxman and C. Walsh, *J. Biol. Chem.* **257,** 10446 (1982).

[80] Y. Fujii-Kuriyama, Y. Mizukami, K. Kawajiri, K. Sogawa, and M. Muramatsu, *Proc. Natl. Acad. Sci. U.S.A.* **79,** 2793 (1982).

[81] A. Kumar, C. Raphael, and M. Adesnik, *J. Biol. Chem.* **258** (in press).

[82] P.-M. Yuan, D. E. Ryan, W. Levin, and J. E. Shively, *Proc. Natl. Acad. Sci. U.S.A.* **80,** 1169 (1983).

On the basis of recently obtained protein sequence data for rabbit phenobarbital-induced cytochrome *P*-450 (LM_2), a model for the transmembrane disposition of cytochrome *P*-450 has been proposed in which five transmembrane segments located within the amino terminal half of the molecule are depicted and are connected by short polar segments exposed on each side of the membrane.[83] This model requires the formation of intrahelical as well as interhelical ion pairs to reduce the prohibitively high free energy required for the insertion of the transmembrane segments into the hydrophobic phase. A large hydrophilic segment near the carboxy terminus is presumed to remain exposed on the cytoplasmic side of the membrane. In this context it is worth noting that immunoelectron microscopic studies[84] have failed to detect antigenic sites for phenobarbital-induced *P*-450 on the luminal side of the perinuclear ER cisterna, although such sites were easily detectable on the cytoplasmic surface of the ER membrane.

Ribophorins

Ribophorins (I and II, M_r 65×10^3 and 63×10^3, respectively) are transmembrane glycoproteins[38] which are characteristic of the rough ER[85–87] and appear to participate in the process by which nascent polypeptides synthesized in bound polysomes are inserted into the membranes.[1,4,25] *In vitro* translation of mRNA extracted from free and bound polysomes of actively dividing rat hepatocytes in culture demonstrated that ribophorins are made exclusively on bound polysomes. The primary translation products of the messengers for the two ribophorins, obtained from cultured hepatocytes or from regenerating livers, comigrated with the respective mature proteins but had slightly higher apparent molecular weights (2 kilodaltons) than the unglycosylated forms immunoprecipitated from cells treated with tunicamycin. This suggests that the polypeptides undergo cotranslational removal of an amino terminal insertion signal.[32]

Kinetic and pulse–chase experiments using untreated cultured rat hepatocytes and cells treated with tunicamycin[32] showed that ribophorins undergo cotranslational core glycosylation in the ER, but are not sub-

[83] F. S. Heinemann and J. Ozols, *J. Biol. Chem.* **257,** 14988 (1982).

[84] S. Matsuura, Y. Fujii-Kuriyama, and Y. Tashiro, *J. Cell Biol.* **78,** 503 (1978).

[85] G. Kreibich, C. M. Freienstein, B. N. Pereyra, B. L. Ulrich, and D. D. Sabatini, *J. Cell Biol.* **77,** 488 (1978).

[86] G. Kreibich, B. L. Ulrich, and D. D. Sabatini, *J. Cell Biol.* **77,** 464 (1978).

[87] E. Marcantonio, R. C. Grebenau, D. D. Sabatini, and G. Kreibich, *Eur. J. Biochem.* **124,** 217 (1982).

jected to electrophoretically detectable posttranslational modifications such as proteolytic cleavage or extensive trimming of mannose residues and terminal glycosylation of oligosaccharides. In addition, the oligosaccharides of the mature proteins are sensitive to endoglycosidase H and insensitive to endoglycosidase D.[32] Because trimming of most of the mannose residues from core oligosaccharides and terminal glycosylation take place in the Golgi apparatus,[cf.7] these observations indicate that ribophorins, once inserted into the ER, are permanently retained in this organelle and never reach the Golgi apparatus.

Since ER as well as plasma membrane proteins and, presumably, integral membrane proteins of other organelles, such as the Golgi apparatus and lysosomes, are all initially inserted into the ER membrane during synthesis on membrane-bound polysomes, a major question which remains to be answered concerns the mechanism controlling the different fates of these classes of polypeptides. The features of the membrane polypeptides synthesized on bound polysomes which serve as sorting signals and are responsible for their retention in the ER or passage toward the Golgi apparatus and the plasma membrane may reside in any of the three distinct domains (luminal, cytoplasmic, and intramembranous) that could exist in integral membrane proteins. Such sorting signals may be common to large subclasses of proteins destined for the same cellular membrane and may be recognized by components of a sorting apparatus that recognizes subcellular addresses and affects the distribution of protein products. Recently, it has been shown[88] by using DNA-mediated transfection procedures with genes encoding truncated polypeptides that a Semliki Forest virus envelope glycoprotein inserted in the ER and destined to the plasma membrane does not require a cytoplasmically exposed segment located near the carboxy terminal end for its exit from the ER. In addition, in several plasma membrane glycoproteins such as influenza A hemagglutinin[89,90] and Semliki Forest virus E1 protein,[91] and in the glycoprotein of avian sarcoma virus,[92] only very short peptide segments are exposed on the cytoplasmic face of the membrane. On the other hand, a common fcature of the ER membrane proteins appears to be the expo-

[88] H. Garoff, personal communication.

[89] W. Min Jou, G. Threlfall, M. Verhoeyen, R. Devos, E. Saman, R. Fang, D. Huylebroeck, W. Fiers, C. Barber, N. Carey, and S. Emtage, *Cell,* **19,** 683 (1980).

[90] M. J. Gething, J. M. White, and M. D. Waterfield, *Proc. Natl. Acad. Sci. U.S.A.* **75,** 2737 (1978).

[91] A.-M. Frischauf, K. Simons, H. Garoff, H. Lehrach, and H. Delius, *Nature (London)* **288,** 236 (1980).

[92] A. P. Czernilofsky, A. D. Levinson, H. E. Varmus, J. M. Bishop, E. Tisher, and H. M. Goodman, *Nature (London)* **287,** 198 (1980).

sure of a substantial portion of the polypeptide on the cytoplasmic aspect of the membrane. It may therefore be speculated that such segments contain information necessary for the retention of the polypeptide within the ER. Because ribophorins are glycosylated and undergo cotranslational cleavage of a peptide segment and several plasma membrane proteins are not cleaved proteolytically (e.g., retinal opsin[93] and band 3 of red cell membranes[94]) or glycosylated (e.g., the catalytic subunit of Na^+,K^+-ATPase[95]), it appears that neither retention of the polypeptide signal for cotranslational insertion nor the absence of carbohydrate, which are characteristic of several ER membrane proteins, constitute features that account for their segregation in the ER. Because functionally related polypeptides may form oligomeric complexes soon after they are inserted in the ER, not all individual polypeptides need to contain a feature that is recognized directly by a cellular sorting apparatus.

Acknowledgments

This work was supported in part by grants GM 21971, GM 20277, and GM/ES 30701.

[93] Y. Schechter, Y. Burstein, R. Zemell, E. Ziv, F. Kantor, and D. Papermaster, *Proc. Natl. Acad. Sci. U.S.A.* **76,** 2654 (1979).

[94] E. L. Sabban, D. D. Sabatini, V. T. Marchesi, and M. Adesnik, *J. Cell Physiol.* **104,** 261 (1980).

[95] K. J. Swedner and S. M. Goldin, *New England J. Med.* **302,** 777 (1980).

[43] Membrane Induction by Drugs

By GUSTAV DALLNER and JOSEPH W. DEPIERRE

It was first discovered almost two decades ago that certain xenobiotics can induce a selective proliferation of the endoplasmic reticulum in the hepatocytes of experimental animals. It has also been found that treatment of animals with thyroxine can selectively increase the levels of certain mitochondrial proteins, whereas administration of hypolipidemic drugs and certain other substances causes increases in the levels of a number of different hepatic parameters, in particular peroxisomes.

All three of these systems provide unique opportunities for investigating the processes of membrane biogenesis and degradation. Upon exposure to the drug, accumulation of specific membrane components commences and continues until exposure is terminated or until a maximal

ISBN 0-12-181996-5

TABLE I
FACTORS INFLUENCING THE INDUCTION OF MEMBRANE COMPONENTS BY DRUGS AND OTHER XENOBIOTICS

Factor	Influence
Animals	
Species	Most types of induction studies are performed using particularly susceptible species.
Strain	Very great variations in, e.g., induction by plasticizers and sensitivity to the tumorigenic effects of 2-acetylaminofluorene.
Sex	Degree of induction is influenced; e.g., hypolipidemic agents are more effective in male rats.
Age	Response is often dependent on age; e.g., many xenobiotics are more effective inducers of drug-metabolizing enzymes in young (60–90 g) rats, whereas plasticizers induce more effectively in mature rats.
State of health	Infections are common in animal houses and can reduce the extent of induction.
State of nutrition	Animals are generally starved overnight before preparation of microsomes in order to remove glycogen, and it is not clear what effect this may have on induction. For example, microsomal glucose-6-phosphatase activity is doubled upon such starvation.
Environment	
Diet	Fat content and composition can modulate induction.
Drinking water	Cation content is important.
Temperature and humidity	May at least in part explain seasonal variations. Elevation of these factors often decreases induction.
Cage contents	Rats and other experimental animals eat wood chips, especially when starved, and these chips often contain preservatives that can uncouple oxidative phosphorylation or induce drug-metabolizing enzymes.
Frequency of excrement removal	Ammonia formed in the excrement of experimental animals can affect the extent of induction.
Cleaning agents used	Volatile components of cleaning agents can be inducers.

state of induction is reached. The factors that determine this maximal state of induction have not as yet been identified. After cessation of drug administration the proliferated membranes are selectively removed until the original control situation is reached once again.

Here we will discuss the use of drug induction in studies of membrane biogenesis and degradation. It is important to remember that a large number of different factors are known to affect the quantitative aspects of induction (Table I). Differences in these parameters may at least partially explain differences in the extent of induction obtained in various laboratories, which are sometimes very large.

The Endoplasmic Reticulum

General

Since the initial discovery that administration of phenobarbital and 3-methylcholanthrene to rats induces drug metabolism by the cytochrome *P*-450 system localized on the endoplasmic reticulum,[1,2] a large number of other xenobiotics have also been found to induce this system (Table II).[3,4] It has been found that microsomal epoxide hydrolase[5] and UDPglucuronyltransferases[6] are induced by certain of these xenobiotics as well. Finally, even a number of cytoplasmic enzymes involved directly and indirectly in drug metabolism—including glutathione *S*-transferases,[7] glutathione reductase,[8] and DT-diaphorase[9]—are induced by such treatment.[3] Increases in the liver weight and number of hepatocytes have also been reported after treatment of rats with inducers of drug-metabolizing enzymes (see DePierre *et al.*[3] and references therein).

Induction of Protein Components

Most investigators have directed their attention toward the increases in many different forms of cytochrome *P*-450, NADPH–cytochrome *P*-450 reductase, microsomal epoxide hydrolase, and at least two different forms of UDPglucuronyltransferase after administration of different xenobiotics to experimental animals. These increases, which generally reach their maximal extent after 3–5 days of treatment, can be manyfold

[1] L. Ernster and S. Orrenius, *Fed. Proc., Fed. Am. Soc. Exp. Biol.* **24,** 1190 (1965).
[2] A. H. Conney, *Pharmacol. Rev.* **19,** 317 (1967).
[3] J. W. DePierre, J. Seidegård, R. Morgenstern, L. Balk, J. Meijer, and A. Åström, *in* "Mitochondria and Microsomes" (C. P. Lee, G. Schatz, and G. Dallner, eds.), p. 585. Addison-Wesley, Reading, Massachusetts, 1981.
[4] R. W. Estabrook and E. Lindenlaub, eds., "The Induction of Drug Metabolism." Schattauer, Stuttgart, 1978.
[5] F. Oesch, *Xenobiotica* **3,** 305 (1973).
[6] G. J. Dutton, *Biochem. Pharmacol.* **24,** 1835 (1975).
[7] W. B. Jakoby, *Adv. Enzymol.* **46,** 383 (1978).
[8] I. Carlberg, J. W. DePierre, and B. Mannervik, *Biochim. Biophys. Acta* **677,** 140 (1981).
[9] C. Lind and L. Ernster, *Biochem. Biophys. Res. Commun.* **56,** 392 (1974).

TABLE II
COMMON INDUCERS OF DRUG-METABOLIZING ENZYMES ON THE ENDOPLASMIC RETICULUM (ER)[a]

Inducer	Structure	Dose and administration	Morphology of the ER	ER proteins induced	Effect on ER lipids
Phenobarbital	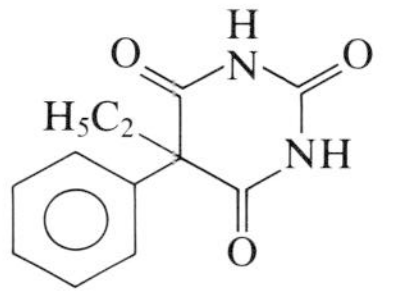	80 mg/kg body wt., i.p. injection in isotonic saline once daily for 5 days	Extensive proliferation	Cyt *P*-450, NADPH-cyt *P*-450 reductase, epoxide hydrolase, cyt b_5, UDP-glucuronyltransferase	Increase 2 to 2.5-fold in amount; small change in composition
3-Methylcholanthrene	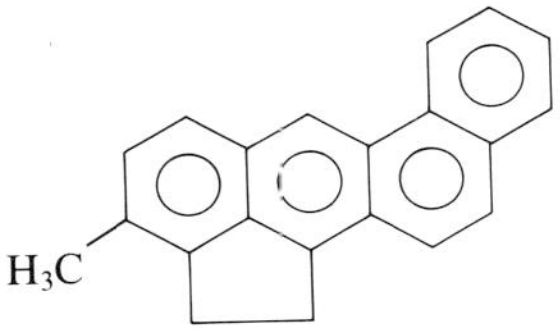	20 mg/kg body wt., i.p. injection in corn oil once daily for 3–5 days	No change	Cyt *P*-450, cyt b_5, UDPglucuronyltransferase	No change
trans-Stilbene oxide	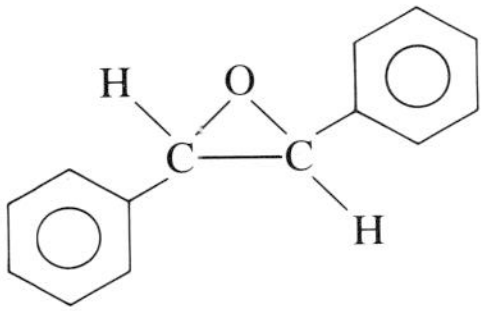	400 mg/kg body wt., i.p. injection in corn oil once daily for 5 days	No change or slight proliferation	Cyt *P*-450, NADPH-cyt *P*-450 reductase, epoxide hydrolase, cyt b_5, UDP-glucuronyltransferase	Small increase in amount, small change in composition
2-Acetylaminofluorene	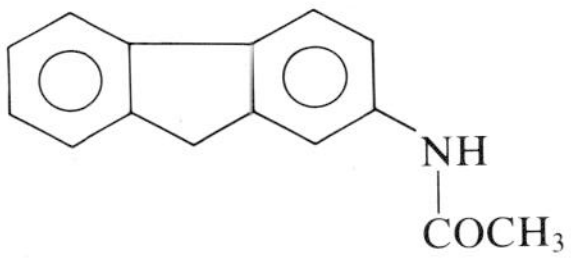	50 mg/kg body wt., i.p. injection in polyethylene glycol 300 once daily for 5 days, or 0.05% in the diet for 1 week	?	Cyt *P*-450, NADPH-cyt *P*-450 reductase, epoxide hydrolase, cyt b_5, UDP-glucuronyltransferase	No change in amount

[a] For relevant references, see text footnotes 1–3, 10, and 11.

(Table II).[1,3,10,11] In addition, different xenobiotics can induce different isozymes of cytochrome *P*-450 and can exhibit individual patterns with respect to the relative induction of cytochrome *P*-450-catalyzed reactions and other enzymes involved in drug metabolism.[3]

A number of experiments indicate that an increased *de novo* synthesis of protein is involved in these induction processes.[12-19] However, there is some disagreement concerning the role of decreased protein breakdown.[14,15] The next step will be to use induction to investigate the process of insertion of proteins into the membrane of the endoplasmic reticulum.

Induction of Phospholipid Components

Much less attention has been paid to the effect of inducers of drug-metabolizing enzymes on the phospholipids of the endoplasmic reticulum. It has long been known that treatment of rats with phenobarbital causes an extensive proliferation of the endoplasmic reticulum as seen in the electron microscope.[1,20] Most of this proliferation is accounted for by hypertrophy of the smooth endoplasmic reticulum,[20] and the phenomenon reflects the fact that phenobarbital treatment causes a 2- to 2.5-fold increase in the phospholipid content of this organelle.[1,21-24] In addition, phenobarbital may cause small changes in the microsomal phospholipid composition,[25-28] as well as in the fatty acid content of these phospholipids.[29]

[10] J. Seidegård, R. Morgenstern, J. W. DePierre, and L. Ernster, *Biochim. Biophys. Acta* **586,** 10 (1979).
[11] A. Åström and J. W. DePierre, *Biochim. Biophys. Acta* **673,** 225 (1981).
[12] A. H. Conney and A. G. Gilman, *J. Biol. Chem.* **238,** 3682 (1963).
[13] D. W. Nebert and H. V. Gelboin, *J. Biol. Chem.* **245,** 160 (1970).
[14] R. T. Schimke, R. Granschow, D. Doyle, and I. M. Arias, *Fed. Proc., Fed. Am. Soc. Exp. Biol.* **27,** 1223 (1968).
[15] Y. Kuriyama, T. Omura, P. Siekevitz, and G. E. Palade, *J. Biol. Chem.* **244,** 2017 (1969).
[16] R. Druyan and A. Kelly, *Biochem. J.* **129,** 1095 (1971).
[17] M. A. Correia and U. A. Meyer, *Proc. Natl. Acad. Sci. U.S.A.* **72,** 400 (1975).
[18] C. Rajamanickam, M. R. Satyanarayana Rao, and G. Padmanaban, *J. Biol. Chem.* **250,** 2305 (1975).
[19] D. A. Haugen, M. J. Coon, and D. W. Nebert, *J. Biol. Chem.* **251,** 1817 (1976).
[20] W. Stäubli, R. Hess, and E. R. Weibel, *J. Cell Biol.* **42,** 92 (1969).
[21] S. Orrenius and J. L. E. Ericsson, *J. Cell Biol.* **28,** 181 (1966).
[22] J. L. Holtzman and J. R. Gilette, *J. Biol. Chem.* **243,** 3020 (1968).
[23] L. C. Eriksson and G. Dallner, *FEBS Lett.* **29,** 351 (1973).
[24] J. W. DePierre and L. Ernster, *FEBS Lett.* **68,** 219 (1976).
[25] H. U. Schulz and H. Staudinger, *Hoppe-Seyler's Z. Physiol. Chem.* **351,** 184 (1970).
[26] C. D. Cooper and G. Feuer, *Can. J. Physiol. Pharmacol.* **50,** 568 (1972).
[27] S. C. Davison and E. D. Wills, *Biochem. J.* **140,** 461 (1974).
[28] K. Ishidate and Y. Nakazawa, *Biochem. Pharmacol.* **25,** 1255 (1976).
[29] M. S. Ilyas, F. A. de la Iglesia, and G. Feuer, *Toxicol. Appl. Pharmacol.* **44,** 491 (1978).

To date, no other inducer of drug-metabolizing enzymes has been found to cause such changes in the phospholipids of endoplasmic reticulum membranes. 3-Methylcholanthrene induction affects neither the microsomal phospholipid content per gram of liver nor the composition of these phospholipids.[30]

Administration of *trans*-stilbene oxide to rats results in an approximately 25% increase in the microsomal phospholipid content and small changes in this phospholipid composition and fatty acid content.[31] Few other inducers have been investigated in these terms.

Phenobarbital has been shown to cause a proliferation of endoplasmic reticulum membranes at least partially by slowing down the degradation of phospholipids.[32] A number of studies on the effects of inducers of drug-metabolizing enzymes on different enzymes involved in triglyceride and phospholipid synthesis have appeared.[32,33]

Studies of Membrane Degradation after Induction

Few investigators have as yet directed attention to the return of components of the endoplasmic reticulum to the control situation after cessation of drug treatment. This is unfortunate, since such studies would be expected to give insight into the process of membrane degradation. Two independent investigations—one morphometric[34] and one biochemical[24]—suggest that the proliferated endoplasmic reticulum is removed by autophagic vacuoles after cessation of phenobarbital treatment.

Methodology

Induction of Drug-Metabolizing Enzymes. This can be obtained effectively using the doses, routes of administration, and time periods of treatment catalogued in Table II. There is often a lag time of about 1 day before induction of different enzymes can be detected assaying activities.[10,11]

Microsomes. These can be prepared effectively from the liver of induced rats using routine procedures. Indeed, in several cases the recovery of membranes from the endoplasmic reticulum in the microsomal fraction from induced liver has been found to be greater than the corresponding recovery from untreated animals.[10,24] Recovery and contamination in the microsomal fraction should be checked, especially if one

[30] J. W. DePierre and G. Lundqvist, unpublished observations.

[31] Y. Suzuki, J. W. DePierre, and L. Ernster, *Biochim. Biophys. Acta* **601,** 532 (1980).

[32] L. C. Eriksson, *Acta Pathol. Microbiol. Scand., Suppl.* **239,** 1 (1973).

[33] K. K. Kohli, B. N. Gupta, P. W. Albro, and J. D. McKinney, *Chem.-Biol. Interact.* **36,** 117 (1981).

[34] R. P. Bolender and E. R. Weibel, *J. Cell Biol.* **56,** 746 (1973).

TABLE III
INDUCERS OF PEROXISOMES

Substance	Dosage	Reference[a]
1. Clofibrate: ethyl α-p-chlorophenoxy-isobutyrate	0.5–2.0% in diet, or 500 mg/kg orally, or 200 mg/kg sub-cutaneous injection	40, 41
Nafenopin: 2-methyl-2-[p-(1,2,3,4-tetrahydro-1-naphthyl)phenoxy]propionic acid	0.12–0.25% in diet	42
AT-308: 3-[4-(1-ethoxycarbonyl-l-methyl-ethoxy)phenyl]-5-(3-pyridyl)-1,2,4-oxadiazole	0.05–0.25% in diet	43
2. Wy-14,643: [4-chloro-6-(2,3-xylidino)-2-pyrimidinylthio]acetic acid	0.12–0.25% in diet	44
Tibric acid: 2-chloro-5-(3,5-dimethyl-piperidinosulfonyl)benzoic acid	0.12–0.25% in diet	44
LK-903: α-methyl-p-myristyroxycinnamic acid-1-monoglyceride	0.05% in diet	45
3. Acetylsalicylic acid	0.5% in diet	46
4. DEHP: di(2-ethylhexyl) phthalate	2% in diet	47
5. High-fat diet		
Corn oil	30% of dietary calories	48
Rapeseed oil	30% of dietary calories	48
Hydrogenated marine oil	30% of dietary calories	48
Linol salad oil	30% of diet	49
Soybean oil	15% of diet	50
6. Vitamin E deficiency	—	51

[a] Numbers refer to text footnotes.

wishes to investigate changes in the phospholipid content and composition of the endoplasmic reticulum after induction.

Induction. This can be ascertained simply by assaying the activities of a number of drug-metabolizing enzymes (Table II). However, an increase in activity does not necessarily reflect an increase in the amount of enzyme protein (which, by definition, must occur if the process is truly an induction). Increases in protein amount can be demonstrated semiquantitatively by SDS disc gel electrophoresis[11,35] or quantitatively using immunochemical techniques.[36]

[35] R. K. Murray, R. N. Sharma, J.-G. Joly, R. G. Cameron, and E. Farber, *Biochem. Soc. Trans.* **7,** 32 (1979).

[36] C. Guthenberg, R. Morgenstern, J. W. DePierre, and B. Mannervik, *Biochim. Biophys. Acta* **631,** 1 (1980).

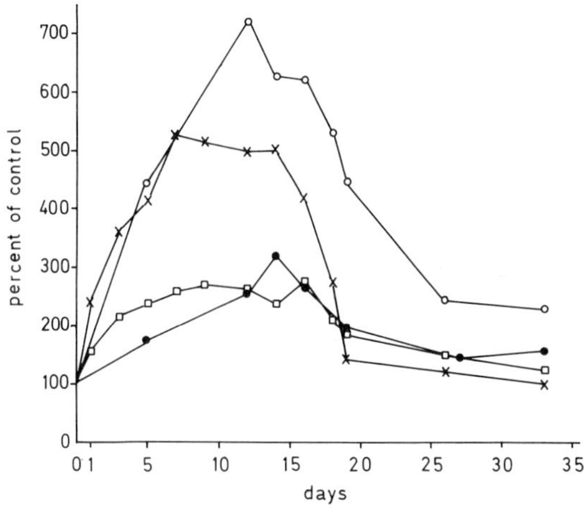

FIG. 1. Induction of drug-metabolizing systems by 3-methylcholanthrene and subsequent return to control levels. The rats were injected on days 0, 1, 2, 3, and 4. Each point represents an average value for 3–9 different animals. The symbols used are as follows: ×, cytochrome *P*-450; □, cytochrome b_5; ●, glutathione *S*-transferase(s); ○, DT-diaphorase. For more details, see DePierre *et al.*[37]

Degradation. After cessation of drug treatment, there is also a lag time of 1 day[10,11] or more[17] (Fig. 1) before return to normal begins,[37] which is perhaps due at least in part to a depot effect.

Peroxisomes

Induction of peroxisomes in liver and kidney has become of great interest for several reasons.[38,39] It turns out that (*a*) these particles are regulators of lipid metabolism; (*b*) hypolipidemic drugs and environmental pollutants (plasticizers) are potent inducers of peroxisomes; and (*c*) peroxisome induction may be associated with carcinogenesis.

Many hypolipidemic drugs and chemical reagents have been demonstrated to be inducers of peroxisomes, and a selection of them is listed in Table III.[40–51] The dosage given is approximate, since it varies greatly

[37] J. W. DePierre, G. Lundqvist, and L. Ernster, *Acta Chem. Scand.* **B36,** 497 (1982).

[38] P. B. Lazarow, *in* "International Cell Biology 1980–1981" (H. G. Schweiger, ed.), p. 633. Springer-Verlag, Berlin and New York, 1981.

[39] J. K. Reddy, J. R. Warren, M. K. Reddy, and N. D. Lalwani, *Ann. N.Y. Acad. Sci.* **386,** 81 (1982).

[40] R. Hess, W. Stäubli, and W. Reiss, *Nature (London)* **208,** 856 (1965).

depending on the type, strain, and sex of the animals used. The most commonly used inducer by far is clofibrate (available as a drug), which is effective in the rat, mouse, dog, and hamster, but not in the guinea pig, chicken, rabbit, or monkey.[52] Male rats respond more effectively than females to low doses. Drugs that induce peroxisomes have effects on many organs and tissues, but proliferation of this organelle has been shown to date only in the liver and kidney.[53]

Certain drugs chemically related to clofibrate (but not available commercially) are better inducers of peroxisomes. Induction of this organelle cannot be related at present to a common chemical structure, since the agents in group 2, Table III, which possess hypolipidemic activity and inductive capacity are chemically unrelated to clofibrate. An important group of inducers are plasticizers, since these materials occur in considerable quantities as pollutants in the air, earth, water, and food. The most commonly used plasticizer, both industrially and in research, is di(2-ethylhexyl) phthalate (DEHP). Like most of the other inducers, the effect of DEHP is apparent after administration for a few days in the diet (2%), and maximal induction is obtained after 2 weeks. Naturally, lower doses are less effective and a 100-fold lower dose (0.02%) requires 1 year of administration to cause an appreciable effect on peroxisomal membranes and enzymes.[54] The explanation for this low dose effect may be accumulation of the plasticizer in tissues.

One physiologically important type of peroxisome induction is that caused by dietary factors. High-fat diets, particularly those containing an excess of C_{22} fatty acid, greatly stimulate the biosynthesis of proteins participating in the oxidation of long-chain fatty acids.

---

[41] M. T. Kahonen, *Biochim. Biophys. Acta* **428,** 690 (1976).

[42] J. K. Reddy, D. L. Azarnoff, D. J. Svoboda, and J. D. Prasad, *J. Cell Biol.* **61,** 344 (1974).

[43] M. Arakawa, H. Miyajima, H. Matsumura, M. Izukawa, and Y. Imai, *Biochem. Pharmacol.* **27,** 167 (1978).

[44] J. K. Reddy and T. P. Krishnakantha, *Science* **190,** 787 (1975).

[45] H. Hayashi, S. Hino, F. Yamasaki, T. Watanabe, and T. Suga, *Biochem. Pharmacol.* **30,** 1817 (1981).

[46] M. Gotoh, C. Griffin, and Z. Hruban, *Virchows Arch. B* **17,** 279 (1975).

[47] D. E. Moody and J. K. Reddy, *Toxicol. Appl. Pharmacol.* **45,** 497 (1978).

[48] R. Z. Christiansen, E. Christiansen, and J. Bremer, *Biochem. Biophys. Acta* **573,** 417 (1979).

[49] H. Ishii, N. Fukumori, S. Horie, and T. Suga, *Biochim. Biophys. Acta* **617,** 1 (1980).

[50] C. E. Neat, M. S. Thomassen, and H. Osmundsen, *Biochem. J.* **186,** 369 (1980).

[51] J. K. Reddy, N. D. Lalwani, A. S. Dabholkar, M. K. Reddy, and S. A. Qureshi, *Biochem. Int.* **3,** 41 (1981).

[52] D. Svoboda, H. Grady, and D. Azarnoff, *J. Cell Biol.* **35,** 127 (1967).

[53] N. D. Lalwani, M. K. Reddy, M. M. Mark, and J. K. Reddy, *Biochem. J.* **198,** 177 (1981).

[54] A. Ganning and G. Dallner, unpublished (1982).

Morphological Appearance

In the case of liver, which is enlarged about 50% because both the number and volume of the hepatocytes are increased, the increase of peroxisomes is dramatic (5- to 20-fold), and they outnumber the mitochondria.[55] In control liver the peroxisomes are uniform in appearance, but in the induced state these particles exhibit considerable variation both in size and shape, which is disadvantageous with respect to fractionation. The homogeneous matrix is now granulated, varies in density, and may contain elongated and tubular structures. The core structure is smaller or disappears as a result of the decrease in urate oxidase. It also appears that the membranes are more easily damaged than in the control tissue.

Cellular Changes in Hepatocytes

In spite of the difficulties in isolating induced peroxisomes, the recovery (in milligrams of protein) after fractionation in most cases exceeds that from control tissue severalfold (Table IV).[39,52,56–69] The specific activities of all enzymes participating in the peroxisomal β-oxidation of fatty acids are increased 2- to 6-fold.[56] This is also the case for the two fatty acyltransferases.[58] The specific activities of both catalase and urate oxidase are substantially decreased.[39] With respect to mitochondria, the particular inducer used is of great importance, since clofibrate increases the number and/or volume of these structures to a moderate extent that depends greatly on the conditions of induction.[60] In contrast, plasticizers may increase the number and volume of mitochondria 2- to 3-fold, but the variations are also great in this case.[61–63] The production of new mitochondria is not a compensation for damaged ATP synthesis, since the

[55] D. Svoboda, D. Azarnoff, and J. Reddy, *J. Cell Biol.* **40,** 734 (1969).
[56] T. Hashimoto, *Ann. N.Y. Acad. Sci.* **386,** 5 (1982).
[57] Y. Shindo and T. Hashimoto, *J. Biochem.* (*Tokyo*) **84,** 1177 (1978).
[58] M. A. Markwell, L. L. Bieber, and N. E. Tolbert, *Biochem. Pharmacol.* **26,** 1697 (1977).
[59] E. L. Appelkvist and G. Dallner, *Biochim. Biophys. Acta* **617,** 156 (1980).
[60] A. R. L. Gear, A. D. Albert, and J. M. Bednarek, *J. Biol. Chem.* **249,** 6495 (1974).
[61] F. P. Bell and P. J. Gillies, *Lipids* **12,** 581 (1977).
[62] T. Yanagita, S. Kuzuhara, N. Enomoto, T. Shimada, and M. Sugano, *Biochem. Pharmacol.* **28,** 3115 (1979).
[63] A. Ganning and G. Dallner, *FEBS Lett.* **30,** 77 (1981).
[64] I. Eggens, U. Brunk, and G. Dallner, *Exp. Mol. Pathol.* **32,** 115 (1980).
[65] B. Borreback, H. Osmundsen, and J. Bremer, *IRCS Libr. Compend.* **7,** 181 (1979).
[66] S. Skrede and O. Halvorsen, *Eur. J. Biochem.* **98,** 223 (1979).
[67] G. Fleischer, D. K. F. Meijer, W. G. Levine, Z. Gaitman, R. Gluck, and I. M. Arias, *Biochem. Biophys. Res. Commun.* **67,** 1401 (1975).
[68] F. P. Bell, *Lipids,* **11,** 769 (1976).
[69] Y. Shindo, T. Osumi, and T. Hashimoto, *Biochem. Pharmacol.* **27,** 2683 (1978).

TABLE IV

CELLULAR CHANGES IN HEPATOCYTES AFTER *in Vivo* ADMINISTRATION OF INDUCERS OF PEROXISOMES

Parameter	Change[a]	References[b]
Peroxisomes		
Protein and phospholipid	Increase severalfold	52
β-Oxidation of fatty acids (acyl-CoA oxidase, enoyl-CoA hydratase,3-hydroxyacyl-CoA dehydrogenase, 3-ketoacyl-CoA thiolase)	Increase 2- to 6-fold	56
Palmitoyl-CoA synthase	Increases 2- to 3-fold	57
Carnitine acetyltransferase	Doubles	58
Carnitine octanoyltransferase	Triples	58
Catalase	Decreases 10–40%	39
Urate oxidase	Decreases 40%	39
Fatty acid binding protein	Increases several fold	59
Mitochondria		
Protein and phospholipid	Increase 30–40% with clofibrate; increase 200–300% with plasticizers	60–63
β-Oxidation of fatty acids	Doubles	56
Carnitine acetyltransferase	Increases 10- to 30-fold	58
Carnitine octanoyltransferase	Triples	58
Carnitine palmitoyltransferase	Increases 3- to 5-fold	58
Dehydrogenases and respiratory enzymes	No or moderate change	63
Respiratory control and oxidative phosphorylation	No change	63
Microsomes		
Protein and phospholipid	Increase 10–30%	63, 64
NADPH–cytochrome *c* reductase	Increases 40–60%	63, 64
Cytochrome *P*-450	Increases 40–60%	63, 64
Cytochrome b_5	Increases 0–50%	63, 64
NADH-cytochrome *c* reductase	No change	63, 64
DT-diaphorase	Increases 0–60%	63, 64
Aminopyrine and naphthalene hydroxylation	Increase 0–100%	63, 64
Nucleoside diphosphatase	Decreases 50% or increases 100%	63, 64
ATPase and glucose-6-phosphatase	No change	63, 64
Cytoplasm		
Acyl-CoA hydrolase	Increases severalfold	65
Pantothene kinase	Doubles	66
Fatty acid binding protein	Triples	67
Whole homogenate		
Sterol and squalene synthesis	Decrease 75%	68
CoA and carnitine	Increase 5- to 6-fold	69
Acyl-CoA and acetylcarnitine	Increase 4-fold	69
Long-chain acyl-CoA and acylcarnitine	Increase 50%	69

[a] Specific activities or amounts on protein basis compared with the control.

[b] Numbers refer to text footnotes.

isolated mitochondrial fraction from induced liver has a completely normal respiratory control and P/O ratio.[63] Mitochondria isolated from treated liver exhibit doubled β-oxidation and trebled oxidation of carnitine-activated medium and long-chain fatty acids, whereas carnitine acetyltransferase increases 10- to 30-fold.[56,58]

The endoplasmic reticulum does not proliferate in a significant manner, but an increase in the two electron transport enzymes participating in drug hydroxylation indicates metabolism of the inducer.[63,64] The increase of acyl-CoA hydrolase in the soluble cytoplasm is important, since it may interfere with the measurement of transferase activities.[65] The pools of various activated fatty acids, CoA, and carnitine are elevated in the homogenate.[69] When sterol and squalene synthesis were monitored, they were found to be greatly decreased.[68]

Isolation of Peroxisomes and Peroxisomal Membranes

Peroxisomes are extremely sensitive to hydrostatic pressure and easily damaged during gradient centrifugation. Isolation procedures were worked out using a specially designed zonal rotor,[70] various sucrose gradients,[71] and Percoll gradients in a vertical rotor.[72] The isolation procedures available are, however, far from being optimal because of the very low recovery, particle damage, and relatively high contamination.

The recovery of peroxisomes from a liver homogenate may be increased by stabilization of the membranes at the start of fractionation by low concentrations of bifunctional cross-linking reagents such as glutaraldehyde.[73]

The livers are homogenized in a Teflon–glass homogenizer by applying three strokes at 885 rpm. To remove cell debris, nuclei, and the majority of mitochondria, centrifugation is performed at 2800 *g* for 15 min. The supernatant after this centrifugation contains about half the total liver peroxisomes, but only 10–20% of the mitochondria.

In order to stabilize peroxisomal membranes, 1 m*M* glutaraldehyde in 50 m*M* cacodylate buffer, pH 7.4, is added to the 2800 *g* supernatant and incubated for 30 min in an ice-water bath. This supernatant (13 ml) is then layered over 6 ml of undiluted Percoll (density 1.13 g/ml) and 18 ml of 40% Percoll in 0.25 *M* sucrose (density 1.08 g/ml) adjusted to pH 7.4 with HCl, and the discontinuous gradient is centrifuged in a VTi 50 rotor (Beckman) at 50,000 *g* for 30 min.

[70] F. Leighton, B. Poole, H. Beaufay, P. Baudhuin, J. W. Coffey, S. Fowler, and C. de Duve, *J. Cell Biol.* **37,** 482 (1968).

[71] N. E. Tolbert, this series, Vol. 52, p. 493.

[72] H. Osmundsen and C. E. Neat, *FEBS Lett.* **107,** 81 (1979).

[73] E. L. Appelkvist, U. Brunk, and G. Dallner, *J. Biochem. Biophys. Methods* **5,** 203 (1981).

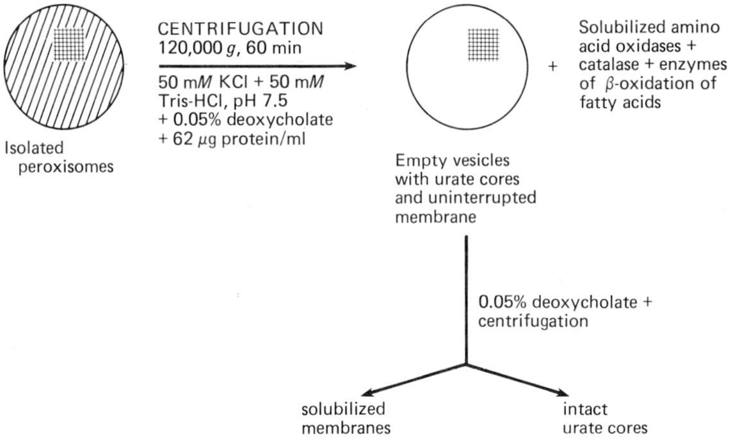

FIG. 2. Procedure for removal of peroxisomal content without destroying vesicle structure. From Appelkvist *et al.*[73]

The band appearing on the top of the undiluted Percoll is removed and mixed with 40% Percoll in 0.25 *M* sucrose to give a volume of 9 ml. Centrifugation is performed in a 50 Ti rotor at 60,000 *g* for 30 min. in order to develop a self-generating gradient. This results in distribution of the material in several bands on the gradient. The lowest visible band, about 2 cm from the bottom and consisting of about 2 ml, is collected and diluted with 0.25 *M* sucrose to a volume of 9 ml. After centrifugation in the 50 Ti rotor at 25,000 *g* for 30 min, the peroxisomes are sedimented and the pellet is resuspended in 1 ml of 0.25 *M* sucrose. The resulting peroxisomes (14% of the total) contain some mitochondrial contamination, but are practically free from microsomes.

In a modification of this isolation procedure, 13 ml of the glutaraldehyde-treated 2800 *g* supernatant is layered over 6 ml of 2.1 *M* sucrose and 18 ml of 1.75 *M* sucrose in a tube fitted to the VTi rotor.[73] After centrifugation at 160,000 *g* for 60 min, the fraction appearing between the 1.75 *M* and 2.1 *M* sucrose layers is removed and designated as the peroxisomal fraction. This fraction is contaminated with microsomes, but almost free of mitochondria.

In all studies involving peroxisome induction, the organelle must be isolated for analysis. Increased amounts of peroxisomal protein can be obtained by fractionation from induced liver, but the great variation in size causes an extra loss of particles in comparison with the control. This poor recovery may create a problem if it turns out that peroxisomes exhibit a heterogeneity similar to that of the microsomal fraction and the isolated fraction represents a special subgroup of the total population.

The peroxisomal membrane, which possesses a high phospholipid : protein ratio (0.55), is not readily available for study, since it makes up only 20% of the total organelle protein. The deoxycholate–KCl procedure for removing the content of microsomes,[74] which is suitable for introducing surface probing reagents into the vesicle lumen,[75] proved to be effective for peroxisomes also (Fig. 2). By applying 0.05% deoxycholate, the peroxisomal membranes are made permeable to macromolecules without destroying the closed vesicular structure.[73] The content enzymes—oxidases, catalase, and enzymes of β-oxidation—are obtained in soluble form.

Mitochondria

Thyroid Hormones and Membrane Synthesis

Mitochondrial membranes, especially inner membranes, and a number of mitochondrial functions can be induced by administration of thyroid hormones.[76] To obtain chronic hyperthyroidism rats are injected subcutaneously with 5–20 μg of 3,5,3′-triiodo-L-thyronine or 10–40 μg of L-thyroxine every day or every second, third, or fourth day during a period of 1–3 weeks. When the acute effects of these hormones are studied, the same or larger doses of thyroid hormones are injected, usually to thyroidectomized rats. Total thyroidectomy can be performed surgically or, much more simply, by administration to young rats (60–80 g) of three doses of 50–60 μCi of carrier-free ^{131}I at intervals of 1 week, the last dose being at least 6 weeks before the experiment.

Thyroid hormones increase both the number and size of the mitochondria and lead to a sizable increase of inner membrane volume (Table V).[76–87] There is a general increase in the translation and transcription of

[74] G. Kreibich, P. Debey, and D. D. Sabatini, *J. Cell Biol.* **58,** 436 (1973).

[75] O. S. Nilsson and G. Dallner, *J. Cell Biol.* **72,** 568 (1977).

[76] J. R. Tata, L. Ernster, O. Lindberg, E. Arrhenius, S. Pedersen, and R. Hedman, *Biochem. J.* **86,** 408 (1963).

[77] T. DeLeo, A. Barletta, S. DiMeo, *Life Sci.* **8,** 747 (1969).

[78] W. L. Wooten and J. Cascarano, *J. Bioenerg. Biomembr.* **12,** 1 (1980).

[79] M. N. Gadaleta, N. DiReda, G. Bove, and C. Saccone, *Eur. J. Biochem.* **51,** 495 (1975).

[80] B. D. Nelson, V. Joste, A. Wielburski, and U. Rosenqvist, *Biochim. Biophys. Acta* **608,** 422 (1980).

[81] D. B. Roodyn, K. B. Freeman, and J. R. Tata, *Biochem. J.* **94,** 628 (1965).

[82] B. Kadenbach, *Biochem. Z.* **344,** 49 (1966).

[83] M. B. Weinberg and M. F. Utter, *J. Biol. Chem.* **254,** 9492 (1979).

TABLE V
CELLULAR CHANGES AFTER *in Vivo* ADMINISTRATION OF THYROID HORMONES

Organ	Change	References[a]
Liver	Morphology	
	Increase in number and size of mitochondria, enrichment of cristae	77
	Mitochondria	
	Increase in RNA and DNA	78
	Increase in RNA polymerase	79
	Increase in mRNA for mitochondrial proteins	79
	Increase in amino acid incorporation	80
	Increase in cytochromes *a*, *b*, $c + c_1$	81
	Increase in respiration and phosphorylation	76
	Increase in α-glycerophosphate, succinate, and isocitrate dehydrogenases	78, 82
	Increase in cytochrome oxidase and pyruvate carboxylase	76, 83
	Microsomes	
	Increase in RNA : protein ratio	76
	Increase in amino acid incorporation	76
	Increase in glucose-6-phosphatase	76
	Increase in NADPH-cytochrome *c* reductase	76
	Increase in albumin synthesis	84
	Nuclei	
	Increase in the DNA-dependent RNA polymerase	85
	Cytosol	
	Increase in malic enzyme	86
Skeletal muscle	Administration of thyroid hormones results in a 3-fold increase in the total mitochondrial volume and a 2-fold increase in respiration and phosphorylation	87
	Thyroidectomy results in a 3-fold increase in the total mitochondrial volume, but no increase in respiration or phosphorylation	87

[a] Numbers refer to text footnotes.

mitochondrially synthesized peptides and many mitochondrial enzymes. The mitochondrial changes are most striking, but, as expected, several other cellular functions are greatly influenced.

Induction of mitochondria by hormonal action is not restricted to the

[84] C. G. Lewallen, J. E. Rall, and M. Berman, *J. Clin. Invest.* **38,** 88 (1959).
[85] C. C. Widnell and J. R. Tata, *Biochim. Biophys. Acta* **72,** 506 (1963).
[86] E. J. Wilson and W. C. McMurray, *Biochem. Biophys. Res. Commun.* **93,** 179 (1980).
[87] R. Gustafsson, J. R. Tata, O. Lindberg, and L. Ernster, *J. Cell Biol.* **26,** 555 (1965).

liver, but occurs to a large extent also in skeletal muscle.[87] Kidney and heart muscle also respond similarly.[82]

Thyroid Hormones and Membrane Degradation

In contrast to what is observed after induction of the endoplasmic reticulum by phenobarbital (see above), there is a dissociation of membrane and enzyme degradation of induced skeletal muscle mitochondria, which speaks against an autophagic degradation mechanism in this case.[88] Transitions from hyperthyroid to normal and from normal to hypothyroid states cause a decrease in enzyme level, but no change, or even an increase, in the total mitochondrial membrane volume.

Phthalate Ester Induction

As described in the section on peroxisomes, DEHP induces mitochondria also.[61–63] Table IV lists a few functions that are found to be induced in isolated liver mitochondria after 2 weeks' treatment of rats with 2% DEHP in the diet. A detailed analysis of the enzyme pattern after such induction has not yet been performed, and, thus, the nature of the newly synthesized mitochondria is not known.

Acknowledgments

The work from the authors' laboratories cited here was supported by grants from the Swedish Medical and Natural Science Research Council and NIH Grant No. 1 RO1 CA 26261-02 awarded by the National Cancer Institute, DHEW, Bethesda, Maryland.

[88] L. Ernster, *Fed. Proc., Fed. Am. Soc. Exp. Biol.* **24,** 1222 (1965).

[44] Preparation of Microsomal β-Glucuronidase and Its Membrane Anchor Protein, Egasyn

By Aldons J. Lusis

Acid hydrolases such as β-glucuronidase are typically located in lysosomes, but in certain tissues of mammals substantial amounts of the enzyme are also present in the membranes of the endoplasmic reticulum. This has been demonstrated using both tissue fractionation and histochemical staining reactions. The microsomal enzyme is firmly bound to

ISBN 0-12-181996-5

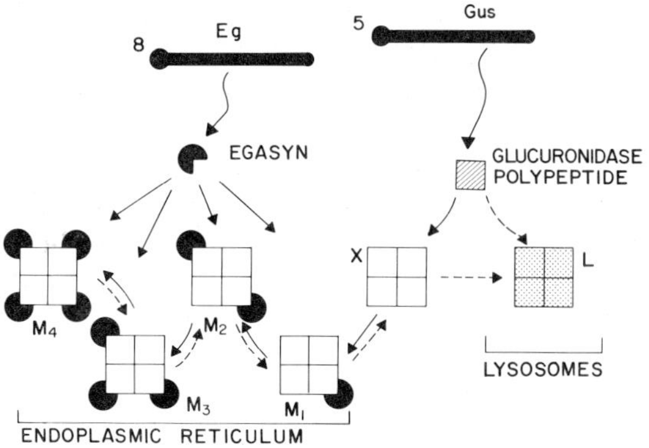

FIG. 1. Model for intracellular localization of mouse β-glucuronidase. Six intracellular forms of β-glucuronidase have been identified. The enzyme in lysosomes consists of a single form, termed L, that is a tetramer of identical subunits, each of molecular weight 71,000. The microsomal enzyme, on the other hand, occurs primarily as a series of higher molecular weight complexes, termed M1, M2, M3, and M4; these consist of a glucuronidase core of slightly different structure, X, complexed with one to four chains of a second polypeptide, egasyn. The X and L subunits are derived from a common structural gene, *Gus*, on chromosome 5, and differ by modification. Egasyn is an integral membrane glycoprotein of molecular weight 64,000. A mutation at the *Eg* gene on chromosome 8, present in strain YBR, results in the absence of immunoreactive egasyn and the inability to maintain β-glucuronidase in endoplasmic reticulum membranes. This and other evidence suggest that egasyn stabilizes the membrane binding of β-glucuronidase (see text). Reprinted from Lusis and Paigen.[4]

membrane and detergents are required to extract it, while the enzyme in lysosomes is released upon rupture of the lysosomal membrane.[1-3]

A combination of biochemical and genetic approaches suggests a model for the localization of β-glucuronidase in mouse tissues[4] (Fig. 1). Genetic studies indicate that the enzyme at both intracellular sites is derived from the same structural gene. The β-glucuronidase in lysosomes occurs as a free tetramer of molecular weight about 280,000, whereas most of the enzyme in microsomes is complexed with an integral membrane protein, egasyn. The complexes can be extracted intact from membrane using Triton X-100, and the egasyn isolated from these complexes is a glycoprotein of molecular weight 64,000 that is structurally distinct from glucuronidase. The microsomal and lysosomal glucuronidase tetramers

[1] C. B. C. de Duve, R. Pressman, R. Gianetto, R. Wattiaux, and F. Appelmans, *Biochem. J.* **60,** 604 (1955).

[2] K. Paigen, *Exp. Cell Res.* **25,** 286 (1961).

[3] W. H. Fishman, S. S. Goldman, and R. DeLellis, *Nature (London)* **213,** 457 (1967).

[4] A. J. Lusis and K. Paigen, *Isozymes: Curr. Top. Biol. Med. Res.* **2,** 63 (1977).

(termed X and L, respectively) differ by covalent modification, the microsomal form being slightly larger and less negatively charged than the lysosomal form, but the functional significance of these differences is not known.

Several observations suggest that egasyn serves to stabilize the binding of β-glucuronidase to membrane. Most convincing is the finding that a mouse strain that lacks egasyn is unable to maintain β-glucuronidase in endoplasmic reticulum membrane.[4–7] In addition, among tissues and during development the levels of β-glucuronidase in microsomes are correlated with the levels of egasyn. For example, liver and kidney contain 20–50% microsomal glucuronidase and have relatively high levels of egasyn, whereas spleen contains only trace levels of microsomal glucuronidase and lacks detectable egasyn. Thus, egasyn appears to function as a membrane anchor for the highly polar glucuronidase tetramer.[4]

Other mammals that have been examined, including rats, rabbits, and humans, also have substantial amounts of microsomal β-glucuronidase in liver.[4] A membrane protein that may be analogous to egasyn has been isolated from rat liver microsomes.[8] The protein was extracted from microsomes with 2% Triton X-100, and it specifically binds β-glucuronidase. Unlike the mouse microsomal β-glucuronidase complexes, the complex between rat glucuronidase and the binding protein is dissociated by relatively low concentrations of Triton X-100. Rabbit liver microsomal glucuronidase is extracted by Triton X-100 as a series of complexes with molecular weights of roughly 280,000, 330,000, 370,000, 420,000, and 500,000; thus, as in mice, the forms increase stepwise in molecular weight increments of roughly 55,000. The rabbit β-glucuronidase complexes are considerably more stable than those of mice and rats, although they can be dissociated using 8 *M* urea.[9]

β-Glucuronidase Purification

Assay. Mammalian β-glucuronidase is capable of hydrolyzing a wide variety of natural and synthetic β-D-glucuronides. A variety of simple fluorometric and spectrophotometric assays of the enzyme, using artificial substrates such as the glucuronides of 4-methylumbelliferone and *p*-nitrophenol, have been developed. In terms of simplicity and sensitivity the fluorometric assay, employing 4-methylumbelliferyl-β-D-glucuronide as the substrate, is the procedure of choice.

[5] R. E. Ganschow and K. Paigen, *Proc. Natl. Acad. Sci. U.S.A.* **58,** 938 (1967).
[6] A. J. Lusis, S. Tomino, and K. Paigen, *Biochem. Genet.* **15,** 115 (1977).
[7] A. J. Lusis, S. Tomino, and K. Paigen, *J. Biol. Chem.* **251,** 7753 (1976).
[8] L. D. Strawser and O. Touster, *J. Biol. Chem.* **254,** 3716 (1979).
[9] R. T. Dean and M. Messer, *Comp. Biochem. Physiol.* **54B,** 107 (1976).

I utilize the following assay protocol:[7] The sample is incubated at 37° with 0.4 m*M* 4-methylumbelliferyl-β-D-glucuronide (Sigma Chemical Co.), 0.1% Triton X-100 (Sigma), and 0.1 *M* sodium acetate, pH 4.6, in a total volume of 0.1 ml. The reaction is stopped after 30 min by immersion in an ice bath followed by the addition of 1.0 ml of 0.1 *M* sodium carbonate. The fluorescence is measured with an Aminco fluorometer (catalog No. 4-7439) equipped with a Corning 7-60 excitation filter (peak wavelength 360 nm) and Kodak 2A emission filter (passing wavelengths above 415 nm), or an instrument with similar capability. One unit of activity is defined as that amount of enzyme that will hydrolyze 1 μmol of substrate per hr at 37°. At very low protein concentrations, the inclusion of 0.1% albumin during the incubation helps to stabilize the enzyme. If necessary, the sensitivity of the assay can be increased by prolonging the incubation period to 24 hr or more.

Electrophoresis. The multiple forms of mouse β-glucuronidase (L, X, M) are resolved using a nondenaturing polyacrylamide gel electrophoresis system containing Tris–glycine, pH 8.1, and 7% acrylamide. Glucuronidase activity is visualized by staining with naphthol-AS-BI-glucuronide (Sigma) as the substrate.[10,11]

The most sensitive technique for analyzing charge variation of the glucuronidase tetramer is isoelectric focusing. I have had best results using a gel system containing 6% acrylamide, 6 *M* urea, 1% pH 3.5–10 Ampholines and 1% pH 5–7 Ampholines.[11] Under these conditions the glucuronidase tetramer remains intact and the gel can be stained for enzyme activity, but the M form complexes between glucuronidase and egasyn are disrupted.

Lysosomal (L Form) β-Glucuronidase. Lysosomal β-glucuronidase has been purified by conventional fractionation from several mammalian sources including bovine liver, human liver, rat liver and preputial gland, and mouse liver, kidney, and urine.[11,12] Rat preputial gland[13] and mouse urine[11] are especially rich sources, permitting rapid and simple purification of milligram quantities of the enzyme.

Antisera. High-titer, monospecific antibodies to β-glucuronidase can be raised using rabbits or goats.[14] Rabbit antiserum is prepared by inoculating outbred rabbits at multiple subcutaneous sites with 0.2 mg of purified L-form glucuronidase in Freund's complete adjuvant. Four weeks later, booster injections are given with 0.2 mg of enzyme in incomplete adjuvant. Serum is taken about 1 week after the boost.

[10] R. T. Swank and K. Paigen, *J. Mol. Biol.* **77,** 371 (1973).
[11] A. J. Lusis and K. Paigen, *J. Biol. Chem.* **253,** 7336 (1978).
[12] O. Touster, this series, Vol. 50, p. 488.
[13] R. K. Keller and O. Touster, *J. Biol. Chem.* **250,** 2739 (1975).
[14] S. Tomino, K, Paigen, D. Tulsiani, and O. Touster, *J. Biol. Chem.* **250,** 8503 (1975).

Antibody Affinity Chromatography. Mouse glucuronidase can be readily purified from liver, kidney, and other tissues by antibody affinity chromatography.[11] Both microsomal (X) and lysosomal (L) forms of the enzyme can be purified, since immunologically they are not discernibly different.

The protocol is as follows: Immunoglobin is purified from antiserum by repeated precipitation with ammonium sulfate at 45% saturation. The precipitate is dissolved in 0.02 *M* Tris, 0.15 *M* NaCl, pH 7.4 (TBS buffer), and dialyzed against the same buffer. The immunoglobin is coupled to cyanogen bromide-activated Sepharose 4B in 0.1 *M* $NaHCO_3$, 0.5 *M* NaCl, and afterward the gel is washed extensively with coupling buffer and with 0.1 *M* sodium acetate, 0.5 *M* NaCl, pH 4.6. The gel is placed in a column and washed with 0.1 *M* sodium acetate, pH 5.2, containing 8 *M* urea to remove any loosely bound material. It is then equilibrated with TBS buffer, and the sample is applied in TBS buffer. Tissue homogenates, extracted with detergents such as Triton X-100 and centrifuged at high speed to remove particulate material (see below), are suitable for chromatography. For isolation of L-form enzyme, however, we have found it useful partially to purify the enzyme using heat treatment at acid pH prior to chromatography.[11] After application of the sample, the column is washed with at least 5 column volumes of (*a*) TBS buffer, (*b*) 0.02 *M* sodium acetate, pH 5.2, and (*c*) 0.02 *M* Tris, 0.3 *M* NaCl, pH 7.4. Glucuronidase is then eluted by applying 0.1 *M* sodium acetate, pH 5.2, containing 8 *M* urea to the column at a rate of about 4 column volumes per hour. The glucuronidase is normally eluted within about 3 column volumes, and the urea is removed by dialysis. Although glucuronidase is relatively resistant to inactivation by urea, dialysis is normally begun shortly after elution of the enzyme. The capacity of the gel can be estimated using a small column and increasing amounts of enzyme; it is possible to bind several hundred units of enzyme (2400 units per milligram of protein) per milliliter of gel. Normally, nearly all of the enzyme binds to the column, and recovery of activity averages 60–80%. After affinity chromatography, preparations contain only traces of contaminants, and these can be removed by chromatography on a 2.6 × 55 cm column of Ultrogel Ac-44 equilibrated with TBS. When stored at 4° in TBS containing 0.1% sodium azide, the affinity gel is stable for more than a year. Similar antibody affinity chromatography procedures have been reported for the isolation of glucuronidase from rat tissue.[15,16]

Purification of Microsomal Glucuronidase from Mouse Liver or Kidney. Tissues are homogenized in a Polytron homogenizer with 10 volumes

[15] J. W. Owens, K. L. Gammon, and P. D. Stahl, *Arch. Biochem. Biophys.* **166**, 258 (1975).

[16] M. Himeno, Y. Nishimura, H. Tsuji, and K. Kato, *Eur. J. Biochem.* **70**, 349 (1976).

of 0.02 *M* imidazole, pH 7.4. The homogenate is centrifuged at high speed (100,000 *g*, 60 min), and the supernatant solution (containing the bulk of the L-form enzyme) is removed. The pellet is suspended in 0.02 *M* imidazole, pH 7.4 (a volume equal to that of the original homogenate) using a Polytron homogenizer. The solution is recentrifuged at high speed, and the pellet (crude membrane) is suspended in an equal volume of 0.02 *M* imidazole, pH 7.4, containing 2% Triton X-100. This solution is centrifuged (100,000 *g*, 1 hr), and the supernatant solution, containing more than 90% of the activity, is carefully decanted. Glucuronidase is purified from this solution using antibody affinity chromatography as described above, except that after application of enzyme the column is first washed with several volumes of TBS containing 0.1% sodium deoxycholate (Sigma) to remove egasyn associated with glucuronidase. Enzyme yields average about 40% of the original membrane-bound activity. Kidneys from androgen-treated mice are the richest source of microsomal glucuronidase. The purified preparations contain about 40% L-form enzyme and about 60% X-form enzyme, judging from staining intensity after polyacrylamide gel electrophoresis. It is possible to obtain enzyme preparations with a greater proportion of X-form enzyme by fractionation and more extensive washing of membrane preparations; however, about 5–10% of the total L-form enzyme is tightly associated with membrane and can be extracted only with detergent.[10,17]

The X and L electrophoretic forms of glucuronidase can be separated on the basis of charge by DEAE-Sephadex chromatography at pH 6.9 (Fig. 2). Fractions obtained after chromatography are examined by electrophoresis at pH 8.1, and those containing enzyme approximately corresponding in mobility to the X or L forms observed in homogenates are pooled (these are termed the X and L fractions). When subjected to electrophoresis in the presence of sodium dodecyl sulfate, the X fraction migrates as a single band corresponding to an apparent molecular weight of about 74,000. The bulk of the L fraction migrates as a band of molecular weight 71,000, not discernibly different in mobility from L-form enzyme isolated from the lysosomal component of liver and kidney. When subjected to isoelectric focusing in polyacrylamide gels, the X fraction can be resolved into seven forms of about equal staining intensity, ranging from p*I* 6.0 to p*I* 6.6. The L fraction also contains several forms, ranging from p*I* 5.7 to p*I* 5.9, and it closely resembles the enzyme isolated from the lysosomal component of liver and kidney. The size difference between X and L β-glucuronidase may result from differences in carbohydrate content, since after treatment with endoglycosidase H both forms comigrate on SDS gels with an apparent molecular weight of about 70,000. Endogly-

[17] A. J. Lusis and K. Paigen, unpublished data (1977).

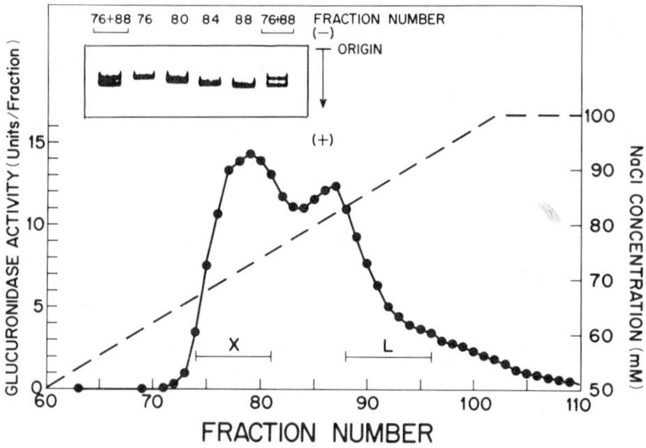

FIG. 2. Separation of X and L mouse β-glucuronidase using DEAE-Sephadex chromatography. Glucuronidase was isolated from crude microsomes of pooled livers and kidneys of 100 male C57BL/6J mice using antibody affinity chromatography. The enzyme (13.6 ml, containing 260 units of enzyme activity) was then dialyzed against 0.02 *M* imidazole, pH 6.9, and applied to a 0.6 × 10 cm column of DEAE-Sephadex A-25, equilibrated with 0.02 *M* imidazole, pH 6.9. Essentially all activity bound to the column. The column was washed with several volumes of equilibration buffer, and the activity was eluted using a linear salt gradient, from 0 to 0.1 *M* NaCl, in equilibration buffer (total volume 200 ml) at a rate of about 20 ml/hr. Fractions of 2.4 ml were collected, glucuronidase activity was determined, and aliquots from peak fractions were subjected to polyacrylamide gel electrophoresis at pH 8.1, followed by staining for glucuronidase activity. The inset shows the pattern of glucuronidase activity after electrophoresis of fractions 76, 80, 84, 88, and mixtures of 76 and 88 (30-μl aliquots were used). Fractions 74–81, corresponding approximately to X enzyme in electrophoretic mobility, and fractions 88–90, corresponding approximately to L enzyme in mobility, were pooled for further characterization (see text).

cosidase treatment has little effect on the charge of glucuronidase.[17] The charge difference, on the other hand, may be due in part to a C-terminal peptide that can be removed from X-form enzyme by protease treatment (Fig. 3). The proteases tested did not discernibly affect the apparent molecular weight of glucuronidase as judged by electrophoresis in sodium dodecyl sulfate, suggesting that the peptide removed from X by protease treatment is short.[17]

Rat Microsomal β-Glucuronidase. Rat liver microsomal β-glucuronidase can be isolated by similar procedures. As in the mouse, the rat microsomal enzyme appears to be slightly larger and less acidic than the lysosomal enzyme.[15,16] Chemical analyses suggest that the two forms differ in carbohydrate and amino acid composition.[18,19]

[18] D. R. P. Tulsiani, H. Six, and O. Touster, *Proc. Natl. Acad. Sci. U.S.A.* **75,** 3080 (1978).

[19] M. Himeno, Y. Nishimura, K. Takahashi, and K. Kato, *J. Biochem.* (*Tokyo*) **83,** 511 (1978).

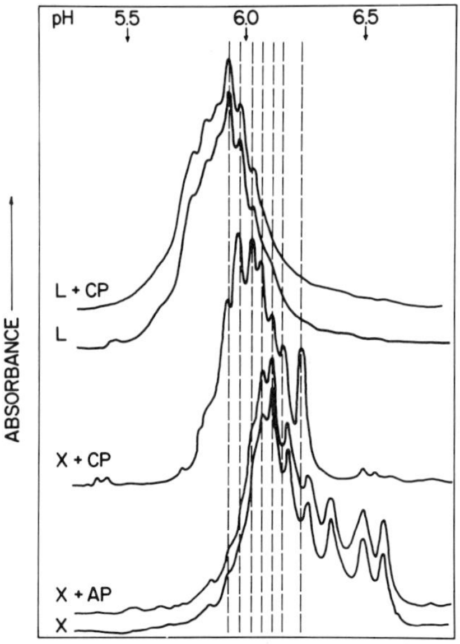

FIG. 3. Effect of protease treatment on the isoelectric focusing pattern of X- and L-form glucuronidase. Isolated glucuronidase preparations (Fig. 2) were treated with proteases and subjected to isoelectric focusing as described.[11] The gels were stained for glucuronidase activity and scanned at 550 nm. The pH gradient along the gel is shown at the top. About 0.5 unit of enzyme activity was applied to each slot, and protease treatment had no effect on total activity. Shown are absorbance tracings for control X form glucuronidase (X), X form treated with aminopeptidase M (X + AP), X form treated with carboxypeptidase Y (X + CP), control L form (L), and L form treated with carboxypeptidase Y (L + CP). Trypsin and chymotrypsin treatment gave the same results as carboxypeptidase Y.

Isolation and Assay of Egasyn

Isolation from Mouse Tissues. In the mouse, microsomal glucuronidase–egasyn complexes can be extracted intact from membranes with Triton X-100.[10] The microsomal glucuronidase can then be specifically precipitated using anti-glucuronidase antibody,[20] and egasyn can be selectively solubilized from such immunoprecipitates by disrupting the glucuronidase–egasyn complex with heating at 50° or with 0.1% sodium deoxycholate.[7] It also is feasible to isolate egasyn by binding microsomal glucuronidase complexes to anti-glucuronidase affinity columns and selectively eluting egasyn by passing 0.1% sodium deoxycholate in TBS through the columns.[17]

[20] S. Tomino and K. Paigen, *J. Biol. Chem.* **250,** 116 (1975).

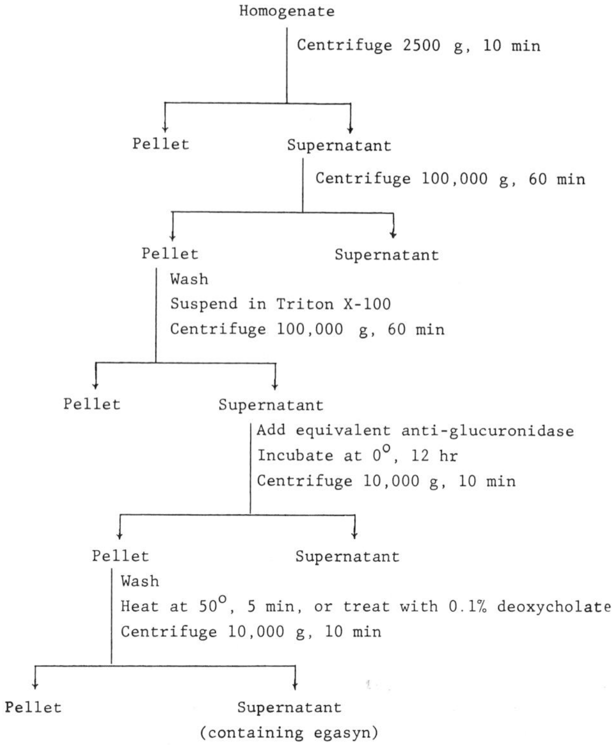

FIG. 4. Isolation of egasyn by release from glucuronidase immunoprecipitates.

The purification of egasyn by specific release from immunoprecipitated microsomal β-glucuronidase is summarized in Fig. 4. All steps are carried out at 0–4° to minimize disruption of egasyn–glucuronidase complexes. The protocol is as follows:

1. The tissues are removed and homogenized in 10 volumes of ice-cold 0.02 *M* imidazole, 0.25 *M* sucrose, pH 7.4, using a Waring blender (4 min at low speed). I normally use as a source the pooled livers and kidneys from 50–100 mice, representing 75–150 g wet weight of tissue. Livers and kidneys are used, since they are relatively rich sources of microsomal glucuronidase.[10] Most of the common strains of mice have similar levels of microsomal enzyme, but C3H/HeJ, AKR/J, and CBA/J have very low levels of β-glucuronidase and YBR/J lacks egasyn. Male or testosterone-induced female mice have elevated levels of kidney enzyme and, thus, are preferable to untreated female mice for egasyn purification.[4]

2. The debris is removed by low-speed centrifugation (2500 *g*, 10 min). The supernatant is centrifuged at high speed (100,000 *g*, 60 min), and the pellet is washed by suspension in the original volume of 0.02 *M* imidazole, 0.25 *M* sucrose, pH 7.4, in a Waring blender, followed by recentrifugation at high speed.

3. The washed high-speed pellet, containing the crude membrane fraction, is suspended in the original volume of 0.02 *M* imidazole, 0.25 *M* sucrose, pH 7.4. Triton X-100 is added to a final concentration of 2%. This suspension is allowed to stand on ice for about 1 hr, and then the solution is centrifuged at 100,000 *g* for 60 min. The clear supernatant, containing over 80% of the microsomal β-glucuronidase activity, is carefully decanted.

4. The high-speed Triton X-100 extract is diluted by addition of an equal volume of 0.02 *M* Tris-HCl, 0.3 *M* NaCl, pH 7.4, and its glucuronidase activity is determined. An equivalent amount of anti-glucuronidase antibody is added, and the mixture is allowed to stand overnight on ice or at 4°. Our original procedure[7] utilized $F(ab)_2$ fragment of anti-glucuronidase IgG, but similar results can be obtained using intact antibody.

5. The immunoprecipitate is collected by centrifugation (10,000 *g*, 10 min) and washed three times with several milliliters of TBS buffer. The immunoprecipitate is then suspended in a small volume of TBS (about 0.2 ml per 100 g of tissue), and the egasyn is quantitatively released by heating at 50° for 5 min or by the addition of sodium deoxycholate (0.1% final concentration). The egasyn is separated from the remaining immunoprecipitate by centrifugation (10,000 *g*, 10 min). When mouse liver is used, the final yield of egasyn is 1–2 μg per gram of tissue.

6. The purity of isolated egasyn should be examined by polyacrylamide gel electrophoresis in sodium dodecyl sulfate.[7] Most preparations exhibit a single protein band corresponding to molecular weight 64,000.

Glucuronidase Binding Protein from Rat Tissues. A protein that specifically binds rat β-glucuronidase is extracted from rat liver microsomes using 2% Triton X-100. Unlike the mouse glucuronidase–egasyn complex, the rat glucuronidase-binding protein complex is sensitive to relatively low concentrations of detergent (0.25% Triton X-100), and thus, it is not possible to recover intact complexes after extraction of tissues with detergents. The binding protein extracted from rat microsomes can be isolated on the basis of its ability to adhere to columns containing immobilized β-glucuronidase, and it resembles mouse egasyn in certain respects.[8]

Antiserum against Egasyn. Rabbits are inoculated intradermally at multiple sites on the back with 50 μg of purified mouse egasyn emulsified in Freund's complete adjuvant. Booster injections containing the same mixture are given at 2-week intervals. Antibody is detected by Ouchterlony double diffusion (with a few micrograms of purified egasyn as the antigen) or by the ability to bind microsomal glucuronidase in Triton X-100 extracts (see below). Rabbits are bled after maximal titer is obtained (this required about 2 months in my experience). The immunoglobin fraction can be partially purified from serum by precipitation with ammonium sulfate at 40% saturation. The antibody preparations that I obtained are specific for egasyn and show no cross-reactivity with either L- or X-form mouse glucuronidase; however, they do bind the M-form complexes containing both glucuronidase and egasyn.[7]

A sensitive method for detecting antibody to egasyn is to expose microsomal glucuronidase to antibody and examine the subsequent ability of glucuronidase–egasyn complexes to migrate in polyacrylamide gels: Fresh homogenates of mouse liver are made 2% in Triton X-100 and centrifuged at high speed. Electrophoresis of the supernate on 7% polyacrylamide gels at pH 8.1, followed by staining for β-glucuronidase with naphthol-AS-BI-glucuronide substrate, gives a band corresponding to lysosomal (L form) glucuronidase and 4 bands corresponding to the microsomal (M form) complexes containing glucuronidase and egasyn. Incubation with anti-egasyn antibody (overnight at 4°) results in the removal of M-form, but not L-form, enzyme upon electrophoresis.[7]

Labeling with ^{125}I. To a test tube are added, in order, 1 mCi of carrier-free $Na^{125}I$ (in 10 μl); 0.5 *M* sodium phosphate, pH 7.5 (10 μl); 10 μg of egasyn isolated using the heat-release procedure described above (20 μl); and 50 μg of chloramine-T in water (10 μl). The mixture is shaken briefly, and 250 μg of sodium metabisulfite (in 0.5 ml) and 100 μg of KI (in 100 μl) are added. Unreacted iodine is separated from the labeled egasyn by passing the mixture through a column of Sephadex G-25 (0.7 cm × 25 cm) equilibrated with 0.15 *M* NaCl, 0.02 *M* Tris, 0.1% Triton X-100, and 0.1% bovine serum albumin, pH 7.5. Fractions of 0.6 ml are collected, and the peak fractions eluting in the void volume, which contain about 30% of the total counts, are pooled.

When examined by polyacrylamide gel electrophoresis in the presence of sodium dodecyl sulfate, the labeled product contains a significant amount of activity that migrates slightly behind the tracking dye in addition to a peak that migrates at the position of an egasyn marker. This low molecular contamination, possibly a breakdown product resulting from the iodination procedure, is only partially removed by exhaustive dialysis; however, the contamination can be removed by chromatography on a 0.5-ml concanavalin A–Sepharose column equilibrated with 0.15 *M* NaCl, 0.02 *M* Tris, 0.1% Triton X-100, 0.1% bovine serum albumin, pH 7.5. The sample is applied slowly to the column, and the column is then washed with several volumes of equilibration buffer. This removes the low molecular weight contamination, which does not bind to the concanavalin A. The labeled egasyn is then eluted by percolating 5 ml of equilibration buffer containing 0.2 *M* methyl-α-mannoside through the column. The ^{125}I-labeled egasyn is stored frozen until use in the radioimmunoassay. As judged by electrophoresis and immunoreactivity, it is stable for at least 6 weeks.

Titration of Labeled Egasyn with Anti-Egasyn Antibody. Titrations and radioimmunoassays can be performed in small glass tubes (0.6 cm × 5 cm). For titrations, antibody to egasyn is added at varying dilutions to about 10,000 cmp of ^{125}I-labeled egasyn, in a final volume of 100 μl. All dilutions are made in TBS containing 0.1% Triton X-100 and 0.1% bovine serum albumin. The mixtures are incubated for 1 hr at 37° and then 24 hr at 4°. Carrier control rabbit serum (40 μl of 4% serum) and antibody to rabbit IgG (30 μl) are added, and the mixtures are incubated for 1 hr at 37° and 4 hr at 4°. The immunoprecipitate pellets are collected by centrifugation (10,000 *g*, 10 min) and washed twice with dilution buffer. Using our antibody preparations, 80–90% of the label is precipitated at high levels of antibody (10-fold dilution) and less than 3% of the label is precipitated in the absence of anti-egasyn antibody.[7]

Radioimmunoassay of Egasyn. For the radioimmunoassay, a concentration of antibody that corresponds to about 50% binding of the ^{125}I-labeled egasyn is used. The antibody is first incubated with sample, ^{125}I-labeled egasyn is then added, and after further incubation the immune complexes are precipitated by the addition of goat antiserum raised against rabbit IgG. The egasyn content of samples is estimated by their ability to displace radioactivity from the precipitate to the supernatant fraction. Standard curves are obtained by the addition of unlabeled isolated egasyn. Most of the antigenic sites on egasyn in homogenates of liver are masked after extraction with Triton X-100 and become immunoreactive only after treatment with 1% sodium deoxycholate (Fig. 5). The reason for this is unclear but it is, therefore, necessary to carry out the radioimmunoassay in the presence of deoxycholate.[7] The assay described here is a liquid-phase procedure with double-antibody precipitation; other variations have not been tested. Thus, it may be possible to improve the sensitivity or simplicity of the assay, for example, with solid-phase immunoassay.

The protocol for the assay is as follows:

1. Anti-egasyn IgG and the sample are incubated in TBS containing 1% sodium deoxycholate (100 μl final volume) for 1 hr at 37° and 24 hr at 4°. The amount of antibody used is sufficient to precipitate about 50% of the labeled egasyn (see below) in the absence of competing antigen and

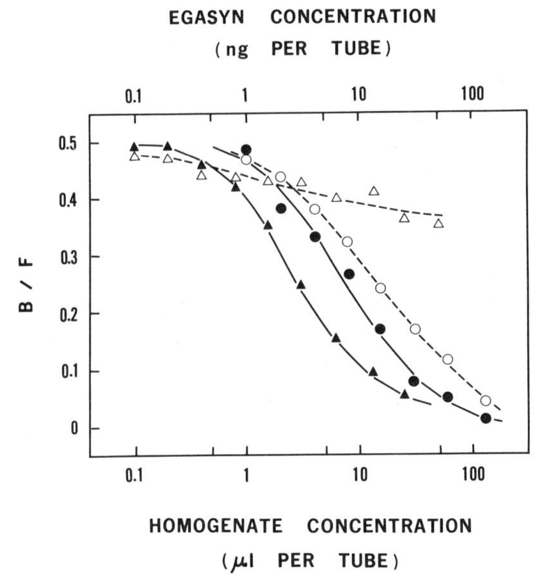

FIG. 5. Radioimmunoassay of mouse egasyn: effect of deoxycholate. Egasyn was assayed in the absence (○, △) or the presence (●, ▲) of 1% sodium deoxycholate as described in the text. The standard (○, ●) was a preparation of egasyn purified from mouse liver and kidney by release from immunoprecipitates of microsomal glucuronidase. The sample (△, ▲) was a 10% homogenate of strain C57BL/6J liver extracted with Triton X-100 and centrifuged at high speed. Plotted on the ordinate is the ratio of antibody-bound ^{125}I (B) to free ^{125}I (F).

represents a 500-fold dilution of the antiserum. Samples are diluted in TBS containing 0.1% bovine serum albumin and 1% Triton X-100. The particulate matter in homogenates results in some interference with the assay. For this reason, assays are generally performed on high speed supernatant solutions of samples extracted with 2% Triton X-100, which solubilizes both egasyn and the microsomal glucuronidase complexes.[7]

2. To the mixture is then added 5000 cpm for ^{125}I-labeled egasyn (in 50 μl of sample buffer). This is incubated for 1 hr at 37° and 72 hr at 4°.

3. Egasyn bound to antibody is separated from free egasyn by the addition of goat antibody to rabbit IgG (30 μl) and 4% control rabbit serum as carrier (40 μl). This is incubated for 1 hr at 37° and 4 hr at 4°. The mixture is centrifuged at 10,000 *g* for 10 min, and the immunoprecipitate pellet is washed twice with 0.5 ml of sample buffer. The radioactivity in the pellet is measured with a gamma counter. A purified preparation of egasyn (from 1 to 50 ng per tube) is used as the standard. Typical radioimmunoassay results are shown in Fig. 5.

[45] Biosynthesis of Sarcoplasmic Reticulum Proteins

By David H. MacLennan and Stella de Leon

We have studied the biosynthesis of sarcoplasmic reticulum proteins both in cell culture and in *in vitro* systems. In cell culture, the initiation of synthesis of the Ca^{2+},Mg^{2+}-ATPase[1,2] and the intrinsic glycoprotein[3] occurs at about the time of myoblast fusion. Initiation of synthesis of the extrinsic protein, calsequestrin,[2,4] and the high-affinity Ca^{2+} binding protein[5] occurs several hours earlier, well before fusion begins. Calsequestrin and the ATPase are synthesized on membrane-bound polyribosomes[6–8]—calsequestrin with a signal sequence,[8] and the ATPase without one.[8,9] In this chapter, methods are described for *in vivo* and *in vitro* synthetic studies.

Myoblast Cultures

Reagents for Myoblast Tissue Culture

PBS (Ca^{2+} free, phosphate-buffered saline): 140 m*M* NaCl, 2.6 m*M* KCl, 8 m*M* Na_2HPO_4, 1.5 m*M* KH_2PO_4, 0.5 m*M* $MgCl_2$ (autoclave)
SSC: 134 m*M* NaCl, 10 m*M* sodium citrate, pH 7.8
Trypsin solution A: 0.3% trypsin (1 : 250) in PBS (filter)
Trypsin solution B: 0.05% trypsin (1 : 250) in SSC (filter)
Trypsin solution C: 0.1% Difco trypsin (1 : 250) in Ca^{2+},Mg^{2+}-free PBS. Add EDTA to 0.5 m*M* (filter).
Gelatin, 0.1% in H_2O (autoclave)
Trasylol (Boehringer)
DME medium (Dulbecco's modified Eagle's medium[10]). This medium is normally prepared commercially or in media preparation

1 P. C. Holland and D. H. MacLennan, *J. Biol. Chem.* **251,** 2030 (1976).
2 A. O. Jorgensen, V. I. Kalnins, E. Zubrzycka, and D. H. MacLennan, *J. Cell Biol.* **74,** 287 (1977).
3 E. Zubrzycka-Gaarn, K. P. Campbell, D. H. MacLennan, and A. O. Jorgensen, *J. Biol. Chem.* **258,** 4576 (1983).
4 E. Zubrzycka and D. H. MacLennan, *J. Biol. Chem.* **251,** 7733 (1976).
5 M. Michalak and D. H. MacLennan, *J. Biol. Chem.* **255,** 1327 (1980).
6 D. C. Greenway and D. H. MacLennan, *Can. J. Biochem.* **56,** 452 (1978).
7 T. L. Chyn, A. N. Martonosi, T. Morimoto, and D. D. Sabatini, *Proc. Natl. Acad. Sci. U.S.A.* **76,** 1241 (1979).
8 R. A. F. Reithmeier, S. de Leon, and D. H. MacLennan, *J. Biol. Chem.* **255,** 11839 (1980).
9 K. E. Mostov, P. de Foor, S. Fleischer, and G. Blobel, *Nature (London)* **292,** 87 (1981).
10 R. Dulbecco and M. Vogt, *J. Exp. Med.* **99,** 167 (1954).

METHODS IN ENZYMOLOGY, VOL. 96

ISBN 0-12-181996-5

centers according to Dulbecco and Vogt.[10] It can be prepared minus Ca^{2+} and minus methionine for later experimental manipulation. It can also be prepared in 2× concentration for supplementation of the freezing medium.

Methionine, 20 m*M* (filter)

$CaCl_2$, 140 m*M* (autoclave)

Horse serum: commercial source

Fetal calf serum: commercial source

Chick embryo extract: lyophilyzed powder, commercial source

Gentamycin, 10 mg/ml: commercial source

Standard medium: 875 ml of DME medium; 100 ml of horse serum; 5 ml of chick embryo extract (rehydrated from lyophilized stock), 10 ml of 140 m*M* $CaCl_2$, 10 ml of 20 m*M* methionine, 1 ml of Gentamycin (10 mg/ml)

Ca^{2+}-deficient medium: as in standard medium but add 0.45 ml of 140 m*M* $CaCl_2$ and 100 ml of horse serum that has been dialyzed overnight against 10 volumes of Ca^{2+}-free PBS.

Triton buffer: 0.5% Triton X-100, 150 m*M* NaCl, 10 m*M* $NaHPO_4$, pH 7.0, 50 kallikrein-inactivating units (KIU) of Trasylol per milliliter

Fractionation buffer: 1% sodium deoxycholate, 150 m*M* KCl, 10 m*M* Tris-HCl, pH 7, .8, 1 m*M* sodium tetrathionate, 100 KIU of Trasylol and 0.1 m*M* PMSF

Freezing medium: 10 ml of glycerol, 10 ml of fetal calf serum, 25 ml of 2× concentrated DME medium, 5 ml of H_2O

Materials for Myoblast Culture

Trypsinizing flasks with side arms and with caps for side arm and neck

Plastic tissue culture flasks, 75 cm^2 and 25 cm^2

Plastic petri dishes, 150 mm in diameter, gelatin coated. Incubate plates at 4° overnight with 20 ml of sterile 0.1% gelatin solution; aspirate.

Plastic freezing vials, 1 ml

Plastic, 15-ml capped, conical centrifuge tubes

Swinnex filter cartridges, 2.5 cm in diameter, with double layer of lens paper

Dissecting forceps, scissors, sterile beakers

Preparation of Rat Myoblast Cultures. The procedure for primary myoblast preparation and culture is that of Yaffe.[11] Two-day-old rats are

[11] D. Yaffe, *in* "Tissue Culture: Methods and Application" (P. Kruse and M. K. Patterson, eds.), p. 106. Academic Press, New York, 1973.

killed by decapitation, sterilized by dipping into 70% ethanol, and transferred to a sterile petri dish. The skin is dissected around the midsection and peeled back to expose the musculature of the hind legs. Leg muscle is excised in strips about 2 mm in diameter and transferred first to a beaker of sterile PBS and then to a trypsinizing flask containing 40 ml of trypsin solution A. The suspension is stirred gently for 40 min at room temperature, and the supernatant containing released cells is decanted. This supernatant is discarded. The digestion process is repeated three to five more times by the addition of fresh trypsin solution A until the tissue is completely digested. All these supernatants are processed immediately by decantation into 15-ml centrifuge tubes containing 1.5 ml of fetal calf serum. The tubes are mixed and centrifuged for 3 min in a clinical centrifuge, and each pellet is washed in 8 ml of standard medium and finally suspended with a Pasteur pipette in a few milliliters of standard medium. The contents of three tubes are combined into 15 ml and filtered through two layers of lens paper in a Swinney adaptor. The cells are diluted to 4×10^6 cells per milliliter, and 15–20 ml are plated in a 75-cm^2 plastic tissue culture flask to permit selective attachment of fibroblasts. After 1 hr at 37° the supernatant is decanted, cells are counted, and 2×10^7 cells are transferred to 150-mm, gelatin-coated petri dishes. The volume is made to 25 ml with standard medium, and the cells are incubated at 37° in a humidified air atmosphere containing 5% CO_2. The medium is changed at 24 hr and at 48 hr thereafter. The cells should be confluent and aligned by 48 hr and should shortly begin fusing. The addition of Ca^{2+}-deficient medium at 24–48 hr prevents fusion, but not differentiation.

Preparation of Prenatal Rabbit Muscle Cell Culture. Excellent primary cultures can be obtained from prenatal rabbit muscle. Rabbits at 25 days gestational age are obtained by cesarean section and killed by decapitation; cells are isolated and cultured from both back and leg muscle exactly as described for rat muscle except that preplating to remove fibroblasts is carried out for 2 hr.

Culture of Continuous Cell Lines. Cell lines L6 and L8[12] are cultured by plating 10^6 cells in 20 ml in 75-cm^2 tissue culture flasks.[13] The growth medium is DME supplemented with 10% horse serum and 10 μg of Gentamycin per milliliter. The medium is routinely changed at 24 hr and at 72 hr intervals thereafter up to a maximum of about 12 days. Fusion begins between days 4 and 5. The line is carried by transfer, prior to fusion at about 72 hr. If the line is to be carried, it is plated at 2×10^5 cells per 75-cm^2 flask. Cells are washed once in SSC, 2 ml of trypsin solution B are

[12] D. Yaffe, *Proc. Natl. Acad. Sci. U.S.A.* **61,** 477 (1965).

[13] S. K. Ng, J. Rogers, and B. D. Sanwal, *J. Cell Physiol.* **90,** 361 (1976).

added, and the flask is incubated at 37° for 3 min to detach the cells. The detached cells are added to 3 ml of DME medium and 10% horse serum and collected by low speed centrifugation. They are then suspended in DME medium at a concentration of 2×10^5 cells per 15 ml and plated in 75-cm^2 flasks.

The line can also be preserved in liquid nitrogen. Cells at a concentration of 2×10^6 per milliliter in freezing solution are dispensed in 1-ml aliquots into plastic freezing vials and frozen in liquid nitrogen. They are thawed at 37°, diluted with 9 ml of DME medium, and centrifuged; the pellet is suspended in 5 ml of DME medium and plated in 25-cm^2 tissue culture flasks.

Labeling of Cells and Isolation of Labeled Products. In a typical labeling protocol, cell cultures at different stages of differentiation are rinsed once with 10 ml of PBS and incubated for 2 hr with 10 ml of methionine-free DME medium supplemented with 0.1 volume of horse serum and 10 μCi of [^{35}S]methionine per milliliter (1000–1200 Ci/mmol). For labeling periods of 4 hr or 24 hr we have added unlabeled methionine to concentrations of 4 μM or 50 μM, respectively. After labeling, the labeling medium is removed and the dishes are washed 3 times with 10 ml of ice cold PBS. Primary cells are detached at 37° by addition of 8 ml of trypsin solution C. After 3 min, when detachment is complete, 2 ml of fetal calf serum and 200 KIU of Trasylol are added to inhibit proteolytic activity. Cells can be harvested also by scraping them from the flask surface with a rubber policeman. The cells are collected by low speed centrifugation and washed with 8 ml of ice cold PBS containing 1 ml of fetal calf serum and 200 KIU of Trasylol. The washed pellet can be frozen in Dry Ice–ethanol and then stored at −20°. After thawing, the cells from a single plate are dissolved in 1 ml of Triton buffer and centrifuged at 100,000 *g* for 45 min. The supernatant contains all sarcoplasmic reticulum proteins.

We have had difficulty in obtaining immunoprecipitation of the high-affinity calcium-binding protein in cell extracts. This problem can be solved by precipitating a large fraction of the muscle protein, including actin, with ammonium sulfate. Washed cells are dissolved in 2 ml of deoxycholate solution and centrifuged at 100,000 *g* for 60 min. The pellet is discarded, and 55 mg of solid $(NH_4)_2SO_4$ are added per milliliter of supernatant. After 10 min at 0°, the precipitate is removed by centrifugation at 10,000 *g* for 20 min. The supernatant is then made 35% saturated with respect to $(NH_4)_2SO_4$ by the addition of 149 mg of $(NH_4)_2SO_4$ per milliliter. The precipitate is removed by centrifugation, and the supernatant is dialyzed for 20 hr against fractionation buffer minus deoxycholate. This procedure removes 30% of the protein and 20% of the radioactivity from the supernatant, but does not precipitate calsequestrin or the high-affinity Ca^{2+}-binding protein. They can be immunoprecipitated from the fractionated extracts.

Calsequestrin can also be purified from the dialyzed, fractionated extracts by two-dimensional gel electrophoresis.[5] The protein is separated in the Weber and Osborn slab gel electrophoretic system,[14] where it runs with an apparent $M_r = 44,000$. The lane is then cut out and placed on top of the gel in a Laemmli slab gel system.[15] The slice is sealed with 0.1% agarose in Tris-glycine buffer, pH 6.8, and equilibrated for 15 min; electrophoresis is begun in the second dimension. Calsequestrin runs at $M_r = 63,000$ in the Laemmli system and, consequently, falls off of the diagonal formed by the various proteins. It is isolated in a discrete spot by this procedure.

Immunoprecipitation

Preparation of Antisera. Proteins for antibody production have been purified by procedures worked out in our laboratory.[1,3–5] As a final purification step these proteins can be subjected to Laemmli slab gel electrophoresis and the eluted product used as an antigen. Antisera against rat muscle Ca^{2+},Mg^{2+}-ATPase have been routinely prepared by injection of 1 mg of protein at 7-day intervals into adult rabbits. Antisera are collected on day 28 and at intervals thereafter by bleeding 1 week after a booster shot. Antisera against the intrinsic glycoprotein, calsequestrin, and the high-affinity Ca^{2+}-binding protein have been obtained by injecting 500-μg aliquots of rabbit proteins into mature sheep. The second injection is made at 3 weeks, the third at 5 weeks, and bleeding is at 8 weeks.

Reagents for Protein-A Immunoprecipitation[16]

Protein A beads: commercial product (Pharmacia)
Bead buffer A: 50 mM Tris-HCl, pH 7.2, 150 mM NaCl, 1% Triton X-100, 100 KIU of Trasylol per milliliter
Bead buffer B: as for buffer A, but containing 1% sodium deoxycholate and 0.1% SDS
Wash solution A: 100 mM Tris-HCl, pH 7.0, 200 mM LiCl, 0.1% 2-mercaptoethanol
Wash solution B: 1 mM Tris-HCl, pH 7.0, 150 mM NaCl

[14] K. Weber and M. Osborn, *J. Biol. Chem.* **244,** 4406 (1969).
[15] U. K. Laemmli, *Nature (London)* **227,** 680 (1970).
[16] P. E. Branton, N. J. Lassam, J. F. Downey, S. I. Yee, F. L. Graham, S. Mak, and S. T. Bayley, *J. Virol.* **37,** 601 (1981).

Laemmli sample buffer (3×): 6% SDS, 30% glycerol, 3% 2-mercaptoethanol, 0.1875 *M* Tris-HCl, pH 6.8, 0.03% Bromphenol Blue

Immunoprecipitation with Protein A Beads. Protein A beads (1.5 g) will swell to about 3.5 ml in bead buffer A. The beads are suspended twice and allowed to settle briefly in about 35 ml of this buffer to remove "fines." They are then stored in 50 ml of bead buffer A at 4° and vortexed just before use to ensure uniform dispensing.

Aliquots of translation assays (up to 200 μl) are mixed in a 1.5-ml Eppendorf tube with 0.8 ml of bead buffer B, 10 μl of antiserum, and 250 μl of the suspension of washed protein A beads. Up to 1 ml of Triton extracts of cell culture can be incubated with 10–20 μl of antiserum and 350 μl of protein A beads suspension. The tubes are rotated slowly at 4° for 3 hr, and the beads are collected by brief centrifugation in the cold. The beads are then washed 3 times in 1 ml of wash solution A, 3 times in wash solution B and, finally, extracted by boiling in 150 μl of 3× Laemmli buffer. The extract is separated from the beads by centrifugation for 3 min at 12,000 *g* and analyzed on Laemmli slab gels. The entire extract is separated on Laemmli slab gels.

Isolation of RNA

Isolation of Free and Bound Polysomes from Rat Muscle

Reagents for Polysomal Isolation

Wash solution A: 50 m*M* Tris-HCl, pH 7.6, 150 m*M* KCl, 10 m*M* $MgCl_2$, 0.5 m*M* EDTA, heparin, 70 U/ml

Wash solution B: as wash solution A, but minus heparin

Storage solution: as in A, but minus heparin and minus EDTA

Sucrose solution A: 250 m*M* sucrose, 50 m*M* Tris-HCl, pH 7.6, 150 m*M* KCl, 10 m*M* $MgCl_2$, 0.5 m*M* EDTA

Sucrose solution B: 2 *M* sucrose, 50 m*M* Tris-HCl, pH 7.6, 150 m*M* KCl, 10 m*M* $MgCl_2$, 0.5 m*M* EDTA

10% Sodium deoxycholate

10% Triton X-100

Procedure. Polysomal preparations are based on the work of Nihei[17] and of Laga *et al.*[18] Muscle (7–8 g) is excised from the hind legs of 40–50 neonatal rats and placed in 50 ml of cold wash buffer A. The muscle is rinsed three times in cold wash buffer B and then homogenized for 15 sec at low speed and 30 sec at high speed in 5 volumes of sucrose solution A.

[17] T. Nihei, *Biochem. Biophys. Res. Commun.* **43,** 1139 (1971).

[18] E. M. Laga, B. S. Baliga, and H. M. Munro, *Biochim. Biophys. Acta* **213,** 391 (1970).

The homogenate is centrifuged at 10,000 *g* for 10 min, the supernatant is filtered through four layers of gauze, and 5-ml aliquots (of a total of 35 ml) are layered over 3 ml of sucrose solution B and centrifuged for 16–20 hr at 100,000 *g*. The pellet passing through 2 *M* sucrose is considered to be free polysomes. The membrane fraction, collected from the 2 *M* sucrose interface, is gently suspended with a glass–Teflon homogenizer and made 1% in deoxycholate and 0.5% in Triton X-100 in a total of 30 ml of sucrose solution A. The suspension, in 5-ml aliquots, is layered over 3 ml of sucrose solution B and centrifuged for 5 hr at 125,000 *g*. The pellet is considered to be bound polyribosomes. The polysome fractions are gently suspended with a glass rod in 0.3 ml of storage solution and frozen at −70°. The RNA content is estimated by the method of Wool and Cavicchi.[19] The yield of free polysomes is about 5 A_{260} units per gram of muscle, and the yield of bound polysomes is about 0.4 A_{260} units per gram. Translation assays require about 5 A_{260} units.

Isolation of mRNA from Cell Cultures and Neonatal Muscle

Reagents for mRNA Isolation from Cell Cultures

Guanidine–SCN buffer: 50% guanidine thiocyanate (w/v), 0.5% sodium lauroyl sarcosine, 0.7% 2-mercaptoethanol, 0.33% antifoam A emulsion, 25 m*M* sodium citrate, pH 7.0. Warm to dissolve and ultrafilter. Store at room temperature for up to 1 month.

Guanidine-HCl solution: Mix 1 ml of 2 *M* potassium acetate, pH 5.5, and 19 ml of 6 *M* guanidine-HCl.

Oligo(dT) (12–18)

Oligo dT column buffer A: 10 m*M* Tris-HCl, pH 7.5, 500 m*M* NaCl, 30 m*M* EDTA, 0.5% SDS, 0.2% sodium deoxycholate. (This solution is made up also at 2× concentration.)

Column buffer B: 10 m*M* Tris-HCl, pH 7.5, 500 m*M* NaCl, 30 m*M* EDTA

Column buffer C: 10 m*M* Tris-HCl, pH 7.5, 1 m*M* EDTA

Siliconized 30-ml glass tubes

Potassium acetate, 2 *M*

All reagents are made up in diethyl pyrocarbonate-treated H_2O and all glassware is baked at 180° for 2 hr.

Procedure. The isolation of mRNA from tissue culture cells is based on procedures of Chirgwin *et al.*[20] Cell cultures are washed twice with 10 ml of PBS and then incubated at 37° with 4 ml of trypsin solution B until cells begin to detach (3–10 min). The cells are collected in 30-ml

[19] I. G. Wool and P. Cavicchi, *Biochemistry* **6,** 1231 (1967).

[20] J. M. Chirgwin, A. E. Przybyla, R. J. MacDonald, and W. J. Rutter, *Biochemistry* **18,** 5294 (1979).

siliconized glass tubes and centrifuged at 4000 rpm for 10 min at 4°. The cells from 5 plates are suspended in 6 ml of guanidine-SCN solution and homogenized with 10 strokes by hand in a loose-fitting Dounce homogenizer. The pH is adjusted to 5.5 by the addition of 0.025 volume of 1 *M* acetic acid. RNA is precipitated by the addition of 0.35 volume of cold ethanol and storage overnight at −20° and recovered by centrifugation at 5000 rpm for 20 min at −10°. The RNA in the pellet is further purified by two cycles of dissolution in 4 ml of guanidine-HCl solution, precipitation with 0.5 volume of ethanol, incubation at −20° for 4 hr, and centrifugation at 5000 rpm for 20 min at −10°. The final RNA pellet is rinsed twice by gentle addition of 4 ml of 80% ethanol to the surface and brief centrifugation. After careful removal of the ethanolic solution, the pellet is dried *in vacuo,* dissolved in 1 ml of H_2O, heated at 55° for 3 min, and centrifuged for 10 min at 12,000 *g* in an Eppendorf centrifuge. The RNA concentration is estimated by absorbance at 260 nm and adjusted to 10 A_{260} units/ml. RNA can be stored at −20° at this stage after the addition of 0.1 volume of 2 *M* KAc, pH 5.5, and 2 volumes of ethanol.

For purification of poly(A)$^+$ RNA, 10 units of RNA are recovered by centrifugation, washed twice in 80% ethanol, dried *in vacuo,* dissolved in 1 ml of column buffer A and heated at 55° for 3 min. Oligo(dT) columns are prepared in plugged, siliconized Pasteur pipettes or 1-ml syringe casings in the ratio with RNA recommended by the manufacture (usually about 50 mg of oligo(dT)/10 units of RNA). The column is washed successively with buffers A, B, and C and then equilibrated with buffer A. The sample, in buffer A, is applied, and the effluent is immediately reapplied. The column is washed with 5 ml of buffer A, 5 ml of buffer B; then poly(A)$^+$-enriched RNA is eluted with 3 ml of buffer C. The sample is made to the composition of buffer A by the addition of an equal volume of 2× buffer A and reapplied on a fresh oligo(dT) column equilibrated with buffer A (one half the size of the original column). The final eluate is stored after addition of 0.3 volume of 2 *M* KAc and 2 volumes of ethanol. The yield of mRNA varies with the age of the culture. We obtain about 30 units of RNA from five 24-hr rabbit cell culture plates. The first oligo(dT) column gives a 10-fold purification, and the second gives a further 5-fold purification of mRNA.

Isolation of Poly(A)$^+$ RNA from Rat or Rabbit Muscle

Reagents for mRNA Isolation from Muscle

Extraction buffer: 50 m*M* Tris-HCl, pH 7.6, 250 m*M* KCl, 10 m*M* MgCl, 250 m*M* sucrose

Guanidine-NaAc: 6 *M* guanidine-HCl, 20 m*M* sodium acetate, pH 5.5

Procedure. Muscle isolated from the hind and forelimbs of 40–50 neonatal rats or the hind legs and back of 7–10 neonatal rabbits is frozen in liquid nitrogen, where it can be stored without degradation of RNA. The frozen muscle is thawed in 5 volumes of extraction buffer and homogenized in a Waring blender for 15 sec at low speed and 30 sec at high speed. The high ionic strength is required to prevent precipitation of myosin. The homogenate is centrifuged for 10 min at 10,000 *g*, and the supernatant is filtered through four layers of sterile gauze. The supernatant is then made 6 *M* in guanidine-HCl by the addition of solid material, and the pH is adjusted to 5.5 with 1 *M* acetic acid.[21] One half volume of ethanol is added, and the RNA is precipitated during a 30-min incubation at −20°. RNA is collected by centrifugation at 4000 *g* for 20 min at 0°. The RNA is further purified by two cycles of dissolution in 10 ml of guanidine-NaAc, precipitation with 0.5 volume of ethanol for 30 min at −20°, and centrifugation. The pellet is washed twice with 80% ethanol, dried *in vacuo*, and purified on oligo(dT) columns as described previously.

Sucrose Gradient Fractionation of RNA

We have fractionated mRNA, passed once through oligo(dT) columns, on isokinetic 10% to 31.4% sucrose density gradients.[22] RNA is dissolved at a concentration of about 3 A_{260} units in 200 μl of 10 m*M* Tris-HCl, pH 7.5, 1 m*M* EDTA, and 0.2% SDS. The sample is heated to 70° for 3 min, quick-chilled, and immediately loaded onto the gradient that is formed in the presence of 10 m*M* Tris-HCl, pH 7.5, 1 m*M* EDTA, and 0.2% SDS. The sample is centrifuged in a Beckman SW40 rotor at 37,000 rpm for 16 hr at 4°. The gradient is scanned at 260 nm with a density gradient fractionator and absorbance monitor. The tubing and flow cell are flushed with 0.05% diethyl pyrocarbonate before use. The RNA is recovered by ethanol precipitation, and aliquots are translated. Calsequestrin mRNA is found in a narrow trough on the heavy side of the 18 S absorbance peak: ATPase mRNA is found in the lighter half of the 28 S absorbance peak.

Translation System for Muscle mRNA

Reagents Required for Translation Assays. The preparation of rabbit reticulocyte lysate described by Pelham and Jackson[23] is suitable for translation of rabbit muscle mRNA. Rabbit reticulocyte lysates are also commercially available (e.g., Bethesda Research Laboratories) and are of

[21] R. A. Cox, this series, Vol. 12B, p. 120.
[22] K. S. McCarty, Jr. and K. S. McCarty, *Anal. Biochem.* **61,** 165 (1974).
[23] H. R. B. Pelham and R. J. Jackson, *Eur. J. Biochem.* **67,** 247 (1976).

high quality. We now use these preparations routinely in the reaction mixtures recommended by the manufacturer. The basic mixtures are as follows:

mRNA passed twice through an oligo(dT) column, 0.03–0.1 A_{260} unit in 30 μl of H_2O. Heat to 55° for 3 min just before use.

Rabbit reticulocyte lysate in 3.5 m*M* $MgCl_2$, 0.05 m*M* EDTA, 25 m*M* KCl, 0.5 m*M* dithiothreitol, 25 m*M* hemin, 1 m*M* $CaCl_2$, 2 m*M* EGTA, 70 m*M* NaCl, 50 mg of creatine kinase per milliliter

Biosynthetic reaction mixture: 75 m*M* HEPES, pH 7.5, 120 m*M* KCl, 260 m*M* KAc, 33 m*M* creatine phosphate, 19 amino acids, 150 μM each (excluding methionine), 125 μCi of [^{35}S]methionine per 30 μl

Trichloroacetic acid, 5%

Triton X-100, 10%

Methionine, 100 m*M*

Trasylol, 10,000 units/ml

Procedure. In a typical experiment, 30 μl of reaction mixture and 30 μl of reticulocyte lysate are mixed in a 1.5-ml Eppendorf tube. The reaction is started by the addition of 30 μl of mRNA and continued for 1 hr at 30°. The tube is transferred to ice, and 5 μl are transferred to a Whatman No. 3 filter paper disk, which is then dried; up to 10 filters are washed once in 100 ml of 5% trichloroacetic acid and boiled for 10 min in 100 ml of 5% trichloroacetic acid. The filter is washed in 100 ml of ethanol–ether (1 : 1) and then in 100 ml of ether, air dried, and counted for incorporation of methionine. A control minus mRNA is used for background. To quench the translation reaction, 10 μl of methionine and 5 μl of 10% Triton are added, and the sample is incubated at 0° for 10 min. It is then centrifuged at 100,000 *g* for 30 min. The supernatant is used for immunoprecipitation of radioactive sarcoplasmic reticulum proteins using protein A beads.

In vitro processing of sarcoplasmic reticulum proteins has been studied in this translation system using pancreatic microsomes. The procedures for studying processing are described in another chapter in this volume.[24]

[24] G. Blobel, this volume [6].

[46] Plant Vacuoles

By CLARENCE A. RYAN and MARY WALKER-SIMMONS

Because of their large sizes, vacuoles are the most conspicuous of all the membrane-enclosed organelles in plant cells. Their presence has been recognized for a long time, but it has been only in the last few years that the full importance of the functional roles of the vacuole has been appreciated. The vacuole not only plays an important osmotic role in maintaining cell turgor, but also participates in ion and metabolite transport and storage, in sequestering "secondary" defense chemicals, and in intracellular digestion during plant metabolism. The hydrolytic role of plant vacuoles has been likened to that of animal lysosomes, and its identity as a large lysosomal organelle is now generally accepted.[1,2] Mounting evidence, including high-resolution electron microscopy and cytochemical and chemical analyses of vacuolar components, continues to support the concept that vacuoles can function as plant lysosomes.

Biochemical studies of the properties of plant vacuoles have been greatly stimulated and facilitated by the recent availability of purified vacuolar preparations. The first large-scale isolation of plant vacuoles was achieved in 1975.[3] The method was based on the greater stability of the vacuolar membrane with respect to the cell membrane in isolated protoplasts subjected to osmotic shock. A number of variations of the technique have now been published,[4-10] but they are essentially similar to the original. A method for the isolation of vacuoles by mechanically shearing fresh tissues with a special slicing apparatus has also been developed.[11] This method produces highly pure vacuoles, although in low yields, but is limited to tissues that can be readily sliced in the apparatus.

[1] P. Matile, *in* "The Lytic Compartment of Plant Cells" (Cell Biology Monographs, Vol. 1). Springer-Verlag, Berlin and New York, 1975.
[2] F. Marty, *Proc. Natl. Acad. Sci. U.S.A.* **75,** 852 (1978).
[3] G. J. Wagner and H. W. Siegelman, *Science* **190,** 1298 (1975).
[4] H. Lörz, C. T. Harms, and I. Potrykus, *Biochem. Physiol. Pflanz.* **169,** 617 (1976).
[5] M. Walker-Simmons and C. A. Ryan, *Plant Physiol.* **60,** 61 (1977).
[6] J. A. Saunders and E. E. Conn, *Plant Physiol.* **61,** 154 (1978).
[7] M. Nishimura and H. Beevers, *Plant Physiol.* **62,** 1978).
[8] T. Boller and H. Kende, *Plant Physiol.* **63,** 1123 (1979).
[9] R. Kringstad, W. H. Kenyon, and C. C. Black, *Plant Physiol.* **66,** 379 (1980).
[10] A. H. Moskowitz and G. Hrazdina, *Plant Physiol.* **68,** 686 (1981).
[11] R. A. Leigh and D. Branton, *Plant Physiol.* **58,** 656 (1976).

ISBN 0-12-181996-5

Preparation of Vacuoles from Lysed Protoplasts

This method, with various modifications, has been successfully employed with a variety of tissues from many plant genera. Any tissue from which protoplasts can be readily produced can potentially yield isolated vacuoles, although careful manipulation of conditions is often critical to maximize purity and yields.

Preparation of Protoplasts

Cellulysin Digestion of Hippeastrum and Wheat.[3,12,13] Large *Hippeastrum* petals are surface sterilized by washing with 70% ethanol. The epidermis is removed by peeling, when possible, but the epidermis of some tissues cannot be easily removed in this manner and are left intact. The tissues are sliced into 2-mm strips and incubated in a 2% solution (w/v) of cellulysin (Calbiochem) in 0.6 *M* mannitol, adjusted to pH 5.8 with NaOH, and then shaken gently for 11 hr at 18°. For wheat, 2% cellulase (Vega Biochemicals) and 0.5% pectinase (Sigma) in 0.7 *M* mannitol are employed for 4–5 hr at 28°.[13] The enzyme solutions are sterilized before use by passage through a 0.22-μm filter. After incubation, the undigested tissue is removed and the protoplasts are sedimented at 500 *g*, washed three times with 0.7 *M* mannitol in 20 m*M* MES–KOH, pH 5.5.[13] *Hippeastrum* protoplasts are further purified by resuspending in 22% sucrose (w/v), 20 m*M* MES–KOH, pH 5.5, 1 m*M* dithiothreitol (DTT), and centrifuged at 700 *g* for 5 min. The protoplasts float to the surface. Protoplasts collected at the surface are diluted in 0.7 *M* mannitol, 20 m*M* MES–KOH (pH 5.5), and sedimented at 50 *g*.

Cellulysin is partially purified[12] by centrifugation of a 20 g/70 ml solution at 5000 *g* for 5 min. The supernatant is passed through a Sephadex G-25 column (75 × 4.5 cm) equilibrated and eluted with 1 m*M* NaCl. The breakthrough peak is collected and utilized for protoplast isolation.

Pectolyase-Cellulase Digestion of Tobacco, Rice, and Oats. Fully expanded tobacco leaves are cut into 0.5-cm slices without peeling the epidermis.[14] Rice and oat leaves are cut longitudinally. To each 1 g of leaves are added 10 ml of an enzyme solution containing 0.1% Pectolyase Y23 (Kikkoman Shoyu Co., Ltd., Noda, Chibaken, Japan; Kanematsu-Jasho Inc., San Francisco, CA), 2% cellulase Onozuka R10 (Yakult Honsha Co., Nishinomiya, Japan) in 0.7 *M* mannitol, pH 5.5. [Macerozyme R10 (Yakult Honsha Co.) can also be used with cellulase R10 instead of Pectolyase[14a].] The leaves are vacuum infiltrated for 1 min and incubated at 30°

[12] G. Wagner, *Plant Physiol.* **64,** 88 (1979).

[13] G. Wagner, P. Mulready, and J. Cutt, *Plant Physiol.* **68,** 1081 (1981).

[14] T. Nagata and S. Ishii, *Can. J. Bot.* **57,** 1820 (1979).

[14a] G. Kaiser, E. Martinoia, and A. Wiemken, *Z. Pflanzenphysiol.* **107,** 103 (1982).

with occasional gentle swirling for 20 min to 1 hr, when nearly all the cells are converted to protoplasts (determined by examination with a microscope). The digest is filtered through a 50-mesh nylon filter, and the resulting protoplast suspension centrifuged at 100 *g* for 2 min to pellet the protoplasts. The protoplasts are twice washed with 0.7 *M* mannitol, pH 5.5, by centrifuging at 100 *g* and resuspended in 0.7 *M* mannitol, pH 5.5.

Isolation of Vacuoles

Hippeastrum vacuoles are released from washed protoplasts by treatment of washed protoplasts with 0.2 *M* potassium phosphate, pH 8, with the aid of a multibladed stirrer (15 rpm) for 2–4 min.[3,12,13,15,16] Many of the cell cytoplasmic organelles accumulate on the blades, and the remaining suspension is filtered through a plastic screen (1-mm openings) and then through two layers of glass wool. The vacuoles are sedimented by centrifugation (100 *g* for 3 min) and further purified by transferring (with a Pasteur pipette) to the bottom of a 50-ml conical centrifuge tube containing 40 ml of 0.7 *M* mannitol, 1 m*M* HEPES buffer, pH 8, and 0.5 m*M* DTT. The vacuoles float up into the solution and are recovered by centrifugation at 100 *g* for 3 min. A final pellet from *Hippeastrum* petals contained about 10^6 intact vacuoles and represented a yield from protoplasts of about 12%.[16] Isolated vacuoles usually remain intact, and retain pigment, for more than 20 hr when suspended in 0.55 *M* sorbitol, 1 m*M* Tris-MES buffer, pH 8.[3]

Wheat vacuoles are released in 0.17 *M* potassium phosphate.[13,15] Debris and remaining intact protoplasts are removed by centrifugation at 500 *g*. The supernatant containing the vacuole is made 17% (w/v) in sucrose and then overlayed with 18% (w/v) sucrose, 20 m*M* HEPES-NaOH, pH 8.0. During centrifugation at 5000 *g* for 10 min, the vacuoles float to the top and can then be collected from the upper surface.

The purity of the vacuoles varied considerably among preparations and appeared to depend upon both the type of tissue and the species of plant. *Hippeastrum* vacuoles, for example, had little contamination,[3] as did those of castor beans,[7] but vacuoles from *Lycopersicon*[5] contained visible contamination. Contaminating materials on vacuoles from sorghum[6] could be eliminated by including 5–8 mg of serum albumin per milliliter in the lysing buffer. Conditions must be sought not only to minimize contamination from cell debris, but to minimize damage to the tonoplast, prevent leakage of vacuolar components, and efficiently wash away the hydrolytic enzymes employed to isolate the protoplasts. Further puri-

[15] G. J. Wagner, *Plant Physiol.* **67,** 591 (1981).

[16] H. C. Butcher, G. J. Wagner, and H. W. Seigelman, *Plant Physiol.* **59,** 1098 (1977).

fication of vacuoles has been achieved by centrifuging through gradients of Ficoll,[6,8] sucrose,[5,7,15] or silicone oil.[16a]

Preparation of Vacuoles by Mechanical Shearing of Intact Tissues

This method has been applied successfully with beet, radish, and onion roots,[17] but its application with leaf or flower tissues has not been reported.

Fresh beetroot (*Beta vulgaris* L.) tissue (450 g) is cut into slices approximately 5 mm thick and loaded into a specially constructed slicing apparatus.[11] Resulting slices are collected into 1 liter of 1 *M* sorbitol, 5 m*M* EDTA, 0.1 mg of sodium 2-mercaptobenzothiazole per milliliter, 50 m*M* Tris-HCl, pH 7.6, at room temperature. Tissue pieces are separated on Miracloth and resliced with the apparatus into 1 liter of fresh medium. Tissue is again removed, and the two filtrates (1800 ml, total volume) are combined and cooled to 4°. All subsequent operations are carried out at this temperature and are gently performed to minimize vacuolar breakage.

Eight 220-ml aliquots of the filtrate are centrifuged at 2000 *g* for 10 min at 4° (GSA rotor; Sorvall RC2-B centrifuge). The supernatants are discarded, and the pellets are gently resuspended and combined in 16 ml of cold 15% (w/v) Metrizamide (Nyegaard Co., A/S Oslo, Norway), in isolation medium (1.5 *M* sorbitol, 1 m*M* EDTA, 10 m*M* Tris-HCl, pH 7.6). Three 1-ml aliquots of the resuspended 2000 *g* pellet are placed in 15-ml glass centrifuge tubes and overlaid with 5 ml of 10% (w/v) Metrizamide, 2.5 ml of 2.5% Metrizamide, and 2.5 ml of 0% Metrizamide, each in isolation medium. The gradients are centrifuged at 650 *g* for 10 min (HB-4 swinging-bucket rotor, Sorvall RC2-B centrifuge), and fractions are transferred into individual tubes for further analysis. The majority of the vacuoles are recovered from the interface of the 10% Metrizamide and 2.5% Metrizamide. For further concentration, the vacuole-containing bands are diluted with 2 volumes of isolation medium and centrifuged at 650 *g* for 10 min. The pellet containing the vacuoles is gently resuspended in the desired volume of isolation buffer.

The vacuoles are counted with a Levy ultraplane improved Neubauer hemacytometer; 450 g of beet tissue yields about 10^7–10^8 vacuoles, but the efficiency of yield is only about 0.2%.[11,17] The vacuolar half-life is a time-dependent, first-order process that is also pH dependent. The maximum

[16a] M. Nishimura, *Plant Physiol.* **70,** 742 (1982).

[17] R. A. Leigh, D. Branton, and F. Marty, *in* "Plant Organelles" (E. Reid, ed.), p. 69. Wiley, New York, 1979.

half-life of vacuoles prepared in this manner is 24 hr when stored at pH 7.5–8.0 at 20°.[11]

Properties of Intact Vacuoles

Vacuolar Sap

The vacuolar sap contains a variety of components including enzymes, inorganic ions, and organic molecules (some classified as metabolites and others as secondary plant compounds). The array of enzymes and other compounds found in plant vacuoles can differ considerably, depending upon the stage of tissue development or differentiation, the species of plant, or the environment of the whole plant. Enzymes that have been reported in vacuoles of higher plant tissues include acid phosphatase,[5,7,8,16–23b] proteinase (endopeptidase),[7,8,13,19,24–26a] carboxypeptidase,[5,7,26a] nuclease,[7,8,16] α-galactosidase,[8,22] β-glucosidase,[7] α-mannosidase,[8,19,22,23a] β-*N*-acetylglucosaminidase,[8,22] phosphodiesterase,[7,8] Mg^{2+}-dependent pyrophosphatase,[26b] ATPase,[17,21,26c,30,45] acid invertase,[17,27] peroxidase,[18,23b] alliinase,[28] and phytase.[7] The proteinases in the vacuoles have received particular attention because of their possible roles in degradation of cytoplasmic proteins[7,19,24–26] and storage proteins,[45a,45b] and recent evidence strongly supports a role for the degradation of specific chloroplast proteins in the vacuole.[25,26]

[18] K. Grob and P. Matile, *Z. Pflanzenphysiol.* **98,** 235 (1980).
[19] U. Heck, E. Martinoia, and P. Matile, *Planta* **151,** 198 (1981).
[20] F. Sasse, D. Backs-Hüsemann, and W. Barz, *Z. Naturforsch., C: Biosci.* **34C,** 848 (1979).
[21] A. Admon, B. Jacoby, and E. E. Goldschmidt, *Plant Sci. Lett.* **22,** 89 (1981).
[22] A. M. Boudet, H. Canut, and G. Alibert, *Plant Physiol.* **68,** 1354 (1981).
[23] R. A. Leigh and R. R. Walker, *Planta* **150,** 222 (1980).
[23a] R. C. Granstedt and R. C. Huffaker, *Plant Physiol.* **70,** 410 (1982).
[23b] M. Thom, A. Maretzki, and E. Komor, *Plant Physiol.* **69,** 1315 (1982).
[24] M. Nishimura and H. Beevers, *Nature (London)* **277,** 412 (1979).
[25] W. Lin and V. A. Wittenbach, *Plant Physiol.* **67,** 969 (1981).
[26] V. A. Wittenbach, W. Lin, and R. R. Hebert, *Plant Physiol.* **69,** 98 (1982).
[26a] S. P. Waters, E. R. Noble, and M. J. Dalling, *Plant Physiol.* **69,** 575 (1982).
[26b] R. R. Walker and R. A. Leigh, *Planta* **153,** 150 (1981).
[26c] R. R. Walker and R. A. Leigh, *Planta* **153,** 140 (1981).
[27] R. A. Leigh, T. Rees, W. A. Fuller, and J. Banfield, *Biochem. J.* **178,** 539 (1979).
[28] J. E. Lancaster and H. A. Collin, *Plant Sci. Lett.* **22,** 169 (1981).
[29] E. Martinoia, U. Heck, and A. Wiemken, *Nature (London)* **289,** 292 (1981).
[30] W. Lin, G. J. Wagner, H. W. Siegelman, and G. Hind, *Biochim. Biophys. Acta* **465,** 110 (1977).
[31] J. J. M. Hooymans, *Z. Pflanzenphysiol.* **100,** 185 (1980).
[31a] D. Sanders, *Plant Physiol.* **68,** 401 (1981).
[31b] H. Holländer-Czytko and N. Amrhein, *Plant Sci. Lett.* **29,** 89 (1983).

The list of inorganic ions, metabolites, and secondary plant compounds that have been identified within the vacuole is quite large. In the past few years reports have included proteinase inhibitors,[5] nitrate,[29] chloride,[30–31a] oxalic acid,[15] ascorbic acid,[18] shikimic acid,[31b] phenolics,[18] nicotine,[32] gibberellin,[31c,33] capsaicinoids,[34] cardiac glycosides,[35] sugars,[7,10,12,14a,27,31d,36,37] glucosinolates,[38] betacyanin,[11,17,39] anthocyanins,[10,12] amino acids,[12,14a,19,20,22,31b,31e,31f] malic acid,[40,41] cyanoglucosides,[6] and 2-hydroxycinnamic acid.[42] Thus, the central vacuole not only appears to be involved with the temporary storage of primary metabolites of the cell,[43,44] but also provides a compartment for the storage of compounds that may be toxic to the plant. Some of the compartmented chemicals are also toxic to plant pests and may have important roles in natural plant protection.

The Vacuole Membrane: The Tonoplast

The availability of pure vacuoles in large quantities has provided the first opportunity to study in detail the composition of the tonoplast. Isolation of the tonoplast from highly pure vacuoles ensures minimum contamination by the membranes of other organelles and allows the analysis of both lipid and protein components of the vacuolar membrane.

Tonoplast Isolation. Tonoplasts were first purified from isolated *Tulipa* and *Hippeastrum* vacuoles by Lin *et al.*[30] Membranes from *Hippeastrum* vacuoles are obtained[45] by lysing vacuoles in 2% sucrose, 10 m*M*

[31c] J. L. Garcia-Martinez, J. B. Ohlrogge, and L. Rappaport, *Plant Physiol.* **68,** 865 (1981).
[31d] M. Thom, E. Komor, and A. Maretzki, *Plant Physiol.* **69,** 1320 (1982).
[31e] R. A. Leigh, A. Naizir, and R. G. Wyn Jones, *Planta* **153,** 34 (1981).
[31f] G. Alibert, A. Carrasco, and A. M. Boudet, *Biochim. Biophys. Acta.* **721,** 22 (1982).
[32] J. A. Saunders, *Plant Physiol.* **64,** 74 (1979).
[33] J. B. Ohlrogge, J. L. Garcia-Martinez, D. Adams, and L. Rappaport, *Plant Physiol.* **66,** 422 (1980).
[34] H. Fujiwake, T. Suzuki, and K. Iwai, *Plant Cell Physiol.* **21,** 1023 (1980).
[35] W. Löffelhardt, B. Kopp, and W. Kubelka, *Phytochemistry* **18,** 1289 (1979).
[36] S. Doll, F. Rodier, and J. Willenbrink, *Planta* **144,** 407 (1979).
[37] M. Guy, L. Reinhold, and D. Michaeli, *Plant Physiol.* **64,** 61 (1979).
[38] K. Grob and P. Matile, *Plant Sci. Lett.* **14,** 327 (1979).
[39] R. Schmidt and R. J. Poole, *Plant Physiol.* **66,** 25 (1980).
[40] C. Buser and P. Matile, *Z. Pflanzenphysiol.* **82,** 462 (1977).
[41] W. H. Kenyon, R. Kringstad, and C. C. Black, *FEBS Lett.* **94,** 281 (1978).
[42] K. Oba, E. E. Conn, H. Canut, and A. M. Boudet, *Plant Physiol.* **68,** 1359 (1981).
[43] P. Matile, *Naturwissenschaften* **66,** 343 (1979).
[44] P. Matile, *Annu. Rev. Plant Physiol.* **29,** 193 (1978).
[45] G. J. Wagner, *Plant Physiol.* **68,** 499 (1981).
[45a] W. Van der Wilden and M. Chrispeels, *Plant Physiol.* **71,** 82 (1983).
[45b] M. J. Chrispeels, *in* "The Biochemistry of Plants" (N. E. Tolbert, ed.), Vol. 1, p. 390. Academic Press, New York, 1980.

HEPES-NaOH, pH 8, containing 3 mM $MgSO_4$, and 0.5 mM DTT at 4°. The membranes are sedimented by centrifuging for 2 hr at 100,000 g, resuspended in the lysing buffer, applied to a 15 to 50% w/w linear sucrose gradient with a 60% (w/w) sucrose cushion prepared in the above buffer, and centrifuged for 17 hr at 4°. The fraction layering at a density of 1.08 to 1.12 g/ml contains the vacuolar membranes.

Vacuole membranes can also be isolated by lysing beet vacuoles in 10 ml of a 10 mM Tris-HCl, pH 7.6, buffer containing 1 mM EDTA followed by freezing and thawing.[46] The mixture is diluted to 35 ml with the lysing buffer and centrifuged at 39,000 g for 30 min to sediment the membranes. After extensive draining of the supernatant, the membranes can be utilized for analysis. Further washing of the membranes is not recommended, since even water can extract a set of the loosely bound, "peripheral" proteins.

Tonoplast Composition

LIPIDS. Beetroot tonoplasts[46] exhibit a density of about 1.10 g/ml and a phospholipid : protein ratio of 0.7, a high percentage of lipids being polar. Seventeen polar lipids including five classes of glycolipids are present.[46,47] The phospholipids include phosphatidylcholine (54%), phosphatidylethanolamine (24%), phosphatidylglycerol (4%), phosphatidylinositol (4%), phosphatidylserine (1%), and phosphatidic acid (13%). The ratio of phosphatidylcholine to phosphatidylethanolamine of about 2 : 1 in beetroot vacuolar membranes resembles that of the tonoplast of yeast.[48,49] Sphingomyelin, a marker of animal lysosomes, was not found in the tonoplast, and phosphatidylinositol, a major component of yeast tonoplast, is not a major component in the beetroot tonoplast.[46] Phosphatidic acid, which comprises 82% of the membrane phospholipids of rubber tree lutoids[50] (lysosome-like organelles found in latex), is present in the beetroot tonoplast but absent in the yeast tonoplast.[48,49]

PROTEINS. About 15 major protein bands were identified in the beetroot tonoplast,[46,47] and five major and nine minor proteins were found in vacuolar membranes of *Hippeastrum* petals.[45] Most of the beetroot proteins were glycoproteins, presumed to be of the "high mannose" type found in animal lysosomes. However, no carbohydrate is found to be associated with the *Hippeastrum* tonoplast peptides.

[46] F. Marty and D. Branton, *J. Cell Biol.* **87,** 72 (1980).

[47] F. Marty, D. Branton, and R. A. Leigh, *in* "The Biochemistry of Plants" (P. Stumpf and E. Conn, eds.), Vol. 1, p. 625. Academic Press, New York, 1980.

[48] R. Kramer, F. Kopp, W. Niedermeyer, and G. F. Fuhrmann, *Biochim. Biophys. Acta* **507,** 369 (1978).

[49] W. Van der Wilden and P. Matile, *Biochem. Physiol. Pflanz.* **173,** 285 (1978).

[50] J. Dupont, F. Moreau, C. Lance, and J.-L. Jacob, *Phytochemistry* **15,** 1215 (1976).

At present only a few enzymes have been associated with the tonoplast. Mg^{2+}-stimulated ATPase has been detected in tonoplast fractions from *Hippeastrum*,[30,45] beetroot,[17,21] and rubber tree lutoid,[51] though not in a tobacco tonoplast fraction.[52] NADH–cytochrome *c* reductase and NADPH–cytochrome *c* reductase have been identified in isolated beetroot[17] vacuolar membranes, and peroxidase has been shown to be associated with the inner surface of the tonoplast membrane of immature tomato fruit cells in cytochemical studies.[53] None of these enzymes are specific markers of the vacuolar membrane, and to date no enzymes unique to the tonoplast have been found. α-Mannosidase, a component of the yeast tonoplast,[54] was not found in *Hippeastrum* tonoplasts.[45]

Vacuole Biogenesis. The origin of most, if not all, organelle membranes in plants, until recently, had been considered to be from the endoplasmic reticulum.[55] The data to support this concept have been primarily from numerous electron microscopy studies that demonstrated lumenal continuities between the endoplasmic reticulum (ER) and the various other organelles of the plant cell.[2,56] Although mitochondrial and peroxysomal membranes are probably not derived from the ER, evidence continues to implicate the Golgi in vacuolar biogenesis. Direct membrane continuities between ER, Golgi, plasmalemma, and provacuoles have been demonstrated with a 3 MeV electron microscope.[56] A flow process from the ER, through the Golgi and provacuoles to the vacuoles was suggested[56] to provide a pathway for continually supplying the vacuoles with proteins and enzymes. However, in contrast to animal lysosomes,[57–60] biochemical evidence for the mechanism and origins of synthesis and compartmentation of the plant vacuolar enzymes has not been reported. Studies of the synthesis of two vacuolar compartmentated proteinase inhibitor proteins[61] indicate that these proteins are synthesized *in vitro* as preproteins,[62] but their origins of synthesis and mode of compartmentation in the vacuole are unknown. Electron micrographs of the tono-

[51] B. Marin, H. Cretin, and J. D'Auzac, *Physiol. Veg.* **20,** 333 (1982).

[52] D. P. Briskin and R. T. Leonard, *Plant Physiol.* **66,** 684 (1980).

[53] R. L. Thomas and J. J. Jen, *J. Food Biochem.* **4,** 247 (1980).

[54] W. Van der Wilden, P. Matile, M. Schellenberg, J. Meyer, and A. Wiemken, *Z. Naturforsch.*, **28C,** 416 (1973).

[55] D. J. Morré, *Annu. Rev. Plant Physiol.* **26,** 441 (1975).

[56] F. Marty, *J. Histochem. Cytochem.* **28,** 1129 (1980).

[57] A. Hasilik, *TIBS* **5,** 237 (1980).

[58] A. Hasilik and E. F. Neufeld, *J. Biol. Chem.* **255,** 4937 (1980).

[59] A. Kaplan, D. T. Achord, and W. S. Sly, *Proc. Natl. Acad. Sci. U.S.A.* **69,** 2026 (1977).

[60] H. D. Fischer, H. Gonzales-Noriega, W. S. Sly, and D. J. Morré, *J. Biol. Chem.* **255,** 9608 (1980).

[61] C. A. Ryan, *Curr. Top. Cell. Regul.* **17,** 1 (1980).

[62] C. E. Nelson and C. A. Ryan, *Proc. Natl. Acad. Sci. U.S.A.* **77,** 1975 (1980).

plast during inhibitor synthesis do not reveal ribosomes attached to the membrane,[62a] indicating that synthesis may be on the ER, with subsequent transport into the vacuole.

Recent evidence concerning the origin and function of storage protein bodies of seeds indicates that they have the same complements of enzymes usually associated with the vacuolar sap of nonstorage tissues,[8,62b,62c] and should be considered vacuoles.[62d] Therefore, the evidence elucidating protein body biogenesis in seeds may be directly applicable to the general understanding of vacuolar protein synthesis and compartmentation[45b] in other tissues. The proteins that accumulate in protein bodies apparently are synthesized and segregated in the RER, sequestered in the lumen of the ER,[62b,62e–62h] and eventually transported to the protein bodies (vacuoles).[45a] No evidence is yet available concerning the mechanisms of segregation of the vacuole-destined proteins.

The yeast vacuole or lysosomal compartment may provide an insight into the plant system. Like those in higher plants, the vacuole is a large, conspicuous organelle that contains a variety of hydrolytic enzymes.[63–67] *Saccharomyces cerevisiae* synthesizes several enzymes that become compartmented within the vacuole. Among these enzymes is carboxypeptidase Y, a glycoprotein of 61,000 daltons which is synthesized as an enzymically inactive 67,000-dalton precursor.[68] One strain of yeast has been identified in which the precursor accumulates as a result of a mutation that prevents processing.[69] Five other vacuolar enzymes are similarly affected by the mutation. However, two enzymes, acid phosphatase and

[62a] C. A. Ryan and R. Davis, in preparation.

[62b] B. Baumgartner, K. T. Tokuyasu, and M. Chrispeels, *J. Cell Biol.* **79,** 10 (1978).

[62c] W. Van der Wilden, R. M. Herman, and M. J. Chrispeels, *Proc. Natl. Acad. Sci. U.S.A.* **77,** 428 (1980).

[62d] S. Craig, D. J. Goodchild, and C. Miller, *Aust. J. Plant Physiol.* **7,** 329 (1980).

[62e] R. Bollini, W. Van der Wilden, and M. J. Chrispeels, *Plant Physiol.* **55,** 82 (1982).

[62f] D. J. Bowles and H. Kauss, *Biochim. Biophys. Acta* **43,** 360 (1976).

[62g] M. J. Chrispeels, T. J. Higgins, S. Craig, and D. Spencer, *J. Cell Biol.* **93,** 5 (1982).

[62h] W. Van der Wilden, N. R. Gilkes, and M. J. Chrispeels, *Plant Physiol.* **66,** 390 (1980).

[63] E. Cabib, R. Ulane, and B. Bowers, *J. Biol. Chem.* **248,** 1451 (1973).

[64] A. Hasilik, H. Müller, and H. Holzer, *Eur. J. Biochem.* **48,** 111 (1974).

[65] J. Lenney, P. Matile, A. Wiemken, M. Schellenberg, and J. Meyer, *Biochem. Biophys. Res. Commun.* **60,** 1378 (1974).

[66] M. Susani, P. Zimniak, F. Fessl, and H. Ruis, *Hoppe-Seyler's Z. Physiol. Chem.* **357,** 961 (1976).

[67] A. Wiemken, M. Schellenberg, and K. Ureck, *Arch. Microbiol.* **123,** 23 (1979).

[68] A. Hasilik and W. Tanner, *Eur. J. Biochem.* **91,** 567 (1978).

[69] B. A. Hemmings, G. S. Zubenko, A. Hasilik, and E. W. Jones, *Proc. Natl. Acad. Sci. U.S.A.* **78,** 435 (1981).

invertase, of the asparagine *N*-acetylglucosamine type, are not vacuolar, but secreted outside the plasma membrane. These two enzymes were unaffected by the mutation. Apparently the mutation is specifically affecting the proteolytic processing of vacuolar enzymes. The synthesis and glycosylation of carboxypeptidase is thought to occur in the ER,[69] supporting the probable role of the Golgi in the segregation and transport of the enzymes to the vacuole compartment. A carbohydrate-free carboxypeptidase Y recently has been shown to be transferred into the yeast vacuole when glycosylation was blocked by tunicamycin,[70] indicating that the carbohydrate moiety was not an absolute requirement for segregation. Similar experiments have not been performed with plant tissue.

[70] H. Schwaiger, A. Hasilik, K. Von Figura, A. Wiemken, and W. Tanner, *Biochem. Biophys. Res. Comm.* **104,** 950 (1982).

[47] Preparation of a Nuclear Matrix–Pore Complex–Lamina Fraction from Embryos of *Drosophila melanogaster*

By PAUL A. FISHER and GÜNTER BLOBEL

In recent years, the cell nucleus has been among the most widely studied of eukaryotic organelles. In particular, a number of laboratories have focused their attention on the three-dimensional architecture of the nucleus both as pertains to its interactions with other cellular structures at the nucleocytoplasmic interface (the nuclear envelope; for a review, see Franke *et al.*[1]) and to the internal organization of the nuclear contents (the nuclear matrix and associated structures; see Shaper *et al.*[2]). Whereas for the nuclear envelope, there are substantial data at the molecular level regarding both its structure and function, a similar understanding of the nuclear interior has not been forthcoming. Thus, despite the suggestions of some workers in the field,[3] there are as yet no definitive results regarding either the specific molecular composition of the nuclear matrix or the relationship between the nuclear matrix and the nuclear envelope.

[1] W. W. Franke, U. Scheer, G. Krohne, and E.-D. Jarasch, *J. Cell Biol.* **91,** 39S (1981).
[2] J. H. Shaper, D. M. Pardoll, S. H. Kaufmann, E. R. Barrack, B. Vogelstein, and D. S. Coffey, *Adv. Enzyme Regul.* **17,** 213 (1979).
[3] R. Berezney, *J. Cell Biol.* **85,** 641 (1980).

ISBN 0-12-181996-5

One of the major problems encountered in attempting to elucidate the biochemistry of the nuclear matrix has been the difficulty in defining cell fractionation procedures that lead reproducibly to the isolation of an intact or at least recognizable matrix structure. Procedures that are superficially quite similar result in some cases in a subnuclear fraction apparently composed exclusively of peripheral elements (i.e., nuclear envelope),[4–6] and in others, in subnuclear fractions containing substantial internal material (nuclear matrix) as well.[7,8] Even in the hands of a single group of investigators, what seem to be relatively trivial modifications of preparative procedure have been shown nevertheless to produce dramatic morphological alterations in the resultant subnuclear fractions.[9] For reasons largely beyond the scope of this chapter, we sought to establish *Drosophila melanogaster* in our laboratory as a system for the study of nuclear structure in higher eukaryotes. The initial outcome of this endeavor has been the development of a rapid, efficient, and highly reproducible method for the preparation of a *Drosophila* subnuclear fraction that we have demonstrated to be composed of a nuclear matrix surrounded by a peripheral lamina; the latter has also been shown to contain morphologically identifiable nuclear pore complexes.[10] The methodology used to prepare this fraction is the subject of this chapter.

Source Material

The organism *Drosophila melanogaster* has a rather complicated but extremely well characterized life cycle involving, at its outset, a 22-hr period (when reared at 25°) of rapid embryogenesis. The present work was performed exclusively with these embryos and, in general, involved organisms between 4 hr and 20 hr old. Results with embryos in more narrowly defined age ranges have also been obtained and are similar to those results generated with the most heterogeneous populations. The organisms were grown in mass culture, and the embryos were harvested essentially according to Allis *et al.*[11] Embryos were dechorionated in half-strength Clorox according to established procedures[12] and stored at −70° after rapid freezing in liquid N_2. A typical preparation, as outlined below,

[4] R. P. Aaronson and G. Blobel, *Proc. Natl. Acad. Sci. U.S.A.* **72,** 1007 (1975).
[5] N. Dwyer and G. Blobel, *J. Cell Biol.* **70,** 581 (1976).
[6] L. Gerace, A. Blum, and G. Blobel, *J. Cell Biol.* **79,** 546 (1978).
[7] R. Berezney and D. S. Coffey, *Biochem. Biophys. Res. Commun.* **60,** 1410 (1974).
[8] R. Berezney and D. S. Coffey, *J. Cell. Biol.* **73,** 616 (1977).
[9] S. H. Kaufmann, D. S. Coffey, and J. H. Shaper, *Exp. Cell Res.* **132,** 105 (1981).
[10] P. A. Fisher, M. Berrios, and G. Blobel, *J. Cell Biol.* **92,** 674 (1982).
[11] C. D. Allis, G. L. Waring, and A. P. Mahowald, *Dev. Biol.* **56,** 372 (1977).
[12] S. C. R. Elgin and L. E. Hood, *Biochemistry* **12,** 4984 (1973).

was initiated with 50 ml packed volume of frozen embryos. One milliliter of dechorionated embryos contains 4 to 5 × 10^4 organisms.

Buffer Solutions

Buffer A: 50 m*M* Tris-HCl, pH 7.5, 50 m*M* NaCl, 5 m*M* $MgCl_2$, 250 m*M* sucrose, 1 m*M* phenylmethylsulfonyl fluoride (PMSF), 1 m*M* (L-tosylamide-2-phenyl)ethyl chloromethyl ketone (TPCK), 2.5 m*M* *N*-ethyl maleimide (NEM)
Buffer B: 20 m*M* Tris-HCl, pH 7.5, 5 m*M* $MgCl_2$
Buffer C: 10 m*M* Tris-HCl, pH 7.5, 0.1 m*M* $MgCl_2$, 290 m*M* sucrose

Preparation of Purified Nuclei from *Drosophila melanogaster* Embryos

The initial fractionation of *Drosophila* embryos, performed essentially as specified by Elgin and Hood,[12] is described here. All work is performed at 0–4°. This procedure, along with the subfractionation of the purified nuclei, is summarized in Fig. 1.[10] Frozen embryos are thawed directly into 10 embryo volumes of buffer A; the protease inhibitors PMSF, TPCK, and NEM are added to the buffer A solution immediately prior to the start of the purification. The suspended embryos are Dounce homogenized (4 strokes, tight pestle) and filtered through two layers of 120 μm nylon mesh; the filtered crude homogenate (FCH) is centrifuged at 1000 *g* for 10 min. After centrifugation, the postnuclear supernatant (PNS) is removed, and the pellet is resuspended in 5 embryo volumes (i.e., with respect to the original volume of embryos used in the preparation) of buffer A. Centrifugation (1000 *g* for 10 min) is repeated, the supernatant (WS-1) is withdrawn, and the pellet is again suspended in 5 volumes of buffer A. The suspension is centrifuged (1000 *g* for 10 min), the supernatant (WS-2) is withdrawn, and the purified nuclear pellet (N) is suspended in 1 embryo volume of buffer B.

Comments on the Purification of Nuclei

The procedure outlined above for the purification of nuclei from *Drosophila melanogaster* embryos is simple, rapid, and efficient. From 50 ml of frozen embryos, it is possible to purify nuclei in less than 2 hr. By direct count, the yield of nuclei is between 80 and 100%. Consistent with this, greater than 90% of the total embryo DNA is recovered in the final nuclear pellet. In terms of protein, our purified nuclear fraction contains approximately 4 mg per milliliter of embryos, or about 6% of the total

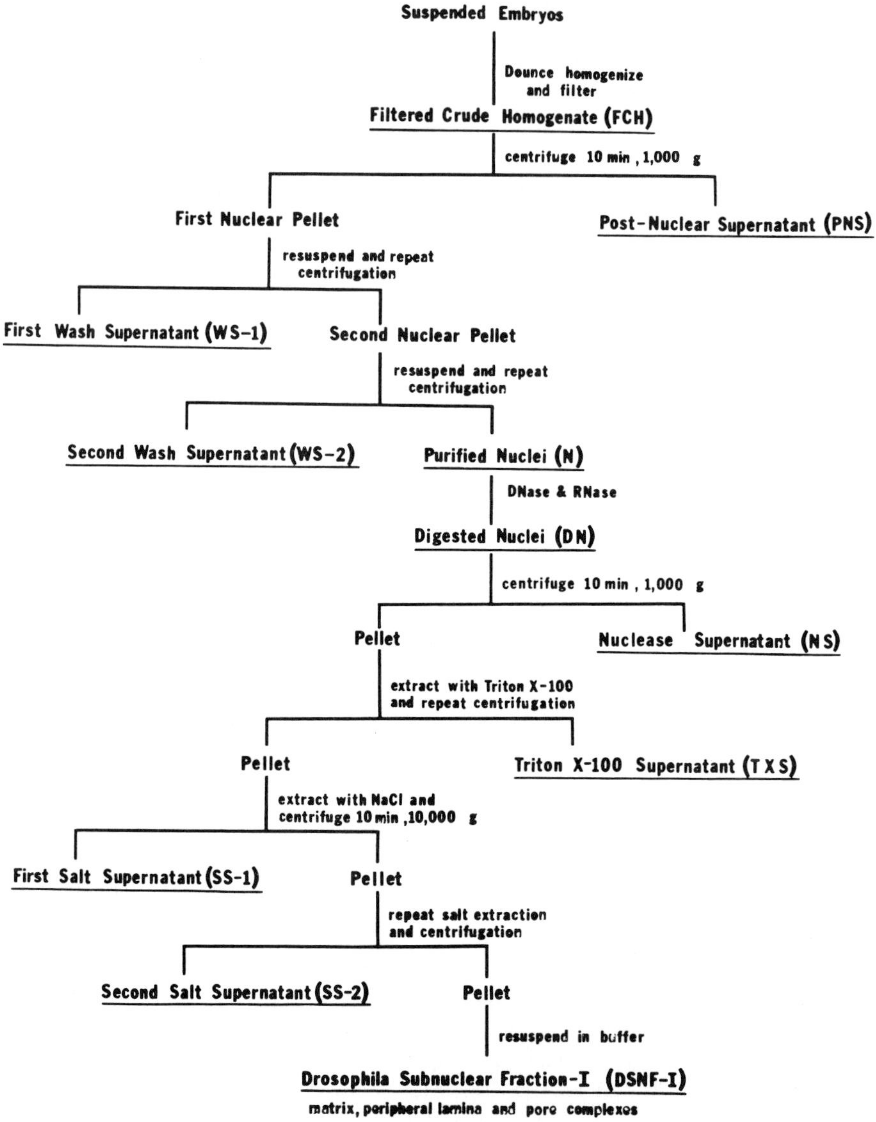

FIG. 1. Flow diagram showing the subcellular fractionation of *Drosophila melanogaster* embryos.[10]

SUBCELLULAR FRACTIONATION OF *Drosophila melanogaster* EMBRYOS

Fraction[a]	Protein (mg)[b]	DNA (μg)[b]	RNA (μg)[b]	Cold-acid soluble nucleotide (μg)[b]
FCH	71.8	625	8440	—
PNS	69.4	93	7168	—
WS-1	0.28	71	274	—
WS-2	0.07	0	163	—
N	4.13	578	536	454
DN	4.06	313	213	1071
NS	0.19	20	19	539
TXS	0.25	19	19	94
SS-1	2.03	222	119	163
SS-2	0.09	18	26	36
DSNF-I[c]	1.64	11	29	80
	(93.2%)	(0.6%)	(1.6%)	(4.5%)

[a] Fraction abbreviations are as defined in the text and Fig. 1.

[b] All values refer to quantities obtained from 1 ml of frozen embryos. A typical preparation was started with 50 ml. (There are approximately 4 to 5 × 10^4 embryos per milliliter.)

[c] Numbers in parentheses represent the relative percentage compositions of each of the DSNF-I components.

embryo protein. (This percentage will vary depending on the developmental stage of the embryo.) A detailed quantitative summary of our fractionation is shown in the table.[10]

In terms of purity, the nuclei obtained from our procedure are free of detectable cytoplasmic contaminants as assessed both by phase contrast light microscopy and transmission electron microscopy. In addition, sodium dodecyl sulfate–polyacrylamide gel electrophoresis (SDS–PAGE) reveals a polypeptide composition of the purified nuclei completely distinct from that of the postnuclear supernatant; this observation has been corroborated and extended by the fact that sedimentation of the purified nuclei through 2.2 *M* sucrose prior to gel analysis does not change the polypeptide pattern observed, as well as by immunochemical comparisons of the nuclear and cytoplasmic fractions.[10] Morphologically, the purified nuclei are quite heterogeneous in size, with diameters ranging from 2 to 5 μm. The vast majority of nuclei appear to be intact and, at the electron microscopic level, the inner nuclear membrane and ribosome-studded outer nuclear membrane are both readily demonstrable.[10]

Subfractionation of Purified *Drosophila* Nuclei

All work is performed at 0–4° unless otherwise indicated. As above, one embryo volume refers to the original amount of embryos from which the nuclear material in question is derived. The procedure and results are summarized in Fig. 1 and the table, respectively. The resuspended purified nuclear pellet (N) is incubated at 37° for 15 min in the presence of 8 μg of RNase A and 10 μg of DNase I per milliliter. (For optimal stability, and thus reproducibility, DNase I is stored at 5 mg/ml, frozen at −20° in 3 m*M* HCl; aliquots are thawed once, and unused portions are discarded. DNase I solutions are never vortexed. RNase A is stored at 20 mg/ml frozen at −20° in 200 m*M* NaCl, 50 m*M* Tris-HCl, pH 7.5. Samples may be frozen and thawed repeatedly.) The nuclease-digested nuclei are then centrifuged at 1000 *g* for 10 min, the nuclease supernatant (NS) is withdrawn, and the remaining pellet is resuspended in 0.9 embryo volume of buffer C. To the resuspended material is added 0.1 embryo volume of 20% (v/v) Triton X-100; the suspension is incubated on ice for 10 min and then centrifuged at 1000 *g* for 10 min. The Triton X-100 supernatant (TXS) is withdrawn, and the pellet is resuspended in 0.45 embryo volume of buffer C plus 0.05 embryo volume of 1 *M* Tris-HCl, pH 7.5. To this resuspended material is added 0.5 embryo volume of 2 *M* NaCl. The suspension is incubated on ice for 10 min and centrifuged at 10,000 *g* for 10 min; the salt supernatant (SS-1) is withdrawn. The pellet is resuspended in buffer C supplemented with 1 *M* Tris-HCl, pH 7.5, as immediately above, brought again to a final NaCl concentration of 1 *M*, incubated on ice for 10 min, and again centrifuged at 10,000 *g* for 10 min. The second salt supernatant (SS-2) is withdrawn, and the final pellet, designated *Drosophila* subnuclear fraction I (DSNF-I) is resuspended in 0.1–1 embryo volume of buffer C. (The amount of buffer C used for the final resuspension depends on the constraints imposed by the subsequent experiments planned for a particular preparation.)

Comments on the Subfractionation of *Drosophila* Embryo Nuclei

As described for the purification of *Drosophila* embryo nuclei, subfractionation of these nuclei is straightforward and may be completed in less than 2 hr. Thus, the total preparative time from 50 ml of frozen embryos to the respective amount of DSNF-I is generally less than 4 hr. As indicated in the table, DSNF-I is greater than 93% by mass protein; the remaining material in the fraction has been shown to consist of small amounts of DNA (0.6%), RNA (1.6%), and cold-acid-soluble nucleotide (4.5%). Quantitatively, DSNF-I represents approximately 2% of the total embryo protein and almost 40% of the purified nuclear protein. As a result, from a typical preparation (50 ml of embryos) it is possible to obtain more than 80 mg of DSNF-I protein.[10]

Detailed SDS–PAGE analysis of DSNF-I[10] reveals a complex polypeptide pattern with major species at molecular weights of 174,000, 74,000, and 42,000; however, the designation of major is somewhat arbitrary in that more than 90% of the total protein in DSNF-I is found in relatively minor species distributed throughout the common molecular weight range (10,000–200,000). Comparison of DSNF-I with the nuclear starting material shows that most of the polypeptides identified in the intact nuclei are retained in DSNF-I; notably absent from DSNF-I, however, are all detectable traces of histone proteins, which are recovered nearly quantitatively in SS-1.

Immunochemical analyses as well as direct examination of polypeptide patterns at each step of the fractionation procedure demonstrate that, in the presence of all three protease inhibitors contained in buffer A, there is little or no proteolysis taking place during our fractionation of *Drosophila* embryos. In contrast, leaving only the NEM out of buffer A results in an approximately 50% loss of DSNF-I protein (appearing most notably in SS-1) and a blurring of the polypeptide band pattern consistent with substantial proteolysis.

Morphological Characteristics of DSNF-I

By phase contrast light microscopy, DSNF-I is difficult to distinguish from the intact nuclear fraction. The sizes and shapes of the parent nuclei appear to be retained in DSNF-I, and nucleoli (or nucleolar remnants) are readily identified. Upon close examination of phase contrast micrographs, it can be seen that the structures observed in DSNF-I are considerably less refractile than the intact nuclei, and this constitutes the only recognizable difference.[10] In contrast to the light microscopic results, transmission electron microscopy reveals substantial differences between the purified nuclear fraction and DSNF-I. Although little or no change is seen in the nuclei after nuclease digestion, Triton X-100 extraction removes all detectable traces of both inner and outer nuclear membranes including outer nuclear membrane-attached ribosomes. Subsequent salt extraction results in a total disruption of chromatin structure, such that the final fraction in the procedure (DSNF-I) appears as a typical nuclear matrix–pore complex–lamina preparation.[10] Pore complexes, a peripheral lamina, nucleolar remnants, and an extensive internal meshwork (matrix) are all readily identified.

Comparison of the Drosophila Nuclear Matrix–Pore Complex–Lamina Fractionation with Vertebrate Nuclear Matrix and Pore Complex–Lamina Preparations

DSNF-I, prepared as described above, appears to be morphologically quite similar to a variety of vertebrate nuclear matrix preparations.[8] However, contrary to what has been reported for the rat liver nuclear matrix,[9] the inclusion of NEM and/or 2-mercaptoethanol in the initial extraction buffer has no adverse effect on the *Drosophila* matrix fraction; as discussed above, NEM or other sulfhydryl blockers appear in fact to be required for optimal DSNF-I preparations. Also, in contrast to results with the rat liver nuclear matrix,[9] addition of high concentrations of RNase A to the purified nuclei prior to DNase digestion is not detrimental to the final recovery of matrix material. At the present time, the bases for this apparent resilience of the *Drosophila* nuclear matrix are not understood. However, possible factors include the gentleness and rapidity with which the purified *Drosophila* nuclei are obtained as well as the small size of the nuclei, which thus minimizes sensitivity to mechanical stresses (e.g., shear forces). Finally, it should be noted that the manner in which we digest *Drosophila* nuclear RNA and DNA is significantly different from the procedure previously used to prepare the rat liver pore complex–lamina fraction.[5,6] In the latter case, two room-temperature nuclease digestions at a final Mg^{2+} concentration of 0.1 mM were used.[13] While chromatin is substantially disrupted under these conditions and DNA would therefore be more accessible to nuclease action, DNase I is less than 1% as active as 0.1 mM as it is at 5 mM Mg^{2+}. Thus, even with chromatin in a stable configuration, the single digestion of *Drosophila* nuclei that we perform at 5 mM Mg^{2+} is extremely effective at fragmenting the nuclear DNA (see the table). It is conceivable that, in fractionating rat liver nuclei, the disruption of chromatin structure induced by the low Mg^{2+} concentration present prior to nuclease digestion may significantly destabilize and/or disrupt any intranuclear matrix structure that is present. Further, the same rationale that predicts that unfolded chromatin is relatively more sensitive to the attack of exogenous nucleases may be invoked to suggest an increased sensitivity of unfolded or exposed matrix structures to the actions of either endogenous or exogenous proteases. In summary, maintenance of chromatin structure through the nuclease digestion step may prove to be crucial to the efficient preparation of a stable nuclear matrix.

[13] R. R. Kay, D. Fraser, and I. R. Johnston, *Eur. J. Biochem.* **30,** 145 (1972).

[48] Proteins of Pore Complex–Lamina Structures from Nuclei and Nuclear Membranes

By GEORG KROHNE and WERNER W. FRANKE

The nuclear envelope (NE) is composed of three prominent architectural components[1,2]: the inner and outer nuclear membrane and the pore walls; the nonmembranous pore complex material; and the nuclear lamina, a thin layer of fibrillar meshwork containing nonmembranous proteins subjacent to the inner nuclear membrane.[3–5] The architectural components of the nuclear pore complexes and the nuclear lamina that interconnects them are the only NE structures resistant to treatments with nucleases, low- and high-salt buffers, and nonionic detergents. Biochemical and immunological analyses have shown that the residual pore complex–lamina structure contains only very few (1–3) major polypeptides, which appear in the molecular weight range of 60,000–80,000 (see Fig. 2).[6–14] The purest pore complex–lamina fractions are obtained when nuclear membranes or whole nuclei are extracted, after removal of chromatin, with buffers containing high salt concentrations and nonionic detergents. This nucleocortical skeleton structure cannot be enriched in comparable purity and structural preservation by other approaches, for example, using buffers containing urea, guanidinium hydrochloride, heparin, or sodium deoxycholate for extraction. Some simple procedures for the isolation of pore complex–lamina structures are described below.

[1] W. W. Franke, *Int. Rev. Cytol., Suppl.* **4,** 71 (1974).
[2] G. G. Maul, *Int. Rev. Cytol., Suppl.* **6,** 75 (1977).
[3] D. W. Fawcett, *Am. J. Anat.* **119,** 129 (1966).
[4] R. R. Aaronson and G. Blobel, *Proc. Natl. Acad. Sci. U.S.A.* **72,** 1007 (1975).
[5] U. Scheer, J. Kartenbeck, M. F. Trendelenburg, J. Stadler, and W. W. Franke, *J. Cell Biol.* **69,** 1 (1976).
[6] L. Gerace, A. Blum, and G. Blobel, *J. Cell Biol.* **79,** 546 (1978).
[7] S. Ely, A. D'Arcy, and E. Jost, *Exp. Cell Res.* **116,** 325 (1978).
[8] G. Krohne, W. W. Franke, S. Ely, A. D'Arcy, E. Jost, *Cytobiologie* **18,** 22 (1978).
[9] G. Krohne, W. W. Franke, and U. Scheer, *Exp. Cell Res.* **116,** 85 (1978).
[10] K. R. Shelton, L. L. Higgins, D. L. Cochran, J. J. Ruffolo, and P. M. Egle, *J. Biol. Chem.* **225,** 10978 (1980).
[11] R. Stick and P. Hausen, *Chromosoma* **80,** 219 (1980).
[12] G. Krohne, M. C. Dabauvalle, and W. W. Franke, *J. Mol. Biol.* **151,** 121 (1981).
[13] R. Stick and G. Krohne, *Exp. Cell Res.* **138,** 319 (1982).
[14] G. G. Maul and N. Avdalovic, *Exp. Cell Res.* **130,** 229 (1980).

ISBN 0-12-181996-5

I. Nuclear Envelopes from *Xenopus laevis* Oocytes

If not indicated otherwise, the biological material should be kept at 4–8°, usually by keeping the solutions on crushed ice.

A. Preparation of Nuclear Envelopes from Manually Isolated Oocyte Nuclei

Solutions

1. Storage medium for oocytes (modified barth medium): 88 m*M* NaCl, 1.0 m*M* KCl, 0.9 m*M* $MgSO_4 \cdot 7H_2O$, 0.33 m*M* $Ca(NO_3)_2 \cdot 4H_2O$, 0.41 m*M* CaCl, 2.4 m*M* $NaHCO_3$, benzylpenicillin and streptomycin sulfate, at 10 mg/1000 ml, 10 m*M* *N*-2-hydroxyethylpiperazine *N'*-2-ethanesulfonic acid (HEPES); adjusted with NaOH to pH 7.6
2. Nuclear isolation medium (NIM): 83 m*M* KCl, 17 m*M* NaCl, 10 m*M* Tris-HCl (pH 7.2)
3. Nuclear envelope isolation medium (NEM): 83 m*M* KCl, 17 m*M* NaCl, 10 m*M* $MgCl_2$, 10 m*M* Tris-HCl (pH 7.2)

In some experiments solutions 2 and 3 contained 2 m*M* dithioerythreitol without considerable changes of the results.

Animals and Removal of Ovary

Full-grown clawed toads (*Xenopus laevis*) of both sexes can be purchased from The South African Snake Farm, Fish Hoek 7975, Cape Province, South Africa. Animals are kept at 20–22° in water tanks (100–400 liters) at a day–night rhythm of 12 : 12 hr. They are fed twice weekly, once with chopped bovine liver and heart and once with fly larvae (*Musca domestica*). Pieces of ovaries are removed as described by Gall.[15] Females are anesthetized for 10–30 min with 0.1% *m*-aminobenzoic acid ethyl ester-methansulfonate (MS222; Sandoz, Basel, Switzerland). A small incision (ca. 1.0 cm) is made in the ventral body wall, either on the left or the right side, and a piece of ovary of desired size is pulled out using watchmaker forceps. The piece of ovary is snipped off and placed in storage medium (solution 1) at 18°. For the preparations described here, the oocytes can be kept for 2–4 days in this medium. The incision is closed with two stitches, using small curved surgical needles and surgical suture (catgut size 2/0). Operated individuals are placed separately and are not fed for 4–7 days. Animals can be operated on 4 or 5 times.

Manual Isolation of Nuclei and Nuclear Envelopes

The isolation is performed under a dissecting microscope. A small

[15] J. G. Gall, *Methods Cell Physiol.* **2,** 37 (1966).

piece of ovary is transferred into a small petri dish (it should be siliconized to prevent sticky adherence of the nuclear envelopes) containing nuclear isolation medium (solution 2). Vitellogenic oocytes of 1.0–1.2 mm diameter (stages V and VI[16]) are suitable for isolation. One oocyte is picked with a pair of watchmaker forceps (Dumont No. 5) at the transition zone between the pigmented and the white half in such a way that a fold of the follicular epithelium, together with the oocyte surface, is formed. This fold is grasped with a second pair of forceps. Then both forceps are pulled in opposite directions and the ooplasm will squeeze out. Somewhere within the masses of yolky ooplasm, the translucent nucleus ("germinal vesicle") is seen. The nucleus is gently pushed away from the bulk of the yolk, using the forceps. It is then freed from the adherent yolk by sucking it up 3–6 times into a pipette with an inner bore diameter of 0.7–0.8 mm.

The purified oocyte nucleus is washed once with fresh NIM and transferred, for preparation of NE,[17] into solution 3. Within 30–60 sec the originally transparent nuclear content gels and forms an opaque and compact aggregate from which the NE detaches and appears as a clearly separated "ghost."[17] By sucking the nucleus into a pipette with an inner bore diameter of 0.3–0.4 mm (nuclear diameter shortly after isolation is 0.4–0.5 mm; nuclear diameter 2 min after incubation in NEM is 0.5–0.7 mm), the NE is locally broken and the gelled ball of nucleoplasm slips out. The isolated NE is cleaned in fresh NIM again by repeated excursions through a small pipette and finally transferred into a Eppendorf reaction tube (1.5 ml size). Usually batches of the nuclear envelopes of 5–10 collected nuclei are cleaned collectively. Portions of 20–50 NEs are pooled and pelleted by centrifugation for 4 min at 8000 *g* in a laboratory centrifuge (Eppendorf Gerätebau, Hamburg, Federal Republic of Germany). The supernatant is discarded, and the pellet is stored at −20°.

Preparation of the Pore Complex–Lamina

The pore complex–lamina is prepared from pellets of fresh isolated NEs by incubation for 20–30 min with 300 μl of 10 m*M* Tris-HCl (pH 7.4) containing 1.0 *M* KCl and 1% Triton X-100. Under these conditions all components of the NE, except the nuclear pore complexes and the nuclear lamina, are solubilized, and the pore complex–lamina structures can be collected by pelleting (4 min, 8000 *g*). Pellets obtained are washed once with 10 m*M* Tris-HCl (pH 7.4).

The morphology and purity of manually isolated NEs has been described by Scheer,[17] and the pore complex–lamina structures of NEs from *Xenopus* oocytes by Krohne *et al.*[9,12] The protein composition of total

[16] J. N. Dumont, *J. Morphol.* **136,** 153 (1972).

[17] U. Scheer, *Z. Zellforsch. Mikrosk. Anat.* **127,** 127 (1972).

NEs and the pore complex-lamina fraction obtained therefrom is compared in slots b and c of Fig. 2 and has been described in detail by Krohne *et al.*[9,12]

B. Large-Scale Isolation of Nuclear Envelopes from Mass-Isolated Oocyte Nuclei

The isolation procedure developed by Scalenghe *et al.*[18] has been used with *some* modifications.

Solutions

1. Medium OR 2 without Ca^{2+} (1000 ml): 87 m*M* NaCl, 2.5 m*M* KCl, 1.0 m*M* $MgCl_2$, 1.0 m*M* $Na_2HPO_4 \cdot 2H_2O$, 5 m*M* HEPES, 100,000 units (60 mg) of penicillin, 0.5 g of polyvinylpyrrolidone; adjusted with NaOH to pH 7.8
2. Collagenase solution (300–350 ml: 0.2% collagenase type I No. C0130, Sigma, St. Louis, Missouri) in medium OR 2 without Ca^{2+}. The collagenase solution can be used 2 or 3 times and is stored at −20°
3. Modified Barth solution (MBS 5000 ml): Solution 1 described in Section I,A but adjusted to pH 7.4 and without streptomycin sulfate.
4. Pronase solution (40–50 ml): Per milliliter of packed oocytes (measured after the collagenase treatment) 1 ml of Pronase solution is needed (528 PUK units per milliliter of MBS; Pronase, nuclease free, No. 537088, Calbiochem, La Jolla, California) and should be prepared before use
5. Bovine serum albumin (BSA) solution: 16 g of BSA in 800 ml of MBS
6. Eagle's lysis medium (200 ml): 180 ml of distilled water, 1.94 g of Eagle's minimum essential medium without $NaHCO_3$ (No. 10-101-26/1F-121A; Flow Laboratories, Rockville, Maryland), 1.0 ml of 1.0 *M* Tris-HCl (pH 7.4), 0.4 ml of Nonidet P-40 (NP-40). To this solution 0.336 g of $NaHCO_3$ is added and brought to 200 ml with distilled water (prepare fresh before use)
7. Nuclear isolation medium (NIM, 1500 ml): 83 m*M* KC1, 17 m*M* NaCl, 2 m*M* $MgCl_2$, 10 m*M* Tris-HCl (pH 7.4), 0.1 *M* sucrose, 2.5 m*M* dithioerythreitol (DTE), 60 mg of phenylmethylsulfonyl fluoride (PMSF; solubilized in 0.5 ml of methanol and added to the 1500 ml immediately before use)
8. NIM without sucrose (500 ml): NIM (solution 7) with 20 mg of PMSF and without sucrose and DTE

[18] F. Scalenghe, M. Buscaglia, M. Steinheil, and M. Crippa, *Chromosoma* **66,** 229 (1978).

Isolation of Oocytes and Oocyte Nuclei

Ovaries from 3–5 *Xenopus laevis* females are removed and collected in MBS (solution 3); 60–70 ml of packed ovary material are needed for an efficient mass isolation. Ovaries are washed 3 or 4 times with MBS (300–400 ml per wash) to remove adherent blood and are cut into small pieces comprising 50–200 vitellogenic oocytes each. Then ovaries are washed two more times with MBS followed by two washes with medium OR 2 (solution 1). The exact volume of the packed ovaries is then determined. Per milliliter of ovary mass, 5 ml of collagenase solution (solution 2) are added and incubated at 28–29° in a slowly shaking water bath until the majority of the full grown oocytes float freely (ca. 2–3 hr). The digestion is then continued for another 15–20 min at 20°. The collagenase solution is decanted and stored at −20°. The freed oocytes are then washed 5–7 times with 200–300 ml of MBS (solution 3). Vitellogenic oocytes are enriched as follows. Oocytes mixed with MBS are allowed to sediment for 5–10 sec, then the supernatant containing most of the previtellogenic oocytes is decanted. Nondigested pieces of ovaries are removed with a wide-bore pipette and are discarded.

The volume of the packed oocytes is determined (ca. 40–50 ml), and equal amounts (20–25 ml) are transferred into 50-ml Falcon plastic tubes (Becton Dickinson, Rutherford, New Jersey). The same volume of Pronase solution (solution 4) is added per tube (20–25 ml). The tubes are put on ice in a horizontal position, oocytes are mixed with the Pronase solution by gently rotating the tube and are incubated for 30 min, with occasional rotating (every 5 min). Then tubes are brought into a vertical position, the Pronase solution is removed by aspiration (water-sucking pipette), and the oocytes are transferred into a 300-ml beaker and put on ice. The oocytes are washed 8 times (200 ml per wash) with the BSA solution (solution 5) to remove and inactivate all Pronase activity. Oocytes thus obtained appear as flattened pancake-shaped cells with high plasticity.

For lysis, oocytes are filled into a special plastic syringe (Fig. 1) in the following way. Piston P is removed and valve B is closed. The oocytes are introduced in chamber A, then 2.5 volumes of lysis medium (solution 6) are added (as reference volume, the volume of packed oocytes determined after the collagenase digestion should be taken). Then chamber A is closed with piston P, the syringe is brought into a vertical position with collecting chamber C to the top. Valve B is opened, and piston P is adjusted so that no air cushion is left in chamber A. Then valve B is closed and the syringe is kept in ice in a horizontal position. Lysis is obtained by gently rotating the syringe. Free oocyte nuclei start to appear in great amounts after 20–30 min of lysis. The protein concentration and viscosity of this suspension are so high that the nuclei float to the surface and are

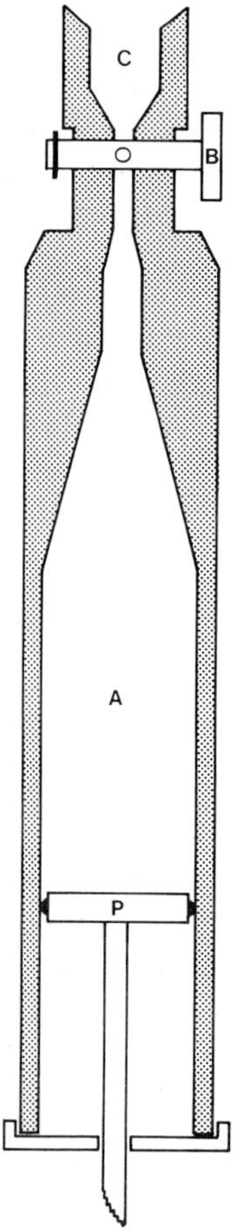

FIG. 1. Schematic drawing of the special syringe designed for the collection of *Xenopus* oocyte nuclei (slightly modified from Scalenghe *et al.*[18] Chamber A has an internal diameter of 40 mm and a length of 24 cm. The hole in valve B has a diameter of 5 mm, and the internal diameter of the collecting chamber is 18 mm. The volume can be adjusted with piston P (40 mm diameter).

recognized as small dark dots. Optimal conditions for the flotation of the nuclei are attained at a 1 : 2.5 ratio of oocyte mass to lysis medium. When several hundred nuclei are floated, the syringe is brought into a vertical position; 1–2 min later, valve B is opened and the lysis medium enriched in nuclei is carefully pushed into the collecting chamber C until it is filled. The syringe is fixed in this position and valve B is closed when the majority of nuclei are floating at the surface of the solution in chamber C.

For further purification of oocyte nuclei a very simple step is employed: 40–60 30-ml Corex tubes are filled with 25 ml of solution 7 (NIM) and kept in ice. Each tube is carefully overlaid with 2.5 ml of solution 8 (NIM without sucrose). The tubes are prepared during the Pronase digestions and lysis of oocytes in the syringe (see above).

Nuclei are collected from the surface of chamber C in aliquots of 200 μl using Eppendorf pipettes with tips from which the front part has been cut off to enlarge the bore diameter (bore diameter: 2–3 mm). To the upper part of each tube 400 μl of nuclear suspension are applied and slightly mixed with the 2.5 ml of solution 8. The oocyte nuclei sediment slowly by gravity while most of the contaminating material floats to the upper part of the tube. When most of the nuclei are sedimented, the supernatant is sucked off, leaving 2–4 ml of overlay medium covering the nuclear sediment.

When the majority of the nuclei present in chamber C have been collected, the residual lysis medium is sucked back into chamber A, valve B is closed, and the syringe is rotated in a horizontal position for 3–5 min. When a large number of nuclei have floated to the surface, the syringe is brought again into a vertical position, chamber C is filled, and the nuclei are collected. This procedure can be repeated 3–5 times.

Manual Isolation of Nuclear Envelopes

The sedimented nuclei of one or two Corex tubes are collected with a Eppendorf pipette tip (inner bore diameter 3 mm) and transferred into a small petri dish containing NE isolation medium (NEM, solution 3 of Section I,A). Batches of 20–50 nuclei are sucked all at once into a fine-bore pipette (inner bore diameter 0.3–0.4 mm). The envelope ghosts freed from the gelled nuclear contents are washed once or twice with fresh NIM (solution 2 of Section I,A) and are collected in 1.5-ml Eppendorf reaction tubes. Within 2 hr 1000–1500 nuclear envelopes can be isolated. NEs obtained with this procedure are slightly contaminated by cytoplasmic components (mainly yolk material). Very pure pore complex–laminae can be prepared when the NE suspension is mixed with the same volume of 10 m*M* Tris-HCl buffer (pH 7.4) containing 2.0 *M* KCl and 2% (w/v) Triton X-100. After a 10-min incubation, the pore complex–lamina structures are pelleted (4 min, 8000 *g*) and washed once with 10 m*M* Tris-HCl of pH 7.2. The protein compositions of pore complex–lamina fractions prepared from mass-isolated oocyte nuclei (see Fig. 2, slot c) and from hand-isolated nuclei[12] are identical.

Preparation of Karyoskeletal Residues Highly Enriched in Pore Complex–Lamina Structures

The sedimented nuclei from 40–50 Corex tubes are collected in the smallest volume possible (15–20 ml). Nuclei are homogenized by vortexing and then pelleted (3000 *g*, 10 min). Sediments containing yolk, pigment granules, nuclear envelopes, and nucleoli are homogenized in 500–1000 μl of extraction buffer (10 m*M* Tris-HCl, pH 7.4, 1.0 *M* KCl, 1% (w/v) Triton X-100). After 10 min of incubation, pigment material, pore complex–lamina structures, and residual nucleoli are pelleted (8000 *g*, 4 min), and this extraction is repeated once more. The final pellets are washed twice with 10 m*M* Tris-HCl buffer of pH 7.2. Since the pigment material is not solubilized in lysis buffers used for electrophoresis all soluble protein represents karyoskeletal material and can be used directly for electrophoresis. The residual material of extracted oocyte nuclei contains two major polypeptides (Fig. 2, slot d) the pore complex–lamina protein[12] (denoted by an arrowhead) and a skeletal protein of molecular weight 145,000 (denoted by a bar) present in oocyte nucleoli.[13,19,20]

II. Nuclear Pore Complex–Lamina Structures from Erythrocytes of *Xenopus laevis* and of Chicken

Solutions

1. Medium A (1500 ml): 0.4 *M* sucrose, 3% (w/v) purified gum arabic, 3–4 m*M* *n*-octanol (2 drops per 100 ml), 10 m*M* Tris-HCl (pH 7.2)
2. Medium A + heparin (200 ml): 200 ml of medium A, 2 ml of heparin 5000 (Braun, Melsungen, Federal Republic of Germany)
3. Medium A + ions (300 ml): to 300 ml of medium A, add 0.1 *M* NaCl and 2 m*M* $MgCl_2$
4. 2.1 *M* sucrose medium (200 ml): 2.1 *M* sucrose, 0.1 *M* NaCl, 2 m*M* $MgCl_2$, 10 m*M* Tris-HCl (pH 7.2)
5. 2.6 *M* sucrose medium (200 ml): 2.6 *M* sucrose, 0.1 *M* NaCl, 2 m*M* $MgCl_2$, 10 m*M* Tris-HCl (pH 7.2)
6. DNase buffer, pH 8.5 (200 ml): 10 m*M* Tris-HCl, pH 8.5, 2 m*M* $MgCl_2$
7. DNase buffer, pH 7.5 (200 ml): 10 m*M* Tris-HCl (pH 8.5), 2 m*M* $MgCl_2$
8. Tris–H_2O (200 ml): distilled water adjusted with Tris base to pH 7.5
9. High-salt–Triton buffer (200 ml): 1.0 *M* KCl, 1% (w/v) Triton X-100, 10 m*M* Tris-HCl (pH 7.4)
10. 10 m*M* Tris (200 ml): 10 m*M* Tris-HCl (pH 7.4)

Inclusion of 2 m*M* DTE in these solutions does not result in gross changes of the fractions obtained.

Purification of Gum Arabic

Commercial gum arabic (Merck, Darmstadt, Federal Republic of Germany) is purified as follows: 400 g of gum are solubilized by heating in

[19] G. Krohne, R. Stick, J. A. Kleinschmidt, R. Moll, W. W. Franke, and P. Hausen, *J. Cell Biol.* **94,** 749 (1982).

[20] W. W. Franke, J. A. Kleinschmidt, H. Spring, G. Krohne, C. Grund, M. F. Trendelenberg, M. Stoehr, and U. Scheer, *J. Cell Biol.* **90,** 289 (1981).

1500 ml of distilled water, filtered through two layers of nylon gauze, and then centrifuged for 2 hr at 32,000 *g* at 40°. The supernatant is slowly mixed with 5 liters of ethanol (technical grade) and the gum arabic precipitates; 15 min later the ethanol is filtered through a sieve, and the precipitated gum is washed once more with 3 liters of ethanol (technical grade), followed by two washes with each 2.5 liters of methanol (technical grade), and a final wash with 3 liters of diethyl ether (analytical grade). The purified gum arabic is then dried for 10 hr in an exsiccator over $CaCl_2$ and then for 4 hr at 60° in an oven. From 400 g of commercial gum, ca. 200 g of purified gum arabic can be obtained.

Isolation and Extraction of Erythrocyte Nuclei

All isolation steps are carried out at 4°, and all buffers contain 0.5 m*M* PMSF. The blood of two decapitated frogs or one decapitated chicken (4–8 weeks old) is collected, under gentle swirling, in 200 ml of medium A + heparin (solution 2) and then centrifuged at low speed in portions of 50–60 ml (350 *g*, 10 min). Each pellet is gently resuspended by shaking in 100 ml of medium A (solution 1). The resuspended cells of all sediments are mixed, and the cell density is controlled by light microscopy using a phase-contrast objective (40×) and an ocular (10×). Cell concentration optimal for the subsequent homogenization is attained when 20–40 erythrocytes are seen in the visual field of the microscope (set up as above). At higher cell densities the plasma membrane of the homogenized cells tend to form aggregates with the nuclei. Then the cells are homogenized in portions of 50 ml with a rotating blade device 4–5 times for 10 sec each at position setting 7 (Fa. Buehler, Tübingen, Federal Republic of Germany). The efficiency of breakage is controlled with the light microscope. Nuclei are pelleted (1500 *g*, 10 min) and are resuspended in ca. 30 ml of solution 3, with the aid of a motor-driven Potter–Elvehjem glass–Teflon homogenizer. The suspension is adjusted to 1.9 *M* sucrose by mixing with 2.6 *M* sucrose (final volume ca. 80–90 ml); 15–16 ml of the 1.9 *M* sucrose are layered on top of a 20-ml cushion of 2.1 *M* sucrose (36-ml centrifuge tubes). After centrifugation (swinging buckets; 38,000 *g*, 75 min) the pellets, highly enriched in nuclei, are washed once with medium A + ions and pelleted again (1500 *g*, 10 min). The nuclear sediment is resuspended by homogenization in 10 ml of DNase buffer (pH 8.5) and incubated for 15 min at 22° with 50 μg of deoxyribonuclease I (DNase I; Worthington Biochemicals, Freehold, New Jersey). The residual nuclear structures are pelleted by centrifugation at 1500 *g* for 10 min, resuspended in 20 ml of Tris-H_2O, and, after 5 min of incubation, pelleted again (1500 *g*, 10 min). The nuclear residues are then resuspended in 10 ml of DNase buffer (pH 7.5), and incubated for 15 min at 22° with 50 μg of DNase I.

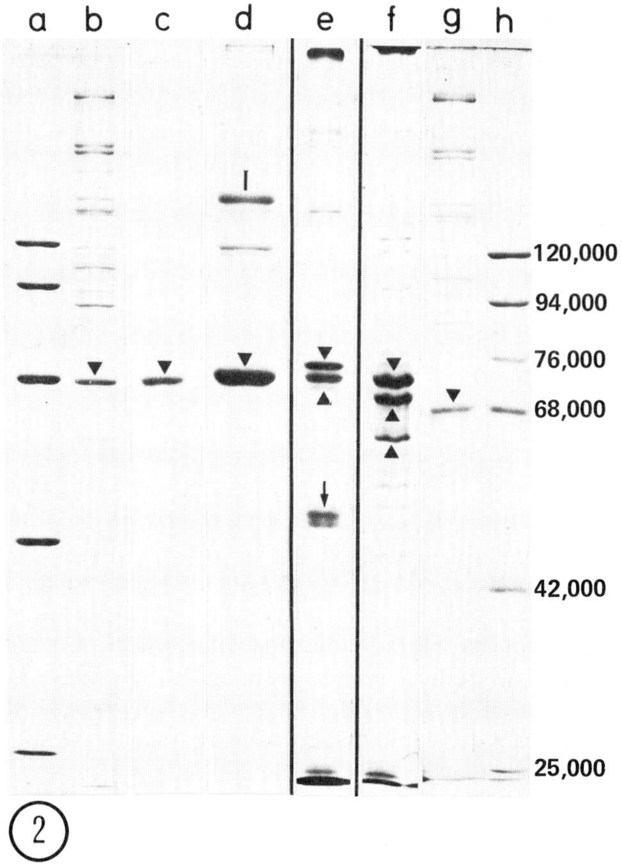

FIG. 2. Sodium dodecyl sulfate–polyacrylamide (10%) gel electrophoresis of isolated nuclear envelopes (NE) and pore complex–lamina fractions after staining with Coomassie Blue. The typical pore complex–lamina proteins are denoted by arrowheads. Slot a: Reference proteins are, from top to bottom, β-galactosidase (120,000), phosphorylase *a* (94,000), bovine serum albumin (68,000), α-actin from rabbit skeletal muscle (42,000), and chymotrypsinogen (25,000). Slot b: 250 nonextracted NEs prepared from manually isolated *Xenopus* oocyte nuclei. Slot c: Pore complex-lamina fraction from ca 350 NEs prepared from mass-isolated *Xenopus* oocyte nuclei. Slot d: Total karyoskeletal residue material of 600–700 mass-isolated *Xenopus* oocyte nuclei. The bar denotes a skeletal protein of M_r 145,000 present in the amplified nucleoli.[19,20] Slot e: Pore complex–lamina fraction from *Xenopus* erythrocyte nuclei. The arrow denotes a degradation product of the pore complex–lamina proteins identified by antibody binding in "immunoblot" experiments (cf. Stick and Krohne[13]). Slots f-h: Comparison of the pore complex-lamina proteins from rat liver nuclei (f) and of *Xenopus* oocyte nuclei (g, 120 nonextracted NEs) with reference proteins (h: same as in a, but including transferrin of M_r 76,000). The apparent M_r values of the pore complex–lamina polypeptides as estimated in this gel system are 68,000 (*Xenopus* oocyte); 72,000 and 68,000 (*Xenopus* erythrocyte); 74,000, 72,000, and 62,000 (rat liver). For corresponding two-dimensional gel electrophoretic comparison, see Shelton *et al.*[10] and Krohne *et al.*[12]

They are pelleted again (1500 *g*, 10 min), rehomogenized in 50–70 ml of high salt–Triton buffer and extracted for 30 min. The insoluble nuclear ghost residues, consisting almost exclusively of pore complex–lamina structures are pelleted (38,000 *g*, 30 min), washed once more with 10 m*M* Tris-HCl, pelleted again (38,000 *g*), and stored at −20°.

The morphology of the extracted erythrocyte nuclei has been described in both species, *Xenopus*[12] and chicken.[10] The protein composition of the pore complex–lamina structure from *Xenopus* erythrocytes is documented in Fig. 2, slot e (for further biochemical and immunological data see Krohne *et al.*[12] and Stick and Krohne[13]).

III. Nuclear Pore Complex–Lamina Structures from Nuclear Membranes of Rat Liver

Solutions

1. Medium A (1000 ml): 0.44 *M* sucrose, 2 m*M* $MgCl_2$, 70 m*M* KCl, 2% (w/v) purified gum arabic
2. Medium A without gum arabic (200 ml)
3. DNase buffer (100 ml): 0.3 *M* sucrose, 0.2 m*M* $MgCl_2$, 10 m*M* Tris-HCl (pH 7.4)
4. 2.1 *M* sucrose medium (300 ml): 2.1 *M* sucrose, 2 m*M* $MgCl_2$, 10 m*M* Tris-HCl (pH 7.4)
5. 2.6 *M* sucrose medium (200 ml): 2.6 *M* sucrose, 2 m*M* $MgCl_2$, 10 m*M* Tris-HCl (pH 7.4)
6. 1.3 *M* sucrose medium (200 ml): 1.3 *M* sucrose, 10 m*M* Tris-HCl (pH 7.4)
7. 1.5 *M* KCl extraction medium (300 ml): 1.5 *M* KCl, 10 m*M* Tris-HCl (pH 7.4)
8. High-salt–Triton buffer (200 ml): 1.5 *M* KCl, 1% (w/v) Triton X-100, 10 m*M* Tris-HCl (pH 7.4)
9. Tris-H_2O (200 ml): see Section II

All solutions contain 2 m*M* DTE, except solution 3.

Procedure

The whole preparation should be carried out at 4°, and the solutions are 0.5 m*M* in respect to PMSF. Nuclei and nuclear membranes are isolated from livers of young rats (Sprague–Dawley) using modifications of the method of Franke *et al.*[21,22] The livers from 7 animals (50–70 g of

[21] W. W. Franke, B. Deumling, B. Ermen, R. D. Jarasch, and H. Kleinig, *J. Cell Biol.* **46,** 379 (1970).

[22] W. W. Franke, T. W. Keenan, J. Stadler, R. Genz, E. D. Jarasch, and J. Kartenbeck, *Cytobiologie* **13,** 28 (1976).

liver) are minced with scissors in 200 ml of medium A and washed once with fresh medium A; the suspension is forced, using a press, through a fine metal sieve (mean pore size: 0.5–1 mm). Most of the connective tissue material is held back in the sieve. To the homogenate, medium A is added up to a total of 300 ml, and the cells are disrupted with the aid of a motor-driven Potter–Elvehjem glass–Teflon homogenizer (highest setting). The homogenate is filtered through 4–6 layers of nylon gauze (No. 7-200; Swiss Silk Bolting Cloth Mfg, Zurich, Switzerland) further to remove residual connective tissue. The filtrate is centrifuged at 850 *g* (12 min), and the pellet containing nuclei, intact cells and parts of the endoplasmic reticulum is resuspended with a loose-fitting Dounce homogenizer in 30 ml of medium A. The homogenate is adjusted with 2.6 *M* sucrose to 2.0 *M* (final volume ca. 80–90 ml); 15–16 ml of the nuclear homogenate in 2.0 *M* sucrose medium is layered on top of a 20-ml cushion of 2.1 *M* sucrose and centrifuged for 1 hr in swinging buckets (110,000 *g*). The nuclear sediment is resuspended with the aid of a loose-fitting Dounce homogenizer in ca. 30 ml of medium A without gum arabic. The nuclei are pelleted (1000 *g*, 10 min) are resuspended with a Dounce homogenizer in 20 ml of Tris-H_2O. The swollen nuclei are sedimented after 5 min (3000 *g*, 10 min), resuspended (Dounce homogenizer) in 20 ml of DNase buffer, and incubated with 100 μg of DNase I (Worthington) at 30°. Then the same amount of DNase I is added, and after a 10-min incubation the reaction is stopped by the addition of 8 volumes (160 ml) of ice-cold 1.5 *M* KCl. The nuclei are extracted for 1 hr and then collected by centrifugation (swinging buckets; 110,000 *g*, 60 min). The pellet is resuspended with a homogenizer of the Potter–Elvehjem type in a total of 5 ml of medium A without gum arabic, adjusted with 2.6 *M* sucrose medium to 2.1 *M* (final volume 22–24 ml), and filled into ultracentrifuge tubes (ca 12 ml per 36-ml tube). The 2.1 *M* sucrose-containing suspension is overlaid by a sucrose gradient ranging from 2.1 *M* to 1.3 *M* sucrose and is centrifuged in swinging buckets at 110,000 *g* for 10–14 hr. After centrifugation a band with a buoyant density in the range of 1.180–1.195 is visible in the centrifuge tubes. This band contains the nuclear membrane material. The membrane material is collected by pipetting, pelleted (100,000 *g*, 1 hr), and extracted with high-salt–Triton buffer for 30 min. The residual pore complex–lamina material is pelleted (100,000 *g*, 30 min), washed once with 10 m*M* Tris-HCl (pH 7.4), and finally collected by centrifugation (100,000 *g*, 30 min). The purity and structural preservation of the nuclear membranes before and after extraction has been described in detail[5,21]; for the protein composition of the pore complex–lamina fraction, see Fig. 2, slot f and Krohne *et al.*[9,12] For a simple and rapid isolation procedure see also Kaufmann *et al.*[23]

[23] S. H. Kaufmann, W. Gibson, and J. H. Shaper, *J. Biol. Chem.* **258,** 2710 (1983).

[49] Subcellular Fractionation and Immunochemical Analysis of Membrane Biosynthesis of Photoreceptor Proteins

By DAVID S. PAPERMASTER

Throughout adult life, vertebrate photoreceptors renew the special light-sensitive membranes of the outer segment, a unique compartment of rod and cones in the retina. In the rod outer segment (ROS) the membranes are assembled as a stack of 1000 disks enveloped by a plasma membrane. This plasma membrane is continuous with the plasma membrane of the rest of the cell. Each day, new disks are formed at the base of the ROS and are progressively displaced by new disk formation on subsequent days.[1] In *Xenopus laevis* toads, 80 disks are formed each day at the base of the ROS.[2] These disks are the size of human erythrocytes (7 μm) yet the membranes are composed almost entirely of the visual pigment rhodopsin and a large intrinsic membrane protein, M_r 290,000 in frogs and cattle.[3,4]

Despite this prodigious feat of membrane assembly and morphogenesis in the outer segment, it contains none of the cell's apparatus for membrane glycoprotein or lipid biosynthesis. The outer segment compartment is separated from the rest of the cell by a connecting cilium. Beneath the cilium most of the cell's mitochondria are clustered in the cytoplasm of the inner segment. Between this region and the nucleus, the cell's rough endoplasmic reticulum and Golgi apparatus are found. Like other compartmentalized and polarized cells, the photoreceptors pose the interesting problem of the mechanism for sorting out and selectively delivering the special molecules of visual sensitivity to the outer segment while maintaining the metabolism and membrane structure of the remainder of the cell, including a synaptic terminal at the opposite end. Cell elongation is prevented by a circadian shedding of the tips of the outer segments.[5,6] This extraordinary commitment to synthesis of a relatively simple membrane in a unique compartment is nearly without equal in biology.

Rhodopsin biosynthesis and disk formation can be divided into four stages: (*a*) synthesis in RER and Golgi; (*b*) vectorial transport of opsin on

[1] R. W. Young, *J. Cell Biol.* **33,** 61 (1967).
[2] J. C. Besharse, J. G. Hollyfield, and M. E. Rayborn, *J. Cell Biol.* **75,** 507 (1977).
[3] D. S. Papermaster and W. J. Dreyer, *Biochemistry* **13,** 2438 (1974).
[4] D. S. Papermaster, C. A. Converse, and M. Zorn, *Exp. Eye Res.* **23,** 105 (1976).
[5] M. M. LaVail, *Science* **194,** 1071 (1976).
[6] S. Basinger, R. Hoffman, and M. Matthes, *Science* **194,** 1074 (1976).

ISBN 0-12-181996-5

vesicles (?cisternae) from the Golgi to the base of the connecting cilium and insertion into the apical plasma membrane of the inner segment; (*c*) transport out the plasma membrane of the connecting cilium; and (*d*) disk morphogenesis at the base of the ROS.[7] At no time during biosynthesis or disk assembly has a soluble precursor of the intrinsic membrane proteins of the ROS, rhodopsin, and the large protein been observed.

In order to evaluate the structure of the cell components involved in transporting opsin and the large intrinsic membrane protein from the inner to the outer segment, techniques of subcellular fractionation and high resolution immunocytochemistry at the EM level have been employed. The procedure for retinal subcellular fractionation is outlined here and full details and the rationale for the experimental conditions and results have been reported.[4,8] The related immunocytochemical and immunochemical procedures are also included in this volume.[9,10]

Materials and Methods

Stock Solutions. All sucrose solutions are weight/weight so that they may be conveniently prepared on a top-loading balance.

Sucrose, 42% freshly prepared
Tris-acetate, 1.0 *M*, pH 7.4
$MgCl_2$, 0.1 *M*
Ethylenediaminetetraacetic acid disodium salt (Na_2EDTA), 0.2 *M*, pH 7.4
Radiolabeled precursor amino acid or sugar

Equipment. All equipment required for isolation of ROS[11] is also used in these experiments. A 10-μl Hamilton syringe is carefully sharpened on a carborundum stone to elongate the point to reduce deformation of the eye during injection of labeled precursors.

Density Gradient Solutions. See Papermaster[11] for preparation of sucrose solutions with a density of 1.10, 1.11, 1.13, 1.15, 1.19 g/ml containing 10 m*M* Tris-acetate and 0.1 m*M* $MgCl_2$.

Homogenizing media

Medium 1. 34% sucrose containing 65 m*M* NaCl, 5 m*M* Tris-acetate, pH 7.4, and 0.2 m*M* $MgCl_2$

[7] For a review of the evidence supporting this outline, see D. S. Papermaster and B. G. Schneider, *in* "Cell Biology of the Eye" (D. McDevitt, ed.), p. 475. Academic Press, New York, 1982.
[8] D. S. Papermaster, C. A. Converse, and J. Siu, *Biochemistry* **14,** 1343 (1975).
[9] C. A. Converse and D. S. Papermaster, this volume [19].
[10] B. G. Schneider and D. S. Papermaster, this volume [38].
[11] D. S. Papermaster, this series, Vol. 81, p. 48.

Medium 2. 0.25 *M* sucrose containing 50 m*M* NaCl, 10 m*M* Tris-acetate, pH 7.4, and 1 m*M* $MgCl_2$

Procedure

Subcellular Fractionation. Frogs (*Rana* sp., >9 cm body length) or toads (juvenile *X. laevis,* 5–10 g, or adult *Bufo marinus,* >300 g) are housed at 20° under incandescent dim light on a 12-hr light : 12-hr dark light cycle. Prior to the isolation of retinal membranes, the radiolabeled precursor is prepared in a suitable carrier and injected intravitreally. Amino acids ([^{35}S]methionine, [^{3}H]- or [^{14}C]leucine, or phenylalanine) or sugars ([^{14}C]- or [^{3}H]mannose) are dried under an N_2 stream and redissolved in 50 m*M* NaCl at 10-fold greater concentration. For long-term incorporation (>1 day), no dark adaptation is needed at the time of injection, but for short incorporations frogs or toads are dark-adapted overnight and all operations are completed under dim red light (Kodak 1A filter, 7.5 W bulb) with the aid of an infrared image converter (Varo, Inc., Garland, Texas). Frogs and toads are anesthetized by immersion for 15–25 min in 0.1% MS222 (Calbiochem, Eastman). When paralyzed, 2.5 μCi of ^{14}C- or 10–50 μCi of ^{35}S- or ^{3}H-labeled amino acid or mannose solution (5–10 μl total volume) is injected in each eye by carefully inserting the needle of a 10-μl Hamilton syringe behind the lens into the vitreous space. The injection is aided by lifting the upper eyelid and pointing the needle toward the back of the eye while entering the globe just behind the corneal–scleral junction to avoid any damage to the anterior chamber and collapse of the globe.

After dark adaptation and incorporation for varying time periods (30, 60, 90 min; 2 or 3 hr; 3 or 9 days), the animals are sacrificed by decapitation (small-animal guillotine, Harvard Apparatus) and the anterior third of the eye is sliced away with a sharp razor blade just behind the ora serrata to eliminate any anterior attachments of the retina to the sclera. Dissection of the anterior segment is completed with a fine scissors. The retina and black pigment epithelium are teased away by rimming with a pointed forceps (Dumont No. 5) and removed from the eye cup by tearing the optic nerve. The tissue is suspended in cold 34% sucrose homogenizing medium in a petri dish. The pigment epithelium is teased away from the nearly transparent retina and discarded. The retina is transferred to fresh homogenizing medium on ice (0.3 ml per retina). The homogenizing and subcellular fractionation are diagrammed in Fig. 1.

Retinas are homogenized by five passes through a No. 15 trochar by gentle aspiration in a 10-ml syringe to shear off ROS. The homogenate is overlaid by 1 ml of 1.10 g/ml sucrose solution containing 10 m*M* Tris-

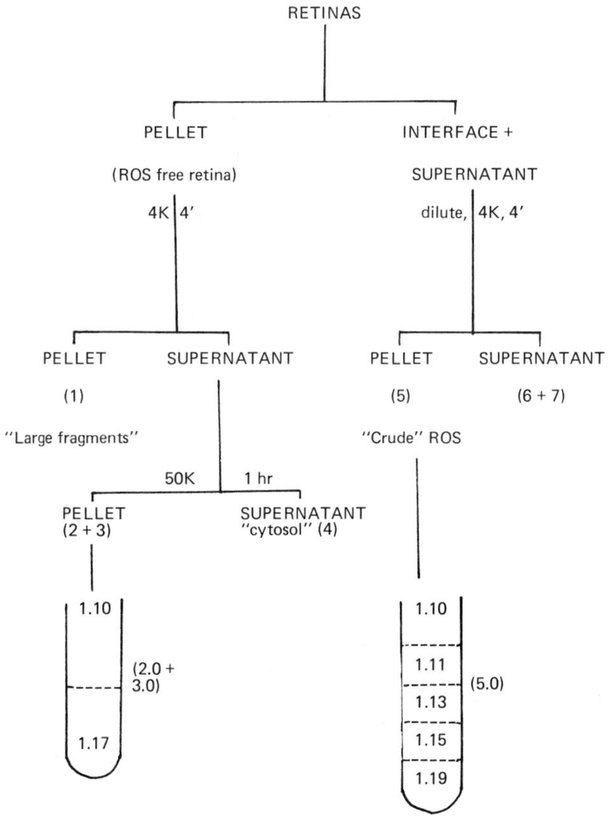

FIG. 1. A simplified protocol of retinal subcellular fractionation modified from Papermaster *et al.*[4,8] After separation of the rod outer segment (ROS) from the rest of the retina in the first step, rehomogenized retinas are further subfractionated by differential and density gradient centrifugation. In this protocol, the fractions 2.0 and 3 of the original procedure are combined to reduce losses for subsequent study.

acetate and 0.1 m*M* $MgCl_2$ and centrifuged at 20,000 rpm (Beckman JA 20 or Sorvall SS34 rotors) for 20 min to float the ROS to the interface. This crude ROS fraction can be further homogenized and the ROS isolated in high purity as described previously.[11] The retinal pellet is rehomogenized in 0.25 *M* sucrose homogenizing medium (0.2 ml per retina) with five passes of a motor-driven loose Teflon–glass homogenizer (Thomas Tissue Grinder, size AA). At the end of the homogenization, sufficient 0.2 *M* EDTA is added to bring the final concentration to 2 m*M* and the homogenate is centrifuged at 4000 rpm to yield a pellet (fraction 1, nuclei and large fragments). This pellet contains about 85% of the retinal tissue, but less than 10% of the newly synthesized opsin. The supernatant can be treated in several ways depending on the next goal of the experiment. Further subfractionation can be obtained by differential centrifugation steps at 15,000 rpm for 15 min (pellet 2, intermediate fragments) and at 50,000 rpm (pellet 3, microsomes, supernatant fraction 4, cytosol). The pellets can also be further subfractionated on sucrose gradients or other suitable media.

Because of the small amounts of material and lack of evidence that these techniques can adequately separate apart from other membranes the small amounts of the membranes that carry ROS proteins, we have simplified the procedure by directly taking the supernatant of the first centrifugation (4000 rpm) and overlaying it on a solution of a 1.17 g/ml sucrose solution containing 10 m*M* Tris-acetate and 0.1 m*M* $MgCl_2$. The sample is centrifuged (13.5 ml, SW40 rotor or 5 ml, SW50.1 rotor) at 100,000 g_{av} for 1 hr. Dense contaminating particles (melanosomes, nuclear fragments) sediment through the barrier, while all membranes of interest (as well as mitochondria and synaptosomes) collect at the interface and are concentrated conveniently for further study. This fraction corresponds to a combined fraction 2.0 and 3 of previous descriptions of the fractionation.[4,8] If not used immediately, samples may be frozen at $-196°$ in plastic vials (Nunc, Vangard Inc.).

Electrophoretic Analysis. The membrane fractions at each time period of incorporation can be compared on a single slab gel of polyacrylamide. The procedure of Fairbanks *et al.*[12] at 8 V/cm (10 cm dye migration) has satisfactorily resolved the many proteins of each subcellular fraction. Sample volumes up to 200 μl in each slot may be run if the slot is large enough to keep the height of the sample below 4 mm (1 cm × 3 mm thick). The wider slot and thicker gel reduces band curvature caused by diffusion into the edges of the slot. Usually the fractions are dissolved in 2.5% SDS–2.5% mercaptoethanol–4% sucrose at room temperature to avoid heat-induced aggregation of opsin. Samples of 75–200 μg are run in each slot. High-temperature denaturation may be necessary to evaluate fraction 1. While still on the glass plate, the gels are marked at 0, 2, 4, and 8 cm and at the Pyronin Y dye front with a syringe dipped in India ink containing [^{14}C]leucine to indicate migration distance. Other buffer-gel combinations are useful depending on the molecular weight range of interest. For example, to study the biosynthesis of the large intrinsic ROS membrane protein, we lowered the acrylamide concentration from 5.6% to 4% to permit greater entry of the 290,000 M_r protein 2 cm into the gel.[8] The radiolabeled proteins are detected by drying the gel[13] and autoradiog-

[12] G. Fairbanks, T. Steck, and D. F. H. Wallach, *Biochemistry* **10,** 2606 (1971).

[13] G. Fairbanks, C. Levinthal, and R. H. Reeder, *Biophys. Biochem. Res. Commun.* **20,** 393 (1965).

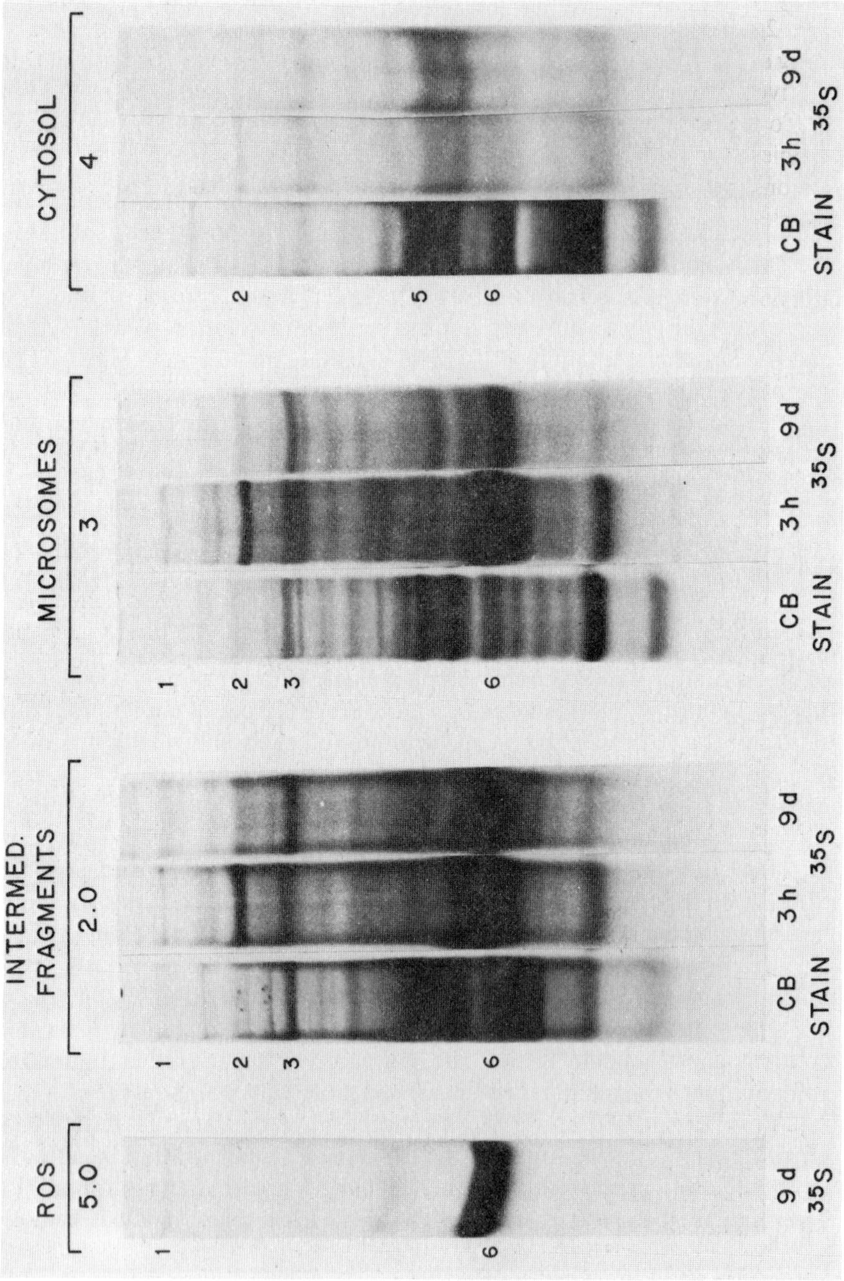
ROS
5.0
INTERMED. FRAGMENTS
2.0
MICROSOMES
3
CYTOSOL
4
9d 35S
CB STAIN
3h 35S
9d 35S
CB STAIN
3h 35S
9d 35S
CB STAIN
3h 35S
9d 35S
1
2
3
5
6

raphy on Kodak RP/R2 or Kodak AR X-Omat X-ray film for 2–13 weeks (Fig. 2). Exposure times can be shortened by using enhancers and sensitizing screens and exposing films at −70°. Semiquantitative analysis of the relative radioactivity in each band can be evaluated by densitometry (Fig. 3). To establish that the film response range has not been overexposed, samples of increasing concentration of membranes should be compared to demonstrate approximate linearity of density and amount of membranes sampled.

Immunochemical Analysis of Membrane and Cytosol Fractions

In order to evaluate any proteins sharing antigenic determinants with mature opsin or the large intrinsic protein, rabbit antibodies were prepared to both proteins and shown to be monospecific.[4,8,14] The retina fractions can be evaluated by two-dimensional immunoelectrophoresis[9] or by the more recently developed procedures of electrophoretic transfer to nitrocellulose.[15] Using the two-dimensional electrophoretic technique, we observed the presence of immunoprecipitable, newly synthesized opsin and large ROS protein only in membrane fractions, predominantly in the postnuclear fractions 2.0 and 3, which are combined in the method described here. Only small amounts were recovered in the large fragment fraction 1, and no significant opsin was detected in cytosol fractions 4 and 7. Nonspecific immunoprecipitation of indifferent proteins (e.g., asparaginase, M_r 34,000 with anti-asparaginase) precipitated only 6% of the opsin precipitated by specific anti-opsin.

This approach to retinal subcellular fractionation must still be considered an early stage in the development of the study of retinal membrane biosynthesis. Until procedures for subfractionation of smooth membrane vesicles with similar biophysical properties become available, it at least

[14] D. S. Papermaster, B. G. Schneider, M. A. Zorn, and J. P. Kraehenbuhl, *J. Cell Biol.* **77**, 196 (1978).

[15] J. Reiser and G. Stark, this volume [14].

FIG. 2. Coomassie Blue-stained electrophoretic patterns of proteins in membranous and cytosol subcellular fractions 2.0, 3, and 4 of retina compared to autoradiographs of fractions obtained at various intervals after intravitreal injection of radiolabeled [^{35}S]methionine. Opsin migrates at 6 cm in outer segments but as a slightly wider band in the fractions from inner segments. This may reflect the relative degree of processing of the oligosaccharide side chains on newly synthesized opsin. The large protein at 1 cm is prominently labeled by 3 hr in fractions 2.0 and 3 and is also transferred to the rod outer segment (ROS). Neither of these ROS proteins is found among the radiolabeled proteins of the cytosol fraction 4. From Papermaster *et al.*,[4] with permission.

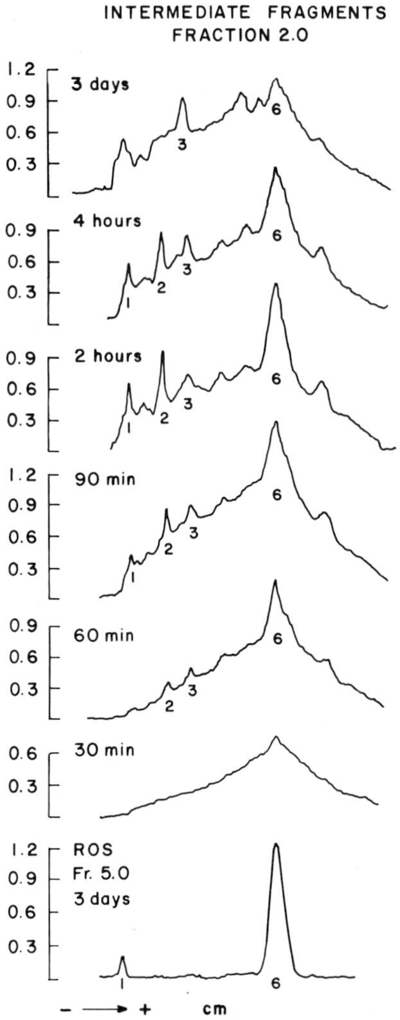

FIG. 3. Densitometric scans of autoradiographs of electrophoretically separated ^{14}C-labeled retinal proteins in fraction 2.0 (prepared according to a slightly more complex protocol[8]). Opsin (6 cm) and the large protein (1 cm) become rapidly labeled and are maximally labeled by 90 min to 2 hr. Despite the incorporation of label into other proteins in this fraction, only the two intrinsic ROS proteins, opsin (6 cm) and the large protein (1 cm) are finally transported in significant amounts to the rod outer segment (ROS). The identity of both the proteins was also established immunochemically by two-dimensional immunoelectrophoresis.[9] Modified from Papermaster *et al.*,[8] with permission.

ensures a rapid recovery of the inner segment membranes in a reasonable yield and at a relative 10 to 50-fold enrichment as a starting material for further study. Its simplicity and rapidity reduce degradative artifacts.

Acknowledgments

Research was supported in part by the NIH Grants EY-00845, GM-21714, and EY-03239, an American Cancer Society grant BC 129, and the Veterans Administration.

[50] Transducin and the Cyclic GMP Phosphodiesterase of Retinal Rod Outer Segments

By LUBERT STRYER, JAMES B. HURLEY, and BERNARD K.-K. FUNG

Flow of Information in the Cyclic Nucleotide Cascade of Vision

Cyclic GMP appears to be important in visual excitation.[1,2] Electrophysiological studies have shown that cyclic GMP depolarizes the plasma membrane of the outer segment of retinal rod cells within milliseconds after being injected intracellularly.[3] The injected cyclic GMP also increases the latency of the hyperpolarization elicited by a subsequent light flash. The concentration of cyclic GMP in retinal rod outer segments (ROS) is unusually high, about 70 μM in the dark. Moreover, the level of cyclic GMP in freshly detached ROS decreases markedly within 100 msec after illumination.[4] These experiments suggest that cyclic GMP keeps sodium channels in the plasma membrane open in the dark and that light closes these channels by lowering the concentration of cyclic GMP. Further support for this hypothesis comes from the finding that ROS contain a light-activated phosphodiesterase specific for cyclic GMP.[5] In disk membrane suspensions, a single photolyzed rhodopsin can lead to hydrolysis of 4×10^5 cyclic GMP per second.[6] This high gain is achieved by the

[1] See W. H. Miller, ed., "Current Topics in Membranes and Transport," Vol. 15. Academic Press, New York, 1981, for an excellent discussion of the cyclic nucleotide cascade and other facets of visual excitation.

[2] See L. Packer, ed., this series, Vol. 81, for numerous valuable articles on experimental methods for the study of retinal rod outer segments.

[3] W. H. Miller and G. D. Nicol, *Nature (London)* **280,** 64 (1979).

[4] M. L. Woodruff and D. Bownds, *J. Gen. Physiol.* **73,** 629 (1979).

[5] J. S. Pober and M. W. Bitensky, *Adv. Cyclic Nucleotide Res.* **11,** 265 (1979).

[6] R. Yee and P. A. Liebman, *J. Biol. Chem.* **253,** 8902 (1978).

ISBN 0-12-181996-5

activation of hundreds of phosphodiesterase molecules following the absorption of a single photon.

How does a single photolyzed rhodopsin molecule activate a large number of phosphodiesterase molecules? Photolyzed rhodopsin does not interact directly with the phosphodiesterase. Rather, a third protein, called transducin, is the amplified signal carrier in this activation process.[7] The flow of information in the cyclic nucleotide cascade is

$$\underset{1}{R^*} \rightarrow \underset{\sim 500}{\text{T-GTP}} \rightarrow \underset{\sim 500}{\text{PDE}^*}$$

where R* denotes photolyzed rhodopsin, T-GTP denotes the GTP complex of transducin, and PDE* denotes the active form of the phosphodiesterase. The role of R* is to catalyze the exchange of GTP for GDP bound to transducin. About 500 T-GTP are formed per absorbed photon.[8] T-GTP then activates the phosphodiesterase by relieving an inhibitory constraint imposed by one of the subunits of the phosphodiesterase. This enzyme returns to the inactive dark state when the bound GTP is hydrolyzed to GDP by the endogenous GTPase activity of transducin.[9-11] Reconstitution experiments have shown that (*a*) T-GTP can be formed in the absence of phosphodiesterase; and (*b*) the phosphodiesterase can be activated by T-GTP in the absence of photolyzed rhodopsin.[7] The separability of these steps *in vitro* shows that information flows from photolyzed rhodopsin to T-GTP and then to the phosphodiesterase.

Light-Activated Amplification Cycle

The major steps in our proposed reaction scheme[7] (Fig. 1) follow.

1. In the dark, nearly all the transducin is in the T-GDP form, which does not activate the phosphodiesterase.
2. Photolyzed rhodopsin encounters T-GDP by diffusion in the plane of the disk membrane and forms an R*-T-GDP complex. R* is known to have high affinity for T-GDP.
3. GTP rapidly exchanges for GDP in the R*-T-GDP complex.
4. T-GTP rapidly dissociates from this complex and diffuses to the phosphodiesterase. R* released in this step binds another T-GDP and thus acts catalytically. The rapid and highly amplified formation of T-GTP requires that R* have high affinity for T-GDP and

[7] B. K.-K. Fung, J. B. Hurley, and L. Stryer, *Proc. Natl. Acad. Sci. U.S.A.* **78,** 152 (1981).
[8] B. K.-K. Fung and L. Stryer, *Proc. Natl. Acad. Sci. U.S.A.* **77,** 2500 (1980).
[9] W. E. Robinson and W. A. Hagins, *Biophys. J.* **17,** 196a (1977).
[10] G. Wheeler and M. W. Bitensky, *Proc. Natl. Acad. Sci. U.S.A.* **74,** 4238 (1988).
[11] W. Godchaux, III and W. F. Zimmerman, *J. Biol. Chem.* **254,** 7874 (1979).

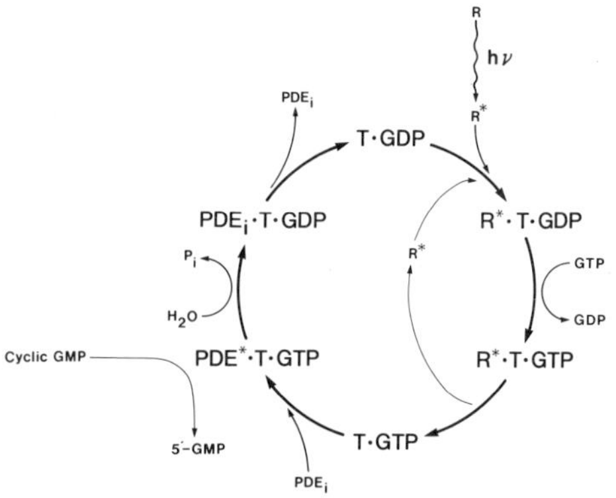

FIG. 1. Proposed light-activated amplification cycle. Information flows from photolyzed rhodopsin (R*) to transducin (T-GTP) and then to the phosphodiesterase (PDE*).

low affinity for T-GTP, as demonstrated by the finding that transducin is released from bleached disk membranes at low ionic strength upon addition of GTP.[12]

5. T-GTP (specifically its α-subunit, as discussed below) activates the phosphodiesterase by relieving an inhibitory constraint. Many molecules of cyclic GMP are hydrolyzed by the activated phosphodiesterase (PDE*). Optical studies have shown that these species of the transducin cycle are formed in times of a hundred milliseconds or less, which means that they are kinetically competent to participate in visual excitation.[13,14]
6. The GTPase activity of transducin slowly converts T-GTP to T-GDP, which inactivates the phosphodiesterase. The dark state is also restored by an ATP-dependent reaction[15] that deactivates R*. The molar ratio of phosphodiesterase to transducin to rhodopsin in ROS is estimated to be of the order 1 : 10 : 70.

Properties of Transducin

Transducin, a peripheral membrane protein, is a complex of α (39 kilodaltons), β (36 kilodaltons), and γ (~10 kilodaltons) subunits.[7,12,16] The

[12] H. Kuhn, *Nature (London)* **283,** 587 (1980).
[13] H. Kuhn, N. Bennett, M. Michel-Villaz, and M. Chabre, *Proc. Natl. Acad. Sci. U.S.A.* **78,** 6873 (1981).
[14] N. Bennett, *Eur. J. Biochem.* **123,** 133 (1982).
[15] P. A. Liebman and E. N. Pugh, Jr., *Nature (London)* **287,** 734 (1980).
[16] W. Baehr, E. A. Morita, R. J. Swanson, and M. L. Applebury, *J. Biol. Chem.* **257,** 6452 (1982).

binding site for GTP or analogs of GTP is on the α-subunit. The dissociation constant of the complex of transducin and guanosine 5′-[β,γ-imido] triphosphate (p[NH]ppG), a nonhydrolyzable analog of GTP, is 2×10^{-7} *M*. The binding of GTP or p[NH]ppG weakens the association of transducin with the disk membrane and promotes the dissociation of the α-subunit from the β- and γ-subunits, which stay together. The complex of T_α with p[NH]ppG fully activates phosphodiesterase bound to unilluminated disk membranes, showing that T_α-GTP is the information carrier between photolyzed rhodopsin and the phosphodiesterase. The β- and γ-subunits of transducin are not required for the activation of the phosphodiesterase. Rather, they participate in the R*-catalyzed exchange of GTP for GDP bound to the α subunit. It seems likely that T_α-GTP encounters the phosphodiesterase by diffusing in the aqueous space between disks. In the presence of physiological concentrations of GTP, T-GTP is formed from T-GDP within 100 msec after the production of R* by a light flash. In contrast, the GTPase reaction, which brings transducin back to the dark state, is slow. Transducin hydrolyzes a molecule of bound GTP in times of seconds to a minute.

Homology between Transducin and the G Protein

The activation of adenylate cyclase by hormones such as epinephrine resembles the activation of the phosphodiesterase by light.[7,17] The G protein in the adenylate cyclase system[18,19] has a role like that of transducin in ROS. The hormone–receptor complex triggers GTP–GDP exchange in the G protein.[20] The GTP complex of the G protein then activates adenylate cyclase until the bound nucleotide is hydrolyzed to GDP.[21] The G protein, like transducin, serves as an amplified signal carrier. Recent studies suggest that transducin and the G-protein are members of the same family of signal-amplifying proteins. Cholera toxin catalyzes the ADP-ribosylation of the α-subunit of transducin, as it does for the α-subunit of the G protein.[22] Another similarity is that ADP-ribosylation inhibits the GTPase activity of both transducin and the G protein. Furthermore, components from the cyclic GMP cascade of retinal ROS can substitute for cognate elements of the adenylate cyclase system of a vari-

[17] T. Shinozawa, I. Sen, G. Wheeler, and M. Bitensky, *J. Supramol. Struct.* **10,** 185 (1979).
[18] M. Rodbell, *Nature (London)* **284,** 17 (1980).
[19] E. M. Ross and A. G. Gilman, *Annu. Rev. Biochem.* **49,** 533 (1980).
[20] D. Cassel and Z. Selinger, *Proc. Natl. Acad. Sci. U.S.A.* **75,** 4155 (1978).
[21] T. Pfeuffer, *FEBS Lett.* **101,** 85 (1979).
[22] M. E. Abood, J. B. Hurley, M.-C. Pappone, H. R. Bourne, and L. Stryer, *J. Biol. Chem.* **257,** 10540 (1982).

ety of tissues.[23] For example, transducin containing bound p[NH]ppG can serve in place of the G protein in activating adenylate cyclase.

Properties of the Phosphodiesterase

The phosphodiesterase, also a peripheral membrane protein, consists of three kinds of subunits: α (88 kilodaltons), β (84 kilodaltons), and γ ($\sim$11 kilodaltons).[24,25] A striking feature of this enzyme is the switching of its catalytic activity from a low level in the dark to a very high level after illumination.[5] It can also be activated by limited tryptic digestion, which suggests that the enzyme in the dark state is subject to an inhibitory constraint. The identity of this inhibitor has been established.[25] Trypsin-activated phosphodiesterase consists only of α- and β-subunits. The γ-subunit is very rapidly destroyed by trypsin. Thus, the catalytic activity of this enzyme resides in its α- or β-subunit or in both. The specific activity of trypsin-activated phosphodiesterase is 1.2 mmol/min mg^{-1}, which corresponds to a turnover number of 3700 sec^{-1}. The K_m value for cyclic GMP is 78 μM. The γ-subunit, isolated by gel filtration in acidic media, inhibits the catalytic activity of trypsin-activated phosphodiesterase more than 99% by lowering its maximal velocity without appreciably affecting its K_m. The γ-subunit has a very high affinity for ($\alpha\beta$). The dissociation constant of this complex is 0.13 nM. Thus, the cyclic GMP phosphodiesterase from ROS consists of distinct regulatory and catalytic subunits. The phosphodiesterase is kept in check in the dark by the inhibitory action of its γ-subunit. This inhibition is relieved by the binding of T_α-GTP formed on illumination. Activated phosphodiesterase has great catalytic prowess. Its k_{cat}/K_m ratio of 4.7×10^7 M^{-1} sec^{-1} approaches the diffusion-controlled limit, indicating that this enzyme has nearly achieved kinetic perfection in the course of evolution.

Purification of the Transducin Complex[7]

Transducin is purified by taking advantage of the tight binding of T-GDP to bleached ROS membranes at low ionic strength.[12] The phosphodiesterase and other peripheral membrane proteins are released from the membrane under these conditions. Addition of GTP then leads to the formation of T-GTP, which dissociates from the membrane. Rod outer segments prepared from 200 bovine retinas by sucrose density centrifuga-

[23] M. W. Bitensky, M. A. Wheeler, M. M. Rosenick, A. Yamazaki, P. J. Stein, K. R. Halliday, and G. L. Wheeler, *Proc. Natl. Acad. Sci. U.S.A.* **79,** 3408 (1982).

[24] W. Baehr, M. O. Devlin, and M. L. Applebury, *J. Biol. Chem.* **254,** 11669 (1979).

[25] J. B. Hurley and L. Stryer, *J. Biol. Chem.* **257,** 11094 (1982).

tion are suspended in 80 ml of ice-cold buffer A and immediately exposed to room light for 10 min. Buffer A (extraction buffer) consists of 5 m*M* Tris-HCl, 0.5 m*M* $MgCl_2$, 0.1 m*M* phenylmethylsulfonyl fluoride, and 1 m*M* dithiothreitol, at pH 7.5. The bleached membranes are then sedimented at 45,000 *g* for 30 min. The pellet is washed with buffer A four more times to remove the phosphodiesterase. Transducin bound to these membranes is then eluted by the addition of 50 ml of buffer A containing 0.1 m*M* GTP. The suspension is centrifuged at 45,000 *g* for 30 min, and the supernatant is collected. This extraction is repeated twice. The combined supernatants containing 8 mg of protein are centrifuged at 45,000 *g* for 30 min to remove a small amount of residual ROS membrane. This solution is loaded on a column, 15 $\times$ 0.7 cm in diameter, of hexylagarose operated at a flow rate of 20 ml/hr. This column is washed with buffer B (hexylagarose chromatography buffer), which consists of 10 m*M* 4-morpholinepropanesulfonic acid (MOPS), 2 m*M* $MgCl_2$, 0.1 m*M* phenylmethylsulfonyl fluoride, and 1 m*M* dithiothreitol, at pH 7.5. Protein is eluted by the stepwise addition of 75 m*M* NaCl and then 300 m*M* NaCl. About 5.5 mg of transducin are obtained from the 300 m*M* NaCl fractions. Sodium dodecyl sulfate (SDS)–polyacrylamide gel electrophoresis shows that these fractions contain three polypeptides, designated at T_α (39 kilodaltons), T_β (36 kilodaltons), and T_γ ($\sim$10 kilodaltons).

Properties of the Transducin Complex

Transducin purified in this way exhibits GTPase activity and amplified binding of GTP (or of p[NH]ppG, a nonhydrolyzable analog of GTP) in the presence of ROS containing photolyzed rhodopsin.[7] GTPase activity is assayed by measuring the formation of radioactive orthophosphate. The assay mixture contains 10 μM [γ-^{32}P]GTP, 300 m*M* NaCl, and 10 μM photolyzed rhodopsin in reconstituted membranes in buffer B. Reconstituted membranes containing hydroxyapatite-purified rhodopsin and phosphatidylcholine at a molar ratio of 1 : 125 are prepared by dialysis. A 20-μl aliquot of the solution to be assayed is added to 20 μl of this mixture. After 5 min at 23°, the reaction is stopped by adding perchloric acid containing potassium phosphate. Orthophosphate is then precipitated by adding ammonium molybdate. This precipitate is collected by filtration and washed six times. The radioactivity retained by the filter is counted. The recovery of orthophosphate is about 85%.

Photolyzed rhodopsin in reconstituted membranes catalyzes the binding of p[NH]ppG to purified transducin. A 0.4-ml aliquot of 1.2 mg/ml transducin is incubated with 0.8 ml of reconstituted membranes containing 75 μM rhodopsin. After 12 hr in the dark at 4°, the membranes are pelleted at 45,000 *g* for 30 min. About 73% of the transducin becomes bound to the membranes. The pellet is resuspended in buffer D (reconstitution assay buffer), which consists of 60 mM KCl, 30 mM NaCl, 2 mM $MgCl_2$, 10 mM MOPS, 1 mM dithiothreitol, and 0.1 mM phenylmethylsulfonyl fluoride, at pH 7.5, to give a rhodopsin concentration of 3 μM. Aliquots (300 μl) of these membranes are added to an equal volume of 20 μM [^{3}H]p[NH]ppG (containing 600,000 dpm) and immediately subjected to different doses of illumination. After 30 min at 22°, the amount of tritiated nucleotide bound to transducin is determined by filtering the suspension through a 0.45 μm Millipore HA filter, washing the membrane five times with 3 ml of buffer D, and counting the radioactivity bound to the membrane. This filter retains nucleotide bound to transducin, but not free nucleotide; the recovery is about 65%. The amount of GTP analog bound under these conditions is half-maximal when the degree of photolysis is 0.027%, which corresponds to the catalyzed uptake of 71 p[NH]ppG molecules per photolyzed rhodopsin molecule.

Purification and Properties of the α-Subunit of Transducin

The complex of transducin with p[NH]ppG is formed by incubating 1 mg of purified transducin per milliliter in 4 ml of buffer B containing 300 mM NaCl with 0.15 mM tritiated nucleotide and 5 μM photolyzed rhodopsin in reconstituted membranes for 12 hr on ice. The membranes are then removed by centrifugation at 45,000 g for 15 min. The supernatant is diluted with 36 ml of buffer B and applied to a 10 × 0.7 cm hexylagarose column operated at a flow rate of 10 ml/hr. The column is washed with buffer B to remove unbound p[NH]ppG. Transducin containing bound p[NH]ppG is eluted by 300 mM NaCl in buffer B. These fractions are pooled and concentrated. This solution (0.3 ml of 1.4 mg/ml protein) is loaded on a 30 × 0.75 cm TSK-125 (Bio-Rad) gel filtration column equilibrated with buffer C, which consists of 20 mM MOPS, 100 mM Na_2SO_4, 1 mM dithiothreitol, 0.1 mM phenylmethylsulfonyl fluoride, and 1.5 mM $MgCl_2$, at pH 7.2. High-pressure liquid chromatography is carried out at a pressure of 100 psi (4.22 × 10^5 Pa) and a flow rate of 0.2 ml/min. The elution profile exhibits two peaks (Fig. 2). The first contains T_α and the second contains T_β and T_γ. All of the radioactive p[NH]ppG emerges with T_α.

Activation of the Phosphodiesterase by the α-Subunit of Transducin

The effect of these fractions on phosphodiesterase activity can be assayed in the following way. The assay solution consists of 0.1 ml of 2 mM cyclic GMP, cyclic [^{3}H]GMP (250,000 dpm), and 5′-[^{14}C]GMP (10,000 dpm) in buffer C. 5′-[^{14}C]GMP serves to monitor the recovery of

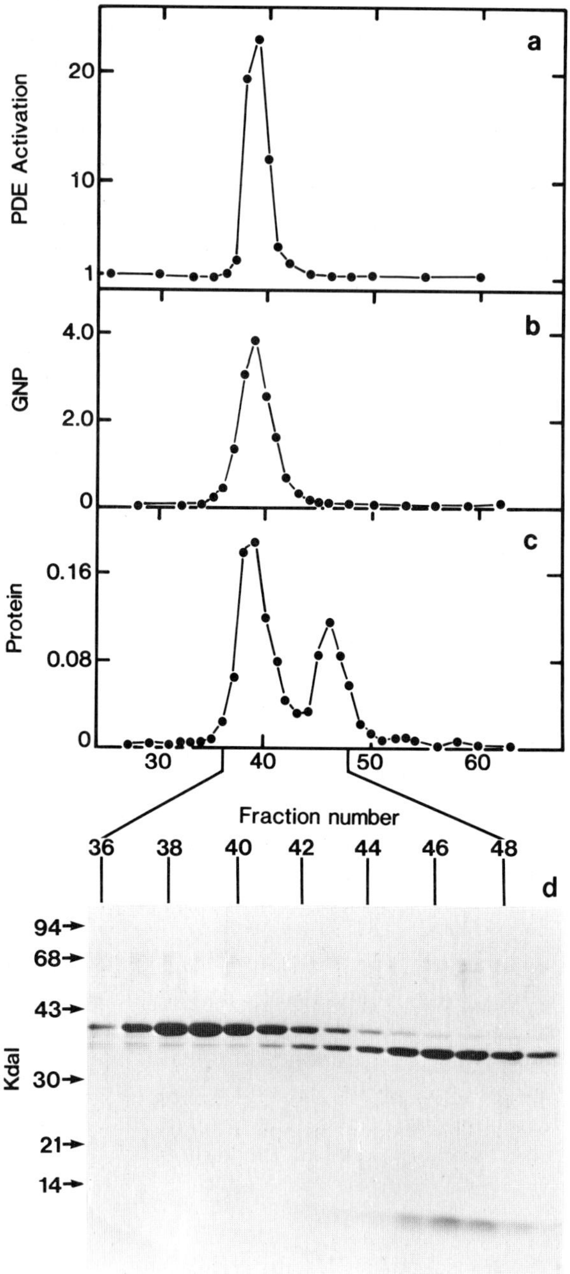

FIG. 2. High-pressure liquid chromatography of transducin. (a) Stimulation of phosphodiesterase activity of unilluminated rod outer segment membranes by addition of aliquots of column fractions. (b) p[NH]ppG concentration (μM). (c) Protein concentration (mg/ml). (d) Sodium dodecyl sulfate–polyacrylamide gel electrophoresis pattern of peak fractions.

the 5′-[^{3}H]GMP product of the phosphodiesterase reaction. Each assay is initiated by the addition of cyclic GMP. After 2 min on ice, the reaction is terminated by adding 0.1 ml of 20 m*M* nonradioactive cyclic GMP and boiling the mixture for 2 min. The 5′-[^{3}H]GMP product is converted to [^{3}H]guanosine by adding 0.5 ml of 0.2 mg/ml snake venom (*Ophiophagus hannah*) and incubating the mixture at 30° for 15 min. This mixture is applied to a 0.7 × 2 cm DEAE–Sephadex column, and the [^{3}H]guanosine is eluted with 2 ml of distilled H_2O. ^{3}H and ^{14}C in the eluent are counted.

Phosphodiesterase bound to unilluminated ROS membranes is activated by high-pressure liquid chromatography fractions containing p[NH]ppG bound to T_α, but not by fractions containing T_β and T_γ (Fig. 2). Phosphodiesterase activity increases nearly linearly with the amount of T_α-p[NH]ppG added up to an asymptotic value. This limiting value is 2.2-fold as high as the catalytic activity obtained by activation with light in the presence of p[NH]ppG. In fact, the maximal activity attainable by the addition of T_α-p[NH]ppG is nearly the same as that obtained by activation of the phosphodiesterase with trypsin. As discussed below, it seems likely that the addition of T_α-p[NH]ppG relieves the same inhibitory constraint on catalytic activity as does the proteolytic action of trypsin.

Preparation of ADP-Ribosylated α-Subunit of Transducin[22]

A freshly bleached suspension of ROS (at a concentration of 5 mg/ml rhodopsin) is incubated with 50 μg of cholera toxin per milliliter activated with dithiothreitol, 1 m*M* [^{32}P]NAD (50 Ci/mol), and 5 μM p[NH]ppG in 20 ml of buffer E (ADP-ribosylation buffer). This buffer consists of 200 m*M* potassium phosphate at pH 6.8, 10 m*M* thymidine, 4 m*M* $MgCl_2$, and 12 m*M* arginine HCl. After 3 hr at 30°, this reaction mixture is diluted 10-fold with buffer D and centrifuged at 40,000 *g* for 30 min. The pellet is washed twice with buffer D. Transducin, extracted with buffer A as described above, is applied to a 1.5 × 6 cm column of ω-aminooctyl-agarose.[26] This column is washed with 40 ml of buffer B and eluted with an NaCl gradient in this buffer. The ^{32}P-labeled ADP-ribosylated α-subunit of transducin elutes at 220 m*M* NaCl, whereas the β and γ subunits emerge at 280 m*M* NaCl. The concentrations of NaCl needed to elute these subunits may vary with the batch of ω-aminooctylagarose.

Purification of the Phosphodiesterase[24,25]

Retinal ROS prepared from 200 bovine retinas are suspended in 200 ml of buffer A and homogenized by manually passing a Teflon–glass homog-

[26] B. K.-K. Fung, *Biochemistry* (in press).

enizer through the suspension three times. This suspension is then exposed to incandescent light for 10 min and centrifuged at 48,000 *g* for 20 min. The phosphodiesterase is partially solubilized by this procedure, whereas transducin remains bound to the disk membranes because of the high affinity of T-GDP for R*. This extraction is repeated, and the two supernatants are pooled. This extract (containing about 95 mg of protein) is applied to a 1.5 × 6 cm hexylagarose column and washed with 50 ml of buffer B followed by a linear gradient of 0 to 400 m*M* NaCl in buffer B. The flow rate is 15 ml/hr; 5-ml fractions are collected. Fractions containing phosphodiesterase activity are pooled, concentrated, and dialyzed against buffer C. The concentrated enzyme (0.24 ml of 2.6 mg/ml) is then injected into a 0.75 × 50 cm Varian 3000 SW high-performance liquid chromatography (HPLC) gel filtration column operated at a flow rate of 1 ml/min. Fractions containing phosphodiesterase activity are pooled, concentrated to 5 mg/ml, dialyzed against buffer B, and then stored at −20° in 50% glycerol. The catalytic activity and subunit composition of the phosphodiesterase remain constant for at least 6 months under these storage conditions.

Preparation of Trypsin-Activated Phosphodiesterase[25]

Trypsin-activated phosphodiesterase is of interest because it exhibits maximum catalytic activity and can be used to assay the inhibitory action of the γ-subunit. Trypsin-activated enzyme can be prepared in two ways. In the first method, purified phosphodiesterase is dialyzed against buffer D to remove phenylmethylsulfonyl fluoride. Aliquots (50 μl of 1 mg/ml) of dialyzed phosphodiesterase are passed over a 0.2 × 1 cm column of insolubilized trypsin (trypsin-agarose) equilibrated with buffer D devoid of phenylmethylsulfonyl fluoride. The degree of activation can be controlled by varying the flow rate. In the second method, phosphodiesterase stored in 50% glycerol is diluted 10-fold with buffer D devoid of phenylmethylsulfonyl fluoride. Trypsin (TPCK-treated) is added to a concentration of 50 μg/ml. After 4 min on ice, bovine pancreatic trypsin inhibitor is added to a concentration of 250 μg/ml to stop the proteolysis. After concentrating the reaction mixture, 0.5 mg is injected into a Varian 3000 SW HPLC gel filtration column operated at a flow rate of 1 ml/min. Fractions containing phosphodiesterase activity are pooled, concentrated to 1 mg/ml in buffer B, and stored in 50% glycerol at −20°.

Purification of the γ Regulatory Subunit of the Phosphodiesterase[25]

The γ-subunit of the phosphodiesterase can be isolated by taking advantage of the fact that it is stable in acid, in contrast to the α and β

subunits. The purification of this regulatory subunit is facilitated by first labeling it with ^{14}C. Purified phosphodiesterase (0.9 ml of 2.2 mg/ml) is dialyzed against 100 m*M* sodium borate, pH 8.3, containing 1 m*M* $MgCl_2$. [^{14}C]Ethyl acetimidate (500 μCi of 485 μCi/mg) is added at 30°. The reaction is stopped after 40 min by addition of 0.9 ml of 0.2 *M* glycine at pH 8, and the mixture is dialyzed. The catalytic activity of phosphodiesterase labeled in this way is virtually the same as that of native enzyme. The ^{14}C-labeled phosphodiesterase (0.8 mg containing 2.3×10^5 dpm) is applied with 3 mg of native enzyme on a 1.5×100 cm column of Sephadex G-100 equilibrated with 2 *M* acetic acid at pH 2.1. The column is eluted at a flow rate of 9 ml/hr; 1.2-ml fractions are collected. Most of the radioactivity emerges in the void volume, followed by a small peak of radioactivity 27 fractions later. The location of the inhibitory subunit in the column profile is ascertained by assaying the effect of a 0.8-μl aliquot of each fraction on the catalytic activity of 50 ng of trypsin-activated phosphodiesterase. Nearly all the inhibitory activity appears in the second peak of the radioactivity profile. Autoradiograms of SDS–polyacrylamide gels of the inhibitory fractions display a single band corresponding to the γ-subunit of the enzyme. The major peak of protein in the void volume consists of denatured α- and β-subunits.

Acknowledgments

The work carried out in the authors' laboratories was supported by research grants from the National Eye Institute. B. K.-K. F. was a Jane Coffin Childs Fellow, and J. B. H. was a Helen Hay Whitney Fellow.

[51] Avian Salt Gland: A Model for the Study of Membrane Biogenesis

By RUSSELL J. BARRNETT, JOSEPH E. MAZURKIEWICZ, and JOHN S. ADDIS

General Considerations

A large amount of membrane is used in the construction of all cells. These membranes, especially the plasma membrane, are diverse, complex, and specialized but contain entities that make them identifiable by certain biochemical and/or structural features. The task to produce, main-

ISBN 0-12-181996-5

tain, and repair even the plasma membrane alone is a very large and continuing process; it is even greater during growth and differentiation.

Although the constituent components, protein and lipid, of the plasma membranes from a variety of cell types have been partially studied or elucidated (see Chapters 20–33, this volume), the organization of these components,[1] and the mechanism and site(s) of their synthesis and assembly[2] into the membrane structure, though currently investigated, are not yet clearly understood. On an *a priori* basis several mechanisms for the biogenesis of plasma membranes may be postulated. The constituent molecules or components may be synthesized at the same or different sites in the cell and assembled either *in situ* or at some central site (multistep) from which the newly assembled constituents move, either as vesicles or subunits to fuse with the preexisting plasma membrane. In the latter case, the Golgi complex has been a suggested site for assembly of the plasma membrane constituents, and the endoplasmic reticulum the site of synthesis of these.[3]

Most studies of plasma membrane biogenesis have been either indirect or based on cell fractionation experiments. Direct studies of the site of synthesis of plasma membrane components and of their assembly using whole cells have been sparse, except for some studies on surface receptors and specialized structures of cell surfaces. One initial problem in the study of plasma membrane biogenesis is the selection of suitable tissue and of circumstances in which the plasma membrane proliferates in response to an experimental stimulus that can be physiological, pharmacological, developmental, or pathological in nature.

Adaptation of the Plasma Membrane of the Salt Gland

The avian salt gland represents an excellent model since it undergoes hypertrophy, hyperplasia, and secretory cell differentiation in response to osmotic stress.[4,5] Several main morphological features concerning organization of the gland are discussed below. The center of the gland is occu-

[1] G. I. Nicholson, G. Poste, and H. J. Tae, *in* "Dynamic Aspects of Cell Surface Organization" (G. Poste and G. I. Nicholson, eds.), p. 1. Elsevier/North-Holland Biomedical Press, Amsterdam, 1977.

[2] D. D. Sabatini, G. Keibich, T. Morimoto, and M. Adesnik, *J. Cell Biol.* **92,** 1 (1982).

[3] J. D. Morré, *in* "Synthesis, Assembly and Turnover of Surface Components" (G. Poste and G. I. Nicholson, eds.), p. 1. Elsevier/North-Holland Biomedical Press, Amsterdam, 1977.

[4] R. A. Ellis, C. C. Goertemiller, R. A. DeLellis, and Y. H. Kablotsky, *Dev. Biol.* **8,** 236 (1963).

[5] M. Peaker and J. L. Linzell, "Salt Gland in Birds and Reptiles." Cambridge Univ. Press, London and New York, 1975.

pied by a central duct into which secretory tubules empty. These tubules begin as a blind ending just under the peripheral connective tissue capsule or its internal branches and contain *peripheral cells* that are undifferentiated and are the only epithelial cells of the tubule that are capable of mitosis. [^{3}H]Thymidine is taken up by these; under control circumstances, labeled nuclei reach the duct level in 12 weeks.[6] Adjacent to the peripheral cells, which occupy a small area, the tubules contain a small zone of *partially differentiated* cells that show some plications of the lateral surface. The remainder of the tubule is occupied by *principal cells,* which show moderate infoldings and plicae of the lateral, but not the basal, portions of the plasma membrane[7,8] (Fig. 1).

On osmotic stress, especially in ducklings that have not previously been subjected to stress, basal infoldings and plicae markedly appear in the principal cells, and the lateral plicae are also increased.[7] Thus, changes in the amount of basolateral, but not of apical, membrane are readily initiated and maintained (adaptation) simply by administration of a hypertonic solution (usually saline, 1%, or seawater) for drinking over a period of time. Other routes and solutions are reviewed.[5] Subsequent reversal (deadaptation) occurs by a substitution of water for saline. The magnitude of the adaptive changes ultimately is quite large, since in the fully stressed secretory cell, the amount of basolateral membrane exceeds the apical membrane by 1000×[8] (Fig. 2).

The ease with which the phenomena can be induced or reversed makes the salt gland an attractive system for studying the cellular and molecular events underlying the process of plasma membrane biogenesis, despite the fact that relatively few such studies have occurred. During membrane amplification or adaptation, parallel morphological, cytochemical, and biochemical changes occur, some of which are covered here. In fact, the cytochemical approach at the electron microscope level, accompanied by biochemical justification of the findings and procedures, is one of the major themes of this review. Once the membrane components are identified and the magnitude of change of these markers is elucidated, "where" becomes a major question concerning synthesis and assembly.

Use of Tissue

Although some details refer to use of the intact gland from the intact animal for the comparison of the states of adaptation (salt-stressed), deadaptation (destressed) and control ducklings, cytochemical studies

[6] R. A. Ellis, *Am. Zool.* **5,** 648 (1965).

[7] S. A. Ernst and R. A. Ellis, *J. Cell Biol.* **40,** 305 (1969).

[8] H. Komnick and E. Knirprath, *Cytobiologie* **1,** 228 (1970).

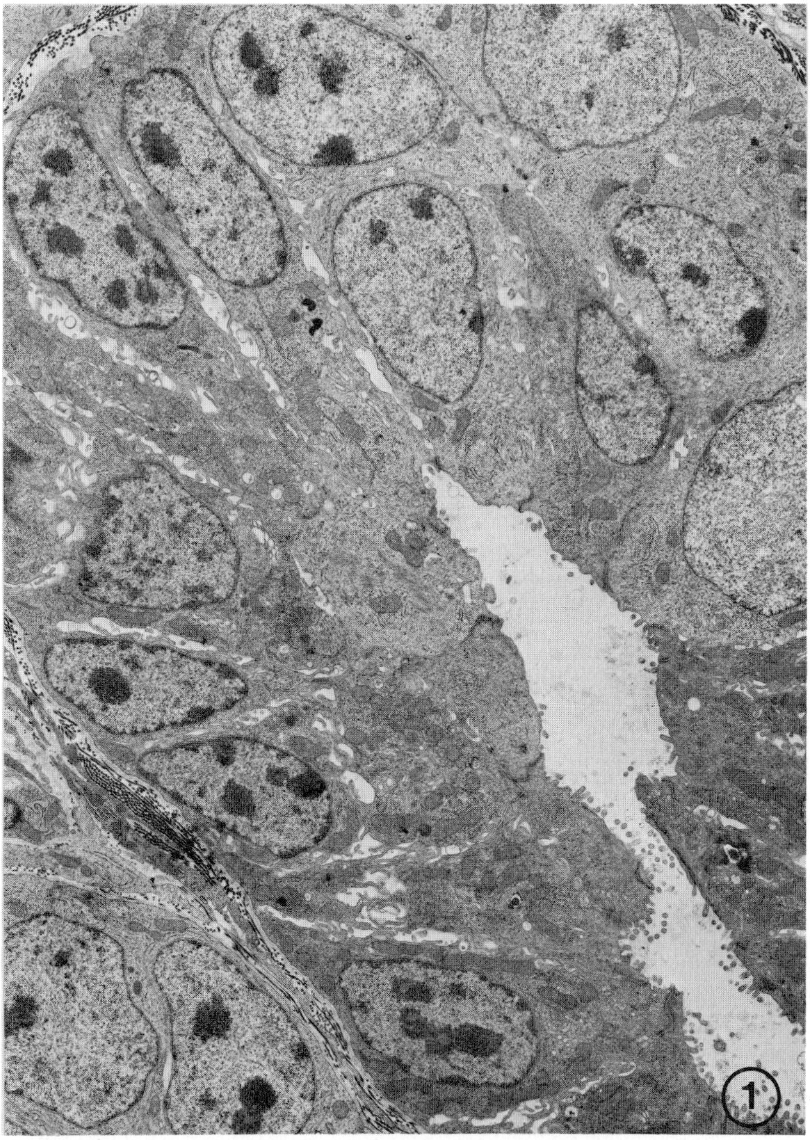

FIG. 1. Base of control tubule showing, in upper right, primitive principal cells; the remainder of the field contains partially differentiated cells with some increase in the lateral border plicae, but not in the base. ×5000.

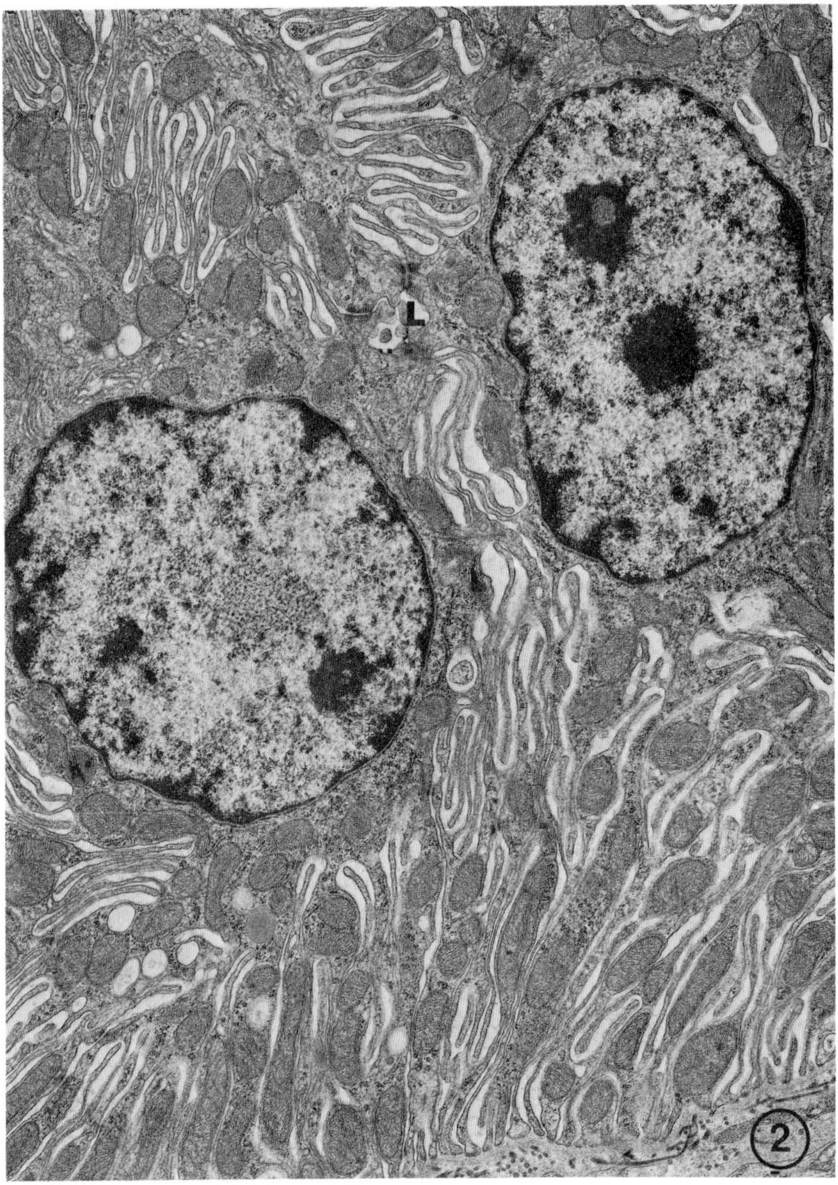

FIG. 2. Principal cells of adapted animal showing crowded lateral plications and large basal infoldings and plications (bottom). Note mitochondria in relation to basal infoldings. Apical surface marking the lumen (L) shows small surface and contains a few microvilli. ×11,200.

(referred to later) require sections (fixed or frozen) or small blocks of tissue (<1 mm^3). In dealing with the intact gland, especially for biochemical studies where localization studies are less possible than with cytochemistry, it should be noted that the gland is replete with connective tissue elements (capsule as well as strands proceeding into the center) and capillaries, both of which respond to adaptation. In fact, prior to the initial time to establish *de novo* increases in a variety of enzyme activities during adaptation, the gland turns bright red from pink, presumably because of increased blood flow.[9] In addition, though principal cells are the major cellular element, these are diluted in the total mass by peripheral cells, fibroblasts and other connective tissue cells, myoepithelial cells, duct cells, and endothelial cells and other elements of the total vascular bed. Thus, after decapitation and rapid removal of the gland, the capsule is usually removed by dissection before the gland is weighed.

In addition to the above, "minilobules," which are realistically organotypic cultures of the avian salt gland, may be produced and used[10]; these may yield interesting studies because many events are complicated and do not lend themselves to whole-gland study. Minilobules are prepared by dissociation by modification of the method of Amsterdam and Jamieson.[11]

1. The gland is removed, trimmed, and placed in culture medium M 199, pinned so that it can be injected in multiple sites with 3 ml of enzyme solution [180 units of crude collagenase (Worthington) per milliliter in medium M 199; the use of purified collagenase showed few signs of disruption and required excessive mechanical manipulation, and addition of hyaluronidase improved the dissociation, but the viability and morphology were poor].
2. Glands are incubated in 5 ml of enzyme solution, gassed with 95% air/5% CO_2 while shaking vigorously for 30–45 min at 37°.
3. Glands are reinjected with fresh enzyme solution and incubated as in step 2 above for 45 min (glands swelled and became friable).
4. Glands are pulled repeatedly by suction into a 10-ml disposable pipette to break tissue into small aggregates.
5. Tissue aggregates are washed in M 199 and gently centrifuged in M 199.
6. Aggregates are placed on strips of 45 μm Millipore membranes stretched across wells of a Microtiter plate; wells are filled with M 199 to touch tissue and are placed in incubators at 37° in a

[9] R. Frange, K. Schmitt-Nielsen, and M. Robinson, *Am. J. Physiol.* **195,** 321 (1958).
[10] J. E. Mazurkiewicz and R. J. Barrnett, *J. Cell Sci.* **48,** 75 (1981).
[11] A. Amsterdam and J. D. Jamieson, *J. Cell Biol.* **63,** 1037 (1974).

humidified atmosphere of 95% air/5% CO_2. Alternatively, aggregates are incubated submerged in medium in a 50-ml flask, gassed as above, and shaken in a water bath at 37°. Under the conditions of culture, M 199 contained 100 mg of gentamycin per milliliter.

These procedures allowed morphological and radiolabeled precursor ([^{3}H]leucine) incorporation studies as well as immunoprecipitation of Na^+,K^+-ATPase.[10] Essentially, tissue aggregates were homogenized (0.25 *M* sucrose, 50 m*M* Tris, pH 7.5) and centrifuged (500 *g*, 15 min, 4°) twice; the supernatants were collected (400 μl) for sample protein determination and liquid scintillation (LS) counts on trichloroacetic acid precipitates. The supernatant was recentrifuged (47,000 *g*, 60 min, 4°) to obtain a pellet, which was resuspended in buffer for protein determination, LS count (trichloroacetic acid precipitate), and immunoprecipitation followed by sodium dodecyl sulfate–polyacrylamide gel electrophoresis (SDS–PAGE) plus slice and LS count or by LS count alone.

Significantly, minilobules could be cultured as long as 72 hr. Incorporation of [^{3}H]leucine into total protein was linear for 90 min of continuous exposure. After 24 hr of culture, a 30-min pulse followed by variable chase showed that more leucine was incorporated into the proteins of previously stressed ducklings than in controls; most of the incorporation occurred in 2 hr, and a plateau was reached in 4 hr, with only a minimal rise in 20 hr. Further analysis using immunoprecipitation of Na^+,K^+-ATPase showed that, after a 20-min pulse, total protein peaked in less than 1 hr but that of Na^+,K^+-ATPase reached a continuous linear rise over a 4-hr period. SDS–PAGE profiles of the antibody precipitate of the membrane fraction of stressed animals in which minilobules labeled with [^{3}H]leucine for 20 min and chased for 24 hr indicated that Na^+,K^+-ATPase was labeled at least six times more than most other proteins.[10]

Finally, it is clear that other functions may be demonstrated in preparations of suspensions of dissociated cells, which has been attempted.[12–13] We have essentially used a modification of one of these[13] on the glands of stressed and control ducklings. The basic protocol is essentially the use of glands minced into 0.5–1 mm^3 cubes that were subjected to: (*a*) first enzymic digestion in Eagle's MEM containing collagenase (250 units/ml), hyaluronidase (500 units/ml), and chymotrypsin (25 units/ml) for 60 min at 37°; (*b*) chelation of divalent cations in Ca^{2+},Mg^{2+}-free KRB with 1 m*M* EDTA for 15 min at 37°; (*c*) replacement of cations with Eagle's MEM containing 1.8 m*M* Ca^{2+}, 0.8 m*M* Mg^{2+} for 5 min at 37°; (*d*) second

[12] F. E. Hossler and M. P. Sarras, *Scanning Electron Microsc.* **2,** 155 (1980).

[12a] I. G. Thompson, *Am. Zool.* **17,** 878 (1977).

[13] S. R. Hootman and S. A. Ernst, *Am. J. Physiol.* **238,** C124 (1980).

enzymic digestion—same as (*a*) above; (*e*) rinses with Ca^{2+},Mg^{2+}-free KRB, twice for 2 min at 24°; (*f*) dissociation by pipetting, flushing 20 times through a siliconized pipette in KRB medium free of Ca^{2+} and Mg^{2+} containing 2.0 mg of soybean trypsin inhibitor per milliliter and deoxyribonuclease I (15 units/ml); (*g*) filtration through 25 μm nylon mesh in medium described in (*f*) above; (*h*) centrifugation for 10 min at 150 *g* to remove debris in Eagle's MEM with 4% BSA; (*i*) resuspension in Eagle's MEM containing 18 m*M* Ca^{2+} and 0.8 m*M* Mg^{2+}. The morphology of such isolated cells was well preserved and recovery of cells was between 24% and 32%. These freshly isolated cells from control and adapted ducklings showed retention of Na^+,K^+-ATPase activity and an O_2 consumption that was stimulated by methacholine and inhibited by ouabain.[13]

However, further culture of these cells should be undertaken with some cautions, based on fine-structural examination of cells during 4 days of culture, especially during the first 24 hr.[14] Fibroblasts are not eliminated and become plentiful as culture time continues. More important, though the isolated secretory cells are comparable to those in the intact gland on direct isolation, during culture they tend to undergo changes compatible with dedifferentiation. The entire area of plication of the basolateral border of most cells at variable times in the first 24 hr becomes capped in a smaller peripheral zone, which subsequently invaginates into the interior of the cells. This pinches off and fuses with lysosomes, so that the cell is left with a regular border (probably apical in nature) and contains sparse microvilli and one or more very large vacuoles, containing debris. After several days, the isolated cells tend to clump, and only at this time do a few invaginations and evaginations appear where cells are joined. With such severe morphological changes, including loss of the basolateral border, the biochemical change must be great and is under further investigation. These findings are already supported by observations on dissociated urinary bladder cells of the frog.[14a]

Enzyme Markers of the Basolateral Membrane

Hokin[15] first demonstrated the presence of Na^+,K^+-ATPase in the salt gland, and this was followed by much significant work, referred to briefly below. Mg^{2+}-ATPase is also present[16] and though its level is twice that of Na^+,K^+-ATPase in the newborn, it only doubles on adaptation whereas

[14] J. Merchant and R. J. Barrnett, unpublished results (1982).
[14a] T. Fujimoto and K. Ogawa, *J. Histochem. Cytochem.* **31,** 131 (1983).
[15] M. R. Hokin, *Biochim. Biophys. Acta* **77,** 108 (1963).
[16] G. L. Fletcher, I. M. Stainer, and W. M. Holmes, *J. Exp. Biol.* **45,** 372 (1967).

Na^+,K^+-ATPase increases fivefold.[17] An anion-activated ATPase attributed to the membrane[5] on the basis of work in other tissues[18] clearly occurs only in mitochondria of the salt gland[19] and, like succinate dehydrogenase, increases during adaptation as mitochondrial profiles increase in size and number. This may be related to the fact that glycolytic enzymes [aspartate aminotransferase (glutamic–oxaloacetic transaminase), phosphofructokinase, hexokinase, and G-6-P dehydrogenase] also increase during adaptation.[20] Other enzymes that have been reported, but are insufficiently studied, include carbonate dehydratase (carbonic anhydrase),[21] cholinesterase,[22] and alkaline phosphodiesterase.[23]

Na^+,K^+-ATPase deserves special attention here because of the high content of this integral protein in the salt gland, its localization to the basolateral membrane, its relationship to glandular function,[16] and its response to adaptation and deadaptation of the salt gland.[24–28] The enzyme has received much attention, especially its biochemistry and biophysical nature, which has been plentifully reviewed.[29–31] That Na^+,K^+-ATPase appears to be the major marker suggests the use of the salt gland as a model system for investigating the pathway traversed by such newly synthesized integral plasma membrane proteins through the cell.

Preparation of Plasma Membrane and Golgi Membrane Fractions

One approach to delineation of the intracellular pathway followed by Na^+,K^+-ATPase would be to quantitate radiolabeled Na^+,K^+-ATPase

[17] A. M. Levine and R. J. Barrnett, unpublished results (1974).

[18] B. Simon, R. Kinne, and G. Sachs, *Biochim. Biophys. Acta* **282,** 293 (1972).

[19] J. S. Addis and R. J. Barrnett, *J. Cell Biol.* **78,** 371a (1977).

[20] I. M. Stainer, M. Ensor, D. M. Phillips, and W. N. Holmes, *Comp. Biochem. Physiol.* **37,** 273 (1970).

[21] S. L. Bonting, L. L. Caravaggio, M. R. Canady, and N. M. Hawkins, *Arch. Biochem. Biophys.* **106,** 49 (1963).

[22] B. Ballantyne, *Cytobios* **9,** 39 (1974).

[23] J. S. Addis and R. J. Barrnett, unpublished results (1982).

[24] S. A. Ernst, C. C. Goertemiller, Jr., and R. A. Ellis, *Biochim. Biophys. Acta* **135,** 682 (1967).

[25] D. J. Stewart, E. W. Semple, G. T. Swart, and A. K. Sen, *Biochim. Biophys. Acta* **419,** 150 (1976).

[26] F. E. Hossler, M. P. Sarras, and R. J. Barrnett, *J. Cell Sci.* **31,** 179 (1977).

[27] S. A. Ernst and J. W. Mills, *J. Cell Biol.* **75,** 74 (1977).

[28] S. A. Ernst, *J. Histochem. Cytochem.* **20,** 13 (1972a).

[29] A. Schwartz, G. E. Lindenmayer, and J. C. Allen, *Pharmacol. Rev.* **27,** 3 (1975).

[30] P. L. Jorgensen, *Physiol. Rev.* **60,** 846 (1980).

[31] J. C. Skou and J. G. Nørby, eds., "Na,K-ATPase: Structure and Kinetics." Academic Press, New York, 1979.

that has been immunoprecipitated from subcellular fractions prepared at intervals after pulse-labeling. In order to evaluate the feasibility of this approach, we developed a fractionation scheme for salt gland[32] based upon the procedure of Hodson and Brenchley[33] and have utilized it to identify a non-plasma membrane-associated species of Na^+,K^+-ATPase.[34,34a] The fractionation procedure is described below. In brief, after a gentle homogenization in hypertonic medium and the removal of large particulates from the homogenate by filtration and low speed centrifugation, Golgi and plasma membranes are separated in a single multistep sucrose gradient and then further enriched by differential centrifugation (see flow diagram, Fig. 3).

The homogenization medium contained 0.5 *M* sucrose (RNase-free, Schwarz-Mann)–89.5 m*M* K_2HPO_4–10.5 m*M* citric acid, pH 7.1 (4°)–1.0% (w/v) dextran (type 200C, Sigma)–1 m*M* $MgCl_2$–1 mg of bovine serum albumin per milliliter (Fraction V, Armour). Step gradient solutions: 0.7 *M*, 1.01 *M*, and 1.28 *M* sucrose in 89.5 m*M* K_2HPO_4–10.5 m*M* citric acid, pH 7.1 (4°).

Fractionation

The salt glands of ducklings salt-stressed for 3 days are excised, trimmed, and weighed in a tared beaker containing ice-cold homogenization medium. The glands are minced finely with razor blades and homogenized (1 g wet weight per 10 ml of homogenization medium) with a Polytron PT-20 (three 10-sec pulses at 1750 rpm followed by two 5-sec pulses at 2250 rpm). The homogenate, after filtration through a single layer of cheesecloth, is centrifuged for 10 min at 480 *g* (Sorvall SS34 rotor). The resulting supernatant is removed and saved. The pellet is then resuspended in a starting volume of homogenization medium with the Polytron (one 5-sec pulse at 2100 rpm), filtered through two layers of cheesecloth, and recentrifuged for 10 min at 480 *g*. The supernatants are combined, and 6.0-ml aliquots are layered over sucrose step gradients comprised of 4.0 ml each of 1.28 *M*, 1.01 *M*, and 0.7 *M* sucrose buffered with K_2HPO_4–citric acid (total volume = 18.0 ml). Gradients are centrifuged for 120 min at 88,000 *g* (Beckman SW27.1 rotor).

After centrifugation, the interface bands are removed with a Pasteur pipette and diluted with two parts of 0.25 *M* sucrose. The diluted 0.7 *M*/1.01 *M* sucrose interface is centrifuged for 20 min at 8400 *g* (SW27.1

[32] J. S. Addis, W. D. Merritt, and R. J. Barrnett, *J. Cell Biol.* **79,** 236a (1978).

[33] S. Hodson and G. Brenchley, *J. Cell Sci.* **20,** 167 (1976).

[34] J. S. Addis, J. E. Mazurkiewicz, and R. J. Barrnett, *J. Cell Biol.* **83,** 273a (1979).

[34a] J. S. Addis, W. D. Merritt, J. E. Mazurkiewicz, and R. J. Barrnett, in preparation.

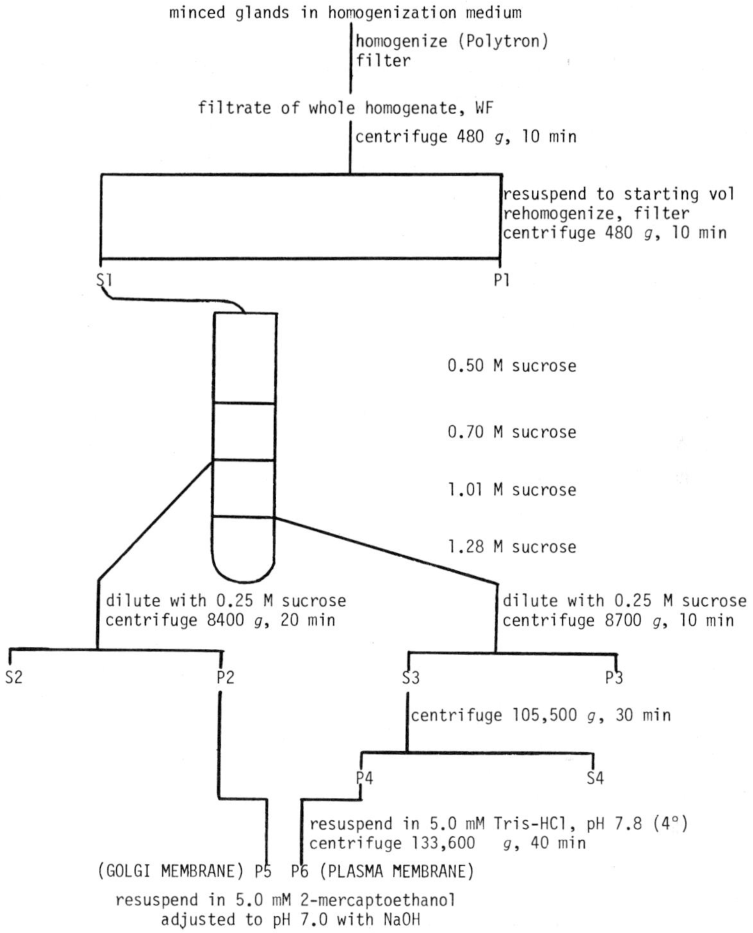

FIG. 3. Flow diagram for separate isolation of elements of the Golgi apparatus and plasma membrane.

rotor). The diluted 1.01 *M*/1.28 *M* sucrose interface is centrifuged for 10 min at 8700 *g* (SS34 rotor), and the resulting supernatant centrifuged for 30 min at 105,500 *g* (Beckman 50 Ti rotor). This latter pellet and the pellet obtained from the 0.7 *M*/1.01 *M* sucrose interface are resuspended in 5 m*M* Tris-HCl, pH 7.8, and centrifuged for 40 min at 133,600 *g* (50 Ti rotor). The pellets are then resuspended in 5 m*M* 2-mercaptoethanol, pH 7.0, divided into aliquots, and frozen at −20°.

One of the major difficulties in fractionating the salt gland is the large amount of connective tissue present. Overhomogenization should be avoided if adequate yields of Golgi membranes are to be obtained. The problem can be partially alleviated by using 3-day-stressed ducklings between 2 and 4 weeks of age and by mincing the tissue finely. The first series of Polytron pulses serves as an additional mince. An alternative approach is to treat the glands with collagenase prior to homogenization.[35] Five to ten ducklings are used per fractionation, depending on the age of the birds and the capacity of the rotor used for the gradient (the SW27 rotor has also been used with comparable results). The salt stress period (3 days) was selected to provide maximal gland size at a time of net Na^+,K^+-ATPase accumulation (2-day salt-stressed ducklings have also been used).

Enzyme Assays

Na^+,K^+-ATPase (EC 3.6.1.3) activity is determined by measuring inorganic phosphate released during hydrolysis of ATP. Samples are preincubated for 15 min at 25° in 0.01 mg of Tween 80 (Sigma) per milliliter in the presence of 2.5 m*M* Na_2ATP (vanadate-free, Sigma)–1 m*M* ethylenediaminetetraacetate (EDTA, Sigma)–30 m*M* Tris-HCl, pH 7.5.[36] A stock solution of Tween 80 is prepared daily, the pH of which is normally between 6 and 7. Aliquots from the preincubation mixture are used to initiate the Na^+,K^+-ATPase reaction. The reaction medium consists of 2.5 m*M* Na_2ATP–5 m*M* $MgCl_2$–100 m*M* NaCl–10 m*M* KCl–1 m*M* EDTA–±1 m*M* ouabain (Sigma).[37] The reaction is terminated by the addition of 0.5 ml of cold 20% trichloroacetic acid, and inorganic phosphate is measured.[38] Tween 80, at the appropriate concentration, is included with the standards.

UDPgalactose : ovomucoid galactosyltransferase (EC 2.4.1.38, glycoprotein β-D-galactosyltransferase) activity is determined using a slight modification of the method of Bretz and Staubli.[39]

Succinate dehydrogenase (EC 1.3.99.1) activity is measured by following the reduction of *p*-iodo-*p*-nitrotetrazolium violet (Sigma) according to the method of Pennington.[40]

Esterase (EC 3.1.1.1, carboxylesterase) activity is assayed using the procedure of Lazarow and de Duve[41] except that samples are incubated at 37° in the presence of 0.1% Triton X-100 (Sigma).

[35] J. S. Addis and R. J. Barrnett, unpublished observations (1980).
[36] J. R. Brotherus, P. C. Jost, O. H. Griffith, and L. E. Hokin, *Biochemistry* **18,** 5043 (1979).
[37] B. E. Hopkins, H. Wagner, Jr., and T. W. Smith, *J. Biol. Chem.* **251,** 4365 (1976).
[38] S. L. Bonting, *in* "Membranes and Ion Transport" (E. E. Bittar, ed.), Vol. 1, p. 257. Wiley, New York, 1970.
[39] R. Bretz and W. Staubli, *Eur. J. Biochem.* **77,** 181 (1977).
[40] R. Pennington, *Biochem. J.* **80,** 649 (1961).
[41] P. B. Lazarow and C. de Duve, *J. Cell Biol.* **59,** 507 (1973).

TABLE I
PARTIAL ENZYMIC CHARACTERIZATION OF GOLGI MEMBRANE (G) AND PLASMA MEMBRANE (P) FRACTIONS[a]

Enzyme	Specific activity[b]		Enrichment (X)[c]		Recovery (%)[c]	
	G	P	G	P	G	P
UDPgalactose : ovomucoid galactosyltransferase	567.4 ± 92.0	36.4 ± 18.9	71.6	4.6	13.1	5.1
Na^+,K^+-ATPase	43.1 ± 13.5	247.1 ± 17.5	1.5	8.6	0.3	11.9
Esterase	691.0 ± 55.6	193.8 ± 11.9	1.7	0.5	0.3	0.5
Succinate dehydrogenase (succinate-iodonitro-tetrazolium reductase)	1.28 ± 0.46	1.90 ± 0.37	3.1	4.5	0.2	2.2

[a] Fractions were prepared from the salt glands of ducklings salt-stressed for 3 days, using the fractionation scheme described in the text. Fractions from three runs were pooled.
[b] Specific activities are expressed as means ± standard deviations in the following units: galactosyltransferase—nanomoles of galactose per hour per milligram of protein; Na^+,K^+-ATPase—micromoles of P_i per hour per milligram of protein; esterase—nanomoles of α-naphthol per hour per milligram of protein; succinate dehydrogenase—micromoles of formazan dye per hour per milligram of protein.
[c] Expressed relative to the low-speed supernatant, S1 (Fig. 1).

Protein is determined by the Markwell *et al.* procedure[42] using bovine serum albumin (fraction V, Armour) as the standard.

Observations

The distribution of enzymatic activities in the Golgi membrane (G) and plasma membrane (P) fractions are shown in Table I. UDPgalactose : ovomucoid galactosyltransferase activity is enriched more than 70-fold in G relative to activity of the low speed supernatant (Fig. 3). Galactosyltransferase activity is also moderately enriched (4.5-fold) in P. Na^+,K^+-ATPase activity is enriched approximately 8-fold in P and shows little enrichment (1.1-fold) in G. Succinate dehydrogenase activity is moderately enriched in both fractions (3- to 5-fold). The distribution of esterase activity, which parallels that of NADPH–cytochrome *c* reductase in the fractionation scheme,[35] is slightly enriched in G and depressed in P. Most esterase activity as well as succinate dehydrogenase activity is recovered in the pellet at the bottom of the gradient (data not shown).

[42] M. K. Markwell, S. M. Haas, L. L. Bieber, and N. E. Tolbert, *Anal. Biochem.* **87,** 206 (1978).

TABLE II
ENZYMIC ACTIVITY AND IMMUNOCHEMICAL QUANTITATION OF Na^+,K^+-ATPase IN GOLGI MEMBRANE (G) AND PLASMA MEMBRANE (P) FRACTIONS[a]

Fraction	Enzymic activity[b]	Immunochemical quantitation[c]
G	50.5 ± 2.7	0.98 ± 0.04
P	362.5 ± 8.1	2.24 ± 0.26
G/P	0.14	0.44 (+215.1%)

[a] Fractions were prepared from the salt glands of 3-day salt-stressed ducklings. Fractions from 3 runs were pooled.

[b] Activity is expressed as μmoles P_i/hr/mg protein.

[c] Determined by measuring peak areas beneath precipitin arcs in crossed immunoelectrophoretic gels; expressed in arbitrary units on the basis of protein recovered in 4.68×10^6 g/min supernatant.

Biochemical measurements of Na^+,K^+-ATPase have been supplemented with immunochemical quantitation utilizing antibody directed against the holoenzyme of Na^+,K^+-ATPase (Table II). If all the enzymic activity in G were due to contamination of the fraction by plasma membrane, then the ratio, G/P, should be the same after both modes of analysis. The fact that the ratios differ suggests that a species of Na^+,K^+-ATPase is present in G that differs from the one in P.

Cytochemistry of Na^+,K^+-ATPase

Ultrastructural cytochemical studies of Na^+,K^+-ATPase have been performed on a variety of tissues considered to function in sodium transport. Since the methods involve the hydrolytic portion of the enzyme, the following formulas are self-explanatory.

$$AB \rightleftharpoons A + B \tag{1}$$

$$B + R \rightarrow BR \tag{2}$$

where AB is the ester substrate, A + B are the products of hydrolysis, R is the capture reagent, and BR is the colored (light microscopy) or electron-opaque (electron microscopy) insoluble final product, deposited at the site of hydrolysis.

Cytochemical methods used are listed in Table III[43–51] as developed historically and include some virtues and faults encountered. Clearly the approach of Ernst in which biochemical studies[27] preceded the cytochemical ones[43] were a turning point, as well as the use of pNPP as substrate which had been known biochemically earlier.[52] Pb^{2+} as capture reagent was avoided because of enzyme inhibition and the metal-induced hydrolysis of substrate. The use of DMSO (25%) in the incubating medium which changed the pH optimum to 9 was the next advance[44] and finally, the use of Pb citrate with DMSO at pH 9 presented a one-step reaction in which the Pb^{2+} effect on the substrate and enzyme was slight.[45] A similar one-step method, simultaneously developed, in which ATP could be used as substrate for Na^+,K^+-ATPase activity at the light microscope level had Pb^{2+} chelated with citrate as the capture reagent.[53] The notion of chelation of capture reagents may be an important step in localization for many cytochemical methods and was referred to earlier[54] in another context.

The detailed method of Mayahara *et al.*[45] thus presents the best repeatable results and essentially uses tissue (1 mm^3) fixed in 2% formaldehyde and 0.5% glutaraldehyde in 0.1 *M* cacodylate buffer for 1 hr. Tissue was stored overnight in buffer (pH 7.2) containing 0.25 *M* sucrose and 10% DMSO. The standard incubation medium contained 1.0 *M* glycine–KOH buffer, pH 9.0, 2.5 ml; 1% lead citrate (in 50 m*M* KOH), 4.0 ml; DMSO (25% v/v), 2.5 ml; 0.1 *M* pNPP (Mg^{2+} salt), 1 ml; Bromotetramisole, 1.87 mg or Levamisole, 6.02 mg; final pH 8.8. Controls included replacement of K^+ with Na^+ (from KOH to NaOH in glycine buffer and also in lead citrate solution) and the use of 10 m*M* ouabain in cacodylate buffer for 30 min prior to incubation. Clearly all the methods using pNPP demonstrate a K^+ dependent, ouabain-sensitive phosphatase activity, and cytochemical reviews of Na^+,K^+-ATPase exist.[49,50]

[43] S. A. Ernst, *J. Histochem. Cytochem.* **20,** 23 (1972b).

[44] L. Guth and R. W. Albers, *J. Histochem. Cytochem.* **22,** 320 (1974).

[45] H. Mayahara, R. Fujimoto, T. Ando, and K. Ogawa, *Histochemistry* **67,** 125 (1980).

[46] M. Wachstein and E. Meisel, *Am. J. Clin. Pathol.* **27,** 13 (1957).

[47] J. S. Charnock, H. A. Trebilcock, and J. R. Casley-Smith, *J. Histochem. Cytochem.* **20,** 1069 (1972).

[47a] J. S. Britten and M. Blank, *Biochim. Biophys. Acta* **159,** 160 (1968).

[48] J. E. Mazurkiewicz, F. E. Hossler, and R. J. Barrnett, *J. Histochem. Cytochem.* **26,** 1042 (1978).

[49] J. A. Firth, *Histochem. J.* **10,** 253 (1978).

[50] S. A. Ernst and S. R. Hootman, *Histochem. J.* **13,** 397 (1981).

[51] M. Borgers, *J. Histochem. Cytochem.* **21,** 812 (1973).

[52] H. Bader and A. K. Sen, *Biochim. Biophys. Acta* **118,** 116 (1966).

[53] J. Chayen, G. T. B. Frost, R. A. Dodds, L. Bitensky, J. Pitchfork, P. H. Baylis, and R. J. Barrnett, *Histochemistry* **71,** 533 (1981).

[54] G. G. Berg, D. Lyon, and M. Campbell, *J. Histochem. Cytochem.* **20,** 39 (1972).

TABLE III
CYTOCHEMICAL METHODS[a] FOR Na^+,K^+-ATPASE

Method	Substrate	Incubation ingredients	Buffer	pH	Reagents	Fixation
A: Wachstein and Meisel[46]	ATP	Mg^{2+}	Tris or maleate	7.2	Pb^{2+}	Formalin
B: Charnock, *et al.*[47]	ATP	Ouabain, Na^+,Mg^{2+}	Trisglycylglycine	7.6	Cs^+	Unfixed, isolated RBC ghosts
C: Ernst[43]	pNPP	Mg^{2+},K^+	Tris-HCl	9.0	Sr^{2+} (see notes)	Paraformaldehyde
D: Guth and Albers[44]	pNPP	K^+, Mg^{2+}, DMSO	2-Amino-2-methyl-1-propanol	9.0	Co^{2+} (see notes)	Frozen section acetone
E: Mayahara *et al.*[45]	pNPP Mg salt	DMSO, bromo-tetramisole	Glycine-KOH	8.8	Pb^{2+} citrate in KOH	Formalin and glutaraldehyde
F: Mazurkiewicz *et al.*[48]	Oxidized ouabain coupled with H11P	Na^+, Mg^{2+}, ATP	Tris-HCl	7.5	Peroxidase method for H11P (see notes)	Paraformaldehyde

[a] Notes and comments. Asterisks (*) denote personal observations.

Method A: Although some Mg-ATPase might be demonstrated, Pb^{2+} spontaneously hydrolyzes ATP and inhibits Na^+,K^+-ATPase. The method is quite controversial.[49,50]

Method B: ATP is hydrolyzed; Cs^+ is substituted for K^+ and binds to the unreleased P_i in the presence of ouabain. The method is unconfirmed on other tissues; it is not satisfactory for salt gland.* T1 also replaced K,[47a] but produces insufficient density with this method.*

Method C: This was the first sound and widely used method. High pH is used for precipitation of SrP_i, which is converted to PbP_i by second-step incubation in $Pb(NO_3)_2$. Conversion of Sr^+ to Pb^+ produces some diffusion and large deposits; not necessary for SrP_i; though soluble in OsO_4, not soluble in alcoholic OsO_4.* Alkaline phosphatase also hydrolyzes pNPP,[51] and inhibitor must be used.[45]

Method D: High pH is used because of changes in pH optimum caused by DMSO. It is a poor method with great diffusion of final product because of three steps: P_i liberated is actually precipitated as MgP_i, which is converted to CoP_i in different solutions.* There is a final conversion of CoP_i to CoS.

Method E: In our hands, this has been the most effective, simplest, and most workable method. Deposits of final product are relatively small and not diffuse. Pb citrate does not significantly hydrolyze pNPP or inhibit enzyme.*

Method F: Peroxidase activity of H11P is detected with 3,3′-diaminobenzidine, imidazole, and H_2O_2; long incubation is required because of weak activity with H11P. This is first visual method to use inhibitor instead of substrate. A dense product is formed by reaction with OsO_4 during fixation for electron microscopy.

What is important is that the localization by appropriate methods reveal at a fine structural level that Na^+,K^+-ATPase activity is completely restricted to the basolateral membrane of the secretory cell of the salt gland (Fig. 4). Deposits of final product have a spotty distribution on the cytoplasmic side of the membrane, only a few sites in the membrane showing areas of confluent deposits over a small area. If the α and β subunits of Na^+,K^+-ATPase were assembled at a central area such as the Golgi (see Morré[3] for review) and transported to the membrane by Golgi vesicles, one might expect to demonstrate activity if the local molecular (especially phospholipid) environment were correct. Unfortunately, this was never the case cytochemically, and the events must be more complex.

Cytochemistry Involving Ouabain

Radiolabeled ouabain can be utilized to localize the Na^+,K^+-ATPase by autoradiography in stressed and unstressed salt gland.[26,27] This compound binds specifically to the enzyme and thus quantifies the number of enzyme molecules. Conditions for optimal binding of [^{3}H]ouabain to avian salt gland homogenates (fresh and fixed) used an assay system modified from Hoffman.[55]

After tissue was obtained in the manner described, the salt glands, both fresh and fixed (formalin), were homogenized in a glass homogenizer (20 mg wet weight per milliliter of Tris–sucrose buffer) and assayed for [^{3}H]ouabain binding. Homogenates (4–10 mg wet weight of tissue) were shaken for 15 min at 37° in incubation medium (0.4–2.0 ml) containing 3 mM ATP, 6 mM $MgCl_2$, 100 mM NaCl, 50 mM Tris-HCl (pH 7.5) at 37° and [^{3}H]ouabain (New England Nuclear, 0.1 Ci/mmol). The reaction was started by addition of homogenate and stopped by an ice bath. Unbound [^{3}H]ouabain was removed by four sequential centrifugations (5 min each, 3500 g_{av}), and resuspended in incubation medium above without ATP and ouabain. Washed pellets were resuspended in toluene–Triton X-100 scintillation fluid with a motor-driven pestle and counted. All of the unbound [^{3}H]ouabain was removed during the first two washes.[26]

Although paraformaldehyde inhibited enzyme activity approximately 30%, it increased ouabain binding over fresh tissue by 43%, and conditions of binding required ATP and Mg^{2+} but not Na^+. Reduced binding (76%) occurred in the presence of 10 mM K^+. In the absence of ATP, Mg^{2+} and the presence of K^+ and EDTA, 5% was bound, and that was considered to be nonspecific. Optimal binding occurred in 15 min at 37° at a concentration of 10^{-6} M, which inhibited the enzyme activity 100% (K_i

[55] J. F. Hoffman, *J. Gen. Physiol.* **54,** 343 (1969).

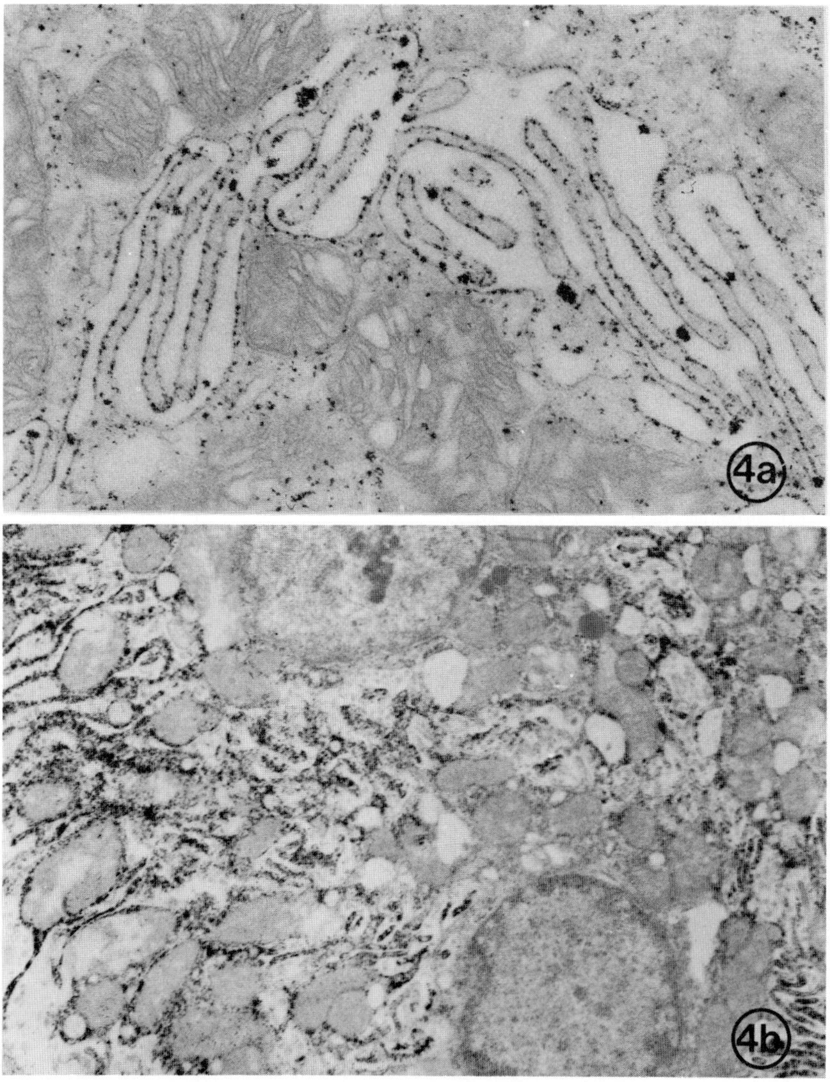

FIG. 4. (a) Ouabain-sensitive pNPPase according to the method of Mayahara *et al.*[45] showing $Pb_3(PO_4)_2$ deposits restricted to the basal and lateral surfaces of the principal cells of a 3-day-adapted duckling. ×9000. (b) Higher magnification of the lateral area of the principal cell of similarly adapted duckling cytochemically reacted for pNPPase with a chelated Pb capture reagent that we have modified for electron microscopy from the method of Chayen *et al.*[53] With the exception of a few extraneous deposits, the reaction product is clearly on the inner aspect of the plasma membrane. ×36,500.

is 1.7×10^7).[26] It is interesting that with extended time of incubation K^+ increases the time for binding, but not the total amount bound.[27]

From these results the incubation conditions and controls for autoradiography were determined. Tissue could be prepared in two ways. First, frozen sections mounted on slides were fixed in formalin, washed, and flooded with [^{3}H]ouabain (10^6; 0.1 Ci/mmol), minced in cold buffer, and dipped into Ilford L 4 photographic emulsion; radiograms were developed 3 days to 3 weeks later.[26] Alternatively,[27] using a modification introduced by Stirling,[56] thick vibratome sections were cut of fresh tissue, incubated with [^{3}H]ouabain; sections were freeze dried, fixed with OsO_4 vapor, and embedded in plastic resin. Sections 1 μm thick were mounted on slides and coated with Kodak NTB-2 emulsion for radiograms. The latter method[27] provides better resolution with light microscopy. No electron microscopy studies have been performed, and, even if they were successful, there is an inherent limit of resolution making such a method less effective for localization than the cytochemical ones mentioned, but more effective quantitatively.

Ouabain binding sites were restricted to the basolateral membranes of the secretory cells with no luminal localization. Partially and fully specialized secretory cells were labeled, whereas the mitotically active peripheral cells were not. There was a gradient of increasing gain density from the blind end of the tubule to the ductal end, with the highly specialized cells at the ductal end being the most heavily labeled. Comparison of ouabain binding and biochemical determination of Na^+,K^+-ATPase activity during salt stress and destress parallel each other. However, the former method was considered more advantageous, especially since the binding sites quantitatively measure the number of molecules, whereas enzymic activity, which is more sensitive to fixatives, does not.[26] These findings lead to the impression that ouabain binding is simple, rapid, and reproducible, and that a potential cytochemical method utilizing it was worthwhile since it was based on specific inhibitor rather than a substrate reaction.

The cytochemical method[48] was based on the oxidation of ouabain (100 mg containing 5 μCi of [^{3}H]ouabain) in distilled H_2O containing 100 mg of $NaIO_4$ for 18 hr at 24° in a light-tight container.The resultant mixture was chromatographed on Dowex AG-1 anion exchanger to separate the oxidized form from the unoxidized form. The eluates were pooled, concentrated on a rotary evaporator, and stored at 4° until used. Heme undecapeptide (H11P) (which has peroxidase activity) was prepared from horse heart cytochrome *c* (500 mg) by digestion with pepsin (15 mg) in 40

[56] D. E. Stirling, *J. Cell Biol.* **53,** 704 (1972).

ml of 0.1 *N* HCl for 24 hr at 24°. Reaction was stopped by titration to pH 8.5 with NaOH; after addition of $(NH_4)_2SO_4$ a precipitate was produced that is soluble in NH_4HCO_3. H11P was isolated by gel filtration on a Sepharose G-25 column and lyophilized. Homogeneity was monitored by ascending paper chromatography on Whatman No. 1 paper and the sample proved to have an amino acid composition of 2 Val, 3 Glu, 1 His, 1 Lys, 1 Ala, 2 Cys, and 1 Thr. H11P (5 mg) was conjugated to oxidized ouabain (5.8 mg) by a Schiff reaction in bicarbonate buffer, pH 9.6, (mole ratio of reactants, ouabain : H11P, 4 : 1). The reactants were mixed at 24° for 18 hr, followed by dropwise addition of 0.1 *M* $NaBH_4$ and allowed to react for 30 min at 24°. Unreacted borohydride was decomposed by addition of HCl followed by titration to pH 8.6 with NaOH. The ouabain–H11P conjugate was isolated by gel filtration on Sephadex G-25, equilibrated and eluted with NH_4HCO_3, pH 8.6.

The resultant ouabain-H11P inhibited Na^+,K^+-ATPase activity 87% at a concentration of 10^{-5} *M* under conditions described above for ouabain. Its peroxidase activity was localized by incubation of formalin-fixed sections with diaminobenzidine as hydrogen donor and H_2O_2 as the substrate in Tris-HCl buffer, pH 9, containing imidazole. The final product, which was colored, reacted with OsO_4 during refixation and preparation of tissues for electron microscopy. Controls with no reaction occurred if the tissues were first incubated with unlabeled ouabain, or if the reagents were allowed to react in a medium lacking ATP and Mg^{2+} and containing K^+ and EDTA. It is interesting that the localization of the oxidized ouabain sites were fine-textured deposits distributed as focal patches on the outer aspects of only the basolateral membrane.[48] Currently, we have prepared the biotin derivative of oxidized ouabain,[57] which may prove to be a better reagent for localization when allowed to react with avidin-ferritin. The biotin derivative also holds promise for the isolation of the basolateral membrane with a Sepharose–avidin column or a column containing the antibody to biotin.

Immunocytochemical Localization of Na^+,K^+-ATPase in Salt Gland Tissue

Although the cytochemical methods described above can be used to localize Na^+,K^+-ATPase in salt gland, the detection depends upon biological activity (turnover of substrate or binding of specific inhibitor). Such methods have limitations if one is interested in determining the intracellular pathway(s) for synthesis and assembly of the multisubunit Na^+,K^+-

[57] R. J. Barrnett and J. Russo, unpublished results (1983).

ATPase, because nonconforming or disassembled forms of the enzyme will not be detected. Immunocytochemical techniques, on the other hand, may be applied to such studies because antisera raised to purified Na^+,K^+-ATPase will contain populations of different antibodies that could bind to many domains in the molecule, irrespective of enzymic activity. Although an inactive or nonfunctioning enzyme cannot be differentiated from an active-functioning enzyme by this technique, a different question can be addressed, and the disassembled subunits may be capable of detection. The technique has the capability of high resolution when used with electron microscopy.

Antisera to Na^+,K^+-ATPase from various sources have been raised in several species of animals.[29,30,58,59] Recently, monoclonal antibodies have been made to the α-subunit of lamb kidney Na^+,K^+-ATPase.[59a] We have raised rabbit antibody to duck salt gland Na^+,K^+-ATPase and have applied it immunocytochemically to salt gland tissue.[60] Na^+,K^+-ATPase was purified from adult, salt-stressed ducks by the method of Jorgensen[61] as modified by Hopkins *et al.*;[62] 600 μg of purified Na^+,K^+-ATPase holoenzyme in Freund's complete adjuvant was injected into the front and rear toepads of adult rabbits, followed by bimonthly intramuscular boosters of 170 μg in Freund's incomplete adjuvant. The presence of precipitating antibody was detected after 16 weeks by Ouchterlony double diffusion in 1% agarose containing PBS and 1% Triton X-100. Two bands of identity were present in Ouchterlony test plates between purified Na^+,K^+-ATPase and Triton X-100 solubilized total microsomal membrane preparations from salt glands. By crossed-rocket immunoelectrophoresis,[63] two peaks were present with R_fs identical to those of the catalytic and glycoprotein subunits in the first-dimension SDS–polyacrylamide gel electrophoresis: a major peak at the catalytic subunit was approximately 10 times taller than the peak at the glycoprotein subunit.

When this antiserum was used in an immunofluorescent method on cryostat sections of 3-day-stressed glands of ducklings, the greatest fluorescence was associated with the epithelium of the secretory tubules. A gradient of increasing fluorescence was seen progressing from virtually unlabeled peripheral cells to highly fluorescent principal cells. On the basis of these findings we examined the localization by electron micros-

[58] H. Koepsell, *J. Membr. Biol.* **44,** 85 (1978).
[59] G. D. Schellenberg, I. V. Pech, and W. L. Stahl, *Biochim. Biophys. Acta* **649,** 691 (1981).
[59a] W. J. Ball, A. Schwartz, and L. Lessard, *Biochim. Biophys. Acta* **719,** 413 (1982).
[60] J. E. Mazurkiewicz and R. J. Barrnett, *J. Cell Biol.* **79,** 216a (1978).
[61] P. L. Jorgensen, *Biochim. Biophys. Acta* **356,** 36 (1974).
[62] B. E. Hopkins, H. Wagner, Jr., and T. W. Smith, *J. Biol. Chem.* **251,** 4365 (1976).
[63] N. H. Chua and F. Blomberg, *J. Biol. Chem.* **254,** 215 (1979).

copy using a peroxidase-labeled antibody technique; only the basal and lateral surface membranes of principal cells were labeled, not apical.[60]

Thus antibodies provide a molecular probe capable of detecting subtle differences in content and distribution of enzyme on the surfaces of cells and, when applied at the electron microscopic level,[64] should be capable of demarcating the intracellular sites along the synthetic pathway. A time course cannot be deduced by these methods—only identification of the compartments that contain the enzyme. Using antibodies directed specifically to each subunit, shared and discrete synthetic compartments may be demonstrated. Combining the morphological methods with the biochemical methods (such as described for the Golgi-enriched fractions), sites and time frames for the steps in the synthetic pathways could be determined. In this regard, the efficient labeling of salt gland tissue slices coupled with the resolution of the major subunit of Na^+,K^+-ATPase with specific anti-alpha antibodies, now currently in progress,[64a] may well represent a precise tool by which the biosynthesis of this important enzyme may be studied.

Cytochemical Localization of Enzymes Involved in Phospholipid Biosynthesis

General Remarks

The lipid constituents of membranes have been studied less in comparison to proteins during membrane biogenesis although they account for approximately 40% of membrane composition. Although more attention has been given to the lipids of organelles, especially the endoplasmic reticulum, much less is known of the plasma membrane except for erythrocytes and prokaryotic cells. Some of the biogenesis has been reviewed,[3] but there is disagreement as to the subcellular sites of distribution of phospholipid synthesizing enzymes and the phospholipid content of the membranes of different compartments. The latter is due to species differences, or rapid phospholipid turnover the exchange of them between membranes[65] and lipid asymmetry.[66]

A series of studies by Karlsson *et al.*[67–70] showed that in the duckling, during adaptation when Na^+,K^+-ATPase increases approximately three-

[64] P. K. Nakane and B. G. Pierce, Jr., *J. Histochem. Cytochem.* **14,** 929 (1966).

[64a] J. Merchant and R. J. Barrnett, unpublished results (1983).

[65] J. C. Kader, *in* "Dynamic Aspects of Cell Surface Specialization" (C. Poste and G. L. Nicholson, eds.), p. 217. Elsevier/North-Holland Biomedical Press, Amsterdam, 1977.

[66] J. A. F. Opdenkamp, *Annu. Rev. Biochem.* **48,** 47 (1979).

[67] K. A. Karlsson, B. E. Samuelsson, and G. O. Steen, *Biochim. Biophys. Acta* **176,** 429 (1969).

fold, phospholipids generally increased 1.3- to 1.6-fold in comparison to control. The one exception was an increase in sulfatides that parallels that of Na^+,K^+-ATPase, and the two were originally considered to be related, but this was later denied. More recently,[71] it was shown that the increase in phospholipid on adaptation was 1.7-fold, the largest quantity being in phosphatidylcholine, phosphatidylethanolamine, and sphingomyelin. The largest present increase on adaptation occurred in phosphatidylserine.

Na^+,K^+-ATPase is known to require phospholipid for catalytic activity, but in the isolated and purified enzyme or reconstituted enzyme there is serious controversy as to the exact phospholipids and the amounts.[72] For example, many workers indicate the significance of phosphatidylserine, but evidence was presented against the notion by conversion of it enzymically to phosphatidylethanolamine[73] or complete substitution by phosphatidylcholine[74] without loss of enzyme activity.

Since many studies regarding membrane biogenesis contained little noncontroversial information on the phospholipids, it might be due to the biochemical approach. Thus we reasoned that if, during large amplification of the basolateral membrane of the secretory cells of the duckling's gland during adaptation, we could investigate the activity of the enzymes concerned with acyl lipid biosynthesis (see Fig. 5), find peaks of activity, show increased incorporation of label, the test conditions of cytochemical reagents on the enzyme activity, we could develop cytochemical methods to demonstrate the sites of the enzymes (see asterisks in Fig. 5).

Acyltransferases

The cytochemistry of acyltransferases with biochemical justification of each step in the procedure was approached first by developing fine-structural methods for carnitine transferase in heart muscle,[75] and the demonstration in absorptive cells of the small intestine of the difference in localization of the α-glycerophosphate (phospholipid) and the monoglyc-

[68] K. A. Karlsson, B. E. Samuelsson, and G. O. Steen, *J. Membr. Biol.* **5,** 169 (1971).

[69] K. A. Karlsson, B. E. Samuelsson, and G. O. Steen, *Eur. J. Biochem.* **46,** 243 (1974).

[70] K. A. Karlsson, B. E. Samuelsson, and G. O. Steen, *Biochim. Biophys. Acta* **356,** 376 (1974).

[71] D. L. Stewart, E. W. Semple, G. T. Swart, and H. K. Sen, *Biochim. Biophys. Acta* **419,** 150 (1976).

[72] B. Raelofsen and L. L. M. Van Deenen, *Eur. J. Biochem.* **40,** 245 (1973).

[73] J. J. H. H. Depont, A. Van Prooijen-Van Eeden, and S. L. Bonting, *Biochim. Biophys. Acta* **323,** 487 (1973).

[74] S. Hilden and L. Hokin, *Biochem. Biophys. Res. Commun.* **69,** 521 (1976).

[75] J. A. Higgins and R. J. Barrnett, *J. Cell Sci.* **6,** 29 (1970).

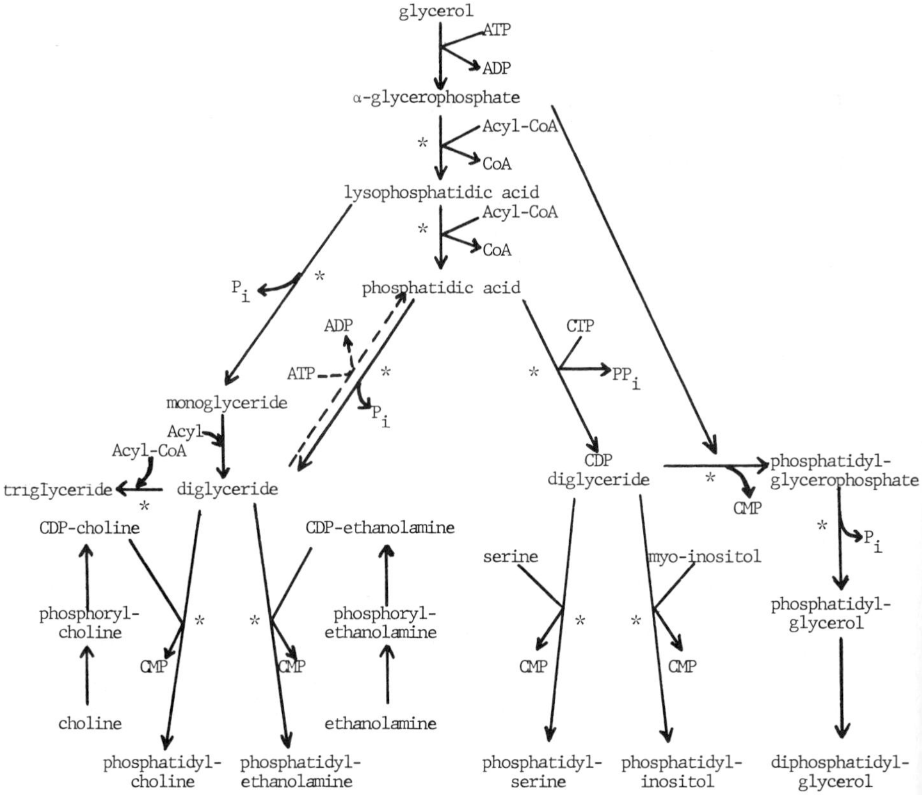

FIG. 5. Part of the general scheme of acyl lipid biosynthesis as deduced largely from animal cells and bacteria [H. Van Den Bosch, *Annu. Rev. Biochem.* **43,** 243 (1974)]. Base exchanges and secondary redistributions of acyl groups are not indicated, nor are the phosphodihydroxyacetone pathway from glycolysis and galactosyl diglycerides from diglyceride. Enzymes at asterisks (*) can be potentially demonstrated cytochemically.

eride (triglyceride) pathways.[76] Essentially, the methods depend on testing the effect of various concentrations of various fixatives, including glutaraldehyde,[77] on the enzymic activity in homogenates and slices of tissue in the hope of finding a fixative that will retain enzyme activity while retaining fine-structural integrity. This was followed by assaying the contents of the cytochemical medium, including capture reagents that react to create a dense final product from one of the reaction products of the enzymic reaction. Essentially, the reaction scheme of the methods was the following:

[76] J. A. Higgins and R. J. Barrnett, *J. Cell Biol.* **50,** 102 (1971).

[77] D. Sabatini, K. Bensch, and R. J. Barrnett, *J. Cell Biol.* **17,** 19 (1963).

$$\alpha\text{-Glycerophosphate} + \text{palmityl-CoA} \rightarrow \text{lysophosphate} + \text{CoASH} \quad (1)$$

$$\text{CoASH} + \text{ferricyanide} \rightarrow \text{CoA—S—S—CoA} + \text{metal ferrocyanide} \quad (2)$$

In reaction (2) metal ferrocyanide precipitated at the site of release of CoA, which was oxidized by ferricyanide. A variety of metal salts, depending on the acyltransferase, could be used, but in most cases the enzyme activity was least affected by Mn^{2+} and unaffected by ferricyanide. In order to eliminate endogenous reactions, the fixed and washed tissue blocks or sections were preincubated in ferricyanide before regular incubation. Controls consisted of elimination of α-glycerophosphate, palmityl-CoA, or Mn^{2+} from the medium or the use of enzyme inhibition during regular incubation.

This system proved to be very valuable in studying the biogenesis of smooth endoplasmic reticulum (SER) of hepatic cells on administration of phenobarbital,[78] which causes a proliferation of SER. In the normal liver, there was distinctly more acyltransferase activity in the rough microsomes (RER), both in fractions and intact cells, biochemically and cytochemically. Phenobarbital administration caused the enzyme activity to peak in the SER within 12 hr, and thereafter activity remained high in this fraction, both cytochemically and biochemically. At the same time the activity in RER fell to a low level, at which it remained. This initial response was unaffected by either puromycin or actinomycin D, suggesting either activation of the enzyme in the SER or translocation from RER so that, during biogenesis of the SER, phospolipid synthesis and assembly occurred *in situ,* and initially the phospholipid-to-protein ratio of the SER changed. In another study with similar methods, choline *O*-acetyltransferase was localized entirely to the membrane-limited synaptic vesicles in the myoneural junction of muscles in the newt.[79]

Salt Gland Studies

With the assurances of these results, the acyltransferases of the salt gland were investigated.[80] Using very young ducklings on salt stress, acyltransferase activity increased 6-fold, reaching a peak on day 6 of adaptation. The initial increase in the first 24 hr was 2.5-fold, and the first significant increase that could be detected was in 4 hr. [This was in contrast to Na^+,K^+-ATPase and Mg-ATPase activities (see preceding section), which increased at later times.] This early response was inhibited by puromycin and actinomycin D, indicating *de novo* synthesis. After day 6, activity fell to a level at 10 days of stress that was only 3-fold that of

[78] J. A. Higgins and R. J. Barrnett, *J. Cell Biol.* **55,** 282 (1972).
[79] M. Fiegenson and R. J. Barrnett, *Brain Res.* **119,** 855 (1972).
[80] A. M. Levine, J. A. Higgins, and R. J. Barrnett, *J. Cell Sci.* **11,** 855 (1972).

controls. With the conditions and ingredients of the cytochemical incubation, over 80% of labeled α-glycerophosphate appeared in the phospholipid fraction, and 70% of that appeared in phosphatidic acid, presumably because of a lack of endogenous substrates in fixed tissue for the complete synthesis of phospholipids. Under the same conditions, 35% of the acyltransferase activity was retained in the fixed tissue. The cytochemical methods used fixed (1% formaldehyde and 1.25% glutaraldehyde in cacodylate buffer, pH 7.4) and unfixed tissues of control and adapted (1–10 days of salt stress) ducklings. Slices of chopped tissues were preincubated for 0.5 hr in potassium ferricyanide (15 mg/ml), rinsed in buffer, and transferred to the incubating medium containing potassium ferricyanide (1.5 mg/ml), $MnCl_2$ (5 mg/ml), palmityl-CoA (2 mg/ml), and α-glycerophosphate (5 mg/ml) in cacodylate buffer (0.05 *M*) containing 4.5% sucrose at pH 7.4 for 1–2 hr at 37°. After incubation, tissues were rinsed in buffer, refixed in 1% OsO_4, dehydrated in acetone, and embedded in Epon. Sections cut on an LKB Ultrotome were viewed under the electron microscope.

Surprisingly, the cytochemical localization of acyltransferase activity was restricted to some, but not all, of the cisternal elements of the Golgi apparatus (Fig. 6). Clearly, many Golgi areas containing numerous elements were found in comparison to controls, and these were localized to the apex of the cells, often close to the lateral border. In addition, it was evident that Golgi cisternae pinched off their lateral ends, creating unreactive vesicles, some of which fused by formation of omega figures to amplify the plasma membrane. On this basis we suggested that the Golgi was the site of synthesis of phospholipids and potentially the site of assembly of at least some of the enzymic proteins (the β fragment of Na^+,K^+-ATPase is heavily glycosylated,[30,81] presumably in the Golgi). Thus, after the synthesis of phospholipid and presumably the assembly of proteins, the events seem to mimic the mechanism of secretion, except that the Golgi vesicles do not export the products, but deliver them to the basolateral plasma membrane.

We are fortified in this position, since examination of the process of myelinization of the trigeminal nerve in neonatal rats indicated that acyltransferase activity also occurred in the Golgi.[82] However, in this circumstance, vesicles that pinched off cisternae contained enzymic activity restricted to the interior of the limiting membrane so that, on fusion with the plasma membrane, enzyme activity was localized to the exterior of

[81] J. A. Perrone, J. F. Hackney, J. F. Dixon, and L. E. Hokin, *J. Biol. Chem.* **250,** 4178 (1975).

[82] F. M. Benes, J. A. Higgins, and R. J. Barrnett, *J. Cell Biol.* **57,** 613 (1973).

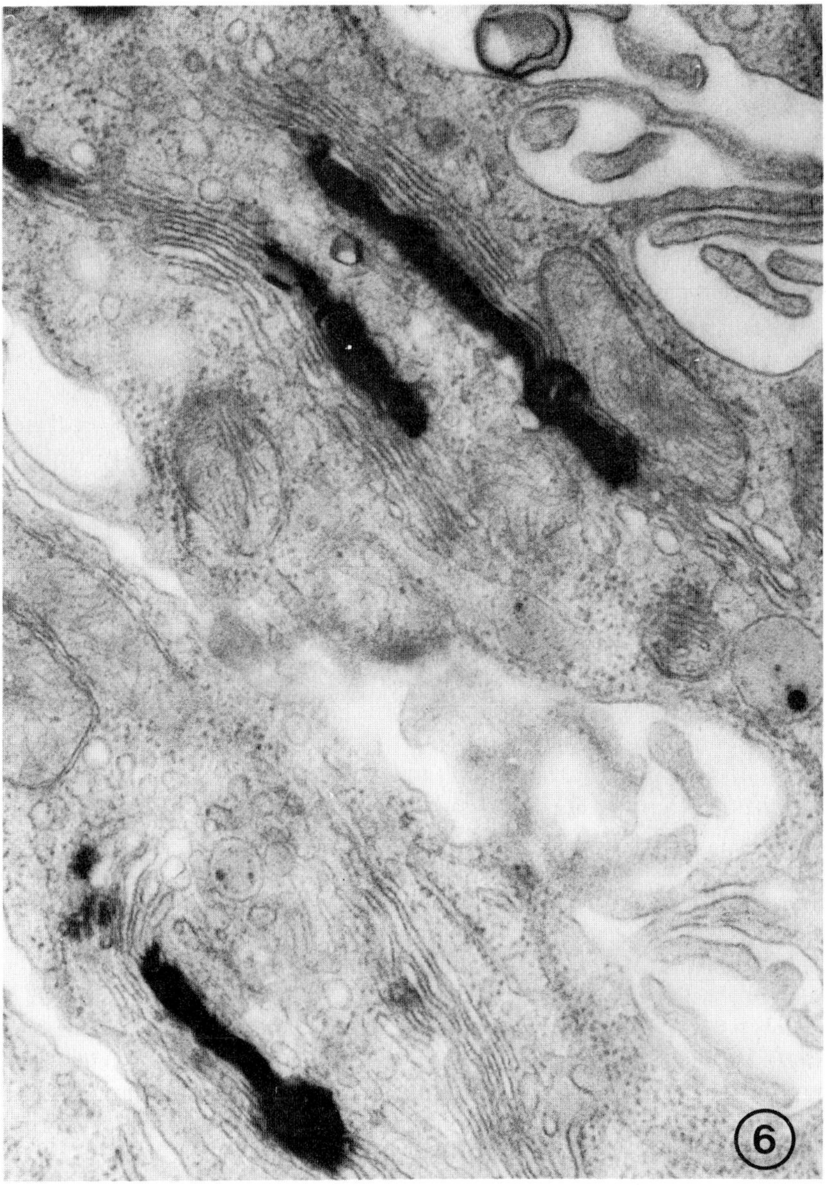

FIG. 6. Acyltransferase activity demonstrated in partially adapted duckling. Activity is completely confined to some cisternae of the Golgi apparatus. Small vesicles that pinch off the end of the cisternae, close to the plasma membrane, consistently show no activity. ×29,000.

that membrane. This activity was lost as the Schwann cell process wound around the axon. On the other hand, in the terminal stages of blastulae of *Drosophila* embryos, which are syncitia and in each of which 3500 cells are separated within a period of approximately 1 hr by the formation of plasma membrane, activity was restricted to elements of the ER, not of the Golgi.[83] Similarly, Golgi localization was not obtained in studies of the cell cycle and thylakoid membrane biogenesis in *Chlamydomonas*.[83a,83b]

Further Studies

In order to fortify or change the implication of the above studies of the adapting salt gland, isolation of the Golgi, along with biochemical studies, is required (see earlier section). In addition, there is the problem of the site of synthesis of acyltransferase on free or bound ribosomes and their route to the Golgi. Cytochemically, it would also be helpful to develop methods for the detection of phosphatidic acid phosphatase (PAPase) and of the enzyme responsible for the synthesis of final phospholipids, e.g., phosphatidylcholine (PC). These are currently in progress,[84] and initial results have been obtained.

Concerning PAPase, the method simply depends on the capture of P_i by a metal salt capture reagent, since the enzyme shows little sensitivity to glutaraldehyde.[84] Some difficulties are the solubility of the substrate, especially in the presence of capture reagent, which appears to be overcome by the simultaneous use of DMSO, chelating agents, nonionic detergents, and the use of pH optimum for the capture of product rather than enzyme optimum. Luckily the latter changes to alkaline levels in the presence of DMSO. For PC, the activity of phosphorylcholine transferase (cholinephosphate cytidylyltransferase) is doubled at 2 and 6 days of stress.[84] The cytochemical method in development is based on our method for cyclic 3′,5′-nucleotide phosphodiesterase.[85] Essentially, phosphorylcholine transferase transfers choline to diglyceride from CDP-choline, and the by-product of the reaction is CMP, which can be localized in the presence of snake venom 5′-nucleotidase as the P_i liberated is captured by metal salt. Sufficient enzyme activity (65%) is retained after fixation as well as incubation in the presence of nucleotidase and metal salts. Successful completion of this method would inspire similar

[83] C. A. Heckman, S. J. Friedman, P. J. Skehan, and R. J. Barrnett, *Dev. Biol.* **59,** 9 (1977).

[83a] D. R. Janero and R. J. Barrnett, *J. Cell Biol.* **91,** 126 (1981).

[83b] C. C. Jelsema, A. S. Michaels, C. C. Jelsema, and R. J. Barrnett, *J. Ultrastruct. Res.* **82,** 35 (1983).

[84] R. J. Barrnett, unpublished results (1983).

[85] N. T. Florendo, R. J. Barrnett, and P. Greengard, *Science* **173,** 745 (1971).

methods, changing only the substrates for phosphatidylethanolamine, phosphatidylserine, and phosphatidylinositol. There appears to be a cycle involving the latter enzyme, which is readily stimulated by acetylcholine in the salt gland.[86] Creation of these cytochemical methods holds additional promise for a host of other problems concerned with phospholipid synthesis in addition to membrane biogenesis.

An additional approach involving lipids includes the use of metal-containing analogs of biological compounds that could be used biochemically as the natural compound and could be localized by energy analysis of the metal probe. In this regard, we have synthesized an analog of palmitic acid in which dimethyl tin replaced carbon-12 of the fatty acid. An analog of phosphatidylcholine was prepared by combining the tin lipid with egg lysophosphatidylcholine and studied by vesicle formation, nuclear magnetic resonance, agarose gel filtration, X-ray diffraction, and differential scanning calorimetry indicating similarity with the natural compound.[87] The tin lipid was then used in a culture medium of *Acholeplasma laidlawii,* and 49% of the label was incorporated into the phospholipid of the plasma membrane and could be detected in the membrane by energy-dispersive X-ray microanalysis in an analytical scanning electron microscope. More recent studies involve the synthesis of Sb-choline[88] for similar studies on the synthesis[89] and localization of acetylcholine. Interestingly, the Sb-choline serves as a substrate of choline kinase, thus potentially relating it to studies of phosphatidylcholine.

Deadaptation

While it is well established that the plasma membranes of eukaryotic cells exist in a dynamic state in which both the lipid and protein components of membranes of organs undergo continued synthesis and turnover,[90] the mechanisms of these heterogeneous phenomena remain obscure. For the salt gland, although no studies have been reported, such a dynamic state can be envisioned for the unstressed control duckling, despite a slight growth and minimal increase in Na^+,K^+-ATPase with time. A different dynamic state of higher amplitude can be envisioned for the duckling subjected to prolonged and continued osmotic stress, in

[86] M. R. Hokin and L. E. Hokin, *J. Gen. Physiol.* **50,** 793 (1966).

[87] S. B. Andrews, J. W. Faller, R. J. Barrnett, and V. Mizuhira, *Biochim. Biophys. Acta* **506,** 1 (1978).

[88] E. M. Meyer, D. L. White, S. B. Andrews, R. J. Barrnett, and J. R. Cooper, *Biochem. Pharmacol.* **30,** 3005 (1981).

[89] E. M. Meyer, R. J. Barrnett, and J. R. Cooper, *J. Neurochem.* **39,** 321 (1982).

[90] D. Doyle and H. Bauman, *Life Sci.* **24,** 592 (1979).

which the response to stress has practically plateaued (the level of Na^+,K^+-ATPase remains at a much higher level than the control). However, during the process of adaptation (1–6 days of salt stress), the process of amplification is so large and continuous that synthesis and assembly simply swamps out turnover, and biochemical and morphological observations support the notion. This does not mean that the response to stress is not a combination of large synthesis and small turnover. The same can be stated for deadaptation; the obvious destruction of basolateral plasma membranes and other cellular components swamp out a concept of construction. In fact, the slope of the loss of plasma membrane Na^+,K^+-ATPase is steeper than the gain in Na^+,K^+-ATPase during adaptation.

Deadaptation, like adaptation of the secretory cells, presents an exaggerated physiological phenomenon, and is simply produced by substituting drinking water for saline in the diet of the fully adapted duckling. This phenomenon has been little studied, and only two morphological studies accompanied by several biochemical observations exist.[91,92] Clearly, from our own work, deadaptation is accompanied by surface degradation at the basolateral borders, shedding from these sites as well as from the apices, as well as apical exocytosis, and endocytic internalization of the basolateral border involving the lysosomal system (including multivescular bodies) and the Golgi[92] (Fig. 7). Cell death was not a prominent feature morphologically, and this is supported by the finding that although the gland lost 37% of its protein during deadaptation, only 5% of the glandular DNA was lost.[93] Furthermore, DNA systems did not increase during or as a result of deadaptation,[94] arguing against incurred cell turnover as an explanation of reduction in cell surface.

The end point of deadaptation was a principal cell that appeared somewhat similar to the end stage of the partially differentiated state in which the basal infoldings were lost and the lateral plicae were shorter, blunter, and much less numerous. These changes occurred simultaneously with a loss of Na^+,K^+-ATPase. Peak of the lysosomal system, accompanied by an increase in acid phosphatase, only occurred early in deadaptation after which the levels fell to below those of the adapted cell.[92] Morphological evidence of degradation and shedding suggest the potential activity of a protease and/or phospholipase that has not yet been studied. In all, this exaggerated state deserves further investigation, since some of the phenomena may simply be one-sided amplification of the lost portion of the

[91] F. E. Hossler, M. P. Sarras, and E. R. Allen, *Cell Tissue Res.* **188,** 299 (1978).
[92] J. A. Addis, P. R. Eager, and R. J. Barrnett, *Am. J. Anat.* (in press).
[93] W. N. Holmes and D. J. Stewart, *J. Exp. Biol.* **48,** 509 (1968).
[94] F. E. Hossler, *Cell Tissue Res.* **226,** 531 (1982).

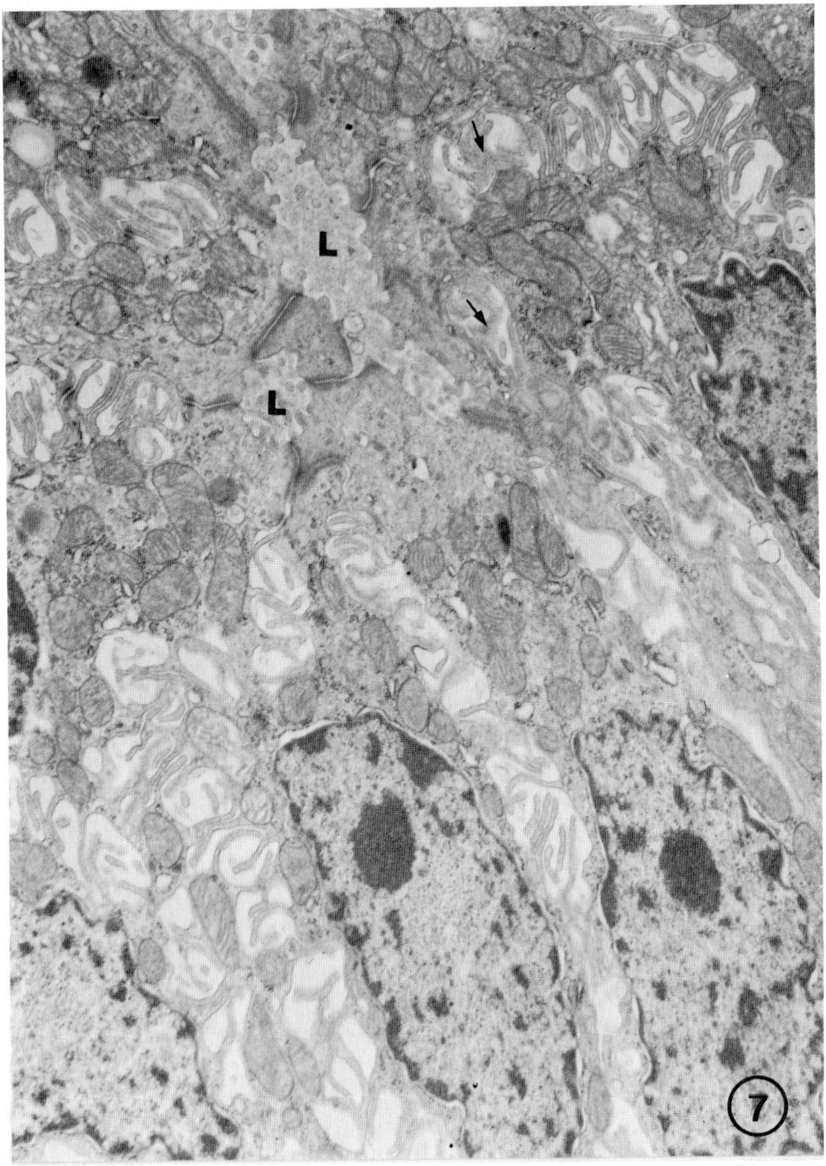

FIG. 7. Principal cells of duckling during period of deadaptation. Lumina (L) show debris, and lateral plicae are irregular, blunted, and fused (arrows) (compare with Fig. 2). ×9000.

dynamic state, which can be studied more easily than the dynamic state itself. No studies on control of deadaptation exist.

Regulation of Membrane Biogenesis into Salt Gland

At present this has been the least investigated area in biogenesis in general, including the salt gland. Given that crossed perfusion from a salt-stressed animal to the head of a control animal does not initiate secretion,[95] that *in vitro* activation of ouabain-sensitive respiration[96] or Na^+,K^+-ATPase activity does not depend on raised osmolality,[97] and that, if the secretory nerve of the gland is severed on one side, response to salt load occurs only on the innervated side,[98,99] the major implication is that innervation (parasympathetic) is required for adaptation. These studies also imply that neither raised plasma tonicity or circulating hormonal factors are sufficient to initiate adaptation including Na^+,K^+-ATPase biosynthesis or plasma membrane biogenesis, but that neural elements are needed for the regulation, though secondary effects of hormones cannot be completely excluded.[100]

Nerve endings in the gland are complicated and occur at the base of the secretory cells and near blood vessels. The former endings show several varieties; ones with only clear vesicles (presumably cholinergic), ones with only large dense-core vesicles (catecholamines), and ones with mixed clear and small dense-core vesicles.[101] Given the above results, and the excitation-secretion coupling of the gland,[102] the mechanism of amplification and secretion may both be correlated. Thus, it is likely that neural control, while probably including acetylcholine, the muscarinic binding receptors of which have been described,[103] may be quite complicated, as is being revealed in salivary glands.[104] Cholinomimetics have been shown to stimulate ouabain-sensitive respiration[105] and pNPPase activity[94] in gland slices in a manner that seems to involve Ca^{2+}-dependent cyclic GMP, and activation of the inositol cycle occurs rapidly.[86,106,107] Based on

[95] A. Hanawell, J. L. Linzell, and M. Peaker, *J. Physiol.* (*London*) **226,** 453 (1972).
[96] R. Borut and K. Schmitt-Nielsen, *Am. J. Physiol.* **204,** 573 (1963).
[97] O. J. Stewart and A. K. Sen, *Am. J. Physiol.* **240,** C207 (1981).
[98] J. B. Pitt and A. D. Hally, *J. Anat.* **114,** 303 (1973).
[99] A. Hanawell and M. Peaker, *J. Physiol.* (*London*) **248,** 193 (1975).
[100] D. Bellamy and J. F. Phillips, *J. Endocrinol.* **36,** 97 (1966).
[101] R. J. Barrnett, personal observation.
[102] G. D. V. van Rossan, *Biochim. Biophys. Acta* **126,** 338 (1966).
[103] S. R. Hootman and S. A. Ernst, *J. Cell Biol.* **91,** 781 (1981).
[104] T. Barka, *J. Histochem. Cytochem.* **28,** 836 (1980).
[105] D. J. Stewart, J. Sax, R. Funk, and A. K. Sen, *Am. J. Physiol.* **237,** C200 (1979).
[106] M. R. Hokin and L. E. Hokin, *J. Biol. Chem.* **239,** 2116 (1964).
[107] M. R. Hokin and L. E. Hokin, *J. Gen. Physiol.* **50,** 793 (1966).

the above, it may be a logical point to investigate the neural regulation of the biogenesis of Na^+,K^+-ATPase and/or phospholipids *in vitro,* related to total protein and RNA synthesis. For this work, it may be wise to follow stringently the α subunit of Na^+,K^+-ATPase. *In vitro* slices or minilobules may well be the best working model, using glands from partially adapted ducklings to show that neural agents may increase the incorporation of label into the unit, over those of partially adapted controls.

Acknowledgment

Much of our work reported here was supported by NIH Grant AM-03688.

Section II

Targeting: Selected Techniques to Study Transfer of Newly Synthesized Proteins into or across Membranes (Eukaryotic Cells)

[52] Control of Intracellular Protein Traffic

By GÜNTER BLOBEL

A eukaryotic cell contains billions of protein molecules. These are steadily being synthesized and degraded. At homeostasis, a given species of protein is represented by a characteristic number of molecules that is kept constant within a narrow range. Very little is known about the cell's accounting procedures, i.e., how it balances and controls biosynthesis and biodegradation.

An important aspect of biosynthesis[1] as well as biodegradation[2,3] is the intracellular topology of proteins. Many protein species spend their entire life in the same compartment in which they are synthesized, others have to be fully translocated across the hydrophobic barrier of *one,* or in some cases, *two* distinct cellular membranes in order to reach the intracellular compartment or extracellular site where they exert their function. Numerous protein species have to be integrated asymmetrically into distinct cellular membranes. Asymmetric integration requires in many instances *partial* translocation, i.e., selective transfer of one or several *distinct hydrophilic* or *charged segments* of the polypeptide chain across the hydrophobic barrier of one or two intracellular membranes. Following complete or partial translocation subpopulations may undergo further "posttranslocational" traffic. The need for posttranslocational traffic arises from the fact (see below) that not all the cellular membranes are competent to translocate proteins. Translocation-incompetent membranes (and the compartments they enclose) therefore receive their constitutive integral membrane proteins from translocation-competent donor membranes. Posttranslocational traffic may proceed in bulk or by receptor-mediated processes, and it may be unidirectional (in which case the protein ends up as a permanent resident of a particular cellular membrane) or may follow a cyclic pattern between distinct cellular membranes (e.g., recycling of receptors). The collective term "topogenesis" has been introduced[1] to encompass protein translocation (partial or complete) across membranes as well as subsequent posttranslocational protein traffic.

Not included in these processes that define topogenesis are distinct traffic patterns that may be required for protein degradation. Theoretical

[1] G. Blobel, *Proc. Natl. Acad. Sci. U.S.A.* **77,** 1496 (1980).

[2] G. Blobel, *in* "Limited Proteolysis in Microorganisms" (G. N. Cohen and H. Holzer, eds.), p. 167. Library of Congress Catalog Card No. 78-600168, U.S. Dept. of Health, Education, and Welfare, Washington, D.C., 1979.

[3] G. Blobel, *Ciba Found. Symp.* [N.S.] **75,** 274 (1980).

ISBN 0-12-181996-5

considerations on the topology of protein degradation have been presented elsewhere.[2,3] In essence, these considerations argue for the existence of three (animal cells) or even four (plant cells) distinct and separate compartments for protein degradation, each containing a distinct set of proteases. The information for the lifespan of a given protein might be largely encoded in its surface structure. For example, the presence of a Glu residue on the surface of a given protein and the presence in this compartment of an endoprotease cleaving after a Glu residue would put this protein at a high risk for proteolytic attack and for a relatively short half-life. Implicit in this proposal is the notion that the distinct present-day half-life of a given species of protein evolved by selection that ultimately affected not only the structure (primarily the surface structure) of the protein, but also the development (or elimination) of specific proteinases in the various compartments.[2,3]

Topogenic Sequences

Detailed proposals have been made also for protein topogenesis.[1] The essence of these proposals is that the information for intracellular protein topogenesis resides in *discrete* "topogenic" sequences that constitute a permanent or transient part of the polypeptide chain. The repertoire of distinct topogenic sequences was predicted to be relatively small because many different proteins would be topologically equivalent, i.e., targeted to the same intracellular address. Four types of topogenic sequences were distinguished.[1]

1. Signal sequences initiate translocation of proteins across specific membranes and are decoded by protein translocators that, by virtue of their signal sequence-specific domain and their location in distinct cellular membranes, effect unidirectional translocation of proteins across specific cellular membranes.
2. Stop-transfer sequences interrupt the translocation process that was previously initiated by a signal sequence and, by excluding a distinct segment of the polypeptide chain from translocation, yield asymmetric integration of proteins into translocation-competent membranes.
3. Sorting sequences act as determinants for posttranslocational traffic of subpopulations of proteins, originating in translocation-competent donor membranes (and compartments) and leading to translocation-incompetent receiver membranes (and compartments).
4. Insertion sequences interact with the lipid bilayer directly and thereby anchor a protein to the hydrophobic core of the lipid bilayer.

Translocation of Proteins across Membranes

Translocation is understood here as transport of an entire polypeptide chain across one (or two) membrane(s), proceeding unidirectionally from the protein biosynthetic compartment. Not considered here will be ectopically synthesized proteins (e.g., toxins, such as the colicins or diphtheria toxins) although their entry into cells may also require complete or partial translocation of polypeptide chains across a membrane, either the plasma membrane directly or an intracellular membrane, following uptake by endocytosis. Most of these ectopically synthesized proteins acquire their ability to interact with the lipid bilayer only outside of the cell in which they are synthesized (e.g., by proteolytic activation).

Hypothetical models for intracellular protein translocation must deal with two essential tenets that appear to underlie the observed phenomenology of this process. First, the permeability barrier of the membrane appears to be reversibly modified for the passage of each translocated polypeptide chain while being maintained for other solutes. Second, the species of protein to be translocated as well as the type of membrane across which a given protein is translocated are highly specific. Both of these tenets can be readily satisfied by postulating that protein translocation is a receptor-mediated process in which specificity is achieved by a "signal" sequence in the protein to be translocated and by a signal-sequence-specific translocation system that is restricted in its location to a distinct cellular membrane.[1]

Biological Membranes Endowed with Protein Translocation Systems

Several signal-sequence-specific translocation systems have been postulated to exist.[1] Table I lists the biological membranes or membrane pairs that have been proposed to be endowed each with *one* signal-sequence-specific translocation system (in multiple copies) that is able to decode the information of *one* type of signal sequence. Two modes of translocation have been distinguished, a cotranslational and a posttranslational mode. In cotranslational translocation[1,4] the passage of the polypeptide chain across the membrane appears to be strictly coupled to translation, whereas in posttranslational translocation[1,5] the polypeptide can traverse the membrane posttranslationally uncoupled from its synthesis.

The conjecture was made,[1] based on possible evolutionary relationships between various cellular membranes (see Fig. 4), that the contem-

[4] C. M. Redman and D. D. Sabatini, *Proc. Natl. Acad. Sci. U.S.A.* **56,** 608 (1966).

[5] B. Dobberstein, G. Blobel, and N.-H. Chua, *Proc. Natl. Acad. Sci. U.S.A.* **74,** 1082 (1977).

TABLE I

CELLULAR MEMBRANES PROPOSED TO BE ENDOWED WITH A TRANSPORT SYSTEM (TRANSLOCATOR) FOR THE UNIDIRECTIONAL TRANSLOCATION OF NASCENT OR NEWLY SYNTHESIZED PROTEINS[a,b]

Mode of translocation	Membrane	Code
Cotranslational	a. Prokaryotic plasma membrane	PPM
	b. Inner mitochondrial membrane	IMM
	c. Thylakoid membrane	TKM
	d. Rough endoplasmic reticulum	RER
Posttranslational (across one membrane)	e. Outer mitochondrial membrane	OMM
	f. Outer chloroplast membrane	OCM
	g. Peroxisomal membrane	PXM
Posttranslational (across two membranes)	h. Mitochondrial envelope	MEN
	i. Chloroplast envelope	CEN

[a] Taken from Blobel.[1]

[b] Each of the translocation-competent membranes listed here (a–i) is proposed to contain only *one* distinct "translocator" (in multiple copies). Each translocator responds to *one* type of signal sequence. Translocation can proceed across a *single* membrane (a–g), or *two* membranes (h, i), cotranslationally (a–d), or posttranslationally (e–i). Suggested abbreviations for these translocation-competent membranes might serve as useful codes. For example, a signal sequence (Si) addressed to the rough endoplasmic reticulum (RER), to the chloroplast envelope (CEN), etc., might be designated Si(RER), Si(CEN), etc. Likewise, a particular signal receptor (SiR) or signal peptidase (SiP) could be classified as SiR(RER), SiR(CEN), or SiP(RER), SiP(CEN), etc.

porary cotranslational translocation systems (Table I, a–d) were derived from a common ancestral system and that they might be highly conserved. A high degree of conservation has indeed been demonstrated for the RER translocation system within the animal and plant kingdoms.[6] Moreover, it has been demonstrated that a signal sequence of a eukaryotic protein addressed to the RER translocation system can be decoded by its putative analog in the prokaryotic plasma membrane.[7,8] The existence of two other cotranslational translocation systems, namely those in the inner mitochondrial membrane and in the thylakoid membrane, has been postulated[1] because of the presence of membrane-bound polysomes in thylakoid membranes[9] and in the inner mitochondrial membrane.[10] These cotranslational translocation systems are most likely involved in

[6] B. Dobberstein and G. Blobel, *Biochem. Biophys. Res. Commun.* **74,** 1675 (1977).

[7] K. S. Talmadge, S. Stahl, and W. Gilbert, *Proc. Natl. Acad. Sci. U.S.A.* **77,** 3369 (1980).

[8] K. Talmadge, J. Kaufman, and W. Gilbert, *Proc. Natl. Acad. Sci. U.S.A.* **77,** 3988 (1980).

[9] N.-H. Chua, G. Blobel, P. Siekevitz, and G. E. Palade, *Proc. Natl. Acad. Sci. U.S.A.* **70,** 1554 (1973).

[10] R. Kuriyama and D. Luck, *J. Cell Biol.* **59,** 776 (1973).

partial translocation, i.e., translocation only of a distinct segment of the nascent chain (not of the entire polypeptide) and therefore function in the integration of membrane proteins (see below).

Posttranslational translocation systems have been postulated[1] for translocation of cytoplasmically synthesized proteins across a *single* membrane (peroxisomal, outer mitochondrial, outer chloroplast membrane) or across *two* membranes (outer and inner membranes of mitochondria and chloroplasts).

Evidence for the existence of a posttranslational translocation system in the peroxisomal membrane rests on the demonstration that liver catalase and uricase (two enzymes located in the peroxisome) are synthesized by free ribosomes, not by membrane-bound ribosomes.[11] Conclusive evidence for the existence of posttranslational translocation systems in the outer mitochondrial membrane[12] and across both outer and inner membranes of chloroplasts[5,13,14] and mitochondria[15] was first derived from data of *in vitro* translation and translocation experiments that were subsequently confirmed by numerous laboratories. The existence of a posttranslational translocation system in the outer chloroplast membrane analogous to that in the outer mitochondrial membrane has not yet been demonstrated.

Signal Sequences

The existence of a "signal sequence" for translocation across the RER was first postulated on theoretical grounds.[16] Subsequently, cell-free synthesis of secretory proteins showed them to be synthesized as larger precursors.[17–20] *In vitro* translocation experiments provided pivotal evidence that the sequence extension present in these precursors functions as a "signal sequence" in translocation.[21–23] Thereafter, signal sequences

[11] B. M. Goldman and G. Blobel, *Proc. Natl. Acad. Sci. U.S.A.* **75,** 5066 (1978).
[12] M. L. Maccecchini, Y. Rudin, and G. Schatz, *J. Biol. Chem.* **254,** 7468 (1979).
[13] P. E. Highfield and R. J. Ellis, *Nature (London)* **271,** 420 (1978).
[14] N.-H. Chua and G. W. Schmidt, *Proc. Natl. Acad. Sci. U.S.A.* **75,** 6110 (1978).
[15] M. L. Maccecchini, Y. Rudin, G. Blobel, and G. Schatz, *Proc. Natl. Acad. Sci. U.S.A.* **76,** 343 (1979).
[16] G. Blobel and D. Sabatini, *Biomembranes* **2,** 193 (1971).
[17] C. Milstein, G. G. Brownlee, T. M. Harrison, and M. B. Mathews, *Nature (London), New Biol.* **239,** 117 (1972).
[18] D. Swan, H. Aviv, and P. Leder, *Proc. Natl. Acad. Sci. U.S.A.* **69,** 1967 (1972).
[19] I. Schechter, D. J. McKean, R. Guyer, and W. Terry, *Science* **188,** 160 (1974).
[20] A. Devillers-Thiery, T. Kindt, G. Scheele, and G. Blobel, *Proc. Natl. Acad. Sci. U.S.A.* **72,** 5016 (1975).
[21] G. Blobel and B. Dobberstein, *J. Cell Biol.* **67,** 835 (1975).
[22] G. Blobel and B. Dobberstein, *J. Cell Biol.* **67,** 852 (1975).
[23] E. Szczesna and I. Boime, *Proc. Natl. Acad. Sci. U.S.A.* **73,** 1179 (1976).

were discovered, by similar *in vitro* approaches, for translocation across the prokaryotic plasma membrane,[24–26] the chloroplast envelope,[5,13,14] the two mitochondrial membranes,[15] and the outer mitochondrial membrane.[12]

Translocation is not always accompanied by cleavage of the signal sequence, and there are now numerous examples for uncleaved signal sequences. Further, the signal sequence is not always located at the NH_2 terminus,[27,28] and there may be more than one signal sequence in a polypeptide.[1,28]

The complete primary structure is known for the signal sequence addressed to (*a*) the RER (numerous examples, see compilation by Steiner *et al.*[29]); (*b*) the prokaryotic plasma membrane (numerous examples, see compilation by Emr *et al.*[30]); and (*c*) the chloroplast envelope (so far only one example, Schmidt *et al.*[31]).

As expected on evolutionary grounds[1] and as demonstrated experimentally,[7,8] the signal sequence addressed to the RER membrane is similar to that addressed to the prokaryotic plasma membrane. At present it is not obvious, at least not from the primary structure of the numerous examples, what features of the signal sequence constitute a consensus structure for the receptor (see below). Elegant experiments with mutants (see review by Emr *et al.*[30]) and with amino acid analogs[32] have shown that replacement in the signal sequence of hydrophobic residues by charged or hydrophilic residues interferes with translocation.

As expected, the primary structure of the signal sequence addressed to the chloroplast envelope[31] differs dramatically from that addressed to the RER or to the prokaryotic plasma membrane. However, the primary structure of more examples would need to be elucidated before one could recognize features of a consensus structure for the corresponding receptor(s) of the chloroplast envelope translocation system.

It should be emphasized that a signal sequence was postulated to be involved only in the initiation of chain translocation.[1,21] Implicit in this

[24] S. Inouye, S. Wang, J. Sekizawa, S. Halegoua, and M. Inouye, *Proc. Natl. Acad. Sci. U.S.A.* **74,** 1004 (1977).

[25] H. Inouye and J. Beckwith, *Proc. Natl. Acad. Sci. U.S.A.* **74,** 1440 (1977).

[26] C. N. Chang, G. Blobel, and P. Model, *Proc. Natl. Acad. Sci. U.S.A.* **75,** 361 (1978).

[27] V. R. Lingappa, J. R. Lingappa, and G. Blobel, *Nature (London)* **281,** 117 (1979).

[28] H. Garoff, A. M. Frischauf, K. Simons, H. Lehrach, and H. Delius, *Nature (London)* **288,** 236 (1980).

[29] D. F. Steiner, P. S. Quinn, S. J. Chan, J. Marsch, and H. S. Tager, *Ann. N.Y. Acad. Sci.* **343,** 1 (1980).

[30] S. D. Emr, M. N. Hall, and T. J. Silhavy, *J. Cell Biol.* **86,** 701 (1980).

[31] G. W. Schmidt, A. Devillers-Thiery, H. Desruisseaux, G. Blobel, and N.-H. Chua, *J. Cell Biol.* **83,** 615 (1979).

[32] G. Hortin and I. Boime, *Proc. Natl. Acad. Sci. U.S.A.* **77,** 1356 (1980).

postulate was that the rest of the polypeptide chain must be compatible with the translocation machinery (see "stop-transfer" sequences below); for example, a polyleucine or a nonsecretory protein[33] linked to a signal sequence may not be compatible with translocation.

Mechanisms of Translocation

Until recently, the postulated translocation machinery[1,21] remained largely undefined, so much so, that it was deemed unnecessary.[34–37] Only after the development of an *in vitro* translocation system[22] that was able to reproduce translocation across the ER membrane (isolated in the form of closed microsomal vesicles) with apparent fidelity, did it become possible to assay and to characterize the ER's translocation activity *in vitro*. Two approaches were taken to dissect the membrane translocation activity: salt extraction[38,39] and limited proteolysis.[40,41] Both approaches yielded membrane vesicles that were largely translocation inactive; translocation activity, however, could be restored by readdition of the salt extract or the tryptic extract. These findings provided an assay for the purification of the active components of the salt extract[39] and of the proteolytic extract.[42] The purified active component of the proteolytic extract consisted of an apparently single polypeptide chain (M_r 60,000),[42] whereas the purified active component of the salt extract was shown to be an 11 S protein of M_r ~250,000 that consisted of six polypeptide chains that could not be separated from each other by a variety of nondenaturing procedures.[39,39a]

Studies on the role of the 11 S protein in the translocation process revealed that it is involved in the recognition of the signal sequence, and therefore it was termed signal recognition protein (SRP).[43] When SRP is present in the cell-free translation system, in the absence of salt-extracted microsomal membranes, it was found to inhibit selectively only the translation of mRNA for secretory protein (bovine prolactin), but not of mRNA for cytosolic proteins (α and β chain of rabbit globin).[43] Moreover,

[33] F. Moreno, A. V. Fowler, M. Hall, T. J. Silhavy, I. Zabin, and M. Schwartz, *Nature (London)* **286,** 356 (1980).
[34] G. von Heijne, and C. Blomberg, *Eur. J. Biochem.* **97,** 175 (1979).
[35] W. Wickner, *Annu. Rev. Biochem.* **48,** 23 (1979).
[36] J. Garnier, P. Gaye, J. C. Mercier, and B. Robson, *Biochimie* **62,** 231 (1980).
[37] D. M. Engelman and T. A. Steitz, *Cell* **23,** 411 (1981).
[38] G. Warren and B. Dobberstein, *Nature (London)* **273,** 569 (1978).
[39] P. Walter and G. Blobel, *Proc. Natl. Acad. Sci. U.S.A.* **77,** 7112 (1980).
[39a] P. Walter and G. Blobel, this volume, p. 682.
[40] P. Walter, R. C. Jackson, M. M. Marcus, V. R. Lingappa, and G. Blobel, *Proc. Natl. Acad. Sci. U.S.A.* **76,** 1795 (1979).
[41] D. Meyer and B. Dobberstein, *J. Cell Biol.* **87,** 498 (1980).
[42] D. Meyer and B. Dobberstein, *J. Cell Biol.* **87,** 503 (1980).
[43] P. Walter, I. Ibrahimi, and G. Blobel, *J. Cell Biol.* **91,** 545 (1981).

SRP was found to bind with a relatively low affinity (apparent K_d 5 × 10^{-5}) to ribosomes, but was shown to bind with a 6000-fold higher affinity (apparent K_d 8 × 10^{-9}) when ribosomes are engaged in the translation of mRNA for secretory proteins.[43] Most interestingly, this high-affinity binding of SRP caused a site-specific and signal sequence-induced arrest of chain elongation.[44] The elongation-arrested peptide of nascent preprolactin was shown to be ~70 amino residues long.[44] Because the signal sequence of nascent bovine preprolactin comprises 30 residues,[45] and because about 40 residues of the nascent chain are buried (protected from proteases) in the large ribosomal subunit, [46,47] it was concluded[44] that it is the signal sequence of the nascent chain (fully emerged on the outside of the large ribosomal subunit) that causes high-affinity binding of SRP, which in turn modulates translation and causes arrest in chain elongation.

Most strikingly, elongation arrest is released upon binding of the elongation-arrested ribosome to salt-extracted microsomal membranes (K-RM) resulting in chain elongation and translocation into the microsomal vesicle.[44] Binding of the translating ribosome to K-RM occurs only in the presence of SRP.[48] Further, treatment of K-RM with low concentrations of trypsin abolishes SRP-mediated binding of the translating ribosome to K-RM.[48] This latter finding suggests that besides SRP (which could be considered a peripheral membrane protein) integral membrane proteins are required for translocation to proceed. It is the hydrophilic cytoplasmic domain of one of these integral membrane proteins (severed by proteolytic enzymes in such a manner that it retains reconstitutability to its parent molecule[40–42,49]) that has recently been purified.

Taken together, these data provide the strongest support to date for the most pivotal postulate of the signal hypothesis,[1,21] namely, that protein translocation across the ER is a receptor-mediated process. These data thus definitively rule out alternative hypotheses that have postulated that chain translocation across the ER occurs spontaneously, without the mediation by proteins.[35–37,50] They also rule out translocation models that, although relying on the participation of specific proteins, have postulated a primary interaction of the signal sequence (because of its hydrophobic

[44] P. Walter and G. Blobel, *J. Cell Biol.* **91,** 557 (1981).

[45] N. L. Sasavage, J. H. Nilson, S. Horowitz, and F. M. Rothman, *J. Biol. Chem.* **257,** 678 (1982).

[46] L. I. Malkin and A. Rich, *J. Mol. Biol.* **26,** 329 (1967).

[47] G. Blobel and D. D. Sabatini, *J. Cell Biol.* **45,** 130 (1970).

[48] P. Walter and G. Blobel, *J. Cell Biol.* **91,** 551 (1981).

[49] D. I. Meyer, D. Louvard, and B. Dobberstein, *J. Cell Biol.* **92,** 579 (1982).

[50] M. S. Bretscher, *Science* **181,** 622 (1973).

nature) with the lipid bilayer.[34,51] Thus, the initial events that lead to translocation and provide for its specificity are protein–protein (signal sequence plus ribosome-SRP), *not* protein–lipid (signal sequence–lipid bilayer) interactions.

The ability of SRP to arrest chain elongation and the finding that microsomal membranes release this arrest is of teleological interest. If this mechanism also operates *in vivo* it would provide the cell with a means to stop the synthesis of secretory proteins (some of which might be harmful if completed in the cytosol) unless sites on the ER are available so that translocation and segregation into the intracisternal space are ensured. These sites in the microsomal membranes could consist of several integral membrane proteins that might form an ensemble undergoing cyclic disassembly and reassembly for each chain translocation event.[1,21] Signal peptidase and core sugar transferase might, as integral membrane proteins, participate in the formation of this ensemble or might be transiently associated with it. Other components of this ensemble might be the so-called ribophorins,[52,53] although their involvement in protein translocation remains to be demonstrated.

Because of evolutionary considerations (see below), and because of the documented mechanistic similarity of protein translocation across the prokaryotic plasma membrane[7,8,26,30,54–56] to that across the ER, our conjecture is[1,56] that there is only *one, cotranslational* translocation system in the bacterial plasma membrane and, moreover, that this system will be essentially similar if not identical to that in the ER. However, it should be noted that this view has been challenged and that a posttranslational mode of translocation across the bacterial plasma membrane has been postulated.[35,57]

The discovery of SRP has permitted us to add more detail to and to expand the previously proposed translocation models. The postulated ribosome receptor and signal-sequence receptor for the cotranslational translocation system were envisioned to be integral membrane proteins.[1] Because SRP (presumably a peripheral membrane protein) is, at least in part, endowed with these postulated receptor properties and because ad-

[51] J. M. DiRienzo, K. Nakamura, and M. Inouye, *Annu. Rev. Biochem.* **47,** 481 (1978).

[52] G. Kreibich, B. C. Ulrich, and D. D. Sabatini, *J. Cell Biol.* **77,** 464 (1978).

[53] G. Kreibich, C. M. Freienstein, P. N. Pereyra, B. C. Ulrich, and D. D. Sabatini, *J. Cell Biol.* **77,** 488 (1978).

[54] W. P. Smith, P. C. Tai, R. C. Thompson, and B. D. Davis, *Proc. Natl. Acad. Sci. U.S.A.* **74,** 2830 (1977).

[55] L. L. Randall, S. J. S. Hardy, and L. G. Josefson, *Proc. Natl. Acad. Sci. U.S.A.* **75,** 1209 (1978).

[56] C. N. Chang, P. Model, and G. Blobel, *Proc. Natl. Acad. Sci. U.S.A.* **76,** 1251 (1979).

[57] D. Koshland and D. Botstein, *Cell* **20,** 749 (1980).

ditional, integral membrane proteins are required for translocation (translocation activity of trypsinized K-RM cannot be restored by SRP), our present cotranslational translocation model[44] is in detail, not in principle, more complex than was previously envisioned.[1]

The discovery of SRP likewise suggests modifications of our models for posttranslational translocation. The latter has been envisioned to be in principle similar to cotranslational translocation except that the existence of only signal sequence receptors (again as integral membrane proteins), but not of ribosome receptors, was envisioned.[1] If signal-sequence-specific SRP analogs exist also for the various posttranslational translocation systems, and if in turn SRP-specific receptors in various organelle membranes were to control import into organelles, one could envision a cytoplasmic pool of translocation-primed complexes consisting of an SRP analog plus a protein to be imported. Such translocation-primed complexes might explain how those integral membrane proteins of mitochondria and choloroplasts that are synthesized in the cytoplasm are kept in a water-soluble form in the cytoplasm prior to import into these organelles.

Integration into Membranes

Many integral membrane proteins (IMPs) require selective translocation of one or more hydrophilic segment(s) of the polypeptide chain in order to acquire their characteristic asymmetric orientation. How could a selective translocation of discrete segment(s) of the polypeptide chain be accomplished?

In considering theoretical solutions to this problem, an arbitrary definition of possible modes of orientation of the polypeptide chain of IMPs with respect to the hydrophobic core and the hydrophilic environment of the lipid bilayer was proposed.[1] IMPs were classified as monotopic, bitopic, and polytopic (see Fig. 1). The polypeptide chain of monotopic IMPs exhibits unilateral topology; i.e., each molecule possesses hydrophilic domain(s) exposed to the hydrophilic environment on only one side of the membrane. The polypeptide chain of bitopic and polytopic IMPs is bilateral in nature, containing two or multiple hydrophilic domains, respectively, exposed on opposite sides of the membrane.

It was proposed[1] that all these orientations could be accomplished by invoking, in addition to the signal sequence, only two additional types of topogenic sequences, termed "stop-transfer sequences" and "insertion sequences." The *stop-transfer sequence* was proposed to contain the information to interrupt the chain translocation process that was initiated by a signal sequence.

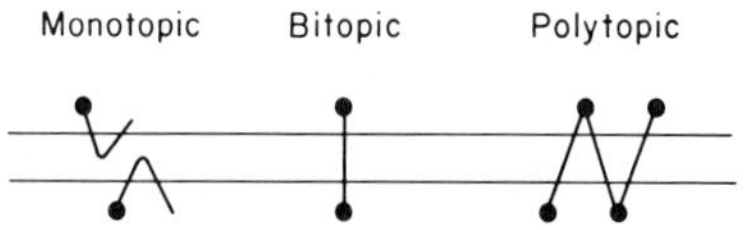

FIG. 1. Classification of integral membrane proteins (IMPs) as monotopic, bitopic, and polytopic. The hydrophobic boundary of the lipid bilayer is indicated by two parallel lines. Filled circles on polypeptide chains indicate major hydrophilic domains. The hydrophilic domain of an individual monotopic IMP is exposed only on one side of the lipid bilayer. A hydrophobic domain is indicated to anchor the polypeptide chain to the hydrophobic core of the lipid bilayer. A monotopic IMP may contain several hydrophilic and hydrophobic segments alternating with each other (not indicated here). However, all hydrophilic domains are unilaterally exposed. The polypeptide chain of bitopic IMPs spans the lipid bilayer once and contains a hydrophilic domain on each side of the membrane. In variants of bitopic IMPs (not indicated), the bilateral hydrophilic domains could be further subsegmented by interspersed hydrophobic domains that are capable of monotopic integration. The polypeptide chain of polytopic IMPs spans the membrane more than once and contains multiple hydrophilic domains on both sides of the membrane. The existence of polytopic IMPs remains to be demonstrated. Two structurally monotopic IMPs located on opposite sides of the membrane could interact via their hydrophobic anchorage domains and form a functionally bilateral ensemble. Taken from Blobel.[1]

Because translocation of the polypeptide chain could be expected to proceed sequentially and asymmetrically in both cotranslational and posttranslational translocation, stop-transfer sequences would be effective means for asymmetric integration of certain IMPs by either modes of translocation (see Table I). There could be as many translocator-specific stop-transfer sequences as there are translocator-specific signal sequences. On the other hand, there could be only one stop-transfer sequence addressed to one component common to all translocators.

The sequence features that constitute a stop-transfer sequence remain to be defined. The stop-transfer sequence may not simply be that stretch of ~25 primarily hydrophobic residues which is found as the transmembrane segment of bitopic IMPs and which might be envisioned to act as a stop-transfer sequence by virtue of being nonpermissive with the translocation process. There are, e.g., viral bitopic IMPs that possess stretches of at least 28 hydrophobic residues in their ectoplasmic domain.[58,59] Since this domain is translocated, it is clear that a long stretch of hydrophobic residue per se is not sufficient to stop the translocation process.

[58] A. Scheid, M. C. Graves, S. M. Silver, and P. W. Choppin, *in* "Negative Strand Viruses and the Host Cell" (B. W. J. Mahy and R. D. Barry, eds.), p. 181. Academic Press, New York, 1978.

[59] M. J. Gething, J. M. White, and M. D. Waterfield, *Proc. Natl. Acad. Sci. U.S.A.* **75,** 2737 (1978).

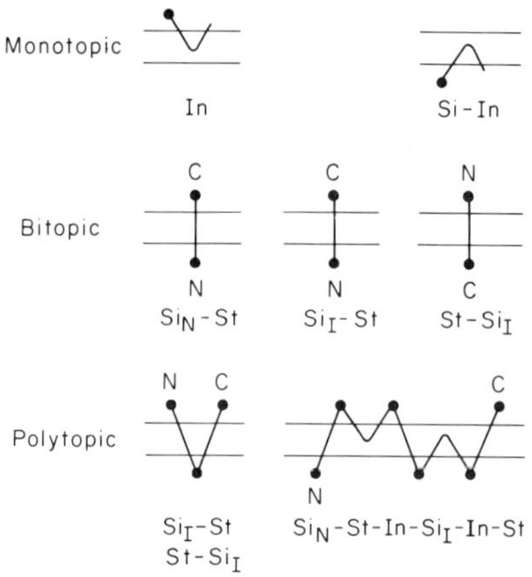

FIG. 2. Program of topogenic sequences for the asymmetric integration into membranes of some representative examples of monotopic, bitopic, and polytopic integral membrane proteins (IMPs). The hydrophobic boundary of the lipid bilayer is indicated by two parallel lines, with the upper line facing the protein biosynthetic compartment. Solid circles represent major hydrophilic domains, which, when indicated, contain an amino (N) or carboxy (C) terminus of the polypeptide chain. Topogenic sequences are: insertion sequence (In), signal sequence (Si), and stop-transfer sequences (St). Si_N and Si_I indicate amino-terminal and internal signal sequences, respectively. Examples given here (except for monotopic IMP at upper left) are for cotranslational integration into RER. Similar programs are conceivable also for cotranslational integration into PPM, IMM, and TKM as well as for post-translational integration into PXM, OMM, OCM, IMM [using Si(MEN)], and ICM/TKM [using Si(CEN)]. An attempt has been made to list topogenic sequences in order of their location along the polypeptide chain starting from the amino terminus. The problems encountered in predicting the order relate to uncertainties as to the order of chain translocation. In particular, in the case of an internal signal sequence (Si_I) there are several possibilities depending on the order of translocation.[27] The orientation of a polytopic IMP such as that indicated at the lower right is entirely hypothetical and is illustrated here only to indicate how such a polypeptide chain could be integrated into the membrane by a program of multiple topogenic sequences. Taken from Blobel.[1]

The *insertion sequence* functions to anchor a protein monotopically to the hydrophobic core of the lipid bilayer. Insertion would be spontaneous and not mediated by specific proteins. It would not be accompanied by the translocation across the membrane's lipid bilayer of large charged segments of the polypeptide chain. The latter can be achieved only by a signal sequence in a receptor-mediated process.

As is the case for the stop-transfer sequence, the structural features of an insertion sequence remain to be defined. It is conceivable that there are several unique insertion sequences that can distinguish lipid composition and therefore insert only into specific membranes. On the other hand, the specificity of insertion into a distinct membrane may be largely dictated by protein–protein interaction (i.e., by an affinity of a protein to be inserted to another IMP).

Although the precise orientation of the polypeptide backbone with respect to the lipid bilayer is unknown for most species of IMPs, the proposed hypothetical schemes using any of the postulated topogenic sequences either alone or combined (Fig. 2) can explain any one orientation by what essentially are a limited number of highly redundant mechanisms.[1] It is clear from these examples (Fig. 2) that the integration of most proteins into the membrane requires a signal sequence and a translocator, except for one subgroup of monotopic IMPs (see Fig. 2, upper left example). Thus, most IMPs can be integrated directly only into translocation-competent membranes. Because the translocators themselves are likely to consist of IMPs (see above) that require translocation for their integration into the membrane, it follows that Virchow's paradigm on the ontogeny of cells could be extended to membranes and paraphrased to *omnis membrana e membrana*.

Information about the mechanism of integration can be derived from assays that mimic the *in vivo* situation as closely as possible. Isolation of an IMP with detergents and its subsequent reconstitution into lipid vesicles,[60] while important for functional studies, cannot yield such information because it is improbable that detergents (either free or bound to proteins) are used by the cell to integrate its IMPs into membranes.

The first example of IMP integration into membranes (RER) under physiological conditions, in an *in vitro* translocation system (developed for *in vitro* translocation of secretory proteins[22]), was that of a bitopic viral IMP, the glycoprotein G of vesicular stomatitis virus (VSV). It was shown[61] that this protein is synthesized with a signal sequence that is addressed to the ER translocation system and is functionally identical to that of a secretory protein (shown by competition experiments). This *in vitro* translocation system also reproduced the bitopic asymmetric orientation of G with fidelity; the amino-terminal portion of newly synthesized G was translocated into the microsomal vesicles (protected by added proteolytic enzymes), whereas its carboxy-terminal portion remained untranslocated and therefore accessible to proteolytic enzymes.[61] We have shown[62] that integration of IMPs into the RER also requires SRP, as was expected, based on results of the earlier competition experiments.[61]

The finding that SRP causes a signal sequence-induced arrest in chain elongation[48] should be useful for mapping the location (NH_2-terminal or internal) of a signal sequence in those IMPs that contain an uncleaved

[60] Y. Kagawa and E. Racker, *J. Biol. Chem.* **246,** 5474 (1971).

[61] V. R. Lingappa, F. N. Katz, H. F. Lodish, and G. Blobel, *J. Biol. Chem.* **253,** 8667 (1978).

[62] D. J. Anderson, P. Walter, and G. Blobel, *J. Cell Biol.* **93,** 501 (1982).

signal sequence.[63,63a] The same approach should be useful also for mapping the location of multiple signal sequences.

Together with the rapidly accumulating information on the primary structure of a variety of IMPs and on their precise topology in the membrane, SRP and the *in vitro* translocation system can also be expected to yield detailed information on the mechanism of integration of those IMPs with other than a simple bitopic orientation.

Phylogeny of Membranes, Protein Translocation, and Compartments

How then could biological membranes with their characteristic asymmetry of proteins have evolved if their assembly depended on the development of a protein translocation system that, because it was made up in part of IMPs, was itself dependent for its assembly on a protein translocation system?

In an attempt to retrace the "phylogeny" of membranes[1] one could distinguish between precellular and cellular stages of evolution. Starting with lipid vesicles (Fig. 3) the first step in the precellular evolution of biological membranes may have been monotopic integration of proteins into the outer leaflet of lipid vesicles via insertion sequences. Such vesicles could have functioned as capturing devices to collect, on their outer surface, components involved in replication, transcription, and translation as well as metabolic enzymes present in the surrounding medium (Fig. 3A). In this way, much of the precellular evolution and assembly of macromolecular complexes (such as the ribosome) may have proceeded on the surface of these vesicles rather than within vesicles. By vesicle fusion, larger vesicles containing a synergistic assortment of functions could have evolved, resulting essentially in the formation of "inside-out cells"[1] (Fig. 3A and B). Concurrent with the evolution of such inside-out cells could have been the development of mechanisms for the translocation of proteins, thus providing the opportunity to segregate proteins, to colonize (with monotopic IMPs) the interior leaflet of the vesicle's lipid bilayer, and to integrate bitopic IMPs. Toward this end, the ribosome–membrane junction could have been remodeled and the insertion sequence could have evolved into a signal sequence so as to achieve first a cotranslational mode of translocation. The development of the stop-transfer sequence (perhaps as a variant of the signal sequence) to integrate bitopic IMPs may have concluded the precellular evolution of the cotranslational mechanism for the assembly of membranes.

The posttranslational mode of translocation may have evolved from

[63] S. Bonatti and G. Blobel, *J. Biol. Chem.* **254,** 12261 (1979).

[63a] I. Schechter, Y. Burstein, K. R. Zemell, E. Ziv, F. Kantor, and D. S. Papermaster, *Proc. Natl. Acad. Sci. U.S.A.* **76,** 2654 (1979).

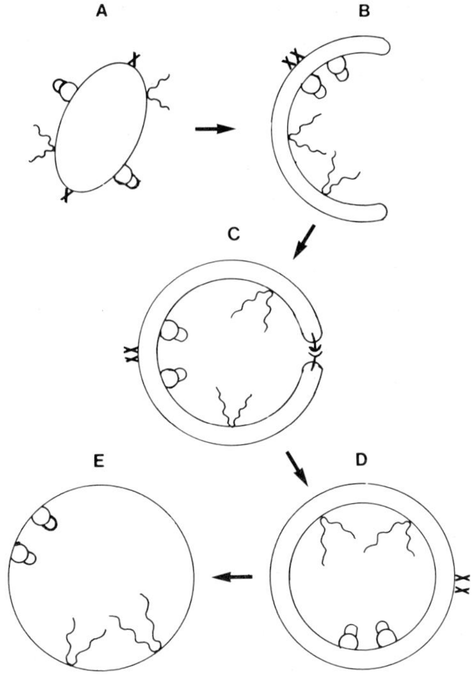

FIG. 3. Schematic illustration of various theoretical stages of precellular evolution on the surface of vesicles culminating in the formation of a primordial cell. (A) Vesicles containing monotopic IMPs (not indicated) are able to bind various macromolecules (X) and macromolecular complexes, among them chromatin and ribosomes; (B) nonrandom distribution of bound components on the vesicle surface and beginning invagination; (C) formation of a "gastruloid" vesicle, perhaps able to open and to close via protein–protein interaction of bitopic IMPs at the orifice; (D) fusion at the orifice, resulting in a primordial cell delimited by two membranes; (E) loss of the outer membrane; D could have evolved into gram-negative bacteria and E into gram-positive bacteria and eukaryotic cells (see Fig. 4). Taken from Blobel.[1]

the cotranslational mode by transposing the information that might be contained in a ribosomal protein and adding it to the signal sequence for cotranslational translocation. The integration of bitopic IMPs into the lipid bilayer permitted the development of transport systems and signaling systems. This set the stage for evolution to continue within a closed system (the primordial cell) effectively sealed from some of the hazards of the surrounding medium by the lipid bilayer but able to communicate with the outside via the lipid bilayer-integrated transport and signaling systems. The primordial cell (Fig. 3D) may have possessed two membranes, a plasma membrane delimiting the newly generated endoplasmic compartment, and an outer membrane enclosing a periplasmic space that represents the remnant of the intravesicular space of the inside-out cell. Subsequent elimination of the outer membrane would have yielded a cell with only one membrane (Fig. 3E), the plasma membrane, and one compartment, the endoplasmic compartment. All other biological membranes could have originated either directly or indirectly from this primordial plasma membrane.

The membranes of eukaryotic cells could be traced to two distinct sources (Fig. 4). One would be the cell's own primordial plasma membrane, generating by invagination various "orthoplasmic" membranes that delimit a new intracellular compartment, the ectoplasmic compartment (Fig. 4, left side). The other source (based on the theory of endosymbiosis[64]) would be the plasma membrane of a foreign symbiotic cell (at a "prenuclear" stage of evolution), which after being interiorized would give rise to "xenoplasmic" membranes delimiting a xenoplasmic subcompartment within the ectoplasmic compartment (Fig. 4, right side).

Posttranslocational Pathways

The nonrandom removal of distinct membrane functions from a pluripotent primordial plasma membrane during evolution would generate a number of highly differentiated intracellular membranes that lack a translocator and that are physically not continuous (at least not permanently) with translocation-competent membranes. These translocation-incompetent membranes (or the subcompartments they enclose) therefore must receive their translocation-dependent, constitutive IMPs (or segregated proteins) from translocation-competent membranes (or subcompartments).

The most significant donor membrane (subcompartment) is the RER, which probably supplies translocation-dependent proteins to essentially all orthoplasmic membranes and ectoplasmic subcompartments.[65] Each of the receiving membranes presumably contains a set of IMPs that are permanent residents (either constitutive to a particular receiving membrane or shared by several other orthoplasmic membranes) and a set of proteins in transit [either on their way to their permanent residence or cycling between orthoplasmic membranes (e.g., carrier proteins, see below)].

The information for posttranslocational traffic could reside in one (or several) discrete segments of the polypeptide chain. Proteins with an identical travel objective could share this information. These sequences,

[64] L. Margulis, "Origin of Eukaryotic Cells." Yale Univ. Press, New Haven, Connecticut, 1970.

[65] G. Palade, *Science* **189,** 347 (1975).

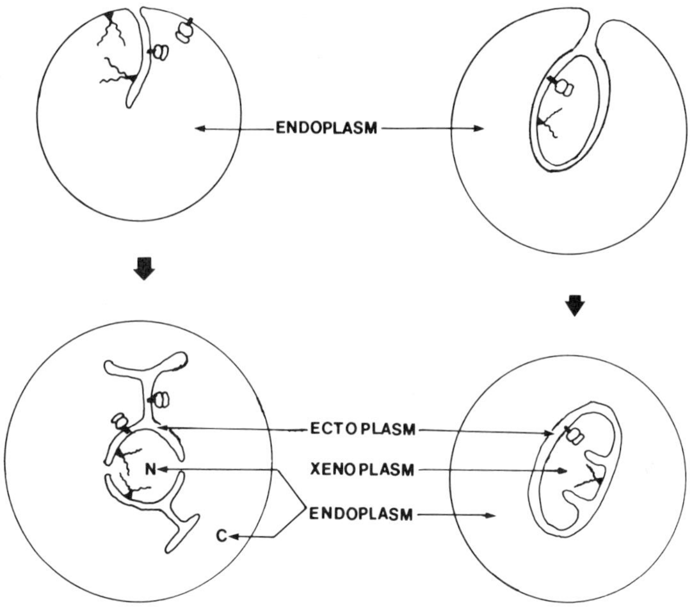

FIG. 4. Schematic illustration of the evolution of intracellular membranes and compartments. (Left side) Aggregation of certain membrane functions in the plane of the pluripotent plasma membrane. Nonrandom removal of these functions from the plasma membrane by invagination and fission results in the formation of a nuclear envelope (pore complexes omitted) continuous with the endoplasmic reticulum (rough and smooth) and generates an ectoplasmic compartment. The endoplasmic compartment is thereby subdivided into nucleoplasm (N) and cytoplasm (C). Note, however, that N and C remain connected via nuclear pores that do not have a membranous barrier. Other intracellular membranes that are distinct from the endoplasmic reticulum, such as lysosomal, peroxisomal, and Golgi complex membranes, also could have developed by invagination from the plasma membrane or could be outgrowths of the endoplasmic reticulum. (Right side) Symbiotic capture of another cell, generating an additional xenoplasmic compartment. Green plant cells have two such xenoplasmic compartments (mitochondrial matrix and chloroplast stroma). Only the inner mitochondrial membrane and the inner chloroplast membrane (including derived thylakoid membrane) would be of xenoplasmic origin, whereas the outer mitochondrial and chloroplast membranes would be of orthoplasmic origin, like all other cellular membranes. The proposed terminology may be useful for describing the precise topology of IMPs (see Fig. 1). For example, monotopic IMPs of the thylakoid membrane may be exposed ectoplasmically (i.e., toward the intradisk space) or xenoplasmically (i.e., toward the stroma); bitopic IMPs of the outer mitochondrial membrane have an ectoplasmic and an endoplasmic domain, etc. Taken from Blobel.[1]

termed "sorting sequences," would therefore constitute another group of topogenic sequences. Sorting sequences may be required not only for proteins that leave the RER, but also for those that need to stay there.

It is possible, however, that individual proteins may be able to reach

TABLE II
ALTERNATE-CHOICE PROGRAMS OF TOPOGENIC SEQUENCES FOR TOPOLOGICALLY EQUIVALENT PROTEINS[a,b]

Membrane	Bitopic IMPs	Content proteins
Peroxisomal	Si(PXM)-St	Si(PXM)
	Si(RER)-St-So	Si(RER)-So
Inner mitochondrial	Si(IMM)-St	Si(MEN)
	Si(MEN)-St	
Thylakoid	Si(TKM)-St	Si(TKM)
	Si(CEN)-St-So	Si(OCM)-So

[a] Taken from Blobel.[1]

[b] Abbreviations are as in Table I; St, stop-transfer sequence; So, sorting sequence. Listed are programs only for bitopic IMPs and content proteins that are not integral membrane proteins. Alternative programs analogous to those shown for the peroxisomal membrane are theoretically possible also for the outer membrane of mitochondria and chloroplasts, whereby the "content" proteins would correspond to proteins that are located in the ectoplasmic compartment (intermembrane space) of mitochondria and chloroplasts (see Fig. 4). Likewise, a program analogous to that shown for the inner mitochondrial membrane is conceivable also for the inner membrane of chloroplasts. For the corresponding "content" proteins in the xenoplasmic compartment there most likely is no alternative program of topogenic sequences: proteins are synthesized either within the xenoplasmic compartment or imported via Si(MEN) or Si(CEN). The alternative programs for bitopic IMPs in the thylakoid membrane are similar to those in the inner chloroplast membrane, except that sorting sequences may be required for the program Si(CEN)-St to distinguish between those bitopic IMPs that remain in the inner membrane and those that continue (by invagination) to become residents of TKM. By the same token, one of the programs [Si(OCM)-So] for the corresponding "content" proteins in the intradisk space is based on the possibility that this space communicates transiently with the ectoplasmic space of chloroplasts.

their target without a sorting sequence(s). They could do this merely by association with another protein (piggybacking) that is endowed with a sorting sequence(s). Likewise, sorting sequences (as defined here) may not be needed for the nonrandom distribution of proteins within physically continuous membranes. Protein–protein interactions to form large ensembles with a decreased rate of diffusion in the plane of the membrane and possibly anchored by cytoskeletal elements could be responsible for the regional differences that are characteristic of continuous membranes.

Decoding of the information contained in the sorting sequences should be effected by specific proteins. For sorting sequences of bilateral IMPs, the effector may be represented by a few distinct peripheral membrane proteins. For sorting sequences of soluble proteins, such as lysosomal enzymes, the effector may be represented by a bilateral IMP that functions as a carrier protein shuttling back and forth between the donor and a receiver compartment. Its ectoplasmic domain may be able to bind reversibly to the sorting sequence(s) of lysosomal enzymes, and its endoplasmic domain may contain a sorting sequence for a cyclic traffic pattern between the donor (RER) and receiver compartments [the latter could be represented by a distinct portion of the Golgi apparatus from which primary lysosomes develop].[66] A defect in the carrier could result in secretion of all lysosomal enzymes.

The need for sorting arose from the use of only one translocator for topologically different proteins. The reverse—namely, the potential to use more than one translocator for topologically equivalent proteins—may have arisen when certain membranes (see Table I) acquired a posttranslational translocator. For example, there could be two programs of topogenic sequences for peroxisomal proteins (Table II), both for the "content" proteins of the peroxisome and for those constitutive of the peroxisomal membrane (exemplified by bitopic IMPs).

Both programs indicated in Table II for the integration of bitopic IMPs into the inner mitochondrial membrane (or the inner membrane of chloroplasts) and into the thylakoid membrane are likely to exist.

Finally, one could conceive of "pleiotopic" proteins that are similar in structure and function but different in topology. Pleiotopic proteins[1,67] could have arisen by the loss or acquisition of a topogenic sequence(s). Such processes may be important (*a*) for achieving dichotomy in the posttranslocational pathway of proteins (e.g., secretory and lysosomal proteins); or (*b*) for achieving either export or retention via binding to membranes (e.g., secreted or membrane-bound form of IgM heavy chains[68]); or (*c*) for achieving either export or retention as a cytoplasmic

[66] A. B. Novikoff, *Proc. Natl. Acad. Sci. U.S.A.* **73,** 2781 (1976).

[67] G. Blobel, P. Walter, C. N. Chang, B. M. Goldman, A. H. Erickson, and V. R. Lingappa, *Symp. Soc. Exp. Biol.* **33,** 9 (1979).

[68] J. Rogers, P. Early, C. Carter, K. Calame, M. Bond, L. Hood, and R. Wall, *Cell* **20,** 303 (1980).

protein[69]; or (*d*) for diversifying the organellar distribution of proteins (e.g., some proteins that may occur both within peroxisomes and the mitochondrial matrix); or (*e*) for anchoring polymeric structures in the membrane (e.g., free and membrane-bound forms of cytoskeletal proteins).

[69] D. Perlman, H. O. Halvorson, and L. E. Cannon, *Proc. Natl. Acad. Sci. U.S.A.* **79,** 781 (1982).

[53] Signal Recognition Particle: A Ribonucleoprotein Required for Cotranslational Translocation of Proteins, Isolation and Properties

By PETER WALTER and GÜNTER BLOBEL

Microsomal membranes (RM)[1] lose their ability to translocate nascent secretory,[2,3] lysosomal,[4] and membrane[5] proteins upon salt extraction if they are assayed in the wheat germ cell-free translation system. From the salt extract, a ribonucleoprotein [termed signal recognition particle (SRP)] was purified to homogeneity and shown to be composed of six different polypeptide chains[3] and one 7 S RNA molecule.[6] Addition of the crude salt extract or purified SRP to salt-extracted RM (K-RM) restored their ability not only to catalyze translocation of secretory[2,3] and lysosomal[4] proteins across the microsomal membrane, but also to catalyze the asymmetric integration of transmembrane proteins[5] into the membrane. The function of SRP in the translocation process was shown to involve the recognition of the signal sequence of these proteins in their nascent state,[7] the specific binding of polysomes synthesizing these proteins to RM vesicles,[8] and the initiation of the translocation event. In the absence

[1] P. Walter and G. Blobel, this volume [6].
[2] G. Warren and B. Dobberstein, *Nature* (*London*) **273,** 569 (1978).
[3] P. Walter and G. Blobel, *Proc. Natl. Acad. Sci. U.S.A.* **77,** 7112 (1980).
[4] A. H. Erickson, P. Walter, and G. Blobel, in preparation.
[5] D. J. Anderson, P. Walter, and G. Blobel. *J. Cell Biol.* **93,** 501 (1982).
[6] P. Walter and G. Blobel, *Nature* (*London*) **299,** 691 (1982).
[7] P. Walter, I. Ibrahimi, and G. Blobel, *J. Cell Biol.* **91,** 545 (1981).
[8] P. Walter and G. Blobel, *J. Cell Biol.* **91,** 551 (1981).

ISBN 0-12-181996-5

of RM vesicles, SRP modulates translation by reversibly arresting the elongation of nascent secretory proteins after their signal peptide and about 40 additional amino acids have been polymerized.[9] This translation arrest can be released if K-RM vesicles are added, resulting in completed and efficiently translocated secretory protein. The arrest-releasing activity of K-RM vesicles has been localized to a 72,000-dalton integral membrane protein, termed SRP-receptor,[10,11] or docking protein.[12]

Here we describe the purification of SRP which, as a component of the translocation machinery, has proved to be a valuable tool for studying the mechanism of protein translocation across the membrane of the endoplasmic reticulum.[7–9] Purified SRP has also been used as an affinity probe to purify SRP-receptor.[11] Finally, it has been helpful as an additional supplement in *in vitro* translation systems containing RM, where it can boost the efficiency of *in vitro* protein translocation to virtually 100%.

Properties of SRP

The SRP is composed of six different polypeptide chains (molecular weights: 72,000, 68,000, 54,000, 19,000, 14,000, and 9000) and one molecule of RNA (identified as 7 SL RNA, which is about 300 nucleotides long, has been sequenced,[13,14] and has a sedimentation coefficient of 11 S). The particle is negatively charged and interacts tightly with DEAE-ion exchange resins.[3] It also interacts (quite selectively) with hydrophobic-ion exchange resins (see footnote 3 and below). Taken together with the fact that SRP's activity is stabilized by very low concentrations of nonionic detergents,[3] this suggests that SRP contains exposed hydrophobic domains. The activity of SRP is dependent on free sulfhydryl group(s).[3,15] To stabilize SRP, we therefore include small amounts of nonionic detergent (Nikkol) and reducing agent (DTT) in most buffers.

SRP is relatively stable to nucleolytic breakdown. We normally observe little (<20%) or no degradation of SRP–RNA, even when taking no special precautions to keep solutions free of nuclease contamination. Preliminary experiments indicate that Mg^{2+} ions are important for SRP's structural integrity (see also below).

[9] P. Walter and G. Blobel, *J. Cell Biol.* **91,** 557 (1981).
[10] R. Gilmore, G. Blobel, and P. Walter, *J. Cell Biol.* **95,** 463 (1982).
[11] R. Gilmore, P. Walter, and G. Blobel, *J. Cell Biol.* **95,** 470 (1982).
[12] D. I. Meyer, E. Krause, and B. Dobberstein, *Nature (London)* **297,** 647 (1982).
[13] E. Ullu, S. Murphy, and M. Melli, *Cell* **29,** 195 (1982).
[14] W. Y. Li, R. Reddy, D. Henning, P. Epstein, and H. Busch, *J. Biol. Chem.* **257,** 5136 (1982).
[15] R. C. Jackson, P. Walter, and G. Blobel, *Nature (London)* **284,** 174 (1980).

We routinely freeze SRP preparations in small aliquots in liquid nitrogen. It can be stored at −80° for at least 1 year, and there is no apparent loss of activity upon three cycles of freezing and thawing.

Solutions

A stock solution of 1.0 *M* triethanolamine was adjusted to pH 7.5 at room temperature with acetic acid and, as such, is referred to as TEA. A stock solution of 4.0 *M* KOAc was adjusted to pH 7.5 at room temperature with acetic acid. Stock solutions of 2.5 *M* sucrose and 1.0 *M* $Mg(OAc)_2$ were not further adjusted. All stock solutions mentioned above were filtered through a 0.45 μm Millipore filter, except the sucrose solution which was filtered through a 1.2 μm Millipore filter. Dithiothreitol (DTT) was kept in small aliquots as a 1.0 *M* stock solution at −20° and diluted into the buffers immediately before use. The nonionic detergent Nikkol (octaethylene glycol dodecyl ether) was purchased directly from Nikko Chemical Corp. in Tokyo. It was chosen because it is available in a chemically pure form and does not absorb UV light at 260 nm or 280 nm (thereby allowing RNA and protein to be followed by optical absorbance). We keep the detergent as a 20% (w/v) stock solution at 4°. In its properties, Nikkol resembles Triton X-100, which can also be used to stabilize SRP activity (however, we have not yet tested to use Triton in the hydrophobic chromatography steps).

Buffer I: 50 m*M* TEA, 500 m*M* KOAc, 5 m*M* $Mg(OAc)_2$, 1 m*M* DTT
Buffer II: 250 m*M* sucrose, 50 m*M* TEA, 1 m*M* DTT
Buffer III: 50 m*M* TEA, 1.0 *M* KOAc, 10 m*M* $Mg(OAc)_2$, 1 m*M* DTT, 0.05% Nikkol
Buffer IV: 12 m*M* TEA, 250 m*M* KOAc, 2.5 m*M* $Mg(OAc)_2$, 1 m*M* DTT, 0.01% Nikkol
Buffer V: 50 m*M* TEA, 350 m*M* KOAc, 3.5 m*M* $Mg(OAc)_2$, 1 m*M* DTT, 0.01% Nikkol
Buffer VI: 50 m*M* TEA, 600 m*M* KOAc, 6.0 m*M* $Mg(OAc)_2$, 1 m*M* DTT, 0.01% Nikkol

Salt Extraction of RM

We have prepared salt extracts from either column-washed RM or EDTA-stripped RM preparations.[1] SRP can be purified from either preparation, but column-washed RM yield consistently cleaner preparations.

To prepare the salt extract from column-washed RM we aimed to keep ribosomes (and SRP) from unfolding and therefore included magnesium ions in our extraction buffers. This allowed us to obtain a ribosome-free salt extract by removing intact, coextracted ribosomes in an additional spin after the initial extraction.[3]

All preparations were carried out at 0–4°. Ten milliliters of an ice-cold salt solution [1.5 *M* KOAc, 15 m*M* $Mg(OAc)_2$] were slowly added to 20 ml of column-washed RM (at a concentration of 1000 eq/ml). The mixture was incubated on ice for 15 min. The membranes were sedimented for 1 hr at 120,000 g_{av} through a cushion of 0.5 *M* sucrose in buffer I. The ratio of load to cushion was about 3 to 1. The supernatant fraction (including the upper half of the cushion) was carefully removed from the top with a Pasteur pipet. To remove ribosomes the supernatant was then recentrifuged for 3.5 hr at 200,000 g_{av} over a cushion of 0.5 *M* sucrose in buffer I. The ratio of load to cushion was again about 3 to 1. The supernatant was carefully removed from the top (again including the upper half of the cushion), yielding a postribosomal salt-extract fraction. Including the upper half of the cushions in the corresponding supernatant fractions allowed essentially complete recovery of SRP in spite of its relatively high sedimentation coefficient.

The membrane pellet from the first centrifugation step was resuspended in 20 ml of buffer II by manual homogenization in a Dounce homogenizer (pestle A, 2–3 strokes). The suspension was then sedimented (1 hr at 120,000 g_{av}) through a cushion of 0.5 *M* sucrose in buffer II and finally resuspended in 20 ml of buffer II to yield the salt-extracted rough microsome fraction (K-RM) at a concentration of 1000 eq/ml.

The extraction of EDTA-stripped RM was carried out identically to the extraction of column-washed RM, except that the membranes were removed from the salt extract by centrifugation for 1.5 hr at 200,000 g_{av} through a cushion of 0.35 *M* sucrose in buffer I. The second centrifugation step was omitted since ribosomes had been unfolded and extracted in the previous EDTA washes. A K-RM fraction prepared from EDTA-stripped RM was functionally indistinguishable from K-RM prepared from column-washed RM.

Fractionation of the Salt Extract and Hydrophobic-Ion Exchange Chromatography

The salt extracts prepared from column-washed RM or EDTA-stripped RM were treated identically. Chromatography of the salt extracts on ω-aminoalkylagarose resins[3,16,17] resulted in a 50-fold enrichment[3] of SRP.

[16] S. Shaltiel, this series, Vol. 34, p. 126.

[17] S. Shaltiel, S. P. Adler, D. Purich, C. Caban, P. Senior, and E. R. Stadtman, *Proc. Natl. Acad. Sci. U.S.A.* **72,** 3397 (1975).

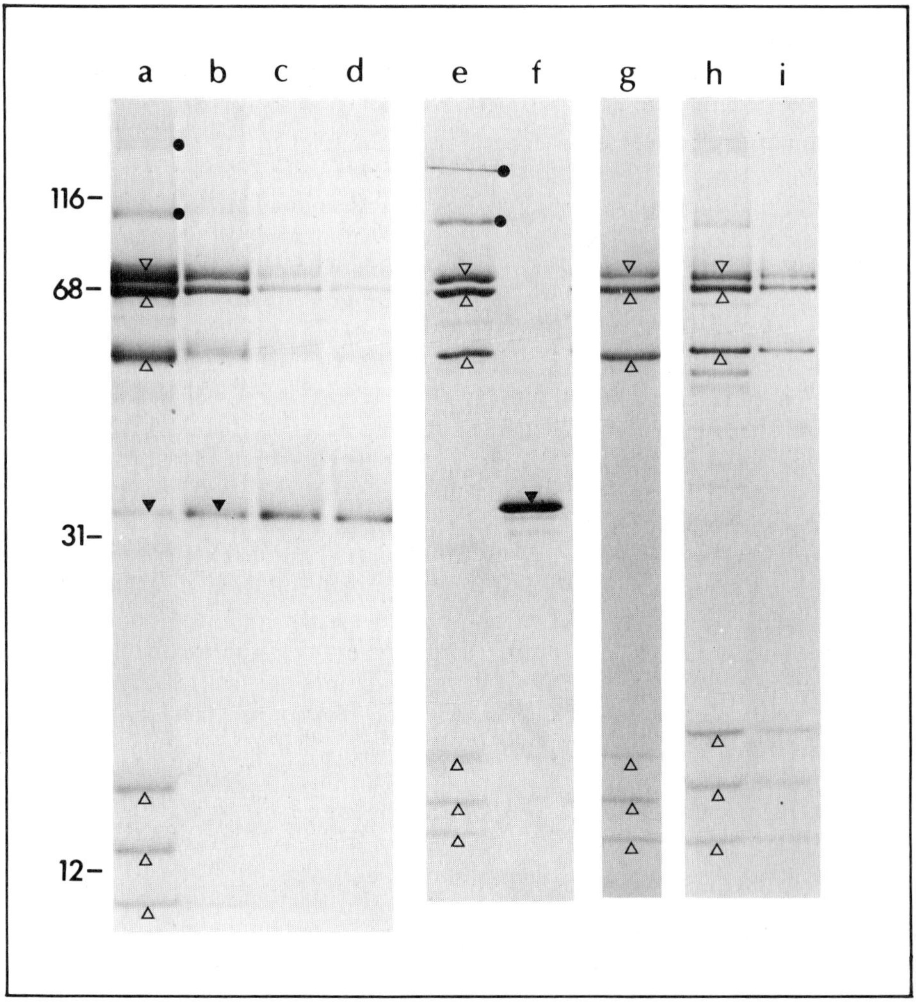

FIG. 1. Polypeptide profile in SDS–PAGE of different SRP preparations. Fractions obtained during the purification of SRP were TCA precipitated and their polypeptide composition visualized by Coomassie Blue staining after PAGE in SDS on 10–15% acrylamide gradient gels. The fractions were derived from either EDTA-stripped RM (lanes a–f) or column-washed RM (lanes g–i). Bands corresponding to polypeptides that have been designated as SRP polypeptides are marked with an open arrowhead. The low molecular weight polypeptides were weakly stained. The contaminating bands marked with dots or solid arrowhead were only obtained when EDTA-stripped RM was used as the starting material (lanes a–f, see text for discussion). The molecular weight standards were: cytochrome *c*, 12,000; DNase, 31,000; bovine serum albumin, 68,000; β-galactosidase, 116,000.

SRP was eluted from an ω-aminopentylagarose column and fractions of one column volume were collected. A 250-μl aliquot of fraction 1 (lane a), fraction 3 (lane b), fraction 5

All preparations were carried out at 0–4°. A 2-ml column of ω-aminopentylagarose (Sigma, 5.7 μmol of 1.5 diaminopentane coupled per milliliter of CNBr-activated agarose) was prewashed with 10 ml of 2 *M* KOAc (to exchange the counterion to acetate), washed with 10 ml of water, and equilibrated with 20 ml of buffer I. It is useful to layer a small amount of Sephadex G-15 on top of the ω-aminopentylagarose bed to prevent protein from binding to the very top surface, which, upon step elution, might become exposed to air. A 35–40 ml salt-extract fraction (obtained from 20 ml of a starting RM fraction) was passed over the column at a flow rate of 6 ml/hr. The column was then washed with 20 ml of buffer I and eluted with a step of buffer III. As soon as buffer III appeared in the eluent (detected by conductivity or the abrupt change in drop size due to the presence of detergent) 0.5-ml fractions were collected. Eluting protein was monitored by absorbance measurements at 260 nm or 280 nm and SRP activity was followed in the translocation assay. Typically, the bulk of the SRP eluted in 1–4 column volumes.

In Fig. 1 (lanes a–d) the elution profile of SRP from the ω-aminopentylagarose column is shown. Note that the salt extract was obtained from EDTA-stripped RM. Polypeptides of individual fractions were displayed by SDS–PAGE. SRP activity and SRP polypeptides (open arrowheads) eluted immediately as the elution buffer was applied to the column. Two contaminating polypeptides of high molecular weight (about 90,000 and 200,000, marked with dots) were also observed in these active fractions. Another distinct contaminant (about 34,000 daltons, marked with solid arrowhead) eluted delayed. If the pool of eluted material (Fig. 1, lanes a–c) was concentrated by DEAE-Sepharose chromatography and then further fractionated by sucrose-gradient centrifugation (see below), two distinct peaks, one at 7 S (Fig. 1, lane f) and one at 11 S (Fig. 1, lane e), were resolved. Only the 11 S peak showed SRP activity, but both peaks appeared to be ribonucleoproteins. A 5 S RNA could be extracted from the 7 S peak and 7 SL RNA (SRP-RNA) was extracted from the 11 S

(lane c), and fraction 7 (lane d) is displayed on the gel. Fractions 1–5 were pooled, concentrated on DEAE–Sepharose, and further fractionated by sucrose-gradient centrifugation. Two symmetrical peaks were obtained, one having a sedimentation coefficient of 7 S and one of 11 S (data not shown). The material displayed in lanes e (comprising the 11 S peak) and f (comprising the 7 S peak) corresponds to 250 μl of the pooled ω-aminopentylagarose eluate.

The eluate of ω-aminopentylagarose columns of two different SRP preparations starting from *column-washed RM* is shown in lanes g and h. Whereas the preparation shown in lane g appeared homogeneous, the material shown in lane h contained several contaminating bands and therefore required further purification by sucrose-gradient centrifugation. The 11 S peak of the gradient was collected and is shown in lane i.

peak. The 7 S RNP represents the known complex of ribosomal 5 S RNA and ribosomal protein L5 (35,000 daltons). It was previously shown[18] that this RNP could be specifically extracted from the large ribosomal subunit by EDTA treatment, consistent with our observation that this contaminant was only observed when the salt extract was prepared from EDTA-treated RM (rather than column-washed RM).

For comparison, the polypeptides of an ω-aminopentylagarose eluate prepared from column-washed RM are displayed in Fig. 1, lane g. Note that the 34,000-dalton polypeptide as well as the high molecular weight contaminants are absent. We feel that upon EDTA treatment of RM a "partial unfolding" of SRP takes place, thus allowing the high molecular weight contaminants to become unspecifically associated with the (highly charged) particle, although at present we have no experimental proof for this rationalization. In general, preparations obtained from column-washed RM are more homogeneous.

Chromatography on ω-aminoalkylagarose resins appears to involve both hydrophobic and ion-exchange effects.[16,17] Buffer III causes elution of SRP from the column by interfering with both types of interactions. The increase in ionic strength causes a weakening of ionic interactions, whereas the inclusion of small amounts of detergent interferes with hydrophobic interactions. Whereas an increase in ionic strength was required to elute SRP, omission of detergent led only to peak broadening.

The fact that the interaction of SRP with the ω-aminopentylagarose is indeed due to a combination of ionic and hydrophobic interactions is demonstrated by the data shown in Fig. 2. It can be seen that SRP under the ionic conditions of the salt extract (buffer I) does not interact with resins containing ω-aminopropyl- or ω-aminobutylagarose, but, as one increases the number of methylene groups in the alkyl chain (and thereby the hydrophobicity of the resin) to the length of ω-aminopentyl-, -hexyl-, or -octylagarose, binding does occur. If, however, the salt concentration in the salt extract is lowered by dilution to 200 m*M*, SRP also binds to ω-aminobutylagarose [indicating that the previously high salt (500 m*M*) prevented it from binding by interfering with presumably ionic interactions]. By analogy, SRP can be prevented from binding to ω-aminopentylagarose by increasing the salt concentration to 1.0 *M*. From these data we designed the elution conditions reflected in the composition of buffer III. Polypeptide profiles of eluted material bound at 200 m*M* salt to ω-aminobutylagarose were more heterogeneous than the material obtained from the ω-aminopentylagarose columns. Attempts to elute active SRP from ω-aminohexylagarose columns failed.

[18] G. Blobel, *Proc. Natl. Acad. Sci. U.S.A.* **68,** 1881 (1971).

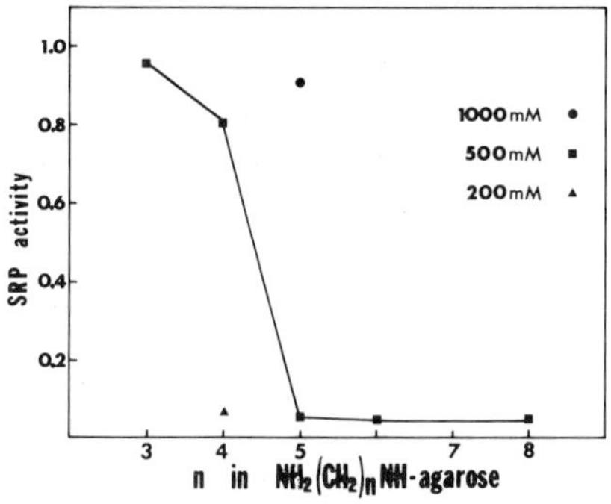

FIG. 2. Behavior of SRP on ω-aminoalkylagarose columns. Fractions of a postribosomal salt extract (500 μl) (ionic conditions of buffer I) were passed over 50-μl columns of ω-aminoalkylagarose of varying hydrophobicity ranging from ω-aminopropylagarose ($n = 3$) to ω-aminooctylagarose ($n = 8$). The material flowing through the column (i.e., not bound) was collected and assayed for SRP activity (■). Separate aliquots were made 1.0 *M* in KOAc prior to passage over the ω-aminopentylagarose column (●) or diluted to 200 m*M* KOAc prior to passage over the ω-aminobutylagarose column (▲). SRP activity is expressed as a fraction of the SRP activity of the starting material.

It is worth noting that *two* distinct RNPs were purified (Fig. 1, lanes e and f) by this chromatographic technique from the complex mixture of proteins in the salt extract prepared from EDTA-stripped RM. There are also reports from other laboratories[19] where chromatography on ω-aminoalkylagarose resins was used as a key step in the purification of RNPs. Thus, it is conceivable that RNPs share structural characteristics that can be recognized by these resins. An intriguing alternative interpretation would be that these interactions are not (or only to a small degree) of hydrophobic nature, but rather that RNPs bind to these resins on the basis of a sterically controlled ionic interaction. The increased affinity of the RNPs to resins with a higher number of methylene groups could therefore be due not (or not only) to an increased hydrophobicity of the matrix, but rather to an increased accessibility of the positively charged amino groups to the negatively charged RNA backbone in the particles. This interpretation is consistent with the fact that an *increase* in ionic strength caused weakening of the interactions and thus caused elution.

Further Purification of SRP

Signal recognition particles can be further purified by sucrose gradient centrifugation[3,8] under high salt conditions. This procedure removes all

[19] M. Hinterberger, I. Pettersson, and J. A. Steitz, *J. Biol. Chem.* **258**, 2604 (1983).

contaminating polypeptide chains that occasionally coelute in the hydrophobic-ion exchange chromatography (even if the salt extract was prepared from column-washed RM). An example is given in Fig. 1, lanes h–i. The relatively heterogeneous SRP preparation obtained after ω-aminopentylagarose chromatography (lane h) is purified to homogeneity (lane i). Prior to sucrose-gradient centrifugation it is helpful to concentrate SRP further by chromatography on DEAE–Sepharose Cl-6B.

For concentration by DEAE–chromatography, the eluate of the ω-aminopentylagarose column was diluted with three parts of ice-cold water to reduce the ionic strength. The solution (derived from 20 ml of RM) was passed over a 0.4-ml DEAE–Sepharose 6B-Cl column which was equilibrated in buffer IV. The column was washed with 2 ml of buffer V and eluted with a step of buffer VI. Signal recognition particle elutes immediately and can be completely recovered in a 0.4-ml fraction (the appearance of buffer VI in the eluate is monitored by conductivity, the appearance of SRP by optical absorbance at 260 nm and 280 nm).

For sucrose-gradient centrifugation, the DEAE-concentrated SRP solution is layered directly on top of a 5–20% linear sucrose gradient in buffer I (containing 0.01% Nikkol). If the eluate of the ω-aminopentylagarose column was used without prior DEAE concentration, it was first diluted with one part of water to approximate the ionic conditions of the gradient. We either layered 250–500 μl on a 12.5-ml gradient (Beckman SW 40 rotor, 40,000 rpm, 20 hr) or 100–200 μl on a 5-ml gradient (Beckman SW 50.1 rotor, 50,000 rpm, 6 hr). The gradients were fractionated using an ISCO gradient fractionator with a continuous absorbance monitor, and the 11 S peak was collected.

Translocation Assay

To assay SRP for its activity, we used a wheat germ cell-free translation system.[20] It was programmed with bovine pituitary mRNA (coding mainly for the secretory protein prolactin) and supplemented with K-RM. Signal recognition particles reconstitute the translocation activity of K-RM which is monitored by the cotranslational conversion of preprolactin to prolactin.[3]

Bovine pituitary RNA (0.2 A_{260} units per 25 μl of translation mix) was translated in a staphylococcal nuclease-treated wheat germ system (6 μl of wheat germ S23 and 25 μCi [^{35}S]Met per 25 μl of translation mix). All translations were supplemented with human placental RNase inhibitor[21]

[20] A. H. Erickson and G. Blobel, this volume [3].

[21] P. Blackburn, *J. Biol. Chem.* **254,** 12484 (1979).

at a final concentration of 0.01 A_{280} units/ml from a 100-fold concentrated stock solution (kept at $-80°$) and a cocktail of selected protease inhibitors (which do not interfere with protein synthesis) from a 200-fold concentrated stock solution (kept at $-20°$) at the following final concentrations: pepstatin A 0.1 μg/ml, chymostatin 0.1 μg/ml, antipain 0.1 μg/ml, leupeptin 0.1 μg/ml, trasylol 10 units/ml. The ions that were added with the wheat germ extract and different amounts of SRP were taken into account and compensated for to yield final ion concentrations of 150 m*M* KOAc and 2.0 m*M* $Mg(OAc)_2$ in all translations. The nonionic detergent Nikkol was present in all translations at a final concentration of 0.002% to stabilize SRP activity. At this detergent concentration, the translocation activity of RM is unaffected. Furthermore, complete protection of the translocated polypeptide chain in posttranslational proteolysis assays[1,22] is maintained. To achieve breakdown of the lipid bilayer and expose signal peptidase, Nikkol concentrations in excess of 0.04% are required. Membranous fractions were always added last after all of the other components were mixed and the detergent uniformly diluted.

A typical assay for SRP activity contained K-RM at a concentration of 40 eq/ml (1 μl of K-RM/25 μl of translation mix). The eluate of the ω-aminopentylagarose column was diluted with one part of water and 0.5–3 μl of this solution was assayed in 25 μl of translation mix. After concentration by DEAE–chromatography or after sucrose-gradient centrifugation, the amount of SRP added to the translocation assay was adjusted according to the concentration or dilution the sample underwent. Usually, the highest amount of SRP that can be added to the translocation assay is limited by the salt and detergent added with the SRP solution. However, 3 μl of the ω-aminopentylagarose eluate assayed in a 25-μl translation mix constitutes a saturating amount, i.e., complete (>90%) translocation of nascent secretory proteins is usually obtained.

Instead of the K-RM fraction, the translation system can also be supplemented with RM or EDTA-stripped RM (at about 40 eq/ml) and SRP can be added to boost its translocation activity (try 1 μl of ω-aminopentylagarose eluate per 25 μl of translation mix as a starting point).

[22] G. Scheele, this volume [7].

[54] Proteins Mediating Vectorial Translocation: Purification of the Active Domain of the Endoplasmic Reticulum Docking Protein

By DAVID I. MEYER and BERNHARD DOBBERSTEIN

Since the discovery that vectorial translocation of nascent secretory proteins can be accomplished *in vitro* using isolated rough microsomes,[1] efforts have been underway to dissect out the membrane components that mediate such transport. To date, two approaches have been taken, and two protein components have been isolated. The first approach utilized high-salt treatment to dissociate an essential component from rough microsomes (RM). Salt-washed microsomes were no longer able to translocate nascent secretory proteins synthesized in the wheat germ cell-free system[2]; however, they retained their ability to translocate proteins in the reticulocyte lysate system.[3] This latter activity could be subsequently proteolytically released from membranes by elastase and high-salt treatment.[4] In each of these cases, the material removed from the microsomes could be added back to the membranes rendered inactive by the treatment, thereby restoring activity in the appropriate cell-free system.

Subsequent studies determined that a protein complex of M_r 250,000 (called signal recognition protein, SRP) was the component liberated by high-salt treatment and required for activity in wheat germ lysate.[5] Elastase and high salt removed a 60 kilodalton domain (from a 72 kilodalton membrane protein) from salt-washed RM, rendering the membranes inactive in reticulocyte lysate.[4] The relationship between the two components and their respective activities in the two translation systems is shown in Fig. 1. It is now known that the 72 kilodalton membrane protein is required for the functional association of SRP.[6] Moreover, evidence has been accumulated demonstrating that SRP may well exert its effect as a soluble component of the cytoplasm.[6] As the 72 kilodalton membrane protein represents the site at which the ribosomal complex + SRP first makes contact with the endoplasmic reticulum (ER), it has been termed the "docking" protein (DP).

[1] G. Blobel and B. Dobberstein, *J. Cell Biol.* **67,** 852 (1975).
[2] G. Warren and B. Dobberstein, *Nature (London)* **273,** 569 (1978).
[3] D. Meyer and B. Dobberstein, *J. Cell Biol.* **87,** 498 (1980).
[4] D. Meyer and B. Dobberstein, *J. Cell Biol.* **87,** 503 (1980).
[5] P. Walter and G. Blobel, *Proc. Natl. Acad. Sci. U.S.A.* **77,** 7112 (1980).
[6] D. Meyer *et al., Nature (London)* **297,** 647 (1982).

ISBN 0-12-181996-5

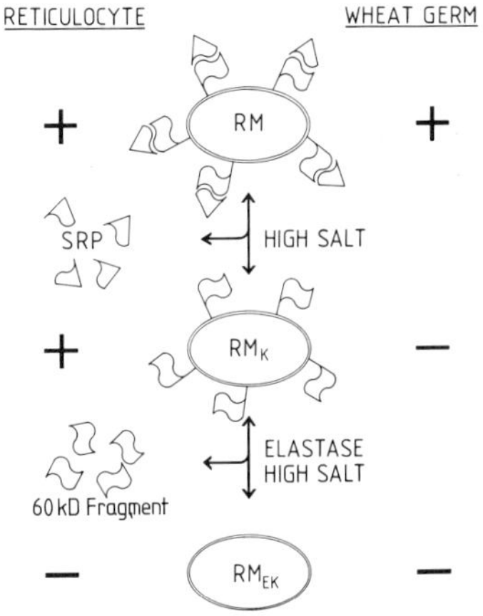

FIG. 1. Dissection and reconstitution of rough microsomes (RM) using protease and high salt. When RM are treated with 0.5 *M* KCl, translocation activity is lost in wheat germ (due to the salt-mediated removal of SRP), and RM_K are generated that are nonetheless active in reticulocyte lysate. Treatment of salt-washed microsomes (RM_K) with elastase and high salt yields RM_{EK} that are inactive in either system. This is due to the loss of a 60 kilodalton peptide fragment from the docking protein (DP_f). Reconstitution can be achieved sequentially by readdition of DP_f to RM_{EK} (thus restoring activity in reticulocyte lysate) followed by readdition of SRP to RM_{EK}–DP_f membranes. In this way full reconstitution is achieved in both cell-free systems. From Meyer *et al.*[6]

Presented here are the procedures needed to isolate and purify the active domain of the docking protein from isolated rough microsomes.

Assay System

The general principle of using dog pancreas microsomes to reconstitute translocation of secretory proteins[7] must be slightly modified to allow the addition of isolated factors to the system, as well as the membranes.

Components

Rabbit reticulocyte lysate[8]
Potassium acetate, 1 *M*
Magnesium acetate, 50 m*M*
Rough microsomes (or microsomes + isolated components) in 0.25 *M* sucrose, 20 m*M* HEPES, 50 m*M* KCl[7]
mRNA for a secretory protein, e.g., IgG light chain, preprolactin, human placental lactogen[9]
Translation grade [^{35}S]methionine (specific activity >1000 Ci/mmol)

Procedure. Assays are carried out in heat-sterilized microfuge tubes (1.5 ml) at 37° in a final volume of 25 μl. The mixture contains 1 μl of mRNA (0.1–0.5 μg of poly(A)-rich RNA), 4 μl of [^{35}S]Met (~40–50 mCi), 0.5 μl of MgAc (50 m*M*), and 4.5 μl of a cocktail consisting of spermidine, creatine phosphate, dithiothreitol (DTT) and GTP.[8] Five microliters of a

[7] P. Walter and G. Blobel, this volume [6].
[8] T. Hunt and R. J. Jackson, this volume [4].
[9] P. M. Lizardi, this volume [3].

membrane solution or a mixture of inactive membranes preincubated with fractions to be tested (see below) are added, and the reaction is initiated by the addition of 10 μl of nuclease-treated reticulocyte lysate. After 1 hr the samples are cooled and processed for PAGE and fluorography.[10] Typically 2–4 μl of the incubation mixture are sufficient.

Quantitation of Activity

The processing of nascent secretory proteins involves cotranslational cleavage of the signal sequence on the luminal side of microsomal vesicles. The processed nascent chains are known to have been translocated across the membrane and sequestered within microsomal vesicles because they are resistant to proteolytic attack.[1] Thus the cleavage of the signal sequence is a direct measure of the translocation capability of a membrane. In other words, the ratio of processed secretory protein to the total amount synthesized (authentic + preprotein) can be expressed as percentage of processing.

Values for percentage of processing can be obtained in two ways. Fluorograms can be scanned by densitometric methods and the intensities of the bands of preprotein and authentic protein compared. It is important in this case that the grains of the X-ray film being used are not saturated, i.e., that the film is developed while photon adsorption is occurring linearly. A more precise measurement is obtained when the respective bands are located on the polyacrylamide gel and are cut out and counted directly. Localization of bands can be accomplished fluorometrically (as above) or by staining using appropriate standard proteins.

Whichever method is used, it is essential to take into account alterations in the specific radioactivity that may have occurred during processing. For example, IgG light chain contains five methionine residues in the preprotein and only three in the authentic light chain. Accurate values for percentage of processing must therefore reflect such a change.

Routinely, 2.5 μl of inactive RM (see below for method of preparation) are incubated with 2.5 μl of the fraction being tested. As the fraction usually has a [KCl] = 0.5 *M*, and the RM are in 50 m*M* KCl, the final [KCl] of the 5-μl sample is 275 m*M*. It is therefore imperative that the cell-free translation system be able to accommodate this addition of KCl. This is easily accomplished by omitting KAc from the translation mixture. The final [KCl] when the 5 μl of RM + sample are added to the 20 μl from the translation system will be about 55 m*M*, very near the optimum.

[10] W. M. Bonner, this volume [15].

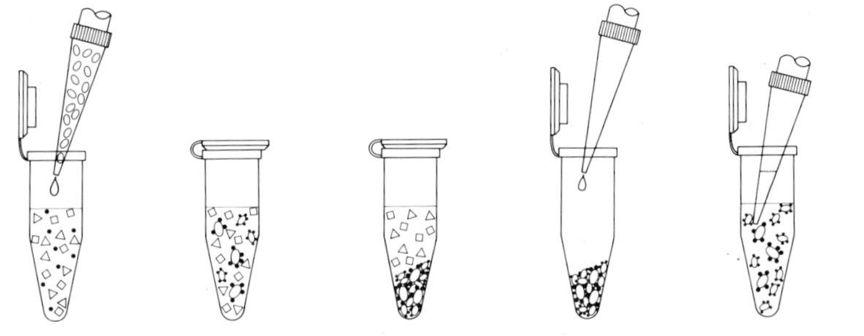

FIG. 2. Membrane-affinity factor isolation assay. Inactive rough microsomes are added to samples in order to adsorb activity-restoring factors. (1) In this example, salt-washed rough microsomes (RM_K), which are inactive in wheat germ lysate, are added to a solution to be tested for signal recognition protein (SRP) activity under appropriate salt conditions. Elastase-treated, high-salt washed membranes (RM_{EK}) could be used in the same way to test for the presence of docking protein fragment (DP_f) in a given sample. (2) Suspensions are incubated for 30 min at 0°. (3) The membranes are pelleted by centrifugation at 12,000 *g* for 20 min. (4) The appropriate cell-free translation system is added to resuspend the membranes. (5) Translations are assayed for the translocation activity of the reconstituted membranes.

An exquisitely sensitive method for assaying factors involved in translocation has been devised. It makes use of the inactive membrane's ability to adsorb the required components from solution and thus reconstitute activity. In the traditional assay as described, the inactive membranes as well as the fraction to be tested are added to the translation system. As there is only a limited volume of exogenous material that one can add to a cell-free translation, reconstitution is largely dependent on the concentration of the activity-restoring factor in the fraction being tested.

The affinity-based assay (see Fig. 2) allows the sample of membranes, typically 2.5 μl at A_{280} = 60 units/ml, to absorb out the active factor from a sample up to 100 times larger than is normally added to a translation. Such a method also allows one to assay for active factors even when there are components present in the fraction that are inhibitory to *in vitro* protein synthesis (such as protease, ribonuclease, or competing initiation factors).

Method. Inactive RM (2.5 μl) at A_{280} = 60 units/ml in membrane buffer are added to the fraction being tested (typically 20–200 μl). The suspension is incubated for 30 min at 0° in a 1.5-ml microfuge tube. The sample is then centrifuged for 20–30 min at 12,500 *g* in a microfuge, and the supernatant is immediately removed with a syringe needle coupled to an aspirator. The pellet, which is barely visible, is then suspended in the complete cell-free translation system and assayed for reconstitution as indicated above. Such a microbinding assay can be used to measure the presence or levels of SRP in cellular extracts,[11] or, as in this case, for following the purification of DP_f.

Purification of Docking Protein

All procedures are carried out at 0° unless specified otherwise.

Step 1. Preparation of Stripped Rough Microsomes. Dog pancreas rough microsomes, prepared by the method described by Blobel,[7] are first stripped of ribosomes by a combination of EDTA and high-salt treatment. The pelleted RM are suspended to a concentration of 60 A_{280}/ml (measured in 2% SDS) in 0.25 *M* sucrose, 20 m*M* HEPES, pH 7.5, 50 m*M* KCl, 5 m*M* DTT (membrane buffer). Using concentrated stock solutions, the [EDTA] is brought to 10 m*M* and [KCl] to 0.5 *M*. The suspension is incubated on ice for 15 min and then layered over a 0.5 volume cushion of 0.75 *M* sucrose, 20 m*M* HEPES, pH 7.5, and 0.5 *M* KCl in polycarbonate centrifuge tubes. The material is then centrifuged for 90 min at 105,000 g_{av}. After centrifugation three fractions are obtained: (*a*) a tight pellet consisting largely of polysomes; (*b*) the cushion into which the RM have migrated, which has a milky appearance; and (*c*) a clear, EDTA-rich upper phase. The upper phase is carefully removed by aspiration and the milky layer is decanted from above the pellet. The milky solution is diluted with two volumes of 20 m*M* HEPES in order to reduce the sucrose concentration to 0.25 *M*. This membrane-containing suspension is then centrifuged for 60–90 min at 105,000 g_{av} to pellet the stripped RM.

Step 2. Detergent Extraction of Stripped RM. Preliminary studies demonstrated that protease-high salt treatment of RM yielded, in addition to others, two proteins of M_r 60,000 whose characteristics were similar enough to complicate purification procedures. One protein is the fragment of the docking protein (DP_f), the other a contaminant having no relevant activity. Fortunately, this contaminant can be removed from RM by detergent treatment at low salt concentrations, a condition that allows the docking protein to remain associated with the detergent-insoluble material.

The pellet of stripped RM is suspended to a concentration of A_{280} = 50 units/ml in membrane buffer containing 40 μg of phenylmethylsulfonyl fluoride (PMSF) per milliliter. Triton X-100, 20% (w/v), is added until a

[11] P. Walter and G. Blobel, this volume [53].

final Triton concentration of 1% is achieved. The membrane suspension is gently agitated and then layered over a cushion of 0.5 volume of 0.5 *M* sucrose, 20 m*M* HEPES, 50 m*M* KCl. Centrifugation is carried out for 90 min at 105,000 g_{av}. The Triton phase, as well as the cushion are aspirated off, and the pellet is resuspended in 0.5 *M* KCl, 20 m*M* HEPES, pH 7.5, 5 m*M* DTT to a concentration of A_{280} = 50 units/ml.

Step 3. Elastase Extraction of Triton-Insoluble Membrane Material. A 1% solution of elastase (from pig pancreas) is treated with an equal volume of 1% Trasylol to inactivate nonelastolytic proteolytic activity. The elastase–Trasylol mixture is then added to the membrane solution to a final elastase concentration of 1 μg/ml. The suspension is then vigorously shaken at 0° for 1 hr in a reciprocally shaking ice-water bath. The proteolysis is stopped by the addition of PMSF to a final concentration of 200 μg/ml. The insoluble material is removed by centrifugation for 2 hr at 105,000 g_{av}, and the resulting supernatant represents the crude elastase extract (EE).

Step 4. Purification of DP_f from EE Using Ion-Exchange Chromatography. The fragment released from the docking protein by elastase and high salt is basic in character. For this reason, cation exchangers such as carboxymethyl-Sephadex represent an effective means of purification. The EE is first dialyzed into 200 m*M* KCl 20 m*M* HEPES, 1 m*M* DTT. A column is prepared of CM-Sephadex equilibrated with the same buffer. Routinely 1–2 ml of packed CM-Sephadex is more than sufficient to adsorb DP_f from 10 A_{280} units of EE. The dialyzed EE is applied to the column, equipped with a UV monitor set to 280 nm, and the column is washed with the equilibration buffer until no further UV-absorbing material comes off of the column. The DP_f is then eluted by the addition of 350

PURIFICATION OF ACTIVE FACTOR FROM ROUGH MICROSOMES

Elastase extract (+RM_i)	Specific activity[a]	Total protein[b]
Crude	4.4	27
After CM-Sephadex	13.3	2.0
After Sephadex G-150	320	0.05

[a] Specific activity is expressed as the reciprocal of the amount of material (in A_{280}) × 10^{-3} needed to convert 50% of IgG light chain precursor to authentic light chain in one 25-μl cell-free translation.

[b] A_{280} units.

m*M* KCl, 20 m*M* HEPES, pH 7.5, 1 m*M* DTT. The peak thus collected is designated CM-bound EE.

Step 5. Gel Filtration of CM-Bound EE. The products obtained from ion exchange lend themselves well to further purification by gel filtration. Since DP_f has an M_r of 60,000 and the major contaminants are of high M_r (>100,000), Sephadex G-150 can be used to purify DP_f to near homogeneity. Typically a 1 × 25 cm column is packed with fine Sephadex G-150 and equilibrated in 0.5 *M* KCl, 20 m*M* HEPES, pH 7.5, 1 m*M* DTT. The sample is then applied in a volume of about 1 ml and eluted at a flow rate of 1.5 ml/hr. Fractions of 0.5 ml are collected and assayed for activity and protein composition.

Preparation of Inactive Rough Microsomes

When DP_f is prepared from extracts of RM the elastase concentration is crucial. If too much is used, the DP_f liberated by the protease is also degraded by it.[3] It was determined that the optimal concentration for release of active DP_f is not sufficient, in a reasonable period of time, to remove all the DP activity from the membrane. Thus it is necessary to prepare the inactive RM separately from DP_f. Stripped RM (see below) are suspended in membrane buffer to a concentration of A_{280} = 60 units/ml. Protease, either trypsin at 5 μg/ml or elastase at 10 μg/ml, is added, and the sample is incubated in an ice-water bath for 60 min with continual agitation. The reaction is stopped by the addition of PMSF to a final concentration of 200 μg/ml. The membranes are pelleted by centrifugation at 105,000 g_{av} for 2 hr and then resuspended to a concentration of A_{280} = 60 units/ml. Such treatment should yield membranes that are capable of translocation to levels of less than 5%. Addition of crude elastase extract (see below) should enable full reconstitution of activity in the reticulocyte lysate system.

Analysis of Purification

At each stage in the purification scheme, activity measurements and an analysis of protein composition should be undertaken. The most convenient way to express the activity of a given sample is to perform a titration of the fraction, holding the membrane concentration constant, and determine the dilution (or protein concentration) at which 50% processing can be reconstituted in a standardized assay (see Quantitation of Activity, above). The results of such a purification analysis are shown in the table.

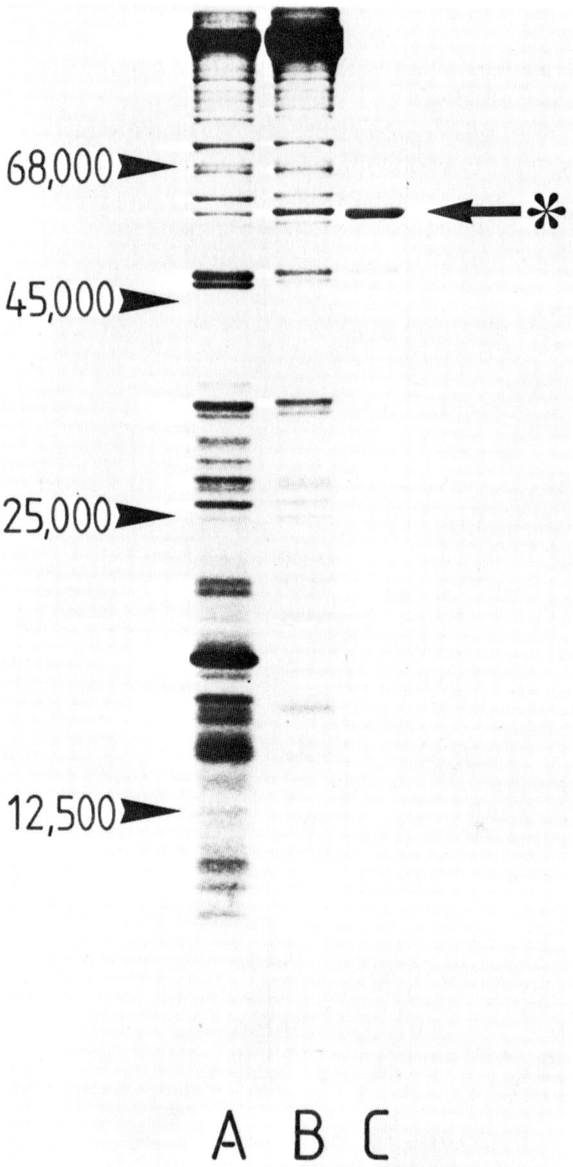

FIG. 3. Purification of material bound to CM-Sephadex by gel filtration. Elastase extract derived from Triton X-100-50 m*M* KCl-extracted microsomes was bound to CM-Sephadex at 200 m*M* KCl and eluted at 350 m*M* KCl. The eluted material was concentrated and further fractionated on Sephadex G-150, as described. Elastase extract fractions containing translocation/processing activity were separated on polyacrylamide gels and stained with Coomassie Blue. Lanes: A, elastase–high-salt extract of microsomes (EE); B, material eluted from CM-Sephadex at 350 m*M* KCl; C, translocation–processing activity-containing fraction obtained by gel filtration on Sephadex G-150. Asterisk-arrow indicates M_r 60,000 protein. From Meyer and Dobberstein.[4]

The protein composition of fractions is best determined by polyacrylamide gel electrophoresis and staining with either Coomassie Blue or silver. In the case of Coomassie Blue, samples have to be concentrated by trichloroacetic acid precipitation prior to gel electrophoresis. Use of the far more sensitive silver staining procedure[12] enables visualization of band patterns without prior concentration or precipitation of samples. A summary of a typical purification is seen in Fig. 3.

[12] C. R. Merril, D. Goldman, and M. L. Van Keuren, this volume [17].

[55] Biosynthesis of Glyoxysomal Proteins

By H. KINDL and C. KRUSE

Glyoxysomes and other microbodies contain a limited number of constituent proteins; they can, as organelles, easily be surveyed. Enzymes for fatty acid β-oxidation and glyoxylate cycle, catalase, some enzymes of urate degradation and glycolate pathway are housed in glyoxysomes.[1–5] The limited number of the components, the single membrane surrounding the organelle, and the fact that the proteins are encoded only by nuclear genes, render glyoxysomes very suitable for studies on the biosynthesis and assembly of an organelle.

Analysis of the Protein Components of Glyoxysomes

Investigations of the suborganellar location of the glyoxysomal enzymes establish that, in general, a preferential occurrence and organization has to be considered.[6–8] It is, therefore, practicable to separate the organellar components into three fractions: matrix, peripheral membrane proteins, and integral membrane proteins. Each of these fractions con-

[1] N. E. Tolbert, *Annu. Rev. Plant Physiol.* **22,** 45 (1971).

[2] H. Beevers, *Annu. Rev. Plant Physiol.* **30,** 159 (1979).

[3] S. E. Frederick, P. J. Gruber, and E. H. Newcomb, *Protoplasma* **84,** 1 (1975).

[4] B. Gerhardt, "Microbodies/Peroxisomen pflanzlicher Zellen." Springer-Verlag, Berlin and New York, 1978.

[5] A. H. C. Huang, T. Moore, and R. N. Trelease, "Microbodies." Academic Press, New York, 1983.

[6] C. Bieglmayer, J. Graf, and R. Ruis, *Eur. J. Biochem.* **37,** 553 (1973).

[7] A. H. C. Huang and H. Beevers, *J. Cell Biol.* **58,** 379 (1973).

[8] W. Köller and H. Kindl, *Arch. Biochem. Biophys.* **181,** 236 (1977).

ISBN 0-12-181996-5

tains only a few proteins. The performance of only a few operations, followed by electrophoretic analysis, leads to an overview. This kind of easily standardized analysis, combined with more selective procedures for the cytosolic fractions, is a valuable tool for allowing the sequence of steps during organelle biosynthesis. For this reason, it is not always necessary to exploit monospecific antibodies for the determination of conversion of radioactivity from precursor amino acids into specific proteins of the glyoxysomes.

Glyoxysomes are obtained by isopycnic density gradient centrifugation.[6–8] Sucrose is the preferred gradient material, but metrizamide[9] or Percoll[10] can also be used. For many purposes it is advantageous to apply onto the gradient a homogenate, freed from cell debris by filtering it through Miracloth, rather than a fraction obtained by differential centrifugation. It may also be advisable to use a combination of flotation and sedimentation.[11]

Fractions with glyoxysomes, characterized by marker enzymes, can be disintegrated by osmotic shock.[6–8] The matrix components are released, and can be separated from the membranes, which are sedimented by centrifugation. Then, the membranes still containing their peripheral proteins are subjected to purification by gradient flotation. Finally, the peripheral proteins can be removed from the membranes by solubilization with 100 m*M* $MgCl_2$.[8]

Procedure 1

Buffer A: 50 m*M* Tris-HCl, pH 7.5

Buffer B: 10 m*M* $MgCl_2$, 2 m*M* glyoxylate, in buffer A

Three milliliters of a suspension of glyoxysomes in 53% (w/w) sucrose are diluted with 6 ml of buffer A and treated in a Potter homogenizer. The matrix enzymes discharged from the organelles are separated from the membranes by centrifugation at 150,000 *g* for 40 min. The membranes obtained as a pellet are resuspended in 1 ml of 55% (w/w) sucrose in buffer A and then layered within a sucrose gradient consisting of 2 ml of 58% sucrose, 1 ml suspension of glyoxysomal membranes in 55% sucrose, and 34 ml of a gradient ranging from 52 to 30% sucrose in buffer A. Centrifugation is in an SW27 rotor (Beckman) at 27,000 rpm for 15 hr. Fractionation of the gradient yields a band at 44% (w/w) sucrose with the membranes containing all the peripheral proteins (Fig. 1).

[9] A. Hüttermann and G. Wendberger-Schieweg, *Biochim. Biophys. Acta* **453,** 176 (1976).

[10] W. M. Becker, H. Riezman, E. M. Weir, D. E. Titus, and C. J. Leaver, *Ann. N.Y. Acad. Sci.* **386,** 329 (1982).

[11] B. Ludwig and H. Kindl, *Hoppe-Seyler's Z. Physiol. Chem.* **357,** 177 (1976).

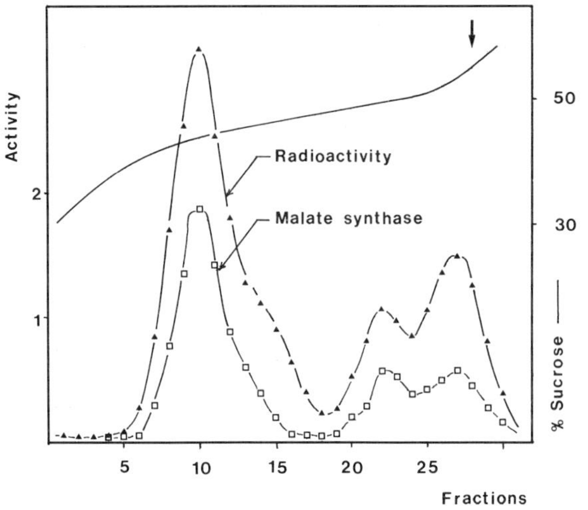

FIG. 1. Flotation of glyoxysomal membranes in a nonlinear sucrose gradient. The position where the crude membranes are applied is indicated with an arrow. The profiles of enzyme activities are almost identical for both malate synthase and citrate synthase. When radioactive glycerol is used to label the phospholipids of membranes *in vivo,* the radioactivity profile is also an indicator of membranes.

Membrane suspension (3 ml) is mixed with 3 ml of 200 mM $MgCl_2$ and shaken for 15 min. The stripped membranes are removed by centrifugation at 150,000 g for 1 hr. In the supernatant, malate synthase, citrate synthase, and malate dehydrogenase are the predominant proteins.

The matrix fraction, as well as the fraction with peripheral proteins, consists of a few proteins only; the differences in M_r or pI (see the table)[8,12–28] are generally sufficient to isolate the proteins by a one-step

[12] S. Schiefer, W. Teifel, and H. Kindl, *Hoppe-Seyler's Z. Physiol. Chem.* **357,** 163 (1976).
[13] H. Ruis, *Can. J. Biochem.* **57,** 1122 (1979).
[14] J. Frevert and H. Kindl, *Eur. J. Biochem.* **92,** 35 (1978).
[15] J. E. Lamb, H. Riezman, W. M. Becker, and C. J. Leaver, *Plant Physiol.* **62,** 754 (1978).
[16] J. Frevert and H. Kindl, *Eur. J. Biochem.* **107,** 79 (1980).
[17] J. Frevert and H. Kindl, *Hoppe-Seyler's Z. Physiol. Chem.* **361,** 537 (1980).
[18] R. A. Walk, S. Michaeli, and B. Hock, *Planta* **136,** 211 (1977).
[19] W. Behrends, U. Rausch, H. G. Löffler, and H. Kindl, *Planta* **156,** 566 (1982).
[20] D. W. Rehfeld and N. E. Tolbert, *J. Biol. Chem.* **248,** 4803 (1972).
[21] M. W. Kerr and D. Groves, *Phytochemistry* **14,** 359 (1975).
[22] H. Kindl, S. Schiefer, and H.-G. Löffler, *Planta* **148,** 199 (1980).
[23] H. Kindl, *Ann. N.Y. Acad. Sci.* **386,** 314 (1982).
[24] T. Osumi and T. Hashimoto, *Biochem. Biophys. Res. Commun.* **83,** 479 (1979).
[25] S. Pawar and H. Schulz, *J. Biol. Chem.* **256,** 3894 (1981).
[26] W. Köller and H. Kindl, *FEBS Lett.* **88,** 83 (1978).

MICROBODY PROTEINS PURIFIED TO HOMOGENEITY

	Molecular properties					
Enzyme	Subunit M_r	M_r	pI	Localized within microbodies	Other forms[a] localized	Leading references
Catalase	53,500[b]	250,000	6	Matrix	—	12, 13
Isocitrate lyase	64,000	260,000	6	Matrix	—	14, 15
Multifunctional protein	75,000	75,000[c]	9	Matrix[d]	Absent in mitochondria[e]	16
Thiolase	45,000	90,000	—	Matrix	—	17
Malate synthase	63,000	550,000	9	Membrane, peripheral	Cytosol[f]	8
Citrate synthase	45,000	100,000[g]	9	Membrane, peripheral	Cytosol,[f] mitochondria[h]	8
Malate dehydrogenase	34,000	70,000[i]	9	Membrane, peripheral	Mitochondria[j] Cytosol (3 forms)	8, 18
erine–glyoxylate aminotransferase	47,000	185,000	5,5	Matrix	—	19, 20
Glycolate oxidase	43,000	700,000[k]	9	Matrix	—	19, 21
-Hydroxypyruvate reductase	46,000	95,000	6	Matrix	In mitochondria of algae	19

[a] Small pools in cytosol are not considered.
[b] Additional form with subunit M_r 57,000 was found.[22,23]
[c] There are great similarities with the respective protein in liver peroxisomes[24] and *Escherichia coli*.[25]
[d] At least 10% of the enzyme was found to be associated with the membrane.
[e] Enzymes of β-oxidation in mitochondria are constructed in another way.
[f] Highly aggregated forms appear in the cytosol.[26]
[g] A tetramer of M_r 200,000 is also detectable.[8]
[h] Mitochondrial citrate synthase: similar M_r, pI = 6.1.[27]
[i] Hexamers were also found.[8]
[j] Mitochondrial malate dehydrogenase: subunit M_r 37,000, pI = 5.4.[28]
[k] In several cases, dimers and tetramers were reported instead of 16-mers.[19]

procedure. Irrespective of such further separations, the mere electrophoresis in the presence of sodium dodecyl sulfate (SDS) permits an analysis and overview, based on the knowledge of subunit M_r (see the table) of the majority of glyoxysomal protein components and facilitated by the fact that most proteins are composed of identical subunits.[28a,b]

[27] C. Schnarrenberger, H. Zehler, and H.-H. Fitting, *Hoppe-Seyler's Z. Physiol. Chem.* **361,** 328 (1980).
[28] B. Hock and C. Gietl, *Ann. N.Y. Acad. Sci.* **386,** 350 (1982).
[28a] H. Kindl, *Int. Rev. Cytol.* **80,** 193 (1982).
[28b] H. Kindl, *Comprehensive Biochem.*, in press.

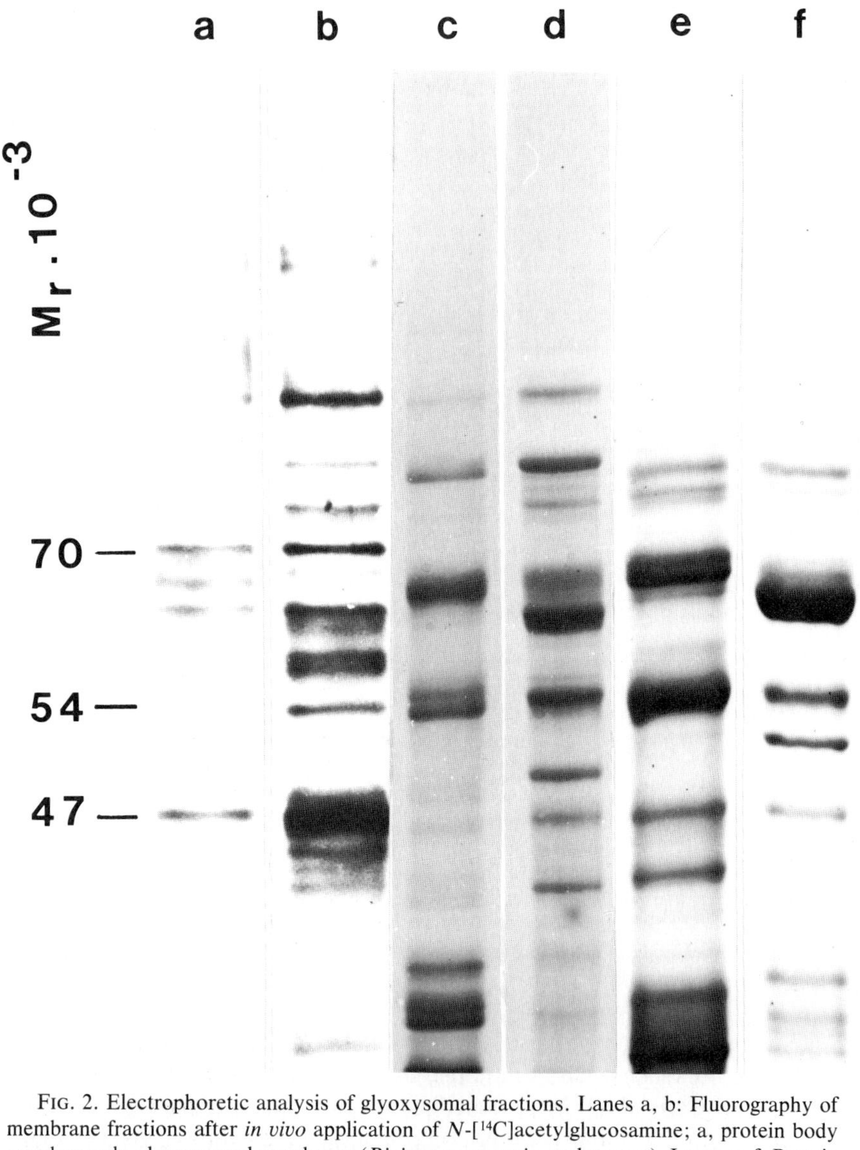

FIG. 2. Electrophoretic analysis of glyoxysomal fractions. Lanes a, b: Fluorography of membrane fractions after *in vivo* application of *N*-[^{14}C]acetylglucosamine; a, protein body membrane; b, glyoxysomal membrane (*Ricinus communis*, endosperm). Lanes c–f: Protein stains; c, matrix of glyoxysomes (*R. communis*). Lane d: Peripheral membrane proteins of glyoxysomes (*R. communis*). Lane e: Matrix of glyoxysomes (*Cucumis sativus*, cotyledons). Lane f: Peripheral membrane proteins of glyoxysomes (*C. sativus*). In lanes c and e, bands in the range of M_r 33,000–37,000 are contaminations originating from protein bodies.

This kind of analysis does not apply to glycoproteins being detected in the matrix and membranes of glyoxysomes as minor components only. They can be visualized after feeding of *N*-[^{14}C]acetylglucosamine *in vivo*[29,30] (Fig. 2).

Monomeric and Oligomeric Forms of Glyoxysomal Enzymes: Detection in the Cytosol

Synthesis and transport of organellar proteins can, in principle, proceed either via the endoplasmic reticulum (ER) or the cytosol. It is, therefore, reasonable to inquire whether pools, in fact precursor pools, of glyoxysomal proteins external to the organelle can be detected, e.g., in the luminal site of the ER or in the cytosol.

First we demonstrate the occurrence in the cytosol of a highly aggregated form of malate synthase.[8,31,32] This species is found at the stage of germination shortly before maximal glyoxysome assembly is observed.[33] It is, therefore, because of the time of its appearance, a likely candidate as precursor[34] of the glyoxysomal malate synthase.

Besides these aggregates, it is necessary to search for monomeric forms which are, in contrast to the oligomers in the organelle, expected to be the primary products of protein synthesis.

Procedure 2

Grinding medium: 150 m*M* Tris-HCl, pH 7.5, 10 m*M* KCl, 1 m*M* EDTA, 1 m*M* $MgCl_2$ 15% (w/w) sucrose

A homogenate is prepared by chopping the cotyledons of 2.5-day-old seedlings with razor blades on a chilled polystyrene block; 15 ml of cold grinding medium were used for 20 g of cotyledons. The homogenate from 60 g of cotyledons is squeezed through Miracloth and layered onto a sucrose gradient in a zonal rotor (14 Ti, Beckman). The total gradient should contain (in buffer A) a linear gradient ranging from 20 to 60% (w/w) sucrose (510 ml), 60 ml of the homogenate, and 80 ml of an overlay (5% sucrose). After the run at 47,000 rpm for 45 min, the gradient is fractionated into 10-ml fractions. A typical result is shown in Fig. 3.

Peak I of Fig. 3 (gradient supernatant) consists of soluble proteins and contains monomeric 5 S malate synthase and octameric 19 S malate syn-

[29] U. Bergner and W. Tanner, *FEBS Lett.* **131,** 68 (1981).

[30] A. Sturm, U. K. A. Kara, and H. Kindl, unpublished.

[31] H. Kindl, W. Köller, and J. Frevert, *Hoppe-Seyler's Z. Physiol. Chem.* **361,** 465 (1980).

[32] W. Köller and H. Kindl, *Hoppe-Seyler's Z. Physiol. Chem.* **361,** 1437 (1980).

[33] E. Gonzalez and H. Beevers, *Plant Physiol.* **57,** 406 (1977).

[34] J. M. Lord and L. Bowden, *Plant Physiol.* **61,** 266 (1978).

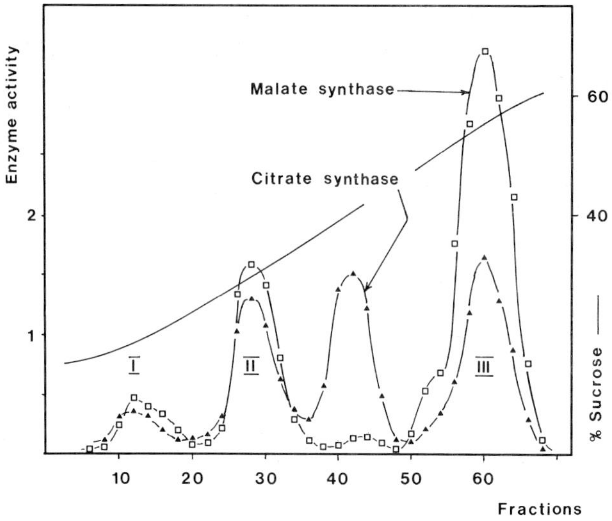

FIG. 3. Separation of three pools of malate synthase.

thase. Peak II includes ER vesicles and the form of malate synthase that is characterized by a sedimentation value of approximately 100 S. Peak III represents the glyoxysomal fraction.

To purify the cytosolic 5 S malate synthase, protein of peak I is concentrated by precipitation with ammonium sulfate (90% saturation). The precipitated protein is dissolved in 10 ml of buffer B, put onto an Ultrogel AcA-44 column (v_t = 400 ml) and run in the presence of Mg^{2+}-glyoxylate (buffer B). Fractionation yields two peaks with malate synthase activity: at M_r 60,000–70,000 (monomer) and at M_r 550,000 (octamer). The catalase peak in between (M_r 250,000) can be exploited as internal standard. The fraction with monomeric malate synthase is then applied onto a Blue Dextran-Sepharose column (v_t = 7 ml). The bulk of proteins is removed by washing the column with buffer B, and a linear gradient of KCl (0 to 300 m*M*) is used to elute malate synthase. The presence of Mg^{2+}-glyoxylate prevents the enzyme from aggregation. In a final step, the malate synthase fraction is bound to phenyl-Sepharose (v_t = 12 ml) and eluted by applying a combined KCl-ethylene glycol gradient. A procedure as outlined in Fig. 4, standardized with all known glyoxysomal proteins and tested with proteins imaginable as main contaminants, permits not only a good separation of hydrophobic glyoxysomal, but also a purification of cytosolic 5 S malate synthase. Malate synthase is the component with the highest affinity toward the hydrophobic material of the column.

The fractions of peak II are diluted with the equal volume of buffer A, and the aggregated malate synthase can be sedimented at 150,000 *g* for 1 hr. The pellet is dissolved in buffer A and sedimented, by zonal centrifugation, as 100 S form (Fig. 5). The peak fraction is diluted with buffer A, and the malate synthase is pelleted as before. The aggregated form is then shifted into the 19 S form (octamer) by dissolving in buffer B. Further purification, as the 19 S form, is achieved by sedimentation velocity centrifugation (Fig. 5) in the presence of Mg^{2+}-glyoxylate (buffer B).

As evidenced by a great number of experiments with different systems, the ER is implicated as site of posttranslational or cotranslational modification, e.g., en bloc glycosidation. It is also a proper location of integration and sorting out of precursors destined for organelles. It was for this reason that intensive studies were performed to characterize the ER in fatty acid degrading cells, and at this very stage when glyoxysome biosynthesis takes place. Especially the relationship between ER and the highly aggregated forms of citrate synthase and malate synthase, both

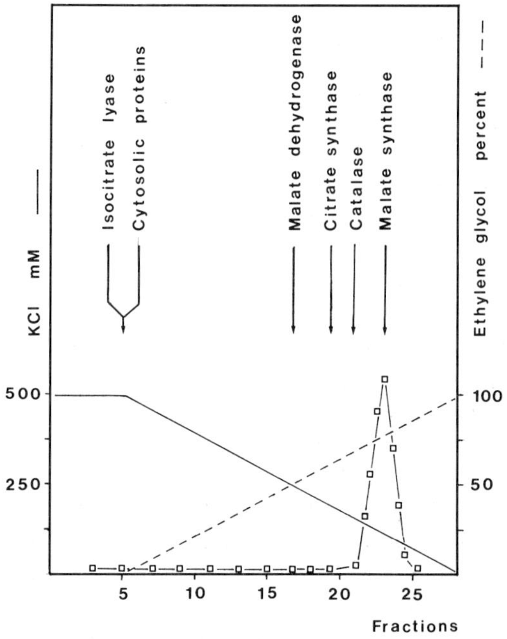

FIG. 4. Chromatography on phenyl-Sepharose. The samples are loaded onto the column equilibrated with Tris-HCl buffer, pH 8.0, containing 500 m*M* KCl. After washing with the same buffer, the proteins are eluted with a decreasing concentration of KCl and an increasing concentration of ethylene glycol.

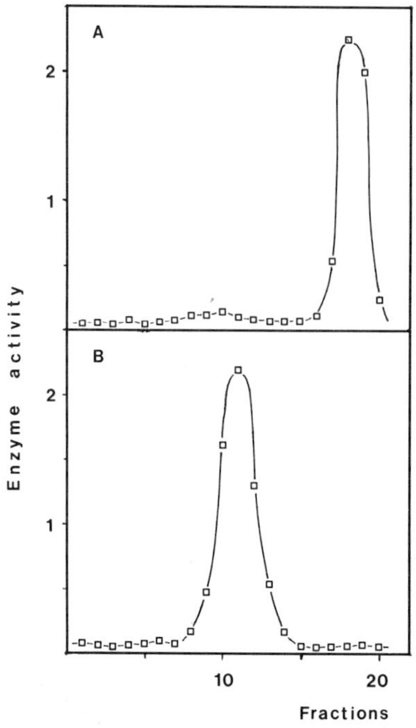

FIG. 5. Sedimentation velocity centrifugation of two different forms of malate synthase: (A) 100 S form; (B) 19 S (octameric) form. Centrifugation on a sucrose gradient (procedure 3) was at 40,000 rpm (SW40 rotor, Beckman) for 10 hr. The gradient was prepared with buffer A in case A, and with Mg^{2+}-glyoxylate (buffer B) in case B.

exhibiting sedimentation values similar to those of the smooth ER vesicles, was subject to examinations with different approaches: sedimentations and flotations[26] in order to purify the ER, and immunological studies aimed at the effective but gentle separation of ER from putative precursor forms of citrate synthase and malate synthase. One of the best procedures to isolate the vesicles of ER is floating them in a sucrose gradient.[23,26] Under these conditions, contaminating highly aggregated malate synthase and citrate synthase will not float up along with the ER, but rather sediment. That is in contrast to malate synthase and citrate synthase associated with the glyoxysomal membrane (Fig. 1). These results[23,31,32] should be appropriate to resolve the controversy concerning a location of malate synthase and citrate synthase in the luminal surface of the ER.[33,35]

Kinetic Studies *in Vivo*

The possibility of cotranslational transport of precursors into the ER and subsequent segregation was suggested by the findings of a malate

[35] T. Kagawa and E. Gonzalez, *Plant Physiol.* **68,** 845 (1981).

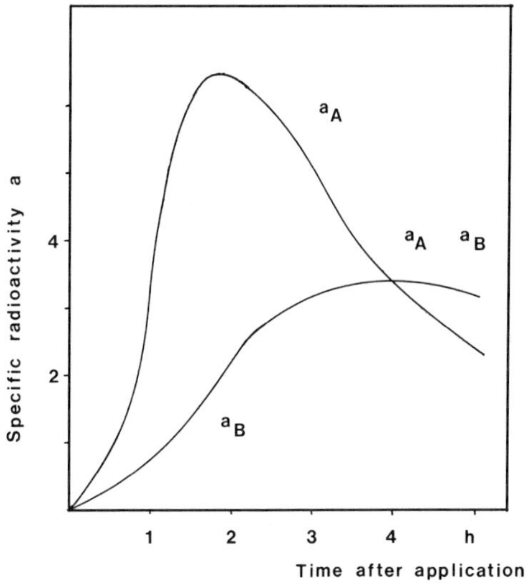

FIG. 6. Time course of specific radioactivity of precursor A and product B.

synthase assigned to the ER[33] and by kinetic studies *in vivo* demonstrating a quantitative transformation[34] of this pool into the pool of glyoxysomal malate synthase. Other investigators,[23] however, found also pools of soluble forms in the cytosol and showed that these cytosolic forms exhibited the highest specific radioactivity when radioactive amino acids were fed *in vivo* to cotyledons[32] or endosperm.[31] These cytosolic pools were established as precursors.

A reliable procedure to prove a precursor–product relationship is to apply the criterion of Zilversmit *et al.*[36] It states that the change of specific radioactivity, a, of the product B during time $(da_B)/(dt)$ is proportional to the difference of specific radioactivity of precursor A and specific radioactivity of product B.

$$da_B/dt = k(a_A - a_B)$$

At the maximum of specific radioactivity in the product pool (a_B), this curve is intersected by the descending curve of the specific radioactivity of the precursor pool (a_A). A theoretical curve is given in Fig. 6.

$$\text{Maximum:} \quad da_B/dt = 0 \qquad \text{for } a_A = a_B$$

Determining *specific* radioactivities of isolated proteins and precursors can pose problems, especially if very small pools are to be measured. In the case of glyoxysomes, an additional dilemma is posed by the rupture

[36] D. B. Zilversmit, C. Entenman, and M. C. Fishler, *J. Gen. Physiol.* **26,** 325 (1943).

of the organelles giving rise to too high values for the amount of protein in the "cytosolic" fraction (gradient supernatant). In the case of soluble matrix enzymes, therefore, a precursor pool can be demonstrated unequivocally only when very pronounced differences in the specific radioactivities between the cytosolic pool and the organelle pool are found. That the specific radioactivities in the cytosolic pools were significantly higher than in any other pools during the first 2.5 hr of application of radioactive amino acids was demonstrated for the multifunctional protein (enoyl-CoA hydratase–3-hydroxyacyl-CoA dehydrogenase)[31,37] and isocitrate lyase.[31,38]

In the case of malate synthase, however, a good fit of data with the criterion of Zilversmit *et al.* was achieved.[23] The pool of cytosolic forms (the sum of 5 S and 19 S forms) proved to be precursors of the glyoxysomal enzyme.

The criterion of Zilversmit *et al.* clearly shows that the highly aggregated form of cytosolic malate synthase, detectable in cotyledons at the stage of maximal glyoxysome biosynthesis, is not on the pathway to the glyoxysomes. Probably it is already modified and would, therefore, not be suitable for import into the organelle.

It is essential to our understanding of the steps in biosynthesis that we can recognize whether oligomerization is intrinsically correlated with import at the organelle membrane or can already take place in the cytosol. Kinetic studies and analysis of the cytosolic precursor forms, as described under procedure 3, revealed that, upon short pulses of radioactive amino acids, almost exclusively 5 S forms of isocitrate lyase or malate synthase were present. With increasing time, the portion of oligomeric forms, e.g., 19 S malate synthase, became significant.

Studies Demonstrating Synthesis and Import *in Vitro*

Poly(A) mRNA coding for plant microbody proteins can be prepared from fatty acid degrading tissues, e.g., endosperm (*Ricinus communis*) or cotyledons (Cucurbitaceae), but also from greening cotyledons and green leaves. Poly(A) containing RNA was also isolated from *Neurospora crassa* grown on acetate and used to direct the synthesis of glyoxysomal proteins.[39] Both reticulocyte lysates[38-40] and the wheat germ system[41]

[37] J. Frevert, H.-H. Gerdes, and H. Kindl, unpublished.

[38] L. M. Roberts and J. M. Lord, *Eur. J. Biochem.* **119,** 43 (1981).

[39] H. Desel, R. Zimmermann, M. Janes, F. Miller, and W. Neupert, *Ann. N.Y. Acad. Sci.* **386,** 377 (1982).

[40] C. Kruse, J. Frevert, and H. Kindl, *FEBS Lett.* **129,** 36 (1981).

[41] E. M. Weir, H. Riezman, J.-M. Grienenberger, W. M. Becker, and C. J. Leaver, *Eur. J. Biochem.* **112,** 469 (1980).

were found to be suitable for *in vitro* translations. For most of the *in vitro* translated microbody proteins, significant differences could not be detected when their M_r was compared with the subunit M_r of the mature enzyme in the organelle.[10,23,38] Precursors of glyoxysomal proteins do not seem to have signal sequences that are split off. A special feature shows malate dehydrogenase[28] that is synthesized at the ribosomes as M_r 41,000 species and is processed at the organelle membrane in more than one step to an M_r 34,000 form. In the case of catalase,[22,23] too, forms with higher M_r than the mature form were detected, but the question remained open whether the change in M_r is linked with the process of import.

Malate synthase and isocitrate lyase remain, after *in vitro* translation, in the monomeric 5 S form.[40,42] Oligomerization can take place both at the glyoxysomal membrane during import or, under certain circumstances, in the cytosol by a process not yet understood. Most likely, a minimal modification, e.g., including acetylation, and a partial refolding of the peptide chain are prerequisites of the oligomerization.

In vitro synthesized malate synthase (*Cucumis sativus*), and isocitrate lyase (*Neurospora crassa*)[42] are distinguished by their amphipathic character. The 5 S malate synthase can be coated with phospholipid and binds very easily to glyoxysomal membranes. If glyoxysomal membranes are tested as acceptors and added posttranslationally, a high percentage of de novo synthesized malate synthase is found to be taken up by the membrane. The integration of the protein gets at least to the stage where externally added proteinase K can only partly split the M_r 63,000 monomer, i.e., to an M_r 48,000 fragment not accessible to further degradation. Although free malate synthase can be degraded to the M_r 48,000 fragment by trypsin, proteinase K would completely degrade, under these conditions, any unprotected malate synthase. It is, therefore, likely that in the former case the malate synthase becomes partly inaccessible because of the protecting membrane.

Given the methodology of preparing glyoxysomes and isolating distinct components, the combination with an efficient translation system[43] provides the opportunity to design tests for demonstrating the import of newly synthesized proteins into the organelles. Probably, the most critical point in these experiments is to use glyoxysomes that are intact and have retained their capacity to take up specifically external components. It seems to be an adequate strategy to use a crude organelle preparation that contains an optimal amount of intact glyoxysomes. We therefore prepare a mixture of organelles all sedimenting similarly to glyoxysomes and separate the glyoxysomes only after the import has taken place. This proce-

[42] R. Zimmermann and W. Neupert, *Eur. J. Biochem.* **112,** 225 (1980).

[43] R. J. Jackson and T. Hunt, this volume [4].

dure permits one to test simultaneously the selectivity of the import, viz. when subsequently other fractions than glyoxysomes are also assayed for the enzyme assumed to be taken up by the glyoxysomes, e.g., malate synthase.

Import in glyoxysomes of malate synthase and malate dehydrogenase was successful. The highest radioactivity of imported malate synthase was determined in the octamer form. The import was dependent on the amount of glyoxysomes, and the maximal efficiency was approximately 50%. The other portion of radioactive malate synthase was detected in membranes of destroyed glyoxysomes and in the soluble fraction. Most results obtained so far reveal that only the proteins destined for glyoxysomes are imported by these organelles. This was demonstrated with malate dehydrogenase,[28] only the glyoxysomal isoenzyme being taken up, but the other multiple forms, one mitochondrial and three distinct cytosolic forms, not being transported. A dependence upon ATP could not be found for the import.

Transfer of isocitrate lyase synthesized *in vitro* into a particle fraction isolated from acetate-grown cells of *Neurospora crassa* was achieved. As a control, no transfer was found to occur when similar particles prepared from noninduced cells were used.[42]

Procedure 3

Preliminary Work. Poly(A) mRNA is prepared from total RNA or from polysomal fractions as described elsewhere.[41,44,44a] The gradients used for sedimentation velocity centrifugation consist of the following sucrose solutions (w/w) in buffer A: 1 ml 50%, 1 ml 35%, 1 ml 30%, 2 ml 27%, 2 ml 25%, 1 ml 20%, 2.5 ml 15%.

Preparation of Monomeric Precursors

a. Two microliters of poly(A) mRNA (0.06 μg/μl), 4 μl of [^{35}S]methionine (1.6 MBq; 30 TBq/mmol), and 16 μl of reticulocyte lysate are incubated for 1 hr at 35°. The translation is terminated by addition of 2 μl of cycloheximide solution (1 mg/ml), and the polysomes are removed by centrifugation in an airfuge at 100,000 *g* for 30 min. Then 2 μl of a 100 m*M* solution of L-methionine are added.

b. 10 MBq of [^{35}S]methionine (30 TBq/mmol) are applied to 1 g of excised cotyledons from 3-day-old seedlings. After 1 hr, a homogenate is prepared as described (Procedure 2) and is separated on a sucrose gradient (15 to 60%, w/w) using an SW27 rotor (Beckman) for 2 hr at 27,000

[44] H. H. Gerdes, W. Behrends, and H. Kindl, *Planta* **156,** 572 (1982).
[44a] P. M. Lizardi, this volume [2].

rpm. The gradient supernatant and the 100 S fractions (compare Fig. 3) are used for further operations.

Fractions either prepared according to (a) or (b) are subjected to sedimentation velocity centrifugation in an SW40 rotor (Beckman). The run is for 20 hr at 40,000 rpm and 4°. After fractionation, 10 μg of inactive isocitrate lyase are added to each fraction, and the immunoprecipitations are performed with 60 μl of anti-isocitrate lyase antibodies. After several washings the precipitates are prepared for gel electrophoresis. After immunoprecipitation the respective supernatants of each fraction were mixed with 10 μg of inactive malate synthase and 15 μl of anti-malate synthase antibodies. The procedures of precipitation, washing of the precipitates, and electrophoresis (9% of acrylamide) are described in detail in this volume [8]. It is important that the bands with a molecular weight of 60,000 migrate at least 5 cm.[44]

Procedure 4

Preliminary Work

Seedlings: Germinate for 3 days at 26° on vermiculite.

Grinding medium: Prepare as in Procedure 2.

Gradient: linear sucrose gradient ranging from 20 to 60%, prepared in 50 mM Tris-HCl, pH 7.5, containing 1 mM EDTA, 10 mM KCl, 1 mM $MgCl_2$.

Poly(A) mRNA: prepared from cotyledons of 2.5-day-old seedlings. The final solution contained 88 A_{260} units/ml.

Import Experiment. Translation is started by incubation of 10 μl of poly(A) mRNA (A_{260} = 22/ml) with 80 μl of reticulocyte lysate (The Radiochemical Centre, Amersham, U.K.) and 20 μl of [^{35}S]methionine (8 MBq; 30 TBq/mmol) in a 2.5-ml Eppendorf cup. Incubation is at 35° for 1 hr. During the time period of incubation, the organelle fraction is prepared: 10 g of cotyledons of 3-day-old seedlings are chilled (5°) and placed on a polystyrene block (trough 6 × 4 cm, 1 cm deep, chilled); 10 ml of grinding medium (0°) are added, and cotyledons are homogenized by using razor blades. The suspension is squeezed through Miracloth and centrifuged for 5 min at 2000 g. The resulting supernatant is again centrifuged for 25 min at 9000 g. The pellet is resuspended in 300 μl of grinding medium (organelle suspension). During the last centrifugation of the organelle suspension, the translation mixture is treated with cycloheximide (final concentration: 0.1 mg/ml) and centrifuged in an airfuge at 100,000 g for 20 min. Subsequently, the sucrose concentration is brought up to 15% (w/w); 200 μl of the organelle suspension are added to this stopped translation system. Import is at 25° for 60 min. Finally, the mixture is allowed to react with proteinase K (final concentration: 0.3 mg/ml for 30 min at 30°), stopped with PMSF,[40] and then applied onto the gradient.

Centrifugation is for 2 hr in an SW27 rotor (Beckman) at 27,000 rpm. At the beginning of the run, low acceleration is used. After the run and the fractionation (1.2 ml), fractions 1–3 are labeled gradient supernatant, fractions 15–17 are mitochondria, and fractions 22–25 are designated as glyoxysomes.

In these fractions, the soluble enzymes were freed by osmotic shock and the membranes were removed by sedimentation. The peripheral enzymes of the membrane were solubilized by adding 100 m*M* $MgCl_2$. The soluble enzyme preparations thus obtained (matrix enzymes, fractions with peripheral enzymes) were used for immunoprecipitation. After the usual procedures, the proteins were separated by SDS–gel electrophoresis.

Other Forms of Plant Microbodies

Glyoxysomal proteins are present not only in germinating seeds when degradation of fatty acids and gluconeogenesis dominate, but also in dry seeds,[45,46] and even at the last stage of seed development.[47,48] On the basis of molecular and immunological properties, these proteins cannot be distinguished from the respective glyoxysomal proteins during germination. That here, too, the biosynthesis proceeds via cytosolic precursors was demonstrated in the case of malate synthase, isocitrate lyase, multifunctional protein, and catalase.[22,48]

At least two different forms of microbodies were observed during seed development: at a late stage, a ripe form very similar or identical with glyoxysomes; and an earlier form of more simple microbodies with catalase and the multifunctional protein, but lacking malate synthase and isocitrate lyase.[22,47,48]

The biosynthesis of leaf peroxisomes, which contain catalase, malate dehydrogenase, and enzymes of the glycolate pathway (glycolate oxidase, serine : glyoxylate aminotransferase, and hydroxypyruvate reductase) was found to be under the control of phytochrome.[49] Poly(A) mRNA from green leaves of *Lens culinaris* appeared to be appropriate for study of the biosynthesis of catalase *in vitro*.

[45] W. Köller, J. Frevert, and H. Kindl, *Hoppe-Seyler's Z. Physiol. Chem.* **360,** 167 (1979).
[46] J. A. Miernyk, R. N. Trelease, and J. S. Choinski, *Plant Physiol.* **63,** 1068 (1979).
[47] J. S. Choinski, Jr., R. N. Trelease, and D. C. Doman, *Planta* **152,** 428 (1981).
[48] J. Frevert, W. Köller, and H. Kindl, *Hoppe-Seyler's Z. Physiol. Chem.* **361,** 1557 (1980).
[49] P. Schopfer, D. Bajracharya, R. Bergfeld, and H. Falk, *Planta* **113,** 73 (1976).

Quite incomplete is our knowledge concerning the transition of glyoxysomes into leaf peroxisomes.[28a] Electron microscopic studies concerning the association of microbodies with lipid bodies and chloroplasts[49] as well as histochemical staining procedures[50] hinted at the possibility that a transition form of organelle has to be taken into account during the first phase of greening of formerly fatty acid-degrading cotyledons. Results of biochemical studies with these subjects (e.g., cucumber, sunflower) were in accord with the morphological data. Glyoxysomal enzymes were found to be synthesized de novo at a stage of greening when the microbodies behave already as fully competent leaf peroxisomes.[51,52] Although unequivocal evidence is lacking, it is very likely that, apart from normal turnover, no new organelles are necessarily to be synthesized, but rather the already existing form of microbodies is gradually changed in content by uptake of precursors from the cytosolic pool.[23] The latter, being dependent on the formation and presence of mRNA species, is altered qualitatively during greening. Modulation in the synthesis of mRNA available in the cytosol[41] is, therefore, the decisive quality that governs the synthesis of glyoxysomes, glyoxyperoxisomes,[28a] leaf peroxisomes, or even "uricosomal" microbodies.[53]

In vitro studies did not give clues for the assumption that a change in selectivity during import or cytosolic precursors by glyoxysomes may be the controlling step. Experiments *in vivo* performed with cotyledons shortly before and after illumination have demonstrated (*a*) that the rate of de novo synthesis of malate synthase and isocitrate lyase did not cease abruptly after illumination of the cotyledons; and (*b*) that catalase and the multifunctional protein,[37,54] both being components also of leaf peroxisomes, were not synthesized de novo at an increased rate comparable with the formation and import of the peroxisomal enzymes of the glycolate pathway. If we postulate, during the transition from glyoxysomes to leaf peroxisomes in the first phase of illumination on fatty acid-degrading cotyledons, the de novo synthesis of a distinct organelle (peroxisome) we would expect to find an enhanced rate of de novo synthesis for all components of this new organelle, namely glycolate oxidase *and* catalase.

Acknowledgments

This work has been supported by investigations of the Deutsche Forschungsgemeinschaft (SFB 103, A8) and of the Fonds der Chemischen Industrie.

[50] J. J. Burke and R. N. Trelease, *Plant Physiol.* **56,** Suppl., 473 (1975).
[51] W. Köller and H. Kindl, *Z. Naturforsch., C: Biosci.* **33C,** 962 (1978).
[52] U. Franzisket and B. Gerhardt, *Plant Physiol.* **65,** 1081 (1980).
[53] Y.-N. Hong and P. Schopfer, *Planta* **152,** 325 (1981).
[54] B. Gerhardt, *FEBS Lett.* **126,** 71 (1981).

[56] *In Vitro* Processing of Plant Preproteins

By F. A. BURR and B. BURR

A cell-free system has been developed for the translation, processing, and packaging of plant preproteins.[1] We have used this method to demonstrate the synthesis, processing, and deposition of zein polypeptides and α-amylase. Comparable systems exist for the *in vitro* processing of preproteins in bacteria,[2] animals,[3] mitochondria,[4,5] and chloroplasts.[6] The present method uses a nuclease-treated rough endoplasmic reticulum (RER)-enriched preparation derived from developing maize endosperm in combination with a wheat germ translation system.[7]

Maize endosperm was found to be a superior source for RER active in processing. In contrast, membranes isolated from either young shoots or nucellar tissue of maize were far less efficient.[1] Comparison of membrane preparations from shoot, nucellus, and endosperm by sodium dodecyl sulfate (SDS)–polyacrylamide gel electrophoresis (PAGE) furthermore showed that the electrophoretic patterns of shoot and nucellus RER greatly resembled each other, but both were quite distinct from endosperm membranes.[1] This could imply that there is a developmental differentiation of the endosperm membranes into a type that is specialized to function in protein processing and sequestration, i.e., a "secretory-type" of RER.

In vivo, zein is synthesized in the maize endosperm on polyribosomes of the RER cisternae that subsequently enlarge to form 1–2 μm globular protein bodies.[8] Zein synthesis therefore occurs on the membrane surface, and the preprotein is transported across the membrane layers and into the RER intracisternal space. The protein is not shunted to another organelle for additional processing, nor is it secreted—it simply remains within the protein body membrane until seed germination, at which time it is degraded *in situ*. We know that the processing steps for the prezeins occur during translation, that is, they are cotranslational, for both message and membranes must be present at the same time in order for pro-

[1] F. A. Burr and B. Burr, *J. Cell Biol.* **90,** 427 (1981).

[2] C. N. Chang, G. Blobel, and P. Model, *Proc. Natl. Acad. Sci. U.S.A.* **75,** 361 (1978).

[3] D. Shields and G. Blobel, *J. Biol. Chem.* **253,** 3752 (1978).

[4] M.-L. Maccecchini, Y. Rudin, G. Blobel, and G. Schatz, *Proc. Natl. Acad. Sci. U.S.A.* **76,** 343 (1979).

[5] K. A. Sevarino and R. O. Poyton, *Proc. Natl. Acad. Sci. U.S.A.* **77,** 142 (1980).

[6] N.-H. Chua and G. W. Schmidt, *Proc. Natl. Acad. Sci. U.S.A.* **75,** 6110 (1978).

[7] B. Burr, F. A. Burr, I. Rubenstein, and M. N. Simon, *Proc. Natl. Acad. Sci. U.S.A.* **75,** 996 (1978).

[8] B. Burr and F. A. Burr, *Proc. Natl. Acad. Sci. U.S.A.* **73,** 515 (1976).

METHODS IN ENZYMOLOGY, VOL. 96

ISBN 0-12-181996-5

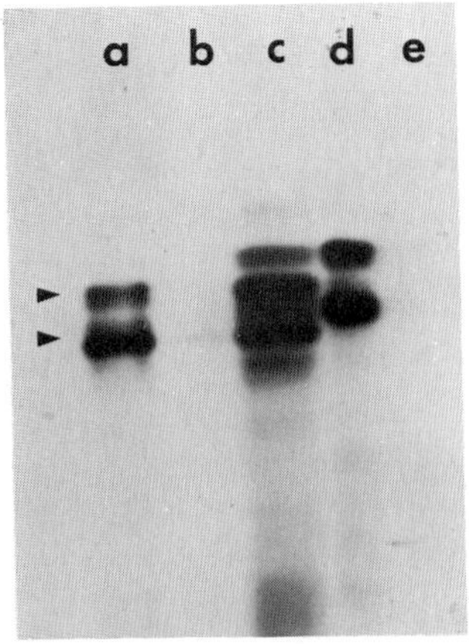

FIG. 1. Fluorogram of SDS–PAGE showing prezeins and processed zeins made *in vitro*: Lanes: a, Untreated membranes; b, membranes after treatment with micrococcal nuclease; c, treated membranes incubated with zein mRNA; d, zein mRNA only; e, synthesis mixture with no additions. Arrowheads at left indicate positions in the gel of authentic zein that were determined by Coomassie Blue staining.

cessing to take place. The processing of prezeins requires an exopeptidase that removes an approximately M_r 2000 presequence from the heavier class of M_r 24,500 preproteins and approximately M_r 1000 from the light class of M_r 20,000 prezeins producing the two mature classes of M_r = 22,500 and 19,000 (Fig. 1). There is also apparently an addition of a glucose residue to the molecule.[9] Other modifications are implicated as suggested by the large diversity of polypeptides generated under processing conditions (Fig. 2). Thus for Illinois High Protein, the particular strain being used, the unprocessed prezeins (mRNA only, no treated membranes) yield only five major spots on two-dimensional isoelectric focusing (IEF)–SDS–polyacrylamide gels, whereas the processed zeins (mRNA + treated membranes) separate into 14+ spots. The processed zein spots correspond exactly to the positions of authentic zein polypeptides whose positions were determined by Coomassie Blue staining. The zeins from each strain of maize have a characteristic and different two-dimensional IEF-SDS–gel pattern.[1,10] The particular pattern produced *in vitro* is known to be dependent solely on the source of the membranes (unpublished).

This technique will, it is hoped, be useful in confirming or distinguishing cloned sequences for plant preproteins, particularly in cases where a number of closely related proteins, such as the zeins,[11] are involved. The cloned material would be employed as a hybridization probe under stringent conditions to select homologous messages that would then be trans-

[9] F. A. Burr and B. Burr, *in* "The Plant Seed: Development, Preservation, and Germination" (I. R. Rubenstein, ed.), p. 27. Academic Press, New York, 1979.

[10] G. Hagen and I. Rubenstein, *Plant Sci. Lett.* **19,** 217 (1980).

[11] B. Burr, F. A. Burr, T. P. St. John, M. Thomas, and R. W. Davis, *J. Mol. Biol.* **154,** 33 (1982).

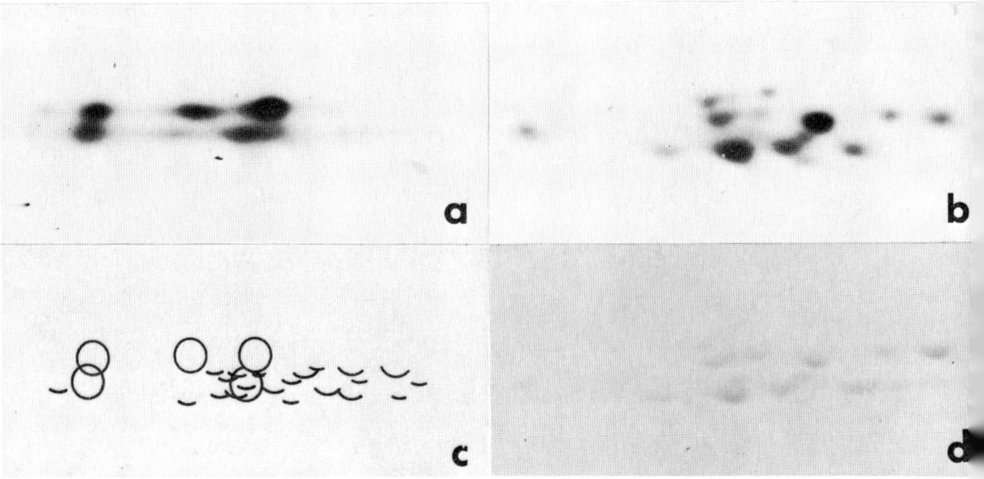

FIG. 2. Two-dimensional isoelectric focusing–SDS-PAGE of unprocessed and processed zeins synthesized *in vitro*. (a) Fluorogram of unprocessed products made by zein mRNA; (b) fluorogram of processed zein produced by zein mRNA in the presence of treated membranes; (c) diagram showing the relationship of unprocessed zein spots seen in (a), indicated by circles, to authentic zein that was electrophoresed in the same gel; (d) same gel as (b) stained with Coomassie Blue to show positions of authentic zein markers.

lated and processed. Although we designed the method to examine the prezeins, we have also been able to show that it can be used to translate and process another maize preprotein, the precursor to α-amylase. This enzyme is made in another tissue of the seed (aleurone) than zein and at another time in development (seed germination), but evidently it still can be processed by the RER of the developing endosperm.

The procedures for the isolation and preparation of treated maize RER membranes can probably be readily adapted to other plant species. We would suggest, however, based on our results with shoot and nucellar membranes, that when selecting the type of tissue to be used the investigator should keep in mind that a highly active "secretory-type" membrane system is probably required.

Material for Membrane Preparation

Nonpigmented kernels, such as Illinois High Protein, are preferred to colored kernels. Developing ears 15–18 days after pollination are harvested, husked, and held in wet ice until use. The endosperms are dissected from the kernels by first cutting the tops off a row of kernels with a razor blade then running a minispatula around the kernel just inside the pericarp to free the endosperm. The endosperm separates very easily from the embryo. The pieces of the endosperm are dropped immediately into a Dewar flask containing liquid nitrogen. Frozen endosperms may be stored at −85° for at least a year with no detectable loss of activity.

Isolation of Membranes

Routinely, 10–20 g of frozen endosperms are put into a 100-ml VirTis homogenizer flask (VirTis Co., Inc., Gardiner, New York), and 1 ml of cold grinding buffer is added per gram of tissue. The grinding buffer, or PB (protein body) buffer, is 50 m*M* Tricine, pH 8.0, 100 m*M* KCl, 5 m*M* $MgCl_2$, 1 m*M* EDTA, 1 m*M* dithiothreitol, 2 m*M* adenosine-2′,3′-monophosphoric acid, 20% sucrose (w/v). The tissue is ground in an ice bath at position 25–30 on the VirTis 45 until a smooth brei is produced, generally in 2–3 min. The homogenate is filtered through two layers of freezer-chilled cheesecloth and one layer of Miracloth, then centrifuged at 3000 *g* for 10 min. The supernatant is decanted carefully into a clean tube and dispensed onto ice-chilled sucrose gradients. The gradients are made up in Beckman SW41 nitrocellulose tubes (Beckman Instruments, Inc., Spinco Division, Palto Alto, California) that contain an 8-ml linear gradient of 35 to 70% sucrose (w/v) in PB buffer overlayered by a 1.5-ml pad of 22.2% sucrose (w/v) in the same buffer. A maximum of 3.0 ml of homogenate can be loaded on each gradient. The gradients are centrifuged at 35,000 rpm for 2 hr at 0°. After centrifugation five bands are seen (Fig. 3). The most prominent band will be the bottommost or protein body band.[1]

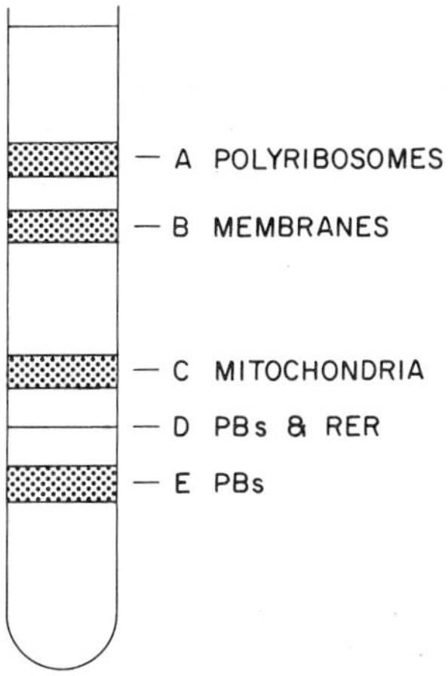

FIG. 3. Diagram of the bands seen in the sucrose density gradients of endosperm homogenates after centrifugation. Each band is identified as to its major component(s).[1]

The RER band is the second band from the top of the gradient and is about 2.5 cm from the surface. (If yellow kernels are used, the band is usually quite yellow from the carotenoids and xanthophylls.) Care should be taken to keep the gradients well chilled on ice until the bands are collected.

A silanized Pasteur pipette that has been bent to form a right angle about 0.5 cm from the tip is lowered into the gradient and used to suck up the band. The pooled membranes are diluted with 2 volumes of 10 m*M* HEPES 4-(2-hydroxyethyl-1-piperazineethane sulfonate), pH 7.6, 2 m*M* $MgCl_2$, and centrifuged at 100,000 *g* for 30 min. The pellets are taken up in 20% glycerol (v/v), 5 m*M* HEPES, pH 7.6, 1 m*M* calcium acetate, to 50 A_{260} units/ml and stored at −85°. Membrane preparations retain full activity for months. It must be emphasized that care should be taken to try and maintain all steps of the procedure at 0–4°.

Preparation of Treated Membranes

Before setting up a synthesis and processing reaction, the membranes are treated with micrococcal nuclease to destroy endogenous zein messages. The treatment apparently affects only mRNA, for electron micrographs of fixed and sectioned RER still show polyribosomal structures associated with the membranes, and gel electrophoresis indicates that no detectable degradation of the rRNAs has occurred in response to nuclease treatment.[1] This is in contrast to EDTA-treated RER that has been described as being stripped of polyribosomes. These nuclease-treated membranes are therefore referred to as "treated membranes" rather than "stripped membranes."

Membranes removed from low-temperature storage are thawed in ice water then treated with *Streptococcus aureus* (micrococcal) nuclease (P-L Biochemicals, Inc., Milwaukee, Wisconsin) at 10–15 units/ml for 10 min in a 20° water bath. The reaction is brought to 4 m*M* EGTA and left on ice for 10 min before centrifuging at 100,000 *g* for 30 min. The pellet is resuspended in 20% glycerol (v/v), 3 m*M* HEPES, pH 7.6, 1 m*M* magnesium acetate, to 100 A_{260} units/ml and kept on ice.

In Vitro Protein Synthesis and Processing

Wheat germ (General Mills) is enriched for embryos by using only the material retained on a No. 25 Tyler sieve. The procedure of Roberts and Patterson[12] is followed except that the extract (S-30) is not preincubated prior to desalting on Sephadex G-25 and potassium acetate is substituted

[12] B. E. Roberts and B. M. Paterson, *Proc. Natl. Acad. Sci. U.S.A.* **70,** 2330 (1973).

for potassium chloride. Desalted extracts are aliquoted in 0.4-ml polyethylene disposable tubes, frozen in liquid nitrogen, and stored at −85°.

The final reaction mixture was adapted from the translation system of Davies and Kaesberg[13] and includes 20 μ*M* L-[^{3}H]leucine, 66 m*M* potassium acetate, 4.4 m*M* magnesium acetate, 4 m*M* dithiothreitol, 20 μg of pyruvate kinase per milliliter, 30 μg of spermine per milliliter, 20 A_{260} units of S-30/ml, and 1/10 of the final reaction volume of "10X -Leu" [25 m*M* ATP, 3.75 m*M* GTP, 50 m*M* phosphoenolpyruvate, 0.25 m*M* each of 19 unlabeled amino acids (all of the 20 common amino acids except leucine), and 0.2 *M* HEPES, adjusted to pH 7.6 with potassium hydroxide]. The reaction volume is typically 25 μl and includes 0.3 μg of poly(A)$^+$ RNA and 4–5 μl of treated membranes at 100 A_{260} units/ml. Incubation is at 25°C for 90 min.

After the incubation period the external polypeptides that have not been taken up and included by the membranes are digested by adding 0.04 μg of proteinase K per microliter (EM Laboratories, Inc., Elmsford, New York) and leaving the tube at 4° for 45 min. Protease activity is terminated by adding 1 μl of a saturated solution of phenylmethylsulfonyl fluoride (Calbiochem-Behring Corp., San Diego, California) in 95% ethanol per 25-μg reaction. After 10 min on ice, the reactions are spun and the pellets are taken up in sample buffer for electrophoresis.

[13] J. W. Davies and P. Kaesberg, *J. Virol.* **12,** 1434 (1973).

[57] Biogenesis of Peroxisomal Content Proteins: *In Vivo* and *in Vitro* Studies

By PAUL B. LAZAROW

In our investigations of the biogenesis of rat liver peroxisomal proteins, we have found it useful to combine two approaches: analysis of events as they take place inside living cells, by means of cell biology methods; and reconstruction of these events *in vitro*, using the tools of molecular biology.

Biogenesis *in Vivo*

We have investigated the site of synthesis, properties, and intracellular transport of proteins destined for the peroxisomes. The basic experi-

ISBN 0-12-181996-5

mental approach involved: (*a*) pulse-labeling of liver proteins with radioactive amino acids; (*b*) fractionation of livers after various "chase" times; (*c*) isolation of peroxisomal proteins with antibodies; and (*d*) measurement of radioactivity in peroxisomal proteins by sodium dodecyl sulfate–polyacrylamide gel electrophoresis (SDS–PAGE) followed by fluorography or liquid-scintillation spectrometry. The subcellular location of newly synthesized peroxisomal proteins was determined by comparing the radioactivity distributions with marker enzyme analyses.

Pulse Labeling. In order to study intracellular transport cleanly, it is essential to label newly synthesized proteins with radioactive amino acids in a time that is short compared to the transport times. Injection of isotopes into the peritoneal cavity, while convenient, results in gradual absorption of the radioactivity, which is unsatisfactory for these purposes. Therefore, we inject the isotope into the portal vein.[1]

Rats are anesthetized with Nembutal at an intraperitoneal dose of 40 mg/kg. If necessary, the Nembutal anesthesia is supplemented with ether, applied with a filter paper nose cone.

The abdomen is opened, and the portal vein is exposed by gently pushing aside the intestines and turning back the overlying liver lobe. The radioactive amino acid is injected through a 25-gauge needle, taking care not to tear the portal vein during the injection. We have used 0.5 to 5 mCi of [^{3}H]leucine or of [^{35}S]methionine. Under these conditions, the isotope goes first directly to the liver; whatever isotope is not taken up and incorporated into liver proteins on the first pass is then diluted into the rest of the rat, and little returns to the liver during the next hour. We observe that approximately 10% of injected [^{3}H]leucine is incorporated into liver proteins, and the half-time of this incorporation is about 2.5 min.[1] This procedure thus gives reasonable "pulse–chase" labeling kinetics.

One-Step Cell Fractionation. In order to determine the subcellular location of newly synthesized proteins, we developed a simple procedure that substantially, although incompletely, resolves many of the major subcellular structures.[2] In particular, cell sap, microsomes, and peroxisomes are well separated.

The liver is excised from the anesthetized rat at the desired time after labeling, dropped into a tared beaker of ice-cold 0.25 *M* sucrose, and weighed. The liver is homogenized (by means of one stroke with a Potter–Elvehjem homogenizer turning at 1000 rpm) with ~2 volumes of 0.25 *M* sucrose (containing 0.1% ethanol to protect catalase activity). A postnuclear supernatant is prepared by differential centrifugation, rehomogeniz-

[1] P. B. Lazarow and C. de Duve, *J. Cell Biol.* **59,** 491 (1973).
[2] P. B. Lazarow and C. de Duve, *J. Cell Biol.* **59,** 507 (1973).

ing and resedimenting the nuclear fraction twice according to de Duve *et al.*[3] The combined supernatants are brought to a final volume of 5 ml per gram of starting liver.

This postnuclear supernatant is layered in the automatic rotor of Beaufay,[4] over a sucrose gradient that increases linearly in density from 1.10 to 1.27 and a sucrose cushion of 1.32 g/cm^3 (see Lazarow and de Duve[2] for details). It is centrifuged at 4° for ~15 min at 35,000 rpm to give

$$W = \int_0^t \omega^2 \, dt = 1.4 \times 10^{10} \text{ sec}^{-1}$$

(measured with an Integrator Accessory available from Beckman Instruments or Sorvall). These conditions were chosen, on the basis of a computer simulation of the sedimentation behavior of peroxisomes, as the minimal time to bring peroxisomes essentially to equilibrium at the bottom of the gradient.[2] Under these short centrifugation conditions, the mitochondria also reach density equilibrium, but the microsomal vesicles do not, improving their separation from peroxisomes.

The positions of the organelles in the gradient are determined by measuring the distribution of their characteristic marker enzymes according to de Duve.[5] Typical results are illustrated in Fig. 1 (top 4 panels). Although complete separation of organelles is not accomplished, sufficient analytic resolution is achieved that an unknown component may be assigned provisionally to one of the organelles if its distribution matches that of the marker enzyme.

This rapid fractionation of a postnuclear supernatant in a sucrose gradient may be carried out in commercially available rotors. Fowler obtained good resolution in the Sorvall TV865B vertical rotor.[6] One milliliter of rat liver postnuclear supernatant (containing 15 mg of protein) was layered over a 15-ml linear sucrose gradient (1.05 to 1.28 g/cm^3) and a 1-ml cushion (1.32 g/cm^3). Centrifugation was for 40 min at 50,000 rpm at 10°.

Satisfactory results may also be obtained in a Beckman SW39 swinging-bucket rotor. Postnuclear supernatant (0.5 ml) is layered over a 4.6-ml linear sucrose gradient (1.10 to 1.27 g/cm^3) and a 0.3 ml 1.32 density cushion and centrifuged for $W = \int \omega^2 \, dt = 1.5 \times 10^{11}$ sec^{-1}.[2] In the absence of an integrator, the time integral of the squared angular velocity

[3] C. de Duve, B. C. Pressman, R. Gianetto, R. Wattiaux, and F. Appelmans, *Biochem. J.* **60,** 604 (1955).

[4] F. Leighton, B. Poole, H. Beaufay, P. Baudhuin, J. W. Coffey, S. Fowler, and C. de Duve, *J. Cell Biol.* **37,** 482 (1968).

[5] C. de Duve, *Science* **189,** 186 (1975).

[6] Stanley D. Fowler, Department of Pathology, University of South Carolina School of Medicine (personal communication, 1981).

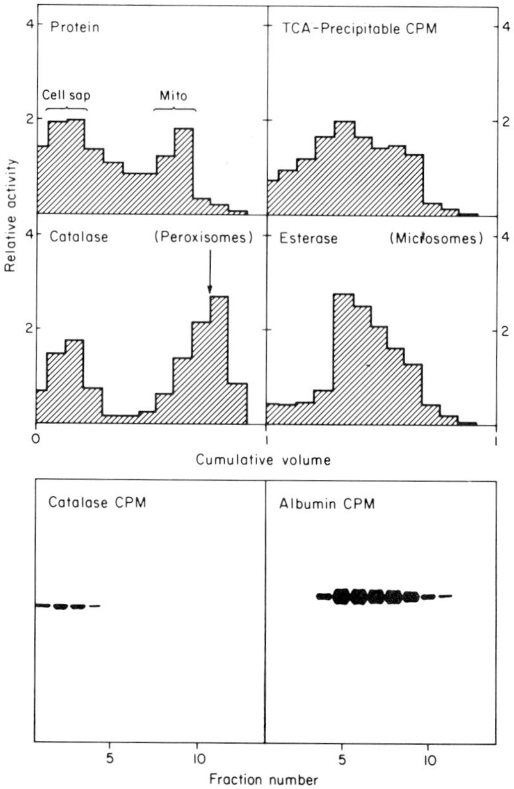

FIG. 1. Subcellular location of newly synthesized catalase and albumin after pulse-labeling *in vivo* with [^{35}S]methionine. A liver postnuclear supernatant was fractionated by centrifugation into a sucrose gradient.[2] The top 4 panels show the distributions of marker enzymes (sedimentation from left to right). The location of mitochondria was determined by measuring cytochrome oxidase (not shown). The lower left panel is a fluorogram of an SDS–PAGE analysis of the immunoprecipitates of catalase from each of the fractions. The lower right panel similarly shows [^{35}S]albumin. From Lazarow *et al.*[18]

may be calculated stepwise from revolutions per minute and time (in seconds) by

$$W = \sum_{0}^{t} \omega^2 \Delta t = \sum_{0}^{t} (2\pi \cdot \text{rpm}/60)^2 \Delta t$$

The g-min integral is related to W by

$$g\text{-min} = (W \times R_{\text{av}})/(981 \times 60)$$

where R_{av} is the radial distance of the middle of the gradient during centrifugation.[4] See de Duve *et al.*[7] for a thorough theoretical treatment of the sedimentation of particles in density gradients.

Isolation of Peroxisomal Proteins by Immunochemistry and SDS–PAGE; Radioactivity Measurements. Catalase is isolated by direct immunoprecipitation (with added carrier if there is less than 2–3 μg of catalase) in the presence of 0.15 *M* NaCl, 10 m*M* phosphate buffer, pH 7, 1% sodium deoxycholate, 1% Triton X-100, and 10 m*M* unlabeled leucine or methionine.[8] After incubation for 1 hr at 25°, the immunoprecipitates are collected in 400-μl microcentrifuge tubes according to Taylor and Schimke[9] by centrifugation (10,000 rpm for 5 min in a Sorvall HB-4 rotor) through layers of 0.5 *M* and 1.0 *M* sucrose, both containing salt, buffer, detergent, and amino acid as above. After freezing of the microcentrifuge tube in Dry Ice–acetone, the tip containing the immunoprecipitate is cut off with a razor blade and placed in a 1.5-ml microcentrifuge tube.

Residual 1 *M* sucrose overlying the immunoprecipitate is removed with Kleenex by capillary action, after which the pellet is dissolved in 5 μl of 0.1 *M* NaOH, agitating as required with an Eppendorf micropipette tip. SDS sample buffer is added (40 μl of 1.25 × normal strength), and the NaOH is neutralized with 5 μl of 0.1 *N* HCl. The samples are boiled, alkylated, and subjected to SDS–polyacrylamide gel electrophoresis in 7 to 15% gradient slab gels as usual.[8] The procedure of sedimenting the immunoprecipitates through sucrose layers[9] eliminates the need to wash the pellets and, in our experience, yields very clean immunoprecipitates. Residual radiochemical contaminants are removed by the subsequent SDS–PAGE.

The radioactivity in catalase is either visualized by fluorography[10] (as in Fig. 1, lower panel) or quantitated.[11] In the latter case, the band is cut out of the SDS–gel and incubated overnight at 60° with 0.5 ml of 10% H_2O_2; its radioactivity is measured by liquid scintillation spectrometry in the presence of 10 ml of Formula 963 Aqueous Counting Cocktail (New England Nuclear).

These procedures have been generalized for the study of many peroxisomal proteins.[12,12a] A polyclonal goat antibody against peroxisomal proteins was covalently linked to Sepharose and used to isolate eight or more peroxisomal proteins. These were then separated and analyzed by SDS–PAGE, followed by fluorography or quantitation of radioactivity. This

[7] C. de Duve, J. Berthet, and H. Beaufay, *Prog. Biophys. Biophys. Chem.* **9,** 325 (1959).

[8] M. Robbi and P. B. Lazarow, *Proc. Natl. Acad. Sci. U.S.A.* **75,** 4344 (1978).

[9] J. M. Taylor and R. T. Schimke, *J. Biol. Chem.* **248,** 7661 (1973).

[10] R. A. Laskey and A. D. Mills, *Eur. J. Biochem.* **56,** 335 (1975).

[11] M. Robbi and P. B. Lazarow, *J. Biol. Chem.* **257,** 964 (1982).

[12] L. Wong and P. B. Lazarow, *Fed. Proc., Fed. Am. Soc. Exp. Biol.* **40,** 1616 (1981).

[12a] P. B. Lazarow, M. Robbi, Y. Fujiki, and L. Wong, *Ann. N.Y. Acad. Sci.* **386,** 285 (1982).

approach permits the comparative analysis of many peroxisomal proteins simultaneously.

Results Obtained. Figure 1 illustrates the type of data obtained with these procedures for the case where a rat was pulse-labeled with [^{35}S]methionine for 8 min. The fluorogram in the panel at the lower left represents a three-dimensional plot of catalase radioactivity: the abscissa gives the position in the sucrose gradient, hence, together with the marker enzyme analyses, subcellular localization; the ordinate gives subunit molecular mass by SDS–PAGE; and the third dimension is intensity of the fluorographic image, giving the amount of radioactivity. The data show that newly synthesized catalase is found in soluble form, together with the cell sap proteins, but not in microsomes. Albumin, included as a control (lower right), is found exclusively in microsomes.

The fourth dimension in the experimental protocol is time: a series of experiments done at different chase intervals after the pulse labeling permit the analysis of the intracellular route and timetable by which catalase reaches peroxisomes. We found that catalase enters peroxisomes from a soluble pool as an apomonomer with a half-time of 14 min; inside the organelle it acquires heme and aggregates to its active tetrameric form.[2] Other peroxisomal proteins also enter the organelle *in vivo* from soluble pools, but at different rates.[12,12a] None of these proteins appear to be made as larger precursors.

Experiments *in Vitro*

Total rat liver polysomal RNA has been translated *in vitro* in both the wheat germ and the reticulocyte lysate cell-free protein-synthesizing systems,[8] by standard procedures described elsewhere in this volume. In the case of the wheat germ translations, we include the pure placental ribonuclease inhibitor purified by Blackburn *et al.*,[13,14] which has been shown to improve the translation of long, minor mRNAs.[15] It is effective in improving the translation of catalase mRNA.[8]

The translation product of catalase mRNA, isolated immunochemically from each of the protein-synthesizing systems, was found to be indistinguishable from the subunit of mature peroxisomal catalase by the criterion of SDS–PAGE.[8] The primary catalase translation product, labeled at its amino terminus using formyl-[^{35}S]Met-tRNA$_f^{Met}$, was also indistinguishable in size from the subunit of the intraperoxisomal enzyme.[11] These results demonstrate that catalase, unlike most other organelle pro-

[13] P. Blackburn, G. Wilson, and S. Moore, *J. Biol. Chem.* **252,** 5904 (1977).

[14] P. Blackburn, *J. Biol. Chem.* **254,** 12484 (1979).

[15] G. Scheele and P. Blackburn, *Proc. Natl. Acad. Sci. U.S.A.* **76,** 4898 (1979).

teins studied to date, is not synthesized as a larger precursor. The information directing catalase to the peroxisome may reside anywhere within its structure.

One potential pitfall in this type of experiment deserves comment. We found that catalase readily undergoes a limited proteolysis during its purification, such that its molecular mass decreases by 4000 daltons.[8] If the translation product were just compared with this purified enzyme, one would come to the erroneous conclusion that it is made as a larger precursor. We found that immunoprecipitation of *in vivo* labeled catalase from purified peroxisomes avoided this proteolytic artifact. Inhibitors of proteases may also be included during enzyme purification and/or immunoprecipitations.

Other Methods

Peptide Mapping. The fact that two molecules comigrate during SDS–PAGE does not demonstrate identity. Subtle structural modifications might be required for packaging an enzyme into an organelle, or might occur after packaging, preventing egress. Such modifications, especially of a large protein, might not be detected by SDS–PAGE. To look for such possible alterations to catalase, we carried out peptide mapping experiments, using both the one-dimensional limited proteolysis technique of Cleveland *et al.*[16] and two-dimensional mapping of tryptic peptides (see Robbi and Lazarow[11] for details and references). Mature peroxisomal catalase was compared both to its *in vivo*-labeled extraperoxisomal precursor and to the cell-free translation product. No differences were found, indicating that within the limits of the methods, catalase apparently undergoes no covalent modification of its primary structure during its packaging into peroxisomes.

Cytochemistry. Because an early theory of peroxisome biogenesis holds that peroxisomes form by budding from the endoplasmic reticulum, we have looked intensively for connections between the two structures. We carried out electron microscopic cytochemistry for catalase, the principal peroxisomal enzyme, and for glucose-6-phosphatase, an endoplasmic reticulum enzyme facing the ER lumen. We expected that cytochemical reaction product characteristic of one organelle might be found within the lumen of the other at points of continuity. No evidence of such connections could be found.[17] Based on these results and considerable other data, we hypothesized that peroxisomes may be interconnected with each

[16] D. W. Cleveland, S. G. Fischer, M. W. Kirschner, and U. K. Laemmli, *J. Biol. Chem.* **252,** 1102 (1977).

[17] H. Shio and P. B. Lazarow, *J. Histochem. Cytochem.* **29,** 1263 (1981).

other to form a "peroxisomal reticulum," rather than being connected to the ER.[18]

Membrane Composition. A new procedure has been devised for the isolation of organelle membranes as flat, unsealed sheets, free of contents and peripheral membrane proteins.[19] Using this procedure, the polypeptide composition of the peroxisomal membrane was found to be highly dissimilar from endoplasmic reticulum or mitochondrial membranes.[20]

Conclusions

A variety of methods have been employed to investigate the biogenesis of rat liver peroxisomal proteins. The *in vivo* studies, the *in vitro* experiments and the peptide mapping all indicate that catalase enters peroxisomes relatively slowly, posttranslationally from the cytosol, without modification of its primary structure. Less complete data suggest that the same conclusion likely applies to several other peroxisomal content proteins, except that the rate of packaging varies. All our results thus far are compatible with the hypothesis that new peroxisomes are formed from preexisting peroxisomes or the "peroxisomal reticulum." A more complete review of recent work from this laboratory appears in the proceedings of the International Conference on Peroxisomes and Glyoxysomes.[12a]

Acknowledgments

This work was supported by NIH Grant AM 19394, by NSF Grant PCM 8008713, and by an Established Fellowship from the New York Heart Association.

[18] P. B. Lazarow, H. Shio, and M. Robbi, *in* "31st Mosbach Colloquium. Biological Chemistry of Organelle Formation" (T. Bucher, W. Sebald, and H. Weiss, eds.), p. 187. Springer-Verlag, New York, 1980.

[19] Y. Fujiki, A. L. Hubbard, S. Fowler, and P. B. Lazarow, *J. Cell Biol.* **93,** 97 (1982).

[20] Y. Fujiki, S. Fowler, H. Shio, A. L. Hubbard, and P. B. Lazarow, *J. Cell Biol.* **93,** 103 (1982).

[58] Studies of Lysosomal Enzyme Biosynthesis in Cultured Cells

By RACHEL MYEROWITZ, APRIL R. ROBBINS, RICHARD L. PROIA, G. GARY SAHAGIAN, CHRISTINA M. PUCHALSKI, and ELIZABETH F. NEUFELD

In the course of their biosynthesis, lysosomal enzymes must be equipped with signals for translocation to lysosomes, the specialized organelles in which they will ultimately function. To that purpose, the newly formed precursor polypeptides undergo numerous posttranslational modifications, such as N-glycosylation followed by processing and phosphorylation of the high-mannose oligosaccharide chains; the polypeptide moieties undergo limited proteolysis. Lysosomal enzymes may also be secreted, or they may be internalized by endocytosis.

Methods for studying the natural history of lysosomal enzymes in intact cells are determined primarily by the low abundance of these interesting proteins. After biosynthetic labeling, a lysosomal enzyme may have to be purified 1000 to 10,000-fold. This is accomplished by immunoprecipitations followed by polyacrylamide gel electrophoresis. Structural studies may be performed after isolating radiolabeled polypeptides from the gel.

Labeling of Cells

The growth characteristics of the cells under study must be considered in planning the protocols. Human diploid skin fibroblasts, which grow slowly, are labeled at high density, whereas rapidly growing Chinese hamster ovary (CHO) cells are labeled at a density such that the cells will be subconfluent at harvest. Cells are grown at pH 7.0 in Eagle's minimum essential medium supplemented with nonessential amino acids, antibiotics, and 10% fetal bovine serum (5% for CHO cells), at 35° in 5% CO_2. It is convenient to plate them into 100-mm petri dishes for labeling experiments. Sterile conditions are maintained throughout the labeling procedures.

Labeling with Radioactive Amino Acids. The concentration of unlabeled amino acid in the medium is reduced in order to maximize isotope incorporation into the enzymes. Human fibroblasts, which can withstand relatively long periods of amino acid depletion, are first rinsed in Waymouth MAB 87/3 medium with added antibiotics, but free of leucine. The

ISBN 0-12-181996-5

cells are incubated for 1 hr in 5 ml of the same medium per 100-mm petri dish, to deplete the intracellular leucine pool, and then labeled for periods of 3 hr or longer in 4 ml of that medium, supplemented with 5% dialyzed fetal bovine serum and 50–100 μCi (1 nmol) per milliliter of [^{3}H]leucine.[1] By contrast, CHO cells do not tolerate amino acid starvation. Therefore, after rinsing in deficient medium, the cells are labeled for 1 hr in medium containing a reduced level of the amino acid of interest (leucine, 1.7 μg/ml; methionine, 1 μg/ml).[2] The labeling medium used for CHO cells is otherwise similar to the growth medium supplemented with dialyzed fetal bovine serum. Labeling is terminated by addition of unlabeled amino acid to a concentration double that of complete medium, or by replacement with unlabeled, complete medium; further incubation is carried out for the desired period of chase.

Labeling with Radioactive Mannose. To label with [2-^{3}H]mannose, the glucose concentration of the medium should be reduced, as glucose competes with mannose for entry and phosphorylation. However, complete depletion of glucose may result in a reduction of the number of mannose residues from nine to five in oligosaccharide chains transferred to newly made proteins.[3] Therefore, medium with low glucose is recommended; for human fibroblasts, Hasilik and Neufeld used Waymouth MAB 87/3 medium in which the glucose concentration was reduced to 1 m*M*,[4] whereas for CHO cells Robbins and Myerowitz used growth medium with dialyzed fetal bovine serum, with glucose reduced to 0.14 m*M* and with added 15 m*M* Tricine.[2]

Labeling may be terminated by addition of glucose to the concentration normally present in the medium or by replacement with complete medium.

Labeling with Radioactive Phosphate. Waymouth MAB 87/3 medium is prepared free of Na_2HPO_4 and KH_2PO_4. Minor components of the medium and dialyzed fetal bovine serum may contribute some inorganic phosphate, so that the actual P_i concentration of the labeling medium is 25–50 μ*M*. Carrier-free $^{32}PO_4$ or $^{33}PO_4$ are used in amounts of 250–500 μCi per 100-mm plate.[4] Labeling is terminated by addition of phosphate to a concentration of 3.6 m*M*. Since phosphatases are not inhibited in low-phosphate medium, secreted proteins may be dephosphorylated partially or completely; thus incubation in such medium should be brief, and the structure of phosphorylated polypeptide or carbohydrate residues should be interpreted with caution.

[1] A. Hasilik and E. F. Neufeld, *J. Biol. Chem.* **255,** 4937 (1980).
[2] A. R. Robbins and R. Myerowitz, *J. Biol. Chem.* **256,** 10623 (1981).
[3] J. I. Rearick, A. Chapman, and S. Kornfeld, *J. Biol. Chem.* **256,** 6255 (1981).
[4] A. Hasilik and E. F. Neufeld, *J. Biol. Chem.* **255,** 4946 (1980).

Labeling in the Presence of NH_4Cl. The presence of 10 m*M* NH_4Cl in the labeling medium inhibits the translocation of newly synthesized precursor enzymes to lysosomes and induces an almost quantitative secretion of these enzymes.[1,5] As 10 m*M* NH_4Cl is not toxic to the cells, incubation may be carried out for as long as 24 hr to maximize the yield of labeled precursor enzymes, using any of the labeling procedures described above. However, it may be desirable to omit fetal bovine serum if its presence would interfere with subsequent use of the labeled secretions[6]; in such cases, Waymouth MAB 87/3 medium is supplemented with 1 mg of bovine serum albumin per milliliter. (Most commercial preparations of bovine serum albumin are contaminated with lysosomal enzymes; the lot used should be pretested if the catalytic activity of the secreted enzymes is of importance to the study.)

Harvest of Labeled Media and Cells

Medium is collected from the plates and freed of floating cells by centrifugation. The pH is maintained by addition of NaP_i, pH 6.0, to 10 m*M*. Solid $(NH_4)_2SO_4$ is added to 70% saturation; the precipitate is collected, dialyzed against PBS-A (0.15 *M* NaCl–10 m*M* NaP_i, pH 6.0). After dialysis, the samples are adjusted to 0.6 ml, and 2 μl of 20% Triton X-100 are added. These samples may be stored frozen.

Cell monolayers are rinsed with PBS-B (0.14 *M* NaCl, 24 m*M* Na_2HPO_4, 2.7 m*M* KCl, 1.5 m*M* KH_2PO_4). Cells may be detached by trypsinization, washed with 0.15 *M* NaCl containing 1% fetal bovine serum to inactivate residual trypsin, and stored frozen in 50 m*M* Tris-HCl, pH 7.0 (0.5 ml for cells from one dish). After thawing, the enzymes are extracted by sonication; the suspensions are made 0.08% in Triton X-100 and 0.03% in protamine sulfate. After 10 min on ice, the extracts are centrifuged at 40,000 *g* for 1 hr.

A mild detergent solution may be used to extract lysosomal enzymes instead of trypsinization. After rinsing the cell monolayer with PBS-B, a solution consisting of 0.15 *M* NaCl, 0.01 *M* Tris-HCl, pH 7.4, 0.02% w/v NaN_3, 1% v/v Nonidet P-40 (0.5–1.0 ml per dish) is added and the dishes are gently rocked. The detergent extract is treated with protamine sulfate as above.

In some instances harsher methods may be required to extract lysosomal enzymes.[6a] After trypsinization and washing as above, cells are dis-

[5] A. Gonzalez-Noriega, J. H. Grubb, V. Talkad, and W. S. Sly, *J. Biol. Chem.* **85,** 839 (1980).

[6] A. Frisch and E. F. Neufeld, *J. Biol. Chem.* **256,** 8242 (1981).

[6a] R. L. Proia and E. F. Neufeld, *Proc. Natl. Acad. Sci. U.S.A.* **79,** 6360 (1982).

solved by heating at 100° for 5 min in a solution containing Tris-HCl, pH 7.4, 0.15 *M* NaCl, and 1% sodium dodecyl sulfate (0.3 ml per dish). The extract is cooled and diluted to 1 ml with a solution of Tris-HCl, pH 7.4, and 0.15 *M* NaCl containing 2% Nonidet P-40 and 10 mg of bovine serum albumin per milliliter. The cell extracts are then treated as above with protamine sulfate.

Isolation of Labeled Enzymes with Antisera

This step must provide extensive purification of the radioactive enzyme of interest, sometimes several thousandfold. Availability of highly specific antisera and close attention to the details of collection and washing of the immune precipitates are essential. Practice is important, but even experienced operators may find a significant number of precipitates to be contaminated. Two procedures are described, the first depending on direct precipitation of the enzyme–antibody complex, and the second on precipitation of the complex with *Staphylococcus aureus* bearing protein A.

Direct Immunoprecipitation. After centrifugation of medium concentrates and protamine-treated cell extracts, the clear supernatants are carefully removed with finely drawn Pasteur pipettes and placed in 1.5 ml of Eppendorf plastic tubes. The samples are now ready for immunoprecipitation and receive (*a*) immunoprecipitation buffer (4% Triton X-100, 0.4 *M* Tris-HCl, pH 7.0, and 1.6 *M* KCl, except for detergent extracts, which receive KCl only); (*b*) carrier enzyme if necessary; and (*c*) antiserum. The amount of immunoprecipitation buffer added to each sample equals one-quarter the total volume of the completed mixture. Samples low in endogenous enzyme activity, such as concentrates of medium or extracts of mutant cells, receive 100–500 ng of purified or partially purified carrier enzyme to facilitate the immunoprecipitation. Enough antiserum is added to precipitate the enzyme of interest, the amount depending on the concentration of enzyme in the sample and the antiserum titer. After all additions have been made, the samples are allowed to stand at room temperature for 30 min and then at 4° for at least 12 hr. Some antisera require a 24-hr period in the cold for maximal precipitation of the enzyme.

It is often desirable, for reasons of economy or expediency, to immunoprecipitate several enzymes from labeled cells or media. In such cases, immunoprecipitation buffer and carrier enzymes (if needed) are added to the samples. The first enzyme selected for immunoprecipitation is that requiring the smallest volume of antiserum; the second, that requiring the next larger amount; and so on. After precipitation of the first enzyme–antibody complex, the mixtures are centrifuged for 5 min at

15,000 *g* in an Eppendorf Model 5412 centrifuge. The pellets are washed and prepared for electrophoresis as described below. The supernatant fluids are treated with a 10% suspension of fixed *Staphylococcus aureus* bearing protein A that is just sufficient to remove all of the added immunoglobulin, and are clarified by centrifugation; this prevents contamination of the second enzyme immunoprecipitate with the first. The second antiserum is then added to the clarified supernatant, and the procedure is repeated until all desired enzymes have been immunoprecipitated.

Collection of the immunoprecipitates is carried out at 4°. The precipitates are sedimented in an Eppendorf Model 5412 centrifuge at 15,000 *g* for 5 min. The supernatant fluids are withdrawn gingerly, using finely drawn out Pasteur pipettes; it is important not to disturb or aspirate the pellets, which may be barely visible. Pellets are washed[7] with 1 ml of a solution containing 1.0 *M* KCl, 50 m*M* Tris-HCl, pH 7.4, and 1.2% Triton X-100, and they are resedimented. As much of the supernatant fluid as possible is then withdrawn from each tube without disturbing the pellets. A second wash of the pellets with 50 m*M* Tris-HCl, pH 7.4, containing 0.1 *M* NaCl, is followed by a final rinse with 1 ml of cold acetone. The acetone wash removes lipids from the immunoprecipitates and increases bulk by precipitation of salts from residual buffer. This facilitates the complete removal of supernatant liquid from the pellets. Pellets are allowed to air dry for several minutes and are then solubilized by heating at 100° for 5 min in 50 μl of gel sample buffer (0.125 *M* Tris-HCl, pH 6.8, 1–3% w/v sodium dodecyl sulfate, 10 m*M* dithiothreitol, 10% v/v glycerol, and 1 drop of 0.2% w/v Bromphenol Blue). In spite of careful washing, nonspecific sticking of radioactive material may occur. This problem can be remedied in some cases by including a wash with a solution containing 0.6 *M* NaCl, 10 m*M* Tris-HCl, pH 8.6, 0.1% w/v sodium dodecyl sulfate, and 0.05% v/v Nonidet P-40,[8] followed by a wash with PBS-B prior to the acetone rinse.

Immunoprecipitation with Staphylococcus aureus. This procedure is used when the enzyme–antibody complex does not precipitate (whether because of the nature of the antibody or because of low enzyme concentration). The protein A on the surface of the fixed bacteria interacts with IgG, so that the enzyme–antibody complex becomes bound to an insoluble support.

Clarified cell extracts (see above) should be freed of fibronectin and of other cellular components that have an affinity for the bacterial cell walls and would otherwise interfere with purification of lysosomal enzymes.

[7] J. W. Roberts and C. W. Roberts, *Proc. Natl. Acad. Sci. U.S.A.* **72,** 142 (1975).

[8] G. Kaplan, H. Plutner, I. Mellman, and J. C. Unkeless, *Exp. Cell Res.* **133,** 103 (1981).

Goat antiserum directed against human fibronectin (10 μl) is added to extract from each plate, and the mixtures are placed on ice. After 30 min, 100 μl of a 10% w/v suspension of *S. aureus* cells is added; the mixtures are allowed to stand on ice for 15 min longer and are centrifuged for 2 min in an Eppendorf 5412 centrifuge. The supernatant solutions are treated with a similar amount of *S. aureus* suspension an additional two times. The suspension of *S. aureus* should be pelleted and suspended in buffer before use to remove small fragments of bacteria.

For the immunoprecipitation of labeled enzymes, the preabsorbed cell extracts are mixed with a volume of antiserum sufficient to bind the antigen completely. After the mixtures have stood for 16 hr at 4°, *S. aureus* (10 μl per microliter of antiserum) suspension is added. After 15 min in the cold, the antigen–antibody complexes bound to the bacteria are pelleted and washed four times with 0.5 ml of a solution containing 0.6 *M* NaCl, 0.01 *M* Tris-HCl, pH 8.6, 0.1% w/v sodium dodecyl sulfate, and 0.05% v/v Nonidet P-40. The final wash is performed with 0.5 ml of 0.15 *M* NaCl–0.01 *M* Tris-HCl, pH 7.4; at that time the pellets are also transferred to clean tubes. The labeled lysosomal enzymes are eluted by resuspending the bacterial pellet in 50 μl of gel sample buffer (as above but without dithiothreitol) and heating for 5 min at 100°. The bacteria are removed by centrifugation, and the proteins in the supernatant solutions are reduced by heating for 5 min at 100° in the presence of dithiothreitol.

Gel Electrophoresis of Immunoprecipitated Enzymes

The solubilized immunoprecipitates may be stored frozen, but should be reheated and centrifuged immediately before use. They are subjected to polyacrylamide gel electrophoresis under denaturing conditions in slab gels 10 cm long, 14 cm wide, and 1.5 mm thick, with 20 wells (Bio-Rad, Model 220). The gels and electrode buffers are prepared essentially as described by Laemmli,[9] except that the ratio of acrylamide to methylene-bisacrylamide in the separating gel is changed from 30 : 0.8 to 30 : 0.312.[1] The concentration of acrylamide is 5.7% in the stacking gel and 12.5% in the separating gel. Sodium dodecyl sulfate is omitted from the anode buffer, and the electrophoresis is performed with circulating coolant at about 10°.

The resolution of some polypeptides appears to depend on alkyl sulfates other than dodecyl. For example, we have found that the α- and β-chain precursors of β-hexosaminidase (apparent M_r = 67,000 and 63,000, respectively) are separable if one uses a preparation containing a mixture

[9] U. K. Laemmli, *Nature (London)* **227,** 680 (1970).

of alkyl sulfates (66% dodecyl, 26% tetradecyl, and 6% hexadecyl; Sigma Chemical Co.) but not in other, presumably purer, preparations of sodium dodecyl sulfate.

Enhanced resolution of peptides of M_r less than 15,000, which may be produced in the processing of lysosomal enzymes, can be achieved in 15% acrylamide–8 *M* urea gels.[10] Inclusion of 8 *M* urea reduces the effective pore size of the gel without undesirable effects such as shrinkage or brittleness. Electrophoresis should be carried out at 20–24° to prevent precipitation of urea.

It is frequently necessary to extract radioactive polypeptides from the gel and subject them to some enzymic or chemical modification and a second electrophoresis. This can be done provided the gel is neither dried nor fixed. To locate radioactive bands, the samples are applied to the slab gel in a relatively broad band in a central portion of the gel, and nonradioactive molecular weight markers are applied to lanes near the edges. After electrophoresis, the marker lanes are cut out, briefly stained and destained, and lined up with the lanes containing the radioactive samples. Regions of the gel estimated to include the radioactive bands of interest are excised and cut transversely into 1-mm slices; a small fragment taken from one end of each slice is solubilized for determination of radioactivity. The remainder of each slice is stored frozen pending the results of the counting. This procedure has been used to treat electrophoretically isolated polypeptides with endohexosaminidase H[4] or with V 8 protease,[11] and to demonstrate that some fragments of lysosomal enzymes remain associated during electrophoresis unless they are separated by heating in the presence of dithiothreitol.[12]

Detection of Radioactive Bands

^{3}H-, ^{35}S-, and ^{33}P-labeled polypeptides are detected by the fluorographic procedure of Bonner and Laskey,[13] using preflashed film.[14] The films are exposed at −70° until the polypeptides of interest are revealed (usually 1–5 days). ^{32}P-labeled polypeptides do not require impregnation of the gel with a fluor, but can be detected by autoradiography with an intensifying screen behind the film. Independent detection of two isotopes used in double-label experiments is possible, as is the quantitation of

[10] R. T. Swank and K. D. Munkres, *Anal. Biochem.* **39,** 462 (1971).
[11] R. Myerowitz and E. F. Neufeld, *J. Biol. Chem.* **256,** 3044 (1981).
[12] C. M. Puchalski and E. F. Neufeld, *Fed. Proc., Fed. Am. Soc. Exp. Biol.* **40,** 1551 (1981).
[13] W. M. Bonner and R. A. Laskey, *Eur. J. Biochem.* **46,** 83 (1974).
[14] R. A. Laskey and A. D. Mills, *Eur. J. Biochem.* **56,** 335 (1975).

radioactivity by densitometry or by excision of the bands, solubilization, and counting.[4,15]

Detection of Phosphorylated Mannose

The presence of phosphorylated mannose residues on oligosaccharides of the high mannose type characterizes newly formed lysosomal enzymes. Recognition of these residues by a specific receptor (see Volume 98 [25]) is required for transport of the enzymes, endogenous as well as endocytosed, to lysosomes. The phosphorylated oligosaccharides can exist in many isomeric forms, with one to three phosphate groups, and at times with the phosphate in diester linkage. Readers are referred to the original literature for methods of resolving and characterizing these complicated structures.[16,17,18] However, it is relatively simple to ascertain the presence or the absence of phosphorylated mannose on biosynthetically labeled enzymes.[4]

The enzymes are isolated by immunoprecipitation after labeling the cells with [2-^{3}H]mannose and/or $^{32}PO_4$. The procedure is carried out on a sufficient number of dishes (up to 10) to ensure an adequate amount of product for analysis. Hydrolysis may be carried out directly on the intact enzymes or on oligosaccharides released from the enzymes by digestion with endohexosaminidase H; the starting material may be immunoprecipitated enzyme (an aliquot of which has been checked for purity by polyacrylamide gel electrophoresis), or polypeptide(s) isolated from the electrophoresis gels. Acid hydrolysis is carried out in 4 *M* trifluoroacetic acid, in sealed or Teflon-stoppered tubes, at 105° for 3 hr. A small amount of mannose 6-phosphate carrier (0.2 μmol) should be added to each tube. The acid is removed by lyophilization.

The mannose 6-phosphate in the hydrolyzate may be identified by paper electrophoresis at pH 5.3 (1% v/v pyridine, 0.4% v/v acetic acid) and by release of [^{3}H]mannose, demonstrated chromatographically, upon treatment with alkaline phosphatase. The best evidence is supplied by conversion to 6-phosphogluconate in the presence of phosphomannose isomerase, phosphoglucoisomerase, glucose-6-phosphate dehydrogenase and NADP. This must be done with ^{32}P-labeled material, since the ^{3}H at the 2 position of mannose is lost during isomerization. The 6-phosphogluconate is identified by electrophoresis in 30 m*M* NH_4HCO_3, pH 8.0.

[15] P. G. Robey and E. F. Neufeld, *Arch. Biochem. Biophys.* **213,** 251 (1982).

[16] A. Varki and S. Kornfeld, *J. Biol. Chem.* **255,** 10847 (1980).

[17] A. Hasilik, U. Klein, A. Waheed, G. Strecker, and K. von Figura, *Proc. Natl. Acad. Sci. U.S.A.* **77,** 7074 (1980).

[18] M. Natowicz, J. U. Baenziger, and W. S. Sly, *J. Biol. Chem.* **257,** 4412 (1982).

[59] Inhibitors of Lysosomal Function

By PER O. SEGLEN

A number of inhibitors have been found to be useful in the study of lysosomal function (Table I).[1] The weak-base (lysosomotropic) amines and the proteinase inhibitors are the most lysosome-specific of these, but the autophagy-inhibitory purines and some inhibitors that affect lysosomal function indirectly (microtubule poisons and protein synthesis inhibitors) will also be considered. The natural regulators of autophagic-lysosomal activity, i.e., the amino acids[2–4] and hormones,[2,5–7] will not be dealt with here.

Weak-Base Amines

Ammonia and the weakly basic alkylamines are (with few exceptions) lysosomotropic in the sense that, at low concentrations, they selectively accumulate in the lysosomes, partly because of an initially low lysosomal pH, partly because of continuous pumping of protons into the lysosome.[8–11] It was originally thought that the selective effect of a weak-base amine on lysosomal function would be due to its ability to accumulate specifically in this organelle, the actual inhibition being ascribed to some other property of the amine molecule, e.g., proteinase inhibition.[8] However, the discovery that ammonia could inhibit lysosomal protein degradation in intact cells,[12,13] without directly affecting lysosomal enzyme

[1] P. O. Seglen, P. B. Gordon, B. Grinde, A. Solheim, A. L. Kovács, and A. Poli, *Acta Biol. Med. Ger.* **40,** 1587 (1981).
[2] C. M. Schworer and G. E. Mortimore, *Proc. Natl. Acad. Sci. U.S.A.* **76,** 3169 (1979).
[3] P. O. Seglen, P. B. Gordon, and A. Poli, *Biochim. Biophys. Acta* **630,** 103 (1980).
[4] A. L. Kovács, B. Grinde, and P. O. Seglen, *Exp. Cell Res.* **133,** 431 (1981).
[5] G. E. Mortimore and C. E. Mondon, *J. Biol. Chem.* **245,** 2375 (1970).
[6] A. Poli, P. B. Gordon, P. E. Schwarze, B. Grinde, and P. O. Seglen, *J. Cell Sci.* **48,** 1 (1981).
[7] F. J. Ballard, S. E. Knowles, S. S. C. Wong, J. B. Bodner, C. M. Wood, and J. M. Gunn, *FEBS Lett.* **114,** 109 (1980).
[8] C. de Duve, T. de Barsy, B. Poole, A. Trouet, P. Tulkens, and F. van Hoof, *Biochem. Pharmacol.* **23,** 2495 (1974).
[9] S. Ohkuma and B. Poole, *Proc. Natl. Acad. Sci. U.S.A.* **75,** 3327 (1978).
[10] S. Ohkuma and B. Poole, *J. Cell Biol.* **90,** 656 (1981).
[11] B. Poole and S. Ohkuma, *J. Cell Biol.* **90,** 665 (1981).
[12] P. O. Seglen, *Biochem. Biophys. Res. Commun.* **66,** 44 (1975).
[13] P. O. Seglen and A. Reith, *Exp. Cell Res.* **100,** 276 (1976).

ISBN 0-12-181996-5

TABLE I
MAJOR CLASSES OF LYSOSOME INHIBITORS

Inhibitor class	Examples
1. pH-elevating agents	
Weak-base amines	Ammonia, methylamine, propylamine, chloroquine, amantadine, procaine
Ionophores	Monensin, nigericin
2. Proteinase inhibitors	
Peptides	Leupeptin, antipain, chymostatin, pepstatin, E-64, Ep-475, Z-Phe-Ala-CHN_2
Phosphate analog	Vanadate
3. Autophagy inhibitors	
Purine derivatives	3-Methyladenine, puromycin aminonucleoside, 6-dimethylaminopurine riboside
Amino acids	AsN, GlN, Leu, His, Phe, Tyr, Trp
Hormones	Insulin, EGF, IGF, serum
4. Fusion inhibitors	
Cytoskeleton poisons	Vinblastine, colchicine
Lectins and polymers	Concanavalin A, chlorite-oxidized amylose
5. Protein synthesis inhibitors	Cycloheximide

activity,[14,15] suggested that the lysosome-inhibitory effect might be due to a neutralization of the lysosomal acidity by the weak base.[13] This hypothesis has now been verified by experiments that directly demonstrate an elevation of the lysosomal pH in ammonia-treated cells.[9–11]

A large number of weak-base alkylamines were subsequently found to affect lysosomal function[9–11,16,17] (Tables II and III). While their primary mechanism of action is basically the same (Fig. 1), individual properties of the different amines may exert a significant influence on some of the many steps leading up to the biological responses eventually observed.

Cellular Uptake of Amines (Fig. 1A)

The cellular uptake of weak-base amines is distinctly biphasic (Fig. 1C), the consensus being that the early, rapid phase represents uptake across the plasma membrane and the second, slow phase represents accumulation in the lysosomes.[9] The effect on lysosomal pH has been shown

[14] D. J. Reijngoud, P. S. Oud, J. Kás, and J. M. Tager, *Biochim. Biophys. Acta* **448,** 290 (1976).
[15] P. O. Seglen and P. B. Gordon, *J. Biol. Chem.* **256,** 7699 (1981).
[16] G. Carpenter and S. Cohen, *J. Cell Biol.* **71,** 159 (1976).
[17] P. O. Seglen and P. B. Gordon, *Mol. Pharmacol.* **18,** 468 (1980).

TABLE II
EFFECT OF VARIOUS ALKYLAMINES (AT 10 mM) ON LYSOSOMAL SWELLING AND PROTEOLYTIC ACTIVITY

Compound	% Inhibition of endogenous hepatocytic protein degradation[a]	Lysosomal swelling: Hepatocytes[17]	Lysosomal swelling: Macrophages[10]	pK[b]
Unsubstituted aliphatic monoamines				
Ammonia	63	+	+	9.2
Methylamine	62	+	+	10.6
Ethylamine	64	+	+	10.7
Propylamine	69	+		
tert-Butylamine	63	+	+	10.7
Dimethylamine	63	+	+	10.8
Diethylamine	62	+	+	11.0
Trimethylamine	60	+	+	9.8
Triethylamine	63	+	+	10.7
Diamines and polyamines				
Ethylenediamine (1,2-diaminoethane)	17	±	−	6.8, 9.9
1,3-Diaminopropane	2	−		
Putrescine (1,4-diaminobutane)	0	−		
Cadaverine (1,5-diaminopentane)	6	−	±	7.0, 10.3
Spermidine	13	−	−	10.9
Spermine	16	−	−	8.0, 8.9
N,N-Diethylaminoethylamine	70	+		
N,N′-Dimethylethylenediamine	59	+		
Amino alcohols and chloroamines				
Tris (hydroxymethyl)methylamine	9	±	−	8.1
3-Amino-1-propanol	18	±		
2-Aminoethanol	35	+		
2-Methylaminoethanol	55	+		
2-Dimethylaminoethanol	61	+		
Isopropanolamine	59	+		
2-Amino-1-butanol	62	+		
N,N-Dimethyl-3-chloropropylamine	64	+		
Miscellaneous				
Imidazole	56	+	+	7.1
Guanidine	21	−	−	10.0, 10.9

[a] Under the conditions of measurement used, 65–70% inhibition would correspond to a virtually complete inhibition of lysosomal degradation.[17]

[b] pK values were taken from Ohkuma and Poole.[10]

TABLE III
ALKYLAMINES AFFECTING LYSOSOMAL FUNCTION AT LOW CONCENTRATIONS (1 m*M* OR BELOW)

	References[a]		
Compound	Lysosomal swelling	Inhibition of protein degradation	Interference with receptor recycling
Acridine orange	—	—	22
Amantadine	—	—	27
4-Aminopyridine	10	—	—
4-Aminoquinaldine	10	22	22
Atropine	10	—	—
Chloroquine	10	22, 24, 25	22, 27, 28
Cocaine	—	16	—
Dansylcadaverine	—	22	22, 29
Dibucaine	—	22	22
4-Dimethylaminopyridine	10	—	—
N,*N*-Dimethylbenzylamine	10	—	—
Ephedrine	10	—	—
Eserine	10	—	—
Lidocaine	10	16, 22	—
Mecamylamine	10	—	—
Neutral red	10, 23	26	—
Procaine	10	16, 22	28
Propranolol	10	—	—
Quinine	10	—	27
Tetracaine	—	22	22
N,*N*,*N*′,*N*′-Tetramethylethylenediamine	—	—	28

[a] Numbers indicate text footnotes.

to correlate with the concentration of neutral base rather than with the total base concentration (different ratios being produced by varying the extracellular pH)[11]; i.e., the protonated amine is presumably not significantly taken up by the (mammalian) cells, excluding participation of ammonium transport systems like those found in bacteria and plants.[18] A considerable part of the amine uptake probably takes place by simple physical diffusion, some uptake being demonstrable even at 0° in the case of methylamine.[19] Uptake by *diffusion* would explain the general exclusion of the hydrophilic protonated form of amines as well as the poor effect (due to low total uptake) of those amines that are very hydrophilic

[18] D. Kleiner, *Biochim. Biophys. Acta* **639,** 41 (1981).
[19] A. E. Solheim and P. O. Seglen, *Biochem. J.* **210,** 929 (1983).

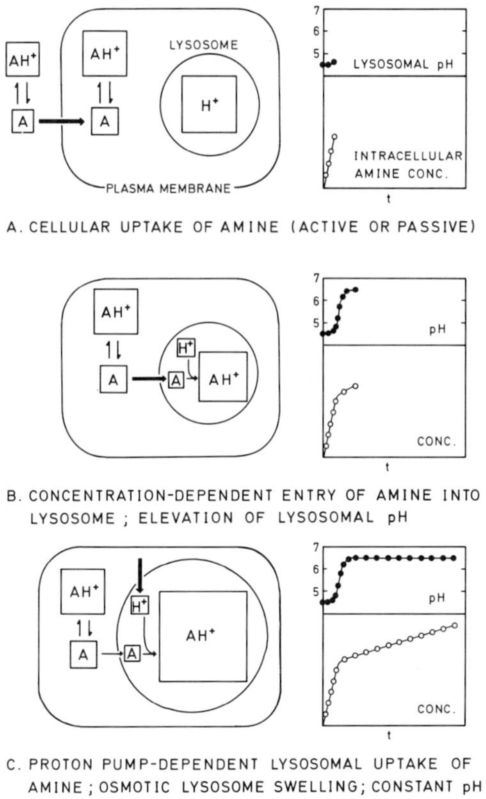

FIG. 1. Mechanism of action of weak-base amines. A, Neutral form of the amine; AH^+, protonated form. Extralysosomal protons are not shown.

even in their neutral form.[11] Examples of the latter in Table II are the unblocked diamines and polyamines, guanidine, Tris (although a lysosomotropic effect of Tris can be obtained at concentrations in excess of 100 m*M*[20]), and to some extent the polar straight-chain amino alcohols (2-aminoethanol and 3-amino-1-propanol). Increasing the lipophilicity of the amino alcohols by the addition of alkyl groups to the amino nitrogen (2-methyl- and 2-dimethylaminoethanol) or as side chains (isopropanolamine and 2-amino-1-butanol) increases the lysosome-inhibitory efficiency correspondingly; blocking of one (*N*,*N*-diethylaminoethylamine) or both (*N*,*N*′-dimethylethylenediamine) amino groups similarly increases the

[20] M. Peterlik and D. Kerjaschki, *Lab. Invest.* **40,** 313 (1979).

activity of the diamines (Table II) (cf. also the rapid uptake[21] and lysosome-inhibitory action[22] of dansylcadaverine in contrast to cadaverine.

In addition to physical diffusion, carrier-mediated uptake of methylamine (as well as of ammonia and triethylamine) by a *passive transport* mechanism (facilitated diffusion) has been demonstrated.[19] *Active (energy-dependent) transport* of amines across the plasma membrane has not been directly documented, but the highly concentrative uptake of several aromatic amines (chloroquine, atropine, propranolol)[10] is more reasonably explained by specific active transport than by nonspecific (pH-dependent) lysosomal accumulation. The fact that certain amines *stimulate* the uptake of others[10] is furthermore incompatible with a pH-dependent uptake mechanism, but can be explained as an osmotic expansion of the total fluid space of the cell. We will therefore assume that the ability of aromatic amines to affect lysosomal function at low extracellular concentrations (i.e., in the micromolar range, in contrast to the simple aliphatic alkylamines, which are maximally effective at 10 m*M* or higher) is due to active (concentrative) uptake across the plasma membrane (Table III).[10,16,22–29] The fact that proteolysis in isolated lysosomes can be inhibited by chloroquine only at concentrations several orders of magnitude higher than those effective on intact cells[29a] supports this view. Whether active transport across the lysosomal membrane also occurs is not known.

Entry of Amines into Lysosomes (Fig. 1B)

A weak-base amine that enters the lysosome in its neutral form will tend to become protonated in the acidic interior of the lysosome. At very low amine concentrations, the resulting equilibrium shift may cause a relative accumulation of protonated amine intralysosomally with only a moderate increase in lysosomal pH; however, at high amine concentrations the acid capacity of the lysosome is rapidly exhausted, and the lysosomal pH is elevated to a new equilibrium value (Fig. 1B). The extent

[21] P. J. A. Davies, D. R. Davies, A. Levitzki, F. R. Maxfield, P. Milhaud, M. C. Willingham, and I. H. Pastan, *Nature (London)* **283,** 162 (1980).

[22] H. Tolleshaug, T. Berg, and K. Holte, *Biochim. Biophys. Acta* **714,** 114 (1982).

[23] A. Bulychev, A. Trouet, and P. Tulkens, *Exp. Cell Res.* **115,** 343 (1978).

[24] M. Wibo and B. Poole, *J. Cell Biol.* **63,** 430 (1974).

[25] P. O. Seglen, B. Grinde, and A. E. Solheim, *Eur. J. Biochem.* **95,** 215 (1979).

[26] A. L. Kovács and P. O. Seglen, *Acta Biol. Med. Ger.* **41,** 125 (1982).

[27] D. W. H. Riches and D. R. Stanworth, *Biochem. J.* **188,** 933 (1980).

[28] G. N. Sando, P. Titus-Dillon, C. W. Hall, and E. F. Neufeld, *Exp. Cell Res.* **119,** 359 (1979).

[29] F. Van Leuven, J.-J. Cassiman, and H. Van den Berghe, *Cell* **20,** 37 (1980).

[29a] J. Ahlberg, L. Marzella, and H. Glaumann, *Lab. Invest.* **47,** 523 (1982).

of pH elevation has been shown to be a function of the concentration of neutral base extracellularly.[9] It must necessarily be an even more direct function of the concentration of neutral base intracellularly, since only the neutral (nonprotonated) form of the molecule is capable of altering the lysosomal pH. The ratio neutral : protonated amine extralysosomally is given by the cytosol pH and the dissociation constant (pK) of the amine:

$$\log (AH^+/A) = pK - pH$$

An amine with a high pK is relatively strongly basic and will therefore have a high tendency to neutralize the lysosomes; on the other hand, the fraction of the amine present in nonprotonated form at the near-neutral cytosol pH will be correspondingly low, and its entry into the lysosome may well be rate-limiting. Because of these opposing tendencies there is no clear-cut correlation between the pK and the pH-elevating ability of an amine, unless the effect is expressed as a function of the calculated *neutral* base concentration.[11] The unsubstituted monoamines presented in Table II cover a pK range from 9.2 (ammonia) to 11.0 (diethylamine), yet their anti-proteolytic potency is strikingly similar. Of course pK has to be kept within certain limits: an amine with a very low pK will be too acidic to be able to elevate the lysosomal pH; while at a very high pK the equilibrium concentration of the neutral form will be too low to be effective (besides, the amine may be too hydrophilic to be taken up by the cell in both cases). Within such limits (approximately pK 6 to pK 11), however, lipohilicity and concentrative uptake would seem to be much more important than pK.

It has been suggested that the rate of back diffusion of protonated base from lysosomes to cytosol might account for differences in the potency of individual amines.[9,11] However, while back diffusion might moderately aid in the removal of protons, it would have the opposite effect on lysosomal swelling (see below), thus violating the excellent correlation actually observed (Table II). A significant role for back diffusion is, furthermore, incompatible with the continued amine accumulation that takes place long after the establishment of a stable, elevated pH value.[9,11] A possible exception is tributylamine, which is very lipophilic even in its protonated form (it kills hepatocytes by detergent action); this amine elevates macrophage pH without inducing lysosomal swelling.[10,11]

Lysosomal Swelling (Fig. 1C)

The normally low intralysosomal pH is generated partly by a Gibbs–Donnan equilibrium[30] depending on the presence of acidic groups such as

[30] D.-J. Reijngoud, *Trends Biochem. Sci.* **3,** 178 (1978).

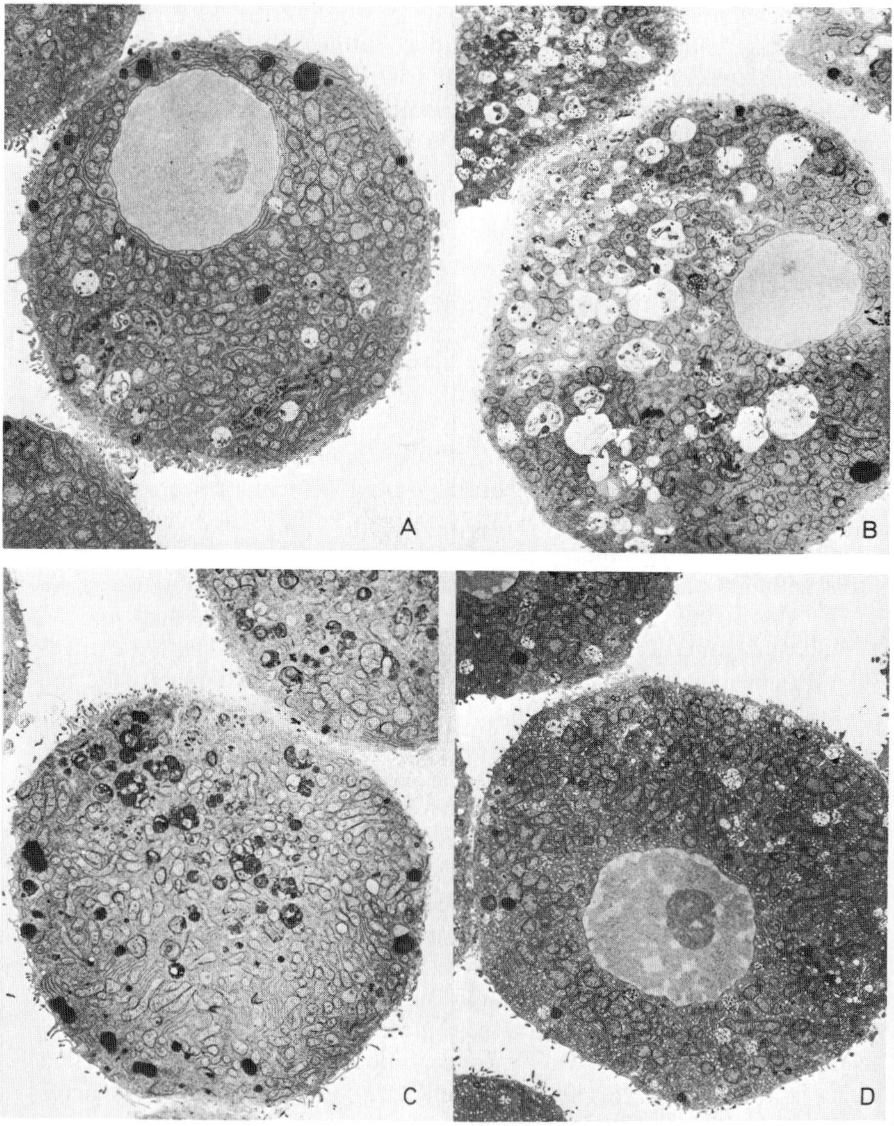

FIG. 2. Ultrastructure of hepatocytes treated with various inhibitors of lysosomal function. (A) Control; (B) propylamine (10 mM); (C) leupeptin (0.5 mM); (D) vinblastine (0.1 mM). ×3200. From Kovács *et al.*[73]

sialic acid[31] in the lysosome wall, partly by the activity of an energy-dependent proton pump.[9,32,33] In the amine-neutralized lysosome, proton pumping apparently proceeds at a high rate for several hours, but the entering protons are now immediately trapped by reaction with the amine. The net result is a continuous accumulation of protonated amine, at constant pH, rather than an acidification of the lysosome[9,11] (Fig. 1C). The amine accumulation is accompanied by an osmotic influx of water, resulting in an energy-dependent,[19] visible swelling of the lysosome (Fig. 2). Provided that the protonated form of the amine is reasonably nondiffusible (which is nearly always the case), lysosomal swelling may serve as a convenient indication of the ability of an amine to enter lysosomes.[10,17] It should be stressed, however, that the proton pump-dependent amine accumulation takes place *after* the amine has produced its biologically relevant effect, which is elevation of the lysosomal pH. Amine accumulation and lysosomal swelling should therefore be regarded as epiphenomena, as yet of no proven biological significance.

The source of the extra membrane required for lysosomal volume expansion is not known. Morphological reorganization of the lysosomal system from small vesicles and tubuli (often interconnected[34]) to larger vacuoles with a higher volume-to-surface ratio may contribute, but the magnitude of the swelling (e.g., a sevenfold volume increase within 1 hr[13]) indicates that a perturbation of cellular membrane flow must take place, causing net accumulation of membrane in the lysosomal compartment.[35]

Side Effects of Amines

In addition to their pH-elevating properties, the various weak-base amines may have individual side effects that affect lysosomal function as well as other cellular processes (see Fig. 3). *Protease inhibition* is a well-known side effect of *chloroquine*.[24] In addition this drug inhibits phospholipases,[36] steroid synthesis,[37] and basal DNA synthesis[38]; it binds to the cell surface[39] and to membrane phospholipids[40]; and it is capable of alter-

[31] H. R. Adhikari and U. K. Vakil, *Biochim. Biophys. Acta* **633,** 465 (1980).
[32] J. L. Mego, *FEBS Lett.* **107,** 113 (1979).
[33] D. L. Schneider, *J. Biol. Chem.* **256,** 3858 (1981).
[34] A. B. Novikoff and W.-Y. Shin, *Proc. Natl. Acad. Sci. U.S.A.* **75,** 5039 (1978).
[35] P. O. Seglen and P. B. Gordon, *FEBS Lett.* **105,** 345 (1979).
[36] Y. Matsuzawa and K. Y. Hostetler, *J. Biol. Chem.* **255,** 5190 (1980).
[37] M. Ascoli and D. Puett, *J. Biol. Chem.* **253,** 7832 (1978).
[38] A. C. King, L. Hernaez-Davis, and P. Cuatrecasas, *Proc. Natl. Acad. Sci. U.S.A.* **78,** 717 (1981).
[39] S. DiDonato, U. N. Wiesmann, and R. N. Herschkowitz, *Biochem. Pharmacol.* **26,** 7 (1977).
[40] A. Harder, S. Kovatchev, and H. Debuch, *Hoppe-Seyler's Z. Physiol. Chem.* **361,** 1847 (1980).

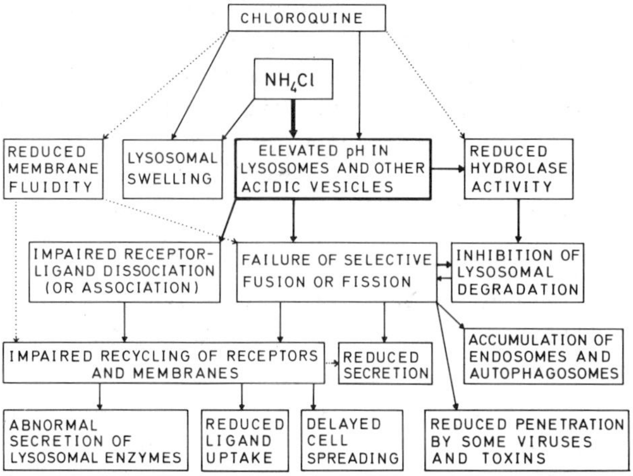

FIG. 3. Primary and secondary effects of weak-base amines on cellular functions. Thick arrows: major effects; dotted arrows; atypical or uncertain effects. Ammonia (NH_4Cl) is representative of the weak-base amines in general, whereas chloroquine is included to show some of the more important side effects.

ing *membrane fluidity*,[41] and is demonstrably toxic in several cell types.[25,42] Local anesthetics (e.g., lidocaine and dibucaine) have a similar structure and share several of the chloroquine effects,[22,38,43] whereas ammonia has none of the side effects mentioned. Ammonia may therefore be one of the most "pure" pH-elevating agents among the weak-base amines, except in the liver (and kidney), where it is very actively metabolized[12,44] and where nonmetabolizable amines such as propylamine[1,3,15] or methylamine[13,25] may be preferable. Both ammonia and the short-chain alkylamines, however, have been found to inhibit both surface receptor clustering[45] and internalization of receptor–ligand complexes.[21,45a] The latter effect has been ascribed to inhibition of the enzyme transglutaminase,[21] whereas an increase in receptor mobility may contribute to the former.[46]

[41] T. Berg and H. Tolleshaug, *Biochem. Pharmacol.* **29,** 917 (1980).

[42] G. Livesey, K. E. Williams, S. E. Knowles, and F. J. Ballard, *Biochem. J.* **188,** 895 (1980).

[43] K. Suzuki and T. Kono, *J. Biol. Chem.* **254,** 9786 (1979).

[44] S. Grisolia, R. Baguena, and F. Mayor, eds., "The Urea Cycle." Wiley, New York, 1976.

[45] F. R. Maxfield, M. C. Willingham, P. J. A. Davies, and I. Pastan, *Nature (London)* **277,** 661 (1979).

[45a] R. B. Dickson, R. Schlegel, M. C. Willingham, and I. Pastan, *Exp. Cell Res.* **140,** 215 (1982).

[46] Y. Yarden, M. Gabbay, and J. Schlessinger, *Biochim. Biophys. Acta* **674,** 188 (1981).

Long-chain alkylamines (octylamine, dodecylamine) may cause cell lysis[17] by virtue of their detergent properties.[47]

The major biological effect of the amines—i.e., the inhibition of lysosomal degradation processes—will have secondary metabolic consequences. For example, the reduced supply of amino acids from degraded proteins may lead to an inhibition of protein synthesis, unless compensated for by the addition of amino acids to the medium.[17]

Biological Effects of Weak-Base Amines: Interference with Receptor–Ligand Dissociation and the Recycling of Receptors and Membranes

Proteins and other ligands destined for delivery to lysosomes generally associate with their receptor at neutral pH (e.g., at the cell surface) and dissociate from the receptor at low pH (i.e., intralysosomally). The receptor is then usually (but not always) free to leave the lysosome (by membrane flow) for another round of ligand transport, while the ligand stays behind. If dissociation is prevented by an amine-induced elevation of the lysosomal pH, the result may be either (*a*) that the receptor–ligand complex returns intact from the lysosome, i.e., failing to deposit the ligand intralysosomally; (*b*) that the receptor–ligand complex becomes trapped in the lysosome, thus effectively interrupting receptor recycling; or (*c*) that receptors that are normally degraded rather than recycled are protected against degradation.

Deposition Failure

This failure occurs in at least the two known cases described below.

Unscheduled Secretion of Newly Synthesized Lysosomal Enzymes. Newly synthesized lysosomal hydrolase precursors are equipped with a recognition marker, 6-phosphomannose, that ensures tight binding to a membrane receptor serving to transport the enzymes from the Golgi apparatus to the lysosomes.[48,49] In the acidic intralysosomal environment, the enzymes dissociate from their receptor, the phosphomannosyl marker is hydrolytically cleaved to ensure irreversibility of dissociation, and the large molecular weight precursors are proteolytically processed to form the mature lysosomal hydrolase species.[49,50]

[47] R. A. Firestone, J. M. Pisano, and R. J. Bonney, *J. Med. Chem.* **22,** 1130 (1979).

[48] A. Gonzalez-Noriega, J. H. Grubb, V. Talkad, and W. S. Sly, *J. Cell Biol.* **85,** 839 (1980).

[49] H. D. Fischer, A. Gonzalez-Noriega, W. S. Sly, and D. J. Morré, *J. Biol. Chem.* **255,** 9608 (1980).

[50] A. Hasilik and E. F. Neufeld, *J. Biol. Chem.* **255,** 4946 (1980).

In the presence of weak-base amines, intralysosomal dissociation and processing are inhibited. Instead of being delivered to the lysosomes, the hydrolase precursors, still attached to their receptors, are redirected to the cell surface, where some of them dissociate; i.e., an unscheduled *secretion* of lysosomal enzymes can be measured.[27,48–50] Intralysosomal dissociation failure likewise accounts for the ability of amines to prevent successful capture of the exogenous hydrolases used for replacement therapy of fibroblasts with I-cell disease (which have recognition marker-deficient endogenous hydrolases).[28]

Transferrin-Mediated Iron Uptake. Iron uptake is similarly disturbed by amines at the lysosomal level; elevation of the lysosomal pH prevents dissociation of the receptor–transferrin–iron complex, and results in a futile recycling of the complex between the lysosome and the cell surface rather than in the normal intralysosomal deposition of iron.[51,52]

Receptor Trapping

Trapping may occur when amines are allowed to interfere with the receptor-mediated endocytosis of proteins that utilize recycling receptors; this seems to be the case with many plasma proteins. A gradual intracellular accumulation of receptor–ligand complexes, eventually resulting in a depletion of cell surface receptors and an inhibition of ligand uptake, has been demonstrated in the case of asialoglycoproteins,[53] mannosylneoglycoproteins,[54] α_2-macroglobulin,[29,55] and low-density lipoprotein (LDL).[56] A considerable fraction of the receptors recycle even in the absence of ligand and can be trapped by amines[53–56]; i.e., recycling and receptor–ligand dissociation can be inhibited independently by a high pH.

Surface receptors are generally endocytosed in *coated pits/vesicles*[57] and subsequently transferred to *receptosomes*[58] and other kinds of endosomes.[59,60] Recent evidence indicates that receptors may recycle at such a prelysosomal level,[54,61,62] even in association with the ligand (so-called

[51] E. H. Morgan, *Biochim. Biophys. Acta* **642,** 119 (1981).
[52] M. Karin and B. Mintz, *J. Biol. Chem.* **256,** 3245 (1981).
[53] H. Tolleshaug and T. Berg, *Biochem. Pharmacol.* **28,** 2919 (1979).
[54] C. Tietze, P. Schlesinger, and P. Stahl, *J. Cell Biol.* **92,** 417 (1982).
[55] J. Kaplan and E. A. Keogh, *Cell* **24,** 925 (1981).
[56] S. K. Basu, J. L. Goldstein, R. G. W. Anderson, and M. S. Brown, *Cell* **24,** 493 (1981).
[57] J. L. Goldstein, R. G. W. Anderson, and M. S. Brown, *Nature (London)* **279,** 679 (1979).
[58] M. C. Willingham and I. Pastan, *Cell* **21,** 67 (1980).
[59] D. A. Wall, G. Wilson, and A. Hubbard, *Cell* **21,** 79 (1980).
[60] J. A. McKanna, H. T. Haigler, and S. Cohen, *Proc. Natl. Acad. Sci. U.S.A.* **76,** 5689 (1979).
[61] H. J. Geuze, J. W. Slot, G. J. A. M. Strous, H. F. Lodish, and A. L. Schwartz, *J. Cell Biol.* **92,** 865 (1982).
[62] M. J. Geisow, *Nature (London)* **295,** 649 (1982).

diacytosis).[63] Endosomes have been shown to be acidic[64]; i.e., it is very likely that they are the main site of a receptor–ligand dissociation, and that amines may cause receptor trapping in endosomes as well as in lysosomes.

Receptor Protection

Receptor protection due to the inhibition of receptor degradation by weak-base amines can be expected in those cases where the receptor is normally degraded in the lysosomes along with the ligand. The latter is the case with many hormone receptors subject to "down-regulation" (i.e., an immediate decrease in the number of surface receptors after exposure to hormone, due to internalization of the hormone–receptor complex and slow receptor renewal),[65] such as epidermal growth factor (EGF),[60,66,67] human chorionic gonadotropin (hCG),[68] human growth hormone (hGF),[69] acetylcholine,[70] prolactin,[71] and insulin in some cell types,[72] but adequate documentation of the effect of amines is available only in the case of EGF and insulin. In the presence of amines, receptor degradation is inhibited, and the total number of cell-associated receptors increases with time as new receptors are supplied by de novo protein synthesis.[66,72] Morphological analysis indicates that amines prevent EGF-receptor dissociation at the level of endosomes (multivesicular bodies).[60] However, the number of receptors on the cell *surface* decreases after exposure to hormones even in the presence of amines; i.e., amines do not prevent down-regulation,[60,66,67,72] and, since no augmentation of the biological response, e.g., to EGF, can be detected,[38,67] the receptor protection offered by the amines would seem to be nonfunctional.

Pinocytosis (fluid-phase endocytosis) is strongly inhibited by weak-base amines in some cell types,[42,55] and the internalization of certain ligand–receptor complexes may also show a time-dependent inhibition,[55] indicating that factors necessary for membrane internalization may be-

[63] H. Tolleshaug, P. A. Chindemi, and E. Regoeczi, *J. Biol. Chem.* **256,** 6526 (1981).

[64] B. Tycko and F. R. Maxfield, *Cell* **28,** 643 (1982).

[65] M. A. Lesniak and J. Roth, *J. Biol. Chem.* **251,** 5720 (1976).

[66] A. C. King, L. Hernaez-Davis, and P. Cuatrecasas, *Proc. Natl. Acad. Sci. U.S.A.* **77,** 3283 (1980).

[67] H. J. Michael, S. Bishayee, and M. Das, *FEBS Lett.* **117,** 125 (1980).

[68] A. Amsterdam, A. Nimrod, S. A. Lamprecht, Y. Burstein, and H. R. Lindner, *Am. J. Physiol.* **236**(2), E129 (1979).

[69] F. C. Kosmakos and J. Roth, *in* "Protein Turnover and Lysosome Function" (H. L. Segal and D. J. Doyle, eds.), p. 763. Academic Press, New York, 1978.

[70] P. Libby, S. Bursztajn, and A. L. Goldberg, *Cell* **19,** 481 (1980).

[71] L.-M. Houdebine, J. Djiane, and H. Clauser, *C. R. Hebd. Seances Acad. Sci.* **289,** 679 (1979).

[72] A. Green and J. M. Olefsky, *Proc. Natl. Acad. Sci. U.S.A.* **79,** 427 (1982).

come trapped in endosomes and lysosomes. Alternatively, the accumulation of membrane in the latter may reduce the rate of membrane recycling, as suggested for the inhibition of *cell spreading* by the weak-base amines[35]; however, the persistent secretion of lysosomal enzymes[27] and the recycling of iron-transferrin–receptor complexes[51,52] indicates that trans-lysosomal membrane flow is not *generally* impeded.

Dose-Response Characteristics. It is a frequent observation that the amine concentrations needed for inhibition of receptor–ligand dissociation and/or receptor recycling are considerably higher than those required for inhibition of protein degradation.[27,53,56] The reason for this is not known, but it is possible either that degradation is more sensitive to pH elevation, or that endosomes (where receptor–ligand dissociation is most likely to occur) are more amine-resistant than lysosomes (where degradation takes place).

Inhibition of Vacuolar Membrane Fusion and Related Processes

Berg and Tolleshaug[41] found that asialofetuin-containing endosomes accumulated in the presence of ammonia, suggesting an inhibition of *endosome–lysosome fusion*. The accumulation of autophagosomes often observed after treatment with weak-base amines could similarly reflect a reduced *autophagosome–lysosome fusion*.[73] During particle phagocytosis in macrophages, a direct inhibition of *phagosome–lysosome fusion* by ammonia was observed.[74] The inhibition was apparently due to an elevation of the *lysosomal* pH, the phagosomal pH being unaffected.[75] (The ability of more lipophilic amines paradoxically to *stimulate* phagosome–lysosome fusion[74] cannot be explained at present.)

Membrane invagination, the process by which the vesicles inside *multivesicular bodies* are formed, is inhibited by weak-base amines.[60] These endosomes are apparently normally acidic, as indicated by their swelling after amine treatment.[60]

In contrast to the above-mentioned processes, *pinosome–lysosome fusion* was not inhibited by the pH-elevating ionophore monensin.[76] The ability of this type of fusion to proceed independently of pH may account for the recruitment of the extra membrane required for amine-induced lysosome swelling.

Virus and Toxin Entry. Infection by certain enveloped animal viruses depends on adsorptive endocytosis of the virus and subsequent fusion

[73] A. L. Kovács, A. Reith and P. O. Seglen, *Exp. Cell Res.* **137,** 191 (1982).

[74] A. H. Gordon, P. D'Arcy Hart, and M. R. Young, *Nature* (*London*) **286,** 79 (1980).

[75] M. J. Geisow, P. D'Arcy Hart, and M. R. Young, *J. Cell Biol.* **89,** 645 (1981).

[76] T. Berg, unpublished results.

between the viral membrane and the membrane of endosomes or lysosomes, the fusion reaction being triggered by the low pH in these vacuoles.[77,78] The fusion is inhibited by the pH-elevating action of weak-base amines; this may be the mechanism of action of amantadine, an inhibitor of influenza virus infection.[77]

Diphtheria toxin is similarly taken up by adsorptive endocytosis, and its toxic A subunit penetrates the cell membrane at the low pH in endosomes or lysosomes. Weak-base amines prevent this penetration and offer complete protection against the toxic effect of diphtheria toxin.[79,80] Both with viruses and diphtheria toxin a direct entry through the surface plasma membrane can be induced by lowering the extracellular pH, in which case amines have no effect, as expected.[76,81] Weak-base amines also offer protection against the toxic lectin modeccin, but in this case the amine block cannot be circumvented by lowering the extracellular pH, indicating a mechanism different from that operating with diphtheria toxin.[80,81a] With the toxins abrin and ricin, the toxic effects are *potentiated* by amines, possibly due to inhibition of toxin degradation in the lysosomes.[81a,82]

Secretion. In the adrenal gland, catecholamines (which are weak-base amines) are sequestered in acidic chromaffin granules by an ATP-dependent mechanism, probably involving protein pumping. Addition of ammonia elevates the intragranular pH and inhibits catecholamine accumulation.[83] The possibility should be considered that the neurotoxic symptoms produced by ammonia *in vivo* may be due to interference with the normal secretion of neurotransmitters from storage granules in the nerve endings.

In hepatocytes, ammonia was found to delay protein secretion and induce swelling of lipoprotein-containing secretory vesicles, suggesting that a certain intravesicular acidity might be required for normal transport and/or fusion of these vesicles with the plasma membrane.[84]

Inhibition of Lysosomal Degradation

Mechanism of Inhibition. Inhibition of lysosomal protein degradation is the most consistent and well-established effect of the weak-base

[77] K. S. Matlin, H. Reggio, A. Helenius, and K. Simons, *J. Cell Biol.* **91,** 601 (1981).
[78] J. White, J. Kartenbeck, and A. Helenius, *J. Cell Biol.* **87,** 264 (1980).
[79] K. Kim and N. B. Groman, *J. Bacteriol.* **90,** 1552 (1965).
[80] K. Sandvig, S. Olsnes, and A. Pihl, *Biochem. Biophys. Res. Commun.* **90,** 648 (1979).
[81] K. Sandvig and S. Olsnes, *J. Cell Biol.* **87,** 828 (1980).
[81a] K. Sandvig and S. Olsnes, *J. Biol. Chem.* **257,** 7504 (1982).
[82] K. Sandvig and S. Olsnes, *J. Biol. Chem.* **256,** 9068 (1981).
[83] R. G. Johnson, N. J. Carlson, and A. Scarpa. *J. Biol. Chem.* **253,** 1512 (1978).
[84] P. O. Seglen and A. Reith, *Biochim. Biophys. Acta* **496,** 29 (1977).

amines,[1,12,13,17] yet the mechanism of inhibition is not clearly understood. An elevation of the lysosomal pH will provide suboptimal working conditions for the lysosomal enzymes, most of which have acid pH optima, and a considerable inhibition by amines of the proteolytic activity of intact, isolated lysosomes can accordingly be observed.[14] In addition, some aromatic amines have been found to inhibit lysosomal proteases directly,[24,43] but this is the exception rather than the rule. In intact cells, the inhibition of receptor–ligand dissociation[60] and endosome–lysosome fusion[41] may play important parts in preventing the degradation of proteins taken up by endocytosis. A reduced rate of autophagosome–lysosome fusion[73] or of invagination of the lysosomal membrane (cf. the inhibition of multivesicular body development[60]) may play a similar role in endogenous protein degradation. The lack of accumulation of undegraded material in the lysosomes of amine-treated cells, contrasting with the picture seen in, e.g., leupeptin-treated cells[1,73] (Fig. 2), may indeed indicate that intralysosomal proteolysis is *not* the rate-limiting step during amine inhibition.

Endogenous Protein Degradation. Weak-base amines have been found to inhibit endogenous protein degradation at least partially in most tissues and cell types examined, e.g., hepatocytes[12,17,25,85] and hepatoma cells,[86] fibroblasts,[87] macrophages,[88,89] temperature-sensitive mouse L cells,[90] yolk sac,[43] heart,[91] skeletal muscle,[92,93] and mammary gland.[94] The degree of inhibition varies from 10–20% in well-nourished cultured cells ("basal" conditions)[87–90] to 70–80% in hepatocytes in an amino acid-free ("step-down") medium.[17,25] This variability can be explained by differences in the relative contribution by different pathways of degradation. It is now generally accepted that cellular proteins can be degraded either by a lysosomal, amine-sensitive pathway, or by a nonlysosomal, amine-resistant pathway[25,86,87] (Fig. 4). Within the former pathway, amino acid-

[85] M. F. Hopgood, M. G. Clark, and F. J. Ballard, *Biochem. J.* **164,** 399 (1977).

[86] S. E. Knowles and F. J. Ballard, *Biochem. J.* **156,** 609 (1976).

[87] J. S. Amenta, T. J. Hlivko, A. G. McBee, H. Shinozuka, and S. Brocher, *Exp. Cell Res.* **115,** 357 (1978).

[88] R. T. Dean, *Biochem. J.* **180,** 339 (1979).

[89] B. Poole, S. Ohkuma, and M. J. Warburton, *Acta Biol. Med. Ger.* **36,** 1777 (1977).

[90] J. Sparkuhl and R. Sheinin, *J. Cell. Physiol.* **105,** 247 (1980).

[91] K. Wildenthal, J. R. Wakeland, J. M. Ord, and J. T. Stull, *Biochem. Biophys. Res. Commun.* **96,** 793 (1980).

[92] A. B. Jenkins, M. Whittaker, and P. J. Schofield, *Biochem. Biophys. Res. Commun.* **86,** 1014 (1979).

[93] M. Mayer, M. Chaouat, R. Hadar, S. Nissan, and O. Z. Lernau, *J. Cell. Physiol.* **109,** 525 (1981).

[94] S. M. Russell, C. J. Wilde, D. A. White, H. R. Hasan, and R. J. Mayer, *Acta Biol. Med. Ger.* **40,** 1397 (1981).

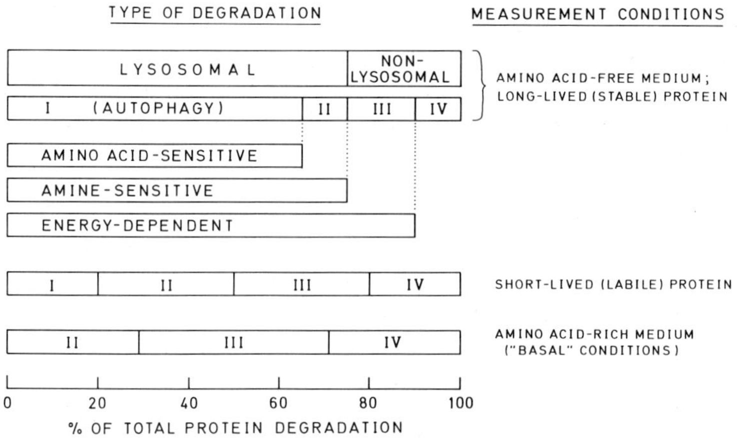

FIG. 4. Pathways of protein degradation in rat hepatocytes. The relative contributions by the various types of degradation have been estimated from data given in references cited in footnotes 1, 3, 4, 6, 15, 17, 25, and 104. Long-lived protein was labeled for 24 hr and short-lived protein for 1 hr before the degradation of radioactive protein was measured.

and hormone-sensitive degradation (I),[1,95] probably representing autophagy,[2,4,96] can be distinguished from degradation resistant to these regulatory agents (II). The nature of the latter is not known; it could represent, e.g., the degradation of secretory proteins already sequestered in vesicles (*crinophagy*), or proteins take up by invaginations of the lysosomal membrane (*lysophagy* or microautophagy).[1,95,96]

The nonlysosomal pathway (which may represent the sum of proteolytic activities in the cytosol,[97–99] mitochondria, [100–102] membranes, etc.[103]) can be tentatively divided into energy-dependent (chymostatin-sensitive) degradation (III) and energy-independent degradation (IV).[1,25,104]

[95] J. S. Amenta and S. C. Brocher, *Exp. Cell Res.* **126,** 167 (1980).

[96] H. Glaumann, J. L. E. Ericsson, and L. Marzella, *Int. Rev. Cytol.* **73,** 149 (1981).

[97] I. A. Rose, J. V. B. Warms, and A. Hershko, *J. Biol. Chem.* **254,** 8135 (1979).

[98] G. N. De Martino and A. L. Goldberg, *J. Biol. Chem.* **254,** 3712 (1979).

[99] G. N. De Martino, *Arch. Biochem. Biophys.* **211,** 253 (1981).

[100] J. Soler, J. Timoneda, D. de Arriaga, and S. Grisolia, *Biochem. Biophys. Res. Commun.* **97,** 100 (1980).

[101] M. Mori, S. Miura, M. Tatibana, and P. P. Cohen, *Proc. Natl. Acad. Sci. U.S.A.* **77,** 7044 (1980).

[102] M. Desautels and A. L. Goldberg, *Proc. Natl. Acad. Sci. U.S.A.* **79,** 1869 (1982).

[103] A. J. Barrett and J. K. McDonald, "Mammalian Proteases: A Glossary and Bibliography," Vol. 1. Academic Press, New York, 1980.

[104] B. Grinde and P. O. Seglen, *Biochim. Biophys. Acta* **632,** 73 (1980).

In amino acid-deprived hepatocytes, all modes of degradation are fully operative (totaling 4–5% per hour), and as much as 70–80% of the degradation (I + II) may be amine-sensitive. In the presence of amino acids ("basal" conditions) autophagy (I) is suppressed, and only 20–30% (II) of the remaining degradation (1.5% per hour) is now amine-sensitive.

Short-lived protein, which represents a small fraction (0.2%) of the total liver protein, but contributes significantly to overall protein turnover (15–20%),[105] can be selectively measured by using a brief (1 hr) isotope labeling period.[25,104] Only about one-half of the short-lived protein is degraded by the lysosomal pathway,[104] mostly by amino acid-resistant mechanism(s) (II), which are not necessarily identical to those used for long-lived protein.

The rationale for using weak-base amines to quantitate lysosomal degradation is twofold: first, the pH-elevating effect of the amines will *specifically* affect proteolysis that is dependent on the participation of an *acidic* cell compartment; this may in fact serve as a useful *definition* of a lysosomal pathway. Proteinases located on the outer (cytosolic) surface of the lysosomes,[106] or normally intralysosomal enzymes operating in the cytosol,[107,108] would by this definition belong to the nonlysosomal pathway. Second, the amines appear to inhibit the lysosomal degradation of endogenous protein essentially *completely*, as indicated by the lack of quantitatively significant additive effect of lysosomal proteinase inhibitors, such as leupeptin,[25,85] antipain,[104] vanadate,[15] epoxysuccinyl peptides[109] and peptidyl diazomethanes.[110] Autophagy inhibitors like the amino acids,[3,4] insulin,[6] 3-methyladenine,[111] and dimethylaminopurine riboside,[112,113] as well as the microtubule poisons colchicine and vinblastine,[114] are likewise nonadditive. Taken together, these observations provide strong support for the assumption that the amine-resistant degradation, being refractory to all these inhibitors of lysosomal function, must be nonlysosomal.

[105] G. E. Mortimore and N. J. Hutson, *Acta Biol. Med. Ger.* **40,** 1577 (1981).

[106] E. Melloni, S. Pontremoli, F. Salamino, B. Sparatore, M. Michetti, and B. L. Horecker, *Proc. Natl. Acad. Sci. U.S.A.* **78,** 1499 (1981).

[107] K. Tanaka, N. Ikegaki, and A. Ichihara, *Arch. Biochem. Biophys.* **208,** 296 (1981).

[108] M. A. McElligott, F. J. Roisen, K. S. Keaton, D. F. Triemer, Q.-S. Li, A. C. St. John, G. Yorke, and J. W. C. Bird, *Acta Biol. Med. Ger.* **40,** 1333 (1981).

[109] B. Grinde, *Biochim. Biophys. Acta* **701,** 328 (1982).

[110] B. Grinde, *Biochim. Biophys. Acta* **757,** 15 (1983).

[111] P. O. Seglen and P. B. Gordon, *Proc. Natl. Acad. Sci. U.S.A.* **79,** 1889 (1982).

[111a] P. B. Gordon and P. O. Seglen, *Exp. Cell Res.* **142,** 1 (1982).

[112] A. L. Kovács and P. O. Seglen, *Biochim. Biophys. Acta* **676,** 213 (1981).

[113] P. B. Gordon and P. O. Seglen, *Arch. Biochem. Biophys.* **217,** 282 (1982).

[114] B. Grinde and P. O. Seglen, *Hoppe-Seyler's Z. Physiol. Chem.* **362,** 549 (1981).

Degradation of Exogenous (Endocytosed) Proteins. The degradation of several proteins ingested by endocytosis (EGF,[16] LDL,[56] acetylated LDL,[115] hCG,[37] and denatured albumin[42]) has been shown to be completely inhibited by weak-base amines, indicating involvement of the lysosomal pathway exclusively. In cases where a partial inhibition is observed,[41,116,117] only a fraction of the ingested protein may have reached the lysosomes[118,119]; alternatively, the protein may be degraded partly in lysosomes and partly elsewhere, e.g., in endosomes[76] or after regurgitation to the extracellular medium.[43] It is therefore possible that the *lysosomal* degradation of exogenous protein in general may be completely inhibited by amines, although the evidence is not as strong as in the case of endogenous protein.

Ionophores

Monovalent carboxylic ionophores like *nigericin* and *monensin* are known primarily as inhibitors of Golgi function. They alter the morphology and interrupt the membrane flow through this organelle, causing, e.g., a rapid inhibition of secretion.[120–122] A more slowly developing inhibition of pinocytosis may be the result of receptor trapping or of membrane trapping in the Golgi apparatus.[122]

Carboxylic ionophores elevate the lysosomal pH,[9] and since they also induce lysosomal swelling,[122] it is likely that they effect a K^+/H^+ exchange rather than an H^+ efflux.[123] Continuous operation of the proton pump would thereby result in an osmotically effective accumulation of K^+ intralysosomally. The ionophores inhibit the degradation of endocytosed proteins like LDL[56] and asialofetuin,[76] but they inhibit the recycling of the receptors for these proteins even more effectively and to a greater extent than do the weak-base amines. The inhibition of degradation may therefore result from an arrest of the transport of endocytosed protein through the Golgi region rather than from the direct action of the carboxylic ionophores upon the lysosome.

[115] T. J. C. Van Berkel, J. F. Nagelkerke, and J. K. Kruijt, *FEBS Lett.* **132,** 61 (1981).
[116] C.-H. Florén and Å. Nilsson, *Biochem. J.* **174,** 827 (1978).
[117] L. Ose, I. Røken, K. R. Norum, and T. Berg, *Exp. Cell Res.* **130,** 127 (1980).
[118] A. D. Attie, R. C. Pittman, and D. Steinberg, *Proc. Natl. Acad. Sci. U.S.A.* **77,** 5923 (1980).
[119] V. Herzog, *Trends Biochem. Sci.* **6,** 319 (1981).
[120] A. Tartakoff and P. Vassalli, *J. Cell Biol.* **79,** 694 (1978).
[121] K. Oda and Y. Ikehara, *Biochem. Biophys. Res. Commun.* **105,** 766 (1982).
[122] D. K. Wilcox, R. P. Kitson, and C. C. Widnell, *J. Cell Biol.* **92,** 859 (1982).
[123] B. C. Pressman, *Annu. Rev. Biochem.* **45,** 501 (1976).

Proteinase-Inhibitory Peptides

A limited number of peptidic proteinase inhibitors have been found to be capable of inhibiting lysosomal protein degradation in intact cells. These include the aspartic proteinase inhibitor pepstatin, the serine proteinase inhibitor chymostatin, and several cysteine proteinase inhibitors (antipain, leupeptin, E-64, Ep-475, and the diazomethylketone Z-Phe-Ala-CHN_2). Since each inhibitor affects only one group of lysosomal enzymes, none of them are quite as effective as the weak-base amines; however, the fact that lysosomal enzymes work sequentially rather than independently makes it possible to inhibit as much as 80% of the amine-sensitive degradation with a single inhibitor like leupeptin.[104]

One important difference between the weak-base amines and the proteinase inhibitors is that the latter specifically inhibit proteolysis, without interfering with other (pH-dependent) lysosomal processes like receptor–ligand dissociation, receptor recycling, or toxin entry.[81] Leupeptin has been found to inhibit endosome-lysosome fusion,[124] but this effect is probably secondary to the accumulation of undigested material intralysosomally, easily visible in the electron microscope.[1,73,125]

Pepstatin, chymostatin, leupeptin, and antipain have been isolated from the culture filtrates of actinomycetes (mostly *Streptomyces* species) by Umezawa and co-workers.[126,127] The epoxysuccinyl peptide E-64 has been isolated from the fungus *Aspergillus japonicus* by Hanada *et al.*[128]; Ep-475 is a closely related synthetic analog. The diazomethane Z-Phe-Ala-CHN_2 was synthesized by Green and Shaw.[129]

Pepstatin. Pepstatin (*N*-isovalerylvalylvalylstatylalanylstatine, M_r 686) contains the unusual amino acid statine (α-hydroxy-β-isobutyl-γ-aminobutyrate) and is active against several aspartic proteinases, including pepsin, renin and the lysosomal enzyme cathepsin D.[103] It is relatively insoluble in water, and is therefore usually dissolved in dimethyl sulfoxide (DMSO).[88,107,130] (Incorporation of pepstatin into liposomes[131] does not seem to be more effective than dissolution in DMSO[130].) Pepstatin is

[124] H. Tolleshaug and T. Berg, *Exp. Cell Res.* **134,** 207 (1981).

[125] N. N. Aronson, P. A. Dennis, and W. A. Dunn, *Acta Biol. Med. Ger.* **40,** 1531 (1981).

[126] T. Aoyagi and H. Umezawa, *in* "Proteases and Biological Control" (E. Reich, D. B. Rifkin, and E. Shaw, eds.), p. 429. Cold Spring Harbor Lab., Cold Spring Harbor, New York, 1975.

[127] T. Aoyagi and H. Umezawa, *Acta Biol. Med. Ger.* **40,** 1523 (1981).

[128] K. Hanada, M. Tamai, M. Yamagishi, S. Ohmura, J. Sawada, and I. Tanaka, *Agric. Biol. Chem.* **42,** 523 (1978).

[129] G. D. J. Green and E. Shaw, *J. Biol. Chem.* **256,** 1923 (1981).

[130] W. F. Ward, B. L. Chua, J. B. Li, H. E. Morgan, and G. E. Mortimore, *Biochem. Biophys. Res. Commun.* **87,** 92 (1979).

[131] R. T. Dean, *Nature (London)* **257,** 414 (1975).

taken up very slowly by cells (probably by endocytosis), and may require 20–30 hr of incubation to produce maximal inhibition.[88,107] Although pepstatin eventually inhibits intracellular cathepsin D activity strongly, its effect on overall endogenous protein degradation (15–20% inhibition) is considerably smaller than that of the thiol proteinase inhibitors.[107] It is, however, additive to the latter,[107,130] and pepstatin may therefore be useful in combination with other proteinase inhibitors, despite its insolubility and slow action.

Chymostatin. Chymostatin, *N*-[(1-carboxy-2-phenylethyl)carbamoyl]-α-(2-iminohexahydro-4-pyrimidyl)-glycylleucylphenylalaninal (M_r 605), is a strong inhibitor of chymotrypsin and chymotrypsin-like serine proteinases like the mast cell chymases, neutrophile cathepsin G, and seminin[103]; however, it also effectively inhibits the lysosomal cysteine proteinases cathepsin B, H, and L[126,132] and the cytosolic Ca^{2+}-activated cysteine proteinase from muscle.[133] Chymostatin can be dissolved in DMSO or in 0.1 *M* HCl, which is subsequently neutralized; the inhibitor should be used at a final medium concentration of 100–200 μg/ml.

Chymostatin inhibits endogenous protein degradation markedly in muscle[133] and yolk sac[134] and strongly in hepatocytes,[104,135,136] most of the latter effect being due to an approximately 75% inhibition of lysosomal proteolysis.[135] The lysosomal degradation of an exogenous protein (asialofetuin) was also partially inhibited.[136] However, in addition chymostatin inhibits the energy-dependent part of the nonlysosomal (amine-resistant) degradation of short-lived protein.[104] The inactivation of two individual short-lived enzymes, tryptophan oxygenase and tyrosine aminotransferase, was inhibited uniquely by chymostatin, methylamine and leupeptin being inactive.[135] Chymostatin is so far the only known proteinase inhibitor capable of effective inhibition of nonlysosomal (as well as lysosomal) degradation. In hepatocytes it inhibits after a lag of only 10 min; its effect is rapidly reversible.[104] At the concentrations needed for maximal inhibition in hepatocytes it does, however, inhibit protein synthesis some 20%, in contrast to the cysteine proteinase inhibitors.[104]

Leupeptin. Leupeptin is *N*-propionyl- and *N*-acetylleucylleucylargininal in an approximately 3:1 ratio (average M_r 438, or 473 as the chloride). It is a strong inhibitor of cysteine proteinases and inhibits the

[132] P. Bohley, H. Kirschke, J. Langner, S. Riemann, B. Wiederanders, S. Ansorge, and H. Hanson, *in* "Protein Turnover and Lysosome Function" (H. L. Segal and D. J. Doyle, eds.), p. 379. Academic Press, New York, 1978.

[133] P. Libby and A. L. Goldberg, *Biochem. J.* **188,** 213 (1980).

[134] S. E. Knowles, F. J. Ballard, G. Livesey, and K. E. Williams, *Biochem. J.* **196,** 41 (1981).

[135] B. Grinde and R. Jahnsen, *Biochem. J.* **202,** 191 (1982).

[136] N. T. Neff, G. N. DeMartino, and A. L. Goldberg, *J. Cell. Physiol.* **101,** 439 (1979).

lysosomal cathepsins B, H, L, N, S, and T as well as the nonlysosomal Ca^{2+}-dependent proteinase II.[103,132,137] It also inhibits a number of serine proteinases, including trypsin, plasmin, tissue kallikrein, ribosomal serine proteinase, and trypsin-like proteinases of uterus and muscle.[103]

Leupeptin dissolves easily in aqueous solutions; for maximal effect it should be used at a concentration of 100 μg/ml.

The inhibitory effect of leupeptin on endogenous protein degradation varies considerably between different tissues, probably reflecting variability in the relative contribution by the lysosomal degradation pathway. In rat liver cells, the degradation of endogenous[25,104] as well as exogenous[124] protein is strongly inhibited; the former up to 70%, representing an estimated 80% inhibition of lysosomal (amine-resistant) degradation and a slight (15%) inhibition of nonlysosomal degradation.[104] The inactivation of short-lived liver enzymes was not detectably inhibited by leupeptin.[135] In the yolk sac, the degradation of denatured albumin (which is 100% inhibitable by amines, i.e., entirely lysosomal) was likewise inhibited up to approximately 80% by leupeptin, the latter being more effective than either antipain or chymostatin (pepstatin and several other proteinase-inhibitory peptides were inactive).[134] Endogenous protein degradation in the yolk sac was, however, inhibited only 20% by leupeptin,[134] in agreement with the lower relative contribution by the lysosomal pathway (45%)[42] to this type of degradation. Leupeptin has furthermore been found to inhibit endogenous protein degradation 30% in mouse macrophages,[88] 30–40% in hepatoma cells,[86] 20–40% in the heart,[91,130] 10–20% in chick skeletal muscle,[138,139] and 35–60% in rat skeletal muscle.[140]

Leupeptin inhibits rapidly, with no detectable lag.[25] The mode of uptake is unknown, but the rapidity of its transfer from liver sinusoids to bile, apparently in an unmodified form, suggests that it is readily transported across cellular membranes.[125] Its effect is rapidly reversible in the yolk sac,[134] but not in hepatocytes.[25] Leupeptin has no effect on protein synthesis[104] or ATP levels.[134] Its lack of toxicity has made it interesting as a possible therapeutic agent,[126] and, e.g., its ability partially to inhibit protein degradation in dystrophic muscle[139,140] has been reported to result in some functional improvement.[141]

The accumulation of undegraded material in the lysosomes of leupeptin-treated hepatocytes[1,73,124,125] has been found to increase the density of

[137] E. Gohda and H. C. Pitot, *J. Biol. Chem.* **256,** 2567 (1981).

[138] P. Libby and A. L. Goldberg, *J. Cell. Physiol.* **107,** 185 (1981).

[139] B. A. Wolitzky, H. L. Segal, and M. S. Hudecki, *Exp. Cell Res.* **137,** 295 (1982).

[140] P. Libby and A. L. Goldberg, *Science* **199,** 534 (1978).

[141] J. H. Scher, A. Stracher, S. A. Shafiq, and J. Hardy-Staskin, *Proc. Natl. Acad. Sci. U.S.A.* **78,** 7742 (1981).

the lysosomes in sucrose gradients,[124,125] and to impair their ability to fuse with endosomes.[124] The marked accumulation of autophagosomes[73] suggests that fusion of lysosomes with the latter may likewise be impaired. The inhibition of fusion is incomplete and develops slowly; it is probably a secondary consequence of lysosomal "constipation,"[124] which does not contribute to the rapid inhibitory effect of leupeptin.

Antipain. Antipain, [1-carboxy-2-phenylethyl]carbamoylarginylarginylvalylargininal (M_r 605), is closely related to leupeptin, both containing the amino aldehyde argininal terminally. Like leupeptin, antipain is easily soluble in water and is maximally effective around 100 μg/ml. Antipain is a stronger inhibitor of trypsin and cathepsin A,[126] but a weaker inhibitor of plasmin,[126] cathepsin B,[126] and overall endogenous protein degradation in hepatocytes,[104,132,136] hepatoma cells,[86] and yolk sac.[134] The effect of antipain is, on an average, 75% of the leupeptin effect, and not additive to the latter.[104] Although antipain may have biological effects not shared by leupeptin,[142] it is less useful (less effective) than leupeptin as a general inhibitor of lysosomal protein degradation.

E-64 and Ep-475. E-64 (L-*trans*-epoxysuccinylleucylagmatine, M_r 341) and its synthetic derivative Ep-475 (L-*trans*-epoxysuccinylleucylamidoisopentane, M_r 298) bind strongly and irreversibly to most cysteine proteinases, and can actually be used as titrants of the latter.[143] They react very rapidly with cathepsins B and L, Ep-475 being the most potent, and somewhat less rapidly with cathepsin H, E-64 being more potent.[143] The epoxides are very specific, having no effect on noncysteine proteinases (unlike, e.g., leupeptin and antipain). E-64 is water-soluble, whereas Ep-475 has to be dissolved in DMSO.

E-64 and Ep-475 inhibited endogenous protein degradation in hepatocytes to a similar maximal extent (45%), at approximately 100 μg/ml and 10 μg/ml, respectively.[109] Although the inhibition was smaller than with leupeptin, the epoxides acted exclusively upon lysosomal degradation; i.e., there was no effect on the amine-resistant proteolysis. The onset of inhibition was rapid (10 min lag), and the effect is presumably irreversible.[143] No inhibition of protein synthesis could be detected.[109] The effects of E-64 and Ep-475 were significantly additive, consistent with their different spectra of cathepsin inhibition, reaching a total inhibition of 60% of the endogenous degradation.[109] The high lysosome specificity would seem to make a mixture of these epoxides a very attractive alternative to leupeptin for the inhibition of lysosomal proteolysis.

Z-Phe-Ala-CHN$_2$. This diazomethane (benzyloxycarbonylphenyl-

[142] M. Mori and P. O. Cohen, *Proc. Natl. Acad. Sci. U.S.A.* **75,** 5339 (1978).
[143] A. J. Barrett, A. A. Kembhavi, and K. Hanada, *Acta Biol. Med. Ger.* **40,** 1513 (1981).

alanylalanyldiazomethane, M_r 394) and other peptidyl diazomethanes (diazomethylketones) are specific inhibitors of thiol proteinases.[129,144] They act irreversibly, forming covalent complexes with the enzymes. Z-Phe-Ala-CHN_2 appears to be the most suitable of these inhibitors for treatment of intact cells. It is sparingly water-soluble, but can be dissolved in DMSO.

In macrophages, Z-Phe-Ala-CHN_2 was taken up very slowly, apparently by endocytosis. After 24 hr a 20–30% inhibition of endogenous protein degradation was obtained in the concentration range 5–40 μg/ml. The effect was additive to that of pepstatin.[145]

In rat hepatocytes, Z-Phe-Ala-CHN_2 exerted its effect rapidly, and at 40 μg/ml it produced a 50% inhibition of endogenous protein degradation within 2 hr.[110] Most of this effect would be on lysosomal degradation, as expected, but a small (15%) inhibition of nonlysosomal degradation was also indicated. New diazomethanes are being developed, and it is yet too early to evaluate their utility as lysosome inhibitors, especially in terms of side effects (inhibition of protein synthesis, etc.).

Vanadate

Vanadate (sodium orthovanadate, $Na_3VO_4 \cdot 14H_2O$, M_r 436) is best known by its ability to inhibit various ATPases and other phosphate-transferring enzymes,[146,147] probably by acting as a phosphate analog. However, it has also been found to stimulate glucose metabolism[148] fatty acid synthesis,[149] and DNA synthesis[150,151] and to inhibit lipolysis[152] and protein degradation[15] in intact cells. Certain bacterial and mitochondrial proteinases are directly inhibited by vanadate.[102,152a]

The stabilization of isolated steroid receptors by vanadate, molybdate, and tungstate is probably due to protection against proteolytic attack.[153–155] Although several other transition-state metal oxides shared the

[144] H. Kirschke and E. Shaw, *Biochem. Biophys. Res. Commun.* **101,** 454 (1981).

[145] E. Shaw and R. T. Dean, *Biochem. J.* **186,** 385 (1980).

[146] J. J. Grantham, *Am. J. Physiol.* **239,** F97 (1980).

[147] I. G. Macara, *Trends Biochem. Sci.* **5,** 92 (1980).

[148] G. R. Dubyak and A. Kleinzeller, *J. Biol. Chem.* **255,** 5306 (1980).

[149] L. Agius and W. J. Vaarties, *Biochem. J.* **202,** 791 (1982).

[150] C. Hori and T. Oka, *Biochim. Biophys. Acta* **610,** 235 (1980).

[151] G. Carpenter, *Biochem. Biophys. Res. Commun.* **102,** 1115 (1981).

[152] H. Degani, M. Gochin, S. J. D. Karlish, and Y. Schechter, *Biochemistry,* **20,** 5795 (1981).

[152a] F. S. Larimore, L. Waxman, and A. L. Goldberg, *J. Biol. Chem.* **257,** 4187 (1982).

[153] L. K. Miller, F. E. B. Tuazon, N. I. U. En-Mei, and M. R. Sherman, *Endocrinology* **108,** 1369 (1981).

[154] H. Nishigori and D. Toft, *Biochemistry* **19,** 77 (1980).

[155] T. Hazato and A. Murayama, *Biochem. Biophys. Res. Commun.* **98,** 488 (1981).

antiproteolytic properties of vanadate, the latter was unique in having very little effect on protein synthesis.[156]

Vanadate dissolves easily in water, giving colorless acidic solutions that, upon neutralization, turn yellow. The chemistry of the vanadates is complex, and the solution probably contains an equilibrium mixture of several vanadium–oxygen complexes of different molecular weights.[147] Although vanadate is quite effective in the micromolar range when tested on isolated enzymes or in homogenates,[153,156] up to 10 mM is required for maximal inhibition of protein degradation in intact hepatocytes, at least partly owing to a slow uptake.[15,156] At this concentration, alterations in cell morphology are observed.[15]

Vanadate inhibits hepatocytic protein degradation 50–60%, most of its effect being on lysosomal proteolysis, but a slight inhibition of nonlysosomal degradation is also indicated.[15] A direct 80% inhibition of lysosomal enzyme activity in an acidified homogenate at pH 5 has been demonstrated.[15] Nothing is yet known about the specificity of inhibition of the mammalian proteinases by vanadate.

In the electron microscope, the lysosomes of vanadate-treated cells can be seen to be filled with undigested material (cf. leupeptin), and autophagosomes are accumulated in large numbers, suggesting a reduced fusion with the "constipated" lysosomes.[156] In accordance with this interpretation, vanadate has been found to inhibit endosome–lysosome fusion strongly.[76]

When radioactivity is measured in vanadate-containing samples by liquid scintillation counting, corrections must be made for color quenching and bioluminescence, the latter being particularly prominent in the presence of sucrose.

3-Methyladenine: an Inhibitor of Autophagy

3-Methyladenine (3-MeAde, 6-amino-3-methylpurine, M_r 149) is not a direct lysosome inhibitor; however, being a specific inhibitor of autophagy, it shuts down the major source of material delivered to the lysosomes.[111] 3-MeAde has accordingly been found to inhibit 60% of the endogenous protein degradation in hepatocytes, affecting exclusively the lysosomal degradation pathway. Like the physiological regulators, the amino acids, 3-MeAde inhibits the primary sequestration of cytoplasm, i.e., the formation of autophagosomes.[111,111a]

Several other purine derivatives (e.g., 6-dimethylaminopurine riboside and puromycin aminonucleoside) have been found to inhibit autophagic/

[156] D. O'Reilly and P. O. Seglen, unpublished results.

lysosomal protein degradation[1,112,113]; however, 3-MeAde is unique in having no inhibitory effect on protein synthesis. It is nontoxic *in vivo,* and may have a certain therapeutic potential under conditions where autophagy contributes to excessive tissue breakdown. Its tissue specificity has not yet been investigated.

3-MeAde can be dissolved in hot water up to a concentration of about 6 mg/ml, it is maximally effective around 1 mg/ml (5–10 mM). 3-MeAde diffuses rapidly into the cells, even at 0°. It is fully active after a lag phase of 20–30 min; its effect is readily reversible.[111]

Microtubule Poisons

Vinblastine has long traditions as an "autophagy inducer"[96,157,158]; however, it is now believed that the characteristic accumulation of autophagosomes seen in vinblastine-treated cells is a secondary effect of reduced autophagosome–lysosome fusion.[26,73] Both vinblastine and *colchicine* are in fact effective *inhibitors* of cellular protein degradation, apparently acting exclusively upon the lysosomal pathway.[114] Progressive inhibition is observed over a wide concentration range, and although degradation is significantly inhibited at the micromolar concentrations known to disrupt microtubuli, maximal inhibition is not reached before the appearance of toxic effects (inhibition of protein synthesis and eventually cell death) at much higher concentrations. A 60–70% inhibition of the lysosomal pathway can be achieved without significant inhibition of protein synthesis at 0.5 mM colchicine, whereas the vinblastine concentration must be kept below 0.1 mM.[114]

Despite their very general effects on the cell, the microtubule poisons may be useful as inhibitors of lysosomal function, acting differently from the other lysosome drugs. However, it should be stressed that the exact mechanism by which they inhibit the lysosomal degradation pathway is not at all understood.

Lectins and Polymers

Certain lectins and polyanions may be used to inhibit lysosomal fusions "from within." These substances are taken up by endocytosis and may eventually enter the lysosomes. By binding to the inside of the endosomal and/or lysosomal membrane, they reduce the ability of these vacuoles to fuse.

[157] A. U. Arstila, I. J. M. Nuuja, and B. F. Trump, *Exp. Cell Res.* **87,** 249 (1974).
[158] J. Kovács, G. Réz, and A. Kiss, *Cytobiologie* **11,** 309 (1975).

Concanavalin A has been shown to inhibit phagosome–lysosome fusion in macrophages.[159] In a study of isolated hepatocytes, Berg and co-workers[160] found that a 15-min preincubation with concanavalin A (100–200 μg/ml) inhibited endosome–lysosome fusion, and thereby strongly suppressed the degradation of the exogenous, endocytosed protein asialofetuin. The lectin accumulated only in the endosomes, and therefore obviously inhibited fusion from the *endosomal* side. (In contrast, a lectin from *Wistaria floribunda* was found to accumulate in the *lysosomes* of mouse fibroblasts.)[161]

Chlorite-oxidized amylose (COAM) and several other polyanions have been found to inhibit phagosome–lysosome fusion in macrophages.[162,163] COAM accumulates in the lysosomes, and apparently inhibits fusion from the *lysosomal* side.[75]

Protein Synthesis Inhibitors

Inhibitors of protein synthesis have been found to suppress endogenous protein degradation in several cell types,[86,95,109,164,165] apparently by an action upon the lysosomal pathway.[86,164] The inhibition seen in hepatocytes treated with *cycloheximide* is evident only after a 1–hr lag; before this time the autophagic-lysosomal pathway is fully operative and sensitive to regulation by amino acids.[6,112] The mechanism of inhibition is not known; it may, e.g., reflect the decay of a short-lived degradation-promoting protein.[166] A regulatory function of deacylated tRNA has been proposed,[167] but is disputable.[166,168,169] Long-term treatment with cycloheximide may result in a different type of reduced lysosome function, based on a relative decrease in the amounts of lysosomal enzymes.[170]

Puromycin inhibits protein degradation by two mechanisms additional to the one shared with cycloheximide: (*a*) by a purine effect specifically

159 P. J. Edelson and Z. A. Cohn, *J. Exp. Med.* **140,** 1364 (1974).
160 H. Tolleshaug, M. Abdelnour, and T. Berg *Biochem. J.* **190,** 697 (1980).
161 R. D. Poretz, R. E. Triemer, A. C. St. John, M. Merion, R. J. Kuchler, D. Cryan, J. H. Carter, and J. W. C. Bird, *Exp. Cell Res.* **128,** 133 (1980).
162 M. J. Geisow, G. H. Beaven, P. D'Arcy Hart, and M. R. Young, *Exp. Cell Res.* **126,** 159 (1980).
163 P. D'Arcy Hart and M. R. Young, *Exp. Cell Res.* **118,** 365 (1979).
164 J. S. Amenta, M. J. Sargus, and F. M. Baccino, *Biochem. J.* **168,** 223 (1977).
165 A. N. Neely, B. P. Nelson and G. E. Mortimore, *Biochim. Biophys. Acta* **338,** 458 (1974).
166 D. Epstein, S. Elias-Bishko, and A. Hershko, *Biochemistry* **14,** 5199 (1975).
167 O. A. Scornik, M. L. S. Ledbetter, and J. S. Malter, *J. Biol. Chem.* **255,** 6322 (1980).
168 C. P. Stanners, T. M. Wightman, and J. L. Harkins, *J. Cell. Physiol.* **95,** 125 (1978).
169 J. M. Gunn, *Exp. Cell Res.* **117,** 448 (1978).
170 J. S. Amenta, M. J. Sargus, and F. M. Baccino, *J. Cell. Physiol.* **97,** 267 (1978).

on autophagic-lysosomal degradation; this inhibition is more rapid in onset than the cycloheximide effect, and is observed also with puromycin aminonucleoside and several other purines (cf. 3-MeAde); (*b*) by a direct inhibition of lysosomal activity, also affecting the degradation of the exogenous protein asialofetuin.[11] The latter effect is not shared by other purines or by cycloheximide; it may possibly be related to the aminopeptidase-inhibitory properties of puromycin.[171]

Cycloheximide is thus clearly the protein synthesis inhibitor of choice, but even this drug is at best a very indirect inhibitor of lysosomal function. It may, however, help to elucidate the mechanisms by which the activity of the lysosomal pathway is controlled.

[171] H. P. Schnebli, M. A. Phillipps, and R. K. Barclay, *Biochim. Biophys. Acta* **569,** 89 (1979).

[60] Biosynthesis of Lysosomal Enzymes

By MELVIN G. ROSENFELD, GERT KREIBICH, DAVID D. SABATINI, and KEITARO KATO

The study of lysosomal biogenesis promises to cast light on the discriminating mechanisms that direct the subcellular traffic of proteins. In several respects, lysosomal hydrolases resemble secretory proteins, and indeed in many cell lines a substantial proportion of these proteins are exported into the extracellular medium.[1] It may be inferred from this fact that lysosomal enzymes, like genuine secretory proteins, are synthesized in membrane-bound ribosomes and at least initially follow the secretory route.[2,3]

This notion was originally based on cytochemical observations[4,5] demonstrating the presence of hydrolases in the lumina of the endoplasmic reticulum, Golgi, and GERL membrane systems. It is also supported by the glycoprotein character common to lysosomal enzymes and many secretory proteins, as well as by the functional analogy that can be made

[1] S. Hickman, L. J. Shapiro, and E. F. Neufeld, *Biochem. Biophys. Res. Commun.* **57,** 55 (1974).
[2] C. de Duve and R. Wattiaux, *Annu. Rev. Physiol.* **28,** 435 (1966).
[3] L. D. Strawser and O. Touster, *Rev. Physiol., Biochem. Pharmacol.* **87,** 169 (1980).
[4] A. B. Novikoff, *Proc. Natl. Acad. Sci. U.S.A.* **73,** 2781 (1976).
[5] D. F. Bainton and M. G. Farquhar, *J. Cell Biol.* **45,** 54 (1970).

ISBN 0-12-181996-5

between the secretory discharge that takes place on the surface of the cell and the release of lysosomal content into phagocytic vacuoles.

We have shown that mRNAs for the lysosomal enzymes β-glucuronidase and cathepsin D are segregated intracellularly in membrane-bound polysomes and provided evidence indicating that the nascent lysosomal polypeptides contain amino-terminal insertion signals that may be functionally analogous to those determining the cotranslational insertion of secretory polypeptides into the endoplasmic reticulum membranes.[6]

This section describes procedures used for the purification of the lysosomal enzymes β-glucuronidase and cathepsin D. Methods are also detailed for the isolation of mRNA and the *in vitro* translation and immunoprecipitation procedures used to study the biosynthesis of these lysosomal proteases.

β-Glucuronidase

Enzymic Assay

Enzymic activity is assayed by incubating samples for 20 min at 37° in a final volume of 1 ml containing 0.2 *M* sodium acetate buffer, pH 4.5, 0.1% bovine serum albumin, and 10 m*M* *p*-nitrophenol β-D-glucosiduronic acid (Sigma). The reaction is stopped by adding 2.0 ml of 0.1 *M* NaOH, 3 ml of water, and absorbance at 400 nm is measured. One unit is the activity that catalyzes the release of 1 μmol of *p*-nitrophenol per minute from the substrate.[7]

Purification Procedure

β-Glucuronidase is purified from the rat preputial gland, which has been shown to be a particularly rich source of this enzyme.[8] Preputial glands (2.5 g) from 45 female rats (150 g) are dissected and homogenized in 80 ml of 0.1 *M* Tris-acetate buffer, pH 7.8, using a Polytron tissue homogenizer (Brinkmann Instruments). The homogenate is centrifuged at 95,500 *g* for 1 hr at 4° and the crude supernatant fraction is brought to 20% saturation by addition of solid ammonium sulfate at 4°.[9] The precipitate formed is discarded and the supernatant adjusted to 65% saturation. This precipitate is collected by centrifugation at 15,000 *g* for 20 min and dis-

[6] M. G. Rosenfeld, G. Kreibich, D. Popov, K. Kato, and D. D. Sabatini, *J. Cell Biol.* **93,** 135 (1982).

[7] M. Himeno, H. Ohhara, Y. Akakaway, and K. Kato, *J. Biochem.* (*Tokyo*) **77,** 427 (1975).

[8] G. A. Levvy, A. McAllan, and C. A. Marsh, *Biochem. J.* **69,** 22 (1958).

[9] A. A. Green and W. L. Hughes, this series, Vol. 1, p. 67.

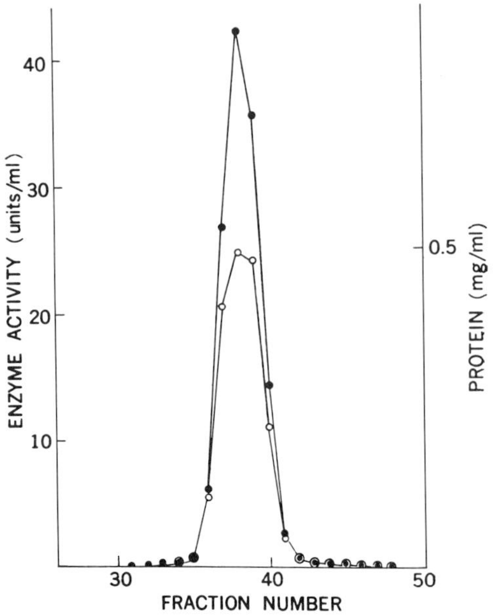

FIG. 1. Gel filtration on Sephadex G-200 of β-glucuronidase obtained by ethanol fractionation. Sample solution (10 ml) was applied to a 2.5 × 90 cm column of Sephadex G-200 equilibrated with 5 m*M* Tris-acetate buffer, pH 7.8. Fractions (5 ml) were collected at a flow rate of 5 ml/hr. β-Glucuronidase activity (●); protein (○).[7]

solved in 50 ml of 0.02 *M* Tris-acetate, pH 7.8. Insoluble material is removed by centrifugation.

One volume containing approximately 6 mg of protein per milliliter is mixed with 0.4 volume of ice-cold ethanol and kept on ice for 10 min. After centrifugation at 15,000 *g* for 30 min the pellet is discarded and the supernatant is treated with 0.8 volume of ice-cold ethanol. After 10 min of incubation at 0°, the mixture is centrifuged as before, and the supernatant is discarded. The precipitate is dissolved in 10 ml of 0.02 *M* Tris-acetate, pH 7.8, and insoluble material is removed by brief centrifugation.

The enriched extract (5.2 mg protein per milliliter) is loaded onto a Sephadex G-200 column (2.5 × 90 cm) equilibrated and eluted with a 0.005 *M* Tris-acetate buffer, pH 7.8. The enzyme activity elutes as a symmetrical peak just after the void volume. A typical chromatography is shown in Fig. 1. This procedure results in a 20-fold enrichment of the β-glucuronidase activity over the total homogenate with a recovery of approximately 60%. The specific activity of the electrophoretically homogeneous enzyme is approximately 85 units per milligram of protein.

A summary of this purification procedure is shown in Table I.

TABLE I
PURIFICATION OF β-GLUCURONIDASE FROM RAT PREPUTIAL GLAND[7]

Procedure	Protein (mg)	Total activity (units)	Specific activity (units/mg)	Purification ratio	Recovery (%)
Crude extract	768.50	4008	5.21	1	100
Ammonium sulfate fractionation	321.30	3823	11.90	2.28	95.4
Ethanol fractionation	52.82	3343	63.28	12.14	83.4
Gel filtration on Sephadex G-200	29.61	2535	85.61	16.42	63.2

Cathepsin D

Enzymic Assay

Cathepsin D activity is determined by a modification of the method of Anson.[10] To the reaction mixture, containing 0.5 ml of 0.1 *M* sodium acetate buffer, pH 3.8, and 1.0 ml of buffered hemoglobin solution (2.5%, w/v, in 0.1 *M* sodium acetate buffer, pH 3.8), 5–100 μl of enzyme solution are added, and the final volume is adjusted to 2.0 ml with distilled water. After 40 min incubation at 37°, the reaction is stopped by rapid cooling of the test tubes to 0° followed by the addition of 2.0 ml of ice-cold 5% trichloroacetic acid solution. After 10 min at 0°, the samples are centrifuged; the peptides liberated in the supernatant (1.0 ml) are measured by the Folin–Lowry reaction.[11] The enzyme activity is determined with reference to a tyrosine standard curve. Units are expressed as micrograms of tyrosine solubilized per minute.[12]

Purification Procedure

This procedure describing the purification of cathepsin D from rat spleen has also been used for purification of the enzyme from rat liver.[12] Rats are sacrificed by decapitation, and the spleens are removed, minced, and homogenized in a Waring blender at top speed for 1 min in ice-cold distilled water (1 ml per gram of tissue). The homogenate is brought to 5 volumes with water and centrifuged at 1000 *g* for 10 min. The pellet,

[10] M. L. Anson, *J. Gen Physiol.* **22,** 79 (1939).
[11] O. H. Lowry, N. J. Rosebrough, A. L. Farr, and R. J. Randall, *J. Biol. Chem.* **193,** 265 (1951).
[12] K. Yamamoto, N. Katsuda, M. Himeno, and K. Kato, *Eur. J. Biochem.* **95,** 459 (1979).

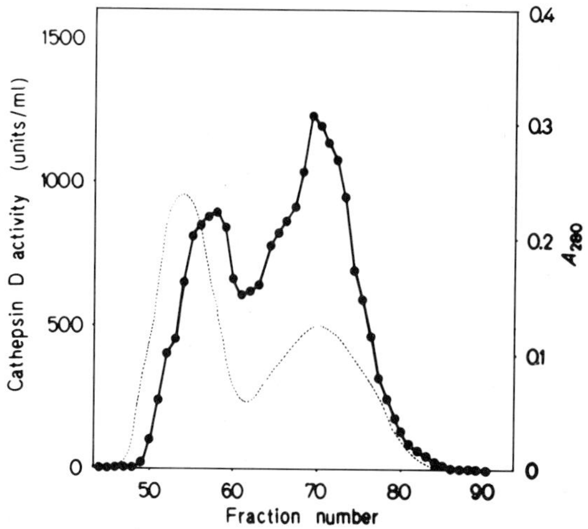

FIG. 2. Gel filtration on Sephadex G-100 of the cathepsin D-enriched fraction. The enzyme solution after concanavalin A–Sepharose 4B affinity chromatography was applied to a Sephadex G-100 column, 2.5 × 90 cm, equilibrated with 0.02 *M* sodium phosphate buffer, pH 7.0. The flow rate was 6 ml/hr, and fractions of 3 ml were collected. ●, Cathepsin D activity; ---, protein.[12]

resuspended in the same volume of 0.02 *M* sodium phosphate buffer, pH 7.0, is centrifuged again as above.

The pooled supernatants are centrifuged at 17,000 *g* for 30 min, and the final supernatant is adjusted to 65% saturation by addition of solid ammonium sulfate while stirring at 4°. After 30 min the precipitate is sedimented (17,000 *g* for 20 min), dissolved in water, and dialyzed for 12 hr against water. Insoluble material is removed by centrifugation (17,000 *g* for 20 min) and the supernatant is adjusted to pH 3.5 with 1 *M* acetic acid. After addition of one-fifth volume of 0.5 *M* sodium acetate buffer, pH 3.5, containing 5 *M* sodium chloride, the precipitate formed is removed by centrifugation (17,000 *g* for 20 min). The supernatant is saved, and the pellet is resuspended in 0.1 *M* sodium acetate buffer, pH 3.5, containing 1 *M* NaCl. The supernatant obtained (105,000 *g* for 30 min) is pooled together with the first one and applied to a pepstatin–Sepharose 4B column (3.2 × 6.0 cm) equilibrated with 0.1 *M* sodium acetate, pH 3.5, containing 1 *M* NaCl. The column is washed with this buffer until the absorbance at 280 nm returns to the baseline. Elution is carried out with 0.1 *M* Tris-HCl, pH 8.6, containing 1 *M* NaCl. Samples corresponding to the protein peak are pooled, concentrated, and dialyzed against 0.02 *M* sodium phosphate, pH 7.0.

The enzyme preparation is applied to a concanavalin A (Con A)–Sepharose column (2.5 × 5.0 cm), equilibrated with 0.02 *M* sodium phosphate, pH 7.0, containing 1 *M* NaCl. After thorough washing of the column, the enzyme is eluted with 0.2 *M* methyl α-D-glucoside. The eluted fraction is concentrated and dialyzed against 0.02 *M* sodium phosphate, pH 7.0, and applied to a Sephadex G-100 column (2.5 × 90 cm) equilibrated with 0.02 *M* sodium phosphate, pH 7.0. The column is eluted with the same buffer (6 ml/hr), and fractions of 3 ml are collected (see Fig. 2). Two peaks of proteolytic activity that hydrolyze bovine hemoglobin at pH 3.5 are found. The first peak, containing approximately 40% of the total activity, corresponds to a cathepsin E-like enzyme. Cathepsin D activity is recovered in the second peak. These samples are pooled, concentrated, and dialyzed against 0.005 *M* sodium phosphate, pH 7.1.

The enriched fraction is applied to a column of DEAE–Sephacel (1.6 × 12 cm) equilibrated with 0.005 *M* sodium phosphate, pH 7:0. The column is washed with the same buffer, and the enzyme is eluted stepwise with an NaCl-containing buffer (for details see legend to Fig. 3). Two peaks of enzyme activity are obtained by elution with 0.05 *M* NaCl. The major peak (cathepsin D-I) represents 70–80% of the total activity, and

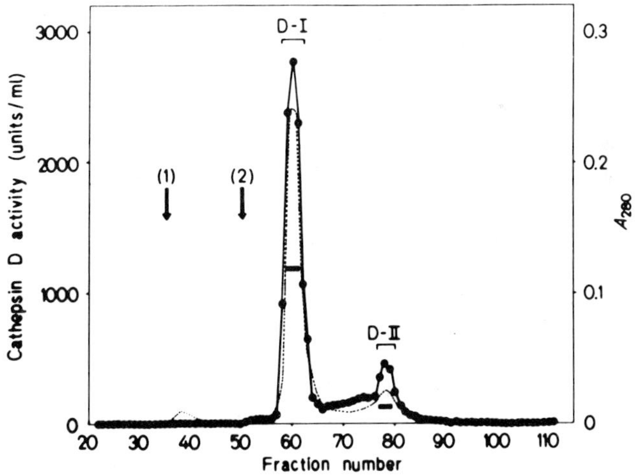

FIG. 3. DEAE–Sephacel column chromatography of the cathepsin D-enriched fraction. The enzyme fraction after gel filtration on Sephadex G-100 was applied to a DEAE-Sephacel column, 1.6 × 12 cm, equilibrated with 0.005 *M* sodium phosphate buffer, pH 7.0. Elution was carried out with the same buffer. The arrows indicate the change of sodium chloride concentrations: (1) 0.02 *M*, (2) 0.05 *M*. The flow rate was 6 ml/hr, and fractions of 3 ml were collected. ●, cathepsin D activity; ---, protein.[12]

the minor peak (cathepsin D-II) represents 10–15%. Any contaminating cathepsin E-like activity is eluted at 0.3 *M* NaCl.

A summary of this purification procedure is shown in Table II.

Preparation of Antibodies

Antibodies to β-glucuronidase and cathepsin D are raised in rabbits. Antigens (0.5 mg) are emulsified in 0.5 ml of 0.05 *M* sodium phosphate buffer, pH 7.0, with an equal volume of Freund's complete adjuvant and injected intramuscularly on days 0, 14, and 28. The animals are bled 14 days after the last injection. Antibodies are affinity purified by conventional procedures.[13]

In Vitro Synthesis of Lysosomal Enzymes

Isolation of mRNA for in Vitro Synthesis of Lysosomal Enzymes

The following procedure for the preparation of mRNA combines features of the methods of Liu *et al.*[14] and Ulrich *et al.*[15] It has been used successfully for the isolation of mRNA from about 5.0 g of tissue, but smaller quantities can be processed using rotors with smaller capacity.

A stock solution of 5 *M* quanidine thiocyanate (Fluka) containing 0.5% sarcosyl is made in distilled water by stirring under low heat followed by filtration through a 0.45 μm Millipore filter. 2-Mercaptoethanol (100 m*M* final) is added just prior to use and the pH is adjusted to pH 5.0 with acetic acid, if necessary.

Rats are sacrificed by cervical dislocation and the tissue to be extracted (preputial gland, spleen, liver) is removed and placed in the homogenization solution (5% w/v). The sample is homogenized with a Polytron homogenizer (Brinkmann Instruments) using a sterilized probe to avoid RNase contamination. If the homogenate is viscous, more of the thiocyanate solution is added and the mixture is rehomogenized. The homogenate is poured into sterile Corex tubes and spun at 10,000 *g* for 5 min. When preputial gland tissue is extracted, a sebaceous material which

[13] R. J. Mayer and J. H. Walker, *in* "Immunological Methods in the Biological Sciences" (J. E. Treherne and P. H. Rubery, eds.), p. 18. Academic Press, New York, 1980.

[14] C. P. Liu, D. L. Slate, R. Gravel, and F. H. Ruddle, *Proc. Natl. Acad. Sci. U.S.A.* **76,** 4503 (1979).

[15] A. Ulrich, J. Shine, J. Chirgrin, R. Piatet, E. Fischer, U. Rutter, and A. M. Goodman, *Science* **196,** 1313 (1977).

TABLE II
PURIFICATION OF CATHEPSIN D FROM RAT SPLEEN[a]

Procedure	Protein (mg)	Total activity (munits)	Specific activity (units/mg)	Purification ratio	Recovery (%)
Homogenate	88,995	801.9	9	1	100
Spleen extract	56,794	752.9	13	1.4	94
Ammonium sulfate	18,808	685.9	36	4.0	86
Acid treatment	5,836	486.5	83	9.2	61
Pepstatin–Sepharose 4B	115	428.2	3732	414.6	53
Concanavalin A–Sepharose 4B	60	350.3	5838	658.7	44
Sephadex G-100	28	186.7	6668	740.9	23 (38)
DEAE-Sephacel					
Peak I (cathepsin D-I)	10	93.4	9526	1058.4	12 (20)
Peak II (cathepsin D-II)	2	18.7	8118	902.0	2 (3)

[a] The starting material was 370 g (wet weight) spleen. The proteolytic activity was determined under the conditions described in the text. The first six lines contain the combined activities of cathepsin D- and E-like enzymes, whereas the last three lines refer to cathepsin D only. Values in parentheses are those corrected by calculating the quantity of cathepsin D in the homogenate to be 60%.[12]

is found at the top of the tube is removed before further processing of the supernatant.

Using sterile tubes for the SW27 rotor (Beckman Instruments), 25 ml of the supernatant is layered on top of a 5-ml cushion (0.2 volume) of 5.7 *M* $CsCl_2$, containing 0.1 *M* EDTA, pH 7.2. After centrifugation for 18 hr at 113,000 *g* (25°), the supernatant is poured off. The glassy pellet at the bottom of the tube comprises the total RNA. The pellet is washed gently with sterile distilled water and dispersed into small pieces with a glass rod. Sterile distilled water (4 ml) is added to each tube and the suspensions are pooled. After adjusting to 0.2 *M* NaCl the RNA is precipitated by addition of 2 volumes of absolute ethanol and overnight storage at −20°. The pellet obtained after centrifugation for 20 min at 12,000 *g* is once again dispersed as described before. Most of the RNA is now soluble, but total solubilization may require gentle heating (50°). The RNA solution is centrifuged at 5000 *g* for 5 min to remove any particulate matter prior to a second ethanol precipitation. Total RNA obtained in this manner is subjected to standard oligo(dT)-cellulose chromatography to obtain poly(A)-containing mRNA.[16]

[16] H. Aviv and P. Leder, *Proc. Natl. Acad. Sci. U.S.A.* **69,** 1408 (1972).

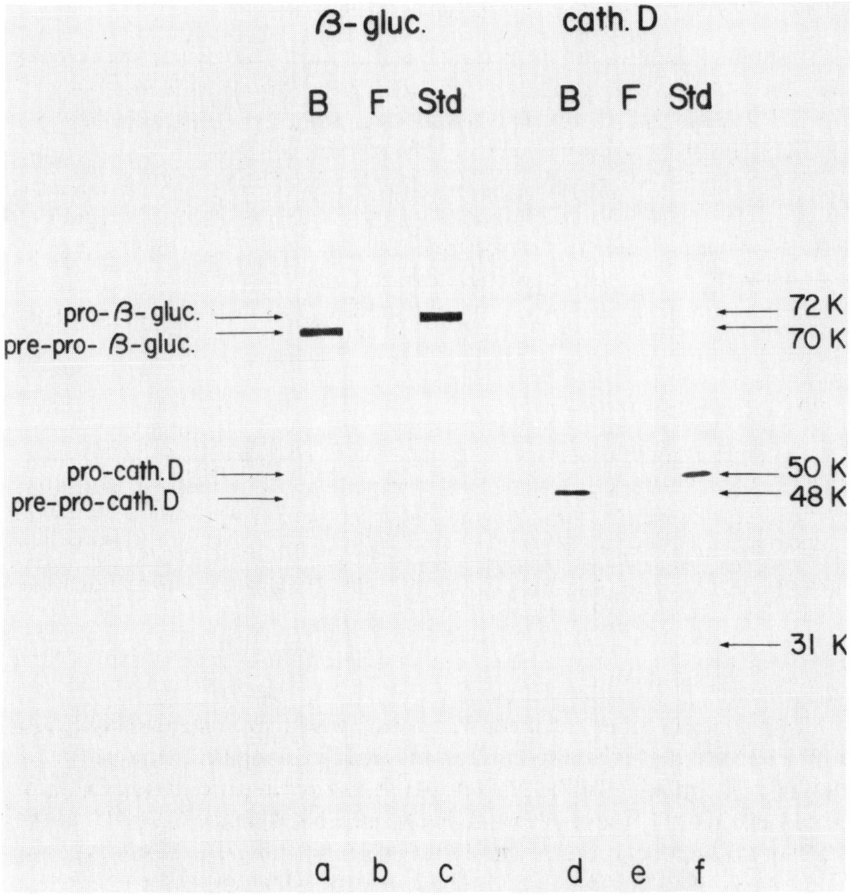

FIG. 4. *In vitro* synthesis of lysosomal enzymes. Messenger RNA samples isolated from free (lanes b and e) or membrane-bound polysomes (lanes a and d) prepared from the mouse spleen (cathepsin D) or rat preputial gland (β-glucuronidase), were used to program wheat germ cell-free translation systems (0.06 A_{260} unit of mRNA in 50 μl). Samples (0.2 ml) containing approximately 10^7 cpm of incorporated [^{35}S]methionine were used for immunoprecipitation of *in vitro* synthesized polypeptides with specific antibodies against cathepsin D or β-glucuronidase (10 μg of IgG for each). The immunoprecipitates were analyzed by SDS–polyacrylamide gel electrophoresis (10%) followed by fluorography (72 hr) of the dried slab gel. Precursors of β-glucuronidase and cathepsin D (lanes c and f), which are included as standards, were labeled in rat liver hepatocytes (C1-9) for 1 hr with [^{35}S]methionine, recovered by immunoprecipitation, and analyzed by gel electrophoresis.[6]

Isolation of Free and Membrane-Bound Polysomes

Free and membrane-bound polysomes are prepared from mouse spleen or rat preputial gland by the procedure of Ramsey and Steele[17] with the following modification. The tissue is gently homogenized with a Polytron homogenizer (Brinkmann Instruments) in 3 volumes (w/v) of 50 m*M* HEPES, pH 7.4, 250 m*M* KCl, 5 m*M* $MgCl_2$, and 3 m*M* DTT. In addition to cell sap, the polysome gradients are supplemented with 500 μg/ml of heparin (Sigma) as an additional RNase inhibitor. All glassware and solutions used in this preparation are sterilized.

Total RNA and mRNA are prepared from the isolated polysomes as described in the preceding section.

Cell-Free Protein Synthesis

A wheat germ extract is prepared as described by McMullen *et al.*,[18] except that the wheat germ is first frozen with liquid nitrogen and then pulverized with mortar and pestle at liquid nitrogen temperature. The cell-free translation system contains 12.5 μCi of [^{35}S]methionine (1000 Ci/mmol) and 0.03 A_{260} unit of mRNA in a total volume of 25 μl and incubation is carried out for 90 min at 25°.

Immunoprecipitation

Immunoprecipitation is carried out essentially as described by Goldman and Blobel.[19] Aliquots from translation mixtures are boiled in 2% SDS, centrifuged for 2 min at 15,000 *g* and diluted fourfold with solution A containing 50 m*M* Tris-HCl 7.4, 2.5% Triton X-100, 190 m*M* NaCl, 2 m*M* EDTA, 0.02% sodium azide, and 10 units of Trasylol per milliliter (Mobay Chemicals). The samples are then incubated with affinity-purified immunoglobulins (10 μg/ml) at room temperature for 1 hr, and at 4° for 14 hr before addition of protein A–Sepharose beads (Sigma). After incubation for 3 hr at room temperature, the beads are sedimented and washed three times with solution A containing 0.2% SDS. The immunoreactive material is extracted from the beads by boiling for 2 min in a solution of 10% SDS, 1 *M* dithiothreitol, 2 m*M* EDTA, and analyzed on polyacrylamide gels followed by fluorography.[20]

[17] J. C. Ramsey and W. J. Steele, *Biochemistry* **15,** 1704 (1976).

[18] M. McMullen, P. Shaw, and P. E. Mantin, *J. Mol. Biol.* **132,** 679 (1979).

[19] B. M. Goldman and G. Blobel, *Proc. Natl. Acad. Sci. U.S.A.* **75,** 5066 (1978).

[20] R. A. Laskey and A. D. Mills, *Eur. J. Biochem.* **56,** 335 (1975).

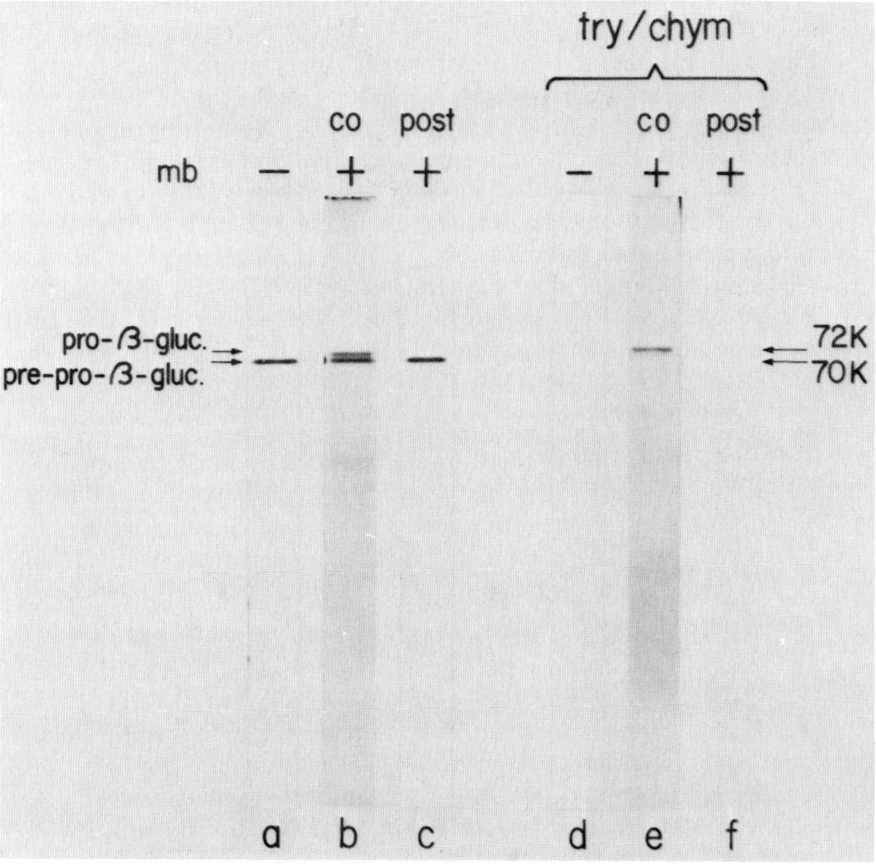

FIG. 5. Cotranslational translocation and processing of β-glucuronidase by microsomal membranes. Messenger RNA samples isolated from the rat preputial gland were translated in wheat germ cell-free systems (0.06 A_{260} unit of RNA in 50 μl final volume). The *in vitro* synthesized β-glucuronidase was purified by immunoprecipitation (10 μg of anti-β-glucuronidase IgG) from translation mixtures that contained no microsomal membranes (lanes a and d) or to which dog pancreas microsomes (2 A_{260} units/ml) were added before (lanes b and e) or after (lanes c and f) translation was completed. After protein synthesis, aliquots (100 μl) of the translation mixtures were subjected to protease treatment (lanes d, e, f; 80 μg/ml each of trypsin and chymotrypsin) for 3 hr at 4°. All samples were then treated for immunoprecipitation with specific antibody, and immunoprecipitates were analyzed by SDS–polyacrylamide gel electrophoresis (10%) and fluorography (72 hr).[6]

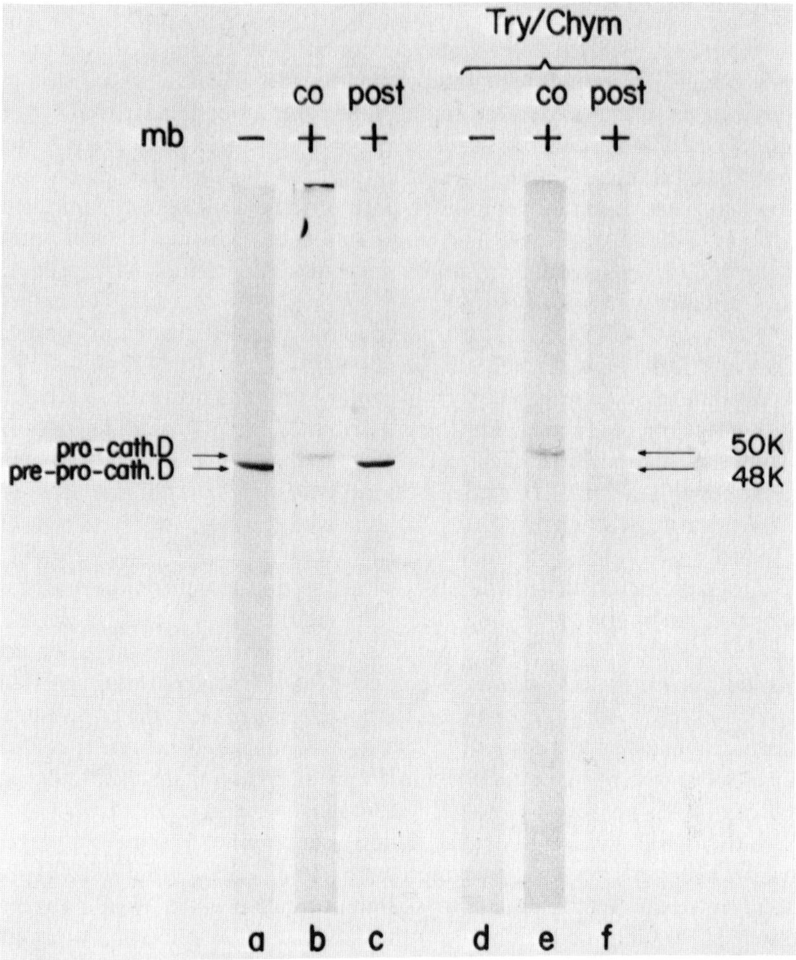

FIG. 6. Cotranslational processing and vectorial discharge of cathepsin D. Messenger RNA samples isolated from the mouse spleen were translated in wheat germ cell-free systems (0.06 A_{260} units of RNA in 50 μl final volume). The *in vitro* synthesized cathepsin D was purified by immunoprecipitation (10 μg of anti-cathepsin D IgG) from translation mixtures that contained no microsomal membranes (lanes a and d) or to which dog pancreas microsomes (2 A_{260} units/ml) were added before (lanes b and e) or after (lanes c and f) translation was completed. After protein synthesis, aliquots (100 μl) of the translation mixtures were subjected to protease treatment (lanes d, e, f; 80 μg/ml each of trypsin and chymotrypsin) for 3 hr at 4°. All samples were then treated for immunoprecipitation with specific antibody and immunoprecipitates were analyzed by SDS–polyacrylamide gel electrophoresis (10%) and fluorography (72 hr).[6]

Interpretation of Results

In vitro translation experiments with mRNAs from free and membrane-bound polysomes obtained from the rat preputial gland and the mouse spleen were carried out to determine the subcellular site of synthesis of these lysosomal hydrolases. Figure 4 shows that only mRNA samples extracted from bound polysomes directed the synthesis of polypeptides that were immunoprecipitated with specific antibodies against each of the lysosomal enzymes. The polypeptide brought down with anti-β-glucuronidase IgG from translation mixtures programmed by preputial gland mRNA was slightly smaller (M_r 70,000; pre-pro-β-glucuronidase, Fig. 4, lane a) than the mature polypeptide purified from the preputial gland or the *in vivo* labeled rat hepatocyte enzyme (M_r 72,000; pro-β-glucuronidase, Fig. 4, lane c), from which it could, however, be clearly resolved in the same gels. On the other hand, the polypeptide immunoprecipitated with antibodies against cathepsin D from systems programmed with mRNA from spleen-bound polysomes was much larger (M_r 48,000; pre-pro cathepsin D, Fig. 4, lane d) than the mature protein (M_r 31,000, Fig. 4, lane f).

The segregation of translatable mRNAs for β-glucuronidase and cathepsin D in bound polysomes implies that the corresponding polypeptides are cotranslationally inserted into ER membranes. A direct demonstration of this insertion was achieved in cell-free translation systems supplemented with microsomal membranes stripped of ribosomes,[21] which also allowed a study of the cotranslational modifications that affect these lysosomal polypeptides (Figs. 5 and 6). When translation was carried out in the presence of membranes (lane b of Figs. 5 and 6), but not when these were added after translation was completed (lane c of Figs. 5 and 6), polypeptides that migrated with slightly lower mobility (corresponding to M_r of 72,000 for pro-β-glucuronidase and 50,000 for procathepsin D) than the products synthesized in the absence of membranes were generated.

To establish whether the cotranslationally modified lysosomal polypeptides had been transferred into the microsomal lumen or incorporated into the membranes, their accessibility to exogenous trypsin and chymotrypsin was determined. It was found (lane e, Figs. 5 and 6) that the modified products were completely resistant to the attack of the proteases, which were added after translation in the presence of membranes had been completed. On the other hand, the unmodified polypeptides present in the same mixtures or in mixtures that received membranes posttranslationally, were completely digested (lanes d and f of Figs. 5 and

[21] D. Shields and G. Blobel, *J. Biol. Chem.* **253,** 3753 (1978).

6). Furthermore, it was determined that the sizes of the cotranslationally modified lysosomal polypeptides were unaffected by the proteases unless detergent was added, in which case digestion proceeded to yield acid-soluble fragments, indicating that the polypeptide had been completely transferred across the membrane and segregated into the microsomal lumen, as is the case with secretory proteins.[22]

[22] G. Blobel and B. Dobberstein, *J. Cell Biol.* **67,** 835 (1975).

[61] Applications of Amino Acid Analogs for Studying Co- and Posttranslational Modifications of Proteins

By GLEN HORTIN and IRVING BOIME

Synthesis of proteins often requires modification of the primary translation products, and numerous types of covalent modification of peptide chains have been observed. These changes can occur on the protein during its synthesis (cotranslational) or after release from the ribosome (posttranslational). Understanding the assembly of mature proteins requires definition of the mechanisms of these reactions and of the structural determinants that limit them to particular sites in the cellular environment. One approach to this question is to incorporate amino acid analogs into protein and to examine the consequent effects on processing. There are several examples of such applications of amino acid analogs. Baltimore and Jacobson[1] used them to inhibit the proteolytic cleavage of viral polyproteins and thus establish that all the peptides were derived from a single precursor. Hydroxyproline synthesis in collagen was probed with proline analogs,[2] clarifying the mechanism of this hydroxylation reaction. Proteolytic cleavagc of prosomatostatin, proinsulin,[3] and pro-opiomelanocortin[4,5] were inhibited by analogs of arginine and lysine. Asparagine-linked glycosylation was inhibited by analogs of threonine[6] and asparagine,[7] yielding insights into the specificity and mechanism of this reaction. Analogs of

[1] M. F. Jacobson and D. Baltimore, *Proc. Natl. Acad. Sci. U.S.A.* **61,** 77 (1968).
[2] A. A. Gottlieb, Y. Fujita, S. Udenfriend, and B. Witkop, *Biochemistry* **4,** 2507 (1965).
[3] B. D. Noe, *J. Biol. Chem.* **256,** 4940 (1981).
[4] P. Crine and E. Lemieux, *J. Biol. Chem.* **257,** 832 (1982).
[5] H. Hoshina, G. Hortin, and I. Boime, *Science* **217,** 63 (1982).
[6] G. Hortin and I. Boime, *J. Biol. Chem.* **255,** 8007 (1980).
[7] G. Hortin, I. Boime, A. M. Stern, B. Miller, and R. H. Abeles, *J. Biol. Chem.* **257,** 4047 (1983).

ISBN 0-12-181996-5

leucine and threonine have been useful for probing the process by which secretory proteins traverse the membrane of the endoplasmic reticulum[8–10] and for examining the specificity of the protease that excises amino-terminal prepeptides from most secretory proteins.[11] The diversity of these examples illustrates the versatility of this approach. Virtually any process dependent on protein structure can be examined in this manner.

Use of amino acid analogs has several characteristics that are often advantageous.

1. Proteins are modified cotranslationally, permitting analysis of modification reactions that occur on nascent peptide chains.
2. Subtle modifications of protein structure can be introduced because most analogs are structurally very similar to the amino acids they supplant.
3. Protein can be labeled with new heteroatoms such as selenium, chlorine, and fluorine. Fluorine, in particular, has been useful as a probe for nuclear magnetic resonance (NMR) spectroscopy of proteins.[12]
4. They can be used in any system—intact organisms, organ culture, cell culture, and cell-free systems.

Owing to the great variety of applications of this approach several general considerations are presented here rather than a detailed protocol with restricted applicability.

Selection of Amino Acid Analogs

A large number of amino acid analogs have been reported to be incorporated into protein. Some of these compounds are listed in the table. Analogs probably can be found to replace all the amino acids with the possible exception of glycine, alanine, and serine. A large number of potential analogs have not been studied thoroughly. Numerous additional amino acid antagonists are listed in other references.[13–15] Some analogs listed in the table may be effective only in particular systems. For exam-

[8] G. Hortin and I. Boime, *Proc. Natl. Acad. Sci. U.S.A.* **77,** 1356 (1980).
[9] G. Hortin and I. Boime, *J. Biol. Chem.* **255,** 7051 (1980).
[10] P. Walter, I. Ibrahimi, and G. Blobel, *J. Cell Biol.* **91,** 545 (1981).
[11] G. Hortin and I. Boime, *Cell* **24,** 453 (1981).
[12] B. D. Sykes, H. I. Weingarten, and M. J. Schlesinger, *Proc. Natl. Acad. Sci. U.S.A.* **71,** 469 (1974).
[13] K. Dittmer, *Ann. N.Y. Acad. Sci.* **52,** 1274 (1950).
[14] W. Shive and C. G. Skinner, *in* "Metabolic Inhibitors" (R. M. Hochster and J. H. Quastel, eds.), p. 2. Academic Press, New York, 1963.
[15] G. L. Igloi, F. von der Haar, and F. Cramer, *Biochemistry* **17,** 3459 (1978).

AMINO ACID ANALOGS INCORPORATED INTO PROTEIN

Amino acid	Analog	Concentration[a] (mM)	References[b]
Arginine	Canavanine	1	19
Asparagine	*threo*-3-Fluoroasparagine	0.2	7, 20
Aspartic acid	*threo*-3-Fluoroaspartic acid	0.1	20
	erythro-3-Fluoroaspartic acid	5	20
Histidine	2-Fluorohistidine	0.2	21
	1,2,4-Triazole-3-alanine	180	22
	2-Methylhistidine	—	23
	3-Amino-1,2,4-triazoylalanine	—	24
Isoleucine	4-Thiaisoleucine	1	25
	O-Ethylthreonine	—	26
	O-Methylthreonine	80	27
	4-Fluoroisoleucine	0.5	28, 29
	Alloisoleucine	—	30
Leucine	*threo*-3-Hydroxyleucine	0.3	8
	5,5,5-Trifluoroleucine	15	16
Lysine	Thialysine (*S*-2-aminoethylcysteine)	0.5	31
	6-*C*-Methyllysine	—	31
	5-Hydroxylysine	8	31
	trans-4,5-Dehydrolysine	—	32
	2,6-Diamino-4-hexynoic acid	—	32
	4-Oxalysine	—	32
	4-Selenalysine	—	33
Methionine	Ethionine	2	34
	Selenomethionine	—	35
	cis-Crotylglycine	—	36
	Norleucine	—	37
Phenylalanine	2-Fluorophenylalanine	0.3	38
	3-Fluorophenylalanine	1	38
	4-Fluorophenylalanine	0.5	39
	β-2-Thienylalanine	4	38
	β-3-Thienylalanine	—	40
	2,5-Dihydrophenylalanine	—	41
	3-Phenylserine	4	42
	3-Furyl-3-alanine	—	43
	m-Tyrosine	2	29
Proline	3,4-Dehydroproline	10	44
	Azetidine-2-carboxylic acid	10	45
	cis-4-Fluoroproline	—	2
	trans-4-Fluoroproline	—	2
	4-Thioproline	15	46
	4-Selenaproline	—	47
	cis-4-Hydroxyproline	20	48
Threonine	3-Hydroxynorvaline	1	9, 49

(continued)

AMINO ACID ANALOGS INCORPORATED INTO PROTEIN (*continued*)

Amino acid	Analog	Concentration[a] (m*M*)	References[b]
Tryptophan	7-Azatryptophan	Inactive	50
	4-Fluorotryptophan	—	51
	5-Fluorotryptophan	—	51
	6-Fluorotryptophan	0.5	52
	Tryptazan	—	53
	4-Methyltryptophan	—	54
Tyrosine	*m*-Fluorotyrosine	0.5	12, 55
	3,4-Dihydroxyphenylalanine	5	56
Valine	2-Amino-3-chlorobutyric acid	0.1	57
	Cyclobutyglycine	—	58
	Cyclopropylglycine	—	59
	Penicillamine	Inactive	60
	Alloisoleucine	70	30

[a] Concentration resulting in 50% inhibition of incorporation of the corresponding amino acid in the Krebs II ascites cell-free system described in Fig. 1.
[b] Numbers refer to text footnotes.

ple, trifluoroleucine was incorporated by bacteria,[16] but not by a eukaryotic system.[17] As a result, when a particular analog is applied to a new system, it is necessary to assess critically whether the analog is incorporated into protein and to establish its effective concentration.

Selection of Analog Concentration

Selection of amino acids for incorporation into protein depends on the selectivity of aminoacyl-tRNA synthetases. These enzymes possess an extraordinary ability to discriminate among structurally similar amino acid substrates.[15,18] Consequently, amino acid analogs generally are linked to tRNA much less efficiently than their corresponding amino acids. As an example, Fig. 1 compares the competitive inhibition of [^{3}H]threonine incorporation by threonine and a threonine analog. Assuming that only the L-isomer is active, the analog is 50 times less effective. As a result, to achieve efficient incorporation of the analog, high concentrations must be used, and the naturally occurring amino acid should be depleted as much as possible. (Concentrations of analogs resulting in 50% inhibition are listed in the table.[2,7–9,12,16,19–60])

[16] O. M. Rennert and H. S. Anker, *Biochemistry* **2,** 471 (1963).
[17] O. M. Rennert and H. S. Anker, *Nature* (*London*) **203,** 1256 (1964).
[18] P. R. Schimmel and D. Söll, *Annu. Rev. Biochem.* **48,** 601 (1979).
[19] P. F. Kruse, P. B. White, H. A. Carter, and T. A. McCoy, *Cancer Res.* **19,** 122 (1959).

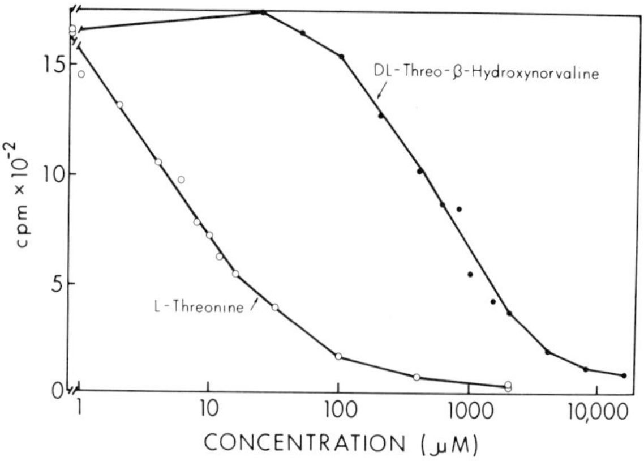

FIG. 1. Inhibition of [^{3}H]threonine incorporation by L-threonine or by DL-*threo*-β-hydroxynorvaline. Varying amounts of these amino acids were added to the Krebs II ascites cell-free system containing 4 μM [^{3}H]threonine (6 Ci/mmol). After incubation for 30 min, reactions were assayed for incorporation of label into tRNA and protein by spotting 20-μl aliquots on Whatman 3 MM filter paper disks, rinsing the disks four times for 10 min in 5% trichloroacetic acid, and counting the dry filters.[64]

[20] A. M. Stern, B. M. Foxman, A. H. Tashjian, Jr., and R. H. Abeles, *J. Med. Chem.* **25,** 544 (1982).

[21] D. C. Klein, J. L. Weller, K. L. Kirk, and R. W. Hartley, *Mol. Pharmacol.* **13,** 1105 (1977).

[22] A. P. Levin and P. E. Hartman, *J. Bacteriol.* **86,** 820 (1963).

[23] S. Schlesinger and M. J. Schelsinger, *J. Biol. Chem.* **244,** 3803 (1969).

[24] A. K. Williams, S. T. Cox, and R. G. Eagon, *Biochem. Biophys. Res. Commun.* **18,** 250 (1965).

[25] T. J. McCord, D. C. Howell, D. L. Tharp, and A. L. Davis, *J. Med. Chem.* **8,** 290 (1965).

[26] C. B. Hiremath, G. Olson, and C. Rosenblum, *Biochemistry* **10,** 1096 (1971).

[27] M. E. Smulson and M. Rabinowitz, *Arch. Biochem. Biophys.* **124,** 306 (1968).

[28] H. Gershon, L. Shanks, and D. D. Clarke, *J. Pharm. Sci.* **67,** 715 (1978).

[29] G. Hortin and I. Boime, unpublished result.

[30] R. B. Loftfield, L. I. Hecht, and E. A. Eigner, *Biochim. Biophys. Acta* **72,** 383 (1963).

[31] M. Rabinowitz and K. Tuve, *Proc. Soc. Exp. Biol. Med.* **100,** 222 (1959).

[32] E. M. Lansford, N. M. Lee, and W. Shive, *Arch. Biochem. Biophys.* **119,** 272 (1967).

[33] C. DeMarco, V. Busiello, M. DiGirolamo, and D. Cavallini, *Biochim. Biophys. Acta* **454,** 298 (1976).

[34] M. Levine and H. Tarver, *J. Biol. Chem.* **192,** 835 (1951).

[35] D. B. Cowie and G. N. Cohen, *Biochim. Biophys. Acta* **26,** 252 (1957).

[36] C. G. Skinner, J. Edelson, and W. Shive, *J. Am. Chem. Soc.* **83,** 2281 (1961).

[37] D. B. Cowie, G. N. Cohen, E. T. Bolton, and H. deRobichon-Szulmajster, *Biochim. Biophys. Acta* **34,** 39 (1959).

[38] W. L. Fangman and F. C. Niedhardt, *J. Biol. Chem.* **239,** 1939 (1964).

[39] R. Munier and G. N. Cohen, *Biochim. Biophys. Acta* **21,** 592 (1956).

[40] B. A. Samal, L. E. Frazier, G. Monto, A. Slesers, Z. Hruban, and R. W. Wissler, *Proc. Soc. Exp. Biol.* **112,** 442 (1963).

Assay of tRNA charging as shown in Fig. 1 is a useful approach for estimating the minimal concentration of amino acid analog required in cell-free systems to achieve efficient incorporation of the analog. If the analog is incorporated into protein, it necessarily competes with a naturally utilized amino acid for aminoacylation of tRNA. However, it should be kept in mind that some compounds not linked to tRNA, such as amino alcohols[61] and adenylates of amino acids,[62] also inhibit aminoacylation of tRNA. Thus, this assay does not demonstrate incorporation of an analog into protein.

As a control for nonspecific inhibition of aminoacylation, amino acid analogs should be assayed for effects on the charging of other amino acids. In the case of the threonine analog in Fig. 1, the highest concentration used did not block incorporation of methionine into aminoacyl-tRNA. Cell-free translation systems appear to tolerate concentrations of greater than 100 mM of many analogs, even though translational efficiency is very dependent on the concentration of some ions such as potassium and magnesium.[63] Neutral amino acids have an acidic isoelectric point so that solutions need to be adjusted with a base such as NaOH or

[41] M. J. Pine, *Antimicrob. Agents Chemother.* **7,** 601 (1975).

[42] J. Janacek, J. Chaloupka, K. Veres, and M. Havranek, *Nature* (*London*) **184,** 1895 (1959).

[43] J. Janacek, *Folia Microbiol.* (*Prague*) **12,** 132 (1967).

[44] J. Rosenbloom and D. J. Prockop, *J. Biol. Chem.* **245,** 3361 (1970).

[45] T. Takeuchi and D. J. Prockop, *Biochim. Biophys. Acta* **175,** 142 (1969).

[46] V. Busiello, *Biochim. Biophys. Acta* **564,** 311 (1979).

[47] V. Busiello, *Biochim. Biophys. Acta* **606,** 347 (1980).

[48] J. Rosenbloom and D. J. Prockop, *J. Biol. Chem.* **246,** 1549 (1971).

[49] P. Christner, A. Carpousis, M. Harsch, and J. Rosenbloom, *J. Biol. Chem.* **250,** 7623 (1975).

[50] A. B. Pardee, V. G. Shore, and L. S. Prestidge, *Biochim. Biophys. Acta* **21,** 406 (1956).

[51] D. T. Browne, G. L. Kenyon, and G. D. Hegemon, *Biochem. Biophys. Res. Commun.* **39,** 13 (1970).

[52] A. B. Pardee and L. S. Prestidge, *Biochim. Biophys. Acta* **27,** 330 (1958).

[53] A. B. Pardee and L. S. Prestidge, *Biochim. Biophys. Acta* **21,** 406 (1956).

[54] R. D. Mosteller and C. Yanofsky, *Fed. Proc., Fed. Am. Soc. Exp. Biol.* **29,** 598 (1970).

[55] R. S. Schweet and E. H. Allen, *J. Biol. Chem.* **233,** 1104 (1958).

[56] R. Calendar and P. Berg, *Biochemistry* **5,** 1690 (1966).

[57] M. Freundlich, *Science* **157,** 823 (1967).

[58] T. H. Porter, S. C. Smith, and W. Shive, *Arch. Biochem. Biophys.* **179,** 266 (1977).

[59] W. M. Harding and M. L. DeShazo, *Arch. Biochem. Biophys.* **118,** 23 (1967).

[60] E. Lodemann, P. Ulrich, and A. Wacker, *Biochim. Biophys. Acta* **474,** 210 (1977).

[61] R. B. Loftfield, *Prog. Nucleic Acid Res. Mol. Biol.* **12,** 87 (1972).

[62] D. Cassio, F. Lemoine, J. Waller, E. Sandrin, and R. A. Boissonnas, *Biochemistry* **6,** 827 (1967).

[63] S. Daniels-McQueen, D. McWilliams, S. Birken, R. Canfield, T. Landefeld, and I. Boime, *J. Biol. Chem.* **253,** 7109 (1978).

tris(hydroxymethyl)aminomethane before they are used. If an analog is an acidic or basic amino acid or it occurs as a hydrochloride salt, the maximum concentration may be limited by the counterions required for neutralization.

Determination of the appropriate analog concentration for use with intact cells presents greater difficulty owing to the added complications of uptake into cells, amino acid metabolism, and endogenous amino acid pools. Competition assays as in Fig. 1 are of less use in these cases than in cell-free systems. In practice, several analog concentrations should be tried, and the maximal concentration that does not cause toxicity or excessive inhibition of protein synthesis should be used.

Demonstration of Amino Acid Analog Incorporation into Protein

Incorporation of an amino acid analog into protein can be shown most directly by use of radioactively labeled analogs. If incorporation of the label occurs, the labeled protein should be hydrolyzed and the hydrolyzate analyzed to ensure that the analog was not converted to different compounds before being incorporated. Other studies have detected the incorporation of unlabeled analogs by amino acid analysis of a protein hydrolyzate.

More indirect evidence for the incorporation of amino acid analogs is provided by slight alteration of the electrophoretic mobility of proteins in polyacrylamide gels containing sodium dodecyl sulfate.[8] This can be examined very easily, but it is not complete proof of incorporation. Mobility could be changed by altered protein processing. Also, if an analog induces starvation for a particular amino acid, other amino acids may be substituted for it in protein—a phenomenon that has been termed stuttering.[64]

Analysis of the Effects of Amino Acid Analogs on Protein Processing

If an amino acid analog affects the processing of a protein, the most basic issue is whether the effect results from incorporation of the analog into the protein of interest. Several criteria, besides demonstration that the analog is capable of incorporation into protein, are helpful.

1. The effective analog concentration equals the concentration required for analog incorporation.
2. The effect should be blocked specifically by the amino acid replaced by the analog.

[64] J. Parker, J. W. Pollard, J. D. Friesen, and C. P. Stanners, *Proc. Natl. Acad. Sci. U.S.A.* **75,** 1091 (1978).

3. The effect should appear with brief exposure to the analog. If cells are exposed to analogs for many hours, the protein processing mechanism may be altered, and the levels of many enzymes may change because incorporation of analogs generally increases the turnover of proteins and increases their susceptibility to proteases.[65]

Acknowledgments

We thank Drs. Robert Abeles, Herman Gershon, Robert Handschumacher, Kenneth Klein, Theodore Otani, Upendra Pandit, and Marco Rabinowitz for supplying amino acid analogs for our studies.

[65] A. Goldberg and A. C. St. John, *Annu. Rev. Biochem.* **45,** 747 (1976).

[62] Quantitative Assay for Signal Peptidase[1]

By ROBERT C. JACKSON

Signal peptidase is the enzyme responsible for removing the signal peptide portion of nascent presecretory and presumably also prelysosomal and premembrane proteins during their cotranslational translocation across the rough-endoplasmic reticulum (RER) membrane. In many respects signal peptidase is a unique protease. It is an integral membrane protein whose substrate is not a full-length protein, but, rather, an incomplete polypeptide chain that is engaged in the translocation process. These unique properties of signal peptidase were the source of several obstacles to its assay.

Signal peptidase activity was first detected by a cotranslational, or translocation-dependent assay.[2] In this assay the entire series of events that occurs at the RER membrane during translocation is reconstituted *in vitro*. The assay involves the translation of mRNA encoding a secretory protein in a cell-free *in vitro* translation system containing microsomal membranes. Polysomes bearing nascent presecretory proteins bind to elements on the cytoplasmic surface of the microsomes, and the nascent presecretory proteins are translocated across the microsomal membrane. During translocation across the membrane they become available to sig-

[1] This work was supported by United States Public Health Service Grant GM26763.
[2] G. Blobel and B. Dobberstein, *J. Cell Biol.* **67,** 852 (1975).

ISBN 0-12-181996-5

nal peptidase and are cleaved. In this assay signal peptidase activity is coupled to both the translation and translocation of the presecretory protein; consequently, it is impossible to examine the signal peptidase reaction independently of these events. Furthermore, the translocation-dependent assay utilizes microsomal signal peptidase. Unless signal peptidase could be solubilized from the microsomal membrane and assayed in its soluble form, it could not be purified or characterized. Finally, the translocation-dependent assay is critically dependent on *in vitro* protein synthesis. The assay is, therefore, restricted to ionic conditions and detergent concentrations that are compatible with *in vitro* translation. These restrictions severely limit the flexibility of the translocation-dependent assay for enzyme purification purposes.

These obstacles to the assay of signal peptidase were partially removed by the introduction of a posttranslational or translocation-independent assay.[3] In this assay the cleavage of full-length [^{35}S]preprolactin to [^{35}S]prolactin by detergent-solubilized signal peptidase proceeds in the absence of *in vitro* translation and polypeptide translocation. Products of the reaction are separated by polyacrylamide gel electrophoresis in sodium dodecyl sulfate and detected by autoradiography or fluorography.

Sequence analysis of the prolactin molecule produced in the translocation-independent assay substantiated that the amino-terminal sequence of the product is identical to that of authentic bovine prolactin,[3] thereby confirming the fidelity of posttranslational cleavage. Several other presecretory proteins, including pregrowth hormone,[3] prepromellitin,[4,5] and human preplacental lactogen,[6,7] have also been shown to serve as substrates in the translocation-independent assay. However, not all full-length presecretory proteins are processed with equal efficiency; several are very poor substrates. This substrate-dependent variability was not altogether unexpected, since the full-length substrates utilized in the translocation-independent assay are not equivalent to the nascent substrates processed *in vivo*. Folding of the full-length presecretory molecule may obscure the cleavage site; hence, it is not surprising that the proportion of processable substrate molecules depends upon the folding pattern and, therefore, upon the primary sequence of a given presecretory protein.

[3] R. Jackson and G. Blobel, *Proc. Natl. Acad. Sci. U.S.A.* **74,** 5598 (1977).

[4] G. Kreil, G. Mollay, R. Kaschnitz, L. Haiml, and U. Vilas, *Ann. N.Y. Acad. Sci.* **343,** 338 (1980).

[5] R. Kaschnitz and G. Kreil, *Biochem. Biophys. Res. Commun.* **83,** 901 (1978).

[6] A. W. Strauss, M. Zimmermann, I. Boime, B. Ashe, R. A. Mumford, and A. W. Alberts, *Proc. Natl. Acad. Sci. U.S.A.* **76,** 4225 (1979).

[7] A. W. Strauss, M. Zimmerman, R. A. Mumford, and A. W. Alberts, *Ann. N.Y. Acad. Sci.* **343,** 168 (1980).

Despite these improvements in the assay, inactivation of signal peptidase during chromatographic procedures remained problematic, until it was recognized that detergent-solubilized signal peptidase specifically requires phosphatidylcholine as a cofactor and can be inactivated by delipidation during chromatographic separations.[8] The additional discovery that activity can be restored to delipidated signal peptidase by readdition of detergent-solubilized phosphatidylcholine allows chromatographic separations to be performed in the absence of phospholipid.[8] Recognition of the phosphatidylcholine requirement has allowed us to define the phospholipid and detergent conditions required for signal peptidase activity and to develop a quantitative assay, which is described below.

Principle

[^{35}S]Prolactin produced by posttranslational cleavage of [^{35}S]preprolactin, in an assay containing 1.0% Triton X-100 and 2.5 m*M* phospholipid, is detected by polyacrylamide gel electrophoresis in sodium dodecyl sulfate (SDS) followed by autoradiography and is quantitated by counting the portion of the gel containing the prolactin band or by densitometry of the autoradiograph.

Reagents

Ethanol-extracted soybean phospholipid, 37.5 m*M* (30 mg/ml) in 40 m*M* triethanolamine-HCl, pH 7.5
Triton X-100, 20% (v/v)
[^{35}S]Methionine-labeled bovine pituitary translation products, synthesized in a wheat germ cell-free translation system
Canine pancreatic signal peptidase in 1.0% (v/v) Triton X-100, 40 m*M* triethanolamine-HCl, pH 8.1, 0.02% NaN_3 (w/v)

Preparation of Reagents

Signal Peptidase. Although signal peptidase can be extracted from canine pancreatic rough microsomes with several detergents, including sodium deoxycholate, Triton X-100, Nikkol (octaethyleneglycol dodecyl ether), and octyl glucoside, the assay described here is designed for use with enzyme prepared with Triton X-100; therefore, only those procedures resulting in Triton X-100-solubilized enzyme are described. Delipidated signal peptidase is prepared essentially as described by Jackson and White,[8] with the exception that the Sepharose CL-6B column is equili-

[8] R. Jackson and W. R. White, *J. Biol. Chem.* **256,** 2545 (1981).

brated in 1.0% Triton X-100 instead of 0.2% deoxycholate. All procedures are conducted on ice or in a cold room at 4°, except as otherwise noted. Canine pancreatic rough microsomes are prepared as described by Shields and Blobel.[9] The rough microsomes are either used immediately or frozen in liquid nitrogen and stored at −80° for future use. The rough microsomes are resuspended with a Teflon pestle homogenizer in 50 m*M* NaCl, 40 m*M* triethanolamine-HCl, pH 8.1, to a concentration of 70 A_{280} units/ml (measured in 1.0% SDS). Sodium deoxycholate (10%, w/v, decolorized with activated charcoal and recrystallized three times from aqueous acetone) is added to a final concentration of 0.7%, and the solubilized membranes are centrifuged at 100,000 *g* for 4 hr at 4°. The resultant deoxycholate extract can be frozen in liquid nitrogen and stored at −80° or used immediately. The deoxycholate extract is delipidated by gel filtration chromatography on a column of Sepharose CL-6B. A 3.0-ml aliquot of delipidation buffer (2.0% deoxycholate, 0.02% NaN_3, 40 m*M* triethanolamine-HCl, pH 8.1) is loaded onto a column (1.6 × 100 cm) containing 200 ml of Sepharose CL-6B equilibrated with 40 m*M* triethanolamine-HCl, pH 8.1, 1.0% Triton X-100, 0.02% NaN_3. The column is pumped at a flow rate of 10 ml/hr, until the delipidation buffer enters the column. Meanwhile the deoxycholate concentration of the signal peptidase extract is increased to 2.0% by the addition of an appropriate volume of 10% deoxycholate. A 9.0-ml aliquot of this sample is loaded on to the column, and the column is eluted with equilibration buffer at 10 ml/hr. Fractions (3.0 ml each) containing signal peptidase activity (assessed qualitatively[8]) are combined, frozen in liquid nitrogen, and stored at −80°. Samples of signal peptidase can be stored in this manner for at least 1 month without loss of activity. Signal peptidase preparations containing endogenous RER phospholipids can be prepared by directly solubilizing rough microsomes or ethylenediaminetetraacetic acid (EDTA)-stripped microsomes[3] in 1.0% Triton X-100.

[^{35}S]Methionine-labeled bovine pituitary translation products are prepared in a wheat germ cell-free translation system,[10,11] as previously described,[3] with two exceptions: The cell-free system is supplemented with placental ribonuclease inhibitor[12] and total pituitary RNA, prepared from bovine pituitaries, by phenol–chloroform–isoamyl alcohol extraction[13] and lithium chloride precipitation,[14] is used in place of poly(A)$^+$ pituitary

[9] D. Shields and G. Blobel, *J. Biol. Chem.* **253,** 3753 (1978).

[10] R. Roman, J. D. Brooker, S. N. Seal, and A. Marcus, *Nature* (*London*) **260,** 359 (1976).

[11] B. Dobberstein and G. Blobel, *Biochem. Biophys. Res. Commun.* **74,** 1675 (1977).

[12] G. Scheele and P. Blackburn, *Proc. Natl. Acad. Sci. U.S.A.* **76,** 4898 (1979).

[13] H. Aviv and P. Leder, *Proc. Natl. Acad. Sci. U.S.A.* **69,** 1408 (1972).

[14] R. E. Rhoads, *J. Biol. Chem.* **250,** 8088 (1975).

RNA. Stocks of bovine pituitary translation products can be stored for at least 2 months at −80° before use. Trichloroacetic acid precipitable radioactivity was determined by spotting 3 μl of the translation products on Whatman 3 MM filter paper disks. The disks were processed as described by Mans and Novelli[15] and counted in toluene-Liquifluor (New England Nuclear, Boston, Massachusetts).

Ethanol-Extracted Soybean Phospholipid. This phospholipid is prepared from soybean asolectin (Associated Concentrates, Woodside, New York) as described by Kagawa and Racker,[16] except that the acetone wash procedure is repeated three times. Five grams of the acetone-insoluble phospholipid pellet is dissolved in 50 ml of anhydrous ether, and 125 ml of absolute ethanol are added with vigorous stirring. Stirring is continued for 2 hr under nitrogen. After centrifugation (700 *g* for 10 min) the supernatant is evaporated under reduced pressure and the residue of phospholipid is dissolved in 10 ml of chloroform–methanol (4 : 1, v/v) and stored at −20°. Phospholipid prepared by this procedure is composed primarily of phosphatidylcholine and phosphatidylethanolamine.[17] Liposomes are prepared as follows. An aliquot of the phospholipid is dried with a stream of nitrogen and lyophilized for 2 hr to remove the last traces of organic solvent. The dried phospholipid is dispersed in 40 m*M* triethanolamine-HCl, pH 8.1, 0.02% NaN_3 by vortex mixing. Liposomes are prepared from the dispersed phospholipid by sonication (under nitrogen) in a Branson 220 sonic bath until a pale opalescent solution is produced. Stock solutions of liposomes are routinely stored at −80° and resonicated before use.

Assay Procedure

A stock substrate mixture sufficient for assaying the desired number of samples is premixed and stored on ice. For instance, for the experiment illustrated in Fig. 1 (12 assay tubes) 350 μl of substrate mix, containing 1.0% Triton X-100, 40 m*M* triethanolamine-HCl, pH 8.1, 5.0 m*M* soybean phospholipid, and 2.10×10^7 cpm of pituitary translation products, was prepared by mixing 174 μl of [^{35}S]Met-labeled pituitary translation products, containing 1.21×10^5 trichloroacetic acid-precipitable counts per minute, 17.5 μl of 20% Triton X-100, 97.8 μl of distilled water, 14 μl of 1.0 *M* triethanolamine-HCl, pH 8.1, and 46.7 μl of 37.5 m*M* soybean phospholipid, which had been dispersed by brief sonication (approximately 10 sec) in a Branson B-220 sonic bath before use. Each assay tube

[15] R. J. Mans and G. D. Novelli, *Arch. Biochem. Biophys.* **94,** 48 (1971).

[16] Y. Kagawa and E. Racker, *J. Biol. Chem.* **246,** 5477 (1971).

[17] C. Miller and E. Racker, *J. Membr. Biol.* **26,** 319 (1976).

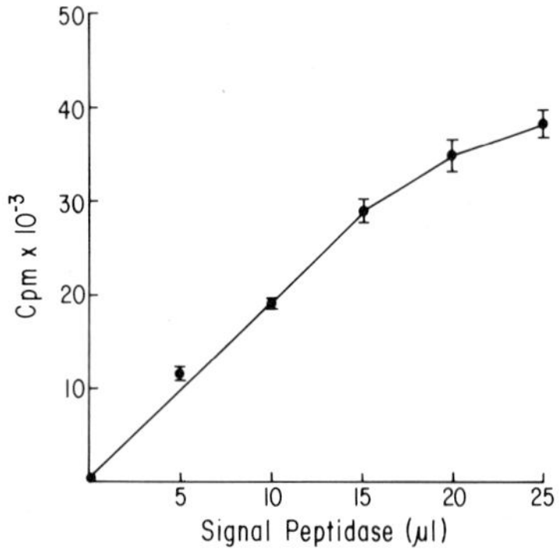

FIG. 1. Quantitation of the translocation-independent assay for signal peptidase. The indicated amounts (0–25 μl) of a sample of delipidated signal peptidase were assayed in the presence of 2.5 m*M* ethanol-extracted soybean phospholipid. The [^{35}S]prolactin produced by the cleavage reaction was separated from other labeled proteins by electrophoresis in 15% polyacrylamide gels containing 0.1% SDS and detected by autoradiography. The amount of [^{35}S]prolactin produced was quantitated by excising and counting the portion of the gel containing the prolactin band. Radioactivity in the prolactin region of control samples (no enzyme) was subtracted from each experimental sample. Bars represent the range of duplicate samples.

receives 25 μl of this substrate mix and 25 μl of a solution containing the indicated amount of delipidated signal peptidase in 1% Triton X-100, 40 m*M* triethanolamine-HCl, pH 8.1. After 90 min of incubation at 25°, the reaction is terminated by the addition of 25 μl of concentrated (3×) gel

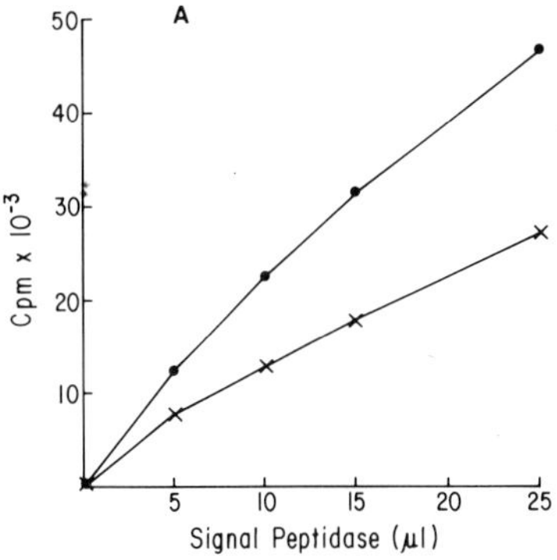

FIG. 2. The concentration of [^{35}S]preprolactin in the assay is rate limiting. The indicated amounts (0–25 μl) of a sample of delipidated signal peptidase were assayed in the presence of 2.5 m*M* ethanol-extracted soybean phospholipid. Each assay tube received either 1.5×10^6 cpm of translation products (●——●) or 0.75×10^6 cpm of translation products (×——×). The [^{35}S]prolactin produced by the cleavage reaction was separated from other labeled proteins by electrophoresis in 15% polyacrylamide gels containing 0.1% SDS and detected by autoradiography (panel B, see next page). The amount of [^{35}S]prolactin produced was quantitated by excising and counting the portion of the gel containing the prolactin band. The percentage of [^{35}S]preprolactin converted to [^{35}S]prolactin was determined by excising and counting the [^{35}S]preprolactin band as well as the [^{35}S]prolactin band. By this method it was determined that 15.1% of the total available [^{35}S]preprolactin in the sample receiving 25 μl of signal peptidase and 1.5×10^6 cpm of translation products was processed to [^{35}S]prolactin. PL, prolactin; pPL, preprolactin.

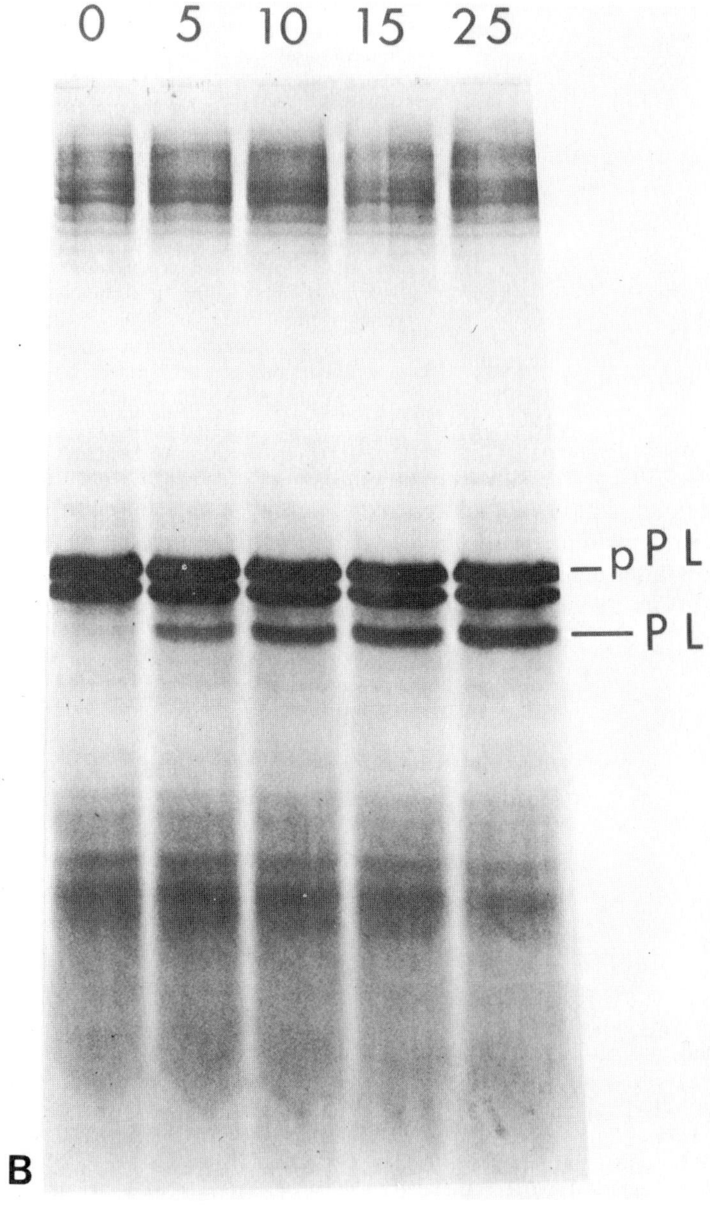

FIG. 2B.

electrophoresis sample buffer containing 8% (w/v) SDS, 60 m*M* dithiothreitol, 12 m*M* EDTA, 48% sucrose, 0.24 *M* Tris-HCl, pH 8.8, and 0.04% Bromphenol Blue. The samples are heated in a boiling water bath for 3 min, cooled to room temperature, and alkylated by the addition of 10 μl of 1.0 *M* iodoacetamide, followed by a 20-min incubation at 37°. If the Bromphenol blue indicator dye in any of the samples turns yellow owing to hydrolytic release of HI from iodoacetamide, the sample pH is restored to the correct range by adding 1.0 *M* Tris base (1–2 μl) until the indicator dye turns blue.

[^{35}S]Met-prolactin is separated from [^{35}S]Met-preprolactin and other labeled translation products by slab gel electrophoresis in SDS.[18] Our gels contain 15% acrylamide and 0.4% bisacrylamide. They are 1.2 mm thick and are cast between 26.5 × 35.5 cm glass plates. Each gel contains 20 sample wells (1.5 cm deep × 1.0 cm wide). Each sample (85 μl) is loaded into a single well. Wells not receiving a sample are loaded with 85 μl of sample buffer (2.7% SDS, 4 m*M* EDTA, 0.08 *M* Tris-HCl, pH 8.8, 14% sucrose, 0.013% Bromphenol Blue). Gels are run at 23 mA for 15 hr. The gels are fixed, and free [^{35}S]methionine is removed by washing each gel three times (1 hr per wash) in 500 ml of a solution containing 50% methanol, 10% acetic acid, and 5% glycerol. The washed gels are dried onto Whatman 3 MM filter paper. The border of filter paper around the dried gel is marked in several places with radioactive ink (approximately 25,000 cpm of ^{35}S per microliter) and an autoradiograph of the dried gel is produced by exposing the gel to an 11 × 14 in sheet of Kodak X-RP-1 X-ray film for 2 days. The amount of [^{35}S]prolactin produced in the assay is quantitated either by directly counting the portion of the gel containing the prolactin band or by densitometry of the autoradiograph.

For direct quantitation, the autoradiograph is placed on a light box and the dried gel is positioned over it, so that the ink spots on the border of the dried gel correspond exactly with their replicas on the autoradiograph. The portion of the gel corresponding to the prolactin band (see Fig. 2B) on the underlying autoradiograph is traced in pencil and excised with scissors. The prolactin band from each lane is deposited into a 20-ml scintillation vial, hydrated with 100 μl of distilled water for approximately 5 min, and heated with 1.0 ml of 90% NCS tissue solubilizer (Amersham Corp., Arlington Heights, Illinois) at 55° for 6 hr. Samples are cooled to room temperature, acidified by the addition of 50 μl of glacial acetic acid, and counted in 10 ml of Aquasol-2 (New England Nuclear, Boston, Massachusetts). Samples that receive no enzyme, but are otherwise identical to the experimental samples, are used to control for background radioactivity.

[18] U. K. Laemmli, *Nature (London)* **227,** 680 (1970).

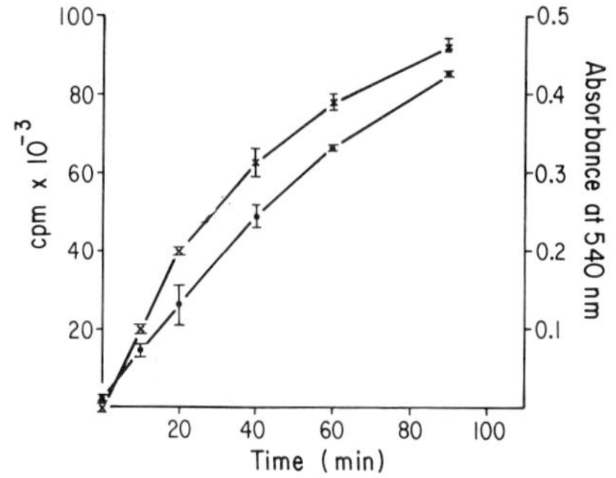

FIG. 3. Production of [^{35}S]prolactin as a function of time of reaction. Duplicate aliquots (50 μl each) of a reaction mixture (700 μl) containing delipidated signal peptidase (350 μl), bovine pituitary translation products (2.08 $\times$ 10^7 cpm), 1.0% Triton X-100, 2.0 mg/ml of ethanol-extracted soybean phospholipid, and 40 mM triethanolamine-HCl, pH 8.1, were removed at the indicated times and assayed for the production of [^{35}S]prolactin. The amount of [^{35}S]prolactin produced was determined by densitometry of a 2-day autoradiograph (×——×) or by excising and counting the portion of the gel containing the prolactin band (●——●). Bars represent the range of duplicate samples.

Radioactivity in the prolactin region of control samples (generally about 20,000–30,000 cpm) is subtracted from the radioactivity of each experimental sample.

Alternatively, the assay is quantitated by densitometrically scanning each lane of the autoradiograph at 540 nm on a Gelman DCD-16 densitometer. The relative amount of processed prolactin produced is proportional to the area under the prolactin peak (or to the maximum absorbance of the prolactin peak, provided that the peak widths at half height are identical, which is usually the case). Both methods of quantitating the assay yield comparable results (see Fig. 3).

Characteristics of the Signal Peptidase Reaction

Utilizing a qualitative assay for signal peptidase, we have previously shown that signal peptidase can be inactivated by delipidation and that activity can be restored to the delipidated enzyme by readdition of phospholipid.[8] However, although phospholipid was required for signal peptidase activity, excess phospholipid proved to be inhibitory.[8] Furthermore, the optimal phospholipid concentration was shown to increase with increasing detergent concentration. (In these previous experiments a mixture of detergents, sodium deoxycholate, and Nikkol, was used. The deoxycholate concentration remained constant while the Nikkol concentration was varied.)

In designing a quantitative assay for signal peptidase we chose to use Triton X-100 as the detergent, because it is both readily available and compatible with most protein purification techniques. The optimal concentration of ethanol-extracted soybean phospholipid to be used in an assay containing 1.0% Triton X-100 was determined to be 2.5 m*M* (i.e., approximately 2.0 mg/ml; see Fig. 4); consequently, we maintain a final phospholipid concentration of 2.5 m*M* in all assays. Samples containing endogenous phospholipid, e.g., detergent-solubilized RER or EDTA-stripped RER, represent an unresolved problem, because the relative efficacies of soybean phospholipid and endogenous RER phospholipid have not as yet been determined. However, since the phospholipid optimum is broad (Fig. 4), a small amount of additional endogenous RER phospholipid probably does not significantly affect signal peptidase activity.

The concentration of [^{35}S]preprolactin in the assay can be estimated from the specific activity of the [^{35}S]methionine used, the number of methionines in preprolactin, and proportion of the total radioactivity in [^{35}S]preprolactin (assumed to be ≤50%, see Fig. 2B), to be on the order of 1 n*M*. This low substrate concentration has several consequences for the signal peptidase assay. First, since the substrate concentration is below

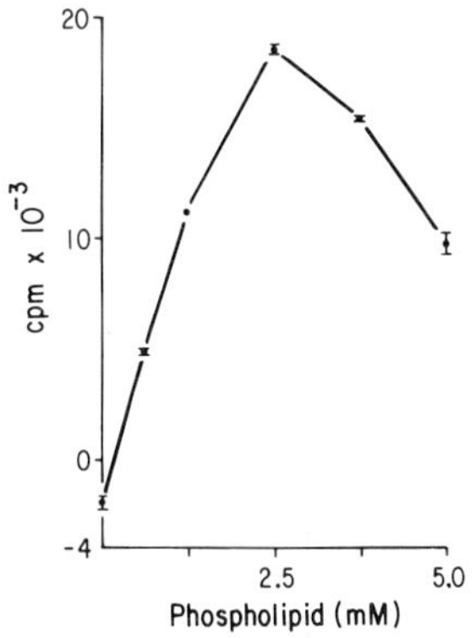

FIG. 4. Determination of the optimal concentration of phospholipid for expression of signal peptidase activity. Canine pancreatic signal peptidase was prepared by deoxycholate extraction of rough microsomes and delipidated by gel filtration chromatography in Sepharose CL-6B equilibrated with 1.0% Triton X-100. Signal peptidase was assayed, in duplicate, by the translocation-independent assay, using [^{35}S]preprolactin as substrate. Assays contained 1.0% Triton X-100, 1.5 × 10^6 cpm [^{35}S]Met-labeled bovine pituitary translation products, 40 m*M* triethanolamine-HCl, pH 8.1, and the indicated concentration of ethanol-extracted soybean phospholipid, in a total volume of 50 μl. The [^{35}S]prolactin produced by the cleavage reaction was quantitated by excising and counting the portion of the gel containing the prolactin band (●——●). Bars represent the range of duplicate samples.

that required to saturate signal peptidase, any decrease in the number of substrate counts per minute in the assay results in a proportional decrease in the number of product counts per minute accumulated. In Fig. 2A, it is shown that a twofold decrease in the amount of [^{35}S]Met-labeled translation products added to the assay results in a similar decrease in the number of counts per minute of [^{35}S]prolactin produced. This situation, wherein the rate of reaction and the amount of product accumulated are directly proportional to the substrate concentration, occurs only when the substrate concentration is much less than the Michaelis constant (i.e., $[S] \ll K_m$).

A second consequence of the limiting amount of substrate in the assay is that the rate of reaction decreases as the substrate is utilized; therefore the accumulation of [^{35}S]prolactin as a function of reaction time deviates from linearity (see Fig. 3). However, since only a small fraction (approximately 15%, see legend to Fig. 2) of the available [^{35}S]preprolactin is converted to [^{35}S]prolactin by 25 μl of signal peptidase during a 90-min incubation, the deviation from linearity is tolerably small. Consequently, we incubate samples for 90 min in order to increase the total [^{35}S]prolactin accumulation and enhance the sensitivity of the assay. As shown in Figs. 1 and 2, this longer incubation time does not seriously affect the linearity of the assay over the range of enzyme concentrations employed.

Finally, the substrate-limiting conditions of the assay make it imperative that each assay tube receive an identical number of substrate counts per minute. When utilizing a single preparation of bovine pituitary translation product this represents no problem, since when the substrate concentration is much less than K_m, the Michaelis–Menten equation reduces to a form in which the fraction of substrate converted to product is independent of the initial substrate concentration. Practically, this means that the results obtained with a particular preparation of translation products should not change as the substrate decays. On the other hand, the results obtained with two different preparations of translation products cannot be compared since the proportion of radioactivity in [^{35}S]preprolactin may change (e.g., owing to the use of different preparations of pituitary RNA). Any change in the proportion of radioactivity in [^{35}S]preprolactin will affect the results obtained by altering the number of counts per minute of [^{35}S]preprolactin added to the assay. Problems of this sort can be avoided by comparing only signal peptidase activities obtained in a particular experiment with a single preparation of bovine pituitary translation products.

[63] Fatty Acid Acylation of Eukaryotic Cell Proteins

By MILTON J. SCHLESINGER

There are numerous examples of proteins and enzymes whose structure and function involve an intimate association with lipid. For most of these proteins, the lipid fraction can be dissociated relatively easily and protein obtained lipid-free. For others, the lipid moiety is tightly complexed and strong denaturants are needed to remove lipid, leading in many cases to denatured enzymes. There is a third group of proteins for which the lipid moiety can be removed only under conditions that cleave covalent bonds. The latter proteins can be divided into two general types. The first type has long-chain fatty acids esterified to glycerol that is linked in a thioether bond to a cysteinyl residue of the protein. Proteins of this kind are found, thus far, only in prokaryotic organisms and are localized to the outer membranes of cells.[1] The second type has long-chain fatty acids esterified to amino acids that are part of the polypeptide backbone of the protein. Proteins of this kind are localized to the membranes of eukaryotic cells or their viruses.[2] It is the latter group that is to be described in some detail in this section.

Many of the data pertaining to fatty acid acylation of eukaryotic cell membrane proteins come from studies with those glycoproteins that form the external spikes on enveloped RNA animal viruses. This is because virus-infected tissue culture cells provide a particularly advantageous system for examining events that occur during synthesis, transport, and maturation of membrane-bound glycoproteins. Infection of cells with many of these viruses establishes within a few hours a virtually synchronous synthesis of millions of molecules of a very few virus-coded proteins that are acted upon by cellular enzymes involved in the normal processing and transporting of cellular membrane proteins (see this volume [36]). There is abundant evidence that the events required to synthesize and translocate virus-specified membrane-bound glycoprotein are identical to those occurring when uninfected cells synthesize glycoproteins and translocate them through intracellular organelles to the plasma membrane.[3] But, when compared to events during normal cell growth in which there might be ~100 different membrane proteins undergoing processing asynchron-

[1] M. Inouye, *Biomembranes* **10,** 141 (1979).

[2] M. J. Schlesinger, *Annu. Rev. Biochem.* **50,** 193 (1981).

[3] H. F. Lodish, W. A. Braell, A. L. Schwartz, G. J. A. M. Strous, and A. Zilberstein, *Int. Rev. Cytol., Suppl.* **12,** 247 (1981).

ISBN 0-12-181996-5

FATTY ACID ACYLATED PROTEINS

Protein[a]	Reference[b]
Sindbis virus glycoproteins PE2*, E1*, E2*	4
Semliki Forest virus glycoproteins E1*, E2*	5
Vesicular stomatitis virus glycoprotein G*	6
Influenza virus hemagglutinin HA_2^*	6
Fowl-plague virus hemagglutinin HA_2	5
New Castle disease virus fusion glycoprotein F1	5
Corona virus glycoprotein E2	5
Brain myelin proteolipoprotein	7, 2
Human transferrin receptor	8
Human histocompatability antigen*	9
Membrane-associated tissue-culture cell proteins	10
Avian and murine virus "sarc" proteins	28
Butyrophilin and xanthine oxidase of milk lipid globule	29
Human red blood cells and rat tissues	30
Catalytic subunits cAMP-dependent protein kinase[c]	26
Murine retrovirus proteins[c]	27

[a] Asterisks indicate proteins having characteristics described in detail in the section Site and Mechanism of Fatty Acid Acylation.
[b] Numbers refer to text footnotes.
[c] Identified as myristic acid acylated in peptide bond to the protein's amino terminus.

ously through a cell growth cycle of 24 hr, the one or two "synchronously" made virus-membrane proteins constitute about a 100- to 1000-fold increase in "sensitivity" for detecting processing events. It is, undoubtedly, this feature of the virus tissue culture system that has permitted the relatively straightforward study of fatty acid acylated proteins.

Distribution of Fatty Acid Acylated Proteins

The proteins that have been reported to contain covalently bound fatty acid are listed in the table.[2,4–10] Most are virus-associated, but it is noteworthy that an important cell surface receptor (for transferrin) is among those listed. The evidence for fatty acid bound to the human histocompa-

[4] M. F. G. Schmidt, M. Bracha, and M. J. Schlesinger, *Proc. Natl. Acad. Sci. U.S.A.* **76,** 1687 (1979).
[5] M. F. G. Schmidt, *Virology* **116,** 327 (1982).
[6] M. F. G. Schmidt and M. J. Schlesinger, *Cell* **17,** 813 (1980).
[7] G. Sherman, and J. Folch-Pi, *Biochem. Biophys. Res. Commun.* **44,** 157 (1971).
[8] M. B. Omary, and I. S. Trowbridge, *J. Biol. Chem.* **256,** 4715 (1981).
[9] H. L. Ploegh, H. T. Orr, and J. L. Strominger, *Cell* **24,** 287 (1981).
[10] M. J. Schlesinger, A. I. Magee, and M. F. G. Schmidt, *J. Biol. Chem.* **255,** 10021 (1980).

tability antigen has not been published in detail. In my laboratory, we detected covalently bound fatty acid in about 30–40 different proteins extracted from membranes of chicken embryo fibroblasts. These membranes were from intracellular organelles as well as the cell surface. A few of these protein bands appear to be common to several other kinds of tissue culture cells, but none have been assigned to a protein of known function (this volume [30]). The covalently bound fatty acids in brain myelin proteolipoprotein were first reported some 30 years ago[11] and subsequently carefully quantitated and detected in tryptic peptide fragments of the protein.[12]

Detection of Fatty Acid Acylated Proteins

The procedures for detecting and identifying protein-bound fatty acids have relied primarily on incorporating high specific radioactive labeled fatty acids into growing tissue culture cells and analyzing proteins from these cells by electrophoresis in sodium dodecyl sulfate (SDS)–polyacrylamide gels. In a typical experiment,50–100 μCi of fatty acid ([9, 10(n)-^{3}H]palmitic acid, 10–30 Ci/mmol, available from Research Products International Corp., Mt. Prospect, Illinois, stored in absolute ethanol at 4° at 5 μCi/μl; note that the isotope should be discarded after 6 weeks in storage) are added to a 35 mm^2 to 100 mm^2 tissue culture dish (10^6 to 10^7 cells) for 4–6 hr at 37°. Serum that contains large amounts of fatty acid should be delipidated. Medium is removed, and cells are washed twice with cold sterile saline solution, once with cold sterile H_2O, and scraped with 0.5 ml of cold sterile H_2O into conical tubes, then centrifuged to form a pellet. The supernatant solution is removed and the cells are stored at −70°. Extracts are prepared by resuspending cells in 0.5 ml of water and sonifying for 30 sec. These extracts are centrifuged for 5 min at 180,000 *g* in a Beckman airfuge or an equivalent rotor. The supernatant fraction is discarded, and the pellets are solubilized in a buffer consisting of 2% sodium dodecyl sulfate (SDS), 10% glycerol, 0.1 *M* Tris-HCl, pH 6.8, and 0.1% Bromphenol Blue.

Radioactivity is determined on a 2-μl sample and 1 to 10 × 10^6 cpm are loaded onto a single lane of a slab gel[13] consisting of 10% polyacrylamide in 0.1% SDS and Tris buffers.[14] Electrophoresis is generally performed at 120 V (constant) for about 3 hr or until the tracking dye has run out the

[11] J. Folch-Pi and M. B. Lees, *J. Bibl. Chem.* **191,** 807 (1951).

[12] J. Jolles, J. L. Nussbaum, F. Schoentgen, P. Mandel, and P. Jollès, *FEBS Lett.* **74,** 190 (1977).

[13] M. S. Reid and R. Bieleski, *Anal. Biochem.* **22,** 374 (1968).

[14] U. K. Laemmli, *Nature (London)* **227,** 680, (1970).

bottom of the gel. The gel is processed *directly* for fluorography using DMSO-PPO.[15] Attempts to stain the gel for protein or fix gels in methanol–acetic acid prior to fluorography may lead to loss of bound fatty acid. Use of other reagents for fluorography also decreases the amounts of bound fatty acid. Of the order of 100–200 cpm of ^3H-labeled fatty acid in a single protein band can be detected after a 4- to 5-day exposure of the film in cassettes stored at −70°. Detection of low levels of radioactivity can be increased by flashing the film prior to exposure.[15] Accurate quantitation (to ±5%) of the radioactivity in protein bands is obtained by orienting the fluorogram over the dried gel and carefully excising the gel with a scalpel. The paper is removed from the gel, and the latter is placed in a vial with scintillation fluid, allowed to swell for 6 hr, and counted. Even a double-labeled band can be measured by this procedure.

Staining of proteins in acrylamide gels previously impregnated with PPO and fluorographed can be carried out as follows.

The dried gel is rehydrated by soaking for 5–10 min in 10% methanol–5% acetic acid. The paper backing is removed, and the gel is transferred to a solution of 0.001% Coomassie Brilliant Blue R250 in 10% methanol–5% acetic acid. The protein bands preferentially and rapidly take up the dye. After 30–60 min the gel is destained briefly in 10% methanol–5% acetic acid and washed twice in water for 30 min each time.

Electrophoresis of protein in gels effectively separates the small amount of labeled protein-bound fatty acid (estimated for a single protein to range from 0.005% to 0.1% of the total lipid label) from the bulk lipid. Because of the extremely low yield of radioactivity into protein-bound lipid and the limited capacity of the gels for protein mass, it is useful to remove nonmembrane proteins (cytoplasmic material) by differential centrifugation prior to gel analysis. In some cases, it may be helpful to perform a partial membrane fractionation by sucrose gradient centrifugation and analyze the membrane fraction only.

If appropriate antibodies are available, one can remove most of the extraneous proteins and a substantial fraction of bulk lipid by preparing antigen–antibody complexes. The procedures are described as follows: Cells previously labeled (as above) with fatty acid are washed and lysed with small volumes (~0.5 ml) of buffer consisting of 1% Triton X-100, 1% sodium deoxycholate, 0.1% SDS, 0.15 *M* NaCl, and 0.05 *M* Tris-Cl, pH 7.2. A general protease inhibitor, e.g., phenylmethanesulfonyl fluoride (0.1 m*M*), is generally included in the lysis buffer. Antibody is added and, after 1-hr incubation at 4°, the antigen–antibody complex is precipitated by adding a preparation of washed *Staphylococcus aureus* (Cowan) ac-

[15] W. M. Bonner and R. A. Laskey, *Eur. J. Biochem.* **46,** 83 (1974).

cording to procedures described by Cullen and Schwartz.[16] The bacteria–Ab-Ag complex is washed twice by vigorous vortexing with the lysis buffer and once with lysis buffer containing *M* NaCl. A brief (15 sec) centrifugation (15,000 *g* in a microfuge) is used to repellent the material after the washes. The final pellet is suspended in 50–100 μl of a gel loading buffer consisting of 2% SDS, 10% glycerol, 5% 2-mercaptoethanol, 0.1% Bromphenol Blue, and 0.6 *M* Tris, ph 6.8, and boiled for 10 min. Insoluble material is removed by microfuge centrifugation and the protein analyzed by SDS–polyacrylamide gel electrophoresis and fluorography as described above.

Analysis of the Fatty Acid

Fatty acid can be rapidly released from protein by treating the sample with mild alkali (0.2 *N* KOH) in absolute methanol at ambient temperatures for 15–30 min (depending on the amounts of material analyzed). When performing this hydrolysis on material separated in gels, it is important to treat samples *before* the gels have been dried or heated. For analyzing protein not well separated from bulk lipid, (i.e., a membrane fraction or a virus preparation) the protein should be thoroughly delipidated by organic solvent extraction, for example, using sequentially chloroform–methanol (2 : 1), chloroform-methanol (1 : 2), chloroform-methanol, H_2O (1 : 1 : 0.3), and acetone. After delipidation, the hydrolysis reaction is carried out. The released fatty acid is extracted with heptane or pentane and analyzed quantitatively by gas chromatography[4] or qualitatively by thin-layer chromatography. The reaction with methanolic KOH converts the fatty acid to the methyl ester, and it can be analyzed without further conversion to an ester with diazomethane.

Protein-bound fatty acid can also be released by incubating protein with *M* hydroxylamine (pH 8.0). The time required (at 23°) for complete release of the fatty acid to the hydroxamate ester depends upon the particular protein and can range from minutes to hours.

Site and Mechanism of Fatty Acid Acylation

The virus glycoproteins that contain fatty acids form external spikes on virions. For many of these proteins, a portion of the carboxyl-terminal region spans the lipid bilayer and the actual terminus is inside the lipid bilayer (this volume [40]). The major part of the glycoprotein that is external to the membranes of vesicular stomatitis virus (VSV), Sindbis

[16] S. E. Cullen and B. D. Schwartz, *J. Immunol.* **117,** 136 (1976).

virus, and fowl plague virus can be removed quantitatively by appropriate proteases under conditions that leave the membrane-embedded portion of the protein (called the root) still associated with the virion. This root portion of the glycoprotein contains the covalent bound fatty acid.[5,17–20] Small peptide fragments of VSV-G with esterified fatty acid have been purified in low yields and shown to be enriched in serine.[6] We do not know yet precisely which amino acid(s) in the protein sequence have acylated fatty acids. There are, however, certain common features in the amino acid sequences of the membrane-spanning regions of several proteins known to contain acylated fatty acids. All these proteins (noted by asterisks in the table) contain a positively charged amino acid (lysine or arginine) on that part of the chain that lies on the internal (cytoplasmic) side of the membrane; all have a sequence of 20–24 hydrophobic residues in the region spanning the bilayer, and all have within 10–12 residues of the membrane spanning segment amino acids that could be acylated.

No information is yet available on the enzyme(s) responsible for acylating the membrane proteins. *In vivo* studies carried out with VSV- and Sindbis virus-infected cells (this volume [40]) show that the glycoproteins are acylated *after* synthesis of the polypeptide chains has been completed, but before substantial processing occurs to the protein-bound oligosaccharides.[21,22] Based on these data, a site in the cell's smooth endoplasmic reticulum or "early" (cis) Golgi complex has been suggested as the cellular site of acylation (this volume [36]). The donor for fatty acid could be newly synthesized long-chain fatty acyl-CoA, fatty acylcarnitine, or fatty acid from neutral or phospholipid. There are data indicating that exchange of fatty acid can occur between membrane lipid and protein.[23]

The function of fatty acid in membrane proteins is unknown. It has been suggested that fatty acid acylation might be important for proper intracellular transport of proteins destined to be localized to the plasma membrane. There are, however, several virus membrane proteins[5] (the neuraminidase of influenza virus, the HN protein of Newcastle disease

[17] M. J. Schlesinger, A. I. Magee, and M. F. G. Schmidt, *in* "The Replication of Negative Strand Viruses" (D. L. Bishop and R. W. Compans, eds.), p. 673. Elsevier, Amsterdam, 1981.

[18] W. A. Petrie, Jr. and R. R. Wagner, *Virology* **107,** 543 (1980).

[19] J. Capone, F. Toneguzzo, and H. P. Ghosh, *J. Biol. Chem.* **257,** 16 (1982).

[20] C. M. Rice, J. R. Bell, M. W. Hunkapiller, E. G. Strauss, and J. H. Strauss, *J. Mol. Biol.* **154,** 355 (1982).

[21] M. F. G. Schmidt and M. J. Schlesinger, *J. Biol. Chem.* **255,** 3334 (1980).

[22] W. G. Dunphy, E. Fries, L. J. Urgani, and J. Rothman, *Proc. Natl. Acad. Sci. U.S.A.* **78,** 7453 (1981).

[23] M. B. Omary and I. S. Trowbridge, *J. Biol. Chem.* **256,** 12888 (1981).

virus, and the E1 protein of corona virus) that do appear at the cell surface membrane but are not acylated. Results of studies using the antibiotic cerulenin showed that VSV G protein acylation can be blocked, yet transport of G to the cell surface is normal.[24] In addition, the G protein from a VSV New Jersey serotype (Ogden) does not become acylated, yet is incorporated efficiently into biologically active virions.[25]

The fatty acid acylated VSV-G (from an Indiana serotype) differs from the nonacylated New Jersey serotype G form in several amino acids in that part of the polypeptide embedded in the lipid bilayer.[25] Thus, appropriate modifications to the protein's structure can circumvent any possible essential need for fatty acid.

All of the fatty acid-bound proteins described thus far are postulated to contain O-acyl ester linkages, based on their lability to alkali and hydroxylamine. There is now unambiguous evidence, however, that long-chain fatty acids are found also in peptide bonds of eukaryotic cell proteins—acylating the amino terminus of two proteins.[26,27] The proteins with this kind of amino terminal modification are considered to be cytoplasmic but are localized in function to cellular membranes. For these polypeptides, fatty acid would facilitate binding to the lipid bilayer. A similar role can be envisaged for fatty acid bound to various tumor virus "sarc" proteins.[28] Indeed, even for the transmembranal proteins the bound lipid could enhance stability between lipid and proteins. But, in addition to providing an anchor, fatty acid should influence the kinds of lipid in the bilayer surrounding the protein and significantly affect protein–lipid interactions.

[24] M. J. Schlesinger and C. Malfer, *J. Biol. Chem.* **257,** 9887 (1982).

[25] C. J. Gallione and J. K. Rose, *J. Virol.* **46,** 162 (1983).

[26] S. A. Carr, K. Biemann, S. Shoji, D. C. Parmelee, and K. Titani, *Proc. Natl. Acad. Sci. U.S.A.* **79,** 6128.

[27] L. E. Henderson, H. C. Krutzsch, and S. Oroszlan, *Proc. Natl. Acad. Sci. U.S.A.* **80,** 339.

[28] B. M. Sefton, I. S. Trowbridge, J. A. Cooper, and E. M. Scolnick, *Cell* **31,** 465 (1982).

[29] T. W. Keenan, H. W. Heid, J. Stadler, E. D. Jarasch, and W. W. Franke, *Eur. J. Cell Biol.* **26,** 270 (1982).

[30] G. V. Marinetti and K. Cattieu, *Biochim. Biophys. Acta* **685,** 109 (1982).

[64] Yeast Secretory Mutants: Isolation and Characterization

By RANDY SCHEKMAN, BRENT ESMON, SUSAN FERRO-NOVICK, CHARLES FIELD, and PETER NOVICK

The yeast cell surface consists of at least three layers: the cell wall, which contains mannoproteins and structural polysaccharides (β-1,3- and β-1,6-linked glucan), a periplasm that contains mannoproteins, and a plasma membrane. Most of the soluble secreted enzymes, such as invertase and acid phosphatase, are located in the periplasm or in the cell wall, where they are accessible to low molecular weight substrates.[1,2] Certain smaller nonglycosylated proteins, such as α-factor and killer toxin, are secreted through the cell wall into the culture medium.[3,4]

Secretion is correlated topologically to the region of cell surface growth. Invertase and acid phosphatase are secreted into the bud portion of a growing cell, which corresponds to the point of cell surface addition during most of the division cycle.[5,6] The correlation between secretion and budding is best accounted for by an exocytotic mechanism of surface growth. Thus, secretory vesicles may fuse with the inner surface of the bud and deliver mannoproteins to the periplasm and membrane precursors to the plasma membrane. The available cytologic evidence strongly supports this notion. Electron microscopic thin-section and freeze-fracture views show 50–100 nm vesicles that fuse with the bud plasma membrane.[7] Histochemical staining of cells secreting acid phosphatase has shown enzyme-specific staining of the bud-localized vesicles, the endoplasmic reticulum (ER), and a Golgi-like organelle.[2]

Although the yeast secretory process appears to resemble the mechanism used by plant and animal cells, one striking difference is the low level of secretory organelles revealed by standard electron microscope thin-section analysis (Fig. 1A). This low level is consistent with a rapid transit time for export of invertase,[8] and a low level of invertase export precursors.[9] The small internal pool of secretory precursors provides a

[1] W. N. Arnold, *J. Bacteriol.* **112,** 1346 (1972).
[2] W. A. M. Linnemans, P. Boer, and P. F. Elbers, *J. Bacteriol.* **131,** 638 (1977).
[3] W. Duntze, V. MacKay, and T. R. Manney, *Science* **168,** 1472 (1970).
[4] D. R. Woods and E. A. Bevan, *J. Gen. Microbiol.* **51,** 115 (1968).
[5] J. S. Tkacz and J. O. Lampen, *J. Bacteriol.* **113,** 1073 (1973).
[6] C. Field and R. Schekman, *J. Cell Biol.* **86,** 123 (1980).
[7] H. Moor, *Arch. Mikrobiol.* **57,** 135 (1967).
[8] P. Novick, S. Ferro, and R. Schekman, *Cell* **25,** 461 (1981).
[9] P. Novick and R. Schekman, *Proc. Natl. Acad. Sci. U.S.A.* **76,** 1858 (1979).

METHODS IN ENZYMOLOGY, VOL. 96

ISBN 0-12-181996-5

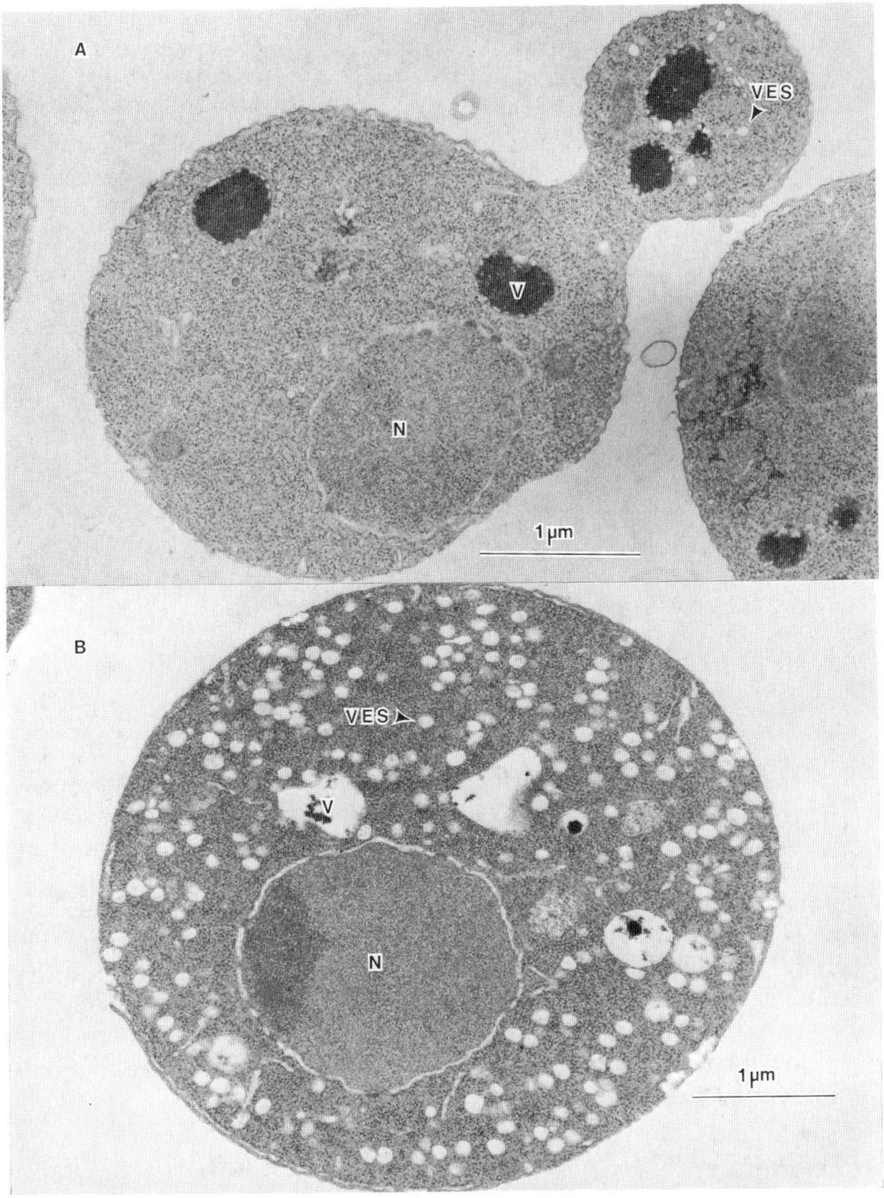

FIG. 1. Thin-section micrographs of cells grown in YPD medium: (A) HMSF 12 (*sec6*-1) grown at 25°; (B) HMSF 93 (*sec8*-4) shifted to 37° for 2 hr. N, nucleus; V, vacuole; VES, vesicle.

sensitive experimental system for the evaluation of mutants that block the secretory pathway and cause an accumulation of secretory enzymes and organelles.

Given the possibility that the secretory process contributes generally to yeast cell surface growth, we have assumed that secretory mutants would be lethal. Thus we have restricted our analysis to temperature-sensitive growth mutants. Secretory mutants are defined as those strains that fail to export active invertase and acid phosphatase, but continue to synthesize protein under restrictive growth conditions. The first mutants (*sec1*-1 and *sec2*-1) were found by direct screening of randomly selected temperature-sensitive mutants.[9] Mutant *sec1* cells stop enlarging and dividing at the nonpermissive temperature (37°), yet macromolecular synthesis continues at a near normal rate for at least 3 hr. This situation produces dense cells that can be separated from normal cells on a Ludox density gradient.[10] Density gradient enrichment has now been used to isolate over 200 secretory mutants.

Procedures for *sec* Mutant Isolation and Classification

Selection Procedure

Mutagenesis

X2180-1A cells (Yeast Genetic Stock Center, University of California, Berkeley) are grown to stationary phase (cell suspension $A_{600} \geq 30$ units; measured in a 1-cm quartz cuvette at 600 nm in a Zeiss PMQII spectrophotometer) in YPD medium (1% yeast extract, 2% peptone, and 2% glucose). Mutagenesis is performed with 3% ethylmethane sulfonate on cells washed and resuspended in sterile 0.05 *M* potassium phosphate (pH 7.0). After 60 min at 25°, a condition that allows 60% survival, cells are diluted into sterile 5% sodium thiosulfate, centrifuged, and resuspended in YPD medium. Nitrous acid mutagenesis has also been employed with a similar distribution of mutant alleles among the *sec* complementation groups.[10]

Density Enrichment

After mutagenesis the cells are grown for 16 hr at 25° to allow recovery and expression of the mutant alleles. A culture containing about 25 A_{600} units of cells is incubated for 3 hr at 37°, and the cells are collected on a

[10] P. Novick, C. Field, and R. Schekman, *Cell* **21**, 205 (1980).

nitrocellulose filter, washed, and resuspended in 0.5 ml of water. The cell sample is layered on 12.5 ml of a mixture containing Wickerham's salts[11] and 60% (v/v) of a purified stock suspension of Ludox AM, in a Falcon 17 × 100 mm polypropylene tube. Ludox is a commercial colloidal silica preparation manufactured by DuPont; samples (1 qt) may be obtained gratis from the manufacturer. Toxic substances may be removed from Ludox by treatment with charcoal, and the pH of the eluate can be adjusted to neutrality with Dowex particles.[12] After centrifugation in a Sorvall SS34 rotor (22,000 *g*, 20 min, 4°), the tube is punctured at the bottom and 1-ml fractions are collected. The A_{600} of the fractions is measured and corrected for the A_{600} of corresponding fractions from a cell-free gradient, and the densest 2% of the cells (1.5 A_{600} unit total) is diluted 2-fold with water. The cells are then centrifuged, the pellet is resuspended in 1 ml of water and diluted 400-fold, and 0.1-ml portions are spread on 200 YPD plates.

Screening the Mutants

Isolation of Temperature-Sensitive (ts) Clones

The YPD plates are incubated at room temperature (20–25°) for 2 days, and colonies are replica-plated onto two YPD plates each; one replica is left at room temperature, the other at 37°. After 26 hr the replicas are compared and colonies that formed at room temperature, but not at 37°, are picked from the master plate. About 15% of the colonies from a density enrichment exhibit ts growth.

Screen for Defective Secretion

The ts mutants are replica plated onto phosphate-free minimal medium plates (agar plates containing Wickerham's medium[11] in which potassium chloride is used in place of potassium phosphate) to derepress the synthesis of acid phosphatase. Copies are incubated at 25° and 37° for 10 hr and then stained for secreted phosphatase activity.[13]

Clones that show ts secretion of phosphatase are screened for conditional secretion and internal accumulation of invertase. Cultures grown at 25° in YP + 5% glucose medium are shifted to 37° for 30 min, after which 2 A_{600} units of cells are sedimented for 1.5 min in a clinical centrifuge. The cell pellets are then resuspended in 2 ml of fresh YP + 0.1% glucose

[11] L. J. Wickerham, *J. Bacteriol.* **52,** 293 (1946).
[12] C. A. Price and E. L. Dowling, *Anal. Biochem.* **82,** 243 (1977).
[13] P. E. Hansche, V. Beres, and P. Lange, *Genetics* **88,** 673 (1978).

medium, and cultures are incubated at 37° for an additional 90 min. The cells are sedimented again, and the pellets are resuspended in 10 m*M* sodium azide at 0°. External (cell wall-bound) invertase is assayed at 37° in whole cell samples using the two-stage incubation described by Goldstein and Lampen.[14] Internal invertase activity is measured in spheroplast lysates. Washed cells (1 A_{600} unit) are converted to spheroplasts in 1 ml of a mixture containing 1.4 *M* sorbitol, 0.1 *M* potassium phosphate (7.5), 0.5 m*M* sodium azide, 20 m*M* 2-mercaptoethanol, and 50 units of lyticase[15] (a purified zymolytic enzyme). After 45 min at 37° the spheroplasts are centrifuged at 10,000 *g* for 5 min, and the pellets are resuspended in 0.5 ml of 1% Triton X-100. Invertase assays are the same as for external activity, except that residual 2-mercaptoethanol present in the spheroplast lysate is eliminated with 0.4 m*M* *N*-ethylmaleimide added after the first assay incubation.

At this stage in the screening procedure, the mutants fall into three groups: those that secrete normally, those that fail to secrete at 37° and accumulate invertase intracellularly, and mutants that fail to secrete or accumulate invertase activity. The second group, referred to as class A *sec* mutants, is distinct from the third group, which consists of protein synthesis mutants and mutants defective in the formation of active secretory enzymes (class B mutants, 16).

Classification of the Mutants

Genetic Complementation Tests

The basis for, and techniques of, genetic complementation in yeast were described by Fink.[17] Briefly, the complementation test involves the formation of diploids from haploid mutant strains that display a common phenotype (e.g., temperature-sensitive growth). When the mutations are allelic (in the same complementation group), diploids display the mutant phenotype; when mutations are not allelic, diploids usually display wild-type behavior.

In practice, two techniques may be used to obtain mutants in the two mating types needed for the complementation test. One approach is to perform the selection procedure with parent strains of both mating types. The resulting mutants from both selections are then crossed with each other to obtain allelic groupings. Alternatively, mutants derived from a single parent strain can be crossed with a wild-type strain of the opposite

[14] A. Goldstein and J. O. Lampen, this series, Vol. 42, p. 504.
[15] J. H. Scott and R. Schekman, *J. Bacteriol.* **142,** 414 (1980).
[16] S. Ferro-Novick, C. Field, and R. Schekman, *J. Cell Biol.* (in press).

mating type. The heterozygous diploids are then sporulated. If the selected mutations are not linked to the mating-type locus, mutant spores of each mating type will be obtained. By these procedures 23 complementation groups have been found among 190 class A *sec* mutants. Similar analysis with 275 mutants that did not accumulate invertase yielded 34 complementation groups.[16]

Protein Synthesis and Cytoplasmic Invertase Production at 37°

Accumulation of secretory forms of invertase by the class A *sec* mutants implies continued protein synthesis at 37°. In order to detect other mutations that specifically affect secretory enzyme formation, representative alleles from each of the complementation groups with mutants that fail to secrete or accumulate derepressed invertase activity are screened for protein synthesis at 25° and 37°. Cells are grown in sulfate-free minimal medium (Wickerham's medium[11] in which chloride salts replace all sulfate salts) supplemented with 0.1 m*M* ammonium sulfate and 2% glucose. Cells from a log-phase culture (1.5 A_{600} units) are sedimented, resuspended in 1.5 ml of fresh medium, and incubated at37° for 30 min. Label (0.1 μCi of carrier-free $^{35}SO_4^{2-}$) is added to 1 ml of each sample, and incubation is continued for 90 min; the remaining 0.5 ml is used to normalize the incorporation rate to cell absorbance. Incubations are stopped with 1 ml of cold 20% trichloroacetic acid, and incorporation is measured in samples collected onto Whatman GF/A filters. Mutants that show a threefold or greater difference between the rates (normalized to wildtype) of protein synthesis and invertase secretion are considered candidates for class B *sec* mutants.[16]

Protein synthesis may also be measured by the production of a cytoplasmic enzyme activity. A convenient assay is the synthesis of cytoplasmic invertase. This enzyme is translated from an mRNA distinct from the one that encodes the secreted form, although both mRNAs are transcribed from the same gene.[18,19] Unlike the secreted enzyme, the cytoplasmic form is not sensitive to glucose repression.[20] Production of the cytoplasmic form is monitored as described above in the section Screen for Defective Secretion, except cells are incubated for the entire period at 37° in YP + 5% glucose medium. Mutants that produce the cytoplasmic invertase but fail to produce active forms of the secreted enzyme are considered to be class B *sec* mutants.[16]

[17] G. R. Fink, this series, Vol. 17A, p. 59.
[18] D. Perlman and H. Halvorson, *Cell* **25,** 525 (1981).
[19] M. K. Grossman and F. K. Zimmerman, *Mol. Gen. Genet.* **175,** 223 (1979).
[20] G. Gallili and J. O. Lampen, *Biochim. Biophys. Acta* **475,** 113 (1977).

Cytologic and Biochemical Characterization of the *sec* Mutants

Electron Microscopy

Standard Thin-Section Analysis

The procedure of Byers and Goetsch[21] has been used for EM sample preparation.

Wild-type cells show occasional enrichment of secretory vesicles (50–70 nm diameter) in the bud and short thin tubules of endoplasmic reticulum are seen apposed to the inner surface of the plasma membrane or in continuity with the nuclear membrane (Fig. 1A). Of the 23 class A *sec* groups, 22 are found to accumulate one or more of three distinct secretory organelles.[10] The most common phenotype, with 10 members, is accumulation of membrane-bounded vesicles (80–100 nm in diameter; Fig. 1B) that are distributed throughout the cell. Another phenotype, with 9 members, is accumulation of ER (Fig. 2A). The accumulated ER contacts the plasma membrane and extends into the cytoplasm, where multiple connections with the nuclear membrane are seen. The lumen of both the ER and the nuclear membrane is wider than the corresponding wild-type structure. Several members of this class produce small vesicles (40–60 nm diameter) in addition to exaggerated ER. A third, more complex phenotype is displayed by mutants in two complementation groups. *sec7* Mutants reveal different organelles depending on the glucose concentration in the growth medium at 37°.[8] In normal YPD medium (2% glucose), structures that we have called Berkeley bodies are seen (Fig. 3A). The Berkeley body, although varied in form, appears to consist of two curved membranes with an enclosed electron-transparent lumen. In some sections the Berkeley body is closed to form a toroid; in other sections it is open at one end to form a cup. When *sec7* cells are incubated in medium containing low glucose (0.1%), instead of Berkeley bodies, stacks of membrane-bounded lamellae, reminiscent of higher eukaryotic Golgi bodies, are seen (Fig. 3B). Of the two structures, the Golgi body appears to be the functional intermediate because *sec7* will secrete accumulated invertase only when cells are returned to 25° in a low glucose medium.[8] Two of the class B *sec* mutants (*sec53, sec59*) produce aberrant, fragmented ER at 37° (Fig. 2B). This may correlate with the accumulation of membrane-bound secretory protein precursors in these mutants (see section on immunoprecipitation, below).

[21] B. Byers and L. Goetsch, *J. Bacteriol.* **124,** 511 (1975).

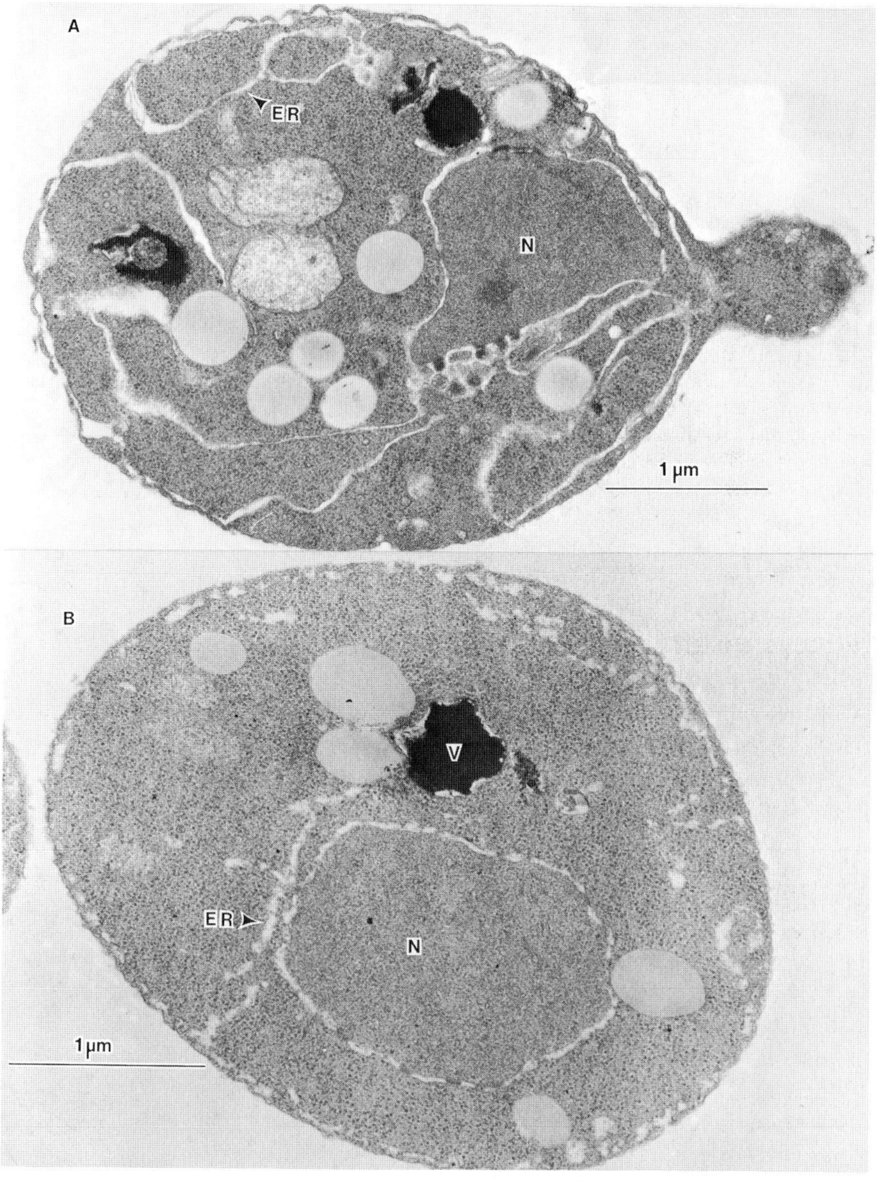

FIG. 2. Thin-section micrographs of cells grown in YPD medium: (A) HMSF 174 (*sec16-2*) shifted to 37° for 2 hr; (B) SF 402-4D (*sec59-1*) shifted to 37° for 3 hr. N, nucleus; ER, endoplasmic reticulum; V, vacuole.

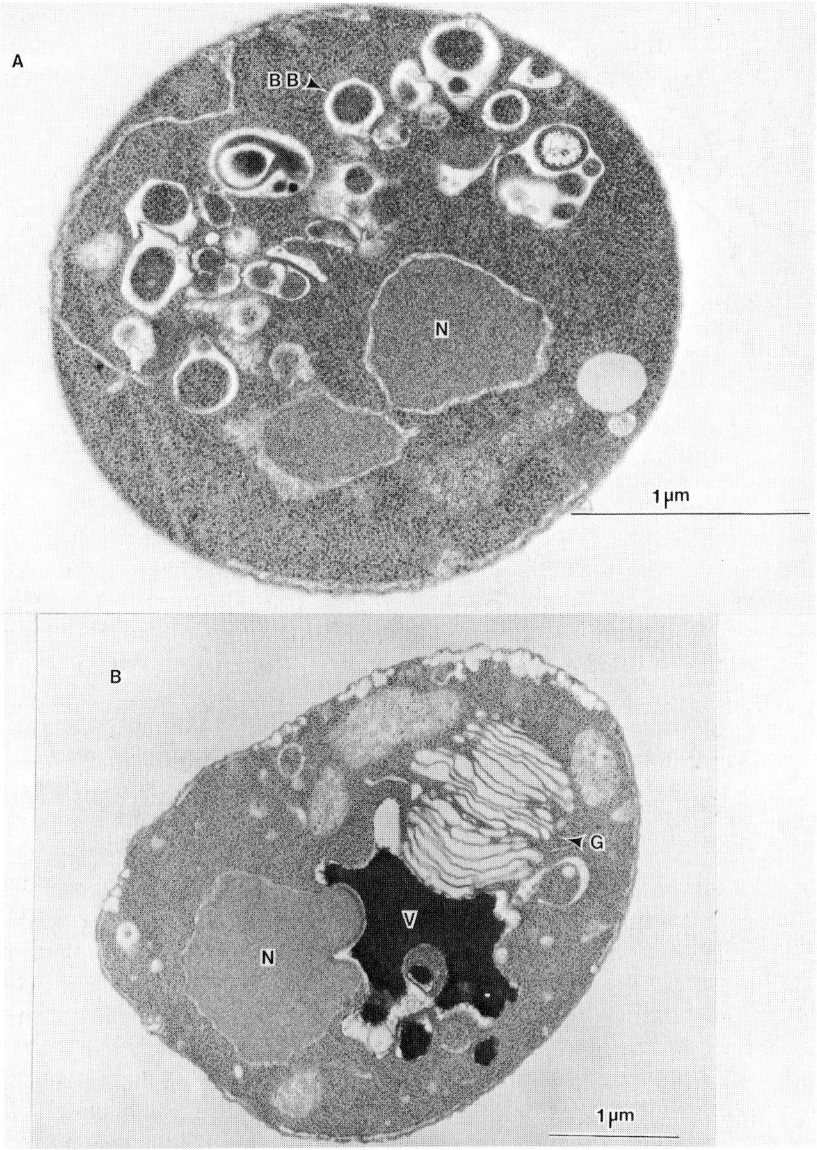

FIG. 3. Thin section micrographs of HMSF 6 (*sec7*-1) shifted to 37° for 2 hr. (A) Grown in YP 2% glucose medium; (B) grown in YP 0.1% glucose medium. N, nucleus; BB, Berkeley bodies; G, Golgi bodies; V, vacuole.

Thin-section electron microscopy has been used to evaluate the ontogenetic relationship among the organelles that accumulate in the *sec* mutants.[8] If these organelles represent stages in the passage of secretory proteins along a linear pathway, a double mutant will accumulate the organelle corresponding to the earliest block. Haploid double mutants are produced by sporulating diploids obtained from a cross of the two mutant parent strains. If the two mutations are unlinked, spores containing both mutations will be produced. The double mutants may be distinguished from single mutants by the complementation test: Double mutants will fail to complement both parent mutants. This analysis has been performed with double *sec* mutants in which each member represented a distinct phenotype. The results demonstrate the following order of events: ER → Golgi → vesicle → cell surface.

Histochemical Staining to Localize Accumulated Acid Phosphatase

A histochemical staining procedure, first developed to detect acid phosphatase activity in spheroplasts of wild-type cells,[22] may be used to assess the intracellular location of phosphatase accumulated in the class A *sec* mutants. For this purpose, *sec* mutant strains are constructed that contain a constitutively expressed phosphatase gene (*pho80*). Under these conditions sufficient enzyme activity accumulates to allow dense staining of accumulated organelles.[9,23]

sec pho80 Cultures are grown in phosphate-depleted YPD medium (phosphate removal by Ca^{2+} precipitation[24]) and transferred to 37° for 90 min. Cells (200 A_{600} units) are collected by filtration on nitrocellulose filters, washed once with 15 ml of 0.1 *M* cacodylate, 5 m*M* $CaCl_2$ (pH 6.0), and resuspended in 6 ml of the same buffer containing 3% glutaraldehyde. After 20 min at 0°, the cells are washed with cacodylate-calcium buffer and resuspended in 5 ml of 50 m*M* cacodylate (pH 7.4) containing 15,000 units of lyticase. Cells are collected after 45 min at 25°, washed with 50 m*M* sodium acetate (pH 5.5), and resuspended in 25 ml of 50 m*M* acetate buffer containing, per milliliter, 1 mg of lead nitrate, 2 mg of *p*-nitrophenyl phosphate, 0.5 mg of nystatin, and 2.5% DMSO. The histochemical reaction mixture is incubated for 1 hr at 30°, and the cells are postfixed in glutaraldehyde followed by osmium tetroxide and uranyl acetate; the poststaining procedure is omitted.[21] Dehydration and embedment are performed as for standard thin-section analysis.[21]

[22] H. J. M. van Rijn, W. A. M. Linnemans, and P. Boer, *J. Bacteriol.* **123,** 1144 (1975).
[23] B. Esmon, P. Novick, and R. Schekman, *Cell* **25,** 451 (1981).
[24] G. M. Rubin, *J. Biol. Chem.* **248,** 3860 (1973).

Electrophoretic Analysis of Accumulated Forms of Invertase

Gel Electrophoresis in Nondenaturing Buffers

Three forms of invertase may be distinguished by native gel electrophoresis. The forms, which are revealed with an enzyme activity stain,[25] are (*a*) a rapidly migrating sharply focused species that corresponds to cytoplasmic invertase[19]; (*b*) an intermediate mobility, discrete species that corresponds to a transit form sequestered within the ER[26,27]; and (*c*) a low mobility, heterogeneous form that corresponds to the secreted enzyme.[19] The mobility differences derive from varying levels of glycosylation. The cytoplasmic invertase has no carbohydrate, whereas the ER transit form has 9 or 10 core oligosaccharides of discrete size for each of the two protein subunits of the enzyme.[28,29] The secreted enzyme has considerably more carbohydrate in the form of outer chain units added in variable length to the core oligosaccharides. These differences allow a rapid test to distinguish the forms of invertase that are made when the secretory pathway is blocked at one of three different points. Class B *sec* mutants will produce only cytoplasmic invertase activity at 37°, ER-blocked mutants will accumulate the ER transit form, and mutants blocked at or after the Golgi step (at which point the outer-chain carbohydrate is added[23]) will accumulate a mature secretory form of invertase.

Cultures (5 ml) of wild-type and *sec* mutant cells are grown in YP + 5% glucose medium at 25°. Cells are collected by centrifugation, resuspended in 2 ml of YP + 0.1% glucose medium, and incubated for 2 hr at 37°. Derepressed cells (10 A_{600} units) are centrifuged at 1000 *g* for 2 min, and the pellets are washed once and resuspended in 0.1 ml of 50 m*M* Tris-HCl (pH 7.5). Samples are mixed with 2-mercaptoethanol (40 m*M*) and lyticase (400 units) and incubated for 30 min at 37°. Debris is removed by centrifugation at 12,000 *g* for 10 min. Soluble fractions (10 μl; ~0.05 unit of invertase; 1 unit = 1 μmol of glucose released from sucrose per minute) are mixed with 5 μl of 50% glycerol with a trace of Bromphenol Blue, and are applied to the gel for electrophoresis at 4°. Native gel electrophoresis is performed on 4.5% polyacrylamide slabs with system F described by Rodbard and Chrambach.[30] Invertase is localized in gels after incubation

[25] J. Meyer and P. Matile, *Biochem. Physiol. Pflanz.* **166,** 377 (1974).
[26] P. Babczinski and W. Tanner, *Biochim. Biophys. Acta* **538,** 426 (1978).
[27] P. Babczinski, *Biochim. Biophys. Acta* **614,** 121 (1980).
[28] R. B. Trimble and F. Maley, *J. Biol. Chem.* **252,** 4409 (1977).
[29] L. Lehle, R. E. Cohen, and C. E. Ballou, *J. Biol. Chem.* **254,** 12209 (1979).
[30] D. Rodbard and A. Chrambach, *Anal. Biochem.* **40,** 95 (1971).

in 0.1 *M* sodium acetate (pH 5.1) containing 0.1 *M* sucrose for 10–90 min at 37° by the method of Gabriel and Wang.[31]

Immunoprecipitation and SDS–Gel Electrophoresis of Radiolabeled Invertase

The structure of invertase and of other exported proteins accumulated at different *sec* blocks is best evaluated by immunoprecipitation and electrophoresis in a denaturing buffer. This is particularly true for the class B mutants, which accumulate immunoreactive but enzymatically inactive forms of invertase.[16]

Preparation of Affinity-Purified Invertase Antibody. Invertase may be purified from strain FH4C by the method of Goldstein and Lampen.[14] Invertase-specific antiserum is obtained by injecting New Zealand white rabbits intradermally with 500 μg of pure invertase in complete Freund's adjuvant. Carbohydrate-specific antibodies are removed by adsorbing the serum with invertase-negative mutant cells (*SUCO* strain; 13 A_{600} units of cells per milliliter of serum) three times. The adsorbed serum is mixed with ammonium sulfate to 50% saturation, and the precipitated Ig fraction is collected by centrifugation, resuspended, and dialyzed against PBS buffer. The dialyzed material is further purified by adsorption to and elution from invertase conjugated to Sepharose 4B.[16] The titer of a typical affinity-purified fraction is 1 : 2 when the antibody (2.9 mg/ml) is tested by the Ouchterlony method with an invertase solution at 0.5 mg/ml.

Radiolabel and Invertase Extraction. Cells are grown in sulfate-free minimal medium containing 0.1 m*M* ammonium sulfate and 5% glucose. An aliquot (6 A_{600} units) is centrifuged, and cells are resuspended in 3 ml of fresh minimal medium containing 0.05 m*M* ammonium sulfate, 0.1% glucose, and 300 μCi of $^{35}SO_4^{2-}$. For mutants that do not express the ts phenotype immediately at 37°, cultures are incubated at the high temperature for as long as 30 min prior to the addition of radiolabel. After the labeling period at 37° (usually 30 min to 1 hr), cells are sedimented, washed with 10 m*M* sodium azide, and resuspended in 0.4 ml of spheroplasting medium [1.4 *M* sorbitol, 23 m*M* potassium phosphate (pH 7.5), 2 m*M* $MgCl_2$, 10 m*M* sodium azide, 40 m*M* 2-mercaptoethanol, and 50 units of lyticase per A_{600} unit of cells). Spheroplasts formed during a 30-min incubation at 30° are sedimented at 3000 *g* for 10 min, and the pellet is dissolved in 0.2 ml of 1% Triton X-100. The sample is diluted with an equal volume of 2× (phosphate-buffered saline [PBS: 0.02 *M* NaCl, 12.5 m*M* potassium phosphate (pH 7.6)] containing 2 m*M* each of phenyl-

[31] O. Gabriel and S. F. Wang, *Anal. Biochem.* **27,** 545 (1969).

methylsulfonyl fluoride (PMSF) and tosyllysine chloromethyl ketone (TLCK). The lysate is centrifuged at 12,000 *g* for 10 min; the supernatant solution is removed and sedimented again at 100,000 *g* for 90 min. These conditions are suitable for extraction of invertase that accumulates in a soluble form.

An alternative procedure, which is a modification of one developed to detect unstable protein precursors in *Escherichia coli*,[32] has been devised for yeast. This technique involves rapid lysis in SDS and has allowed the recovery of immunoreactive forms of invertase from two class B *sec* mutants (*sec53, sec59*) which accumulate precursors that are sensitive to proteolysis and insoluble in Triton X-100. Also, a more sensitive detection of invertasc is provided by the use of a high copy number plasmid that contains an insert of the invertase gene (*SUC2*).[33] About 30-fold more invertase is produced when a wild-type strain is transformed with this plasmid.

Mutant strains containing the invertase plasmid are grown in minimal medium under conditions where the plasmid selectable marker (*URA*) is maintained (minimal medium normally has no uracil). Samples (1.9 A_{600} units) are incubated at 37° for 30 min. The cells are sedimented and resuspended in 0.63 ml of minimal medium as above except that 750 μCi of $^{35}SO_4^{2-}$ is used for labeling during a 30-min further incubation at 37°. Labeled cells are collected and washed with 10 m*M* sodium azide, and the cell pellets are mixed with 0.15 g of glass beads (0.3–0.5 mm) and 30 μl of 1% SDS, 50 m*M* 2-mercaptoethanol, 1 m*M* PMSF, and 1 m*M* TLCK. Cells are broken by vortexing for 90 sec at 40° and then immediately heated in a boiling water bath for 3 min. Lysates are diluted with 0.8 ml of PBS containing 2% Triton X-100 and centrifuged at 100,000 *g* for 90 min.

Immune Precipitation and SDS–Gel Electrophoresis. The final soluble fractions are mixed with 1–5 μl of affinity-purified antibody (2.9 mg/ml) in an Eppendorf centrifuge tube and incubated at 0° for 16–18 hr. The immune complex is precipitated with fixed *Staphylococcus* A cells (10% suspension [w/v]). The *Staphylococcus* A cells (Enzyme Center, Boston, Massachusetts) are washed twice with equal volumes of PBS containing Triton X-100 (0.5%) and bovine serum albumin (1 mg/ml), and 25 μl are added to the immunoprecipitates. The mixture is kept at 0° for 30 min and centrifuged for 1 min in a Beckman microfuge B, and the pellet is washed twice with PBS–Triton–BSA buffer. The precipitate is transferred to a new Eppendorf tube, washed twice more, and mixed with 50 μl of sample buffer [45 m*M* Tris-HCl (pH 6.8), 9% glycerol, 1.7% SDS, 1% 2-mercap-

[32] K. Ito, P. J. Bassford, and J. Beckwith, *Cell* **24,** 707 (1981).

[33] M. Carlson and D. Botstein, *Cell* **28,** 145 (1982).

toethanol, and 0.01% Bromphenol Blue]. After heating in a boiling water bath for 3 min, samples are centrifuged in a microfuge to remove the *Staphylococcus* A cells. Solubilized fractions are subjected to electrophoresis on SDS–polyacrylamide (7.5% or 10%) slab gels, stained for protein, and treated with EnHance (New England Nuclear, Boston, Massachusetts), as described by the manufacturer. Dried gels are allowed to expose Kodak X-Omat R film XR-5 for 1 or more days at −70°.

[65] Secretory Mutants in the Cellular Slime Mold *Dictyostelium discoideum*

By RANDALL L. DIMOND, DAVID A. KNECHT, KEVIN B. JORDAN, ROBERT A. BURNS, and GEORGE P. LIVI

During the past two decades the cellular slime mold *Dictyostelium discoideum* has become a popular organism for biochemical and genetic studies of development and gene expression. Its popularity as a "model system" reflects the relative simplicity of its life cycle and its accessibility for experimentation. The organism grows as free-living single-celled amoebae that have no cell wall or rigid cuticular structures. The ease with which large numbers of cells can be cultured and manipulated synchronously makes it ideal for biochemical studies, while its small genome size and normal haploid condition permit genetic studies that are often difficult or impossible in other eukaryotes. The amoebae grow with a generation time of approximately 4 hr when grown on bacteria or 8 hr during axenic growth in a rich broth medium. (For a review of the life cycle and developmental attributes of the organism, the reader is referred to the excellent monograph by Loomis.[1])

The lysosomal system is a prominent feature of *D. discoideum* cellular ultrastructure. In addition to its role in phagocytosis, digestion, and autodegradation, it also functions in secretion. We have shown that under normal physiological conditions of growth or development a number of lysosomal enzymes are secreted in substantial amounts. We have also developed a suspension assay system in which a number of lysosomal enzymes are very efficiently secreted, resulting in greater than 50% of

[1] W. F. Loomis, "*Dictyostelium discoideum:* A Developmental System." Academic Press, New York, 1975.

ISBN 0-12-181996-5

each enzyme appearing extracellularly within a few hours.[2] Using this assay we have shown that the ability to secrete lysosomal enzymes is not a constitutive property of the organism. Amoebae grown in axenic broth culture are the most efficient in the secretion of these lysosomal enzymes, whereas vegetative amoebae grown on bacteria are relatively inefficient in enzyme secretion. Secretory ability is significantly enhanced during the first few hours of development but drops throughout the developmental phase, resulting in very little secretion during culmination.[3] In addition to these effects of cell differentiation on the secretory ability, the amount of enzyme secreted is also regulated in response to the cellular environment. The presence of external amino acids is particularly effective in inhibiting secretion while other changes in the medium composition also influence secretory behavior.[3–5]

Isolation of Secretory Mutants

We have taken advantage of the fact that significant amounts of lysosomal enzymes are secreted during axenic growth in order to isolate mutants that accumulate less extracellular enzyme during growth than do wild-type strains. Our initial screenings have been simplified by the use of *p*-nitrophenyl and 4-methylumbelliferyl derivatives of sugars which serve as substrates for the lysosomal enzymes and to which the *D. discoideum* plasma membrane is impermeable.

Mutagenesis and Screening Procedures. Amoebae are grown in TM medium[6] to a titer of 1 to 2 $\times$ 10^6 cells/ml. The cells are mutagenized with *N*-methyl-*N*′-nitro-*N*-nitrosoguanidine by the procedure of Dimond *et al.*[7] The concentration of the mutagen and time of exposure is optimized for each batch of mutagen to produce a consistent survival frequency of about 0.1%. Freshly mutagenized cells are inoculated into 96-well tissue culture trays using the multiple pipette shown in Fig. 1. The cell titer is adjusted so that each well on the average receives one viable cell. The multiple pipette can be used not only for inoculating trays, but also for adding assay reagents or removing culture supernatants. The trays are incubated at 21° in a moist chamber for 3–4 weeks until most of the colonies approach stationary phase (>10^6 cells/ml). In order to screen for

[2] R. L. Dimond, R. A. Burns, and K. B. Jordan, *J. Biol. Chem.* **256,** 6565 (1981).

[3] R. A. Burns and R. L. Dimond, submitted for publication.

[4] R. A. Burns, G. P. Livi, and R. L. Dimond, *Dev. Biol.* **84,** 407 (1981).

[5] E. V. Crean and E. F. Rossomando, *J. Gen. Microbiol.* **110,** 315 (1979).

[6] S. J. Free and W. F. Loomis, *Biochimie* **56,** 1525 (1974).

[7] R. L. Dimond, M. Brenner, and W. F. Loomis, Jr., *Proc. Natl. Acad. Sci. U.S.A.* **70,** 3356 (1973).

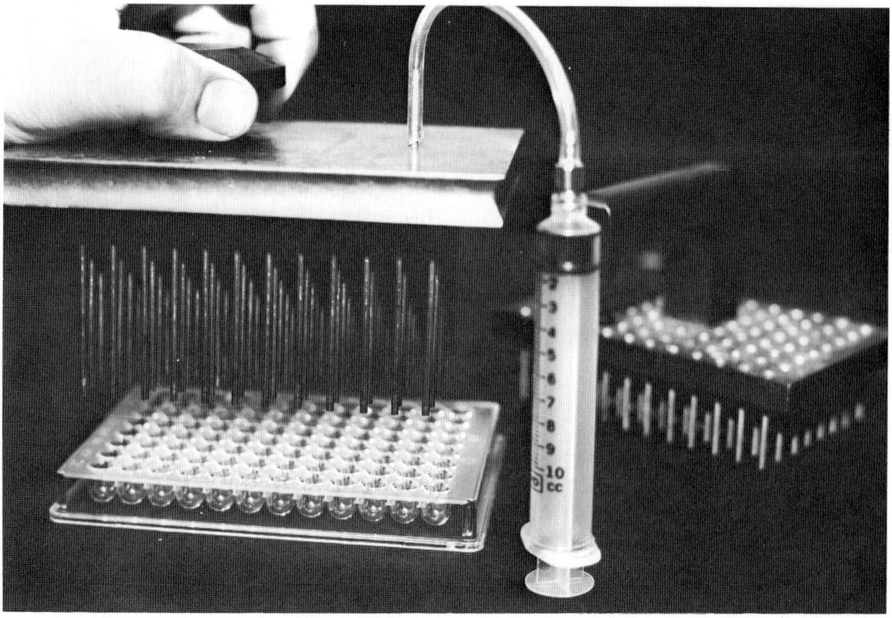

FIG. 1. Inoculation and replication equipment. The multiple pipette used for inoculating 96-well tissue culture trays is shown at the left. The replicator used for replica plating the trays is at the right.

multiple enzymes, the initial trays are replica plated. This is done by gently vibrating the trays on a vortex mixer, dipping a sterile replicator (Fig. 1) into the cell suspensions and then sequentially inoculating two trays previously filled with fresh medium. After growth, replicate clones are assayed for the presence of a particular enzyme in the extracellular medium by adding the appropriate substrate in 10 m*M* phosphate buffer, pH 6.5, using the multiple pipette. Enzymes should be assayed before most of the colonies enter stationary phase, at which time the medium turns dark yellow, making screening difficult. Enzymic hydrolysis of the substrate generates *p*-nitrophenol, which is yellow above pH 6. Using the pH 6.5 buffer, positive wells turn dark yellow within a short period of time at 30–35° depending on the particular enzyme assayed. For example, extracellular *N*-acetylglucosaminidase can be visualized after 30 min whereas α-mannosidase detection requires several hours of incubation. Alternatively, 1 *M* Na_2CO_3 can be added to quickly raise the pH and intensify the color. Once wells containing putative secretory mutants are identified, replica clones can be obtained from the original trays. Half of the contents of each well are diluted into 2 ml of fresh TM medium in a test tube and incubated on a reciprocating shaker at 21°, and half is plated in association with *Klebsiella pneumoniae* on nutrient agar plates.[8]

These primary screening procedures are designed to recognize both regulatory and structural gene mutants for the particular enzyme screened as well as the desired secretory mutants. The 2-ml TM cultures are used to identify the secretory mutants in a secondary screening. After growth of the cultures to 2 to 5 $\times$ 10^6 cells/ml, the cells are separated from the culture supernatant by centrifugation (3000 rpm, 3 min, table-top centrifuge). Cellular and supernatant fractions are assayed for enzymic activity, and the results are plotted as a histogram of the ratio of extracellular to cellular activities of each putative mutant. We have routinely selected those strains differing by a factor of 5 from the mean ratio for further analysis.

Characterization of Mutants Defective in Secretory Ability

We have taken advantage of the increased secretory ability of amoebae grown in axenic broth culture to define optimal conditions for the study of lysosomal enzyme secretion.[2,4] In our standard secretion assay, these amoebae are washed once and resuspended in a low ionic strength nonnutrient buffer to minimize inhibitory effects of ionic constituents or nutrients. Since under these conditions there is very little or no lysosomal enzyme synthesis, the enzymes act as physical markers for the vesicles that contain them and their secretion reflects the functioning of those vesicles. Furthermore, enzyme secretion in *Dictyostelium* is an active process requiring an intact energy-generating system.[2] One group of enzymes (including *N*-acetylglucosaminidase, α-mannosidase, β-glucosidase-1, α-glucosidase-1, and β-galactosidase-1) is very efficiently secreted, 50–80% of the total cellular activity of each enzyme becoming extracellular within a few hours. Since all these enzymes have similar or identical secretion kinetics they may come from the same lysosomal vesicles. In contrast, only 15–40% of the cellular acid phosphatase is secreted. Differences in secretion kinetics along with the differential effects of inhibitors have led us to suggest that the secretable acid phosphatase activity may be localized within different vesicles than are the secretable glycosidases. Other lysosomal enzymes, including β-galactosidase-2 and a number of proteases,[9] are very poorly secreted.

Although some of these lysosomal enzymes are involved in digestion in food vacuoles, we have shown that they are not released in any signifi-

[8] M. Sussman, *Methods Cell Physiol.* **2,** 397 (1966).

[9] M. J. North, *J. Gen. Microbiol.* **128,** 1653 (1982).

cant amounts during the egestion of indigestible material from terminal phagosomes.[2] Thus, there are likely to be three classes of lysosomal vesicles in *Dictyostelium*; those involved in the secretion of the efficiently secreted glycosidases; those responsible for the secretion of acid phosphatase; and others that are not involved in secretion.

Procedure. Amoebae are grown in TM medium until log or early stationary phase of growth, harvested by centrifugation in a refrigerated centrifuge (Sorvall, 3000 rpm, 3 min, SS34 rotor), and washed once in cold 10 m*M* potassium phosphate buffer, pH 6.0. This washing step is important since any residual amino acids trapped in the pellet will partially inhibit the secretory response. The cells are resuspended in cold buffer at 10^8 cells/ml and diluted 1 : 10 into 20 ml of prewarmed buffer in a 125-ml stoppered Erlenmeyer flask. Although the volume of buffer used is not critical, the cell density is important with many strains. The secretion flasks are incubated for 6 hr at 24° to 27° with gentle mixing throughout the experiment. We use a reciprocating shaker, set at 43 oscillations per minute with a 2.75 cm stroke, but gentle gyratory mixing is also quite adequate. Violent mixing results in a decreased secretory response. During the first 30 min the cells clump into groups of 2–5 cells, which eventually disperse into single cells that remain throughout the rest of the experiment. At the time of inoculation and for subsequent sampling, the flasks are vigorously swirled and 2 ml of the suspension are removed. Cellular and supernatant fractions are separated by centrifugation, the cells are resuspended in 2 ml of cold distilled water, and both fractions are frozen at −20° until assayed. The cells are lysed by the addition of Triton X-100 to 0.1%, whereas supernatant fractions remain untreated, and both are assayed as previously described.[2]

Data Analysis. Direct assay of the cellular extracts provides an accurate estimate of the cellular enzymic activity. However, the supernatant activities are corrected to reflect secretion during the experiment by subtracting the amount of enzyme activity in the t_0 supernatant from all supernatant activities. This should never be more than a few percent of the total enzyme activity. Cell lysis during the secretion experiment can be detected by assaying the amount of total protein or β-galactosidase-2 activity (a poorly secreted enzyme) in supernatant fractions. The lesser of these two measurements gives a maximal estimate of cell lysis.

Simply plotting the amount of enzyme released by the cells is not a particularly useful way of comparing data because there is a large variation in the total amount of any one enzyme between different strains and cultures of the same strain. Since the enzymes simply act as markers for the vesicles that contain them, it is more useful to express the data as the percentage of total enzyme that is extracellular at each time point. This

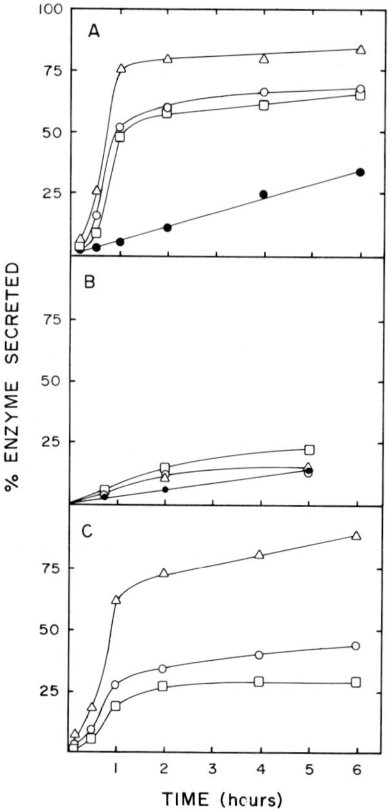

FIG. 2. Secretion of lysosomal enzymes from *Dictyostelium discoideum*. Amoebae from exponentially growing cultures were used in the standard secretion assay and analyzed as described in the text. *N*-Acetylglucosaminidase (○—○); α-mannosidase (□—□); β-glucosidase (△—△); acid phosphatase (●—●). (A) Wild-type strain, Ax3; (B) a secretory mutant affecting the secretion of all glycosidases; (C) a secretory mutant that is normal for β-glucosidase.

percentage is fairly reproducible (SD is about 14%) in the wild-type strain, Ax3, even though the amount of enzyme secreted varies as much as 10-fold. A typical secretion profile for the wild type (Fig. 2A) shows the complex secretion kinetics characteristic of the glycosidases as compared to the linear secretion kinetics of acid phosphatase. In contrast, the secretory mutants show a number of different variations from this pattern. For example, some mutants are deficient in the secretion of all the efficiently secreted glycosidases (Fig. 2B), whereas other mutants indicate much greater heterogeneity. Figure 2C shows a mutant that is affected in the secretion of *N*-acetylglucosaminidase and α-mannosidase but which is normal for the secretion of β-glucosidase.

Another useful way of analyzing the data is to plot the kinetics of release of each enzyme during the standard secretion assay. The amount of extracellular enzyme at each time point is normalized to the amount at t_6. Figure 3A presents this type of analysis for the wild-type strain again showing the two kinetic classes of enzymes, the efficiently secreted glycosidases with complex kinetics and acid phosphatase with linear kinetics. Figure 3B presents a similar analysis for a mutant that alters the kinetics of secretion of the glycosidases much more than the actual percentage of enzyme released from the cells.

Characterization of Mutants Defective in the Regulation of Secretion

Dictyostelium discoideum regulates the secretion of lysosomal enzymes in response to the nutrient environment it encounters. The standard secretion assay has been designed to relieve cells from environmental influences as much as possible in order to reflect the basic secretory ability of the cells. Some mutants which appear normal in the standard secretion assay but which are abnormal in secretion during axenic growth may be specifically affected in the regulation of the secretory system rather than in basic secretory mechanisms. This type of mutant can be characterized by quantitating secretion during axenic growth in broth culture when the secretory system is strongly inhibited by the presence of nutrients in the medium.[4]

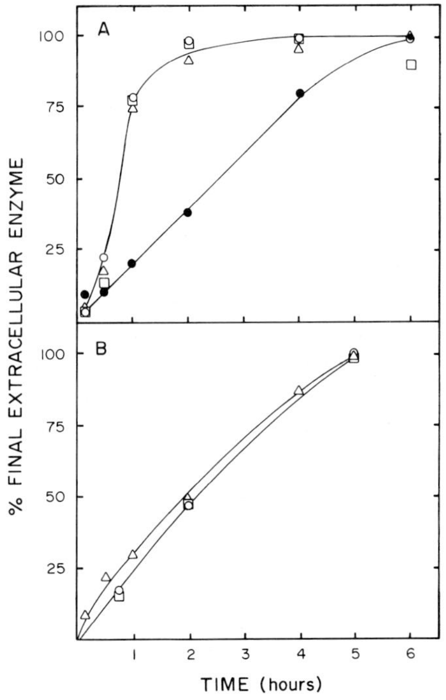

FIG. 3. Kinetics of enzyme secretion. The kinetics of secretion of lysosomal enzymes in the standard secretion assay have been analyzed as described in the text. (A) Wild-type strain, Ax3; (B) secretory mutant. The symbols are the same as in Fig. 2.

Procedure and Data Analysis. It is important that strains be fully adapted to growth in TM medium[6] before any data are collected. This can be accomplished by growing small volume cultures to late exponential phase before inoculating the cultures to be analyzed for secretion. This also allows an accurate determination of growth rate and stationary phase titer (1 to 2 $\times$ 10^7 cells/ml for wild type). Our standard protocol involves inoculating 100-ml cultures in 250-ml Erlenmeyer flasks at 5 $\times$ 10^4 cells/ml with cells from exponentially growing cultures. The cultures are grown at 24° with gyratory shaking. Once the cultures have reached 5 $\times$ 10^5 cells/ml, 2-ml samples are taken twice a day until stationary phase is approached. The cells are separated from the supernatant by centrifugation (3000 rpm, 3 min, GLC2 centrifuge) and the pellets are washed in 10 m*M* potassium phosphate buffer, pH 6.5, and resuspended in cold distilled water. This washing step is particularly important at low titer, when the small pellets can be significantly contaminated with protein or enzyme from the medium. The samples are frozen at −20° until assayed.

In the wild-type strain, Ax3, the secretion of glycosidases is not constant during exponential growth but varies in a complex manner, as does the total amount of enzyme per cell. As cultures approach the end of the growth phase, the variation between cultures in the amount of enzyme per cell and the percentage of enzyme appearing extracellularly increases significantly. Unlike the standard secretion assay, the percentage of total enzyme appearing extracellularly during growth is influenced by the pattern of cellular accumulation of the enzyme. Thus, it is important to distinguish secretory mutants from mutants that primarily alter the regulation of the enzyme and as a result affect the percentage found extracellularly. One approach is to restrict the analysis to data obtained from the culture at relatively low titers. During this time the amount of enzyme per cell is more consistent in both the wild-type and the enzyme-regulatory mutants we have seen to date. Table I compares the secretion during growth of the wild-type strain and several mutants that predominantly affect the regulation of secretion and, as a result, secrete either more or less enzyme than the wild type under comparable conditions. An alternative method for separating the effects of altered gene regulation from the secretory components of these data is to approximate the amount of enzyme secreted per cell per hour over the interval between samplings normalized to the average amount of enzyme per cell over the interval. We have written basic programs that are easily adapted to most microcomputers for this analysis.

Analysis of secretion during growth in media other than TM may aid in the characterization of the specific regulatory defects present in different secretory mutants. We have shown that the secretion of lysosomal en-

TABLE I
MUTANTS AFFECTED IN THE SECRETION OF *N*-ACETYLGLUCOSAMINIDASE DURING AXENIC GROWTH[a]

Strain	% Enzyme extracellular
Ax3	39 ± 5
HMW500	71
HMW501	84
HMW502	14
HMW503	25
HMW504	64
HMW505	27
HMW506	24
HMW507	72

[a] Data for the wild type are presented with the standard error of the mean. Strains were sampled at titers between 5 and 10 × 10^5 cells/ml.

zymes in the wild-type strain is very sensitive to minor modifications of the complex medium.[4] We have also shown that secretion is substantial in the defined medium developed by Franke and Kessin.[10,11] The use of the defined medium allows a more precise analysis of those components that affect the secretion process. Use of these media may help to reveal the nature of the defect present in some secretory mutants.

Isoelectric Focusing Methods for Detecting Defects in Modification of Very Negatively Charged Enzymes

The standard secretion assay provides information about the stability of lysosomal enzymes, since there is no appreciable enzyme synthesis during the experiment. In the wild type the total amount of enzyme activity remains constant (±25%) over the entire 6-hour period. Some of the secretory mutants, however, show significant loss of enzyme activity under these conditions (Table II). The fact that the strain known to be deficient in the modification of lysosomal enzymes, M31,[12,13] shows the same defect in stability indicates that some of our secretory mutants may also alter the modification of these enzymes. Since these lysosomal en-

[10] J. Franke and R. Kessin, *Proc. Natl. Acad. Sci. U.S.A.* **74,** 2157 (1977).
[11] G. P. Livi, unpublished observations.
[12] S. J. Free, R. T. Schimke, H. Freeze, and W. F. Loomis, *J. Biol. Chem.* **253,** 4102 (1978).
[13] S. J. Free and R. T. Schimke, *J. Biol. Chem.* **253,** 4107 (1978).

TABLE II
STABILITY OF ENZYMES IN THE STANDARD SECRETION ASSAY

Strain	% Activity lost by 6 hr[a]	
	N-Acetylglucosaminidase	α-Mannosidase
Ax3	0	0
HMW508	10	36
HMW509	38	50
HMW510	6	31
HMW511	23	30

[a] Sum of cellular and extracellular enzyme activity.

zymes have extremely low isoelectric points due to extensive phosphorylation of amino acids[14] and phosphorylation and sulfation of carbohydrate moieties,[15] we have used low pH nonequilibrium isoelectric focusing gels to determine whether or not our secretory mutants alter the modification of certain enzymes.

Preparation of Isoelectric Focusing Gels. Agarose-EF (0.132 g) (LKB, Rockville, Maryland) is added to 15.2 ml of water containing 1.65 g of sorbitol or sucrose and placed in a steam oven. While the agar is melting, a 10 × 12.5 cm piece of Gelbond film (FMC Corporation, Rockland, Maine) is cut and placed on a wetted glass plate in a 65° oven. When the agarose is in solution and free of bubbles, it is equilibrated to 65° and ampholytes are added. For *D. discoideum* lysosomal enzymes the following amounts of Serva ampholytes from 40% stock solutions are added to the agarose solution: 0.25 ml, pH 2–4; 0.75 ml, pH 3–5; 0.15 ml, pH 4–6. The glass plate with Gelbond is removed from the oven, placed on a level surface, and the agarose solution is distributed evenly over the surface of the Gelbond with a prewarmed pipette. The gel is allowed to solidify at room temperature (within 30 min).

Electrophoresis. A Pharmacia (Piscataway, New Jersey) flat-bed electrophoresis apparatus is cooled to approximately 5° with circulating water. The gel, supported by the Gelbond, is placed on the apparatus with the short dimension (10 cm) between the electrodes. The wicks (either paper from LKB or sponge from Pharmacia) are saturated with 1 *M* H_3PO_4 or 0.2 *N* HCl for the acid wick and 2% pH 6–8 ampholytes (LKB)

[14] Dr. Hudson Freeze, personal communication.
[15] H. H. Freeze and A. L. Miller, *Mol. Cell. Biochem.* **35,** 17 (1980).

for the basic wick. The wicks are positioned at the very edge of the gel and the electrodes are aligned in contact with the wicks throughout their length, with the (+) electrode at the acid end and the (−) electrode at the basic end. If the electrodes are not in contact throughout the length of the wicks, the samples will not run orthogonal to the electrodes. Electrophoresis is maintained with constant wattage corresponding to an initial voltage of 180 V. Initial conditions have to be adjusted for the apparatus and ampholyte mixtures used to prevent overheating and drying out. A 10–15-min prerun establishes the pH gradient prior to sample loading.

At the end of the prerun, sample wells are formed in the agar within 1 cm of the basic wick in a region of neutral pH using a 200-μl glass micropipette with suction. If the wells are positioned farther from the wick, acid precipitates are formed that interfere with the focusing of acidic proteins. The correct position for sample application can also be judged by spotting a hemoglobin mixture on the gel before the prerun and positioning the wells near the resolved hemoglobin band. Samples are usually prepared in water with 1% Triton X-100. However, there is no significant difference in focusing if samples of higher ionic strength are used. Samples containing equal units (0.1–2 units per lane) of the enzyme of interest in a total volume of 5–10 μl are loaded into the wells. Larger volumes can be loaded using the plastic overlay system of Pharmacia. LKB paper wicks are not used because they have been found to retain significant amounts of enzyme. Bromphenol Blue is loaded with some samples as a tracking dye and in the outside lands of the gel to prevent edge effects. Electrophoresis is continued for 150 min or until the Bromphenol Blue tracking dye approaches the acidic wick and begins to turn yellow. The enzymes migrate slightly slower than the dye.

Enzymic Staining of the Gel. Our staining conditions are adapted from those developed by Hayashi.[16] Enzyme activity is visualized using the appropriate naphthyl derivative of the sugar substrate. The substrates we prefer are naphthol-AS-BI-*N*-acetyl-β-D-glucosaminide for *N*-acetylglucosaminidase; 1-naphthyl α-D-mannopyranoside for α-mannosidase; and α-naphthyl phosphate for acid phosphatase. The gel is stained in 100 ml of a freshly prepared solution containing 0.1 *M* sodium acetate, pH 5.0, (plus 0.01 *M* EDTA for acid phosphatase) containing 0.02% of the naphthyl sugar derivative and 0.1% Fast Garnet GBC salt. The substrates are first dissolved in a small volume of dimethyl sulfoxide or methanol before addition to the staining buffer. The action of the enzyme liberates a naphthol product which combines with the Fast Garnet GBC salt and

[16] M. Hayashi, *J. Histochem. Cytochem.* **13,** 355 (1965).

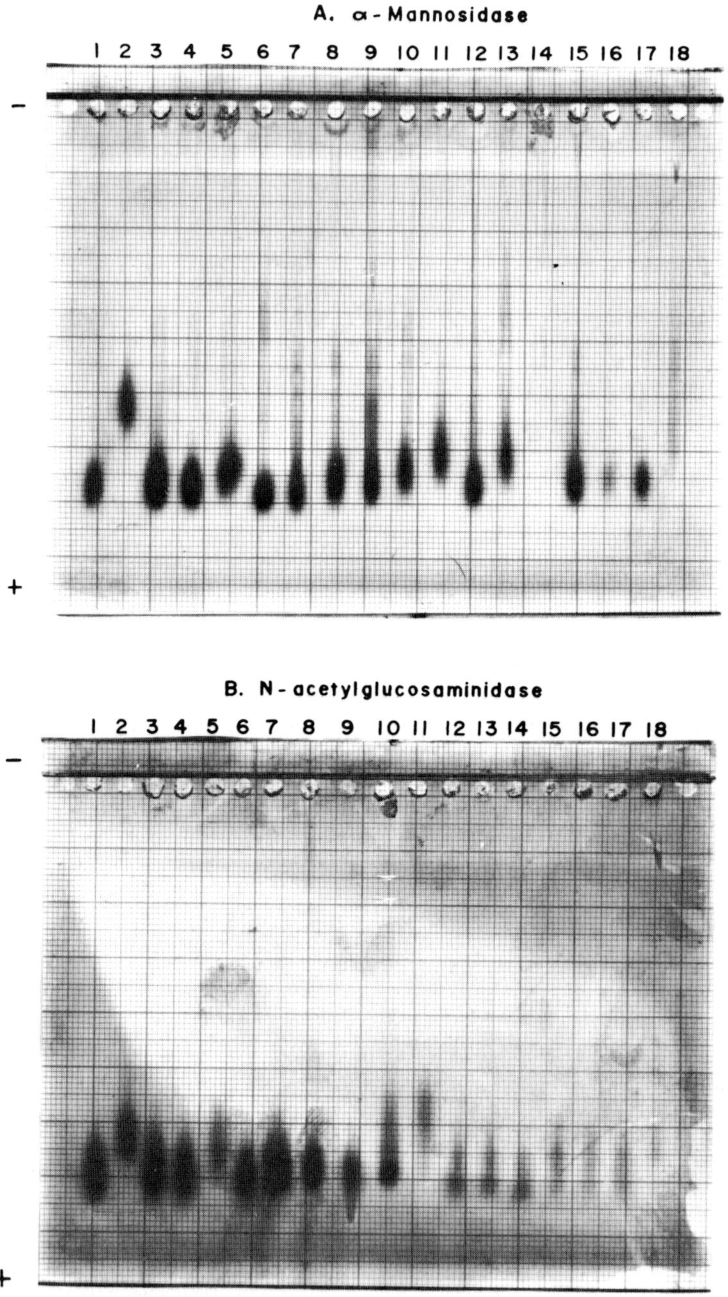

FIG. 4. Nonequilibrium isoelectric focusing of enzymes from wild-type and mutant strains. Lanes 1 and 3: wild type; lane 2: strain M31; other lanes were loaded with cellular extracts of other mutants.

forms an insoluble colored precipitate. The gel is incubated in the staining solution at 30° until sufficiently stained (usually 2–5 hr), and then rinsed in water and air-dried to a thin film.

If more sensitive assays are required for a particular application (as is usually the case for the enzyme β-glucosidase), then fluorescent 4-methylumbelliferyl derivatives of the sugars can be used as substrates. While this procedure has the advantage of higher sensitivity it also has a significant disadvantage since the product of the reaction, 4-methylumbelliferone, is not insoluble and diffuses away from the region of enzyme activity. In this procedure a piece of filter paper saturated with the acetate buffer containing 0.02% of the 4-methylumbelliferyl sugar derivative is placed on the surface of the gel. Enzyme activity is visualized under UV illumination using a hand-held light source (366 nm). Enzyme activity appears as a light-blue spot on a dark-blue background. The stained gel can be photographed using a Tiffen UV1 filter after removing the filter paper and placing the gel on a UV light box fitted with long-wave 366 nm bulbs. The photographic materials should be assembled before staining, since the gels must be photographed quickly before the reaction product diffuses significantly.

This nonequilibrium isoelectric focusing gel system is sufficient to identify at least some of the modificational alterations that affect isoelectric points of lysosomal enzymes (Fig. 4). The only well characterized modificational defect involving lysosomal enzymes in *D. discoideum* is the one caused by the mutation at the *mod A* locus in strain M31.[12,13] This mutation causes a rather dramatic shift in the position of lysosomal enzymes in this gel system (Fig. 4) resulting from the loss of several phosphate and sulfate groups per enzyme molecule. We are not sure how sensitive the gel system is to small changes in the isoelectric point of these enzymes. Although we have been able to detect changes in enzyme modification that cause a much smaller change in p*I* than the M31 mutation, the chemical nature of these alterations has not yet been determined.

Summary

Our initial studies have shown that the cellular slime mold *Dictyostelium discoideum* is a particularly suitable organism for the study of lysosomal enzyme secretion. During appropriate stages in the life cycle, secretion is prominent for a number of lysosomal enzymes. The methods described here have been developed to investigate various aspects of the secretion process. Moreover, our evidence that regulation of the secretory system is influenced by environmental changes and by cell differentiation indicates that this organism may be useful for studying the functional regulation of this organellar system. To a large degree these types of studies have been limited in the past due to the lack of an appropriate experimental system.

The ability to isolate secretory mutants affecting the secretion of lysosomal enzymes adds another dimension to investigations using *D. discoideum*. In our initial attempts we have been successful in isolating a variety of different types of mutants that alter the secretion of one or more lysosomal enzymes. While the results are in agreement with the results of our physiological investigations, they also indicate much more heterogeneity in the lysosomal system than we had previously suspected. The indications that many of our secretory mutants may also affect modification of the enzymes is also intriguing. This observation may also help to explain the fact that many of these strains are defective in normal development. Together with the immunological methods available in this organism for studying posttranslational modification,[17] the mutants may be valuable in deciphering the relationship between modification of lysosomal enzymes and their proper localization and secretion from the cell. Thus, *Dictyostelium discoideum* may become as useful for the study of some questions of cell biology as it has been for development.

[17] D. A. Knecht, R. C. Mierendorf, Jr., and R. L. Dimond, this series, Vol. 98 [14].

Author Index

Numbers in parentheses are footnote reference numbers and indicate that an author's work is referred to although his name is not cited in the text.

A

Aaronson, R. P., 590
Aaronson, R. R., 597
Ab, G., 35
Abbott, R. E., 397
Abdelnour, M., 763
Abeles, R. H., 777, 779(7, 20), 780(7, 20), 781
Abood, M. E., 620, 625(22)
Abrahamson, D. R., xlv
Achord, D. T., 587
Ackerman, E. J., 152, 154(10), 156(10)
Acuto, O., 389, 393(22)
Adamani, A. M., 268(15), 269
Adamietz, P., 25
Adams, D., 585
Adams, L. D., 238
Adams, P. R., 332
Adamson, S. D., 50, 69(1)
Addis, J. A., 656
Addis, J. S., 635, 636, 638, 639(35)
Adelman, M. A., 3, 22
Adelman, M. R., 521, 528, 529
Adesnik, M., xl, xlii(60), 32, 121, 123(2), 124, 125(2, 39), 126(2), 127(39, 47, 51), 128(39), 130(2), 140(47), 141(47), 258, 496, 530, 531(1), 532(1), 533(30), 535(30), 537(5), 538(30, 56), 539(5, 30, 56), 540(1), 542, 628
Adhikari, H. R., 745
Adler, S. P., 685, 688(17)
Admon, A., 584, 587(21)
Agius, L., 760
Agostini, B., 150
Agre, P., 324
Ahlberg, J., 742
Ainsworth, S. K., 490
Akakaway, Y., 765, 766(7)
Akaza, T., 386
Albers, R. W., 641
Albert, A. D., 551, 552(60)
Albert, A. W., 14, 19(17), 20(17), 785
Albert, J., 488
Albro, P. W., 547
Algranati, I. D., 295
Alibert, G., 584, 585(22)
Allen, A. K., 166
Allen, E. H., 780(55), 782
Allen, E. R., 656
Allen, J. C., 635, 647(29)
Allington, W., 311
Allis, C. D., 590
Alperti, G., 512
Alterman, L., 124, 125(39), 127(39), 128(39), 532, 533(30), 535(30), 538(30), 539(30)
Amar-Costesec, A., 531, 532, 533
Amenta, J. S., 752, 753, 763(95)
Ames, B. W., 337
Ames, G., 306
Aminoff, D., 437
Amrhein, N., 584, 585(31b)
Amsterdam, A., 632, 749
Anderson, C. R., 333
Anderson, C. W., 63, 77, 81(17)
Anderson, D. J., xxxviii, xxxix, 50, 85, 112, 348, 369, 371(5), 374, 375, 376(12), 377, 675, 682
Anderson, J. J., 131
Anderson, R. G. W., xxxiv, xlvii, 748, 750(56), 755(56)
Andersson, L. C., 125, 127(49), 282, 283, 290(16), 295(18), 295(16, 18)
Ando, T., 641, 642(45), 644(45)
Andrews, E. P., 269
Andrews, S. B., 655
Anfinsen, C. B., 531
Anker, H. S., 779(16), 780
Anson, M. L., 767
Ansorge, S., 757, 758(132), 759(132)
Anstee, D. J., 281, 286
Aoyagi, T., 756, 757(126), 758(126), 759(126)
Appelkvist, E. L., 551, 552(59), 553, 554
Appelmans, F., 558, 723
Applebury, M. L., 619, 621, 625(24)
Arai, M., 75, 78(11), 79
Arakawa, M., 548(43), 549(43), 550
Argos, P., xli
Arias, I. M., 546, 551, 552(67)

Armes, L. G., 539
Armitage, I. M., 268, 277
Arnold, W. N., 802
Aronson, N. N., 756, 758(125), 759(125)
Arpin, M., 130, 131(80), 140(79), 142(79), 145(79, 80), 149(79, 80)
Arrhenius, E., 555, 556(76)
Arstila, A. U., 762
Ascoli, M., 745, 755(37)
Ash, J. F., 510
Ashe, B., 785
Ashwell, G., 157
Åström, A., 532, 544, 545(3, 11), 546(3), 547(11), 548(11), 549(11)
Atkinson, P. H., 159, 182(16)
Attie, A. D., 755
Audalovic, N., 597
Auffray, C., 35
Aula, P., 461
Austen, B. M., 405
Austin, B. M., 122
Autilio-Gambetti, 238
Aviv, H., 15, 31, 98, 122, 285, 326, 382, 667, 771, 787
Avrameas, S., 482
Awerbuch, T. E., 160
Axelrod, D., xxxiv
Axelrod, J., xxxv
Azarnoff, D. L., 548(42), 549(42), 550, 551(52), 552(52)

B

Babczinski, P., 812
Bablanian, R. H., 436
Baccino, F. M., 763
Bachrach, H. L., 443, 447(4)
Bachur, N. R., 167
Backs-Hüsemann, D., 584, 585(20)
Bader, H., 641
Baehr, W., 619, 621, 625(24)
Baenziger, J. U., 158, 736
Baer, E. A., 488
Baglioni, C., 15, 40, 50, 67
Baguena, R., 746
Bailey, J., 33
Bainton, D. F., xlviii, 764
Bajracharya, D., 714, 715(49)
Baker, R. M., 160, 161(24), 169, 183
Balaspiri, L., 78
Balcarova, J., 481
Baliga, B. S., 575
Balk, L., 532, 544, 545(3), 546(3)
Balkow, K., 69, 70(16)
Ball, W. J., 647
Ballantyne, B., 635
Ballard, F. J., 737, 746, 749(42), 752, 754(85), 755(42), 757, 758(42, 86, 134), 759(86, 134), 763(86)
Ballas, L. M., xxxiii
Ballou, C. E., 812
Balsamo, J., 157
Baltimore, D., 260, 285, 452, 777
Banauch, D., 393
Banik, N., 379
Bantle, J. A., 31
Bar, R. S., 423
Baratti, J., 407, 414
Barbarese, E., 385
Barber, C., 126, 127(65), 435, 496, 541
Barclay, R. K., 764
Barcley, A. N., 268
Bard, E., 535, 538(56), 539(56)
Barenholz, Y., xxxiii
Barka, T., 658
Barletta, A., 555, 556(77)
Barlow, J. J., 30
Bar-Nun, S., 122, 124, 125(39), 127(39), 128(39), 134, 531, 532, 533(30), 535(30), 538(30, 56), 539(30, 56), 540(4)
Barrack, E. R., 589
Barrett, A. J., 753, 756(103), 757(103), 758(103), 759
Barrnett, R. J., li, 632, 633(10), 634, 635, 636, 638, 639(35), 641, 643(26), 644(53), 645(26, 48), 646(48), 647, 648(60), 649, 650, 651, 652, 654, 655, 656, 658
Bartlett, S. G., xliii, 40, 140, 141(115)
Barz, W., 584, 585(20)
Base, K. K., 152
Basha, M., 191
Basinger, S., 609
Bassford, P. J., 814
Basu, S. K., 748, 750(56), 755(56)
Baudhuin, P., 553, 723, 725(4)
Baum, P. R., 63, 77, 81(17)
Bauman, H., 655
Baumbach, G. A., 191
Baumgartner, B., 588
Bause, E., 268
Baxt, B., 443, 447(4)

Baxter, J. D., 369
Bayley, H., 150, 388(18), 389, 391(15), 392(15)
Bayley, S. T., 574
Baylis, P. H., 641, 644(53)
Bayly, H., 280
Beaufay, H., 531, 532, 533, 553, 723, 725(4)
Beaven, G. H., 763
Becher, P., 245
Becht, H., 435
Becker, W. M., 134, 135(98), 701, 702, 703(15), 710, 711(10), 712(41), 715(41)
Beckwith, J., 668, 814
Bednarek, J. M., 551, 552(60)
Beevers, H., 580, 582(7), 583(7), 584(7), 585(7), 700, 701(7), 705, 708(33), 709(33)
Behal, F. J., 406
Behrends, W., 702, 703(19), 712
Bell, F. P., 551, 552(61, 68), 553(68), 557(61)
Bell, J. R., 800
Bell, R. M., xxxiii, 531
Bellamy, D., 658
Benajiba, A., 406, 412, 413(16), 414(16, 17)
Benda, P., 361
Bendayan, M., 148, 469
Bender, W. W., 397
Benes, F. M., 652
Bengtsson, S., 447, 448(19)
Bennet, C. D., 14, 19(17), 20(17)
Bennet, G., 419, 420(38)
Bennett, J., 311
Bennett, N., 619
Bennett, V., 115, 313, 315, 316(5), 317(5), 318, 319, 320, 321(6), 322(3, 6), 323, 324(5, 6)
Bensch, K., 650
Bercovici, T., 392
Beres, V., 805
Berezney, R., 589, 590, 596(8)
Berg, C., 397
Berg, G. G., 641
Berg, P., 496, 780(56), 782
Berg, R. A., 531
Berg, T., 742, 746(22), 748, 750, 751(76), 752(41), 755(41, 76), 756, 758(124), 759(124), 761(76)
Berger, E., 469, 484(20)
Berger, E. G., xxxvi
Bergeron, J. J. M., xlvi, xlvii, 135, 144(100), 532(37), 533, 534(37)
Bergfeld, R., 714, 715(49)
Bergmann, J. E., xlix
Bergner, U., 705
Berk, W., 443
Berman, M., 555(84), 556
Berman, P. W., 332, 352(5)
Bernadac, A., 407, 410, 422(9)
Berrios, M., 590, 591(10), 592(10), 593(10), 595(10)
Berthet, J., 531, 533, 725
Berthold, W., 378
Besharse, J. C., 609
Bevan, E. A., 802
Beyreuther, K., 268, 277
Bhaduri, S., 152
Bieber, L. L., 167, 551, 552(58), 553(58), 639
Bieger, W., 105, 110
Bieglmayer, C., 700, 701(6)
Bieleski, R., 797
Bielinska, M., 93
Biemann, K., 796(26), 801
Bigler-Meier, B., 127, 129(67), 387, 391(9), 400(9, 13), 401(9), 405(9)
Billups, L. C., 448
Binastein, M., 32
Binder, M., 479
Binder, R., 38
Bird, J. W. C., 754, 763
Birken, S., 782
Bishayee, S., 749
Bishop, J. M., 126, 541
Bitensky, L., 641, 644(53)
Bitensky, M. W., 617, 618, 620, 621(5)
Bittner, M., 193, 198(7), 206
Black, C. C., 580, 585
Black, S. D., 150, 536, 537(62)
Blackard, W., 423
Blackburn, P., 45, 95, 98, 514, 690, 726, 787
Blank, M., 641
Blattler, D. P., 216
Blattner, F. R., 126
Blaustein, J., 166
Blobel, B., 348
Blobel, G., xxxi, xxxvii, xxxviii(41, 42), xxxix, xl, xlii(59), xliii(77), liii, 3, 6, 7, 16, 22(2), 38, 41, 43(4), 45(9), 48, 50, 84, 85(2), 86(5, 30, 31), 87(12), 91(15), 93(6), 104, 111, 112, 114, 115, 116(1), 119, 122, 123(16, 17), 124, 127(43), 128(43), 130, 131, 134, 136(30), 140(16), 141(111), 143(30), 146(71), 147(71), 148, 285, 359, 369, 371(5), 374, 375, 376(12), 377, 378,

385, 418, 419, 513, 514, 515(12), 517(11), 518(11), 521, 526, 528, 529(16), 531, 532(6), 570, 579, 590, 591(10), 592(10), 593(10), 595(10), 596(5, 6), 597, 663, 644(1, 2, 3), 665(1), 666, 667(5), 668(1, 5, 15, 21), 669(1, 21, 22), 670(1, 21, 40, 43), 671(1, 21, 26), 672(1, 48), 673, 674(27), 675(22, 48), 676(1), 677, 679, 680, 681(1), 682, 683(3, 7, 8), 684(1), 685(3), 688, 689(3, 8), 690(3), 691(1), 692, 693, 694(1), 696, 716, 773, 776, 777, 778, 784, 785, 787(3)
Blödorn, J., 434, 435(2), 436(2)
Bloemendal, H., 151
Bloemers, H. P. J., 38
Bloj, B., xxxiv
Blomberg, C., xxxix, 405, 669, 671(34)
Blomberg, F., 244(6), 245, 248, 647
Blout, E. R., 78
Blum, A., 590, 596(6), 597
Blumenfeld, O. O., 268(15), 269
Bock, R. M., 338
Bode, B., 445, 450(15)
Bodmer, W. F., 325
Bodner, J. B., 737
Boer, P., 802, 811
Bohley, P., 757, 758(132), 759(132)
Böhni, P. C., 200, 201(16)
Boime, I., 14, 93, 667, 668, 777, 778, 779(7, 8, 9, 29), 780(7–9), 782, 785
Boissonnas, R. A., 782
Bole, D. G., 75
Bolender, R. P., 547
Boller, T., 580, 583(8), 584(8), 588(8)
Bollini, R., 588
Bolton, A., 322, 385
Bolton, E. T., 779(37), 780(37), 781
Bon, S., 358
Bonatti, S., 512, 513, 515(12), 517(11), 518, 676
Bond, M., 126, 681
Bonicel, 414, 418(19)
Bonner, D. M., 77
Bonner, M., 105, 110(15)
Bonner, W. M., 49, 190, 217, 219, 221(9), 289, 327, 515, 694, 735, 798
Bonney, R. J., 747
Bonting, S. L., 635, 638, 649
Bossman, A., 156
Booth, A., 417
Booth, A. G., 392
Borgers, M., 641
Borgese, N., xliv, 135, 144(101), 512, 526, 528(14), 529, 530(14), 532(36), 533, 534(36), 535(44, 45)
Borreback, B., 551, 552(65), 553(65)
Borsook, H., 133
Borut, R., 658
Bosch, F. X., 434, 435, 440
Botelho, L. H., 537
Botstein, D., 671, 814
Boudet, A. M., 584, 585(22)
Boulanger, P., 449
Boulikas, T., 235
Bourguignon, L. Y. W., li
Bourne, H. R., 620, 625(22)
Bourrillon, R., 167
Bove, G., 555, 556(79)
Bowden, L., 705, 709(34)
Bowen, B., 193, 206, 212(3)
Bowers, B., 588
Bowles, A. C., 440(41), 441
Bownds, D., 617
Boyles, J., xlviii
Bracha, M., 796, 799(4)
Brackmann, K., 110, 212
Bradley, A., 216
Braell, W. A., 114, 118(7), 125, 127(48), 140(48), 257, 258(1–3), 261(1), 266(3), 795
Brand, R., 469, 482(25), 483(25), 484(25)
Branfield, J., 584
Branton, D., xlviii, 313, 315, 316(5), 317(5), 324(5), 580, 583(11), 584(11, 17), 585(11, 17), 586, 587(17)
Branton, P. E., 574
Braun, H., 386, 387(7)
Braun, P., 385
Brawerman, G., 15, 24, 25(3)
Bremer, J., 548(48), 549(48), 550, 551, 552(65), 553(65)
Brenchley, G., 636
Brennan, M. D., 93
Brenner, M., 816
Bresnik, E., 532
Bretscher, M. S., xlvii, 150, 397, 670
Bretz, H., xxxv
Bretz, R., xxxv, 638
Briggs, H. R. C., 212
Briles, E. B., 160, 161(27), 181(27)
Brinkley, S. A., 93
Briskin, D. P., 587
Brisson, O., 153
Britten, J. S., 641

Brocher, S. C., 752, 753, 763(95)
Brodbeck, U., 354
Broglie, R., 30
Brooker, J. D., 787
Brophy, E. J., 417
Brosseau, M., 532, 533(31)
Brotherus, J. R., 638
Brown, A., 311
Brown, C., 281
Brown, D. D., 32
Brown, D. T., 445, 450(15)
Brown, J. L., 151
Brown, M. S., xxxiv, xlvii, xlviii, 748, 750(56), 754(56)
Brown, W. R. A., 268
Browne, D. T., 780(51), 782
Brownlee, G. G., xxxvii, 122, 667
Bruder, G., xliv, 533
Bruggeman, A., 35
Brümmer, W., 393
Brunette, D. M., 183
Brunk, U., 551, 552(64), 553(64), 554(73)
Brunner, J., 386, 387(7), 388(19, 20), 389, 391(8, 19), 392(20), 393, 395(28), 396(28), 397(28), 398(28), 399(19, 28), 406(28)
Buell, D. N., 423, 426
Bulychev, A., 742
Burge, B. W., 456
Burger, M. M., 171
Bürgi, R., 393, 395, 396, 397(28), 398, 399(28), 406(28)
Burk, R. R., 27
Burke, B., 469, 482(25), 483(25)
Burke, J. J., 715
Burness, A. T. H., 447
Burnette, W. N., 191, 193, 198(8), 206, 212(7)
Burns, A. T. H., 32
Burns, R. A., 816, 818(2, 4), 819(2), 821(4), 823(4)
Burnstein, Y., 152, 676, 749
Burr, B., 240, 716, 717(1), 719(1)
Burr, F. A., 240, 716, 717(1), 719(1)
Burridge, K., 191
Burstein, U., 542
Bursztajn, S., 350, 749
Buscaglia, M., 600, 602(18)
Busch, H., 683
Buser, C., 585
Busiello, V., 779(33, 46, 47), 780(33, 46, 47), 781, 782
Butcher, H. C., 582, 584(16)
Butow, R. A., xlii, xliii(74), xlv(74), 194, 198(10)
Bye, J., 496
Byers, B., 808, 811(21)
Byus, C., 28

C

Caban, C., 685, 688(17)
Cabib, E., 588
Caen, J. P., 269
Caillibot, V., 160, 161(19, 20), 163(19, 20), 173(20)
Calame, K., 126, 681
Calendar, R., 780(56), 782
Caliguiri, L. A., 439
Cameron, R. G., 548
Campbell, E. A., 53, 66(8)
Campbell, K. P., 570, 574(3)
Campbell, M., 641
Campbell, P. N., 122
Canady, M. R., 635
Cancedda, R., 512, 513, 517(11), 518(11)
Candelas, G., 52
Canfield, R., 782
Cannon, L. E., 295, 682
Canut, H., 584, 585(22)
Capecchi, M. R., 505
Capone, J., 800
Caravaggio, L. L., 635
Card, D. J., 350
Carey, N. H., 126, 127(65), 435, 496, 541
Carignan, P., 148
Carlberg, I., 544
Carlemalm, E., 469, 493
Carlson, M., 814
Carlson, N. J., 751
Carne, T., 94, 96(1), 97(1), 98(1), 100(1), 102(1), 104(1), 139, 140(107), 141(107)
Carpenter, G., 738, 742(16), 755(16), 760
Carpousis, A., 779(49), 780(49), 782
Carr, S. A., 796(26), 801
Carrasco, A., 585
Carrol, J. S., 232, 233(13)
Carroll, R. J., 122
Carson, D. D., 92
Carson, J., 80, 385
Carter, C., 126, 681
Carter, H. A., 779(19), 780
Carter, J. H., 763
Carver, J., 182

Carver, J. P., 159, 163(17), 178
Casals, J., 455
Cascarano, J., 555, 556(78)
Cashmore, A. R., xlii
Casley-Smith, J. R., 641, 642(47)
Cassel, D., 620
Cassiman, J.-J., 742, 748(29)
Cassio, D., 782
Castro, J., 150
Cattieu, K., 796(30), 801
Caulfield, M. P., 80
Cavallini, D., 779(33), 780(33), 781
Cavicchi, P., 576
Cawthon, M. L., 31
Cerasuolo, A., 512
Cerda, J. J., 389
Chabre, M., 619
Chaloupka, J., 779(42), 780(42), 782
Chamberlin, J. P., 218
Chambers, D. A., 82
Chambon, P., 153
Chan, S. J., 122, 538, 668
Chang, C. N., xxxvii, xxxviii(42), 668, 671(26), 681, 716
Changeux, J. P., 339, 367
Chaouat, M., 752
Chapman, A., 183, 730
Charnock, J. S., 641, 642(47)
Chatterjee, N. K., 152
Chayen, J., 641, 644
Chen, C., 167
Chen, M., 215
Chen, T. R., 165
Cheng, H. L., 126
Cheong, L. C., 82
Cherry, R. J., 281
Chi, M. M-Y., 124
Childs, G., 74, 78(3)
Chin, D. J., xxxiv
Chin, S. J., 152, 154(10), 156(10)
Chindemi, P. A., 749
Ching, E., 115
Chirgrin, J., 770
Chirgwin, J. M., 28, 98, 266, 578
Chiu, T. H., 333
Choinski, J. S., Jr, 714
Choppin, P. W., 434, 435(3), 436(3), 439(23), 442, 448, 673
Chowdhry, V., 388, 391(16), 392(16)
Chrambach, A., 812
Chrispeels, M. J., 584(45a, 45b), 585, 588(45a, 45b)
Christen, P., 387, 400(14), 403(14), 404(14)
Christiansen, E., 548(48), 549(48), 550
Christiansen, L., 387, 391(9), 400(9, 11), 401(9), 405(9)
Christiansen, R. Z., 548(48), 549(48), 550
Christner, P., 779(49), 780(49), 782
Chua, B. L., 756, 757(130), 758(130)
Chua, N. H., xlii(79), xliii, liii, 30, 40, 130, 140, 141(115), 244(6), 245, 248, 647, 665, 666, 667(5), 668(5, 14), 716
Chubb, I. W., 354
Chui, D., 324
Chung, A., 152
Chyn, T. L., 11, 24(16), 124, 128(36), 147(36), 570
Chyu, J. Y., 354
Claesson, L., 325, 328(10)
Clark, M. G., 752, 754(85)
Clarke, D. D., 779(28), 780(28), 781
Clarke, D. H., 455
Claudio, T., 371
Clausen, T., 451
Clauser, H., 749
Clegg, C. S., 466
Clegg, J. B., 282
Clegg, J. C. S., 205, 512
Clemens, M. J., 70
Clements, P. R., 336, 423, 424(4), 425(4)
Cleve, H., 286
Cleveland, D. W., 36, 186, 191(8), 223, 225, 727
Cochran, D. L., 597, 606(10)
Cocoran, J. J., 184
Coffey, D. S., 589, 590, 596(8, 9)
Coffey, J. W., 553, 723, 725(4)
Cohen, C., 313
Cohen, C. M., xlviii
Cohen, G. N., 779(35, 37, 39), 780(35, 37, 39), 781
Cohen, J., 214
Cohen, M., 77
Cohen, N., 50
Cohen, P. O., 759
Cohen, P. P., 753
Cohen, R. E., 812
Cohen, S., xlviii, 738, 742(16), 748, 769(60), 750(60), 752(60), 755(16)
Cohen, S. N., 193, 198(4), 215
Cohn, Z. A., 433, 763
Colas, B., 420, 421(42), 422(42)
Coleman, R. A., xxxiii, 531
Coleman, R. M., xxxiii

Collier, R. J., 160
Collin, H. A., 584
Colman, D. R., 125, 127(51), 382, 385(11)
Compans, R. W., 435, 439(14)
Conboy, J. G., xlii(78), xliii, xlv(84)
Conn, E. E., 580, 582(6), 583(6), 585(6)
Conner, G., 84
Conney, A. H., 535, 538(57), 544, 545(2), 546
Connolly, D. T., 424, 425(17)
Converse, C. A., 245, 248(8), 249, 251, 253(5), 256, 609, 610(4), 612(4, 8), 613(4, 8), 615(4, 8, 9), 616(8, 9)
Cook, J. S., 57, 336, 337(23)
Coon, M. J., 150, 536, 537(62), 539, 546
Cooper, C. D., 546
Cooper, J. A., xli, 796(28), 801
Cooper, J. R., 655
Cooper, P. D., 445, 447(12), 450(12), 451(12), 452(12)
Coppock, S. S., 251
Corcoran, D., 135, 144(104), 150(104), 531, 533(20)
Cordy, A., 311
Correia, M. A., 546, 549(17)
Corte, G., xliv
Côté, C., 203(19), 204
Cotmore, S. F., 279(36), 280
Coudrier, E., 482
Cowan, N. J., 36
Coward, J. E., 444
Cowie, D. B., 779(35, 37), 780(35, 37), 781
Cowman, A. P., 215
Cox, R. A., 578
Cox, S. T., 779(24), 780(24), 781
Craig, S., 588
Cramer, F., 778, 780(15)
Cramer, J. A., 277
Crane, R. K., 389
Crawford, L. V., 214
Crean, E. V., 816
Cretin, H., 587
Crine, P., 777
Crippa, M., 600, 602(18)
Crkvenjakov, R., 28
Cronan, J. E., Jr., liii
Crowell, R. L., 443, 444(2), 446(2), 447(2, 3, 8, 16), 448, 449(2, 3, 16, 30, 31), 450(20), 451(32, 40), 452(44)
Crowther, R. A., xlvii
Crumpton, M. J., 325
Cryan, D., 763
Cuatrecasas, P., 286, 428, 745, 746(38), 749(38)
Cullen, S. E., 799
Curthoys, N. P., 392
Curtis, D. R., xli
Cutt, J., 581
Czako-Graham, M., 122, 134, 520, 521(4), 531, 537, 540(4)
Czernilofsky, A. P., 541
Czichi, U., 122, 531, 540(4)
Czombos, J., 78

D

Dabauvalle, M. C., 597, 599(12), 600(12), 603(12), 604(12), 606(12), 607(12), 608(12)
Dabholkar, A. S., 548(51), 549(51), 550
Dahlberg, A. E., 36, 37, 193, 198(6), 206
Dahr, W., 268, 277, 286
Dailey, H. A., 135, 144(104, 105), 150(104, 105)
Dale, R. M. K., 167
Dales, S., lii, 450
Dalling, M. J., 584
Dallner, G., xxix, li(1), 533, 546, 550, 551, 552(59, 63, 64), 553(63, 64), 554(73), 555, 557(63)
Daniels, C. J., 75, 131
Danielsen, E. M., 406, 418
Daniels-McQueen, S., 782
Darby, N. B., 437
D'Arcy, A., 597
D'Arcy Hart, P., 750, 763(75)
Darnell, J. E., 26, 32, 239, 445, 447(13), 451(13)
Das, M., 749
Das, R. C., 87, 93
Date, T., xliv
Daum, G., xliv, xlv(86), 200, 201(16)
D'Auzac, J., 587
Davidson, N., 33
Davies, D. R., 742, 746(21)
Davies, P. J. A., 742, 746(21)
Davis, A. L., 779(25), 780(25), 781
Davis, A. R., 435
Davis, B. D., xl, 78, 671
Davis, D. C., 535
Davis, R., 588
Davis, R. W., 717
Davison, A. N., 378

Davison, S. C., 546
Dean, R. T., 559, 752, 756(88), 757(88), 758(88), 760
de Arriaga, D., 753
de Barsy, T., 737
Debby, P., 147
Debey, P., 555
Debuch, H., 745
Decastel, M., 167
deDuve, C., 532, 553, 558, 638, 722, 723(2), 724(2), 725(4), 737, 764
Deeley, R. G., 32
deFonbrune, P., 505
DeFoo, P., 128, 146(71), 147(71)
deFoor, P., 570
Degani, H., 760
DeGiuli, G., 512
de la Iglesia, F. A., 546
DeLellis, R. A., 558, 628
DeLeo, T., 555, 556(77)
de Leon, S., 570
Delius, H., 126, 466, 496, 512, 541, 668
DeLorenzo, F., 531
DeMarco, C., 779(33), 780(33), 781
DeMartino, G. N., 753, 757, 759(136)
DeMeyts, P., 423, 426
Demoss, J. A., 79
Dennis, P. A., 756, 758(125), 759(125)
DePierre, J. W., 532, 533, 544, 545(3, 10, 11), 546(3), 547(10, 11, 24), 548(11), 549(10, 11)
Depont, J. J. H. H., 649
deRobichon-Szulmajster, H., 779(37), 781
Desautels, M., 753, 760(102)
Desch, F., 532, 533(33), 534(33), 535(33), 536(33), 537(33)
Desel, H., 134, 135(99), 710
DeSena, J., 452
DeShazo, M. L., 780(59), 782
Desnuelle, P., 407, 414, 415(11), 418(19)
Desruisseaux, H., 130, 668
Detraz, M., 347
Deuchler, J., 157
Deumling, B., 607, 608(21)
Devillers-Thiery, A., 85, 93(6), 130, 667, 668
Devlin, M. O., 621, 625(24)
Devos, R., 126, 126(65), 496, 541
Devreotes, P. N., 336, 340, 342, 343, 344, 345(30), 348(31), 349(18), 350(30), 424
Dickson, R. B., 746
DiDonato, S., 745
Dieckert, J. P., 308
Dieckert, J. W., 308
DiGirolamo, M., 779(33), 780(33), 781
DiMeo, S., 555, 556(77)
Dimond, R. L., 816, 818(2, 4), 819(2), 821(4), 823(4), 828
Dingman, C. W., 36
DiReda, N., 555, 556(79)
DiRienzo, J. M., 74, 75(2), 76(2), 77(2), 78(2), 80(2), 671
Dittmer, K., 778
Dixon, J. F., 652
Djiane, J., 749
Dobberstein, B., xxxvii, xxxviii(41), xxxix, xl(45), 6, 16, 84, 85(2), 86(32, 33), 87(17), 91(32), 93, 114, 118(6), 122, 123(16, 17), 124(46), 125, 128(46), 140(16), 143(26), 295, 325, 326(5), 327(5), 328(10), 466, 515, 517, 532, 665, 666, 667(5), 668(5, 21), 669(21, 22), 670(21, 41, 42), 671(21), 675(22), 682, 683, 692, 694(1), 698(3), 699, 777, 784, 787
Dodds, R. A., 641, 644(53)
Dokow, M., 387, 400(14), 403(14), 404(14)
Dolci, E. D., xli
Doll, S., 585
Doman, D. C., 714
Donahue, T. F., 183
Donohue, A. M., 14, 19(17), 20(17)
Dorval, G. K., 460
Douglas, M. G., 133, 194, 198(10)
Douglas, R., 126
Doumeng, C., 406
Dounce, R., xxxv
Dowling, E. L., 805
Downey, J. F., 574
Doyle, D., 546, 655
Drachman, D. B., 332, 352(8)
Dreyer, W. J., 609
Drickamer, L. K., xl, 125, 126(53), 150(53)
Druyan, R., 546
Dryer, R., 157
Drzeniek, Z., 276, 437
Dubyak, G. R., 760
Duguid, J. R., 122
Dulbecco, R., 570, 571
Dumont, J. N., 599
Dunau, M. L., 230
Dunbar, J., 299
Dunn, W. A., 756, 758(125), 759(125)

Dunphy, W. G., 800
Duntze, W., 802
Duodock, B., 6
Dupont, J., 586
Dupuis, D., 269
Dutton, A. H., 468, 470(15), 478(15), 479(33)
Dutton, G. J., 544
Dwyer, N., 590, 596(5)
Dyckes, D. F., 166

E

Eager, P. R., 656
Eagon, R. G., 779(24), 780(24), 781
Earhart, C. F., 80
Earles, B. J., 92
Early, P., 126, 681
Ebeling, W., 393
Ebert, M. H., 230(8, 9), 231, 233(8), 235(8), 238
Edelson, J., 779(36), 780(36), 781
Edelson, P. J., 763
Edidin, E., 422
Edmonds, M., 31
Efron, D., 38
Eggens, I., 551, 552(64)
Eggers, H. J., 436, 445, 450(15)
Egle, P. M., 597, 606(10)
Ehrenfried, G., 231, 234(11)
Ehrenreich, J. H., xlvi
Eibl, H., 112
Eigner, E. A., 779(30), 780(30), 781
Einarson, B., 371
Ekblom, P., 461
Elbers, P. F., 802
Elder, K., 434
Elgin, S. C. R., 590, 591
Elias-Bishko, S., 763
Elicieri, G. L., 164
Ellis, R. A., 628, 629, 635
Ellis, R. J., 667, 668(13)
Ellis, S. A., 629
Elson, E. L., xxxiv
Ely, S., 597
Emmerling, M. R., 361
Emr, S. D., xxxvii, xxxviii(43), 668, 671(30)
Emtage, J. S., 435, 496
Emtage, S., 126, 127(65), 496, 541
Emura, J., 133
Engelberg, A., 27, 98
Engelman, D. M., xl, 405, 417, 418, 419(35), 669, 670(37)
En-Mei, N. I. U., 760, 761(153)
Enomoto, N., 551, 552(62), 557(62)
Enterman, C., 709
Epstein, D., 763
Epstein, P., 683
Erickson, A. H., xxxvii, xxxviii(42), 38, 41(4), 84, 111, 116(1), 681, 682, 690
Ericsson, J. L. E., 546, 753, 762(96)
Ericsson, L., 152, 155(8)
Erikson, A. H., 124
Eriksson, L. C., 546, 547
Erlich, H. A., 193, 198(4), 215
Ermen, B., 607, 608(21)
Ernst, S. A., 633, 634(13), 635, 641(27), 642(27), 645(27), 658
Ernster, L., 536, 544, 545(1, 10), 546(1), 547(10, 24), 549(10), 555(87), 556(76), 557(87)
Esch, F. S., 126, 496
Eshdat, Y., 256, 260, 262(9), 321
Esmon, B., 811
Estabrook, R. W., 544
Estola, T., 454
Etzler, M. E., 166
Everitt, E., 449
Eytan, G., lii

F

Fairbanks, G., 246, 247(15), 252, 271, 300, 301(7), 302(7), 613
Faivre-Bauman, A., 361
Falk, H., 714, 715(49)
Faller, J. W., 655
Fambrough, D. M., xxxi, 332, 333, 336, 339, 340(18), 341(20), 342(18), 343(18), 344, 345(30), 347, 348(31), 349(18, 20), 350(20, 30, 34, 35), 352(4, 6), 354, 355, 357(8), 359(7, 8), 361(7, 8), 362, 363(7), 364, 367, 424
Fambrough, W., 126
Fang, R., 126, 127(65), 496, 541
Fangman, W. L., 779(38), 780(38), 781
Fantus, L. G., 423
Farber, E., 548
Farley, R. A., 150
Farquhar, M. G., xlv, xlvi(93), xlvii, xlix, 764

Farr, A. L., 7, 767
Farrell, P. J., 66, 69, 70(16)
Faulk, W. P., 468
Faust, J. R., xxxiv
Fawcett, D. W., 597
Fayet, G., 407, 422(9)
Feigner, P. L., xxxiii
Feldman, R. A., 7, 13(12), 23(12), 24(12), 127
Felsted, R. L., 167
Feman, E. M., 50
Fenger, T. W., 444
Fenton, W. A., xliii, xlv(84)
Fenwick, M. L., 445, 447(12), 450(12), 451(12), 452(12)
Feracci, H., 406, 407, 408, 412(5), 413, 414(5), 418(19), 422(9)
Ferro-Novick, S., 802, 806, 808(8), 811(8), 813(17)
Ferson, T., 183
Fessl, F., 588
Feuer, G., 546
Fiegenson, M., 651
Field, C., 802, 804, 806, 808(10), 813(17)
Fiers, W., 126, 127(65), 496, 541
Fiete, D., 158
Filip, C., 80
Fink, A., xliv, 533
Fink, G. R., 807
Finkelstein, D., 194, 198(10)
Finne, J., 158, 171
Firestone, R. A., 747
Firth, J. A., 641, 642(49)
Fischer, E., 770
Fischer, H. D., 587, 747, 748(49)
Fischer, S. G., 186, 191(8), 223, 727
Fishbein, J., 313
Fisher, P. A., 590, 591(10), 592(10), 593(10), 595(10)
Fishler, M. C., 709
Fishman, W. H., 558
Fitting, H.-H., 702(27), 703
Fitzmaurice, L., 126
Fleck, A., 7
Fleischer, B., xxxiii, xxxv, xxxvi
Fleischer, G., 551, 552(67)
Fleischer, S., xxxiii, xxxv, 128, 146(71), 147(71), 570
Fletcher, G., 80
Fletcher, G. L., 634, 635(16)
Fleming, P. J., 135, 144(104), 150(104)
Florén, C.-H., 755
Florendo, N. T., 654
Folch-Pi, J., 384, 796, 797
Fong, S. L., xli
Fonnum, F., 355
Fowler, A. V., 669
Fowler, S. D., 533, 553, 723, 725(4), 728
Fox, C. F., 213
Foxman, B. M., 779(20), 780(20), 781
Fraker, P. J., 190
Francis, T., Jr., 435
Frange, R., 632
Frank, G., 387, 391(8)
Franke, J., 823
Franke, U., 183
Franke, W. W., xliv, 533, 589, 597, 599(9, 12), 600(9, 12), 603(12), 604(12), 606(12, 19, 20), 607(12), 608(5, 9, 12, 21), 796(29), 801
Franssen, H. J., 38
Franzisket, U., 715
Fraser, D., 596
Frazier, L. E., 779(40), 780(40), 781
Frederick, S. E., 700
Free, S. J., 816, 822(6), 823, 827
Freedman, R. B., 89
Freeman, K. B., 555, 556(81)
Freeze, H., 823, 824, 827
Freisenstein, C. M., 86, 520, 528, 540, 671
French, J. S., 150
Frenoy, J.-P., 167
Freundlich, M., 780(57), 782
Frevert, J., 702, 703(14, 16, 17), 705, 708(31), 709(31), 710(31), 711(40), 714, 715(37)
Frey, A. B., xli, xliv(67), xlv(67), 124, 125(40), 127(40), 135(40), 144(40), 147(40), 382, 385(11), 532, 533(33), 534(33), 535(33), 536(33), 537(33)
Friedman, S. J., 654
Frielle, T., 392
Fires, E., 800
Firesen, J. D., 781(64), 783
Frisch, A., 731
Frischauf, A. M., 126, 466, 496, 512, 541, 668
Froehner, S. C., 373
Frost, G. T. B., 641, 644(53)
Fry, D. M., 354, 359(9), 361(9), 363(9)
Fruton, J. S., 133
Fuhrmann, G. F., 586
Fujii-Kuriyama, Y., 124, 539, 540

Fujiki, Y., 725, 728
Fujimoto, K., 183
Fujimoto, R., 641, 642(45), 644(45)
Fujimoto, T., 634
Fujita, Y., 777, 779(2), 780(2)
Fujiwake, H., 585
Fukuda, M., 260, 262, 321
Fukumori, N., 548(49), 549(49), 550
Fuller, W. A., 584
Fung, B. K.-K., 618, 619(7), 620(7), 621(7), 622(7), 625
Funk, R., 658
Furthmayr, H., 268, 269, 271(8), 272(2, 8), 273(8), 274(8), 275(8, 9), 276(10), 277(9, 24), 279(1, 8, 36), 280(1, 2, 10), 281, 488

G

Gabbay, M., 746
Gabel, C. A., xxxvi, xli(38)
Gabriel, O., 813
Gadaleta, M. N., 555, 556(79)
Gaetani, S., 7, 13(12), 23(12), 24(12), 127, 135, 144(101), 532(36), 533, 534(36)
Gage, L. P., 32
Gagnon, J., 152, 155(8), 531
Gahmberg, C. G., 125, 127(49), 282, 283, 286, 290(16), 295(18), 297(16, 18), 298
Gaitman, Z., 551, 552(67)
Galardy, R. E., 277
Gall, J. G., 598
Galliland, D. G., 160
Gallili, G., 807
Gallione, C. J., 801
Gallyas, F., 230
Gambetti, P., 238
Gammon, K. L., 561, 563(15)
Ganapathi, R., 184
Gander, J. E., 80
Ganning, A., 550, 551, 552(63), 553(63), 557(63)
Ganschow, R. E, 559
Garan, H., 397
Garavito, M., 469, 493
Garcia-Martinez, J. L., 585
Gardner, J. M., 336, 339, 340(18), 341(20), 342(18), 343(18), 345, 347, 349(18, 20), 350(20), 424
Garner, F., 216
Garnier, J., 669, 670(36)
Garoff, H., xxxviii, xl(45), 114, 118(6), 124(46), 125, 126, 128(46), 295, 325, 326(5), 327(5), 466, 467(1), 480, 496, 497, 511(11), 512, 517, 541, 668
Garten, W., 435, 436, 440, 443
Gasior, E., 50
Gasser, S. M., xliv, xlv(86)
Gavin, J. R., III, 423, 426
Gaye, P., 669, 670(36)
Gear, A. R. L., 551, 552(60)
Geiger, B., 468, 470(15), 478(15)
Geisow, M. J., 748, 750, 763(75)
Gelboin, H. V., 546
Genz, R., 607
Gerace, L., 590, 596(6), 597
Gerald, C., 135, 144(103), 150(103)
Gerard, C., 533, 534(40)
Gerdes, H.-H., 710, 712, 715(37)
Gerhardt, B., 700, 715
Gershon, H., 779(28), 780(28), 781
Gershoni, J. M., 198
Gesteland, R. F., 63, 77, 81(17)
Gething, M.-J., 434, 481, 496, 541, 673
Geuze, H. J., 468, 470, 479, 748
Ghersa, P., 387, 400(14), 403(14), 404(14)
Ghosh, B. K., 389
Ghosh, H. P., 140, 800
Ghrayeb, J., 538
Gianetto, R., 558, 723
Gibson, J., 199(15), 200, 202(15), 204(15)
Gibson, W., 198, 608
Gielen, W., 277
Gierer, A., 442
Gietl, C., 703
Gilbert, C. S., 70, 71(19)
Gilbert, J. M., 378
Gilbert, W., 666, 668(7, 8), 671(7, 8)
Gilead, Z., 116
Gillette, J. R., 535, 546
Gillies, P. J., 551, 552(61), 557(61)
Gilman, A. G., 546, 620
Gilmore, R., 86(30, 31), 93, 532, 683
Ginsberg, M., 269
Girard, M., 24
Girardet, M., 281
Girol, D., 531
Gitler, C., 150, 392
Glass, J., 74, 75(4), 78(4), 79(4), 81(4), 82(4), 83(4)

Glass, W. F., 212
Glaumann, H., 742, 753, 762(96)
Glenney, J. R., xlviii
Glenney, P., xlviii
Glick, M. C., 439
Glisin, V., 28
Glover, J. S., 393
Gluck, R., 551, 552(67)
Gochin, M., 760
Godchaux, W., III, 618
Goding, J. W., 114
Godschaux, W., 50, 69(1)
Goertemiller, C. C., Jr., 628, 635
Goetsch, L., 808, 811(21)
Gohda, E., 758
Goldberg, A. L., 350, 749, 753, 757, 758, 759(136), 760(102), 784
Goldberg, D., xxxvi, xli(38)
Goldberger, R. F., 32, 531
Goldfine, I. D., 423
Goldin, S. M., 542
Goldman, B. M., xxxvii, xxxviii(42), 85, 114, 134, 385, 526, 667, 681, 773
Goldman, D., 189, 230(8–10), 231, 232(10), 233(8, 10), 234, 235(8, 10), 237(17), 238, 700
Goldman, D. W., 150, 280
Goldman, S. S., 558
Goldschmidt, E. E., 584, 587(21)
Goldstein, A., 806, 813
Goldstein, I. J., 158, 167(6)
Goldstein, J. L., xxxiv, xlvii, xlviii, 748, 750(56), 755(56)
Gomatos, P. J., 480
Gonzalez, E., 705, 708(33), 709(33)
Gonzalez, F. J., 124, 125(41), 127(41), 135, 144(102), 532, 533(34, 35), 535(34), 536(34), 537(35), 538
Gonzalez-Noriega, A., 731, 747, 748(48, 49)
Gonzalez-Noriega, H., 587
Goodchild, D. J., 588
Goodman, A. M., 770
Goodman, H. M., 126, 205, 541
Goodman, J. M., xliv
Gorden, P., 423
Gordon, A. H., 750
Gordon, J., 110, 193, 197(3), 198(3), 206, 212(4)
Gordon, J. I., 32
Gordon, P. B., 737, 738, 739(17), 745(17), 746(1, 3, 15), 747(16), 750(35), 752(1, 17), 753(1, 2, 6, 15, 17), 754(3, 6, 15), 756(1), 758(1), 761(111, 111a), 762(1, 111, 113), 763(6)
Gorsky, J. P., 531
Gorvel, J. P., 406, 410, 418(15), 418(15), 420(15), 422(15)
Gosh, H. P., xxxviii, xl(48)
Gosser, L. B., 452
Gossran, R., 406
Gotoh, M., 548(46), 549(46), 550
Gottlieb, A. A., 777, 779(2), 780(2)
Gottlieb, C., 161
Goudie, R. B., 481
Grady, H., 550, 551(52), 552(52)
Graf, J., 700, 701(6)
Graham, F. L., 574
Granschow, R., 546
Granstedt, R. C., 584
Grantham, J. J., 760
Gratzer, W., 299, 317
Gravel, R., 382, 770
Graves, M. C., 673
Gray, W. R., 398
Grebenau, R. C., 520, 521(4), 523(5), 526(5), 537, 540
Green, A. A., 749, 765
Green, G. D. J., 756, 760(129)
Green, H., 164, 447
Green, J., 482, 484(45)
Green, M., 110, 116, 212
Green, N. M., 150
Green, P., 50
Green, R. W., 286
Greenawalt, J. W., 149
Greenberg, J. R., 25
Greene, L. A., 357, 363(14)
Greengard, P., 654
Greenway, D. C., 124, 570
Greenwood, F. C., 393, 526
Gregor, I., xlii, 131, 199(15), 200, 202(15), 204(15)
Grey, H. M., 126
Grienenberger, J.-M., 710, 712(41), 715(41)
Griffin, C., 548(46), 549(46), 550
Griffith, J. M., 468
Griffith, O. H., 638
Griffiths, G. W., 468, 469, 482, 483(25), 484(25, 45), 491, 492(15), 493(15)
Grinde, B., 737, 742, 746(1, 25), 752(1, 25), 753(1, 4, 6, 25), 754(6, 25, 104), 756(1, 104), 757(104), 758(1, 25, 104, 135),

759(104, 109), 760(110), 762(1, 114), 763(6, 109)
Grinstein, S., 386, 394
Grisolia, S., 746, 753
Grob, K., 584, 585(18)
Grollman, A. P., 15
Groman, N. B., 751
Groot, G. S. P., 197, 198(12), 200(12)
Grossman, A., xliii
Grossman, A. R., xlii, 40, 140, 144(115)
Grossman, M. K., 807, 812(19)
Groves, D., 702, 703(21)
Grubb, J. H., 731, 747, 748(48)
Gruber, M., 35
Gruber, P. J., 700
Grund, C., 604, 606(20)
Guengerich, F. P., 150, 538
Guenthner, T. M., xli, xliv(67), 532, 533(33), 534(33), 535(33), 536(33), 537(33)
Guidotti, G., 125
Gum, J. R., 150
Gumbiner, B., xlvi
Gunn, J. M., 737, 763
Gunsalus, R. P., 131
Gunthner, T. M., 124, 125(40), 127(40), 135(40), 144(40), 147(40)
Gupta, B. N., 547
Gupta, N. K., 152
Gupta, R. S., 164
Gustafson, R., 555(87), 556, 557(87)
Guth, L., 641
Guthenberg, C., 548
Guthrie, J. P., 77
Guttman, N., 452
Guy, M., 585
Guyer, R., 667
Guzelian, P., 423

H

Haas, S. M., 167, 639
Habener, J. F., 122, 152, 154(7), 156(7)
Hackett, A. J., 445, 450(14)
Hackney, J. F., 652
Hadar, R., 752
Haddad, A., 419, 420(38)
Hagen, G., 717
Hagins, W. A., 618
Hahn, L. H. E., xxx
Hahn, W. E., 30, 31
Haigh, W. B., Jr., 299, 305
Haigler, H. T., xlviii, 748, 749(60), 750(60), 752(60)
Haimi, L., 785
Hakimi, J., 182
Hakomori, S.-i., 283
Halegoua, S., 74, 75(1), 80(1), 81(1), 83, 405, 537, 668
Hall, C. W., 378, 742, 748(28)
Hall, G. E., 184, 185(5)
Hall, M. N., xxxvii, xxxviii(43), 668, 669, 671(30)
Hall, Z. W., 354
Hallermayer, G., 130
Hallewell, R. A., 435, 496
Halliday, K. R., 621
Hally, A. D., 658
Halvorsen, O., 551, 552(66)
Halvorson, H. O., 338, 682, 807
Hamada, A., 268
Hamaguchi, H., 286
Hamilton, T. A., 215
Hamum, C., 130, 140(79), 141(79), 145(79), 149(79)
Hanada, K., 756, 759
Hanawell, A., 658
Hancock, R., 235
Haneishi, T., 79
Hansche, P. E., 805
Hanson, H., 757, 758(132), 759(132)
Harder, A., 745
Harder, M. E., 77
Harding, W. M., 780(59), 782
Hardy, S. J. S., 81, 83, 671
Hardy-Staskin, J., 758
Hargrave, P. A., xli
Harkins, J. L., 763
Harmey, M. A., 130
Harms, C. T., 580
Harnik, V. M., 532, 533(32), 540(32), 541(32)
Harpaz, N., 158
Harris, E. D., 126
Harrison, T. M., xxxvii, 122, 667
Harsch, M., 779(49), 780(49), 782
Harter, D. H., 448
Harter, M. L., 116
Hartley, R. W., 779(21), 780(21), 781
Hartman, P. E., 779(22), 780(22), 781
Hartzell, H. C., 333, 343, 345, 350(34)
Hasan, H. R., 752

Hashimoto, K., 456, 458, 459(8, 10), 460(11), 461(11), 463(11), 465(10, 11)
Hashimoto, T., 551, 552(56, 57, 69), 553(56, 69), 702, 703(24)
Hasilik, A., 587, 588, 589(69), 730, 731(1), 736(4), 747, 748(50)
Hasselbach, W., 150
Haugen, D. A., 539, 546
Hauri, H-P., 127, 129(67), 387, 400(12, 13, 14), 403(14), 404(14), 418, 419(30), 422
Hausen, P., 597, 604, 606(19)
Hauser, H., 386, 387(7), 391(8)
Hauser, P., 533
Havranek, M., 779(42), 780(42), 782
Hawkins, H. C., 89
Hawkins, N. M., 635
Hay, H. J., 435
Hayashi, H., 548(45), 549(45), 550
Hayashi, M., 824
Hayes, C. E., 158, 167(6)
Häyry, P., 281
Hazato, T., 760
Heath, E. C., 87, 93
Hebert, R. R., 584
Hecht, L. I., 779(30), 780(30), 781
Heck, U., 584, 585(19, 29)
Heckman, C. A., 654
Hedman, K., 465
Hedman, R., 555, 556(76)
Hegemon, G. D., 780(51), 782
Heid, H. W., 796(29), 801
Heidmann, T., 367
Heinemann, F. S., 540
Heinrich, P. G., 131
Heinrikson, R. L., 122
Heitzmann, H., 488
Helenius, A., 407, 414(12), 459, 480, 481, 510, 751
Helenius, A. H., xlv
Hemmings, B. A., 588, 589(69)
Henderson, D., 112
Henderson, L. E., 313, 796(27), 801
Henderson, R., xli
Hendler, R. W., 122
Hennache, B., 449
Hennig, B., 140, 141(113)
Henning, D., 683
Heppel, L. A., 81
Herbert, E., 50, 69(1)
Herbert, P., 53, 66(8)
Herman, R. M., 588
Hermolin, J., 157
Hernaez-Davis, L., 745, 746(38), 749(38)
Herrera, F., 50
Herschkowitz, R. N., 745
Hershey, J. W. B., 193, 198(8a)
Hershko, A., 753, 763
Herzog, V., 521, 755
Hess, R., 536, 546, 548(40), 549
Heuyerjans, J. H., 38
Hewgwill, D., 325
Hickey, E. D., 15, 40, 67
Hickman, S., 764
Hiem Yap, S., 5
Higgins, J. A., 649, 650, 651, 652
Higgins, L. L., 597, 606(10)
Higgins, T. J., 588
Highfield, P. E., 667, 668(13)
Hilden, S., 649
Hilz, H., 25, 31
Himeno, M., 561, 562(16), 563, 765, 766(7), 767, 768(12), 769(12), 771(12)
Hind, G., 584, 585(30), 587(30)
Hino, S., 548(45), 549(45), 550
Hinterberger, M., 689
Hirashima, A., 74, 75, 78(3), 81(5, 6, 7), 82(5), 83(6), 84(6)
Hirata, F., xxxv
Hiremath, C. B., 779(26), 780(26), 781
Hirschberg, C. B., 160
Hirst, G. K., 441
Hiti, A. C., 435
Hiwada, K., 407, 419
Hizuka, N., 423
Hlivko, T. J., 752
Hnilica, L. S., 212
Hock, B., 702(28), 703(18), 712(28)
Hodson, S., 636
Hoffer, J. G., 348
Hoffman, J. F., 643
Hoffman, R., 609
Hofmann, T., xxxviii, xl(48)
Hogg, N., 214
Hokin, L. E., 638, 649, 652, 655, 658(86)
Hokin, M. R., 634, 655, 658(86)
Holland, J. J., 445, 448, 452
Holland, P. C., 570, 574(1)
Holländer-Czytko, H., 584, 585(31a)
Hollenberg, M. D., 447
Hollyfield, J. G., 609

Holmes, D. S., 499
Holmes, W. M., 634, 635(16)
Holmes, W. N., 656
Holte, K., 742, 746(22)
Holtzman, J. L., 546
Holzer, H., 588
Homandberg, A., 133
Homma, M., 434
Hong, Y.-N., 715
Honjo, T., 126
Honma, K., 268
Hood, L. E., 126, 348, 371, 590, 591, 681
Höök, M., xxxi
Hoopes, R. R., 531
Hootman, S. R., 633, 634(13), 641, 642(50), 658
Hooymans, J. J. M., 584, 585(31)
Hopgood, M. F., 752, 754(85)
Hopkins, B. E., 638, 647
Hoppeler, H., 484
Horecker, B. L., 754
Hori, C., 760
Horie, S., 548(49), 549(49), 550
Hornsberger, M., 468
Horowitz, M. I., 158
Horowitz, S., 670
Horst, M. N., 191
Hortin, G., 668, 777, 778, 779(7, 8, 9, 29), 780(7–9, 29), 781
Hoshina, H., 777
Hossler, F. E., 633, 635, 641, 643(26), 645(26, 48), 646(48), 656, 658(94)
Hostetter, K. Y., 745
Houdebine, L.-M., 749
House, L. H. W., 481
Housman, D., 152
Hovsépian, S., 407, 422(9)
Howard, B. H., 496
Howard, I. K., 166
Howe, C., 444
Howe, J. G., 193, 198(8a)
Howell, D. C., 779(25), 780(25), 781
Howell, K. E., xliv
Howell, R. R., 22
Hoyer, B. H., 452
Hrazdina, G., 580, 585(10)
Hruban, Z., 548(46), 549(46), 550, 779(40), 780(40), 781
Hsu, C. J., 256, 300, 305, 306
Hu, A. S. L., 338
Huang, A. H. C., 700, 701(7)
Huang, M. T., 15
Huang, R. T. C., 437
Hubbard, A. L., 532, 728, 748
Hubbard, L. M. L., 392
Hubbard, S. C., 531, 532(7), 541(7)
Hudecki, M. S., 758
Huffaker, R. C., 584
Hughes, M. A., xxxiii
Hughes, W. L., 765
Hung, P. P., 215
Hunkapiller, M. W., 348, 371, 800
Hunt, L., 158
Hunt, T., 50, 52, 53, 66(8), 69, 70(16), 113, 152, 693, 711
Hunter, A. R., 152
Hunter, E., 454
Hunter, F., 311
Hunter, W., 322, 385
Hunter, W. M., 393, 526
Hurley, J. B., 618, 619(7), 620(7), 621(7), 622(7), 625(22, 25), 626(25)
Hussain, M., 78
Hutson, N. J., 754
Hüttermann, A., 337, 701
Huylebroeck, D., 126, 127(65), 496, 541
Hyman, R., 160, 182(29), 183

I

Ibrahimi, I., 41, 45(9), 86, 123, 669, 670(43), 682, 683(7), 778
Ichihara, A., 754, 756(107), 757(107)
Ichihara, S., 78
Igloi, G. L., 778, 780(15)
Ikegaki, N., 754, 756(107), 757(107)
Ikehara, Y., 755
Ilyas, M. S., 546
Imai, Y., 124, 548(43), 549(43), 550
Inouye, H., 668
Inouye, K., 133
Inouye, M., 50, 74, 75(1, 2, 4), 76(2), 77(2), 78(2, 3, 4), 79(4), 80(1, 2), 81(4, 5, 6, 7), 82(4, 5), 83(4, 6), 84(6), 405, 537, 668, 671, 795
Inouye, S., 75, 83, 668
Inukai, M., 75, 78(11), 79
Irimura, T., 257, 258(4)

Irving, R. A., xxxviii, xl(48)
Ishidate, K., 546
Ishii, H., 548(49), 549(49), 550
Ishii, S., 581
Isowa, Y., 133
Isselbacher, K. J., 387, 400(12), 418, 419(30), 422(30)
Ito, A., xliv
Ito, K., xliii, 75, 814
Ito, T., 407, 417
Ivatt, R. J., 531, 532(7), 541(7)
Ivy, J., 183
Iwai, K., 585
Izukawa, M., 548(43), 549(43), 550

J

Jackson, R., 785, 786, 787(3, 8), 792(8)
Jackson, R. C., 85, 86(5), 87, 91(15), 123, 136(30), 143(30), 531, 532(6), 669, 670(40), 683
Jackson, R. J., 6, 15(9), 16, 44, 50, 51, 52(4), 53, 66(8), 69(8), 70(16), 90, 95, 100, 113, 152, 285, 514, 578, 693, 711
Jackson, R. L., 150
Jackson, R. W., 79
Jacob, J.-L., 586
Jacobs-Lorena, M., 152
Jacobson, M. F., 777
Jacoby, B., 584, 587(21)
Jacoby, R., 94, 96(1), 97(1), 98(1), 100(1), 102(1), 104(1), 106, 109(16), 110, 139, 140(107), 141(107)
Jahnsen, R., 757, 758(135)
Jakoby, W. B., 544
Jakubowicz, T., 50
James, M., 134, 135(99)
Jamieson, G. A., 268
Jamieson, J. C., 182
Jamieson, J. D., 468, 469(8), 485, 632
Janacek, J., 779(42, 43), 780(42, 43), 782
Janero, D. R., li, 531, 654
Janes, M., 710
Jarasch, E.-D., xliv, 533, 589, 796(29), 801
Jarasch, R. D., 607, 608(21)
Jaussi, R., 387, 400(14), 403(14), 404(14)
Jelsema, C. C., 654
Jen, J. J., 587
Jeng, Y., 116
Jenkins, A. B., 752
Jenny, B., 354
Jockush, B. M., 469
John, K., 305
Johnson, C. A., 126
Johnson, C. D., 353, 355, 356, 357(2)
Johnson, R. G., 751
Johnston, F. B., 39
Johnston, I. R., 596
Jokinen, J., 125, 127(49)
Jokinen, M., 282, 290(16), 295(18), 297(16, 18)
Joklik, W. K., 50, 445, 447(13), 451(13)
Jolles, J., 797
Jollès, P., 797
Joly, J.-G., 548
Jones, E. W., 588, 589(69)
Jones, P. P., 325
Jordan, F. I., 232, 233(13)
Jordan, K. B., 816, 818(2), 819(2)
Jordan, W. S., 448
Jorgensen, A. O., 570, 574(3)
Jørgensen, P. L., 128, 150, 635, 647(30), 652(30)
Josefson, L. G., 671
Jost, E., 597
Jost, P. C., 638
Joste, V., 555, 556(80)
Joyard, J., xxxv
Juszczak, E., xli

K

Kääriäinen, L., 125, 127(49), 282, 285, 295(18), 297(18), 453, 456, 458, 459(8, 10), 460(11), 461(11), 463(11), 465(10, 11), 480
Kabat, D., 31
Kablotsky, Y. H., 628
Kadenbach, B., 555, 556(82), 557(82)
Kader, J. C., 648
Kaesberg, P., 721
Kagawa, T., 708
Kagawa, Y., 675, 788
Kahane, I., 150
Kahlenberg, A., 147
Kahn, C. R., 423
Kahonen, M. T., 548(41), 549(41), 550
Kaiser, G., 581, 585(14a)

Kalnins, V. I., 570
Kanaya, T., 133
Kantor, F., 542
Kaplan, A., 587
Kaplan, G., 731
Kaplan, J., 748, 749(55)
Kara, U. K. A., 705
Karcher, D., 230
Karhi, K. K., 125, 127(49), 282
Karin, M., 748, 750(52)
Karlin, A., 332, 339
Karlish, S. J. D., 150, 760
Karlsson, K. A., 648(68–70), 649
Karnovsky, M. J., 490
Karnovsky, M. L., 531
Kartenbeck, J., xliv, 533, 597, 607, 608(5), 751
Kás, J., 738, 752(14)
Kasamatsu, H., 193, 198(8b)
Kaschnitz, R., 785
Kasper, C. B., 124, 125(41), 127(41), 135, 144(102), 531, 532, 533(34, 35), 535(34), 536(34), 537(35), 538
Kato, K., xxxviii, 124, 561, 563(16), 765, 766(7), 767, 768(12), 769(12), 771(12), 773(6), 774(6), 775(6)
Katsuda, N., 767, 768(12), 769(12), 771(12)
Katz, A. N., 140, 141(111)
Katz, B., 333
Katz, F., 124(45), 125, 128(45), 512
Katz, F. N., xxxviii, 84, 124, 127(42), 128(42, 43), 378, 418, 675
Katz, R. N., 289
Kaufman, J., 666, 668(8), 671(8)
Kaufman, S. H., 423, 424(3), 589, 590, 596(9)
Kauss, H., 588
Kawajiri, K., 539
Kay, R. R., 596
Keaton, K. S., 754
Keenan, T. W., 607, 796(29), 801
Kehry, M., 126
Keibich, G., 628
Keil, W., 440
Keim, P. S., 122
Keller, R. K., 560
Kelley, D. E., 25
Kelly, A., 546
Kelly, R. B., xlvi
Kembhavi, A. A., 759
Kemp, D. J., 215
Kemper, B., 122, 152, 154(7), 156(7)
Kempf, T., 225
Kende, H., 580, 583(8), 584(8), 588(8)
Kennedy, E. P., xxxiv
Kenny, A. J., 129, 417
Kenny, J. A., 392, 406, 407(4)
Kenyon, G. L., 780(51), 782
Kenyon, W. H., 580, 585
Keogh, E. A., 748, 749(55)
Keränen, S., 285, 456, 458, 459(8, 10), 465(10)
Kerenyi, L., 230
Kerjaschki, D., 741
Kerr, I. M., 50, 70, 71(19)
Kerr, M. W., 702, 703(21)
Kessin, R., 823
Kessler, M., 386, 389, 393
Kim, K., 751
Kim, Y. S., 417
Kindl, H., 700, 701(8), 702(8), 703(8, 12, 14, 16, 17, 22, 23, 26), 705(8), 708(23, 26, 31, 32), 709(23, 31, 32), 710(23, 31), 711(22, 23, 40), 712, 714(22), 715(23, 37)
Kindt, T., 667
King, A. C., 745, 746(38), 749(38)
Kinne, R., 635
Kirby, K. S., 24, 25(2), 30
Kirk, K. L., 779(21), 780(21), 781
Kirkpatrick, F. H., 245, 256
Kirsch, D. R., 230
Kirschke, H., 757, 758(132), 759(132), 760
Kirschner, M. W., 36, 186, 191(8), 223, 225, 727
Kirsh, K., 419, 420, 422(39)
Kishida, Y., 531
Kiss, A., 762
Kissonerghis, A.-M., 265, 325
Kitson, R. P., 755
Kjellen, L., xxxi
Klein, D. C., 779(21), 780(20, 21), 781
Klein, J., 325, 326
Klein, U., 736
Kleiner, D., 740
Kleinig, H., 607, 608(21)
Kleinschmidt, J. A., 604, 606(19, 20)
Kleinzeller, A., 760
Klenk, H.-D., 434, 435(2), 436(2, 6), 437, 438, 439(12, 24), 440, 443
Klip, A., 386, 394
Knecht, D. A., 828

Knipe, D. M., 124, 127(42), 128(42), 260
Knirprath, E., 629
Knowland, J. S., 217
Knowles, J. R., 388, 391(15), 392(15)
Knowles, S. E., 737, 746, 749(42), 752, 755(42), 757, 758(42, 86, 134), 759(86, 134), 763(86)
Knowles, W. J., 300, 305, 306(3, 4), 309(4)
Knutson, V. P., 423, 424(5), 425(5, 12), 432(18)
Kobata, A., 158, 268, 277, 281
Koelle, G. B., 353, 359, 360
Koennecke, I., 440(42), 441
Koepsell, H., 647
Koerner, T. A. W., 268
Kohama, T., 436
Kohler, H., 260, 261(8), 321
Kohler, R. E., 78
Kohli, K. K., 547
Kohno, K., 289
Kokubu, T., 407, 417
Kolansky, D., xliii, xlv(84)
Köller, W., 700, 701(8), 702(8), 703(8, 26), 705(8), 708(26, 31, 32), 709(31, 32), 710(31), 714, 715
Komnick, H., 629
Komor, E., 584, 585
Kondor-Koch, C., 496, 497, 511(11)
Koningsberg, W., 313
Kono, T., 746, 752(43), 755(43)
Kopp, B., 585
Kopp, F., 586
Koppel, D. E., xxxiv
Korant, B. D., 447, 451(21)
Korb, H., 130
Kordowicz, M., 268
Korner, A., 50
Kornfeld, R., xxxv, 158
Kornfeld, S., xxxv, xxxvi, xli(38), 158, 159(14), 161(14), 183(14), 289, 730, 736
Korzeniowski, D., 538
Koshland, D., 671
Kosmakos, F. C., 749
Kovács, A. L., 737, 742, 744, 746(1), 750, 752(1, 73), 753(1, 4), 754(4), 756(1, 73), 758(1, 73), 759(73), 762(1, 26, 73, 112), 763(112)
Kovács, J., 762
Kovacs, K., 78
Kovatchev, S., 745
Koziarz, J. J., 260, 261(8), 321
Kraehenbuhl, J. P., xxxi, 244, 468, 469(8), 485, 488(1), 491, 492, 493(15), 495(2), 615
Krah, D. L., 448, 449(31)
Kramer, R., 586
Krangel, M. S., 295, 325
Krauhs, E., 225
Kraus, J. P., 23
Krause, E., xxxix, 86(32), 91(32), 93, 123, 532, 683
Kreibich, G., xxxviii, xxxix, xl, xli, xlii(60), xliv(67), xlv(67), xlix(35), 86, 121, 122, 123(2), 124, 125(2, 39, 40), 126(2), 127(39, 40, 51), 128(39), 130(2), 134, 135(40), 144(40), 147(31, 40), 258, 382, 384, 385(11), 461, 496, 520, 521(1, 4), 523(1, 5), 526(5), 528(14), 529(14), 530(14), 531(1), 532(1), 533(30, 31, 32, 33), 534(33), 535(30, 33), 536(33), 537(5, 33), 538(30), 539(5, 30), 540(1, 4, 25, 32, 38), 541(32), 555, 671, 765, 773(6), 774(6), 775(6)
Kreike, J., 197, 198(12), 200(12)
Kreil, G., 785
Kringstad, R., 580, 585
Kirshan, A., 184
Krishnakantha, T. P., 548(44), 549(44), 550
Krohne, G., 589, 597, 599, 600, 603(12), 604(12, 13), 606(19, 20), 607(12), 608
Kruczinna, R., 440
Krüger, J., 277
Kruijt, J. K., 755
Krupp, M. N., 423, 424(8), 425(8, 17), 427(8), 428(8), 429(8), 430(8), 431, 432
Kruppa, J., 3, 528, 529
Kruse, C., 710, 711(40)
Kruse, P. F., 779(19), 780
Krusius, T., 158
Krutzsch, H. C., 796(27), 801
Krystosek, A., 31
Kubelka, W., 585
Kubo, R. T., 126
Kucharzewska, T., 50
Kuchler, R. J., 763
Kuhn, H., 619, 621(12)
Kuhn, L. C., xxxi
Kumar, A., 539
Kunicki, T., 269
Kunin, C. M., 448, 449
Kunkel, H. G., 325
Kupferer, P., 193, 198(7), 206

Kuriyama, R., 666
Kuriyama, Y., 536, 546
Kuter, D. J., 256
Kuwahara, S., 533
Kuzuhara, S., 551, 552(62), 557(62)
Kvist, S., 325, 328(10)
Kyte, J., 127

L

Labbe, R. F., 58
Labrecque, D., 122
LaCelle, P., 256
Lacorbiere, M., 160
Laemmli, U. K., 77, 81(16), 186, 191(8), 193, 194, 206, 208, 212(3), 214, 223, 239, 248, 262, 307, 327, 402, 404, 574, 727, 734, 791, 797
Laga, E. M., 575
Lagone, J. J., 212
Lalley, P. A., 183
Lalwani, N. D., 548(51), 549(51), 550, 551(39), 552(39)
Lamb, J. E., 702, 703(15)
Lamb, R. A., 436, 439(23)
Lampén, J. O., 80, 802, 806, 807, 813
Lamprecht, S. A., 749
Lancaster, J. E., 584
Lance, C., 586
Land, B. R., 333
Landau, B. J., 444, 447(8), 450(20)
Landefeld, T., 782
Landfried, K., 286
Landicks, R., 131
Landwall, P., 286
Lane, D. P., 214
Lane, M. D., 336, 423, 424(3, 4, 5, 8), 425(4, 5, 8, 11, 12, 17), 427(8), 428(8, 11), 429(8), 430(8), 431, 432(18), 531
Lang, H., 393
Lange, P., 805
Langner, J., 757, 758(132), 759(132)
Lansford, E. M., 779(32), 780(32), 781
Larimore, F. S., 760
Laskey, R. A., 105, 110(15), 217, 219, 220, 289, 327, 515, 725, 735, 773, 798
Laskowski, M., Jr., 133
Lassam, N. J., 574
La Torre, J., 25
Laurell, C.-B., 245
Laurila, P., 459, 461
LaVail, M. M., 609
Laver, W. G., 481
Lazarow, P. B., 549, 638, 722, 723(2), 724(2), 725, 726(8, 11, 12), 727(8), 728
Lazarowitz, S. G., 434, 435(3), 436(3)
Lazdunski, C., 415
Leaver, C. J., 134, 135(98), 701, 702, 703(15), 710, 711(10), 712(41), 715(41)
Leavitt, R. D., 167
Leblond, C. P., 419, 420(38)
Ledbetter, M. L. S., 763
Leder, P., 15, 31, 98, 122, 285, 326, 382, 667, 771, 787
Lee, C. Y., 333
Lee, J. S., 325
Lee, N. M., 779(32), 780(32), 781
Lee, S. G., 215
Lee, Se Y., 15
Lees, M. B., 384, 796
Legocki, R. P., 197, 198(11), 212, 213(12)
Lehle, L., 812
Lehrach, H., 126, 466, 496, 512, 541, 668
Lehtovaara, P., 285
Leigh, R. A., 580, 583(11), 584(11, 17), 585(11, 17), 586, 587(17)
Leighton, F., 553, 723, 725(4)
Leikola, J., 286
Lemay, A., 256
Lemieux, E., 777
Lemoine, F., 782
Lennarz, W. J., xxxv, 92
Lenney, J., 588
Leon, M. A., 166
Leonard, K., 481
Leonard, R. T., 587
Leonardi, C. L., 49
Lernau, O. Z., 752
Leskes, A., xxix, lii(2)
Lesniak, M. A., 749
Lessard, L., 647
Leterrier, F., 358
Leuthard, P., 217
Levin, A. P., 779(22), 780(22), 781
Levin, W., 537, 538, 539
Levine, A. M., 635, 651
Levine, J., xlviii
Levine, M., 779(34), 781
Levine, W., 532, 533(31)
Levine, W. G., 551, 552(67)
Levinson, A. D., 126, 227

Levinson, J. R., 193, 198(4), 215
Levinson, A. D., 541
Levinthal, C., 613
Levitzki, A., 742, 746(21)
Levvy, G. A., 765
Lewallen, C. G., 555(84), 556
Lewin, A. S., xlii, 131
Lewis, W. H., 160, 161(26)
Leyold, K., 393
Li, E., 161, 289
Li, J. B., 756, 757(130), 758(130)
Li, Q.-S., 754
Li, W. Y., 683
Libby, P., 350, 749, 757, 758
Lichtenberg, D., xxxiii
Liebman, P. A., 617, 619
Lillien, J., 157
Lim, L., 378
Lin, W., 193, 198(86), 584, 585(30), 587(30)
Lind, C., 544
Lindberg, O., 555(87), 556(76), 557(87)
Linden, D. C., 336, 435
Lindenlaub, E., 544
Lindenmayer, G. E., 635, 647(29)
Lindner, H. R., 749
Lindstrom, J. M., 371
Ling, N. C., 126, 496
Ling. V., 160, 161(24)
Lingappa, J. R., 122, 668, 674(27)
Lingappa, V. R., xxxvii, xxxviii(42), 84, 85, 86, 91(15), 93(6), 122, 123, 125, 127(43), 128(43), 136(30), 140, 141(111), 143(30), 378, 418, 668, 669, 670(40), 674(27), 675, 681
Linkhart, T. A., 354, 359(9), 361(9), 363(9)
Linnemans, W. A. M., 802, 811
Linzell, J. L., 628, 629(5)
Lippman, A., 535, 538(56), 539(56)
Lisowska, E., 276
Litman, D., 256, 306
Little, H. G., 406
Little, M., 225
Liu, C. P., 382, 770
Livesey, G., 746, 749(42), 755(42), 757, 758(42, 134), 759(134)
Livi, G. P., 816, 818(4), 821(4), 823(4)
Livingston, J. M., 423, 428
Lizardi, P. M., 27, 30, 38, 44, 52, 98, 369, 693, 712
Lockwood, D. H., 423
Lodemann, E., 780(60), 782
Lodish, H. F., xxxviii, xl, xlix, 50, 84, 114, 118(7), 124(44, 45), 125, 127(42, 43, 44, 48), 128(42, 44, 45), 140(48), 141(111), 152, 257, 258(1–3), 260, 261, 265, 266(3), 285, 289, 325, 378, 418, 512, 675, 748, 795
Loeb, J. N., 22
Loffelhardt, W., 585
Löffler, H. G., 702, 703(19, 22), 711(22), 714(22)
Loftfield, R. B., 779(30), 780(30), 781, 782
Lohmeyer, J., 440
Lomedico, P. T., 43
Loomis, W. F., 815, 822(6), 823, 827
Lonberg-Holm, K., 443, 444, 445, 446, 447(3, 6), 449(3), 451(21), 452
Lord, J. M., 705, 709(34), 710, 711(38)
Lörz, H., 580
Louvard, D., xlvi, 86, 244(4), 245, 407, 410(13), 414(13), 415(11), 417, 469, 482(25), 483(25), 484(25, 45), 510, 670
Lowenthal, A., 230
Lowry, O. H., 7, 124, 767
Loyda, Z., 406
Lozzio, B. B., 281
Lozzio, C. B., 281
Lu, A. Y. H., 532, 535(21), 536(21), 538
Luck, D., 666
Luck, D. J. L., 244(7), 245
Ludwig, B., 200, 701
Lundqvist, G., 547, 549
Lusis, A. J., 558, 559(4), 560(7), 561(11), 562, 563(17), 564(7, 11, 17), 565(4), 566(7), 567(7), 568(7), 569(7)
Luskey, K. L., xxxiv
Lüthi, U., 218
Lux, S., 305
Lyon, D., 641

M

McAda, P. C., 133
McAllan, A., 765
Macara, I. G., 760, 761(147)
McBee, A. G., 752
McBurney, M. W., 164
McCarty, K. S., 578
McCarty, K. S., Jr., 578

McCauley, J., 434
Maccecchini, M. L., 140, 667, 668(12, 15), 716
McClintock, P. R., 448
McCord, T. J., 779(25), 780(25), 781
McCoy, T. A., 779(19), 780
McCullough, G., 311
McDevitt, H. O., 193, 198(4), 215, 325
McDonald, J. K., 753, 756(103), 757(103), 758(103)
MacDonald, R. J., 28, 98, 266, 578
McDonough, J., 157
McDowell, J. H., xli
McDowell, M., 285
McDowell, M. J., 50
McElligott, M. A., 754
McGarrity, G., 165
McGeady, M. L., 451, 452(44)
McGuire, E. J., 297
Macino, G., 418
McIsaac, R. S., 360
Mackall, J. C., 423, 424(3)
McKanna, J. A., xlviii, 748, 749(60), 750(60), 752(60)
MacKay, V., 802
McKean, D. J., 667
McKinney, J. D., 547
McLaughlin, C. A., 531
McLaughlin, C. S., 50
McLean, I. W., 344
McLean, J. D., 485
MacLennan, D. H., 124, 128, 150, 570, 574(1, 3, 4, 5)
McMullen, M., 514, 773
McMurray, W. C., 555(86), 556
McPherson, I. A., 437
McWilliams, D., 782
Madin, S. H., 437
Madri, J. A., 488
Maeda, T., 74, 75(4), 78(4), 79(4), 81(4), 82(4), 83(4)
Maestracci, D., 389
Magee, A. I., 796, 800
Mahaboob, S., 191
Mahdavi, V., 52
Mahowald, A. P., 93, 590
Maizel, J. V., 16, 105, 248
Maizel, J. V., Jr., 239, 241, 525
Mak, S., 574
Malcolm Neumeier, M., 420
Maley, F., 289, 812
Malfer, C., 801
Malinoff, H. L., xxx
Malkin, L. I., 670
Malter, J. S., 763
Mandel, B., 445, 450(10), 451(36), 452
Mandel, G., xliii
Mandel, P., 797
Mankovitz, R., 183
Mannervik, B., 548
Manney, T. R., 802
Manning, D. J., 245
Mans, R. J., 6, 18, 47, 101, 788
Mansfield, E. H., 191
Mantin, P. E., 773
Marcantonio, E. E., 122, 520, 523(5), 526(5), 531, 532, 533(32), 540(4, 25, 32), 541(32)
Marchesi, V., 125, 127(47), 140(47), 141(47)
Marchesi, V. T., xlviii, 150, 256, 260, 266(9), 268, 269, 271, 272, 275(9), 277(9, 24), 279(36), 280, 281, 286, 297, 298, 299, 300, 303(4), 305, 306(3, 4), 309(4), 317, 321, 542
Marcker, K. A., 152
Marcu, K. B., 81
Marcus, A., 38, 787
Marcus, K., 6
Marcus, M., 123, 136(30), 143(30)
Marcus, M. M., 86, 91(15), 669, 670(40)
Maretzki, A., 584, 585
Margoliash, E., 130, 131(80), 140(79), 142(79), 145(79, 80), 149(79, 80)
Margulis, L., 678
Marin, B., 587
Marinetti, G. V., 796(30), 801
Marks, M. M., 550
Markwell, M. A. K., 167, 213, 551, 552(58), 553(58), 639
Maroney, P. A., 40
Maroux, S., 406, 407(4), 408, 410(13), 412(5), 413(16), 414(5, 13, 16, 17), 415(11), 417, 418(15, 19), 419(15), 420(15), 421(42), 422(9, 15, 42)
Marsch, J., 668
Marsh, C. A., 765
Marsh, J., 122, 538
Marsh, M., xlv
Martial, J. A., 369
Martin, E., 158
Martin, R. G., 337

Martin, T. M., 514
Martinoia, E., 581, 584, 585(14a, 19, 29)
Martire, G., 512
Martonosi, A. N., 11, 24(16), 124, 128(36), 147(36), 570
Marty, F., 580, 583, 584(17), 585(17), 586, 587(2, 17)
Marzella, L., 742, 753, 762(96)
Maschio, F., 535, 538(56), 539(56)
Mason, T. L., xlii, 131
Massoulié, J., 353, 358
Masters, B. S. S., 536
Mathews, M. B., xxxvii, 122, 667
Mathias, A. P., 30
Matile, P., 580, 584, 585(18, 19), 586, 587, 812
Matlin, K. S., 510, 751
Matsumura, H., 548(43), 549(43), 550
Matsuura, S., 130, 131(80), 140(79), 142(79), 145(79, 80), 149(79, 80), 540
Matsuzawa, Y., 745
Matthaei, J. H., 82
Matthes, M., 609
Matthew, M. B., 15, 50
Maul, G. G., 597
Max, S. R., 354
Maxfield, F. R., 742, 746(21), 749
Maxwell, F., 30
Maxwell, I. H., 30, 31
Mayahara, H., 641, 642(45), 644
Mayer, M., 752
Mayer, R. J., 752, 770
Mayor, F., 746
Mazauskas, C., 160, 182(29), 183
Mazurkiewicz, J. E., 632, 633(10), 636, 641, 645(48), 646(48), 647, 648(60)
Meakin, G., 311
Medrano, L., 447
Mego, J. L., 745
Meijer, D. K. F., 551, 552(67)
Meijer, J., 532, 544, 545(3), 546(3)
Meisel, E., 641
Mekada, E., 160
Meldolesi, J., xliv, 512, 533, 535(44, 45)
Melli, M., 683
Melloni, E., 754
Mellman, I., 731
Mellman, I. S., xlv
Mendecki, J., 15
Mendez, B., 369
Merchant, J., 634, 648
Mercier, J. C., 669, 670(36)
Merion, M., 763
Merlie, J. P., 348, 371
Merrill, C. R., 189, 230(8–10), 231, 232(10), 233(1, 8, 10), 234, 235(1, 8, 10), 237(17), 238(1, 2), 700
Merritt, W. D., 636
Messer, M., 559
Metaxas, M. N., 268
Metaxas-Bühler, M., 268
Metz, H., 393
Meunier, J. L., 339
Meyer, D. I., xxxix, 86(32, 33), 87(17), 91(32), 93, 123, 143(26), 532, 669, 670(41, 42), 683, 692, 698(3), 699
Meyer, E. M., 655
Meyer, J., 587, 588, 812
Meyer, U. A., 546, 549(17)
Michael, H. J., 749
Michaeli, D., 585
Michaeli, S., 702, 703(18)
Michaels, A., li
Michaels, A. S., 654
Michalak, M., 570, 574(5)
Michel-Villaz, M., 619
Michetti, M., 754
Mierendorf, R. C., 828
Miernyk, J. A., 714
Migala, A., 150
Mihara, K., xlii, xliii(77), 131, 148, 534
Mikawa, R., 124
Miledi, R., 333
Milhaud, P., 742, 746(21)
Milikowski, C., 184
Miljanich, G. P., 397
Miller, A. L., 824
Miller, B., 777
Miller, C., 588, 788
Miller, F., 134, 135(99), 521, 710
Miller, J. H., 75, 79(14), 81(14)
Miller, L. K., 760, 761(153)
Miller, R. S., 427
Miller, W. H., 617
Mills, A. D., 217, 219, 220, 725, 735, 773
Mills, J. W., 635, 641(27), 643(27), 645(27)
Milstein, C., xxxvii, 122, 295, 667
Mindich, L., liii
Min Jou, W., 126, 127(65), 496, 541
Minna, J. D., 183
Mintz, B., 748, 750(52)
Mitchell, D., 311
Miura, S., 133, 149(88), 753

Miyajima, H., 548(43), 549(43), 550
Miyata, T., 126
Mizuhira, V., 655
Mizukami, Y., 539
Mizushima, S., 78
Model, P., 668, 671(26), 716
Moehring, J. M., 160
Moehring, T. J., 160
Mohandas, N., 305
Mok, W., 134, 526, 528(14), 529(14), 530(14)
Moldave, K., 50
Mole, J., 311
Moll, R., 604, 606(19)
Mollay, G., 785
Molnar, J., 418
Mondon, C. E., 737
Monis, B., 406
Monto, G., 779(40), 780(40), 781
Moody, D. E., 548(47), 549(47), 550
Moog, F., xxix
Moor, H., 802
Moore, A. E., 435
Moore, J., 533
Moore, S., 726
Moore, T., 700
Moreau, F., 586
Morein, B., 459
Moreno, F., 669
Morgan, E. H., 748, 750(51)
Morgan, H. E., 756, 757(130), 758(130)
Morgenstern, R., 532, 544, 545(3, 10), 546(3), 547(10), 548, 549(10)
Mori, K., 133
Mori, M., 133, 149(88), 753, 759
Morihara, K., 133
Morimoto, T., xl, xlii(60), 7, 11, 13(12), 23(12), 24(12, 16), 121, 123(2), 124, 125(2), 126(2), 127(50, 51), 128(36), 130, 131(80), 140(79), 142(79), 145(79, 80), 147(36), 149(79, 80), 258, 496, 530, 531(1), 532(1), 537(5), 539(5), 540(1), 570, 628
Morita, E. A., 619
Morita, T., 131, 149(88)
Moroney, P. A., 67
Morré, D. J., 533, 587, 747, 748(49)
Morré, J., xliv
Morre, J. D., 628, 648(3)
Morris, C. F., 193, 198(7), 206
Morris, N. R., 220
Morris, R., 230
Morrison, M., 150, 269
Morrow, J. S., 299, 300, 303(4), 305, 306(3, 4), 309(4)
Mortimore, G. E., 737, 753(2), 754, 756, 757(130), 758(130), 763
Mosbach, E. H., lii
Moscarello, M., 461
Moskowitz, A. H., 580, 585(10)
Moss, W., 183
Mosteller, R. D., 780(54), 782
Mostov, K. E., 43, 128, 146(71), 147(71), 570
Mueller, T. J., 150, 269
Mulder, J., 35
Muller, H., 588
Müller, M., 389, 393(22)
Mullet, J. E., 40, 140, 141(115)
Mulligan, R. C., 122, 496, 497
Mullinex, K. P., 32
Mulready, P., 581
Mumford, R. A., 785
Mundry, K. W., 442
Munier, R., 779(39), 780(39), 781
Munkres, K. D., 239, 242(5), 735
Munro, H. N., 7, 575
Murachi, T., 315
Muramatsu, M., 539
Muramatsu, T., 158
Murayama, A., 760
Murayama, J., 268
Murer, H., 389, 393(22)
Murokami, T., 315
Murphy, D. B., 325
Murphy, S., 683
Murray, R. K., 548
Musinski, J. F., 126
Mutafschiev, S., 414
Myerowitz, R., 730, 735
Myllylä, G., 286
Mysels, K. J., 245

N

Nack, E., 134
Nagai, Y., 434, 436(6), 439(24), 443
Nagata, T., 581, 584(1)
Nagelkerke, J. F., 755
Naizir, A., 585
Nakai, S., 126
Nakajima, M., 79
Nakajo, S., 268

Nakamura, K., 435, 671
Nakamura, Y., 268
Nakane, P. K., 344, 648
Nakaya, K., 268
Nakazato, H., 31
Nakazawa, Y., 546
Narasimhan, S., 158
Nash, B., 127, 129(66)
Nass, M. K., 459
Natowicz, M., 736
Natowitz, M. R., 124
Nayak, D. P., 435
Neat, C. E., 548(50), 549(50), 550, 553
Nebert, D. W., 546
Neely, A. N., 763
Neff, N. T., 757, 759(136)
Negishi, M., 124, 125(39), 127(39), 128(39), 532, 533(30), 535(30), 538(30), 539(30)
Neher, E., 333
Nelson, B. D., 555, 556(80)
Nelson, B. P., 763
Nelson, C. E., 587
Nelson, J., 311
Nelson, N., xlii, 131, 200
Nemes, P. P., 397
Neu, H. C., 81
Neuberger, A., 166
Neufeld, E. F., 157, 587, 730, 731(1), 735, 736(4), 747, 748(28, 50), 760(15), 761(15), 764
Neuhoff, V., 235
Neupert, W., 130, 134, 135(99), 140, 141(113), 710, 711, 712(42)
Neville, D. M., 208, 239, 248
Neville, D. M., Jr., 423, 426
Newcomb, E. H., 700
Nexø, E., 447
Ng, S. K., 572
Nicholson, G. I., 628
Nicol, G. D., 617
Nicolson, G. L., xxx, 160, 166
Nieberg, P. S., 354, 359(9), 361(9), 363(9)
Niedermeyer, W., 586
Niedhardt, F. C., 779(38), 780(38), 781
Nielsen, J. B., 80
Niemann, H., 440
Nigg, E. A., 281
Nihei, T., 375
Nikaido, K., 306
Nilson, J. H., 670
Nilsson, Å., 755
Nilsson, K., 282
Nilsson, O. S., 533, 555
Nimrod, A., 749
Nirenberg, M. W., 82
Nishi, Y., 386
Nishida, G., 58
Nishigori, H., 760
Nishimura, M., 580, 582(7), 583(7), 583, 584(7, 16a), 585(7)
Nishimura, Y., 561, 562(16), 563
Nishizawa, E. E., 238
Nissan, S., 752
Noble, E. R., 584
Noe, B. D., 777
Nørby, J. G., 635
Nordlie, R. C., 531
Nordling, S., 286, 461
Norén, O., 387, 391(9), 392, 400(9, 11), 401(9), 405(9), 406, 418
Norman, A. W., 469
North, M. J., 818
Norton, W. T., 382
Norum, K. R., 755
Notkins, A. L., 448
Novelli, G. D., 6, 18, 47, 101, 788
Novick, P., 802, 804(9), 808(8, 10), 811(8, 9)
Novikoff, A. B., 681, 745, 764
Nurden, A. T., 269
Nussbaum, J. L., 797
Nutley, M. A., 251
Nuuja, I. J. M., 762

O

Oakley, B. R., 230
Oba, K., 585
O'Brien, W. A., 160
Oda, K., 755
Oesch, F., xli, xliv(67), xlv(67), 124, 125(40), 127(40), 135(40), 144(40), 147(40), 535, 544
O'Farrell, P. H., 184, 185(4), 205, 232, 236, 306, 307(12)
O'Farrell, P. Z., 205
Ogata, S., 158
Ogawa, K., 634, 641, 642(45), 644(45)
Ogura, H., 436, 439(24)
Oh, T. H., 354

Ohad, I., lii
Ohashi, A., 199, 200, 202(15), 204(15)
Ohhara, H., 765, 766(7)
Ohkuma, S., 737, 738(9–11), 739, 740(11), 741(11), 742(10), 743(9–11), 745(9, 10, 11), 752, 755(9), 764(11)
Ohlrogge, J. B., 585
Ohmori, N., 133
Ohmura, S., 756
Ohuchi, M., 434
Ojakian, G., 523
Oka, T., 133, 760
Okada, Y., xli, xliv(67), xlv(67), 124, 125(40), 127(40), 134, 135(40), 144(40), 147(40), 160, 532, 533(33), 534(33), 535(33, 42), 536(33), 537(33)
O'Keefe, E., 115
Okita, R. J., 536
Okumura, T., 268
Oldberg, A., xxxi
Olden, K., 191
Olefsky, J. M., 749
Olsen, B. R., 531
Olsen, R. W., 339
Olsnes, S., 159, 751, 756(81)
Olson, G., 779(26), 780(26), 781
Omary, M. B., 796, 800
Omary, O. B., xli
Omura, T., xxx, li(5), 147, 150(120), 531, 533(19), 534, 535(19, 42), 536, 546
O'Neill, B., 386
Opdenkamp, J. A. F., 648
Oppermann, H., 227
Orci, L., 469
Ord, J. M., 752, 758(91)
O'Reilly, D., 761
Orlich, M., 434, 435(2), 436(2)
Oroszlan, S., 313, 796(27), 801
Orr, H. T., 126, 295, 325, 796
Orrenius, S., 536, 544, 545(1), 546(1)
Orringer, E., 324
Osawa, M., 79
Osawa, T., 257, 258(4)
Osborn, M., 239, 241(4), 574
Osborn, M. J., 80, 265
Ose, L., 755
Osmundsen, H., 548(50), 549(50), 550, 551, 552(65), 553(65)
Osumi, T., 551, 552(69), 553(69), 702, 703(24)
Ott, P., 354
Oud, P. S., 738, 752(14)
Ovchinhikov, Y. A., xli
Owen, M. J., 325
Owens, J. W., 561, 563(15)
Oxender, D. L., 75, 131
Ozoles, J., 135, 144(103), 150(103)
Ozols, J., 533, 534(40), 540

P

Pace, N. R., 79
Packer, L., 617
Padmanaban, G., 546
Pagès, J. M., 415
Paigen, K., 558, 559(4), 560(7), 561(11), 562(10), 563(17), 564(7, 10, 11, 17), 565(4, 10), 566(7), 567(7), 568(7), 569(7)
Painter, R. G., 256, 269
Palade, G. E., xxix, xxx, xxxii, xxxv, xli, xliv, xlv, xlvi, xlvii, xlviii(94), xlix, l(14), li(1, 5), lii(2, 118), liii, 84, 147, 150(120), 198, 357(17), 359, 536, 546, 666, 678
Palmiter, R. D., 51, 71, 122, 142, 152, 155, 156, 157(18), 531
Paluch, U., 130
Papahadjopoulos, D., liv
Papasozonenos, S. C., 238
Papermaster, D. S., 244(3, 5), 245, 248(8), 249, 251, 253(5), 256, 468, 485, 488(1), 491, 492(14), 494(3), 495(2, 3, 6, 11), 542, 609, 610(4), 612(11), 613(4, 8), 615(4, 8, 9), 616(9), 676
Pappone, M.-C., 620, 625(22)
Pardee, A. B., 780(50, 52, 53), 782
Pardoll, D. M., 589
Parfett, C. L. J., 160, 161(26), 182
Parisi, E., 80
Parker, J., 781(64), 783
Parmelee, D. C., 796(26), 801
Pastan, I. H., 452, 468, 742, 746(21), 748
Paterson, B. M., 38, 720
Patrick, J., 332, 352(5)
Pawar, S., 702, 703(25)
Peacock, A. C., 36
Peaker, M., 628, 629(5), 658
Pearse, B. M. F., xlvii
Pech, I. V., 647
Pedersen, S., 555, 556(76)

Pekas, D. J., 126
Pelham, H. R. B., 44, 51, 52(4), 60, 64, 65, 67, 90, 285, 514, 578
Pelham, H. R. E., 6, 15(9), 16
Pelham, R. B., 95, 100
Penke, B., 78
Pennington, R., 638
Penttinen, K., 458, 460(11), 461(11), 463(11), 465(11)
Pereyra, P. N., 86, 520, 533, 540(38), 671
Perez, M., 160
Perlman, D., 682, 807
Perrelet, A., 469
Perrone, J. A., 652
Perry, R. P., 25
Persson, R., 449
Pesonen, M., 453, 465
Pestana, A., 531
Peter, J., 78
Peterlik, M., 741
Peterson, D., 308
Peterson, P. A., 325, 328(10)
Peterson, R. F., 301
Peterson, R. L., 79
Petrie, W. A., Jr., 800
Petrini, G., xliv
Pettersson, I., 689
Pettersson, R. F., 459
Pfefferkorn, E. R., 456
Pfeuffer, T., 620
Philipson, L., 285, 444, 447(3, 6), 448(19), 449(3), 450(20), 451
Phillipps, M. A., 764
Phillips, J. F., 658
Phillips, J. N., 245
Piatet, R., 770
Pidard, D., 269
Pierce, B. G., 648
Pietrini, G., 533, 535(45)
Pihl, A., 159, 751
Pine, M. J., 779(41), 780(41), 782
Piperno, G., 244(7), 245
Pirkola, A., 286
Pirtle, R., 84
Pisano, J. M., 747
Pitchfork, J., 641, 644(53)
Pitot, H. C., 531, 758
Pitt, J. B., 658
Pittman, R. C., 755
Place, A., 313
Platt, T., 313
Plesset, J., 50
Ploegh, H. L., 126, 295, 325, 796
Plummer, T. H., Jr., 289
Plutner, H., 731
Pober, J. S., 280, 617, 621(5)
Podleski, T. R., 333
Poduslo, S. E., 382
Poehling, H., 235
Poli, A., 737, 746(1, 3), 752(1), 753(1, 3, 6), 754(3, 6), 756(1), 758(1), 762(1), 763(6)
Pollack, L. R., 336, 337(23)
Pollard, J. W., 781(64), 783
Ponstingl, H., 225
Pontremoli, S., 754
Poole, B., 553, 723, 725(4), 737, 738(9–11), 739, 740(11), 741(11), 742(10), 743(9–11), 745(9–11, 24), 752(24), 755(9), 764(11)
Poole, R. J., 585
Popov, D., xxxviii, 124, 765, 773(6), 774(6), 775(6)
Poretz, R. D., 763
Porter, A. G., 435, 496
Porter, M., 266
Porter, T. H., 780(58), 782
Poste, G., liv, 628
Potrykus, I., 580
Potter, V. R., 3, 7, 22
Potts, J. J., Jr., 122
Potts, J. T., Jr., 152, 154(7), 156(7)
Powell, J. T., 158
Powers, D., 313
Powever, F., 148
Poyton, R. O., 716
Prasad, J. D., 548(42), 549(42), 550
Preiser, H., 389
Pressman, B. C., 184, 723, 755
Pressman, R., 558
Prestidge, L. S., 780(50, 52, 53), 782
Price, C. A., 805
Prieels, J.-P., 171
Prockop, D. J., 531, 779(44, 45, 48), 780(44, 45, 48), 782
Prohaska, R., 268
Proia, R. L., 731
Przybyla, A. E., 28, 98, 266, 578
Puchalski, C. M., 735
Puett, D., 745, 755(37)
Pugh, E. N., Jr., 619

Puglia, K. V., 268(15), 269
Pumplin, D. W., 332, 352(6)
Purich, D., 685, 688(17)
Purvis, B. J., 423

Q

Quaroni, A., 127, 129(67), 387, 400(12, 13), 418, 419(30), 420, 422(30)
Quay, S. C., 75
Queen, C., 84
Quigley, M., 499
Quillen, M., 184
Quinn, P. S., 122, 152, 154(10), 156(10), 482, 484(45), 538, 668
Quiroga, M., 225
Qureshi, S. A., 548(51), 549(51), 550

R

Raaka, B. M., 424
Rabinowitz, M., 779(27, 31), 780(27, 31), 781
Rachubinski, R. A., 135, 144(100), 532(37), 533, 534(37)
Racine, L., 468, 469(8), 491, 492(15), 493(15)
Racker, E., 675, 788
Radcliffe, C. W., 79
Raelofsen, B., 649
Raftery, M. A., 348, 371
Rafto, S., 373
Raivio, K. O., 461
Rajamanickan, C., 546
RajBhandary, U. L., 152
Rall, J. E., 555(84), 556
Ralston, G. B., 299
Ramos, B., 321
Ramos, R., 125, 126(52), 150(52)
Ramsey, J. C., 3, 11(5), 22, 378, 379, 530, 771
Randall, L. L., xliii, 81, 83, 671
Randall, R. J., 7, 767
Randerath, K., 218
Ranki, A., 281
Raphael, C., 539
Rappaport, L., 585
Raschke, C., 126
Rash, J. E., 345, 350(34)
Ratzkin, B., 215
Rausch, U., 702, 703(19)
Rauvala, H., 158
Ravazzola, M., 469
Ray, B., 161
Ray, P. N., 164
Rayborn, M. E., 609
Rearick, J. I., 730
Reddy, G., 260, 261(8), 321
Reddy, J. K., 548(42, 44, 47, 51), 549(42, 44, 47, 51), 550, 551(39), 552(39)
Reddy, M. K., 548(51), 549(51), 550, 551(39), 552(39)
Reddy, R., 683
Redline, R., 531, 533(20)
Redman, C. M., 121, 123(1), 665
Redman, K., 157
Reed, B. C., 336, 423, 424(3, 4), 425(4, 11), 428(11)
Reed, R. E., 427
Reeder, R. H., 613
Rees, T., 584
Reggio, H., 244(4), 245, 482, 510, 751
Regoeczi, E., 749
Rehfeld, D. W., 702, 703(20)
Reid, G. A., 200, 204
Reid, M. S., 797
Reijngoud, D. J., 738, 743, 752(14)
Reik, L., 532, 533(31), 538
Reilly, P., 491, 495(11)
Reinhold, L., 585
Reiser, J., 191, 193, 198(2), 206, 207, 212(2), 213(2), 214
Reiss, W., 548(40), 549
Reith, A., 737, 738(13), 744(73), 745(13), 746(13), 750, 751, 752(13, 73), 756(73), 758(73), 759(73), 762(73)
Reithmeier, R. A. F., 570
Reitman, M. L., xxxvi, xli(38), 158, 159(14), 161(14), 183(14)
Reizman, H., 134, 135(98)
Remacle, J., 533
Renart, J., 193, 198(2), 206, 212, 213
Rennert, O. M., 779(16), 780
Réz, G., 762
Rhee, S. H., xxxviii, xl(48)
Rho, H. M., 116
Rhoads, R. E., 787
Rice, C. M., 512, 800
Rice, R. H., 166
Rich, A., 122, 152, 154(7), 156(7), 670
Richards, F. M., 488
Riches, D. W. H., 742, 748(27), 750(27)

Rick, W., 393
Rickli, E. E., 127, 129(67), 387, 391(9), 400(9, 13), 401(9), 405(9)
Riedel, H., 496, 497, 511(11)
Rieger, F., 357, 361, 363(14)
Riemann, S., 757, 758(132), 759(132)
Riezman, H., 702, 703(15), 710, 711(10), 712(41), 715(41)
Rindfrey, H., 393
Ritchie, A. K., 345, 350(34)
Roa, J. K. M., xli
Robbi, M., 531, 724(18), 725, 726(8, 11), 727(8)
Robbins, A. R., 730
Roberts, B. E., 38, 720
Roberts, C. W., 731
Roberts, J. W., 733
Roberts, L. M., 710, 711(38)
Roberts, R. M., 191
Roberts, S., 378
Robertson, H. D., 70
Robey, P. G., 736, 760(15), 761(15)
Robinson, M., 632
Robinson, P. J., 114, 118(6), 295, 325, 326(5), 327(5)
Robinson, W. E., 618
Robson, B., 669, 670(36)
Rodbard, D., 812
Rodbell, M., 620
Roden, L., 158
Rodewald, R., xlv
Rodier, F., 585
Rodkery, J. A., 14, 19(17), 20(17)
Rodriquez-Boulan, E., 127, 461, 520, 523, 533, 540(38)
Roesing, T. G., 448, 451(32)
Rogers, G., 93
Rogers, J., 126, 572, 682
Rogers, M. J., 150, 531, 533(20), 535
Roisen, F. J., 754
Røken, I., 755
Roll, F. J., 488
Rolleston, F. S., 122
Roman, R., 787
Romano, E. L., 468
Romano, M., 468
Ron, E. Z., 78
Ronner, P., 386, 390(5)
Ronnett, G. V., 336, 423, 424(4, 5), 425(4, 5, 12), 432(18)
Roodyn, D. B., 555, 556(81)
Ropson, I., 313
Rose, D. J., 245, 256
Rose, I. A., 753
Rose, J. K., 126, 496, 801
Rosebrough, N. J., 7, 767
Roseman, M. A., xxxiii
Roseman, S., xxxv, xxxvi(34), 297
Rosenberg, L. E., xlii(78), xliii, xlv(84), 23
Rosenberg, M., 84
Rosenbloom, J., 779(44, 48, 49), 780(44, 48, 49), 782
Rosenblum, C., 779(26), 780(26), 781
Rosenbusch, J. P., 386
Rosenfeld, M. G., xxxviii, 122, 124, 134, 531, 532, 533(32), 540(4, 32), 541(32), 765, 773(6), 774(6), 775(6)
Rosenick, M. M., 621
Rosenqvist, U., 555, 556(80)
Ross, E. M., 620
Rosset, J., 468
Rossomando, E. F., 816
Roth, J., xxxvi, 423, 426, 469, 479, 484(20), 749
Rothman, F. M., 670
Rothman, J. E., xl, 84, 124(44), 125, 127(42, 43, 44), 128(42, 43, 44), 289, 378, 418, 800
Rothstein, A., 386
Rott, R., 200, 434, 435(2), 436(2, 6), 437, 438, 439(12), 440, 443
Rotundo, R. L., xxxi, 345, 347, 350(35), 354, 355, 357(8), 359(7, 8), 361(7, 8), 362, 363(7), 364
Rougeon, F., 35
Roychoudhury, R., 215
Rubenstein, I., 716, 717
Rubenstein, P., 157
Rubin, G. M., 811
Rubin, H., 30, 153
Rubin, R. W., 49, 184, 185(5)
Rucinsky, T., 93
Ruddle, F. H., 382, 770
Rudin, U., 140
Rudin, Y., 140, 667, 668(12, 15), 716
Rueckert, R. R., 443, 447(1)
Ruffolo, J. J., 597, 606(10)
Ruis, H., 588
Ruis, R., 700, 701(6), 702, 703(13)
Rulli, D., 190

Russell, R. L., 353, 355, 356, 357(2)
Russell, S. M., 752
Russo, J., 646
Rutherford, T. R., 282
Rutter, R. W., 98
Rutter, U., 770
Rutter, W. J., 28, 225, 266, 578
Rutz, R., 157
Ryan, C. A., 580, 582(5), 583(5), 584(5), 585(5), 587, 588
Ryan, D. E., 532, 533(31), 537, 538, 539
Ryan, J., 423
Ryan, T. E., 150

S

Sabatini, D. D., xxxvii, xxxviii, xxxix, xl, xli, xlii(60), xliv(67), xlv(67), xlix(35), 3, 11, 22(2), 24(16), 86, 104, 121, 122, 123(1, 2), 124, 125(2, 39, 40), 126(2), 127(39, 40, 47, 50, 51), 128(36, 39), 130, 131(80), 134, 135(40), 140(47, 79), 141(47), 142(79), 144(40), 145(79, 80), 147(31, 36, 40), 149(79, 80), 258, 382, 384, 385(11), 461, 496, 520, 521(1, 4), 523(1, 5), 526(5), 528(14), 529(14, 16), 530(14), 531(1), 532(1), 533(30, 32, 33), 534(33), 535(30, 33), 536(33), 537(5, 33), 538(30), 539(5, 30), 540(1, 4, 25, 32, 38), 541(32), 542, 555, 570, 628, 650, 665, 667, 670, 671, 765, 773(6), 774(6), 775(6)
Sabban, E. L., 125, 127(47, 51), 140(47), 141(47), 542
Saccone, C., 555, 556(79)
Sachs, G., 635
Sachs, L., 172
Sadnik, I., 50
Sage, H. J., 166, 286
Sakakibara, R., 534
Sakakibara, S., 133
Sakmann, B., 333
Saksela, E., 281
Salamino, F., 754
Salpeter, E. E., 333
Salpeter, M. M., 333
Samal, B. A., 779(40), 780(40), 781
Saman, E., 126, 127(65), 496, 541
Sambrook, J., 496
Sammons, D. W., 238
Samuels, H. H., 424
Samuelsson, B. E., 648(68–70), 649
Sandeen, G., 30, 153
Sandelin, K., 454
Sanders, D., 584, 585(31a)
Sando, G. N., 742, 748(28)
Sandrin, E., 782
Sandvig, K., 751, 756(81)
Sanwal, B. D., 572
Sarama, J., 165
Saraste, J., 456, 458, 459(8, 10), 460(11), 461(11), 463(11), 465(10, 11)
Sargus, M. J., 763
Saris, C. J. M., 38
Sarras, M. P., 633, 635, 643(26), 645(26), 656
Sarvas, M., 481
Sasame, H. A., 535
Sasavage, N. L., 670
Sasse, F., 584, 585(20)
Sato, K., 534
Sato, M., 133
Sato, R. F., 148, 531, 533(19), 534, 535(19)
Sato, T., 75
Satyanarayana Rao, M. R., 546
Saunders, J. A., 580, 582(6), 583(6), 585(6)
Sawada, J., 756
Sax, J., 658
Scalenghe, F., 600, 602
Scarpa, A., 751
Schachter, H., xxxv, xxxvi(34), 158, 461
Schaeffer, B. E., liv
Schäfer, W., 435
Schaffer, F. L., 445, 450(14)
Schaffner, W., 185
Schatz, G., xlii, xliii(74), xliv, xlv(74, 86), 130, 131, 140, 192, 199(15), 200, 201(16), 202(15), 203(19), 204(15), 667, 668(12, 15), 716
Schechter, D., 397
Schechter, I., 122, 152, 667, 676
Schechter, Y., 542, 760
Scheckman, R., 470
Scheele, G., 45, 94, 95, 96(1), 97(1), 98, 100(1), 102, 104(1), 105, 106, 109(16), 110(13), 139, 140(107), 141(107), 667, 691, 726, 787
Scheer, U., 589, 597, 599(9), 600(9), 604, 608(5, 9)
Scheffer, R. C. T., 468
Scheid, A., 434, 436(5), 442, 673

Schekman, R., 802, 804(9), 806, 808(8, 10), 811(8, 9), 813(17)
Schellenberg, G. D., 647
Schellenberg, M., 587, 588
Schelsinger, M. J., 779(23), 781
Scher, J. H., 758
Scherrer, K., 25
Schfritz, D. A., 5
Schick, M. J., 245
Schiefer, S., 702, 703(12, 22), 711(22), 714(22)
Schimke, R. T., xxx, 14, 546, 725, 823, 827
Schimmel, P. R., 780
Schirrmacher, V., 459
Schlegel, R., 746
Schlesinger, M. J., xli, 512, 515, 778, 780(12, 23), 781, 795, 796(2), 799(4), 800(6), 801
Schlesinger, P., 748
Schlesinger, S., 512, 515, 779(23), 780(23), 781
Schlessinger, J., xxxiv, 746
Schmelzer, E., 131
Schmidt, G. W., xli(79), xliii, 40, 130, 140, 141(115), 667, 668(14), 716
Schmidt, M. F. G., xli, 435, 439(8), 796, 799(4), 800(5, 6)
Schmidt, R., 585
Schmitt-Nielsen, K., 632, 658
Schmitz, J., 389
Schnaitman, C., 149
Schnarrenberger, C., 702(27), 703
Schnebli, H. P., 764
Schneider, B. G., 244(3), 245, 485, 488(1), 491, 494(3), 495(2, 3, 11), 610, 615
Schneider, D. L., 745
Schoentgen, F., 797
Schofield, P. J., 752
Scholtissek, C., 435, 438, 439(12), 440(41), 441
Schopfer, P., 714, 715(49)
Schor, S., li, lii(118)
Schrenk, W. J., 215
Schulman, H. M., 50, 69(2)
Schulman, S., 422
Schulte, T. H., 277
Schulz, H. U., 546, 702, 703(25)
Schwaiger, H., 589
Schwartz, A., 635, 647(29)
Schwartz, A. L., 257, 258(1), 261(1), 748, 795
Schwartz, B. D., 799
Schwartz, M., 669
Schwarz, R. T., 440
Schwarze, P. E., 737, 752(6), 754(6), 763(6)
Schweet, R. S., 780(55), 782
Schworer, C. M., 737, 753(2)
Scolnick, E. M., 796(28), 801
Scornik, O. A., 532, 763
Scott, J. H., 806
Seal, S. N., 787
Sebbane, R., 348
Sedman, S. A., 230(8), 231, 233(8), 235(8)
Sefton, B. M., xli, 496, 512, 796(28), 801
Sefton, M., 126
Segal, H. L., 758
Seglen, P. O., 737, 738(13), 739(17), 740, 742(19), 744(73), 745(13, 17, 19), 746(1, 3, 12, 13, 15, 25), 746(17), 750(35), 751, 752(1, 12, 13, 17, 25, 73), 753(1, 3, 4, 6, 15, 17, 25), 754(3, 4, 6, 15, 25, 104), 756(1, 73, 104), 757(104), 758(1, 25, 73, 104), 759(73, 104), 761(111, 111a), 762(1, 26, 73, 111, 112, 113, 114), 763(6, 112)
Segrest, J. P., 150
Seidegard, J., 532, 544, 545(3, 10), 546(3), 547(10), 549(10)
Sekizawa, J., 75, 79(9), 86, 668
Seligman, A. M., 406
Selinger, Z., 620
Semenza, G., 127, 129(67), 386, 387, 388(19, 20), 389, 390(5), 391(8, 9, 19), 392(20), 393(22), 394, 395(28), 396(28), 397(28), 398(28), 399(19, 28), 400(9, 10, 11, 13, 14), 401(9), 403(14), 404(14), 405(9), 406(28)
Sémériva, M., 417
Semple, E. W., 635, 649
Sen, A. K., 635, 641, 658
Sen, H. K., 649
Sen, I., 620
Senior, P., 685, 688(17)
Setlow, B., 531, 533(20)
Seto, J. T., 437
Sevarino, K. A., 716
Shafig, S. A., 758
Shaltiel, S., 685, 688(16, 17)
Shamoo, A. E., 128, 150
Shanks, L., 779(28), 780(28), 781
Shaper, J. H., 589, 590, 608
Shapiro, A. L., 241
Shapiro, D. L., 277, 280, 297

Shapiro, L. J., 764
Sharma, R. N., 548
Sharma, R. S., 531
Sharon, N., 166
Shaw, E., 78, 756, 760(129)
Shaw, P., 514, 773
Sheetz, M. P., 256
Sheinin, R., 752
Shelanski, M. L., 357, 363(14)
Shelton, K. R., 597, 606
Sherman, F., 162
Sherman, G., 796
Sherman, J., 125, 127(50, 51)
Sherman, M. R., 760, 761(153)
Shields, D., 52, 85, 91, 122, 140, 141(109), 285, 514, 521, 716, 776, 787
Shifrin, S., 230, 233(1), 235(1), 238(1)
Shimada, T., 551, 552(62), 557(62)
Shimizu, K., 75, 78(11)
Shimshick, E. J., 452
Shin, S.-I., 183
Shin, W.-Y., 745
Shindo, Y., 551, 552(57, 69), 553(69)
Shine, J., 770
Shinozawa, T., 620
Shinozuka, H., 752
Shio, H., 724(18), 727, 728
Shive, W., 778, 779(32, 36), 780(32, 36, 58), 781
Shively, J. E., 539
Shohet, S., 305
Shoji, S., 796(26), 801
Shore, G. C., 3, 122, 148
Shore, V. G., 780(50), 782
Siak, J.-S., 444, 448, 449(30), 452
Sibley, R., 126
Sidransky, H., 3
Siegelman, H. W., 580, 581(3), 582(3), 584(16, 30), 587(30)
Siekevitz, P., xxix, xxx, xlvi, li(1, 5), lii(2, 118), liii, 147, 150(120), 521, 536, 546, 666
Sigler, P. B., 152, 154(10), 156(10)
Sigrist, H., 386, 390(5), 392
Sigrist-Nelson, K., 392
Silbert, D. F., 77
Silhavy, T. J., xxxvii, xxxviii(43), 668, 669, 671(30)
Silver, S. M., 673
Silverberg, M., 277
Silverstein, S. C., 433, 532
Simon, B., 635
Simon, M. N, 716
Siminovitch, L., 160, 161(19–21), 162, 163(19–21), 164, 173(20), 183
Simons, K., 124(46), 125, 126, 128(46), 407, 414(12), 459, 466, 467(1), 480, 481, 493, 496, 510, 512, 517, 541, 668, 751
Simpson, R. W., 441
Singer, I. I., xxxi
Singer, S. J., xxx, xlix, 256, 420, 468, 470(15), 478(15), 479(33), 485, 493, 510
Singh, J., 321
Singh, M. K., 260, 261(8)
Siu, J., 244(5), 245, 610, 612(8), 613(8), 615(8), 616(8)
Siuta-Mangano, P., 531
Six, H., 563
Sjöström, H., 387, 391(9), 392, 400(9, 11), 401, 405, 406, 418
Skehan, P. J., 654
Skehel, J. J., 434, 436, 496
Skinner, A. M., 161
Skinner, C. G., 778, 779(36), 780(36), 781
Skou, J. C., 635
Skovbjerg, H., 406
Skrede, S., 551, 552(66)
Skultelsky, E., 488
Slate, D., 382
Slate, D. L., 770
Slesers, A., 779(40), 780(40), 781
Slot, J. W., 468, 470, 479, 748
Slusarenko, M., 214
Sly, W. S., 124, 587, 731, 736, 747, 748(48, 49)
Small, B., xlix, 124(45), 125, 128(45), 512
Small, J. V., 476, 479(31)
Small, M., 423
Smilowitz, H., 347
Smith, A., 434
Smith, A. D., 354
Smith, A. E., 152
Smith, D. W. E., 52
Smith, F. G., 7
Smith, M., 379
Smith, P., 157
Smith, S. C., 780(58), 782
Smith, T. W., 638, 647
Smith, W. P., 671
Smulson, M. E., 779(27), 780(27), 781

Smyth, D. G., 136
Snider, M. D., 266
Soderberg, K., 497, 511(11)
Sofer, W., 313
Sogawa, K., 539
Soler, J., 753
Solheim, A. E., 737, 740, 742(19), 745(19), 746(1, 25), 752(1, 25), 753(1, 25), 754(25), 756(1), 758(1, 25), 762(1)
Solioz, M., 203(19), 204
Söll, D., 780
Somerville, L. L., 235
Sonderegger, P., 387, 400(14), 403(14), 404(14)
Sparatore, B., 754
Sparkuhl, J., 752
Spatz, L., xliv, 150, 531, 533(20), 534(39), 535
Speck, J. C., Jr., 190
Speicher, D. W., 300, 305, 306(3, 4), 309(4)
Spencer, D., 588
Spencer, L. A., 160
Spiess, M., 386, 387, 388(19), 389, 391(9), 399(19), 400(9), 401(9), 405(9)
Spring, H., 604, 606(20)
Sprinzl, M., 130
Stadler, J., 597, 607, 608(5), 796(29), 801
Stadtman, E. R., 685, 688(17)
Staehelin, T., 3, 110, 191, 193, 197(3), 198(3), 206, 212(4)
Stahl, P. D., 561, 563(15), 748
Stahl, S., 666, 668(7), 671(7)
Stahl, W. L., 647
Stainer, I. M., 634, 635(16)
Stanley, P., 159, 160, 161(19–23, 25, 38), 162(23, 25), 163(15, 17, 19–23, 25, 38), 172(22), 173(15, 20, 22, 23), 174(22), 178, 179(22, 23), 181(22, 23, 25), 182(16), 183(38)
Stanley, W. M., Jr., 153, 154(15)
Stanners, C. P., 164, 763, 781(64), 783
Stanworth, D. R., 742, 748(27), 750(27)
Stark, G. R., 191, 193, 198(2), 206, 212(2), 213(2)
Starnes, W. L., 406
Staros, J., 392
Stäubli, W., 536, 546, 548(40), 549, 638
Staudinger, H., 546
Stavarek, S., 160
Steck, G., 217
Steck, T. L., 125, 126(52), 146, 150(52), 246, 247(15), 252(15), 260, 261, 271, 300, 301(7), 302(7), 321, 613
Stedman, J. D., 219, 221(9)
Steele, W. J., 3, 11(5), 22, 378, 379, 530, 771
Steen, G. O., 648(68–70), 649
Steers, E., Jr., 531
Stein, M. D., 166
Stein, P. J., 621
Stein, S., 313
Steinbach, H., 277
Steinberg, D., 755
Steinberg, J., 193, 206, 212(3)
Steiner, D. F., 122, 152, 154(10), 156(10), 538, 668
Steinheil, M., 600, 602(18)
Steinman, R. M., 433
Steitz, J. A., 689
Steitz, T. A., xl, 405, 417, 418, 419(35), 669, 670(37)
Stellwag, E. J., 37, 193, 198(6), 206
Stenbuck, P., 315, 316, 319, 321(6), 322(3), 323, 324(6)
Stenman, S., 459
Stephens, R., 308, 311
Stern, A. M., 777, 779(7, 20), 780(7, 20), 781
Stern, H., 39
Stevens, C. F., 332, 333
Stewart, D. J., 635, 656, 658
Stewart, D. L., 649
Stewart, M. L., 15
Stewart, O. J., 658
Stewart, P. S., 128, 150
Stick, R., 597, 604(13), 606(19), 607
Stirling, D. E., 645
St. John, A. C., 754, 763, 784
St. John, T. P., 717
Stoehr, M., 604
Stoker, M. G. P., 437
Stominger, J. L., 126
Stone, G. C., 184, 185(5)
Storelli, C., 389, 393(22)
Stotz, E., 7
Stracher, A., 758
Strader, C. D., 348, 371
Strair, R. K., 5
Strapazon, E., 125, 126(52), 150(52), 321
Strauss, A. W., 14, 19, 20, 785
Strauss, E. G., 456, 512, 800
Strauss, J. H., 456, 512, 800

Strawser, L. D., 559, 566(8), 764
Strecker, G., 736
Strittmatter, P., xliv, 135, 144(104, 105), 147, 150(104, 105, 119), 531, 533(20), 534(39), 535
Strobel, H., 150
Strominger, J. L., 295, 325
Strous, G. J. A. M., 257, 258(1), 261(1), 748, 795
Struck, D. K., xxxv
Stryer, L., 618, 619(7), 620(7), 621(7), 622(7), 625(22, 25), 626(25)
Student, A. K., 423, 424(3)
Studier, F. W., 242
Stull, J. T., 752, 758(91)
Sturgess, J. M., 461
Sturm, A., 705
Suchard, S. J., 184
Sudo, T., 159, 163(17), 173(15)
Suga, T., 548(45, 49), 549(45, 49), 550
Sugano, M., 551, 552(62), 557(62)
Sugawara, K., 116
Suissa, M., xliii
Summers, D. F., 239
Sunderland, C. A., 268
Sunders, G. F., 43
Susani, M., 588
Sussman, M., 818
Suzuki, E., 32
Suzuki, K., 746, 752(43), 755(43)
Suzuki, T., 585
Suzuki, Y., 32, 315, 547
Svensson, U., 449
Svoboda, D. J., 548(42), 550, 551(52), 552(52)
Swan, D., 122, 667
Swank, R. T., 239, 242(5), 560, 562(10), 564(10), 565(10), 735
Swanson, R. J., 619
Swart, G. T., 635, 649
Swedner, K. J., 542
Switzer, R. C., 230, 233(1), 235(1), 238(1, 2)
Sykes, B. D., 778, 780(12)
Symington, J., 110, 212
Szczesna, E., 667

T

Tabas, J., 289
Tae, H. J., 628
Tager, H. S., 122, 538, 668
Tager, J. M., 738, 752(14)
Tai, P. C., xl, 671
Tajima, S., 534
Takahashi, K., 563
Takatsuki, A., 289
Takeishi, K., 84
Takeshita, K., 268, 277(9)
Takesue, S., 534
Takesue, Y., 386
Takeuchi, M., 75, 78(11)
Takeuchi, T., 779(45), 780(45), 782
Talbot, D., 308
Talens, L. T., 440
Talkad, V., 731, 747, 748(48)
Talmadge, K. S., 666, 668(7, 8), 671(7, 8)
Tamai, M., 756
Tamm, I., 436, 439
Tamura, G., 289
Tamura, R., 386
Tanaka, I., 756
Tanaka, K., 754, 756(107), 757(107)
Tannenbaum, C., 386
Tanner, M. J. A., 286
Tanner, W., 588, 589, 705, 812
Tarentino, A. L., 289
Tarlow, D. M., 427
Tarone, G., 260, 262(9), 321
Tarr, G. E., 537
Tartakoff, A., 347, 755
Tarver, H., 779(34), 780(34), 781
Tashiro, Y., 124, 521, 540
Tashjian, A. H., 779(20), 780(20), 781
Tata, J. R., 3, 122, 555(85, 87), 556(76, 81), 557(87)
Tate, E. A., 336, 337(23)
Tate, S. S., 127, 129(66)
Tatibana, M., 133, 149(88), 753
Taylor, G. M., 468
Taylor, J. M., 14, 24, 43, 725
Taylor, P. B., 357, 363(14)
Teifel, W., 702, 703(12)
Tennekoon, G., 425, 432(18)
Tepperman, K., 350
Ternynck, T., 482
Terry, W., 667
Tharp, D. L., 779(25), 780(25), 781
Thibodeau, S. N., 155
Thines-Sempoux, D., 531
Thom, M., 584, 585
Thomas, D. B., 281

Thomas, M., 717
Thomas, P. E., 532, 533(31), 538
Thomas, R. L., 587
Thomassen, M. S., 548(50), 550
Thompson, I. G., 633
Thompson, L. H., 169, 183
Thompson, R. C., 78, 671
Thompson, T. E., xxxiii
Thorens, B., 469
Thorley-Lawson, D. A., 150
Threlfall, G., 126, 127(65), 435, 496, 541
Tietz, C., 748
Till, J. E., 183
Timoneda, J., 753
Timonen, T., 281
Tischer, E., 126
Tisher, E., 541
Titani, K., 796(26), 801
Titus, D. E., 134, 135(98), 701, 711(10)
Titus-Dillon, P., 742, 748(28)
Tkacz, J. S., 802
Tochino, Y., 133
Toft, D., 760
Toggenburger, G., 386
Tokuyasu, K. T., xlix, 420, 468, 469, 470(15, 22), 472, 478(15, 30), 479(33), 493, 588
Tolbert, N. E., 167, 551, 552(58), 553(58), 639, 700, 702, 703(20)
Tolleshaug, H., 742, 746(22), 748, 749, 750, 751(41), 755(41), 756, 758(124), 759(124), 763
Tomaselli, H., 305
Tomino, S., 559, 560(7), 564(7), 566(7), 567(7), 568(7), 569(7)
Tomita, M., 150, 268, 271, 275(9), 277, 281
Toneguzzo, F., xxxviii, xl(48), 140, 800
Torian, B., 452
Toselli, P. A., 448, 451(32)
Touster, O., 559, 560, 563, 566(8), 764
Towbin, H., 110, 191, 193, 197(3), 198(3), 206, 212
Trachsel, H., 69, 70(16)
Trebilcock, H. A., 641, 642(47)
Trelease, R. N., 700, 714, 715
Trendelenburg, M. F., 597, 604, 606(20), 608(5)
Triemer, D. F., 754
Triemer, R. E., 763
Trimble, R. B., 812
Trouet, A., 533, 737, 742
Trowbridge, I. S., xli, 158, 159(14), 160, 161(14), 182(29), 183(14), 796(28), 800, 801
Trowsdale, J., 325
Trump, B. F., 762
Tse, T. P. H., 43
Tseng, L. F., 333
Tsuji, H., 561, 562(16)
Tsuji, T., 257, 258(4)
Tsuzuki, H., 133
Tuazon, F. E. B., 760, 761(153)
Tucker, P. W., 126
Tuite, M. F., 50
Tulkens, P., 737, 742
Tullis, R. H., 30, 153
Tulsiani, D. R. P., 560, 563
Tuszynski, G., 313
Tuve, K., 779(31), 780(31), 781
Tycko, B., 749
Tyler, J., xlviii
Tzagoloff, A., 418
Tzartos, S. J., 371

U

Uchida, T., 160
Udenfriend, S., 777, 779(2), 780(2)
Uhlenbruck, G., 281
Ukamara, S., 472, 478(30)
Ulane, R., 588
Ullu, E., 683
Ulmanen, I., 125, 127(49), 282, 285, 295(18), 299(18)
Ulrich, A., 770
Ulrich, B. C., 86, 671
Ulrich, B. L., xxxix, xlix(35), 384, 520, 521(1), 523(1), 540
Ulrich, P., 780(60), 782
Umezawa, H., 756, 757(126), 758(126), 759(126)
Ungewickell, E., 299, 317
Unkeless, J. C., 731
Unwin, P. W. T., xli
Ureck, K., 588
Urgani, L. J., 800
Utter, M. F., 555, 556(83)

V

Vaarties, W. J., 760
Vaessen, R. T. M. J., 197, 198(12), 200(12)

Vakil, U. K., 745
Valentine, R. C., 481
Valenzuela, P., 225, 369
Venaman, V., 165
Van Berkel, T. J. C., 755
Van Deenen, L. L. M., 649
Van den Berghe, H., 742, 748(29)
Van Den Bosch, H., 650
van der Ende, A., 35
van der Ley, P. A., 468
Van der Wilden, W., 584(45a), 585, 586, 587, 588(45a)
van Diggelen, O. P., 183
van Eenbergen, J., 38
van Hoof, F., 737
Van Keuren, M. L., 189, 230(10), 231, 232(10), 233(10), 234, 235(10), 238(2), 700
Van Leuven, F., 742, 748(29)
Vannier, C., 407, 415(11)
Van Prooijen-Van Eeden, A., 649
van Rijn, H. J. M., 811
van Rossan, G. D. V., 658
Van Slyke, K., 216
Van Soom, G., 230
Varki, A., xxxvi, xli(38), 736
Varmus, H. E., 126, 227, 541
Vassalli, P., 347, 755
Veres, K., 779(42), 780(42), 782
Verheecke, P., 230
Verhoeyen, M., 126, 127(65), 496, 541
Verma, D. P. S., 135, 144(100), 197, 198(11), 212, 213(12), 532(37), 533, 534(37)
Verney, E., 3
Vigny, M., 358, 361
Vilas, U., 785
Villa-Komaroff, L., 50, 285, 512
Villinger, W., 469, 493
Vincent, A., 332
Virtanen, I., 458, 459, 460(11), 461(11), 463(11), 465(11)
Vlasuk, G. P., 538
Voelker, D. R., xxxiv
Vogel, H. J., 77
Vogelstein, B., 589
Vogt, M., 570, 571
von Bonsdorff, C.-H., 456, 458, 459(8, 10), 465(10), 480
von der Haar, F., 778, 780(15)
von Figura, K., 589, 736
von Heijne, G., xxxix, 405, 669, 671(34)

W

Wachstein, M., 641
Wacker, A., 780(60), 782
Wacker, H., 127, 129(67), 386, 387, 391(8, 9), 400(9, 11, 13, 14), 401(9), 403, 404, 405(9)
Wada, H., 534
Wagner, G. J., 580, 581(3), 582(12, 13), 583(15), 584(13, 16, 45), 585(12, 15, 30), 586(45), 587(30, 45)
Wagner, H., Jr., 638, 647
Wagner, R. R., 800
Wagstaff, W., 286
Waheed, A., 736
Wakeland, J. R., 752, 758(91)
Wald, S. E., 126
Walk, R. A., 702, 703(18)
Walker, C. R., 354, 359(9), 361(9), 363(9)
Walker, J. H., 770
Walker, R. R., 584
Walker-Simmons, M., 581, 582(5), 583(5), 584(5), 585(5)
Wall, D. A., 748
Wall, E. J., 232, 233(13)
Wall, R., 126, 681
Wallace, R., 308
Wallach, D. F. H., 246, 247(15), 252(15), 271, 300, 301(7), 302(7), 613
Waller, J., 782
Walsh, C., 539
Walsh, K. A., 152, 155(8), 531
Walter, P., xxxi, xxxvii, xxxviii(42), xxxix, 41, 45(9), 48, 86(30, 31), 87(12), 91, 93, 119, 123, 136(30), 143(30), 374, 376(12), 531, 669, 670(40, 43), 672(48), 675(48), 681, 682, 683(3, 7, 8), 684(1), 685(3), 688(3, 8), 690(3), 691(1), 692, 693, 696, 778
Waltz, F. G., Jr., 538
Wang, J. K., xli
Wang, K., 235
Wang, S., 75, 81(6), 83(6), 84(6), 668
Wang, S. F., 813
Warburton, M. J., 752
Ward, C. W., 435
Ward, W. F., 756, 757(130), 758(130)
Wardale, J., 207
Waring, G. L., 590
Warms, J. V. B., 753
Warren, G., xxxviii, xl(45), 86, 114, 118(6),

123, 244(4), 245, 295, 325, 326(5), 327(5), 469, 482(25), 483(25), 484(25, 45), 493, 669, 682, 692
Warren, J. R., 549, 551(39), 552(39)
Warren, L., 313, 439
Warren, T. G., 93
Wartiovaara, J., 459
Waser, P. G., 218
Wasniowska, K., 276
Wasserkrug, H., 406
Wasteney, H., 133
Watanabe, K., 133
Watanabe, T., 548(45), 549(45), 550
Waterfield, M. D., 434, 496, 541, 673
Waters, S. P., 584
Watkins, P. A., 427
Watson, D., 160
Wattiaux, R., 558, 723, 764
Waxman, D. J., 539
Waxman, L., 760
Weatherall, D. J., 282
Webb, W. W., xxxiv
Weber, K., xlviii, 112, 239, 241(4), 256, 313, 574
Weber, L. A., 15, 40, 50, 67
Weber, P., 126
Weeks, D. P., 38
Weibel, E. R., 468, 476, 491, 492(14), 536, 546, 547
Weinberg, C. B., 354
Weinberg, M. B., 555, 556(83)
Weiner, A., 313
Weingarten, H. I., 778, 780(12)
Weintraub, H., 193, 206, 212(3)
Weir, E. M., 134, 135(98), 701, 710, 711(10), 712(41), 715(41)
Weiser, M. M., 419, 420, 422(39)
Weissmann, C., 185
Welch, W. J., 126, 496, 512
Weller, J. L., 779(21), 780(21), 781
Welsh, I., 460
Wendberger-Schieweg, G., 701
Wendlberger, G., 337
West, S. B., 532, 535(21), 536(21), 538(21)
Westheimer, F. H., 388, 391(16), 392(16)
Wheeler, G. L., 618, 620, 621
Wheeler, M. A., 621
White, D. A., 752
White, D. L., 397, 655
White, G., 280
White, J., 751
White, J. M., 541, 673
White, J. O., 378
White, M., 299
White, P. B., 779(19), 780
White, W. R., 86, 786, 787(8), 792(8)
Whiteley, N. M., 397, 446
Whitmore, C. F., 164, 183
Whittaker, M., 752
Wibo, M., 531, 533, 742, 745(24), 752(24)
Wicha, M. S., xxx
Wickerham, L. J., 805, 807
Wickner, W., xl, xliii(62), xliv, 418, 669, 670(35), 671(35)
Widnell, C. C., 555(85), 556, 755
Wiederanders, B., 757, 758(132), 759(132)
Wiegers, U., 25, 31
Wielburski, A., 555, 556(80)
Wiemken, A., 581, 584, 585(14a, 29), 587, 588, 589
Wieringa, B., 35
Wieslander, L., 36
Wiesmann, U. N., 745
Wightman, T. M., 763
Wigzell, H., 460
Wilchek, M., 488
Wilcockson, J., 27
Wilcox, D. K., 755
Wilde, C. J., 752
Wildenthal, K., 752, 758(91)
Wiley, D. C., 436
Wilkinson, P. C., 481
Willard, M., xlviii
Willenbrink, J., 585
Williams, A. F., 268
Williams, A. K., 779(24), 780(24), 781
Williams, B. R. G., 70, 71(19)
Williams, C. H., 150
Williams, K. E., 746, 749(42), 755(42), 757, 758(42, 134), 759(134)
Williams, P. B., 126
Williams, R. J., 245
Williamson, R., 30
Willingham, M. C., 452, 468, 742, 746(21), 748
Wills, E. D., 546
Wilson, B. W., 354, 359(9), 361(9), 363(9)
Wilson, D. L., 184, 185(5), 190
Wilson, E. J., 555(86), 556
Wilson, G., 726, 748
Wilson, I. A., 436
Wilson, J. N., 450
Wilson, J. R., 158
Wilson, K. J., 386

Wiman, K., 325, 328(10)
Winchester, R. J., 325
Winzler, R. J., 126, 281
Wirth, D. F., 124(45), 125, 128(45), 512
Wirtz, K. W. A., xxxiii
Wisner-Provost, A., 410, 418(15), 419(15), 420(15), 422(15)
Wissler, R. W., 779(40), 780(40), 781
Witkop, B., 777, 779(2), 780(2)
Wittenbach, V. A., 584
Wold, W., 116
Wolfe, L., 305
Wolitzky, B. A., 758
Wöllert, W., 435, 439(12)
Wollman, Y., 172
Wolska-Mitaszko, B., 50
Wong, L., 725, 726(12)
Wong, S. S. C., 737
Wong, Y.-S., 256
Woo, S. L. C., 85, 122
Wood, C. M., 737
Woodruff, M. L., 617
Woods, D. R., 802
Wool, I. G., 576
Wooten, W. L., 555, 556(78)
Work, T. S., 50
Wright, J. A., 160, 161(26), 182
Wu, H. C., 161
Wulff, J. L., 80
Wyn, Jones, R. G., 585
Wynne, D., 488

Y

Yaffe, D., 571, 572
Yamada, K. M., xxx, 191
Yamada, S. S., 468
Yamagishi, M., 756
Yamamoto, K., 767, 768(12), 769(12), 771(12)
Yamasaki, F., 548(45), 549(45), 550
Yamawaki-Kataoka, Y., 126
Yamazaki, A., 621
Yanagita, T., 551, 552(62), 557(62)
Yanofsky, C., 131, 780(54), 782
Yarden, Y., 746
Yasumura, M., 84
Yasunobu, K. T., 539
Yee, R., 617
Yee, S. I., 574
Yokoyama, M., 417
Yorke, G., 754
Yoshima, H., 268, 277, 281
Young, M. R., 750, 763(75)
Young, N. M., 166
Young, R. W., xxxvi, 609
Yphantis, D., 308
Yu, J., 146
Yu, P., 308
Yuan, P.-M., 539
Yura, T., 75
Yurchenco, P. D., 269

Z

Zabin, I., 669
Zajac, I., 450, 451(40)
Zaldivar, J., 225
Zambrano, F., xxxvi
Zarbowsky, H., 305
Zehler, H., 702(27), 703
Zemell, K. R., 152, 542, 676
Zerinilofsky, A. P., 126
Ziegler, A., 295
Ziemiecki, A., 466
Zilberstein, A., 257, 258(1), 261(1), 266, 795
Zilversmit, D. B., xxxiii, xxxiv, 709
Zimmerman, F. K., 807, 812(19)
Zimmerman, G., 130
Zimmerman, S. B., 30, 153
Zimmerman, W. F., 618
Zimmermann, M., 785
Zimmermann, R., 130, 134, 135(99), 140, 141(113), 710, 711, 712(42)
Zimriak, P., 588
Ziomek, C. A., 422
Ziv, E., 542, 676
Zoccoli, M. A., 531
Zomzley, C. E., 378
Zorn, M. A., 244, 248, 485, 488(1), 495(2), 609, 610(4), 612(4), 613(4), 615(4)
Zubay, G., 82
Zubenko, G. S., 588, 589(69)
Zuber, H., 387, 391(8)
Zubrzycka, E., 570, 574(3, 4)
Zucker, W. V., 50, 69(2)
Zunick, M., 535, 538(56), 539(56)
Zurawski, G., 131
Zweig, S. E., 256
Zwergel, E. E., 427

Subject Index

A

Abelson virus, transforming protein, xli
Abrin
  cytotoxicity, 159
  potentiation, 751
Acetonyl-CoA, 157
2-Acetylaminofluorene
  induction of epoxide hydrolase biosynthesis, 537
  induction of ER protein components, 545
Acetylcholine, receptors
  biosynthesis, 341–350
    early events, 348
    in vitro approach, 367, 368
  cell-free synthesis, 369, 370
  degradation, 350–352
  glycosylation variants, 376
  heavy isotope-labeled
    density shift, 338–340
    separation from normals, 337, 338
  heavy isotopic labeling, 335–341
  intracellular transport, 341–347, 350–352
    effects of inhibitors, 345–347
  measurement, methods, 332–334
  in membrane recycling, evidence against, 348–350
  metabolism, regulation, 352
  subunits
    synthesized *in vitro*
      immunoprecipitation, 370, 371
      integration into microsomal membranes, 374–378
      primary translation product, electrophoretic analysis, 371–374
Acetylcholinesterase
  active site-directed inhibitors in study of, 359–361
  biosynthesis and intracellular transport, 347
  cell surface, transport and accumulation, 364, 365
  cell surface activity, assay, 354–357
  cell surface versus intracellular, 354, 355, 357–359
  intracellular
    recovery in culture, after DFP treatment, 361–363
    transport, 363, 364
  molecular forms
    assembly, 366
    localization, in culture, 365, 366
  newly synthesized, orientation, 359, 360
  properties, 353, 354
  secretion, 363, 364
  subcellular distribution, 354–361
  synthesis, 361–363
N-Acetylglucosaminidase, in yeast secretory mutants, 817, 818, 820, 823–825
Acid phosphatase
  accumulation in yeast secretory mutants, localization, 811
  secretion, in yeast, 802
  in yeast secretory mutants, 820, 825
Acridine orange, lysosome-inhibitory effects, 740
Acyltransferase
  activity, localization, species differences, 652–654
  cytochemical localization, 649–651
Adenosine triphosphatase
  Ca^{2+}
    in vitro insertion into microsomal membrane
      demonstration by alkali treatment, 146, 147
      demonstration by deoxycholate solubilization, 147
    membrane insertion, 128
    of sarcoplasmic reticulum, synthesis, 124

Ca^{2+}, Mg^{2+}
antisera, preparation, 574
immunoprecipitation, 574, 575
mRNA, sucrose gradient fractionation, 578
synthesis, 570
F_1
β-subunit, identification and quantitation, 199, 200
γ-subunit, antiserum, characterization, 203, 204
Na^+, K^+
antibody, 647, 648
assay, in avian salt gland fractions, 638–640
of avian salt gland basolateral membrane, 634, 635, 643, 644
catalytic subunit, 542
cytochemistry, 640–643
with ouabain, 643–646
heavy chains, heavy isotope labeling, 336
immunocytochemical localization, in avian salt gland, 646–648
intracellular pathway, study, 648
localization, 127
marker for basolateral membrane, 420, 421
phospholipid requirement, 649
α-subunit, membrane insertion, 128, 129
β-subunit, 127
synthesis, 125
ADP-ATP translocator, 130
Agglutinin
Lens culinaris, CHO cells resistant to, 163
wheat germ
CHO cells resistant to, 163
probe for glycoproteins transferred to nitrocellulose, 205
Albumin
bovine serum
embedding technique, for retinal membrane protein study, 485–495
lysosomal enzymes in, 731
marker protein, for reinitiation of protein synthesis *in vitro,* by bound polysomes or rough microsomes, 14–21
silver staining, linearity, 236, 237
subcellular location, in rat liver, 724, 726
Alkylamine
affecting lysosomes at low concentrations, 740
effects
on cellular functions, 746, 747
on lysosomal proteolytic activity, 739, 740, 745
on lysosomal swelling, 739, 740, 743–745
Alloisoleucine, 779, 780
Alphavirus
mutants, with defects in structural proteins, identification, 458
ts mutants, induction, 456
Amantadine
action, mechanism, 751
lysosome-inhibitory effects, 740
Amine
cellular uptake of, 738–742
effects
on cellular functions, 745–747
on endogenous protein degradation, 752–754
on exogenous protein degradation, 755
on receptor-ligand dissociation, 747–749, 752
on toxin entry, 751
on virus penetration of cell, 750, 751
entry into lysosomes, 742, 743
inhibition of lysosomal degradation, 751–755
inhibition of vacuolar membrane fusion, 750
weak-base
compared to proteinase inhibitors, 756
inhibitors of lysosomal function, 737–755
Amino acid
analogs
in analysis of protein processing, 777–784
advantages, 778
applications, 777, 778
concentration, selection, 780–783

effects on protein processing, 783, 784
incorporation into protein, demonstration, 783
selection, 778–780
heavy
from algae, 336, 337
for heavy-isotope density-shift study, 425, 426
Amino alcohol, lysosomal-inhibitory effects, 739, 741
ω-Aminoalkylagarose resin, signal recognition particle chromatography on, 685–689
2-Amino-1-butanol, lysosome-inhibitory effects, 739, 741
2-Amino-3-chlorobutyric acid, 780
2-Aminoethanol, lysosome-inhibitory effects, 739, 741
Aminopeptidase, 406–423
anchor peptide, 407
amino acid composition, 413
integration in membrane, 416, 417
molecular weight determination, 411–414
position in molecule, 414
preparation, 410
antibody
in localization of hydrophilic domain, 415, 416
obtaining, 410
in purification of enzyme, 409, 410
biosynthesis, relation to enzyme integration in membrane, 417–423
detergent form, molecular weight, 411
homology, 406
hydrophilic domain
immunochemical localization, 414–416
molecular weight, 411
molecular weight determination, 411–414
newly synthesized
immunoprecipitation, 422
incorporation into basolateral membrane, 421–423
incorporation into brush-border membrane, 421–423
solubilization, 422
proteolysis, 407
structure, 411, 412, 417, 418
relation to integration in brush border membrane, 407–417
Aminopeptidase A, 406
pig, molecular weight, 411
Aminopeptidase M
hog kidney
inhibitors, 138
pH optimum, 138
specificity, 138
mouse submaxillaris gland
inhibitors, 138
pH optimum, 138
specificity, 138
Aminopeptidase N, 406
antibody, precipitation of enterocyte basolateral membranes, 421, 422
pig, molecular weight, 411
rabbit
anchor peptide, amino acid composition, 413
detergent form, purification, 407–409
molecular weight, 411
N-terminal sequence, 414, 415
protease form, purification, 407–410
Aminophenyl thioether paper, preparation, 207
3-Amino-1-propanol, lysosome-inhibitory effects, 739, 741
4-Aminopyridine, lysosome-inhibitory effects, 740
4-Aminoquinaldine, lysosome-inhibitory effects, 740
3-Amino-1,2,4-triazoylalanine, 779
Ammonia
effects
on cellular function, 746
on protein secretion, 751
inhibitor of lysosomal function, 737–739, 743
Amoeba, *see* Slime mold
Amylase
nascent, biological activity, 110
resistance to posttranslational proteolysis, 101–103
synthesis, cell-free vs. cellular, 96, 97

α-Amylase, precursor, processing, in maize, 718
Amylose, chloride-oxidized, lysosome-inhibitory effects, 763
Anans sativus, bromelain, 137
Angiotensin, protease-catalyzed synthesis, 133
Anemia, hemolytic, spectrin alterations in, 305
Ankyrin, 257
  cytoskeletal association, 313
  fragment containing spectrin-binding site, purification, 320, 321
  ^{125}I-labeled, binding to inverted erythrocyte vesicles, 324
  purification, 318–320
  radiolabeling, 322, 323
  storage, 320
Antibody
  aminopeptidase, 410, 415
  cathepsin D, 770
  egasyn, 566–568
  β-glucuronidase, 770
  incubation with, in immunoprecipitation, 113, 114
  invertase, 813
  labeling of intracellular antigens, 467–469
  to myelin-specific proteins, preparation, 385
  Na^+, K^+-ATPase, 647, 648
  N-terminal amino acids, analysis, 397
  against retinal proteins
    labeling with, 488–491
    monoclonal vs. polyclonal, choice of, 493, 494
    preparation, for use in immunocytochemistry, 488
  ribophorin, preparation, 525, 526
  against Semliki Forest virus spike proteins
    labeling of Semliki Forest virus injected cells, 482, 483
    preparation, 480–482
Antigen, *see also* Histocompatibility antigen
  detection, in immune complex mixture, 198–200
  intracellular
    labeling, 267–269
    study, techniques, 485, 486
  membrane, two-dimensional immunoelectrophoresis, 244–257
Antipain
  isolation, 756
  lysosome-inhibitory effects, 759
Antiserum, characterization, 200–205
Apocytochrome *c*
  addressing signal, 145
  CNBr fragment containing addressing signal, subfragmentation by protease, 132
  primary addressing signal, 130, 131
  subfragment, in protease-catalyzed synthesis of chimeric polypeptides, 133, 134
Aquacide II, 486
Arginine, analog, 779
Asialofetuin, degradation, inhibition, 755, 757, 763, 764
Asparagine, analog, 779
Aspartic acid, analogs, 779
Atropine, lysosome-inhibitory effects, 740, 742
Autophagy, 753
  inhibitors, 738, 754, 761
Autoradiography, effect of silver staining, 234
Axon fragments, myelinated
  fractionation from rat brain, 378
  protein purification from, 382
7-Azatryptophan, 780
Azetidine-2-carboxylic acid, 779

B

Bacillus polymyxa, neutral protease, 138
Bacillus subtilis, subtilisin, 137
Bacillus thermoproteolyticus, thermolysin, 139
Bacterioopsin, xli
Band III protein, *see* Protein, band III
Beetroot
  tonoplast
    composition, 586, 587
    isolation, 586
  vacuoles, preparation, 583
Benzyloxycarbonylalanine chloromethyl ketone, inhibition of *E. coli* membrane protein biosynthesis, 78, 79

Beta vulgaris, see Beetroot
Bovine serum albumin, *see* Albumin
Brain, rat, protein fractionation, 378–385
Brainstem, subcellular fractionation, for RNA extraction, 379–382
Bromelain, *Anans sativus*
 inhibitors, 137
 pH optimum, 137
 specificity, 137
Brush border membrane
 action of papain on, 386, 387
 aminopeptidase incorporation
  kinetics, 422, 423
  relation to enzyme biosynthesis, 417–423
  relation to enzyme structure, 407–417
 sucrase-isomaltase, 386–406
 vesicles, amidination with DAP, 393, 394
α-Bungarotoxin
 properties, 334
 in study of acetylcholine receptors, 333, 334, 338–345, 348–351
tert-Butylamine, lysosome-inhibitory effects, 739
BW284c51, acetylcholinesterase inhibitor, 360, 365

C

Ca^{2+}-binding protein, high-affinity, in sarcoplasmic reticulum, 570
 antisera, 574
Cadaverine, lysosome-inhibitory effects, 739
Calsequestrin, 124
 antisera, preparation, 574
 immunoprecipitation, 574, 575
 mRNA, sucrose gradient fractionation, 578
 purification, 574
 synthesis, 570
Canavanine, 779
Carbamol phosphate synthetase, posttranslational insertion, 131
Carbonate dehydratase, silver-stained, quantitative analysis, 236, 237
Carboxypeptidase A, bovine pancreas
 inhibitors, 138
 pH optimum, 138
 specificity, 138
Carboxypeptidase B, bovine pancreas
 inhibitors, 138
 pH optimum, 138
 specificity, 138
Carboxypeptidase Y, 588, 589
 yeast
  inhibitors, 137
  pH optimum, 137
  specificity, 137
Casein, 38
Catalase
 biosynthesis, *in vitro,* 711
 localization, 703
 of microbody, 714, 715
 molecular properties, 703
 peroxisomal
  cytochemistry, 727
  in vitro translation, 726, 727
  peptide mapping, 727
 silver-stained, quantitative analysis, 236, 237
 subcellular location, in rat liver, 724–726
 synthesis, 134
Catecholamine, secretion, effects of amines, 751
Cathepsin, inhibition, 759
Cathepsin C, bovine spleen
 inhibitors, 137
 pH optimum, 137
 specificity, 137
Cathepsin D, 38, 124
 antibody, preparation, 770
 assay, 767
 cotranslational processing and vectorial discharge, 775, 776
 immunoprecipitation, 773, 776
 purification, 767–771
 synthesis, *in vitro,* 770–777
Cell
 amine uptake, 739–742
 animal, lectin-resistant, 157
  biochemical characterization, 181, 182
  cloning, 178
  cross-reactivities, 171

D_{10} values, 178
genetic characterization, 179–181
glycosylation-defective, 160
identification of isolates, 175, 176
mixed lectin selections, 172, 173
multiple mutations, 171
nomenclature, 161–163
phenotype testing, 176–178
picking colonies, 176
sequential lectin selection protocol, 173, 174
single lectin selections, 171, 172
types other than CHO, 182, 183
baby hamster kidney
Semliki Forest virus-infected, labeling, 479–482
for viral replication, 437
chick embryo
preparation, 454, 455
rough microsomes, 23, 24
for Semliki Forest virus propagation, 454
Chinese hamster, lectin-resistant, survival curves, 169, 170
Chinese hamster ovary
culture, 164, 165
labeling
in presence of NH_4Cl, 731
with radioactive amino acids, 729, 730
with radioactive mannose, 730
with radioactive phosphate, 730
lectin-resistant, 160
nomenclature, 163
selection protocols, 171–182
mutagenesis, 165, 166
choices, for study of histocompatibility antigens, 326
compartment, evolution, 677–679
culture
viral inoculation, 437, 438
virus-infected, radioactive labeling, 436
virus replication in, 437–440
cultured, labeling, 729–731
embedding, for microtomy, 467–469
eukaryotic
DNA injection into, by micropipette, 499, 504–510
expression of viral membrane proteins, 496–511
evolution, 676, 677
fixation, for immunocytochemistry, 470
growth, for virus receptor assay, 446
HeLa
lysate, 50
plasma membrane solubilization, 448
human fibroblast, labeling
in presence of NH_4Cl, 731
with radioactive amino acids, 729, 730
with radioactive mannose, 730
human fibroblast, labeling, with radioactive phosphate, 730
K562 line, 281
culture, 283
glycophorin A in, 282
mRNA, isolation, 284, 285
pulse-chase labeling with [^{35}S]methionine, 289
radioactive metabolic labeling, 283, 284
source, 282, 283
labeled, harvest, for lysosomal enzyme extraction, 731, 732
lines
L6, culture, 572, 573
L8, culture, 572, 573
for viral replication, 437
liver, monolayer culture, 426, 427
lysate
bacterial, 50
Chinese hamster ovary, 50
mouse Krebs ascites, 50
yeast, 50
Madin–Darby canine kidney, xlvi, 436, 437
media, labeled, harvest, for lysosomal enzyme extraction, 731, 732
microinjected, immunofluorescence, 499–501, 510, 511
mouse lymphoma, lectin-resistant, 182, 183
mouse myeloma, ribosome preparation, 528
muscle, culture, for acetylcholine receptor study, 334, 335
nucleus, 589

primordial, 677, 678
proteinase K-SDS digestion, 26–28
Raji line, HLA-DR antigen, 326–330
rat liver, fractionation, 8–12, 21–23
rat myoblast, lectin-resistant, 182, 183
spleen
erythropoietic,
labeling with [^{35}S]methionine, 259
preparation, 258, 259
membrane, preparation, 263, 264
microsomes, preparation, 264
mRNA, preparation, 266, 267
pulse-labeled, chymotrypsin digestion, 264
spreading, inhibition, by amines, 750
types, used in heavy-isotope density-shift study of receptors, 425
virus-infected
cytoplasmic membrane preparation, 439
plasma membrane, isolation, 439, 440
Chlamydomonas
acyltransferase activity, localization, 654
membrane assembly, li–liii
Chlorella pyrenoidosa, isolation of heavy amino acids from, 425
Chloroamine, lysosome-inhibitory effects, 739
Chloroplast
posttranslational insertion of proteins into, 129–134
protein translocation in, xlii, xliii, lv, 666–668
Chloroquine
effects on cellular functions, 745, 746
lysosome-inhibitory effects, 740, 742, 745, 746
CHO cell, *see* Cell, Chinese hamster ovary
Cholesterol, synthesis, xxxiv
Choline, analog, 655
Choline *O*-acetyltransferase, localization, 651
Chymostatin
isolation, 756
lysosome-inhibitory effects, 757
Chymotrypsin
amounts, for proteolysis in peptide mapping, 225
in posttranslational proteolysis, 101–105, 121
properties, 136
Sendai virus mutant activated by, 442
α-Chymotrypsin, bovine pancreas
inhibitors, 137
pH optimum, 137
specificity, 137
Chymotrypsinogen, synthesis, 96
Citrate synthase
localization, in endoplasmic reticulum, 707, 708
of microbody, 703
Clathrin, in coated vesicles, xlvii, xlix
Clofibrate, peroxisome inducer, 548, 550
Clostridium histolyticum, clostripain, 138
Clostripain, *C. histolyticum*
inhibitor, 138
pH optimum, 138
specificity, 138
Coated pit, xlvii–xlix
Coated vesicle, xlvii, xlviii
Cocaine, lysosome-inhibitory effects, 740
Colchicine, lysosome-inhibitory effects, 762
Colloidal gold, 468, 478, 479
Concanavalin A
carbohydrate binding by, 158, 159
CHO cells resistant to, 163
lysosome-inhibitory effects, 763
probe
for cell organelle visualization, 461, 462
for glycoproteins transferred to nitrocellulose, 205
Corona virus, glycoprotein, 796
Coxsackievirus A, 443
attachment, 445
Coxsackievirus B, 443
receptor, 444
solubilization, 448
Coxsackievirus B3
A particles, 452
binding, to solubilized receptors, assay, 449

C particles, 452
receptors, quantitation, 447
Crinophagy, 753
cis-Crotylglycine, 779
Cyanogen bromide, fragmentation, of rough microsome translation products, 19–21
Cyclic guanosine monophosphate, in visual excitation, 617, 618
Cyclic guanosine monophosphate phosphodiesterase
activation by transducin α-subunit, 623–625
properties, 621
purification, 625, 626
γ regulatory subunit, 621
purification, 626, 627
trypsin-activated, preparation, 626
in visual excitation, 617, 618
Cyclic nucleotide cascade, of vision, information flow in, 617, 618
Cyclobutyglycine, 780
Cycloheximide
inhibitor of viral glycoprotein synthesis, 482, 484
lysosome-inhibitory effects, 763, 764
Cyclopropylglycine, 780
Cytochrome b_5
hydrophobic domain, in binding to microsomal membrane, 144, 145
induction in ER, by drugs, 545
membrane insertion, xliv, 134, 135
mRNA, localization, 534
posttranslational insertion into ER membranes, 533–535
verification, by high salt treatment, 147
subfragment, in protease-catalyzed synthesis of chimeric polypeptide, 134
Cytochrome *c*
horse heart, cyanogen bromide cleavage products, molecular weights, 240–242
newly synthesized, proteolysis, by endogenous protease, 142, 143
posttranslational transfer into mitochondria, study, 132
Cytochrome c_1, yeast mutants lacking, immunological screening for, 200, 202
Cytochrome *c* oxidase, activity, in rat liver microsome preparation, 12, 13
Cytochrome oxidase, posttranslational insertion, 131
Cytochrome *P*-450, 538–540
amino acid sequencing, 539
biosynthesis, 538, 539
cDNA clones, 538, 539
cotranslational insertion into ER membrane, 535
induction in ER, by drugs, 544–546
localization, 127
membrane insertion, 128
synthesis, 124, 125
transmembrane disposition, 540
Cytochrome *P*-450 reductase, cotranslational insertion into ER membrane, 535

D

Dansylcadaverine, lysosome-inhibitory effects, 740
DEHP, *see* Di(2-ethylhexyl) phthalate
trans-4,5-Dehydrolysine, 779
3,4-Dehydroproline, 779
Deoxycholate, 112
stepwise solubilization of microsomal membranes, 147
Deoxynucleic acid
fragments
ligation, in low gelling temperature agarose, 502, 503
separation, in low gelling temperature agarose, 502
introduction to eukaryotic cell, by micropipette, 499, 504–510
plasmid, extraction, 503, 504
in polyacrylamide gels, staining, 235–237
processing, in presence of low gelling temperature agarose, 499–503
removal, from RNA extraction, 30
Diamine, lysosome-inhibitory effects, 739, 741
2,6-Diamino-4-hexynoic acid, 779
1,3-Diaminopropane, lysosome-inhibitory effects, 739

Diazobenzyloxymethyl paper
with affinity ligand, protein transfer to, 214, 215
protein transfer to, 206
protein transfers, reuse, 213, 214
Diazomethane, 759, 760
Diazophenyl thioether paper
identification of proteins bound to, 212–215
protein transfer to, 207, 209–212
Dibucaine
effects on cellular functions, 746
effect on posttranslational proteolysis, 101–104
lysosome-inhibitory effects, 740
molecular structure, 104
Dictyostelium discoideum, see Slime mold
Diethylamine, lysosome-inhibitory effects, 739
N,N-Diethylaminoethylamine, lysosome-inhibitory effects, 739, 741
Di(2-ethylhexyl)phthalate
inducer of mitochondria, 557
peroxisome inducer, 548, 550
Diethyl pyrocarbonate, 24, 25
Digitonin, for demonstration of protein transfer to mitochondria, 148, 149
2,5-Dihydrophenylalanine, 779
3,4-Dihydroxyphenylalanine, 780
Diisopropyl fluorophosphate, inhibitor of acetylcholinesterase, 359–363, 366
3-Dimethyl-2-[(acetimidoxyethyl)-ammonio]propanesulfonic acid, 393
Dimethylamine, lysosome-inhibitory effects, 739
2-Dimethylaminoethanol, lysosome-inhibitory effects, 739, 741
6-Dimethylaminopurine riboside, 761
4-Dimethylaminopyridine, lysosome-inhibitory effects, 740
N,N-Dimethylbenzylamine, lysosome-inhibitory effects, 740
N,N-Dimethyl-3-chloropropylamine, lysosome-inhibitory effects, 739
N,N'-Dimethylethylenediamine, lysosome-inhibitory effects, 739, 741
Dimethyl sulfoxide, 216
cautions with, 217
2,5-Diphenyloxazole, scintillant, for fluorography, 216–218
Diphosphatidylglycerol, synthesis, site, xxxiii
Diphtheria toxin, cell penetration, inhibition, by amines, 751
DMSO, *see* Dimethyl sulfoxide
Docking protein, xxxix, 86, 123, 683, 692
active domain, assay, 693–696
purification, 696–698
analysis, 698–700
Drosophila melanogaster, embryo
acyltransferase activity, localization, 654
fractionation, 590–592
for nuclear matrix study, 590, 591
nuclei, purification, 591–593
subcellular fractionation, 592–595
Drug, *see also* specific drug
induction of membrane, 542–557

E

E-64
isolation, 756
lysosome-inhibitory effects, 759
Echothiophate, acetylcholinesterase inhibitor, 361, 366
Echovirus, 443
Egasyn, 558, 559
antiserum, 566, 567
^{125}I-labeled, titration with anti-egasyn antibody, 568
isolation, from mouse tissue, 564–566
labeling, with ^{125}I, 567, 568
radioimmunoassay, 568, 569
Elastase, Sendai virus mutants activated by, 442, 443
Electron microscopy, *see also* Immune electron microscopy
of antibody-labeled retinal proteins, 491–494
of yeast secretory mutants, 808–811
Electrophoresis, *see also* Immunoelectrophoresis
gel, solutions, 223, 224
identification of proteins, limits, 222
polyacrylamide gel
detection, of protein-protein association, 304
gel preparation, for fluorography, 216

low molecular weight slab gel
apparatus, 242, 243
conditions, 244
gel preparation, 242, 243
for resolution of oligopeptides, 239–244
sample preparation, 243, 244
staining, 244
silver staining methods, 230–239
artifacts, 238, 239
destaining procedures, 233
and gel thickness, 238
handling gel, 239
image intensification, 232, 233
precipitate on gel surfaces with, 238
protein detection, 238
recycling for increased sensitivity, 232
sensitivity, 233, 237, 238
water for, 237, 239
of spectrin oligomers, 298–304
apparatus, 300
buffer, 300
drying, 304
gel loading, 302
gel preparation, 300, 301
handling, 304
quantitative analysis, 304
running, 302
slab gel preparation, 301
staining, 302–304
of proteins, 205, 206
SDS–PAGE, 208
of glycophorins, 272, 273
of membrane proteins, 192–195
two-dimensional, analytical, of spectrin peptides, 309
two-dimensional polyacrylamide gel, of membrane proteins, 184–192
autoradiography, 190
electroblot, 191
fluorography, 189, 190
immunolocalization, 191
lectin-binding studies, 190, 191
peptide mapping, 191, 192
photography, 189
silver stain, 189
staining, 189
two-dimensional, preparative, of spectrin peptides, 306, 307
Embedding media
low-temperature, 469
for retina thin sections, 492, 493
water-soluble, 469
Embedding technique, using bovine serum albumin, 485–487
Encephalomyocarditis virus, binding to host cells, 448
Endo-β-*N*-acetylglucosaminidase H
digestion of acetylcholine receptor subunits, 376–378
source, 376–378
Endocytosis, xlv, liv
Endoglycosidase H
digestion, of band III protein immunoprecipitants, 265, 266
treatment, of glycophorin A, 287–289
Endopeptidase, proline-specific, from *F. meningosepticum*
inhibitors, 137
pH optimum, 137
specificity, 137
Endoplasmic reticulum
hepatocyte, protein biosynthesis, 530–542
induction of drug metabolizing enzymes on, 544–547
methodology, 547–549
induction of phospholipid components, by drugs, li, 545–547
induction of protein components, li, 544–546
membrane
degradation, after cessation of drug treatment, 547, 549
polypeptides inserted cotranslationally into, 535, 536
polypeptides inserted posttranslationally into, 533–535
posttranslational insertion of proteins into, 135
protein insertion, xlv, xlvi, 121–129
protein translocation across, 670, 671
protein transport in, xxxvi–xl, xliv,
rough
maize endosperm, in study of preproteins, 716

membrane, protein translocation, 84–87, 94–111
posttranslocational pathways in, 678–681
protein translocation across, 666–668
retinal, labeling, 494, 495
ribophorins, 520, 521
Semliki Forest virus spike protein transport through, 482–485
smooth, biogenesis, in hepatocyte, 651
visualization, in virus-infected cells, 463
Endoproteinase lysine-C, *Lysobacter enzymogenes*
inhibitors, 137
pH optimum, 137
specificity, 137
Endosome, receptor trapping, caused by amines, 748, 749
EnHance, 217, 218
Enterocyte basolateral membrane
aminopeptidase incorporation, kinetics, 421–423
brush-border hydrolases in, 420–422
precipitation, by specific anti-aminopeptidase N antibodies, 421, 422
Ep-475, 756
lysosome-inhibitory effects, 759
Ephedrine, lysosome-inhibitory effects, 740
Epon embedding technique, 468, 469
Epoxide hydrolase, 537, 538
biosynthesis, drug induced, 537
cDNA clones, 538
cotranslational insertion into ER membrane, 535, 536
microsomal, induction in ER, by drugs, 544, 545
L-*trans*-Epoxysuccinylleucylagmatine, *see* E-64
L-*trans*-Epoxysuccinylleucylamidoisopentane, *see* Ep-475
Erythrocyte
anion transport protein, 257–267
band III, *see* Protein, band III
fatty acid acylated, 796
ghosts, preparation, 314, 315
glycophorins, 268–280
infrastructure, xlviii
inside-out vesicles, spectrin-depleted, preparation, 315, 316
lacking glycophorins, 280
membrane, band 3, *see* Protein, band III
membrane proteins, release, by NaI treatment, 147, 148
membrane skeleton
proteins, 305, 313–324
radiolabeling, for binding studies, 322, 323
structure, 298, 299
methionine-labeled, preparation, 259
radioactive cell surface labeling, 283
receptors, for picornaviruses, 447
trypsin digestion, for release of cell surface glycopeptides, 274–276
X. laevis
nuclei
extraction, 605–607
isolation, 605
pore complex-lamina structure, 604–607
Escherichia coli
lipoprotein, synthesis, inhibitors, 78, 79
membrane biogenesis, 74–84
phenethyl alcohol treated, semi-*in vitro* protein synthesis, 80, 81
tRNA acylating enzyme, 153, 156
tRNA formylating enzyme, 153
Eserine, lysosome-inhibitory effects, 740
Esterase, assay, in avian salt gland, 638, 639
Ethionine, 779
Ethylamine, lysosome-inhibitory effects, 739
Ethylenediamine, lysosome-inhibitory effects, 739
O-Ethylthreonine, 779
Eukaryote, membrane biogenesis, xxix, xxx, xxxiii, xxxv
Exocytosis, xlv, liv

F

Fatty acid
acylated proteins
characteristics, 800
detection, 797–799

distribution, 796, 797
acylation, xli, liii
of eukaryotic cell proteins, 795–801
mechanism, 799, 800
site, 799, 800
analog, in lipid biogenesis study, 655
analysis, 799
Ferritin, 468, 479
silver-stained, quantitative analysis, 236, 237
Fibroblast
chick embryo, *see* Cell, chick embryo
human embryonal, organelle visualization with lectin probes, 461, 462
Ficin, *F. carica*
inhibitors, 138
pH optimum, 138
specificity, 138
Ficus carica, ficin, 138
Flavobacterium meningosepticum, proline-specific endopeptidase, 137
threo-3-Fluoroasparagine, 779
erythro-3-Fluoraspartic acid, 779
threo-3-Fluoroaspartic acid, 779
Fluorography
of agarose gels, 220
definition, 215
effect of silver staining, 234
film image, intensifying, on X-ray film, 221, 222
intensifying screen, 215, 216
for isotope detection, in PAGE, 189, 190, 215–220
with PAGE
exposure of gel to film, 218
flashing film to linearize its response, 219
gel dryers, 218
gel preparation, 216
quantitation, 219, 220
radioactive ink scribing on gel, 219
for peptide mapping visualization, 229
procedures
PPO-DMSO, 217, 218
types, 215, 216
scintillants, 215–218
of solid porous supports, 221
of sucrase-isomaltase precursor synthesized *in vitro,* 402–404
2-Fluorohistidine, 779
4-Fluoroisoleucine, 779
2-Fluorophenylalanine, 779
3-Fluorophenylalanine, 779
4-Fluorophenylalanine, 779
cis-4-Fluoroproline, 779
trans-4-Fluoroproline, 779
4-Fluorotryptophan, 780
5-Fluorotryptophan, 780
6-Fluorotryptophan, 780
m-Fluorotyrosine, 780
5-Fluorouracil, mutagenesis of influenza A virus, 440, 441
Fodrin, xlviii
Foot-and-mouth disease virus, 443
attachment, 445
Fowl plague virus
chemical mutagenesis, 440, 441
glycoprotein, fatty acid acylation, 799, 800
Frozen thin section tissue preservation, 469–479
3-Furyl-3-alanine, 779
Fusion protein, paramyxovirus, 434
posttranslational cleavage, 434
protease sensitivity mutants, 442, 443

G

Galactolipid, synthesis, xxxv
β-Galactosidase-2, in yeast secretory mutants, 818, 819
Globin, 38, 51, 52
Globomycin, effect on *E. coli* membrane protein biogenesis, 78, 79
Glucocerebrosidase, 38
Glucose-6-phosphatase, lii
cytochemistry, in peroxisome, 727
β-Glucosidase, in yeast secretory mutants, 818, 820, 827
β-Glucuronidase, 38, 124
antibody, preparation, 770
antibody affinity chromatography, 561
antisera, preparation, 560
assay, 559, 560, 765
cotranslational translocation and processing, 774, 776

electrophoresis, 560
immunoprecipitation, 773, 776
intracellular localization, 557, 558
lysosomal, 557–559
purification, 560
microsomal, 557–559
purification, from mouse liver or kidney, 561–563
rat, 563
protease treatment, 563, 564
purification, 559–564, 765-767
species differences, 559
synthesis, *in vitro,* 770–777
Glucuronidase binding protein, from rat, 566
γ-Glutamyltransferase
cleavage by organelle-specific protease, 129
intracellular transport, 127
Glutaraldehyde, fixative, for immunocytochemistry, 470
Glycolate oxidase, of microbody, 703, 715
Glycophorin, 127
A, 268
antiserum, preparation, 286
biogenesis *in vitro,* 281–289
cell-free synthesis, 291–296
endoglycosidase H treatment, 287–289
immune precipitation, 287, 288
isolation, 271–273
in K562 cells, 282
lectin-Sepharose affinity chromatography, 286, 287
N-glycosylation, inhibition, 289–292
O-glycosylation, 295, 297, 298
orientation, 280
phosphate in, demonstration, 291, 293
polyacrylamide slab gel electrophoresis, 288, 289
precursor, identification, 288, 289
purity, 276, 277
structure, 278, 279
sulfate in, demonstration, 291, 294
B, 268
isolation, 272–274
purity, 277
C, 268
isolation, 272–274
crude fraction, preparation, 269–271
function, 268, 269
glycosylated (cell surface) domains, preparation, 274–276
isolation, 269–276
properties, 277–279
red cells lacking, 280
sources, 268
synthesis, 125
Glycoprotein, *see also* Glycophorin
brush-border, integration in membrane, 419, 420
canine pancreatic, synthesis, 96
core glycosylation, in microsomal membranes, 92
integral membrane, synthesis, 296, 297
intrinsic, in sarcoplasmic reticulum, 570
antisera, 574
myelin-associated, 383
myxovirus
activity, plaque assay, 436
biosynthesis, 434–443
carbohydrate side chains
radiolabeling, 438
types, 435
functions, 434
hemadsorption assay, 436, 437
hemagglutination titration, 436
hemolysis assay, 437
incorporated into virion, labeling, 438, 439
localization in cell, 439
by immune electron microscopy, 440
processing, 434, 435
synthesis, 434
transport, influenza virus mutants with defects in, 440–442
types, 434
secretory, interaction with cell membrane, xxx, xxxi
Semliki Forest virus, spike
intracellular transport, 482–485
protein micelle preparation from, 480, 481

synthesis, 795
sites, xxxv, xxxvi
transferred to nitrocellulose sheets, identification, 205
vesicular stomatitis virus, 257, 258, 261
viral envelope, synthesis, 124, 125
virus
fatty acid acylation, 795, 796, 799, 800
intracellular localization with organelle-specific lectins, 461–465
intracellular transport, immunoelectron microscopic study, 466–485
root, 800
virus membrane
biosynthesis *in vitro,* 512–519
transport, 453–465, 541, 542
Glycoprotein G, integration into membrane, 675
Glyoxysome
enzyme
monomeric form, detection in cytosol, 705–708
oligomeric form, detection in cytosol, 705–708
fractions, electrophoresis, 704
membrane, flotation, on sucrose gradient, 701, 702
proteins
biosynthesis, 700–715
in vitro, 710–714
kinetic studies *in vivo,* 708–710
import, *in vitro,* 711–714
protein components, analysis, 700–705
transition to leaf peroxisome, 714, 715
Golgi apparatus
acyltransferase activity in, in avian salt gland, 652–654
in aminopeptidase integration in membrane, 419, 420, 422
visualization, in virus-infected cells, 463
Golgi cisternae, Semliki Forest virus spike glycoprotein transport through, 482–485
Golgi complex
protein transport in, xxxvi, xxxvii
visualization, with lectin probes, 461, 462
G protein, xlix, 127
in adenylate cyclase system, homology with transducin, 620, 621
Guanidine, lysosome-inhibitory effects, 739, 741
Gum arabic, purification, 604, 605

H

Harvey sarcoma virus, transforming protein, xli
protocol, 429–432
of insulin receptor transit to and from cell surface, 431–433
of receptors, cell types used, 425
Heavy-isotope density-shift study
of insulin, 423–433
heavy amino acids used, 424–426
principle, 424
Heavy isotope labeling, of acetylcholine receptors, 335–341
Helical hairpin hypothesis, xxxix, xl, 405, 419
Helix pomatia agglutinin, probe, for cell organelle visualization, 461, 462
Hemagglutinin
fatty acid acylated, 796
influenza virus, 434
posttranslational cleavage, 434
primary structure analysis, types used for, 435, 436
transport, influenza virus mutants with defects in, 440–442
Hemagglutinin-neuraminidase, paramyxovirus, 434
posttranslational cleavage, 434
Hemin, control of protein initiation, in reticulocyte lysates, 50, 69, 70
Hepatocyte
changes, after *in vivo* administration of hepatocyte inducers, 551–553
differentiation, xxix, li, lii
endoplasmic reticulum
electron transport chains, 531, 532
enzyme systems, 531
lipid biosynthesis, 531

proteins, biosynthesis, 530–542
function, effects of ammonia, 751
smooth endoplasmic reticulum, induction by phenobarbital, 651
treated with lysosomal inhibitors, 744
Hexokinase, yeast, antiserum, characterization, 203, 204
Hippeastrum
cellulysin digestion, 581
tonoplast
composition, 586, 587
isolation, 585, 586
vacuoles, isolation, 582
Histidine, analogs, 779
Histocompatibility antigen, 796, 797
assembly, 325–331
choice of cells, for study of, 326
classes, 325
H-2
assembly, 328
immunocharacterization, 326, 327
HLA-DR
assembly *in vivo,* 328–330
immunocharacterization, 326, 327
oligomeric, characterization, 328
immunoprecipitation, protocol, 327
membrane-spanning segment, size determination, 330, 331
mRNA coding for, isolation, 326
nomenclature, 325
Hormone, receptors, down regulation, 749
Hydrolase, brush-border, 420, 421
presence in enterocyte basolateral membranes, 420–422
threo-3-Hydroxyleucine, 779
5-Hydroxylysine, 779
3-Hydroxynorvaline, 779
cis-4-Hydroxyproline, 779
β-Hydroxypyruvate reductase, of microbody, 703
Hypnorm, 53

I

Imidazole, lysosome-inhibitory effects, 739
Immune electron microscopy
frozen thin-section tissue preservation, 469–479
localization of viral glycoprotein in infected cell, by indirect ferritin-antibody labeling technique, 440
Immunocytochemistry
of retinal membrane protein biosynthesis, 485–495
study of Semliki Forest virus spike protein transport, 466–485
Immunoelectrophoresis, two-dimensional
fenestrated gel washer, 252, 253
first dimension, 246–248
of membrane antigens, 244–257
photography, 252, 253
principle, 245, 246
procedure, 246–253
results, interpretation, 253–257
second-dimension electrophoresis, 251, 252
second-dimension gel, preparation, 248–251
solutions, 246, 247
staining, 253–255
Immunofluorescence, of proteins expressed from cloned cDNA, 499–501, 510, 511
Immunoprecipitation, 397
direct, of cell-free translation products, 114
indirect, 111–120
cleaning immunoprecipitates, 118, 119
dilution of detergent-solubilized translation reaction mixture, 113
elution of sample, from immunoadsorbent, 117, 118
immunoadsorbents, 114, 115
incubation
with immunoadsorbent, 114, 115
with antibody, 113, 114
solubilization of translation reaction mixture, 112, 113
specificity
checking, 118
lack of, 119, 120
washing immunoprecipitates, 115–117
of mRNA translation products, 49, 50
Imposil, 468, 479

Influenza virus
hemagglutination titration, 436
hemagglutinin, 796
intracellular transport, 127
mutants, with defects in glycoprotein transport, 440–442
infectivity assay, 441, 442
plaquing efficiency, 441, 442
replication, cell lines for, 437
strains, for glycoprotein biosynthesis study, 435
surface glycoproteins, 434
Influenza A virus, chemical mutagenesis, 440, 441
Insertion sequence, 664, 672, 674, 675
Insulin
cell responsiveness, 423, 424
^{125}I-labeled
binding to cell surface, determination, 427, 428
binding to soluble receptor, determination, 428, 430, 431
for heavy-isotope density-shift study, 426
protease-catalyzed synthesis, 133
receptor, 423, 424
active, definition, 424
cell-surface, labeling, 432
extraction-solubilization, for density-gradient centrifugation, 427
functional definition, 424
half-life, 430–432
heavy isotope labeling, 336
inactivation, 430–432
isopycnic banding in CsCl density gradients, 428, 429
synthesis, kinetics, 430–432
transit time to and from cell surface, 431–433
turnover, kinetics, 430–432
Invertase
accumulated in yeast secretory mutants, electrophoretic analysis, 812–815
antibody, affinity-purified, preparation, 813
extraction, from yeast, 813, 814
forms, 812
radiolabeled, immunoprecipitation and SDS-PAGE, 813–815
secretion, in yeast, 802
synthesis, in yeast, in mutant screening, 807
Ionophore, effects, on lysosomal function, 738, 755
Isocitrate lyase
import *in vitro,* into glyoxysomes, 712
microbody, 703, 714
synthesis, in glyoxysome, 710, 711
Isoleucine, analogs, 779
Isopropanolamine, lysosome-inhibitory effects, 739, 741
Isotopic dilution technique, for peptide molecular weight determination, 411–414

K

K562 cells, *see* Cell, K562
Kidney, peroxisome induction in, 548–555
Knife
for obtaining frozen thin sections, 471, 472
checking, 475, 476
optimal glass, 472–474
making, 474, 475

L

Lactate dehydrogenase, silver-stained, quantitative analysis, 236, 237
L-cell, lysate, 50
Lectin, 158, *see also* specific lectin
binding studies, in polyacrylamide gel, 190, 191
cytotoxicity, 159
effects of sera, 167, 168
survival curves, 168–170
variation, 167
L. culinaris, isolation, 286
lysosome-inhibitory effects, 762, 763
organelle-specific, in intracellular localization of virus glycoproteins, 461–465
pea, carbohydrate binding, 158
radiolabeled, probe for glycoproteins transferred to nitrocellulose, 205
Ricinus communis, labeling of viral spike protein in infected cells, 483, 484

for selection of resistant cell mutants, 166, 167
specificity, assay, 167
storage, 167
wheat germ, isolation, 286
Wistaria floribunda, accumulation in lysosomes, 763
Lectin-Sepharose affinity chromatography, of glycophorin A, 286, 287
Lectin-Sepharose column, preparation, 285, 286
Lens culinaris
lectin, 286
poly(A) mRNA, 714
Leucine, analogs, 779
Leucine aminopeptidase, hog kidney
inhibitor, 138
pH optimum, 138
specificity, 138
Leupeptin
isolation, 756
lysosome-inhibitory effects, 754, 756–759
Lidocaine
effects on cellular function, 746
lysosome-inhibitory effects, 740
Lipase, canine pancreatic, synthesis, 96
Lipid
association with protein, types, 795
polar, xxx
concentration differences among cell membranes, xxxiii, xxxiv
site of synthesis, xxxiii, xxxiv
Liver, *see also* Hepatocyte
cell fractionation, 722–725
cellular changes, after thyroid hormone administration, 556
peroxisome induction in, 548–555
rat
cell fractionation, 8–12, 21–23
homegenizing solution, volume, 10
high-speed supernatant, 3–5, 9, 11, 12, 23, 24
Sephadex G-100 fraction, 5, 12
LKB knife maker, 472–475
Loop mechanism, 405, 419, 537
Lutoid, rubber tree, tonoplast, composition, 586, 587
Lycopersicon, vacuoles, 582
Lysine, analogs, 779
Lysobacter enzymogenes, endoproteinase lysine-C, 137
Lysophagy, 753
Lysosome
amine entry into, 742, 743
amine uptake, mechanism, 739–742
autophagosome fusion, inhibition, 750, 752, 762
degradation, inhibition, 751–755
effects of proteinase inhibitors, 754, 756–760
effects of vanadate, 760, 761
endosome fusion, inhibition, 750, 752, 761
enzymes
biosynthesis, 764–777
in vitro, 729–736, 770–777
gel electrophoresis, 734, 735
labeled
direct immunoprecipitation, 732, 733
immunoprecipitation with *S. aureus,* 733, 734
phosphorylated mannose on, detection, 736
radioactive bands, detection, 735, 736
secretion, in slime mold, 815, 818, 820, 821, 827, 828
unscheduled secretion, 747, 748
inhibition, by ionophores, 755
inhibitors, 737–764
classes, 737, 738
phagosome fusion, inhibition, 750
pinosome fusion, 750
proteolytic activity, effects of alkylamines on, 739, 740, 745
receptors
protection, by amines, 749, 750
recycling, alkylamines affecting, 740, 747–750
trapping, caused by amines, 748, 749
receptor-ligand dissociation in, effects of amines on, 747–749, 752
swelling
effects of alkylamines on, 739, 740, 743–745
effect of ionophores, 755

transferrin-mediated iron uptake, effects of amines on, 748

M

Madin–Darby bovine kidney cell line, for viral replication, 436, 437
Madin–Darby canine kidney cell line, for viral replication, 436, 437
Maize
  endosperm
    membrane
      isolation, 719, 720
      preparation, material, 718, 719
      secretory type, 716, 718
      treated, preparation, 720
    protein synthesis and processing *in vitro,* 720, 721
    source of rough endoplasmic reticulum active processing, 716
  leaf, mRNA extraction, 29, 30
Malate dehydrogenase
  biosynthesis, *in vitro,* 711
  import *in vitro,* into glyoxysomes, 712
  of microbody, 703
Malate synthase
  detection in cytosol, 705–708
  import *in vitro,* into glyoxysomes, 712
  of microbody, 703, 714
  synthesis, in glyoxysome, 710, 711
Mannose
  phosphorylated, detection, on lysosomal enzymes, 736
  radioactive, cell labeling with, 730
α-Mannosidase, in yeast secretory mutants, 817, 818, 820, 824, 825
Mecamylamine, lysosome-inhibitory effects, 740
Membrane
  assembly
    modes, l–lii
    topography, lii, liii
  bacterial
    biogenesis, 74–84
      in vitro, 81–84
      in vivo, 75–79
      semi-*in vitro* systems, 79–81
    avian salt gland model, 627–659
    Semliki Forest virus as model, 467
  biogenesis, xxix, xxx
  brush border, sucrase-isomaltase, 386–406
  chemistry, xxx, xxxi
  components, xxx
    lateral diffusion, liii, liv
    sites of synthesis, xxxiii–xxxvi
  continuity
    functional, xxxii, xxxiii
    spatial, xxxii
    temporal, xxxii
  cytoplasmic, of viral infected cell, preparation, 439
  endosperm, protein processing *in vitro,* 716–721
  enterocyte basolateral, *see* Enterocyte basolateral membrane
  E. coli, perturbation, 75–77
  fluidity, significance, liii, liv
  fluid mosaic model, xxx, xxxi
  function, xxxi, xxxii
    as self template, liv, lv
  gradients across, xxxi, xxxii
  induction, by drugs, 542–557
    factors affecting, 543
  inner mitochondrial, topogenic sequences, 680
  invagination, inhibition by amines, 750
  microsomal
    from dog pancreas
      in immunoprecipitation, 119
      preparation, 87
    integration of acetylcholine receptor subunits into, 374–378
    nuclease treatment, 95–97
    posttranslational proteolysis assay, 92, 93, 101–105
    preparation, 84–93
    protein insertion, verification
      alkali treatment, 146, 147
      deoxycholate solubilization, 147
    rebinding of labeled ribosomes, 528–530
    recycling, xlvi
    salt extract
      fractionation, 685–689
      hydrophobic ion exchange chromatography, 685–689

salt-extracted, protein translocation in, 670, 682, 683
salt extraction, 684, 685
sources, 93
specificity, liv
control, xlvi–xlix
turnover, xxix
types, xxix, xxx
use in wheat germ mRNA translation system, 45
mitochondrial, and thyroid hormone, 555–557
permeability, 397
peroxisomal
composition, 728
isolation, 553–555
phylogeny, 676–678, 679
protein, *see* Protein
protein translocation across, 665
rough endoplasmic reticulum, protein translocation, 84–87, 94–111
spleen cell, preparation, 263, 264
translocation-competent, 665–667
Metabolic labeling, with heavy isotopes, 335–341
Methane sulfonyl fluoride, inhibitor of acetylcholinesterase, 359
Methionine
analogs, 779
radiolabeled
effect on protein synthesis in rabbit reticulocyte lysate, 68, 69
storage, 68
Methionine aminopeptidase, action, 150, 151
3-Methyladenine, autophagy inhibitor, 761, 762
Methylamine, lysosome-inhibitory effects, 739
2-Methylaminoethanol, lysosome-inhibitory effects, 739, 741
3-Methylcholanthrene, inducer of ER protein components, 545
2-Methylhistidine, 779
6-*C*-Methyllysine, 779
O-Methylthreonine, 779
4-Methyltryptophan, 780
Microinjection
cell preparation for, 507, 508
chamber, 508
of cloned cDNA, into eukaryotic cells, 499, 504–510
instruments, 505
materials, 505, 506
by micropipette, into cell nuclei, 499, 504–510
principle, 504, 505
procedure, 508–510
sample, 507, 508
solutions, 506
Micropipette, preparation, 506, 507
Microsome
membrane, *see* Membrane, microsomal
rough
activity, 11
ascites, 93
bovine pituitary, 93
chemical analysis, 7, 8
chicken oviduct, 93
column washing, 89
core glycosylation activity, 92
in different translation systems, 90, 91
from dog pancreas
crude preparation, 88, 89
isolation, 94, 95
Drosophila melanogaster embryo, 93
EDTA stripping, 89, 90, 96, 97
enzyme assay, 7, 8
fractionation from rat brain, 378, 381
fractions, ribophorin content correlation with ribosome binding capacity, 526–530
glycosylation capacity, 376
heavy, crude fraction, 8, 11
inactive, preparation, 698
inverted
biochemical characterization, 525
recovery of ribophorins in, 523–525
isolated
free ribosome contamination, 13, 17
intactness, 13–21
mRNA, intactness, 13–17, 23

properties, 12–21
protein synthesis, *in vitro,* 15–21
purity, 12, 13
light, 5
crude fraction, 8, 11
RNA to protein ratio, 528
nuclease treatment, 90
partial solubilization, and isolation of ribophorins, 521–523
preparation, 3–24
from cells lacking RNase inhibitor, 24
from chicken embryonic muscle, 23, 24
from rat brain, 24
for translation in cell-free system, 9, 10
rat liver, 93
RNA to protein ratio, 528
salt-extracted, translocation activity, 692, 693
storage, 11
stripped, 6
detergent extraction, 696, 697
preparation, 696
translation mixture, for *in vitro* system, 6, 7
translation products, cyanogen bromide fragmentation, 19–21
translocation ability, 143, 144
transmembrane glycoproteins, 520
treated with Triton X–100, smooth vesicles obtained from, 524, 525
use in translation systems, 91, 92
smooth
crude fraction, 8, 10, 11
RNA to protein ratio, 528
in starved rats, 5
spleen cell, preparation, 264
Microtubule poison, lysosome-inhibitory effects, 762
Mitochondria
hepatocyte, changes, after peroxisome induction, 551–553
induction, by DEHP, 557
membrane, synthesis, and thyroid hormones, 555–557
outer membrane, recognition of cytoplasmically synthesized proteins, 145
posttranslational insertion of proteins into, 129–134
protein transfer to, verification, 148, 149
protein translocation in, xlii, xliii, lv, 666–668
yeast, subfractionation, immunological test for purity, 200, 201
Modeccin, 751
Monensin, 465
Myasthenia gravis, 332, 352
Myelin
crude fraction, 380, 381
fragments, proteolipid protein, purification, 382–384
membrane proteins
electrophoretic patterns, 383
purification, 382–384
proteolipoprotein, 796
Myelin basic protein
large
electrophoretic pattern, 383
immunoprecipitation, 385
small
electrophoresis, 383–385
electrophoretic pattern, 383
immunoprecipitation, 385
synthesis, 385
Myoblast
culture, 570–574
labeling, 573
prenatal rabbit muscle, culture, 572
rat, culture, 571, 572
Myoglobin, horse muscle, cyanogen bromide cleavage products, molecular weights, 240–242
Myxovirus
culture, 436
glycoproteins, biosynthesis, 434–443
replication, in cell cultures, 437–440

N

Neuraminidase
activity assay, 437
influenza virus, 434

unit of activity, 437
Neurospora crassa
poly(A) mRNA, in study of microbody biosynthesis, 710, 711
proteolipid, 418
Neutral red, lysosome-inhibitory effects, 740
Newcastle disease virus
fusion glycoprotein, 796
glycoprotein, 436
hemagglutination titration, 436
mutants, with altered protease sensitivity of fusion protein, 443
protein biosynthesis study, cell line for, 437
Nicotinamide adenine dinucleotide-cytochrome b_5 reductase
membrane insertion, 135
mRNA, localization, 534
posttranslational insertion into ER membranes, 533–535
transport, xliv
Nicotinamide adenine dinucleotide phosphate-cytochrome *P*-450 reductase, 536, 537
biosynthesis, drug induced, 537
cDNA clones, 538
induction in ER, by drugs, 544, 545
localization, 127
synthesis, 124, 125
transport, xliv
Nitrocellulose sheet
detection of immune complexes on, 195–197, 198–200
electrophoretic transfer of proteins to, 193, 195, 196, 206
efficiency, 198
filter replica, reuse, 197, 213
incubation solution, reuse, 197
Nonidet P-40, 112
Norleucine, 779
Nuclear envelope, 589
architectural components, 597
X. laevis oocyte
large-scale isolation, 600–604
manual isolation, 598, 599, 603
preparation, 598–604
SDS-PAGE, 606
Nuclear matrix, 589
chromatin structure, effect of destabilization, 596
vertebrate, comparison to *Drosophila* fraction, 596
Nuclear matrix-pore complex-lamina fraction, *Drosophila*
comparison with vertebrate preparations, 596
morphology, 595
preparation, 591–595
Nuclease
micrococcal
Ca^{2+} requirement, 51
removal of endogenous mRNA from rabbit reticulocyte lysate, 51
staphylococcal, treatment of rough microsome, 90, 95–97
treatment of rabbit reticulocyte lysate, 57–59, 62, 63
Nucleus, *X. laevis* oocyte, manual isolation, 598, 599

O

Oat, pectolyase-cellulase digestion, 581, 582
Oligopeptide, low molecular weight, resolution, by slab gel system, 239–244
Oligosaccharide, high mannose, phosphorylation, xxxvi
OmpA protein, *see* Outer membrane, *E. coli,* protein A
Opsin, xli, 542
biosynthesis, analysis, 495
electrophoretic analysis, 614, 615
transport to ROS, immunochemical analysis, 615, 616
Ornithine transcarbamylase, posttranslational processing, 149
Ouabain, in Na^+, K^+-ATPase cytochemistry, 643–646
Outer membrane, *E. coli,* protein A, 75, 79, 80
Ovalbumin, silver-stained, quantitative analysis, 236, 237
4-Oxalysine, 779

P

Palmitic acid, analog, 655
Papain
  action, on brush border membrane, 386, 387
  amounts, for proteolysis in peptide mapping, 225
  P. carica
    inhibitors, 138
    pH optimum, 138
    specificity, 138
Paramyxovirus
  protease activation mutants, 442, 443
  replication, cell lines for, 437
  surface glycoproteins, 434
Penicillamine, 780
Pepsin, porcine gastric mucosa
  inhibitor, 139
  pH optimum, 139
  specificity, 139
Pepstatin
  isolation, 756
  lysosome-inhibitory effects, 756, 757
Peptidase, organelle-specific, 133
Peroxisomal reticulum, 727, 728
Peroxisome
  biosynthesis, 714, 715
  enzymes, 714
  inducers, 548
  induction, 548–555
    morphological effect on liver, 551
  isolation, 553–555
  membrane
    isolation, 553–555
    topogenic sequences, 680, 681
  posttranslational insertion of proteins into, 135
  protein translocation in, 666, 667
  rat liver
    catalase, peptide mapping, 727
    cytochemistry, 727, 728
    fractions, radioactivity measurements, 725, 726
    membrane composition, 728
    protein biogenesis, 721–728
      in vitro, 726, 727
      in vivo, 721–726
    protein isolation
      by immunochemistry, 725
      by SDS-PAGE, 725
PHA-L, sugar specificity, 158
Phenethyl alcohol, membrane perturbation, 75–77, 79–81
Phenobarbital
  effect on ER lipids, 545, 546
  increase of hepatic microsomal cytochrome *P*-450, 538
  induction of acyltransferase in hepatocyte, 651
  induction of ER membrane biosynthesis, in rats, 536
  induction of ER proteins, 544–546
Phenylalanine, analogs, 779
3-Phenylserine, 779
Phosphate
  analog, lysosome inhibitor, 738, 760, 761
  radioactive, cell labeling with, 730
Phosphatidic acid phosphatase, detection, in avian salt gland, 654
Phosphatidylcholine
  analog, 655
  synthesis, in avian salt gland, 649, 654
Phosphatidylethanolamine, synthesis, in avian salt gland, 649, 655
Phosphatidylinositol, 586
  cycle, in avian salt gland, 655
Phosphatidylserine, synthesis, in avian salt gland, 649, 655
Phosphodiesterase, *see* Cyclic guanosine monophosphate phosphodiesterase
Phospholine, acetylcholinesterase inhibitor, 361
Phospholipid
  biosynthesis, enzymes involved in
    cytochemical localization, 648–654
    drug induction, 545–547
  in signal peptidase reaction, 792, 793
  soybean, ethanol-extracted, preparation, 788
  synthesis, site, xxxiii
  transport, xxxiii
6-Phosphomannose, lysosomal enzyme marker, 747
Phosphorylase *b,* silver-stained, quantitative analysis, 236, 237
Phosphorylcholine transferase, activity, in avian salt gland, 654

Photolysis, of membrane protein, for topological study, 388–392
Photoreceptor
membrane biosynthesis, 609
proteins, electrophoretic analysis, 613–615
structure, 609
Phytohemagglutinin, CHO cells resistant to, 163
survival curves, 169, 170
Picornavirus
A particles, 451, 452
attachment
assays of receptors for, 445–448
to cellular site, kinetics, 446, 447
to erythrocytes, 447, 448
to solubilized membranes, assay, 448, 449
to specific receptor, 444, 445
to subcellular fractions, assay, 448, 449
cell penetration, assay, 450
cellular receptor unit, 444, 445
eclipse
assay, 450–452
definition, 445, 450
infection, events in, 444, 445
receptors, 443, 444
quantitation, 447
types, 443
uncoating, assay, 450–452
Pinocytosis, inhibition, by amines, 749
Plant
microbody, *see also* specific microbody proteins, purification, 703
preproteins, *in vitro* processing, 716–721
Plant vacuole, *see* Vacuole
Plasma membrane, 627, 628
avian salt gland, adaptation of, 628–631
bacterial, protein translocation across, 671
Band 3 protein, 125
prokaryotic, protein translocation in, 668
proteins
localization, 127
synthesis, 124, 125
of virus-injected cell, isolation, 439, 440
Plasmid, small-scale preparation, 503, 504
Plasticizer, inducer of peroxisomes, 549, 550
Platelet, glycophorins, 268, 269
Poliovirus
A particle, 452
receptors
quantitation, 447
solubilization, 448
specificity, 443
Polyamine, lysosome-inhibitory effects, 739, 741
Polyanion, lysosome-inhibitory effects, 762, 763
Polypeptide
chimeric, protease-catalyzed synthesis, 133, 132
free, primary addressing signal, 130
nascent
cotranslational insertion, 122–129
insertion into ER membrane, by vectorial discharge, 123, 126
Polysaccharide, inhibition of mRNA translation in reticulocyte lysates, 71
Polysome
free, xxxv
fractionation from rat brain, 378, 381
isolation, 771–773
from rat muscle, 575, 576
mRNA, intactness, 13–16
protein biosynthesis on, 81, 82
rat liver, preparation, 8, 9
isolation, from *E. coli,* 81, 82
membrane-bound, xxxv
isolated, intactness, 13–21
isolation, 771–773
from rat muscle, 575, 576
isolation of ribophorins in association with, 521–523
membrane polypeptides synthesized on, sorting mechanism, 541, 542
membrane protein biosynthesis and processing on, 82, 83
mRNA, intactness, 13–17, 23

preparation, 3–24, 83
MgCl$_2$ concentration, 11, 12
protein synthesis *in vitro,* 15–21
rat liver
preparation, 9
translation system, *in vitro,* 6, 7
segregation from free polysomes, 122–125
translation products, cyanogen bromide fragmentation, 19–21
rat liver
activity, 11
quantitative recovery as pellets, 12
storage, 11
Pore complex-lamina fraction, rat liver, comparison to *Drosophila* fraction, 596
Pore complex-lamina structure, 597
rat liver, 607, 608
SDS-PAGE, 606
X. laevis erythrocyte, 604–607
SDS-PAGE, 606
X. laevis oocyte
karyoskeletal residues enriched in, preparation, 603, 604
preparation, 599, 600, 606
SDS-PAGE, 606
Porin, transfer to mitochondria, verification, 148
Precursor-product relationship, criterion, 709
Procaine, lysosome-inhibitory effects, 740
Procarboxypeptidase, A + B, synthesis, cell-free vs. cellular, 96, 97
Proelastase, synthesis, 96
Prokaryote, membrane biogenesis, xxix, xxxvii, liii
Prolactin, 38
labeled, in signal peptidase assay, 785, 786
Proline, analogs, 779
Propranolol, lysosome-inhibitory effects, 740, 742
Propylamine, lysosome-inhibitory effects, 739
Protease, 136–139
in construction of chimeric polypeptides, 133, 134
digestion of protein in presence of SDS
for peptide analysis, 222–229
useful ranges, 224, 225
endogenous
activity, in membrane insertion study, 142, 143
inhibition, 149
inhibitor, added to mRNA translation system, 45
in intracellular transport of membrane proteins, 128, 129
membrane insertion study, 121–150
proteolytic conditions, 136–143
neutral, *B. polymyxa*
pH optimum, 138
specificity, 138
nonspecific, *S. griseus*
inhibitors, 138
pH optimum, 138
specificity, 138
organelle-specific, in membrane protein intracellular transport, 129
posttranslational protection from, 92, 93, 101–105
serine
specificities, 137
synthesis, 97
S. aureus V8
inhibitors, 137
pH optimum, 137
specificity, 137
in study of transfer of mitochondrial and chloroplast proteins, 132
Protein
assembly into bilayer, procedures, xlii–xlv
associated with endoplasmic reticulum membrane, segregation from proteins discharged into ER lumen, 126
band III, xl, 257–267, 542
antiserum, production, 263
autoradiography, 262
CH38 fragment, 260, 261, 264
purification, 263
CH65 fragment, 261, 264
cytoplasmic domain
purification, 321–323

radiolabeling, 322, 323
cytoskeletal association, 313
gel electrophoresis, 262
immunoprecipitation, 265
immunoprecipitants, endoglycosidase H digestion, 266
membrane insertion, 258
protease fragments, preparation, 260, 261
purification, for use as immunogen, 262, 263
synthesis, 258, 266
transmembrane configuration, 257, 261
TR41 fragment, 261
TR52 fragment, 260, 261
purification, 263
biogenesis
in *E. coli,* 75
in hepatocyte endoplasmic reticulum, 530–542
bound to DPT paper, identification, 212–215
cotranslational
posttranslational proteolysis assay, 92, 93
translocation, 84, 85
dansylation, 394, 397–399
degradation
endogenous, 752–754, 757–760
exogenous, 755
in dried gel slices, digestion procedure, 227–229
electrophoretic transfer to DPT paper
at pH 6.5, room temperature, and low voltage, 211, 212
at pH 9.2, freezing temperature, and high voltage, 209, 210
endosperm, *in vitro* processing, 716–721
eukaryotic cell, fatty acid acylation, 795–801
fatty acid acylated
characteristics, 800
detection, 797–799
distribution, 796, 797
glyoxysomal, biosynthesis, 700–715
heavy isotope labeling, 336, 337
in hydrated gel slices, digestion procedure, 226, 227
immunodetection, by fractionation in gels and transfer to paper, 205–215
insertion signal
interior permanent, 125
N-terminal permanent, 125
permanent, 125
transient, 123
integral membrane
classification, 672, 673
integration into membranes, 672
interaction with cell membrane, xxx, xxxi
intracellular traffic, control, 663–682
isolation, from modified (amidinated) membranes, 397
lysosomal
biosynthesis, in cell culture, 729–736
translocation, xlii
membrane
amino-terminal amino acid residue cleavage, 129
classifications, xxx, xxxi
electrophoretic transfer to nitrocellulose sheet, 193, 195, 196, 206
efficiency, 198
E. coli, biogenesis
effect of lipid fluidity, 75–78
inhibitors, 78, 79
in vitro, 81–84
using purified mRNA, 83, 84
fatty acylation, xli
halt transfer signal, 126
immunodetection, after SDS-PAGE, 192–205
insertion
cotranslational, 121–129
posttranslational, 121, 129–135
in vitro insertion, verification, 145–149
newly synthesized, interaction with organelle membranes, study, 143–145
peripheral, xlix
synthesis
in vitro, RNase contamination, 98–100
sites, xxxv

traffic, xxxv–xxxvii
regulation, xxxvii–xlii
translation products, proteolysis, termination, 141
two-dimensional polyacrylamide gel electrophoresis, 184–192
analysis techniques, 189–192
procedure, 186–189
sample preparation, 184–186
mitochondrial
cytoplasmically synthesized, recognition by outer membrane, 145
precursor, digestion by organelle-specific protease, 132, 133
multifunctional, of microbody, 703, 710, 714
myelin basic, *see* Myelin basic protein
myelin membrane, purification, 382–384
myelin-specific, biosynthesis, 378–385
NH_2 terminus
acetylation, 151
prevention, 156, 157
labeling, with *N*-formyl[^{35}S]methionine, 154–156
relationship to primary translation products, 151
N-terminal amino group derivatization, quantitation, 397–399
nuclear pore complex-lamina structure, SDS-PAGE, 606
peptide analysis, low molecular weight slab gel system, 239–244
peptide mapping
after electrophoresis, 191, 192
in one dimension, 222–229
sensitivity, 229
visualization, 229
peroxisomal content, biogenesis, 721–728
pleiotopic, 681, 682
posttranslocational pathways, 678–682
posttranslocational traffic, xlii, xliii, 663
presecretory
biological activity, 110, 111
proteolytic processing, gel electrophoretic analysis, 105–110
primary translation products, identifying, 150–157
purified, peptide mapping, 224–226
rabbit reticulocyte
analysis, 63–66
content, 63
of rod outer segment, localization, 494, 495
sarcoplasmic reticulum, biosynthesis, 570–579
SDS–PAGE, 208
secretory
apparent molecular weight, determination, 105, 106
biological activity, analysis, 110, 111
compared to lysosomal enzymes, 764, 765
conformation, determination, 106–110
intracellular transport, 127–129
nascent
gel electrophoresis, 105–110
vectorial discharge, 123, 126, 127
resistance to posttranslational proteolysis, 101–103
synthesis, cell-free vs. cellular, 96, 97
translocation, xlii
sedimentation constant, in heavy isotope labeling, 337, 338
silver-stained
density versus concentration, 234–237
density variation with variable loading and differential staining, 235, 236
quantitative analysis, 234, 235
solubilization, for indirect immunoprecipitation, 112, 113
src, peptide mapping, 228, 229
synthesis, 121, 122
inhibitors, lysosome-inhibitory effects, 763, 764
in vitro, 38, 111, 112
reinitiation, 15–22
mRNA specific stimulation, 48

in rough microsome translation mixture, 15–21
inhibition, 15–18
topogenesis, 663, 664
transfer
to cyanogen bromide activated paper, 215
to mitochondria, verification, 148, 149
to paper, methods, 206
translocation, xxxvi, xxxix, xl, xliii, xliv, 663, 665
cotranslational, 665, 666
mechanisms, 669–672
partial, 667
posttranslational, 665–667
evolution, 676, 677
quantitation of activity, 694–696
transmembrane, asymmetric assembly, xl, xli
vesicular transport, xlv, xlvi
control, xlvi
Protein A, 468
^{125}I-labeled, radioimmune assay with, in screening for transport-defective virus, 460
immunoadsorbent, 114, 115
immunoprecipitation of lysosomal enzymes with, 733, 734
Protein acetyltransferase, action, 151
Proteinase, lysosomal, inhibitors, 738, 754, 756–760
Proteinase K
SDS digestion, of cells, 26–28
Tritirachium album
inhibitors, 137
pH optimum, 137
specificity, 137
Protein kinase, cAMP-dependent, catalytic subunits, 796
Proteolipid protein, myelin
antibody preparation, 385
electrophoresis, 383–385
purification, 382–384
synthesis, 385
Protoplast, lysed, vacuole preparation from, 581–583
Pulse-chase labeling, 335
Pulse labeling, 335
of liver proteins, 722
Puromycin, lysosome-inhibitory effects, 763, 764
Puromycin aminonucleoside, 761
Putrescine, lysosome-inhibitory effects, 739
Pyropoikilocytosis, hereditary, spectrin alterations in, 305

Q

Quinine, lysosome-inhibitory effects, 740

R

Rabbit
anemia, induction, 54, 55
bleeding, 55, 56
Rabbit reticulocyte lysate
in vitro translation assay, 100, 101
mRNA translation in, 50–74
nuclease-treated
assay of mRNA, 59–63
gel-filtered, preparation, 72–74
preparation, 57–59
nuclease treatment, 51, 57–59, 62, 63
preparation, 53–57, 100
required components, 53
tRNA supplementation, 52
Rana pipiens, retina, large protein of rod outer segment, 494, 495
Reagent, labeling, permeability measurements, 396, 397
Receptosome, 748
Reticulocyte, tRNA, 51, 52
Reticulocyte lysate, *see also* Rabbit reticulocyte lysate
amino acid pools, 67–69
amino acid requirement, 67–69
composition, 66
K^+ optima, 66, 67
Mg^{2+} optima, 66, 67
potassium acetate vs. potassium chloride with, 67
properties, 66–71
protein synthesis, control of initiation, 69, 70
Retina
antibody labeling technique
choice of method, 492–495

control of background labeling, 490, 491
protocol, 488–490
reagent preparation, 488
cones, labeling, 489
cytosol fraction, immunochemical analysis, 615–617
embedding, in bovine serum albumin, 486, 487
labeled with anti-opsin and ferritin conjugates, quantitation of ferritin densities, 492
membrane fraction, immunochemical analysis, 615–617
opsin biosynthesis, analysis, 495
preparation, for subcellular fractionation, 611, 612
rod outer segment, 489, 492, 493, 495
disk formation, 609, 610
large protein, 609, 613, 615
immunochemical analysis, 615, 616
localization, 494, 495
in visual excitation, 617, 618
subcellular fractionation, 610–613
thin section, for membrane biosynthesis study, 485
X. laevis, disk formation, 609
Retrovirus, murine, proteins, 796
Rhinovirus
attachment, 445
receptors, quantitation, 447
Rhodopsin, 609, 610
photolyzed, in information flow in vision, 618, 619
Ribonuclease
A, protease-catalyzed synthesis, 133
contamination, in *in vitro* protein synthesis, 98–100
inhibitor
added to mRNA translation system, 45
from human placenta, isolation, 98–100
precautions against, in RNA extraction, 24, 25
in rough microsome preparation, 3, 4, 11, 12, 23, 24
Ribonucleic acid
double-stranded, inhibition of mRNA translation in reticulocyte lysate, 70, 71
extraction, 24–31
from brainstem fraction, 382
contaminant removal, 30, 31
DNA removal, 30
ethanolic sodium perchlorate, 27, 28
exogenous nucleases in, precautions against, 24, 25
guanidinium thiocyanate-CsCl, 28, 29
phenol, 25–27
from problem tissues, 30, 31
fractionation, 31–38
messenger
assay, with nuclease-treated rabbit reticulocyte lysate, 59–63
cell-free translation
amino acids, 41, 42
background radioactivity, 47
components of, 41–46
controls, 47, 48
endogenous stimulation, 47, 48
energy-generating system, 41–43
immunoprecipitation of proteins from, 111–120
immunoprecipitation of specific products, 49, 50
incorporation of radiolabel into total protein, 47
nuclease treatment, efficiency, 48
products, analysis, 47–50
reaction temperature, 46
reaction termination, 46
reaction time, 46
scale of reaction, 45, 46
standard net stimulation, 48
stimulation of radiolabel incorporation into protein, 47
systems, 50
wheat germ compensating buffer, 43
in wheat germ system, 38–50, 120

cell-free translation, in rabbit reticulocyte lysate, 50–74, 61, 62, 100, 101
assay of radioactivity incorporation, 61
product analysis by SDS-PAGE, 63–66
centrifugation, 32
in formamide, 32
coding for histocompatibility antigen, isolation, 326
concentration, for cell-free translation system, 44
electroblotting, 36–38
extraction
from corn leaf, 29, 30
from lepidopteran larval fat body, 30
gel electrophoresis, 33–35
gel filtration, 32, 33
isolation
from cell culture, 576, 577
for lysosomal enzyme synthesis, 770, 771
from neonatal muscle, 576, 577
from pancreatic tissue, 97, 98
from subcellular fractions, 31
K562 cell
isolation, 284, 285
translation *in vitro,* 285
muscle
sucrose gradient fractionation, 578
translation system for, 578, 579
poly(A), in study of plant microbody, 710, 714
preparation, 24–38
purified, membrane biosynthesis using, 83, 84
recovery from gels, 35–38
removal, from rough microsomes, 90, 95–97
size fractionation, 32–38
spleen cell, preparation, 266, 267
in translatable rough microsome preparation, intactness, 13–17
poly(A)$^+$
isolation, from muscle, 577, 578
purification, 577
poly(A)-containing, preparation, 31, 32
recovery, in rat liver microsome preparation, 12, 13
26 S, 512–514
tissue, extraction, 25–31
tobacco mosaic virus, for testing mRNA assay with nuclease-treated lysate, 61, 62
Torpedo californica electroplax, isolation, 368, 369
transfer
acylating enzyme, 153, 156
added to cell-free mRNA translation system, 45
aminoacylation, inhibition, with amino acid analogs, 780–783
formylating enzyme, 153
isolation, 152, 153
methionyl, labeled, preparation, 153–155
N-formylmethionyl
in identifying primary translation products, 151–156
preparation, 153, 155
in reticulocytes, 51, 52
whole-cell, extraction, 25–31
Ribophorin, xxxix, xlvii, 86
biosynthesis, 540
content, of microsomal fraction, correlation with ribosome binding capacity, 526–530
functions, 521
hepatocyte, 540–542
immunoprecipitation, 526, 527
properties, 520
recovery, in inverted rough microsomes, 523–525
segregation in ER, mechanism, 541, 542
Ribophorin I
antibody, preparation, 525, 526
molecular weight, 520
Ribophorin II
antibody, preparation, 525, 526
molecular weight, 520
Ribosome, xxxv
binding, to microsomal fractions, correlation to ribophorin content, 526–530
^{3}H-labeled, preparation, 528
receptors, 670–672

Ribulose-1,5-bisphosphate carboxylase, posttranslational insertion, 130
Rice, pectolyase-cellulase digestion, 581, 582
Ricin
CHO cells resistant to, 163
cytotoxicity, 159
potentiation, 751
sugar specificity, 158
Ricinus communis, agglutinin I, labeling of viral spike protein in infected cells, 483, 484
Rough endoplasmic reticulum, *see* Endoplasmic reticulum
Rough microsome, *see* Microsome
Rous sarcoma virus, transforming protein, xli

S

Salt gland, avian
acyltransferases, cytochemical localization, 651–654
basolateral membrane, enzyme markers, 634, 635
deadaptation, 629, 635, 655–658
dissociated cells, preparation, 633, 634
fractionation, 636–638
Golgi membrane fraction
enzymic characterization, 638–640
preparation, 635–638
membrane biogenesis, regulation, 658, 659
minilobules, 632, 633
model of membrane biogenesis, 627–659
Na^+, K^+-ATPase
assay, 638–640
biosynthesis, 658, 659
cytochemistry, 640–646
immunoprecipitation, 633
localization, 635, 643, 644
immunocytochemical, 646–648
loss, during deadaptation, 656
plasma membrane, 628–631
plasma membrane fraction
enzymic characterization, 638–640
preparation, 635–638
radiolabeling, 633
structure, 628, 629
use of tissue, 629–634
Sarcoplasmic reticulum, proteins
biosynthesis, 570–579
immunoprecipitation, 574, 575
Seed, microbodies, 714, 715, *see also* Glyoxysome
4-Selenalysine, 779
4-Selenaproline, 779
Selenomethionine, 779
Semliki Forest virus, 466
cell infection, sequence of events, 466, 467
envelope glycoprotein, 124
glycoproteins, 466, 796
infected cells
double fluorescence with antibodies and organelle-specific lectins, 463–465
frozen thin sections, labeling, 479–482
isolation, from cell monolayer culture, 479, 480
membrane protein, expression system, 497–499
mutagen treatment, 456, 457
mutants
with defects in structural proteins, identification, 458
isolation, 453
selection, 457
transport defect, reversibility, 461
transport-defective, screening for, 458–461
stock, preparation, 455
wild-type, cloning, 455, 456
Sendai virus, 436
hemagglutination titration, 436
intracellular transport, 127
mutants, activated by chymotrypsin and elastase, isolation, 442, 443
protein biosynthesis study, cell line for, 437
Sephadex, ribosome binding, 72
Sephadex G-25, for gel-filtered lysates, 72
Sephadex G-50, for gel-filtered lysates, disadvantages, 72
Serine-glyoxylate aminotransferase, of microbody, 703

Signal peptidase, 85, 86, 92, 123
assay, 784–794
translocation-dependent, 784, 785
translocation-independent, 785
procedure, 788–792
quantitation, 789–792
delipidated, restoration of activity, 786, 792
preparation, 786–788
properties, 784
reaction, characteristics, 792–794
Signal recognition particle, *see* Signal recognition protein
Signal recognition protein, xxxi, xxxviii, xxxix, xliii, 86, 123, 669–672, 692
function, 682, 683
properties, 683, 684
purification, 684–690
receptor, 86, 683
stability, 683
storage, 684
structure, 682
translocation assay, 690, 691
Signal segment, 122, 123
Signal sequence, xxxvii, xxxviii, xliii, 664, 665, 667–669, 674
function, study, 496
localization, 675, 676
Silver stain
photochemical, 231
for polyacrylamide gel electrophoresis, 230–239
in two-dimensional protein electrophoresis, 189
for visualization of peptide maps, 229
Silver staining
artifacts, 238, 239
DNA, in polyacrylamide gels, 235–237
effect
on autoradiography, 234
on fluorography, 234
proteins, colored patterns, 238
sensitivity, 233, 237, 238
Simian virus, intracellular transport 127
Sindbis virus
C protein, 512, 515–517, 519
envelope glycoprotein, 124
glycoproteins, 512–519
fatty acid acylation, 796, 799, 800
6K peptide, 512
mutants, transport-defective, screening for, 458–461
protein B_1, 515–517
protein synthesis
in rabbit reticulocyte lysate, 514
in wheat germ S_{23} translation system, 514
proteins, synthesized *in vivo* and *in vitro*
concanavalin A-Sepharose affinity chromatography, 519
core-glycosylation, 518, 519
posttranslational assays, 517–519
SDS-PAGE analysis, 515–517
26 S RNA, 512
preparation, 513, 514
translation products, 512, 513
stock, 455
structural proteins, labeling *in vivo,* 514, 515
transport-defective mutants, 456
Skeletal muscle, cellular changes, after thyroid hormone administration, 556
Slime mold
lysosomal enzymes, secretion, 818, 820, 821, 827, 828
lysosomal system, 815, 816
mutagenesis, 816
mutants, screening procedures, 816–818
secretory mutants, 815–828
defective in regulation of secretion, characterization, 821–823
defective in secretory ability, characterization, 818–821
isolation, 816–818
modification of enzymes, determination by isoelectric focusing, 823–827
stability of enzymes in standard secretion assay, 823, 824
secretory system model, 827, 828
Smooth endoplasmic reticulum, *see* Endoplasmic reticulum
Smooth microsome, *see* Microsome
Sodium deoxycholate, solubilization of HeLa cell plasma membrane, 448

Sodium dodecyl sulfate, detergent, for immunoprecipitation, 112
Sodium salicylate, scintillant, for fluorography, 218
Sorghum, vacuoles, 582
Sorting sequence, 664, 678–681
Spark chamber, detection of isotopes on solid supports, 221
Spectrin, xlviii
  alterations, in hemolytic anemias, 305
  chemical domains, isolation, 305–313
  cytoskeletal association, 313, 314
  ^{125}I-labeled, binding to inverted erythrocyte vesicles, 324
  intermediate-sized peptides
    generation, 306
    preparative two-dimensional electrophoresis, 306, 307
  isolation, 306
  oligomers
    detection, by PAGE, 298–304
    nondenaturing PAGE, procedure, 300–304
  peptides
    electrophoretic elution, 309–313
    identification, 307–309
  properties, 299
  purification, 316–318
  radiolabeling, 322, 323
  self-association, 298, 299
  two-dimensional immunoelectrophoresis, 254–256
Spermidine, lysosome-inhibitory effects, 739
Spermine, lysosome-inhibitory effects, 739
Sphingomyelin, synthesis
  in avian salt gland, 649
  site, xxxiv
Staphylococcus aureus
  bearing protein A, immunoprecipitation of lysosomal enzymes, 733, 734
  protease, 137
    amounts, for proteolysis in peptide mapping, 225
    solution proteolysis of tubulin, 225, 226
Statine, 756
trans-Stilbene oxide
  effect on ER lipid content, 545, 547
  inducer of ER protein components, 545
  induction of epoxide hydrolase biosynthesis, 537
Stop-transfer sequence, xl, 664, 672–674, 676
Streptomyces griseus, nonspecific protease, 138
Stuttering, 783
Subtilisin
  amounts, for proteolysis in peptide mapping, 225
  B. subtilis
    inhibitors, 137
    pH optimum, 137
    specificity, 137
Succinate dehydrogenase, assay, in avian salt gland, 638, 639
Sucrase-isomaltase
  of brush border membrane, 386–406
  activity, 386
  anchoring peptide, 387, 390, 391
  biosynthesis, 405, 406
  cleavage by organelle-specific protease, 129
  complex
    assembly, 399
    topology, 399
  dansylamino acids, extraction, 394, 395
  isolation, from DAP-labeled brush border membrane, 394
  localization, 127
  N-terminal amino acid of isomaltase-anchoring segment, orientation, 392–399
  N-terminal amino groups, amidination at, quantitation, 394–396, 398
  one-chain, two active sites precursor (Pro-SI), 400, 401
  photolysis, 388–392
    apparatus, 389
    reagent choice, 390–392
  precursor, 387, 400, 418
    amino-terminal sequences, 400, 401
    cell-free *in vitro* translation, 400–404

subunit embedded in lipid bilayer, hydrophobic photolabeling, 388–392
subunits, molecular weights, 386, 387
translation product, peptide map, 404

T

Tetracaine
effect on posttranslational proteolysis, 101–105
lysosome-inhibitory effects, 740
molecular structure, 104
N, N, N′,N′-Tetramethylethylenediamine, lysosome-inhibitory effects, 740
Thermolysin, *B. thermoproteolyticus*
inhibitor, 139
pH optimum, 139
specificity, 139
4-Thiaisoleucine, 779
Thialysine, 779
β-2-Thienylalanine, 779
β-3-Thienylalanine, 779
Thiolase, microbody, 703
4-Thioproline, 779
Threonine, analog, 779
Thylakoid membrane
assembly, li–liii
topogenic sequences, 680, 681
Thyroid hormone
inducer of mitochondrial membranes, 555–557
and mitochondrial membrane degradation, 557
Tissue
eyecup
embedded in BSA, sectioning, 487, 488
embedding, 486, 487
fixation, 486
freezing, 471
frozen thin sections
contrast staining, 478, 479
embedding, 478, 479
immunolabeling, 477, 478
obtaining, 471
knife used, 471, 472
optimal glass knife, 472–476
retrieval, 476, 477
mounting, 471
preparation, for mRNA extraction, 26
preservation, frozen-section technique, 469–479
section, thickness, and interference color, 476, 477
sucrose infusion, 470, 471
Tobacco, pectolyase-cellulase digestion, 581, 582
Tonoplast
composition, 586, 587
isolation, 585, 586
Topogenesis
definition, 663
sequences, 664, 672–675
for topologically equivalent proteins, 680, 681
Torpedo californica, electroplax
RNA, isolation, 368, 369
for study of acetylcholine receptor, 367, 368
Transcytosis, xlv
Transducin
complex
properties, 622, 623
purification, 621, 622
function, 618, 619
homology with G protein, 620, 621
properties, 619, 620
α-subunit
activation of phosphodiesterase, 623–625
ADP-ribosylated preparation, 625
properties, 623
purification, 623
subunits, 619, 620
Transferrin receptor, 796
Translocation, *see* Protein, translocation
1,2,4-Triazole-3-alanine, 779
Triethylamine, lysosome-inhibitory effects, 739
5,5,5-Trifluoroleucine, 779
Trifluoromethyl-(*m*-[^{125}I]iodophenyl) carbene, 388
3-Trifluoromethyl-3-(*m*-[^{125}I]iodophenyl)-diazirine, in topological study of sucrase-isomaltase, 388–392
Trigeminal nerve, neonatal rat, acyltransferase activity, 652–654

Trimethylamine, lysosome-inhibitory effects, 739
Tris (hydroxymethyl)methylamine, lysosome-inhibitory effects, 739, 741
Tritirachium album, proteinase K, 137
Triton X–100, 112
Trypsin
  in posttranslational proteolysis, 101–105, 121
  properties, 136, 137
  in study of transfer of mitochondrial and chloroplast proteins, 132
Trypsin inhibitor, silver-stained, quantitative analysis, 236, 237
Trypsinogen
  nascent, activity, 110
  synthesis, 96
Tryptazan, 780
Tryptophan, analogs, 780
Tubulin, purified, peptide mapping, 225, 226
Tunicamycin
  effect on glycophorin A synthesis, 289–292
  inhibitor of N-glycosylation of proteins, 289, 290
Tyrosine, analogs, 780
m-Tyrosine, 779

U

UDPgalactose: ovomucoid galactosyltransferase, assay, in avian salt gland fractions, 638, 639
UDPglucuronyltransferase, induction in ER by drugs, 544, 545
Urate oxidase, synthesis, 134

V

Vaccinia virus, membrane assembly, lii
Vacuole
  biogenesis, 587–589
  function, 580
  isolation, 580
  membrane, 585–587
  preparation
    from lysed protoplasts, 581–583
    by mechanical shearing of tissues, 583, 584
  properties, 584–589
  purification, 582, 583
  sap, 584, 585
Valine, analogs, 780
Vanadate, lysosome-inhibitory effects, 738, 760, 761
Vesicular stomatitis virus
  envelope glycoprotein, 124, 125
  glycoprotein, 257, 258, 261
    fatty acid acylation, 799–801
  glycoprotein G, integration into membrane, 675
  G protein, xlix, 127, 620, 621
  intracellular transport, 127
Vinblastine, lysosome-inhibitory effects, 762
Virion attachment protein, 444
Virus, *see also* specific virus
  cell penetration, inhibition, by amines, 750, 751
  eclipse, 445, 450
  for glycoprotein study, 435, 436
  infection, in membrane-bound glycoprotein study, 795
  infectivity
    plaque assay, 436, 447
    radioactive assay, 447
  membrane protein, expression
    from cloned cDNA in eukaryotic cell, 496–511
    system, 497–499
  neutral red-labeled, 450, 451
  saturation experiment, 447
  stock
    preparation, 455
    storage, 455
  tumor, sarc proteins, 796, 801
  uncoating, 450
  wild-type, cloning, 455, 456
Vision, light-activated amplification cycle, 618, 619

W

Wheat
  cellulysin digestion, 581
  vacuoles, isolation, 582
Wheat germ
  compensating buffer, for mRNA translation system, 43

extract
concentration, for cell-free mRNA translation, 44
for *in vitro* translation system, preparation, 39, 40
S23, 41, 43
flotation, 39
homogenate, centrifugation, 41
homogenization, 40
mRNA translation system, 38–50
source, 39
supernatant, gel filtration, 41
Wheat germ agglutinin, probe, for visualization of cell organelles, 461, 462
Wolfgram region, 383

X

Xenopus laevis
erythrocyte, pore complex-lamina structures, 604–607
oocyte
isolation, 601
nuclear envelope, preparation, 598–604
nuclei, isolation, 601–603
ovary, removal, 598
retina
embedded in BSA, 489
labeled with anti-opsin and ferritin conjugates, quantitation of ferritin densities, 492
rod cell, Golgi region, labeling, 494, 495
source, 598

Y

Yeast
cell surface
growth, 802, 804
layers, 802
genetic complementation test, 806, 807
invertase, extraction, 813, 814
mitochondria, proteolipid, 418
mitochondrial subfractionation, immunological test for purity, 200, 201
mutagenesis, in *sec* complementation groups, 804
mutants
classification, 806, 807
cytoplasmic invertase production, 807
lacking cytochrome c_1, immunological screening of, 200, 202
protein synthesis, screening, 807
screening, 805, 806
for defective secretion, 805, 806
selection, 804, 805
proteins, antisera, characterization, 201–205
secretory mutants, 802–815
acid phosphatase accumulation, localization by histochemical staining, 811
cytology, 808–811
density enrichment, 804, 805
double, 811
electron microscopy, 808–811
secretory organelles, 802, 803
secretory process, 802–804
temperature-sensitive clones, isolation, 805
tonoplast, composition, 586, 587
vacuole, 588, 589

Z

Zein
preproteins, processing, 717, 718
synthesis, *in vivo,* 716, 717
Z-Phe-Ala-CHN_2
lysosome-inhibitory effects, 759, 760
synthesis, 756